AQA
A Level Further Maths

Year 2

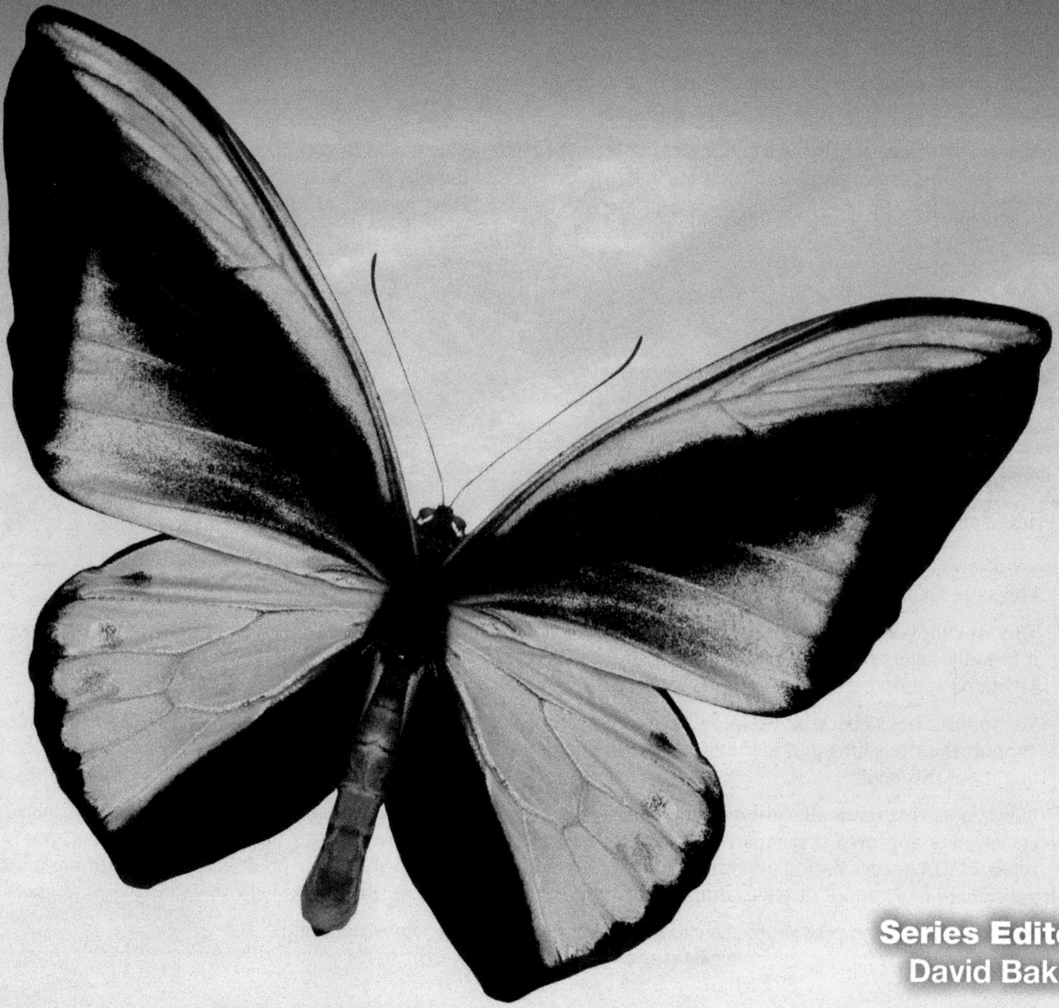

Series Editor
David Baker

Authors
Brian Jefferson, David Bowles, Eddie Mullan, Garry Wiseman,
John Rayneau, Katie Wood, Mike Heylings, Rob Wagner

OXFORD
UNIVERSITY PRESS

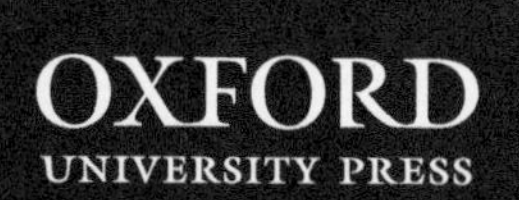

OXFORD
UNIVERSITY PRESS

Great Clarendon Street, Oxford, OX2 6DP, United Kingdom

Oxford University Press is a department of the University of Oxford.

It furthers the University's objective of excellence in research, scholarship, and education by publishing worldwide. Oxford is a registered trade mark of Oxford University Press in the UK and in certain other countries.

First published in 2018

British Library Cataloguing in Publication Data
Data available

978 0 19 841293 9

10 9 8 7 6 5

Paper used in the production of this book is a natural, recyclable product made from wood grown in sustainable forests.
The manufacturing process conforms to the environmental regulations of the country of origin.

Printed and bound by CPI Group (UK) Ltd, Croydon, CR0 4YY

Acknowledgements

Authors
Brian Jefferson, David Bowles, Eddie Mullan, Garry Wiseman, John Rayneau, Katie Wood, Mike Heylings, Rob Wagner

Series editor
David Baker

Editorial team
Dom Holdsworth, Ian Knowles, Matteo Orsini Jones, Felicity Ounsted

With thanks also to Sue Lyons, Geoff Wake, Matt Woodford, Deb Dobson, Katherine Bird, Linnet Bruce, Keith Gallick, Laurie Luscombe and Amy Ekins-Coward for their contribution.

Index compiled by Milla Hills

Although we have made every effort to trace and contact all copyright holders before publication, this has not been possible in all cases. If notified, the publisher will rectify any errors or omissions at the earliest opportunity.

Cover: Shutterstock

p1, **p25**, **p42**, **p45**, **p76**, **p178**(b), **p183**, **p204**, **p207**, **p242**, **p268**, **p271**, **p292**(m), **p292**(b), **p328**(b), **p351**(t), **p370**, **p373**, **p399**, **p418**, **p423**, **p452**, **p457**, **p476** Shutterstock; **p21** New York Public Library/Science photo library; **p79**, **p121** iStockphoto; **p127** Mikephotos/ Dreamstime; **p178**(t), **p245**, **p292**(t), **p297**, **p328**(t), **p328**(m) iStockphoto; **p351**(b), **p355**, **p373** iStockphoto.

Message from AQA

This student book has been approved by AQA for use with our qualification. This means that we have checked that it broadly covers the specification and we are satisfied with the overall quality. We have not, however, reviewed the MyMaths and InvisiPen links, and have therefore not approved this content.

We approve books because we know how important it is for teachers and students to have the right resources to support their teaching and learning. However, the publisher is ultimately responsible for the editorial control and quality of this book.

Please note that mark allocations given in assessment questions are to be used as guidelines only: AQA have not reviewed or approved these marks. Please also note that when teaching the AQA A Level Maths course, you must refer to AQA's specification as your definitive source of information. While the book has been written to match the specification, it cannot provide complete coverage of every aspect of the course.

Full details of our approval process can be found on our website: www.aqa.org.uk

Contents

MECHANICS

STATISTICS

DISCRETE

About this book

This book has been specifically created for those studying the AQA 2017 Further Mathematics A Level. It has been written by a team of experienced authors and teachers, and it's packed with questions, explanation and extra features to help you get the most out of your course.

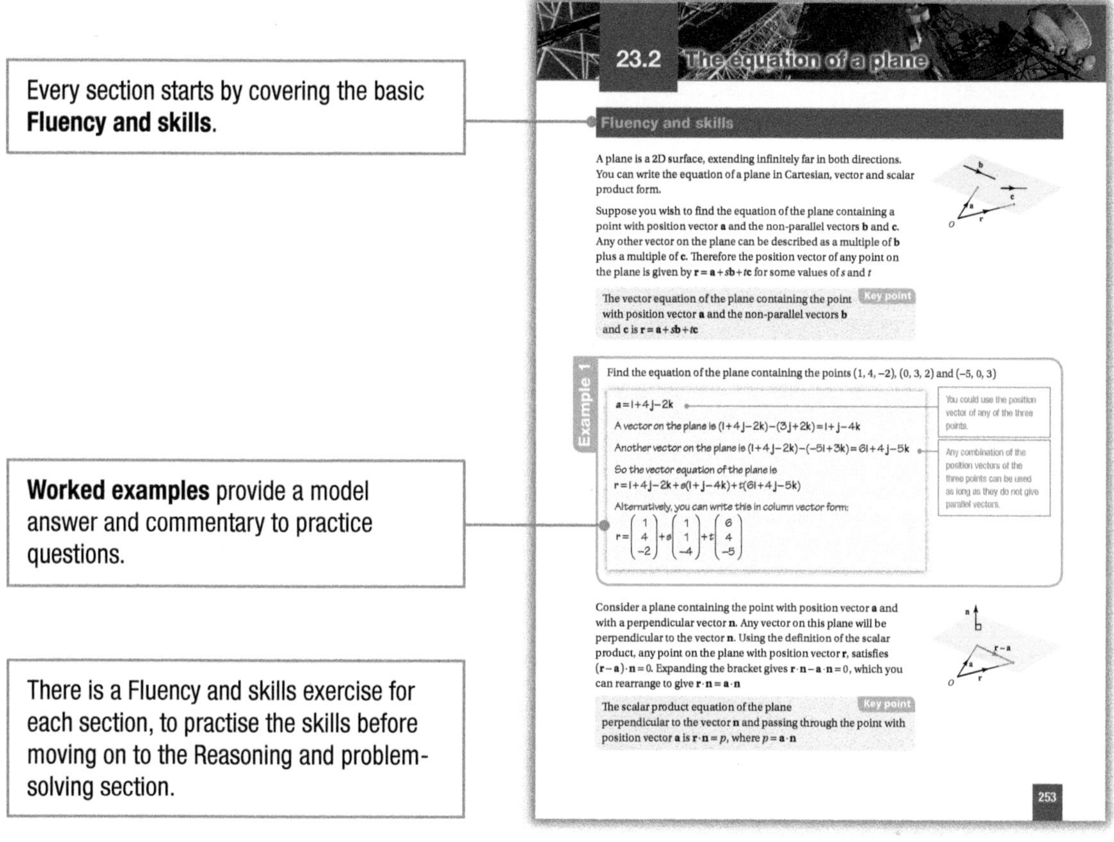

Every section starts by covering the basic **Fluency and skills**.

Worked examples provide a model answer and commentary to practice questions.

There is a Fluency and skills exercise for each section, to practise the skills before moving on to the Reasoning and problem-solving section.

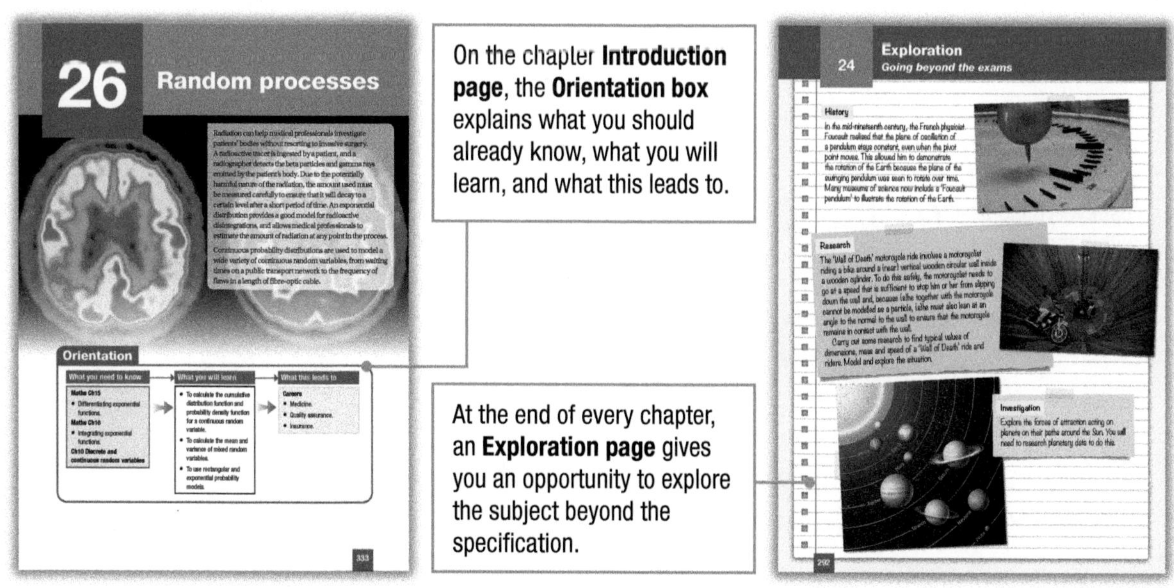

On the chapter **Introduction page**, the **Orientation box** explains what you should already know, what you will learn, and what this leads to.

At the end of every chapter, an **Exploration page** gives you an opportunity to explore the subject beyond the specification.

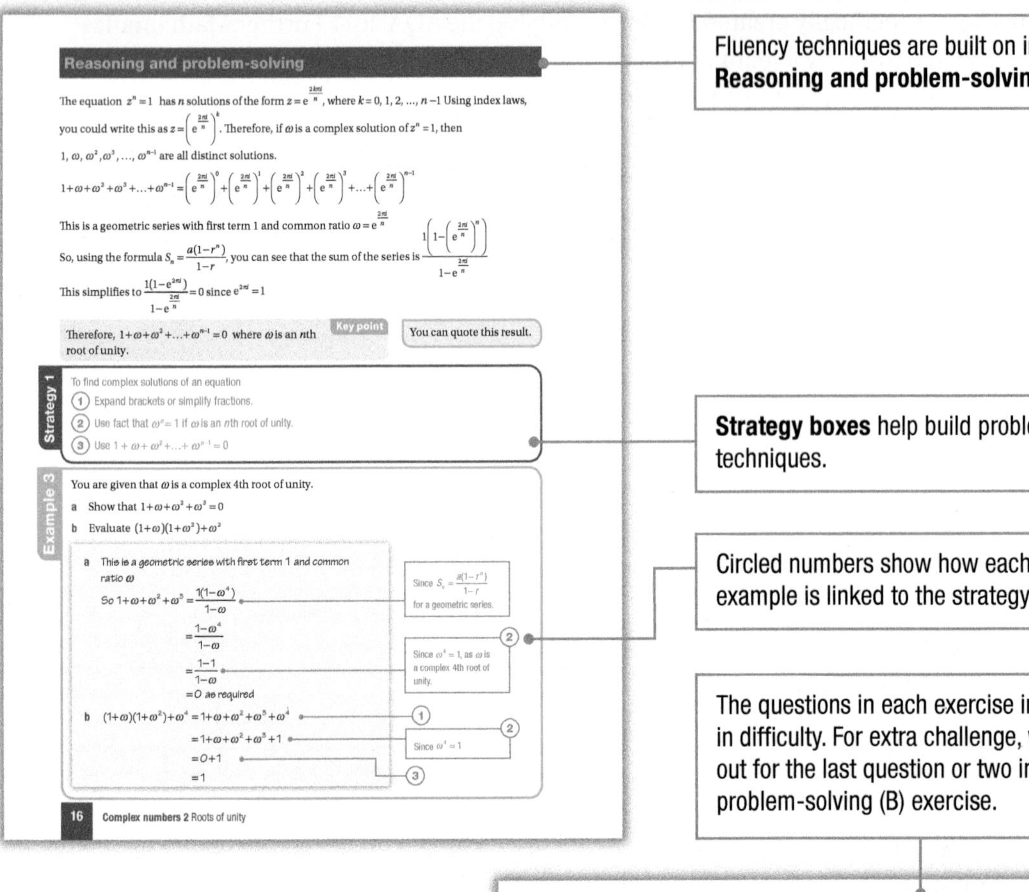

Reasoning and problem-solving

The equation $z^n = 1$ has n solutions of the form $z = e^{\frac{2k\pi i}{n}}$, where $k = 0, 1, 2, ..., n-1$ Using index laws, you could write this as $z = \left(e^{\frac{2\pi i}{n}}\right)^k$. Therefore, if ω is a complex solution of $z^n = 1$, then $1, \omega, \omega^2, \omega^3, ..., \omega^{n-1}$ are all distinct solutions.

$$1+\omega+\omega^2+\omega^3+...+\omega^{n-1} = \left(e^{\frac{2\pi i}{n}}\right)^0 + \left(e^{\frac{2\pi i}{n}}\right)^1 + \left(e^{\frac{2\pi i}{n}}\right)^2 + \left(e^{\frac{2\pi i}{n}}\right)^3 + ... + \left(e^{\frac{2\pi i}{n}}\right)^{n-1}$$

This is a geometric series with first term 1 and common ratio $\omega = e^{\frac{2\pi i}{n}}$

So, using the formula $S_n = \frac{a(1-r^n)}{1-r}$, you can see that the sum of the series is $\frac{1\left(1-\left(e^{\frac{2\pi i}{n}}\right)^n\right)}{1-e^{\frac{2\pi i}{n}}}$

This simplifies to $\frac{1(1-e^{2\pi i})}{1-e^{\frac{2\pi i}{n}}} = 0$ since $e^{2\pi i} = 1$

Key point: Therefore, $1+\omega+\omega^2+...+\omega^{n-1} = 0$ where ω is an nth root of unity.

You can quote this result.

Strategy 1

To find complex solutions of an equation

1. Expand brackets or simplify fractions.
2. Use fact that $\omega^n = 1$ if ω is an nth root of unity.
3. Use $1+\omega+\omega^2+...+\omega^{n-1} = 0$

Example 3

You are given that ω is a complex 4th root of unity.

a Show that $1+\omega+\omega^2+\omega^3 = 0$

b Evaluate $(1+\omega)(1+\omega^2)+\omega^2$

a This is a geometric series with first term 1 and common ratio ω

So $1+\omega+\omega^2+\omega^3 = \frac{1(1-\omega^4)}{1-\omega}$ — Since $S_n = \frac{a(1-r^n)}{1-r}$ for a geometric series.

$= \frac{1-\omega^4}{1-\omega}$ (2)

$= \frac{1-1}{1-\omega}$ — Since $\omega^4 = 1$, as ω is a complex 4th root of unity.

$= 0$ as required

b $(1+\omega)(1+\omega^2)+\omega^4 = 1+\omega+\omega^2+\omega^3+\omega^4$ (1)

$= 1+\omega+\omega^2+\omega^3+1$ (2) Since $\omega^4 = 1$

$= 0+1$ (3)

$= 1$

16 Complex numbers 2 Roots of unity

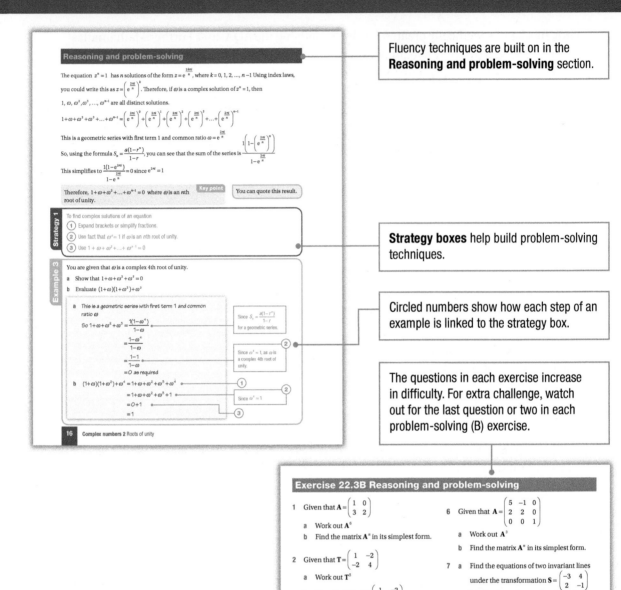

Fluency techniques are built on in the **Reasoning and problem-solving** section.

Strategy boxes help build problem-solving techniques.

Circled numbers show how each step of an example is linked to the strategy box.

The questions in each exercise increase in difficulty. For extra challenge, watch out for the last question or two in each problem-solving (B) exercise.

Exercise 22.3B Reasoning and problem-solving

1 Given that $\mathbf{A} = \begin{pmatrix} 1 & 0 \\ 3 & 2 \end{pmatrix}$

a Work out $\mathbf{A}^6$

b Find the matrix $\mathbf{A}^n$ in its simplest form.

2 Given that $\mathbf{T} = \begin{pmatrix} 1 & -2 \\ -2 & 4 \end{pmatrix}$

a Work out $\mathbf{T}^5$

b Show that $\mathbf{T}^n = 5^{n-1}\begin{pmatrix} 1 & -2 \\ -2 & 4 \end{pmatrix}$

3 Given that $\mathbf{M} = \begin{pmatrix} 5 & 2 \\ 2 & 5 \end{pmatrix}$

a State matrices $\mathbf{U}$ and $\mathbf{D}$ such that $\mathbf{M} = \mathbf{UDU}^{-1}$

b Work out the eigenvalues of $\mathbf{M}^3$

c Write down the eigenvectors of $\mathbf{M}^3$

6 Given that $\mathbf{A} = \begin{pmatrix} 5 & -1 & 0 \\ 2 & 2 & 0 \\ 0 & 0 & 1 \end{pmatrix}$

a Work out $\mathbf{A}^3$

b Find the matrix $\mathbf{A}^n$ in its simplest form.

7 a Find the equations of two invariant lines under the transformation $\mathbf{S} = \begin{pmatrix} -3 & 4 \\ 2 & -1 \end{pmatrix}$

b Which of these lines is also a line of invariant points? Explain your answer.

8 The matrix $\mathbf{M}$ is given by $\mathbf{M} = \begin{pmatrix} a & b \\ b & a \end{pmatrix}$ where a and b are constants and $b \neq 0$

Find the eigenvalues and eigenvectors of $\mathbf{M}$

9 The matrix $\mathbf{M} = \begin{pmatrix} -15 & 24 \\ -8 & 13 \end{pmatrix}$ represents a

Assessment sections at the end of each chapter test everything covered within that chapter.

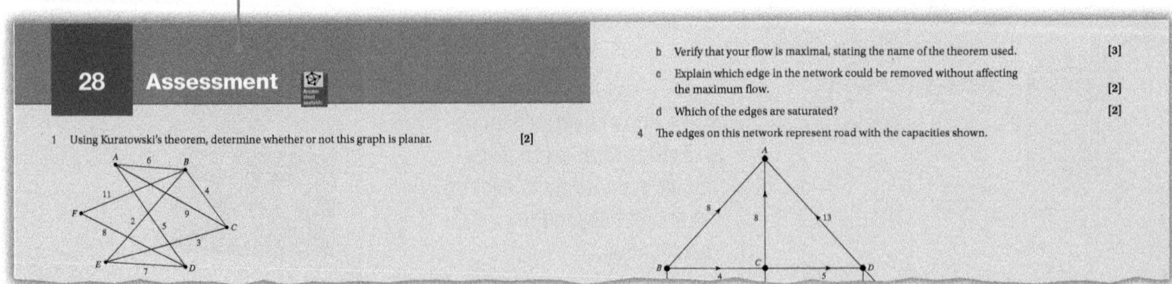

28 Assessment

1 Using Kuratowski's theorem, determine whether or not this graph is planar. [2]

b Verify that your flow is maximal, stating the name of the theorem used. [3]

c Explain which edge in the network could be removed without affecting the maximum flow. [2]

d Which of the edges are saturated? [2]

4 The edges on this network represent road with the capacities shown.

16 Complex numbers 2

Liquids such as water, and gases such as air, are known as fluids. In many ways, the flow of water can be treated the same as the flow of air. The study of such flow is known as fluid dynamics, where complex functions are used to model flow. Aerodynamics is the application of fluid dynamics to the flow of air. Hydrodynamics is the application of fluid dynamics to the flow of liquids. The scientific principles and the underpinning mathematics of fluid dynamics are important in many areas. For example, in the design of vehicles that move in gases and liquids. However, understanding gas and liquid flow is also important when planning how water and gas will reach homes and businesses.

An aircraft that is full of passengers, their luggage, and other cargo can lift off from a runway primarily as a result of the motion of the wing through the air. The mathematics of fluid dynamics allows this motion of the aircraft relative to the air, a fluid, to be analysed in detail. The mathematics relies on complex numbers to provide insight into the flow of the air. It is used by aeronautical engineers when they are designing the wings of an aircraft.

Orientation

What you need to know	What you will learn	What this leads to
Ch1 Complex numbers 1	• How to use exponential form. • How to use de Moivre's theorem. • How to use roots of unity.	**Careers** • Electrical engineering. • Aeronautical engineering. • Mechanical engineering.

16.1 Exponential form

Fluency and skills

A complex number $z = a + bi$ can be expressed in modulus-argument form as $z = r(\cos\theta + i\sin\theta)$ where $r = |z|$ and $\theta = \arg z$

Im

$z = a + bi$

b

r

θ

Re

O

a

The first few terms of the series expansions of $\cos\theta$ and of $\sin\theta$ are

See Ch2.5
For a reminder of Maclaurin expansions.

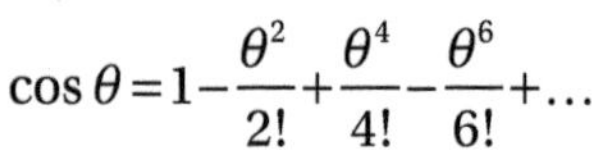

$$\cos\theta = 1 - \frac{\theta^2}{2!} + \frac{\theta^4}{4!} - \frac{\theta^6}{6!} + \ldots$$

$$\sin\theta = \theta - \frac{\theta^3}{3!} + \frac{\theta^5}{5!} - \frac{\theta^7}{7!} + \ldots$$

Therefore $z = r\left[\left(1 - \frac{\theta^2}{2!} + \frac{\theta^4}{4!} - \frac{\theta^6}{6!} + \ldots\right) + i\left(\theta - \frac{\theta^3}{3!} + \frac{\theta^5}{5!} - \frac{\theta^7}{7!} + \ldots\right)\right]$

$$= r\left(1 + i\theta - \frac{\theta^2}{2!} - \frac{\theta^3 i}{3!} + \frac{\theta^4}{4!} + \frac{\theta^5 i}{5!} - \frac{\theta^6}{6!} - \frac{\theta^7 i}{7!} + \ldots\right)$$

$$= r\left(1 + i\theta + \frac{(i\theta)^2}{2!} + \frac{(i\theta)^3}{3!} + \frac{(i\theta)^4}{4!} + \frac{(i\theta)^5}{5!} + \frac{(i\theta)^6}{6!} + \frac{(i\theta)^7}{7!} + \ldots\right)$$

since $(i\theta)^2 = -\theta$, $(i\theta)^3 = -i\theta$, $(i\theta)^4 = \theta^4$ and so on.

This is the expansion of $e^{i\theta}$

Key point

The formula $re^{i\theta} = r(\cos\theta + i\sin\theta)$ is known as **Euler's formula.**

Key point

So, using Euler's formula, you can write the complex number z in **exponential form** as $z = re^{i\theta}$ where $r = |z|$ and $\theta = \arg z$, $-\pi < \theta \le \pi$

Example 1

Write $2e^{\frac{3\pi i}{4}}$ in the form $a + bi$

To get from exponential form to $a + bi$ form, you need to first convert to modulus–argument form.

$$2e^{\frac{3\pi i}{4}} = 2\left(\cos\left(\frac{3\pi}{4}\right) + i\sin\left(\frac{3\pi}{4}\right)\right)$$

Use Euler's formula.

$$= 2\left(-\frac{\sqrt{2}}{2} + \frac{\sqrt{2}}{2}i\right)$$

$$= -\sqrt{2} + \sqrt{2}i$$

Example 2

Write $z = 3 - i$ in exponential form.

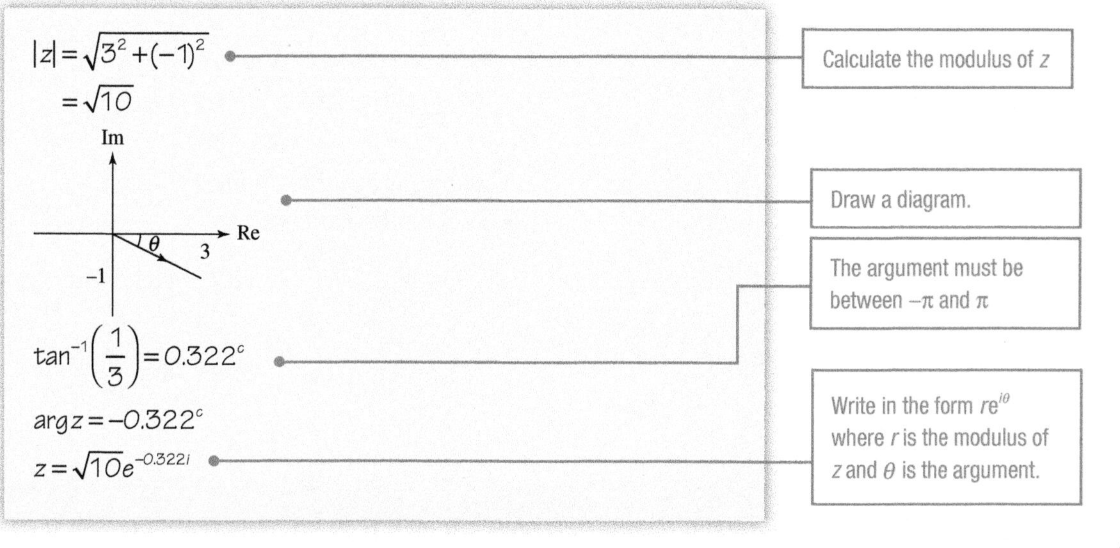

$|z| = \sqrt{3^2 + (-1)^2}$
$= \sqrt{10}$
Calculate the modulus of z
Im
Re
θ
3
–1
Draw a diagram.
The argument must be between $-\pi$ and π
$\tan^{-1}\left(\frac{1}{3}\right) = 0.322^c$
$\arg z = -0.322^c$
$z = \sqrt{10}e^{-0.322i}$
Write in the form $re^{i\theta}$ where r is the modulus of z and θ is the argument.

Calculator

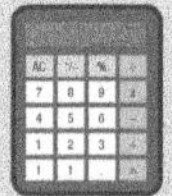

Try it on your calculator

Calculators can be used to convert to and from modulus-argument form.

Find out how to convert $\sqrt{2}e^{-\frac{\pi}{4}i}$ to Cartesian form on your calculator.

$\sqrt{2}\angle -\frac{\pi}{4} \blacktriangleright a+bi$

$1-i$

Exercise 16.1A Fluency and skills

1 Write each of these numbers in exponential form.

a $3+4i$ b $2-i$

c 10 d -5

e $2i$ f $-6i$

g $-5+12i$ h $-4-8i$

i $\sqrt{3}+i$ j $5-5i$

2 Write each of these complex numbers in exponential form.

a $2\left(\cos\left(\frac{\pi}{12}\right)+i\sin\left(\frac{\pi}{12}\right)\right)$

b $4\left(\cos\left(-\frac{2\pi}{3}\right)+i\sin\left(-\frac{2\pi}{3}\right)\right)$

c $3\left(\cos\left(\frac{5\pi}{6}\right)-i\sin\left(\frac{5\pi}{6}\right)\right)$

d $6\left(\cos\left(\frac{\pi}{7}\right)-i\sin\left(\frac{\pi}{7}\right)\right)$

e $\cos\left(\frac{7\pi}{5}\right)+i\sin\left(\frac{7\pi}{5}\right)$

f $\sqrt{2}\left(\cos\left(-\frac{15\pi}{8}\right)+i\sin\left(-\frac{15\pi}{8}\right)\right)$

g $\sqrt{3}\left(\cos\left(-\frac{5\pi}{6}\right)-i\sin\left(-\frac{5\pi}{6}\right)\right)$

h $8\left(\cos\left(-\frac{17\pi}{12}\right)-i\sin\left(-\frac{17\pi}{12}\right)\right)$

3 Write each of these complex numbers in the form $a+bi$

a $2e^{\frac{\pi}{2}i}$ b $7e^{-\frac{\pi}{3}i}$

c $\sqrt{2}e^{\frac{\pi}{4}i}$ d $e^{-\frac{\pi}{6}i}$

e $\sqrt{8}e^{-\pi i}$ f $\sqrt{3}e^{\frac{5\pi}{6}i}$

4 Given that $z=2e^{\frac{\pi}{3}i}$ and $w=3e^{-\frac{\pi}{3}i}$, calculate the value of

a $|zw|$ b $\left|\frac{z}{w}\right|$

c $\arg(zw)$ d $\arg\left(\frac{z}{w}\right)$

5 Given that $z=5e^{\frac{2\pi}{7}i}$ and $w=\frac{1}{5}e^{-\frac{\pi}{7}i}$, calculate the value of

a $|zw|$ b $\left|\frac{z}{w}\right|$

c $\arg(zw)$ d $\arg\left(\frac{z}{w}\right)$

6 Given that $z_1=\sqrt{6}e^{-\frac{\pi}{4}i}$ and $z_2=\sqrt{3}e^{-\frac{5\pi}{6}i}$, calculate the value of

a $|z_1z_2|$ b $\left|\frac{z_1}{z_2}\right|$

c $\arg(z_1z_2)$ d $\arg\left(\frac{z_1}{z_2}\right)$

Reasoning and problem-solving

Using the expansions at the beginning of this section, you can define trigonometric functions in terms of sums of exponentials, in particular:

Key point

$$\cos\theta=\frac{e^{i\theta}+e^{-i\theta}}{2} \text{ and } \sin\theta=\frac{e^{i\theta}-e^{-i\theta}}{2i}$$

Strategy

These results can then be used to prove trigonometric identities

1. Use Euler's formula: $e^{i\theta}\equiv\cos\theta+i\sin\theta$
2. Use the facts that $\cos(-\theta)\equiv\cos\theta$ and $\sin(-\theta)\equiv-\sin\theta$
3. Use $\cos\theta\equiv\frac{e^{i\theta}+e^{-i\theta}}{2}$ or $\sin\theta\equiv\frac{e^{i\theta}-e^{-i\theta}}{2i}$
4. Use index laws.

Example 3

Prove that $\cos\theta=\frac{e^{i\theta}+e^{-i\theta}}{2}$

$e^{i\theta}=\cos\theta+i\sin\theta$ — (1) Start with Euler's formula.

$e^{-i\theta}=\cos(-\theta)+i\sin(-\theta)$

$=\cos\theta-i\sin\theta$ — (2) Since $\cos(-\theta)=\cos(\theta)$ and $\sin(-\theta)=-\sin(\theta)$

$e^{i\theta}+e^{-i\theta}=\cos\theta+i\sin\theta+\cos\theta-i\sin\theta$

$=2\cos\theta$

Therefore $\cos\theta=\frac{e^{i\theta}+e^{-i\theta}}{2}$, as required.

Example 4

Prove that $\sin 2\theta \equiv 2\sin\theta\cos\theta$

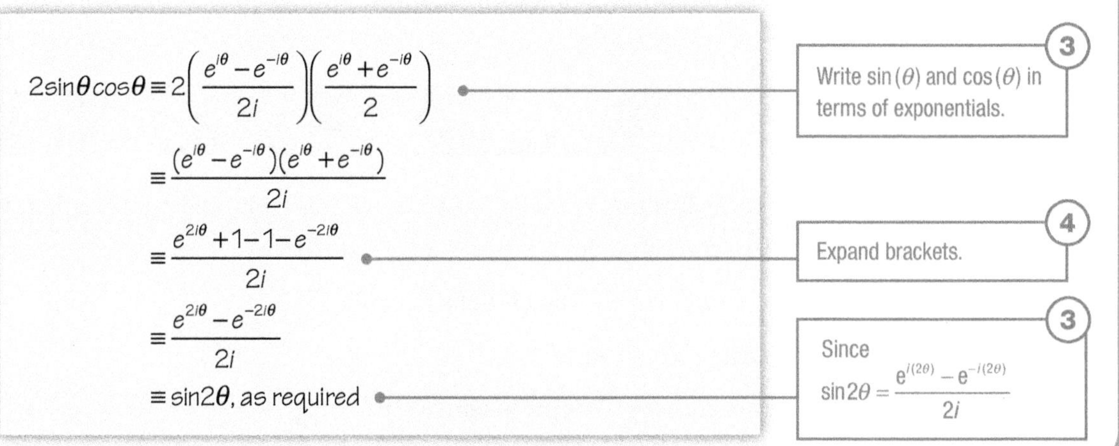

$2\sin\theta\cos\theta \equiv 2\left(\frac{e^{i\theta}-e^{-i\theta}}{2i}\right)\left(\frac{e^{i\theta}+e^{-i\theta}}{2}\right)$
$\equiv \frac{(e^{i\theta}-e^{-i\theta})(e^{i\theta}+e^{-i\theta})}{2i}$
$\equiv \frac{e^{2i\theta}+1-1-e^{-2i\theta}}{2i}$
$\equiv \frac{e^{2i\theta}-e^{-2i\theta}}{2i}$
$\equiv \sin 2\theta$, as required
Write $\sin(\theta)$ and $\cos(\theta)$ in terms of exponentials.
3
Expand brackets.
4
Since $\sin 2\theta = \frac{e^{i(2\theta)}-e^{-i(2\theta)}}{2i}$
3

PURE

Exercise 16.1B Reasoning and problem-solving

1 Use Euler's formula to show that $\sin\theta = \frac{e^{i\theta}-e^{-i\theta}}{2i}$

2 A complex number z has modulus 1 and argument θ

a Show that $z^n + \frac{1}{z^n} = 2\cos(n\theta)$

b Show that $z^n - \frac{1}{z^n} = 2i\sin(n\theta)$

3 Given that $z_1 = r_1e^{\theta_1 i}$ and $z_2 = r_2e^{\theta_2 i}$, show that

a $|z_1z_2| = |z_1||z_2|$ and $\arg(z_1z_2) = \arg z_1 + \arg z_2$

b $\left|\frac{z_1}{z_2}\right| = \frac{|z_1|}{|z_2|}$ and $\arg\left(\frac{z_1}{z_2}\right) = \arg z_1 - \arg z_2$

4 Use $\cos\theta = \frac{e^{i\theta}+e^{-i\theta}}{2}$ and $\sin\theta = \frac{e^{i\theta}-e^{-i\theta}}{2i}$ to show that

a $\sin(A+B) \equiv \sin A\cos B + \sin B\cos A$

b $\cos(A+B) \equiv \cos A\cos B - \sin A\sin B$

5 Use exponentials to show that

a $\cos 2x \equiv \cos^2 x - \sin^2 x$

b $\cos^2 x + \sin^2 x \equiv 1$

6 Use exponentials to show that

a $(\cos\theta + i\sin\theta)^2 \equiv \cos 2\theta + i\sin 2\theta$

b $(\cos\theta + i\sin\theta)^n \equiv \cos(n\theta) + i\sin(n\theta)$

7 Given that $z = 4\left(\cos\left(\frac{\pi}{9}\right) + i\sin\left(\frac{\pi}{9}\right)\right)$ and $w = 3\left(\cos\left(\frac{2\pi}{9}\right) + i\sin\left(\frac{2\pi}{9}\right)\right)$, show that $zw = 6 + 6\sqrt{3}i$

8 Given that $z = 8\left(\cos\left(\frac{5\pi}{12}\right) + i\sin\left(\frac{5\pi}{12}\right)\right)$ show that

$z^2 = -32\sqrt{3} + 32i$

9 The complex number z is such that $|z| = k$ and $\arg(z) = \theta$ for $k > 0$ and $-\pi < \theta \le \pi$

Another complex number is defined as $w = 1 - i$

Find expressions in terms of k and θ for the modulus and the argument of

a zw

b $\frac{z}{w}$

16.2 De Moivre's theorem

Fluency and skills

If you write a complex number in the form $z = r(\cos\theta + i\sin\theta)$, where r is a rational number, then you can see that $z^n = \left[r(\cos\theta + i\sin\theta)\right]^n$. You can write this as $r^n(\cos\theta + i\sin\theta)^n$

Therefore $z^n = r^n(e^{i\theta})^n$ since $e^{i\theta} = \cos\theta + i\sin\theta$ (using Euler's formula).

You can then use index laws to write $r^n(e^{i\theta})^n = r^n e^{in\theta}$

Using Euler's formula again, this becomes $r^n(\cos(n\theta) + i\sin(n\theta))$

Putting these two results together gives

Key point

$\left[r(\cos\theta + i\sin\theta)\right]^n = r^n(\cos(n\theta) + i\sin(n\theta))$, for all integers n, which is known as **de Moivre's theorem.**

You can prove this result using proof by induction.

De Moivre's theorem can be used to simplify powers of complex numbers.

For example, $(\cos\theta + i\sin\theta)^3 = \cos 3\theta + i\sin 3\theta$

$$\frac{1}{\cos\theta + i\sin\theta} = (\cos\theta + i\sin\theta)^{-1} = \cos(-\theta) + i\sin(-\theta)$$

Example 1

Write each of these numbers in the form $a + bi$

a $\left(\cos\frac{\pi}{3} + i\sin\frac{\pi}{3}\right)^4$ **b** $\left(\cos\frac{\pi}{4} - i\sin\frac{\pi}{4}\right)^6$

a $\left(\cos\frac{\pi}{3} + i\sin\frac{\pi}{3}\right)^4 = \cos\frac{4\pi}{3} + i\sin\frac{4\pi}{3}$ — Use de Moivre's theorem.

$= -\frac{1}{2} - \frac{\sqrt{3}}{2}i$ — In the form $a + bi$

b $\left(\cos\frac{\pi}{4} - i\sin\frac{\pi}{4}\right)^6 = \left(\cos\left(-\frac{\pi}{4}\right) + i\sin\left(-\frac{\pi}{4}\right)\right)^6$ — Using $\cos(-\theta) = \cos\theta$ and $\sin(-\theta) = -\sin\theta$

$= \left(\cos\left(-\frac{3\pi}{2}\right) + i\sin\left(-\frac{3\pi}{2}\right)\right)^6$ — Needs to be in modulus–argument form and then you can apply de Moivre's theorem.

$= i$

Example 2

Given the complex number $z=-\sqrt{3}+i$, use de Moivre's theorem to find z^{-2} in the form $a+bi$

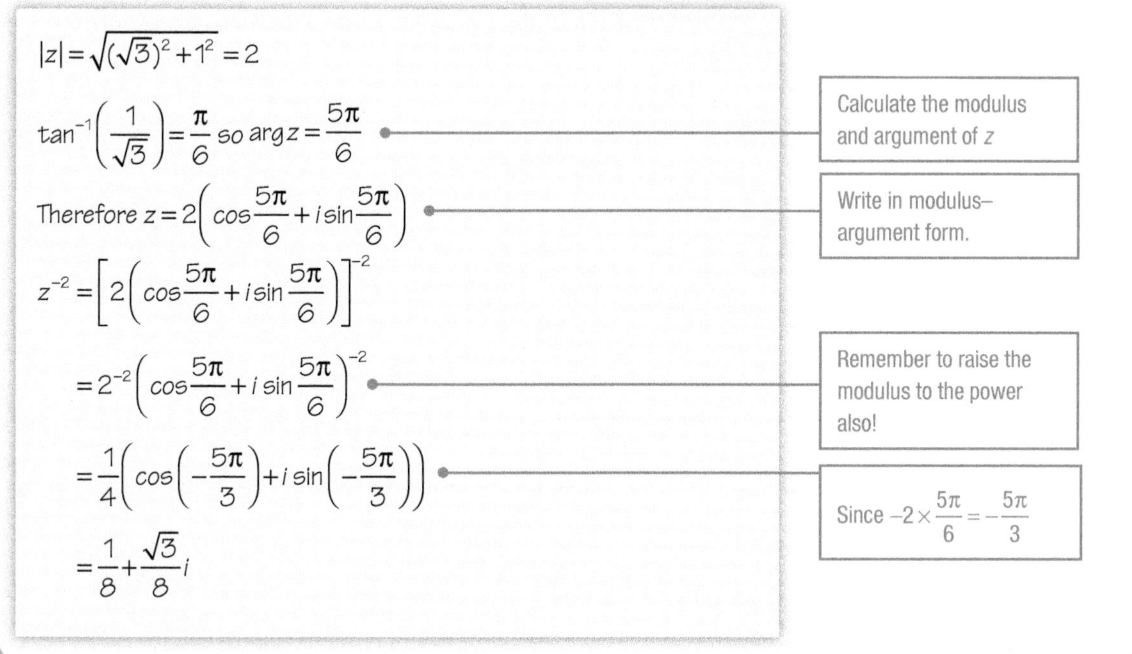

$|z|=\sqrt{(\sqrt{3})^2+1^2}=2$
$\tan^{-1}\left(\frac{1}{\sqrt{3}}\right)=\frac{\pi}{6}$ so $\arg z=\frac{5\pi}{6}$
Calculate the modulus and argument of z
Therefore $z=2\left(\cos\frac{5\pi}{6}+i\sin\frac{5\pi}{6}\right)$
Write in modulus–argument form.
$z^{-2}=\left[2\left(\cos\frac{5\pi}{6}+i\sin\frac{5\pi}{6}\right)\right]^{-2}$
$=2^{-2}\left(\cos\frac{5\pi}{6}+i\sin\frac{5\pi}{6}\right)^{-2}$
Remember to raise the modulus to the power also!
$=\frac{1}{4}\left(\cos\left(-\frac{5\pi}{3}\right)+i\sin\left(-\frac{5\pi}{3}\right)\right)$
Since $-2\times\frac{5\pi}{6}=-\frac{5\pi}{3}$
$=\frac{1}{8}+\frac{\sqrt{3}}{8}i$

PURE

Exercise 16.2A Fluency and skills

1 Express each of these numbers in the form $a+bi$

a $\left(\cos\frac{\pi}{3}+i\sin\frac{\pi}{3}\right)^6$

b $\left(\cos\frac{\pi}{6}+i\sin\frac{\pi}{6}\right)^5$

c $\left(\cos\left(-\frac{\pi}{12}\right)+i\sin\left(-\frac{\pi}{12}\right)\right)^4$

d $\left(\cos\frac{\pi}{14}-i\sin\frac{\pi}{14}\right)^7$

e $\left(\cos\frac{\pi}{4}+i\sin\frac{\pi}{4}\right)^{-3}$

2 Given that $z=3\left(\cos\left(\frac{\pi}{24}\right)+i\sin\left(\frac{\pi}{24}\right)\right)$, express in exact Cartesian form

a z^4 b z^{-6}

3 Given that $z=2\left(\cos\left(\frac{\pi}{12}\right)+i\sin\left(\frac{\pi}{12}\right)\right)$, express in exact Cartesian form

a z^3 b z^{-2}

c z^6 d z^{-4}

4 Given that $z=4\left(\cos\left(\frac{3\pi}{2}\right)+i\sin\left(\frac{3\pi}{2}\right)\right)$, express in exact Cartesian form

a z^2 b z^3

c $\frac{1}{z}$ d $16z^{-4}$

5 Given that $z=1-i$, use de Moivre's theorem to write the following powers of z in the form $a+bi$

a z^3 b z^7

c z^{-5} d z^{-6}

6 Given that $z=3i$, use de Moivre's theorem to write the following powers of z in Cartesian form.

a z^2 b z^{-1}

c z^{-3} d $\frac{3}{z^3}$

7 Given that $z=-\sqrt{3}+i$, use de Moivre's theorem to write the following in Cartesian form.

a z^4 b z^{-3}

c z^{-2} d $\frac{8}{z^6}$

Reasoning and problem-solving

You can use de Moivre's theorem combined with a binomial expansion to prove trigonometric identities involving powers of $\sin\theta$ or $\cos\theta$

Strategy 1

To write a power of $\cos\theta$ or $\sin\theta$ as a series involving $\cos(n\theta)$ or $\sin(n\theta)$

1. Use $\cos\theta \equiv \frac{e^{i\theta}+e^{-i\theta}}{2}$ or $\sin\theta \equiv \frac{e^{i\theta}-e^{-i\theta}}{2i}$
2. Write out the binomial expansion.
3. Use rules of indices to simplify.
4. Group terms together to write the expression in terms of trigonometric functions.

Example 3

Prove that $8\cos^4\theta \equiv \cos 4\theta + 4\cos 2\theta + 3$

$$16\cos^4\theta \equiv (2\cos\theta)^4$$

You want $2\cos\theta$ or $2i\sin\theta$ to start with.

$$\equiv (e^{i\theta}+e^{-i\theta})^4$$

(1) Since $2\cos\theta \equiv e^{i\theta}+e^{-i\theta}$

$$\equiv (e^{i\theta})^4+4(e^{i\theta})^3(e^{-i\theta})+6(e^{i\theta})^2(e^{-i\theta})^2 +4(e^{i\theta})(e^{-i\theta})^3+(e^{-i\theta})^4$$

(2) Write the binomial expansion of $(e^{i\theta}+e^{-i\theta})^4$

$$\equiv e^{4i\theta}+4e^{2i\theta}+6+4e^{-2i\theta}+e^{-4i\theta}$$

(3) Simplify using index rules.

$$\equiv (e^{4i\theta}+e^{-4i\theta})+4(e^{2i\theta}+e^{-2i\theta})+6$$

(4) $e^{4i\theta}+e^{-4i\theta} = 2\cos 4\theta$ and $e^{2i\theta}+e^{-2i\theta} = 2\cos 2\theta$

$$\equiv 2\cos 4\theta + 8\cos 2\theta + 6$$

Remember, this is $16\cos^4\theta$

$$8\cos^4\theta \equiv \frac{2\cos 4\theta+8\cos 2\theta+6}{2}$$

$$\equiv \cos 4\theta + 4\cos 2\theta + 3$$

Strategy 2

To write $\cos(n\theta)$ or $\sin(n\theta)$ in terms of powers of $\cos\theta$ or $\sin\theta$

1. Use de Moivre's theorem.
2. Write out the binomial expansion.
3. Simplify powers of i
4. Equate coefficients of real or imaginary parts.
5. Use $\cos^2\theta+\sin^2\theta \equiv 1$ to write the expression as powers of either $\cos\theta$ or $\sin\theta$

Example 4

Express $\sin 5x$ in the form $A\sin x + B\sin^3 x + C\sin^5 x$ where A, B and C are constants to be found.

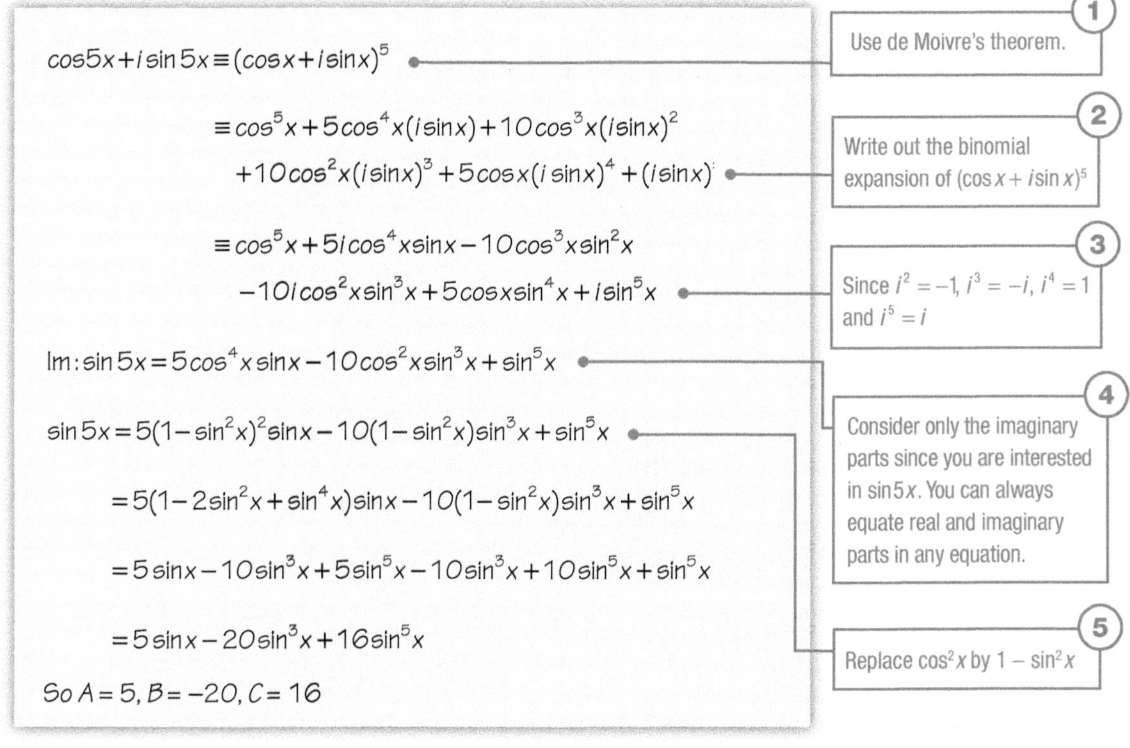

$$\cos 5x + i\sin 5x \equiv (\cos x + i\sin x)^5$$

1. Use de Moivre's theorem.

$$\equiv \cos^5 x + 5\cos^4 x(i\sin x) + 10\cos^3 x(i\sin x)^2 + 10\cos^2 x(i\sin x)^3 + 5\cos x(i\sin x)^4 + (i\sin x)^5$$

2. Write out the binomial expansion of $(\cos x + i\sin x)^5$

$$\equiv \cos^5 x + 5i\cos^4 x\sin x - 10\cos^3 x\sin^2 x - 10i\cos^2 x\sin^3 x + 5\cos x\sin^4 x + i\sin^5 x$$

3. Since $i^2 = -1$, $i^3 = -i$, $i^4 = 1$ and $i^5 = i$

$$\text{Im}: \sin 5x = 5\cos^4 x\sin x - 10\cos^2 x\sin^3 x + \sin^5 x$$

4. Consider only the imaginary parts since you are interested in $\sin 5x$. You can always equate real and imaginary parts in any equation.

$$\sin 5x = 5(1 - \sin^2 x)^2\sin x - 10(1 - \sin^2 x)\sin^3 x + \sin^5 x$$

5. Replace $\cos^2 x$ by $1 - \sin^2 x$

$$= 5(1 - 2\sin^2 x + \sin^4 x)\sin x - 10(1 - \sin^2 x)\sin^3 x + \sin^5 x$$

$$= 5\sin x - 10\sin^3 x + 5\sin^5 x - 10\sin^3 x + 10\sin^5 x + \sin^5 x$$

$$= 5\sin x - 20\sin^3 x + 16\sin^5 x$$

So $A = 5$, $B = -20$, $C = 16$

Strategy 3

To find the sum of a series involving $\sin(r\theta)$ or $\cos(r\theta)$

1. Consider a sum involving $\cos(r\theta) + i\sin(r\theta)$
2. Use de Moivre's theorem.
3. Use the formula for the sum of a geometric series.
4. Use formulae for $\sin(A \pm B)$ or $\cos(A \pm B)$
5. Select only the real or the imaginary parts as required.

Example 5

Show that $\sum_{r=1}^{n}\sin(r\theta)=\dfrac{\sin\left(\frac{(n+1)\theta}{2}\right)\sin\left(\frac{n\theta}{2}\right)}{\sin\left(\frac{\theta}{2}\right)}$

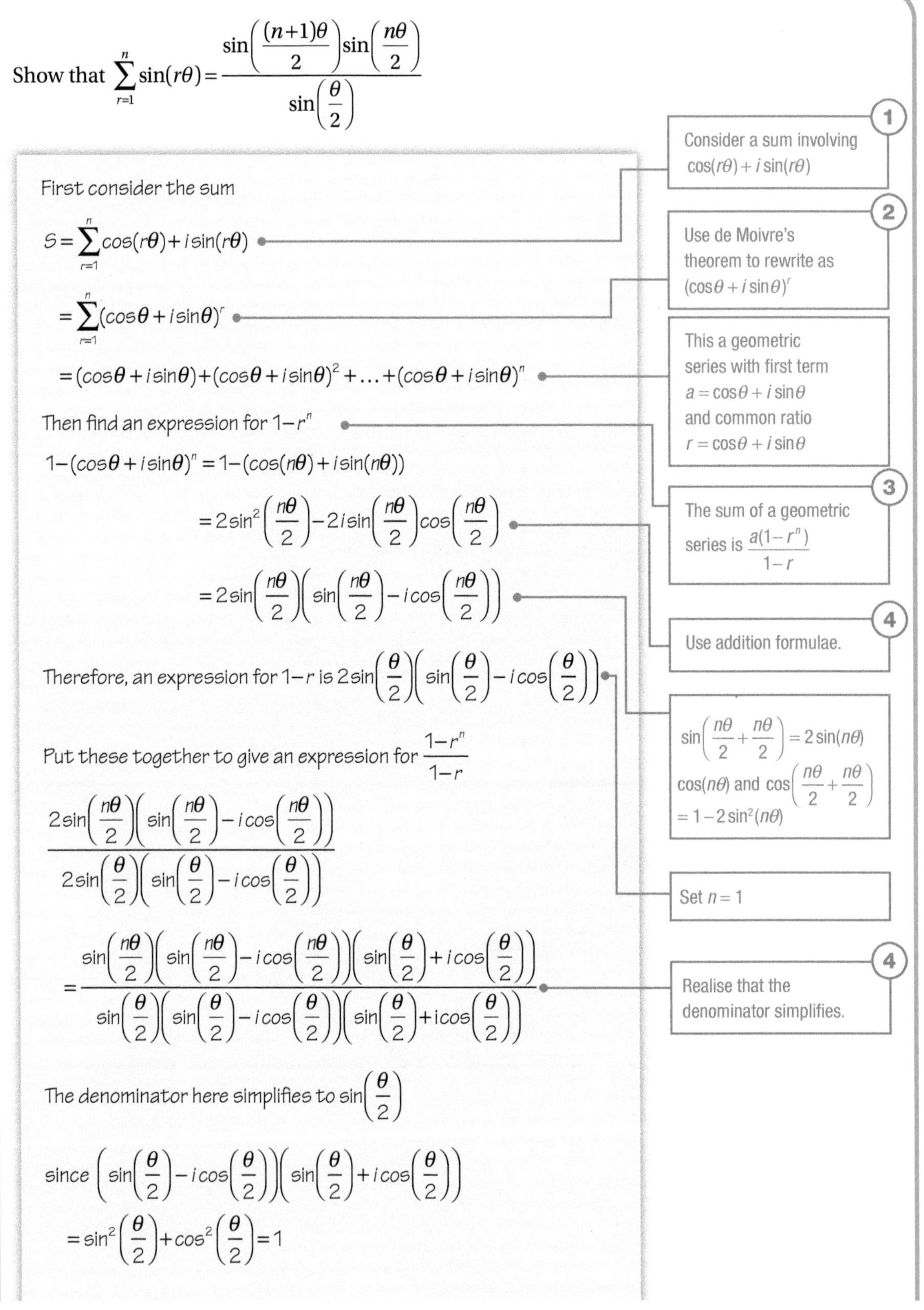

First consider the sum

$$S=\sum_{r=1}^{n}\cos(r\theta)+i\sin(r\theta)$$

$$=\sum_{r=1}^{n}(\cos\theta+i\sin\theta)^r$$

$$=(\cos\theta+i\sin\theta)+(\cos\theta+i\sin\theta)^2+\ldots+(\cos\theta+i\sin\theta)^n$$

Then find an expression for $1-r^n$

$$1-(\cos\theta+i\sin\theta)^n=1-(\cos(n\theta)+i\sin(n\theta))$$

$$=2\sin^2\left(\frac{n\theta}{2}\right)-2i\sin\left(\frac{n\theta}{2}\right)\cos\left(\frac{n\theta}{2}\right)$$

$$=2\sin\left(\frac{n\theta}{2}\right)\left(\sin\left(\frac{n\theta}{2}\right)-i\cos\left(\frac{n\theta}{2}\right)\right)$$

Therefore, an expression for $1-r$ is $2\sin\left(\frac{\theta}{2}\right)\left(\sin\left(\frac{\theta}{2}\right)-i\cos\left(\frac{\theta}{2}\right)\right)$

Put these together to give an expression for $\dfrac{1-r^n}{1-r}$

$$\frac{2\sin\left(\frac{n\theta}{2}\right)\left(\sin\left(\frac{n\theta}{2}\right)-i\cos\left(\frac{n\theta}{2}\right)\right)}{2\sin\left(\frac{\theta}{2}\right)\left(\sin\left(\frac{\theta}{2}\right)-i\cos\left(\frac{\theta}{2}\right)\right)}$$

$$=\frac{\sin\left(\frac{n\theta}{2}\right)\left(\sin\left(\frac{n\theta}{2}\right)-i\cos\left(\frac{n\theta}{2}\right)\right)\left(\sin\left(\frac{\theta}{2}\right)+i\cos\left(\frac{\theta}{2}\right)\right)}{\sin\left(\frac{\theta}{2}\right)\left(\sin\left(\frac{\theta}{2}\right)-i\cos\left(\frac{\theta}{2}\right)\right)\left(\sin\left(\frac{\theta}{2}\right)+i\cos\left(\frac{\theta}{2}\right)\right)}$$

The denominator here simplifies to $\sin\left(\frac{\theta}{2}\right)$

since $\left(\sin\left(\frac{\theta}{2}\right)-i\cos\left(\frac{\theta}{2}\right)\right)\left(\sin\left(\frac{\theta}{2}\right)+i\cos\left(\frac{\theta}{2}\right)\right)$

$$=\sin^2\left(\frac{\theta}{2}\right)+\cos^2\left(\frac{\theta}{2}\right)=1$$

1 Consider a sum involving $\cos(r\theta)+i\sin(r\theta)$

2 Use de Moivre's theorem to rewrite as $(\cos\theta+i\sin\theta)^r$

This a geometric series with first term $a=\cos\theta+i\sin\theta$ and common ratio $r=\cos\theta+i\sin\theta$

3 The sum of a geometric series is $\dfrac{a(1-r^n)}{1-r}$

4 Use addition formulae.

$\sin\left(\frac{n\theta}{2}+\frac{n\theta}{2}\right)=2\sin(n\theta)$

$\cos(n\theta)$ and $\cos\left(\frac{n\theta}{2}+\frac{n\theta}{2}\right)$
$=1-2\sin^2(n\theta)$

Set $n=1$

4 Realise that the denominator simplifies.

(*Continued on the next page*)

PURE

The numerator expands to give

$$\sin\left(\frac{n\theta}{2}\right)\left[\left(\cos\left(\frac{n\theta}{2}\right)\cos\left(\frac{\theta}{2}\right)+\sin\left(\frac{n\theta}{2}\right)\sin\left(\frac{\theta}{2}\right)\right)\right.$$
$$\left.+i\left(\cos\left(\frac{\theta}{2}\right)\sin\left(\frac{n\theta}{2}\right)-\cos\left(\frac{n\theta}{2}\right)\sin\left(\frac{\theta}{2}\right)\right)\right]$$
$$=\sin\left(\frac{n\theta}{2}\right)\left(\cos\left(\frac{(n-1)\theta}{2}\right)+i\sin\left(\frac{(n-1)\theta}{2}\right)\right)$$

4 Use formulae for $\cos\left(\frac{n\theta}{2}-\frac{\theta}{2}\right)$ and $\sin\left(\frac{n\theta}{2}-\frac{\theta}{2}\right)$.

Therefore,

$$S=\frac{(\cos\theta+i\sin\theta)\sin\left(\frac{n\theta}{2}\right)\left(\cos\left(\frac{(n-1)\theta}{2}\right)+i\sin\left(\frac{(n-1)\theta}{2}\right)\right)}{\sin\left(\frac{\theta}{2}\right)}$$

3 Use $S=\frac{a(1-r^n)}{1-r}$

$$\sum_{r=1}^{n}\sin(r\theta)=\frac{\sin\left(\frac{n\theta}{2}\right)\left(\sin\theta\cos\left(\frac{(n-1)\theta}{2}\right)+\sin\left(\frac{(n-1)\theta}{2}\right)\cos\theta\right)}{\sin\left(\frac{\theta}{2}\right)}$$

5 Select the imaginary parts of the sum.

$$=\frac{\sin\left(\frac{(n+1)\theta}{2}\right)\sin\left(\frac{n\theta}{2}\right)}{\sin\left(\frac{\theta}{2}\right)}$$

4 Use formula for $\sin\left(\frac{(n-1)\theta}{2}+\theta\right)$

For $\sum_{r=1}^{n}\cos(r\theta)$ you would need to consider the real part of the sum.

Exercise 16.2B Reasoning and problem-solving

1 Prove each of these identities.

a $2\cos^2\theta \equiv \cos 2\theta + 1$ b $8\sin^3\theta \equiv 6\sin\theta - 2\sin 3\theta$

c $4\sin^4\theta \equiv \frac{1}{2}\cos 4\theta - 2\cos 2\theta + \frac{3}{2}$

2 a Show that $\cos^5\theta \equiv A(10\cos\theta + 5\cos 3\theta + \cos 5\theta)$, where A is a constant to be found.

b Hence find $\int \cos^5\theta \, d\theta$

3 a Show that $\sin^6\theta \equiv B(15\cos 2\theta - 6\cos 4\theta + \cos 6\theta - 10)$, where B is a constant to be found.

b Hence find $\int \sin^6\theta \, d\theta$

4 a Show that $2\sin^3\theta \equiv \frac{3}{2}\sin\theta - \frac{1}{2}\sin 3\theta$

b Hence solve the equation $3\sin\theta - \sin 3\theta = \frac{1}{2}$ for $-\pi \le \theta \le \pi$

5 a Show that $5\cos^4\theta \equiv A\cos 4\theta + B\cos 2\theta + C$, where A, B and C are constants to be found.

b Hence solve the equation $\cos 4\theta + 4\cos 2\theta + 3 = 2$ for $-\pi \le \theta \le \pi$

6 Use de Moivre's theorem to prove the following identities.

a $\sin 2\theta \equiv 2\cos\theta\sin\theta$ b $\sin 3\theta \equiv 3\sin\theta - 4\sin^3\theta$

c $\cos 3\theta \equiv 4\cos^3\theta - 3\cos\theta$ d $\sin 4\theta \equiv 4\cos\theta\sin\theta - 8\cos\theta\sin^3\theta$

7 Prove the identities

a $\cos 6\theta \equiv 32\cos^6\theta - 48\cos^4\theta + 18\cos^2\theta - 1$ b $\sin 6\theta \equiv 2\sin\theta\cos\theta(16\sin^4\theta - 16\sin^2\theta + 3)$

8 a Use de Moivre's theorem to show that $\cos 5\theta \equiv 16\cos^5\theta - 20\cos^3\theta + 5\cos\theta$

b Hence find 3 solutions to the equation $16x^5 - 20x^3 + 5x = 1$

9 a Use de Moivre's theorem to show that $\cos 4\theta \equiv 8\cos^4\theta - 8\cos^2\theta + 1$

b Hence find 4 solutions to the equation $x^4 - x^2 = -\frac{1}{16}$

10 Use de Moivre's theorem to show that $\tan 2\theta \equiv \frac{2\tan\theta}{1-\tan^2\theta}$

11 Use proof by induction to prove that $[r(\cos\theta + i\sin\theta)]^n = r^n(\cos n\theta + i\sin n\theta)$ for all positive integers n

12 a Given that $z = \cos\theta + i\sin\theta$, use de Moivre's theorem to show that $2\cos(n\theta) = z^n + \frac{1}{z^n}$

b Hence show that $4\cos\theta\sin^2\theta \equiv \cos\theta - \cos(3\theta)$

13 a Given that $z = \cos\theta + i\sin\theta$, use de Moivre's theorem to show that $2i\sin(n\theta) = z^n - \frac{1}{z^n}$

b Hence show that $16\sin^3\theta\cos^2\theta \equiv 2\sin\theta + \sin(3\theta) - \sin(5\theta)$

14 Show that $\sum_{r=1}^{n}\cos(r\theta) = \frac{\cos\left(\frac{(n+1)\theta}{2}\right)\sin\left(\frac{n\theta}{2}\right)}{\sin\left(\frac{\theta}{2}\right)}$

16.3 Roots of unity

Fluency and skills

A fundamental rule in maths is that an equation of order n must have n solutions. So, the equation $z^3 = 1$ must have three roots so there must be three cube roots of 1. One of them is real (1) and two are complex. These are called **the cube roots of unity**.

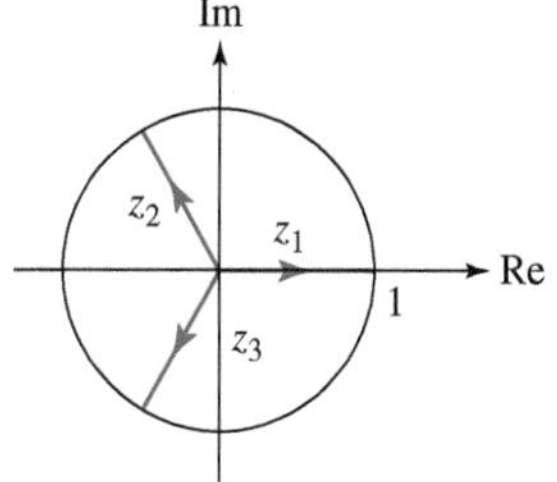

This is easily seen in the case of $z^4 = 1$. Square root both sides to give $z^2 = \pm 1$

If $z^2 = 1$ then $z = 1$ or $z = -1$

If $z^2 = -1$ then $z = \mathrm{i}$ or $z = -i$

So there are four solutions because the equation has order 4
You can use the exponential form of a complex number to solve equations of the form $z^n = 1$ to find the ***n*th roots of unity**.

Example 1

a Find all the solutions to the equation $z^6 = 1$, giving your answers in Cartesian form.

b Illustrate the 6th roots of unity on an Argand diagram.

a $z^6 = e^{2k\pi i}$

Since the modulus is 1 and the argument of any positive real number is 0 or $\pm 2\pi$ or $\pm 4\pi$ and so on, in general, this could be written as $2k\pi$, where k is an integer.

$z = (e^{2k\pi i})^{\frac{1}{6}}$

$= e^{\frac{2k\pi}{6}i}$

Use the index law.

Consider each of the possible values of k from 0 to 5. Going beyond 5 would just repeat the same values again.

$k = 0: z = e^0 = 1$

$k = 1: z = e^{\frac{2\pi}{6}i} = \frac{1}{2} + \frac{\sqrt{3}}{2}i$

$k = 2: z = e^{\frac{4\pi}{6}i} = -\frac{1}{2} + \frac{\sqrt{3}}{2}i$

$k = 3: z = e^{\frac{6\pi}{6}i} = -1$

$k = 4: z = e^{\frac{8\pi}{6}i} = -\frac{1}{2} - \frac{\sqrt{3}}{2}i$

$k = 5: z = e^{\frac{10\pi}{6}i} = \frac{1}{2} - \frac{\sqrt{3}}{2}i$

There should be 6 solutions to the equation.

(*Continued on the next page*)

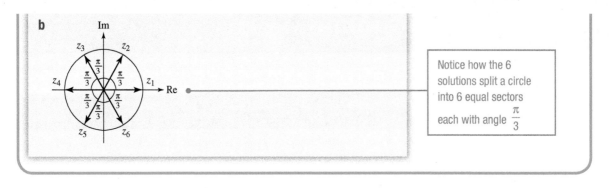

Key point

The equation $z^n = 1$ has n solutions of the form $z = e^{\frac{2k\pi i}{n}}$, where $k = 0, 1, 2, \ldots, n$

This method can be extended to find the nth roots of any complex number.

Example 2

Solve the equation $z^4 = 2\sqrt{3} - 2i$, giving your answers in the form $re^{i\theta}$ where $r > 0$ and $-\pi < \theta \le \pi$

$|2\sqrt{3} - 2i| = \sqrt{(2\sqrt{3})^2 + 2^2} = 4$

$\arg(2\sqrt{3} - 2i) = -\frac{\pi}{6}$

Calculate the modulus and argument.

So in, general, the argument is $-\frac{\pi}{6} + 2k\pi = \frac{(12k-1)\pi}{6}$

$z^4 = 4e^{\frac{(12k-1)\pi}{6}i}$

Write a general term in exponential form.

$z = \left(4e^{\frac{(12k-1)\pi}{6}i}\right)^{\frac{1}{4}}$

$= \sqrt{2}e^{\frac{(12k-1)\pi}{24}i}$

Consider each of the possible values of k from 0 to 3

$k = 0: z = \sqrt{2}e^{-\frac{\pi}{24}i}$

$k = 1: z = \sqrt{2}e^{\frac{11\pi}{24}i}$

$k = 2: z = \sqrt{2}e^{\frac{23\pi}{24}i}$

$k = 3: z = \sqrt{2}e^{\frac{35\pi}{24}i}$

The final solution is not in the interval $-\pi < \theta \le \pi$

So instead write $z = \sqrt{2}e^{-\frac{13\pi}{24}i}$

Since $-\frac{13\pi}{24} = \frac{35\pi}{24}$

Exercise 16.3A Fluency and skills

PURE

1 a Solve the equation $z^3 = 1$, giving your answers in Cartesian form.

b Illustrate the cube roots of unity on an Argand diagram.

2 a Solve the equation $z^4 = 1$, giving your answers in Cartesian form.

b Illustrate the 4th roots of unity on an Argand diagram.

3 a Solve the equation $z^5 = 1$, giving your answers in exponential form

b Illustrate the 5th roots of unity on an Argand diagram.

4 Solve each of these equations, giving your solutions in Cartesian form.

a $z^3 = 8$

b $z^3 = i$

c $z^4 = -49$

d $z^3 = -27i$

e $z^6 = -8$

5 Solve each of these equations, giving your solutions in exponential form.

a $z^4 = -16i$ b $z^5 = 32i$

6 Solve each of these equations, giving your solutions in exponential form

a $z^3 = 4\sqrt{2} + 4\sqrt{2}i$

b $z^3 = -4\sqrt{2} + 4\sqrt{2}i$

c $z^3 = -4\sqrt{2} - 4\sqrt{2}i$

d $z^3 = 4\sqrt{2} - 4\sqrt{2}i$

7 Solve each of these equations, giving your solutions in modulus–argument form with θ given to 2 decimal places.

a $z^4 = 3\sqrt{5} + 6i$

b $z^4 = -3\sqrt{5} + 6i$

c $z^4 = -6 - 3\sqrt{5}i$

d $z^4 = 6 - 3\sqrt{5}i$

8 Solve each of these equations, giving your solutions in exponential form.

a $z^5 + 243 = 0$

b $32z^5 + 1 = 0$

c $z^6 + 64i = 0$

d $4z^4 + 3i = 4$

9 Solve each of these equations, giving your solutions in Cartesian form.

a $z^4 = 8 + 8\sqrt{3}i$

b $z^4 = -8 + 8\sqrt{3}i$

10 Solve the equation $z^{\frac{3}{2}} = (2-2i)^2$, giving your answers in the form $a + bi$

11 a Solve the equation $z^4 = -32 + 32\sqrt{3}i$, giving your solution in Cartesian form.

b Represent the solutions on an Argand diagram.

12 Solve each of these equations, giving your solutions in the form $re^{i\theta}$ where $r > 0$ and $-\pi < \theta \leq \pi$

a $z^3 = 4\sqrt{3} + 4i$

b $z^4 = 3\sqrt{2} - 3\sqrt{2}i$

13 Solve each of these equations, giving your solutions in the form $r(\cos\theta + i\sin\theta)$, where $r > 0$ and $-\pi < \theta \leq \pi$

a $z^3 = \sqrt{2} - \sqrt{6}i$

b $z^6 = -4\sqrt{3} + 4i$

Reasoning and problem-solving

The equation $z^n = 1$ has n solutions of the form $z = \mathrm{e}^{\frac{2k\pi i}{n}}$, where $k = 0, 1, 2, \ldots, n-1$ Using index laws, you could write this as $z = \left(\mathrm{e}^{\frac{2\pi i}{n}}\right)^k$. Therefore, if ω is a complex solution of $z^n = 1$, then $1, \omega, \omega^2, \omega^3, \ldots, \omega^{n-1}$ are all distinct solutions.

$$1+\omega+\omega^2+\omega^3+\ldots+\omega^{n-1} = \left(\mathrm{e}^{\frac{2\pi i}{n}}\right)^0 + \left(\mathrm{e}^{\frac{2\pi i}{n}}\right)^1 + \left(\mathrm{e}^{\frac{2\pi i}{n}}\right)^2 + \left(\mathrm{e}^{\frac{2\pi i}{n}}\right)^3 + \ldots + \left(\mathrm{e}^{\frac{2\pi i}{n}}\right)^{n-1}$$

This is a geometric series with first term 1 and common ratio $\omega = \mathrm{e}^{\frac{2\pi i}{n}}$

So, using the formula $S_n = \dfrac{a(1-r^n)}{1-r}$, you can see that the sum of the series is $\dfrac{1\left(1-\left(\mathrm{e}^{\frac{2\pi i}{n}}\right)^n\right)}{1-\mathrm{e}^{\frac{2\pi i}{n}}}$

This simplifies to $\dfrac{1(1-\mathrm{e}^{2\pi i})}{1-\mathrm{e}^{\frac{2\pi i}{n}}} = 0$ since $\mathrm{e}^{2\pi i} = 1$

Key point

Therefore, $1+\omega+\omega^2+\ldots+\omega^{n-1} = 0$ where ω is an nth root of unity.

You can quote this result.

Strategy 1

To find complex solutions of an equation

1. Expand brackets or simplify fractions.
2. Use fact that $\omega^n = 1$ if ω is an nth root of unity.
3. Use $1+\omega+\omega^2+\ldots+\omega^{n-1} = 0$

Example 3

You are given that ω is a complex 4th root of unity.

a Show that $1+\omega+\omega^2+\omega^3 = 0$

b Evaluate $(1+\omega)(1+\omega^2)+\omega^2$

a This is a geometric series with first term 1 and common ratio ω

So $1+\omega+\omega^2+\omega^3 = \dfrac{1(1-\omega^4)}{1-\omega}$ — Since $S_n = \dfrac{a(1-r^n)}{1-r}$ for a geometric series.

$= \dfrac{1-\omega^4}{1-\omega}$

$= \dfrac{1-1}{1-\omega}$ — (2) Since $\omega^4 = 1$, as ω is a complex 4th root of unity.

$= 0$ as required

b $(1+\omega)(1+\omega^2)+\omega^4 = 1+\omega+\omega^2+\omega^3+\omega^4$ — (1)

$= 1+\omega+\omega^2+\omega^3+1$ — (2) Since $\omega^4 = 1$

$= 0+1$ — (3)

$= 1$

Strategy 2

To solve geometric problems

1. Write z^n in exponential form.
2. Find the roots of an equation.
3. Sketch an Argand diagram.
4. Use Pythagoras' theorem or trigonometry to find lengths and areas.

Example 4

The points *A*, *B* and *C* represent the solutions to the equation $z^3 = 8i$

Calculate the exact area and perimeter of triangle *ABC*

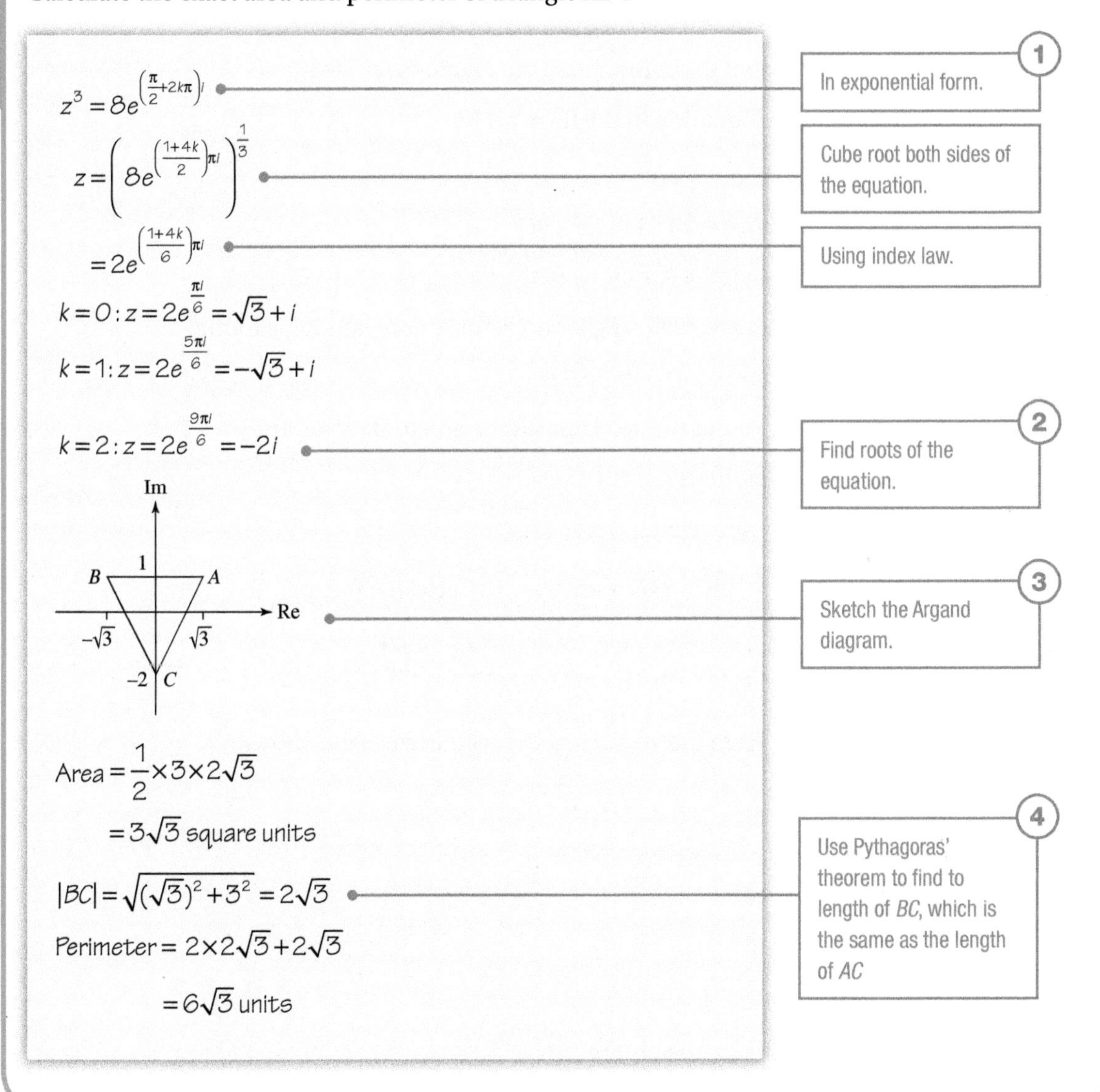

$z^3 = 8e^{\left(\frac{\pi}{2}+2k\pi\right)i}$
In exponential form.
$z = \left(8e^{\left(\frac{1+4k}{2}\right)\pi i}\right)^{\frac{1}{3}}$
Cube root both sides of the equation.
$= 2e^{\left(\frac{1+4k}{6}\right)\pi i}$
Using index law.
$k = 0: z = 2e^{\frac{\pi i}{6}} = \sqrt{3} + i$
$k = 1: z = 2e^{\frac{5\pi i}{6}} = -\sqrt{3} + i$
$k = 2: z = 2e^{\frac{9\pi i}{6}} = -2i$
Find roots of the equation.
Im
Re
B
A
C
1
$-\sqrt{3}$
$\sqrt{3}$
-2
Sketch the Argand diagram.
$\text{Area} = \frac{1}{2} \times 3 \times 2\sqrt{3}$
$= 3\sqrt{3}$ square units
$|BC| = \sqrt{(\sqrt{3})^2 + 3^2} = 2\sqrt{3}$
$\text{Perimeter} = 2 \times 2\sqrt{3} + 2\sqrt{3}$
$= 6\sqrt{3}$ units
Use Pythagoras' theorem to find to length of *BC*, which is the same as the length of *AC*

PURE

Exercise 16.3B Reasoning and problem-solving

1 Given that ω is a complex cube root of unity

a Show that $1+\omega+\omega^2=0$

b Evaluate the following expressions.

i $(1+\omega)^2-\omega$ ii $(1+\omega)(1+\omega^2)$ iii $\omega(\omega+1)$ iv $\frac{2\omega+1}{\omega-1}+\omega$

2 Given that ω is a complex 5th root of unity

a Show that $1+\omega+\omega^2+\omega^3+\omega^4=0$

b Evaluate the following expressions.

i $\omega(1+\omega)(1+\omega^2)$ ii $\frac{\omega^2}{\omega+1}+\omega^3+1$ iii $\omega(1+\omega+\omega^2+\omega^3)$

3 The points *A*, *B* and *C* represent the solutions to the equation $z^3=-27i$

a Find the solutions to the equation in the form $a+bi$

b Calculate the exact

i Area, ii Perimeter of triangle *ABC*

4 The points *A*, *B* and *C* represent the solutions to the equation $z^3=-125$

a Show that the area of triangle *ABC* is $k\sqrt{3}$, where k is a constant to be found,

b Calculate the exact perimeter of triangle *ABC*

5 The points *A*, *B*, *C* and *D* represent the solutions to the equation $z^4=-4i$

a State the name of the quadrilateral *ABCD*

b Calculate the area of the quadrilateral *ABCD*

6 a Find the fourth roots of $\frac{1}{9}$, giving your answers in Cartesian form.

b Show that the points representing these roots form a square.

c Find the area of the square.

7 The points *A*, *B*, *C*, *D*, *E* and *F* are the vertices of the regular hexagon shown.

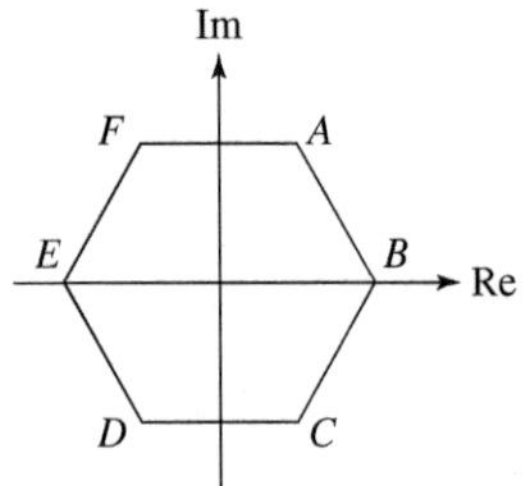

Given that *A*, *B*, *C*, *D*, *E* and *F* are all solutions to $z^n=1$

a State the value of n

b Find the coordinates of *F*

c Calculate the area of the hexagon.

16 Summary and review

Chapter summary

- A complex number $z = a + bi$ can be written
  - in modulus–argument form as $z = r(\cos\theta + i\sin\theta)$
  - in exponential form as $z = re^{i\theta}$

  where $r = |z|$ and $\theta = \arg z$, $-\pi < \theta \leq \pi$
- For complex numbers z and w
  - $|zw| = |z||w|$
  - $\left|\frac{z}{w}\right| = \frac{|z|}{|w|}$
  - $\arg(zw) = \arg z + \arg w$
  - $\arg\left(\frac{z}{w}\right) = \arg z - \arg w$
- Euler's formula is $re^{i\theta} = r(\cos\theta + i\sin\theta)$
- $\cos\theta = \frac{e^{i\theta} + e^{-i\theta}}{2}$ and $\sin\theta = \frac{e^{i\theta} - e^{-i\theta}}{2i}$
- De Moivre's theorem states that $[r(\cos\theta + i\sin\theta)]^n = r^n(\cos(n\theta) + i\sin(n\theta))$ for all integers, n
- The equation $z^n = 1$ has n solutions of the form $z = e^{\frac{2k\pi i}{n}}$, where $k = 0, 1, 2, \ldots, n$
- Equations of the form $z^n = a + bi$ have solutions of the form $z = r^{\frac{1}{n}} e^{\frac{(\theta + 2k\pi)i}{n}}$ where $|z| = r$, $\arg(z) = \theta$ and $k = 0, 1, 2, \ldots, n$
- $1 + \omega + \omega^2 + \ldots + \omega^{n-1} = 0$ where ω is an nth root of unity.

Check and review

You should now be able to...	Review
✔ Write complex numbers in exponential form.	1, 2
✔ Convert complex numbers from exponential to Cartesian form.	3
✔ Use rules to find the argument and modulus of products of complex numbers.	4
✔ Use rules to find the argument and modulus of quotients of complex numbers.	5
✔ Use de Moivre's theorem to simplify powers of complex numbers.	6, 7
✔ Use de Moivre's theorem to prove trigonometric identities.	8
✔ Simplify powers of trigonometric functions using a binomial expansion.	9
✔ Calculate roots of unity.	10
✔ Calculate the nth root of a complex number.	11–12

1 Write these complex numbers in exponential form.

a $-3i$
b $1+i$
c 5
d $-\sqrt{3}-i$
e $\sqrt{2}-i$
f $-1+\sqrt{3}i$

2 Write each of these complex numbers in exponential form

a $3\left(\cos\frac{\pi}{7}+i\sin\frac{\pi}{7}\right)$

b $\sqrt{2}\left(\cos\frac{\pi}{9}+i\sin\frac{\pi}{9}\right)$

c $\sqrt{3}\left(\cos\frac{\pi}{8}-i\sin\frac{\pi}{8}\right)$

d $5\left(\cos\left(-\frac{\pi}{5}\right)-i\sin\left(-\frac{\pi}{5}\right)\right)$

3 Write these complex numbers in the form $a+bi$

a $7e^{\left(\frac{\pi}{2}\right)i}$
b $6e^{\left(\frac{\pi}{3}\right)i}$
c $\sqrt{3}e^{\left(-\frac{\pi}{6}\right)i}$
d $\sqrt{2}e^{\left(\frac{3\pi}{4}\right)i}$
e $\sqrt{6}e^{\left(-\frac{2\pi}{3}\right)i}$
f $2\sqrt{3}e^{\left(\frac{5\pi}{6}\right)i}$

4 Find the argument and modulus of zw in each case.

a $z=3\left(\cos\frac{\pi}{5}+i\sin\frac{\pi}{5}\right)$ and
$w=5\left(\cos\frac{\pi}{7}+i\sin\frac{\pi}{7}\right)$

b $z=\sqrt{2}\left(\cos\left(-\frac{\pi}{8}\right)+i\sin\left(-\frac{\pi}{8}\right)\right)$ and
$w=\sqrt{6}\left(\cos\frac{\pi}{3}+i\sin\frac{\pi}{3}\right)$

c $z=1+\sqrt{3}i$ and $w=3-3i$

d $z=\frac{1}{2}e^{\frac{2\pi}{9}i}$ and $w=4e^{\frac{\pi}{3}i}$

5 Find the argument and modulus of $\frac{z}{w}$ in each case.

a $z=5\left(\cos\frac{\pi}{2}+i\sin\frac{\pi}{2}\right)$ and
$w=10\left(\cos\frac{\pi}{4}+i\sin\frac{\pi}{4}\right)$

b $z=\sqrt{15}\left(\cos\left(-\frac{3\pi}{4}\right)+i\sin\left(-\frac{3\pi}{4}\right)\right)$ and
$w=\sqrt{5}\left(\cos\left(-\frac{\pi}{8}\right)+i\sin\left(-\frac{\pi}{8}\right)\right)$

c $z=-5+5i$ and $w=\sqrt{6}-3\sqrt{2}i$

d $z=16e^{-\frac{2\pi}{11}i}$ and $w=\sqrt{2}e^{\frac{5\pi}{11}i}$

6 Given that $z=8\left(\cos\left(\frac{\pi}{2}\right)+i\sin\left(\frac{\pi}{2}\right)\right)$, write these powers of z in Cartesian form.

a z^2
b z^3

7 Given that $w=-2\sqrt{3}-2i$, express w^2 in the form $a+bi$

8 a Use de Moivre's theorem to show that $\sin 5\theta \equiv 5\sin\theta - 20\sin^3\theta + 16\sin^5\theta$

b Hence find 3 solutions to the equation $5x-20x^3+16x^5=0$
Give your answers to 3 significant figures.

9 Prove that $\cos^3\theta \equiv A(\cos 3\theta + 3\cos\theta)$, where A is a constant to be found.

10 a Calculate the 8th roots of unity. Give your answers in exact Cartesian form.

b Draw the roots on an Argand diagram.

11 Solve these equations, giving your answers in Cartesian form.

a $z^8=16$
b $z^3=i$
c $z^2=-9i$
d $z^6=-125$

12 Solve the equation $z^4=-2\sqrt{2}-2\sqrt{2}i$, giving your solutions in the form $re^{i\theta}$ where $r>0$ and $-\pi<\theta\leq\pi$

Exploration

16

Going beyond the exams

History

Abraham de Moivre was a French mathematician (1667-1754). He was a contemporary and friend of Isaac Newton and Edmund Halley (the astronomer after whom Halley's comet is named). As well as his work on de Moivre's formula, he wrote a major work on probability *The Doctrine of Chances*.

Halley suggested that de Moivre looked at the astronomical world, and so he also worked on the mathematical ideas associated with centripetal force.

Note

Complex numbers are widely used across many areas of engineering, particularly in electrical/electronic engineering. Engineers use 'j' rather than 'i' to represent the imaginary part of a complex number, so a complex number is written, for example, as $1 + j$

Investigation

Investigate the nth roots of unity for $n = 1, 2, 3, 4, 5, \dots$

Use a graph plotting package such as GeoGebra to plot these.

- What do you know about the results?
- How does it relate to what you know about geometry?
- What happens to the distance between roots as the series develops?
- What are the connections with series?
- What are the connections with calculus?

Research

The hyperbolic functions $f(x) = \sinh(x)$, $g(x) = \cosh(x)$ are complex analogues of the circular functions $f(x) = \sin(x)$ and $g(x) = \cos(x)$.

Research how the hyperbolic functions relate to the circular functions, in terms of their relationships with circles and hyperbolas.

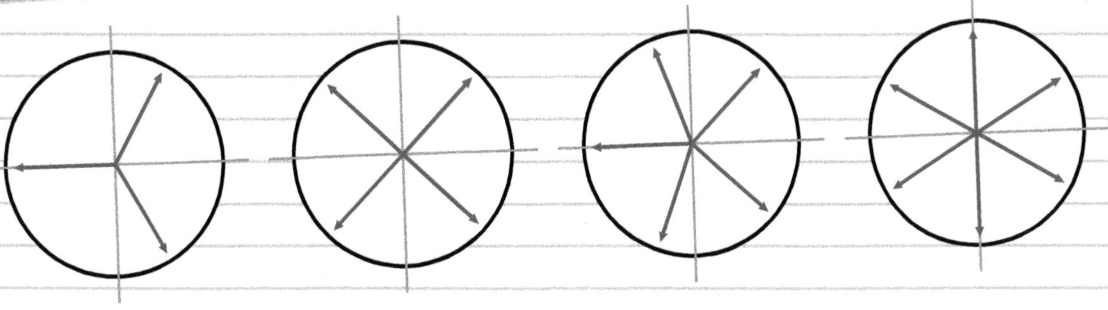

16 Assessment

1 The complex number z is defined by $z=3-\sqrt{3}i$

a Write z in the form $re^{i\theta}$ where r is given as a surd in its simplest form and θ is given as a multiple of π **[4 marks]**

b Hence work out z^4, giving your answer in modulus-argument form. You must show your working. **[2]**

2 The complex number z is defined as $z=-k+k\sqrt{3}i$

a Find the modulus and argument of z in terms of k where appropriate, **[4]**

b Given that $k=2$, use de Moivre's theorem to express z^4 in the form $a+bi$, where a and b are exact constants to be found. **[4]**

3 a Solve the equation $z^3=-i$, giving your answers in the form $e^{i\theta}$, where $-\pi<\theta\leq\pi$ **[4]**

b Sketch the solutions on an Argand diagram. **[2]**

4 Given that $z=-2+2i$

a Write z in modulus-argument form, **[2]**

b Hence use de Moivre's theorem to find z^{-3} in the form $a+bi$, where a and b are constants to be found. **[3]**

5 Express $(1+\sqrt{3}i)^4$ in exact Cartesian form. You must show your working. **[4]**

6 Use de Moivre's theorem to prove that $(\cos\theta-i\sin\theta)^n=\cos(n\theta)-i\sin(n\theta)$ **[4]**

7 a Show that $\sin 5\theta\equiv 5\sin\theta-20\sin^3\theta+16\sin^5\theta$ **[5]**

b Hence solve the equation $5\sin\theta-20\sin^3\theta+16\sin^5\theta=0.5$ for θ in the range $0<\theta\leq 180°$ **[3]**

8 a Prove that $32\cos^5 x=2\cos(5x)+10\cos(3x)+20\cos(x)$ **[5]**

b Hence find $\int\cos^5 x\,dx$ **[2]**

9 a Show that $\sin 3\theta=3\cos^2\theta\sin\theta-\sin^3\theta$ **[4]**

b Hence show that $\tan 3\theta\equiv\dfrac{3\tan\theta-\tan^3\theta}{1-3\tan^2\theta}$ **[4]**

c Find the exact value of $\tan 3\theta$, given that $\tan\theta=-2\sqrt{3}$ **[2]**

10 a Find the value of n such that $(\sqrt{2})^n=8$ **[2]**

b Solve the equation $z^6+8=0$, giving your answers in the form $r(\cos\theta+i\sin\theta)$, where $r>0$ is an exact surd and $-\pi<\theta\leq\pi$ **[5]**

11 Use exponentials to prove that $\cos(2x)=1-2\sin^2 x$ **[4]**

12 A complex number is given as $z=2e^{\frac{\pi}{2}i}$

A second complex number is $w=a+bi$, where $\arg(w)=\frac{\pi}{4}$

Calculate the exact value of

a $\arg(zw)$ b $\arg\left(\frac{z}{w}\right)$

c $\arg(w^n)$ d $\arg\left(\frac{w}{z^2}\right)$ **[7]**

13 a Use de Moivre's theorem to show that

$\cos 7\theta=\cos\theta\,(64\cos^6\theta-112\cos^4\theta+56\cos^2\theta-7)$ **[5]**

b Hence solve the equation $64x^6-112x^4+56x^2=7$
Give your answers to 3 significant figures. **[4]**

14 Solve the equation $z^8+4\sqrt{12}+8i=0$, giving your answers the form $re^{i\theta}$ where $r>0$ and $-\pi<\theta\leq\pi$ **[5]**

15 a Prove that $\sin^4\theta\equiv\frac{1}{8}(\cos(4\theta)-4\cos(2\theta)+3)$ **[4]**

b Hence find $\int 8\sin^4\theta\,d\theta$ **[3]**

16 a Solve the equation $z^4=-6-6\sqrt{3}i$, giving your answers in modulus-argument form, where r is given to 3 significant figures and θ is given as an exact multiple of π **[5]**

b Hence solve the equation $(z-1)^4=-6-6\sqrt{3}i$, giving your answers in Cartesian form to 2 significant figures. **[3]**

17 a Use de Moivre's theorem to show that if $z=\cos\theta+i\sin\theta$, then

$z^n-\frac{1}{z^n}\equiv 2i\sin(n\theta)$ **[4]**

b Hence show that $4\sin^3\theta\equiv 3\sin\theta-\sin(3\theta)$ **[4]**

c Solve the equation $2\sin(3\theta)-6\sin\theta=1$ for θ in the range $0<\theta\leq 2\pi$ **[4]**

18 a Illustrate the solutions to $z^5=1$ on an Argand diagram, **[2]**

b Given that ω is a complex 5th root of unity, evaluate

i $\omega(1+\omega)(1+\omega^2)$ ii $(1+\omega^5)^2$ **[5]**

19 a Use de Moivre's theorem to show that $\cos 5\theta\equiv 16\cos^5\theta-20\cos^3\theta+5\cos\theta$ **[5]**

b Hence find the general form of all the solutions to the equation $\cos 5\theta=5\cos\theta$ **[4]**

20 a Show that $\omega=\cos\left(\frac{2\pi}{7}\right)+i\sin\left(\frac{2\pi}{7}\right)$ is one of the seventh roots of unity, **[3]**

b Write down the other non-real roots of the equation $z^7=1$ in terms of ω **[1]**

c State the value of $\sum_{r=1}^{6}\omega^r$ **[1]**

21 This question refers to the equation $(z-1)^3=1$

a Verify that $z_1=\frac{1}{2}+\frac{\sqrt{3}}{2}i$ is a root of the equation, **[3]**

b Find the other two roots of the equation, **[2]**

c Produce an Argand diagram showing the three roots of the equation, **[2]**

d State the centre and radius of the circle on which the three roots lie. **[2]**

22 a Solve the equation $z^4 + 16 = 0$

Give your answers in the form $a + b\text{i}$ where a and b are real numbers. **[5]**

b The points A, B, C and D in the complex plane represent these four roots.

Find the area of the square $ABCD$ **[2]**

23 a Prove that $\sin 5\theta = 16\sin^5\theta - 20\sin^3\theta + 5\sin\theta$ **[5]**

b Use the result in part **a** to solve the equation $16x^5 - 20x^3 + 5x = 1$ **[4]**

Give your solutions in the form $x = \sin k\pi$, where $0 \le k < 2$

24 a Solve the equation $(z-1)^3 = 8$

Give your solutions in the form $a + bi$, where a and b are real numbers. **[4]**

The points A, B, C in the complex plane represent these three roots.

b Draw the points A, B and C on an Argand diagram. **[2]**

c Calculate the area of the triangle ABC **[3]**

d The triangle ABC is rotated $\frac{\pi}{3}$ radians anticlockwise around the origin.

Find the complex numbers that represent the vertices of the triangle after this rotation. **[4]**

25 a Use de Moivre's theorem to show that $z = \frac{1}{2}(1+i)$ is a solution to the equation $\left(\frac{z-1}{z}\right)^6 = -1$ **[5]**

b Find the other 5 solutions to the equation.

Give your answers in the form $a + bi$ where a and b are real numbers. **[6]**

The root $z = \frac{1}{2}(1+i)$ is represented by the point A in the complex plane.

c Find the coordinates of the image of A following a clockwise rotation of $\frac{\pi}{6}$ radians around the origin. **[3]**

26 a Solve the equation $z^3 = 4 - 4\sqrt{3}i$

Give your answers in the form $re^{i\theta}$, where $r > 0$ and $-\pi < \theta \le \pi$ **[6]**

b The roots of the equation $z^3 = 4 - 4\sqrt{3}i$ are represented by the points A, B and C on an Argand diagram.

Calculate the exact area of the triangle ABC **[3]**

c Show that $\sin\left(\frac{\pi}{9}\right) - \sin\left(\frac{5\pi}{9}\right) + \sin\left(\frac{7\pi}{9}\right) = 0$ **[3]**

27 Use proof by induction to show that de Moivre's theorem $(\cos\theta + i\sin\theta)^n = \cos(n\theta) + i\sin(n\theta)$ holds for all integers n (Hint: consider separately the cases when n is positive or negative.) **[11]**

17 Series

Around the world, nations are increasingly moving away from fossil fuels and towards sources of renewable energy, such as wind power. This helps to minimise pollution and depletion of the world's natural resources. The move has led to rapid growth in the number of modern wind turbines that we now see on our skyline, both on land and out at sea. In some places several turbines are clustered together to form wind farms.

The technology that underpins this 'harvesting of the wind' is developed by engineers from across disciplines and, in particular, brings together the expertise of both mechanical and electrical engineers. The design of this technology relies on the mathematics of series.

Orientation

What you need to know

Ch2 Algebra and series

- The method of differences.

What you will learn

- To use partial fractions to sum series.
- To expand functions using the Maclaurin series.
- To evaluate limits using L'Hopital's rule.

What this leads to

Careers

- Mechanical engineering.
- Electrical engineering.

17.1 Summing series using partial fractions

Fluency and skills

If the general term of a function can be expressed as $f(r+1)-f(r)$, you can find the sum of the series using the **method of differences**.

You can prove $\sum_{1}^{n} r \equiv \frac{n(n+1)}{2}$ using the formula for an arithmetic progression.

Example 1

By considering the identity $2r \equiv r(r+1) - r(r-1)$, use the method of differences to prove that $\sum_{1}^{n} r \equiv \frac{n(n+1)}{2}$

$$2r \equiv r(r+1) - r(r-1)$$

$$\begin{aligned}
\text{Hence } 2\sum_{1}^{n} r &\equiv 2 - 0\\
&+ 6 - 2\\
&+ 12 - 6\\
&\vdots\\
&+ (n-1)(n) - (n-1)(n-2)\\
&+ n(n+1) - n(n-1)
\end{aligned}$$

Write the 'differences' for successive terms vertically and cancel terms wherever possible.

$$2\sum_{1}^{n} r \equiv n(n+1) - 0 \equiv n(n+1)$$

$$\text{So } \sum_{1}^{n} r \equiv \frac{n(n+1)}{2}$$

Collect up the remaining terms and simplify.

In order to express a function as $f(r+1)-f(r)$ you may need to use partial fractions.

You can decompose some functions into their partial fractions.

Start by checking that the degree of the numerator is smaller than the degree of the denominator.

Then split into partial fractions.

Key point

$$\frac{px+q}{(x-a)(x-b)} \equiv \frac{A}{(x-a)} + \frac{B}{(x-b)}$$

You can work out the constants A, B, etc. using substitution or by comparing coefficients (or a combination of these methods).

Example 2

PURE

a Find the partial fractions of $\frac{1}{r(r+2)}$

b Use your answer to find $\sum_1^n \frac{1}{r(r+2)}$

a Let $\frac{1}{r(r+2)} \equiv \frac{A}{r} + \frac{B}{(r+2)}$

So $1 \equiv A(r+2) + Br$

Multiply through by the denominator.

When $r = 0$: $1 = 2A$ so $A = \frac{1}{2}$

When $r = -2$: $1 = -2B$ so $B = -\frac{1}{2}$

Substitute values for r to work out A and B

Hence $\frac{1}{r(r+2)} \equiv \frac{1}{2r} - \frac{1}{2(r+2)} = \frac{1}{2}\left(\frac{1}{r} - \frac{1}{(r+2)}\right)$

b $\sum_1^n \frac{1}{r(r+2)} \equiv \frac{1}{2}\sum_1^n \left(\frac{1}{r} - \frac{1}{(r+2)}\right)$

$$\begin{aligned}
\equiv \frac{1}{2}\Bigg[& 1 - \frac{1}{3} \\
& + \frac{1}{2} - \frac{1}{4} \\
& + \frac{1}{3} - \frac{1}{5} \\
& \downarrow \qquad \downarrow \\
& + \frac{1}{(n-2)} - \frac{1}{n} \\
& + \frac{1}{(n-1)} - \frac{1}{(n+1)} \\
& + \frac{1}{n} - \frac{1}{(n+2)} \Bigg]
\end{aligned}$$

Write the 'differences' for successive terms vertically and eliminate where possible.

This time the terms that cancel are two rows apart.

$$\equiv \frac{1}{2}\left[1 + \frac{1}{2} - \frac{1}{(n+1)} - \frac{1}{(n+2)}\right]$$

$$\equiv \frac{3}{4} - \frac{(n+2)+(n+1)}{2(n+1)(n+2)}$$

$$\equiv \frac{n(3n+5)}{4(n+1)(n+2)}$$

The expression can be factorised and written as a single fraction.

Exercise 17.1A Fluency and skills

1 Express these functions as the sum of their partial fractions.

a $\frac{6x+10}{(x-5)(x+5)}$

b $\frac{-4x}{(x-3)(x-7)}$

c $\frac{7x+56}{(1-x)(x+6)}$

d $\frac{27x+46}{(x-2)(x+2)(x+3)}$

2 $f(x) \equiv \frac{1}{x(x-1)}$

a Show that $f(x) - f(x+1) \equiv \frac{2}{x(x-1)(x+1)}$

b Use the method of differences to find $\sum_{2}^{n} \frac{1}{x(x-1)(x+1)}$

3 Find the sum of each series.

a $\sum_{1}^{n} \frac{1}{r+1} - \frac{1}{r+3}$

b $\sum_{1}^{n} \frac{1}{2r-1} - \frac{1}{2r+1}$

c $\sum_{0}^{n} \frac{1}{2r+1} - \frac{2}{2r+3} + \frac{1}{2r+5}$

4 Use partial fractions to find these sums.

a $\sum_{2}^{n} \frac{1}{r(r-1)}$

b $\sum_{1}^{n} \frac{1}{(r+1)(r+2)}$

c $\sum_{1}^{n} \frac{1}{(r+1)(r+3)}$

d $\sum_{2}^{n} \frac{1}{(r-1)(r+1)}$

Reasoning and problem-solving

Strategy

To sum a series using partial fractions and the method of differences

1. Express the function using partial fractions.
2. Use the method of differences to sum the series.
3. If necessary, consider what happens for large values of n to work out the sum of an infinite number of terms.

Example 3

Express $\dfrac{16}{r(r+2)(r+4)}$ in partial fractions.

Hence show that $\sum_{r=1}^{\infty} \dfrac{16}{r(r+2)(r+4)} = \dfrac{11}{6}$

$$\frac{16}{r(r+2)(r+4)} \equiv \frac{A}{r} + \frac{B}{r+2} + \frac{C}{r+4}$$

$16 \equiv A(r+2)(r+4) + B(r)(r+4) + C(r)(r+2)$

$r=0$: $16 = 8A$ so $A = 2$

$r=-2$: $16 = -4B$ so $B = -4$

$r=-4$: $16 = 8C$ so $C = 2$

So, $\dfrac{16}{r(r+2)(r+4)} \equiv \dfrac{2}{r} - \dfrac{4}{r+2} + \dfrac{2}{r+4}$

$$\begin{aligned}\sum_{r=1}^{n} \frac{16}{r(r+2)(r+4)} \equiv\ & \frac{2}{1} - \frac{4}{3} + \frac{2}{5} \\ & + \frac{2}{2} - \frac{4}{4} + \frac{2}{6} \\ & + \frac{2}{3} - \frac{4}{5} + \frac{2}{7} \\ & + \frac{2}{4} - \frac{4}{6} + \frac{2}{8} \\ & + \frac{2}{5} - \frac{4}{7} + \frac{2}{9} \\ & + \frac{2}{6} - \frac{4}{8} + \frac{2}{10} \\ & \quad\downarrow \quad\quad \downarrow \quad\quad \downarrow \\ & + \frac{2}{n-4} - \frac{4}{n-2} + \frac{2}{n} \\ & + \frac{2}{n-3} - \frac{4}{n-1} + \frac{2}{n+1} \\ & + \frac{2}{n-2} - \frac{4}{n} + \frac{2}{n+2} \\ & + \frac{2}{n-1} - \frac{4}{n+1} + \frac{2}{n+3} \\ & + \frac{2}{n} - \frac{4}{n+2} + \frac{2}{n+4}\end{aligned}$$

$$\sum_{r=1}^{n} \frac{16}{r(r+2)(r+4)} \equiv \frac{2}{1} - \frac{4}{3} + \frac{2}{2} - \frac{4}{4} + \frac{2}{3} + \frac{2}{4} + \frac{2}{n+1} + \frac{2}{n+2} - \frac{4}{n+1} + \frac{2}{n+3} - \frac{4}{n+2} + \frac{2}{n+4}$$

$$\equiv \frac{11}{6} - \frac{2}{n+1} - \frac{2}{n+2} + \frac{2}{n+3} + \frac{2}{n+4}$$

As $n \to \infty$, each of $\dfrac{2}{n+1}, \dfrac{2}{n+2}, \dfrac{2}{n+3}$ and $\dfrac{2}{n+4} \to 0$

Hence $\sum_{1}^{\infty} \dfrac{16}{r(r+2)(r+4)} = \dfrac{11}{6}$

1 Multiply through by the denominator.

Substitute values of r to eliminate factors.

2 Write the 'differences' for successive terms vertically and eliminate where possible.

This time the terms that cancel are several rows apart.

3 Consider what happens for large n

Exercise 17.1B Reasoning and problem-solving

1 a Show that
$r(r+1)(r+2)-(r-1)r(r+1)\equiv 3r(r+1)$

b Hence find $\sum_1^n r(r+1)$

2 a Find the partial fractions of $\frac{1}{4r^2-1}$

b Hence find $\sum_1^n \frac{1}{4r^2-1}$

c Find the sum to infinity of this series.

3 $f(r)=\frac{1}{(r+1)(r+2)}$

a Show that $f(r)-f(r+1)\equiv\frac{2}{(r+1)(r+2)(r+3)}$

b Hence find $\sum_1^n \frac{1}{(r+1)(r+2)(r+3)}$

c Find the sum to infinity of this series.

4 a Simplify the expression $r^2(r+1)^2-r^2(r-1)^2$

b Use your answer to find a formula for $\sum_1^n r^3$

5 a Use partial fractions to find $\sum_2^n \frac{4}{x^2-1}$

b Use your result to find the maximum value of $\sum_2^n \frac{1}{x^2-1}$ for $n\geq 2$

6 The rth term of a series, S, is $\frac{2r-1}{r(r+1)(r+2)}$

Show that $\sum_1^\infty S=\frac{3}{4}$

7 a Show that $\frac{x-1}{x}-\frac{x-2}{x-1}\equiv\frac{1}{x(x-1)}$

b Use your answer to find a formula for $\sum_3^n \frac{1}{x(x-1)}$

c Hence evaluate $\sum_3^\infty \frac{1}{x(x-1)}$

8 Find a formula for $\sum_2^n \ln\left(\frac{r}{r-1}\right)$

9 a Express $\frac{x-3}{(x-1)(x)(x+1)}$ in its partial fractions.

b Hence find $\sum_2^n \frac{x-3}{(x-1)(x)(x+1)}$

c Hence show that the sum to infinity is 0

10 Use partial fractions to show that

$$\sum_1^\infty\left[\frac{1}{r(r+1)(r+2)(r+3)}\right]=\frac{1}{18}$$

11 a Express $\frac{1}{(1-x)(1+2x)}$ in its partial fractions.

b Hence find the first four terms in the binomial expansion of $\frac{\sqrt{1+x}}{(1-x)(1+2x)}$

17.2 Maclaurin series 2

Fluency and skills

A function f(x) can be expanded using the Maclaurin series given that

- f(x) can be expanded as a **convergent** infinite series of terms
- each of the terms in f(x) can be differentiated
- each of the differentiated terms has a finite value when $x = 0$

A convergent series is one where an infinite number of terms has a finite sum.

Key point

The Maclaurin series, or expansion, for f(x) is

$$f(x) \equiv f(0) + xf'(0) + \frac{x^2}{2!}f''(0) + \frac{x^3}{3!}f'''(0) + \frac{x^4}{4!}f''''(0) + \ldots + \frac{x^r}{r!}f^{(r)}(0) + \ldots$$

Here are the range of values of x for which the Maclaurin series is valid for different functions.

Function	Range of x where series is valid
e^x	all values of x
$\sin x$	all values of x
$\cos x$	all values of x
$(1+x)^n$	$-1 < x < 1$ for $n \in \mathbb{R}$
$\ln(1+x)$	$-1 < x \le 1$

You came across the series for these functions in Chapter 2, but now you should be able to derive them yourself.

Example 1

a Explain why a Maclaurin series of $f(x) = \ln(x)$ is not possible.

b Derive the Maclaurin series of $f(x) = \ln(1+x)$

c The first three terms of the series for $\tan x$ are $x + \frac{x^3}{3} + \frac{2x^5}{15}$. Use this series and your answer to part **b** to find the expansion of $\tan(2x)\ln(1+x)$ as far as the term in x^5

a The first constant $f(0) = \ln(0) = -\infty$, which is not finite.

Hence a Maclaurin series of $y = \ln(x)$ is not possible.

b $f(x) = \ln(1+x)$ so $f(0) = \ln(1) = 0$

$f'(x) = (1+x)^{-1} = \frac{1}{1+x}$, so $f'(0) = 1$ — Use the chain rule to differentiate f(x)

$f''(x) = -(1+x)^{-2}$ so $f''(0) = -1$

$f'''(x) = 2(1+x)^{-3}$ so $f'''(0) = 2$

$f''''(x) = -6(1+x)^{-4}$ so $f''''(0) = -6 = -3!$

$f'''''(x) = 24(1+x)^{-5}$ so $f'''''(0) = 4!$ — The pattern is now clear and can be proved, for example, by induction.

(*Continued on the next page*)

$$f(x) \equiv f(0) + xf'(0) + \frac{x^2}{2!}f''(0) + \frac{x^3}{3!}f'''(0) + \frac{x^4}{4!}f''''(0) + \ldots$$

Use the Maclaurin expansion.

$$f(x) = \ln(1+x) \equiv x - \frac{x^2}{2!} + \frac{2x^3}{3!} - \frac{6x^4}{4!} + \frac{24x^5}{5!} - \ldots$$

$$\equiv x - \frac{x^2}{2} + \frac{x^3}{3} - \frac{x^4}{4} + \frac{x^5}{5} - \ldots$$

c $$\tan(2x)\ln(1+x) \equiv \left(2x + \frac{(2x)^3}{3} + \frac{2(2x)^5}{15} + \ldots\right) \times \left(x - \frac{x^2}{2} + \frac{x^3}{3} - \frac{x^4}{4} + \frac{x^5}{5} - \ldots\right)$$

$$\equiv \left(2x + \frac{8x^3}{3} + \frac{64x^5}{15} + \ldots\right)\left(x - \frac{x^2}{2} + \frac{x^3}{3} - \frac{x^4}{4} + \frac{x^5}{5} - \ldots\right)$$

$$\equiv 2x^2 + x^3\left[2x - \frac{1}{2}\right] + x^4\left[\frac{2}{3} + \frac{8}{3}\right] + x^5\left[-\frac{1}{2} - \frac{8}{6}\right] + \ldots$$

You only need to multiply the terms as far as x^5

$$\equiv 2x^2 - x^3 + \frac{10}{3}x^4 - \frac{11}{6}x^5 + \ldots$$

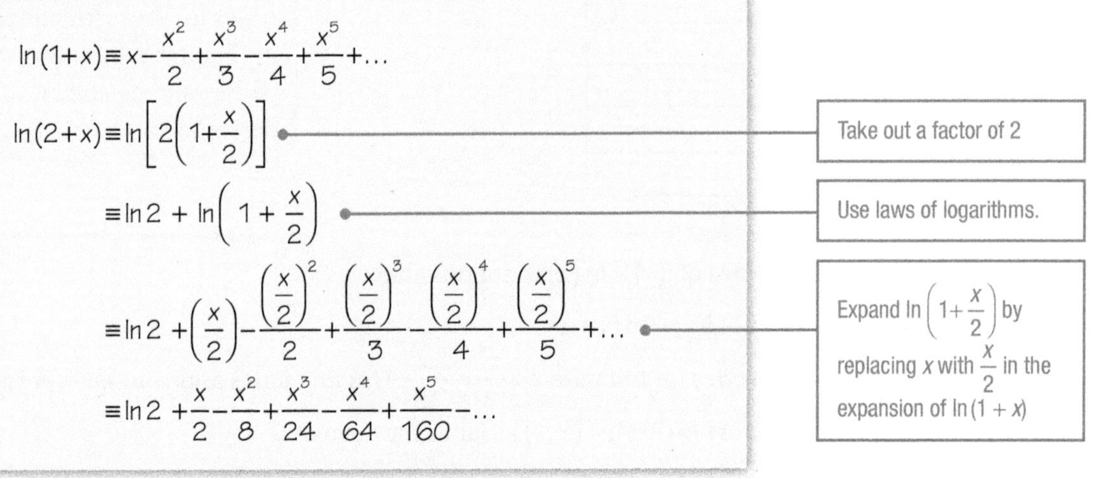

Example 2

Derive the Maclaurin series of $f(x) = \ln(2+x)$ by adapting the series for $\ln(1+x)$ you found in Example 1

$$\ln(1+x) \equiv x - \frac{x^2}{2} + \frac{x^3}{3} - \frac{x^4}{4} + \frac{x^5}{5} + \ldots$$

$$\ln(2+x) \equiv \ln\left[2\left(1 + \frac{x}{2}\right)\right]$$

Take out a factor of 2

$$\equiv \ln 2 + \ln\left(1 + \frac{x}{2}\right)$$

Use laws of logarithms.

$$\equiv \ln 2 + \left(\frac{x}{2}\right) - \frac{\left(\frac{x}{2}\right)^2}{2} + \frac{\left(\frac{x}{2}\right)^3}{3} - \frac{\left(\frac{x}{2}\right)^4}{4} + \frac{\left(\frac{x}{2}\right)^5}{5} + \ldots$$

Expand $\ln\left(1 + \frac{x}{2}\right)$ by replacing x with $\frac{x}{2}$ in the expansion of $\ln(1+x)$

$$\equiv \ln 2 + \frac{x}{2} - \frac{x^2}{8} + \frac{x^3}{24} - \frac{x^4}{64} + \frac{x^5}{160} - \ldots$$

General term

See Ch2.5

For the general terms of some well-known Maclaurin series.

Sequences and series can be represented by a formula for the general term, sometimes called the rth term or nth term. For example, the formula $3r - 2$ represents the sequence of terms 1, 4, 7, 10, ...

Sometimes you will only know the first few terms of a sequence, and need to find the rth term. You do this by looking for patterns which you can write as a formula.

PURE

Example 3

Use your knowledge of these Maclaurin series to write down their general terms.

a e^{2x} **b** $\ln(1+3x)$

a $e^x = 1 + x + \frac{x^2}{2!} + \ldots + \frac{x^r}{r!} + \ldots$

So the general term of the sequence for e^{2x} is $\frac{(2x)^r}{r!}$ or $\frac{2^r x^r}{r!}$.

b $\ln(1+x) = x - \frac{x^2}{2} + \frac{x^3}{3} - \ldots + (-1)^{r+1}\frac{x^r}{r} + \ldots$

So the rth term for $\ln(1+3x)$ is $(-1)^{r+1}\frac{(3x)^r}{r}$ or $(-1)^{r+1}\frac{3^r x^r}{r}$

Example 4

a $f(x) \equiv e^x \cos x$

i Show that $f''(x) = -2e^x \sin x$

ii Find $f'''(x)$ and $f''''(x)$

b Use the Maclaurin series to find the first four non-zero terms in the expansion of $f(x)$

a i $f(x) \equiv e^x \cos x$

$f'(x) = e^x(\cos x - \sin x)$

Use the product rule to differentiate.

$f''(x) = e^x(\cos x - \sin x - \sin x - \cos x) = -2e^x \sin x$

ii $f'''(x) = -2e^x(\sin x + \cos x)$

$f''''(x) = -2e^x(\sin x + \cos x + \cos x - \sin x) = -4e^x \cos x$

b $f(x) \equiv f(0) + xf'(0) + \frac{x^2}{2!}f''(0) + \frac{x^3}{3!}f'''(0) + \frac{x^4}{4!}f''''(0) + \ldots$

$\equiv e^0 \cos 0 + xe^0(\cos 0 - \sin 0) + \frac{x^2}{2!}(-2\, e^0 \sin 0)$

Substitute values for f(0), f′(0), etc.

$+ \frac{x^3}{3!}[-2e^0(\sin 0 + \cos 0)] + \frac{x^4}{4!}(-4e^0 \cos 0) + \ldots$

$f(x) \equiv 1 + x(1) + \frac{x^2}{2!}(0) + \frac{x^3}{3!}(-2) + \frac{x^4}{4!}(-4) + \ldots$

$\equiv 1 + x - \frac{x^3}{3} - \frac{x^4}{6} + \ldots$

Limits

A limiting value, or limit, is a specific value that a function approaches or tends towards as the variable approaches a particular value. This idea is used when differentiating from first principles.

The derivative of a function f(x) is defined as $f'(x)=\lim_{h\to 0}\frac{f(x+h)-f(x)}{h}$

It gives the gradient of the curve at any value of x

In this case $f'(x)$ is the limit as h tends to zero, but you can also find a limit approaching other numbers, for example, 0, 1 or $\frac{\pi}{2}$

You can sometimes easily work out the limit of a function

$$\lim_{n\to 0}(2+e^n)=2+e^0=2+1=3 \qquad \lim_{n\to\frac{\pi}{2}}(1-\cos n)=1-\cos\frac{\pi}{2}=1-0=1$$

In other cases, you may need to manipulate the function to find the limit. For example, when finding $\lim_{n\to\infty}\frac{n+10}{n}$ you must manipulate the expression before you find the limit

$$\lim_{n\to\infty}\frac{n+10}{n}=\lim_{n\to\infty}\frac{n}{n}+\frac{10}{n}=\lim_{n\to\infty}\frac{n}{n}+\frac{10}{n}=1+0=1$$

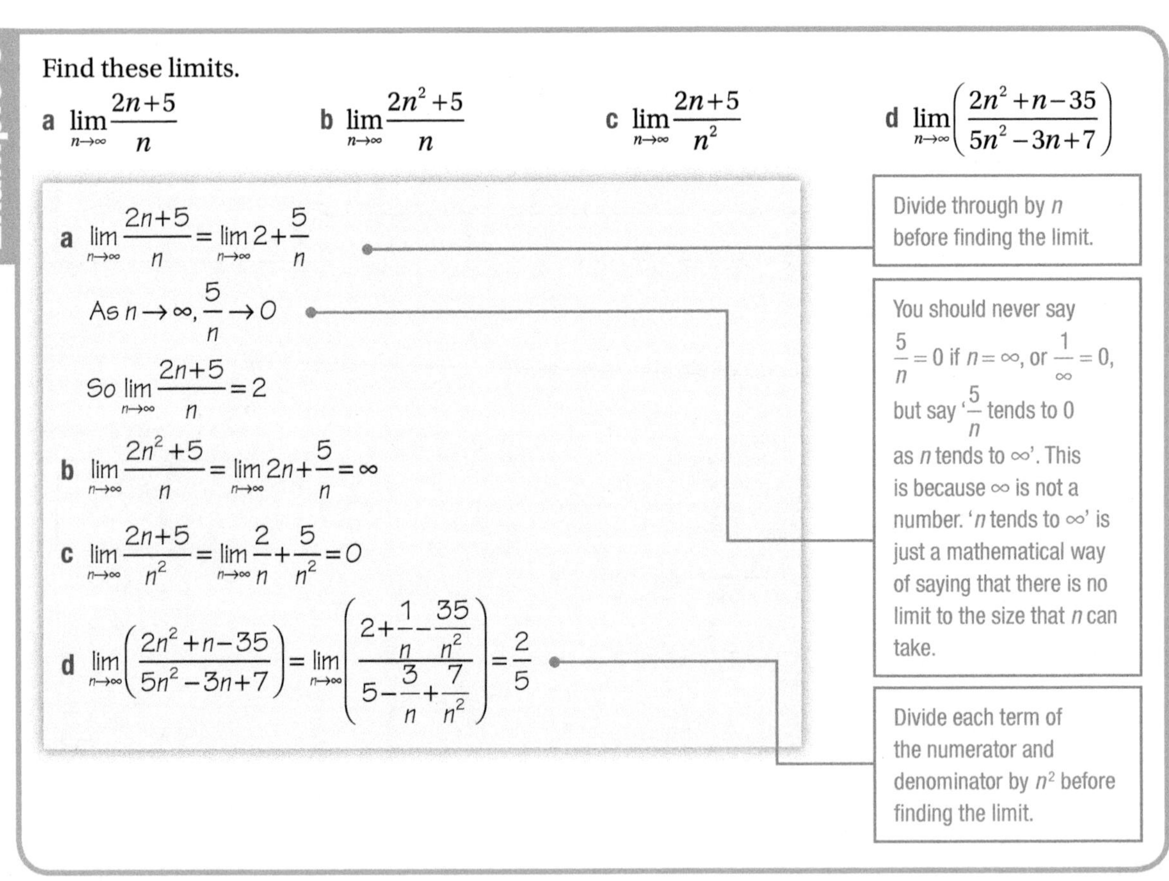

Example 5

Find these limits.

a $\lim_{n\to\infty}\frac{2n+5}{n}$ **b** $\lim_{n\to\infty}\frac{2n^2+5}{n}$ **c** $\lim_{n\to\infty}\frac{2n+5}{n^2}$ **d** $\lim_{n\to\infty}\left(\frac{2n^2+n-35}{5n^2-3n+7}\right)$

a $\lim_{n\to\infty}\frac{2n+5}{n}=\lim_{n\to\infty}2+\frac{5}{n}$

As $n\to\infty,\ \frac{5}{n}\to 0$

So $\lim_{n\to\infty}\frac{2n+5}{n}=2$

b $\lim_{n\to\infty}\frac{2n^2+5}{n}=\lim_{n\to\infty}2n+\frac{5}{n}=\infty$

c $\lim_{n\to\infty}\frac{2n+5}{n^2}=\lim_{n\to\infty}\frac{2}{n}+\frac{5}{n^2}=0$

d $\lim_{n\to\infty}\left(\frac{2n^2+n-35}{5n^2-3n+7}\right)=\lim_{n\to\infty}\left(\frac{2+\frac{1}{n}-\frac{35}{n^2}}{5-\frac{3}{n}+\frac{7}{n^2}}\right)=\frac{2}{5}$

Divide through by n before finding the limit.

You should never say $\frac{5}{n}=0$ if $n=\infty$, or $\frac{1}{\infty}=0$, but say '$\frac{5}{n}$ tends to 0 as n tends to ∞'. This is because ∞ is not a number. 'n tends to ∞' is just a mathematical way of saying that there is no limit to the size that n can take.

Divide each term of the numerator and denominator by n^2 before finding the limit.

L'Hopital's rule

Sometimes when you investigate a limit you are not able to evaluate it in the same way as you did in Example 5

For example $\lim_{x\to 0}\left(\frac{\sin x}{x}\right)$ gives you $\frac{0}{0}$

The expression $\frac{\sin x}{x}$ is known as an **indeterminate form** when $x = 0$ and hence the limit is indeterminate.

Indeterminate forms occur when, for an expression of the form $\frac{f(x)}{g(x)}$, in the limit as $x \to a$,

$$\lim_{x\to a} f(x) = \lim_{x\to a} g(x) = 0$$

or

$$\lim_{x\to a} f(x) = \pm\infty \quad \text{and} \quad \lim_{x\to a} g(x) = \pm\infty$$

The French mathematician, **l'Hopital**, published a rule which allows you to evaluate limits like this. His rule says that when the limit of $\frac{f(x)}{g(x)}$ is indeterminate, it can usually be found by evaluating the limit of $\frac{f'(x)}{g'(x)}$

So $\quad \lim_{x\to 0}\left(\frac{\sin x}{x}\right) = \frac{0}{0} = \lim_{x\to 0}\left(\frac{\cos x}{1}\right) = \frac{1}{1} = 1$

Key point

L'Hopital's rule states that

If $\quad \lim_{x\to a}\frac{f(x)}{g(x)}$ is indeterminate, where a is any real number or $\pm\infty$,

then $\quad \lim_{x\to a}\frac{f(x)}{g(x)} = \lim_{x\to a}\frac{f'(x)}{g'(x)}$

This method can be repeated as many times as necessary if $\lim_{x\to a}\frac{f'(x)}{g'(x)}$, etc. is still indeterminate.

Example 6

Use l'Hopital's rule to evaluate these limits.

a $\lim_{x\to 1}\frac{x^3-1}{x^2-1}$ **b** $\lim_{x\to 0}\frac{(2-2\cos x)}{x^2}$

a $\lim_{x\to 1}\frac{x^3-1}{x^2-1} = \frac{0}{0}$

$\lim_{x\to 1}\frac{f'(x)}{g'(x)} = \lim_{x\to 1}\frac{3x^2}{2x} = \lim_{x\to 1}\frac{3x}{2} = \frac{3\times 1}{2} = \frac{3}{2}$ — Apply l'Hopital's rule.

Hence $\lim_{x\to 1}\frac{x^3-1}{x^2-1} = \frac{3}{2}$

b $\lim_{x\to 0}\frac{(2-2\cos x)}{x^2} = \frac{0}{0}$

$\lim_{x\to 0}\frac{f'(x)}{g'(x)} = \lim_{x\to 0}\frac{2\sin x}{2x} = \lim_{x\to 0}\frac{\sin x}{x} = \lim_{x\to 0}\frac{\sin 0}{0} = \frac{0}{0}$

$\lim_{x\to 0}\frac{f''(x)}{g''(x)} = \lim_{x\to 0}\frac{\cos x}{1} = \lim_{x\to 0}\frac{\cos 0}{1} = 1$ — Sometimes you have to apply l'Hopital's rule more than once.

Hence $\lim_{x\to 0}\frac{(2-2\cos x)}{x^2} = 1$

Exercise 17.2A Fluency and skills

1 Use your knowledge of these Maclaurin series to write down their general terms.

a e^{-x} b $\ln(1-2x)$

c $\sin\left(\frac{x}{5}\right)$ d $(1+3x)^n$

2 Use the Maclaurin expansion to derive a series for $\sin x$. Include the general term.

3 Use the Maclaurin expansion to derive a series for $\cos x$. Include the general term.

4 Find the first three non-zero terms in the Maclaurin expansion of $f(x)=\ln\left(\frac{1-2x}{1+2x}\right)$

5 Derive the first three non-zero terms in the Maclaurin expansion of $f(x)=\cos^2 x$

6 Derive the first three non-zero terms in the Maclaurin expansion of $f(x)=\sqrt{1+3x^2}$

7 Find the first three non-zero terms in the Maclaurin expansion of $f(x)=e^{2x}\sin x$

8 Find the first three non-zero terms in the Maclaurin expansion of $f(x)=e^x\ln(1+x)$

9 Evaluate these limits.

a $\lim\limits_{x\to\infty}(1+e^{-x})$ b $\lim\limits_{x\to 1}\frac{3x+5}{x^2-9}$

c $\lim\limits_{x\to 0}\frac{3\cos x}{\sin x+\cos x}$ d $\lim\limits_{x\to\infty}\frac{x^2+2x+3}{2x^2-5}$

e $\lim\limits_{x\to 0}\frac{x^2+1}{3x^2+2x-4}$ f $\lim\limits_{x\to\infty}\frac{x^2+1}{3x^2+2x-4}$

g $\lim\limits_{x\to\infty}\frac{3x^2+8}{2-3x^2}$ h $\lim\limits_{x\to -1}\frac{3x^2+8}{2-3x^2}$

10 Evaluate these limits.

a $\lim\limits_{x\to 2}\sqrt{\frac{x^3-2x^2+135}{9x^3-12}}$

b $\lim\limits_{x\to\infty}\sqrt{\frac{x^3-2x^2+1}{9x^3+4}}$

c $\lim\limits_{x\to 1}\frac{3x+2}{x^2-9}$

d $\lim\limits_{x\to 0}\frac{x+2}{\sqrt{2x^2-3x+1}}$

e $\lim\limits_{x\to\infty}\frac{x+2}{\sqrt{2x^2-3x+1}}$

11 Use l'Hopital's method to evaluate these limits.

a $\lim\limits_{x\to 0}\frac{\sin^{-1}x}{x}$

b $\lim\limits_{x\to 0}\frac{\tan^{-1}x}{x}$

c $\lim\limits_{x\to\infty}\frac{\ln(1+2x)}{x}$

d $\lim\limits_{x\to\infty}\frac{e^{2x}+1}{e^{2x}-1}$

e $\lim\limits_{x\to 0}\frac{x}{x+1-\sqrt{x+1}}$

f $\lim\limits_{x\to 0}\frac{1-\cos 3x+2x\sin 5x}{6x^2}$

g $\lim\limits_{x\to 0}\frac{e^x\sin 5x}{\ln(2x+1)}$

PURE

Reasoning and problem-solving

When you investigate a limit and end up with an indeterminate value such as $\frac{0}{0}$ or $\frac{\infty}{\infty}$ you can use Maclaurin expansions to evaluate the limit as an alternative to using l'Hopital's rule.

Strategy

To solve problems involving Maclaurin series and limits

1. Derive the Maclaurin series from first principles or adapt a standard Maclaurin series.
2. Substitute the series into the expression.
3. Use the Maclaurin series or l'Hopital's rule to work out the limit of the function.

Example 7

$$f(x) \equiv \frac{4x^4}{3\ln(1+x^2)-3x^2}$$

Adapt the known Maclaurin series for $\ln(1+x)$ in $f(x)$ to calculate $\lim_{x\to 0} f(x)$

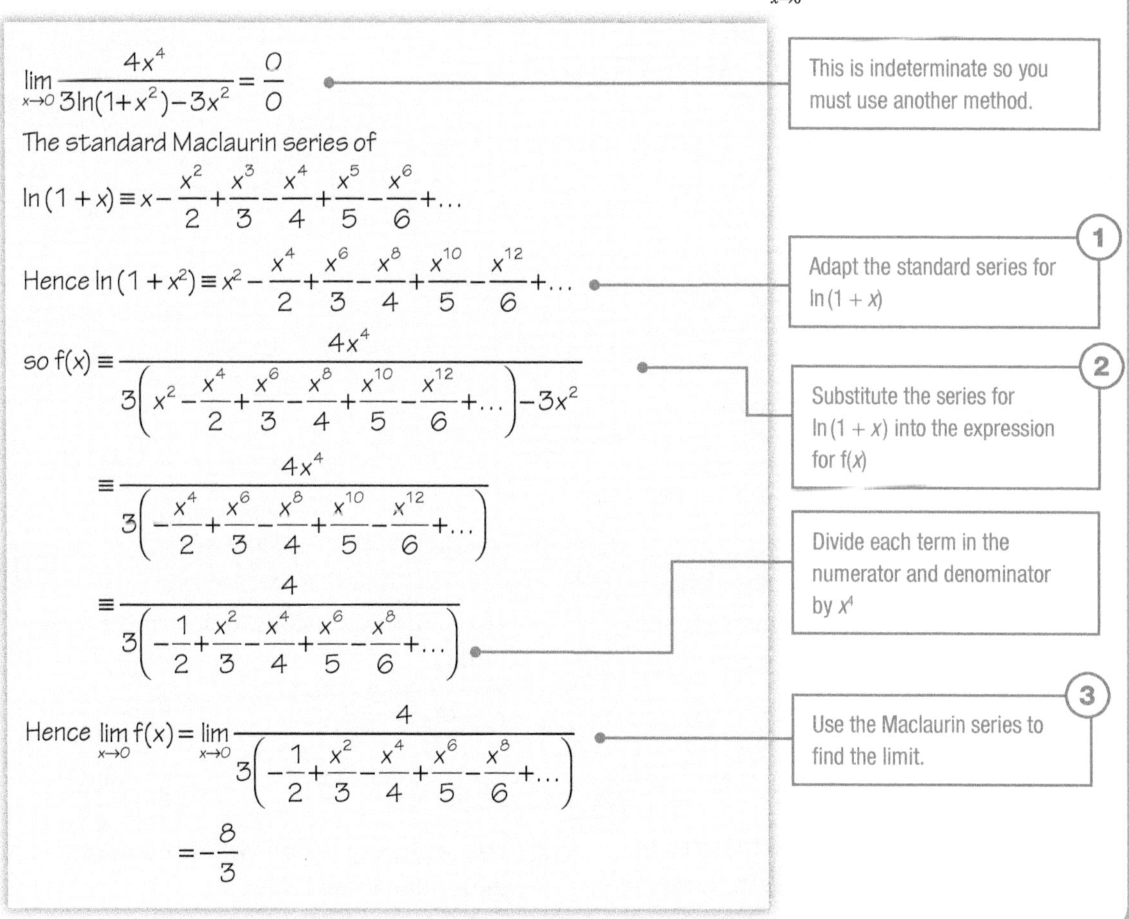

$$\lim_{x\to 0}\frac{4x^4}{3\ln(1+x^2)-3x^2}=\frac{0}{0}$$

This is indeterminate so you must use another method.

The standard Maclaurin series of

$$\ln(1+x)\equiv x-\frac{x^2}{2}+\frac{x^3}{3}-\frac{x^4}{4}+\frac{x^5}{5}-\frac{x^6}{6}+\dots$$

$$\text{Hence } \ln(1+x^2)\equiv x^2-\frac{x^4}{2}+\frac{x^6}{3}-\frac{x^8}{4}+\frac{x^{10}}{5}-\frac{x^{12}}{6}+\dots$$

1 Adapt the standard series for $\ln(1+x)$

$$\text{so } f(x)\equiv\frac{4x^4}{3\left(x^2-\frac{x^4}{2}+\frac{x^6}{3}-\frac{x^8}{4}+\frac{x^{10}}{5}-\frac{x^{12}}{6}+\dots\right)-3x^2}$$

2 Substitute the series for $\ln(1+x)$ into the expression for $f(x)$

$$\equiv\frac{4x^4}{3\left(-\frac{x^4}{2}+\frac{x^6}{3}-\frac{x^8}{4}+\frac{x^{10}}{5}-\frac{x^{12}}{6}+\dots\right)}$$

$$\equiv\frac{4}{3\left(-\frac{1}{2}+\frac{x^2}{3}-\frac{x^4}{4}+\frac{x^6}{5}-\frac{x^8}{6}+\dots\right)}$$

Divide each term in the numerator and denominator by x^4

$$\text{Hence } \lim_{x\to 0}f(x)=\lim_{x\to 0}\frac{4}{3\left(-\frac{1}{2}+\frac{x^2}{3}-\frac{x^4}{4}+\frac{x^6}{5}-\frac{x^8}{6}+\dots\right)}$$

$$=-\frac{8}{3}$$

3 Use the Maclaurin series to find the limit.

Example 8

$$f(x) = \frac{e^{2x} - e^x}{x}$$

a Find the first four terms and the general term of the expansion of f(x)

b Use your expansion to find $\lim_{x \to 0} f(x)$

c Use l'Hopital's rule to confirm your answer to part **b**.

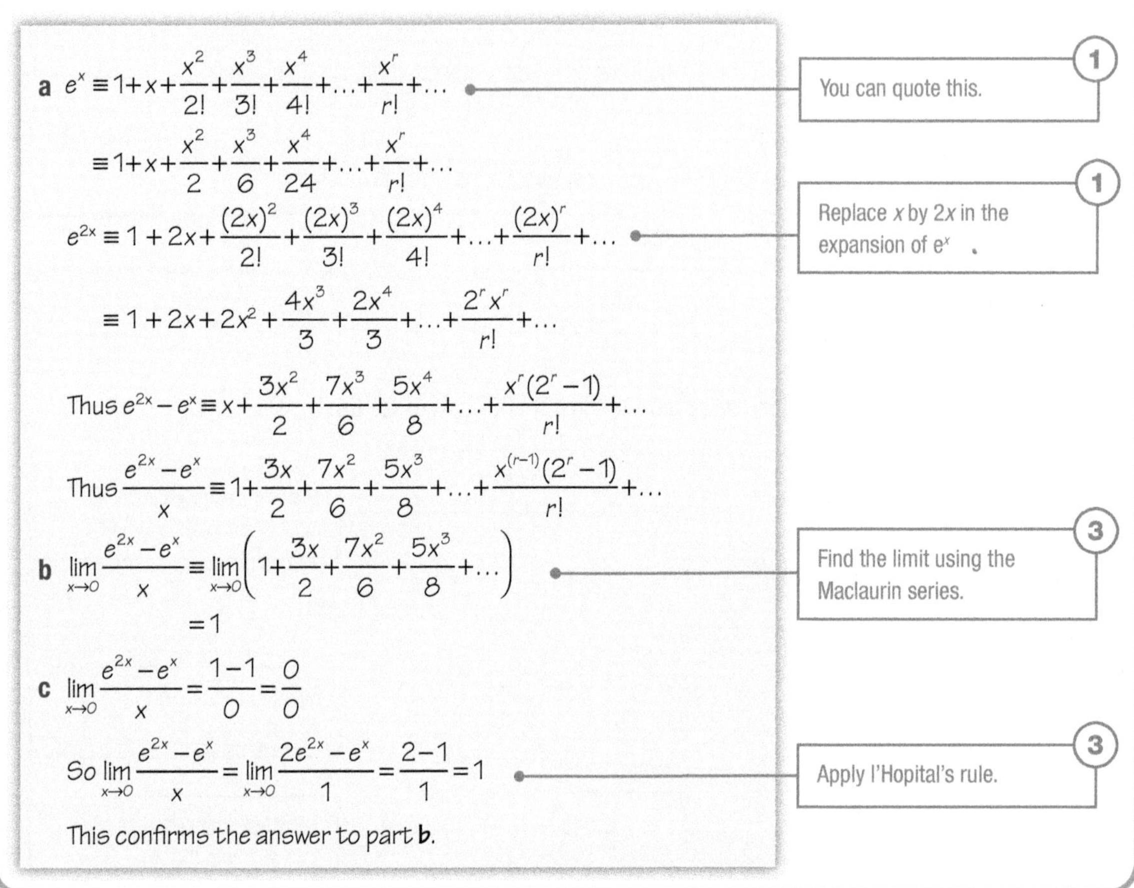

a $e^x \equiv 1+x+\frac{x^2}{2!}+\frac{x^3}{3!}+\frac{x^4}{4!}+...+\frac{x^r}{r!}+...$
$\equiv 1+x+\frac{x^2}{2}+\frac{x^3}{6}+\frac{x^4}{24}+...+\frac{x^r}{r!}+...$
$e^{2x} \equiv 1+2x+\frac{(2x)^2}{2!}+\frac{(2x)^3}{3!}+\frac{(2x)^4}{4!}+...+\frac{(2x)^r}{r!}+...$
$\equiv 1+2x+2x^2+\frac{4x^3}{3}+\frac{2x^4}{3}+...+\frac{2^r x^r}{r!}+...$
Thus $e^{2x}-e^x \equiv x+\frac{3x^2}{2}+\frac{7x^3}{6}+\frac{5x^4}{8}+...+\frac{x^r(2^r-1)}{r!}+...$
Thus $\frac{e^{2x}-e^x}{x} \equiv 1+\frac{3x}{2}+\frac{7x^2}{6}+\frac{5x^3}{8}+...+\frac{x^{(r-1)}(2^r-1)}{r!}+...$
b $\lim_{x\to 0}\frac{e^{2x}-e^x}{x} \equiv \lim_{x\to 0}\left(1+\frac{3x}{2}+\frac{7x^2}{6}+\frac{5x^3}{8}+...\right)$
$=1$
c $\lim_{x\to 0}\frac{e^{2x}-e^x}{x}=\frac{1-1}{0}=\frac{0}{0}$
So $\lim_{x\to 0}\frac{e^{2x}-e^x}{x}=\lim_{x\to 0}\frac{2e^{2x}-e^x}{1}=\frac{2-1}{1}=1$
This confirms the answer to part b.
You can quote this.
1
Replace x by 2x in the expansion of e^x.
1
Find the limit using the Maclaurin series.
3
Apply l'Hopital's rule.
3

Exercise 17.2B Reasoning and problem-solving

1 a Use the binomial theorem to expand $\frac{4}{1+4x}$

b $\int_0^x \frac{4}{1+4y}\,dy = \ln(1+4x)$

Use this and your answer to part **a** to obtain the first four terms in the series of $\ln(1+4x)$

c Check your solution by using the Maclaurin series for $\ln(1+4x)$

2 Use series expansions to determine these limits.

a $\lim_{x\to 0}\frac{(1+2x)^{-3}-1}{x}$ b $\lim_{x\to 0}\frac{x}{\sin 2x}$

c $\lim_{x\to 0}\frac{1-\cos 4x}{x^2}$ d $\lim_{x\to 0}\frac{x\ln(1-x)}{e^{x^2}-1}$

3 Use series expansions where necessary to determine these limits.

a $\lim_{x\to\infty}(x-\sqrt{x^2-4x})$

b $\lim_{x\to\infty}(\sqrt[3]{(x^3-2)}-x)$

c $2\lim_{x\to\infty}\ln\left(\frac{x^4+3}{x^4+2}\right)$

d $\lim_{x\to0}\frac{e^{2x}-3}{4e^{2x}+6}$

e $\lim_{x\to\infty}\frac{e^{2x}-3}{4e^{2x}+6}$

4 a Find the first four non-zero terms in the expansion of $\frac{\ln(1+x)}{1-x}$

b i Find the first four non-zero terms in the expansion of $\frac{\sqrt{(2-x)}}{1+2x}$

ii Write down $\lim_{x\to0}\frac{\sqrt{(2-x)}}{1+2x}$ and confirm the result using your answer from part **i**.

5 a Find the partial fractions of $\frac{3(1-2x)}{(x+2)(1+x)}$

b Hence expand $\frac{3(1-2x)}{(x+2)(1+x)}$ as a series as far as the term in x^3.

c Write down $\lim_{x\to0}\left(\frac{3(1-2x)}{(x+2)(1+x)}\right)$ and confirm the result using your answer from part **b**.

6 a Use standard Maclaurin expansions to find $\lim_{x\to0}\frac{e^x-e^{2x}}{x}$

b Confirm your answer by evaluating $\lim_{x\to0}\frac{e^x-e^{2x}}{x}$ using l'Hopital's rule.

7 a Use standard Maclaurin expansions to find the first three non-zero terms in the expansion of $(2-x)e^{(2-x)}$ and hence find $\lim_{x\to0}(2-x)e^{(2-x)}$

b Without using your answer to part **a** write down $\lim_{x\to0}(2-x)e^{(2-x)}$

8 a Find the first three non-zero terms in the expansion of $\frac{x}{e^x-1}$. Hence find $\lim_{x\to0}\frac{x}{e^x-1}$

b Use l'Hopital's rule to confirm your answer.

9 a Write down the first three non-zero terms of the expansion, in ascending powers of x, of $1+e^{-x}$

b Find the first two non-zero terms in the expansion, in ascending powers of x, of $\ln\left(\frac{1+e^{-x}}{2-3x}\right)$

c Find $\lim_{x\to0}\left(\frac{\ln\left(\frac{1+e^{-x}}{2-3x}\right)}{4x}\right)$

10 a Write down the first three non-zero terms in the expansions of e^{x^2} and $\sin 2x$

b Find the expansion of $\ln\left(\frac{\sin 2x}{2x}\right)$ as far as the term in x^4

c Evaluate $\lim_{x\to0}\frac{\ln\left(\frac{\sin 2x}{2x}\right)}{(e^{x^2}-1)}$

11 Make use of known series expansions to obtain the expansion of $\frac{1}{2}(e^x-e^{-x})$ up to the term in x^5

Hence evaluate $\lim_{x\to0}\frac{6x}{(e^x-e^{-x})}$

12 a Use Maclaurin series to expand $\ln\left(\frac{2+x}{1-x}\right)$ up to the term in x^3

b Write down $\lim_{x\to0}\left(\ln\left(\frac{2+x}{1-x}\right)\right)$ and confirm the result using your answer from part **a**.

13 a Show that

$$\ln(\cos x)\equiv-\frac{x^2}{2}-\frac{x^4}{12}-\frac{x^6}{45}+\ldots$$

b Hence find $\lim_{x\to0}\frac{\ln(\cos x)+x^2}{x^2}$

14 Evaluate $\lim_{x\to\infty}(x-\sqrt{x^2-3})$

[Hint: $A^2-B^2\equiv(A-B)(A+B)$]

17 Summary and review

Chapter summary

- If the general term of a function can be expressed as $f(r+1)-f(r)$, you can find the sum of the series using the method of differences.
- To express a function as $f(r+1)-f(r)$ you may need to use partial fractions.
- To express a function in partial fractions, check that the degree of the numerator is at least one lower than the denominator and then split it up using $\frac{px+q}{(x-a)(x-b)} \equiv \frac{A}{(x-a)} + \frac{B}{(x-b)}$
- You can use substitution, comparing coefficients, or a mixture of the two, to work out the constants A, B, etc.
- The Maclaurin series, or expansion, for a function $f(x)$ is
  $f(x) \equiv f(0) + xf'(0) + \frac{x^2}{2!}f''(0) + \frac{x^3}{3!}f'''(0) + \frac{x^4}{4!}f''''(0) + \ldots + \frac{x^r}{r!}f^{(r)}(0) + \ldots$
- Maclaurin series are valid for specific ranges of x

Function	Range of x where series is valid
e^x	all values of x
$\sin x$	all values of x
$\cos x$	all values of x
$(1+x)^n$	$-1 < x < 1$ for $n \in \mathbb{R}$
$\ln(1+x)$	$-1 < x \leq 1$

- A limiting value, or limit, is a specific value that a function approaches or tends towards as the variable approaches a particular value.
- L'Hopital's rule states that if $\lim_{x \to a} \frac{f(x)}{g(x)}$ is indeterminate, then $\lim_{x \to a} \frac{f(x)}{g(x)} = \lim_{x \to a} \frac{f'(x)}{g'(x)}$
  (where a is any real number, including 0 or $\pm\infty$).
- L'Hopital's rule can be applied to a function repeatedly.
- You can also use Maclaurin expansions to evaluate limits.

Check and review

You should now be able to...	Review Questions
✔ Express a function in terms of its partial fractions.	1, 2
✔ Find the sum of a series using the method of differences.	2, 3
✔ Derive the Maclaurin series for a function and find its range of validity.	4–8
✔ Find the limit of a function.	9
✔ Use l'Hopital's rule.	9

PURE

1 Express $\dfrac{7x-41}{(2x+9)(3x-1)}$ in its partial fractions.

2 a Express $\dfrac{4}{x(x+4)}$ in its partial fractions.

b Hence find $\sum_{1}^{n}\dfrac{4}{x(x+4)}$ and find the sum to infinity.

3 a Show that

$$\frac{1}{r^2}-\frac{2}{(r+1)^2}+\frac{1}{(r+2)^2}\equiv\frac{2(3r^2+6r+2)}{r^2(r+1)^2(r+2)^2}$$

b Use the method of differences to find $\sum_{1}^{n}\dfrac{2(3r^2+6r+2)}{r^2(r+1)^2(r+2)^2}$

c Hence find the sum to infinity.

4 Use known series expansions to find the first three non-zero terms of the Maclaurin series for $e^x\ln(1+2x)$

5 Use your knowledge of standard Maclaurin series to write down the general terms in the expansion of these series.

a $e^{\frac{x}{3}}$ b $\ln(1-x^2)$ c $\cos\left(\dfrac{x}{3}\right)$

d $\sin(4x+5)$ e $\left(1-\dfrac{x}{6}\right)^n$ f xe^x

6 Write down the range of values of x for which these series are valid.

a $(2-x^2)^{-1}$ b $\ln\left(1+\left(\dfrac{x^2}{4}\right)\right)$

7 Write down the range of values of x for which these series are valid.

a $\left(2+\dfrac{x}{3}\right)^{-4}$ b $\ln(1-3x)$ c $\left(2+\dfrac{x}{3}\right)^{-4}(\ln(1-3x))$

8 Use differentiation and the Maclaurin expansion to find the first three non-zero terms in the expansions of these functions.

a $e^{2x}\sin x$ b $\dfrac{4}{(1+2x)}$ c $\dfrac{4}{\ln(2+x)}$

d $\cos 2x-\sin 2x$ e 3^x f $\dfrac{2\cos x}{5+\sin x}$

9 Use l'Hopital's rule to evaluate these limits.

a $\lim\limits_{n\to\infty}\dfrac{4n}{3n+1}$ b $\lim\limits_{n\to\infty}\dfrac{n^2+3}{n}$ c $\lim\limits_{n\to\infty}\dfrac{(2n+1)^2}{(n-2)^2}$

d $\lim\limits_{x\to0}\dfrac{e^x}{4x}$ e $\lim\limits_{x\to0}\dfrac{x^5+8}{2\sin x}$ f $\lim\limits_{n\to0}\dfrac{(8n-4)^2}{3n^2}$

g $\lim\limits_{x\to1}\dfrac{x^2-1}{x-1}$

Exploration

Going beyond the exams

Investigate

The Maclaurin series of some even functions.
The Maclaurin series of some odd functions.
Can you explain what you find?

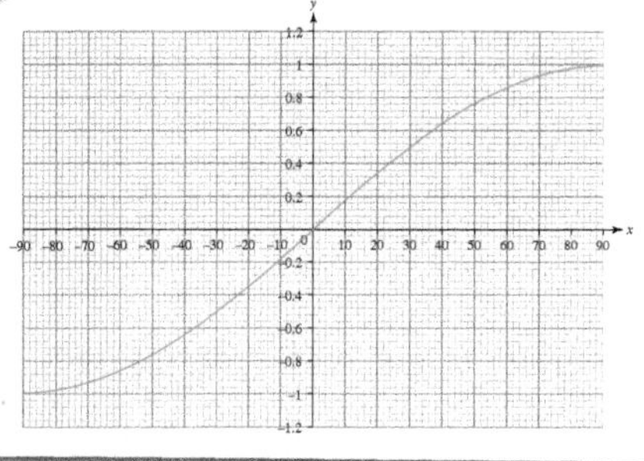

Investigation

The formula for the amount, £P, in a bank account with compound interest for an initial deposit £P_0 is

$$P = P_0\left(1+\frac{r}{n}\right)^{nt}$$

where n is the number of times per year that compound interest is added, r is the rate of interest in that period and t is the number of years.

For continuous compounding, $n \to \infty$, so

$$P = \lim_{n\to\infty} P_0\left(1+\frac{r}{n}\right)^{nt}$$

To evaluate this it is best to proceed by taking natural logarithms of both sides. L'Hopital's rule will be helpful when finding the limit of the resulting expression. Try this to see if you can arrive at the formula used for continuous compounding:

$$P = P_0 e^{rt}$$

Use this to explore the effect of continuous compounding as opposed to adding compound interest over discrete periods during a year.

Research

Use Maclaurin series to explore the validity of trig identities such as $2\sin x\cos x = \sin 2x$

Research further identities such as

$$\sin 3x = 3\sin x - 4\sin^3 x$$

$$\cos 3x = 4\cos^3 x - 3\cos x$$

17 Assessment

1 a Find an expression in terms of n for $\sum_{r=1}^{n} \frac{1}{(r+1)(r+2)}$ [5]

b Hence, find the value of k for which $\sum_{r=11}^{k} \frac{1}{(r+1)(r+2)} = \frac{5}{78}$ [3]

2 Use the method of differences to prove that $\sum_{r=1}^{n} r^2 = \frac{n}{6}(n+1)(2n+1)$ [6]

3 a Give the first four terms of the expansion of $\sin(2x^2)$. Simplify each term to the form Ax^n [4]

b State the range of values of x for which the expansion is valid. [1]

c Use your expansion to find the limit of $\frac{x^2}{\sin(2x^2)}$ as $x \to 0$ [3]

4 a Give the first four terms of the expansion of $\ln(1-5x)$. Simplify each term to the form Ax^n [4]

b State the range of values of x for which the expansion is valid. [1]

c Use your expansion from part **a** to find the limit of $\frac{x}{\ln(1-5x)}$ as $x \to 0$ [3]

5 a Find the expansion of $e^x \sin x$ up to the term in x^5 [4]

b Use your expansion to estimate the value of $e^{0.3} \sin(0.3)$ [2]

6 a Find an expression in terms of n for $\sum_{r=1}^{n} \frac{3}{r(r+1)}$ [5]

b Hence, show that $\sum_{r=n}^{2n} \frac{3}{r(r+1)} = \frac{3(n+1)}{n(2n+1)}$ [3]

c Evaluate $\frac{1}{2} + \frac{1}{6} + \frac{1}{12} + \frac{1}{20} + \ldots + \frac{1}{930}$ [3]

7 a Use the method of differences to prove that $\sum_{r=2}^{n} \frac{1}{r^2-1} = \frac{(3n+2)(n-1)}{4n(n+1)}$ [6]

b Show that $\sum_{r=2}^{\infty} \frac{1}{r^2-1} = \frac{3}{4}$ [3]

8 a Use Maclaurin's theorem to show that $\cos(3x) = 1 - \frac{9}{2}x^2 + \frac{27}{8}x^4 - \ldots$ [4]

b Give the rule for the general term in the expansion of $\cos(3x)$ [2]

9 a Find the expansion of $\sqrt{1-3x}$ up to the term in x^3 [3]

b Hence find the limit of $\frac{2x}{1-\sqrt{1-3x}}$ as $x \to 0$ [3]

10 Find the Maclaurin expansion of $\sec x$ up to the term in x^3 **[5]**

11 Use Maclaurin's theorem to find the expansion of $e^{\sin 2x}$ up to the term in x^2 **[5]**

12 a Find $\lim_{x\to 0}\frac{2x}{e^{3x}-1}$ **[3]**

b Explain why l'Hopital's rule can be used in this case. **[2]**

13 Use l'Hopital's rule to evaluate

a $\lim_{x\to 0}\frac{\sin(x^2)}{3x^2}$ **[3]**

b $\lim_{x\to 1}\frac{\ln\left(x^2\right)}{x-1}$ **[3]**

14 a Show that $\frac{1}{r!}-\frac{1}{(r+1)!}=\frac{r}{(r+1)!}$ **[3]**

b Hence use the method of differences to show that $\sum_{r=1}^{n}\frac{r}{(r+1)!}=1-\frac{1}{(n+1)!}$ **[3]**

c Find an expression for $\frac{n}{(n+1)!}+\frac{n+1}{(n+2)!}+...+\frac{2n}{(2n+1)!}$, expressing your answer as a single fraction. **[3]**

15 a Express $\frac{2}{r(r+1)(r+2)}$ in partial fractions. **[4]**

b Hence use the method of differences to show that $\sum_{r=1}^{n}\frac{1}{r(r+1)(r+2)}=\frac{n(n+3)}{4(n+1)(n+2)}$ **[6]**

c Find $\sum_{r=1}^{\infty}\frac{1}{r(r+1)(r+2)}$ **[3]**

16 Use l'Hopital's rule to evaluate the $\lim_{x\to 0}\frac{3-3\cos x}{2x^2}$

You must explain why l'Hopital's rule is applicable. **[5]**

17 Use l'Hopital's rule to evaluate $\lim_{x\to 0}x^2\ln\left(x^2\right)$

You must explain why l'Hopital's rule is applicable. **[4]**

18 Use l'Hopital's rule to evaluate $\lim_{x\to 0}\frac{x-x^3-\sin x}{x^3}$ **[4]**

19 Use Maclaurin's series to find $\lim_{x\to 0}\left(\frac{x-3\sin x}{2e^x-3x^2-2}\right)$ **[5]**

20 Use partial fractions and the method of differences to show that $\sum_{r=1}^{\infty}\frac{1}{r(r+1)(r+2)}=\frac{1}{4}$ **[7]**

21 Use partial fractions and the method of differences to find the sum to infinity of the sequence given by $u_n=\frac{1}{n^2+9n+20}$ **[7]**

18 Curve sketching 2

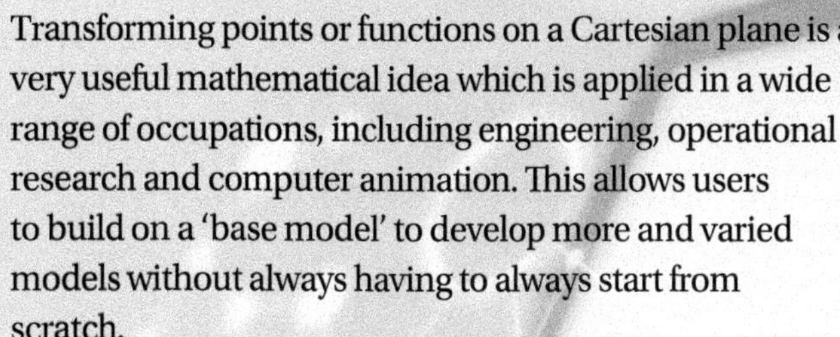

Transforming points or functions on a Cartesian plane is a very useful mathematical idea which is applied in a wide range of occupations, including engineering, operational research and computer animation. This allows users to build on a 'base model' to develop more and varied models without always having to always start from scratch.

Operational researchers use such techniques in their attempts to model complex situations and scenarios to optimise, for example, profits for a company importing and selling high street goods. In their work, as they explore the impact of varying a number of key parameters, ideas of transformations of functions have an important role to play. Although such techniques have been understood for decades, the use of computer technology has allowed such explorations to be carried out at high speed on an hour-by-hour basis.

Orientation

What you need to know	What you will learn	What this leads to
Ch3 Curve sketching 1	• To solve problems involving reciprocal and modulus functions. • To transform graphs of conic sections. • To sketch graphs of inverse hyperbolic functions. • To sketch graphs of reciprocal hyperbolic functions. • To work with graphs with oblique asymptotes.	**Careers** • Operational research. • Engineering. • Computer animation.

18.1 Reciprocal and modulus graphs

Fluency and skills

The graph of $y = \frac{1}{f(x)}$ is related to the graph of $f(x)$ in the following ways.

- $\frac{1}{f(x)}$ is undefined when $f(x) = 0$, so any roots (or zeros) of $y = f(x)$ become vertical asymptotes in $y = \frac{1}{f(x)}$
- As $f(x) \to 0$ from above $\frac{1}{f(x)} \to +\infty$

  as $f(x) \to 0$ from below $\frac{1}{f(x)} \to -\infty$
- The sign of $\frac{1}{f(x)}$ is the same as the sign of $f(x)$
- As $f(x)$ increases, $\frac{1}{f(x)}$ decreases and vice versa
- If $f(x) = \pm 1$ exists, then $\frac{1}{f(x)}$ also equals ± 1 and the graphs intersect
- Minimum points become maximum points and vice-versa (except for where these points are zero where they become asymptotes)
- If $f(x) > 1$, then $0 < \frac{1}{f(x)} < 1$

  and if $f(x) < -1$, then $-1 < \frac{1}{f(x)} < 0$
- If $0 < f(x) < 1$, then $\frac{1}{f(x)} > 1$

  and if $-1 < f(x) < 0$, then $\frac{1}{f(x)} < -1$

The values of x where $f(x)$ crosses the x-axis are called the roots or zeros of the function.

Example 1

$f(x) = 2x^2 + x - 1$

a Sketch the graph of $y = f(x)$

b On the same axes, sketch $y = \frac{1}{f(x)}$

a $y = 2x^2 + x - 1$ is a quadratic curve with intercepts $(0, -1)$, $(-1, 0)$ and $\left(\frac{1}{2}, 0\right)$

$y = 2\left(x + \frac{1}{4}\right)^2 - \frac{9}{8}$

Completing the square shows there is a minimum point at $\left(-\frac{1}{4}, -\frac{9}{8}\right)$

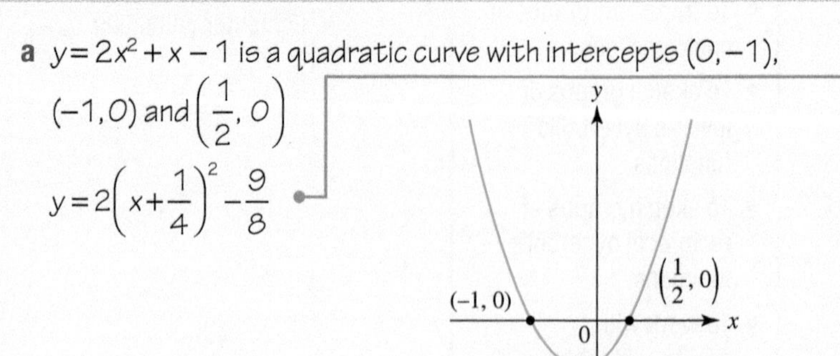

(*Continued on the next page*)

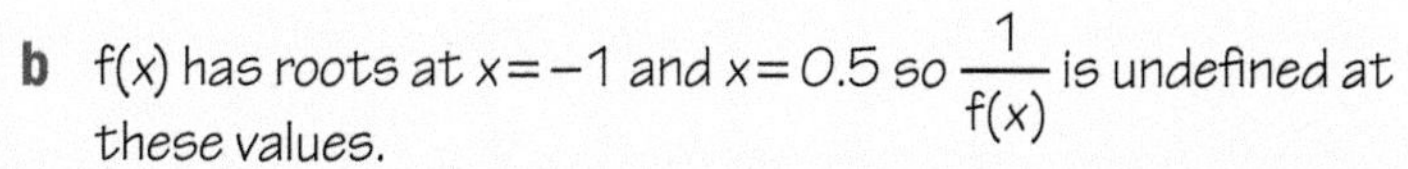

b f(x) has roots at $x=-1$ and $x=0.5$ so $\dfrac{1}{f(x)}$ is undefined at these values.

Hence $x=-1$ and $x=0.5$ are asymptotes for $\dfrac{1}{f(x)}$

> Consider the different features of f(x) and $\dfrac{1}{f(x)}$ in turn.

To find the points of intersection

$f(x) = 1$

$y = 2x^2 + x - 1 = 1$

$2x^2 + x - 2 = 0$

$x = \dfrac{-1 \pm \sqrt{17}}{4} \approx 0.78$ and -1.28

$f(x) = -1$

$y = 2x^2 + x - 1 = -1$

$2x^2 + x = 0$

$x = 0$ or $x = -0.5$

So $y = f(x)$ and $y = \dfrac{1}{f(x)}$ intersect at $(0, -1)$, $(-0.5, -1)$, $(0.78, 1)$ and $(-1.28, 1)$

> $f(x) = \pm 1$ does exist, so solve for x to find where f(x) and $\dfrac{1}{f(x)}$ intersect.

To find the signs and gradient

f(x) has a minimum point at $x=-0.25$, so $x=-0.25$ becomes a maximum point of $\dfrac{1}{f(x)}$

$y = f(x)$ is increasing for $x > -0.25$ and decreasing for $x < -0.25$

So $y = \dfrac{1}{f(x)}$ decreases for $x > -0.25$ and increases for $x < -0.25$

f(x) and $\dfrac{1}{f(x)}$ are both negative for $-1 < x < 0.5$ and positive elsewhere.

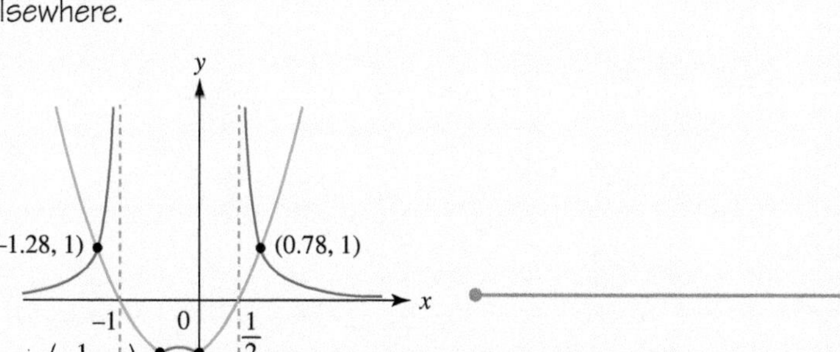

> Use the information to draw the graph of $\dfrac{1}{f(x)}$
>
> You can see that for
>
> $f(x) > 1$, then $0 < \dfrac{1}{f(x)} < 1$
>
> $f(x) < -1$, then $-1 < \dfrac{1}{f(x)} < 0$
>
> $0 < f(x) < 1$, then $\dfrac{1}{f(x)} > 1$
>
> $-1 < f(x) < 0$, then $\dfrac{1}{f(x)} < -1$

Make sure you know how to check your answer using a graphical calculator.

Sometimes you may not be given the function and instead have to work out the shape of the reciprocal graph by just looking at $f(x)$ and using what you know about the relationship between $f(x)$ and $\dfrac{1}{f(x)}$

Example 2

The graph of $y = \mathrm{f}(x)$ is shown.

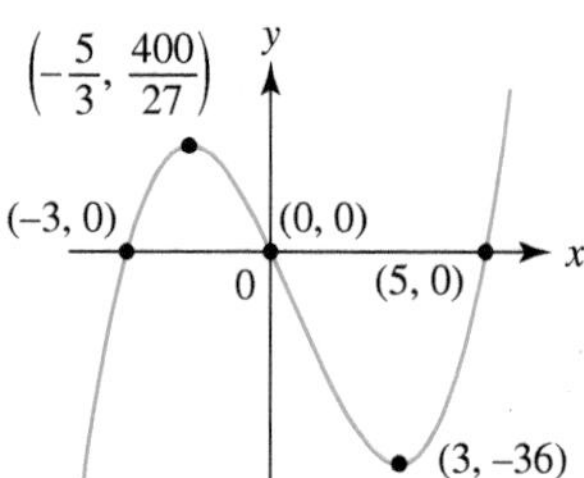

Use this to sketch the graph of $y = \frac{1}{\mathrm{f}(x)}$

Explain all your working.

f(x) has three roots at $x = -3$, $x = 0$ and $x = 5$ so there are three vertical asymptotes where $\mathrm{f}(x) = 0$

The maximum point of f(x) becomes a minimum point of $\frac{1}{\mathrm{f}(x)}$

f(x) at the maximum point is positive so the minimum point is also positive. $\frac{1}{\left(\frac{400}{27}\right)} = \frac{27}{400}$ so the minimum point is at $\left(-\frac{5}{3}, \frac{27}{400}\right)$

The minimum point of f(x) becomes a maximum point of $\frac{1}{\mathrm{f}(x)}$

f(x) at the minimum point is negative so the maximum point is also negative. The value of $\frac{1}{\mathrm{f}(x)}$ is $-\frac{1}{36}$ so the maximum point is at $\left(3, -\frac{1}{36}\right)$.

f(x) is decreasing between the maximum and minimum points so $\frac{1}{\mathrm{f}(x)}$ will increase for these x values.

Elsewhere, f(x) is increasing so $\frac{1}{\mathrm{f}(x)}$ will decrease.

When f(x) approaches zero $\frac{1}{\mathrm{f}(x)}$ will approach infinity,

so $y = \frac{1}{\mathrm{f}(x)}$ will have three vertical asymptotes.

When f(x) approaches infinity in either the positive or negative direction, $\frac{1}{\mathrm{f}(x)}$ will approach zero.

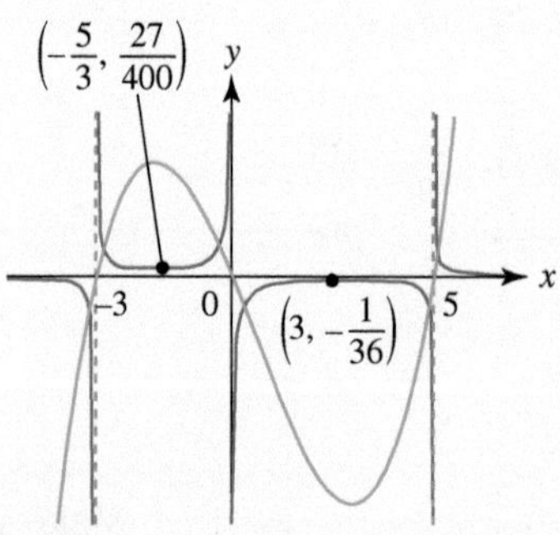

Graphs of the modulus function

$|x|$ is called the **modulus** of x

It is also known as the absolute value of x

The modulus of a real number is always positive. You can think of it as its distance from the origin.

For example, if $x = -3$, then $|x| = 3$

To sketch the graph of $y = |f(x)|$

- Start with a sketch of the graph of $y = f(x)$
- Reflect any negative part of $f(x)$ in the x-axis.

The diagram shows the graphs of $y = x$ (in blue) and $y = |x|$ (in red) drawn on the same axes.

When you carry out the reflection, any minimum turning point below the x-axis will be reflected into a maximum turning point above the x-axis.

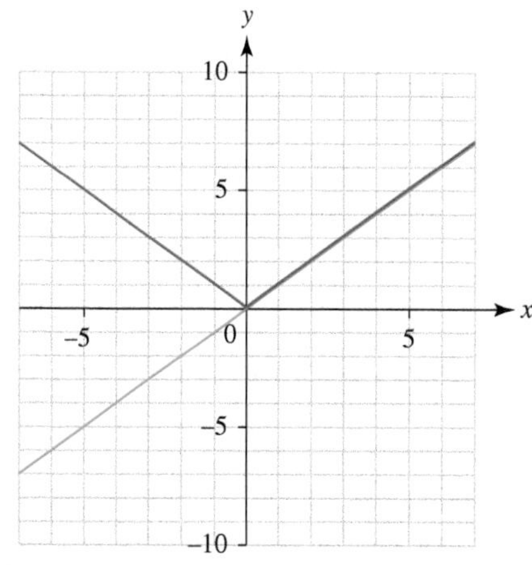

PURE

Example 3

Sketch $y = f(x)$ and $y = |f(x)|$ for these functions.

a $f(x) = 2x - 4$ **b** $f(x) = 2x^2 + 3x - 9$

a $y = 2x - 4$ is a straight line with gradient 2

It crosses the x-axis at $(2, 0)$ and crosses the y-axis at $(0, -4)$

so, $y = |2x - 4|$ crosses the axes at $(2, 0)$ and $(0, 4)$

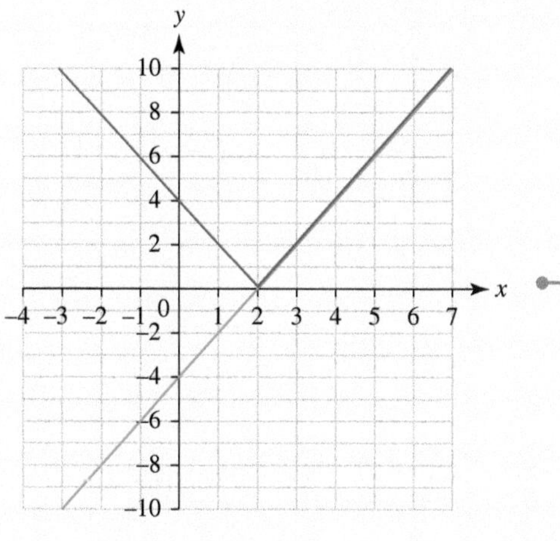

$y = 2x - 4$ is the blue graph and $y = |2x - 4|$ is the red graph.

b $y = 2x^2 + 3x - 9$ is a positive quadratic curve.

It crosses the x-axis at $(-3, 0)$ and $\left(\frac{3}{2}, 0\right)$ and the y-axis at $(0, -9)$

$y = |2x^2 + 3x - 9|$ crosses the x-axis at $(-3, 0)$ and $\left(\frac{3}{2}, 0\right)$ and the y-axis at $(0, 9)$

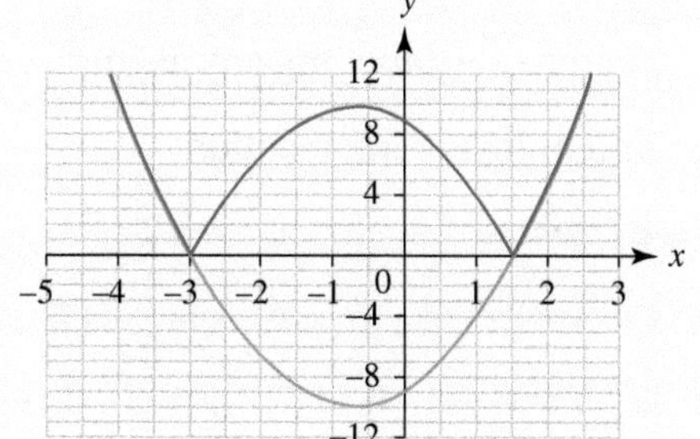

$y = 2x^2 + 3x - 9$ is the blue graph and $y = |2x^2 + 3x - 9|$ is the red graph.

Example 4

The graph of $y=\text{f}(x)$ is shown.

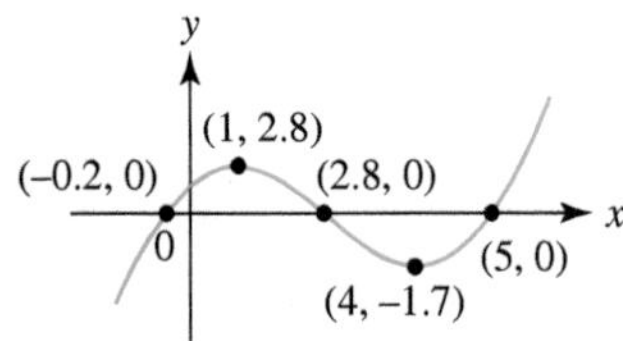

Sketch the graph of $y=|\text{f}(x)|$

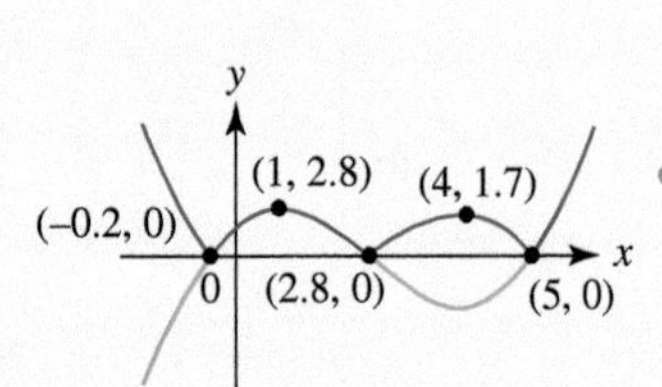

Reflect the negative parts of the graph in the *y*-axis.

The minimum point becomes a maximum point. Coordinate point (4, –1.7) becomes (4, 1.7).

Exercise 18.1A Fluency and skills

1 For each graph $y=\text{f}(x)$

i Copy the graph and sketch on the same axes the graph of $y=\frac{1}{\text{f}(x)}$

ii On a second copy of the graph, sketch $y=|\text{f}(x)|$

a

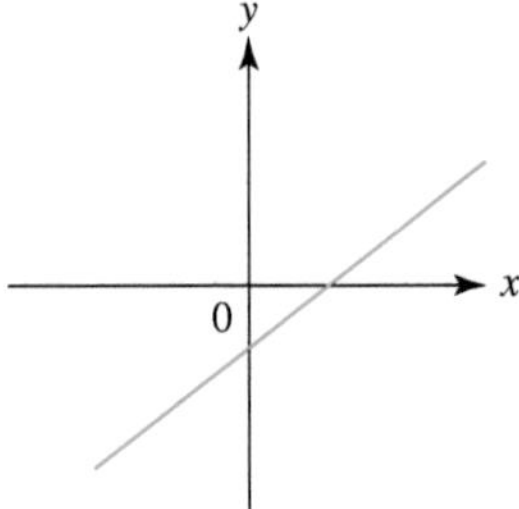

b

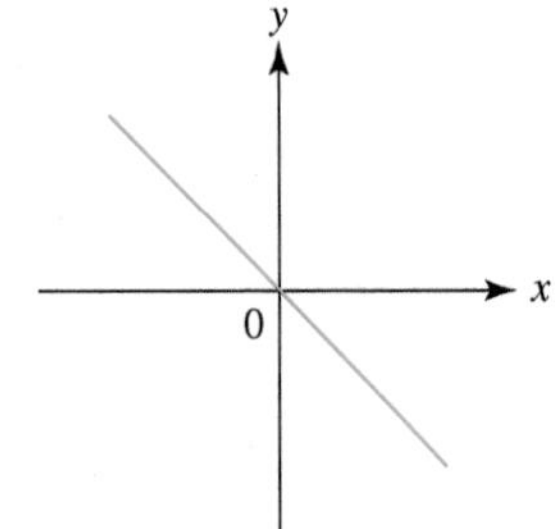

c

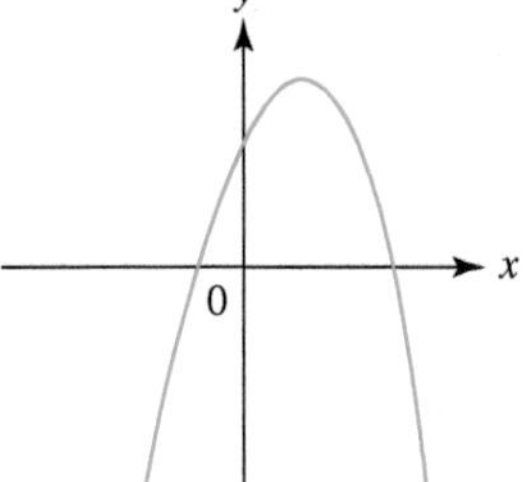

d

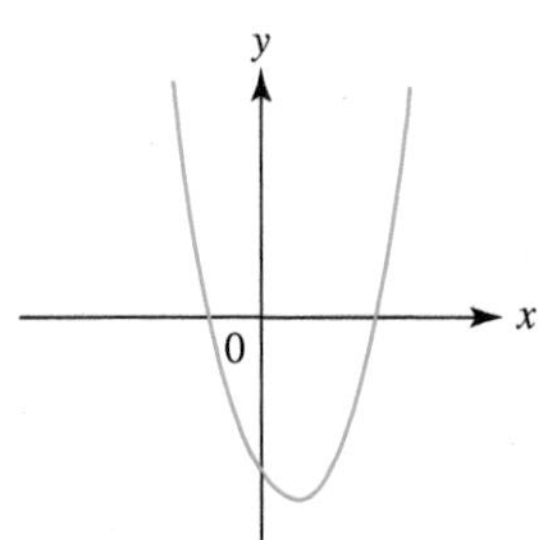

2 For each function

i Sketch $\text{f}(x)$

ii Sketch $\frac{1}{\text{f}(x)}$ labelling any asymptotes and points of intersection.

a $y=x$

b $y=2x+1$

c $y=3x-4$

d $y=\frac{1}{2}x+3$

e $y=x^2-x-6$

f $y=x^2-8x+15$

g $y=x^2+6x+8$

h $y=-x^2-x+12$

3 Sketch $|f(x)|$ for each of the functions in question **2**

4 For each graph $y = f(x)$

i Copy the graph and sketch on the same axes the graph of $y = \frac{1}{f(x)}$

ii On a second copy of the graph, sketch $y = |f(x)|$

a

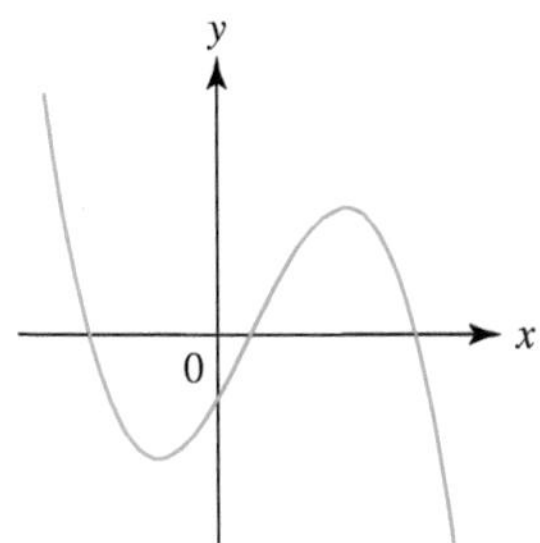

b

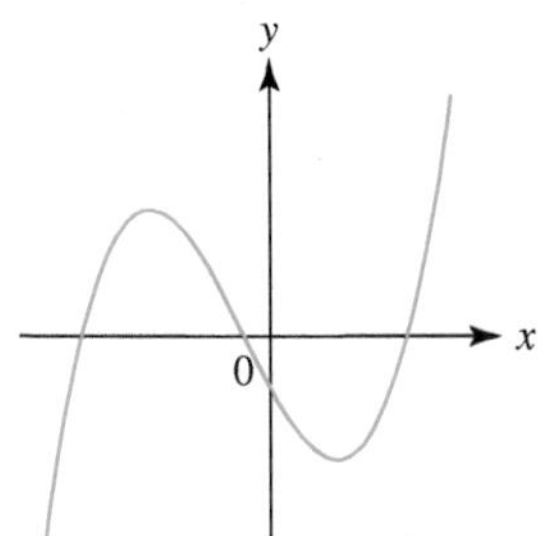

5 For each function

i Sketch $f(x)$

ii Sketch $\frac{1}{f(x)}$ labelling any asymptotes and points of intersection,

iii Sketch $|f(x)|$

a $y = x^3 - 4x$

b $y = x^3 - 2x^2 - 24x$

c $y = x^3 - 3x^2 - 13x + 15$

6 The diagram shows the graph of $y = f(x)$ and the points P, Q and R

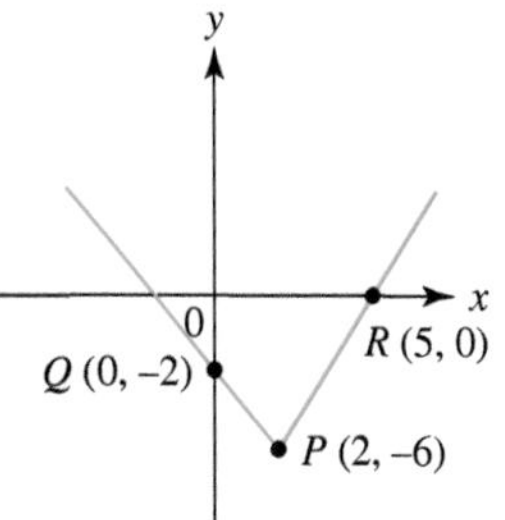

Sketch the graph of $y = |f(x)|$, labelling the transformed points P', Q' and R'

7 The diagram shows the graph of $y = f(x)$ and the points P, Q and R

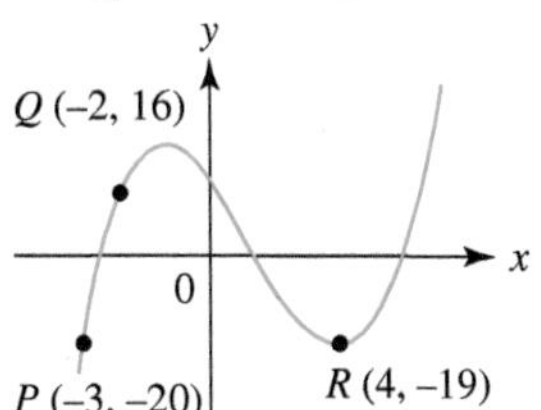

Sketch the graph of $y = |f(x)|$, labelling the transformed points P', Q' and R'

8 The diagram shows the graph of $y = f(x)$ and the points P, Q and R

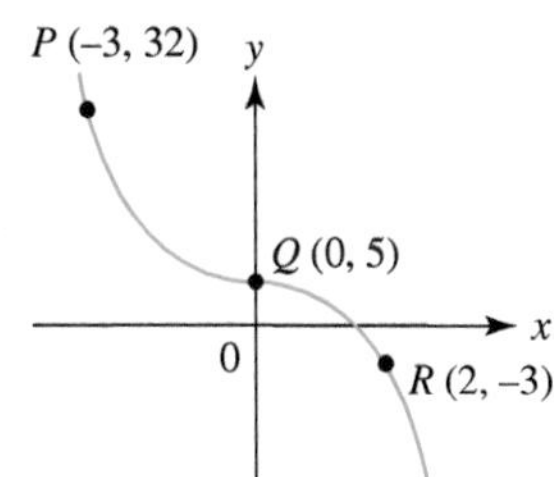

Sketch the graph of $y = |f(x)|$, labelling the transformed points P', Q' and R'

Reasoning and problem-solving

Strategy

To sketch a graph of $y = \frac{1}{f(x)}$ or $y = |f(x)|$

1. If it has not been provided, sketch the graph of $y = f(x)$
2. Consider the key features and transform them appropriately, clearly labelling the final graph.

Example 5

The sketch shows part of the graph of f(x)

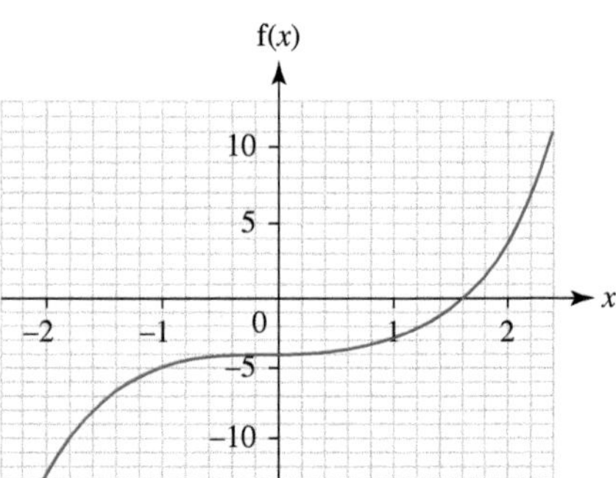

a Find the coordinates of the images of the points (−2, −12), (0, −4) and (2, 4) when f(x) is transformed into |f(x)|

b Sketch the graph of |f(x)| and use it to solve |f(x)| ≤ 4

c Sketch the graph of $\frac{1}{f(x)}$ and use it to estimate the solution to $\frac{1}{f(x)} < 0$

a Under the transformation $f(x) \to |f(x)|$

$(-2, -12) \to (-2, 12)$, $(0, -4) \to (0, 4)$, $(2, 4) \to (2, 4)$

Negative y-coordinates become positive.

All negative parts of the graph are reflected in the x-axis.

b

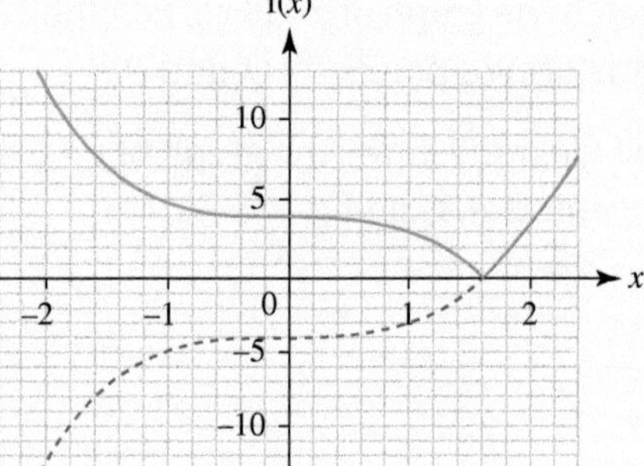

$|f(x)| \leq 4$ when $0 \leq x \leq 2$

The line $y = 4$ intersects with |f(x)| at (0, 4) and (2, 4)

c

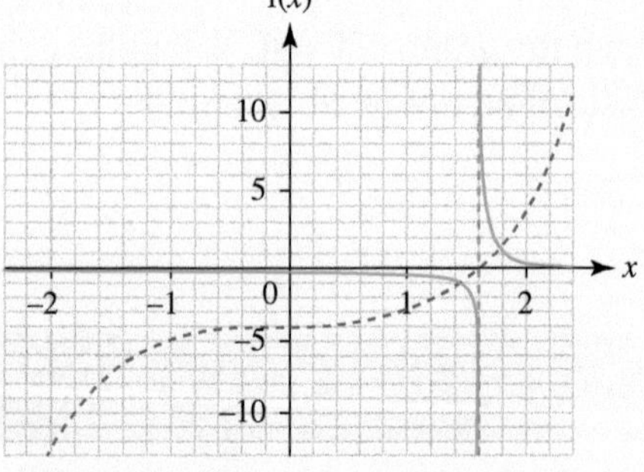

$\frac{1}{f(x)} < 0$ when $x < 1.6$ (approximately)

The asymptote is approximately $x = 1.6$

The point where f(x) = 0 becomes a vertical asymptote.

As f(x) becomes very large, either positive or negative, $\frac{1}{f(x)}$ tends to zero.

Exercise 18.1B Reasoning and problem-solving

1 a Use the relationship between the graphs of $y=\frac{1}{f(x)}$ and $y=f(x)$ to sketch, on the same axes

i $\sin x$ and $\frac{1}{\sin x}$ for $-180° \le x \le 180°$

ii $\cos x$ and $\frac{1}{\cos x}$ for $-180° \le x \le 180°$

iii $\tan x$ and $\frac{1}{\tan x}$ for $-180° \le x \le 180°$

b Use your sketches to write down the interval(s) where

i Both $\sin x$ and $\frac{1}{\sin x} \ge 0$

ii Both $\cos x$ and $\frac{1}{\cos x} \ge 0$

iii Both $\tan x$ and $\frac{1}{\tan x} \ge 0$

2 The diagram shows the graph of $y=f(x)$ and points P, Q and R

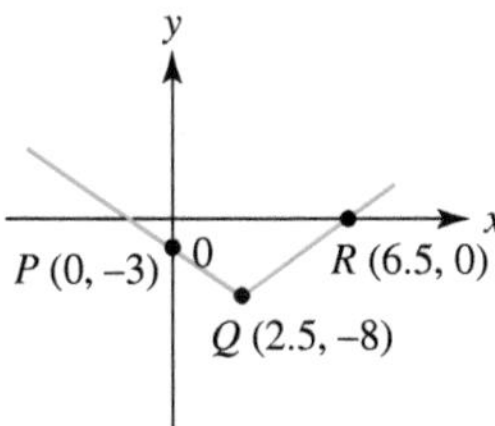

a Sketch the graph of $y=|f(x)|$, labelling the transformed points P', Q' and R'

b $f(x)$ can be written as $a|x+b|+c$

Write down the values of a, b and c

3 The diagram shows $y=f(x)$ and points A and B

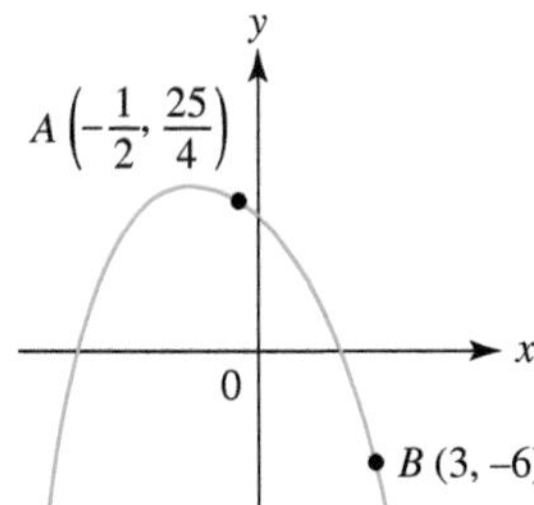

a Write down the coordinates of the points A' and B' when $f(x)$ is transformed into $\frac{1}{f(x)}$

b Sketch the graph of $y=\frac{1}{f(x)}$ and use it to solve the inequality $\frac{1}{f(x)}>0$

4 $f(x)=\frac{2-x}{x-3}$

Sketch the graph of $\frac{1}{f(x)}$ and use it to solve the inequality $\frac{1}{f(x)}<0$

5 A curve has equation $y=\frac{2-x}{1-x}$

a i Write down the equations of the asymptotes to the curve and the points of intersection with the coordinate axes.

ii Sketch the curve, indicating clearly the coordinates of the points of intersection of the curve with the coordinate axes.

b Sketch also the curve $y=\left|\frac{2-x}{1-x}\right|$

c Add the straight line $y=2-x$ to each sketch and hence solve the inequalities $\frac{2-x}{1-x}<2-x$ and $\left|\frac{2-x}{1-x}\right|<2-x$

6 A curve has equation $f(x)=\frac{x}{3+x}$

By drawing suitable curves and lines, solve

a $\frac{x}{3+x}<-x$

b $\left|\frac{x}{3+x}\right|<-x$

18.2 Transformations

Fluency and skills

Key point

An enlargement is equivalent to a stretch by the same scale factor in both the x-direction and the y-direction.

For an enlargement scale factor k replace x with $\frac{x}{k}$ and y with $\frac{y}{k}$

Example 1

An ellipse with equation $\frac{x^2}{4}+\frac{y^2}{9}=1$ is enlarged by scale factor 3

a Find the equation of the transformed curve and state its points of intersection with the axes.

b Sketch the transformed curve.

a $\frac{\left(\frac{x}{3}\right)^2}{4}+\frac{\left(\frac{y}{3}\right)^2}{9}=1$ — Substitute for x and y

$\frac{x^2}{36}+\frac{y^2}{81}=1$ — Simplify to find the equation of the transformed curve.

When $x=0$, $y^2=81 \Rightarrow y=\pm 9 \Rightarrow$ coordinates $(0, 9)$ and $(0, -9)$

When $y=0$, $x^2=36 \Rightarrow x=\pm 6 \Rightarrow$ coordinates $(6, 0)$ and $(-6, 0)$ — Substitute $x=0$ and $y=0$ to find the coordinates of the points of intersection.

b

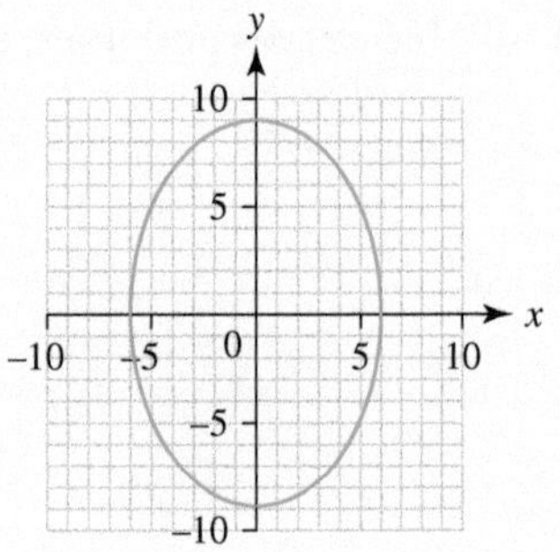

Use the coordinates from part **a** to sketch the curve of the transformed ellipse.

A conic can be rotated through θ radians about the origin. It is convention to describe rotations in an anticlockwise direction. In this course, the rotations are limited to multiples of $\frac{\pi}{2}$

Key point

To rotate a conic by $\frac{\pi}{2}$ radians, replace x by y and y by $-x$

To rotate a conic by π radians, replace x by $-x$ and y by $-y$

To rotate a conic by $\frac{3\pi}{2}$ radians, replace x by $-y$ and y by x

These results come from substituting $\theta = \frac{\pi}{2}$, π and $\frac{3\pi}{2}$ into the general transformation formulae:

$$X = x\cos\theta + y\sin\theta$$
$$Y = -x\sin\theta + y\cos\theta$$

where (X, Y) are the coordinates of the transformed point (x, y).

PURE

Example 2

The hyperbola $\frac{x^2}{9} - \frac{y^2}{16} = 1$ is rotated anticlockwise through $\frac{3\pi}{2}$ radians about (0, 0).

a Write down the equation of the transformed curve.

b Sketch the transformed curve and find the equations of the asymptotes.

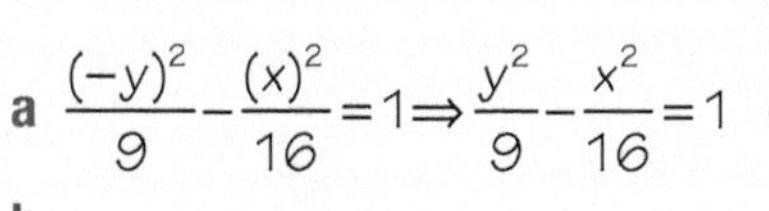

a $\frac{(-y)^2}{9} - \frac{(x)^2}{16} = 1 \Rightarrow \frac{y^2}{9} - \frac{x^2}{16} = 1$ — Substitute for x and y and simplify.

b

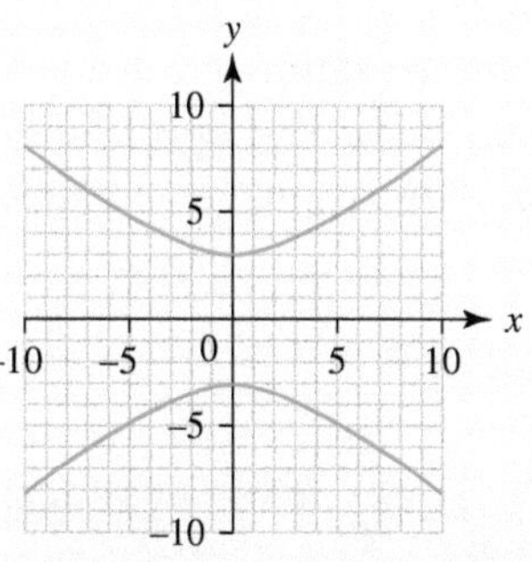

The asymptotes of the original hyperbola are at $y = \pm\frac{4}{3}x$ — Use the result that the asymptotes to a hyperbola are at $y = \pm\frac{b}{a}x$

The asymptotes of the transformed curve are at

$x = \pm\frac{4}{3}(-y) \Rightarrow y = \pm\frac{3}{4}x$ — Apply the same substitution as the one used to transform the conic and rearrange to make y the subject.

Exercise 18.2A Fluency and skills

1 Find the equation of the new curve and sketch its graph when the parabola $y^2 = 12x$ is transformed by

a An enlargement scale factor 2

b An enlargement scale factor $\frac{1}{3}$

c A rotation of $\frac{\pi}{2}$ radians anticlockwise about (0, 0)

d A rotation of π radians anticlockwise about (0, 0)

2 Find the equation of the new curve when the ellipse $\frac{x^2}{16} + \frac{y^2}{25} = 1$ is transformed in the following ways. In each case, sketch the new graph and write down the coordinates of the points where the curve crosses the axes.

a An enlargement scale factor $\frac{1}{2}$

b An enlargement scale factor 3

c A rotation of $\frac{3\pi}{2}$ radians anticlockwise about (0, 0)

d A rotation of π radians anticlockwise about (0, 0)

3 Find the equation of the new curve when the hyperbola $\frac{x^2}{9}-\frac{y^2}{4}=1$ is transformed in the following ways. In each case, sketch the new graph and write down the equations of the asymptotes.

a An enlargement scale factor 4

b An enlargement scale factor $\frac{3}{2}$

c A rotation of $\frac{\pi}{2}$ radians anticlockwise about (0, 0)

d A rotation of $\frac{3\pi}{2}$ radians anticlockwise about (0, 0)

4 Find the equation of the new curve when the rectangular hyperbola $xy = 16$ is transformed in the following ways. In each case, sketch the curve and write down the equations of the asymptotes.

a An enlargement scale factor $\frac{2}{3}$

b An enlargement scale factor $\frac{4}{3}$

c A rotation of π radians anticlockwise about (0, 0)

d A rotation of $\frac{\pi}{2}$ radians anticlockwise about (0, 0)

5 An ellipse with equation $\frac{x^2}{4}+\frac{y^2}{25}=1$ is enlarged by scale factor k

The equation of the transformed curve is $\frac{x^2}{16}+\frac{y^2}{100}=\frac{1}{9}$

Find the value of k

6 A hyperbola with equation $\frac{x^2}{25}-\frac{y^2}{4}=1$ is enlarged by scale factor k

The equation of the transformed curve is $\frac{x^2}{36}-\frac{25y^2}{144}=1$

Find the value of k

7 A parabola with equation $y^2 = 40x$ is rotated anticlockwise through θ radians. The equation of the transformed curve is $x^2 = -40y$

Find the value of θ

Reasoning and problem-solving

You need to be able to find the equation and sketch the graph of a given conic after two or more successive transformations and also identify a sequence of transformations that has been applied to transform one curve into another.

Strategy 1

To find the equation of a conic after two or more transformations

1. Identify the order that the transformations are applied.
2. Transform the original conic using the first transformation.
3. Transform the *new* conic using the next transformation.
4. Repeat step 3 until an equation of the final conic is found.
5. Ensure the final equation is given in the correct form.
6. Sketch the graph if required.

Example 3

The ellipse with equation $x^2+\frac{y^2}{4}=1$ is enlarged by scale factor 2, then rotated by $\frac{\pi}{2}$ radians anticlockwise about (0, 0) before being translated by vector $\begin{pmatrix}3\\2\end{pmatrix}$

Find the equation of the new conic in the form $ax^2+by^2+cx+dy+e=0$ where a, b, c, d and e are constants to be found and sketch the graph.

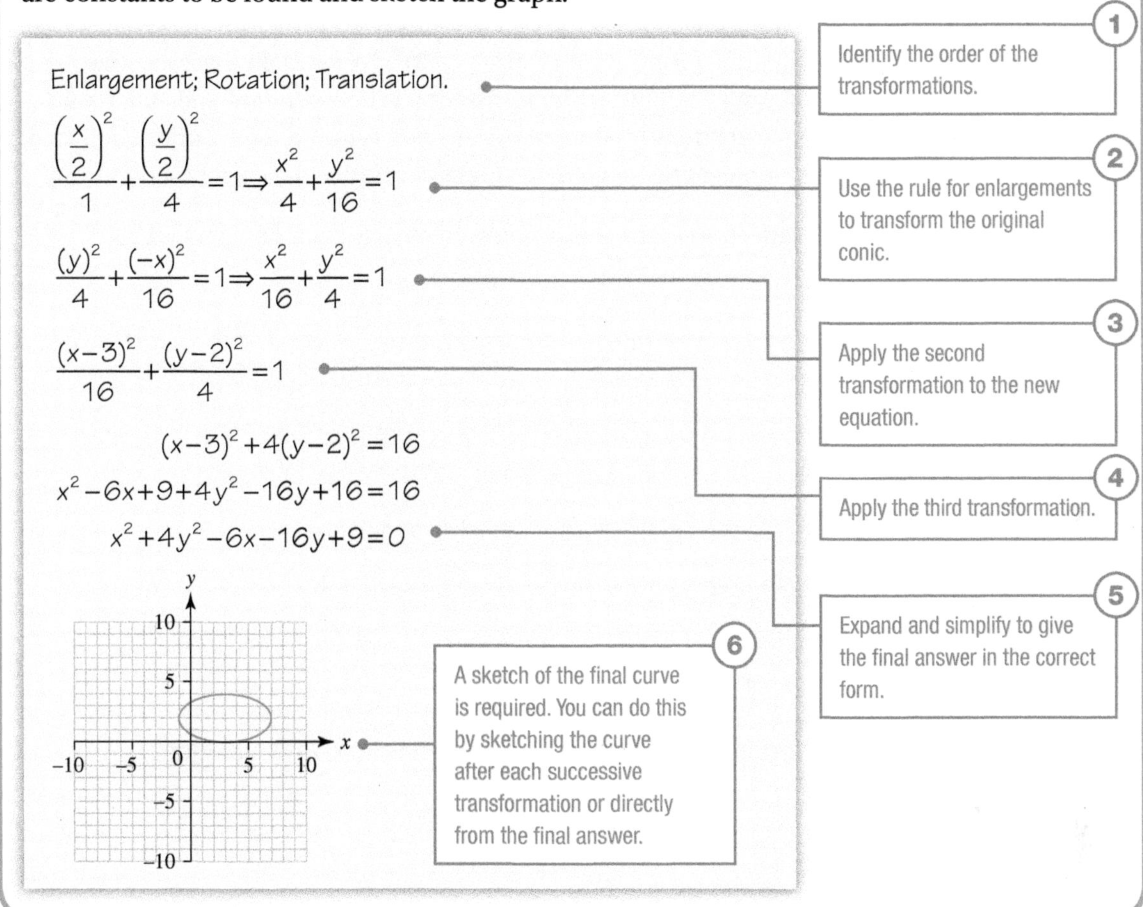

Enlargement; Rotation; Translation.
$\frac{\left(\frac{x}{2}\right)^2}{1}+\frac{\left(\frac{y}{2}\right)^2}{4}=1 \Rightarrow \frac{x^2}{4}+\frac{y^2}{16}=1$
$\frac{(y)^2}{4}+\frac{(-x)^2}{16}=1 \Rightarrow \frac{x^2}{16}+\frac{y^2}{4}=1$
$\frac{(x-3)^2}{16}+\frac{(y-2)^2}{4}=1$
$(x-3)^2+4(y-2)^2=16$
$x^2-6x+9+4y^2-16y+16=16$
$x^2+4y^2-6x-16y+9=0$
y
10
5
-10
-5
0
5
10
x
-5
-10
1 Identify the order of the transformations.
2 Use the rule for enlargements to transform the original conic.
3 Apply the second transformation to the new equation.
4 Apply the third transformation.
5 Expand and simplify to give the final answer in the correct form.
6 A sketch of the final curve is required. You can do this by sketching the curve after each successive transformation or directly from the final answer.

Example 4

A parabola C has equation $y^2=x$

Describe a sequence of two transformations which maps C onto the curve with equation $y^2-6y+x+11=0$

$$y^2-6y+x+11=0$$

$$(y-3)^2-9+x+11=0$$

$$(y-3)^2=-x-2$$

① Rearrange the transformed equation into the same form as the original parabola. $(y-3)^2$ indicates a translation in the positive y-direction of 3 units.

$-x-2$ indicates a positive x-direction translation of 2 units and a reflection in the y-axis in that order.

② Determine the transformations that have taken place.

(*Continued on the next page*)

A possible sequence with two transformations is a translation using vector $\begin{pmatrix} 2 \\ 3 \end{pmatrix}$, followed by a reflection in the y-axis.

③ Determine the order in which the transformations have taken place.

An alternative way of writing the transformed equation is $(y-3)^2 = -(x+2)$. In this case, the order of transformations is different. The y-direction translation is independent and still 'plus 3 units,' but the sequence of transformations on the right-hand side is now a reflection in the y-axis followed by a translation of two units in the *negative* x-direction. Hence the correct sequence of two transformations is now a reflection in the y-axis followed by a translation using vector $\begin{pmatrix} -2 \\ 3 \end{pmatrix}$.

Strategy 2

To find the sequence of transformations that have been used to map one given conic onto another

1. Rearrange the transformed equation into the standard form for the given conic.
2. Determine the transformations that have taken place.
3. Determine, where necessary, the order in which they have taken place.

Example 5

A hyperbola H has equation $x^2 - \dfrac{y^2}{3} = 1$

Describe a sequence of transformations which maps H onto the curve with equation $12x^2 = y^2 + 48x - 36$

$12x^2 = y^2 + 48x - 36$

$12x^2 - 48x - y^2 = -36$

$x^2 - 4x - \dfrac{y^2}{12} = -3$

$(x-2)^2 - 4 - \dfrac{y^2}{12} = -3$

$(x-2)^2 - \dfrac{\left(\dfrac{y}{2}\right)^2}{3} = 1$

① Rearrange the transformed equation into the same form as the original hyperbola.

$(x-2)^2$ indicates a translation in the positive x-direction of 2 units.

$\left(\dfrac{y}{2}\right)^2$ indicates a stretch parallel to the y-axis scale factor 2

② Determine the transformations that have taken place.

A possible sequence is a stretch parallel to the y-axis scale factor 2, followed by a translation in the positive x-direction of 2 units.

③ Determine the order in which the transformations have taken place. Note that in some cases, like this one, the transformations can be applied in *either order*.

Exercise 18.2B Reasoning and problem-solving

1 The hyperbola $\frac{x^2}{4}-\frac{y^2}{9}=1$ is rotated by $\frac{3\pi}{2}$ radians anticlockwise about (0, 0) and then enlarged by scale factor 3

a Find the equation of the new conic in the form $ax^2+by^2=c$ where a, b and c are constants to be found.

b State the equations of the asymptotes.

2 The parabola $y^2=6x$ is enlarged by scale factor 2, translated by vector $\begin{pmatrix}-1\\2\end{pmatrix}$ and then rotated through π radians about (0, 0).

a Find the equation of the new conic in the form $ay^2+by+cx=d$ where a, b, c and d are constants to be found.

b Find the coordinates of the turning point.

c Sketch the graph of the new conic.

3 The ellipse $\frac{x^2}{2}+\frac{y^2}{3}=1$ is translated by vector $\begin{pmatrix}3\\-1\end{pmatrix}$ before being rotated anticlockwise by $\frac{\pi}{2}$ radians about the origin and stretched in the x-direction by scale factor 2

a Show that the equation of the new conic can be written as $x^2+6y^2-4x-36y+c=0$

where c is a constant to be found.

b Given that the line $y=k$ intersects the new conic at two points, find the possible values of k

4 An ellipse C has equation $\frac{x^2}{4}+\frac{y^2}{5}=1$ Describe a sequence of transformations which maps C onto the curve with equation $5x^2+16y^2-20x+32y+16=0$

5 A rectangular hyperbola H has equation $xy=25$

Describe a sequence of transformations which maps H onto the curve with equation $xy-3y-9x=198$

6 An ellipse, E, has equation $\frac{x^2}{9}+\frac{y^2}{16}=1$

a Sketch the ellipse E and state the coordinates of the points where the curve intercepts the coordinate axes.

b Given that the line $y=c-x$ intersects the ellipse at two distinct points, show that $-5<c<5$

c The ellipse E is translated by vector $\begin{pmatrix}a\\b\end{pmatrix}$ to form another ellipse with equation $16x^2+9y^2-128x+18y+d=0$

Find the values of the constants a, b and d

d Hence find the equation for each of the tangents to the ellipse

$16x^2+9y^2-128x+18y+d=0$

which are parallel to the line $y=-x$

7 A hyperbola H has equation $\frac{x^2}{4}-\frac{y^2}{5}=1$

a Show that the line with equation $2x+3y=4$ intersects H at two points and find the exact coordinates of these points.

The hyperbola H is reflected in the line $y=x$

b Show that the line $2x+3y=4$ is now a tangent to the reflected curve.

18.3 Hyperbolic functions 2

Fluency and skills

Here are the definitions of the hyperbolic functions, and their inverses.

See Ch3.5

For a reminder of hyperbolic functions.

Key point

$\sinh x \equiv \frac{1}{2}(e^x - e^{-x})$ $\quad$ $\text{arsinh}\, x \equiv \ln(x+\sqrt{x^2+1})$

$\cosh x \equiv \frac{1}{2}(e^x + e^{-x})$ $\quad$ $\text{arcosh}\, x \equiv \ln(x+\sqrt{x^2-1}),\ x \geq 1$

$\tanh x \equiv \frac{e^x - e^{-x}}{e^x + e^{-x}}$ $\quad$ $\text{artanh}\, x \equiv \frac{1}{2}\ln\left(\frac{1+x}{1-x}\right),\ -1 < x < 1$

The domain of the function $f(x) = \sinh x$ is $x \in \mathbb{R}$ and the range is $f(x) \in \mathbb{R}$, so the domain of the function $f^{-1}(x) = \text{arsinh}\, x$ is $x \in \mathbb{R}$ and the range is $f^{-1}(x) \in \mathbb{R}$

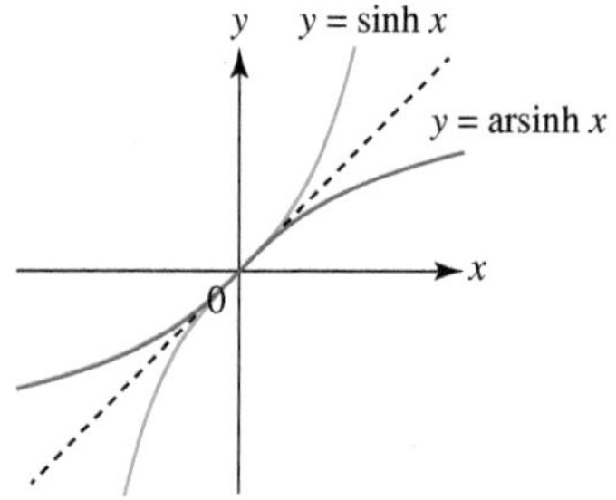

In order to be able to define an inverse you must restrict the domain of $f(x) = \cosh x$ to $x \in \mathbb{R},\ x \geq 0$. The range of $f(x) = \cosh x$ is $f(x) \in \mathbb{R}$, $f(x) \geq 1$, so the domain of $f^{-1}(x) = \text{arcosh}\, x$ is $x \in \mathbb{R},\ x \geq 1$ and the range is $f^{-1}(x) \in \mathbb{R},\ f^{-1}(x) \geq 0$

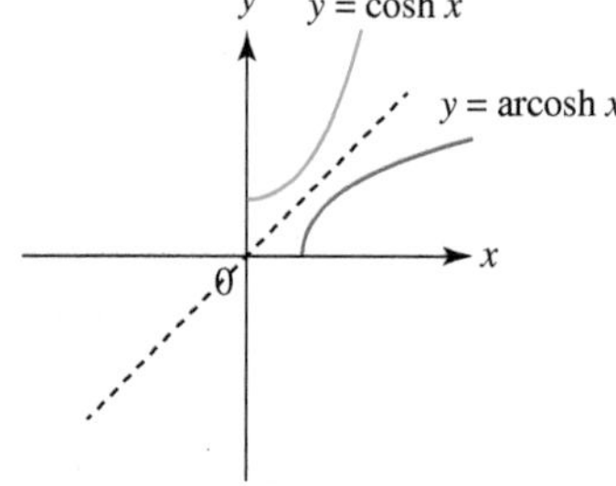

The domain of the function $f(x) = \tanh x$ is $x \in \mathbb{R}$ and the range is $f(x) \in \mathbb{R},\ -1 < f(x) < 1$ since the graph of $y = \tanh x$ has horizontal asymptotes at $y = \pm 1$

Therefore, the graph of $y = \text{artanh}\, x$ will have vertical asymptotes at $x = \pm 1$. So the domain of $f^{-1}(x) = \text{artanh}\, x$ is $x \in \mathbb{R},\ -1 < x < 1$ and the range is $f^{-1}(x) \in \mathbb{R}$

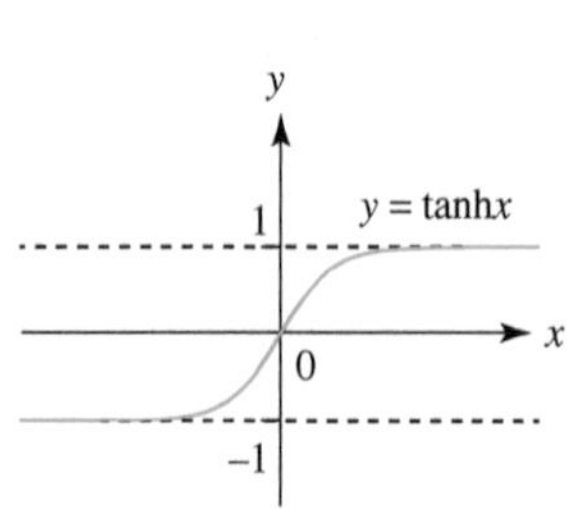

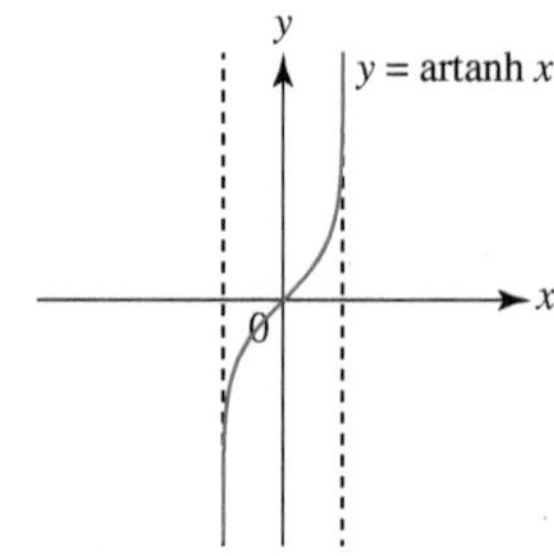

Example 1

Sketch the graph of $y = \text{arcosh}\left(\frac{x}{2}\right)$, and state its domain and range.

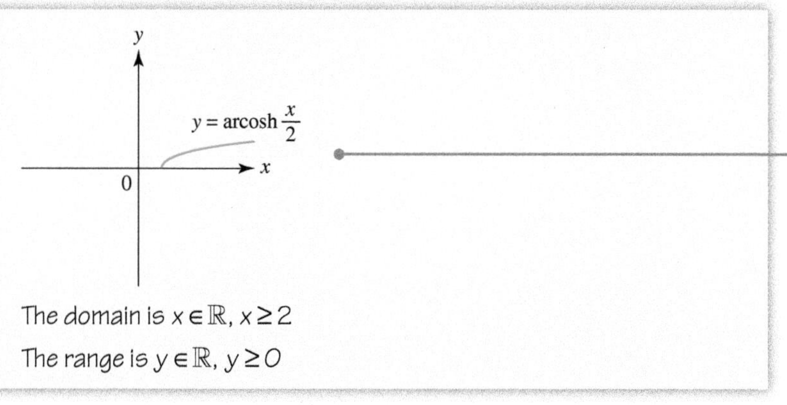

The graph of $y = \cosh^{-1}(x)$ has been stretched by scale factor 2 in the x-direction.

The domain is $x \in \mathbb{R}$, $x \geq 2$

The range is $y \in \mathbb{R}$, $y \geq 0$

There are also reciprocal hyperbolic functions, defined in the same way as with trigonometric functions.

Key point

$$\text{cosech}\, x \equiv \frac{1}{\sinh x} \equiv \frac{2}{e^x - e^{-x}} \quad \text{for } x \in \mathbb{R}, x \neq 0$$

$$\text{sech}\, x \equiv \frac{1}{\cosh x} \equiv \frac{2}{e^x + e^{-x}} \quad \text{for } x \in \mathbb{R}$$

$$\coth x \equiv \frac{1}{\tanh x} \equiv \frac{e^{2x} + 1}{e^{2x} - 1} \quad \text{for } x \in \mathbb{R}, x \neq 0$$

These functions are commonly read as, 'cosetch', 'setch' and 'coth'.

You can sketch the graphs of these reciprocal functions using the techniques covered earlier in this chapter.

See Ch18.1

For a reminder of sketching reciprocal graphs.

The domain of $f(x) = \text{cosech}\, x$ is $x \in \mathbb{R}$, $x \neq 0$

and the range is $f(x) \in \mathbb{R}$, $f(x) \neq 0$

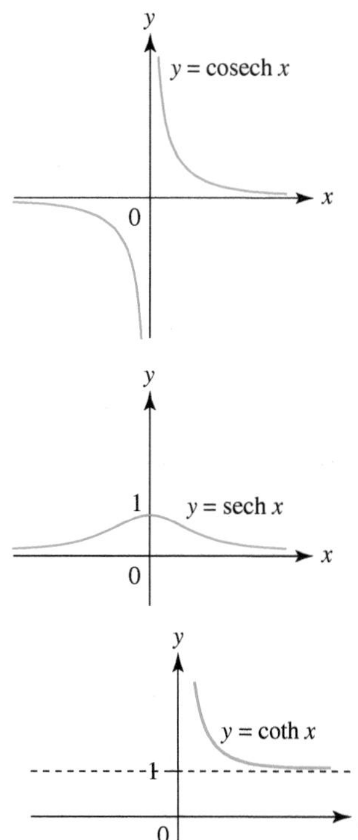

The domain of $f(x) = \text{sech}\, x$ is $x \in \mathbb{R}$,

and the range is $f(x) \in \mathbb{R}$, $0 < f(x) \leq 1$

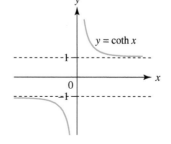

The domain of $f(x) = \coth x$ is $x \in \mathbb{R}$, $x \neq 0$

and the range is $f(x) \in \mathbb{R}$, $f(x) < -1$ or $f(x) > 1$

Example 2

Find the exact value of the expression $\operatorname{sech}(\ln\sqrt{3})$. Show your working.

$$\operatorname{sech}(\ln\sqrt{3}) = \frac{2}{e^{\ln\sqrt{3}} + e^{-\ln\sqrt{3}}}$$
$$= \frac{2}{e^{\ln\sqrt{3}} + e^{\ln\left(\frac{1}{\sqrt{3}}\right)}}$$
$$= \frac{2}{\sqrt{3} + \frac{1}{\sqrt{3}}}$$
$$= \frac{\sqrt{3}}{2}$$

Use definition $\operatorname{sech} x = \frac{2}{e^x + e^{-x}}$ since $-\ln\sqrt{3} = \ln\sqrt{3}^{-1} = \ln\left(\frac{1}{\sqrt{3}}\right)$

Example 3

Find the exact solution to the equation $1 + \coth x = 4$

$1 + \coth x = 4 \Rightarrow \coth x = 3$

Rearrange.

Therefore $\tanh x = \frac{1}{3}$

Since $\tanh x = \frac{1}{\coth x}$

Which gives $x = \frac{1}{2}\ln\left(\frac{1+\frac{1}{3}}{1-\frac{1}{3}}\right)$

$= \frac{1}{2}\ln 2$

$= \ln\sqrt{2}$

Use the definition of the inverse, $\tanh^{-1} x = \frac{1}{2}\ln\left(\frac{1+x}{1-x}\right)$

Exercise 18.3A Fluency and skills

1 Find the exact value of each of these expressions and give your answers in their simplest form. Show all your working and do not use a calculator.

a $\operatorname{sech}(\ln 2)$ b $\operatorname{cosech}(\ln 5)$

c $\coth(\ln\sqrt{2})$ d $\operatorname{sech}(2\ln 4)$

e $\operatorname{cosech}\left(\frac{1}{2}\ln 5\right)$ f $\coth(\ln 3)$

2 Sketch the graph of $y = f(x)$ for each of these functions and state the domain and range.

a $y = \operatorname{arcosh}(x+2)$ b $y = \operatorname{arsinh}\left(\frac{x}{3}\right)$

c $y = 1 + \operatorname{artanh}(x)$ d $y = \operatorname{arcosh}(4x)$

e $y = \operatorname{artanh}(x-2)$ f $y = \operatorname{artanh}(3x)$

3 Sketch the graph of $y = f(x)$ for each of these functions and state the domain and range.

a $y = \operatorname{sech}(2x)$ b $y = \operatorname{cosech}(x+1)$

c $y = 1 + \coth(x)$ d $y = \operatorname{sech}(x) - 2$

e $y = -\operatorname{cosech}(x)$ f $y = 1 - \operatorname{sech}(x)$

4 Solve each of these equations. Give your answers in the form $\ln k$ where k is a constant to be found.

a $\operatorname{cosech} x = 2$ b $\coth x = 3$

c $\operatorname{sech} x = \frac{1}{\sqrt{2}}$ d $2\coth x = 5$

e $1 + \operatorname{cosech} x = -3$ f $5 - 6\operatorname{sech} x = 2$

5 Find the exact solutions to these equations.

a $\operatorname{arsech} x = \ln 7$ b $\operatorname{arcosech} x = 3\ln 2$

c $\operatorname{arcoth} 3x = \ln\sqrt{3}$ d $2\operatorname{arsech} 5x = \ln 16$

6 Find the exact solutions to the equations

a $\operatorname{sech}^2 x = \frac{3}{4}$ b $3\operatorname{cosech}^2 x = 1$

c $\coth^4 x - 9 = 0$

Reasoning and problem-solving

PURE

The identity $\cosh^2 x - \sinh^2 x \equiv 1$ can be used to prove other identities. For example, dividing both sides of $\cosh^2 x - \sinh^2 x \equiv 1$ by $\cosh^2 x$ leads to the identity $1 - \tanh^2 x \equiv \text{sech}^2 x$

Similarly, dividing by $\sinh^2 x$ leads to the identity $\coth^2 x - 1 \equiv \text{cosech}^2 x$

Key point

$\cosh^2 x - \sinh^2 x \equiv 1$

$1 - \tanh^2 x \equiv \text{sech}^2 x$

$\coth^2 x - 1 \equiv \text{cosech}^2 x \quad \text{for } x \neq 0$

The first identity is given in the formula book. You should memorise the other two and be able to derive them from the first.

See Ch3.5 For a reminder of hyperbolic identities.

You can also use the definitions for the hyperbolic functions in terms of exponentials to prove the double angle formulae:

Key point

$\sinh 2x \equiv 2 \sinh x \cosh x$

$\cosh 2x \equiv \cosh^2 x + \sinh^2 x$

$\tanh 2x \equiv \dfrac{2 \tanh x}{1 + \tanh^2 x}$

The first two identities are given in the formula book. You should be able to derive the third identity using the fact that $\tanh(2x) \equiv \dfrac{\sinh(2x)}{\cosh(2x)}$

For example, to prove the double angle formula for $\cosh(2x)$, use the definitions $\sinh x = \dfrac{e^x - e^{-x}}{2}$ and $\cosh x = \dfrac{e^x + e^{-x}}{2}$ to write

$\cosh^2 x + \sinh^2 x$ as $\left(\dfrac{e^x + e^{-x}}{2}\right)^2 + \left(\dfrac{e^x - e^{-x}}{2}\right)^2$

Expand and use index rules to give $\dfrac{e^{2x} + 2 + e^{-2x}}{4} + \dfrac{e^{2x} - 2 + e^{-2x}}{4}$

which simplifies to $\dfrac{2e^{2x} + 2e^{-2x}}{4} = \dfrac{e^{2x} + e^{-2x}}{2}$

which is the definition of $\cosh 2x$

You can also use the identity $\cosh^2 x - \sinh^2 x \equiv 1$ to rewrite the double angle formula for $\cosh 2x$

Key point

Double angle formulae

$\cosh 2x \equiv 1 + 2\sinh^2 x$ or $\cosh 2x \equiv 2\cosh^2 x - 1$

These versions are particularly useful when integrating as you will see in the next chapter.

You will need to use these identities to solve equations involving hyperbolic functions.

Strategy 1

To solve equations involving hyperbolic functions

1. Use $\text{sech}\, x = \dfrac{1}{\cosh x}$, $\text{cosech}\, x = \dfrac{1}{\sinh x}$ or $\coth x = \dfrac{1}{\tanh x}$
2. Use the definitions of $\cosh x$, $\sinh x$ or $\tanh x$ in terms of exponentials.
3. Form a quadratic in e^x then solve.
4. Find the values of x

Example 4

Solve the equation $4\operatorname{sech}x+\tanh x=4$

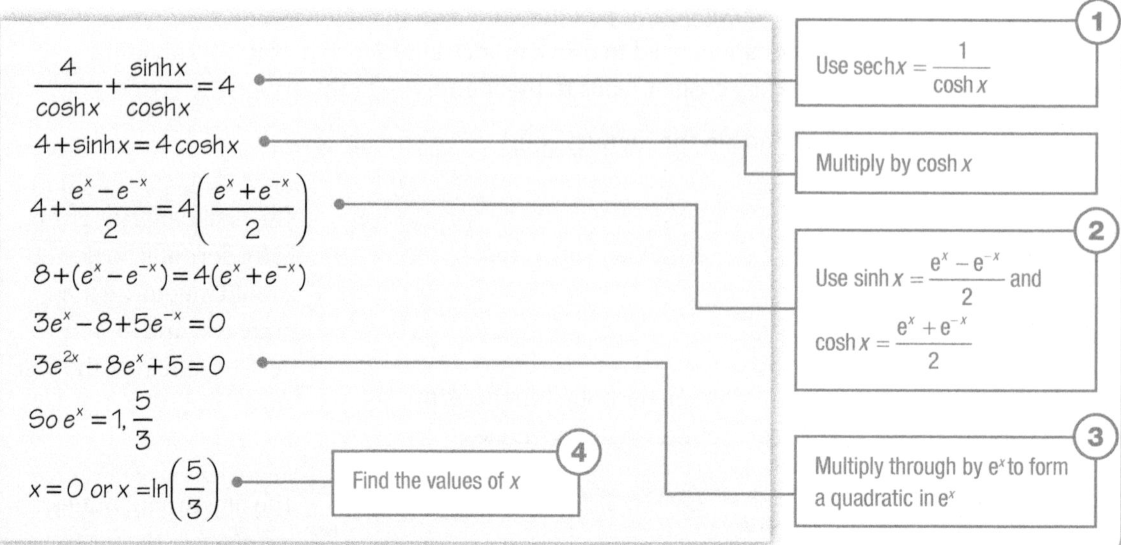

$\frac{4}{\cosh x}+\frac{\sinh x}{\cosh x}=4$
$4+\sinh x=4\cosh x$
$4+\frac{e^x-e^{-x}}{2}=4\left(\frac{e^x+e^{-x}}{2}\right)$
$8+(e^x-e^{-x})=4(e^x+e^{-x})$
$3e^x-8+5e^{-x}=0$
$3e^{2x}-8e^x+5=0$
So $e^x=1, \frac{5}{3}$
$x=0$ or $x=\ln\left(\frac{5}{3}\right)$
1 Use $\operatorname{sech}x=\frac{1}{\cosh x}$
Multiply by $\cosh x$
2 Use $\sinh x=\frac{e^x-e^{-x}}{2}$ and $\cosh x=\frac{e^x+e^{-x}}{2}$
3 Multiply through by e^x to form a quadratic in e^x
4 Find the values of x

Strategy 2

To solve quadratic equations involving reciprocal hyperbolic functions

1. Use identities to write the equation in terms of a single hyperbolic function.
2. Solve the quadratic to find the possible values of the reciprocal hyperbolic function.
3. Use the definitions of the reciprocal hyperbolic functions to find the values of $\sinh x$, $\cosh x$ or $\tanh x$
4. Use the definitions of the inverse hyperbolic functions to find the exact values of x

Example 5

Find the exact solutions to the equation $\coth^2 x-2\operatorname{cosech}x=4$

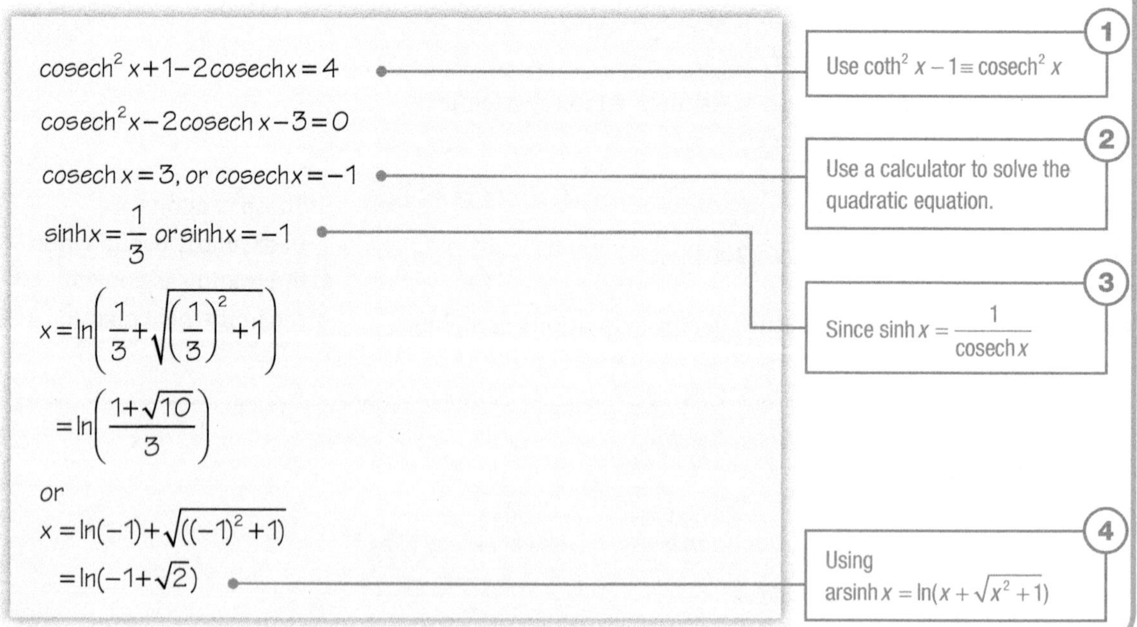

$\operatorname{cosech}^2 x+1-2\operatorname{cosech}x=4$
$\operatorname{cosech}^2 x-2\operatorname{cosech}x-3=0$
$\operatorname{cosech}x=3$, or $\operatorname{cosech}x=-1$
$\sinh x=\frac{1}{3}$ or $\sinh x=-1$
$x=\ln\left(\frac{1}{3}+\sqrt{\left(\frac{1}{3}\right)^2+1}\right)$
$=\ln\left(\frac{1+\sqrt{10}}{3}\right)$
or
$x=\ln(-1)+\sqrt{((-1)^2+1)}$
$=\ln(-1+\sqrt{2})$
1 Use $\coth^2 x-1\equiv\operatorname{cosech}^2 x$
2 Use a calculator to solve the quadratic equation.
3 Since $\sinh x=\frac{1}{\operatorname{cosech}x}$
4 Using $\operatorname{arsinh}x=\ln(x+\sqrt{x^2+1})$

Exercise 18.3B Reasoning and problem-solving

PURE

1 Use definitions in terms of exponentials to prove these identities

a $1+\text{cosech}^2 x \equiv \coth^2 x$

b $\text{cosech}2x \equiv \frac{1}{2}\text{cosech}\, x\, \text{sech}\, x$

c $\coth x \equiv \dfrac{\coth^2\left(\frac{x}{2}\right)+1}{2\coth\left(\frac{x}{2}\right)}$

d $\tanh^2\left(\frac{x}{2}\right) \equiv \dfrac{1-\text{sech}x}{1+\text{sech}x}$

2 a Show that $\cosh x + 2\,\text{sech}\, x = 3$ can be written as $\cosh^2 x - 3\cosh x + 2 = 0$

b Hence, solve the equation $\cosh x + 2\,\text{sech}\, x = 3$

3 a Use the exponential definitions of $\cosh x$ and $\sinh x$ to prove that $\sinh(2x) \equiv 2\sinh x \cosh x$

b Hence, show that $\coth x - \tanh x \equiv 2\,\text{cosech}(2x)$

c Hence, solve the equation $\coth x - \tanh x = -1$

4 Solve the equation $\text{cosech}x - \text{sech}x = \frac{1}{2}e^x$

Give your answer in the form $\ln A$ where A is a constant to be found.

5 a Use the definitions of $\sinh x$ and $\cosh x$ in terms of exponentials to prove that $\sinh^2 x = \frac{1}{2}(\cosh(2x)-1)$

b Hence solve the equation $\sinh(x)+\cosh(2x)=1$

6 a Prove that $\text{arcosech}\, x \equiv \ln\left(\frac{1}{x}+\sqrt{\frac{1}{x^2}+1}\right)$

b Hence, prove that $\text{arcosech}\, x + \text{arcosech}(-x) = 0$ for all $x \in \mathbb{R}$

7 Prove that $\text{arsech}\, x \equiv \pm\ln\left(\frac{1}{x}+\sqrt{\frac{1}{x^2}-1}\right)$

8 Given that $\text{cosech}\, x = \sqrt{2}$, find the exact values of

a $\cosh x$ b $\tanh x$

9 Given that $\text{sech}x = \frac{1}{\sqrt{3}}$, find the exact values of

a $\sinh x$ b $\coth x$

10 a Use the double angle formulae for sinh and cosh to prove that $\tanh(2x) \equiv \dfrac{2\tanh x}{1+\tanh^2 x}$

b Hence show that $\coth(2x) \equiv \frac{1}{2}(\coth x + \tanh x)$

c Solve the equation $\coth x + \tanh x = 4$

11 a Prove that $\dfrac{1}{1+\coth x}+\dfrac{1}{1-\coth x} \equiv -2\sinh^2 x$

b Hence solve the equation $\dfrac{1}{1+\coth x}+\dfrac{1}{1-\coth x} = -\dfrac{9}{2}$

Give your answers as simplified logarithms.

12 a On the same diagram sketch the graph with equations $y=\sinh x$ and $y = 2\,\text{sech}\, x$

b Hence state the number of solutions to the equation $\sinh x = 2\,\text{sech}\, x$

c Solve the equation $\sinh x = 2\,\text{sech}\, x$

13 a On the same diagram sketch the graph with equations $y = 2\coth x$ and $y=\sinh x$. State the equations of any asymptotes.

b Solve the equation $2\coth x = \sinh x$

Give your solutions to 3 significant figures.

14 Solve the equation $\text{cosech}^2 x + 2\,\text{cosech}\, x = 3$ Give your answers as exact logarithms.

15 Find the exact solutions to the equation $\text{sech}^2 x - \tanh^2 x = 1 - 3\,\text{sech}\, x$

16 Solve the equation $6\,\text{sech}^2 x - \tanh x = 4$

Give your answers as logarithms in their simplest form.

17 Find the solution to the equation $3\,\text{cosech}^2 x + \coth^2 x = 4\coth x$

Give your answer in the form $\ln k$, where k is a positive constant to be found.

18 Solve the equation $3\tanh 2x = \coth x$

Give your answers as simplified logarithms.

18.4 Rational functions with oblique asymptotes

Fluency and skills

See Ch3.1
For a reminder on finding asymptotes.

You can sketch the graphs of rational functions with horizontal and/or vertical asymptotes.

In this section, you will look at extending this to include rational functions that have **oblique** (or slant) asymptotes.

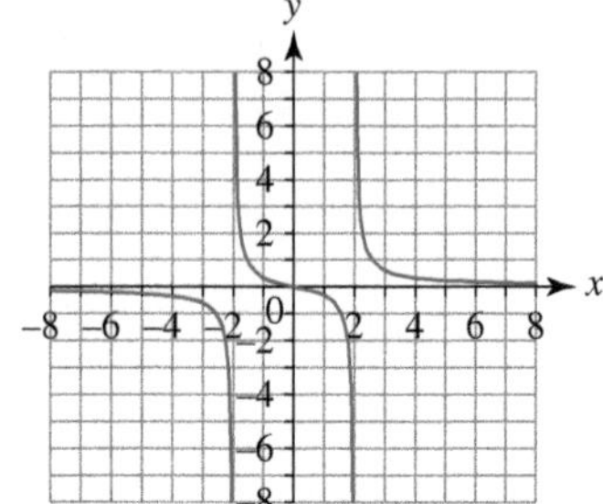

Consider the graph of the function $y = \dfrac{x}{x^2 - 4}$

There are vertical asymptotes at $x = 2$ and $x = -2$ since the function is undefined for these values of x

There is also a horizontal asymptote at $y = 0$ since y tends to zero as $x \to \pm\infty$

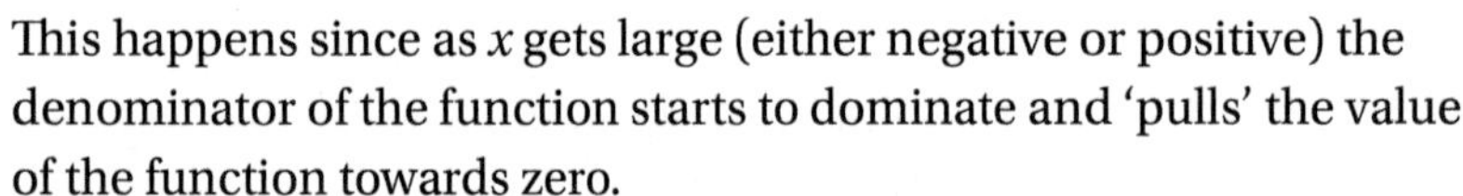

This happens since as x gets large (either negative or positive) the denominator of the function starts to dominate and 'pulls' the value of the function towards zero.

If the rational function is **improper** (or 'top heavy') this logic does not hold. For large x, the **numerator** of the function starts to dominate.

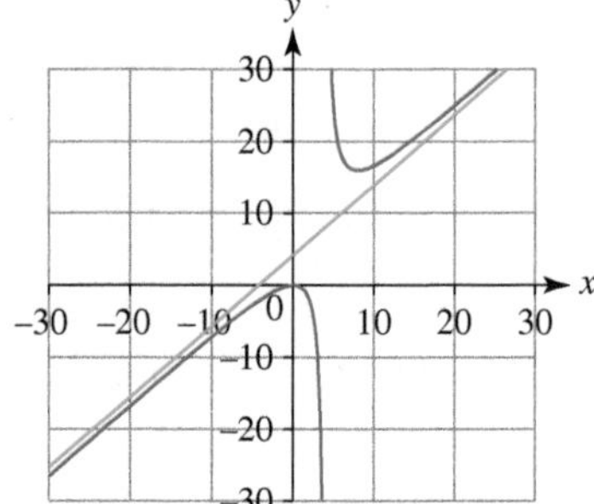

Consider the graph of the function $y = \dfrac{x^2}{x - 4}$

There is a vertical asymptote when $x = 4$ since the function is undefined for this value of x

There is also an oblique asymptote as shown.

Using algebraic long division, you can re-write the function as

$y = x + 4 + \dfrac{16}{x - 4}$

Now, as $x \to \pm\infty$ you can see that the fraction part of the functions tends to zero as in the case of proper rational functions, but that leaves the $x + 4$ ('whole') bit. Hence, as $x \to \pm\infty$, y tends to $x + 4$ and there is an oblique asymptote with equation $y = x + 4$

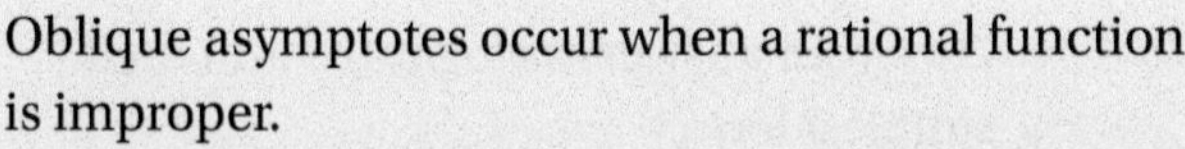

Key point

Oblique asymptotes occur when a rational function is improper.

You find the equation of an oblique asymptote by dividing the function out to find the 'whole' part and setting y equal to this.

Example 1

Find the equations of the asymptotes of the curve $y=\dfrac{x^2-x}{x+1}$

There is a vertical asymptote at $x=-1$

The function is undefined when $x=-1$ since the denominator would be zero.

$$\begin{array}{r} x-2 \\ x+1\overline{)x^2-x} \\ \underline{x^2+x} \\ -2x \\ \underline{-2x-2} \\ 2 \end{array}$$

Use algebraic long division to divide out the function.

Hence $y=x-2+\dfrac{2}{x+1}$

There is an oblique asymptote with equation $y=x-2$

Use the divided out form to identify the equation of the oblique asymptote.

Exercise 18.4A Fluency and skills

1 Find the equations of the asymptotes for each of these curves.

a $y=\dfrac{x^2+2x-12}{x-5}$ b $y=\dfrac{2x^2}{1-x}$

c $y=\dfrac{2-3x^2}{x-1}$ d $y=\dfrac{x^2+x-1}{x-2}$

2 Sketch the graphs of each of these functions. Show clearly any asymptotes.

a $y=\dfrac{x^2-1}{x+3}$ b $y=\dfrac{2x^2-1}{2x+3}$

c $y=\dfrac{x^2-2x+1}{4-x}$ d $y=\dfrac{x^2-4x+3}{3-2x}$

3 Find the equations of the asymptotes for each of these curves. Hence sketch the curves.

a $y=\dfrac{x^3-1}{2x^2-1}$ b $y=\dfrac{x^3-x}{x^2-4}$

Reasoning and problem-solving

You need to be able to find the value of unknowns in a rational function when you are given the equation of the oblique asymptote.

Typically, this will be the values of a and/or b when the numerator is in the form ax^2+bx+c

Strategy

To find the value of unknowns in the rational function when you are given the equation of the oblique asymptote

1. Multiply the asymptote by the denominator of the rational function to form a quadratic expression in the form ax^2+bx+c
2. Read off the values of a and/or b from this quadratic and these will correspond to the values of a and/or b in the rational function.

Example 2

A curve with equation $y=\dfrac{ax^2+bx+1}{x-2}$ has an asymptote $y=2x+1$

a Find the values of a and b

b Write down the equation of the other asymptote.

c Without using calculus, find the coordinates of the turning points.

d Sketch the curve.

a $(2x+1)(x-2)=2x^2-3x-2$

Hence $a=2$ and $b=-3$

The rational function is $y=\dfrac{2x^2-3x+1}{x-2}$

b $x=2$

c The range of allowable values of y can be found by setting $y=k$ and solving for k

$k(x-2)=2x^2-3x-1$

$0=2x^2-3x-kx-1+2k$

$0=2x^2-(3+k)x+(2k-1)$

Solving for k using $b^2-4ac\geq 0$ gives

$k\geq 5+2\sqrt{2}$ or $k\leq 5-2\sqrt{2}$

Hence the curve exists for $y\geq 5+2\sqrt{2}$ and $y\leq 5-2\sqrt{2}$

There will be a minimum point at $y=5+2\sqrt{2}$ and a maximum point at $y=5-2\sqrt{2}$

To find the x-values for these turning points, substitute the y-values into the original function

$(5\pm 2\sqrt{2})(x-2)=2x^2-3x-1$

$\Rightarrow 2x^2-(8\pm 2\sqrt{2})x+(9\pm 2\sqrt{2})=0$

Solving for x in this equation gives the x-values as $\dfrac{4\pm\sqrt{2}}{2}$

Hence the turning points are:

Minimum at $\left(\dfrac{4+\sqrt{2}}{2}, 5+2\sqrt{2}\right)$

Maximum at $\left(\dfrac{4-\sqrt{2}}{2}, 5-2\sqrt{2}\right)$

d

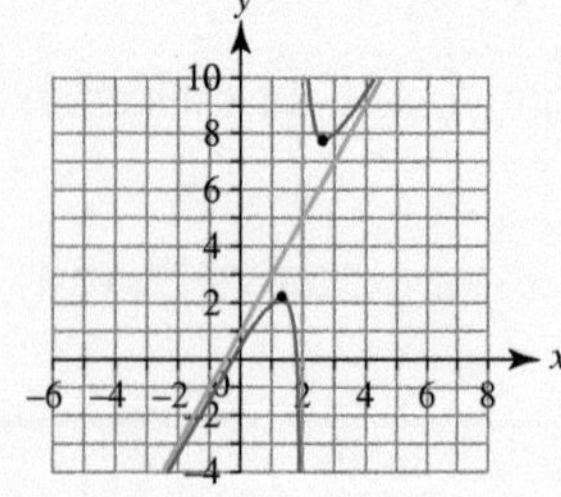

(1) Multiply the asymptote by the denominator of the rational function.

(2) Read off the values of a and b

This is the value of x for which the function is undefined.

For a full review of this technique, see Section 3.2, Example 3

Draw in both asymptotes and use your answer to part **c** to sketch the sections of the graph.

Exercise 18.4B Reasoning and problem-solving

1 A curve with equation $y=\frac{ax^2+bx-1}{x+1}$ has an asymptote $y=4x-2$

a Find the values of a and b

b Write down the equation of the other asymptote.

c Without using calculus, find the coordinates of the turning points.

d Sketch the curve.

2 A curve with equation $y=\frac{ax^2+bx+1}{x-2}$ has an asymptote $y=-x-5$

a Find the values of a and b

b Write down the equation of the other asymptote.

c Without using calculus, find the coordinates of the turning points.

d Sketch the curve.

3 $f(x)=\frac{x^3-1}{2(x^2-1)}$

a Show that $f(x)$ can be written in the form $g(x)+\frac{bx+c}{2(x+1)}$ where b and c are constants to be found.

b Hence write down the equations of the asymptotes to the curve $y=f(x)$

c Find the coordinates of the turning points on the curve $y=f(x)$

d Hence sketch the curve $y=f(x)$

4 $f(x)=\frac{x^3}{x-1}$

a Write $f(x)$ in the form $g(x)+\frac{c}{x-1}$ where c is a constant.

b Hence write down the equations of the asymptotes to the curve $y=f(x)$

c Sketch the curve $y=f(x)$

18 Summary and review

Chapter summary

- Given $y = \mathrm{f}(x)$ you can sketch the graph of $y = \frac{1}{\mathrm{f}(x)}$ by following some simple rules
  - The sign of $\frac{1}{\mathrm{f}(x)}$ is the same as the sign of $\mathrm{f}(x)$
  - Any roots (or zeros) of $y = \mathrm{f}(x)$ become vertical asymptotes in $y = \frac{1}{\mathrm{f}(x)}$
  - As $\mathrm{f}(x) \to 0$ then $\frac{1}{\mathrm{f}(x)} \to \pm\infty$ and vice versa
  - As $\mathrm{f}(x)$ increases, $\frac{1}{\mathrm{f}(x)}$ decreases and vice versa
  - If $\mathrm{f}(x) = \pm 1$ exists then $\frac{1}{\mathrm{f}(x)}$ also equals ± 1 and the graphs intersect
  - Minimum points become maximum points or asymptotes and vice versa
- To sketch the graph of $y = |\mathrm{f}(x)|$
  - Sketch $y = \mathrm{f}(x)$ and reflect any negative parts of the graph in the x-axis
  - Any minimum turning point below the x axis will be reflected to a maximum turning point above the x axis.
- To transform graphs of conic sections
  - Enlarge by scale factor k by replacing x with $\frac{x}{k}$ and y with $\frac{y}{k}$
  - Rotate by $\frac{\pi}{2}$ by replacing x by y and y by $-x$
  - Rotate by π by replacing x by $-x$ and y by $-y$
  - Rotate by $\frac{3\pi}{2}$ by replacing x by $-y$ and y by x
- $\sinh x = \frac{e^x - e^{-x}}{2} \Rightarrow \operatorname{arsinh} x = \ln(x + \sqrt{x^2 + 1})$
- $\cosh x = \frac{e^x + e^{-x}}{2} \Rightarrow \operatorname{arcosh} x = \ln(x + \sqrt{x^2 - 1}),\ x \geq 1$
- $\tanh x = \frac{e^x - e^{-x}}{e^x + e^{-x}} \Rightarrow \operatorname{artanh} x = \frac{1}{2}\ln\left(\frac{1+x}{1-x}\right), -1 < x < 1$
- Graphs of inverse hyperbolic functions

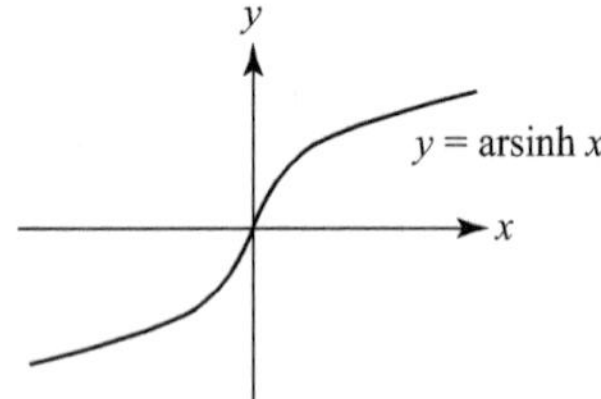

domain is $x \in \mathbb{R}$,
range is $y \in \mathbb{R}$

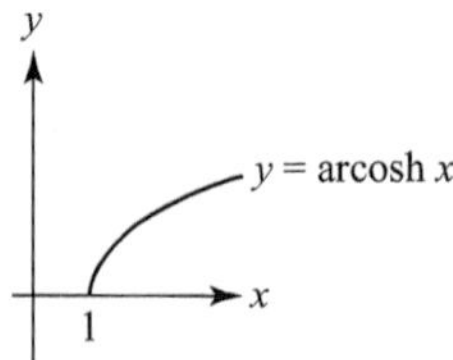

domain is $x \in \mathbb{R},\ x \geq 1$
range is $y \in \mathbb{R},\ y \geq 0$

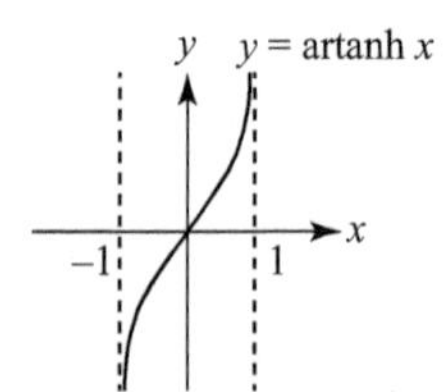

domain is $x \in \mathbb{R}, -1 < x < 1$
range is $y \in \mathbb{R}$

$f(x)$	Domain of $f(x)$	Range of $f(x)$	$f^{-1}(x)$	Domain of $f^{-1}(x)$	Range of $f^{-1}(x)$
$\sinh x$	$x \in \mathbb{R}$	$f(x) \in \mathbb{R}$	$\text{arsinh}\, x$	$x \in \mathbb{R}$	$f^{-1}(x) \in \mathbb{R}$
$\cosh x$	$x \in \mathbb{R}$, $x \geq 0$	$f(x) \in \mathbb{R}, f(x) \geq 1$	$\text{arcosh}\, x$	$x \in \mathbb{R}$, $x \geq 1$	$f^{-1}(x) \in \mathbb{R}, f^{-1}(x) \geq 0$
$\tanh x$	$x \in \mathbb{R}$	$f(x) \in \mathbb{R}$, $-1 < f(x) < 1$	$\text{artanh}\, x$	$x \in \mathbb{R}$, $-1 < x < 1$	$f^{-1}(x) \in \mathbb{R}$

- $\text{cosech}\, x \equiv \dfrac{1}{\sinh x} \equiv \dfrac{2}{e^x - e^{-x}}$
- $\text{sech}\, x \equiv \dfrac{1}{\cosh x} \equiv \dfrac{2}{e^x + e^{-x}}$
- $\coth x \equiv \dfrac{1}{\tanh x} \equiv \dfrac{e^{2x}+1}{e^{2x}-1}$
- Graphs of reciprocal hyperbolic functions

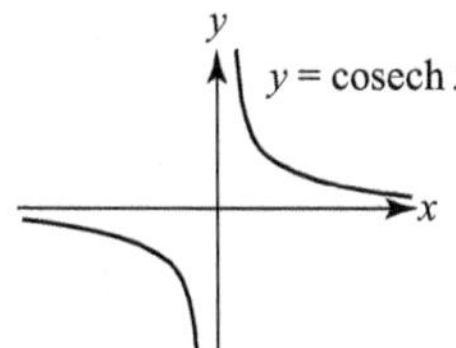

domain is $x \in \mathbb{R}$, $x \neq 0$, range is $y \in \mathbb{R}, y \neq 0$

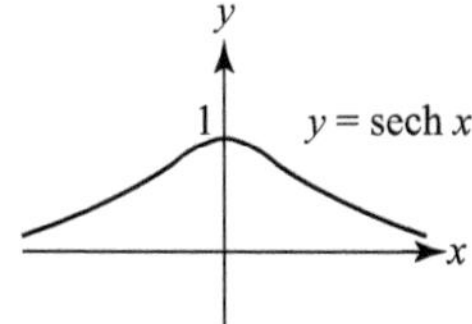

domain is $x \in \mathbb{R}$, range is $y \in \mathbb{R}$, $0 < y \leq 1$

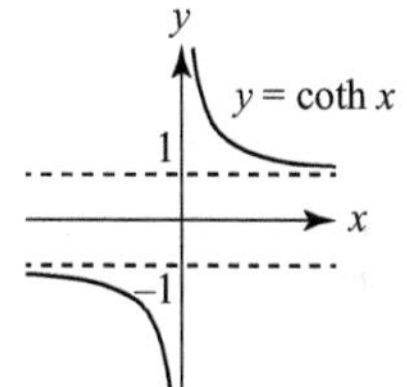

domain is $x \in \mathbb{R}$, $x \neq 0$ range is $y \in \mathbb{R}$, $y < -1$, $y > 1$

$f(x)$	Domain	Range
$\text{cosech}\, x$	$x \in \mathbb{R}, x \neq 0$	$f(x) \in \mathbb{R}, f(x) \neq 0$
$\text{sech}\, x$	$x \in \mathbb{R}$	$f(x) \in \mathbb{R}, 0 < f(x) \leq 1$
$\coth x$	$x \in \mathbb{R}, x \neq 0$	$f(x) \in \mathbb{R}, f(x) < -1$ or $f(x) > 1$

- $\cosh^2 x - \sinh^2 x \equiv 1 \Rightarrow 1 - \tanh^2 x \equiv \text{sech}^2 x$ and $\coth^2 x - 1 \equiv \text{cosech}^2 x$
- Oblique asymptotes occur when a rational function is improper.
- You find the equation of an oblique asymptote by dividing the function out to find the 'whole' part and setting y equal to this.
- To find the value of unknowns in the rational function when you are given the equation of the oblique asymptote
  - Multiply the asymptote by the denominator of the rational function to form a quadratic expression in the form $ax^2 + bx + c$
  - Read off the values of a and/or b from this quadratic and these will correspond to the values of a and/or b in the rational function.

Check and review

You should now be able to...	Review Questions
✔ Use graphical and algebraic methods to find and sketch reciprocal and modulus functions.	1, 3, 4–9
✔ Use graphical and algebraic methods to solve inequalities involving reciprocal and modulus functions.	2, 9
✔ Enlarge graphs of conic sections.	10, 11
✔ Rotate graphs of conic sections.	12, 13
✔ Combine transformations of conic sections.	12–15
✔ Sketch graphs of inverse hyperbolic functions and state each domain and range.	16
✔ Sketch graphs of reciprocal hyperbolic functions and state each domain and range.	17
✔ Know, prove and use identities involving hyperbolic functions.	19–20
✔ Solve equations involving hyperbolic functions.	18–21
✔ Understand and use the concept of an oblique asymptote, including finding the equation of an oblique asymptote.	22–23

1 Sketch the graph of $y=\frac{(x-1)}{(x+4)}$

2 Solve these inequalities graphically, naming any intercepts and vertical and/or horizontal asymptotes.

a $-3<\frac{x+12}{x-4}<9$ **b** $7\geq\frac{x-6}{x-12}\geq-5$

3 Sketch the graphs of

a $y=\frac{x^2+12x+24}{x^2+2x+2}$ **b** $y=\frac{x^2+12x+24}{x^2+2x-3}$

4 Find the localised minimum turning point of the function $f(x)=\frac{x^2-5x}{x^2-5x+5}$ and find a maximum value.

5 The diagram shows the graph of $y=f(x)$ and the points P, Q and R

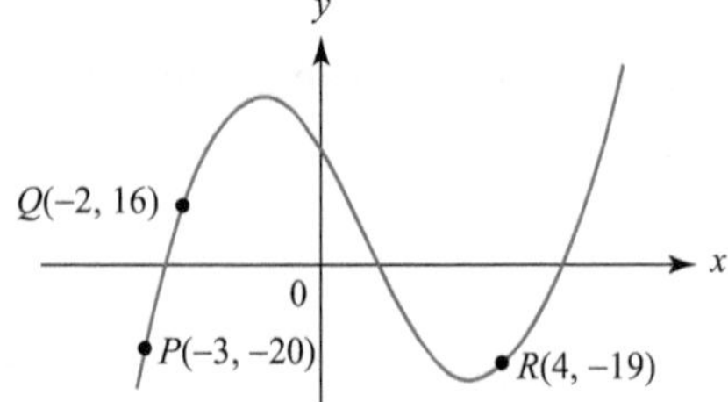

Sketch the graph of $y=|f(x)|$, labelling the transformed points P', Q' and R'

6 Sketch the graphs of $y = |f(x)|$ for these functions.

a $f(x) = x^2 - 2x - 5$

b $f(x) = 2x^2 - 3x - 6$

7 Copy each graph $y = f(x)$ and sketch on the same axes the graph of $y = \dfrac{1}{f(x)}$

a

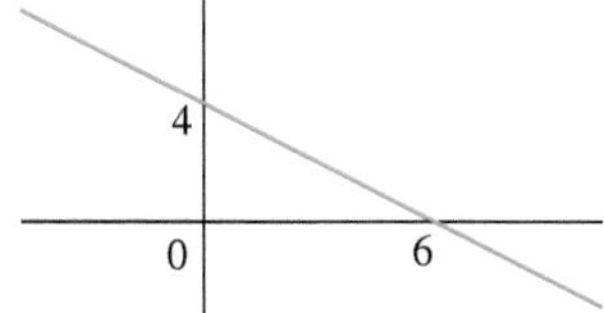

b

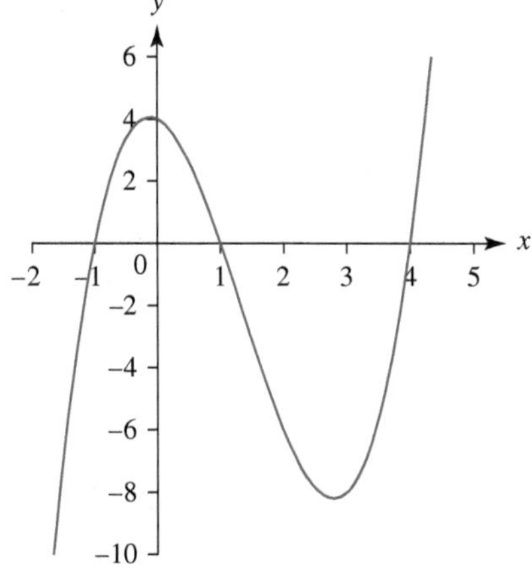

8 For each function

i Sketch $y = f(x)$

ii Sketch $\dfrac{1}{f(x)}$, labelling any asymptotes and points of intersection.

a $y = -3x - 2$

b $y = x^2 + 6x + 5$

c $y = x^3 + 4$

9 A curve has equation $y = \dfrac{4+x}{2-x}$

a i Write down the equations of the asymptotes to the curve and the points of intersection with the coordinate axes.

ii Sketch the curve, indicating clearly the coordinates of the points of intersection of the curve with the coordinate axes.

b Sketch also the curve $y = \left|\dfrac{4+x}{2-x}\right|$

c Add the straight line $y = 2 - \dfrac{3}{2}x$ to each sketch and hence solve the inequalities $\dfrac{4+x}{2-x} > 2 - \dfrac{3}{2}x$ and $\left|\dfrac{4+x}{2-x}\right| > 2 - \dfrac{3}{2}x$

PURE

10 The ellipse with equation $\frac{x^2}{5}+\frac{y^2}{9}=1$ is enlarged by scale factor 3

a Find the equation of the transformed curve.

b Sketch the transformed curve, stating the coordinates of the points of intersection with the coordinate axes.

11 The hyperbola with equation $\frac{x^2}{64}-\frac{y^2}{36}=1$ is enlarged by scale factor $\frac{1}{2}$

a Find the equation of the transformed curve.

b Sketch the transformed curve and find the equations of the asymptotes.

12 The parabola with equation $y^2=18x$ is rotated $\frac{\pi}{2}$ radians anticlockwise about the origin.

a Find the equation of the transformed curve.

The curve is then translated by the vector $\begin{pmatrix}-1\\2\end{pmatrix}$

b Find the equation of the transformed curve.

c Sketch the transformed curve, and state the coordinates of the minimum point.

13 The ellipse with equation $\frac{x^2}{9}+\frac{y^2}{4}=1$ is rotated $\frac{3\pi}{2}$ radians anticlockwise about the origin.

a Find the equation of the transformed curve.

The curve is then translated by vector $\begin{pmatrix}2\\3\end{pmatrix}$

b Find the equation of the transformed curve.

c Sketch the transformed curve.

14 The ellipse with equation $\frac{x^2}{4}+\frac{y^2}{3}=1$ is first enlarged by scale factor 5 then translated by vector $\begin{pmatrix}1\\-4\end{pmatrix}$

a Find the equation of the transformed curve.

b Sketch the transformed curve.

15 The parabola with equation $y^2=7x$ is reflected in the line $y=-x$ then rotated $\frac{\pi}{2}$ radians anticlockwise about the origin.

a Find the equation of the transformed curve.

b Sketch the transformed curve.

16 Sketch the graph of $y=\mathrm{f}(x)$ for each of these functions and state each domain and range.

a $y=1+\operatorname{arsinh}(x)$

b $y=\operatorname{arcosh}(2x)$

c $y=\operatorname{artanh}(x+1)$

17 Sketch the graph of $y = \mathrm{f}(x)$ for this function and state the domain and range.

$y = 3\operatorname{sech}(x)$

18 Solve these equations and give your answer in the form $a \ln b$ where a and b are constants to be found.

a $\coth(x) = 3$

b $\mathrm{e}^x \operatorname{cosech}(x) = -1$

c $\operatorname{cosech} x - \operatorname{sech} x = 2\mathrm{e}^{-x}$

19 a Use exponentials to prove that $\operatorname{cosech}(2x) \equiv \frac{1}{2}\operatorname{cosech}(x)\operatorname{sech}(x)$

b Hence solve the equation $\operatorname{cosech}(x)\operatorname{sech}(x) = -2$

20 Find the exact solutions to each of these equations.

a $\operatorname{cosech}^2 x + 3\operatorname{cosech} x = 4$

b $5\coth x = \operatorname{cosech}^2 x + 1$

21 Show that the equation $\tanh^2 x + \operatorname{sech} x + 5 = 0$ has no real solutions.

22 $\mathrm{f}(x) = \dfrac{x^2 - 2x - 1}{x - 5}$

a Find the equations of the asymptotes of the curve $y = \mathrm{f}(x)$

b Hence sketch the curve $y = \mathrm{f}(x)$

23 A curve with equation $y = \dfrac{ax^2 + bx - 1}{3x - 1}$ has an asymptote $y = x - \dfrac{1}{3}$

a Find the values of a and b

b Write down the equation of the other asymptote.

c Sketch the curve.

PURE

18 Exploration

Going beyond the exams

History

The Cartesian coordinate system is named after the French mathematician René Descartes (1596–1650). It is suggested that he developed the system whilst musing about how to best describe the position of a fly on his ceiling: this led to his development of a system of coordinates which referred to an origin at the corner of the ceiling. This development led to the possibility of describing shapes such as circles and parabolas using equations that describe the relationships between their x and y coordinates, thus linking geometry and algebra.

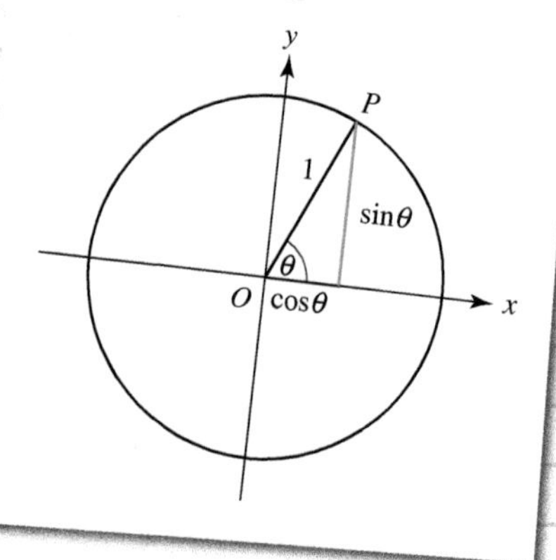

Note

Sine and cosine functions are known as circular functions as they can be defined using the position of a point on a circle. Let a point, P, lie on a unit circle with centre O such that the angle at O measured anticlockwise from the positive x-axis to the line segment OP is θ. As the point moves around the circle, the x- and y-coordinates of P define cosine and sine respectively.

The hyperbolic functions are defined in a similar way using the hyperbola $x^2 - y^2 = 1$ $(x > 1)$ What is similar and what is different?

In particular, notice that the hyperbolic angle is defined in terms of the area enclosed between a line passing through a point on the hyperbola, the hyperbola and the x-axis. (Areas beneath the x-axis are considered negative.)

Explore the graphs of the hyperbolic functions by considering how they can be traced out as the point, P, moves on the hyperbola.

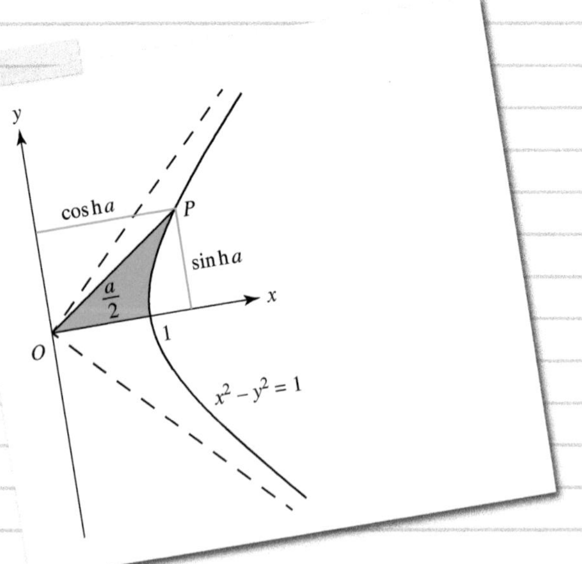

Assessment

1 $f(x) = 2x^2 - 3x - 2$

a Sketch the graph of $y = f(x)$ showing clearly the points where the curve crosses the coordinate axes. **[2 marks]**

b On the same set of axes, sketch the graph of $y = \frac{1}{f(x)}$, labelling clearly any asymptotes. **[3]**

c Calculate the set of values of x for which $f(x) < \frac{1}{f(x)}$ **[6]**

2 The graph of $y = f(x)$ is shown in the diagram.

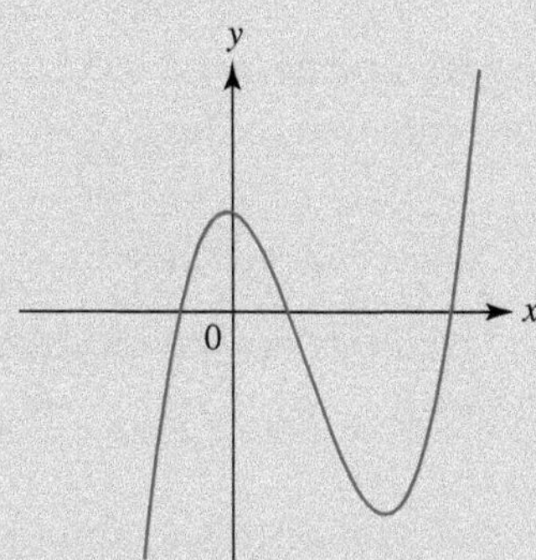

a Sketch the graph of $y = \frac{1}{f(x)}$, showing clearly any asymptotes. **[4]**

b Sketch the graph of $y = |f(x)|$ **[3]**

3 $f(x) = \frac{3 - x}{1 + x}$

a Write down the equations of any asymptotes on the graph of $y = f(x)$ and state the coordinates of any points of intersection with the coordinate axes. **[3]**

b Sketch the graph of $y = |f(x)|$ **[3]**

c Using your sketch or otherwise, find the set of values of x for which $|f(x)| < 3$ **[2]**

4 The ellipse E has equation $\frac{x^2}{4} + \frac{y^2}{2} = 1$

The ellipse is enlarged by scale factor 2 and then translated by vector $\begin{pmatrix} 2 \\ -3 \end{pmatrix}$

a Show that the new conic has equation $x^2 + 2y^2 - 4x + 12y + k = 0$ where k is a constant to be found. **[5]**

b Find the equations of the tangents to the new conic which are parallel to the x-axis. **[4]**

5 A parabola P has equation $y^2 = 16x$

Describe a sequence of transformations that map P onto the curve $x^2 - 2x + 16y + 33 = 0$ **[5]**

6 A hyperbola H has equation $\frac{x^2}{9}-\frac{y^2}{4}=1$

a Show that the line $3y-x=3$ intersects the hyperbola at two points and find the exact coordinates of these points. **[5]**

The hyperbola H is stretched by scale factor 2 in the x-direction and then rotated anticlockwise through $\frac{\pi}{2}$ radians about the origin to produce the hyperbola H'

b Sketch the graph of H' **[3]**

c Write down the equations of the asymptotes to H' **[1]**

7 From the definitions of $\sinh x$ and $\cosh x$ in terms of exponentials,

a Prove that $\cosh^2 x - \sinh^2 x = 1$ **[3]**

b Hence, or otherwise, solve the equation $\operatorname{cosech} x = 2(1+\coth x)$, giving your answer in the form $a\ln b$ where a and b are constants to be found. **[5]**

8 a Use exponentials to show that $2\cosh^2 x = \cosh 2x + 1$ **[3]**

b Hence, or otherwise, solve the equation $5\cosh x = \cosh 2x - 2$, giving your answers in exact logarithmic form. **[4]**

9 a Show that $\sinh x + 3\operatorname{cosech} x = 4$ can be written as $\sinh^2 x - 4\sinh x + 3 = 0$ **[2]**

b Hence find the exact solutions to the equation $\sinh x + 3\operatorname{cosech} x = 4$ **[3]**

10 $f(x) = \operatorname{cosech}(2x+1)$

a Sketch the graph of $y = f(x)$, showing clearly the equations of any asymptotes. **[3]**

b State the domain and range of $f(x)$ **[2]**

c Show that the coordinates of the point where the graph of $y = f(x)$ intersects the y-axis is $\left(0, \frac{2e}{e^2-1}\right)$ **[3]**

11 $f(x)=\frac{2x^2-3}{x+1}$

a Find the equations of the asymptotes of the curve $y = f(x)$ **[3]**

$g(x)=\frac{ax^2+bx-1}{2x-1}$

The curve with equation $y = g(x)$ has the same oblique asymptote as the curve with equation $y = f(x)$

b Find the values of a and b **[2]**

c Hence sketch the curve $y = g(x)$ **[3]**

19 Integration 2

Economics is an area of study and work that has become increasingly important. You can see this by the time and space that is devoted to business and the economy in various parts of the media. Many aspects of the work of economists has become more mathematical, as they develop, and use, a range of mathematical models. These models allow them to understand not only what has happened in the past, but to also predict what may happen in the future.

Aspects of calculus are important in the mathematics used by economists. Methods and techniques associated with integration and differentiation allow economists to consider how measurable quantities change with time. For example, supply and demand, in relation to (number of) sales and pricing, are quantities that can be modelled using mathematical functions. Once the model is set up, calculus can be used to work out optimal conditions.

Orientation

What you need to know	What you will learn	What this leads to
Chapter 4 Integration 1	• How to work with improper integrals. • How to differentiate inverse trigonometric functions. • How to differentiate hyperbolic functions. • How to use partial fractions to help with integration. • How to use a reduction formula. • How to use integration to find areas enclosed by a polar curve. • How to use integration to find lengths and areas.	**Careers** • Economics.

19.1 Improper integrals

Fluency and skills

You can find the value of a definite integral by integrating and then substituting in the limits. This value represents a finite area between the curve and an axis.

Key point

An **improper integral** is a definite integral where either:

- one or both of the limits is $\pm\infty$
- the integrand (expression to be integrated) is undefined at one of the limits of the integral
- the integrand is undefined at some point between the limits of the integral.

It is sometimes possible to calculate the value of an improper integral by replacing a limit of $\pm\infty$ with a variable and then considering what happens as that variable tends to $\pm\infty$

Key point

To evaluate an improper integral with a limit of $\pm\infty$ use

$$\int_a^\infty f(x)dx = \lim_{t\to\infty}\int_a^t f(x)dx \quad \text{or} \quad \int_{-\infty}^a f(x)dx = \lim_{t\to-\infty}\int_t^a f(x)dx$$

If the limit exists, then the improper integral is called **convergent**.

If the limit does not exist, then the improper integral is called **divergent**.

For example, the integral, $\int_1^\infty \frac{1}{x^2}dx$ is an improper integral because one of the limits is infinite. To evaluate it, use

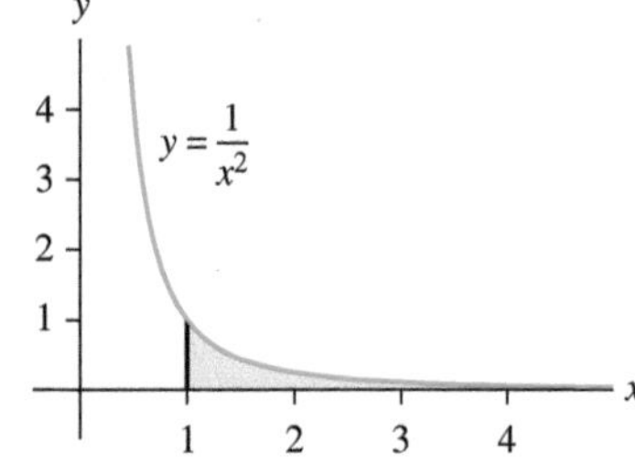

$$\int_1^\infty \frac{1}{x^2}dx = \lim_{t\to\infty}\int_1^t \frac{1}{x^2}dx$$

$$= \lim_{t\to\infty}\left(1-\frac{1}{t}\right) \qquad \text{since } \int_1^t \frac{1}{x^2}dx = \left[-\frac{1}{x}\right]_1^t = 1-\frac{1}{t}$$

$$= 1 \text{ since } \frac{1}{t}\to 0 \text{ as } t\to\infty$$

Therefore, the improper integral $\int_1^\infty \frac{1}{x^2}dx$ is convergent and represents a finite area of 1

However, if you were to try to evaluate the integral $\int_0^\infty x^2dx$ then

$$\lim_{t\to\infty}\int_0^t x^2dx = \lim_{t\to\infty}\left(\frac{t^3}{3}\right) \qquad \text{since } \int_0^t x^2dx = \left[\frac{x^3}{3}\right]_0^t = \frac{t^3}{3}$$

$\frac{t^3}{3}\to\infty$ as $t\to\infty$, therefore the improper integral $\int_0^\infty x^2dx$ is divergent and represents an infinite area.

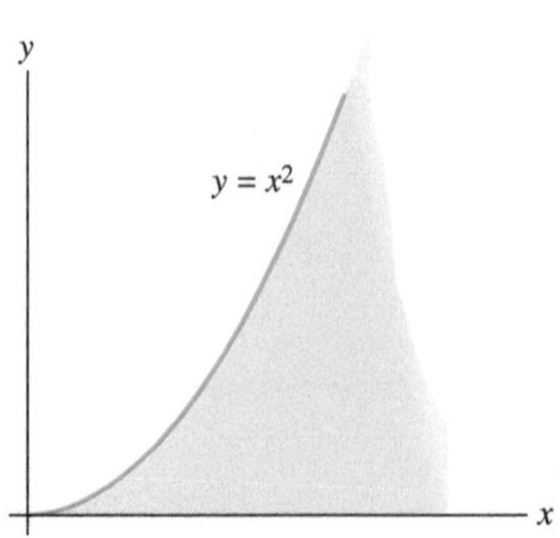

PURE

Example 1

Find the value of $\int_2^{\infty} \frac{2}{x^3}\,dx$

$\int_2^{\infty} \frac{2}{x^3}dx = \lim_{t\to\infty}\int_2^t 2x^{-3}dx$

Replace ∞ with t

$= \lim_{t\to\infty}\left[-x^{-2}\right]_2^t$

$= \lim_{t\to\infty}((-t^{-2})-(-2^{-2})) = \lim_{t\to\infty}\left(-\frac{1}{t^2}+\frac{1}{4}\right) = \frac{1}{4}$ since $\frac{1}{t^2}\to 0$ as $t\to\infty$

If both the limits are ±∞, then the integral needs to be split into two integrals, each with one finite limit. Any point can be chosen for this limit, so just choose a convenient value.

When splitting the integral, you need to use different variables for ∞ and −∞
This is because both parts of the integral must be convergent for the original integral to exist.

Example 2

a Find the value of $\int_{-\infty}^{\infty} \frac{x}{e^{x^2}}\,dx$

b Show that the improper integral $\int_{-\infty}^{\infty} e^x$ is divergent.

a $\int_{-\infty}^{\infty} \frac{x}{e^{x^2}}dx = \lim_{a\to-\infty}\int_a^0 xe^{-x^2}dx + \lim_{b\to\infty}\int_0^b xe^{-x^2}dx$

Choose to split the integral at 0

$= \lim_{a\to-\infty}\left[-\frac{1}{2}e^{-x^2}\right]_a^0 + \lim_{b\to\infty}\left[-\frac{1}{2}e^{-x^2}\right]_0^b$

Since $\frac{d}{dx}(e^{-x^2}) = -2xe^{-x^2}$ using the chain rule.

$= \lim_{a\to-\infty}\left(-\frac{1}{2}e^0 - -\frac{1}{2}e^{-a^2}\right) + \lim_{b\to\infty}\left(-\frac{1}{2}e^{-b^2} - -\frac{1}{2}e^0\right)$

$= \lim_{a\to-\infty}\left(-\frac{1}{2}+\frac{1}{2}e^{-a^2}\right) + \lim_{b\to\infty}\left(-\frac{1}{2}e^{-b^2}+\frac{1}{2}\right)$

Since $e^0 = 1$

$= \left(-\frac{1}{2}\right)+\left(\frac{1}{2}\right)$

Both limits exist, therefore the improper integral is convergent.

Since as $a\to-\infty$, $e^{-a^2}\to 0$ and as $b\to\infty$, $e^{-b^2}\to 0$

So the improper integral $\int_{-\infty}^{\infty}\frac{x}{e^{x^2}}dx = -\frac{1}{2}+\frac{1}{2} = 0$

b $\int_{-\infty}^{\infty} e^x dx = \lim_{a\to-\infty}\int_a^0 e^x dx + \lim_{b\to\infty}\int_0^b e^x dx$

$= \lim_{a\to-\infty}\left[e^x\right]_a^0 + \lim_{b\to\infty}\left[e^x\right]_0^b$

$= \lim_{a\to-\infty}(e^0 - e^a) + \lim_{b\to\infty}(e^b - e^0)$

$= \lim_{a\to-\infty}(1-e^a) + \lim_{b\to\infty}(e^b - 1)$

As $a\to-\infty$, $e^a\to 0$ so $\lim_{a\to-\infty}\int_a^0 e^x dx = 1$

However, as $b\to\infty$, $e^b\to\infty$ so $\lim_{b\to\infty}\int_0^b e^x dx$ does not exist.

Both limits must exist for the improper integral to be convergent.

Therefore, the improper integral $\int_{-\infty}^{\infty} e^x dx$ is divergent.

Another type of improper integral is where the integrand is undefined at one of the limits of the integral. To evaluate these integrals you replace that limit of integration with a variable as before. You then consider what happens as the variable tends to the original value of the limit of integration.

Example 3

Find the value of $\int_0^4 \frac{2}{\sqrt{x}}\,dx$

$$\int_0^4 \frac{2}{\sqrt{x}}\,dx = \lim_{t\to 0}\int_t^4 x^{-\frac{1}{2}}\,dx$$

Replace 0 with t since $\frac{1}{\sqrt{x}}$ is undefined at $x = 0$

$$= \lim_{t\to 0}\left[2x^{\frac{1}{2}}\right]_t^4$$

$$= \lim_{t\to 0}(4 - 2\sqrt{t})$$

$$= 4$$

Since $2\sqrt{t} \to 0$ as $t \to 0$, the value of the integral is 4

When evaluating integrals, you sometimes need to find the limit of $x^k e^{-x}$ as $x \to \infty$ for some value of k. Since $x^k \to \infty$ but $e^{-x} \to 0$ it is not obvious what will happen to the value of $x^k e^{-x}$. You need to know this result:

Key point

For any real number k, $x^k e^{-x} \to 0$ when $x \to \infty$

This can be proved using the series expansion of e^x but you can simply quote it.

Another common limit is that of $x^k \ln x$ as $x \to 0$. Again, as $x^k \to 0$ and $\ln x \to -\infty$ it is not obvious what the result will be, but the previous result can be used to prove the following:

Key point

For any real number k, $x^k \ln x \to 0$ when $x \to 0+$ (this means that x approaches zero from above as x must be positive for $\ln x$ to be defined).

You can also quote this result.

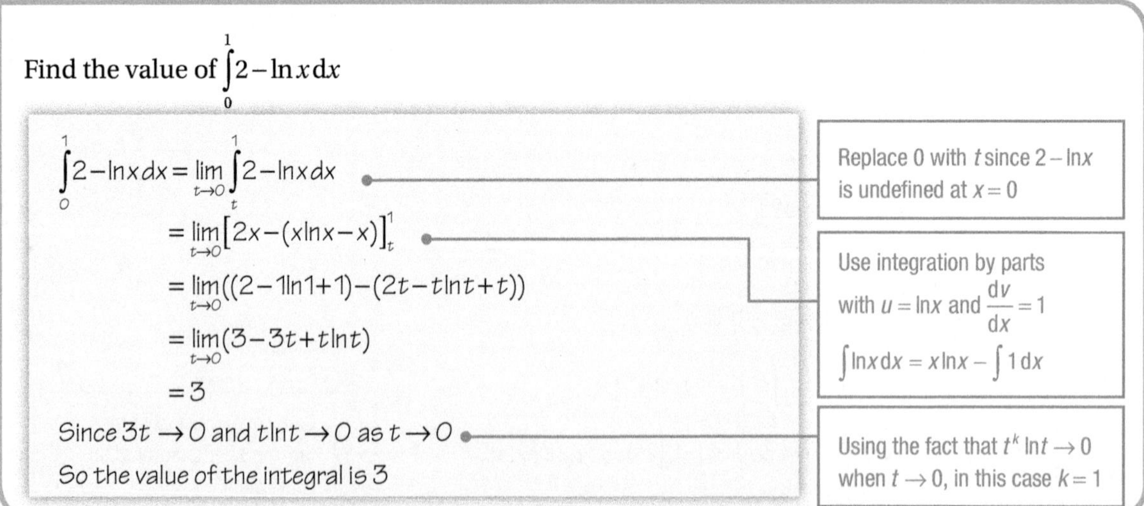

Example 4

Find the value of $\int_0^1 2 - \ln x\,dx$

$$\int_0^1 2 - \ln x\,dx = \lim_{t\to 0}\int_t^1 2 - \ln x\,dx$$

Replace 0 with t since $2 - \ln x$ is undefined at $x = 0$

$$= \lim_{t\to 0}\left[2x - (x\ln x - x)\right]_t^1$$

$$= \lim_{t\to 0}((2 - 1\ln 1 + 1) - (2t - t\ln t + t))$$

Use integration by parts with $u = \ln x$ and $\frac{dv}{dx} = 1$

$\int \ln x\,dx = x\ln x - \int 1\,dx$

$$= \lim_{t\to 0}(3 - 3t + t\ln t)$$

$$= 3$$

Since $3t \to 0$ and $t\ln t \to 0$ as $t \to 0$

So the value of the integral is 3

Using the fact that $t^k \ln t \to 0$ when $t \to 0$, in this case $k = 1$

To evaluate an improper integral where the integrand is undefined at a point between the limits of the integral, you need to split the integral into two parts about the point of discontinuity.

PURE

Example 5

Evaluate the improper integral $\int_{-1}^{e} x\ln|x|\,dx$

You need to identify the point of discontinuity and split the integral at this point.

$$\int_{-1}^{e} x\ln|x|\,dx = \int_{-1}^{0} x\ln|x|\,dx + \int_{0}^{e} x\ln|x|\,dx$$

Since $x\ln|x|$ is undefined at $x = 0$

$$\int_{-1}^{0} x\ln|x|\,dx = \lim_{a\to 0}\int_{-1}^{a} x\ln|x|\,dx$$

$$= \lim_{a\to 0}\left[\frac{1}{2}x^2\ln|x| - \frac{1}{4}x^2\right]_{-1}^{a}$$

Use integration by parts with $u = \ln|x|$ and $\frac{dv}{dx} = x^2$

$$= \lim_{a\to 0}\left(\left(\frac{1}{2}a^2\ln|a| - \frac{1}{4}a^2\right) - \left(\frac{1}{2}(-1)^2\ln|-1| - \frac{1}{4}(-1)^2\right)\right)$$

$$= \lim_{a\to 0}\left(\frac{1}{2}a^2\ln|a| - \frac{1}{4}a^2 + \frac{1}{4}\right)$$

$$= \frac{1}{4} \text{ since } \frac{1}{4}a^2 \to 0 \text{ and } a^2\ln|a| \to 0 \text{ as } a \to 0$$

Using the fact that $a^k \ln a \to 0$ when $a \to 0$ in this case $k = 2$

$$\int_{0}^{e} x\ln|x|\,dx = \lim_{b\to 0}\int_{b}^{e} x\ln|x|\,dx$$

$$= \lim_{b\to 0}\left[\frac{1}{2}x^2\ln|x| - \frac{1}{4}x^2\right]_{b}^{e}$$

$$= \lim_{b\to 0}\left(\left(\frac{1}{2}e^2\ln|e| - \frac{1}{4}e^2\right) - \left(\frac{1}{2}b^2\ln|b| - \frac{1}{4}b^2\right)\right)$$

$$= \lim_{b\to 0}\left(\frac{1}{4}e^2 - \frac{1}{2}b^2\ln|b| + \frac{1}{4}b^2\right)$$

Since $\ln|e| = 1$

$$= \frac{1}{2}e^2 - \frac{1}{4}e^2 \quad \text{since } \frac{1}{4}b^2 \to \text{ and } \frac{1}{2}b^2\ln|b| \to \text{ as } b \to 0$$

Therefore, since both limits exist, the improper integral converges

$$\int_{-1}^{e} x\ln|x|\,dx = \frac{1}{4} + \frac{1}{2}e^2 - \frac{1}{4}e^2 = \frac{1}{4}(1+e^2)$$

Exercise 19.1A Fluency and skills

1 Which of these are improper integrals? Explain your answers.

a $\int_{0}^{5} e^{-x}\,dx$

b $\int_{0}^{2} \ln x\,dx$

c $\int_{1}^{\infty} \frac{1}{x^2}\,dx$

d $\int_{-\infty}^{\infty} \sin x\,dx$

e $\int_{\frac{\pi}{4}}^{\frac{\pi}{2}} \frac{1}{\sin x}\,dx$

f $\int_{0}^{\pi} \tan x\,dx$

2 Evaluate each of these improper integrals.

a $\int_{1}^{\infty} \frac{1}{x^2}\,dx$

b $\int_{2}^{\infty} \frac{3}{x^4}\,dx$

c $\int_{-\infty}^{0} \frac{1}{(1-x)^2}\,dx$

d $\int_{-\infty}^{0} \frac{1}{(2-3x)^2}\,dx$

e $\int_{0}^{\infty} \frac{1}{(x+2)^3}\,dx$

f $\int_{-\infty}^{1} \frac{1}{(x-2)^4}\,dx$

g $\int_{0}^{\infty} xe^{-2x}\,dx$

h $\int_{1}^{\infty} \frac{\ln x}{x^2}\,dx$

3 Find the exact value of this improper integral.

$$\int_3^{\infty}(x-3)e^{-x}dx$$

4 Evaluate this improper integral by first splitting the integral into two parts.

$$\int_{-\infty}^{\infty}\frac{1}{e^{|x|}}dx$$

5 Evaluate each of these improper integrals.

a $\int_0^9 \frac{1}{\sqrt{x}}dx$ b $\int_0^{27} \frac{1}{x^{\frac{1}{3}}}dx$

c $\int_2^4 \frac{1}{\sqrt{x-2}}dx$ d $\int_0^3 \frac{x}{\sqrt{9-x^2}}dx$

e $\int_0^1 \frac{\ln x}{\sqrt{x}}dx$ f $\int_0^{\ln 2} \frac{e^x}{\sqrt{e^x-1}}dx$

g $\int_0^{\frac{\pi}{2}} \frac{\sin x}{\sqrt{\cos x}}dx$ h $\int_0^{\frac{\pi}{12}} \frac{\cos 2x}{\sqrt{\sin 2x}}dx$

6 Find the exact value of each of these improper integrals.

a $\int_0^e x\ln x\,dx$ b $\int_0^e x^2\ln x\,dx$

7 Show that $\int_{1-e}^{1}\ln(1-x)dx=0$

8 By splitting the integral into two parts, find the exact value of

$$\int_{-e}^{e}x^2\ln|x|\,dx$$

Reasoning and problem-solving

Many improper integrals are divergent because the areas they represent are not finite. You can use algebra to show that an improper integral is divergent.

Strategy

To decide whether an improper integral converges or diverges

1. Replace the limit where the integrand is undefined by a variable.
2. Integrate and substitute in the limits.
3. Consider the behaviour of the integral as the variable tends towards the original limit.

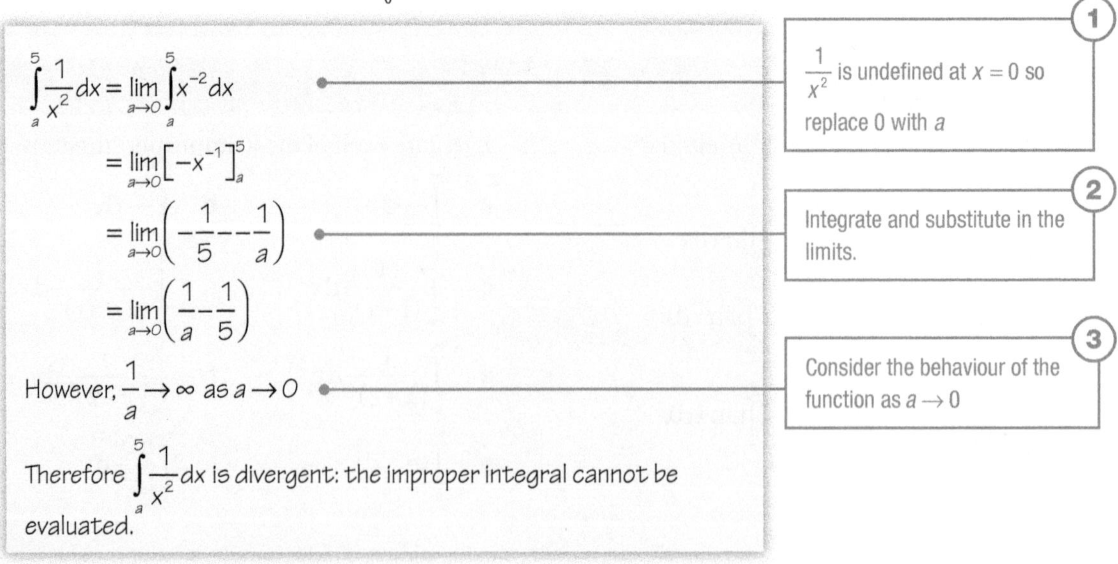

Example 6

Show that the improper integral $\int_0^5 \frac{1}{x^2}dx$ is divergent.

$$\int_a^5 \frac{1}{x^2}dx = \lim_{a\to 0}\int_a^5 x^{-2}dx$$

$$= \lim_{a\to 0}\left[-x^{-1}\right]_a^5$$

$$= \lim_{a\to 0}\left(-\frac{1}{5}-\frac{1}{a}\right)$$

$$= \lim_{a\to 0}\left(\frac{1}{a}-\frac{1}{5}\right)$$

However, $\frac{1}{a}\to\infty$ as $a\to 0$

Therefore $\int_a^5 \frac{1}{x^2}dx$ is divergent: the improper integral cannot be evaluated.

1 $\frac{1}{x^2}$ is undefined at $x=0$ so replace 0 with a

2 Integrate and substitute in the limits.

3 Consider the behaviour of the function as $a\to 0$

PURE

You may need to consider the behaviour of more complex expressions, such as those involving **rational expressions.**

You do this by dividing the numerator and the denominator by the highest power of the variable.

For example, the expression $\frac{x^2+2x}{3x^2-5}$ can be written as $\frac{1+\frac{2}{x}}{3-\frac{5}{x^2}}$ by dividing the numerator and the denominator by x^2

As $x \to \infty, \frac{2}{x} \to 0$ and $\frac{5}{x^2} \to 0$, therefore you can see that $\frac{1+\frac{2}{x}}{3-\frac{5}{x^2}} \to \frac{1}{3}$

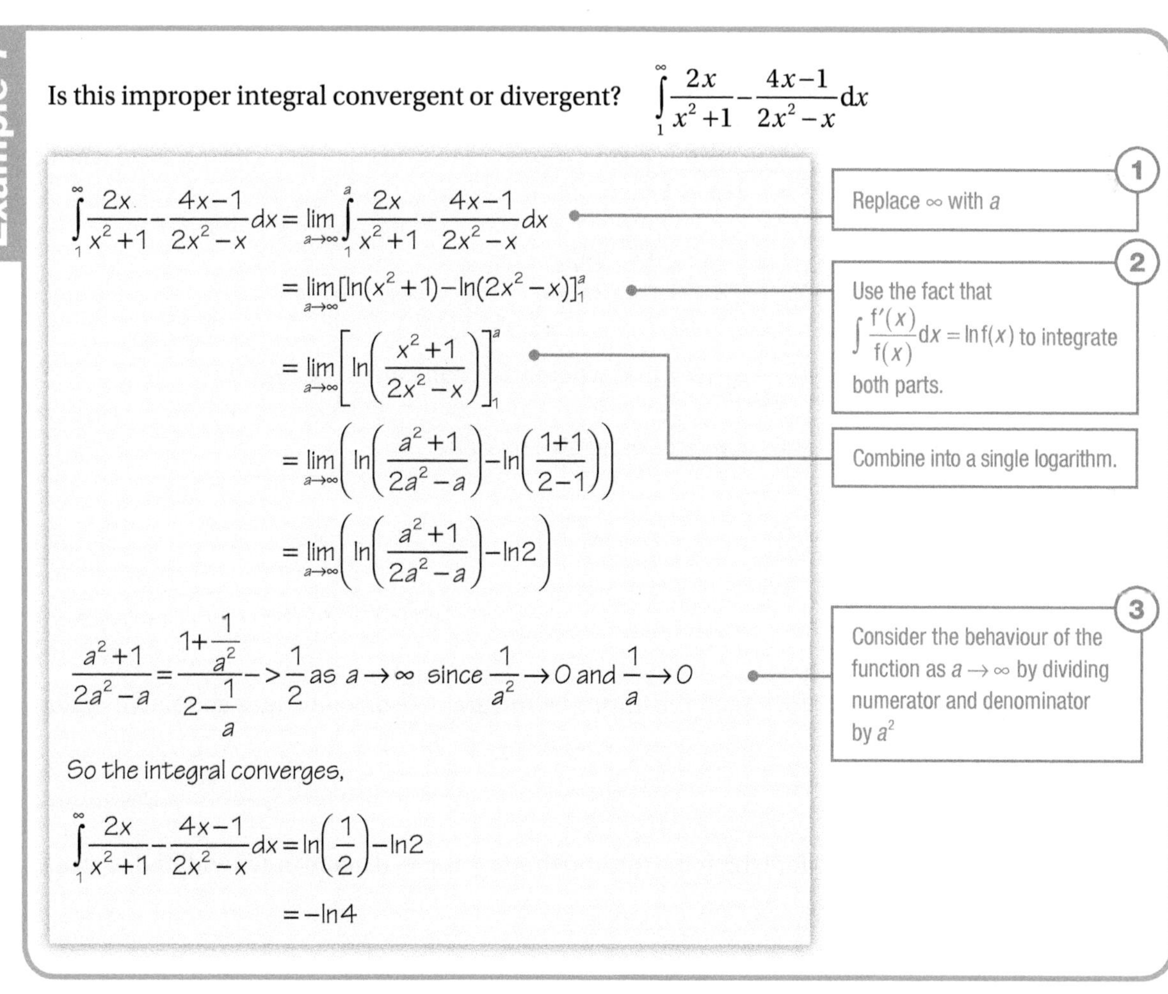

Example 7

Is this improper integral convergent or divergent? $\int_1^\infty \frac{2x}{x^2+1} - \frac{4x-1}{2x^2-x}\,dx$

$$\int_1^\infty \frac{2x}{x^2+1} - \frac{4x-1}{2x^2-x}\,dx = \lim_{a\to\infty} \int_1^a \frac{2x}{x^2+1} - \frac{4x-1}{2x^2-x}\,dx$$

1 Replace ∞ with a

$$= \lim_{a\to\infty} [\ln(x^2+1) - \ln(2x^2-x)]_1^a$$

2 Use the fact that $\int \frac{f'(x)}{f(x)}dx = \ln f(x)$ to integrate both parts.

$$= \lim_{a\to\infty} \left[\ln\left(\frac{x^2+1}{2x^2-x}\right)\right]_1^a$$

Combine into a single logarithm.

$$= \lim_{a\to\infty} \left(\ln\left(\frac{a^2+1}{2a^2-a}\right) - \ln\left(\frac{1+1}{2-1}\right)\right)$$

$$= \lim_{a\to\infty} \left(\ln\left(\frac{a^2+1}{2a^2-a}\right) - \ln 2\right)$$

$$\frac{a^2+1}{2a^2-a} = \frac{1+\frac{1}{a^2}}{2-\frac{1}{a}} -> \frac{1}{2} \text{ as } a \to \infty \text{ since } \frac{1}{a^2} \to 0 \text{ and } \frac{1}{a} \to 0$$

3 Consider the behaviour of the function as $a \to \infty$ by dividing numerator and denominator by a^2

So the integral converges,

$$\int_1^\infty \frac{2x}{x^2+1} - \frac{4x-1}{2x^2-x}\,dx = \ln\left(\frac{1}{2}\right) - \ln 2$$

$$= -\ln 4$$

Exercise 19.1B Reasoning and problem-solving

1 Decide whether or not each of these integrals converges.

If it does converge, find its value. If it diverges, explain why.

a $\int_0^1 \frac{1}{x^4} \mathrm{d}x$ b $\int_1^\infty \frac{1}{x^4} \mathrm{d}x$

c $\int_0^{\frac{\pi}{4}} \tan x \, \mathrm{d}x$ d $\int_0^{\frac{\pi}{2}} \tan x \, \mathrm{d}x$

e $\int_0^\infty \cos x \, \mathrm{d}x$ f $\int_1^\infty \frac{1}{x} \mathrm{d}x$

g $\int_{-\infty}^0 \frac{1}{3-x} \mathrm{d}x$ h $\int_0^7 \frac{1}{\sqrt{7-x}} \mathrm{d}x$

i $\int_0^7 \frac{1}{(7-x)^2} \mathrm{d}x$ j $\int_2^\infty \frac{1}{x-1} - \frac{2}{2x-1} \mathrm{d}x$

k $\int_1^\infty \frac{1}{x} - \frac{2x}{x^2+1} \mathrm{d}x$ l $\int_1^\infty \frac{x}{x^2+1} - \frac{2x}{2x^2+1} \mathrm{d}x$

2 a Show that $\int \frac{x}{x^2+3} - \frac{2}{2x+3} \mathrm{d}x = \frac{1}{2} \ln \frac{(x^2+3)}{4x^2+12x+9}$

b Hence show that $\int_0^\infty \frac{x}{x^2+3} - \frac{2}{2x+3} \mathrm{d}x = \ln k$, where k is a constant to be found.

3 Show that each of these integrals converges and give its value.

a $\int_{-\infty}^0 \frac{6}{3x-2} - \frac{2x}{x^2+4} \mathrm{d}x$

b $\int_{-\infty}^0 \frac{3-x}{x^2-6x+1} - \frac{8}{5-8x} \mathrm{d}x$

4 Find the range of values of p for which the improper integral $\int_0^1 \frac{1}{x^p} \mathrm{d}x$ converges and find its value in terms of p

5 Find the range of values of p for which the improper integral $\int_1^\infty \frac{1}{x^p} \mathrm{d}x$ converges and find its value in terms of p

19.2 Inverse trigonometric functions

Fluency and skills

To differentiate inverse trigonometric functions you need to use the relationship $\frac{dy}{dx}=\frac{1}{\frac{dx}{dy}}$

See Maths Ch15.5

For a reminder of the chain rule.

For example, to differentiate $y=\arcsin x$, first rearrange to give $x=\sin y$

Then you know that $\frac{dx}{dy}=\cos y$ so $\frac{dy}{dx}=\frac{1}{\cos y}$

Now rewrite this in terms of x, using the fact that $\cos y=\sqrt{1-\sin^2 y}=\sqrt{1-x^2}$

Substituting this gives the result $\frac{dy}{dx}=\frac{1}{\sqrt{1-x^2}}$

The derivatives of $\arccos x$ and $\arctan x$ can be derived in a similar way and different functions of x can be used.

Key point

$$\frac{d(\arcsin x)}{dx}=\frac{1}{\sqrt{1-x^2}} \qquad \frac{d(\arccos x)}{dx}=-\frac{1}{\sqrt{1-x^2}} \qquad \frac{d(\arctan x)}{dx}=\frac{1}{1+x^2}$$

Example 1

Differentiate $y=\arctan 3x$ with respect to x

$$3x=\tan y \Rightarrow x=\frac{1}{3}\tan y$$

$$\frac{dx}{dy}=\frac{1}{3}\sec^2 y$$

$$\frac{dy}{dx}=\frac{1}{\frac{1}{3}\sec^2 y}$$

Use $\frac{dy}{dx}=\frac{1}{\frac{dx}{dy}}$

$$=\frac{3}{1+\tan^2 y}$$

Use $1+\tan^2 y=\sec^2 y$

$$=\frac{3}{1+(3x)^2}$$

Write in terms of x, using the fact that $3x=\tan y$

$$=\frac{3}{1+9x^2}$$

Using the fundamental theorem of calculus, you can use these derivatives to obtain the following integration results:

Key point

$\int \frac{1}{\sqrt{1-x^2}} dx = \arcsin x + c$ $\quad \int \frac{1}{\sqrt{1-x^2}} dx = -\arccos x + c$ $\quad \int \frac{1}{1+x^2} dx = \arctan x + c$

These results can also be derived using a suitable substitution.

Here you are given the substitution, but you could work out what to use from your knowledge of the derivative of $\arcsin x$

Example 2

Use the substitution $x = \sin u$ to prove that $\int \frac{1}{\sqrt{1-x^2}} dx = \arcsin x + c$

$\frac{dx}{du} = \cos u$

$\int \frac{1}{\sqrt{1-x^2}} dx = \int \frac{1}{\sqrt{1-\sin^2 u}} \cos u \, du$ — Substitute for x and use $dx = \cos u \, du$

$= \int \frac{\cos u}{\sqrt{\cos^2 u}} du$ — Use the fact that $1 - \sin^2 u = \cos^2 u$

$= \int 1 \, du$

$= u + c$

$= \arcsin x + c$, as required — Write in terms of x using the fact that $x = \sin u$

Exercise 19.2A Fluency and skills

1 Use the technique in the example above to differentiate these expressions with respect to x

a $\arccos x$ b $\arctan x$

c $\arcsin 2x$ d $\arccos 5x$

e $\arctan(x-1)$ f $2\arcsin x$

g $3\arccos\left(\frac{x}{3}\right)$ h $3\arcsin(2-x)$

i $\arccos x^2$ j $x\arcsin x$

2 Prove that $\frac{d(\text{arcsec}\, x)}{dx} = \frac{1}{x\sqrt{x^2-1}}$

3 Prove that $\frac{d(\text{arccosec}\, x)}{dx} = -\frac{1}{x\sqrt{x^2-1}}$

4 Prove that $\frac{d(\text{arccot}\, x)}{dx} = -\frac{1}{x^2+1}$

5 Use the derivatives of $\arccos x$, $\arcsin x$ and $\arctan x$ to find $\frac{dy}{dx}$ in each case.

a $y = e^x \arctan x$ b $y = \arccos(3x^2-1)$

c $y = \sin x \arccos 2x$ d $y = (\arcsin x)^2$

e $y = \arcsin(e^x)$ f $e^{\arctan 2x}$

6 Use the substitution $x = \cos u$ to show that $\int \frac{1}{\sqrt{1-x^2}} dx = -\arccos x + c$

7 Use the substitution $x = \tan u$ to show that $\int \frac{1}{1+x^2} dx = \arctan x + c$

8 Use the substitution $x = \frac{1}{3}\sin u$ to integrate $\int \frac{1}{\sqrt{1-9x^2}} dx$

9 Use the substitution $x = 5\cos u$ to integrate $\int \frac{5}{\sqrt{25-x^2}} dx$

10 Use the substitution $x = 3\tan u$ to integrate $\int \frac{1}{9+x^2} dx$

PURE

Reasoning and problem-solving

You will not always be told what substitution to use so you will need to choose a suitable one. From the questions in Exercise 19.2A, you can see that these substitutions are often suitable.

Key point

- For an integral involving $\sqrt{a^2-x^2}$, try the substitution $x=a\sin u$
- For an integral involving a^2+x^2, try the substitution $x=a\tan u$

You may need to do some rearranging of the integral.

For example, $\int \frac{1}{1+\frac{x^2}{4}}\,dx$ can be written as $\int \frac{4}{4+x^2}\,dx$. You can then see that a suitable substitution is $x=2\tan u$ since the integral involves a^2+x^2, where $a=2$

Sometimes it is useful to complete the square for the denominator first.

For example, $\int \frac{1}{\sqrt{8+2x-x^2}}\,dx$ can be written as $\int \frac{1}{\sqrt{9-(x-1)^2}}\,dx$, so a suitable substitution is $x-1=3\sin u$ since the integral involves $\sqrt{a^2-x^2}$, where 'x' $=x-1$ and $a=3$

Strategy

To find an integral using a trigonometric substitution

1. Rewrite the integrand so it involves either $\sqrt{a^2-x^2}$ or a^2+x^2
2. Choose the correct substitution, either $x=a\sin u$ or $x=a\tan u$
3. Use integration by substitution to work out the integral and then write the answer in terms of x

Example 3

Work out each of these integrals.

a $\int \frac{3}{\sqrt{1-9x^2}}\,dx$ **b** $\int \frac{9}{x^2+6x+25}\,dx$

a $\int \frac{3}{\sqrt{1-9x^2}}\,dx = \int \frac{1}{\sqrt{\frac{1}{9}-x^2}}\,dx$ — (1) Divide numerator and denominator by 3

Let $x=\frac{1}{3}\sin u$

Then $\frac{dx}{du}=\frac{1}{3}\cos u$ — (2) $a^2=\frac{1}{9}$ so $a=\frac{1}{3}$

$$\int \frac{1}{\sqrt{\frac{1}{9}-\left(\frac{1}{3}\sin u\right)^2}}\frac{1}{3}\cos u\,du = \frac{1}{3}\int \frac{\cos u}{\sqrt{\frac{1}{9}-\frac{1}{9}\sin^2 u}}\,du$$

$$=\frac{1}{3}\int \frac{\cos u}{\sqrt{\frac{1}{9}\cos^2 u}}\,du$$ — Use $1-\sin^2 u=\cos^2 u$

(*Continued on the next page*)

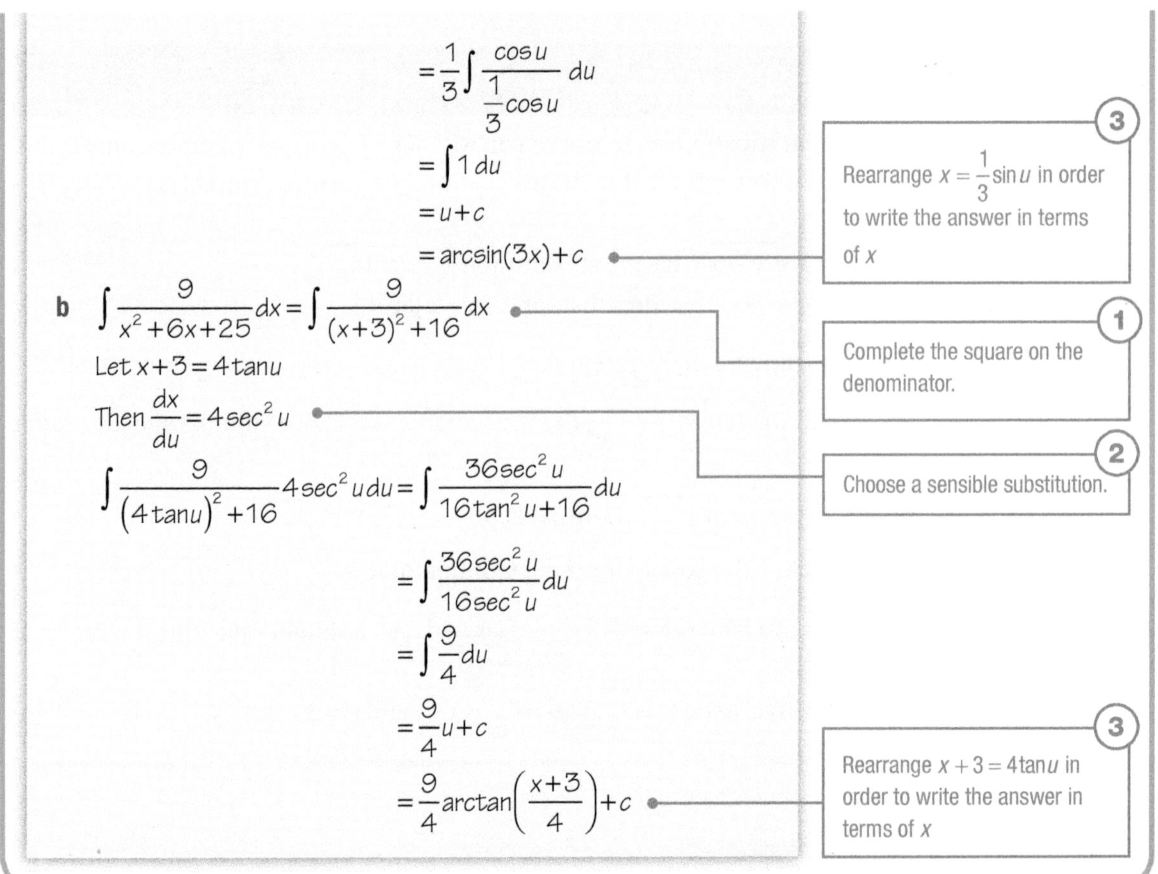

$$= \frac{1}{3}\int \frac{\cos u}{\frac{1}{3}\cos u}\,du$$

$$= \int 1\,du$$

$$= u + c$$

$$= \arcsin(3x) + c$$

(3) Rearrange $x = \frac{1}{3}\sin u$ in order to write the answer in terms of x

b $$\int \frac{9}{x^2+6x+25}dx = \int \frac{9}{(x+3)^2+16}dx$$

(1) Complete the square on the denominator.

Let $x+3 = 4\tan u$

Then $\frac{dx}{du} = 4\sec^2 u$

(2) Choose a sensible substitution.

$$\int \frac{9}{(4\tan u)^2+16}4\sec^2 u\,du = \int \frac{36\sec^2 u}{16\tan^2 u+16}du$$

$$= \int \frac{36\sec^2 u}{16\sec^2 u}du$$

$$= \int \frac{9}{4}du$$

$$= \frac{9}{4}u + c$$

$$= \frac{9}{4}\arctan\left(\frac{x+3}{4}\right) + c$$

(3) Rearrange $x+3 = 4\tan u$ in order to write the answer in terms of x

Exercise 19.2B Reasoning and problem-solving

1 Work out each of these integrals by choosing a suitable substitution.

a $\int \frac{1}{\sqrt{9-x^2}}dx$ b $\int \frac{1}{\sqrt{100-x^2}}dx$

c $-\int \frac{2}{\sqrt{36-x^2}}dx$ d $\int \frac{-4}{\sqrt{8-x^2}}dx$

e $\int \frac{1}{25+x^2}dx$ f $\int \frac{1}{49+x^2}dx$

g $\int \frac{-3}{x^2+2}dx$ h $\int \frac{6}{x^2+6}dx$

i $\int \sqrt{1-x^2}\,dx$ j $\int \sqrt{16-x^2}\,dx$

2 Work out each of these integrals by first rearranging the integrand then choosing a suitable substitution.

a $\int \frac{1}{1+\frac{x^2}{64}}dx$ b $\int \frac{25}{1+25x^2}dx$

c $\int \frac{2}{\frac{x^2}{2}+4}dx$ d $\int \frac{4}{2x^2+6}dx$

e $\int \frac{1}{\sqrt{1-\frac{x^2}{9}}}dx$ f $\int \frac{1}{\sqrt{16-4x^2}}dx$

g $-\int \frac{1}{\sqrt{1-\frac{x^2}{36}}}dx$ h $\int \frac{3}{\sqrt{18-9x^2}}dx$

i $\int \sqrt{1-\frac{x^2}{81}}\,dx$ j $\int \sqrt{36-9x^2}\,dx$

3 Work out each of these integrals.

a $\int \frac{1}{x^2+2x+5}dx$ b $\int \frac{1}{x^2-6x+10}dx$

c $\int \frac{1}{x^2-14x+53}dx$ d $\int \frac{2}{x^2+12x+38}dx$

e $\int \frac{1}{\sqrt{20-8x-x^2}}dx$ f $\int \frac{1}{\sqrt{1+2x-x^2}}dx$

g $-\int \frac{2}{\sqrt{8x-x^2}}dx$ h $\int \frac{1}{\sqrt{4-4x-x^2}}dx$

i $\int \sqrt{12+4x-x^2}\,dx$ j $\int \sqrt{16-6x-x^2}$

4 Calculate the exact value of the integral

$$\int_{-\frac{1}{4}}^{\frac{1}{2}} \frac{1}{\sqrt{8+4x-4x^2}}\,dx$$

19.3 Hyperbolic functions

Fluency and skills

You can use the exponential definition of hyperbolic functions to work out their derivatives.

For example, $\sinh x = \frac{1}{2}(e^x - e^{-x})$ so $\frac{d(\sinh x)}{dx} = \frac{1}{2}(e^x + e^{-x})$ which is the definition of $\cosh x$

Key point

$$\frac{d(\sinh x)}{dx} = \cosh x \qquad \frac{d(\cosh x)}{dx} = \sinh x \qquad \frac{d(\tanh x)}{dx} = \operatorname{sech}^2 x$$

See Ch3.5 For a reminder about hyperbolic functions.

These results can be used to find other derivatives.

$$\frac{1}{\cosh x} = \operatorname{sech} x \qquad \frac{1}{\sinh x} = \operatorname{cosech} x \qquad \frac{1}{\tanh x} = \coth x$$

Example 1

Differentiate $\operatorname{sech} x$ with respect to x

$\operatorname{sech} x = (\cosh x)^{-1}$

$\frac{d(\operatorname{sech} x)}{dx} = -(\cosh x)^{-2}(\sinh x)$ — Use the chain rule.

$= -\frac{\sinh x}{\cosh^2 x}$

$= -\tanh x \operatorname{sech} x$ — Notice the difference in sign compared with the derivative of $\sec x$ which is $\tan x \sec x$

You can also use these derivatives:

Key point

$$\frac{d(\operatorname{sech} x)}{dx} = -\operatorname{sech} x \tanh x \qquad \frac{d(\operatorname{cosech} x)}{dx} = -\operatorname{cosech} x \coth x \qquad \frac{d(\coth x)}{dx} = -\operatorname{cosech}^2 x$$

You can integrate $\sinh x$ and $\cosh x$ using the fundamental theorem of calculus:

Key point

$$\int \sinh x = \cosh x + c, \int \cosh x = \sinh x + c$$

To find the integral of $\tanh x$, use the fact that $\tanh x = \frac{\sinh x}{\cosh x}$. Then

$$\int \tanh x\,dx = \int \frac{\sinh x}{\cosh x}\,dx$$

$$= \ln \cosh x + c \text{ since } \int \frac{f'(x)}{f(x)}\,dx = \ln f(x) + c$$

Example 2

Use the identity $\cosh(2x) \equiv 1 + 2\sinh^2 x$

Work out the exact value of $\int_0^{\ln 3} \sinh^2 x \, dx$

$$\int_0^{\ln 3} \sinh^2 x \, dx = \int_0^{\ln 3} \frac{1}{2}(\cosh(2x) - 1)\, dx$$

Use the fact that $\sinh^2 x = \frac{1}{2}(\cosh(2x) - 1)$

$$= \left[\frac{1}{4}\sinh(2x) - \frac{x}{2}\right]_0^{\ln 3}$$

Integrate and substitute in the limits.

$$= \frac{1}{4}\left(\frac{e^{2\ln 3} - e^{-2\ln 3}}{2}\right) - \frac{1}{2}\ln 3 - 0$$

$$= \frac{e^{\ln 3^2} - e^{\ln 3^{-2}}}{8} - \frac{1}{2}\ln 3$$

Use rules of logarithms to write $e^{2\ln 3}$ as $e^{\ln 3^2} = 9$ and $e^{-2\ln 3}$ as $e^{\ln 3^{-2}} = \frac{1}{9}$

$$= \frac{9 - \frac{1}{9}}{8} - \frac{1}{2}\ln 3$$

$$= \frac{10}{9} - \frac{1}{2}\ln 3$$

You can check this on your calculator, but it will not give an exact answer.

You can differentiate the inverse hyperbolic functions using a similar method as for inverse trigonometric functions in Section 19.2

Example 3

Use the identity $\cosh^2 x - \sinh^2 x \equiv 1$

Differentiate arcosh$x$ with respect to x

Let $y = \text{arcosh}x$

Then $x = \cosh y$

$$\frac{dx}{dy} = \sinh y$$

So $\frac{dy}{dx} = \frac{1}{\sinh y}$

Use the chain rule: $\frac{dy}{dx} = \frac{1}{\frac{dx}{dy}}$

$$= \frac{1}{\sqrt{\cosh^2 y - 1}}$$

Use the identity $\cosh^2 y - \sinh^2 y \equiv 1$

$$= \frac{1}{\sqrt{x^2 - 1}}$$

The derivative of arsinh$x$ can be found in a similar way.

Key point

$$\frac{d(\text{arcosh}x)}{dx}=\frac{1}{\sqrt{x^2-1}} \qquad \frac{d(\text{arsinh}x)}{dx}=\frac{1}{\sqrt{x^2+1}}$$

PURE

Exercise 19.3A Fluency and skills

1 Use the definition of cosh$x$ and sinh$x$ to show that $\frac{d(\cosh x)}{dx}=\sinh x$

2 a Use exponentials to show that $\frac{d(\tanh x)}{dx}=\text{sech}^2 x$

b Hence show that $\frac{d(\coth x)}{dx}=-\text{cosech}^2 x$

3 Use exponentials to show that $\frac{d(\text{cosech}x)}{dx}=-\text{cosech}x\coth x$

4 Differentiate these expressions with respect to x

a $3\text{cosech}x$ b $\coth(x+1)$

c $\sinh 2x$ d $\cosh\left(\frac{x}{3}\right)$

e $\tanh(x-2)$ f $\sinh x^2$

g $\cosh(2x-3)$ h $\sinh^2 x$

i $x\tanh x$ j $x\sqrt{\text{sech}x}$

5 Calculate these integrals.

a $\int \text{sech}^2 x\,dx$ b $\int \coth x\,\text{cosech}x\,dx$

c $\int \cosh 2x\,dx$ d $\int \sinh\left(\frac{x}{3}\right)dx$

e $\int \cosh x\sinh^2 x\,dx$ f $\int \cosh^2 x\,dx$

g $\int \coth 4x\,dx$ h $\int \text{cosech}^2 5x\,dx$

6 Evaluate these integrals. Give the exact answer in each case.

a $\int_0^{\ln 2} \sinh(2x)dx$ b $\int_0^{\ln 3} \cosh(x-\ln 3)$

c $\int_{\ln 2}^{\ln 5} \tanh x\,\text{sech}x\,dx$ d $\int_{\ln 2}^{\ln 3} \text{cosech}^2(2x)dx$

7 Show that $\frac{d(\text{arsinh}x)}{dx}=\frac{1}{\sqrt{x^2+1}}$

8 Show that $\frac{d(\text{artanh}2x)}{dx}=\frac{2}{1-4x^2}$

9 Show that $\frac{d(\text{arcoth}x)}{dx}=\frac{1}{1-x^2}$

10 Show that $\frac{d(\text{arcosech}x)}{dx}=\frac{1}{-x\sqrt{x^2+1}}$

11 Differentiate the following expressions with respect to x

You can quote the derivatives of arsinh$x$ and arcosh$x$

a $\text{arcosh}4x$ b $\text{arsinh}\left(\frac{x}{5}\right)$

c $\text{arcosh}(x^2)$ d $\text{arsinh}\,e^x$

e $\text{arcosh}(\sinh x)$ f $e^x\text{arsinh}(x^2-1)$

Reasoning and problem-solving

You can use the derivatives of $\text{arsinh}\,x$ and $\text{arcosh}\,x$ to choose a suitable substitution to use when integrating.

Key point

- For an integral involving $\sqrt{x^2+a^2}$, try the substitution $x = a\sinh u$
- For an integral involving $\sqrt{x^2-a^2}$, try the substitution $x = a\cosh u$

You may need to rearrange the integral first, using the same techniques that you saw in Section 19.2

Strategy

To find an integral using a hyperbolic substitution

1. Rewrite the integrand so that it involves either $\sqrt{x^2+a^2}$ or $\sqrt{x^2-a^2}$
2. Choose the correct substitution, either $x = a\sinh u$ or $x = a\cosh u$
3. Use integration by substitution to work out the integral and then write the answer in terms of x

Example 4

Work out each of these integrals

a $\int \frac{6}{\sqrt{36+4x^2}}\,dx$ **b** $\int \frac{1}{\sqrt{x^2-8x-20}}\,dx$

a $\int \frac{6}{\sqrt{36+4x^2}}\,dx = \int \frac{3}{\sqrt{9+x^2}}\,dx$ — (1) Divide numerator and denominator by 2

Let $x = 3\sinh u$ — (2) $a^2 = 9$ so $a = 3$

Then $\frac{dx}{du} = 3\cosh u$

$$\int \frac{3}{\sqrt{9+(3\sinh u)^2}}3\cosh u\,du = \int \frac{9\cosh u}{\sqrt{9+9\sinh^2 u}}\,du$$

$$= \int \frac{9\cosh u}{\sqrt{9\cosh^2 u}}\,du$$ — Use $1+\sinh^2 u = \cosh^2 u$

$$= \int \frac{9\cos u}{3\cos u}\,du$$

$$= \int 3\,du$$

$$= 3u + c$$

$$= 3\,\text{arsinh}\left(\frac{x}{3}\right) + c$$ — (3) Rearrange $x = 3\sinh u$ in order to write the answer in terms of x

(*Continued on the next page*)

b $\int \frac{1}{\sqrt{x^2-8x-20}}dx = \int \frac{1}{\sqrt{(x-4)^2-36}}dx$

Complete the square on the denominator. (1)

Let $x-4=6\cosh u$

Then $\frac{dx}{du}=6\sinh u$

Choose a sensible substitution. (2)

$\int \frac{1}{\sqrt{(6\cosh u)^2-36}}6\sinh u\,du = \int \frac{6\sinh u}{\sqrt{36\cosh^2 u-36}}du$

$= \int \frac{6\sinh u}{\sqrt{36\sinh^2 u}}du$

$= \int \frac{6\sinh u}{6\sinh u}du$

$= \int 1du$

$= u+c$

$= \text{arcosh}\left(\frac{x-4}{6}\right)+c$

Rearrange $x-4=6\cosh u$ in order to write the answer in terms of x (3)

Exercise 19.3B Reasoning and problem-solving

1 Work out each of these integrals.

a $\int \frac{1}{\sqrt{x^2+49}}dx$ b $\int \frac{1}{\sqrt{x^2-81}}dx$

c $\int \frac{2}{\sqrt{64+4x^2}}dx$ d $\int \frac{-1}{\sqrt{\frac{x^2}{9}-1}}dx$

e $\int \frac{2}{\sqrt{x^2+6x+25}}dx$

f $\int \frac{1}{\sqrt{x^2-10x+26}}dx$

g $\int \frac{1}{\sqrt{x^2+14x+24}}dx$

h $\int \frac{3}{\sqrt{x^2-24x+44}}dx$

2 Evaluate these integrals, giving your answers to 3 significant figures.

a $\int_0^1 \frac{1}{\sqrt{6+x^2}}dx$ b $\int_6^{12} \frac{\sqrt{27}}{\sqrt{\frac{x^2}{12}-3}}dx$

3 Calculate the exact values of these integrals.

a $\int_0^3 \sqrt{\frac{3}{27+3x^2}}dx$ b $\int_4^5 \frac{1}{\sqrt{3x^2-48}}dx$

4 Integrate each of these expressions with respect to x

a $\frac{x+1}{\sqrt{16+9x^2}}$ b $\frac{3-x}{\sqrt{\frac{x^2}{4}-3}}$

c $\frac{x+4}{\sqrt{16-2x^2}}$ d $\frac{2-5x}{\frac{x^2}{7}+7}$

5 Use integration by parts to show that

$\int \text{arsinh}\,x\,dx = x\,\text{arsinh}\,x - \sqrt{x^2+1}+c$

6 Work out each of these integrals.

a $\int \text{arcosh}\,x\,dx$ b $\int \text{arcoth}\,x\,dx$

7 One of these improper integrals converges for $a>0$ and the other does not.

A: $\int_a^{2a} \frac{1}{\sqrt{x^2-a^2}}dx$ B: $\int_{2a}^{\infty} \frac{1}{\sqrt{x^2-a^2}}dx$

Explain why one doesn't converge and find the exact value of the integral that does converge.

19.4 Partial fractions

Fluency and skills

See Maths Ch12.5
For a reminder on partial fractions.

Rational functions which have linear factors in the denominator can be split into partial fractions to help you to integrate the function.

If the degree of the numerator is the same as or greater than the degree of the denominator, then this is an **improper** fraction.

Key point

A fraction $\frac{\text{f}(x)}{\text{g}(x)}$ where degree of $\text{f}(x) \geq$ degree of $\text{g}(x)$ is called an **improper** fraction and can be written in the form $\text{P}(x)+\frac{\text{Q}(x)}{\text{g}(x)}$

It may then be possible to write $\frac{\text{Q}(x)}{\text{g}(x)}$ in partial fractions. You can use long division to find $\text{P}(x)$ and $\text{Q}(x)$

However, if you prefer, you can consider the degree of the numerator and the denominator to decide on a general form for the quotient.

For example, if you have $\frac{\text{f}(x)}{\text{g}(x)}$ where $\text{f}(x)$ is a quartic (polynomial of degree 4) and $\text{g}(x)$ is a quadratic, then you know that the quotient is of the form Ax^2+Bx+C

Example 1

a Write $\frac{30x^3-13x^2+6x+6}{15x^2+x-6}$ in partial fractions. **b** Hence work out $\int \frac{30x^3-13x^2+6x+6}{15x^2+x-6}\,dx$

a Dividing a cubic by a quadratic will give a linear quotient.

$$\frac{30x^3-13x^2+6x+6}{15x^2+x-6} = Ax+B+\frac{C}{3x+2}+\frac{D}{5x-3}$$

Alternatively, you could use long division.

$$30x^3-13x^2+6x+6 = (Ax+B)(15x^2+x-6)+C(5x-3)+D(3x+2)$$

Multiply both sides by $(3x+2)(5x-3)$

Equating coefficients

$x^3: 30=15A \Rightarrow A=2 \qquad x^2: -13=A+15B \Rightarrow B=-1$

$x: 6=-6A+B+5C+3D \Rightarrow 5C+3D=19$

$1: 6=-6B-3C+2D \Rightarrow 3C-2D=0$

Solve simultaneously to give $C=2, D=3$

$$\frac{30x^3-13x^2+6x+6}{15x^2+x-6} = 2x-1+\frac{2}{3x+2}+\frac{3}{5x-3}$$

b $$\int \frac{30x^3-13x^2+6x+6}{15x^2+x-6}\,dx = \int 2x-1+\frac{2}{3x+2}+\frac{3}{5x-3}\,dx$$

$$= x^2-x+\frac{2}{3}\ln(3x+2)+\frac{3}{5}\ln(5x-3)+c$$

Use the answer from part **a**.

You can also find partial fractions when the denominator includes a quadratic factor which cannot be factorised, for example x^2+5. In these cases, you should use a linear expression $Ax+B$ as the numerator.

Key point

$\frac{f(x)}{(\alpha x^2+\beta)(\gamma x+\delta)}$ can be split into partial fractions of the form $\frac{Ax+B}{\alpha x^2+\beta}+\frac{C}{\gamma x+\delta}$

In some questions you may have to factorise the denominator yourself.

Example 2

Work out $\int \frac{2x+12}{(x+1)(x^2+9)}dx$

$\frac{2x+12}{(x+1)(x^2+9)}=\frac{A}{x+1}+\frac{Bx+C}{x^2+9}$

$2x+12=A(x^2+9)+(Bx+C)(x+1)$

$=Ax^2+9A+Bx^2+Bx+Cx+C$ — Multiply both sides by $(x+1)(x^2+9)$

Equating coefficients

$x^2: 0=A+B$ so $A=-B$ — Or use an alternative method to find the values of A, B and C

$x: 2=B+C$ (equation 1)

$1: 12=9A+C$ (equation 2)

Subtract equation 1 from equation 2 to give: $9A-B=10$

$A=-B \Rightarrow -9B-B=10 \Rightarrow B=-1$ — Solve the three equations simultaneously.

So $A=1$ and $C=3$

$\int \frac{2x+12}{(x+1)(x^2+9)}dx=\int \frac{1}{x+1}+\frac{3-x}{x^2+9}dx$

$=\int \frac{1}{x+1}+\frac{3}{x^2+9}-\frac{x}{x^2+9}dx$ — Split the numerator of the second fraction.

$=\ln(x+1)+\arctan\left(\frac{x}{3}\right)-\frac{1}{2}\ln(x^2+9)+c$ — Use the substitution $x=3\tan u$ to integrate the second fraction.

which can be written as $\ln\left(\frac{x+1}{\sqrt{x^2+9}}\right)+3\arctan\left(\frac{x}{3}\right)+c$

Exercise 19.4A Fluency and skills

1 Work out each of these integrals by first expressing the integrand in partial fractions.

a $\int \frac{5x^2+14x-42}{(x-2)(x+4)}dx$

b $\int \frac{2x^2+31x+115}{x^2+14x+49}dx$

c $\int \frac{5x^3+x^2-46x-24}{x^2-9}dx$

d $\int \frac{8x^3+92x^2+243x+72}{x^3+11x^2+24x}dx$

e $\int \frac{x^4-7x^3-20x^2+14x+46}{(x-9)(x+1)}dx$

f $\int \frac{9x^4}{(x+1)(x-2)^2}dx$

g $\int \frac{12x^4+7x^3-22x-9}{12-7x+12x^2}dx$

2 Work out each of these integrals by first expressing the integrand in partial fractions.

a $\int \frac{9-7x}{(x+3)(x^2+1)}dx$

b $\int \frac{10+8x}{(x-3)(x^2+25)}dx$

c $\int \frac{x+66}{(x^2+36)(6-x)}dx$

d $\int \frac{7x-1}{(1-x)(x^2+2)}dx$

e $\int \frac{36x+6}{(x+1)(x-2)(x^2+9)}dx$

f $\int \frac{79-28x}{(x-1)^2(x^2+16)}dx$

3 Work out each of these integrals by first expressing the integrand in partial fractions.

a $\int \frac{3x+49}{x^3+49x}dx$ b $\int \frac{x-192}{2x^3+128x}dx$

c $\int \frac{10x+9}{3x^3-x^2+12x-4}dx$

4 a Given that x^2+5 is a factor of $x^4-6x^3+14x^2-30x+45$, write the fraction $\frac{20x-46}{x^4-6x^3+14x^2-30x+45}$ in partial fractions.

b Hence work out

$\int \frac{20x-46}{x^4-6x^3+14x^2-30x+45}dx$

Reasoning and problem-solving

You can also use partial fractions when evaluating improper integrals. The integral will only converge if all parts of it converge.

Strategy

To decide whether an improper integral converges when the integrand is a rational function

1. Factorise the denominator as far as possible.
2. Split into partial fractions and establish any points of discontinuity for each fraction.
3. Integrate and consider the behaviour of the integral as the variable tends towards the original limit.

Example 3

a Show algebraically that the improper integral $\int_{\frac{1}{2}}^{1} \frac{x-9}{2x^3-x^2+8x-4}dx$ does not converge.

b Explain whether the improper integral $\int_{1}^{\infty} \frac{x-9}{2x^3-x^2+8x-4}dx$ converges.

a $\frac{x-9}{2x^3-x^2+8x-4}=\frac{x-9}{(2x-1)(x^2+4)}$

(1) x^2+4 cannot be factorised further so will be a quadratic factor.

This can be written in partial fractions: $\frac{x-9}{(2x-1)(x^2+4)}=\frac{A}{2x-1}+\frac{Bx+C}{x^2+4}$

$x-9=A(x^2+4)+(Bx+C)(2x-1)$

Consider coefficients.

$x^2: 0=A+2B$

$x: 1=-B+2C$

$1: -9=4A-C$

Solve simultaneously to give $A=-2, B=1, C=1$

So $\frac{x-9}{(2x-1)(x^2+4)}=\frac{x+1}{x^2+4}-\frac{2}{2x-1}$

(*Continued on the next page*)

$\int_{\frac{1}{2}}^{1} \frac{x+1}{x^2+4}dx$ is a proper integral so can be calculated directly.

However, $\frac{2}{2x-1}$ has a point of discontinuity at $x=\frac{1}{2}$ so $\int_{\frac{1}{2}}^{1}\frac{2}{2x-1}dx$ is an improper integral.

$$\int_{\frac{1}{2}}^{1}\frac{2}{2x-1}dx = \lim_{a\to\frac{1}{2}}\int_a^1 \frac{2}{2x-1}dx$$

$$= \lim_{a\to\frac{1}{2}}\left[\ln(2x-1)\right]_a^1$$

$$= \lim_{a\to\frac{1}{2}}(\ln 1-\ln(2a-1))$$

$$= \lim_{a\to\frac{1}{2}}(-\ln(2a-1))$$

$2a-1\to 0$ as $a\to\frac{1}{2}$ so $\ln(2a-1)\to\infty$

Therefore, the integral does not converge.

b $$\int_1^{\infty}\frac{x-9}{2x^3-x^2+8x-4}dx = \lim_{a\to\infty}\int_1^a \frac{x}{x^2+4}+\frac{1}{x^2+4}-\frac{2}{2x-1}dx$$

$$= \lim_{a\to\infty}\left[\frac{1}{2}\ln(x^2+4)+\frac{1}{2}\arctan\left(\frac{x}{2}\right)-\ln(2x-1)\right]_1^a$$

$$= \lim_{a\to\infty}\left(\left(\frac{1}{2}\ln(a^2+4)+\frac{1}{2}\arctan\left(\frac{a}{2}\right)-\ln(2a-1)\right)-\left(\frac{1}{2}\ln 5+\frac{1}{2}\arctan\frac{1}{2}-\ln 1\right)\right)$$

$$= \lim_{a\to\infty}\left(\frac{1}{2}\ln\left(\frac{a^2+4}{(2a-1)^2}\right)+\frac{1}{2}\arctan\left(\frac{a}{2}\right)-\frac{1}{2}\ln 5-\frac{1}{2}\arctan\frac{1}{2}\right)$$

$$\frac{a^2+4}{(2a-1)^2} = \frac{a^2+4}{4a^2-4a+1}$$

$$= \frac{1+\frac{4}{a^2}}{4-\frac{4}{a}+\frac{1}{a^2}} \to \frac{1}{4} \text{ as } a\to\infty \text{ since } \frac{4}{a},\frac{1}{a^2}\to 0$$

So $\frac{1}{2}\ln\left(\frac{a^2+4}{(2a-1)^2}\right)\to\frac{1}{2}\ln\frac{1}{4}$

$\arctan\left(\frac{a}{2}\right)\to\frac{\pi}{2}$ as $a\to\infty$

Therefore, the improper integral does converge.

$$\int_1^{\infty}\frac{x-9}{2x^3-x^2+8x-4}dx = \frac{1}{2}\ln\frac{1}{4}+\frac{1}{2}\left(\frac{\pi}{2}\right)-\frac{1}{2}\arctan\frac{1}{2}-\frac{1}{2}\ln 5$$

$$= \frac{\pi}{4}-\frac{1}{2}\arctan\frac{1}{2}-\frac{1}{2}\ln 20$$

2 Consider any points of discontinuity for either fraction within the limits given.

3 Consider behaviour as $a\to\frac{1}{2}$

3 So this part of the integral does converge.

3 Consider the graph of $y=\arctan x$, it has a horizontal asymptote at $y=\frac{\pi}{4}$

Exercise 19.4B Reasoning and problem-solving

1 Find the exact value of these improper integrals.

a $\int_0^{\infty} \frac{1}{x^2+7x+12} \mathrm{d}x$

b $\int_1^{\infty} \frac{10}{3-8x-3x^2} \mathrm{d}x$

c $\int_0^{\infty} \frac{7x+26}{x^3+9x^2+24x+20} \mathrm{d}x$

d $\int_{-\infty}^{0} \frac{9}{8x^2-22x+5} \mathrm{d}x$

2 One of these improper integrals converges and the other does not. Explain which does not converge and find the value of the integral that does converge.

A: $\int_1^{\infty} \frac{-2x-12}{2x^3-x^2+6x-3} \mathrm{d}x$

B: $\int_0^{1} \frac{-2x-12}{2x^3-x^2+6x-3} \mathrm{d}x$

3 a Express $\frac{2x+a}{x^3+ax}$ in partial fractions.

b Hence show that the improper integral $\int_0^{a} \frac{2x+a}{x^3+ax}$ does not converge for any $a>0$

4 For what range of values of a (if any) does each improper integral converge? Explain your answers.

a $\int_{-\infty}^{0} \frac{3-a}{(x+a)(x+3)} \mathrm{d}x$

b $\int_0^{\infty} \frac{3-a}{(x+a)(x+3)} \mathrm{d}x$

19.5 Reduction formulae

Fluency and skills

Integration by parts can be used to find integrals of the form $\int u\frac{dv}{dx}dx$ where you know how to find $\frac{du}{dx}$ and v. For example,

$$\int xe^{-x}dx = -xe^{-x} - \int -e^{-x}dx \qquad u = x, \frac{dv}{dx} = e^{-x}$$

$$= -xe^{-x} - e^{-x} + c$$

Key point

The formula for integration by parts is

$$\int u\frac{dv}{dx}dx = uv - \int v\frac{du}{dx}dx$$

You can apply this formula twice, for example,

$$\int x^2e^{-x}dx = -x^2e^{-x} - \int -2xe^{-x}dx \qquad u = x^2, \frac{dv}{dx} = e^{-x}$$

$$= -x^2e^{-x} + 2\int xe^{-x}dx$$

Then apply again to the new integral:

$$= -x^2e^{-x} + 2\left(-xe^{-x} - \int -e^{-x}dx\right) \qquad u = x, \frac{dv}{dx} = e^{-x}$$

$$= -x^2e^{-x} - 2xe^{-x} - 2e^{-x} + c$$

You could use a similar process for higher powers of x but it would be very time-consuming. Instead, you can find a **reduction formula** which relates $I_n = \int x^n e^{-x}dx$ to $I_{n-1} = \int x^{n-2}e^{-x}dx$

$$I_n = \int x^n e^{-x}dx = -x^n e^{-x} - \int -nx^{n-1}e^{-x}dx \qquad u = x^n, \frac{dv}{dx} = e^{-x}$$

$$= -x^n e^{-x} + n\int x^{n-1}e^{-x}dx$$

$$= -x^n e^{-x} + nI_{n-1}$$

You can now use this formula to find, for example, $I_3 = \int x^3e^{-x}dx$

$I_3 = -x^3e^{-x} + 3I_2$, now use the formula again for I_2

$$= -x^3e^{-x} + 3(-x^2e^{-x} + 2I_1)$$

You can continue to use the formula until you have an integrand which is simpler to integrate.

$$I_3 = -x^3e^{-x} - 3x^2e^{-x} + 6I_1$$

$$= -x^3e^{-x} - 3x^2e^{-x} + 6(-xe^{-x} + I_0)$$

$$= -x^3e^{-x} - 3x^2e^{-x} - 6xe^{-x} + 6I_0$$

$I_0 = \int e^{-x}dx = -e^{-x}$ so substitute this to find the solution for I_3

$$I_3 = -x^3e^{-x} - 3x^2e^{-x} - 6xe^{-x} - 6e^{-x} + c$$

Key point

A **reduction formula** for I_n is an equation that relates I_n to I_{n-1} and/or I_{n-2}

It can be repeatedly applied to reduce the integral to one not requiring integration by parts.

You can use a reduction formula for a definite or an indefinite integral.

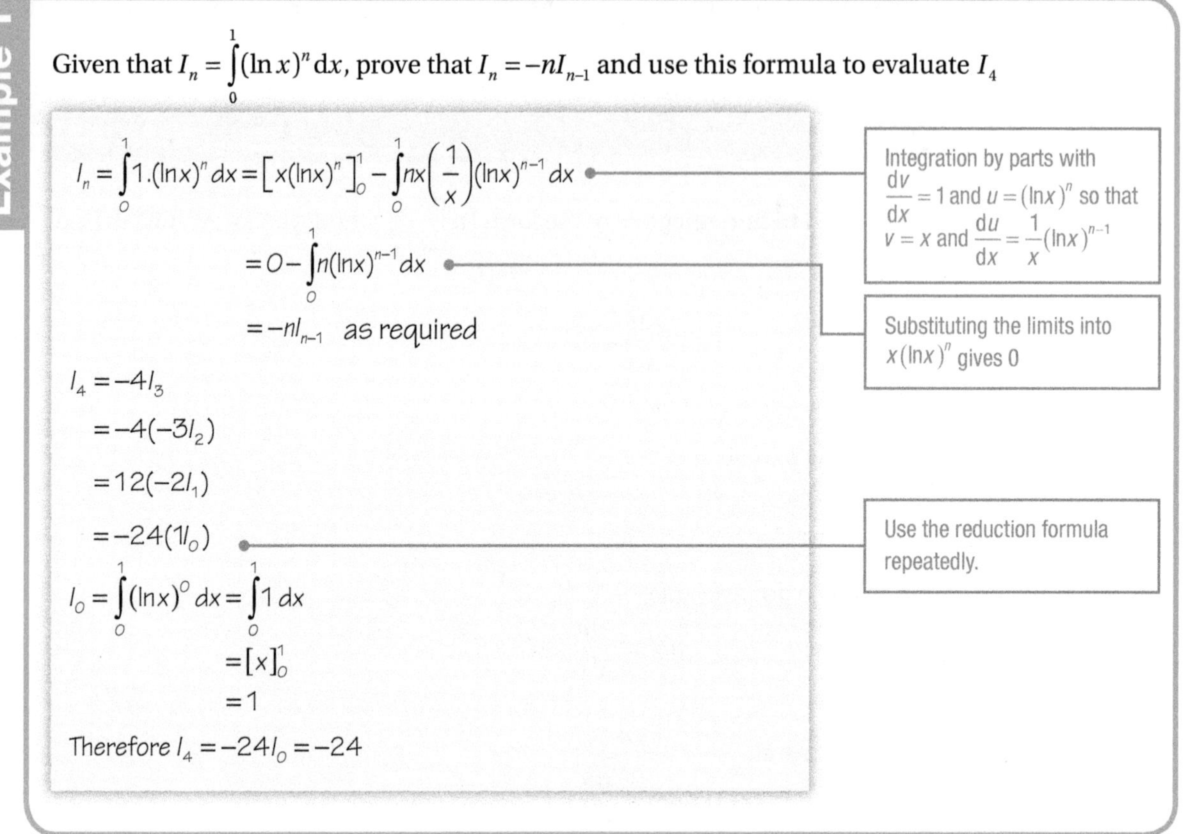

Example 1

Given that $I_n = \int_0^1 (\ln x)^n \, dx$, prove that $I_n = -nI_{n-1}$ and use this formula to evaluate I_4

$I_n = \int_0^1 1.(\ln x)^n dx = \left[x(\ln x)^n\right]_0^1 - \int_0^1 nx\left(\frac{1}{x}\right)(\ln x)^{n-1} dx$

Integration by parts with $\frac{dv}{dx} = 1$ and $u = (\ln x)^n$ so that $v = x$ and $\frac{du}{dx} = \frac{1}{x}(\ln x)^{n-1}$

$= 0 - \int_0^1 n(\ln x)^{n-1} dx$

Substituting the limits into $x(\ln x)^n$ gives 0

$= -nI_{n-1}$ as required

$I_4 = -4I_3$

$= -4(-3I_2)$

$= 12(-2I_1)$

$= -24(1I_0)$

Use the reduction formula repeatedly.

$I_0 = \int_0^1 (\ln x)^0 dx = \int_0^1 1 \, dx$

$= [x]_0^1$

$= 1$

Therefore $I_4 = -24I_0 = -24$

After you have used the integration by parts formula, you may need to rearrange the integrand so that it is clearly in the form of I_{n-1} and/or I_{n-2}

Example 2

Given that $I_n = \int_{-1}^{0} x^n \sqrt{x+1} \, dx$, show that $I_n = -nI_{n-1}$ and use this formula to evaluate $\int_{-1}^{0} x^3 \sqrt{x+1} \, dx$

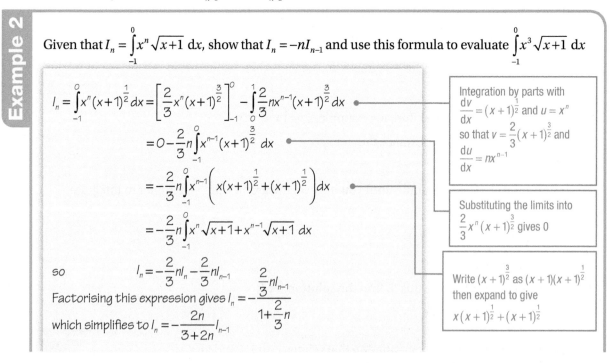

$I_n = \int_{-1}^{0} x^n (x+1)^{\frac{1}{2}} dx = \left[\frac{2}{3}x^n(x+1)^{\frac{3}{2}}\right]_{-1}^{0} - \int_{0}^{1} \frac{2}{3}nx^{n-1}(x+1)^{\frac{3}{2}} dx$

Integration by parts with $\frac{dv}{dx} = (x+1)^{\frac{1}{2}}$ and $u = x^n$ so that $v = \frac{2}{3}(x+1)^{\frac{3}{2}}$ and $\frac{du}{dx} = nx^{n-1}$

$= 0 - \frac{2}{3}n\int_{-1}^{0} x^{n-1}(x+1)^{\frac{3}{2}} dx$

Substituting the limits into $\frac{2}{3}x^n(x+1)^{\frac{3}{2}}$ gives 0

$= -\frac{2}{3}n\int_{-1}^{0} x^{n-1}\left(x(x+1)^{\frac{1}{2}} + (x+1)^{\frac{1}{2}}\right) dx$

Write $(x+1)^{\frac{3}{2}}$ as $(x+1)(x+1)^{\frac{1}{2}}$ then expand to give $x(x+1)^{\frac{1}{2}} + (x+1)^{\frac{1}{2}}$

$= -\frac{2}{3}n\int_{-1}^{0} x^n\sqrt{x+1} + x^{n-1}\sqrt{x+1} \, dx$

so $\quad I_n = -\frac{2}{3}nI_n - \frac{2}{3}nI_{n-1}$

Factorising this expression gives $I_n = -\dfrac{\frac{2}{3}nI_{n-1}}{1+\frac{2}{3}n}$

which simplifies to $I_n = -\frac{2n}{3+2n}I_{n-1}$

(*Continued on the next page*)

So $I_3 = -\frac{6}{9}I_2$

$= -\frac{6}{9}\left(-\frac{4}{7}\right)I_1$

$= -\frac{6}{9}\left(-\frac{4}{7}\right)\left(-\frac{2}{5}\right)I_0$

$= -\frac{16}{105}I_0$

Use the reduction formula repeatedly.

$I_0 = \int_{-1}^{0}(x+1)^{\frac{1}{2}}\,dx = \left[\frac{2}{3}(x+1)^{\frac{3}{2}}\right]_{-1}^{0}$

$= \frac{2}{3}$

Therefore $\int_{-1}^{0} x^3\sqrt{x+1}\,dx = -\frac{16}{105}\left(\frac{2}{3}\right) = -\frac{32}{315}$

Exercise 19.5A Fluency and skills

1 $I_n = \int x^n e^x dx$

a Show that $I_n = x^n e^x - nI_{n-1}$

b Use this formula to find I_4

2 $I_n = \int x^n e^{-\frac{x}{2}} dx$

a Show that $I_n = 2nI_{n-1} - 2x^n e^{-\frac{x}{2}}$

b Use this formula to find $\int x^3 e^{-\frac{x}{2}} dx$

3 $I_n = \int_0^1 x^n e^{3x} dx$

a Show that $I_n = \frac{1}{3}e^3 - \frac{n}{3}I_{n-1}$

b Use this formula to evaluate $\int_0^1 x^4 e^{3x} dx$

4 $I_n = \int x(\ln x)^n\,dx$

a Show that $I_n = \frac{x^2}{2}(\ln x)^n - \frac{n}{2}I_{n-1}$

b Use this formula to find $\int x(\ln x)^2\,dx$

5 $I_n = \int_0^1 \frac{x^n}{\sqrt{1-x}}\,dx$

a Show that $I_n = \frac{2n}{1+2n}I_{n-1}$

b Use this formula to find $\int_0^1 \frac{x^6}{\sqrt{1-x}}\,dx$

6 $I_n = \int x^n \sin x\,dx$

a Show that

$I_n = -x^n\cos x + nx^{n-1}\sin x - n(n-1)I_{n-2}$ for $n > 1$

b Use this formula to find

i $\int x^4 \sin x\,dx$

ii $\int x^3 \sin x\,dx$

7 $I_n = \int_0^{\frac{\pi}{4}} x^n \cos 2x\,dx$

a Show that

$I_n = \frac{1}{2}\left(\frac{\pi}{4}\right)^n - \frac{n(n-1)}{4}I_{n-2}$ for $n > 1$

b Use this formula to evaluate

i $\int_0^{\frac{\pi}{4}} x^6 \cos 2x\,dx$

ii $\int_0^{\frac{\pi}{4}} x^5 \cos 2x\,dx$

Reasoning and problem-solving

Sometimes the integrand is not in the form of the product of two functions. In these cases, you can split up the integral yourself. For example, you can write $\sin^n x$ as $\sin x \sin^{n-1} x$

Strategy

To find and use a reduction formula

1. Split up the function if necessary and integrate.
2. Find an expression for I_n in terms of I_{n-1} and/or I_{n-2}
3. Use the formula repeatedly until you reach I_0 or I_1
4. Calculate I_0 or I_1 and use with your formula to find I_n for the value of n required.

Example 3

Find a reduction formula for $I_n = \int_0^{\pi} \sin^n x\,dx$ and use it to find

a $\int_0^{\pi} \sin^8 x\,dx$ **b** $\int_0^{\pi} \sin^7 x\,dx$

$I_n = \int_0^{\pi} \sin^n x\,dx = \int_0^{\pi} \sin x \sin^{n-1} x\,dx$

$= [-\cos x \sin^{n-1} x]_0^{\pi} - \int_0^{\pi} -(n-1)\cos^2 x \sin^{n-2} x\,dx$

$= (n-1)\int_0^{\pi} \cos^2 x \sin^{n-2} x\,dx$

$= (n-1)\int_0^{\pi} \sin^{n-2} x - \sin^n x\,dx$

$= (n-1)(I_{n-2} - I_n)$

$= (n-1)I_{n-2} - (n-1)I_n$

Rearrange to give $I_n = \frac{n-1}{n} I_{n-2}$

a Use to find I_8

$I_8 = \frac{7}{8} I_6$

$= \frac{7}{8}\frac{5}{6} I_4$

$= \frac{7}{8}\frac{5}{6}\frac{3}{4} I_2$

$= \frac{7}{8}\frac{5}{6}\frac{3}{4}\frac{1}{2} I_0$

$= \frac{35}{128} I_0$

$I_0 = \int_0^{\pi} 1\,dx$

$= [x]_0^{\pi}$

$= \pi$

So $I_8 = \frac{35}{128}\pi$

(1) Split the function.

(1) Integration by parts with $\frac{dv}{dx} = \sin x$ and $u = \sin^{n-1} x$ so that $v = -\cos x$ and $\frac{du}{dx} = (n-1)\sin^{n-2} x \cos x$

Replace $\cos^2 x$ with $1 - \sin^2 x$

(2) This is your reduction formula.

(3) Use the formula repeatedly until I_0 reached.

(4) Substitute value of I_0 into formula for I_8

(*Continued on the next page*)

b Use to find I_7

$I_7 = \frac{6}{7} I_5$

$= \frac{6}{7}\frac{4}{5} I_3$

$= \frac{6}{7}\frac{4}{5}\frac{2}{3} I_1$

$= \frac{16}{35} I_1$

This time, I_1 is reached. (3)

$I_1 = \int_0^{\pi} \sin x \, dx = -[\cos x]_0^{\pi}$

$= -(\cos\pi - \cos 0)$

$= -(-1-1)$

$= 2$

So $I_7 = \frac{16}{35}(2) = \frac{32}{35}$

Substitute value of I_1 into formula for I_7 (4)

PURE

Exercise 19.5B Reasoning and problem-solving

1 $I_n = \int_0^{\frac{\pi}{2}} \cos^n x \, dx$

a Show that $I_n = \frac{n-1}{n} I_{n-2}$ for $n > 1$

b Use this formula to evaluate

i $\int_0^{\frac{\pi}{2}} \cos^5 x \, dx$ **ii** $\int_0^{\frac{\pi}{2}} \cos^6 x \, dx$

2 $I_n = \int_{\frac{\pi}{2}}^{\pi} \sin^n x \, dx$

a Show that $I_n = \frac{n-1}{n} I_{n-2}$

b Use this formula to evaluate

i $\int_{\frac{\pi}{2}}^{\pi} \sin^7 x \, dx$ **ii** $\int_{\frac{\pi}{2}}^{\pi} \sin^8 x \, dx$

3 $I_n = \int \tan^n x \, dx$

a By writing $\tan^n x$ as $\tan^{n-2} x \tan^2 x$, find a reduction formula for I_n

b Use this formula to find

i $\int \tan^6 x \, dx$ **ii** $\int_0^{\frac{\pi}{4}} \tan^7 x \, dx$

4 $I_n = \int_0^1 x^n \sqrt{1+x^2} \, dx$

a Show that $I_n = \frac{2\sqrt{2} - (n-1)I_{n-2}}{2+n}$

b Use this formula to find

i $\int_0^1 x^5 \sqrt{1+x^2} \, dx$ **ii** $\int_0^1 x^4 \sqrt{1+x^2} \, dx$

5 $I_n = \int_0^{\pi} \frac{\cos nx}{\cos x} \, dx$

a Show that $I_n = -I_{n-2}$

Hint: write $nx = (n-1)x + x$ and use the addition formula for cos.

b Use the reduction formula to find

$\int_0^{\pi} \frac{\cos 7x}{\cos x} \, dx$

c Hence write down the value of

i $\int_0^{\pi} \frac{\cos 27x}{\cos x} \, dx$ **ii** $\int_0^{\pi} \frac{\cos 29x}{\cos x} \, dx$

d Show that $\int_0^{\pi} \frac{\cos nx}{\cos x} \, dx$ does not converge when n is an even number.

19.6 Polar graphs and areas

Fluency and skills

See Ch3.3 For a reminder of polar coordinates.

Curves can be defined in polar form.

For example, $r = 2$ is the circle $x^2 + y^2 = 4$ and $\theta = \frac{\pi}{4}$ is a half-line that lies on $y = x$. To sketch polar curves, you work out the value of r for certain values of θ such as $\theta = 0, \frac{\pi}{2}, \pi, \frac{3\pi}{2}, 2\pi$

Example 1

Sketch the curve $r = 3(1 + \cos\theta)$

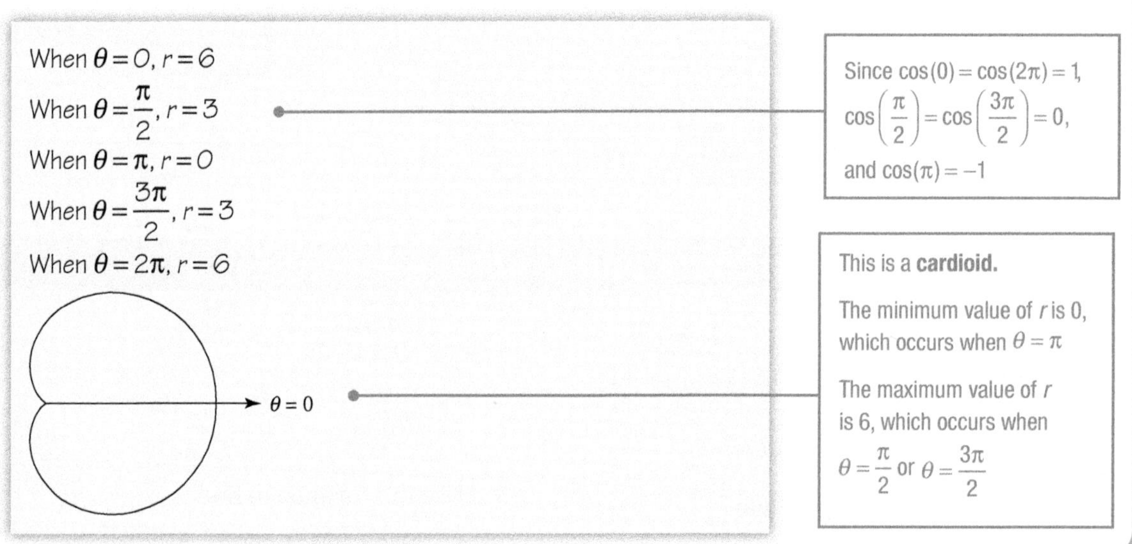

When $\theta = 0, r = 6$
When $\theta = \frac{\pi}{2}, r = 3$
When $\theta = \pi, r = 0$
When $\theta = \frac{3\pi}{2}, r = 3$
When $\theta = 2\pi, r = 6$
$\theta = 0$
Since $\cos(0) = \cos(2\pi) = 1$, $\cos\left(\frac{\pi}{2}\right) = \cos\left(\frac{3\pi}{2}\right) = 0$, and $\cos(\pi) = -1$
This is a **cardioid.**
The minimum value of r is 0, which occurs when $\theta = \pi$
The maximum value of r is 6, which occurs when $\theta = \frac{\pi}{2}$ or $\theta = \frac{3\pi}{2}$

Polar curves of the form $r = a\cos(n\theta)$ or $r = b\sin(n\theta)$ will have n 'loops'. Since r is the radius, it must be positive. So you can work out the position of these loops by considering where r is positive.

Example 2

Sketch the curve $r = 2\cos(3\theta)$

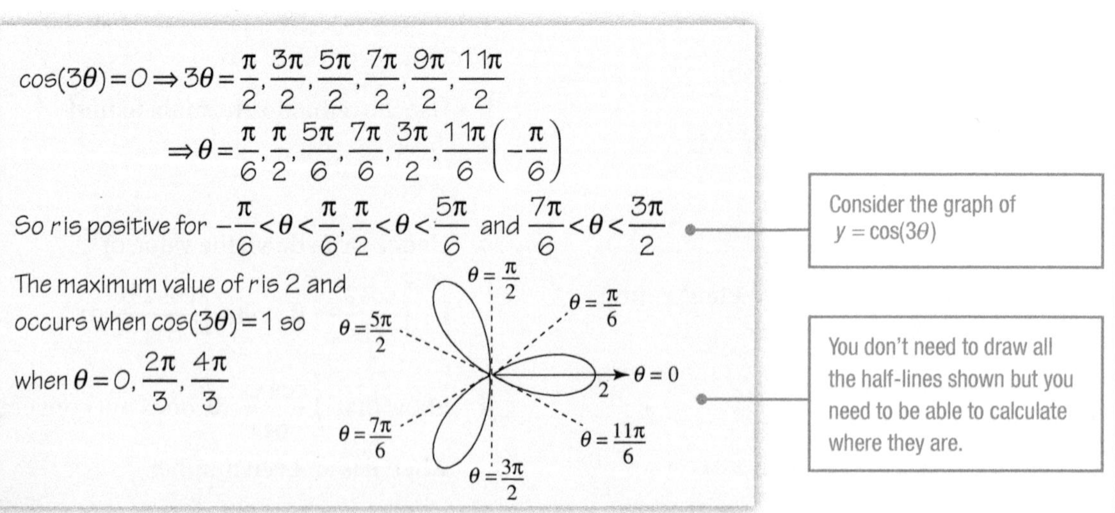

$\cos(3\theta) = 0 \Rightarrow 3\theta = \frac{\pi}{2}, \frac{3\pi}{2}, \frac{5\pi}{2}, \frac{7\pi}{2}, \frac{9\pi}{2}, \frac{11\pi}{2}$
$\Rightarrow \theta = \frac{\pi}{6}, \frac{\pi}{2}, \frac{5\pi}{6}, \frac{7\pi}{6}, \frac{3\pi}{2}, \frac{11\pi}{6}\left(-\frac{\pi}{6}\right)$
So r is positive for $-\frac{\pi}{6} < \theta < \frac{\pi}{6}, \frac{\pi}{2} < \theta < \frac{5\pi}{6}$ and $\frac{7\pi}{6} < \theta < \frac{3\pi}{2}$
The maximum value of r is 2 and occurs when $\cos(3\theta) = 1$ so when $\theta = 0, \frac{2\pi}{3}, \frac{4\pi}{3}$
$\theta = \frac{\pi}{2}$
$\theta = \frac{\pi}{6}$
$\theta = \frac{5\pi}{2}$
$\theta = 0$
2
$\theta = \frac{7\pi}{6}$
$\theta = \frac{11\pi}{6}$
$\theta = \frac{3\pi}{2}$
Consider the graph of $y = \cos(3\theta)$
You don't need to draw all the half-lines shown but you need to be able to calculate where they are.

The area of a sector of a circle is given by the formula area $= \frac{1}{2}r^2\theta$

You can use integration to calculate the area enclosed by a polar curve between two half-lines.

Split the area into n slices, each with an angle of $\delta\theta$

The area of each slice can be approximated as the area of a sector of a circle. The formula for the area of a sector of angle θ from a circle of radius r is $A = \frac{1}{2}r^2\theta$

Therefore, the area of the slice shown is $\delta A \approx \frac{1}{2}[\mathrm{f}(\theta_i)]^2\,\delta\theta$

Taking the limit as $n \to \infty$ of the sum of these areas gives the integral $\int_a^b \frac{1}{2}[\mathrm{f}(\theta)]^2\,\mathrm{d}\theta$, which can be written $\frac{1}{2}\int_a^b r^2\,\mathrm{d}\theta$

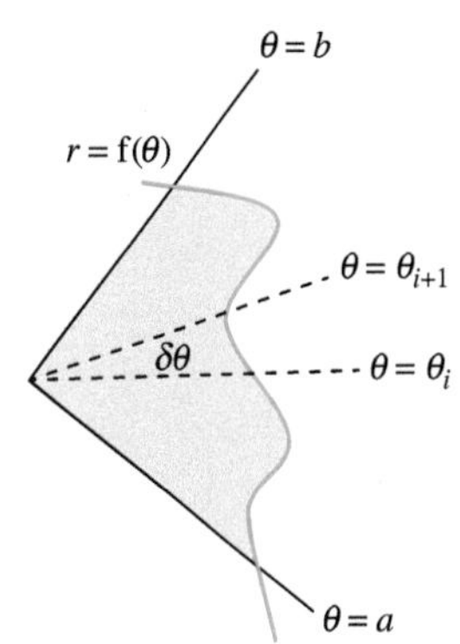

Key point

The area of a sector of a polar curve $r = \mathrm{f}(\theta)$ between the half lines $\theta = a$ and $\theta = b$ is given by $A = \frac{1}{2}\int_a^b r^2\,\mathrm{d}\theta$

Remember that θ must be measured in radians.

Example 3

Calculate the area enclosed by the curve $r = \sin(2\theta)$, where $r > 0$

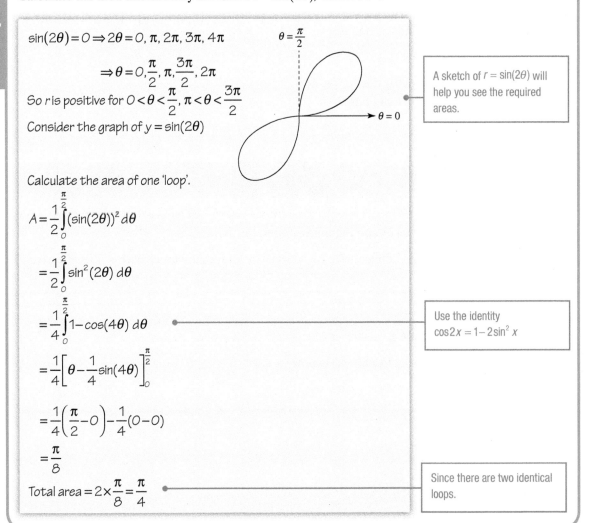

$\sin(2\theta) = 0 \Rightarrow 2\theta = 0, \pi, 2\pi, 3\pi, 4\pi$

$\Rightarrow \theta = 0, \frac{\pi}{2}, \pi, \frac{3\pi}{2}, 2\pi$

So r is positive for $0 < \theta < \frac{\pi}{2}$, $\pi < \theta < \frac{3\pi}{2}$

Consider the graph of $y = \sin(2\theta)$

$\theta = \frac{\pi}{2}$

$\theta = 0$

A sketch of $r = \sin(2\theta)$ will help you see the required areas.

Calculate the area of one 'loop'.

$$A = \frac{1}{2}\int_0^{\frac{\pi}{2}} (\sin(2\theta))^2\,\mathrm{d}\theta$$

$$= \frac{1}{2}\int_0^{\frac{\pi}{2}} \sin^2(2\theta)\,\mathrm{d}\theta$$

$$= \frac{1}{4}\int_0^{\frac{\pi}{2}} 1 - \cos(4\theta)\,\mathrm{d}\theta$$

Use the identity $\cos 2x = 1 - 2\sin^2 x$

$$= \frac{1}{4}\left[\theta - \frac{1}{4}\sin(4\theta)\right]_0^{\frac{\pi}{2}}$$

$$= \frac{1}{4}\left(\frac{\pi}{2} - 0\right) - \frac{1}{4}(0 - 0)$$

$$= \frac{\pi}{8}$$

$$\text{Total area} = 2 \times \frac{\pi}{8} = \frac{\pi}{4}$$

Since there are two identical loops.

Exercise 19.6A Fluency and skills

1 Sketch each of these curves or lines that are given in polar form.

a $r=8$

b $\theta=\frac{\pi}{3}$

c $\theta=\frac{7\pi}{6}$

d $r=2\theta$ for $0\le\theta\le2\pi$

2 For each of these polar equations, where $r>0$

i State the maximum and the minimum values of r

ii Sketch the curve.

a $r=1-\cos\theta$

b $r=2+\sin\theta$

c $r=2(1-\sin\theta)$

d $r=6+3\cos\theta$

e $r=5+\sin\theta$

f $r=\cos(2\theta)$

g $r=a\sin(3\theta)$

h $r=b\cos(4\theta)$

i $r^2=c^2\sin\theta$

3 Calculate the area bounded by the curve with equation $r=4\cos\theta$ and the half lines $\theta=0$ and $\theta=\frac{\pi}{3}$

4 Calculate the area bounded by the curve with equation $r=\theta$ and the half lines $\theta=0$ and $\theta=\frac{\pi}{2}$

5 Calculate the area bounded by the curve with equation $r=2\sin\theta$ and the half lines $\theta=\frac{\pi}{12}$ and $\theta=\frac{5\pi}{12}$

6 Calculate the area bounded by the curve with equation $r^2=4\sin\theta$ and the half lines $\theta=\frac{\pi}{4}$ and $\theta=\frac{3\pi}{4}$

7 Calculate the area enclosed by each of these cardioids.

a $r=3+\cos\theta$

b $r=5+2\sin\theta$

c $r=1+\sin\theta$

d $r=3-2\cos\theta$

8 Calculate the total area enclosed by these curves where $r>0$

a $r=\cos2\theta$

b $r=\sin4\theta$

c $r=4\cos3\theta$

d $r=5\sin3\theta$

e $r^2=2\sin2\theta$

f $r^2=(5+2\sin\theta)$

Reasoning and problem-solving

To find the area between two polar curves, you first need to work out where they intersect. As with Cartesian equations, you do this by solving the equations simultaneously.

You can then calculate the area between the two curves by adding the areas of the different parts.

Strategy

To calculate the area between two polar curves

1. Sketch the curves and solve simultaneously to find their point of intersection.
2. Calculate the area of a sector for each curve using $\frac{1}{2}\int_a^b r^2 d\theta$
3. Add the areas together to find the required area.

Example 4

Find the area bounded by the curves $r=3\cos\theta$ and $r=1+\cos\theta$

To find point of intersection, solve $3\cos\theta = 1+\cos\theta$

$\Rightarrow \cos\theta = \frac{1}{2}$

$\Rightarrow \theta = \pm\frac{\pi}{3}$

① The points of intersection are $\left(\frac{3}{2},\frac{\pi}{3}\right)$ and $\left(\frac{3}{2},-\frac{\pi}{3}\right)$. Draw the half lines through these points.

$\theta=\frac{\pi}{3}$
$r = 1+\cos\theta$
$r = 3\cos\theta$
A_2
A_1
$\theta = 0$
2
3
$\theta=-\frac{\pi}{3}$

The total required area is shaded.

Consider the area labelled A_1 which is enclosed by $1+\cos\theta$ between the half lines $\theta=0$ and $\theta=\frac{\pi}{3}$

$$A_1=\frac{1}{2}\int_0^{\frac{\pi}{3}}(1+\cos\theta)^2\,d\theta$$

② Use the formula $A=\frac{1}{2}\int r^2 d\theta$

$$=\frac{1}{2}\int_0^{\frac{\pi}{3}}1+2\cos\theta+\cos^2\theta\,d\theta$$

$$=\frac{1}{2}\int_0^{\frac{\pi}{3}}1+2\cos\theta+\frac{1}{2}(1+\cos2\theta)d\theta$$

Use the identity $\cos2x = 2\cos^2 x - 1$

$$=\frac{1}{2}\left[\theta+2\sin\theta+\frac{1}{2}\theta+\frac{1}{4}\sin2\theta\right]_0^{\frac{\pi}{3}}$$

$$=\frac{1}{2}\left(\frac{3}{2}\left(\frac{\pi}{3}\right)+\sqrt{3}+\frac{\sqrt{3}}{8}\right)-0$$

$$=\frac{1}{4}\pi+\frac{9}{16}\sqrt{3}$$

Consider the area labelled A_2 which is enclosed by $3\cos\theta$ between the half lines $\theta=\frac{\pi}{3}$ and $\theta=\frac{\pi}{2}$

$$A_2=\frac{1}{2}\int_{\frac{\pi}{3}}^{\frac{\pi}{2}}(3\cos\theta)^2\,d\theta$$

$$=\frac{1}{2}\int_{\frac{\pi}{3}}^{\frac{\pi}{2}}9\cos^2\theta\,d\theta$$

(*Continued on the next page*)

PURE

$$= \frac{9}{4}\int_{\frac{\pi}{3}}^{\frac{\pi}{2}} 1+\cos 2\theta \, d\theta$$

$$= \frac{9}{4}\left[\theta + \frac{1}{2}\sin 2\theta\right]_{\frac{\pi}{3}}^{\frac{\pi}{2}}$$

$$= \frac{9}{4}\left(\frac{\pi}{2}+0\right) - \frac{9}{4}\left(\frac{\pi}{3}+\frac{1}{4}\sqrt{3}\right)$$

$$= \frac{3\pi}{8} - \frac{9}{16}\sqrt{3}$$

$$A_1 + A_2 = \frac{1}{4}\pi + \frac{9}{16}\sqrt{3} + \frac{3\pi}{8} - \frac{9}{16}\sqrt{3}$$

$$= \frac{5}{8}\pi$$

So total area $= 2\times\left(\frac{5}{8}\pi\right) = \frac{5}{4}\pi$

3 This is the top half of the required area.

Exercise 19.6B Reasoning and problem-solving

1 The diagram shows the graphs of $r = \cos\theta + \sin\theta$ and $r = 2 + \sin\theta$ for $r > 0$

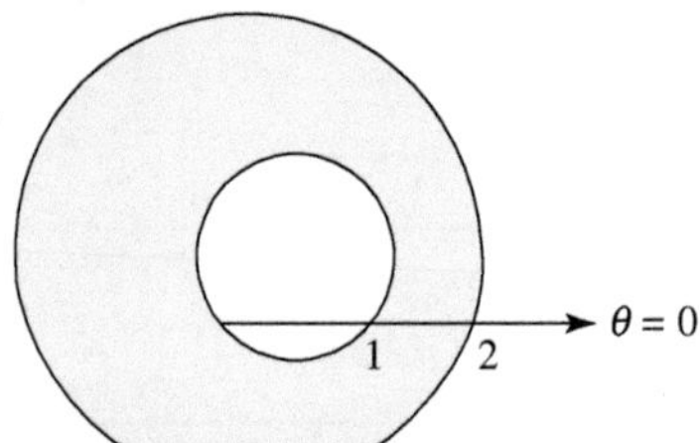

a Prove that the curves do not intersect.

b Calculate the shaded area.

2 The diagram shows the graphs of $r = \sin\theta\sqrt{2\cos\theta}$ and $r = \sin\theta$, for $r > 0$

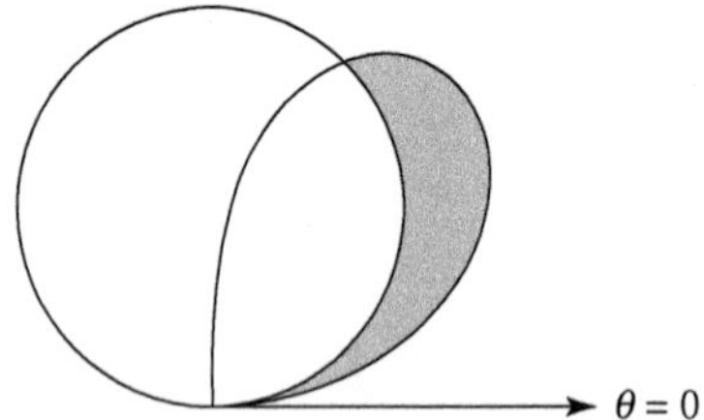

a Find the polar coordinates of the points of intersection of the two curves.

b Calculate the shaded area.

3 The curve shown has polar equation $r = 2 + \cos(2\theta),\ 0 \le \theta \le \frac{\pi}{2}$

At the point A on the curve, $r = 1.5$

Calculate the shaded area.

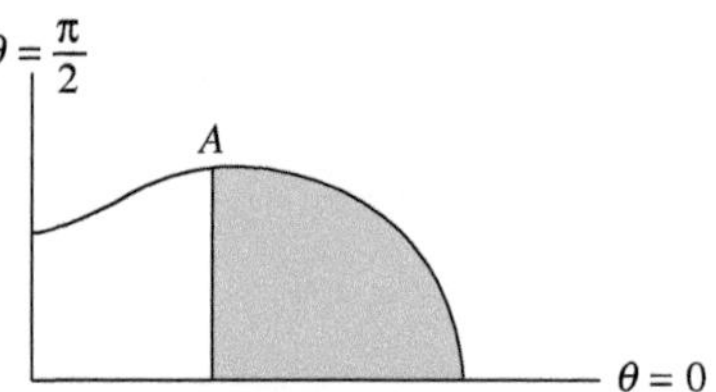

4 Calculate the area bounded by these pairs of curves.

a $r = \cos\theta,\ r = \sqrt{3}\sin\theta$

b $r = 2 - \cos\theta,\ r = 3\cos\theta$

c $r = \sqrt{2} + \sin\theta,\ r = 3\sin\theta$

d $r^2 = 1 - \sin\theta,\ r = \sqrt{2}\sin\theta$

e $r = 5 + 2\cos\theta,\ r = 4$

5 The circle with equation $x^2 + (y-2)^2 = 4$ intersects the curve with polar equation $r = 1 + \sin\theta$ at the points A and B

a Calculate the length of AB

b Calculate the area enclosed by the circle $x^2 + (y-2)^2 = 4$ but not the curve $r = 1 + \sin\theta$

19.7 Lengths and surface areas

Fluency and skills

You can use integration to calculate the length of a curve between two points. Consider two points, A and B, that are close to each other on a curve. If you approximate the length of the curve, s, between these two points by a straight line, then you can use Pythagoras' theorem to see that the length of this line is

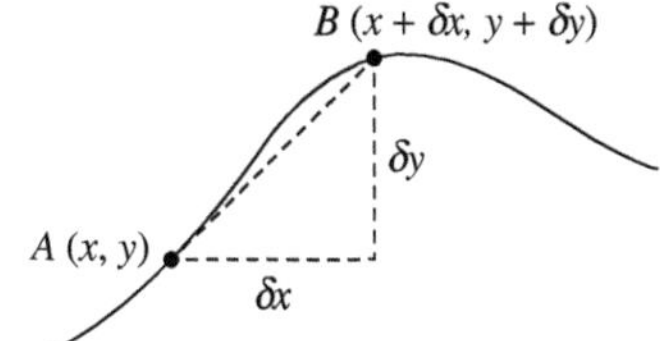

$$(\delta s)^2 = (\delta x)^2 + (\delta y)^2$$

Dividing by $(\delta x)^2$ gives $\frac{(\delta s)^2}{(\delta x)^2} = 1 + \frac{(\delta y)^2}{(\delta x)^2}$, which can be written as

$$\frac{\delta s}{\delta x} = \sqrt{1 + \left(\frac{\delta y}{\delta x}\right)^2}$$

As $\delta x \to 0$, $\frac{\delta s}{\delta x} \to \frac{\mathrm{d}s}{\mathrm{d}x}$ and $\frac{\delta y}{\delta x} \to \frac{\mathrm{d}y}{\mathrm{d}x}$; therefore $\frac{\mathrm{d}s}{\mathrm{d}x} = \sqrt{1 + \left(\frac{\mathrm{d}y}{\mathrm{d}x}\right)^2}$

You can integrate this to find the length of the curve.

$$s = \int \sqrt{1 + \left(\frac{\mathrm{d}y}{\mathrm{d}x}\right)^2}\,\mathrm{d}x$$

Key point

The length of the arc of the curve $y = \mathrm{f}(x)$ from x_1 to x_2 is given by $s = \int_{x_1}^{x_2} \sqrt{1 + \left(\frac{\mathrm{d}y}{\mathrm{d}x}\right)^2}\,\mathrm{d}x$

Example 1

Calculate the length of the curve $y = 2x\sqrt{x}$ between the points (0, 0) and (4, 16)

Write $y = 2x^{\frac{3}{2}}$, then $\frac{\mathrm{d}y}{\mathrm{d}x} = 3\sqrt{x}$

$s = \int_0^3 \sqrt{1 + (3\sqrt{x})^2}\,\mathrm{d}x$ — Use $s = \int_{x_1}^{x_2} \sqrt{1 + \left(\frac{\mathrm{d}y}{\mathrm{d}x}\right)^2}\,\mathrm{d}x$

$= \int_0^3 \sqrt{1 + 9x}\,\mathrm{d}x$

$= \left[\frac{2}{27}(1 + 9x)^{\frac{3}{2}}\right]_0^3$ — You can check this result using differentiation.

$= \frac{2}{27}(1 + 27)^{\frac{3}{2}} - \frac{2}{27}(1 + 0)^{\frac{3}{2}}$

$= \frac{2}{27}(56\sqrt{7} - 1)$

If you are given a curve defined by parametric equations $x = f(t)$, $y = g(t)$, then you need to divide by $(\delta t)^2$ instead.

So $\quad (\delta s)^2 = (\delta x)^2 + (\delta y)^2$ and divide by $(\delta t)^2$ to give

$$\left(\frac{\delta s}{\delta t}\right)^2 = \left(\frac{\delta x}{\delta t}\right)^2 + \left(\frac{\delta y}{\delta t}\right)^2 \text{ which can be written as}$$

$$\frac{\delta s}{\delta t} = \sqrt{\left(\frac{\delta x}{\delta t}\right)^2 + \left(\frac{\delta y}{\delta t}\right)^2}$$

As $\delta t \to 0$, $\frac{\delta s}{\delta t} \to \frac{ds}{dt}$, $\frac{\delta x}{\delta t} \to \frac{dx}{dt}$ and $\frac{\delta y}{\delta t} \to \frac{dy}{dt}$; therefore

$$\frac{ds}{dt} = \sqrt{\left(\frac{dx}{dt}\right)^2 + \left(\frac{dy}{dt}\right)^2}$$

You can then integrate this to find the length of the curve.

$$s = \int \sqrt{\left(\frac{dx}{dt}\right)^2 + \left(\frac{dy}{dt}\right)^2}\, dt$$

Key point

The length of the section of the curve $x = f(t)$, $y = g(t)$ from t_1 to t_2 is given by $s = \int_{t_1}^{t_2} \sqrt{\left(\frac{dx}{dt}\right)^2 + \left(\frac{dy}{dt}\right)^2}\, dt$

Example 2

Calculate the length of the arc of the curve $x = t^2 - 1$, $y = \frac{2}{3}t^3 + 1$ from $t = 0$ to $t = 2$

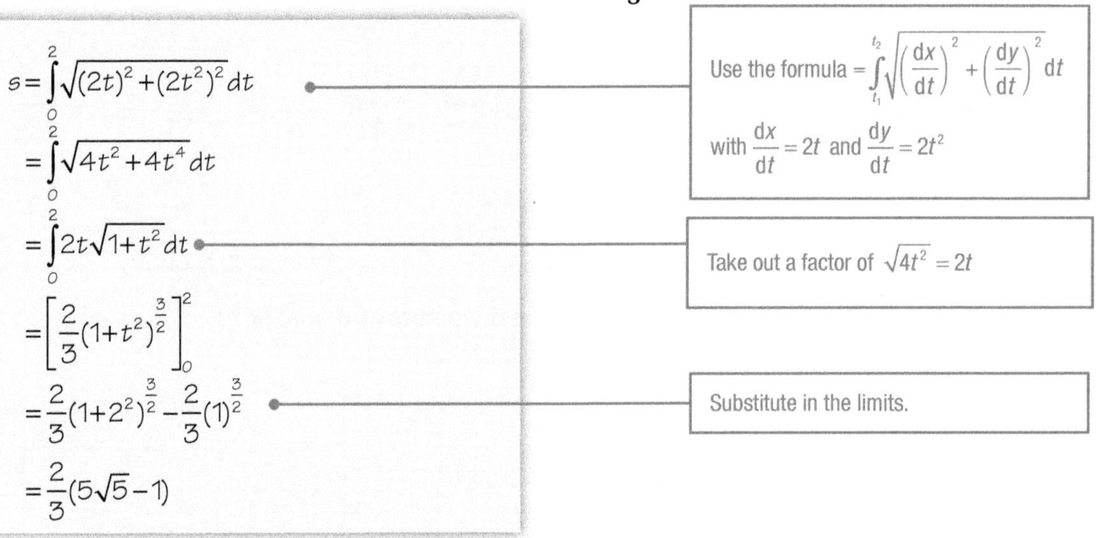

$$s = \int_0^2 \sqrt{(2t)^2 + (2t^2)^2}\, dt$$

Use the formula $= \int_{t_1}^{t_2} \sqrt{\left(\frac{dx}{dt}\right)^2 + \left(\frac{dy}{dt}\right)^2}\, dt$ with $\frac{dx}{dt} = 2t$ and $\frac{dy}{dt} = 2t^2$

$$= \int_0^2 \sqrt{4t^2 + 4t^4}\, dt$$

$$= \int_0^2 2t\sqrt{1+t^2}\, dt$$

Take out a factor of $\sqrt{4t^2} = 2t$

$$= \left[\frac{2}{3}(1+t^2)^{\frac{3}{2}}\right]_0^2$$

$$= \frac{2}{3}(1+2^2)^{\frac{3}{2}} - \frac{2}{3}(1)^{\frac{3}{2}}$$

Substitute in the limits.

$$= \frac{2}{3}(5\sqrt{5} - 1)$$

See Ch4.2 For a reminder of volumes of revolution.

You can use the formula $V = \pi \int_{x_1}^{x_2} y^2\, dx$ to calculate the volume of revolution when $f(x)$ between x_1 and x_2 is rotated through 2π radians around the x-axis.

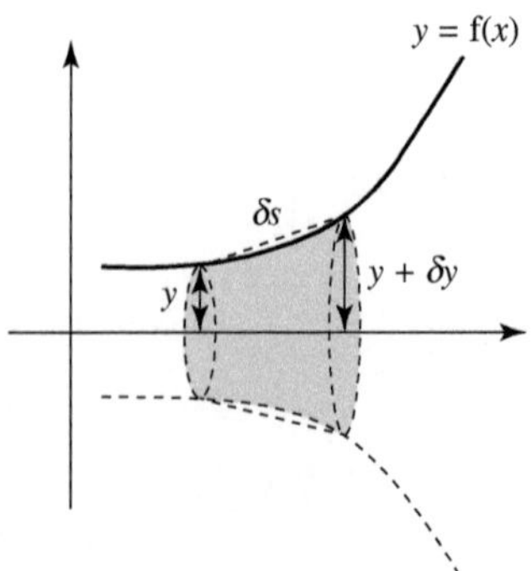

It is also possible to calculate the surface area of revolution. Consider a small section of the curve rotated through 2π radians around the x-axis. The shape can be approximated by a truncated cone (also known as a frustum).

The curved surface area of a truncated cone is given by $S = \pi(r_1 + r_2)l$ where r_1 is the radius of the base, r_2 is the radius of the top and l is the length of the slanted side of the frustum.

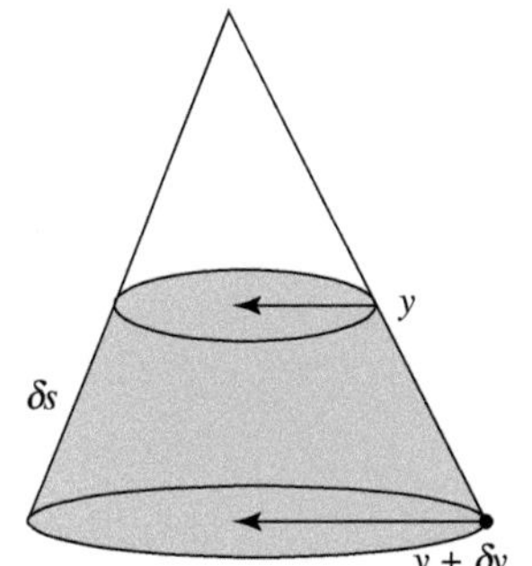

Therefore $\delta S = \pi(y + \delta y + y)\delta s = \pi(2y + \delta y)\delta s$

Divide by δx to give $\frac{\delta S}{\delta x} = \pi(2y + \delta y)\frac{\delta s}{\delta x}$

As $\delta x \to 0$, $\delta y \to 0$, $\frac{\delta s}{\delta x} \to \frac{\mathrm{d}s}{\mathrm{d}x}$ and $\frac{\delta S}{\delta x} \to \frac{\mathrm{d}S}{\mathrm{d}x}$; therefore
$\frac{\mathrm{d}S}{\mathrm{d}x} = 2y\pi\frac{\mathrm{d}s}{\mathrm{d}x}$

You can then integrate both sides with respect to x to find the surface area of revolution.

$$S = \int 2y\pi\,\mathrm{d}s$$

As $\frac{\mathrm{d}s}{\mathrm{d}x} = \sqrt{1 + \left(\frac{\mathrm{d}y}{\mathrm{d}x}\right)^2}$, you can write this as

$$S = \int 2y\pi\sqrt{1 + \left(\frac{\mathrm{d}y}{\mathrm{d}x}\right)^2}\,\mathrm{d}x$$

Key point

The surface area of revolution when the curve $y = \mathrm{f}(x)$ from x_1 to x_2 is rotated through 2π radians around the x-axis is given by $S = 2\pi\int_{x_1}^{x_2} y\sqrt{1 + \left(\frac{\mathrm{d}y}{\mathrm{d}x}\right)^2}\,\mathrm{d}x$

Example 3

The section of the curve with equation $y = x^3$ between $x = 0$ and $x = \frac{1}{2}$ is rotated through 2π radians around the x-axis. For the solid formed, calculate the exact value of the curved surface area.

$y = x^3$ so $\frac{\mathrm{d}y}{\mathrm{d}x} = 3x^2$

$$s = 2\pi\int_0^{\frac{1}{2}} x^3\sqrt{1 + (3x^2)^2}\,\mathrm{d}x$$

Use $2\pi\int_{x_1}^{x_2} y\sqrt{1 + \left(\frac{\mathrm{d}y}{\mathrm{d}x}\right)^2}\,\mathrm{d}x$

$$= 2\pi\int_0^{\frac{1}{2}} x^3(1 + 9x^4)^{\frac{1}{2}}\,\mathrm{d}x$$

$$= 2\pi\left[\frac{1}{54}(1 + 9x^4)^{\frac{3}{2}}\right]_0^{\frac{1}{2}}$$

Substitute in the limits of 0 and $\frac{1}{2}$

$$= 2\pi\left(\frac{1}{54}\left(\frac{125}{64} - 1\right)\right)$$

$$= \frac{61}{1728}\pi$$

See Maths Ch12.3

For a reminder of parametric equations.

You can also find a formula for the surface area of revolution when the curve is defined by parametric equations.

Using $\frac{\mathrm{d}s}{\mathrm{d}t}=\sqrt{\left(\frac{\mathrm{d}x}{\mathrm{d}t}\right)^2+\left(\frac{\mathrm{d}y}{\mathrm{d}t}\right)^2}$ in the formula for surface area of revolution, $S=\int 2y\pi\,\mathrm{d}s$ gives $S=\int 2y\pi\sqrt{\left(\frac{\mathrm{d}x}{\mathrm{d}t}\right)^2+\left(\frac{\mathrm{d}y}{\mathrm{d}t}\right)^2}\,\mathrm{d}t$

Key point

The surface area of revolution when the curve $x=\mathrm{f}(t)$, $y=\mathrm{g}(t)$ from t_1 to t_2 is rotated through 2π radians around the x-axis is given by

$S=2\pi\int_{t_1}^{t_2} y\sqrt{\left(\frac{\mathrm{d}x}{\mathrm{d}t}\right)^2+\left(\frac{\mathrm{d}y}{\mathrm{d}t}\right)^2}\,\mathrm{d}t$, where y is replaced by $\mathrm{g}(t)$

Exercise 19.7A Fluency and skills

1 Show that the length of the arc of the curve $y=\frac{2}{3}x^{\frac{3}{2}}$ between the points $x=0$ and $x=1$ is $A\sqrt{2}+B$, where A and B are constants to be found.

2 Calculate the arc length of the curve $y=x\sqrt{x}$ between (4, 8) and (9, 27), giving your answer to 3 significant figures.

3 Calculate the length of the curve with equation $y=\cosh x$, $\ln 2\le x\le \ln 4$

4 Calculate the length of the arc of the curve $y=\frac{1}{2}\cosh 2x$ between the points $x=0$ and $x=\ln 2$

5 Calculate the length of the arc of the curve $y=(1+x)^{\frac{3}{2}}$ between $(-1, 0)$ and $\left(\frac{1}{3},\frac{8}{9}\sqrt{3}\right)$

6 a Prove that $\frac{\mathrm{d}(\ln(\sec x+\tan x))}{\mathrm{d}x}=\sec x$

b Hence calculate the length of the arc of the curve $y=\ln\left(\frac{1}{2}\cos x\right)$ between $x=0$ and $x=\frac{\pi}{6}$

7 The curve C is defined by the parametric equations $x=t^2$, $y=2t$. The section of C between $t=0$ and $t=2$ is rotated through 360° about the x-axis.

Calculate the surface area of revolution.

8 A curve is given by the parametric equations $x=\frac{t^3}{6}+1$, $y=t^2$, $0\le t\le 3$

Calculate the arc length.

9 A curve C has equation $y=4x^3$, $0\le x\le \frac{1}{2}$

The curve C is rotated through 2π radians around the x-axis. Calculate

a The volume of the solid formed,

b The curved surface area of the solid formed.

10 The arc of the curve with equation $y=\cosh x$ between $x=0$ and $x=\ln 3$ is rotated through 2π radians around the x-axis. Calculate

a The volume of the solid formed,

b The curved surface area of the solid formed.

Reasoning and problem-solving

You may need to use substitution and other integration techniques to calculate arc lengths and surface areas of revolution.

Strategy

To calculate the surface area of revolution

1. Choose the correct formula for arc length or for surface area, depending on whether the equation of the curve is in parametric or Cartesian form.
2. Use a suitable substitution, being sure to apply the correct limits.
3. Use trigonometric or hyperbolic identities.

The hyperbolic identities $\cosh^2 x = \frac{1}{2}(\cosh 2x + 1)$ and $\sinh^2 x = \frac{1}{2}(\cosh 2x - 1)$ are particularly useful for integration.

Example 4

The curve C has equation $y = 2x^2$. Calculate the length of the curve between (0, 0) and $\left(\frac{1}{4}, \frac{1}{8}\right)$

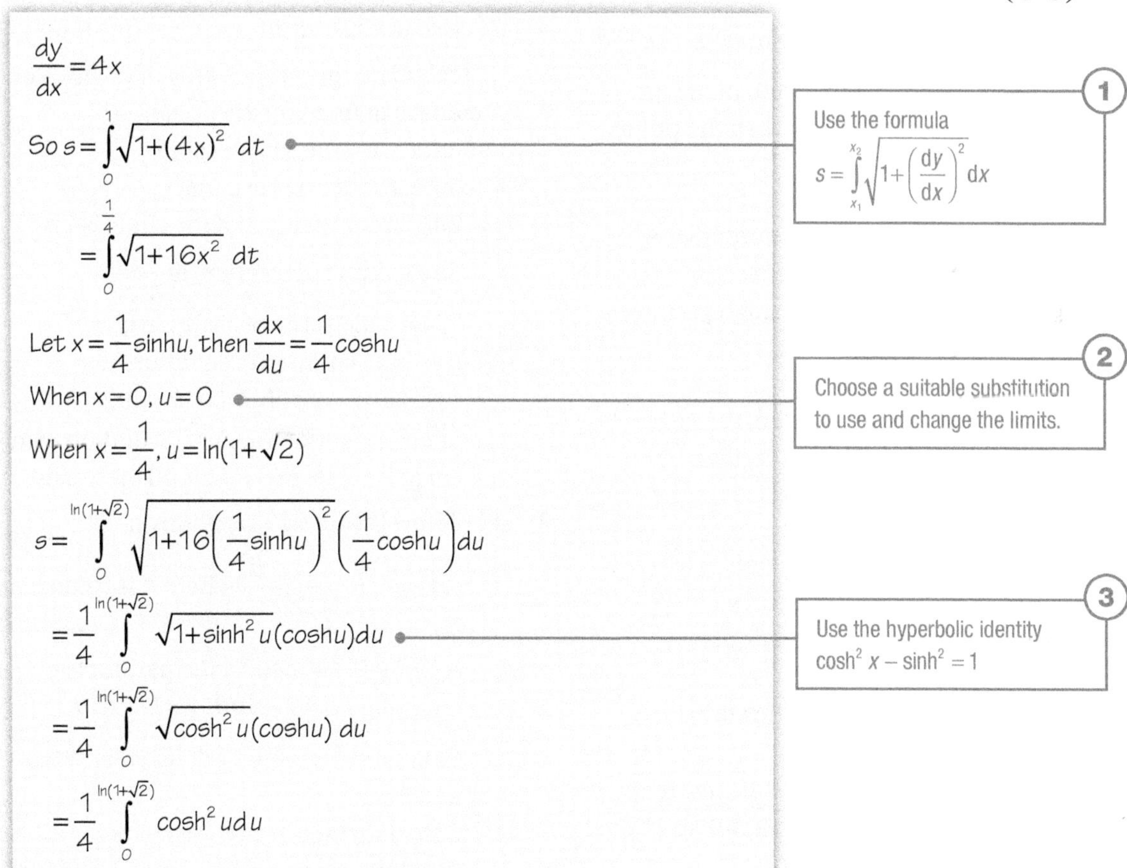

$\frac{dy}{dx} = 4x$

So $s = \int_0^{\frac{1}{4}} \sqrt{1+(4x)^2}\, dt$

$= \int_0^{\frac{1}{4}} \sqrt{1+16x^2}\, dt$

Use the formula $s = \int_{x_1}^{x_2} \sqrt{1+\left(\frac{dy}{dx}\right)^2}\, dx$ (1)

Let $x = \frac{1}{4}\sinh u$, then $\frac{dx}{du} = \frac{1}{4}\cosh u$

When $x = 0$, $u = 0$

When $x = \frac{1}{4}$, $u = \ln(1+\sqrt{2})$

Choose a suitable substitution to use and change the limits. (2)

$s = \int_0^{\ln(1+\sqrt{2})} \sqrt{1+16\left(\frac{1}{4}\sinh u\right)^2}\left(\frac{1}{4}\cosh u\right)du$

$= \frac{1}{4}\int_0^{\ln(1+\sqrt{2})} \sqrt{1+\sinh^2 u}(\cosh u)du$

Use the hyperbolic identity $\cosh^2 x - \sinh^2 = 1$ (3)

$= \frac{1}{4}\int_0^{\ln(1+\sqrt{2})} \sqrt{\cosh^2 u}(\cosh u)\, du$

$= \frac{1}{4}\int_0^{\ln(1+\sqrt{2})} \cosh^2 u\, du$

(*Continued on the next page*)

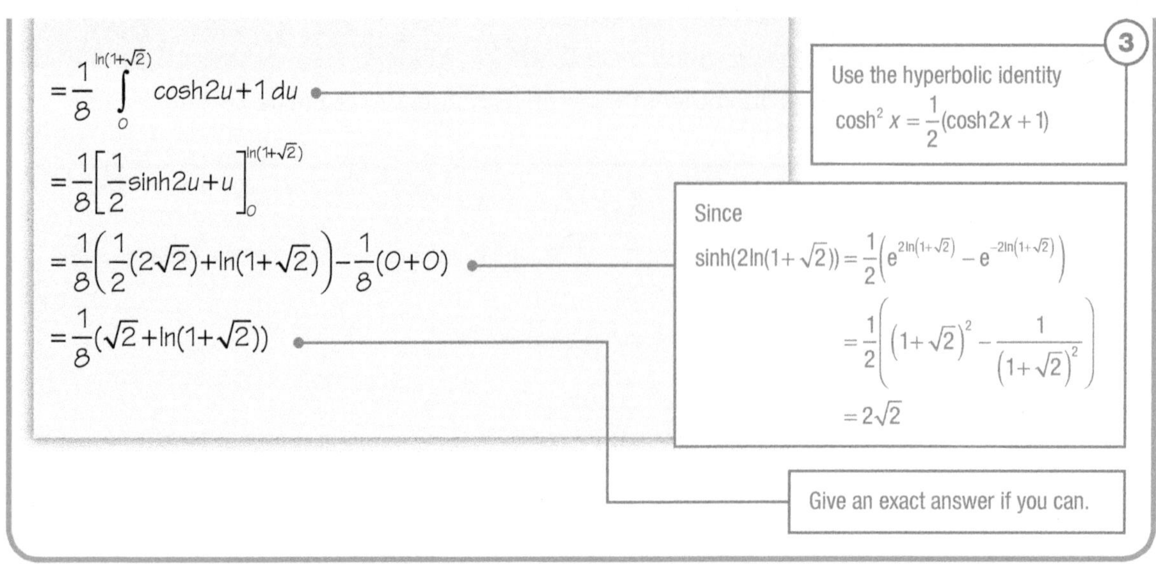

$=\frac{1}{8}\int_0^{\ln(1+\sqrt{2})} \cosh 2u+1\,du$
$=\frac{1}{8}\left[\frac{1}{2}\sinh 2u+u\right]_0^{\ln(1+\sqrt{2})}$
$=\frac{1}{8}\left(\frac{1}{2}(2\sqrt{2})+\ln(1+\sqrt{2})\right)-\frac{1}{8}(0+0)$
$=\frac{1}{8}(\sqrt{2}+\ln(1+\sqrt{2}))$
3
Use the hyperbolic identity
$\cosh^2 x=\frac{1}{2}(\cosh 2x+1)$
Since
$\sinh(2\ln(1+\sqrt{2}))=\frac{1}{2}\left(e^{2\ln(1+\sqrt{2})}-e^{-2\ln(1+\sqrt{2})}\right)$
$=\frac{1}{2}\left((1+\sqrt{2})^2-\frac{1}{(1+\sqrt{2})^2}\right)$
$=2\sqrt{2}$
Give an exact answer if you can.

Exercise 19.7B Reasoning and problem-solving

1 a Use a suitable substitution to show that $\int\sqrt{1+16x^2}\,dx$ $=\frac{1}{8}\text{arsinh}(4x)+\frac{1}{2}x\cosh(\text{arsinh}\,4x)+c$

b Hence calculate the length of the arc of the curve $y=2x^2$ between the points $x=0$ and $x=\frac{1}{4}$

2 A curve is defined by the parametric equations $x=t^2,\ y=2t$

Calculate the exact arc length from $t=0$ to $t=1$

3 A curve has equation $y=2\sqrt{x}$

a Show that the arc length of the curve between $x=0$ and $x=4$ is given by $s=\int_0^4\sqrt{\frac{x+1}{x}}\,dx$

b Use the substitution $x=\sinh^2 u$ to calculate the exact value of s

4 A curve has parametric equations $x=\cos^3 t,\ y=\sin^3 t$

Calculate the arc length of the curve between (1, 0) and (0, 1)

5 The section of the curve $y=2e^{-x}$ between (0, 2) and (ln2, 1) is rotated 2π radians around the x-axis.

a Calculate the volume of revolution.

b Show that the surface area of revolution is given by $4\pi\int_0^{\ln 2}e^{-x}\sqrt{1+4e^{-2x}}\,dx$

c Use the substitution $e^{-x}=\frac{1}{2}\sinh u$ to calculate the surface area of revolution.

6 A circle is defined by the parametric equations $x=r\cos\theta,\ y=r\sin\theta$

Use integration to prove that

a The circumference of the circle is $2\pi r$

b The volume of a sphere is $\frac{4}{3}\pi r^3$

c The surface area of a sphere is $4\pi r^2$

7 A curve has parametric equation $x=\cosh 2\theta,\ y=2\sinh\theta,\ 0<\theta<\ln 2$

Calculate the surface area of revolution when the curve is rotated 360° around the x-axis.

8 a Show that, if $\sin\frac{\theta}{2}$ is positive, $\sqrt{1-\cos\theta}=A\sin\frac{\theta}{2}$, where A is a constant to be found.

b A curve is defined by the parametric equations $x=\theta-\sin\theta,\ y=1-\cos\theta$

Calculate the arc length between $\theta=0$ and $\theta=\pi$

c Calculate the surface area of revolution when the section of the curve between $\theta=0$ and $\theta=\pi$ is rotated 2π radians around the x-axis.

19 Summary and review

Chapter summary

- An improper integral is a definite integral where either:
  - one or both of the limits is $\pm\infty$
  - the integrand (expression to be integrated) is undefined at one of the limits of the integral
  - the integrand is undefined at some point between the limits of the integral.
- To evaluate an improper integral, replace the limit where the integrand is undefined with a variable and then consider what happens to the integral as the variable tends to the original value of the limit.
- If the integral is undefined at more than one point, then split it into two integrals.
- Learn and use these important limits:
  - For any real number k, $x^k e^{-x} \to 0$ when $x \to \infty$
  - For any real number k, $x^k \ln x \to 0$ when $x \to 0+$
- Inverse trigonometric functions can be differentiated using the chain rule:
  - $\frac{\mathrm{d}(\arcsin x)}{\mathrm{d}x} = \frac{1}{\sqrt{1-x^2}}$ $\quad \frac{\mathrm{d}(\arccos x)}{\mathrm{d}x} = -\frac{1}{\sqrt{1-x^2}}$ $\quad \frac{\mathrm{d}(\arctan x)}{\mathrm{d}x} = \frac{1}{1+x^2}$
- Trigonometric functions can be used as substitutions in integration:
  - for an integral involving $\sqrt{a^2 - x^2}$, try the substitution $x = a\sin u$
  - for an integral involving $a^2 + x^2$, try the substitution $x = a\tan u$
- The hyperbolic functions are defined using exponentials:
  - $\sinh x = \frac{1}{2}(e^x - e^{-x})$ $\quad \cosh x = \frac{1}{2}(e^x + e^{-x})$ $\quad \tanh x = \frac{e^x - e^{-x}}{e^x + e^{-x}}$
- Particularly useful hyperbolic identities are:
  - $\cosh^2 x - \sinh^2 x \equiv 1$
  - $\sinh 2x = 2\sinh x \cosh x$ and $\cosh 2x = \cosh^2 x + \sinh^2 x$
- The reciprocal hyperbolic functions are:
  - $\operatorname{cosech} x = \frac{1}{\sinh x}$ $\quad \operatorname{sech} x = \frac{1}{\cosh x}$ $\quad \coth x = \frac{1}{\tanh x}$
- Hyperbolic functions can be differentiated and integrated:
  - $\frac{\mathrm{d}(\sinh x)}{\mathrm{d}x} = \cosh x$ $\quad \frac{\mathrm{d}(\cosh x)}{\mathrm{d}x} = \sinh x$ $\quad \frac{\mathrm{d}(\tanh x)}{\mathrm{d}x} = \operatorname{sech}^2 x$
  - $\frac{\mathrm{d}(\operatorname{sech} x)}{\mathrm{d}x} = -\operatorname{sech} x \tanh x$ $\quad \frac{\mathrm{d}(\operatorname{cosech} x)}{\mathrm{d}x} = -\operatorname{cosech} x \coth x$ $\quad \frac{\mathrm{d}(\coth x)}{\mathrm{d}x} = -\operatorname{cosech}^2 x$
  - $\int \sinh x = \cosh x + c$, $\int \cosh x = \sinh x + c$
- Inverse hyperbolic functions can be differentiated using the chain rule:
  - $\frac{\mathrm{d}(\operatorname{arcosh} x)}{\mathrm{d}x} = \frac{1}{\sqrt{x^2-1}}$ $\quad \frac{\mathrm{d}(\operatorname{arsinh} x)}{\mathrm{d}x} = \frac{1}{\sqrt{x^2+1}}$
- Hyperbolic functions can be used as substitutions in integration:
  - for an integral involving $\sqrt{x^2 + a^2}$, try the substitution $x = a\sinh u$
  - for an integral involving $\sqrt{x^2 - a^2}$, try the substitution $x = a\cosh u$

- A fraction $\frac{f(x)}{g(x)}$ where degree of $f(x) \geq$ degree of $g(x)$ is called an **improper** fraction and can be written in the form $P(x)+\frac{Q(x)}{g(x)}$
- $\frac{f(x)}{(\alpha x^2+\beta)(\gamma x+\delta)}$ can be split into partial fractions of the form $\frac{Ax+B}{\alpha x^2+\beta}+\frac{C}{\gamma x+\delta}$
- The formula for integration by parts is $\int u\frac{dv}{dx}dx = uv - \int v\frac{du}{dx}dx$
- A reduction formula for I_n is an equation that relates I_n to I_{n-1} and/or I_{n-2}. It can be repeatedly applied to reduce the integral to one not requiring integration by parts.
- If a point P has polar coordinates (r, θ), then:
  - r is the length of OP
  - θ is the angle between OP and the initial line.
- A half-line is a straight line that extends infinitely from a point.
- The area of a sector of a polar curve $r = f(\theta)$ between the half lines $\theta = a$ and $\theta = b$ is given by $A = \frac{1}{2}\int_a^b r^2 d\theta$
- The length of the curve $y = f(x)$ from x_1 to x_2 is given by $s = \int_{x_1}^{x_2}\sqrt{1+\left(\frac{dy}{dx}\right)^2}dx$
- The length of the curve $x = f(t)$, $y = g(t)$ from t_1 to t_2 is given by $s = \int_{t_1}^{t_2}\sqrt{\left(\frac{dx}{dt}\right)^2+\left(\frac{dy}{dt}\right)^2}dt$
- The volume of revolution when the curve $y = f(x)$ from x_1 to x_2 is rotated 2π radians around the x-axis is given by $V = \pi\int_{x_1}^{x_2} y^2 dx$
- The surface area of revolution when the curve $y = f(x)$ from x_1 to x_2 is rotated 2π radians around the x-axis is given by $S = 2\pi\int_{x_1}^{x_2} y\sqrt{1+\left(\frac{dy}{dx}\right)^2}dx$
- The surface area of revolution when the curve $x = f(t)$, $y = g(t)$ from t_1 to t is rotated 2π radians around the x-axis is given by $S = 2\pi\int_{t_1}^{t_2} y\sqrt{\left(\frac{dx}{dt}\right)^2+\left(\frac{dy}{dt}\right)^2}dt$

Check and review

You should now be able to...	Review Questions
✔ Recognise and evaluate an improper integral.	1, 2
✔ Decide whether an improper integral converges.	3, 4
✔ Derive and use the derivatives of inverse trigonometric functions.	5, 6
✔ Use trigonometric functions as substitutions in integration.	7
✔ Derive and use hyperbolic identities.	8, 10

✔ Differentiate hyperbolic functions.	9
✔ Differentiate inverse hyperbolic functions.	10, 11
✔ Use hyperbolic functions as substitutions in integration.	12
✔ Write an improper fraction in the form $P(x)+\frac{Q(x)}{g(x)}$	13
✔ Integrate a rational function by first splitting into partial fractions.	13, 14
✔ Derive a reduction formula for indefinite and definite integrals.	15, 16
✔ Use a reduction formula for indefinite and definite integrals.	15, 16
✔ Sketch polar curves.	17
✔ Use integration to calculate the area of a sector of a polar curve.	18
✔ Find the point of intersection between two polar curves.	19
✔ Calculate the area between two polar curves.	19
✔ Calculate the length of a curve given by a Cartesian equation.	20
✔ Calculate the length of a curve given by parametric equations.	21
✔ Calculate the surface area of revolution of a curve given by a Cartesian equation.	22
✔ Calculate the surface area of revolution of a curve given by parametric equations.	23

1 Which of these are improper integrals? Explain your answers.

A: $\int_0^3 \frac{1}{1-x}\,dx$ B: $\int_0^4 \frac{1}{x+2}$ C: $\int_1^\infty \frac{1}{x}\,dx$

2 Evaluate these improper integrals.

a $\int_2^\infty \frac{2}{x^2}\,dx$ b $\int_{-6}^3 \frac{2}{\sqrt{3-x}}\,dx$

3 Show that the integral $\int_0^{2e} x^3 \ln x\,dx$ converges and find its value.

4 Show that the integral $\int_1^\infty \frac{2}{x}\,dx$ does not converge.

5 Prove that $\frac{d(\arccos x)}{dx} = -\frac{1}{\sqrt{1-x^2}}$

6 Differentiate these expressions with respect to x

a $x^2 \arctan x$ b $\arcsin\left(\frac{x^2}{2}\right)$

7 Use a substitution to evaluate each of these integrations.

a $\int \frac{1}{\sqrt{4-x^2}}\,dx$ b $\int \frac{1}{x^2+16}\,dx$

8 Use the definitions of $\sinh x$ and $\cosh x$ to prove these identities.

a $\sinh(A-B) \equiv \sinh A \cosh B - \sinh B \cosh A$

b $\cosh^2 x \equiv \frac{1}{2}(1+\cosh 2x)$

9 Differentiate these expressions with respect to x

a $\cosh 2x$ b $x^2 \sinh x$

10 Show that $\frac{d(\operatorname{artanh} x)}{dx} = \frac{1}{1-x^2}$

11 Differentiate these expressions with respect to x

a $\operatorname{arsinh}(2x^2)$ b $x \operatorname{arcosh}(x-1)$

12 Evaluate these integrals.

a $\int_0^5 \frac{1}{\sqrt{25+x^2}}\,dx$ **b** $\int_3^9 \frac{1}{\sqrt{2x^2-18}}\,dx$

13 **a** Write $\frac{2x^3-x^2-12x+32}{(x-2)(x+3)}$ in the form $P(x)+\frac{Q(x)}{g(x)}$

b Hence evaluate $\int_3^4 \frac{2x^3-x^2-12x+32}{(x-2)(x+3)}\,dx$

14 Calculate the exact value of $\int_0^3 \frac{x^2+5x+15}{(4-x)(x^2+1)}\,dx$

15 $I_n = \int x^n e^{-2x}\,dx$

a Show that $I_n = -\frac{1}{2}x^n e^{-2x} + \frac{n}{2}I_{n-1}$

b Use this formula to find I_3

16 $I_n = \int_0^{\pi} x^n \cos x\,dx$

a Show that $I_n = -n\pi^{n-1} - n(n-1)I_{n-2}$ for $n > 1$

b Use this formula to evaluate

i $\int_0^{\pi} x^6 \cos x\,dx$ **ii** $\int_0^{\pi} x^5 \cos x\,dx$

17 For each polar equation

i Sketch the curves for $r \ge 0$ and $0 \le \theta < 2\pi$

ii State the maximum and the minimum value of r for for $r \ge 0$ and $0 \le \theta < 2\pi$

a $r = 8\cos 2\theta$

b $r = -3\sin\theta$

c $r = \frac{\theta}{4}$

d $r = 7 + 5\sin\theta$

18 Calculate the total area enclosed within the polar curves for $r \ge 0$ and $0 \le \theta < 2\pi$

a $r = 3 + \cos\theta$ **b** $r = 2\sin 4\theta$

19 The graphs of $r = \sqrt{2} + \sin\theta$ and $r = 3\sin\theta$ are shown.

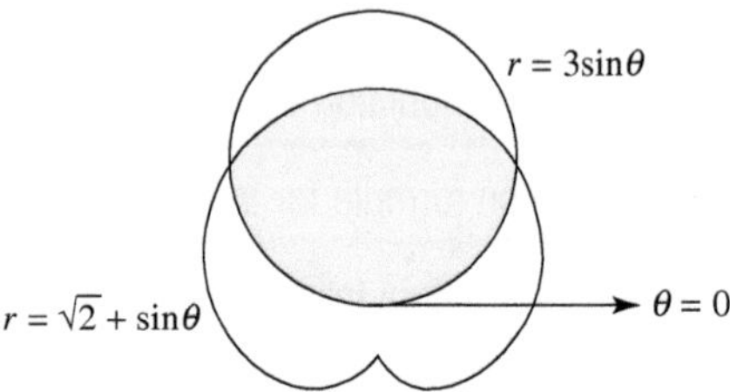

Calculate the shaded area.

20 Calculate the length of the curve $y = (2x+1)^{\frac{3}{2}}$ between the points where $x = 3$ and $x = 5$

21 A curve is given by the parametric equations

$x = 2t,\ y = t^2,\ 0 \le t \le 2$

Calculate the length of the curve.

22 Calculate the surface area of revolution when the arc of the curve $y = \frac{1}{3}x^3$ between $(0, 0)$ and $\left(1, \frac{1}{3}\right)$ is rotated 2π radians around the x-axis.

23 A curve is given by the parametric equations $x = \cos\theta,\ y = \sin\theta,\ \frac{\pi}{2} \le \theta \le \frac{3\pi}{4}$

Show that the surface area of revolution when the curve is rotated 360° around the x-axis is $A\pi$, where A is a constant to be found.

19 Exploration

Going beyond the exams

Investigation

A geometric solid that has finite volume but infinite surface area is called Gabriel's horn.

The solid is formed by rotating the function $f(x) = \frac{1}{x}$ about the x-axis between $x = 1$ and infinity.

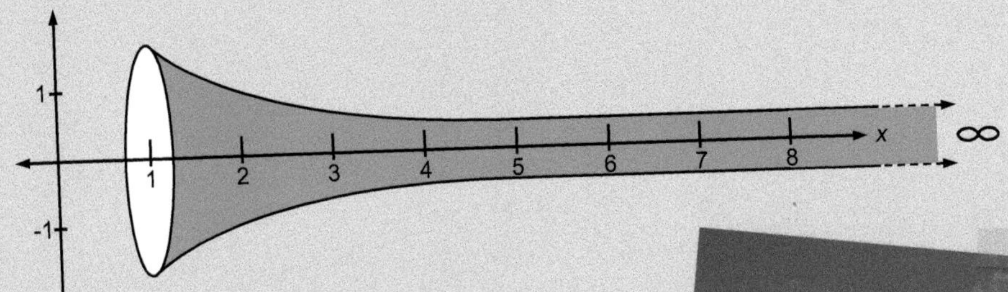

Try this to calculate the finite volume of the solid of revolution.

Research:

- How the formula for the surface area of revolution is derived,
- How the formula for the surface area of revolution leads to the answer of Gabriel's horn having infinite surface area.

Consider the lower bound of your integrals to be a where $0 \le a \le 1$

The results imply that, although the horn could contain a finite amount of paint, this would not be sufficient to paint its surface. This is sometimes known as the painter's paradox.

Investigation

The mean value of a trigonometric wave will be zero over one complete cycle because exactly half of the area enclosed by the function and the horizontal axis is positive and the other half is negative. One way to overcome this problem is to find the root mean square value, that is, the square root of the mean value of the function squared.

Investigate the root mean square of various trigonometric functions such as: $f(x)=A\sin x$ in the interval $0 \le x \le 2\pi$

That is find the value of $\sqrt{\frac{1}{2\pi}\int_0^{2\pi} A^2 \sin^2(\omega t)\,dt}$

Root mean square values have many applications in electrical/electronic engineering where engineers work with the flow of alternating and direct current. Other waveforms such as those illustrated may also be used in these fields of engineering.

Investigate their root mean square values.

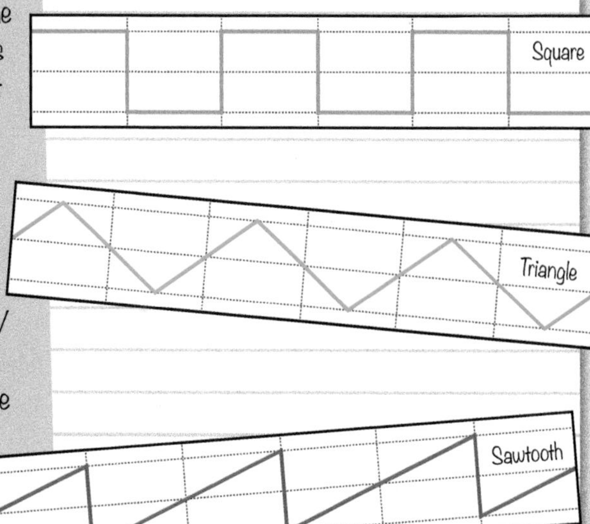

19 Assessment

1 Find $\int \frac{3}{5\sinh x - 4\cosh x}\,dx$ **[10 marks]**

2 a Given $I_n = \int (\ln x)^n\,dx$, show that, for $n \geq 1$, $I_n = x(\ln x)^n - nI_{n-1}$

b Hence evaluate $\int_1^4 (\ln x)^3\,dx$ **[10]**

3

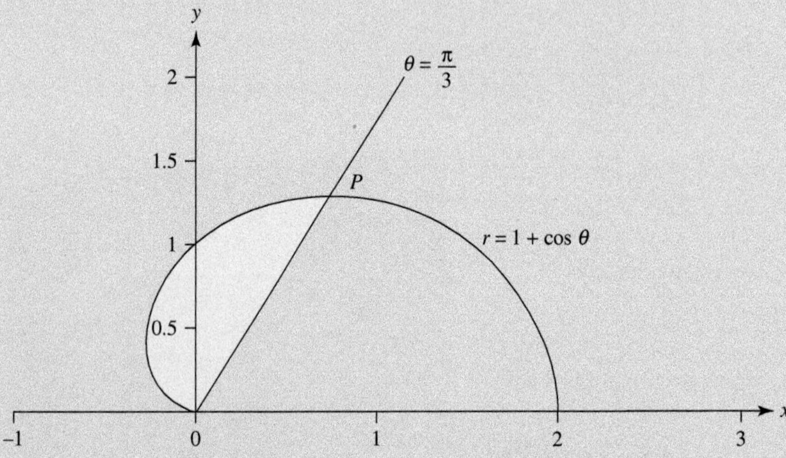

The diagram shows the polar curve with equation $r = 1 + \cos\theta$ for $0 \leq \theta \leq \pi$, along with the line $\theta = \frac{\pi}{3}$. The curve and the line intersect at the origin, O, and at the point P

a Find the coordinates of the point P

b Find the area of the shaded region that is bounded by the line and the curve. **[8]**

4 a Find $\int_0^1 \frac{1}{\sqrt{1-x^2}}\,dx$ b Find $\int_1^2 \frac{1}{\sqrt{x^2-1}}\,dx$ **[12]**

5

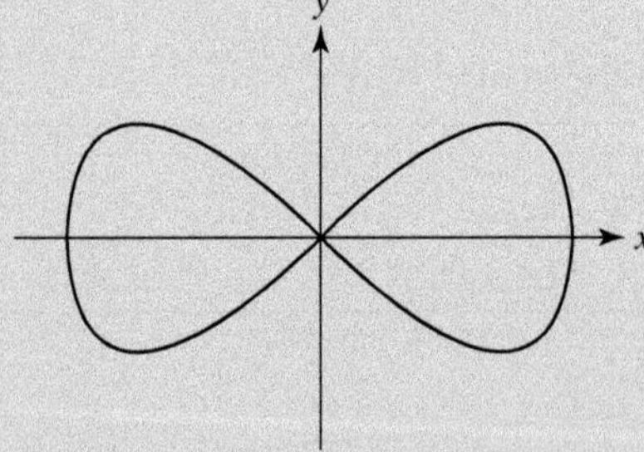

The curve with equation $a^2y^2 = x^2(a^2 - x^2)$ has two loops, as shown in the diagram above.

a Find an expression, in terms of a, for the area of one of the loops.

The curve is rotated through π radians about the x-axis to form two equal solids.

b Find the volume of the solids. **[8]**

6 **a** Given $I_n = \int \tanh^n x\,dx$, show that, for $n \geq 2$, $I_n = I_{n-2} - \dfrac{\tanh^{n-1} x}{n-1}$

b Hence find an expression for $\int \tanh^8 x\,dx$ **[8]**

7 **a** Sketch the curve with polar equation $r = 4\sin 3\theta$, $0 \leq \theta \leq \pi$

b Find the area enclosed by one loop of this curve. **[9]**

8 **a** Given $y = \sinh^{-1} x$, prove that $\dfrac{dy}{dx} = \dfrac{1}{\sqrt{1+x^2}}$

b Show that $\int_0^2 \sinh^{-1} x\,dx = 2\ln(2+\sqrt{5}) + 1 - \sqrt{5}$ **[9]**

9 An arc of a curve is given parametrically by the equations $x = a\cos^3 t$, $y = a\sin^3 t$ for $0 \leq t \leq \dfrac{\pi}{2}$ and $a > 0$

The points A and B on the curve correspond to the values $t = 0$ and $t = \dfrac{\pi}{2}$ respectively.

a Find the length of the arc, AB, of the curve.

b Find the area of the curved surface generated when this arc is rotated through 360° about the x-axis. **[11]**

10 Show that $\int_1^4 \dfrac{2x+1}{\sqrt{x^2-2x+10}}\,dx = 6\sqrt{(2)} + 3\,\text{arsinh}(1) - 6$ **[11]**

11 The region bounded by the curve $y = \sin x^2$, the y-axis and the line $y = 1$ is rotated through 360° about the y-axis. Find the volume of the solid of revolution. **[8]**

12 **a** Find $\int \dfrac{1}{x^2-4x+13}\,dx$ **b** Evaluate $\int_0^\infty x^2 e^{-x}\,dx$ **[15]**

13 The diagram shows the polar curves with equations $r = \sin 2\theta$ and $r = \cos\theta$ for $0 \leq \theta \leq \dfrac{\pi}{2}$
The curves intersect at the origin, O, and at the point P

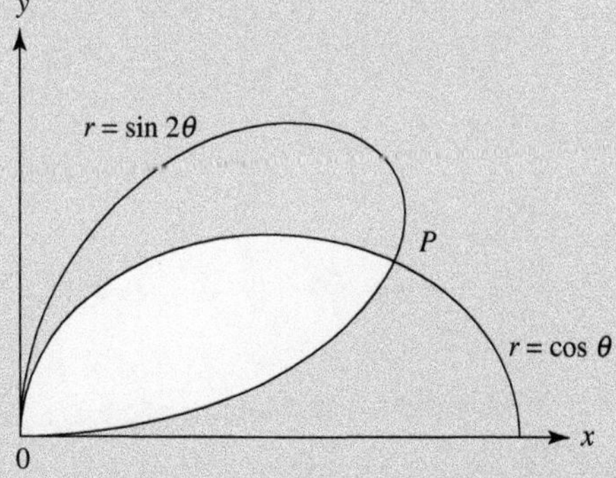

a Find the polar coordinates of the point P

b Show that the area of the shaded region that is bounded by the two curves is given by $\dfrac{\pi}{8} - \dfrac{3\sqrt{3}}{32}$ **[13]**

14 **a** Find $\int_{\frac{\pi}{3}}^{\frac{\pi}{2}} \dfrac{1}{1+\cos\theta}\,d\theta$ **b** Show that $\int_1^3 \dfrac{3x-x^2}{(x+1)(x^2+3)}\,dx = \dfrac{\pi}{2\sqrt{3}} - \ln 2$ **[16]**

15 An arc, L, of a parabola is given parametrically by the equations $x = at^2$, $y = 2at$, for $0 \le t \le 2$

L is rotated through 360° about the x-axis.

a Show that the area of the surface of revolution is given by $8\pi a^2 \int_0^2 t\sqrt{1+t^2}\,dt$

b Find this surface area.

c Show also that the length of L is given by $2a\int_0^2 \sqrt{1+t^2}\,dt$

d Find this length. **[16]**

16 a Given $I_n = \int_0^{\frac{\pi}{2}} \sin^n x\,dx$, show that, for $n \ge 2$, $nI_n = (n-1)I_{n-2}$

The region R is bounded by the curve $y = \sin^4 x$, the line $x = \frac{\pi}{2}$ and the x-axis, between $x = 0$ and $x = \frac{\pi}{2}$

b Find the area of R.

c Find the volume generated when R is rotated through 360° about the x-axis. **[15]**

17 The curve C has polar equation $r = 1 + \cos\theta$, $0 \le \theta \le 2\pi$. The curve D has polar equation $r = 2 - \cos\theta$, $0 \le \theta \le 2\pi$. The two curves intersect at the points P and Q

a Find the polar coordinates of the points P and Q

b Sketch, on one diagram, the graphs of C and D.

c Find the area of the region that is both outside D and inside C. **[13]**

18 a Find $\int \frac{9x^3 - 4x + 6}{9x^2 - 4}\,dx$

b Use the substitution $x = \sin\theta$ to show that $\int_0^{\frac{1}{2}} \frac{24x^2}{\sqrt{1-x^2}}\,dx = 2\pi - 3\sqrt{3}$ **[16]**

19 a Given that $I_n = \int_0^{\frac{\pi}{4}} \frac{\sin^{2n+1} x}{\cos x}\,dx$, show that $I_n = I_{n-1} - \frac{1}{2^{n+1} n}$

b Hence find the exact value of $\int_0^{\frac{\pi}{4}} \frac{\sin^7 x}{\cos x}\,dx$ **[11]**

20

y
P
x
Q

The diagram shows the curves with polar equations $r = a(3 + 2\cos\theta)$ and $r = a(5 - 2\cos\theta)$ for $0 \le \theta \le 2\pi$. The curves intersect at the points P and Q

a Find the coordinates of P and Q

b Show that the area of the shaded region that is bounded by the two curves is given by $\frac{a^2(49\pi - 48\sqrt{3})}{3}$ **[16]**

21 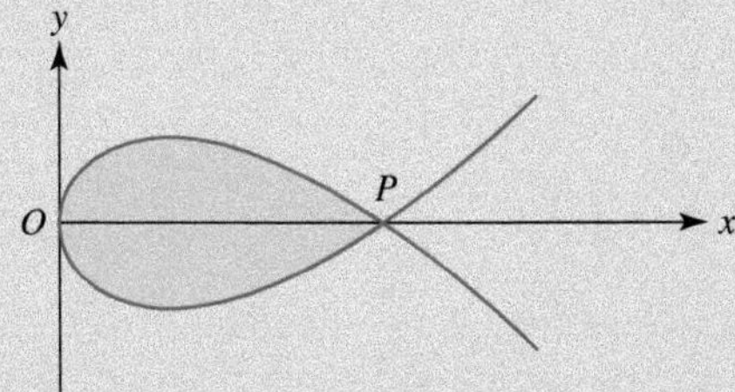

The diagram shows the curve with equation $3y^2 = x(1-x)^2$. The curve meets the x-axis at the origin and at the point P

a Find the coordinates of P

b Show that the perimeter of the closed shaded region between O and P is given by $\int_0^1 \frac{1+3x}{\sqrt{3x}}\,dx$

c Evaluate this perimeter.

The shaded region is rotated through 180° about the x-axis.

d Show that the surface area of the curved surface is $\frac{\pi}{3}$ **[14]**

22 a Find i $\int \frac{1}{\sqrt{12-4x-x^2}}\,dx$ ii $\int \operatorname{artanh} 2x\,dx$

b Use the substitution $x = 4\operatorname{cosech} u$ to show that $\int_1^\infty \frac{1}{x\sqrt{x^2+16}}\,dx = \frac{1}{4}\ln(4+\sqrt{17})$ **[15]**

23 a Given that $I_n = \int_0^1 x^n\sqrt{1-x^2}\,dx$, show that $I_{2n+1} = \frac{2n}{2n+3}I_{2n-1}$ for $n \geq 1$

b Hence show that $\int_0^1 x^{2n+1}\sqrt{1-x^2}\,dx = \frac{2^{2n+1}n!\,(n+1)!}{(2n+3)!}$ **[14]**

24 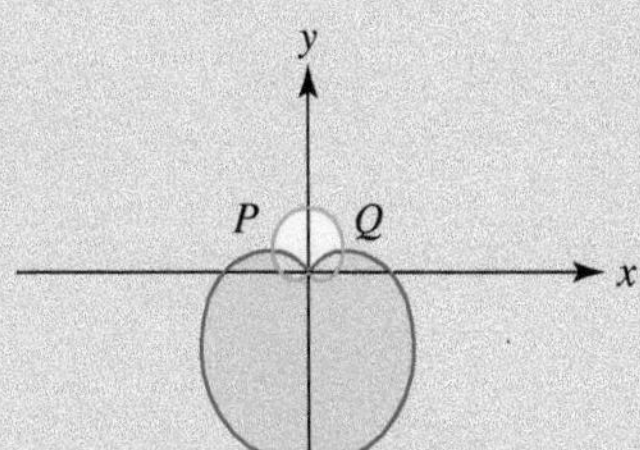

The diagram shows the curves with polar equations $r = 3(1-\sin\theta)$ and $r = 1+\sin\theta$ for $0 \leq \theta \leq 2\pi$
The curves intersect at the pole, and at the points P and Q

a Find the coordinates of P and Q

b Prove that the area that is bounded by the blue shape but not the red shape is $9\sqrt{3} - 4\pi$ **[15]**

25 a Find $\int \frac{1}{\sqrt{x^2-2x+10}}\,dx$

b Show that $\int_{-\infty}^{\infty} \frac{1}{1+x^2}\,dx = \pi$

c What is wrong with the following argument?

Since $\frac{1}{(1+x)^2} > 0$ for all x, clearly $\int_{-a}^{a} \frac{1}{(1+x)^2}\,dx > 0$ for any positive value of a

But $\int_{-a}^{a} \frac{1}{(1+x)^2}\,dx = \left[-\frac{1}{1+x}\right]_{-a}^{a} = -\frac{1}{1+a} + \frac{1}{1-a} = \frac{2a}{1-a^2}$

As $a \to \infty$ $\frac{2a}{1-a^2} \to 0$ so $\int_{-\infty}^{\infty} \frac{1}{(1+x)^2}\,dx = 0$, so we have a contradiction. **[12]**

26 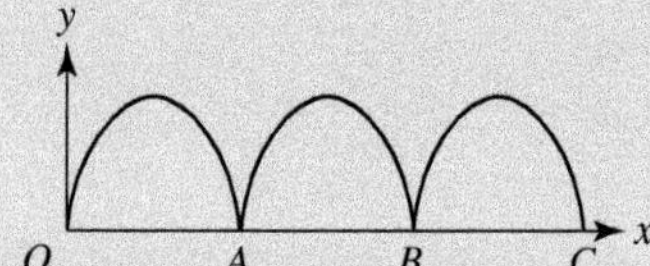

The diagram shows a cycloid with parametric equations $x = \theta - \sin\theta,\ y = 1 - \cos\theta$ for $0 \le \theta \le 6\pi$
The cycloid passes through the origin, and meets the x-axis again at the points A, B and C, as in the diagram.

a Find the coordinates of the points A, B and C

b Show that the length of the arc of the cycloid between points O and C is given by

$$3\sqrt{2}\int_0^{2\pi}\sqrt{1-\cos\theta}\,\mathrm{d}\theta$$

c Find the value of this arc length.

This arc, between O and C, is rotated through 2π radians about the x-axis.

d Show that the surface area of the curved surface generated is 64π **[14]**

27 The diagram shows part of the graph of $y = \dfrac{1}{\sqrt{4-x^2}}$

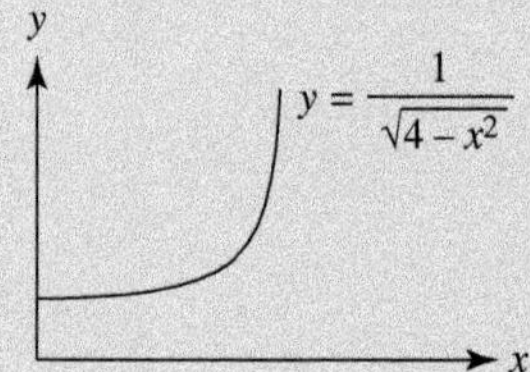

a Find the area of the region bounded by the curve, the x-axis, the y-axis and the line $x = \sqrt{2}$

b Find the volume of the solid generated when this region is rotated through 360° about the x-axis.

c Find also the volume generated when this same region is rotated through 360° about the y-axis. **[19]**

20 Differential equations

Mathematics, and particularly calculus, plays a large part in the science and engineering that underpins space exploration. For centuries, the idea of space travel was only a remote possibility, and the mathematics needed to support the engineering required was only just being developed. The physical principles that these mathematicians worked with are relatively straight forward. For example, rocket launches rely on the application of Newton's third law to the jet engines that are used.

One of the differential equations that space engineers work with is the Tsiolkovsky rocket equation. This is named after the man who developed it, Konstantin Tsiolkovsky, a Russian physicist who was born in 1857 and died in 1935. The equation allows space engineers to calculate how much propellant is necessary to lift a rocket off the ground.

Orientation

What you need to know	What you will learn	What this leads to
Maths Ch16 • Integration and differential equations.	• How to solve first order differential equations. • How to solve second order differential equations. • How differential equations are used to model simple harmonic motion. • How differential equations are used to model damped harmonic motion. • How to work with differential equations that involve more than two variables.	**Careers** • Engineering. • Economics. • Ecology.

20.1 First order equations

Fluency and skills

An equation that involves only a first order derivative, such as $\frac{dy}{dx}$, is called a **first order differential equation**. There are some first order differential equations where the terms involving x can be factorised out from the terms involving y. This method is called **separating the variables**. There are a number of methods for solving first order differential equations. One method involves using an **integrating factor**.

Example 1

Solve the differential equation $x^2\frac{dy}{dx}+2xy=4x^3$, giving your answer in the form $y=f(x)$

$\frac{d}{dx}\left[x^2y\right]=4x^3$ — The LHS can be rewritten as the derivative of a product.

$x^2y=x^4+c$ — Integrate each side.

$y=\frac{x^4+c}{x^2}$ — Rearrange to get an expression for y in terms of x

Key point

An expression that can be integrated by spotting that it is a perfect differential of a product is called an **exact equation**.

Suppose instead you had been asked to solve the differential equation $\frac{dy}{dx}+\frac{2}{x}y=4x$. This is not as easy because you can no longer spot the product rule on the LHS. However, by multiplying through by x^2, you can transform the equation into an exact equation. The multiplier, x^2, is known as the **integrating factor**. You need to know how to find the integrating factor.

First order differential equations that can be solved using an integrating factor can often be written in the form

$$\frac{dy}{dx}+P(x)y=f(x)$$

where $P(x)$ and $f(x)$ are functions involving only the variable x

PURE

To find the integrating factor, first notice that if you differentiate a product in the form $e^{\int P(x)dx}y$

then

$$\frac{d}{dx}\left[e^{\int P(x)dx}y\right]=e^{\int P(x)dx}\frac{dy}{dx}+P(x)e^{\int P(x)dx}y$$

so if you take the equation $\frac{dy}{dx}+P(x)y=f(x)$ and multiply throughout by $e^{\int P(x)dx}$, the equation becomes

$$e^{\int P(x)dx}\frac{dy}{dx}+P(x)e^{\int P(x)dx}y=e^{\int P(x)dx}f(x)$$

The LHS is now an exact equation, and you have

$$\frac{d}{dx}\left[e^{\int P(x)dx}y\right]=e^{\int P(x)dx}f(x)$$

which you can solve by integrating each side.

Key point

The expression $e^{\int P(x)dx}$ is called the integrating factor.

To solve the equation $\frac{dy}{dx}+\frac{2}{x}y=4x$, $P(x)=\frac{2}{x}$, and the integrating factor is $e^{\int\frac{2}{x}dx}=e^{2\ln x}=e^{\ln x^2}=x^2$

Multiplying the equation by x^2 gives you $x^2\frac{dy}{dx}+x^2\frac{2}{x}y=x^2 4x$, i.e. $x^2\frac{dy}{dx}+2xy=4x^3$, and you are back to Example 1

Example 2

Solve the differential equation $\frac{dy}{dx}-y\tan x=3\sin^2 x$, for $0\le x<\frac{\pi}{2}$

This equation is in the form $\frac{dy}{dx}+P(x)y=f(x)$, where $P(x)=-\tan x$ and $f(x)=3\sin^2 x$

$e^{\int P(x)dx}=e^{\int-\tan x dx}=e^{\ln(\cos x)}=\cos x$ — Find the integrating factor.

$\cos x\frac{dy}{dx}-y\cos x\tan x=3\cos x\sin^2 x$ — Multiply through by $\cos x$

Since $\cos x\tan x=\cos x\frac{\sin x}{\cos x}=\sin x$

$\cos x\frac{dy}{dx}-y\sin x=3\cos x\sin^2 x$

$\frac{d}{dx}[y\cos x]=3\cos x\sin^2 x$ — Use the fact it is now an exact equation.

$y\cos x=\sin^3 x+c$ — Integrate each side. You can leave the solution in implicit form.

A solution that includes a constant of integration is called a **general solution**. If you can find a value for the constant you obtain a **particular solution.**

Key point

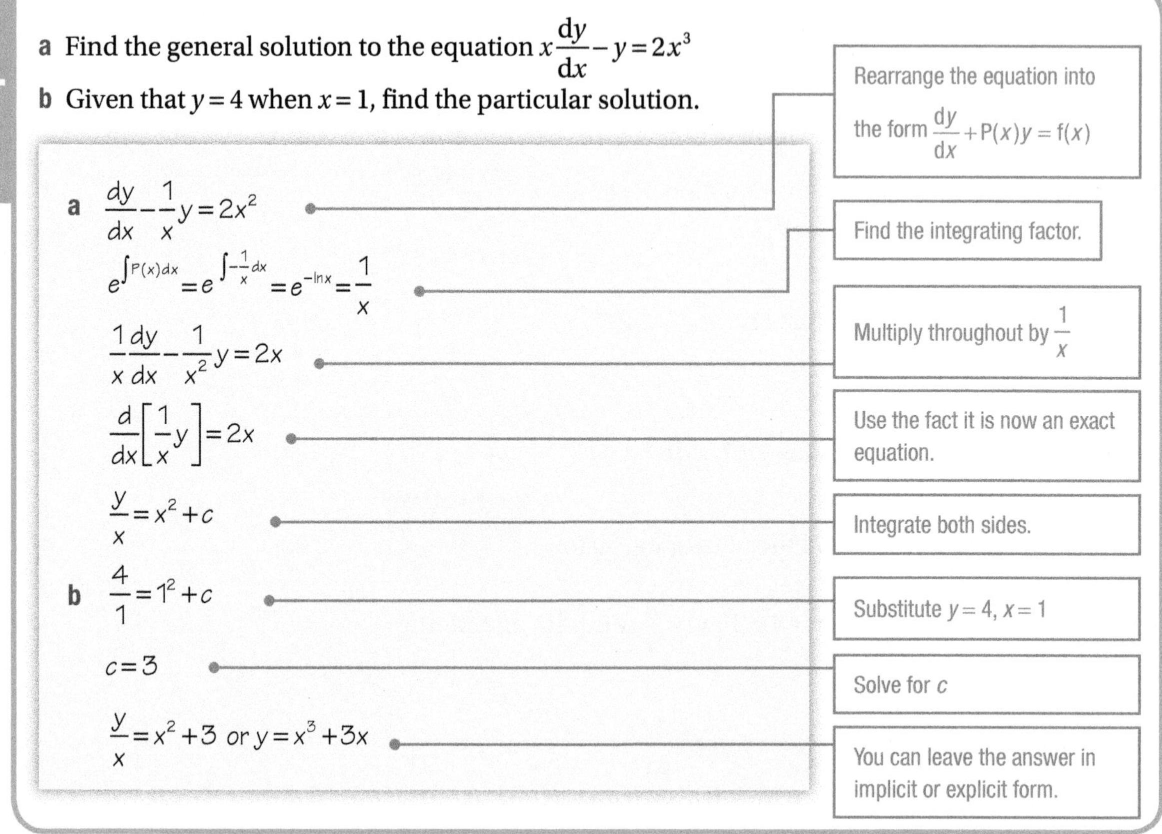

Example 3

a Find the general solution to the equation $x\frac{dy}{dx}-y=2x^3$

b Given that $y=4$ when $x=1$, find the particular solution.

a $\frac{dy}{dx}-\frac{1}{x}y=2x^2$ — Rearrange the equation into the form $\frac{dy}{dx}+P(x)y=f(x)$

$e^{\int P(x)dx}=e^{\int -\frac{1}{x}dx}=e^{-\ln x}=\frac{1}{x}$ — Find the integrating factor.

$\frac{1}{x}\frac{dy}{dx}-\frac{1}{x^2}y=2x$ — Multiply throughout by $\frac{1}{x}$

$\frac{d}{dx}\left[\frac{1}{x}y\right]=2x$ — Use the fact it is now an exact equation.

$\frac{y}{x}=x^2+c$ — Integrate both sides.

b $\frac{4}{1}=1^2+c$ — Substitute $y=4$, $x=1$

$c=3$ — Solve for c

$\frac{y}{x}=x^2+3$ or $y=x^3+3x$ — You can leave the answer in implicit or explicit form.

Sometimes you need to use a given substitution to transform the differential equation into one that can be solved using an integrating factor.

Example 4

a Use the substitution $y=u^2$ to transform the differential equation $\frac{dy}{dx}+x^2y=x\sqrt{y}$ into the differential equation $\frac{du}{dx}+\frac{1}{2}x^2u=\frac{1}{2}x^2$

b Hence use an integrating factor to find the general solution to the differential equation $\frac{dy}{dx}+x^2y=x\sqrt{y}$ in the form $y=f(x)$

a $\frac{dy}{dx}=\frac{dy}{du}\times\frac{du}{dx}=2u\frac{du}{dx}$ — Find an expression for $\frac{dy}{dx}$

Now $\frac{dy}{dx}+x^2y=x\sqrt{y}$ becomes $2u\frac{du}{dx}+x^2u^2=x^2u$ — Substitute for $\frac{dy}{dx}$, y and $\sqrt{y}$

Hence $\frac{du}{dx}+\frac{1}{2}x^2u=\frac{1}{2}x^2$ — Divide through by $2u$

(*Continued on the next page*)

b The integrating factor is $e^{\int \frac{1}{2}x^2\,dx} = e^{\frac{1}{6}x^3}$

Find the integrating factor.

$e^{\frac{1}{6}x^3}\frac{du}{dx} + \frac{1}{2}x^2 e^{\frac{1}{6}x^3}u = \frac{1}{2}x^2 e^{\frac{1}{6}x^3}$

Multiply through by the integrating factor.

Hence $e^{\frac{1}{6}x^3}u = \int \frac{1}{2}x^2 e^{\frac{1}{6}x^3}\,dx$

$= e^{\frac{1}{6}x^3} + c$

Solve the differential equation in u and x

$e^{\frac{1}{6}x^3}\sqrt{y} = e^{\frac{1}{6}x^3} + c$

$y = \left(1 + \frac{c}{e^{\frac{1}{6}x^3}}\right)^2$

Substitute for y and make y the subject.

Exercise 20.1A Fluency and skills

1 Find the general solution to each of the following differential equations.

a $\frac{dy}{dx} + 3y = e^{-x}$

b $\frac{dy}{dx} + 2xy = 8x$

c $\frac{dy}{dx} + y\tan x = \sin^2 x\cos^2 x$

d $\frac{dy}{dx} - \frac{y}{x} = \frac{x}{x-5}$

e $3x\frac{dy}{dx} + y = \sqrt{x}$

f $x\frac{dy}{dx} - y = x^2\ln x$

g $\cos x\frac{dy}{dx} - y\sin x = 4\sec^2 x$

h $\frac{1}{x}\frac{dy}{dx} + 2\frac{(x+1)}{x}y = 2e^{(x-1)^2}$

2 Find the particular solution to each of the following differential equations, giving your answers in the form $y = f(x)$

a $\frac{dy}{dx} - 2y = 2xe^{3x}$, given $y = -1$ when $x = 0$

b $\frac{dy}{dx} + \frac{3}{x}y = \frac{4}{x^2}$, given $y = 5$ when $x = 1$

c $\frac{dy}{dx} + y\cot x = 4\cos^3 x$, given $y = \frac{1}{4}$ when $x = \frac{\pi}{6}$

d $\frac{dy}{dx} - 2y\cos 2x = 2e^{\sin 2x}$, given $y = 1$ when $x = 0$

e $(x+1)\frac{dy}{dx} + 2y = \frac{14}{(x+1)}$, given $y = 3$ when $x = -2$

f $\sin x\frac{dy}{dx} + y\cos x = -3\cos x\sin 2x$, given $y = 2$ when $x = \frac{\pi}{4}$

g $x\frac{dy}{dx} - y = x^3\ln x$, given $y = 2$ when $x = 1$

h $\coth x\frac{dy}{dx} + y = 5e^{5x}\operatorname{cosech} x$, given $y = 4$ when $x = 0$

3 a Find the general solution to the equation $\cos x\frac{dy}{dx} + 2y\sin x = \sin^2 x\cos x$

b Given that $y = \frac{1}{2}$ when $x = \frac{\pi}{4}$, find the particular solution.

4 Find the solution to the differential equation $\frac{dy}{dx} - y\tan x = 3x^2\sec x$, given that $y = 2$ when $x = 0$

5 Find the solution to the differential equation $x\frac{dy}{dx} - 2y = x^3\ln x$, given that $y = 5$ when $x = 1$

6 a Find the general solution to the equation $\frac{dy}{dx} + 2y = 5\sin x$

b Given that $y = 1$ when $x = 0$, find the particular solution.

7 a Given that $u = \frac{1}{y^2}$ and $\frac{dy}{dx} = y + 2xy^3$, show that $\frac{du}{dx} + 2u = -4x$

b Hence find the solution to the differential equation $\frac{dy}{dx} = y + 2xy^3$ given that $y = \frac{1}{2}$ when $x = 0$

8 a Find the general solution of the differential equation $\cos x\frac{dy}{dx} + 2y\sin x = e^x\cos^3 x$

b Find the particular solution for which $y = 5$ at $x = 0$, giving your solution in the form $y = f(x)$

PURE

9 a Find the general solution of the differential equation $\tan x\frac{dy}{dx}+y=12\sin 2x\tan x$

b Hence show that the particular solution to this equation for which $y=2$ at $x=\frac{\pi}{4}$, is given by $y\sin x=8\sin^3 x-\sqrt{2}$

10 a Find $\int x^3 e^x\,dx$

b Hence show that the general solution to the differential equation $x\frac{dy}{dx}+(x+2)y=2x^2$ is given by $x^2 y=2(x^3-3x^2+6x-6)+ce^{-x}$

11 a Show that $\int \sec^3 x\,dx=\frac{1}{2}\left[\sec x\tan x+\ln(\sec x+\tan x)\right]+c$

b Hence show that the general solution to the differential equation

$x\frac{dy}{dx}+(1-x\tan x)y=2\sec^4 x$ is $y=\frac{\sec x\tan x+\ln(\sec x+\tan x)+c}{x\cos x}$

12 Find the general solution of the differential equation $x+(t\ln t)\frac{dx}{dt}=2te^{2t}$, $t>0$

13 a Find the general solution of the differential equation $(1+t)\frac{dx}{dt}+3x=\ln(1+t)$, $t>-1$

b Show that the particular solution for which $x=\frac{8}{9}$ at $t=0$, is given by

$$x=\frac{3\ln(1+t)-1}{9}+\frac{1}{(1+t)^3}$$

14 a Given that $z=y^{\frac{1}{2}}$ and $\frac{dy}{dx}-6y\tan x=3y^{\frac{1}{2}}$, show that $\frac{dz}{dx}-3z\tan x=\frac{3}{2}$

b Hence find the general solution to the differential equation $\frac{dy}{dx}-6y\tan x=3y^{\frac{1}{2}}$ in the form $y=f(x)$

15 a Use the substitution $v=\frac{dx}{dt}$ to transform the differential equation $t\frac{d^2x}{dt^2}-\frac{dx}{dt}=5t^2$ into $\frac{dv}{dt}-\frac{1}{t}v=5t$

b Hence find the general solution of the differential equation $t\frac{d^2x}{dt^2}-\frac{dx}{dt}=5t^2$, giving your answer in the form $y=f(x)$

Reasoning and problem-solving

Strategy

To answer a question that involves first order differential equations

1. Decide whether to separate the variables or use an integrating factor.
2. Find the general solution of the differential equation.
3. Substitute the given values to find the constant, and hence obtain the particular solution.
4. Answer the question in context.

Example 5

A population of bacteria has an initial size of 100. After t hours, the size of the population is P

The connection between P and t can be modelled by the equation $\frac{dP}{dt}=2(50t-P)$

a Solve this equation to show that $P=25(2t+5e^{-2t}-1)$

b Find the size of the population after 24 hours.

c Prove that the number of bacteria never falls below 40

a $\frac{dP}{dt}+2P=100t$ — Rearrange the equation into the form $\frac{dP}{dt}+F(t)P=f(t)$

$e^{\int F(t)dt}=e^{\int 2dt}=e^{2t}$ — Find the integrating factor.

$e^{2t}\frac{dP}{dt}+2e^{2t}P=100te^{2t}$ — Multiply throughout by e^{2t}

$\frac{d}{dt}\left[e^{2t}P\right]=100te^{2t}$ — Use the fact it is now an exact equation.

$e^{2t}P=\int 100t\,e^{2t}\,dt$ — Integrate each side.

$e^{2t}P=50te^{2t}-\int 50\,e^{2t}dt=50te^{2t}-25\,e^{2t}+c$ — Use integration by parts.

$100=-25+c\Rightarrow c=125$ — Substitute $t=0$, $P=100$ to find c

$e^{2t}P=50te^{2t}-25\,e^{2t}+125$

$P=25(2t+5e^{-2t}-1)$

b At $t=24$, $P=25[2(24)+5e^{-2(24)}-1]$ — Substitute $t=24$

$=1175$

c $\frac{dP}{dt}=25(2-10e^{-2t})=0$ — Set $\frac{dP}{dt}=0$

$2-10e^{-2t}=0$ gives $t=\frac{1}{2}\ln 5$ — Solve $\frac{dP}{dt}=0$

$P=25(\ln 5+1-1)=25\ln 5$ — Substitute $t=\frac{1}{2}\ln 5$

By inspection this is a minimum, so the value of P never falls below $25\ln 5\approx 40$

Example 6

A raindrop falls vertically through a cloud. Initially the raindrop is at rest. At time t seconds, the velocity, v m s^{-1} of the raindrop, satisfies the equation $(2+t)\frac{dv}{dt}+v=10(2+t)$

a Solve this equation to find an expression for v in terms of t

b Find the time at which the velocity of the raindrop is 21 m s^{-1}

c Criticise the model.

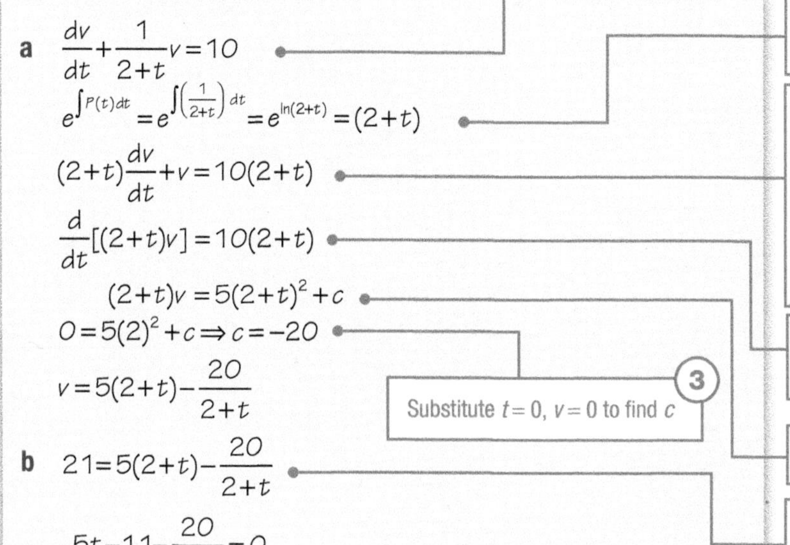

a $\frac{dv}{dt}+\frac{1}{2+t}v=10$

Rearrange the equation into the form $\frac{dv}{dt}+P(t)v=f(t)$ (1)

$e^{\int P(t)dt}=e^{\int\left(\frac{1}{2+t}\right)dt}=e^{\ln(2+t)}=(2+t)$

Decide to use an integrating factor.

$(2+t)\frac{dv}{dt}+v=10(2+t)$

Multiply throughout by $(2+t)$

Note that the differential equation was already in the form $[u'v+uv']$. The integrating factor manipulations are here in case you didn't spot that.

$\frac{d}{dt}[(2+t)v]=10(2+t)$

Use the fact it is now an exact equation.

$(2+t)v=5(2+t)^2+c$

Find the general solution. (2)

$0=5(2)^2+c \Rightarrow c=-20$

Substitute $t=0$, $v=0$ to find c (3)

$v=5(2+t)-\frac{20}{2+t}$

b $21=5(2+t)-\frac{20}{2+t}$

Substitute $v=21$ to get an equation in t

$5t-11-\frac{20}{2+t}=0$

$5t^2-t-42=0$

$(5t+14)(t-3)=0$

So $t=3$

You only need the positive value of t

c Under this model, as t increases, v is unlimited, but, in practice, the raindrop would approach a terminal velocity.

Answer the question in context. (4)

Example 7

A fuel-filled rocket of mass 25 kg burns fuel at a rate of 2 kg s^{-1}. The rocket is initially at rest. After t seconds, the velocity v m s^{-1} of the rocket, satisfies the equation $(25-2t)\frac{dv}{dt}+v=100$ for $t\leq 12$

a Show that $v=100-20\sqrt{25-2t}$

b Find the speed of the rocket when $t=12$

c Sketch a graph of v against t

d What happens for $t>12.5$?

(Continued on the next page)

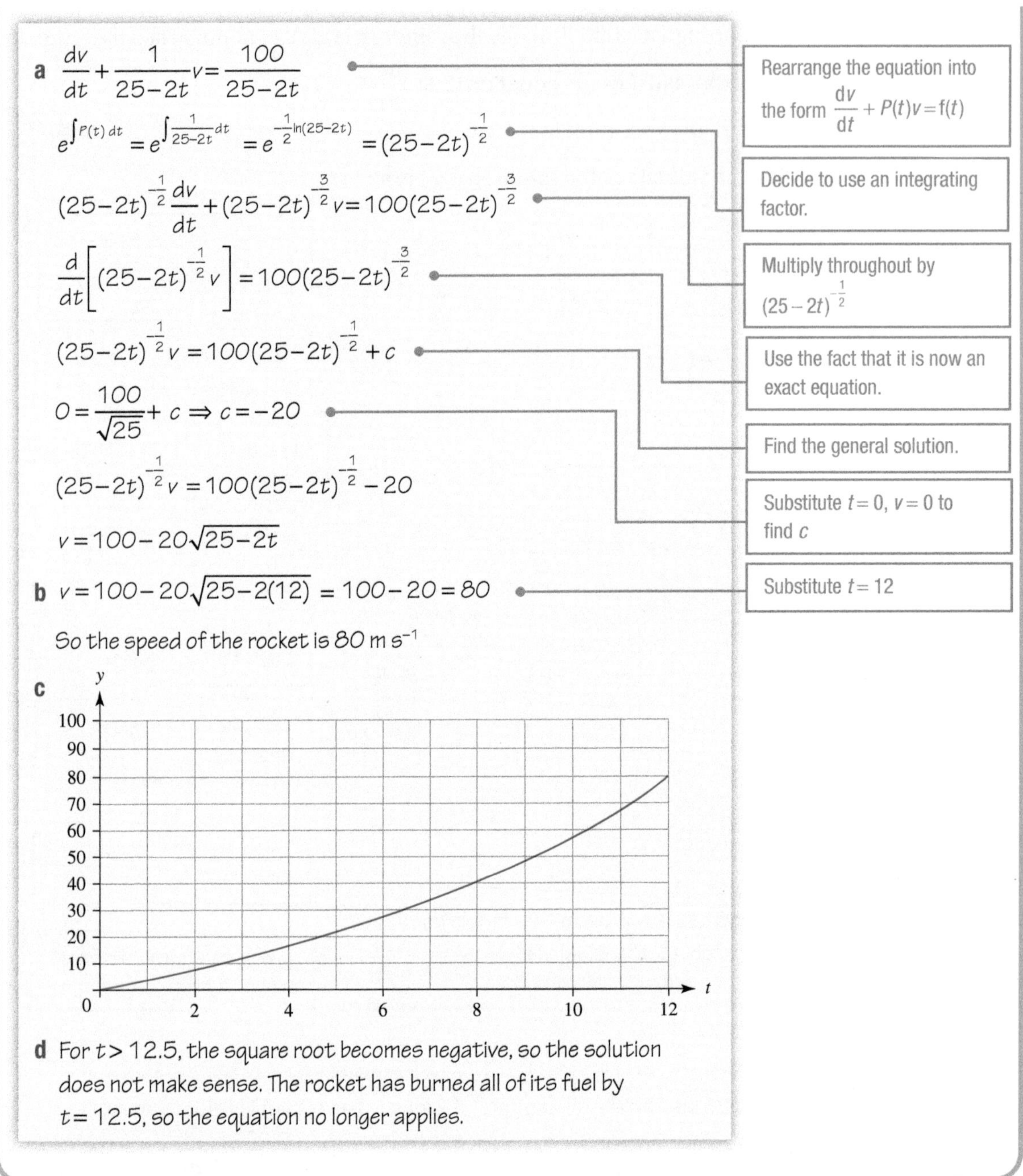

a $\frac{dv}{dt}+\frac{1}{25-2t}v=\frac{100}{25-2t}$ — Rearrange the equation into the form $\frac{dv}{dt}+P(t)v=f(t)$

$e^{\int P(t)\,dt}=e^{\int\frac{1}{25-2t}dt}=e^{-\frac{1}{2}\ln(25-2t)}=(25-2t)^{-\frac{1}{2}}$ — Decide to use an integrating factor.

$(25-2t)^{-\frac{1}{2}}\frac{dv}{dt}+(25-2t)^{-\frac{3}{2}}v=100(25-2t)^{-\frac{3}{2}}$ — Multiply throughout by $(25-2t)^{-\frac{1}{2}}$

$\frac{d}{dt}\left[(25-2t)^{-\frac{1}{2}}v\right]=100(25-2t)^{-\frac{3}{2}}$ — Use the fact that it is now an exact equation.

$(25-2t)^{-\frac{1}{2}}v=100(25-2t)^{-\frac{1}{2}}+c$ — Find the general solution.

$0=\frac{100}{\sqrt{25}}+c\Rightarrow c=-20$ — Substitute $t=0$, $v=0$ to find c

$(25-2t)^{-\frac{1}{2}}v=100(25-2t)^{-\frac{1}{2}}-20$

$v=100-20\sqrt{25-2t}$

b $v=100-20\sqrt{25-2(12)}=100-20=80$ — Substitute $t=12$

So the speed of the rocket is 80 m s^{-1}

c

d For $t>12.5$, the square root becomes negative, so the solution does not make sense. The rocket has burned all of its fuel by $t=12.5$, so the equation no longer applies.

Exercise 20.1B Reasoning and problem-solving

1 A population has an initial size of 40
After t days the size of the population is P
The connection between P and t can be modelled by the equation

$$\frac{dP}{dt}=20t-P$$

a Solve this equation to show that $P=20(t+3e^{-t}-1)$

b Find the size of the population after 2 weeks.

c Prove that the size of the population never falls below 21

2 In an electric circuit the current, I amps at time t seconds is modelled by the equation

$$\frac{dI}{dt}+2I=6$$

Initially $I=8$

a Using an integrating factor, solve this equation to show that $I=3+5e^{-2t}$

b Show that the current never drops below 3 amps.

c Sketch a graph of I against t

3 A population of 8 million bacteria is injected into a body. After t days the size of the population in the body is x million where x and t satisfy the differential equation $\frac{dx}{dt}=t-x+4$

a Show that the size of the population initially starts to decline.

b Find an expression for x in terms of t

c Find the minimum size of the population.

d What happens to the population in the long term?

4 A hailstone falls vertically through a cloud. Initially the hailstone is at rest. At time t seconds, the velocity, $v\,\mathrm{m\,s^{-1}}$, of the hailstone satisfies the equation $\frac{dv}{dt}=10-\frac{2v}{1+2t}$

a Solve this equation to find an expression for v in terms of t

b Find, correct to one decimal place, the time at which the velocity of the hailstone is $24\,\mathrm{m\,s^{-1}}$.

c Criticise the model.

5 As part of an industrial process a chemical is dissolved in a solution. t minutes after the start of the process, the mass, C kg, of the chemical that has dissolved can be modelled by the equation $\frac{dC}{dt}+\frac{C}{20+t}=4$

Initially $C=10$ kg.

a Solve this equation to show that

$$C=2(20+t)-\frac{600}{20+t}$$

b Find the time at which $C=40$

6 Water is draining from a tank. The depth of water in the tank is initially 2 metres, and after t minutes, the depth is x metres. The depth can be modelled by the equation

$$\frac{dx}{dt}=-\frac{1}{2}(x+e^{\frac{1}{2}t}\cos t)$$

a Solve this equation to find an expression for x in terms of t

b Find the depth of the water in the tank after five minutes.

7 A particle is projected vertically upwards with a velocity of $13\,\mathrm{m\,s^{-1}}$. During the motion it absorbs moisture so that when the particle is a distance x metres above the point of projection, its velocity, $v\,\mathrm{m\,s^{-1}}$ is given by the equation $2v\frac{dv}{dx}+\frac{2}{1+2x}v^2=-4$

a Solve this equation to show that $v^2(1+2x)=170-(1+2x)^2$

b Calculate the greatest height reached by the particle.

8 A population of bacteria grows from an initial size of 10 000. After t years, the size of the population is P. The connection between P and t can be modelled by the equation

$$\frac{dP}{dt}=\frac{P}{1+t}+2$$

a Solve this equation to show that $P=2(1+t)[5000+\ln(1+t)]$

b Find the size of the population of bacteria after six years.

9 In fluid dynamics, Bernoulli's equation is used to measure flow. The equation states that

$$\frac{dy}{dx}+py=qy^n$$

where p and q are both functions of x

a Show that the differential equation

$$x\frac{dy}{dx}+4y=x^4y^2 \qquad \text{(I)}$$

is a Bernoulli equation.

b Use the substitution $y=\frac{1}{v}$ to solve equation (I) given that

$$y=-2 \text{ when } x=2$$

PURE

20.2 Second order equations

Fluency and skills

An equation that involves a second order derivative, such as $\frac{d^2y}{dx^2}$, but nothing of higher order, is called a **second order differential equation.** Some second order differential equations can be written in the form

$a\frac{d^2y}{dx^2}+b\frac{dy}{dx}+cy=f(x)$, where a, b and c are constants.

You can solve these types of equation in two stages. The first stage is to solve the related **homogeneous equation**, that is, the equation

$$a\frac{d^2y}{dx^2}+b\frac{dy}{dx}+cy=0$$

Homogeneous means '= 0'

The full formal method to solve this is as follows. You will see later that there is also a much quicker method.

Example 1

Solve the differential equation $\frac{d^2y}{dx^2}+\frac{dy}{dx}-6y=0$

Rewrite the equation as $\frac{d^2y}{dx^2}+3\frac{dy}{dx}-2\frac{dy}{dx}-6y=0$ — This just uses the fact that $\frac{dy}{dx}=3\frac{dy}{dx}-2\frac{dy}{dx}$. This is very much like solving a quadratic equation.

So $\frac{d}{dx}\left[\frac{dy}{dx}+3y\right]-2\left[\frac{dy}{dx}+3y\right]=0$ — Since $\frac{d^2y}{dx^2}+3\frac{dy}{dx}=\frac{d}{dx}\left[\frac{dy}{dx}+3y\right]$

Substitute $u=\frac{dy}{dx}+3y$ to get $\frac{du}{dx}-2u=0$

So $e^{-2x}\frac{du}{dx}-2e^{-2x}u=0$ — Multiplying throughout by the integrating factor, e^{-2x}

$ue^{-2x}=C$ — Integrate each side.

So $u=Ce^{2x}\Rightarrow\frac{dy}{dx}+3y=Ce^{2x}$ — Substitute $u=\frac{dy}{dx}+3y$

$e^{3x}\frac{dy}{dx}+3e^{3x}y=Ce^{5x}$ — Multiply throughout by the integrating factor, e^{3x}

$e^{3x}y=\frac{1}{5}Ce^{5x}+A$ — Integrate both sides.

$y=\frac{1}{5}Ce^{2x}+Ae^{-3x}$ — Rearrange.

The solution is $y=Ae^{-3x}+Be^{2x}$ — Relabel the constant $\frac{1}{5}C$ as B

PURE

You don't need to go through all this working every time. By assuming that solutions exist of the form $y = Ce^{mx}$, where C and m are constants, you can cut out several steps of the working.

If $y = Ce^{mx}$, then $\frac{dy}{dx} = Cme^{mx}$ and $\frac{d^2y}{dx^2} = Cm^2e^{mx}$

If you substitute each of these into the differential equation, then $\frac{d^2y}{dx^2} + \frac{dy}{dx} - 6y = 0$ becomes $Cm^2e^{mx} + Cme^{mx} - 6Ce^{mx} = 0$ which simplifies to $m^2 + m - 6 = 0$

This is called the **auxiliary equation**, and when solving these problems, you can use the solution to this equation to complete the first stage of the solution to the differential equation.

So $(m+3)(m-2) = 0 \Rightarrow m = -3$ or $m = 2$, giving solutions $y = Ce^{-3x}$ and $y = Ce^{2x}$

Since either of these solutions satisfies the homogeneous equation, any linear combination of these solutions will also satisfy the homogeneous equation, and the general solution is, therefore, $y = Ae^{-3x} + Be^{2x}$

As there are two solutions, you need two different constants.

This method only works if the auxiliary equation has real distinct roots. There are two other cases for the roots of the auxiliary equation. The equation could have one repeated root, or the roots could be a complex conjugate pair. These three cases need three different general solutions.

Table 1

Roots of auxiliary equation $ax^2 + bx + c = 0$	**General solution**
Real distinct roots, m_1 and m_2	$y = Ae^{m_1x} + Be^{m_2x}$
Repeated root, m	$y = (Ax + B)e^{mx}$
Complex roots $m \pm in$	$y = e^{mx}(A\cos nx + B\sin nx)$

You will be able to prove why these work in the next exercise.

Example 2

Solve the differential equation $\frac{d^2y}{dx^2} + 7\frac{dy}{dx} + 12y = 0$

The auxiliary equation is $m^2 + 7m + 12 = 0$

Write down and solve the auxiliary equation.

$(m+3)(m+4) = 0$, so $m = -3$ or $m = -4$

The general solution is $y = Ae^{-3x} + Be^{-4x}$

This is 'real distinct roots' from Table 1

Example 3

Solve the differential equation $\frac{d^2y}{dx^2}-8\frac{dy}{dx}+16y=0$

The auxiliary equation is $m^2-8m+16=0$

$(m-4)^2=0$, so $m=4$

The general solution is $y=(Ax+B)e^{4x}$

This is 'repeated root' from Table 1

Example 4

Solve the differential equation $\frac{d^2y}{dx^2}-4\frac{dy}{dx}+13y=0$

The auxiliary equation is $m^2-4m+13=0$

$(m-2)^2=-9$, so $m=2\pm3i$

The general solution $y=e^{2x}(A\cos3x+B\sin3x)$

This is 'complex roots' from Table 1

The second stage is used to solve equations that are not homogeneous.

To solve an equation like $\frac{d^2y}{dx^2}+\frac{dy}{dx}-6y=e^{4x}$, the first step is to solve the related homogeneous equation $\frac{d^2y}{dx^2}+\frac{dy}{dx}-6y=0$

This solution, $y=Ae^{-3x}+Be^{2x}$, is now called the **complementary function.**

You have already solved this equation in Example 1

You know that this gives the answer 'zero' when substituted into

$$\frac{d^2y}{dx^2}+\frac{dy}{dx}-6y$$

For the second stage you need to look for a function that will give e^{4x} when substituted into $\frac{d^2y}{dx^2}+\frac{dy}{dx}-6y$. This function is called the **particular integral**. Adding together the **complementary function** and the **particular integral** gives you the **general solution.**

Key point

The general solution is obtained by adding together the complementary function and the particular integral.

In this example, the particular integral is easy to find. It is likely to be a multiple of e^{4x}, so try $y=ae^{4x}$

$$y=ae^{4x}\Rightarrow\frac{dy}{dx}=4ae^{4x}\text{ and }\frac{d^2y}{dx^2}=16ae^{4x}$$

Substituting gives

$$\frac{d^2y}{dx^2}+\frac{dy}{dx}-6y=16ae^{4x}+4ae^{4x}-6ae^{4x}$$
$$=14ae^{4x}$$

This gives

$14ae^{4x}=e^{4x}$ and $a=\frac{1}{14}$

The **complementary function** is $y=Ae^{-3x}+Be^{2x}$ and the **particular integral** is $y=\frac{1}{14}e^{4x}$

Adding the two together, gives the **general solution**

$$y=Ae^{-3x}+Be^{2x}+\frac{1}{14}e^{4x}$$

Not all particular integrals are that easy to find. The following table offers suggestions as to what to try.

Table 2

Form of f(x)	Form of particular integral
C – constant	c – constant
$Mx+C$	$mx+c$
$P_n(x)$ – a polynomial of degree n	$p_n(x)$ – a polynomial of degree n
Ce^{mx}	ce^{mx}
$A\cos kx+B\sin kx$	$a\cos kx+b\sin kx$
$A\cosh kx+B\sinh kx$	$a\cosh kx+b\sinh kx$

Example 5

Solve the differential equation $\frac{d^2y}{dx^2}-4\frac{dy}{dx}+4y=12x-8$

The auxiliary equation is $m^2-4m+4=0$

$(m-2)^2=0$, so $m=2$

So the complementary function is $y=(Ax+B)e^{2x}$ — This is the 'repeated root' type equation from Table 1

For the particular integral try $y=ax+b$ — This is from Table 2

$\frac{dy}{dx}=a$, and $\frac{d^2y}{dx^2}=0$

So $\frac{d^2y}{dx^2}-4\frac{dy}{dx}+4y=-4a+4(ax+b)=4ax+(-4a+4b)=12x-8$

$4a=12$, and $-4a+4b=-8$ — Equate coefficients.

$a=3$ and $b=1$

So the particular integral is $y=3x+1$

and the general solution is $y=(Ax+B)e^{2x}+3x+1$ — Add together the complementary function and the particular integral to get the general solution.

Exercise 20.2A Fluency and skills

1 Find the general solution to each of the following differential equations.

a $\frac{d^2y}{dx^2}-6\frac{dy}{dx}+8y=0$

b $\frac{d^2y}{dx^2}+8\frac{dy}{dx}+16y=0$

c $\frac{d^2y}{dx^2}-4\frac{dy}{dx}+5y=0$

d $\frac{d^2y}{dx^2}+3\frac{dy}{dx}=0$

e $\frac{d^2y}{dx^2}+\frac{dy}{dx}-12y=0$

f $\frac{d^2y}{dx^2}+y=0$

g $2\frac{d^2y}{dx^2}+5\frac{dy}{dx}-3y=0$

h $4\frac{d^2y}{dx^2}+4\frac{dy}{dx}+5y=0$

2 Find the general solution to each of the following differential equations.

a $\frac{d^2y}{dx^2}+2\frac{dy}{dx}-15y=19-30x$

b $\frac{d^2y}{dx^2}-6\frac{dy}{dx}+9y=3e^{2x}$

c $\frac{d^2y}{dx^2}+2\frac{dy}{dx}+17y=8\cos 3x-6\sin 3x$

d $\frac{d^2y}{dx^2}-16y=32x-48$

e $\frac{d^2y}{dx^2}+3\frac{dy}{dx}-4y=4e^{-3x}$

f $\frac{d^2y}{dx^2}+2\frac{dy}{dx}+y=x^2+4x+7$

g $3\frac{d^2y}{dx^2}-8\frac{dy}{dx}-3y=5e^{2x}$

h $4\frac{d^2y}{dx^2}+4\frac{dy}{dx}+y=5x+18$

3 a Show that the differential equation $\frac{d^2y}{dx^2}-2a\frac{dy}{dx}+a^2y=0$ can be written as $\frac{du}{dx}-a\times u=0$, where a is a constant, and $u=\frac{dy}{dx}-a\times y$

b Solve the differential equation $\frac{du}{dx}-a\times u=0$

c Hence show that $y=(Ax+B)e^{ax}$, where A and B are constants.

4 If the auxiliary equation has complex conjugate roots $m\pm in$, use Euler's formula to deduce that the general solution $y=Ae^{(m+in)x}+Be^{(m-in)x}$ can be expressed as $y=e^{mx}(\alpha\cos nx+\beta\sin nx)$ for constants α and β

Reasoning and problem-solving

Strategy

To answer a question that involves a second order differential equation

1. Write down the auxiliary equation.
2. Find the complementary function.
3. Decide on the form of the particular integral.
4. Find the particular integral.
5. Write down the general solution.
6. Use the information given in the question to find the values of the constants.
7. Answer the question(s) in context.

Example 6

Solve the differential equation $\frac{d^2y}{dx^2}+2\frac{dy}{dx}+10y=13\cos x+16\sin x$, given that $y(0)=1$ and $y'(0)=8$

The auxiliary equation is $m^2+2m+10=0$

$(m+1)^2=-9$, so $m=-1\pm3i$

and the complementary function is $y=e^{-x}(A\cos3x+B\sin3x)$

For the particular integral, try $y=a\cos x+b\sin x$

$\frac{dy}{dx}=-a\sin x+b\cos x$, and $\frac{d^2y}{dx^2}=-a\cos x-b\sin x$

So $\frac{d^2y}{dx^2}+2\frac{dy}{dx}+10y=-a\cos x-b\sin x+2(-a\sin x+b\cos x)$

$+10(a\cos x+b\sin x)$

$=(9a+2b)\cos x+(9b-2a)\sin x$

$=13\cos x+16\sin x$

$9a+2b=13$, and $9b-2a=16$

$a=1$ and $b=2$

So the particular integral is $y=\cos x+2\sin x$

The general solution is $y=e^{-x}(A\cos3x+B\sin3x)$

$+\cos x+2\sin x$

$A+1=1$, so $A=0$

$y=Be^{-x}\sin3x+\cos x+2\sin x$

$\frac{dy}{dx}=-Be^{-x}\sin3x+3Be^{-x}\cos3x-\sin x+2\cos x$

$3B+2=8$, so $B=2$

The particular solution is $y=2e^{-x}\sin3x+\cos x+2\sin x$

1 Write down the auxiliary equation.

2 Find the complementary function.

This is the 'complex roots' type equation from Table 1

3 Decide on the form of the particular integral.

This is from Table 2

Equate coefficients.

Solve for a and b

4 Find the particular integral.

5 Write down the general solution.

Add together the complementary function and the particular integral to get the general solution.

Use $y(0)=1$ to substitute $x=0$, $y=1$

6 Find the values of the constants.

Use $y'(0)=8$ to substitute $x=0$, $y'=8$

7 Answer the question in context.

Key point

When you have found the values of the constants in the general solution, the final answer is called the **particular solution**.

If the form of the particular integral you try matches the complementary function, it will not work. You simply end up with '$0=0$'. As long as this is not the 'repeated root' case in Table 1 you should simply multiply the function you try for the particular integral by x. In the case of the 'repeated root', multiplying x will also give '$0=0$'. In this case, instead of multiplying by x, you multiply by x^2

These cases are illustrated in Example 7

Example 7

Solve the differential equation $\frac{d^2y}{dx^2}-5\frac{dy}{dx}+6y=6e^{2x}$, given that $y(0)=3$ and $y'(0)=-2$

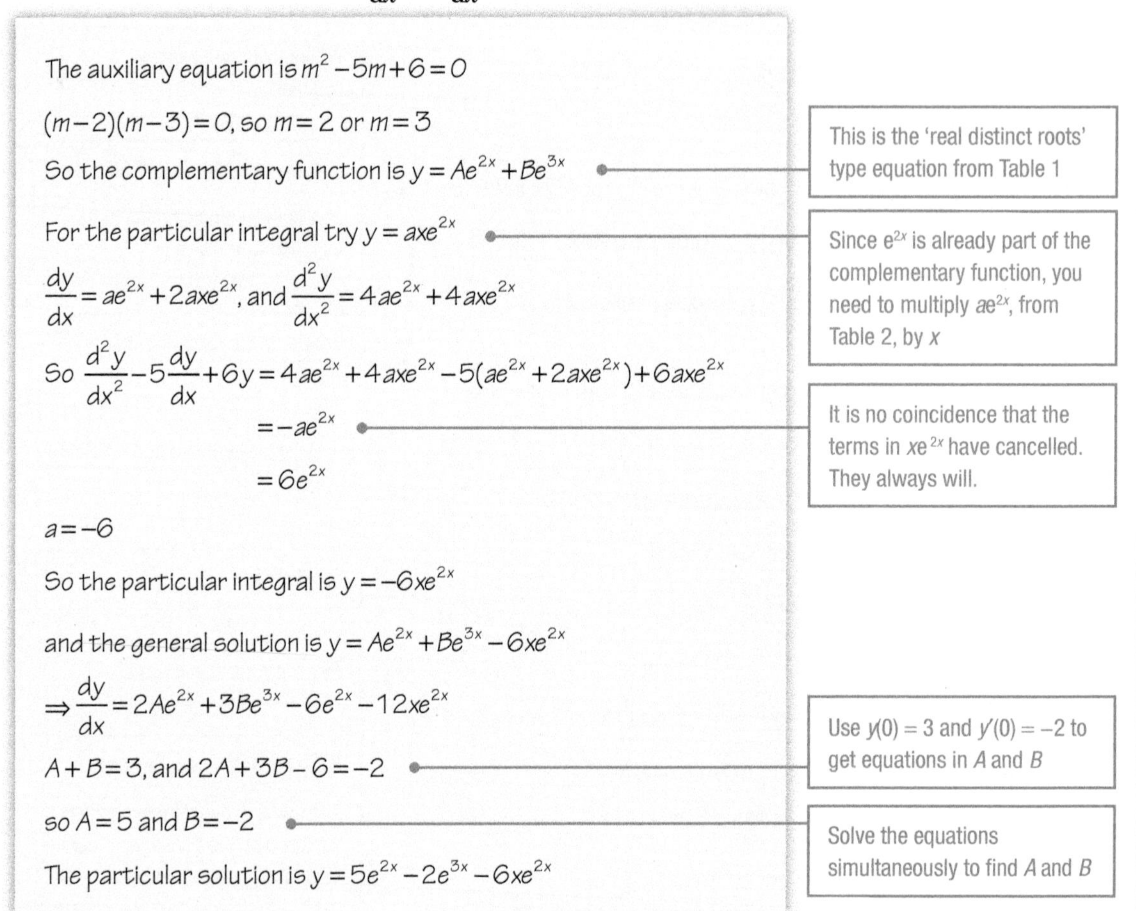

The auxiliary equation is $m^2-5m+6=0$

$(m-2)(m-3)=0$, so $m=2$ or $m=3$

So the complementary function is $y=Ae^{2x}+Be^{3x}$

This is the 'real distinct roots' type equation from Table 1

For the particular integral try $y=axe^{2x}$

Since e^{2x} is already part of the complementary function, you need to multiply ae^{2x}, from Table 2, by x

$\frac{dy}{dx}=ae^{2x}+2axe^{2x}$, and $\frac{d^2y}{dx^2}=4ae^{2x}+4axe^{2x}$

So $\frac{d^2y}{dx^2}-5\frac{dy}{dx}+6y=4ae^{2x}+4axe^{2x}-5(ae^{2x}+2axe^{2x})+6axe^{2x}$

$=-ae^{2x}$

$=6e^{2x}$

It is no coincidence that the terms in xe^{2x} have cancelled. They always will.

$a=-6$

So the particular integral is $y=-6xe^{2x}$

and the general solution is $y=Ae^{2x}+Be^{3x}-6xe^{2x}$

$\Rightarrow\frac{dy}{dx}=2Ae^{2x}+3Be^{3x}-6e^{2x}-12xe^{2x}$

$A+B=3$, and $2A+3B-6=-2$

Use $y(0)=3$ and $y'(0)=-2$ to get equations in A and B

so $A=5$ and $B=-2$

Solve the equations simultaneously to find A and B

The particular solution is $y=5e^{2x}-2e^{3x}-6xe^{2x}$

Exercise 20.2B Reasoning and problem-solving

1 Solve each of the following differential equations subject to the given boundary conditions.

a $\frac{d^2y}{dx^2}-7\frac{dy}{dx}+6y=0$, given that $y(0)=3$ and $y'(0)=8$

b $\frac{d^2y}{dx^2}+4\frac{dy}{dx}+4y=0$, given that $y(0)=1$ and $y'(0)=2$

c $\frac{d^2y}{dx^2}+25y=0$, given that $y(0)=3$ and $y\left(\frac{\pi}{10}\right)=6$

d $\frac{d^2y}{dx^2}+\frac{dy}{dx}-6y=0$, given that $y(0)=1$ and $y'(0)=12$

e $\frac{d^2y}{dx^2}-6\frac{dy}{dx}+9y=0$, given that $y(0)=5$ and $y'(0)=3$

f $\frac{d^2y}{dx^2}-\frac{dy}{dx}-12y=0$, given that $y(0)=12$ and $y'(0)=6$

g $9\frac{d^2y}{dx^2}-12\frac{dy}{dx}+4y=0$, given that $y(0)=12$ and $y'(0)=8$

h $4\frac{d^2y}{dx^2}-4\frac{dy}{dx}+5y=0$, given that $y(0)=6$ and $y'(0)=-1$

2 Solve each of the following differential equations subject to the given boundary conditions.

a $\frac{d^2y}{dx^2}-7\frac{dy}{dx}+10y=12e^x$, given that $y(0)=13$ and $y'(0)=8$

b $\frac{d^2y}{dx^2}-12\frac{dy}{dx}+36y=72$, given that $y(0)=3$ and $y(0)=-2$

c $\frac{d^2y}{dx^2}+4\frac{dy}{dx}+5y=16\cos x$, given that $y(0)=5$ and $y'(0)=-6$

d $\frac{d^2y}{dx^2}+2\frac{dy}{dx}-8y=30-24x$, given that $y(0)=1$ and $y'(0)=0$

e $9\frac{d^?y}{dx^2}-4y=4x^2+2$, given that $y(0)=-8$ and $y'(0)=10$

f $\frac{d^2y}{dx^2}-4\frac{dy}{dx}+5y=\cos 2x+8\sin 2x$, given that $y(0)=9$ and $y'\left(\frac{\pi}{2}\right)=4e^{\pi}-1$

g $4\frac{d^2y}{dx^2}-9y=26+27x-9x^2$, given that $y(0)=8$ and $y'(0)=-9$

h $16\frac{d^2y}{dx^2}-8\frac{dy}{dx}+y=12e^{-\frac{1}{4}x}$, given that $y(0)=10$ and $y'(0)=2$

3 Given that the differential equation $\frac{d^2y}{dx^2}-2\frac{dy}{dx}-8y=12e^{-2x}$ has a particular integral of the form $y=axe^{-2x}$, determine the value of the constant a, and find the general solution of the differential equation.

4 Given that the differential equation $\frac{d^2y}{dx^2}-2\frac{dy}{dx}+y=6e^x$ has a particular integral of the form $y=ax^2e^x$, determine the value of the constant a, and find the general solution of the differential equation.

5 Find the solution of the differential equation $\frac{d^2y}{dx^2}+5\frac{dy}{dx}-6y=21e^x+12$ for which $y=-2$ and $\frac{dy}{dx}=17$ at $x=0$

6 Find the solution of the differential equation $\frac{d^2y}{dx^2}-6\frac{dy}{dx}+9y=34e^{3x}$ for which $y=3$ and $\frac{dy}{dx}=11$ at $x=0$

7 Solve the differential equation $\frac{d^3y}{dx^3}-6\frac{d^2y}{dx^2}+11\frac{dy}{dx}-6y=12e^{4x}$, given that $y(0)=4$, $y'(0)=11$ and $y''(0)=35$

8 a Given $x=e^u$ and $x^2\frac{d^2y}{dx^2}-4x\frac{dy}{dx}+6y=12$, show that $\frac{d^2y}{du^2}-5\frac{dy}{du}+6y=12$

b Hence solve the equation $x^2\frac{d^2y}{dx^2}-4x\frac{dy}{dx}+6y=12$, given that $y(1)=7$ and $y(2)=14$

20.3 Simple harmonic motion

Fluency and skills

Differential equations can be used to model many different situations. In particular, they can be used to describe things that oscillate. This includes the motion of a weight hanging on the end of a spring, moving up and down, or a pendulum swinging from side to side, or the depth of the water in a harbour as the tide moves in and out. Oscillatory motions like these follow a regular, cyclic pattern. If you plot a displacement–time graph of the motion of an oscillatory system, you will often find that it is a sine graph.

Example 1

a Solve the differential equation $\frac{d^2x}{dt^2} = -4x$, given that $x(0) = 4$ and $x\left(\frac{\pi}{4}\right) = -4$

b Sketch the solution, and state its amplitude.

a First rewrite the equation as $\frac{d^2x}{dt^2} + 4x = 0$

The auxiliary equation is $m^2 + 4 = 0$

$m^2 = -4$, so $m = \pm 2i$

So the general solution is $x = A\cos 2t + B\sin 2t$

$A = 4$ — Use $x(0) = 4$ to substitute $t = 0$, $x = 4$

$B = -4$ — Use $x\left(\frac{\pi}{4}\right) = -4$ to substitute $t = \frac{\pi}{4}$, $x = -4$

The particular solution is $x = 4\cos 2t - 4\sin 2t$

$$= 4\sqrt{2}\left(\frac{1}{\sqrt{2}}\cos 2t - \frac{1}{\sqrt{2}}\sin 2t\right)$$

$$= 4\sqrt{2}\cos\left(2t + \frac{\pi}{4}\right)$$ — Use harmonic form.

b

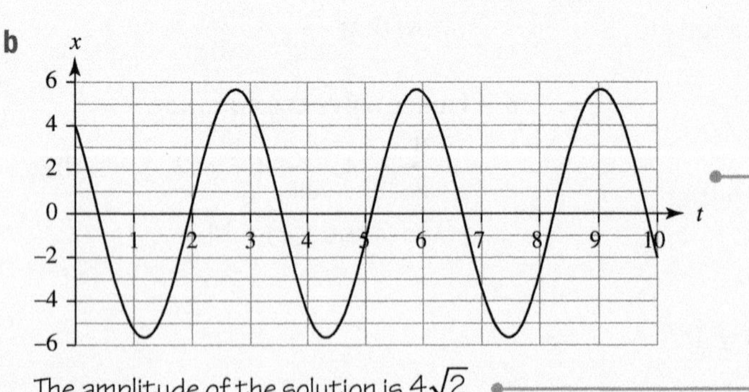

Try plotting this on your graphical calculator.

The amplitude of the solution is $4\sqrt{2}$ — The amplitude is the maximum displacement from the centre of the motion.

Key point

Motion that satisfies a differential equation of the form $\frac{d^2x}{dt^2} = -\omega^2 x$ is called simple harmonic motion (SHM).

There are certain standard results that apply to SHM. These are derived below.

Using $\frac{d^2x}{dt^2} = v\frac{dv}{dx}$ gives $v\frac{dv}{dx} = -\omega^2 x$

$$\frac{d^2x}{dt^2} = \frac{dv}{dt} = \frac{dv}{dx} \times \frac{dx}{dt} = \frac{dv}{dx} \times v$$

So $\int v\,dv = -\int \omega^2 x\,dx$

Separate the variables.

$\frac{v^2}{2} = -\frac{\omega^2 x^2}{2} + c \qquad (1)$

Let the amplitude of the motion be a, then

$\frac{0^2}{2} = -\frac{\omega^2 a^2}{2} + c \Rightarrow c = \frac{\omega^2 a^2}{2}$

Use the fact that when $x = a$ the velocity is zero.

So $\frac{v^2}{2} = -\frac{\omega^2 x^2}{2} + \frac{\omega^2 a^2}{2}$

Substitute for c into (1)

i.e. $v^2 = \omega^2(a^2 - x^2)$ or $v = \omega\sqrt{a^2 - x^2}$

$\frac{dx}{dt} = \omega\sqrt{a^2 - x^2}$

Use $v = \frac{dx}{dt}$

$\int \frac{1}{\sqrt{a^2 - x^2}}\,dx = \int \omega\,dt$

Separate the variables, and integrate.

$\sin^{-1}\left(\frac{x}{a}\right) = \omega t + \varepsilon$

$x = a\sin(\omega t + \varepsilon)$

Use the standard result $\int \frac{1}{\sqrt{a^2 - x^2}}\,dx = \sin^{-1}\left(\frac{x}{a}\right)$, where ε is a constant of integration

The value of ε depends on when the clock starts ticking, i.e. where we measure $t = 0$ from.

- If $t = 0$ when $x = 0$, then $\varepsilon = 0$, and $x = a\sin\omega t$
- If $t = 0$ when $x = a$, then $\varepsilon = \frac{\pi}{2}$, and $x = a\sin\left(\omega t + \frac{\pi}{2}\right) = a\cos\omega t$

The period of a simple $\sin t$ or $\cos t$ function is 2π. Hence the period of $x = a\sin(\omega t + \varepsilon)$ is $\frac{2\pi}{\omega}$

The following results apply to any particle that is moving with simple harmonic motion, i.e. any motion for which $\frac{d^2x}{dt^2} = -\omega^2 x$

Standard results

$$\frac{d^2x}{dt^2} = -\omega^2 x \Rightarrow v = \omega\sqrt{a^2 - x^2}$$

$$x = a\cos\omega t \quad \text{or}$$

$$x = a\sin\omega t$$

$$T = \frac{2\pi}{\omega}$$

where a is the amplitude and T is the period.

It is the starting conditions that determine which form of x to use.

Example 2

A particle moves with SHM defined by the equation $x = -16x$

$\left(\text{This is another way of writing } \frac{d^2x}{dt^2} = -16x\right)$

The amplitude of the motion is 3 metres.

Find

a The period of the motion,

b The maximum speed of the particle.

a $T = \frac{2\pi}{4} = \frac{\pi}{2}$ s — Use $T = \frac{2\pi}{\omega}$ with $\omega = \sqrt{16} = 4$

b $v = 4\sqrt{3^2 - 0^2} = 12\text{ m s}^{-1}$ — Use $v = \omega\sqrt{a^2 - x^2}$ with $\omega = 4$ and $x = 0$

Key point

The maximum speed occurs at the centre of the motion, i.e. when $x = 0$

Example 3

A particle is projected from a point O at time $t = 0$, and performs SHM with O as the centre of oscillation.

The motion is of amplitude 5 metres, and the period is $\frac{\pi}{3}$ s

Find

a The speed of projection,

b The time it takes for the particle to first reach a point that is 4 metres from O

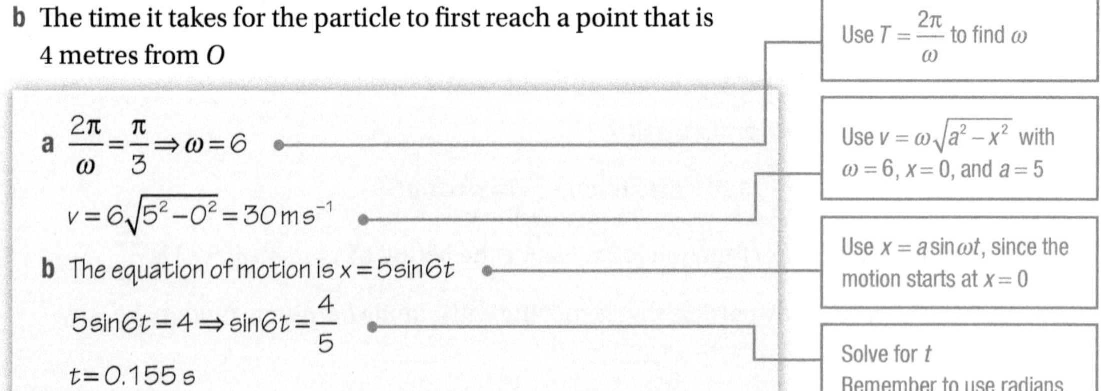

a $\frac{2\pi}{\omega} = \frac{\pi}{3} \Rightarrow \omega = 6$
Use $T = \frac{2\pi}{\omega}$ to find ω
$v = 6\sqrt{5^2 - 0^2} = 30\text{ m s}^{-1}$
Use $v = \omega\sqrt{a^2 - x^2}$ with $\omega = 6$, $x = 0$, and $a = 5$
b The equation of motion is $x = 5\sin 6t$
Use $x = a\sin\omega t$, since the motion starts at $x = 0$
$5\sin 6t = 4 \Rightarrow \sin 6t = \frac{4}{5}$
$t = 0.155$ s
Solve for t
Remember to use radians.

Exercise 20.3A Fluency and skills

1 Solve each of the following differential equations of SHM, subject to the given initial and boundary conditions.

a $\frac{d^2x}{dt^2} = -25x$, given that $x = 7$ when $t = 0$, and $x = 5$ when $t = \frac{\pi}{10}$

b $\frac{d^2x}{dt^2} = -4x$, given that $x = 2\sqrt{2}$ when $t = \frac{\pi}{8}$, and $x = 8$ when $t = \frac{\pi}{4}$

c $\frac{d^2x}{dt^2} = -100x$, given that $x = -5$ and $\frac{dx}{dt} = 30$ when $t = 0$

d $\frac{d^2x}{dt^2} = -x$, given that $x = -1$ and $\frac{dx}{dt} = 9$ when $t = 0$

e $4\frac{d^2x}{dt^2} = -x$, given that $x = \sqrt{2}$ when $t = \frac{\pi}{2}$, and $x = -4\sqrt{2}$ when $t = \frac{3\pi}{2}$

f $9\frac{d^2x}{dt^2} = -x$, given that $x = 7$ when $t = \frac{\pi}{2}$, and $x = \sqrt{3}$ when $t = \pi$

2 Solve each of the following differential equations of SHM, subject to the given initial and boundary conditions.

a $\frac{d^2x}{dt^2} = -16x + 48$, given that $x = 0$ when $t = 0$, and $x = 5$ when $t = \frac{\pi}{8}$

b $\frac{d^2x}{dt^2} = -25x + 100$, given that $x = 5$ when $t = \frac{\pi}{10}$, and $x = -1$ when $t = \frac{\pi}{5}$

c $\frac{d^2x}{dt^2} = -9x + 36$, given that $x = -1$ and $\frac{dx}{dt} = 12$ when $t = 0$

d $\frac{d^2x}{dt^2} = -4x + 12$, given that $x = -3$ and $\frac{dx}{dt} = 8$ when $t = 0$

e $16\frac{d^2x}{dt^2} + x - 1 = 0$, given that $x = 8$ when $t = \pi$, and $x = 4$ when $t = 3\pi$

f $4\frac{d^2x}{dt^2} + 25x + 50 = 0$, given that $x = 0$ when $t = \frac{2\pi}{5}$, and $x = -2\sqrt{2} - 2$ when $t = \frac{\pi}{2}$

3 Solve the differential equation $\frac{d^2x}{dt^2} = -25x$, given that $x(0) = 4$ and $x\left(\frac{\pi}{10}\right) = 3$

Hence sketch the graph of x against t

4 Solve the differential equation $\frac{d^2x}{dt^2} = -9x + 18$, given that $x(0) = 7$ and $x\left(\frac{\pi}{2}\right) = 14$

Hence sketch the graph of x against t

5 A particle moves with SHM defined by the equation $\ddot{x} = -9x$. The amplitude of the motion is 2 metres.

Find

a The period of the motion,

b The maximum speed of the particle.

6 A particle moves with SHM defined by the equation $\ddot{x} = -4x$. The maximum speed of the particle is 12 m s^{-1}

Find

a The period of the motion,

b The amplitude of the motion.

7 A particle moves with SHM defined by the equation $\ddot{x} = -\frac{x}{25}$. When $x = 0.4$ metres, the speed of the particle is 0.06 m s^{-1}.

Find

a The period of the motion,

b The amplitude of the motion,

c The speed of the particle when $x = 0.3$ metres.

8 A particle is projected from a point O and performs SHM with O as the centre of oscillation.

The motion has amplitude 0.9 metres, and the period is π s.

Find

a The speed of projection,

b The time it takes for the particle to first reach a point P, where P is 0.7 metres from O

9 A particle is projected from a point O at time $t=0$ and performs SHM with O as the centre of oscillation.

The motion is of amplitude 0.6 metres, and the period is $\frac{\pi}{2}$ s.

Find

a The speed of projection,

b The time it takes for the particle to first reach a point P, where P is 0.3 metres from O

10 A particle moves with SHM about a centre O The amplitude of the motion is 1.3 metres, and the period is $\frac{\pi}{4}$ s.
Find how far the particle is from O when its speed is $4\ \text{m s}^{-1}$.

11 A man on the end of a bungee performs SHM of period 5π seconds and amplitude 10 metres about a centre O. After passing through O the man passes through a point X which is 3 metres below O, and then he passes through a point Y which is 9 metres below O

Find the time taken for the man to travel from X to Y

12 The head of a piston moves with SHM about a centre O. When the piston-head is 1 metre from O its speed is $7.2\ \text{m s}^{-1}$, and when it is 2.4 metres from O its speed is $3\ \text{m s}^{-1}$.

Find

a The amplitude of the motion,

b The period of the motion.

13 A toy on the end of a spring moves with SHM about a centre O. When the toy is 30 cm from O its speed is $0.8\ \text{cm s}^{-1}$, and when it is 40 cm from O its speed is $0.6\ \text{cm s}^{-1}$.

Find

a The amplitude of the motion,

b The period of the motion.

Reasoning and problem-solving

Strategy

To answer a question involving simple harmonic motion

1. Draw a force diagram.
2. Use Newton's 2nd law to form a second order differential equation.
3. Find the general solution to that differential equation.
4. Use the information given in the question to find the particular solution.
5. Answer the question in context.

Example 4

A particle, P, of mass 0.06 kg moves in a horizontal straight line under the action of a force directed towards a fixed point O. The force varies with the distance of the particle from O, and is equal to $(1.5x)$ N, where x, measured in metres, is the displacement of P from O. P is initially at rest at a displacement of 10 metres from O

a Prove that the motion of P is SHM.

b Find the velocity of P when P is 6 metres from O

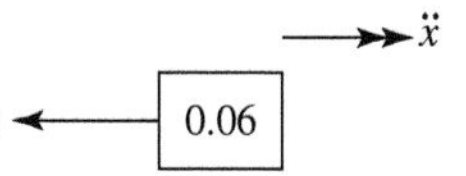

a $-1.5x = 0.06\ddot{x}$ — The minus sign is because the force acts towards O

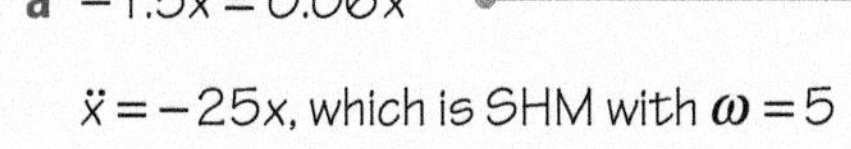

$\ddot{x} = -25x$, which is SHM with $\omega = 5$

b $v^2 = 5^2(10^2 - 6^2)$ — Use $v^2 = \omega^2(a^2 - x^2)$ with $\omega = 5$, $a = 10$, $x = 6$

$v = 40\ \text{m s}^{-1}$

Example 5

One end of a light elastic spring of unstretched length 2 metres is fixed to a point O on a smooth horizontal table. A particle of mass 0.1 kg is attached to the other end of the spring and rests in equilibrium at a point A on the table. The particle is then pulled away from A, in the direction OA, causing the spring to stretch by 20 cm. The force in the spring is directed towards A, and has magnitude $10x$ N, where x, measured in metres, is the displacement of the particle from A. The body is released from rest at time $t = 0$ seconds.

a Show that $\dfrac{d^2x}{dt^2} = -100x$

b Solve this differential equation to show that $x = 0.2\cos(10t)$

c Calculate the distance of the particle from O when $t = \dfrac{\pi}{30}$

d Calculate the speed of the particle when $t = \dfrac{\pi}{30}$

a Apply $F = ma$ to get $-10x = 0.1\dfrac{d^2x}{dt^2}$

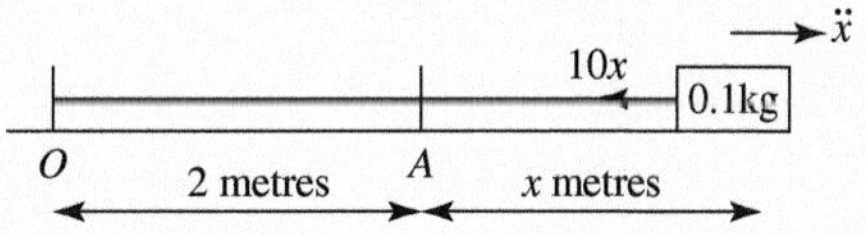

$$\Rightarrow \frac{d^2x}{dt^2} = -100x$$

(Diagram labels: $\ddot{x}$, $10x$, 0.1kg, O, 2 metres, A, x metres)

b The auxiliary equation is

$m^2 + 100 = 0$

$m^2 = -100$, so $m = \pm 10i$

So the general solution is $x = A\cos 10t + B\sin 10t$

Use $x = 0.2$, $\dot{x} = 0$ at $t = 0$ to get $A = 0.2$, $B = 0$

$\Rightarrow x = 0.2\cos 10t$

c $x = 0.2\cos\left(\dfrac{10\pi}{30}\right) = 0.2\cos\left(\dfrac{\pi}{3}\right) = 0.1$ metres — Substitute $t = \dfrac{\pi}{30}$

d $\dfrac{dx}{dt} = -2\sin 10t = -2\sin\left(\dfrac{10\pi}{30}\right) = -\sqrt{3}$ — Differentiate x to get an expression for the velocity and substitute $t = \dfrac{\pi}{30}$

The speed of the particle is $\sqrt{3}\ \text{m s}^{-1}$

PURE

Example 6

A small box of mass m kg is at rest at a point O on a smooth horizontal table. The box is attached to two identical springs. The other ends of the springs are attached to the points A and B, respectively, which are 3 metres apart on a straight line through O.

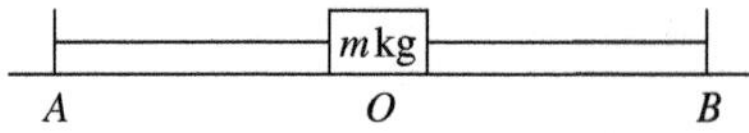

The box is struck so that it moves towards B with an initial velocity of 1.2 m s^{-1}. At a time t seconds after the box is struck, the displacement of the box from O, in the direction OB, is x metres. The force that each of the springs applies to the box, is $(2mx)$ N in the direction BA

a Prove that the motion of the box is SHM.

b Write down the period of the motion.

c Find the minimum distance of the box from B in the subsequent motion.

d State three assumptions that you have made in forming your model.

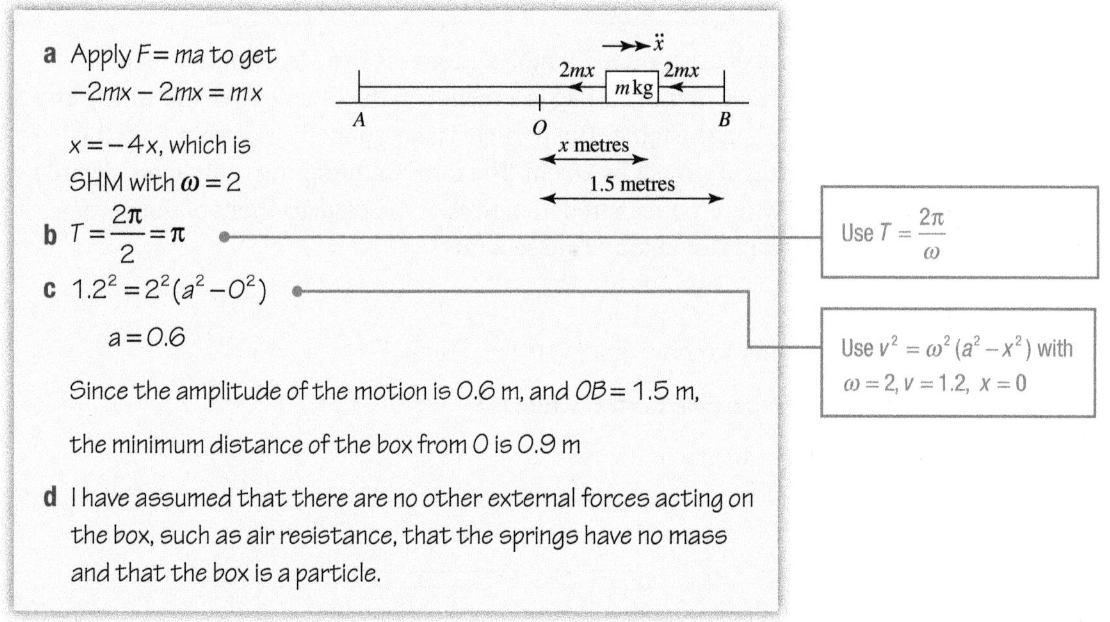

a Apply $F = ma$ to get

$-2mx - 2mx = mx$

$x = -4x$, which is SHM with $\omega = 2$

b $T = \dfrac{2\pi}{2} = \pi$

Use $T = \dfrac{2\pi}{\omega}$

c $1.2^2 = 2^2(a^2 - 0^2)$

$a = 0.6$

Use $v^2 = \omega^2(a^2 - x^2)$ with $\omega = 2$, $v = 1.2$, $x = 0$

Since the amplitude of the motion is 0.6 m, and $OB = 1.5$ m, the minimum distance of the box from O is 0.9 m

d I have assumed that there are no other external forces acting on the box, such as air resistance, that the springs have no mass and that the box is a particle.

Example 7

A particle of mass 0.5 kg rests at a fixed point P, on a smooth horizontal surface, attached to one end on a light elastic spring of natural length 3 metres. The other end of the spring is attached to O, where O is a fixed point on the surface. The particle is pulled a distance 2 metres from P, in the direction OP, and released from rest at time $t = 0$. At time t seconds, the displacement of the particle from O is x metres. The force in the spring is directed towards P and has magnitude $32(x-3)$ newtons.

a Write down the equation of motion of the particle.

b Find an expression for x in terms of t

c Sketch the graph of x against t

d Write down the value of t when the particle is closest to O for the first time.

(*Continued on the next page*)

a $-32(x-3)=0.5\ddot{x}$

$\ddot{x}+64x=192$

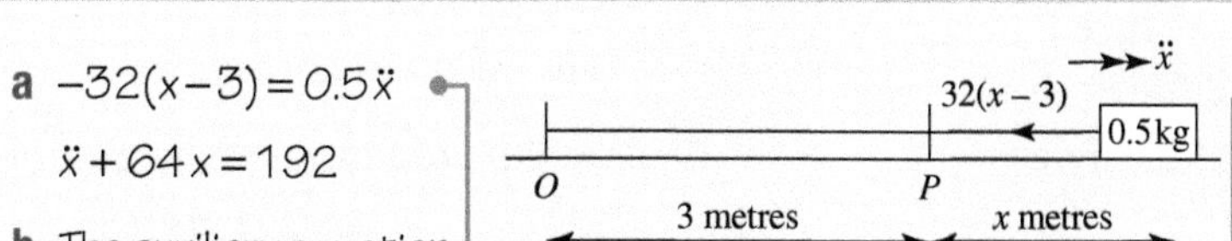

Apply Newton's 2nd law in the direction *OP*

b The auxiliary equation is $m^2+64=0$

$m^2=-64$, so $m=\pm 8i$

So the complementary function is $x=A\cos 8t+B\sin 8t$

By inspection the particular integral is $x=3$

So the general solution is $x=3+A\cos 8t+B\sin 8t$

Use $x=5, \dot{x}=0$ at $t=0$ to get $A=2, B=0$

$x=3+2\cos 8t$

c

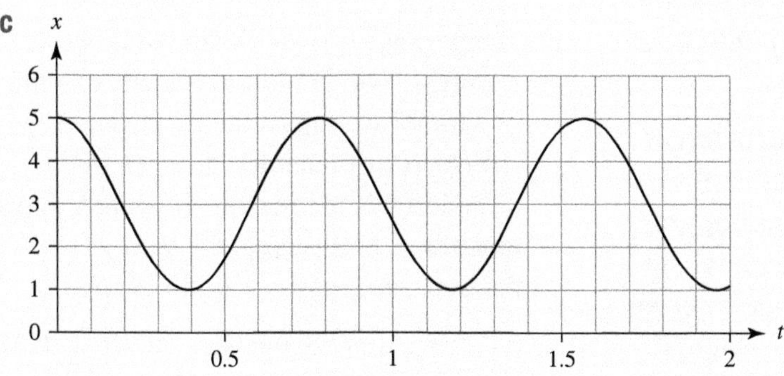

d $\cos 8t=-1$

At the minimum point.

$3+2\cos 8t=1$

$8t=\pi$

So $t=\dfrac{\pi}{8}$

Key point

This motion is still SHM. It is centred on $x=3$ rather than $x=0$

Exercise 20.3B Reasoning and problem-solving

1 A box *B*, of mass 5 kg, moves in a horizontal straight line under the action of a force directed towards a fixed point *O*. The force varies with the displacement of the particle from *O*, and is equal to $20x$ N, where x, measured in metres, is the distance of *B* from *O*. The box is initially at rest at distance of 5 metres from *O*

a Prove that the motion of *B* is SHM.

b Write down the period of the motion.

c Calculate the velocity of the box when it is 3 metres from *O*

2 One end of a light elastic spring of unstretched length 2 metres is fixed to a point *O* on a smooth horizontal table. A body of mass 200 g is attached to the other end of the spring and rests in equilibrium at a point *A* on the table. The body is then pulled away from *A*, a distance 30 cm in the direction *OA*, and released from rest. The force in the spring is directed towards *A*, and

has magnitude $5x$ N, where x, measured in metres, is the displacement of the body from A at time t seconds after the particle is released.

a Show that $\frac{d^2x}{dt^2} = -25x$

b Solve this differential equation to show that $x = 0.3\cos(5t)$

c Calculate the distance of the body from O when $t = \frac{\pi}{15}$

d Calculate also the speed of the body when $t = \frac{\pi}{15}$

3 Two points A and B are 1 metre apart on a smooth horizontal table. The point C lies on AB and is 80 cm from A

A light spring of unstretched length 80 cm has one end fastened to the table at A, and the other end is fastened to a body of mass 3 kg which is held at rest at B. When extended, the force in the spring is directed towards C, and is equal to $12x$ N, where x, measured in metres, is the displacement of the body from C. The body is released from rest at B

a Prove that the motion is SHM.

b Write down the period of the motion.

c Calculate the speed of the body when it is 4 cm from B

d State three assumptions that you have made in forming your model.

4 One end of a light elastic spring of unstretched length 60 cm is fixed to a point O in a smooth horizontal table. A small parcel of mass 100 g is attached to the other end of the spring. The parcel is at rest at the point P on the table, where $OP = 60$ cm. At time $t = 0$ the parcel is struck so that it moves in the direction OP with an initial velocity of $1.2\,\text{m s}^{-1}$. The force in the spring is directed towards P and has magnitude $10x$ N, where x, measured in metres, is the displacement of the parcel from P at time t seconds.

a Show that $\frac{d^2x}{dt^2} = -100x$

b Calculate the amplitude of the motion.

c Calculate the speed of the parcel when it is 6 cm from P

5 A small package of mass 1.5 kg is at rest at a point O on a smooth horizontal table. The package is attached to two identical springs. The other ends of the springs are attached to the points A and B, respectively, which are 2.4 metres apart on a straight line through O

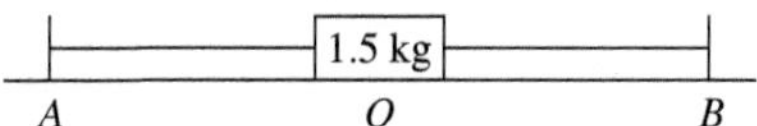

The package is struck so that it moves towards B with an initial velocity of $4\,\text{m s}^{-1}$ At a time t seconds after the package is struck, its displacement from O, in the direction OB, is x metres. The force, directed towards A, that each of the springs applies to the package is of magnitude $48x$ N.

a Prove that the package moves with SHM.

b Write down the period of the motion.

c Calculate the time at which the package is first 0.7 metres from B

d Calculate the speed of the package when it is 0.9 metres from B

6 A light elastic string of unstretched length 4 metres is stretched between two points, A and B, 6 metres apart on a smooth horizontal surface. O is the midpoint of AB A body of mass 0.5 kg is attached to the string at O. The body is pulled 90 cm towards B and released from rest. At a time t seconds after the body is released, its displacement from O is x metres, in the direction OB. The total force, directed towards A, that the string exerts on the body, is $0.32x$ N

a Show that $\frac{d^2x}{dt^2} = -0.64x$

b Solve this differential equation to show that $x = 0.9\cos(0.8t)$

c Calculate also the speed of the body when $t = \dfrac{5\pi}{12}$

d Explain how the model would need to be changed if, instead of pulling the body 90 cm towards B, it had been pulled 120 cm.

7 A small box of mass m kg is at rest in equilibrium at a point O on a smooth horizontal table. The box is attached to two springs. The other ends of the springs are attached to the points A and B which are 4 metres apart, with $AO = 1$ metre and $OB = 3$ metres.

m kg

A O B

The box is pulled 20 cm in the direction OB, and released from rest. At a time t seconds after the box is released, its displacement from O, in the direction OB, is x metres. The spring connected to B applies a force of mx N to the box in the direction BO, and the spring connected to A applies a force of $3mx$ N to the box in the direction OA

a Prove that the box moves with SHM.

b Write down the period of the motion.

c Find the speed of the box when it is 10 cm from O

d State three assumptions that you have made in forming your model.

8 A small box of mass 0.1 kg rests at a fixed point P on a large smooth horizontal table, attached to one end of a light elastic spring of natural length 2 metres. The other end of the spring is attached to O, a fixed point on the table. The box is pulled a distance 0.4 metres from P, in the direction OP and released from rest at time $t = 0$. At time t seconds, the displacement of the box from O is x metres. The force in the spring is directed towards P, and has magnitude $\dfrac{2(x-2)}{5}$ newtons.

a Write down the equation of motion of the box.

b Show that $x = 2 + \dfrac{2}{5}\cos(2t)$

c Calculate the maximum speed of the box.

d Calculate the value of t when the box is closest to O for the first time.

9 A small body of mass 0.3 kg rests at a fixed point A, on a smooth horizontal surface, attached to one end on a light elastic spring of natural length 1.5 metres. The other end of the spring is attached to O, a fixed point on the table. At time $t = 0$ the body is struck so that it moves in the direction OA with an initial velocity of $20\,\text{m s}^{-1}$. The force in the spring is directed towards A, and has magnitude $15(2x-3)$ N, where x, measured in metres, is the displacement of the body from P at time t seconds.

a Show that $\ddot{x} + 100x = 150$

b Find an expression for x in terms of t

c Find $\dfrac{dx}{dt}$

d Write down the maximum speed of the body.

10 A simple pendulum consists of a bob of mass m kg suspended from a light inextensible string of length ℓ metres. The other end of the string is attached to a fixed point O. The bob oscillates to and fro. At time t seconds, the angle between the string and the downward vertical through O is θ radians.

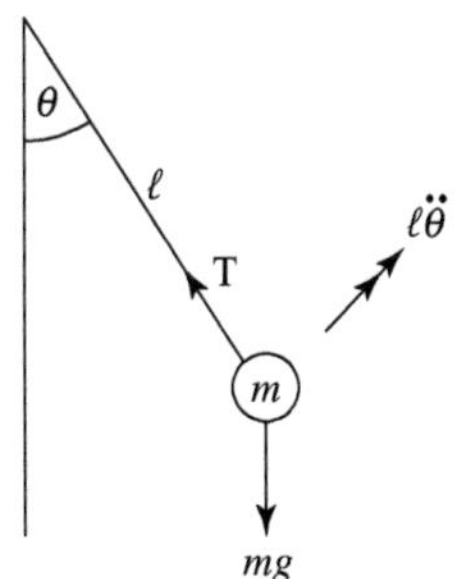

a Show that $-mg\sin\theta = m\ell\ddot{\theta}$

b Using the fact that, for small θ, $\sin\theta \approx \theta$, deduce that this motion is approximately SHM with period $2\pi\sqrt{\dfrac{\ell}{g}}$

20.4 Damped and forced harmonic motion

Fluency and skills

When a particle is attached to an elastic string or spring, and set in motion, with no extra forces acting on it, the motion is simple harmonic. However, in practice, the amplitude of the oscillations decreases with time. In fact, in some cases, there are no oscillations at all. This is due to other external forces such as friction and air resistance.

Example 1

A particle, P, moves along a horizontal line under the action of a force directed towards a fixed point O. The displacement, x metres, of P from its initial position, at a time t seconds from the start, satisfies the differential equation $\frac{d^2x}{dt^2}+2\frac{dx}{dt}+17x=34$

At time $t=0$ the particle is moving through the point $x=2$ with velocity $20\,\text{m s}^{-1}$.

a Solve the differential equation to obtain an expression for x in terms of t

b Sketch the graph of x against t

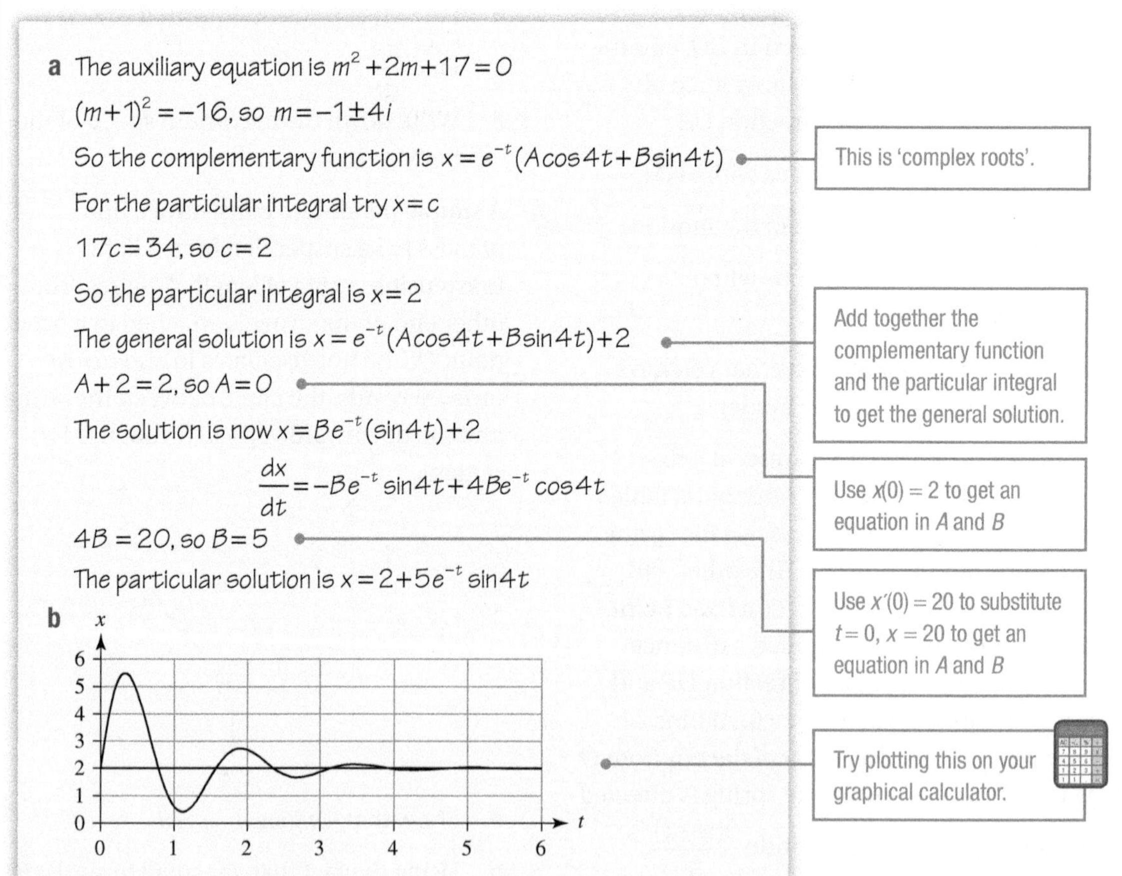

a The auxiliary equation is $m^2+2m+17=0$

$(m+1)^2=-16$, so $m=-1\pm4i$

So the complementary function is $x=e^{-t}(A\cos4t+B\sin4t)$ — This is 'complex roots'.

For the particular integral try $x=c$

$17c=34$, so $c=2$

So the particular integral is $x=2$

The general solution is $x=e^{-t}(A\cos4t+B\sin4t)+2$ — Add together the complementary function and the particular integral to get the general solution.

$A+2=2$, so $A=0$ — Use $x(0)=2$ to get an equation in A and B

The solution is now $x=Be^{-t}(\sin4t)+2$

$$\frac{dx}{dt}=-Be^{-t}\sin4t+4Be^{-t}\cos4t$$

$4B=20$, so $B=5$ — Use $x'(0)=20$ to substitute $t=0$, $x=20$ to get an equation in A and B

The particular solution is $x=2+5e^{-t}\sin4t$

b

Try plotting this on your graphical calculator.

Key point

This is called damped harmonic motion. The graph oscillates about $x=2$. The damping effect is caused by the term $2\frac{dx}{dt}$

In practice, one way of thinking of SHM is a particle moving freely on the end of a spring. Damped harmonic motion is a bit like putting the particle and spring in a tub of treacle. If the treacle is thin, the particle will oscillate much as before, with the amplitude slowly decreasing. If the treacle is thick, there might be no oscillations at all, and the particle will just move slowly towards an equilibrium position.

In general, simple harmonic motion is modelled by the differential equation $\frac{d^2x}{dt^2}=-\omega^2 x$ or $\frac{d^2x}{dt^2}+\omega^2 x=0$

For damped harmonic motion the motion of the particle is subject to a resistive force, which is proportional to the velocity of the particle. This can be modelled by a differential equation of the form $\frac{d^2x}{dt^2}+k\frac{dx}{dt}+\omega^2 x=0$, where $k>0$. The larger the value of k, the stronger is the resistance to motion (i.e. the thicker the treacle). The larger the value of ω^2, the stiffer the spring. So it is the ratio of $k:\omega^2$ that determines whether the particle will oscillate towards the equilibrium position as in Example 1, or approach the equilibrium position without oscillating.

This can be summarised with reference to the auxiliary equation, $m^2+km+\omega^2=0$

Key point

- If $k^2-4\omega^2>0$, $x=Ae^{m_1 t}+Be^{m_2 t}$ This is called heavy damping
- If $k^2-4\omega^2=0$, $x=(At+B)e^{mt}$ This is called critical damping
- If $k^2-4\omega^2<0$, $x=e^{-\frac{kt}{2}}\left(A\sin\frac{\alpha t}{2}+B\cos\frac{\alpha t}{2}\right)$ This is called light damping where $\alpha^2=4\omega^2-k^2$

Example 2

A particle moves along a horizontal line under the action of a force directed towards a fixed point O. The displacement, x metres of the particle from its initial position, at a time t seconds from the start, satisfies the differential equation $\frac{d^2x}{dt^2}+5\frac{dx}{dt}+6x=3$

At time $t=0$ the particle is moving through the point $x=1$ with velocity 3ms^{-1}

a Solve the differential equation to obtain an expression for x in terms of t

b Find the time at which the particle is at rest.

c Sketch the graph of x against t

d Explain whether the type of damping is light, critical or heavy.

a The auxiliary equation is $m^2+5m+6=0$

$(m+2)(m+3)=0$, so $m=-2$ or -3

So the complementary function is $x=Ae^{-2t}+Be^{-3t}$ — This is 'real distinct roots'.

(*Continued on the next page*)

For the particular integral, try $x = c$

$6c = 3$, so $c = \frac{1}{2}$

So the particular integral is $x = \frac{1}{2}$

The general solution is $x = Ae^{-2t} + Be^{-3t} + \frac{1}{2}$

$x = -2Ae^{-2t} - 3Be^{-3t}$

$t = 0, x = 1$ gives $A + B + \frac{1}{2} = 1$

$t = 0, x = 3$ gives $-2A - 3B = 3$

So $A = 4.5$ and $B = -4$

The particular solution is $x = 4.5e^{-2t} - 4e^{-3t} + 0.5$

This is from Table 2

Add together the complementary function and the particular integral to get the general solution.

b $\frac{dx}{dt} = -9e^{-2t} + 12e^{-3t} = 0$

$9e^{t} = 12$

$t = \ln\left(\frac{4}{3}\right) \approx 0.288$

c

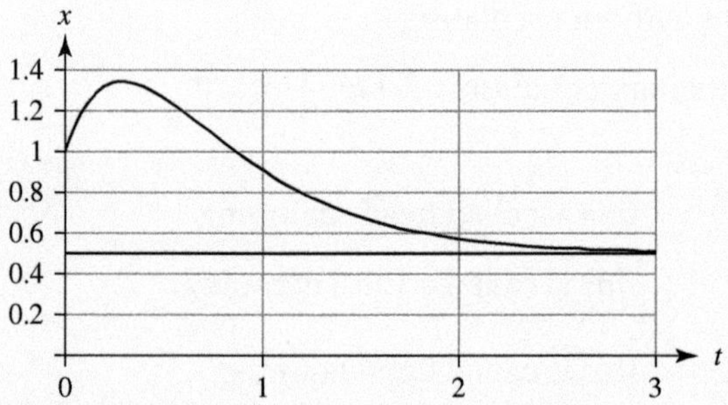

d The auxiliary equation is $m^2 + 5m + 6 = 0$

The discriminant, $\Delta = 5^2 - 4(1)(6) = +1 > 0$

Hence the damping is heavy.

Not all harmonic motion is simple or damped. In the two examples above, one end of the spring has been attached to a fixed point. Suppose, instead, you allow that end to move. It is still possible for the motion of the particle to follow a regular pattern.

Example 3

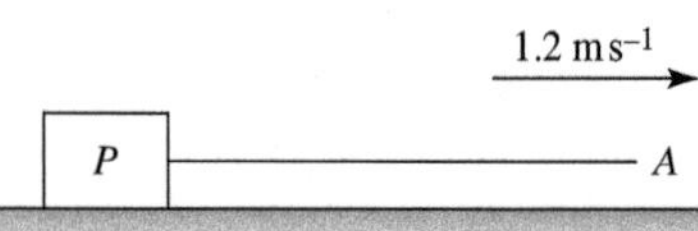

A box, P, is attached to one end of a light elastic string. The box is initially at rest on a smooth horizontal table when the other end of the spring, A, starts to move away from P with constant velocity 1.2 m s^{-1}. The displacement, y metres, of P from its initial position at a time t seconds from the start, satisfies the differential equation $\frac{d^2x}{dt^2} + 25y = 10t$

a Find an expression for y in terms of t

b Calculate the time at which P is next at rest.

c Sketch the graph of y against t

(Continued on the next page)

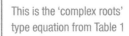

a The auxiliary equation is $m^2 + 25 = 0$

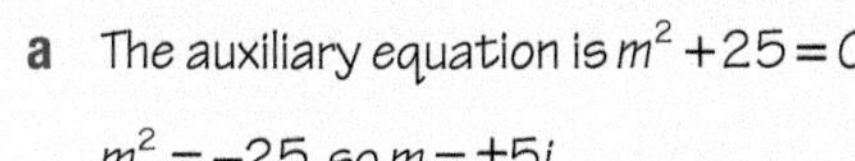

$m^2 = -25$, so $m = \pm 5i$

So the complementary function is $y = A\cos 5t + B\sin 5t$

> This is the 'complex roots' type equation from Table 1

For the particular integral, try $y = at + b$

> This is from Table 2

Then $\dot{y} = a$ and $\ddot{y} = 0$

$25(at + b) = 10t$

giving $a = \dfrac{2}{5}$ and $b = 0$

So the particular integral is $y = \dfrac{2}{5}t$

The general solution is $y = A\cos 5t + B\sin 5t + \dfrac{2}{5}t$

> **1** Find the general solution to the differential equation.

$A = 0$

> Substitute $t = 0$, $y = 0$ because y is the displacement from its starting position.

The solution is now $y = B\sin 5t + \dfrac{2}{5}t$

$\dfrac{dy}{dt} = 5B\cos 5t + \dfrac{2}{5}$

$5B + \dfrac{2}{5} = 0$, so $B = -\dfrac{2}{25}$

> Substitute $t = 0$, $y' = 0$ because the assumption is the box is initially at rest (it is the other end of the spring that starts to move at $1.2\,\text{m s}^{-1}$, and that is not relevant here).

The particular solution is $y = -\dfrac{2}{25}\sin 5t + \dfrac{2}{5}t$

> **2** Use the information given in the question to find the particular solution.

b $\dfrac{dy}{dt} = -\dfrac{2}{5}\cos 5t + \dfrac{2}{5}$

> Differentiate y to get velocity.

$-\dfrac{2}{5}\cos 5t + \dfrac{2}{5} = 0$

> Set velocity = 0

$\cos 5t = 1$

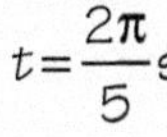

$t = \dfrac{2\pi}{5}\,\text{s}$

c

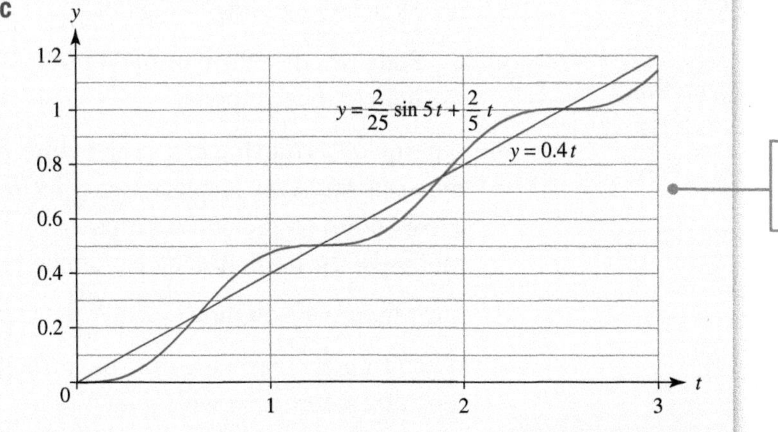

> Try plotting this on your graphical calculator.

As you can see from the diagram, the graph oscillates about the line $y = 0.4t$

PURE

Exercise 20.4A Fluency and skills

1 Solve each of the following differential equations subject to the given initial conditions, and classify each type of damping as heavy, critical or light.

a $\frac{d^2x}{dt^2}+6\frac{dx}{dt}+13x=0$, given that $x=-5$ and $\frac{dx}{dt}=3$ when $t=0$

b $\frac{d^2x}{dt^2}+6\frac{dx}{dt}+8x=0$, given that $x=2$ and $\frac{dx}{dt}=-2$ when $t=0$

c $\frac{d^2x}{dt^2}+4\frac{dx}{dt}+4x=0$, given that $x=-1$ and $\frac{dx}{dt}=5$ when $t=0$

d $\frac{d^2x}{dt^2}+10\frac{dx}{dt}+41x=0$, given that $x=2$ and $\frac{dx}{dt}=-6$ when $t=0$

e $2\frac{d^2x}{dt^2}+10\frac{dx}{dt}+13x=0$, given that $x=6$ and $\frac{dx}{dt}=-8$ when $t=0$

f $9\frac{d^2x}{dt^2}+6\frac{dx}{dt}+x=0$, given that $x=0$ and $\frac{dx}{dt}=6$ when $t=0$

2 Solve each of the following differential equations subject to the given initial conditions.

a $\frac{d^2x}{dt^2}+10\frac{dx}{dt}+9x=-18$, given that $x=0$ and $\frac{dx}{dt}=6$ when $t=0$

b $\frac{d^2x}{dt^2}+8\frac{dx}{dt}+16x=48$, given that $x=5$ and $\frac{dx}{dt}=-3$ when $t=0$

c $\frac{d^2x}{dt^2}+10\frac{dx}{dt}+29x=29$, given that $x=6$ and $\frac{dx}{dt}=-21$ when $t=0$

d $\frac{d^2x}{dt^2}+4\frac{dx}{dt}+5x=20$, given that $x=3$ and $\frac{dx}{dt}=2$ when $t=0$

e $\frac{d^2x}{dt^2}+4\frac{dx}{dt}+4x=12$, given that $x=4$ and $\frac{dx}{dt}=9$ when $t=0$

f $\frac{d^2x}{dt^2}+2\frac{dx}{dt}+2x=10$, given that $x=6$ and $\frac{dx}{dt}=0$ when $t=0$

3 Solve the differential equation $\frac{d^2x}{dt^2}+4\frac{dx}{dt}+3x=6$, given that $x(0)=5$ and $\dot{x}(0)=-1$, and hence sketch the graph of x against t

4 Solve the differential equation $\frac{d^2x}{dt^2}+2\frac{dx}{dt}+5x=20$, given $x(0)=4$ and $\dot{x}(0)=2$, and hence sketch the graph of x against t

5 A particle moves along a horizontal line under the action of a force directed towards a fixed point O. The displacement, x metres of P from O, at a time t seconds from the start, satisfies the differential equation $\frac{d^2x}{dt^2}+4\frac{dx}{dt}+8x=0$. At time $t=0$ the particle is moving through O with velocity $0.8\,\text{m s}^{-1}$.

a Solve the differential equation to obtain an expression for x in terms of t

b Sketch the graph of x against t

c State whether the type of damping is light, critical or heavy.

6 A particle P is moving in a straight line. At time t seconds, the displacement of P from a fixed point on the line is x metres. The motion of the particle can be modelled by the differential equation $\frac{d^2x}{dt^2}+7\frac{dx}{dt}+12x=6$
When $t=0$, the particle is moving through the point where $x=1$ with velocity $2\,\text{m s}^{-1}$

a Solve the differential equation to obtain an expression for x in terms of t

b Calculate the time at which the particle is at rest.

c Sketch the graph of x against t

d State whether the type of damping is light, critical or heavy.

7 A particle, P, is attached to one end of a light elastic spring. The other end of the spring is attached to a fixed point A, and P hangs at rest, vertically below A. At time $t=0$, P is projected vertically downwards with speed u m s^{-1}. The displacement of P downwards from its equilibrium position at time t seconds is x metres. The motion of P can be modelled by the differential equation
$$\frac{d^2x}{dt^2}+8\omega\frac{dx}{dt}+16\omega^2=0,$$
where ω is a constant.

a Find an expression for x in terms of u, t and ω

b Find an expression in terms of ω for the time at which P first comes to rest.

c State whether the type of damping is light, critical or heavy.

8 When an alternating electromotive force of $2\cos 2t$ is applied across an electrical circuit, the current, I amps at a time t seconds satisfies the differential equation
$$\frac{d^2I}{dt^2}+16I=36\sin 2t$$
At $t=0$, $I=0$ and $\frac{dI}{dt}=14$

a Find an expression for I in term of t

b Find the time at which the current first returns to zero.

9 A swing door is fitted with a damping device. The angular displacement, θ radians, of the door from its equilibrium position at time t seconds, is modelled by the differential equation $\frac{d^2\theta}{dt^2}+6\frac{d\theta}{dt}+13\theta=0$

The door starts from rest at an angle of $\frac{\pi}{4}$ to the equilibrium position.

a Find an expression for θ in terms of t

b Sketch the graph of θ against t

c What does the model predict as t becomes large?

10 The differential equation $\frac{d^2x}{dt^2}+\frac{dx}{dt}+\frac{x}{2}=\frac{3}{4}e^{-\frac{1}{2}t}$ describes the motion of a particle along the x-axis, where x, measured in metres, is the displacement of the particle from the origin at time t seconds.

At time $t=0$, $x=5$ and $\frac{dx}{dt}=-\frac{3}{2}$

a Solve this differential equation to get an expression for x in terms of t

b Prove that the particle never reaches the origin.

c Calculate the value of t when the particle first comes to rest.

11 In an experiment, John attaches a small weight, W, to one end of an elastic spring. John holds the other end of the spring, and W hangs at rest at a fixed point A. John then starts to move his end of the spring up and down, causing the weight to oscillate. The vertical displacement, x cm, of W above A, at a time t seconds after John has started to move his end, is given by the differential equation $2\frac{d^2x}{dt^2}+5\frac{dx}{dt}+2x=10\cos t, t\geq 0$

a Solve this differential equation to find an expression for x in terms of t

b Find the distance of W from A at $t=2$, making clear whether W is above A or below A at that time.

12

0.5 m s^{-1}

P A

A box, P, is attached to one end of a light elastic string. The box in initially at rest on a rough horizontal table, when the other end of the spring, A, starts to move away from P with constant velocity 0.5 m s^{-1}.

The displacement, x metres, of P from its initial position, at a time t seconds from the start, satisfies the differential equation
$$\frac{d^2x}{dt^2}+4x=2t+1$$

a Find an expression for x in terms of t

b Calculate the time at which P is next at rest.

c Sketch the graph of x against t

13 A particle, P, moves along a horizontal line under the action of a force directed towards a fixed point O

The displacement, x metres, of P from its initial position, at a time t seconds from the start, satisfies the differential equation

$$\frac{d^2x}{dt^2}+2\frac{dx}{dt}+2x=0$$

At $t=0$, $x=1$ and $\frac{dx}{dt}=0$

a Find an expression for x in term of t

b Find the times at which P is at rest.

c Show that the total distance moved by P will not exceed $\coth\left(\frac{\pi}{2}\right)$

Reasoning and problem-solving

Strategy

To answer a question involving damped or forced simple harmonic motion

1. Draw a force diagram.
2. Use Newton's 2nd law to form a second order differential equation.
3. Find the general solution to that differential equation.
4. Use the information given in the question to find the particular solution.
5. Answer the question in context.

Example 4

A particle, P, of mass 2 kg moves along the positive x-axis under the action of a force directed towards the origin. At time t seconds, the displacement of P from O is x metres, and P is moving away from O with a speed of v m s^{-1}. The force has magnitude $40x$ N. The particle, P, is also subject to a resistive force of magnitude $16v$ N.

a Show that the equation of motion of P is $\frac{d^2x}{dt^2}+8\frac{dx}{dt}+20x=0$

Given also that $x(0)=0$ and $x\left(\frac{\pi}{4}\right)=3e^{-\pi}$

b Solve the differential equation, and find an expression for x in terms of t

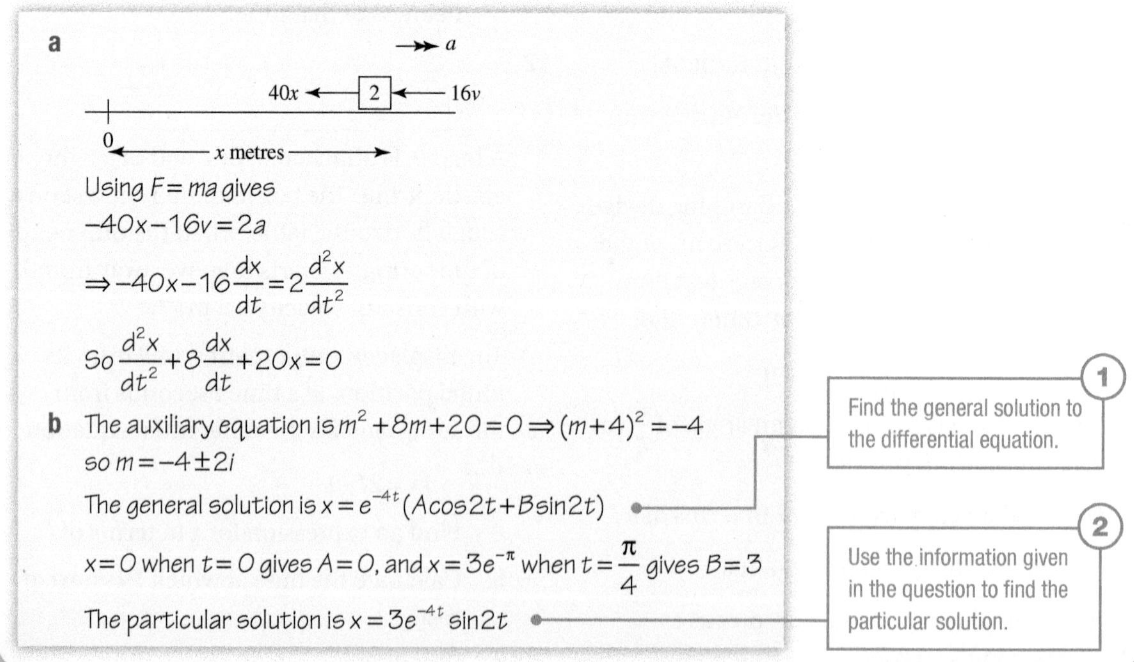

Using $F=ma$ gives

$-40x-16v=2a$

$\Rightarrow -40x-16\frac{dx}{dt}=2\frac{d^2x}{dt^2}$

So $\frac{d^2x}{dt^2}+8\frac{dx}{dt}+20x=0$

b The auxiliary equation is $m^2+8m+20=0 \Rightarrow (m+4)^2=-4$

so $m=-4\pm 2i$

The general solution is $x=e^{-4t}(A\cos 2t+B\sin 2t)$

$x=0$ when $t=0$ gives $A=0$, and $x=3e^{-\pi}$ when $t=\frac{\pi}{4}$ gives $B=3$

The particular solution is $x=3e^{-4t}\sin 2t$

(1) Find the general solution to the differential equation.

(2) Use the information given in the question to find the particular solution.

Example 5

A train of mass m kg runs into the buffers. The buffers can be modelled as an elastic spring that is fixed at one end. At time t seconds after the train first comes into contact with the buffers, the compression of the buffers is x metres, and the magnitude of the force in the spring is given by $m\omega^2 x$ newtons, where ω is a positive constant. The motion of the train is also subject to a resistive force of magnitude mkv newtons, where v m s^{-1} is the speed of the train and k is a positive constant.

a Show that $\frac{d^2x}{dt^2}+k\frac{dx}{dt}+\omega^2 x=0$

b At time $t=0$ the train is travelling with speed U. Given that $k=\frac{10\omega}{3}$, find an expression for x in terms of U, ω and t

c Show that the train comes to instantaneous rest when $t=\frac{3}{4\omega}\ln 3$

d State three assumptions you have made in forming your model.

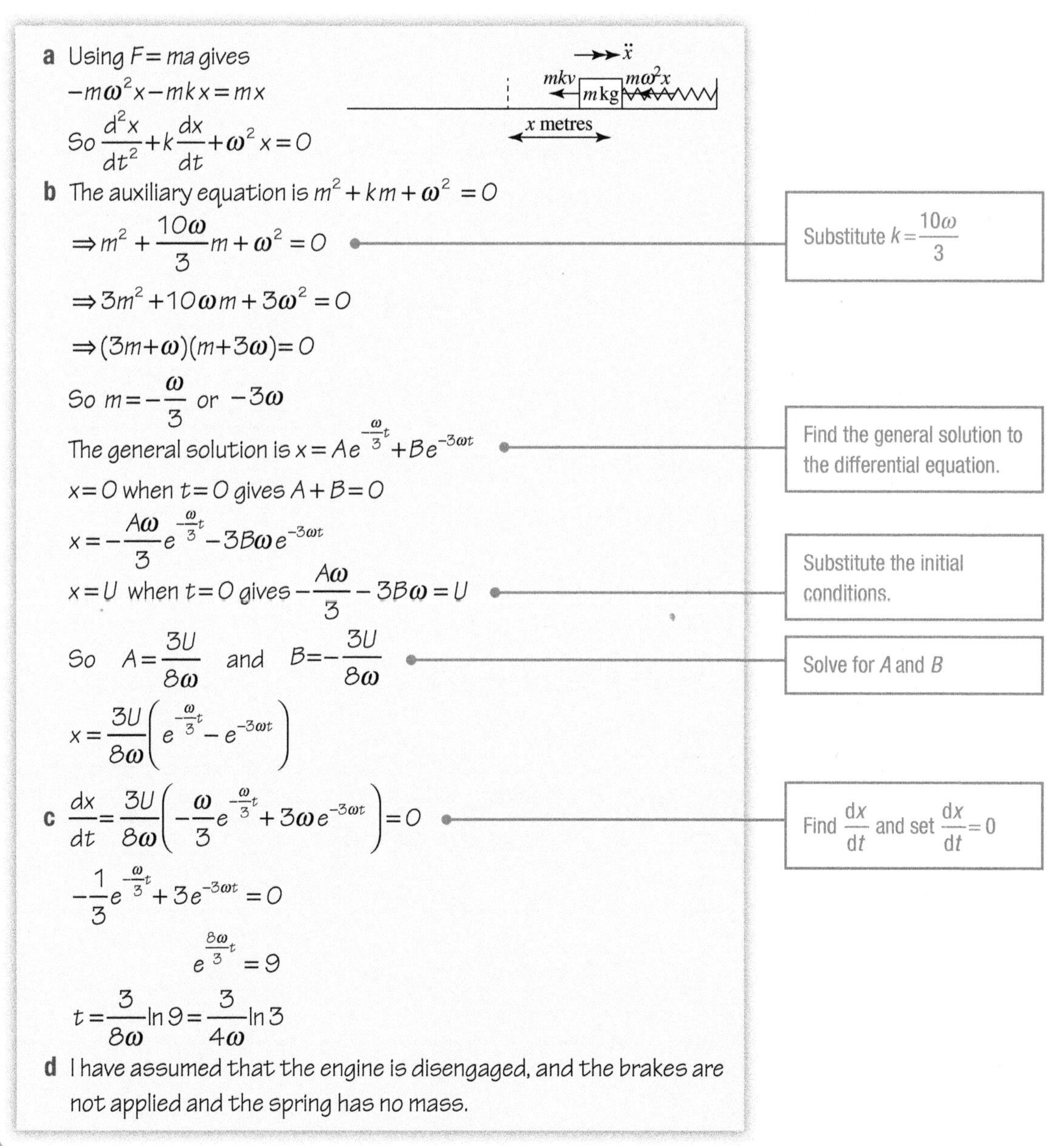

a Using $F=ma$ gives

$-m\omega^2 x-mkx=m\ddot{x}$

So $\frac{d^2x}{dt^2}+k\frac{dx}{dt}+\omega^2 x=0$

b The auxiliary equation is $m^2+km+\omega^2=0$

$\Rightarrow m^2+\frac{10\omega}{3}m+\omega^2=0$ — Substitute $k=\frac{10\omega}{3}$

$\Rightarrow 3m^2+10\omega m+3\omega^2=0$

$\Rightarrow (3m+\omega)(m+3\omega)=0$

So $m=-\frac{\omega}{3}$ or -3ω

The general solution is $x=Ae^{-\frac{\omega}{3}t}+Be^{-3\omega t}$ — Find the general solution to the differential equation.

$x=0$ when $t=0$ gives $A+B=0$

$\dot{x}=-\frac{A\omega}{3}e^{-\frac{\omega}{3}t}-3B\omega e^{-3\omega t}$

$\dot{x}=U$ when $t=0$ gives $-\frac{A\omega}{3}-3B\omega=U$ — Substitute the initial conditions.

So $A=\frac{3U}{8\omega}$ and $B=-\frac{3U}{8\omega}$ — Solve for A and B

$x=\frac{3U}{8\omega}\left(e^{-\frac{\omega}{3}t}-e^{-3\omega t}\right)$

c $\frac{dx}{dt}=\frac{3U}{8\omega}\left(-\frac{\omega}{3}e^{-\frac{\omega}{3}t}+3\omega e^{-3\omega t}\right)=0$ — Find $\frac{dx}{dt}$ and set $\frac{dx}{dt}=0$

$-\frac{1}{3}e^{-\frac{\omega}{3}t}+3e^{-3\omega t}=0$

$e^{\frac{8\omega}{3}t}=9$

$t=\frac{3}{8\omega}\ln 9=\frac{3}{4\omega}\ln 3$

d I have assumed that the engine is disengaged, and the brakes are not applied and the spring has no mass.

Example 6

A box of mass 0.2 kg is attached to the end of a light elastic string. The other end of the spring is attached to a fixed point. The mass falls freely vertically under gravity, and is travelling with a speed of $7\,\text{m s}^{-1}$ when the string becomes taut for the first time. At time t seconds after the string first becomes taut, the extension of the string is x metres. While the string is taut, the three forces that act on the box are an upwards force in the string of magnitude $2x$ newtons, a resistance force of magnitude $0.4\dot{x}$ newtons, and the weight of the box.
Taking $g = 10\,\text{m s}^{-2}$,

a Show that $\ddot{x} + 2\dot{x} + 10x = 10$

b Solve this equation to show that $x = e^{-t}(2\sin 3t - \cos 3t) + 1$

c Find the time at which particle box first comes to rest.

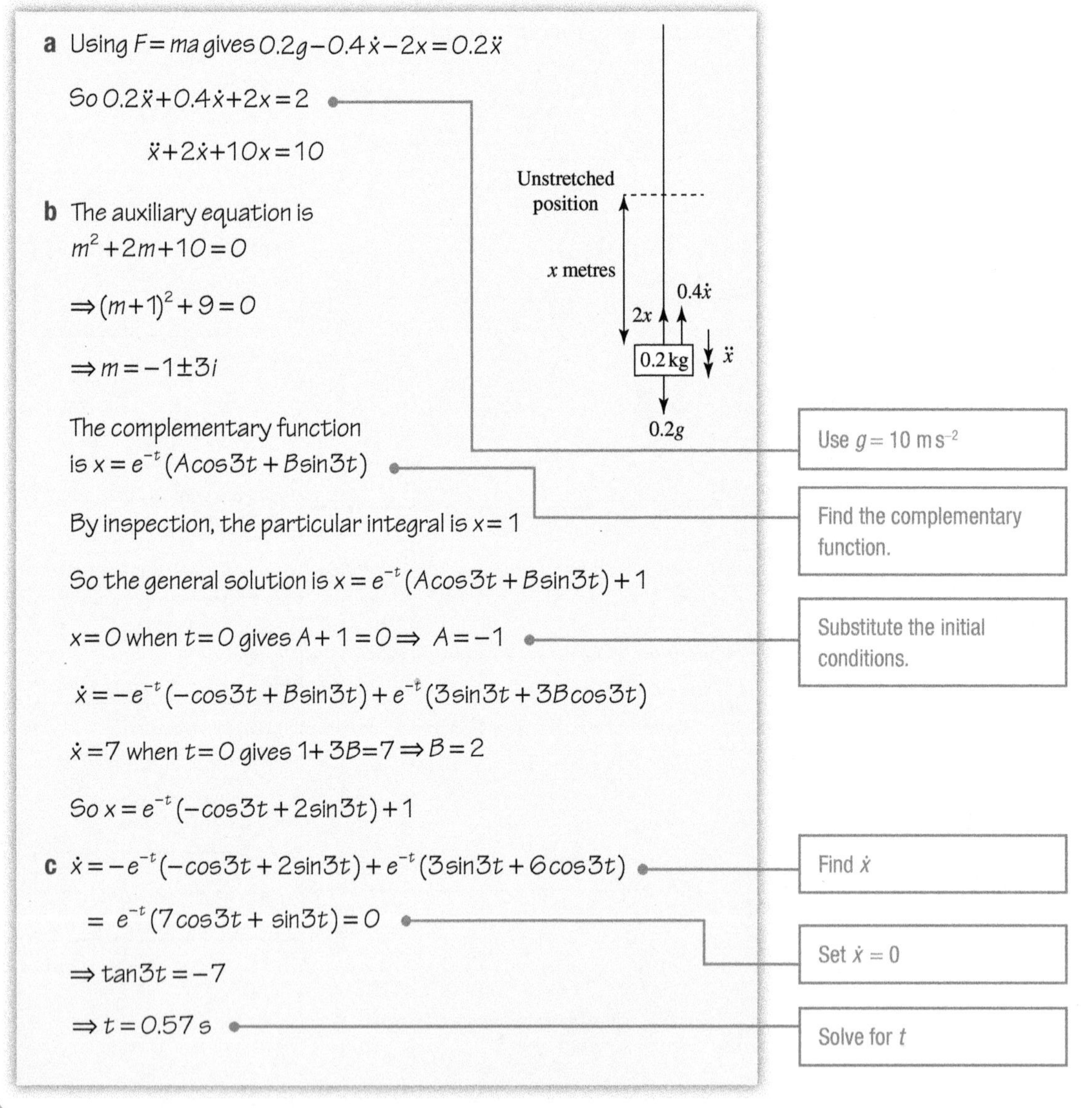

a Using $F = ma$ gives $0.2g - 0.4\dot{x} - 2x = 0.2\ddot{x}$

So $0.2\ddot{x} + 0.4\dot{x} + 2x = 2$

$$\ddot{x} + 2\dot{x} + 10x = 10$$

Use $g = 10\,\text{m s}^{-2}$

b The auxiliary equation is
$m^2 + 2m + 10 = 0$

$\Rightarrow (m+1)^2 + 9 = 0$

$\Rightarrow m = -1 \pm 3i$

The complementary function is $x = e^{-t}(A\cos 3t + B\sin 3t)$

Find the complementary function.

By inspection, the particular integral is $x = 1$

So the general solution is $x = e^{-t}(A\cos 3t + B\sin 3t) + 1$

$x = 0$ when $t = 0$ gives $A + 1 = 0 \Rightarrow A = -1$

Substitute the initial conditions.

$\dot{x} = -e^{-t}(-\cos 3t + B\sin 3t) + e^{-t}(3\sin 3t + 3B\cos 3t)$

$\dot{x} = 7$ when $t = 0$ gives $1 + 3B = 7 \Rightarrow B = 2$

So $x = e^{-t}(-\cos 3t + 2\sin 3t) + 1$

c $\dot{x} = -e^{-t}(-\cos 3t + 2\sin 3t) + e^{-t}(3\sin 3t + 6\cos 3t)$

Find $\dot{x}$

$= e^{-t}(7\cos 3t + \sin 3t) = 0$

Set $\dot{x} = 0$

$\Rightarrow \tan 3t = -7$

$\Rightarrow t = 0.57\,\text{s}$

Solve for t

Exercise 20.4B Reasoning and problem-solving

1 A body B, of mass 9 kg moves in a horizontal straight line. At time t seconds, the displacement of B from a fixed point O on the line, is x metres, and B is moving with a velocity of $\dot{x}$ m s^{-1}. Throughout the motion two forces act on B, a restorative force of magnitude $36x$ newtons directed towards O, and a resistance force of magnitude $45\dot{x}$ newtons. At time $t=0$, $x=-2$ and $\dot{x}=11$

a Show that $\ddot{x}+5\dot{x}+4=0$

b Find an expression for x in terms of t

c Hence state the type of damping that occurs.

d Find the value of t when B passes through O for the first time.

2 A particle P, of mass m kg moves in a horizontal straight line. At time t seconds, the displacement of P from a fixed point O on the line, is x metres, and P is moving with a velocity of $\dot{x}$ m s^{-1}. Throughout the motion two forces act on P, a restorative force of magnitude $25mn^2x$ newtons directed towards O, and a resistance force of magnitude $km\dot{x}$ newtons, where n and k are positive constants.

a Show that $\ddot{x}+k\dot{x}+25n^2x=0$

Given the damping is critical

b Find an expression for k in terms of n

c Find a general expression for x in terms of n and t

3 Two points, A and B, are 2 metres apart on a horizontal table. The point C lies on AB and is 160 cm from A. A light spring of unstretched length 160 cm has one end fastened to the table at A, and the other end fastened to a body of mass m kg which is held at rest at B. When extended, the force in the spring is directed towards C, and is equal to $\frac{4mx}{9}$ N, where x, measured in metres, is the displacement of the body from C

The body is released from rest at B. When in motion, two additional resistive forces act on the body, one a constant force of magnitude $\frac{4m}{45}$ N, and the other a variable force of magnitude $\frac{4m\dot{x}}{3}$ N.

a Suggest two physical properties that might cause the two resistive forces.

b Write down an equation of motion for the body as it moves towards A

c Show that $x=\frac{1}{15}\left[(2t+3)e^{-\frac{2}{3}t}+3\right]$

d With reference to the type of damping, explain what happens as the value of t increases.

4 A particle P of mass 3 kg is moving on the x-axis. At time t seconds the displacement of P from the origin O, is x metres, and the velocity of P is v m s^{-1}. Three forces act on P, namely, a restoring force towards O, of magnitude $15x$ newtons, a resistance to motion of magnitude $6v$ newtons, and a force of $24e^{-t}$ newtons acting in the direction OP

a Show that $\frac{d^2x}{dt^2}+2\frac{dx}{dt}+5x=8e^{-t}$

Given also that at time $t=0$, $x=5$ and $v=3$

b Find an expression for x in terms of t

c Find the value of t for which $x=0$ for the first time.

5 A particle P of mass m kg is attached to one end of a light elastic string of natural length ℓ metres. P is at rest in equilibrium on a smooth horizontal table. The other end of the string is attached to a fixed point, O, on the table. A time-dependent force is then applied to P in the direction OP. At time t seconds after the force is applied, the extension of the string is x metres. The magnitude of the force in the string is $6m\omega^2 x$, where ω is a constant, and the magnitude of the time-dependent force is $m\omega^2 \ell e^{-\omega t}$. The motion of the particle is opposed by a force of magnitude $5m\omega\dot{x}$.

- **a** Write down a differential equation connecting, x, t, ω and ℓ
- **b** Solve this equation to show that $x = \frac{\ell}{2}e^{-\omega t}(1-e^{-\omega t})^2$
- **c** Show that the greatest extension of the string is $\frac{2\ell}{27}$

6 On a water-ride at a theme park, passengers travel down a course in an open tub. At the end of the course, the tub is brought to rest by running into a retarding device. The retarding device is a large elastic spring that is fixed at the far end. At time t seconds after a tub first comes into contact with the spring, the compression of the spring is x metres, and the magnitude of the force in the spring is $m\omega^2 x$ newtons, where m is the combined mass of the passengers and tub, and ω is a positive constant. The motion of the tub is also subject to a resisting force of magnitude $2mk\dot{x}$ newtons, where $\dot{x}$ m s^{-1} is the velocity of the tub and k is a positive constant.

- **a** Show that $\ddot{x} + 2k\dot{x} + \omega^2 x = 0$

At time $t = 0$ the tub is travelling with speed U as it hits the spring.

- **b** Given that $k=\omega$, find an expression for x in terms of U, ω and t
- **c** Hence state the type of damping that occurs.

Health and safety rules state the the tub must come to rest within two seconds of the tub hitting the spring.

- **d** Show that $\omega > \frac{1}{2}$
- **e** State two assumptions that you have made in forming your model.

7 A box of mass 5 kg is attached to the end of a light elastic string. The other end of the string is attached to a fixed point. The box falls vertically under gravity and is travelling with a speed of 3 m s^{-1} when the string becomes taut for the first time. At time t seconds after the string first becomes taut, the extension of the string is x metres. While the string is taut, the three forces that act on the box are an upwards force in the string of magnitude $25x$ newtons, a resistive force of magnitude $10\dot{x}$ newtons, and the weight of the box. Taking g=10 m s^{-2},

- **a** Show that $\ddot{x} + 2\dot{x} + 5x = 10$
- **b** Solve this equation to get an expression for x in terms of t
- **c** Find the time at which the box comes to rest for the first time.

8 A stuntman of mass 80 kg attaches himself to one end of an elastic rope. The other end of the rope is attached to a fixed point on a high bridge. The man throws himself off the bridge and falls vertically under gravity. He is travelling with a speed of $15\,\text{m}\,\text{s}^{-1}$ when the rope becomes taut for the first time. At time t seconds after the rope first becomes taut, the extension of the rope is x metres. While the rope is taut, the three forces that act on the man are an upwards force in the string of magnitude $10x$ newtons, a resistive force of magnitude $40\dot{x}$ newtons, and the weight of the man. Taking $g = 10\,\text{m}\,\text{s}^{-2}$,

a Show that $8\frac{d^2x}{dt^2} + 4\frac{dx}{dt} + x = 80$

b Solve this equation to get an expression for x in terms of t

c Find the time at which the man first comes to rest.

d State two modelling assumption you have made in answering the question.

9

3 ms^{-1} →

5 kg —————— A

A box, B, of mass 5 kg is attached to one end of a light elastic spring of natural length 2 metres and modulus 90 N. The box is initially at rest on a smooth horizontal table, with the spring in equilibrium when the other end of the spring, A, starts to move away from B with constant velocity $3\,\text{m}\,\text{s}^{-1}$. The air resistance acting on B has magnitude $30v$ N, where $v\,\text{m}\,\text{s}^{-1}$ is the speed of B. At time t seconds, the extension of the string is x metres, and the displacement of B from its initial position is y metres.

a Show that

i $x + y = 3t$

ii $\frac{d^2x}{dt^2} + 6\frac{dx}{dt} + 9x = 18$

iii $x = 2 - (2 + 3t)e^{-3t}$

b Describe the motion of the box for large values of t

20.5 Coupled equations

Fluency and skills

So far all of the differential equations you have looked at have involved just two variables. Many differential equations involve more than two variables. Consider, for example, an island populated by foxes and rabbits. You can denote the number of foxes by F, and the number of rabbits by R. Foxes eat rabbits and are known as the **predators**. Rabbits eat grass, and are eaten by the foxes, and are known as the **prey**.

- $\frac{\mathrm{d}R}{\mathrm{d}t}$ depends on R and F. The more rabbits in the population, the more breeding and deaths there are, but the more foxes in the population, the more rabbits are eaten.
- $\frac{\mathrm{d}F}{\mathrm{d}t}$ depends on R and F. The more rabbits in the population, the more food there is for the foxes, but the more foxes in the population, the more breeding.

This gives rise to equations of the type

$$\frac{\mathrm{d}R}{\mathrm{d}t}=\alpha R-\beta F \quad (1) \qquad \text{and} \qquad \frac{\mathrm{d}F}{\mathrm{d}t}=\lambda R+\mu F \quad (2)$$

Key point

Problems that involve equations of this type are called **predator-prey** problems.

To solve equations of this type, first differentiate one of the equations with respect to t

Differentiating equation (1) gives

$$\frac{\mathrm{d}^2R}{\mathrm{d}t^2}=\alpha\frac{\mathrm{d}R}{\mathrm{d}t}-\beta\frac{\mathrm{d}F}{\mathrm{d}t}$$

Now substitute for $\frac{\mathrm{d}F}{\mathrm{d}t}$ from equation (2) to get

$$\frac{\mathrm{d}^2R}{\mathrm{d}t^2}=\alpha\frac{\mathrm{d}R}{\mathrm{d}t}-\beta(\lambda R+\mu F)$$

$$\frac{\mathrm{d}^2R}{\mathrm{d}t^2}=\alpha\frac{\mathrm{d}R}{\mathrm{d}t}-\beta\lambda R-\mu\beta F$$

Now substitute $\beta F = \alpha R - \frac{dR}{dt}$ from equation (1) to get

$$\frac{d^2R}{dt^2} = \alpha\frac{dR}{dt} - \beta\lambda R - \mu\left(\alpha R - \frac{dR}{dt}\right)$$

which rearranges to

$$\frac{d^2R}{dt^2} - (\alpha+\mu)\frac{dR}{dt} + (\beta\lambda + \alpha\mu)R = 0$$

This is a second order differential equation in R and t that you can solve in the usual way.

Example 1

A system of differential equations is given by

$$\frac{dx}{dt} = x + 4y \qquad (1)$$

$$\frac{dy}{dt} = 2x + 3y - 10 \qquad (2)$$

where $(x, y) = (3, 2)$ when $t = 0$

Find expressions for x and y in terms of t

Differentiating equation (2) gives

$$\frac{d^2y}{dt^2} = 2\frac{dx}{dt} + 3\frac{dy}{dt} = 2(x+4y) + 3\frac{dy}{dt}$$

Substitute $\frac{dx}{dt} = x + 4y$ from equation (1)

$$= \frac{dy}{dt} - 3y + 10 + 8y + 3\frac{dy}{dt}$$

Substitute $2x = \frac{dy}{dt} - 3y + 10$ from equation (2)

So $\frac{d^2y}{dt^2} - 4\frac{dy}{dt} - 5y = 10$

$m^2 - 4m - 5 = 0 \Rightarrow (m-5)(m+1) = 0$ so $m = -1$ or $m = 5$

The complementary function is

$y = Ae^{5t} + Be^{-t}$

The particular integral is $y = -2$

and the general solution is $y = Ae^{5t} + Be^{-t} - 2 \qquad (3)$

So $\dot{y} = 5Ae^{5t} - Be^{-t} \qquad (4)$

Differentiate the general solution.

From equation (2), $\dot{y}(0) = 2x(0) + 3y(0) - 10$

Substitute $t = 0$ into equation (2)

$= 2(3) + 3(2) - 10 = 2$

So $5A - B = 2$

Substitute $t = 0$, $\dot{y} = 0$ into equation (4)

and $A + B - 2 = 2$

Substitute $t = 0$, $y = 2$ into equation (3)

giving $A = 1$, $B = 3$

Solve equations simultaneously.

$y = e^{5t} + 3e^{-t} - 2$, and $x = e^{5t} - 6e^{-t} + 8$

Substitute for y into equation (2)

Exercise 20.5A Fluency and skills

1 Solve each of the following systems of differential equations to find expressions for y in terms of t

a $\frac{dx}{dt}=x-2y; \frac{dy}{dt}=x+4y$

b $\frac{dx}{dt}=-3y; \frac{dy}{dt}=3x$

c $\frac{dx}{dt}=5x+4y; 3\frac{dy}{dt}=x+4y$

d $\frac{dx}{dt}=4x-y; \frac{dy}{dt}=6x-3y+2$

e $\frac{dx}{dt}=2x-10y; \frac{dy}{dt}=5x-12y$

f $\frac{dx}{dt}=-3x+4y+\cos t; \frac{dy}{dt}=-2x+y+\sin t$

2 Solve each of the following systems of differential equations, subject to the given boundary conditions, to find expressions for x and y in terms of t

a $\frac{dx}{dt}=2x+4y; \frac{dy}{dt}=x-y$; given that $y(0)=3$ and $\dot{y}(0)=4$

b $\frac{dx}{dt}=2x-y+3; \frac{dy}{dt}=5x-4y$; given that $y(0)=-1$ and $\dot{y}(0)=-8$

c $\frac{dx}{dt}=x-5y; \frac{dy}{dt}=x-3y$; given that $y(0)=3$ and $\dot{y}(0)=-5$

d $\frac{dx}{dt}=x+2y+t+2; \frac{dy}{dt}=-2x-3y+3t$; given that $y(0)=9$ and $\dot{y}(0)=-2$

e $\frac{dx}{dt}=-3x-y-3; \frac{dy}{dt}=2x-y+2$; given that $y(0)=5$ and $\dot{y}(0)=2$

f $\frac{dx}{dt}=7x-9y+3e^{-2t}; \frac{dy}{dt}=4x-5y+e^{-2t}$; given that $y(0)=2$ and $\dot{y}(0)=3$

3 A system of differential equations is given by

$$\frac{dx}{dt}=2x+y \qquad (1)$$

$$\frac{dy}{dt}=x+2y \qquad (2)$$

where $(x, y)=(3, 1)$ when $t=0$

Find expressions for x and y in terms of t

4 A system of differential equations is given by

$$\frac{dx}{dt}=-3x-2y+t \qquad (1)$$

$$\frac{dy}{dt}=2x+y+3t-1 \qquad (2)$$

where $(x, y)=(8, -11)$ when $t=0$

Find expressions for x and y in terms of t

5 A system of differential equations is given by

$$\frac{dx}{dt}=x+2y \qquad (1)$$

$$\frac{dy}{dt}=y-z \qquad (2)$$

$$\frac{dz}{dt}=-x \qquad (3)$$

At $t=0$, $x=0$, $\frac{dx}{dt}=4$ and $\frac{d^2x}{dt^2}=5$

a Show that $\frac{d^3x}{dt^3}-2\frac{d^2x}{dt^2}+\frac{dx}{dt}-2x=0$

b Solve this equation to find an expression for x in terms of t

PURE

Reasoning and problem-solving

Strategy

To solve problems involving a coupled system

1. Define any variables that you need.
2. Use the information in the question to set up the coupled differential equations.
3. Differentiate one of the equations and use the original equations to eliminate the other variable to obtain a second order differential equation.
4. Solve the second order differential equation and hence solve for both variables.
5. Interpret your solution in the context of the question.

Example 2

In a chemical reaction, substance X decays into substance Y, which itself decays.

The rate of decay of X, in grams per hour, is given by twice the amount of substance Y, in grams.

The rate of change of Y, in grams per hour, is given by the amount of substance X, in grams, minus three times the amount of substance Y, in grams.

a Set up two differential equations for x and y, the amounts in grams, of substances X and Y respectively.

b Given that initially $x = 20$ and $y = 0$, solve for x and y at time t hours.

c Prove that there can never be equal amounts of X and Y

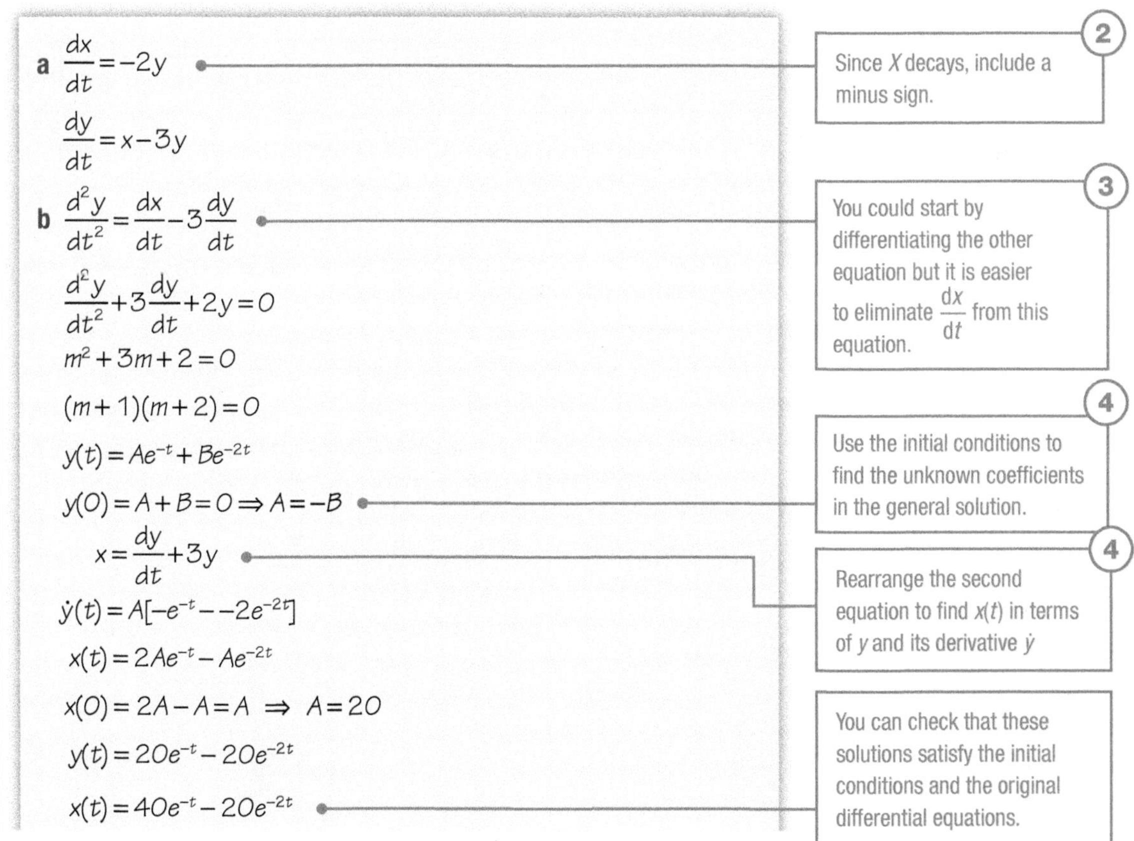

a $\frac{dx}{dt} = -2y$
$\frac{dy}{dt} = x - 3y$
b $\frac{d^2y}{dt^2} = \frac{dx}{dt} - 3\frac{dy}{dt}$
$\frac{d^2y}{dt^2} + 3\frac{dy}{dt} + 2y = 0$
$m^2 + 3m + 2 = 0$
$(m+1)(m+2) = 0$
$y(t) = Ae^{-t} + Be^{-2t}$
$y(0) = A + B = 0 \Rightarrow A = -B$
$x = \frac{dy}{dt} + 3y$
$\dot{y}(t) = A[-e^{-t} - -2e^{-2t}]$
$x(t) = 2Ae^{-t} - Ae^{-2t}$
$x(0) = 2A - A = A \Rightarrow A = 20$
$y(t) = 20e^{-t} - 20e^{-2t}$
$x(t) = 40e^{-t} - 20e^{-2t}$
2 Since X decays, include a minus sign.
3 You could start by differentiating the other equation but it is easier to eliminate $\frac{dx}{dt}$ from this equation.
4 Use the initial conditions to find the unknown coefficients in the general solution.
4 Rearrange the second equation to find $x(t)$ in terms of y and its derivative $\dot{y}$
You can check that these solutions satisfy the initial conditions and the original differential equations.

(Continued on the next page)

c Suppose that at time t there are equal amounts of X and Y

$20e^{-t} - 20e^{-2t} = 40e^{-t} - 20e^{-2t}$

$20e^{-t} = 0$

But, $e^{-t} > 0$ for all t

$\therefore$ there can never be equal amounts of X and Y

5 Answer the question in context.

Example 3

An isolated island supports populations of sparrowhawks and finches.

- The number of sparrowhawks increases at a rate proportional to the number of finches. When there are 64 finches present, the rate of increase of sparrowhawks is 16 per year.
- If there are no sparrowhawks present, then the finch population would increase by 120% per year.
- If there are sparrowhawks present, then, on average, each sparrowhawk kills 1.44 finches per year.

a Set up two differential equations to model the population of sparrowhawks and finches.

Initially there are 75 sparrowhawks and 120 finches.

b Find the number of sparrowhawks and finches t years later.

c What happens to the populations of finches and sparrowhawks in the distant future?

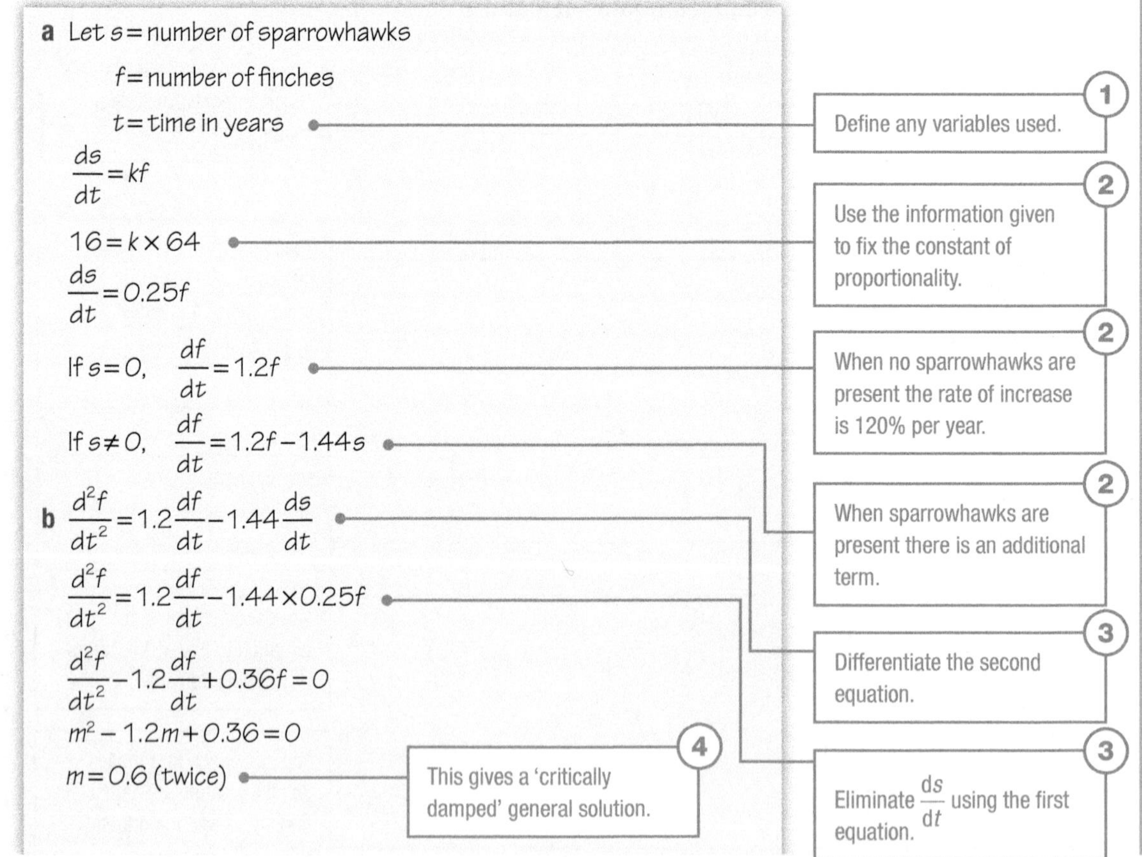

a Let s = number of sparrowhawks

f = number of finches

t = time in years

$\frac{ds}{dt} = kf$

$16 = k \times 64$

$\frac{ds}{dt} = 0.25f$

If $s = 0$, $\frac{df}{dt} = 1.2f$

If $s \neq 0$, $\frac{df}{dt} = 1.2f - 1.44s$

b $\frac{d^2f}{dt^2} = 1.2\frac{df}{dt} - 1.44\frac{ds}{dt}$

$\frac{d^2f}{dt^2} = 1.2\frac{df}{dt} - 1.44 \times 0.25f$

$\frac{d^2f}{dt^2} - 1.2\frac{df}{dt} + 0.36f = 0$

$m^2 - 1.2m + 0.36 = 0$

$m = 0.6$ (twice)

1 Define any variables used.
2 Use the information given to fix the constant of proportionality.
2 When no sparrowhawks are present the rate of increase is 120% per year.
2 When sparrowhawks are present there is an additional term.
3 Differentiate the second equation.
3 Eliminate $\frac{ds}{dt}$ using the first equation.
4 This gives a 'critically damped' general solution.

(*Continued on the next page*)

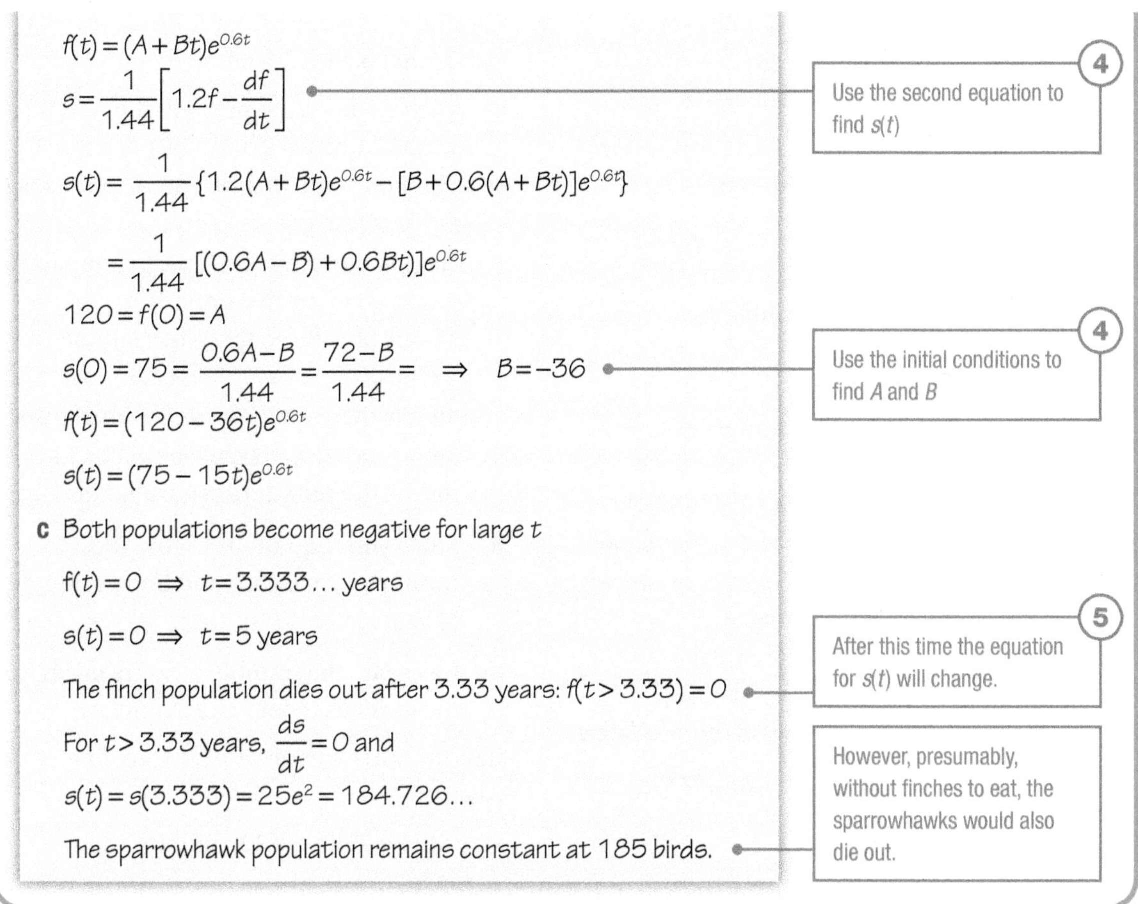

$f(t) = (A + Bt)e^{0.6t}$

$s = \frac{1}{1.44}\left[1.2f - \frac{df}{dt}\right]$

$s(t) = \frac{1}{1.44}\{1.2(A + Bt)e^{0.6t} - [B + 0.6(A + Bt)]e^{0.6t}\}$

$= \frac{1}{1.44}[(0.6A - B) + 0.6Bt)]e^{0.6t}$

$120 = f(0) = A$

$s(0) = 75 = \frac{0.6A - B}{1.44} = \frac{72 - B}{1.44} = \Rightarrow B = -36$

$f(t) = (120 - 36t)e^{0.6t}$

$s(t) = (75 - 15t)e^{0.6t}$

c Both populations become negative for large t

$f(t) = 0 \Rightarrow t = 3.333...$ years

$s(t) = 0 \Rightarrow t = 5$ years

The finch population dies out after 3.33 years: $f(t > 3.33) = 0$

For $t > 3.33$ years, $\frac{ds}{dt} = 0$ and

$s(t) = s(3.333) = 25e^2 = 184.726...$

The sparrowhawk population remains constant at 185 birds.

4 Use the second equation to find $s(t)$

4 Use the initial conditions to find A and B

5 After this time the equation for $s(t)$ will change.

However, presumably, without finches to eat, the sparrowhawks would also die out.

Exercise 20.5B Reasoning and problem-solving

1 In a chemical reaction, substance X changes into substance Y which evaporates away.

- The rate of disappearance of X, in grams per hour, is given by three times the amount of substance Y, in grams.
- The rate of change of Y, in grams per hour, is given by the amount of substance X, in grams, minus four times the amount of substance Y, in grams.

a Set up two differential equations for x and y, the amounts, in grams, of substances X and Y respectively.

b Given that initially $x = 40$ and $y = 0$, solve for x and y at time t hours.

c Find the time at which there is precisely four times as much of substance X as there is of substance Y

2 Two declining populations of mollusc species, X and Y, are competing for supremacy.

- The rate of change of the numbers of species X, in thousands per year, is given by three times the amount the numbers of species Y, in thousands, minus twice the number of species X, in thousands.
- The rate of change of the numbers of species Y, in thousands per year, is given by the number of species X, in thousands, minus four times the number of species Y, in thousands.

a Set up two differential equations to model the numbers of molluscs of species X and Y

b Given that initially there are 7000 molluscs of species X and 5000 molluscs of species Y, find the number of molluscs of species X and Y a number of years later.

c Show that the number of molluscs of species X will always be greater than the number of molluscs of species Y

3 Two species of insect, X and Y, compete for survival in the same environment.

- The rate of change of the number of species X, in millions per month, is given by three times the difference in the numbers of species X and species Y, measured in millions.
- The rate of change of the number of species Y, in millions per month, is given by five times the number of species Y, in millions, minus the numbers of species X, in millions.
- Initially there are 5×10^7 insects of species X and 3×10^7 insects of species Y

a Set up and solve two differential equations for the numbers of insects of species X and Y a number of months later.

b Which species becomes extinct? When does this happen?

4 A chemical reaction occurs between two liquids, A and B

- The rate of change of liquid A, in litres per minute, is given by the amount of liquid B, measured in litres, minus four times the amount of liquid A, measured in litres.
- In addition, liquid A is added at the rate of 7 litres per minute.
- The rate of change of liquid B, in litres per minute, is given by six times the amount of liquid A, measured in litres, minus five times the amount of liquid B, measured in litres.

a Set up two differential equations to model the amounts of liquids A and B

b Given that initially there are 3 litres of liquid A and 14 litres of liquid B, find the amounts of A and B a number of minutes later.

c Show that as the reaction progresses, the ratio of the amount of liquid A to the amount of liquid B tends to a constant ratio. Find this ratio.

5 Gazelles and lions are fighting for survival on an enclosed grass plain.

- When no lions are present, the number of gazelles increases by 100% per year.
- When lions are present, on average, each lion eats one gazelle per year.
- The rate of change of lions, in animals per year, is given by five times the number of lions plus three times the number of gazelles.
- Initially there are 120 gazelles and 80 lions.

a Set up two differential equations to model the numbers of gazelles and lions.

b Solve your equations to find the numbers of gazelles and lions a number of years later.

c Calculate the number of years taken for the gazelles to become extinct.

6 Rabbits and foxes are introduced on to an island.

- Initially there are 100 rabbits and 160 foxes.
- When no foxes are present, the number of rabbits increases by 600% per year.
- When foxes are present, on average, each fox eats one rabbit per year.
- The rate of change of foxes, in animals per year, is given by twice the number of foxes plus three times the number of rabbits.

a Set up and solve two differential equations for the numbers of rabbits and foxes a number of years later.

b Show that, over time, there will be approximately equal numbers of rabbits and foxes.

7 A forest contains bamboo plants and pandas. The pandas like to eat bamboo.

- The rate of change of bamboo plants is proportional to the number of pandas present. Three pandas eat one bamboo plant per year.
- The panda population increases by 20% per year plus 3% of the number of bamboo plants present.

- Initially there are 100 pandas and 2000 bamboo plants.

How many years pass before there are no bamboo plants left in the forest?

8 A small colony of bears feed on fish in a lake.

- When no bears are present, the number of fish would increase at a rate of 20% per year.
- When bears are present, on average, each bear eats a fifth of a fish per year.
- The rate of increase of bears per year is equal to 40% of the number of bears plus 10% of the number of fish.
- Initially there are 4 bears in the colony and 1000 fish in the lake.

Let x represent the number of bears, y represent the number of fish and t represent the time that has passed in years.

a Show that
$$\frac{d^2y}{dt^2}-0.6\frac{dy}{dt}-0.1y=0$$

b Find expressions for x and y in terms of t

c Using the model, how many years pass before there are no fish in the lake?

d Give one criticism of this population model.

9 An Alaskan forest is populated by Canadian lynx and snowshoe hares.

- The rate of increase of lynx per year equals 5% of the number of hares present.
- When no lynx are present, the number of hares would increases at a rate of 120% per year.
- When lynx are present, on average, each lynx eats 1.15 hares per year.
- Initially there are 60 lynx and 500 hares.

a If L represents the number of lynx, H represents the number of hares and t represents the time that has passed in years, explain why
$$\frac{dH}{dt}=1.2H-1.15L$$

b Show that $\frac{d^2H}{dt^2}-1.2\frac{dH}{dt}+0.0575H=0$

c Find expressions for the number of lynx and the number of hares at time t years.

d Give one criticism of this model.

10 An African nature reserve is populated by lions and zebra.

- The rate of increase of lions per year equals 20% of the number of zebra present.
- When no lions are present, the number of zebra would increase at a rate of 130% per year.
- When lions are present, on average, each lion kills 1.1 zebra per year.
- Initially there are 100 lions and 1000 zebra.

a If L represents the number of lions, Z represents the number of zebra and t represents the time that has passed in years, show that
$$\frac{d^2Z}{dt^2}-1.3\frac{dZ}{dt}+0.22Z=0$$

b Find expressions for the number of lions and the number of zebras at time t years.

c Show that, for large values of t, the ratio number of lions : number of zebra $=2:11$

11 A radioactive element X decays into a second radioactive element Y which in turn decays into the stable element Z

- The rate of decay of X equals 0.2 times the amount of X present.
- The rate of decay of Y equals 0.1 times the amount of Y present.
- Initially, there are 100 milligrams of X present and no Y or Z

This process can be modelled by the equations
$$\frac{dx}{dt}=-0.2x \qquad \frac{dy}{dt}=0.2x-0.1y \qquad \frac{dz}{dt}=0.1y$$
where x, y and z are the masses of elements X, Y and Z respectively, measured in milligrams, and t is the time, measured in seconds.

a Explain why the equations take these forms.

b Find expressions for x, y and z at time t

c Prove that $x+y+z=100$ for all t

20 Summary and review

Chapter summary

- An equation that involves only a first-order derivative, such as $\frac{dy}{dx}$, is called a **first order differential equation.**
- An equation of the form $\frac{dy}{dx}=F(x)G(y)$ is solved by **separating the variables** to get $\int F(x)dx=\int\frac{1}{G(y)}dy$
- An equation of the form $\frac{dy}{dx}+P(x)y=f(x)$ is solved by multiplying throughout by an **integrating factor** $e^{\int P(x)dx}$
- A solution with constant(s) is called a **general solution.** Substitute given value(s) to get a **particular solution.**
- An equation that involves a second order derivative, such as $\frac{d^2y}{dx^2}$, but nothing of higher order, is called a **second order differential equation.**
- To solve an equation of the form $a\frac{d^2y}{dx^2}+b\frac{dy}{dx}+cy=f(x)$
    - First solve the **auxiliary equation** $am^2+bm+c=0$. There are three cases for the **complementary function** that come from the solution of the auxiliary equation.
    - Next find the **particular integral.**
    - To get the **general solution,** add together the complementary function and the particular integral.
- Differential equations that include two or more variables linked to a common additional variable are called **coupled equations.**
- To solve a system of coupled equations, use differentiation to change the differential equations into a single equation of higher order connecting just two variables.

Check and review

You should now be able to...	Review Questions
✔ Solve a first order differential equation using an integrating factor.	1
✔ Solve a first order differential equation and interpret the solution in context.	2
✔ Set up a first order differential equation from a model and solve it.	3, 4
✔ Solve a second order differential equation of the form $a\frac{d^2y}{dx^2}+b\frac{dy}{dx}+cy=f(x)$	5
✔ Solve a second order differential equation and interpret the solution in context.	6
✔ Set up a second order differential equation from a model and solve it.	7
✔ Set up and solve a system of coupled equations.	8

1 Find the solution to the differential equation $\frac{dy}{dx}+y\cot x=\cos^3 x$ for which $y=1$ at $x=\frac{\pi}{2}$, giving your answer in the form $y=f(x)$

2 As part of an industrial process, salt is dissolved in a liquid. The mass, S kg, of the salt dissolved, t minutes after the process begins, is modelled by the equation

$$\frac{dS}{dt}+\frac{2S}{300-t}=\frac{1}{2}, 0\le t<300$$

Given that $S=0$ when $t=0$,

a Find S in terms of t

b Calculate the maximum mass of salt that the model predicts will be dissolved in the liquid.

3 At time t minutes, the rate of change of temperature of a cooling liquid is proportional to the temperature, T°C, of that liquid at that time. Initially $T=100$

a Show that $T=100e^{-kt}$, where k is a positive constant.

Given also that $T=25$ when $t=6$,

b Show that $k=\frac{1}{3}\ln 2$

c Calculate the time at which the temperature of this liquid will reach 5 °C.

4 A woman and her bicycle have a combined mass of 120 kg. The woman starts from rest, exerting a constant force of 600 N. When her velocity is v m s^{-1} her motion is subject to a resistive force of $6v^2$ N.

a Show that $100-v^2=20\frac{dv}{dt}$

b Calculate the time taken by the woman to reach a speed of 9 m s^{-1} from rest.

c Calculate the distance travelled by the woman in reaching a speed of 9 m s^{-1} from rest.

5 For the differential equation

$$\frac{d^2y}{dx^2}+4\frac{dy}{dx}+13y=90e^{4x}$$

a Find the general solution,

b Find the particular solution for which $y(0)=6$ and $y'(0)=-3$

6

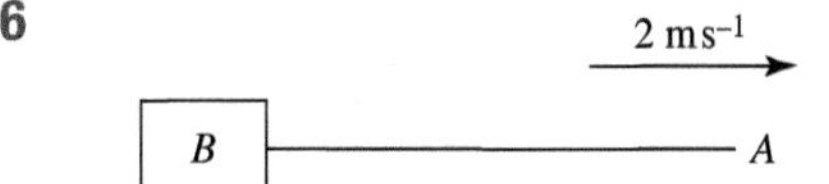

A box, B, is attached to one end of a light elastic string. The box is initially at rest on a smooth horizontal table, when the other end of the string, A, starts to move away from B with constant velocity 2 m s^{-1}. The displacement, y metres, of B from its initial position, at a time t seconds from the start, satisfies the differential equation

$$\frac{d^2y}{dt^2}+16y=8t$$

a Find an expression for y in terms of t

b Calculate the time when B is next at rest.

c Sketch the graph of y against t

7 A particle, P, of mass 4 kg moves along the horizontal x-axis under the action of a force directed towards the origin. At time t seconds, when the displacement of P from the origin is x metres, P is moving with a speed of v m s^{-1} and the force has magnitude $9x$ N. The particle is also subject to a resistive force of $12v$ N.

a Show that $4\frac{d^2x}{dt^2}+12\frac{dx}{dt}+9x=0$

Initially $x=6$ and $\frac{dx}{dt}=-11$

b Show that $x=e^{\frac{3t}{2}}(6-2t)$

c Write down the time when P passes through the origin.

d Find the speed of P at that time.

8 A particle moves in the plane such that at time $t\ge 0$, it has coordinates (x, y)

- The rate of change of x equals the difference in the x- and y-coordinates minus three times the elapsed time.
- The rate of change of y equals the height of the particle above the line $y=4x+3$
- The particle is initially at (2, 4)

a Show that

$$\frac{d^2x}{dt^2}-2\frac{dx}{dt}-3x=3t$$

b Find the position of the particle at time t

20 Exploration

Going beyond the exams

Research

Bungee jumping was first attempted in 1979, by students jumping off the Clifton suspension bridge in Bristol. Their inspiration came from vine jumping, as practised by the islanders of Vanuatu in the Pacific. The bungee cords used in Northern and Southern hemispheres tend to have different construction, with the Northern hemisphere cords giving a harder, sharper bounce than those in the southern hemisphere.

Try modelling the motion of a bungee jumper using differential equations where appropriate. Research the sorts of values you might use in this motion, including different values for the elasticity of the ropes in the different hemispheres.

Investigation

The diagram shows a potential situation that can be developed as a water feature.

Water flows into the top tank and, once it reaches the holes in the tank, it flows into the middle tank. As the water in the middle tank fills and reaches the holes in this tank, it then flows into the lower tank, which starts to fill. Water continues to fill the lower tank until it overflows.

Investigate this situation using differential equations. Develop your own mathematical model. Assume that that the rate of flow of water out of a tank depends on the volume of water in the tank. In other words, $\frac{dV}{dt} = g(V)$.

Start your investigation with a function such as $g(V) = -kV$

Try different functions to develop a design that would be attractive.

Research

Engineers use differential equations when designing a rocket to launch a satellite into space. These equations are derived from Newton's Laws of motion. Important factors that had to be considered when the differential equations were developed included the changing mass of the rocket and the change in the pull of gravity and air resistance as the rocket gets further away from the Earth. Research the use of differential equations to model the motion of a rocket.

20 Assessment

1 Find the general solution to the differential equation $\frac{dy}{dx} = e^{x+y}\cos x$ **[5 marks]**

2 Solve the differential equation $\frac{1}{\ln x}\frac{dy}{dx} + e^{y} = 0$ given that when $x = 1$, $y = 0$ **[6]**

3 Given that $y\frac{dy}{dx} - x\ln x = 0$

a Find the general solution to the equation, **[4]**

b Find the particular solution if when $x = 1$, $y = 4$ **[2]**

4 The number of sparrows nesting in a wooded area is thought to be declining. The rate at which the population is changing is known to depend on the population, *P*, at any given time and on the space available to build nests (200 – *P*)

An initial model predicts that $\frac{dP}{dt} = kP + (200 - P)$, where *k* is a constant of proportionality and *t* is the number of years.

Initially, there were 550 birds in the area.

a Taking the value of *k* to be 5, find the general solution to the differential equation, expressing *P* as an explicit function of *t* **[6]**

b Use the initial conditions to find the particular solution. **[2]**

c What does the model predict will happen to the number of sparrows in the wood after a number of years? **[2]**

d Comment on the suitability of the model for predicting the future population of sparrows. **[1]**

5 Find the general solution to the differential equation $x\frac{dy}{dx} + 2y = x^2$ **[4]**

6 Solve the differential equation $\frac{dy}{dx} - y\tan x = 4x^2$ given that when $x = 0$, $y = 1$ **[6]**

7 Given that $2x\frac{dy}{dx} + y = \ln x$

a Find the general solution to the equation. **[5]**

b Find the particular solution if when $x = e^2$, $y = e^{-1}$ **[2]**

8 A family of differential equations takes the form

$$2\frac{d^2y}{dx^2}+8\frac{dy}{dx}+ky=0 \text{ where } k \text{ is a constant.}$$

Find the general solution to the equation when

a $k=6$ **[4]** b $k=8$ **[4]** c $k=10$ **[4]**

9 Find the particular solution of $2\frac{d^2y}{dx^2}+3\frac{dy}{dx}-2y=0$ given that when $x=0$, $y=5$ and $\frac{dy}{dx}=-5$ **[7]**

10 Find the particular solution of $9\frac{d^2y}{dx^2}-6\frac{dy}{dx}+y=0$ given that when $x=0$, $y=6$ and $\frac{dy}{dx}=4$ **[7]**

11 Find the particular solution of $\frac{d^2y}{dx^2}+2\frac{dy}{dx}+5y=0$ given that when $x=0$, $y=10$ and $\frac{dy}{dx}=2$ **[7]**

12 Find the general solution of $\frac{d^2y}{dx^2}+\frac{dy}{dx}-12y=12x+1$ **[7]**

13 Find the particular solution of $\frac{d^2y}{dx^2}-4\frac{dy}{dx}+13y=40\sin x$ given that when $x=0$, $y=5$ and $\frac{dy}{dx}=-1$ **[7]**

14 The force acting on a compressed spring is given by the equation $m\frac{d^2x}{dt^2}=-\frac{7}{2}\frac{dx}{dt}-(5x-6e^{-t})$, where m kg is the mass of the spring and x m is the displacement from its initial position at time t seconds.

a If $m=0.5$ kg, show that the equation can be written as $\frac{d^2x}{dt^2}+7\frac{dx}{dt}+10x=12e^{-t}$ **[2]**

b Find the particular solution of this equation, given that when $t=0$, $x=0$ and $\frac{dx}{dt}=2$ **[11]**

c Comment on the nature of the damping of the spring. Give a reason for your answer. **[2]**

d Sketch a graph of x against t **[2]**

15 The motion of an air particle when a musical note is played by a wind instrument can be modelled by the equation $\frac{d^2x}{dt^2}+9x-8\sin t=0$, where x m is the displacement of the particle from its rest position.

a Given that when $t=\frac{\pi}{2}$, $x=0$ and $\frac{dx}{dt}=0$, find the particular solution to the differential equation. **[9]**

b Use the formula $\sin A+\sin B=2\sin\left(\frac{A+B}{2}\right)\cos\left(\frac{A-B}{2}\right)$ to rewrite your answer to part **a** and find the times at which the displacement of the particle from its rest position is zero. **[4]**

16 Money placed in a savings account will grow in direct proportion to the amount of money in the bank.

Initially £1000 is placed in the account. At the end of year 1, there is £1005 in the account. Let £A represent the amount after a time t years.

a Form a differential equation to model the situation. **[1]**

b Find the particular solution to the equation. **[6]**

c How much will be in the bank after 5 years? **[1]**

17 The voltage in a circuit containing a resistor and an inductor connected in series with a power supply is given by the equation

$$V=RI+L\frac{dI}{dt}$$

where I amps is the current, V volts is the voltage of the power supply, C farads is the capacitance of the capacitor and R ohms is the resistance.

a Show that the equation can be written as $\frac{dI}{dt}+\frac{RI}{L}=\frac{V}{L}$ **[1]**

b Given that $I=0$ when $t=0$, solve the differential equation to find I as a function of t **[9]**

If $R=50$ ohms, $V=100$ V and $L=10$ H

c Find the value of I when $t=0.1$ seconds. **[2]**

d What happens to the current as t becomes large? **[2]**

e Sketch a graph of I against t **[2]**

18 A particle moves so that at time t seconds, it is x units from the origin.

Its motion is modelled by $100\frac{d^2x}{dt^2}=-25x$

Initially $x=0$ when $t=0$. When $t=\pi$, $x=4$

a Find the particular solution of the equation. **[6]**

b Find the value of x when $t=\frac{\pi}{3}$ **[1]**

19 A damped oscillating system is modelled by the differential equation

$\frac{d^2x}{dt^2}+0.3\frac{dx}{dt}+0.15^2x=0$

a Explain why the damping will be critical. **[3]**

b Given that $x=0.5$ when $t=0$ and $\frac{dx}{dt}=0.425$ when $t=0$, solve the differential equation. **[8]**

c Sketch a graph of x against t **[2]**

20 A particle of mass m attached to a spring is subject to a damping force proportional to its velocity given by $-5m\frac{dx}{dt}$ and a tension in the spring given by $-6mx$. A disturbing force $3m\sin 2t$ is applied to the particle.

a Show that $\frac{d^2x}{dt^2}+5\frac{dx}{dt}+6x=3\sin 2t$ **[2]**

b Find the complementary function of this differential equation. **[3]**

c Find the particular integral. **[8]**

d Find the particular solution given that $x=0$ when $t=0$ and $\frac{dx}{dt}=1$ when $t=0$ **[6]**

21 A reaction vessel contains two chemicals, X and Y

- The rate of increase of X, in grams per minute, is given by the amount of chemical X plus eight times the amount of chemical Y, both in grams.
- The rate of decrease of Y, in grams per minute, is given by the amount of chemical X plus five times the amount of chemical Y, both in grams.
- Chemical X is added to the reaction vessel at a rate of one gram per minute.
- Chemical Y is removed from the reaction vessel at a rate of four grams per minute.

a Set up two differential equations to describe the amounts of chemicals X and Y in the reaction vessel. **[5]**

b Initially the vessel contains one gram of Y and no X
Find the amounts of X and Y at later times. **[17]**

21 Numerical methods

Those working in computer animation need to understand differential equations. Differential equations are used to describe the motion of objects that are being animated. These equations become complex very quickly and so numerical methods are used to 'solve' them. High-powered computers process the equations rapidly and efficiently so that the animation is smooth and effective.

The mathematical and computing methods that underpin the development of computer animation have applications in many fields throughout science and technology. For example, they can be used to model, and therefore understand, the flow of liquids and gases in aero-engines and nuclear reactors. They can also help designers of cars ensure that their designs are aerodynamic.

Orientation

What you need to know	What you will learn	What this leads to
Maths Ch16 • Differential equations. **Maths Ch17** • Numerical integration.	• How to use Simpson's rule. • How to use Euler's method.	**Careers** • Computer animation. • Aerodynamic engineering.

21.1 Numerical integration

Fluency and skills

You find the area under a curve by integrating the equation of the curve between appropriate limits.

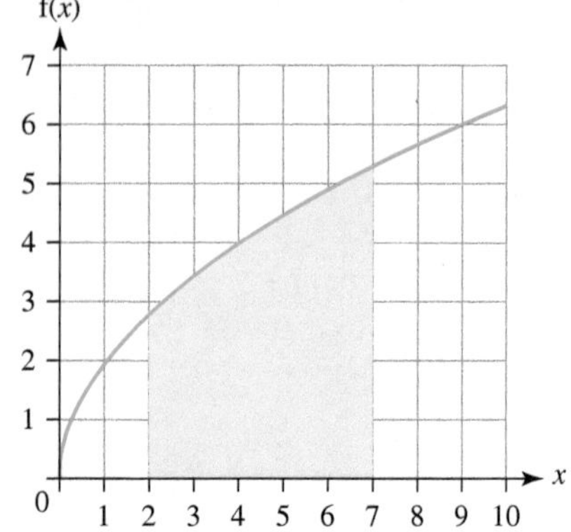

For example, in the top diagram the area in blue, under the function f(x), between the values $x = 2$ and $x = 7$, is found by calculating $\int_2^7 f(x)\,dx$

However, there are many functions which you cannot integrate using straightforward methods. Simple examples are $y = \cos\frac{1}{2}x$, $y = e^{x^2}$ and $y = e^{\frac{1}{x}}$

The same problem arises if the graph is from gathered data and so you do not know the equation of the graph.

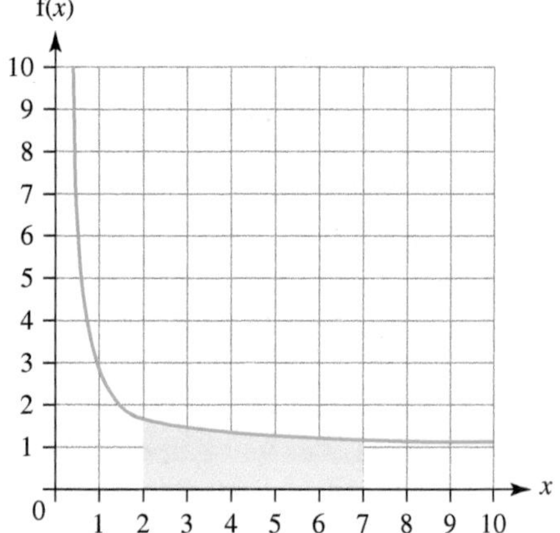

Even if the functions cannot be integrated, they can still be drawn, and so the area under the curve will exist. For example, the middle diagram shows the curve of $f(x) = e^{\frac{1}{x}}$ with the area beneath the curve between $x = 2$ and $x = 7$ shaded.

One way of finding an area like this is to use the **mid-ordinate rule**. Using this rule you divide the area required into vertical rectangles where the height of each rectangle is the y value (ordinate) in the middle of the rectangle. For example, in the function shown in the bottom diagram, the area from $x = 0$ to $x = 10$ is divided into 5 rectangles.

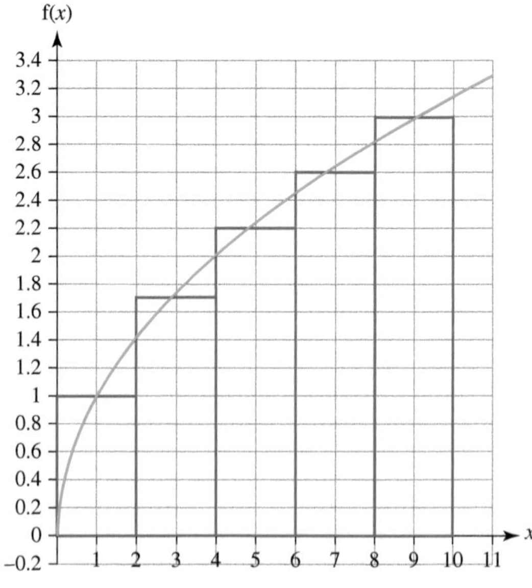

The sum of the areas of all the rectangles is an approximation to the area under the curve between $x = 0$ and $x = 10$, i.e. $\int_0^{10} f(x)\,dx$

At the top of each rectangle, a small area on the left, above the curve, is added into the area, but this is compensated for by the small area on the right, which is left out. This helps to minimise any errors. The sum of the area of the rectangles is often very close to the actual area under the curve. Estimating the values of the ordinates from the graph at $x = 1$, 3, 5, 7 and 9, we can sum the rectangle areas as

$(2 \times 1) + (2 \times 1.7) + (2 \times 2.2) + (2 \times 2.6) + (2 \times 3) = 21$

Key point

The **mid-ordinate rule** states that if you divide the area to be estimated under the function, from a to b, into n rectangles, then the width of each rectangle is $\frac{b-a}{n}$

If you then let the ordinates at the centre of each of these rectangles be $y_1, y_2, \ldots, y_n$, then the estimated area is given by the formula $A \left(\frac{b-a}{n}\right)(y_1+y_2+y_3+ \quad +y_n)$

The more rectangles you use the more accurate your estimate will be.

The function shown is $y=\sqrt{x}$ and, using calculus, $\int_0^{10}\sqrt{x}\,\mathrm{d}x = 21.1$ (3 s.f.).

In the example shown above, y_1, y_2, y_3, y_4 and y_n are the ordinates when $x = 1, 3, 5, 7$ and 9

Example 1

a Draw the graph of $y = 10x - x^2$

b Divide the area between the curve and the x-axis into 10 rectangles and draw them on your graph.

c Use your diagram to estimate the area between the curve and the x-axis.

d Use calculus to find the exact area between the curve and the x-axis.

e Calculate the relative error between your estimate and the exact area.

a

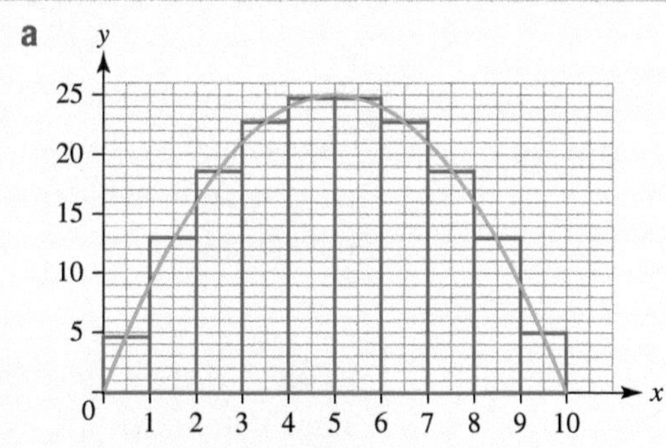

b The width of each rectangle = 1

c The sum of the required ordinates is

$(10\times0.5-0.5^2)+(10\times1.5-1.5^2)+(10\times2.5-2.5^2)$
$+(10\times3.5-3.5^2)+(10\times4.5-4.5^2)+(10\times5.5-5.5^2)$
$+(10\times6.5-6.5^2)+(10\times7.5-7.5^2)+(10\times8.5-8.5^2)$
$+(10\times9.5-9.5^2)$ — Use the mid-ordinate rule.

$= 12.75+18.75+22.75+24.75+22.75+18.7+12.75+4.75$

Hence, $A = 162.75$

d $\int_0^{10}(10x-x^2)\,dx = \left[5x^2 - \frac{x^3}{3}\right]_0^{10}$ — Integrate.

$= \left[5(10)^2 - \frac{(10)^3}{3}\right] - \left[5(0)^2 - \frac{(0)^3}{3}\right]$

$= 166\frac{2}{3}$ — Apply the limits and evaluate.

e Relative error $= \frac{\text{actual error}}{\text{true value}}$

$= \frac{162.75-166.67}{166.67}$

$= -0.0235$ (3 s.f.)

The mid-ordinate rule is a simple formula for estimating the area under a curve. Thomas Simpson, in 1743, published a rule which set out to give a more accurate answer. Simpson's rule minimises the discrepancies at the top of each rectangle by replacing the mid-ordinates with a quadratic curve.

Just as you can always draw a straight line between two points, you can always draw a quadratic curve between three points.

Take three points on this curve, which has equation $y = ax^2 + bx + c$

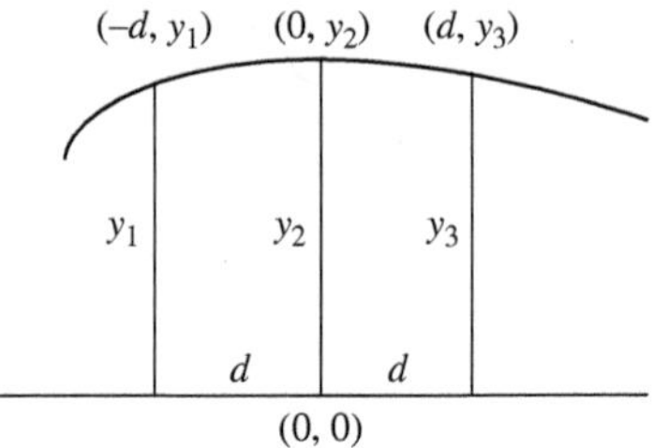

The area under the parabola is given by

$$A = \int_{-d}^{d} (ax^2 + bx + c)\,dx$$

$$= \left[\frac{ax^3}{3} + \frac{bx^2}{2} + cx\right]_{-d}^{d}$$

$$= \frac{2}{3}ad^3 + 2cd$$

Substituting for the three points, you get

$y_1 = ad^2 - bd + c$ (1)

$y_2 = c$ (2)

$y_3 = ad^2 + bd + c$ (3)

You don't have to find the actual equation of the parabola which passes through these three points because that is governed by the values y_1, y_2, y_3 and d which you have used.

From these equations, if you work out $(1) + 4 \times (2) + (3)$, you get

$y_1 + 4y_2 + y_3 = 2ad^2 + 6c$

Multiplying through by $\frac{d}{3}$ you get

$$\frac{d}{3}(y_1 + 4y_2 + y_3) = \frac{d}{3}(2ad^2 + 6c)$$

$$= \frac{2}{3}ad^3 + 2cd$$

Therefore the area under this parabola is $\frac{d}{3}(y_1 + 4y_2 + y_3)$

So an approximation for the area between a curve and the x-axis passing through the points $(-d, y_1)$, $(0, y_2)$ and (d, y_3) is given by the formula $\frac{d}{3}(y_1 + 4y_2 + y_3)$

This is known as **Simpson's rule**.

In this example there are 3 ordinates covering two equally spaced intervals. However, normally you will need the area under a curve for a larger range of ordinates.

The diagram shows the area under a curve divided into eight 'strips' or, more importantly, four **pairs** of strips.

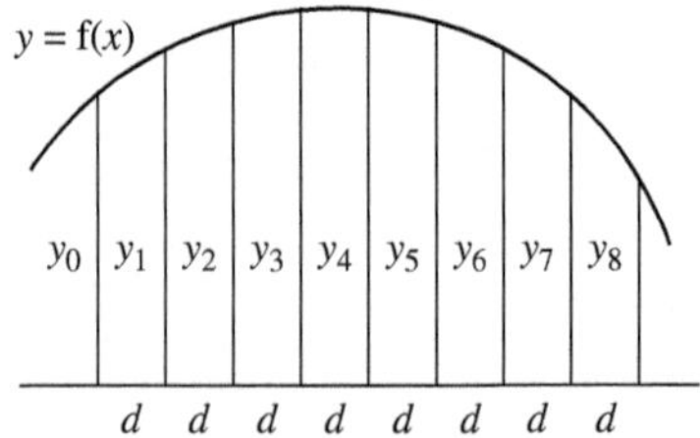

The area under the first two strips is approximated by $\frac{d}{3}(y_0 + 4y_1 + y_2)$

The area under the second two strips is approximated by $\frac{d}{3}(y_2 + 4y_3 + y_4)$

The area under the third two strips is approximated by $\frac{d}{3}(y_4 + 4y_5 + y_6)$

The area under the last two strips is approximated by $\frac{d}{3}(y_6 + 4y_7 + y_8)$

Therefore the area under the whole section shown is approximated by

$$A = \frac{d}{3}(y_0 + 4y_1 + y_2) + \frac{d}{3}(y_2 + 4y_3 + y_4) + \frac{d}{3}(y_4 + 4y_5 + y_6) + \frac{d}{3}(y_6 + 4y_7 + y_8)$$

$$= \frac{d}{3}(y_0 + 4y_1 + 2y_2 + 4y_3 + 2y_4 + 4y_5 + 2y_6 + 4y_7 + y_8)$$

$$= \frac{d}{3}[(y_0 + y_8) + 4(y_1 + y_3 + y_5 + y_7) + 2(y_2 + y_4 + y_6)]$$

This is typically much more accurate than trying to divide the area into just two strips.

You could remember this as $\frac{d}{3} \times$ [(sum of first and last ordinates) $+ 4 \times$ (sum of the odd ordinates) $+ 2 \times$ (sum of the even ordinates)]

Therefore a generalised version of Simpson's rule is

Key point

If the area required is divided into n strips of equal thickness, **where n is even,** then $\int_{x_0}^{x_n} \mathrm{f}(x)\,\mathrm{d}x \simeq \frac{d}{3}[(y_0 + y_n) + 4(y_1 + y_3 + y_5 + \dots + y_{n-1}) + 2(y_2 + y_4 + y_5 + \dots + y_{n-2})]$

PURE

Example 2

a Using Simpson's rule with ten strips, find the approximate area for the function and range given in Example 1

b Comment on your result.

a $d = 1$ — Define d

x	$10x - x^2$	y_x
0	$10(0) - (0)^2$	0
1	$10(1) - (1)^2$	9
2	$10(2) - (2)^2$	16
3	$10(3) - (3)^2$	21
4	$10(4) - (4)^2$	24
5	$10(5) - (5)^2$	25
6	$10(6) - (6)^2$	24
7	$10(7) - (7)^2$	21
8	$10(8) - (8)^2$	16
9	$10(9) - (9)^2$	9
10	$10(10) - (10)^2$	0

$$A = \frac{d}{3}[(y_0 + y_{10}) + 4(y_1 + y_3 + y_5 + y_7 + y_9) + 2(y_2 + y_4 + y_6 + y_8)]$$

$$= \frac{1}{3}[(0+0) + 4(9 + 21 + 25 + 21 + 9) + 2(16 + 24 + 24 + 16)]$$ — Substitute y values.

$$= \frac{1}{3}[(0) + (340) + (160)]$$

$$= 166\frac{2}{3}$$ — Evaluate the area.

Hence $\int_0^{10} (10x - x^2)\,dx \simeq 166\frac{2}{3}$

b The function given in Example 1 is, in fact, a quadratic, so the answer should be exact.

Exercise 21.1A Fluency and skills

1 Use the mid-ordinate rule with the number of rectangles stated to estimate each of these areas to 3 significant figures.

a $\int_1^{11} e^{\frac{1}{x}}\,dx$ with 10 rectangles

b $\int_0^{2\pi} \cos^2 x\,dx$ with 8 rectangles

c $\int_{\frac{\pi}{4}}^{2\pi} |\ln(x)\sin(x)|\,dx$ with 7 rectangles

d $\int_{-2}^{2} e^{x^2}\,dx$ with 8 rectangles

e $\int_1^{11} \left(\frac{x}{1+e^x}\right)dx$ with 10 rectangles

f $\int_{-\frac{\pi}{2}}^{\frac{\pi}{2}} \sqrt{\cos x}\,dx$ with 10 rectangles

g $\int_{-\pi}^{\pi} \left(\frac{\sin x}{x}\right)dx$ with 10 rectangles

h $\int_1^{5} \left(\frac{\ln x}{e^x}\right)dx$ with 8 rectangles

2 Use Simpson's rule with an appropriate number of strips to estimate these areas to 4 s.f.

a $\int_1^{11} e^{\frac{1}{x}}\,dx$ b $\int_2^{4} |\ln(x)\tan(x)|\,dx$

c $\int_{-2}^{2} e^{-x^2}\,dx$ d $\int_1^{4} \left(\frac{-x}{1-e^x}\right)dx$

e $\int_0^{\frac{\pi}{3}} \sqrt{\tan x}\,dx$ f $\int_{\frac{3\pi}{2}}^{\frac{5\pi}{2}} \left(\frac{\cos x}{x}\right)dx$

g $\int_1^{5} \left(\frac{\ln x}{e^{-x}}\right)dx$

3 Use the mid-ordinate rule with an appropriate number of rectangles to estimate these areas to 4 s.f. In each case, also find the exact answer by integration and hence calculate the relative error.

a $\int_8^{12} (x^2-6x)\,dx$ b $\int_0^{\pi} (\sin x)\,dx$

c $\int_1^{6} (-x^3+5x^2+8x-12)\,dx$ d $\int_1^{11} \ln(5x)\,dx$

4 Use Simpson's rule with an appropriate number of strips to estimate these areas to 4 s.f. In each case, also find the exact answer by integration and hence calculate the relative error.

a $\int_2^{4} \sec^2 x\,dx$ b $\int_2^{12} \left(\frac{5}{x}\right)dx$

c $\int_1^{7} \left(\frac{5}{x^{\frac{3}{2}}}\right)dx$ d $\int_0^{10} \left(\frac{16x}{x^2+1}\right)dx$

Reasoning and problem-solving

Strategy

To find an approximate area under a curve using Simpson's rule

1. Divide the required area into an appropriate even number of strips of equal width, d
2. Calculate the values of the ordinates for each x-value.
3. Complete the area approximation by substituting into the formula
$\frac{d}{3} \times$ [(sum of first and last ordinates) + 4 × (sum of odd ordinates) + 2 × (sum of even ordinates)]

Example 3

The cross-section of a vase is circular. The radius r cm, at any point h cm above the base, is shown in this table. The vase is 40 cm high.

h	0	5	10	15	20	25	30	35	40
r	6	5	11	13	15	13	11	5	6

Estimate the volume of the vase in litres to the nearest ml.

$$V \approx \frac{5}{3}[(\pi\times 6^2+\pi\times 6^2)+4(\pi\times 5^2+\pi\times 13^2+\pi\times 13^2+\pi\times 5^2)$$
$$+2(\pi\times 11^2+\pi\times 15^2+\pi\times 11^2)]$$
$$\approx \frac{5\pi}{3}[(72)+4(388)+2(467)]$$
$$\approx 13.394 \text{ litres}$$

3 Substitute into the formula. The volume of the vase is found by finding the area under a graph of circular cross-sectional area plotted against height.

Example 4

The equation of a circle, radius 12, is given by the formula $x^2+y^2=144$

Using Simpson's rule in the first quadrant with 12 equal intervals, find an estimate for π correct to 2 d.p.

Find also the relative error, to 1 s.f., of your estimate from the 2 d.p. value of π

$$x^2+y^2=144 \Rightarrow y^2=144-x^2, \text{ so } y=\sqrt{144-x^2} \text{ and } A=\int_0^{12}(\sqrt{144-x^2})dx$$

x	0	1	2	3	4	5	6	7	8	9	10	11	12
y	12	11.958	11.832	11.619	11.314	10.909	10.392	9.747	8.944	7.937	6.633	4.796	0

$$A \approx \frac{1}{3}[(12+0)+4(11.958+11.619+10.909+9.747+7.937+4.796)$$
$$+2(11.832+11.314+10.392+8.944+6.633)]$$
$$\approx \frac{1}{3}[(12)+(227.861)+(98.23)]$$
$$\approx \frac{1}{3}(338.091)$$
$$\approx 112.697$$

This area is an approximation to a quarter of the area of a circle.

$$\frac{1}{4}\times\pi\times 12^2 \approx 112.697$$
$$36\pi \approx 112.697$$
$$\pi \approx \frac{112.697}{36}$$
$$\approx 3.13$$

$$\text{Relative error} = \frac{3.13-3.14}{3.14} = -0.003$$

Exercise 21.1B Reasoning and problem-solving

1 The table shows the velocity of a particle over 9 seconds.

Time (s)	1	2	3	4	5	6	7	8	9
Velocity ($\mathrm{m\,s^{-1}}$)	4.8	19	44	78	122	176	240	313	397

Use Simpson's rule to calculate how far the particle has travelled over this time.

2 a Find the first three non-zero terms in the expansion of $\cosh x = \left(\frac{e^x + e^{-x}}{2}\right)$

b Use Simpson's rule with this expansion to estimate $\int_0^{0.6} \cosh x \, dx$ to 5 d.p.

3 a Write down the first four terms in the expansion of $\sin x$

b Use Simpson's rule with this expansion to estimate $\int_0^{1} \sin x \, dx$ to 4 d.p.

4 a Prove by integration that the area between the curve $y = \tan x$ and the coordinate axes between $x = 0$ and $x = \frac{\pi}{4}$ is $\ln\sqrt{2}$

b Use Simpson's rule with 8 strips to calculate an approximation to this value and find the relative error.

5 Use the binomial theorem to expand $(1+x^2)^{\frac{1}{3}}$ as far as the term in x^6

Hence find $\int_0^{0.6} (1+x^2)^{\frac{1}{3}} \, dx$ to 4 d.p.

a By integration,

b By using Simpson's rule with six intervals.

6 Use the mid-ordinate rule to estimate $\int_0^{10} \left(\sqrt[3]{1+\frac{x^3}{8}}\right) dx$ to 5 d.p.

7 a Use the mid-ordinate rule and Simpson's rule to estimate $\int_0^{1} \left(\frac{24x}{4+x^2}\right) dx$

b Find $\int_0^{1} \left(\frac{24x}{4+x^2}\right) dx$ by integration.

c Find the relative error for the calculation in part **a**.

8 $\int \frac{1}{x}\,dx = \ln(x)$. Use Simpson's rule with six strips between $x = 1$ and $x = 3$ to find an estimate for $\ln(3)$

9 The coordinates of the centre of mass of a lamina are found by using the formulae

$$\bar{x} = \frac{\int xy\,dx}{\int y\,dx},\ \bar{y} = \frac{\int \frac{y^2}{2}\,dx}{\int y\,dx}$$

A lamina in the shape of the quadrant of the ellipse $\frac{x^2}{16} + \frac{y^2}{9} = 1$ is formed enclosing the curve and the positive x- and y-axes. Use direct integration to calculate, between appropriate limits, $\int xy\,dx$ and $\int \frac{y^2}{2}\,dx$ and Simpson's rule with an appropriate number of strips to calculate $\int y\,dx$ and hence find the coordinates of the centre of mass of the lamina.

PURE

21.2 Euler's method

Fluency and skills

Up to this point, all the differential equations that you have met could be solved. However, almost all first order differential equations cannot be solved using calculus and algebra. When this is the case, you have to use numerical methods such as Euler's method to find approximate solutions to the differential equation.

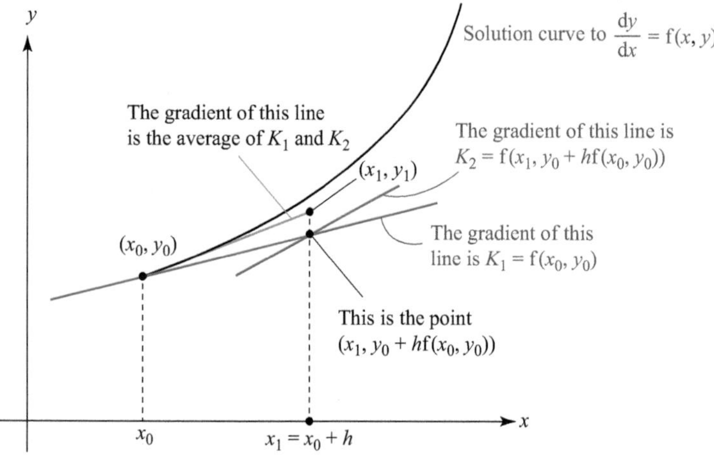

Assume you wish to solve the differential equation $\frac{dy}{dx} = f(x, y)$, where you know one point which lies on the curve, say $y = y_0$ when $x = x_0$

Euler's method enables you to estimate successive points (x_1, y_1), (x_2, y_2), (x_3, y_3), ... which lie approximately on the solution curve of the differential equation.

Since $\frac{dy}{dx} = f(x_0, y_0)$ at the point (x_0, y_0), then the value of $f(x_0, y_0)$ is the gradient of the tangent to the solution curve at that point (shown in red on the diagram). Therefore, if x increases by a small amount, h, then the y value of that straight line will increase by an amount $hf(x_0, y_0)$

This is because on a straight line the change in y is equal to the change in x times the gradient.

If h is small, the solution curve should stay close to the tangent line, and so the point $(x_1, y_1) = (x_0 + h, y_0 + hf(x_0, y_0))$ should be close to the correct solution.

The accuracy is improved the smaller the step length, h and when the start and end points are close.

Clearly, this process can be repeated, so successive approximations to points on the solution curve are given by $x_2 = x_1 + h$, $y_2 = y_1 + hf(x_1, y_1)$; $x_3 = x_2 + h$, $y_3 = y_2 + hf(x_2, y_2)$ and so on.

If $x_{r+1} = x_r + h$ then $y_{r+1} = y_r + hf(x_r, y_r)$ **Key point**

This is called Euler's method, named after the Swiss mathematician, Leonhard Euler (1707–1783).

Euler approximations are above the curve when it is concave and below when it is convex.

Example 1

PURE

a $\frac{dy}{dx} = 4x^3$

Taking (x_0, y_0) as (1, 1) and step length, h, as 0.5, use Euler's method to calculate the approximate value of the curve when $x = 2$

b Find the relative error of this value.

c Change the step length to $h = 0.1$ and find the new approximate value of the curve when $x = 2$ and the resulting relative error.

a

r	x_r	y_r
0	1	1
1	1.5	$1 + 0.5 \times 4 \times 1^3 = 3$
2	2	$3 + 0.5 \times 4 \times 1.5^3 = 9.75$

In each calculation, $y_{r+1} = y_r + hf(x_r, y_r)$.

The relative value required is 9.75

b $\frac{dy}{dx} = 4x^3$ so $F(x) = x^4$

The constant of integration equals 0 due to the point (1, 1) lying on the solution curve.

Hence, when $x = 2$, the accurate value is $2^4 = 16$

Relative error $= \frac{9.75 - 16}{16} = -0.390625$

c

r	x_r	y_r
0	1	1
1	1.1	$1 + 0.1 \times 4 \times 1^3 = 1.4$
2	1.2	$1.4 + 0.1 \times 4 \times 1.1^3 = 1.9324$
3	1.3	$1.9324 + 0.1 \times 4 \times 1.2^3 = 2.6236$
4	1.4	$2.6236 + 0.1 \times 4 \times 1.3^3 = 3.5024$
5	1.5	$3.5024 + 0.1 \times 4 \times 1.4^3 = 4.6$
6	1.6	$4.6 + 0.1 \times 4 \times 1.5^3 = 5.95$
7	1.7	$5.95 + 0.1 \times 4 \times 1.6^3 = 7.5884$
8	1.8	$7.5884 + 0.1 \times 4 \times 1.7^3 = 9.5536$
9	1.9	$9.5536 + 0.1 \times 4 \times 1.8^3 = 11.8864$
10	2.0	$11.8864 + 0.1 \times 4 \times 1.9^3 = 14.63$

The value required is 14.63

Relative error $= \frac{14.63 - 16}{16} = -0.085625$

Find out how to use a spreadsheet, your graphical calculator or computer software to do the calculations in part **c** quickly and simply.

Example 2

$\frac{dy}{dx} = x + 2y - 3xy$ and $y = 2$ when $x = 1$

Use Euler's method to estimate the value of y when $x = 1.5$ using a step length, h, of 0.1

Give your answer to 4 d.p.

r	x_r	y_r
0	1	2
1	1.1	$2 + 0.1 \times (1 + 2 \times 2 - 3 \times 1 \times 2) = 1.9$
2	1.2	$1.9 + 0.1 \times (1.1 + 2 \times 1.9 - 3 \times 1.1 \times 1.9) = 1.763$
3	1.3	$1.763 + 0.1 \times (1.2 + 2 \times 1.763 - 3 \times 1.2 \times 1.763)$ $= 1.60092$
4	1.4	$1.60092 + 0.1 \times (1.3 + 2 \times 1.60092 - 3 \times 1.3 \times 1.60092)$ $= 1.4267452$
5	1.5	$1.4267452 + 0.1 \times (1.4 + 2 \times 1.4267452 - 3 \times 1.4 \times 1.4267452)$ $= 1.252861256$

$y_5 = 1.2529$ (4 d.p.)

The improved Euler method uses the average of the gradients at the initial point (A) and at the end point of the step (B). As before, the curve has function $F(x)$ and the gradient function at any point is given by $f(x)$

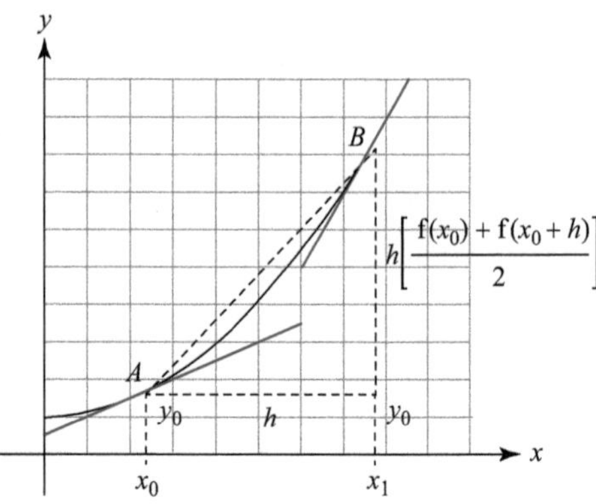

Let A be (x_0, y_0) and B be (x_1, y_1). Therefore, using Euler's method, $x_1 = x_0 + h$ and $y_1 = y_0 + hf(x_0, y_0)$, so, the improved method gives

$$y_1 = y_0 + \frac{hf(x_0, y_0) + hf((x_0 + h), (y_0 + hf(x_0, y_0)))}{2}$$

If you write k_1 as $hf(x_0, y_0)$ and k_2 as $hf(x_0 + h, y_0 + k_1)$, you get the improved Euler formula $y_1 = y_0 + \frac{1}{2}(k_1 + k_2)$. So, in general,

It is helpful to note that, in effect, working out k_2 is just using the ordinary Euler method on the right-hand boundary.

Key point

$$y_{r+1} = y_r + \frac{1}{2}(k_1 + k_2)$$

$k_1 = hf(x_0, y_0)$ and $k_2 = hf(x_0 + h, y_0 + k_1)$

Example 3

a Estimate the value of y when $x = 0.2$ for the differential equation $\frac{dy}{dx} = 3 + xy$ given that $y = 2$ when $x = 0$

b Use a spreadsheet or your graphical calculator or computer software, to do these calculations quickly and simply.

a $\frac{dy}{dx} = 3 + xy$ so $(x_0, y_0) = (0, 2)$ and $h = 0.1$

Step 1

$k_1 = 0.1(3 + (0)(2)) = 0.3$

$k_2 = 0.1f(0.1, 2 + k_1) = 0.1(3 + (0.1)(2 + 0.3))$

$k_2 = 0.1(3.23) = 0.323$

Hence $y_1 = 2 + \frac{1}{2}(0.3 + 0.323)$

$y_1 = 2.3115$

Step 2

$(x_1, y_1) = (0.1, 2.3115)$

$k_1 = 0.1(3 + (0.1)(2.3115)) = 0.323115$

$k_2 = 0.1f(0.2, 2.3115 + k_1) = 0.1f(0.2, 2.634615)$
$= 0.1(3 + (0.2)(2.634615))$

$k_2 = 0.1(3.526923) = 0.3526923$

Hence $y_2 = 2.3115 + \frac{1}{2}(0.323115 + 0.3526820)$

$y_2 = 2.64940365$

$= 2.649$ (3 d.p.).

b

$y_{r+1} = y_r + \frac{1}{2}(k_1 + k_2)$		$k_1 = hf(x_r, y_r)$	$k_2 = hf(x_r + h, y_r + k_1)$		$f(x) = 3 + xy$		$h = 0.1$
r	x	y (formula)	y	k (formula)	k	k (formula)	k
0	0		2				
1	0.1	(D3) + 0.5*((F4) + (H4))	2.3115	0.1*(3 + (B3)*(D3)	0.3	0.1*(3 + (B4)*((D3)+(F3))	0.323
2	0.2	(D4) + 0.5*((F5) + (H5))	2.649404	0.1*(3 + (B4)*(D4)	0.323115	0.1*(3 + (B5)*((D4)+(F4))	0.3526923
3	0.3	(D4) + 0.5*((F5) + (H6))	3.020934	0.1*(3 + (B5)*(D5)	0.352988073	0.1*(3 + (B6)*((D5)+(F5))	0.390071752
4	0.4	(D5) + 0.5*((F6) + (H7))	3.434479	0.1*(3 + (6)*(D6)	0.390628007	0.1*(3 + (B7)*((D6)+(F6))	0.436462463
5	0.5	(D6) + 0.5*((F7) + (H8))	3.899965	0.1*(3 + (B7)*(D7)	0.437379152	0.1*(3 + (B8)*((D7)+(F7))	0.493592897
6	0.6	(D7) + 0.5*((F8) + (H9))	4.429313	0.1*(3 + (B8)*(D8)	0.494998241	0.1*(3 + (B9)*((D8)+(F8))	0.563697784
7	0.7	(D8) + 0.5*((F9) + (H10))	5.03702	0.1*(3 + (B9)*(D9)	0.56575877	0.1*(3 + (B10)*((D9)+(F9))	0.649655012
8	0.8	(D9) + 0.5*((F10) + (H11))	5.7409	0.1*(3 + (B10)*(D10)	0.652591381	0.1*(3 + (B11)*((D10)+(F10))	0.755168888
9	0.9	(D10) + 0.5*((F11) + (H12))	6.563044	0.1*(3 + (B11)*(D11)	0.759271989	0.1*(3 + (B12)*((D11)+(F11))	0.885015466

This is a spreadsheet method for calculating these estimates. Since, after the initial 'programming,' the spreadsheet takes the drudgery out of the calculations, the table goes up to the value of y at $x = 0.9$

An alternative improved Euler method is sometimes referred to as the midpoint formula since it compares the gradient of the chord PQ on a curve F(x, y) with the gradient of the curve f(x, y) at the midpoint of the chord at A, (x_0, y_0).

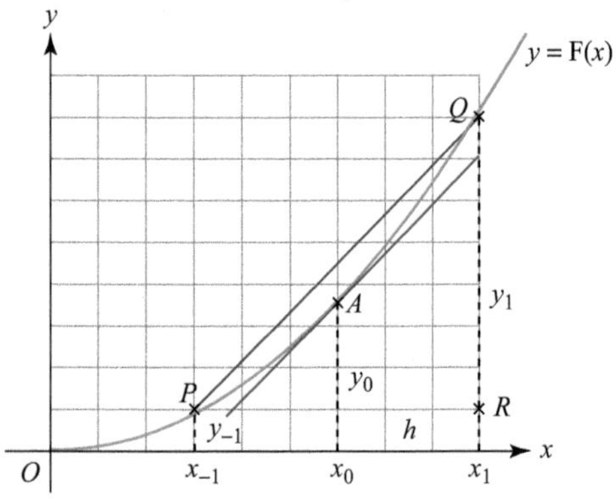

Most questions give you a starting point (x_0, y_0) and a gradient function f(x, y). This means you do not know (x_{-1}, y_{-1}). You will have to use the general Euler formula $y_1 = y_0 + h\mathrm{f}(x_0, y_0)$ to create (x_1, y_1). You can then use $y_{r+1} = y_{r-1} + 2h\mathrm{f}(x_r, y_r)$ to create $y_2 = y_0 + 2h\mathrm{f}(x_1, y_1)$ and so on.

Let P be (x_{-1}, y_{-1}) and let Q be (x_1, y_1) and let $x_0 - x_{-1} = x_1 - x_0 = h$

The gradient of the chord $PQ = \frac{QR}{PR} = \frac{y_1 - y_{-1}}{x_1 - x_{-1}} = \frac{y_1 - y_{-1}}{2h}$

This gradient is taken as an approximation to the gradient of the curve at A,

i.e. $\left(\frac{\mathrm{d}y}{\mathrm{d}x}\right)_0 = \mathrm{f}(x_0, y_0)$

Hence $\mathrm{f}(x_0, y_0) \approx \frac{y_1 - y_{-1}}{2h}$

Rearranging this equation gives $y_1 = y_{-1} + 2h\mathrm{f}(x_0, y_0)$, or in general

Key point

$$y_{r+1} = y_{r-1} + 2h\mathrm{f}(x_r, y_r)$$

Example 4

Use the midpoint method to estimate the value of y when $x = 0.3$ for the differential equation $\frac{\mathrm{d}y}{\mathrm{d}x} = 3 + xy$ given that $y = 2$ when $x = 0$

Use a step length of 0.1

$(x_0, y_0) = (0, 2)$ and $h = 0.1$

$y_1 = 2 + 0.1 \times 3 = 2.3$ — Use $y_1 = y_0 + h\mathrm{f}(x_0, y_0)$ to find y_1

$(x_1, y_1) = (0.1, 2.3)$

$y_2 = 2 + 0.2(3 + 0.1 \times 2.3) = 2.646$ — Use $y_2 = y_0 + 2h\mathrm{f}(x_1, y_1)$ to find y_2

$(x_2, y_2) = (0.2, 2.646)$

$y_3 = 2.3 + 0.2(3 + 0.2 \times 2.646) = 3.00584$

Thus an estimate for y when $x = 0.3$ is 3.01 (3 s.f.)

Exercise 21.2A Fluency and skills

In each of the questions use Euler's method when

1 $\frac{dy}{dx} = 2x + 3y$, $x_0 = 0$, $y_0 = 1$, $h = 0.1$ and the endpoint is at $y(0.1)$

2 $\frac{dy}{dx} = 3 + xy$, $x_0 = 1$, $y_0 = 1$, $h = 0.1$ and the endpoint is at $y(1.2)$

3 $\frac{dy}{dx} = x^2y$, $x_0 = 0$, $y_0 = 1$, $h = 0.1$ and the endpoint is at $y(0.3)$

4 $\frac{dy}{dx} = \cos x$, $x_0 = 0$, $y_0 = 0$, $h = 0.1$ and the endpoint is at $y(0.5)$

5 $\frac{dy}{dx} = \sqrt{x}$, $x_0 = 0$, $y_0 = 5$, $h = 1$ and the endpoint is at $y(10)$

Use the improved Euler method when

6 $\frac{dy}{dx} = \ln(\sqrt{x} + \sqrt{y})$, $x_0 = 1$, $y_0 = 2$, $h = 1$ and the endpoint is at $y(5)$

7 $\frac{dy}{dx} = \sin(x^2)$, $x_0 = 1$, $y_0 = 1$, $h = 0.1$ and the endpoint is at $y(1.4)$

8 $\frac{dy}{dx} = \sin^2 x$, $x_0 = 1$, $y_0 = 1$, $h = 0.1$ and the endpoint is at $y(1.4)$

9 $\frac{dy}{dx} = (x - y)e^{-x}$, $x_0 = 0$, $y_0 = 5$, $h = 1$ and the endpoint is at $y(5)$

10 $\frac{dy}{dx} = ye^{\tan x}$, $x_0 = 0$, $y_0 = 5$, $h = 0.25$ and the endpoint is at $y(1)$

Use the midpoint method on these functions:

11 $\frac{dy}{dx} = 2x - 3y$, $x_0 = 0$, $y_0 = 1$, $h = 0.1$ and the endpoint is at $y(0.3)$

12 $\frac{dy}{dx} = 3x + xy - 4$, $x_0 = 1$, $y_0 = 1$, $h = 0.1$ and the endpoint is at $y(1.3)$

13 $\frac{dy}{dx} = x^2y^2$, $x_0 = 0$, $y_0 = 1$, $h = 0.1$ and the endpoint is at $y(0.3)$

14 $\frac{dy}{dx} = \tan x$, $x_0 = 0$, $y_0 = 0$, $h = 0.1$ and the endpoint is at $y(0.5)$

15 $\frac{dy}{dx} = x^{-\frac{1}{2}}$, $x_0 = 1$, $y_0 = 5$, $h = 1$ and the endpoint is at $y(7)$

Reasoning and problem-solving

Strategy

To find an approximate solution to a differential equation

1. Identify the differentiated function $f(x, y)$, a starting point (x_0, y_0) and a suitable step value, h
2. Calculate $y_1 = y_0 + hf(x_0, y_0)$
3. Continue step 2 using $y_{r+1} = y_r + hf(x_r, y_r)$ until the desired endpoint is reached.
4. Answer the question in context.

Example 5

a Solve the equation $\frac{dy}{dx} = y$ where $y = 1$ when $x = 0$

b Use Euler's method with a step length of 0.1 to find, correct to 5 d.p., an approximate value of e.

c Find the relative error.

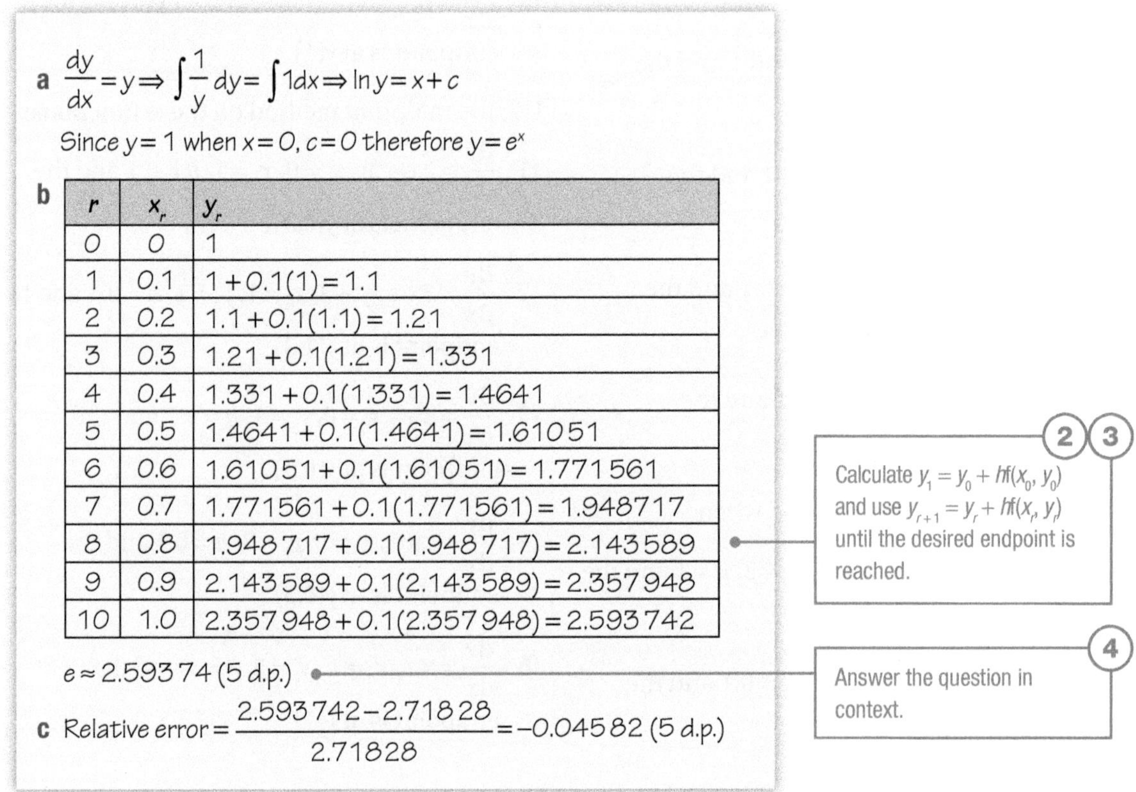

a $\frac{dy}{dx} = y \Rightarrow \int \frac{1}{y} dy = \int 1 dx \Rightarrow \ln y = x + c$

Since $y = 1$ when $x = 0$, $c = 0$ therefore $y = e^x$

b

r	x_r	y_r
0	0	1
1	0.1	$1 + 0.1(1) = 1.1$
2	0.2	$1.1 + 0.1(1.1) = 1.21$
3	0.3	$1.21 + 0.1(1.21) = 1.331$
4	0.4	$1.331 + 0.1(1.331) = 1.4641$
5	0.5	$1.4641 + 0.1(1.4641) = 1.61051$
6	0.6	$1.61051 + 0.1(1.61051) = 1.771561$
7	0.7	$1.771561 + 0.1(1.771561) = 1.948717$
8	0.8	$1.948717 + 0.1(1.948717) = 2.143589$
9	0.9	$2.143589 + 0.1(2.143589) = 2.357948$
10	1.0	$2.357948 + 0.1(2.357948) = 2.593742$

$e \approx 2.59374$ (5 d.p.)

c Relative error $= \frac{2.593742 - 2.71828}{2.71828} = -0.04582$ (5 d.p.)

2 3 Calculate $y_1 = y_0 + hf(x_0, y_0)$ and use $y_{r+1} = y_r + hf(x_r, y_r)$ until the desired endpoint is reached.

4 Answer the question in context.

Example 6

The rate of temperature decrease along a straight copper wire is modelled as $\frac{dT}{dx} = -\frac{1}{5000} - \frac{x}{100}$, where x (cm) is the distance from one end. $T = 100°$ when $x = 0$

a Use Euler's method to estimate the temperature when $x = 10$

Take a step length of 1

b Solve the original equation analytically and hence find the relative error in your estimate.

a

r	x_r	T_r
0	0	100
1	1	99.9998
2	2	99.9896
3	3	99.9694
4	4	99.9392

2 3 Calculate $y_1 = y_0 + hf(x_0, y_0)$ and use $y_{r+1} = y_r + hf(x_r, y_r)$ until the desired endpoint is reached.

(*Continued on the next page*)

5	5	99.899
6	6	99.8488
7	7	99.7886
8	8	99.7184
9	9	99.6382
10	0	99.548

The estimated temperature 10 cm from the end is 99.548°

4 Answer the question in context.

b $\frac{dT}{dx} = -\frac{1}{5000} - \frac{x}{100} \Rightarrow T = \int\left(-\frac{1}{5000} - \frac{x}{100}\right)dx$

$$= -\frac{x}{5000} - \frac{x^2}{200} + C$$

When $T = 100$, $x = 0$, hence $C = 100$

$$T = 100 - \frac{x}{5000} - \frac{x^2}{200}$$

When $x = 10$, $T = 99.498°$

$$\text{Relative error} = \frac{99.548 - 99.498}{99.498} = 0.000\,502\,5 \text{ (4 s.f.)}$$

Example 7

The number of cells in a bacterial culture is governed by the formula $n = 1000e^{0.2t}$, where t is in hours.

a How many cells are there initially in the culture?

b After how many hours, to the nearest hour, have the cells in the culture quadrupled?

c Use an improved Euler method to find an estimate for the number of cells after this time.

a Initially $t = 0$, so $n = 1000e^0 = 1000$

b $n = 1000e^{0.2t}$ so $\frac{n}{1000} = e^{0.2t}$ and hence

$$0.2t = \ln\left(\frac{n}{1000}\right) \text{ or } t = 5\ln\left(\frac{n}{1000}\right)$$

Thus, when $n = 4000$, $t = 5\ln 4 = 6.93$ hours.

Therefore the number of cells has quadrupled in 7 hours, to the nearest hour.

(*Continued on the next page*)

c If $n = 1000e^{0.2t}$, then $\frac{dn}{dt} = 200e^{0.2t}$

Hence using a starting point of (0, 1000) with a step length of 1, you get

1	1	(D3)+0.5*((F4)+(H4))	1222.140276	200*EXP(0.2*(B3))	200	200*(EXP(0.2*(B4)))	244.2805516
2	2	(D4)+0.5*((F5)+(H5))	1493.463021	200*EXP(0.2*(B4))	244.2805516	200*(EXP(0.2*(B5)))	298.3649395
3	3	(D4)+0.5*((F6)+(H6))	1824.857371	200*EXP(0.2*(B5))	298.3649395	200*(EXP(0.2*(B6)))	364.4237601
4	4	(D5)+0.5*((F7)+(H7))	2229.623344	200*EXP(0.2*(B6))	364.4237601	200*(EXP(0.2*(B7)))	445.1081857
5	5	(D6)+0.5*((F8)+(H8))	2724.00562	200*EXP(0.2*(B7))	445.1081857	200*(EXP(0.2*(B8)))	543.6563657
6	6	(D6)+0.5*((F9)+(H9))	3327.845495	200*EXP(0.2*(B8))	543.6563657	200*(EXP(0.2*(B9)))	664.0233845
7	7	(D6)+0.5*((F10)+(H10))	4065.377184	200*EXP(0.2*(B9))	664.0233845	200*(EXP(0.2*(B10)))	811.0399934
8	8	(D6)+0.5*((F11)+(H11))	4966.200423	200*EXP(0.2*(B10))	811.0399934	200*(EXP(0.2*(B11)))	990.6064849
9	9	(D6)+0.5*((F12)+(H12))	6066.468412	200*EXP(0.2*(B11))	990.6064849	200*(EXP(0.2*(B12)))	1209.929493
10	10	(D6)+0.5*((F13)+(H13))	7410.338768	200*EXP(0.2*(B12))	1209.929493	200*(EXP(0.2*(B13)))	1477.81122
11	11		9051.745728		1477.81122		1805.0027
12	12		11056.56472		1805.0027		2204.635276

The number of cells after 7 hours is approximately equal to 4065

Using formulae in a spreadsheet, you can see clearly that the number of cells has quadrupled in 7 hours.

Exercise 21.2B Reasoning and problem-solving

1 The vertical velocity of a hot-air balloon t s after some ballast has been ejected, is given by the approximate formula $\frac{ds}{dt} = (2t - 1)\ \text{m s}^{-1}$.

  a The balloon was 150 m above the ground when the ballast was ejected. Use a step size of 0.5 to estimate its height above the ground after 4 s.

  b Use integration to find the accurate height.

2 A body, falling freely from rest under gravity, encounters air resistance. The acceleration of the body is given approximately by the formula $\frac{dv}{dt} = 10 - 0.5v$. Using a step length of 0.1 estimate the velocity of the body after one second.

3 Water is steadily poured into a hemispherical bowl of radius 10 cm. When the height is h the rate of change of volume against height is given by the formula $\frac{dv}{dh} = \pi(20h - h^2)$. Using a step length of 0.5 estimate the volume when $h = 2$ cm.

4 An iron marble runs along a channel under the influence of a magnet. Its motion at time t s is given by the formula $\frac{ds}{dt} = 3t^2 - 10t + 18$ where $v = 0$ when $t = 0$

  a Use a step length of 1 to find its distance from the start when $t = 6$

  b Use integration to find the accurate distance.

5 The acceleration of a body is governed by the equation $\frac{dv}{dt}=\frac{4}{\sqrt{t}}$ where $v=4$ when $t=1$

a Use a step length of 2 to estimate the velocity of the body when $t=21$

b Use integration to find the accurate velocity and find the relative and percentage errors.

6 For a minute after leaving a signal from rest, the speed of a train is given approximately by the formula $\frac{dv}{dt}=\frac{t(60-3t)}{1000}$ where t is the time in seconds and v is the speed in m s^{-1}.

a Use a step size of 1 to estimate the velocity of the train after 10 s.

b Use integration to find the accurate velocity.

7 A particle is following simple harmonic motion in a straight line. At any time its rate of change of velocity with respect to its distance from the centre of motion is given by the formula $v\frac{dv}{dx}=-x$ and $v=10$ cm s^{-1} when $x=0$ cm.

a Using a step length of 1 estimate the velocity of the particle when $x=5$

b Use integration to find the accurate velocity.

8 The ellipse $\frac{x^2}{16}+\frac{y^2}{9}=1$ has the differential equation $\frac{dy}{dx}=-\frac{9x}{16y}$

a Use an improved Euler method with a step length of 0.5 to estimate a value of y when $x=3$

b Find the relative and percentage errors.

9 The acceleration of a body is governed by the equation $\frac{dv}{dt}=\frac{2}{\sqrt[3]{t}}$ and $v=4$ when $t=1$

Use an improved Euler method with a step length of 2 to estimate the velocity of the body when $t=21$

10 The gradient of a curve at any point (x, y) is given by $\frac{dy}{dx}=\frac{y}{x^2+1}$ and it passes through the point (1, 2)

a Use a step length of 0.2 to estimate the value of y when $x=2$

b Use integration to find the accurate value of y

11 A particle runs along a groove. Its motion at time t s is given by the formula $\frac{ds}{dt}=t^2-5t+8$ and $v=0$ when $t=0$

a Use the midpoint method with a step length of 1 to find its distance from the start when $t=5$

b Use integration to find the accurate distance.

21 Summary and review

Chapter summary

- To find an approximate area under a curve using the mid-ordinate rule:
  - Divide the required area into an appropriate number of vertical rectangles.
  - Calculate and sum the values of the ordinates at the midpoint of each rectangle.
  - Complete the area approximation by multiplying the sum of the ordinates by the width of each rectangle.
- To find an approximate area under a curve using Simpson's rule:
  - Divide the required area into an appropriate even number of strips of equal width, d
  - Calculate the values of the ordinates for each x-value
  - Complete the area approximation by substituting into the formula:
  - $\frac{d}{3} \times$ [(sum of first and last ordinates) + 4 × (sum of odd ordinates) + 2 × (sum of even ordinates)]
- Euler's method states that $x_{r+1} = x_r + h$, $y_{r+1} = y_r + h\mathrm{f}(x_r, y_r)$
- To find an approximate value in an Euler equation:
  - Identify the differentiated function $\mathrm{f}(x, y)$, a starting point (x_0, y_0) and a suitable step value, h
  - Calculate $y_1 = y_0 + h\mathrm{f}(x_0, y_0)$
  - Continue using $y_{r+1} = y_r + h\mathrm{f}(x_r, y_r)$ until the desired endpoint is reached.
- The improved Euler equation is $y_{r+1} = y_r + \frac{1}{2}(k_1 + k_2)$ where $k_1 = h\mathrm{f}(x_r, y_r)$ and $k_2 = h\mathrm{f}(x_r + h, y_r + k_1)$
- The midpoint formula is $y_{r+1} = y_{r-1} + 2h\mathrm{f}(x_r, y_r)$

Check and review

You should now be able to...	Review Questions
✔ Find an approximation for the area under a graph using the mid-ordinate rule.	1
✔ Find an approximation for the area under a graph using Simpson's rule.	2
✔ Find the relative error between an approximation and the accurate answer.	3
✔ Use Euler's step-by-step first order method for equations of the form $y' = \mathrm{f}(x, y)$	4
✔ Use improved Euler methods for equations of the form $y' = \mathrm{f}(x, y)$	5, 6, 7, 8, 9

1 Use the mid-ordinate rule with an appropriate number of rectangles to estimate these areas to 3 s.f.

a $\int_1^{10} e^{\frac{-2}{x}}\,dx$ b $\int_0^6 \ln(1+x^2)\,dx$

c $\int_{\frac{\pi}{4}}^{2\pi} e^{\sin x}\,dx$

2 Use Simpson's rule with an appropriate number of strips to estimate these areas to 3 s.f.

a $\int_1^7 e^{\sin(x^3)}\,dx$ b $\int_0^{\frac{\pi}{3}} \tan^2 x\,dx$

c $\int_1^5 \frac{\sqrt{1+x}}{x^2}\,dx$

3 Questions using Simpson's rule led to these results. Calculate the relative and percentage errors in each case.

a Approximate result 212.6, accurate answer 220

b Approximate result 5.97, accurate answer 5.9

c Approximate result 0.002 3, accurate answer 0.01

d Approximate result 4×10^7, accurate answer 39 450 000

4 Use Euler's method when

a $\frac{dy}{dx} = 10 - 3x$, $x_0 = 0$, $y_0 = 1$, $h = 0.1$, endpoint is at $y(0.4)$

b $\frac{dy}{dx} = 10 - 3y$, $x_0 = 1$, $y_0 = 1$, $h = 0.1$, endpoint is at $y(1.5)$

c $\frac{dy}{dx} = x^2\sqrt{y}$, $x_0 = 3$, $y_0 = 2$, $h = 0.01$, endpoint is at $y(3.05)$

5 Use the improved Euler method when

a $\frac{dy}{dx} = 2x - y + xy$, $x_0 = 3$, $y_0 = 4$, $h = 0.1$, endpoint is at $y(3.6)$

b $\frac{dy}{dx} = \sqrt{xy}$, $x_0 = 0$, $y_0 = 36$, $h = 0.01$, endpoint is at $y(0.06)$

c $\frac{dy}{dx} = \frac{e^{x^2}}{y}$, $x_0 = 1$, $y_0 = 4$, $h = 1$, endpoint is at $y(5)$

6 The mass of a block of 10 g of francium-223 is governed by the formula $m \simeq 10e^{-0.0318t}$, where t is in minutes.

a After t minutes, the mass has halved in size. This is called its 'half life'. Calculate the half life of francium-223.

b Use an improved Euler method to find out an estimate for the half life.

c What is the mass after 12 minutes?

7 The gradient of a curve at any point (x, y) is given by $\frac{dy}{dx} = \frac{y}{1-x^2}$ and it passes through the point (0, 1)

a Use a step length of 0.1 in an improved Euler method to estimate the value of y when $x = 0.6$

b Use integration to find the accurate value of y

8 Use the midpoint method on these functions.

a $\frac{dy}{dx} = 2xy$, $x_0 = 1$, $y_0 = 1$, $h = 0.1$, endpoint is at $y(1.4)$

b $\frac{dy}{dx} = \tan^2 x$, $x_0 = 0$, $y_0 = 0$, $h = 0.2$, endpoint is at $y(1)$

c $\frac{dy}{dx} = 2x + \ln(1+x)$, $x_0 = 1$, $y_0 = 2$, $h = 0.01$, endpoint is at $y(1.05)$

9 The village of Far Far Away has a population, $n = 1000$

The government needs to implement a massive house building project in order to cope with a predicted population increase of rate $\frac{dn}{dt} = 700 \times 1.5^t$ where t is measured in years.

a Use integration to find the accurate population prediction for the population of Far Far Away after 5 years.

b Use a midpoint method to find out an estimate for the population of Far Far Away at this time.

Investigation

In the first part of their fall from an aircraft, sky divers free-fall under gravity. If we assume the acceleration due to gravity has a value of 10 ms^{-2}, then the motion of a sky diver can be modelled by the differential equation $\frac{dv}{dt} = 10$ where v ms^{-1} is the velocity of the sky diver (in this case measured vertically downwards) after time t seconds. This assumes that there is no force due to air resistance, whereas in reality this is not the case.

Experience suggests that the force of air resistance increases as speed increases. Experiments in wind tunnels suggest that forces due to air resistance can often be modelled as being proportional to velocity or the square of velocity.

The motion of sky divers, taking air resistance into account, can be modelled by one of these differential equations:

$$\frac{dv}{dt} = 10 - kv \qquad \frac{dv}{dt} = 10 - cv^2$$

Use the Euler step method to explore what would typically happen for a sky diver if you apply each of these models for air resistance in turn. Assume that a sky diver has a speed downwards of zero at $t = 0$

(You may wish to get started by using a value of $k = 0.1$ and a value of $c = 0.01$)

Try different values of k and c

Explore the resulting terminal velocities and how long it takes to reach these. Research typical values and consider which models might be most appropriate.

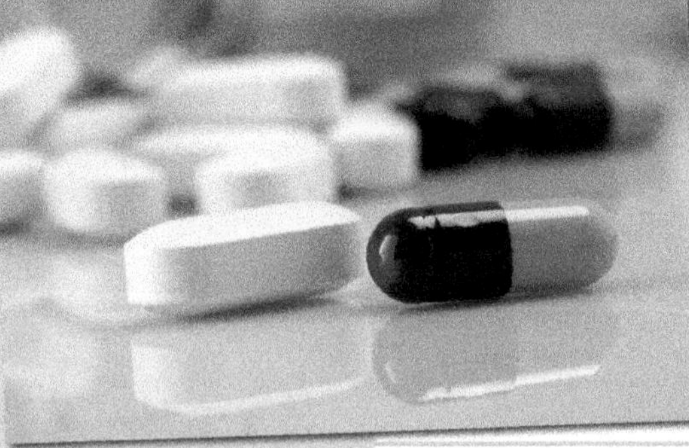

Investigation

Medicinal drugs typically decay exponentially with time in the body.

The differential equation $dA/dt = -kA$ where A is the amount of drug in the body at time t and k is a constant that primarily depends on the body's physiology, can be used to model the rate of change of concentration of the drug in a body.

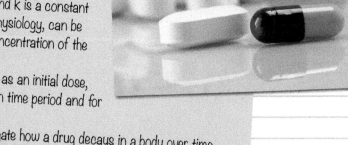

Doctors prescribe drugs to be taken as an initial dose, followed by a repeat dose after a certain time period and for a certain number of days.

Use the Euler step method to investigate how a drug decays in a body over time, and how this is affected by taking repeat doses of the drug every so often. Start by considering the drug to have a half-life of one hour (that is, the drug decays so that its concentration in the body naturally halves every hour).

1 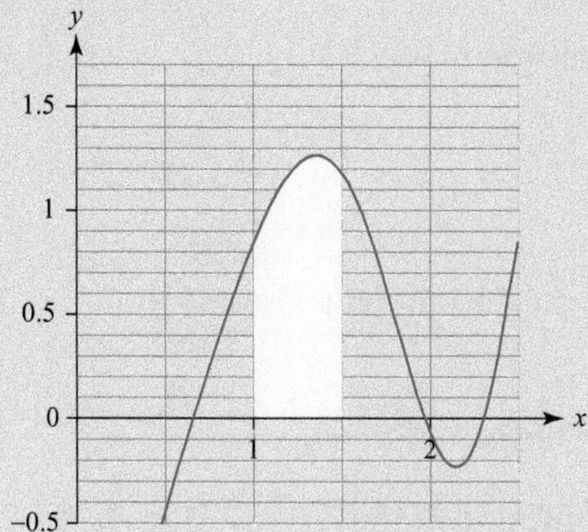

The diagram shows the graph of $y = f(x)$ where $f(x) = \ln x + \sin(x^2)$

The shaded area is bounded by the curve, the x-axis and the lines $x = 1$ and $x = 1.5$

a Use the mid-ordinate rule with five strips to find an estimate for the shaded area. **[5 marks]**

b Given that your estimate in part **a** gives a percentage error of 0.26%, find, correct to four significant figures, the actual area of the shaded region. **[2]**

2 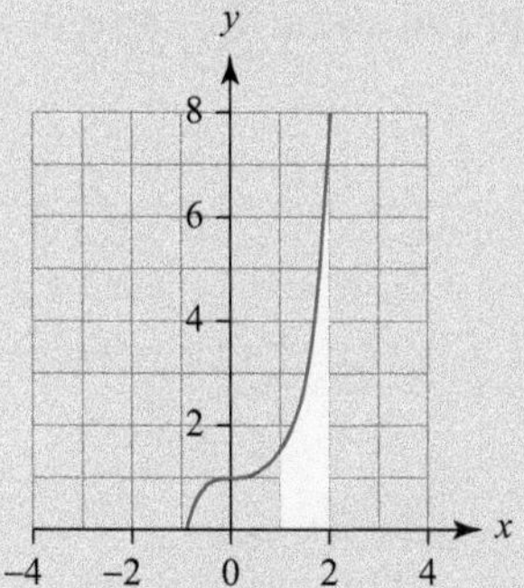

The diagram shows the graph of $y = f(x)$ where $f(x) = \cos(x^2) + x^3$

The shaded area is bounded by the curve, the x-axis and the lines $x = 1$ and $x = 2$

a Use Simpson's rule with six strips to estimate the shaded area. **[5]**

b Suggest how your estimate in part **a** could be improved. **[1]**

3 The diagram shows the design for a new turbine fin. The fin is modelled as the area enclosed by the curve $y = \tan x + x^2$, the x-axis and the lines $x = 0$ and $x = \frac{\pi}{4}$ where each unit on the axes represents 10 cm.

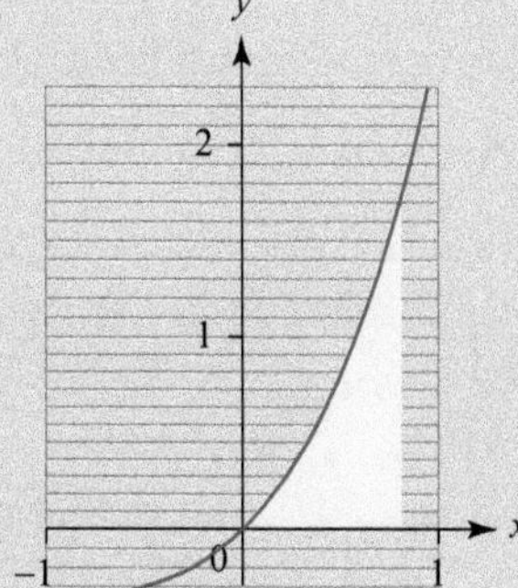

a Use Simpson's rule with five ordinates to estimate the area of the fin, correct to three significant figures. **[4]**

b Find the exact area of the fin, giving your answer in the form $p\pi^3 + q\ln(\sqrt{2})$ where p and q are rational constants to be found. **[4]**

c In light of your answer to part **b**, comment on your answer to part **a**. **[1]**

4 $\frac{dy}{dx} = 3x - 1$

Given that $y = 2$ when $x = 1$

a Use Euler's method with a step length of 0.5 to estimate the value of y when $x = 2$ **[4]**

b Use integration to find the exact value of y when $x = 2$ **[4]**

c Hence find the percentage error in using your approximation in part **a**. **[2]**

5 The value, x hundred pounds, of a particular tradeable stock t hours after it is purchased is modelled by the differential equation $\frac{dx}{dt} = \frac{x^2 - 2t}{2xt - t^2}$

If the stock is worth £500 three hours after it is purchased, use two iterations of Euler's method to estimate, correct to the nearest ten pence, the value of the stock four hours after it is purchased. **[6]**

6 $\frac{dy}{dx} = 2 - x^2 y$

Joanna uses an improved Euler method to estimate the particular solution of the differential equation when $x = 1.5$ given that $y = 1$ when $x = 1$

She uses a strip width of 0.25

a Show that after one iteration, Joanna gets a y-value of $\frac{579}{512}$ **[4]**

b Complete Joanna's method to find her estimated solution, correct to four decimal places. **[4]**

7 A particle moves such that the rate of change of displacement with respect to time has differential equation $\frac{ds}{dt} = t^3 - 4t + 2$

Given that $s = 0$ when $t = 0$,

a Use the midpoint formula with step length 1 to estimate the displacement when $t = 3$ **[5]**

b By solving the differential equation, find the exact value of s when $t = 3$ **[4]**

c Find the relative error in using your approximation in part **a**. **[2]**

8 The velocity, v m s^{-1}, of a particle attached to a vertically hanging spring, at the point where the spring becomes taut, is modelled using the differential equation $\frac{dv}{dx} = \frac{3x - 23}{2v} - 0.01v$ where x is the vertical displacement, in cm, from the point of release. Given that $v = 6$ when $x = 4$, use the midpoint formula with a step length of 0.25 to find the value of v, to 3 decimal places, when $x = 5$ **[6]**

22 Matrices 2

The mathematical techniques that use matrices, eigenvalues and eigenvectors have uses in computer models in many disciplines. For example, engineering, geology, statistics and financial analysis.

One area where such computer models are used extensively is in weather forecasting.

Weather forecasts help people with everyday decisions. They are also useful for those who take part in outdoor activities such as sailing and hill walking. For people who work outdoors, such as farmers and fishermen, it is important that forecasts are accurate for as far into the future as possible. This is only possible if the use of the mathematics of matrices continues to develop.

Orientation

What you need to know	What you will learn	What this leads to
Ch5 Matrices 1	• How to calculate the determinate of a matrix, and the inverse of a matrix. • How to solve a system of linear equations. • How to work out and use eigenvalues and eigenvectors. • How to use matrix diagonalization.	**Careers** • Computer programming. • Meteorology.

22.1 Determinants, inverse matrices and linear equations

Fluency and skills

See Ch5.3
For a reminder of matrices.

You can find the determinant of the 2×2 matrix $\mathbf{A} = \begin{pmatrix} a & b \\ c & d \end{pmatrix}$ either on your calculator or using the fact that $\det(\mathbf{A}) = ad - bc$

In order to find the determinant of a 3×3 matrix, you need to find the **minors** of the top row of elements in your matrix. Given a 3×3 matrix, the minor of an element is the determinant of the 2×2 matrix remaining when the row and column of the element are crossed out.

For example, to find the minor of a in the matrix $\begin{pmatrix} a & b & c \\ d & e & f \\ g & h & i \end{pmatrix}$, you cross out the row and the column involving a then find the determinant of the matrix $\begin{pmatrix} e & f \\ h & i \end{pmatrix}$ that remains.

Key point

Then $\det\begin{pmatrix} a & b & c \\ d & e & f \\ g & h & i \end{pmatrix} = a\begin{vmatrix} e & f \\ h & i \end{vmatrix} - b\begin{vmatrix} d & f \\ g & i \end{vmatrix} + c\begin{vmatrix} d & e \\ g & h \end{vmatrix}$

Example 1

Find the determinant of the matrix $\begin{pmatrix} 3 & 2 & -1 \\ 0 & 4 & -2 \\ -3 & 1 & 5 \end{pmatrix}$

$\det\begin{pmatrix} 3 & 2 & -1 \\ 0 & 4 & -2 \\ -3 & 1 & 5 \end{pmatrix} = 3\begin{vmatrix} 4 & -2 \\ 1 & 5 \end{vmatrix} - 2\begin{vmatrix} 0 & -2 \\ -3 & 5 \end{vmatrix} - 1\begin{vmatrix} 0 & 4 \\ -3 & 1 \end{vmatrix}$

Use the formula $a\begin{vmatrix} e & f \\ h & i \end{vmatrix} - b\begin{vmatrix} d & f \\ g & i \end{vmatrix} + c\begin{vmatrix} d & e \\ g & h \end{vmatrix}$

$= 3(20 - -2) - 2(0 - 6) - 1(0 - -12)$

Take care with negative signs.

$= 66 + 12 - 12$

$= 66$

Matrices can be used to perform transformations. Consider a square with vertices at (0, 0), (1, 0), (0, 1) and (1, 1)

Under the transformation $\mathbf{T} = \begin{pmatrix} a & 0 \\ 0 & a \end{pmatrix}$ the square is enlarged by scale factor a centre the origin. So the vertices of the image are $(0, 0)$, $(a, 0)$, $(0, a)$ and (a, a). The area of the original square is 1 and the area of the image is a^2, which is also the determinant of **T**. This result extends to all linear transformations.

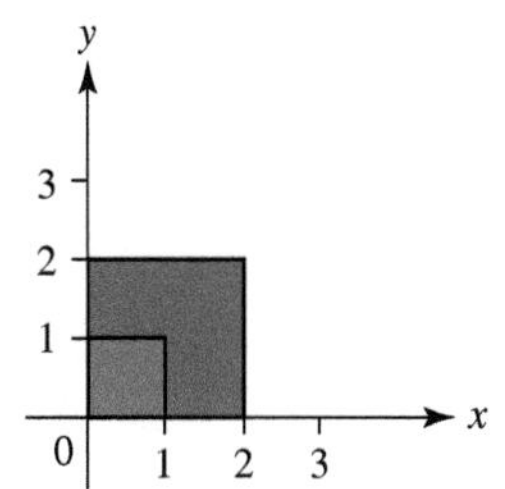

Key point

Under a transformation represented by the 2×2 matrix **T**
area of image = area of original $\times |\det(\mathbf{T})|$

Example 2

The triangle with vertices $(3, -1)$, $(4, 2)$ and $(0, 2)$ is stretched by scale factor 2 parallel to the x-axis and stretched by scale factor -3 parallel to the y-axis. Find the area of the image under this transformation.

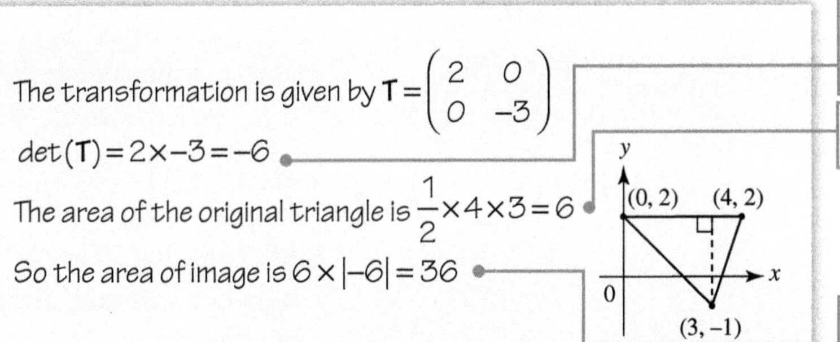

The transformation is given by $\mathbf{T} = \begin{pmatrix} 2 & 0 \\ 0 & -3 \end{pmatrix}$

$\det(\mathbf{T}) = 2 \times -3 = -6$ — A negative determinant indicates that the shape has been reflected.

The area of the original triangle is $\frac{1}{2} \times 4 \times 3 = 6$ — A sketch will help with the area.

So the area of image is $6 \times |-6| = 36$ — Using area of original $\times |\det(\mathbf{T})|$

Key point

If $\det(\mathbf{T}) < 0$ then the transformation represented by **T** involves a reflection.

If $\det(\mathbf{T}) > 0$ then the orientation of the original shape is preserved.

The same principles can be extended to the volumes of shapes.

Key point

Under a transformation represented by the 3×3 matrix **M**
volume of image = volume of original $\times |\det(\mathbf{M})|$

A parallelepiped is a prism where each face is a parallelogram. A cuboid is a special case of a parallelepiped.

Example 3

A cube is transformed by the matrix $\mathbf{M} = \begin{pmatrix} 1 & -2 & -1 \\ 0 & 3 & 2 \\ 2 & 1 & 4 \end{pmatrix}$ to create a parallelepiped.

The volume of the parallelepiped is 36 cm^3.

a Calculate the volume of the original cube.

b Explain whether the transformation **M** involves a reflection.

a $\det(\mathbf{M}) = 1(12-2) - -2(0-4) - 1(0-6) = 8$

$36 \div 8 = 4.5$ — Volume of original $\div |\det(\mathbf{M})|$

So the volume of the original cube is 4.5 cm^3.

b $\det(\mathbf{M}) > 0$, therefore the transformation **M** does not involve a reflection.

You can find the inverse of the 2×2 matrix $\mathbf{A} = \begin{pmatrix} a & b \\ c & d \end{pmatrix}$, either on your calculator or using the rule $\mathbf{A}^{-1} = \frac{1}{\det(\mathbf{A})}\begin{pmatrix} d & -b \\ -c & a \end{pmatrix}$

To find the inverse of a 3×3 matrix, **A**, you first need to find the **matrix of minors**, **M**. You do this by replacing each element in **A** by its minor.

So if $\mathbf{P} = \begin{pmatrix} a & b & c \\ d & e & f \\ g & h & i \end{pmatrix}$ then the matrix of minors is

$\mathbf{M} = \begin{pmatrix} A & B & C \\ D & E & F \\ G & H & I \end{pmatrix}$ where A is the minor of a, B is the minor of b and so on.

You then transpose this matrix (swap rows and columns) and change the sign of alternating elements to give $\begin{pmatrix} A & -D & G \\ -B & E & -H \\ C & -F & I \end{pmatrix}$

Remember, you can only find the inverse of a non-singular matrix. Singular matrices (which have a determinant of zero) do not have an inverse.

The minor of element a is

$A = \det\begin{pmatrix} e & f \\ h & i \end{pmatrix} = ei - fh$

The sign matrix

$\begin{pmatrix} + & - & + \\ - & + & - \\ + & - & + \end{pmatrix}$

indicates the elements of which you need to change the sign.

Key point

The inverse of **P** is $\mathbf{P}^{-1} = \frac{1}{\det(\mathbf{P})}\begin{pmatrix} A & -D & G \\ -B & E & -H \\ C & -F & I \end{pmatrix}$

Example 4

Find the inverse of $\mathbf{A} = \begin{pmatrix} 2 & 1 & -3 \\ 3 & 1 & -2 \\ 0 & 2 & -1 \end{pmatrix}$, given that it is non-singular.

$$\mathbf{M} = \begin{pmatrix} (1\times-1)-(-2\times2) & (3\times-1)-(-2\times0) & (3\times2)-(1\times0) \\ (1\times-1)-(-3\times2) & (2\times-1)-(-3\times0) & (2\times2)-(1\times0) \\ (1\times-2)-(-3\times1) & (2\times-2)-(-3\times3) & (2\times1)-(1\times3) \end{pmatrix}$$

$$= \begin{pmatrix} 3 & -3 & 6 \\ 5 & -2 & 4 \\ 1 & 5 & -1 \end{pmatrix}$$

Find the minor of every element.

$$\det(\mathbf{A}) = (2\times3) - (1\times-3) + (-3\times6)$$

Use the minors from the top row of the table as $\det(\mathbf{A}) = 2\begin{vmatrix} 1 & -2 \\ 2 & -1 \end{vmatrix} - 1\begin{vmatrix} 3 & -2 \\ 0 & -1 \end{vmatrix} - 3\begin{vmatrix} 3 & 1 \\ 0 & 2 \end{vmatrix} = -9$

$$\text{Therefore } \mathbf{A}^{-1} = -\frac{1}{9}\begin{pmatrix} 3 & -5 & 1 \\ 3 & -2 & -5 \\ 6 & -4 & -1 \end{pmatrix}$$

Use $\mathbf{A}^{-1} = \frac{1}{\det(\mathbf{A})}\begin{pmatrix} A & -D & G \\ -B & E & -H \\ C & -F & I \end{pmatrix}$

This could be worked out on a calculator but you do need to know the method.

Exercise 22.1A Fluency and skills

1 Find the determinant of each of these matrices.

a $\begin{pmatrix} 3 & -7 \\ 2 & -4 \end{pmatrix}$ b $\begin{pmatrix} a & 1 \\ -b & 5 \end{pmatrix}$

c $\begin{pmatrix} 6 & 2 & -1 \\ -2 & 4 & 0 \\ 3 & 5 & 1 \end{pmatrix}$ d $\begin{pmatrix} 2a & 3 & a \\ -b & 0 & b \\ 3 & 5 & 1 \end{pmatrix}$

2 Find the values of x for which each of these matrices is singular.

a $\begin{pmatrix} x & -3 \\ 4 & 2 \end{pmatrix}$ b $\begin{pmatrix} -x & x+2 \\ 3x & 4-x \end{pmatrix}$

c $\begin{pmatrix} 1 & 2 & -x \\ x & 4 & -3x \\ 1 & 0 & 2 \end{pmatrix}$ d $\begin{pmatrix} 4 & x & 2 \\ -x & 3 & 3x \\ -2 & x & 1 \end{pmatrix}$

3 A rectangle has area 7 square units.

a Find the area of the image of the rectangle under each of these transformations.

i $\begin{pmatrix} -2 & 3 \\ 1 & -3 \end{pmatrix}$ ii $\begin{pmatrix} 2 & 1 \\ 5 & 2 \end{pmatrix}$

b Explain whether or not each of the transformations in part **a** involves a reflection.

4 A triangle is transformed by the matrix $\begin{pmatrix} 1 & -3 \\ 4 & -2 \end{pmatrix}$. The area of the image is 15 square units.

a Calculate the area of the original triangle.

b Explain whether or not the triangle has been reflected.

5 A cube with volume 5 cube units is transformed by the matrix $\begin{pmatrix} 1 & 3 & 2 \\ -2 & 1 & 2 \\ -1 & 2 & 0 \end{pmatrix}$

a Calculate the volume of the image.

b Explain whether or not the transformation represented by M involves a reflection.

6 Find the inverse of each of these matrices.

a $\begin{pmatrix} 2 & 3 \\ -1 & 0 \end{pmatrix}$ b $\begin{pmatrix} a & 2 \\ a & 2-a \end{pmatrix}$ c $\begin{pmatrix} -2 & -1 & 0 \\ 4 & 0 & 2 \\ 5 & 1 & 3 \end{pmatrix}$

d $\begin{pmatrix} 5 & 1 & 4 \\ 6 & 2 & -3 \\ -2 & -1 & 4 \end{pmatrix}$ e $\begin{pmatrix} -4a & 2 & 1 \\ a & 0 & 2 \\ 3a & -1 & 1 \end{pmatrix}$ f $\begin{pmatrix} a & -2 & 2a \\ 0 & 2 & 3-a \\ -3a & 3 & 1 \end{pmatrix}$

7 Find the inverse of the matrix

$\mathbf{B} = \begin{pmatrix} 1 & -1 & k \\ k & 0 & 1 \\ 0 & k & 1 \end{pmatrix}$ in terms of k

8 A transformation is represented by the matrix $\mathbf{T}=\begin{pmatrix} 1 & 0 & 0 \\ 0 & \frac{\sqrt{2}}{2} & \frac{\sqrt{2}}{2} \\ 0 & -\frac{\sqrt{2}}{2} & \frac{\sqrt{2}}{2} \end{pmatrix}$

The image of the point A under $\mathbf{T}$ is $A'(1,-\sqrt{2},\sqrt{2})$. Use the inverse of $\mathbf{T}$ to find the coordinates of point A

Reasoning and problem-solving

See Ch23.1 For more on equations of planes.

You can use matrices to solve a system of two linear equations in two variables. The solution gives you the point of intersection of two lines.

This method can be extended to three dimensions, where the solution represents the intersection of three planes.

An equation of the form $ax+by+cz=d$ can represent a plane. If you have three different planes then one of these will apply:

- There are no points that lie on all three planes (their equations are said to be inconsistent). This could be because two or three of the planes are parallel or because the three planes form a triangular prism.

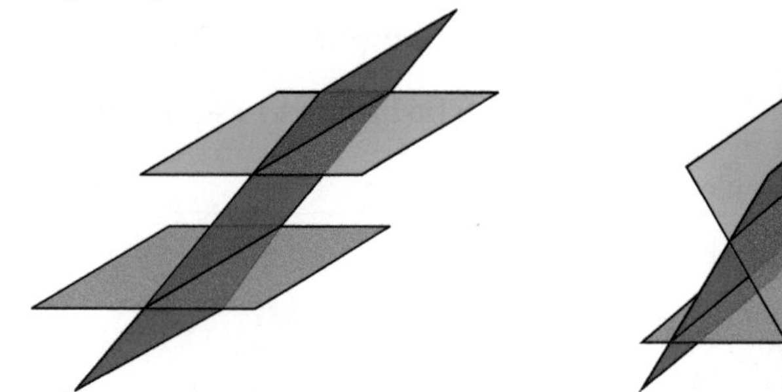

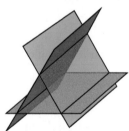

- There are infinitely many solutions as the three planes meet along a line (called a sheaf).

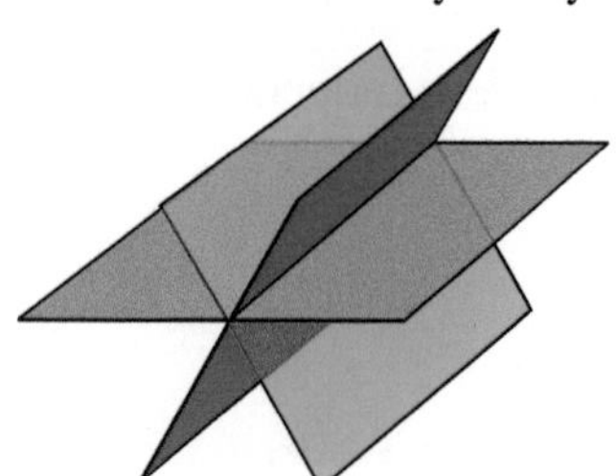

- They intersect at a single point, so there is exactly one solution.

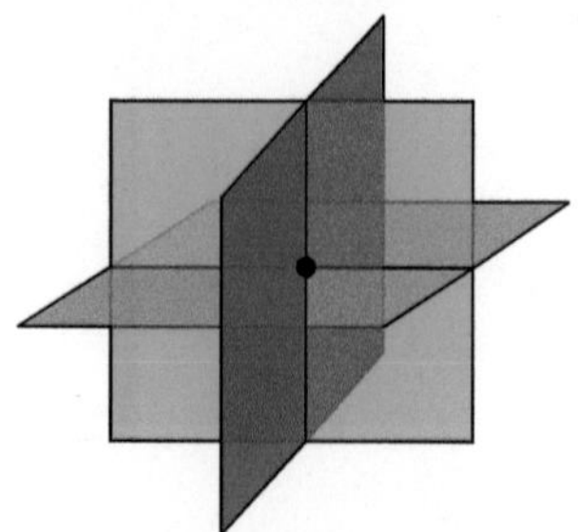

PURE

Strategy 1

To solve problems involving systems of linear equations

1. Rewrite a system of linear equations using matrices.
2. Pre-multiply or post-multiply a matrix by its inverse.
3. Use the fact that $\mathbf{AA}^{-1} = \mathbf{A}^{-1}\mathbf{A} = \mathbf{I}$

Example 5

Given that there is a unique solution, use matrices to solve this system of equations:

$x+y+z=2 \qquad 2x-3y-z=-1 \qquad 3x-2y+2z=11$

Write the equations in the form $\begin{pmatrix}1 & 1 & 1\\ 2 & -3 & -1\\ 3 & -2 & 2\end{pmatrix}\begin{pmatrix}x\\ y\\ z\end{pmatrix} = \begin{pmatrix}2\\ -1\\ 11\end{pmatrix}$ — (1) Rewrite the system of linear equations using matrices.

$$\det\begin{pmatrix}1 & 1 & 1\\ 2 & -3 & -1\\ 3 & -2 & 2\end{pmatrix} = 1(-6-2)-1(4--3)+1(-4--9) = -10$$

Calculate the determinant.

$$\begin{pmatrix}1 & 1 & 1\\ 2 & -3 & -1\\ 3 & -2 & 2\end{pmatrix}^{-1} = -\frac{1}{10}\begin{pmatrix}-8 & -4 & 2\\ -7 & -1 & 3\\ 5 & 5 & -5\end{pmatrix}$$

Find the inverse.

$$-\frac{1}{10}\begin{pmatrix}-8 & -4 & 2\\ -7 & -1 & 3\\ 5 & 5 & -5\end{pmatrix}\begin{pmatrix}1 & 1 & 1\\ 2 & -3 & -1\\ 3 & -2 & 2\end{pmatrix}\begin{pmatrix}x\\ y\\ z\end{pmatrix} = -\frac{1}{10}\begin{pmatrix}-8 & -4 & 2\\ -7 & -1 & 3\\ 5 & 5 & -5\end{pmatrix}\begin{pmatrix}2\\ -1\\ 11\end{pmatrix}$$

(2) Pre-multiply both sides of the original equation by the inverse matrix.

$$\begin{pmatrix}x\\ y\\ z\end{pmatrix} = -\frac{1}{10}\begin{pmatrix}-8 & -4 & 2\\ -7 & -1 & 3\\ 5 & 5 & -5\end{pmatrix}\begin{pmatrix}2\\ -1\\ 11\end{pmatrix}$$

(3) Use $AA^{-1} = A^{-1}A = I$

$$= -\frac{1}{10}\begin{pmatrix}10\\ 20\\ -50\end{pmatrix}$$

So $x=-1$, $y=-2$, $z=5$

Strategy 2

To decide if three planes intersect

1. Check if any of the equations represent parallel planes – if so, they will not intersect.
2. Write two of the variables in terms of the third.
3. If the equations are inconsistent then the planes form a triangular prism.

Example 6

Explain why these systems of equations have no solutions.

a $2x-3y-z=2$
$4x-6y-2z=1$
$12x-18y-6z=3$

b $x+y-z=5$
$2x+2y=10$
$x+y+z=7$

a $4x-6y-2z=1$ is a multiple of $2x-3y-z=2$ except the constant term, therefore these represent parallel planes. Since parallel planes do not intersect, there are no solutions.

b Adding the first and third equations together gives $2x+2y=12$ however, the second equation is $2x+2y=10$ so the equations are inconsistent and there are no solutions.

1. Also, notice that $12x-18y-6z=3$ is a multiple of $4x-6y-2z=1$ so these represent the same plane.
2. None of the planes are the same or parallel so you need to start trying to simplify the problem by eliminating a variable.
3. The planes form a triangular prism.

The system of equations in Example 6b can be written in matrix form as

$$\begin{pmatrix} 1 & 1 & -1 \\ 2 & 2 & 0 \\ 1 & 1 & 1 \end{pmatrix}\begin{pmatrix} x \\ y \\ z \end{pmatrix}=\begin{pmatrix} 5 \\ 10 \\ 7 \end{pmatrix}$$

You can attempt to solve these by pre-multiplying both sides of the equation by the inverse of

$$\begin{pmatrix} 1 & 1 & -1 \\ 2 & 2 & 0 \\ 1 & 1 & 1 \end{pmatrix}$$

However, for this particular system of equations

$$\det\begin{pmatrix} 1 & 1 & -1 \\ 2 & 2 & 0 \\ 1 & 1 & 1 \end{pmatrix}=1(2-0)-1(2-0)--1(2-2)=0$$

This implies that there is either no solution (as in this case) or an infinite number of solutions.

Key point

A system of linear equations represented by the matrix equation $\mathbf{M}\mathbf{x}=x'$ has a unique solution if and only if $\det(\mathbf{M})\neq 0$

Strategy 3

To decide on the nature of a system of simultaneous equations

1. Rewrite the system of linear equations using matrices.
2. Calculate a determinant – a non-zero determinant implies a unique solution.
3. Write two of the variables in terms of the third.
4. Check for inconsistencies.

Example 7

You are given a system of simultaneous equations:

$x+3y-7z=8,\ 2x-2z=-2,\ x-y+z=-4$

a Decide whether there is a unique solution, an infinite number of solutions or no solutions.

b Describe the geometric significance.

a $\begin{pmatrix}1 & 3 & -7\\ 2 & 0 & -2\\ 1 & -1 & 1\end{pmatrix}\begin{pmatrix}x\\ y\\ z\end{pmatrix}=\begin{pmatrix}8\\ -2\\ -4\end{pmatrix}$

$\det\begin{pmatrix}1 & 3 & -7\\ 2 & 0 & -2\\ 1 & -1 & 1\end{pmatrix}=1(0-2)-3(2--2)-7(-2-0)$

$=0$

This implies there is not a unique solution.

Subtracting the third equation from the first equation gives

$4y-8z=12 \Rightarrow y=3+2z$

The second equation gives $x=z-1$

Verify there are no inconsistencies:

$(z-1)+3(3+2z)-7z=8$

$2(z-1)-2z=-2$

$(z-1)-(3+2z)+z=-4$

So there is an infinite number of solutions. These lie on the line with equations $y=3+2z,\ x=z-1$

b The planes meet in a line: they form a sheaf.

1. Rewrite the system of linear equations using matrices.

2. Since the determinant is zero, there must either be no solution or an infinite number of solutions.

3. Here y and x are expressed in terms of z, but you could have chosen any of the three variables to write the other two in terms of.

4. You can verify there are no inconsistencies by substituting back into the original equations.

You will learn more about writing the equation of lines in 3D in Chapter 23

Exercise 22.1B Reasoning and problem-solving

1 Use matrices to find the point of intersection between the three planes in each case.

a $x+y-2z=3$

$2x-3y+5z=4$

$5x+2y+z=-3$

b $4x+6y-z=-3$

$2x-3y=2$

$8y+4z=0$

2 Given that $\mathbf{A}\begin{pmatrix} 2 & 0 & -3 \\ 0 & 1 & 4 \\ -5 & 2 & -1 \end{pmatrix} = \begin{pmatrix} 14 & -9 & 9 \\ 25 & -10 & 5 \\ 9 & 1 & 7 \end{pmatrix}$, find the matrix **A**.

3 Given that $\begin{pmatrix} 5 & -1 & 0 \\ 0 & 2 & -2 \\ 3 & 1 & 4 \end{pmatrix}\mathbf{B} = \begin{pmatrix} -32 & 16 & 15 \\ -4 & 8 & 6 \\ 0 & -12 & -3 \end{pmatrix}$, find the matrix **B**.

4 If $\mathbf{A} = \begin{pmatrix} 8 & 0 & 2 \\ 4 & 3 & -1 \\ 5 & 0 & 2 \end{pmatrix}$ and $\mathbf{AB} = \begin{pmatrix} -2 & 28 & 14 \\ 13 & 10 & -5 \\ -2 & 19 & 11 \end{pmatrix}$, find the matrix **BA**.

5 Show that each of these systems of linear equations are inconsistent and explain the geometric significance.

a $3x-2y+z=7$

$6x-4y+2z=5$

$x+3y+2z=3$

b $-2x+3y-z=4$

$6x-9y+3z=7$

$5x+y+z=3$

6 A cube is transformed by the matrix $\mathbf{T} = \begin{pmatrix} 1 & 0 & k \\ 0 & 2 & -3 \\ 2-k & -1 & k \end{pmatrix}$ to give a parallelepiped.

a State the range of values of k for which the transformation preserves the orientation of the cube.

The point $C(-3, 2, 1)$ is transformed by **T** to $C'(-4, 1, -12)$

b Find the value of k

7 Three planes have equations

$-7x-3y+5z=4$

$x+y+z=0$

$x+2y+4z=1$

a Show that the planes do not have a unique point of intersection.

b Show that a general point on the line is given by $(2\lambda+1, -3\lambda-2, \lambda+1)$

8 For each system of equations, show that the planes that the equations represent form a sheaf.

a
$$\begin{aligned} x+2z&=1 \\ -2x+y+4z&=0 \\ 9x-2y+2z&=5 \end{aligned}$$

b
$$\begin{aligned} 5x-y+z&=0 \\ 2x+y-2z&=7 \\ 36x+11y-24z&=91 \end{aligned}$$

9 Find the values of a for which this system of simultaneous equations does not have a unique solution.

$$\begin{aligned} 3y+az&=4 \\ -x+ay+2z&=1 \\ x+y&=3 \end{aligned}$$

10 A family keeps rabbits, hamsters and fish as pets.

They initially have 17 pets in total and one more rabbit than hamsters.

Two of the fish die. They then have 8 more fish than hamsters.

a Write a matrix equation to represent this situation.

b Solve the matrix equation and state how many of each pet the family had initially.

11 A property developer owns 12 homes which have a mix of two, three and four bedrooms.

The total number of bedrooms in all her houses is 36 and she has the same number of two-bedroom homes as four-bedroom homes.

a Write a matrix equation to represent this situation.

b Show that there is no unique solution to this problem.

c How many possible solutions are there? Explain your answer.

12 The triangle ABC is stretched by scale factor k parallel to the x-axis then reflected in the line $y=x$. The images of vertices A, B and C are given by $A'(4, -6)$, $B'(0, 0)$ and $C'(4, 2)$

a Work out the area of the original triangle in terms of k

b State the range of values of k for which the transformation preserves the orientation of the triangle.

c Find the coordinates of A, B and C in terms of k

d If the coordinates of C are $(-1, 4)$, find the value of k

13 Three planes have equations:

$\Pi_1 : kx - y - z = 3,$

$\Pi_2 : 2x + ky - z = 4,$

$\Pi_3 : 3x + 2y - 2z = 3$

a Show that the planes intersect at a single, unique point for all possible values of k

b Find the coordinates of the point of intersection when $k = 2$

14 **a** Find the value of k for which the equations

$x + 2y + kz = 3$

$x - y - 3z = 0$

$2x + y + z = -3$

do not have a unique solution.

b Assuming k is not equal to the value found in part **a**, find the solution to the equations in terms of k

15 The system of equations

$x - y + z = b$

$y + z = 0$

$x - ay = 4$

does not have a unique solution

a Find the value of a

b Given that the planes with these equations form a sheaf, find the value of b

22.2 Manipulating determinants

Fluency and skills

The determinant of a 3×3 matrix, $\mathbf{A} = \begin{pmatrix} a & b & c \\ d & e & f \\ g & h & i \end{pmatrix}$ was defined as

$\det(\mathbf{A}) = a\begin{vmatrix} e & f \\ h & i \end{vmatrix} - b\begin{vmatrix} d & f \\ g & i \end{vmatrix} + c\begin{vmatrix} d & e \\ g & h \end{vmatrix}$, which uses the top row to multiply out the determinant.

In fact, any row or column can be used. The sign of the terms is given by the sign matrix

$$\begin{pmatrix} + & - & + \\ - & + & - \\ + & - & + \end{pmatrix}$$

For example, using, the second row gives

$$\det(\mathbf{A}) = -d\begin{vmatrix} b & c \\ h & i \end{vmatrix} + e\begin{vmatrix} a & c \\ g & i \end{vmatrix} - f\begin{vmatrix} a & b \\ g & h \end{vmatrix}$$
$$= -bdi + cdh + aei - ceg - afh + bfg$$

which is the same as using the original definition

$$\det(\mathbf{A}) = a\begin{vmatrix} e & f \\ h & i \end{vmatrix} - b\begin{vmatrix} d & f \\ g & i \end{vmatrix} + c\begin{vmatrix} d & e \\ g & h \end{vmatrix}$$
$$= aei - afh - bdi + bfg + cdh - ceg$$

Using, for example, the third column will also give this solution

$$\det(\mathbf{A}) = c\begin{vmatrix} d & e \\ g & h \end{vmatrix} - f\begin{vmatrix} a & b \\ g & h \end{vmatrix} + i\begin{vmatrix} a & b \\ d & e \end{vmatrix}$$
$$= cdh - ceg - afh + bfg + aei - bdi$$

Example 1

a Multiply out $\begin{vmatrix} 5 & 0 & 2 \\ 7 & 1 & 3 \\ 6 & 2 & 4 \end{vmatrix}$ using

i The second column, **ii** The third row,

b Hence write down the determinant of $\begin{pmatrix} 5 & 7 & 6 \\ 0 & 1 & 2 \\ 2 & 3 & 4 \end{pmatrix}$

a $\begin{vmatrix} 5 & 0 & 2 \\ 7 & 1 & 3 \\ 6 & 2 & 4 \end{vmatrix} = -0 + 1\begin{vmatrix} 5 & 2 \\ 6 & 4 \end{vmatrix} - 2\begin{vmatrix} 5 & 2 \\ 7 & 3 \end{vmatrix}$

$= 1(20-12) - 2(15-14)$

$= 8 - 2$

$= 6$

Ensure that you use the correct pattern of signs.

(Continued on the next page)

$$\begin{vmatrix} 5 & 0 & 2 \\ 7 & 1 & 3 \\ 6 & 2 & 4 \end{vmatrix} = 6\begin{vmatrix} 0 & 2 \\ 1 & 3 \end{vmatrix} - 2\begin{vmatrix} 5 & 2 \\ 7 & 3 \end{vmatrix} + 4\begin{vmatrix} 5 & 0 \\ 7 & 1 \end{vmatrix}$$

$$= 6(0-2) - 2(15-14) + 4(5-0)$$

$$= -12 - 2 + 20$$

$$= 6$$

You should get the same answer using any row or column.

b $\begin{pmatrix} 5 & 7 & 6 \\ 0 & 1 & 2 \\ 2 & 3 & 4 \end{pmatrix} = \begin{pmatrix} 5 & 0 & 2 \\ 7 & 1 & 3 \\ 6 & 2 & 4 \end{pmatrix}^T$ so multiplying out $\begin{vmatrix} 5 & 7 & 6 \\ 0 & 1 & 2 \\ 2 & 3 & 4 \end{vmatrix}$

using the second row is the same as multiplying out $\begin{vmatrix} 5 & 0 & 2 \\ 7 & 1 & 3 \\ 6 & 2 & 4 \end{vmatrix}$

using the second column.

Therefore $\det\begin{pmatrix} 5 & 7 & 6 \\ 0 & 1 & 2 \\ 2 & 3 & 4 \end{pmatrix} = 6$

Key point

$|\mathbf{M}| = |\mathbf{M}^{\mathrm{T}}|$ where $\mathbf{M}^{\mathrm{T}}$ is the transpose of matrix **M**.

From Example 1, you can see that there is some advantage to using a row or column with the most zeros in order to simplify the calculation of the determinant. Therefore, you may wish to manipulate a matrix in order to simplify the calculation of the determinant. You can do this using **row and column operations**.

Key point

These row and column operations can be carried out without affecting the value of the determinant

- Adding or subtracting any multiple of a row to another row.
- Adding or subtracting any multiple of a column to another column.

Key point

However, these row and column operations change the sign of the determinant

- Swapping two rows.
- Swapping two columns.

Finally

Key point

- Multiplying a row or a column of a matrix by a scalar will multiply the determinant by that scalar.
- Dividing a row or a column of a matrix by a scalar will divide the determinant by that scalar.

Example 2

a Use row and column operations to show that $\begin{vmatrix} 24 & 20 & -1 \\ 9 & 4 & -2 \\ 3 & 6 & 2 \end{vmatrix} = 6\begin{vmatrix} 0 & 0 & -1 \\ 2 & 5 & 0 \\ 1 & 6 & 2 \end{vmatrix}$

b Evaluate $\begin{vmatrix} 24 & 20 & -1 \\ 9 & 4 & -2 \\ 3 & 6 & 2 \end{vmatrix}$

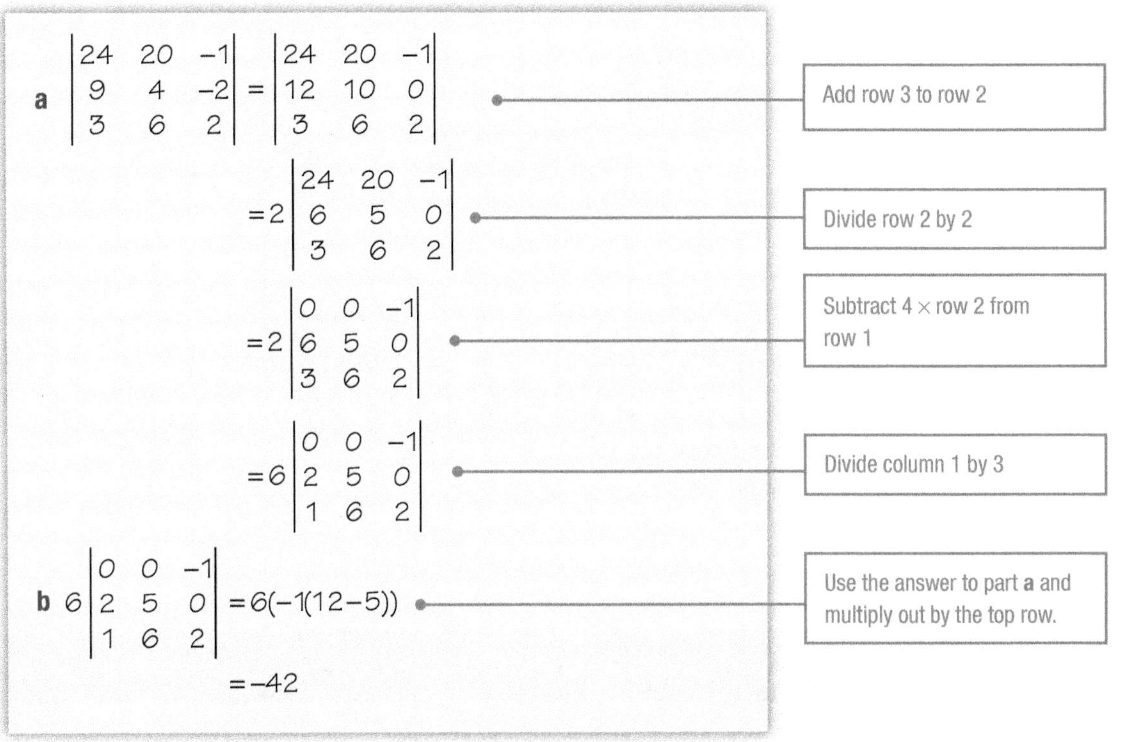

a $\begin{vmatrix} 24 & 20 & -1 \\ 9 & 4 & -2 \\ 3 & 6 & 2 \end{vmatrix} = \begin{vmatrix} 24 & 20 & -1 \\ 12 & 10 & 0 \\ 3 & 6 & 2 \end{vmatrix}$
Add row 3 to row 2
$= 2\begin{vmatrix} 24 & 20 & -1 \\ 6 & 5 & 0 \\ 3 & 6 & 2 \end{vmatrix}$
Divide row 2 by 2
$= 2\begin{vmatrix} 0 & 0 & -1 \\ 6 & 5 & 0 \\ 3 & 6 & 2 \end{vmatrix}$
Subtract 4 × row 2 from row 1
$= 6\begin{vmatrix} 0 & 0 & -1 \\ 2 & 5 & 0 \\ 1 & 6 & 2 \end{vmatrix}$
Divide column 1 by 3
b $6\begin{vmatrix} 0 & 0 & -1 \\ 2 & 5 & 0 \\ 1 & 6 & 2 \end{vmatrix} = 6(-1(12-5))$
$= -42$
Use the answer to part **a** and multiply out by the top row.

PURE

Exercise 22.2A Fluency and skills

1 Multiply out each of these determinants, using the row or column specified; show your working.

a $\begin{vmatrix} 2 & 5 & 1 \\ 4 & 3 & -1 \\ -3 & 2 & 0 \end{vmatrix}$ using the third column

b $\begin{vmatrix} 0 & 14 & 3 \\ 0 & -1 & 2 \\ 8 & 3 & 1 \end{vmatrix}$ using the first column

c $\begin{vmatrix} 8 & 13 & -2 \\ 0 & 1 & 5 \\ 2 & 5 & 7 \end{vmatrix}$ using the second row

d $\begin{vmatrix} 12 & 4 & 6 \\ 8 & -2 & 5 \\ 0 & 9 & 0 \end{vmatrix}$ using the third row

e $\begin{vmatrix} 0 & b & 2a \\ a & a & 1 \\ 0 & b & 0 \end{vmatrix}$ using the first column

f $\begin{vmatrix} b & a & -b \\ 0 & -3 & 0 \\ a & 2 & a \end{vmatrix}$ using the second row.

2 **a** Use one or more column operations to write $\begin{vmatrix} 2 & 1 & -3 \\ 4 & 2 & 1 \\ 1 & 1 & 2 \end{vmatrix}$ as a determinant with at least two zero elements,

b Hence find $\det\begin{pmatrix} 2 & 1 & -3 \\ 4 & 2 & 1 \\ 1 & 1 & 2 \end{pmatrix}$

3 a Use at least 3 column operations to simplify $\begin{vmatrix} 6 & 3 & -3 \\ 9 & 2 & -1 \\ -5 & -7 & -2 \end{vmatrix}$ and calculate its value,

b State the value of $\det\begin{pmatrix} 6 & 9 & -5 \\ 3 & 2 & -7 \\ -3 & -1 & -2 \end{pmatrix}$

4 Use row and column operations to write $\begin{vmatrix} 5 & -2 & 12 \\ -2 & -8 & 24 \\ 4 & 1 & 16 \end{vmatrix}$ as a determinant with at least two zero elements.

5 Given that the determinant of $\begin{pmatrix} a & 2 & -3 \\ 1 & a & 4 \\ 3 & 2 & 5 \end{pmatrix}$ is 128, calculate each of these determinants. Justify your answer in each case.

a $\begin{vmatrix} 1 & a & 4 \\ a & 2 & -3 \\ 3 & 2 & 5 \end{vmatrix}$

b $\begin{vmatrix} a & 1 & 3 \\ 2 & a & 2 \\ -3 & 4 & 5 \end{vmatrix}$

c $\begin{vmatrix} a+1 & 2+a & 1 \\ 1 & a & 4 \\ 3 & 2 & 5 \end{vmatrix}$

d $\begin{vmatrix} a & 2 & -3-4a \\ 1 & a & 0 \\ 3 & 2 & -7 \end{vmatrix}$

e $\begin{vmatrix} a & 6 & -3 \\ 1 & 3a & 4 \\ 3 & 6 & 5 \end{vmatrix}$

f $\begin{vmatrix} a & 1 & 3 \\ 1 & \frac{a}{2} & -4 \\ 3 & 1 & -5 \end{vmatrix}$

g $\begin{vmatrix} -3 & 2a & 2 \\ 4 & 2 & a \\ 5 & 6 & 2 \end{vmatrix}$

h $\begin{vmatrix} a & 8 & -3 \\ 1 & 4a & 4 \\ \frac{3}{2} & 4 & \frac{5}{2} \end{vmatrix}$

Reasoning and problem-solving

Row and column operations can be used to factorise a determinant.

Key point

A determinant can be factorised by using row or column operations to obtain rows or columns with a common factor.

Strategy

To factorise a determinant

1. Take out common factors from rows or columns.
2. Perform row and column operations to create rows or columns where the elements have a common factor.
3. Multiply out the determinant.

Example 3

a Factorise $\begin{vmatrix} a & b & 1 \\ a^2 & b^2 & 1 \\ a^3 & b^3 & 1 \end{vmatrix}$

b Under what conditions on a and b does this system of linear equations have a unique solution?

$ax+by+z=4 \qquad a^2x+b^2y+z=-2 \qquad a^3x+b^3y+z=0$

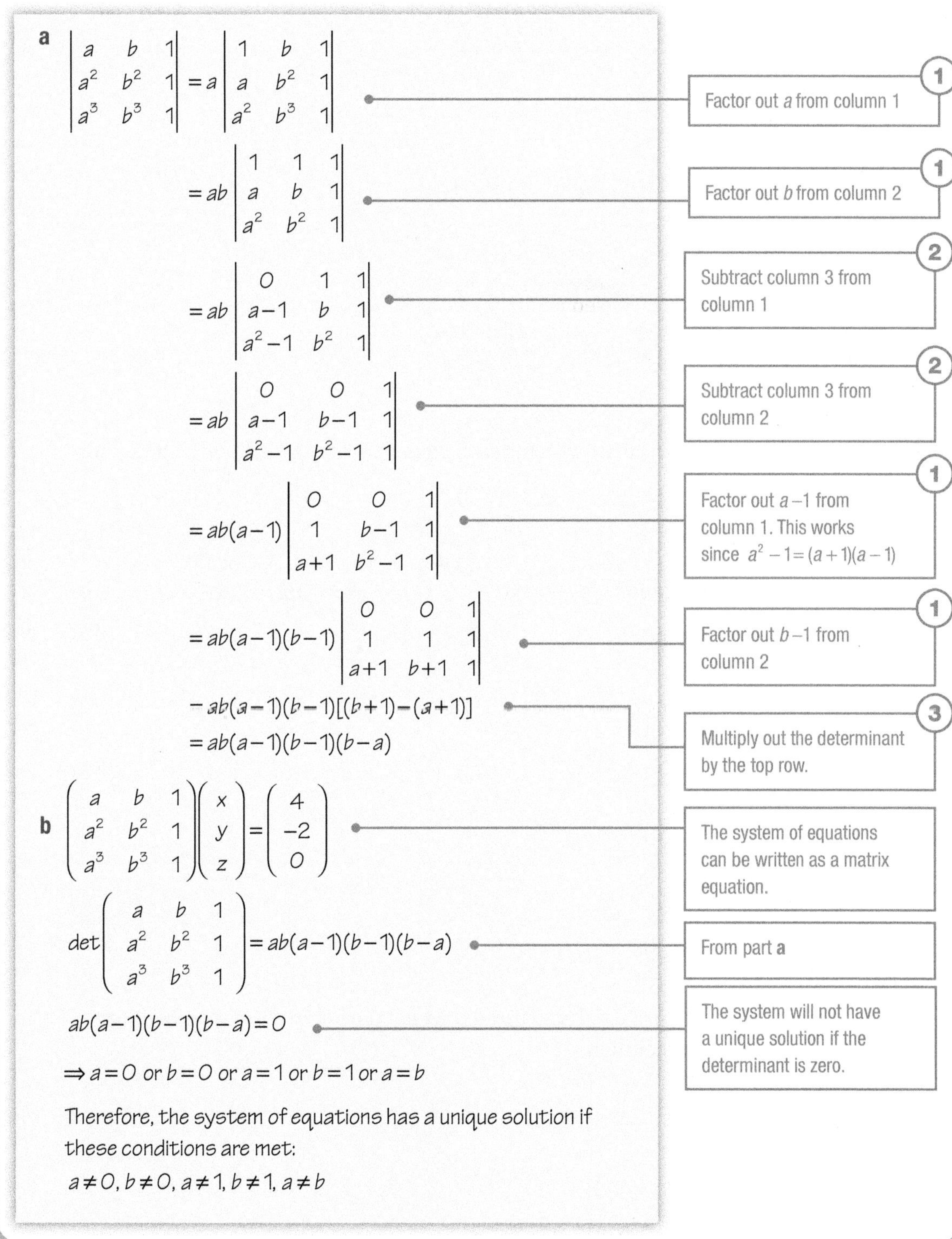

a $\begin{vmatrix} a & b & 1 \\ a^2 & b^2 & 1 \\ a^3 & b^3 & 1 \end{vmatrix} = a\begin{vmatrix} 1 & b & 1 \\ a & b^2 & 1 \\ a^2 & b^3 & 1 \end{vmatrix}$ — Factor out a from column 1 (1)

$= ab\begin{vmatrix} 1 & 1 & 1 \\ a & b & 1 \\ a^2 & b^2 & 1 \end{vmatrix}$ — Factor out b from column 2 (1)

$= ab\begin{vmatrix} 0 & 1 & 1 \\ a-1 & b & 1 \\ a^2-1 & b^2 & 1 \end{vmatrix}$ — Subtract column 3 from column 1 (2)

$= ab\begin{vmatrix} 0 & 0 & 1 \\ a-1 & b-1 & 1 \\ a^2-1 & b^2-1 & 1 \end{vmatrix}$ — Subtract column 3 from column 2 (2)

$= ab(a-1)\begin{vmatrix} 0 & 0 & 1 \\ 1 & b-1 & 1 \\ a+1 & b^2-1 & 1 \end{vmatrix}$ — Factor out $a-1$ from column 1. This works since $a^2-1=(a+1)(a-1)$ (1)

$= ab(a-1)(b-1)\begin{vmatrix} 0 & 0 & 1 \\ 1 & 1 & 1 \\ a+1 & b+1 & 1 \end{vmatrix}$ — Factor out $b-1$ from column 2 (1)

$= ab(a-1)(b-1)[(b+1)-(a+1)]$

$= ab(a-1)(b-1)(b-a)$ — Multiply out the determinant by the top row. (3)

b $\begin{pmatrix} a & b & 1 \\ a^2 & b^2 & 1 \\ a^3 & b^3 & 1 \end{pmatrix}\begin{pmatrix} x \\ y \\ z \end{pmatrix} = \begin{pmatrix} 4 \\ -2 \\ 0 \end{pmatrix}$ — The system of equations can be written as a matrix equation.

$\det\begin{pmatrix} a & b & 1 \\ a^2 & b^2 & 1 \\ a^3 & b^3 & 1 \end{pmatrix} = ab(a-1)(b-1)(b-a)$ — From part **a**

$ab(a-1)(b-1)(b-a)=0$ — The system will not have a unique solution if the determinant is zero.

$\Rightarrow a=0$ or $b=0$ or $a=1$ or $b=1$ or $a=b$

Therefore, the system of equations has a unique solution if these conditions are met:

$a\neq 0, b\neq 0, a\neq 1, b\neq 1, a\neq b$

PURE

Exercise 22.2B Reasoning and problem-solving

1 Use row or column operations to show that

a $\begin{vmatrix} a & b & c \\ b & c & a \\ c & a & b \end{vmatrix} = (a+b+c)(ab+ac+bc-a^2-b^2-c^2)$

b $\begin{vmatrix} a^2 & b^2 & c^2 \\ bc & ca & ab \\ 1 & 1 & 1 \end{vmatrix} = (b-a)(c-a)(c-b)(a+b+c)$

c $\begin{vmatrix} a+b & a+c & b+c \\ c & b & a \\ c^2 & b^2 & a^2 \end{vmatrix} = (b-a)(a-c)(c-b)(a+b+c)$

d $\begin{vmatrix} a & -b & c \\ c-b & a+c & a-b \\ -bc & ac & -ab \end{vmatrix} = (a+b)(c-a)(b+c)(a-b+c)$

2 Use row or column operations to fully factorise the determinant $\begin{vmatrix} 1 & 1 & 1 \\ a & a^2 & a^3 \\ b & b^2 & b^3 \end{vmatrix}$. Show your working clearly.

3 a Use row or column operations to show that $\begin{vmatrix} a & b & c \\ a^2 & b^2 & c^2 \\ a^3 & b^3 & c^3 \end{vmatrix} = abc\begin{vmatrix} 1 & 1 & 1 \\ 0 & b-a & c-a \\ 0 & 0 & (c-b)(c-a) \end{vmatrix}$

b Hence state the values of

i $\begin{vmatrix} 2 & 3 & 4 \\ 4 & 9 & 16 \\ 8 & 27 & 64 \end{vmatrix}$ ii $\begin{vmatrix} 5 & 25 & 125 \\ 1 & 1 & 1 \\ 7 & 49 & 343 \end{vmatrix}$

4 a Show that $a^3-b^3=(a-b)(a^2+ab+b^2)$

b Fully factorise $\begin{vmatrix} a & 1 & a^3 \\ b & 1 & b^3 \\ c & 1 & c^3 \end{vmatrix}$

c Hence find the conditions on a, b and c under which the following system of linear equations has a unique solution:

$ax+y+a^3z=1$

$bx+y+b^3z=2$

$cx+y+c^3z=3$

5 a Fully factorise the determinant $\begin{vmatrix} a^3 & b^3 & c^3 \\ a^2 & b^2 & c^2 \\ 1 & 1 & 1 \end{vmatrix}$

b Given that $ab+ac+bc>0$, find the conditions on a, b and c under which the following system of linear equations has a unique solution:

$a^3x+b^3y+c^3z=0$

$a^2x+b^2y+c^2z=-1$

$x+y+z=2$

6 a Express $\begin{vmatrix} a & b & a+b-1 \\ a+1 & b+1 & 2 \\ b & a & 1 \end{vmatrix}$ as a product of three linear factors,

b Therefore, find the conditions on a and b under which the system of equations

$ax+(a+1)y+bz=0$

$bx+(b+1)y+az=0$

$(a+b-1)x+2y+z=0$

has a unique solution other than $x=y=z=0$

7 $\mathbf{S}=\begin{pmatrix} a^2 & b^2 & a \\ 1 & c & b \\ a & ac & ab \end{pmatrix}$

Prove that matrix **S** is singular for all values of a, b and c

8 $\mathbf{T}=\begin{pmatrix} b & x^2 & x \\ b & a^2 & a \\ x & bx & x \end{pmatrix}$, where $a \neq b$

Given that matrix **T** is singular find the possible values of x in terms of a and b

9 $\mathbf{M}=\begin{pmatrix} a+b & x^2 & x \\ a+x & b^2 & b \\ b+x & a^2 & a \end{pmatrix}$, where $a \neq b$

Given that matrix **M** is singular find the possible values of x in terms of a and b

10 $\mathbf{A}=\begin{pmatrix} a & a^2-a & a^2+b^2+c^2 \\ b & b^2-b & a^2+b^2+c^2 \\ c & c^2-c & a^2+b^2+c^2 \end{pmatrix}$ where $a \neq b \neq c$

Show that matrix **A** is non–singular for all real values of a, b and c

22.3 Eigenvalues and eigenvectors

Fluency and skills

A vector whose direction is maintained under a transformation is known as an **eigenvector**. If the matrix used for the transformation is **A** then when the vector **x** is transformed using **A**, the result is a multiple of **x**.

Key point

Eigenvectors of the square matrix **A** are non-zero vectors that satisfy the equation $\mathbf{Ax} = \lambda\mathbf{x}$.
The scalar λ is known as the **eigenvalue**.

You can rearrange the equation $\mathbf{Ax} = \lambda\mathbf{x}$ to give $\mathbf{Ax} - \lambda\mathbf{Ix} = 0$ since $\mathbf{x} = \mathbf{Ix}$

then factorise to give $(\mathbf{A} - \lambda\mathbf{I})\mathbf{x} = 0$

Since **x** is a non-zero vector, it must be the case that the matrix $\mathbf{A} - \lambda\mathbf{I}$ is singular.
Therefore $\det(\mathbf{A} - \lambda\mathbf{I}) = 0$

You need to write $\lambda\mathbf{x}$ as $\lambda\mathbf{Ix}$ where **I** is an identity matrix with the same dimensions as **A** in order to be able to carry out the subtraction.

Key point

The equation $\det(\mathbf{A} - \lambda\mathbf{I}) = 0$ is the **characteristic equation** of **A** and is used to find the eigenvalues.

Example 1

Find the eigenvalues and corresponding eigenvectors of the matrix $\mathbf{A} = \begin{pmatrix} 3 & -1 \\ 4 & -2 \end{pmatrix}$

$\mathbf{A} - \lambda\mathbf{I} = \begin{pmatrix} 3 & -1 \\ 4 & -2 \end{pmatrix} - \lambda\begin{pmatrix} 1 & 0 \\ 0 & 1 \end{pmatrix}$

$= \begin{pmatrix} 3-\lambda & -1 \\ 4 & -2-\lambda \end{pmatrix}$

$\det(\mathbf{A} - \lambda\mathbf{I}) = (3-\lambda)(-2-\lambda) - -4$

$\lambda^2 - \lambda - 2 = 0$

$(\lambda - 2)(\lambda + 1) = 0$ — Solve the equation $\det(\mathbf{A} - \lambda\mathbf{I}) = 0$ to find the possible values of λ

$\lambda = 2, -1$ — These are the eigenvalues.

$\begin{pmatrix} 3 & -1 \\ 4 & -2 \end{pmatrix}\begin{pmatrix} a \\ b \end{pmatrix} = 2\begin{pmatrix} a \\ b \end{pmatrix}$ — Use $\mathbf{Ax} = \lambda\mathbf{x}$ with $\mathbf{x} = \begin{pmatrix} a \\ b \end{pmatrix}$ and $\lambda = 2$ to find the possible vectors.

$3a - b = 2a \Rightarrow a = b$

$4a - 2b = 2b \Rightarrow 4a = 4b$ — Both equations give $a = b$

(continued on the next page)

So a possible eigenvector corresponding to the eigenvalue 2 is $\begin{pmatrix}1\\1\end{pmatrix}$

$\begin{pmatrix}3 & -1\\4 & -2\end{pmatrix}\begin{pmatrix}a\\b\end{pmatrix}=-1\begin{pmatrix}a\\b\end{pmatrix}$

Repeat process with $\lambda=-1$ to find the possible vectors **x**

$3a-b=-a \Rightarrow 4a=b$

$4a-2b=-b \Rightarrow 4a=b$

Both equations give $4a=b$

So a possible eigenvector corresponding to the eigenvalue -1 is $\begin{pmatrix}1\\4\end{pmatrix}$

Any non-zero multiples of these will also be eigenvectors.

The same process can be used for any square matrix **A**.

You will need to be able to find eigenvalues and eigenvectors of 2×2 and 3×3 matrices.

Example 2

Find the eigenvalues and eigenvectors of the matrix $\mathbf{T}=\begin{pmatrix}2 & 2 & 1\\1 & 1 & 2\\0 & 0 & 2\end{pmatrix}$

$\begin{vmatrix}2-\lambda & 2 & 1\\1 & 1-\lambda & 2\\0 & 0 & 2-\lambda\end{vmatrix}=(2-\lambda)[(2-\lambda)(1-\lambda)-2]$

Use the simplest method to calculate the determinant; in this case using the third row to multiply out.

$(2-\lambda)(\lambda^2-3\lambda)=0$

$\lambda(2-\lambda)(\lambda-3)=0$

$\lambda=0, 2, 3$

Solve $\det(\mathbf{A}-\lambda\mathbf{I})$ to find the eigenvalues.

$\begin{pmatrix}2 & 2 & 1\\1 & 1 & 2\\0 & 0 & 2\end{pmatrix}\begin{pmatrix}a\\b\\c\end{pmatrix}=0\begin{pmatrix}a\\b\\c\end{pmatrix}$

Use $\mathbf{Ax}=\lambda\mathbf{x}$ with $\mathbf{x}=\begin{pmatrix}a\\b\\c\end{pmatrix}$ and $\lambda=0$ to find the possible vectors **x**

$2a+2b+c=0$

$a+b+2c=0$

$2c=0 \Rightarrow c=0$

$\Rightarrow a+b=0$ so $a=-b$

So a possible eigenvector corresponding to the eigenvalue 0 is $\begin{pmatrix}1\\-1\\0\end{pmatrix}$

$\begin{pmatrix}2 & 2 & 1\\1 & 1 & 2\\0 & 0 & 2\end{pmatrix}\begin{pmatrix}a\\b\\c\end{pmatrix}=2\begin{pmatrix}a\\b\\c\end{pmatrix}$

Repeat process with $\lambda=2$ to find the possible vectors **x**

(*continued on the next page*)

$2a+2b+c=2a \Rightarrow c=-2b$

$a+b+2c=2b \Rightarrow a+b+2(-2b)=2b \Rightarrow a=5b$

$2c=2c$

So a possible eigenvector corresponding to the eigenvalue

2 is $\begin{pmatrix} 5 \\ 1 \\ -2 \end{pmatrix}$

$$\begin{pmatrix} 2 & 2 & 1 \\ 1 & 1 & 2 \\ 0 & 0 & 2 \end{pmatrix}\begin{pmatrix} a \\ b \\ c \end{pmatrix}=3\begin{pmatrix} a \\ b \\ c \end{pmatrix}$$

Repeat process with $\lambda=3$ to find the possible vectors **x**

$2c=3c \Rightarrow c=0$

$2a+2b+c=3a \Rightarrow a=2b$

$a+b+2c=3b \Rightarrow a=2b$

So a possible eigenvector corresponding to the eigenvalue

3 is $\begin{pmatrix} 2 \\ 1 \\ 0 \end{pmatrix}$

The eigenvectors are $\begin{pmatrix} 1 \\ -1 \\ 0 \end{pmatrix}$, $\begin{pmatrix} 5 \\ 1 \\ -2 \end{pmatrix}$ and $\begin{pmatrix} 2 \\ 1 \\ 0 \end{pmatrix}$

Key point

A matrix will always satisfy its own **characteristic equation.**

This property is known as the Cayley-Hamilton Theorem.

To prove this for any 2×2 matrix, consider $\mathbf{A}=\begin{pmatrix} a & b \\ c & d \end{pmatrix}$

The characteristic equation is $\det(\mathbf{A}-\lambda\mathbf{I})=0$ which leads to $(a-\lambda)(d-\lambda)-bc=0$

Expanding the brackets gives $\lambda^2-(a+d)\lambda+(ad-bc)=0$

- Now replace λ by matrix **A**

$$\begin{pmatrix} a & b \\ c & d \end{pmatrix}^2-(a+d)\begin{pmatrix} a & b \\ c & d \end{pmatrix}+(ad-bc)\begin{pmatrix} 1 & 0 \\ 0 & 1 \end{pmatrix}$$

$$=\begin{pmatrix} a^2+bc & ab+bd \\ ac+cd & bc+d^2 \end{pmatrix}-\begin{pmatrix} a^2+ad & ab+bd \\ ac+cd & ad+d^2 \end{pmatrix}+\begin{pmatrix} ad-bc & 0 \\ 0 & ad-bc \end{pmatrix}$$

$$=\begin{pmatrix} a^2+bc-a^2-ad+ad-bc & ab+bd-ab-bd \\ ac+cd-ac-cd & bc+d^2-ad-d^2+ad-bc \end{pmatrix}$$

$=\begin{pmatrix} 0 & 0 \\ 0 & 0 \end{pmatrix}$ as required.

Notice that the $(ad-bc)$ term is multiplied by the identity matrix, **I**

Otherwise you would be trying to add a constant to a matrix which isn't possible.

PURE

Example 3

Verify the that the matrix $\mathbf{A}=\begin{pmatrix}1 & 0\\ 2 & 3\end{pmatrix}$ satisfies its characteristic equation.

$$\det\begin{pmatrix}1-\lambda & 0\\ 2 & 3-\lambda\end{pmatrix}=0 \Rightarrow (1-\lambda)(3-\lambda)-0=0$$

This becomes $\lambda^2-4\lambda+3=0$

Replace λ with the matrix $\mathbf{A}$

$$\begin{pmatrix}1 & 0\\ 2 & 3\end{pmatrix}^2-4\begin{pmatrix}1 & 0\\ 2 & 3\end{pmatrix}+3\begin{pmatrix}1 & 0\\ 0 & 1\end{pmatrix}$$

You need to replace 3 by 3**I** where **I** is the 2×2 identity matrix.

$$=\begin{pmatrix}1 & 0\\ 8 & 9\end{pmatrix}-\begin{pmatrix}4 & 0\\ 8 & 12\end{pmatrix}+\begin{pmatrix}3 & 0\\ 0 & 3\end{pmatrix}$$

$$=\begin{pmatrix}1-4+3 & 0-0+0\\ 8-8+0 & 9-12+3\end{pmatrix}$$

$$=\begin{pmatrix}0 & 0\\ 0 & 0\end{pmatrix} \text{ as required}$$

Key point

A square matrix is **diagonal** if all its elements except those on the leading diagonal are zero.

You can convert a square matrix into diagonal form. This can be useful when dealing with powers of matrices.

Key point

A matrix, **M**, can be **diagonalised** by finding **P** and **D** such that $\mathbf{M}=\mathbf{PDP}^{-1}$. It can be shown that:

- **D** is a diagonal matrix with the eigenvalues of M along the leading diagonal
- **P** is a matrix where the columns are the eigenvalues of M

The eigenvectors in the columns of **P** must occur in the same order as their corresponding eigenvalues in **D**

Example 4

Diagonalise the matrix $\mathbf{M}=\begin{pmatrix}2 & 0 & 0\\ 0 & 1 & -2\\ 0 & -2 & 4\end{pmatrix}$ by finding matrices **P** and **D** such that $\mathbf{M}=\mathbf{PDP}^{-1}$

$$\det\begin{pmatrix}2-\lambda & 0 & 0\\ 0 & 1-\lambda & -2\\ 0 & -2 & 4-\lambda\end{pmatrix}=(2-\lambda)[(1-\lambda)(4-\lambda)-4]=0$$

(*continued on the next page*)

$(2-\lambda)(\lambda^2-5\lambda+4-4)=0$

$\lambda(2-\lambda)(\lambda-5)=0 \Rightarrow \lambda=0, 2, 5$

Use the characteristic equation to find the eigenvalues.

$\begin{pmatrix} 2 & 0 & 0 \\ 0 & 1 & -2 \\ 0 & -2 & 4 \end{pmatrix}\begin{pmatrix} a \\ b \\ c \end{pmatrix}=\lambda\begin{pmatrix} a \\ b \\ c \end{pmatrix}$

Use $\mathbf{Mx}=\lambda\mathbf{x}$ to find the eigenvectors.

When $\lambda=0$

$2a=0 \Rightarrow a=0$

$b-2c=0 \Rightarrow b=2c$

$-2b+4c=0 \Rightarrow b=2c$

So eigenvector is $\begin{pmatrix} 0 \\ 2 \\ 1 \end{pmatrix}$

When $\lambda=2$

$2a=2a$

$b-2c=2b \Rightarrow b=-2c$

$-2b+4c=2c \Rightarrow b=c$

Therefore $b=c=0$

So eigenvector is $\begin{pmatrix} 1 \\ 0 \\ 0 \end{pmatrix}$

$\begin{pmatrix} 2 & 0 & 0 \\ 0 & 1 & -2 \\ 0 & -2 & 4 \end{pmatrix}\begin{pmatrix} a \\ b \\ c \end{pmatrix}=\lambda\begin{pmatrix} a \\ b \\ c \end{pmatrix}$

When $\lambda=5$

$2a=5a \Rightarrow a=0$

$b-2c=5b \Rightarrow -2b=c$

$-2b+4c=5c \Rightarrow c=-2b$

So eigenvector is $\begin{pmatrix} 0 \\ 1 \\ -2 \end{pmatrix}$

Therefore $\mathbf{D}=\begin{pmatrix} 0 & 0 & 0 \\ 0 & 2 & 0 \\ 0 & 0 & 5 \end{pmatrix}$ and $\mathbf{P}=\begin{pmatrix} 0 & 1 & 0 \\ 2 & 0 & 1 \\ 1 & 0 & -2 \end{pmatrix}$

Ensure you put the columns of **P** in the correct order.

You could use your calculator to check that $\mathbf{M}=\mathbf{PDP}^{-1}$

Exercise 22.3A Fluency and skills

1 a Find the eigenvalues and eigenvectors for each of these matrices.

i $\begin{pmatrix} 2 & 3 \\ 0 & -4 \end{pmatrix}$ ii $\begin{pmatrix} 3 & -1 \\ 4 & -2 \end{pmatrix}$ iii $\begin{pmatrix} 3 & 0 \\ 3 & 7 \end{pmatrix}$ iv $\begin{pmatrix} 4 & -5 \\ -1 & 0 \end{pmatrix}$

b Show that each of the matrices in part **a** satisfies its characteristic equation.

2 Find the eigenvalues and eigenvectors for each of these matrices.

a $\begin{pmatrix} 1 & 0 & -5 \\ 4 & 5 & 1 \\ 0 & 0 & -3 \end{pmatrix}$ b $\begin{pmatrix} 1 & 3 & 0 \\ 2 & 0 & 1 \\ 0 & 0 & 2 \end{pmatrix}$

c $\begin{pmatrix} 5 & -3 & 0 \\ 0 & 2 & 0 \\ 0 & -5 & -4 \end{pmatrix}$ d $\begin{pmatrix} -7 & 0 & 0 \\ 5 & 4 & 1 \\ 4 & 0 & 9 \end{pmatrix}$

3 Given that $\mathbf{A} = \begin{pmatrix} 1 & 3 & -1 \\ -2 & 3 & 2 \\ 1 & 0 & 2 \end{pmatrix}$

a Show that 3 is the only real eigenvalue of **A** and find the eigenvector corresponding to it.

b Show that **A** satisfies its characteristic equation.

4 Given that $\mathbf{A} = \begin{pmatrix} -2 & 6 & -3 \\ 3 & 2 & 6 \\ 2 & 6 & -3 \end{pmatrix}$

a Show that −7 is an eigenvalue of **A** and find the other two eigenvalues.

b Find the eigenvector corresponding to the eigenvalue −7

c Show that **A** satisfies its characteristic equation.

5 Given $\mathbf{A} = \begin{pmatrix} 2 & 0 \\ 0 & 7 \end{pmatrix}$, find matrices **P** and **D** such that $\mathbf{A} = \mathbf{PDP}^{-1}$

6 Given $\mathbf{B} = \begin{pmatrix} 2 & 6 \\ 1 & -3 \end{pmatrix}$, write **B** in the form $\mathbf{B} = \mathbf{PDP}^{-1}$

7 Given $\mathbf{M} = \begin{pmatrix} -4 & 0 & 0 \\ 5 & 2 & 6 \\ 1 & 6 & -3 \end{pmatrix}$, find matrices **P** and **D** such that $\mathbf{M} = \mathbf{PDP}^{-1}$

8 Given $\mathbf{T} = \begin{pmatrix} 7 & -1 & 0 \\ -1 & 7 & 0 \\ 0 & 0 & -7 \end{pmatrix}$, write **T** in the form $\mathbf{T} = \mathbf{PDP}^{-1}$

Reasoning and problem-solving

You can use the diagonalised form of a matrix to calculate powers of a matrix more easily.

If $\mathbf{A} = \mathbf{PDP}^{-1}$ then $\mathbf{A}^n = (\mathbf{PDP}^{-1})^n$

$$\begin{aligned} &= (\mathbf{PDP}^{-1})(\mathbf{PDP}^{-1})(\mathbf{PDP}^{-1})\ldots(\mathbf{PDP}^{-1}) \\ &= (\mathbf{PD})(\mathbf{P}^{-1}\mathbf{P})\mathbf{D}(\mathbf{P}^{-1}\mathbf{P})\mathbf{D}\ldots(\mathbf{P}^{-1}\mathbf{P})(\mathbf{DP}^{-1}) \\ &= (\mathbf{PD})\mathbf{DD}\ldots\mathbf{D}(\mathbf{DP}^{-1}) \\ &= \mathbf{PD}^n\mathbf{P}^{-1} \end{aligned}$$

Since matrix multiplication is associative.

Since $\mathbf{P}^{-1}\mathbf{P} = \mathbf{I}$

Key point

If $\mathbf{A} = \mathbf{PDP}^{-1}$ then $\mathbf{A}^n = \mathbf{PD}^n\mathbf{P}^{-1}$

You can use this to solve problems involving $\mathbf{A}^n$, since

$$\mathbf{D}^n = \begin{pmatrix} a & 0 & 0 \\ 0 & b & 0 \\ 0 & 0 & c \end{pmatrix}^n = \begin{pmatrix} a^n & 0 & 0 \\ 0 & b^n & 0 \\ 0 & 0 & c^n \end{pmatrix}$$

You can use proof by induction to prove this result.

Strategy

To calculate powers of a square matrix, **A**

1. Diagonalise the matrix using $\mathbf{A} = \mathbf{PDP}^{-1}$
2. Use $\mathbf{A}^n = \mathbf{PD}^n\mathbf{P}^{-1}$ to find powers of **A**
3. Use the fact that $\begin{pmatrix} a & 0 & 0 \\ 0 & b & 0 \\ 0 & 0 & c \end{pmatrix}^n = \begin{pmatrix} a^n & 0 & 0 \\ 0 & b^n & 0 \\ 0 & 0 & c^n \end{pmatrix}$

Example 5

Given that $\mathbf{A} = \begin{pmatrix} 3 & -1 \\ 0 & 5 \end{pmatrix}$, show that $\mathbf{A}^n = \begin{pmatrix} 3^n & \frac{3^n}{2} - \frac{5^n}{2} \\ 0 & 5^n \end{pmatrix}$

$$\det\begin{pmatrix} 3-\lambda & -1 \\ 0 & 5-\lambda \end{pmatrix} = 0 \Rightarrow (3-\lambda)(5-\lambda) = 0$$

$\Rightarrow$ eigenvalues are $\lambda = 3$ and 5

$$\begin{pmatrix} 3 & -1 \\ 0 & 5 \end{pmatrix}\begin{pmatrix} a \\ b \end{pmatrix} = 3\begin{pmatrix} a \\ b \end{pmatrix}$$

$5b = 3b \Rightarrow b = 0$

$3a - b = 3a$

So eigenvector is $\begin{pmatrix} 1 \\ 0 \end{pmatrix}$

$$\begin{pmatrix} 3 & -1 \\ 0 & 5 \end{pmatrix}\begin{pmatrix} a \\ b \end{pmatrix} = 5\begin{pmatrix} a \\ b \end{pmatrix}$$

$5b = 5b$

$3a - b = 5a \Rightarrow b = -2a$

So eigenvector is $\begin{pmatrix} 1 \\ -2 \end{pmatrix}$

$$\text{So } \mathbf{A} = \begin{pmatrix} 1 & 1 \\ 0 & -2 \end{pmatrix}\begin{pmatrix} 3 & 0 \\ 0 & 5 \end{pmatrix}\begin{pmatrix} 1 & \frac{1}{2} \\ 0 & -\frac{1}{2} \end{pmatrix}$$

$$\mathbf{A}^n = \begin{pmatrix} 1 & 1 \\ 0 & -2 \end{pmatrix}\begin{pmatrix} 3 & 0 \\ 0 & 5 \end{pmatrix}^n\begin{pmatrix} 1 & \frac{1}{2} \\ 0 & -\frac{1}{2} \end{pmatrix}$$

$$= \begin{pmatrix} 1 & 1 \\ 0 & -2 \end{pmatrix}\begin{pmatrix} 3^n & 0 \\ 0 & 5^n \end{pmatrix}\begin{pmatrix} 1 & \frac{1}{2} \\ 0 & -\frac{1}{2} \end{pmatrix}$$

$$= \begin{pmatrix} 3^n & 5^n \\ 0 & -2(5^n) \end{pmatrix}\begin{pmatrix} 1 & \frac{1}{2} \\ 0 & -\frac{1}{2} \end{pmatrix}$$

$$= \begin{pmatrix} 3^n & \frac{3^n}{2} - \frac{5^n}{2} \\ 0 & 5^n \end{pmatrix} \text{ as required}$$

Find the eigenvalues using the characteristic equation.

1 Find $\mathbf{P}^{-1}$ then write in the form $\mathbf{A} = \mathbf{PDP}^{-1}$

2 Use $\mathbf{A}^n = \mathbf{PD}^n\mathbf{P}^{-1}$

3 Simplify D^n then multiply the matrices together.

Example 6

The matrix $\mathbf{B}=\begin{pmatrix} a & a \\ b & b \end{pmatrix}$ where $a \neq -b$ and $b \neq 0$

a Find the eigenvalues and corresponding eigenvectors of **B**

b Write **B** in the form $\mathbf{ADA}^{-1}$ for a diagonal matrix **D**

c Hence, show that $\mathbf{B}^n = (a+b)^{n-1}\mathbf{B}$

a $\det\begin{pmatrix} a-\lambda & a \\ b & b-\lambda \end{pmatrix}=0$

$\Rightarrow (a-\lambda)(b-\lambda)-ab=0$

$\Rightarrow ab-(a+b)\lambda+\lambda^2-ab=0$

$\Rightarrow \lambda^2-(a+b)\lambda=0$

$\Rightarrow \lambda(\lambda-(a+b))=0$

$\Rightarrow \lambda=0,\ a+b$

Solve $\det(\mathbf{B}-\lambda\mathbf{I})=0$ to find the eigenvalues.

$\begin{pmatrix} a & a \\ b & b \end{pmatrix}\begin{pmatrix} x \\ y \end{pmatrix}=\begin{pmatrix} 0 \\ 0 \end{pmatrix}$

$ax+ay=0 \Rightarrow x=-y$

So an eigenvector corresponding to the eigenvalue of 0 is $\begin{pmatrix} 1 \\ -1 \end{pmatrix}$

Use $\mathbf{Bv}=0$ to find the eigenvectors corresponding to the eigenvalue of 0

$\begin{pmatrix} a & a \\ b & b \end{pmatrix}\begin{pmatrix} x \\ y \end{pmatrix}=(a+b)\begin{pmatrix} x \\ y \end{pmatrix}$

$ax+ay=ax+bx \Rightarrow ay=bx$

So an eigenvector corresponding to the eigenvalue of $a+b$ is $\begin{pmatrix} a \\ b \end{pmatrix}$

Use $\mathbf{Bv}=(a+b)\mathbf{v}$ to find the eigenvectors corresponding to the eigenvalue of $(a+b)$

b $\mathbf{A}=\begin{pmatrix} 1 & a \\ -1 & b \end{pmatrix}$

A is composed of the eigenvectors of **B**

so $\mathbf{A}^{-1}=\dfrac{1}{a+b}\begin{pmatrix} b & -a \\ 1 & 1 \end{pmatrix}$

and $\mathbf{D}=\begin{pmatrix} 0 & 0 \\ 0 & a+b \end{pmatrix}$

D is a diagonal matrix with the eigenvalues on the leading diagonal.

Therefore, $\mathbf{B}=\dfrac{1}{a+b}\begin{pmatrix} 1 & a \\ -1 & b \end{pmatrix}\begin{pmatrix} 0 & 0 \\ 0 & a+b \end{pmatrix}\begin{pmatrix} b & -a \\ 1 & 1 \end{pmatrix}$

c $\mathbf{B}^n=\dfrac{1}{a+b}\begin{pmatrix} 1 & a \\ -1 & b \end{pmatrix}\begin{pmatrix} 0 & 0 \\ 0 & a+b \end{pmatrix}^n\begin{pmatrix} b & -a \\ 1 & 1 \end{pmatrix}$

$=\dfrac{1}{a+b}\begin{pmatrix} 1 & a \\ -1 & b \end{pmatrix}\begin{pmatrix} 0 & 0 \\ 0 & (a+b)^n \end{pmatrix}\begin{pmatrix} b & -a \\ 1 & 1 \end{pmatrix}$

$=\dfrac{1}{a+b}\begin{pmatrix} 0 & a(a+b)^n \\ 0 & b(a+b)^n \end{pmatrix}\begin{pmatrix} b & -a \\ 1 & 1 \end{pmatrix}$

(*Continued on the next page*)

$$=\frac{1}{a+b}\begin{pmatrix} a(a+b)^n & a(a+b)^n \\ b(a+b)^n & b(a+b)^n \end{pmatrix}$$

$$=\frac{(a+b)^n}{a+b}\begin{pmatrix} a & a \\ b & b \end{pmatrix}$$

$$=(a+b)^{n-1}\mathbf{B}$$

See Ch5.2 For a reminder of invariant line.

Another application of eigenvectors is in finding the direction of an invariant line through the origin.

Key point

The eigenvectors of a matrix **A** determine the direction of the invariant lines through the origin of the transformation represented by **A**

If an eigenvalue of a transformation represented by matrix **A** is $\lambda = 1$, then $\mathbf{Av} = \mathbf{v}$ for all corresponding eigenvectors **v**

Therefore,

Key point

If a transformation given by matrix **A** has an eigenvalue of 1 then the corresponding eigenvectors determine the direction of a line of invariant points though the origin.

If you have a repeated eigenvalue then you will sometimes be able to find two non-parallel corresponding eigenvectors. In these cases, the eigenvalue has a whole plane of associated eigenvectors, found by any linear combination of the two non-parallel eigenvectors. So the transformation represented by **A** has an invariant plane, namely the plane of eigenvectors. In other cases, there will only be one eigenvector associated with the repeated eigenvalue. In these cases the transformation has an invariant line corresponding to this eigenvector.

Key point

If a transformation given by matrix **A** has a repeated eigenvalue and this eigenvalue has two non-parallel corresponding eigenvectors, then **A** has an invariant plane defined by a linear combination of two corresponding non-parallel eigenvectors.

Example 7

The matrix $\mathbf{A} = \begin{pmatrix} 4 & -3 & 3 \\ 6 & -5 & 3 \\ 0 & 0 & -2 \end{pmatrix}$ represents a transformation.

Given that the eigenvalues of **A** are 1, −2 and −2,

a For each eigenvalue, find a full set of eigenvectors,

b Describe the geometric significance of the eigenvectors of **A** in relation to the transformation that **A** represents.

(*continued on the next page*)

a For $\lambda = 1$

$$\begin{pmatrix} 4 & -3 & 3 \\ 6 & -5 & 3 \\ 0 & 0 & -2 \end{pmatrix}\begin{pmatrix} x \\ y \\ z \end{pmatrix} = \begin{pmatrix} x \\ y \\ z \end{pmatrix}$$

Use $\mathbf{Av} = \mathbf{v}$ to find the eigenvectors corresponding to the eigenvalue of 1

$-2z = z \Rightarrow z = 0$

$4x - 3y + 3z = x \Rightarrow 3x - 3y = 0$

Since $z = 0$

$\Rightarrow x = y$

Therefore, the set of eigenvectors corresponding to the eigenvalue 1 is the set of all vectors of the form $\alpha\begin{pmatrix} 1 \\ 1 \\ 0 \end{pmatrix}$

Remember, there are infinitely many eigenvectors, all of which are scalar multiples of each other.

For $\lambda = -2$

$$\begin{pmatrix} 4 & -3 & 3 \\ 6 & -5 & 3 \\ 0 & 0 & -2 \end{pmatrix}\begin{pmatrix} x \\ y \\ z \end{pmatrix} = -2\begin{pmatrix} x \\ y \\ z \end{pmatrix}$$

Use $\mathbf{Av} = -2\mathbf{v}$ to find the eigenvectors corresponding to the eigenvalue of -2

$4x - 3y + 3z = -2x \Rightarrow 6x - 3y + 3z = 0$

$\Rightarrow 2x - y + z = 0$

$6x - 5y + 3z = -2y \Rightarrow 6x - 3y + 3z = 0$

This is the same as the first equation.

$-2z = -2z$

This just tells us that z can be any value.

So, we have the equation of a plane $2x - y + z = 0$

$\begin{pmatrix} 0 \\ 1 \\ 1 \end{pmatrix}$ and $\begin{pmatrix} 1 \\ 2 \\ 0 \end{pmatrix}$

Choose any two, non-parallel vectors on the plane.

Therefore, the eigenvectors corresponding to the eigenvalue -2 are linear combinations of $\begin{pmatrix} 0 \\ 1 \\ 1 \end{pmatrix}$ and $\begin{pmatrix} 1 \\ 2 \\ 0 \end{pmatrix}$,

i.e. $\beta\begin{pmatrix} 0 \\ 1 \\ 1 \end{pmatrix} + \gamma\begin{pmatrix} 1 \\ 2 \\ 0 \end{pmatrix}$

b The eigenvector $\begin{pmatrix} 1 \\ 1 \\ 0 \end{pmatrix}$ gives the direction of a line of invariant points through the origin.

The set of eigenvectors $\beta\begin{pmatrix} 0 \\ 1 \\ 1 \end{pmatrix} + \gamma\begin{pmatrix} 1 \\ 2 \\ 0 \end{pmatrix}$ represents an invariant plane.

PURE

Exercise 22.3B Reasoning and problem-solving

1 Given that $\mathbf{A}=\begin{pmatrix}1 & 0\\ 3 & 2\end{pmatrix}$

a Work out $\mathbf{A}^6$

b Find the matrix $\mathbf{A}^n$ in its simplest form.

2 Given that $\mathbf{T}=\begin{pmatrix}1 & -2\\ -2 & 4\end{pmatrix}$

a Work out $\mathbf{T}^5$

b Show that $\mathbf{T}^n=5^{n-1}\begin{pmatrix}1 & -2\\ -2 & 4\end{pmatrix}$

3 Given that $\mathbf{M}=\begin{pmatrix}5 & 2\\ 2 & 5\end{pmatrix}$

a State matrices **U** and **D** such that $\mathbf{M}=\mathbf{UDU}^{-1}$

b Work out the eigenvalues of $\mathbf{M}^3$

c Write down the eigenvectors of $\mathbf{M}^3$

4 The matrix **B** can be written as

$$\mathbf{B}=\begin{pmatrix}1 & 2\\ -3 & 1\end{pmatrix}\begin{pmatrix}7 & 0\\ 0 & -7\end{pmatrix}\begin{pmatrix}\frac{1}{7} & -\frac{2}{7}\\ \frac{3}{7} & \frac{1}{7}\end{pmatrix}$$

a Write down the vector equations of two invariant lines under the transformation represented by matrix **B**

b State the eigenvalues and eigenvectors of $\mathbf{B}^2$

5 The matrix **R** can be written as

$$\mathbf{R}=\begin{pmatrix}1 & 2 & 1\\ 0 & -3 & 1\\ 1 & 0 & 2\end{pmatrix}\begin{pmatrix}3 & 0 & 0\\ 0 & 2 & 0\\ 0 & 0 & -4\end{pmatrix}\begin{pmatrix}6 & 4 & -5\\ -1 & -1 & 1\\ -3 & -2 & 3\end{pmatrix}$$

a State the eigenvalues and the eigenvectors of $\mathbf{R}^4$

b Write down the vector equations of the invariant lines through the origin of the transformation represented by matrix **R**

6 Given that $\mathbf{A}=\begin{pmatrix}5 & -1 & 0\\ 2 & 2 & 0\\ 0 & 0 & 1\end{pmatrix}$

a Work out $\mathbf{A}^3$

b Find the matrix $\mathbf{A}^n$ in its simplest form.

7 a Find the equations of two invariant lines under the transformation $\mathbf{S}=\begin{pmatrix}-3 & 4\\ 2 & -1\end{pmatrix}$

b Which of these lines is also a line of invariant points? Explain your answer.

8 The matrix **M** is given by $\mathbf{M}=\begin{pmatrix}a & b\\ b & a\end{pmatrix}$

where a and b are constants and $b\neq 0$

Find the eigenvalues and eigenvectors of **M**

9 The matrix $\mathbf{M}=\begin{pmatrix}-15 & 24\\ -8 & 13\end{pmatrix}$ represents a transformation.

a Find the eigenvalues and corresponding full sets of eigenvectors of **M**

b Describe the geometric significance of the eigenvectors of **M** in relation to the transformation **M** represents.

10 The matrix $\mathbf{A}=\begin{pmatrix}3 & -10 & -2\\ 0 & -1 & 0\\ 0 & 2 & 1\end{pmatrix}$ represents a transformation.

a Find the eigenvalues of **A**

b For each eigenvalue, find a full set of eigenvectors.

c Describe the geometric significance of the eigenvectors of **A** in relation to the transformation **A** represents.

11 The matrix $\mathbf{T}=\begin{pmatrix} 3 & -2 & 1 \\ 1 & 0 & 1 \\ -2 & 4 & 0 \end{pmatrix}$ represents a transformation.

a Use row and column operations to find the eigenvalues of $\mathbf{T}$

b For each eigenvalue, find a full set of eigenvectors.

c Describe the geometric significance of the eigenvectors of $\mathbf{T}$ in relation to the transformation $\mathbf{T}$ represents.

12 The matrix $\mathbf{M}$ is given by $\mathbf{M}=\begin{pmatrix} 1 & 1 & 2 \\ 0 & 1 & -1 \\ 0 & 3 & 0 \end{pmatrix}$

a Show that there is only one real eigenvalue of $\mathbf{M}$

b Hence, show that the x–axis is a line of invariant points under the transformation represented by $\mathbf{M}$

13 a Given that the matrix $\mathbf{M}$ can be written $\mathbf{UDU}^{-1}$ where $\mathbf{D}$ is a diagonal matrix, show that $\mathbf{M}^n = \mathbf{UD}^n\mathbf{U}^{-1}$

The 2×2 matrix $\mathbf{A}$ has eigenvalues of $\frac{3}{5}$ and 1 with corresponding eigenvectors $\begin{pmatrix} 1 \\ 0 \end{pmatrix}$ and $\begin{pmatrix} -1 \\ 2 \end{pmatrix}$

b Given that $\mathbf{A}^n \to \mathbf{L}$ as $n \to \infty$, find the matrix $\mathbf{L}$

14 Given the matrix $\mathbf{M}=\begin{pmatrix} 3 & k+3 \\ k-3 & -3 \end{pmatrix}$, where k is a constant and $k \neq \pm 3$

a Find the eigenvalues and the corresponding eigenvectors of $\mathbf{M}$

b Hence, show that $\mathbf{M}^{2n+1} = k^{2n}\mathbf{M}$ for all integers n

PURE

22 Summary and review

Chapter summary

- The determinant of a 3×3 matrix is
$$\det\begin{pmatrix} a & b & c \\ d & e & f \\ g & h & i \end{pmatrix} = a\begin{vmatrix} e & f \\ h & i \end{vmatrix} - b\begin{vmatrix} d & f \\ g & i \end{vmatrix} + c\begin{vmatrix} d & e \\ g & h \end{vmatrix}$$
- Under a transformation represented by a 2×2 matrix $\mathbf{T}$, area of image = area of original $\times\,|\det(\mathbf{T})|$
- Under a transformation represented by a 3×3 matrix $\mathbf{T}$, volume of image = volume of original $\times\,|\det(\mathbf{T})|$
- If $\det(\mathbf{T}) < 0$ then the transformation represented by $\mathbf{T}$ involves a reflection.
- If $\mathbf{P} = \begin{pmatrix} a & b & c \\ d & e & f \\ g & h & i \end{pmatrix}$ then the matrix of minors is $\mathbf{M} = \begin{pmatrix} A & B & C \\ D & E & F \\ G & H & I \end{pmatrix}$ where A is the minor of a, that

  is $\det\begin{pmatrix} e & f \\ h & i \end{pmatrix} = (ei - fh)$, B is the minor of b and so on,

  then the inverse matrix is $\mathbf{P}^{-1} = \dfrac{1}{\det(\mathbf{P})}\begin{pmatrix} A & -D & G \\ -B & E & -H \\ C & -F & I \end{pmatrix}$
- A system of linear equations represented by the matrix equation $\mathbf{Mx} = \mathbf{x}'$ has a unique solution if and only if $\det(\mathbf{M}) \neq 0$
- Effect of row and column operations:
  - adding or subtracting any multiple of a row to another row or any multiple of a column to another column will not affect the determinant
  - Swapping two rows or swapping two columns will change the sign of the determinant
  - Multiplying or dividing a row or a column of a matrix by a scalar will multiply or divide the determinant by that scalar.
- Eigenvectors of the square matrix $\mathbf{A}$ are non-zero vectors that satisfy the equation $\mathbf{Ax} = \lambda\mathbf{x}$.
- λ is a scalar known as the eigenvalue.
- The characteristic equation $\det(\mathbf{A} - \lambda\mathbf{I}) = 0$ can be used to find the eigenvalues.
- A matrix will always satisfy its own characteristic equation.
- The eigenvectors of a matrix $\mathbf{T}$ determine the direction of the invariant lines through the origin of the transformation represented by $\mathbf{T}$
- If a transformation given by matrix $\mathbf{T}$ has an eigenvalue of 1 then the corresponding eigenvectors determine the direction of a line of invariant points though the origin.
- If a transformation given by matrix $\mathbf{T}$ has a repeated eigenvalue then $\mathbf{T}$ has an invariant plane defined by a linear combination of two corresponding non-parallel eigenvectors.
- A square matrix is diagonal if all its elements except those on the leading diagonal are zero.

- A 3×3 matrix, $\mathbf{M}$, can be diagonalised by finding $\mathbf{P}$ and $\mathbf{D}$ such that $\mathbf{M} = \mathbf{PDP}^{-1}$
  - $\mathbf{D} = \begin{pmatrix} \lambda_1 & 0 & 0 \\ 0 & \lambda_2 & 0 \\ 0 & 0 & \lambda_3 \end{pmatrix}$ where λ_1, λ_2, λ_3 are the eigenvalues of $\mathbf{M}$
  - $\mathbf{P} = \begin{pmatrix} a_1 & a_2 & a_3 \\ b_1 & b_2 & b_3 \\ c_1 & c_2 & c_3 \end{pmatrix}$ where $\mathbf{v}_n = \begin{pmatrix} a_n \\ b_n \\ c_n \end{pmatrix}$ is an eigenvector corresponding to the eigenvalue λ_n
- If $\mathbf{M} = \mathbf{PDP}^{-1}$ then $\mathbf{M}^n = \mathbf{PD}^n\mathbf{P}^{-1}$

Check and review

You should now be able to...	Review Questions
✔ Use determinants as scale factors.	1, 4
✔ Understand the implications of a negative determinant.	1, 4
✔ Calculate the determinant of a 3×3 matrix.	2, 3
✔ Find the inverse of a 3×3 matrix.	5
✔ Use matrices to solve systems of linear equations.	6
✔ Identify when a system of 3 simultaneous equations has a unique solution, infinitely many solutions and no solutions, and interpret geometrically.	7
✔ Calculate the determinant of a matrix by multiplying out any row or column.	8
✔ Understand the effect of row and column operations on the determinant of a matrix.	9
✔ Factorise a determinant using row and column operations.	10, 11
✔ Find the eigenvalues and eigenvectors of a 2×2 matrix and a 3×3 matrix.	12, 13, 15
✔ Demonstrate that a matrix satisfies its characteristic equation.	14
✔ Diagonalise a square matrix.	16
✔ Use diagonal form to find the power of a matrix.	17
✔ Understand the geometrical significance of eigenvalues and eigenvectors.	18, 19

1 A triangle has area 8 square units.

a Give the area of each image of the triangle under each of these transformations

i $\begin{pmatrix} 5 & 0 \\ 0 & 5 \end{pmatrix}$ ii $\begin{pmatrix} 2 & 4 \\ 3 & 1 \end{pmatrix}$ iii $\begin{pmatrix} \frac{1}{\sqrt{2}} & -\frac{1}{\sqrt{2}} \\ \frac{1}{\sqrt{2}} & \frac{1}{\sqrt{2}} \end{pmatrix}$ iv $\begin{pmatrix} 2a & 3a \\ 4a & -2a \end{pmatrix}$

b Explain which of the transformations involve(s) a reflection.

2 Work out the determinant of each of these matrices.

a $\begin{pmatrix} 3 & -2 & -1 \\ 4 & 0 & 5 \\ 2 & 1 & -3 \end{pmatrix}$ b $\begin{pmatrix} a & -3 & 2a \\ 2b & 4 & b \\ 5 & 0 & -1 \end{pmatrix}$

3 Calculate the values of x for which each of these matrices are singular.

a $\begin{pmatrix} 2 & -1 & 1 \\ x & 3 & 0 \\ -x & 4 & 2 \end{pmatrix}$ b $\begin{pmatrix} x & 3 & 0 \\ 1 & 2 & 2x \\ 1 & -1 & 2 \end{pmatrix}$

4 A cube is transformed by matrix **T** and the volume of the image is 56 cubic units.

a Give the volume of the original cube when

i $\mathbf{T} = \begin{pmatrix} 3 & 4 & 2 \\ 0 & 3 & 1 \\ -5 & 5 & -1 \end{pmatrix}$

ii $\mathbf{T} = \begin{pmatrix} a & -2 & 4 \\ 3 & 4-a & 2a \\ -2a & 0 & 0 \end{pmatrix}$ for $a > 0$

b Explain which of the transformations involve(s) a reflection.

5 Find the inverse, if it exists, of each of these matrices. If it does not exist, explain why not.

a $\begin{pmatrix} 3 & 2 & -1 \\ 4 & 0 & 5 \\ 2 & 1 & -3 \end{pmatrix}$ b $\begin{pmatrix} 6 & -2 & -3 \\ 1 & 0 & -1 \\ 0 & 2 & -3 \end{pmatrix}$

6 Use matrices to find a solution to the simultaneous equations

$3x - 2y + z = 6$

$x - y + 3z = -23$

$5x - 3y + 2z = 5$

7 Three planes have equations

$6x - 2y + 4z = 3$

$(k+4)x - 3y + z = -5$

$-3x + ky - 2z = 7$

a Find the values of k for which the system of equations does not have a unique solution.

b Describe the arrangement of the three planes for each value of k found in part **a**.

8 Multiply out the determinant $\begin{vmatrix} 3 & -2 & 5 \\ -6 & 4 & 0 \\ 5 & -1 & -3 \end{vmatrix}$ using the 3rd row and show that the solution is the same as the result of multiplying out using the 3rd column.

9 Given that $\begin{vmatrix} a & 3 & 2b \\ -6 & 0 & 4b \\ a & 2 & -5b \end{vmatrix} = 90$, state the values of

a $\begin{vmatrix} a & 2b & 3 \\ -6 & 4b & 0 \\ a & -5b & 2 \end{vmatrix}$ b $\begin{vmatrix} a & 3 & 2b \\ 3 & 0 & -2b \\ a & 2 & -5b \end{vmatrix}$

c $\begin{vmatrix} a & -6 & a \\ 3 & 0 & 2 \\ 2b & 4b & -5b \end{vmatrix}$ d $\begin{vmatrix} 0 & 1 & 7 \\ 3 & 0 & -2 \\ a & 2 & -5 \end{vmatrix}$

10 a Show that

$$\begin{vmatrix} 1 & -1 & 1 \\ a & a^2 & -a^2 \\ b & b^2 & -b^2 \end{vmatrix} = ab(a+1)(b+1)\begin{vmatrix} 1 & -1 \\ 1 & -1 \end{vmatrix}$$

b Hence state the value of $\begin{vmatrix} 1 & -1 & 1 \\ a & a^2 & -a^2 \\ b & b^2 & -b^2 \end{vmatrix}$

11 Fully factorise the determinant $\begin{vmatrix} bc & a & a^2 \\ ac & b & b^2 \\ ab & c & c^2 \end{vmatrix}$

12 Find the eigenvalues and corresponding eigenvectors of

a $\begin{pmatrix} -3 & 5 \\ 4 & -2 \end{pmatrix}$ b $\begin{pmatrix} -15 & 12 \\ 6 & -21 \end{pmatrix}$

13 Find the eigenvalues and corresponding eigenvectors of

a $\begin{pmatrix} 2 & 0 & 0 \\ 2 & 3 & 2 \\ 7 & -4 & -3 \end{pmatrix}$ b $\begin{pmatrix} 4 & 2 & 9 \\ 0 & 9 & 0 \\ 2 & 1 & 7 \end{pmatrix}$

14 a Show that the characteristic equation of the matrix $\mathbf{A} = \begin{pmatrix} a & 3 \\ -2 & 2a \end{pmatrix}$ is

$\lambda^2 - 3a\lambda + 2a^2 + 6 = 0$

b Demonstrate that the matrix **A** satisfies its own characteristic equation.

15 Given that $\mathbf{M} = \begin{pmatrix} -1 & 0 & 2 \\ -1 & 2 & 0 \\ -4 & 0 & 3 \end{pmatrix}$

a Show that $\lambda = 2$ is the only real eigenvalue of **M**

b Find the eigenvector corresponding to $\lambda = 2$

16 $\mathbf{A} = \begin{pmatrix} -3 & -2 & 0 \\ 0 & 2 & 0 \\ 3 & 1 & 0 \end{pmatrix}$

State a matrix **P** and a diagonal matrix **D** such that $\mathbf{A} = \mathbf{PDP}^{-1}$

17 $\mathbf{M} = \begin{pmatrix} 2 & 2 & 0 \\ 1 & 1 & 0 \\ 0 & 2 & 1 \end{pmatrix}$

Show that $\mathbf{M}^n = \begin{pmatrix} 2(3^{n-1}) & 2(3^{n-1}) & 0 \\ 3^{n-1} & 3^{n-1} & 0 \\ 3^{n-1}-1 & 3^{n-1}+1 & 1 \end{pmatrix}$

18 The matrix **A** can be written

$\mathbf{A} = \begin{pmatrix} 3 & 2 \\ -1 & 3 \end{pmatrix}\begin{pmatrix} 1 & 0 \\ 0 & -2 \end{pmatrix}\begin{pmatrix} 3 & 2 \\ -1 & 3 \end{pmatrix}^{-1}$

Write down the vector equation of

a A line of invariant points,

b An invariant line which is not a line of invariant points.

19 a Find the eigenvalues of the matrix

$\mathbf{B} = \begin{pmatrix} -3 & 5 & -10 \\ 10 & -8 & 20 \\ 5 & -5 & 12 \end{pmatrix}$

b For each eigenvalue, find the full set of corresponding eigenvectors.

c Describe the geometrical significance of the eigenvectors of **B** in relation to the transformation it represents.

22 Exploration
Going beyond the exams

Did you know?

Internet search engines use algorithms that rank the importance of pages according to an eigenvector of a weighted link matrix.

Research

There are several alternative methods for solving systems of linear equations. Find out about:

- Cramer's rule
- Cholesky decomposition
- Quantum algorithm for linear systems of equations.

Research

Find the eigenvalues and the eigenvectors of the matrix $\mathbf{A} = \begin{bmatrix} 2 & 1 \\ 1 & 2 \end{bmatrix}$.

Now plot the four points (0,0), (1,0), (0, 1) and (1,1) and draw in the three vectors as shown below.

Transform each of the four points using matrix **A**. Draw in the new vectors created by the transformed points.

For each of the vectors consider the following questions.

- Which vectors have had their direction maintained and why?
- Which vectors have been enlarged, by how much, and why?
- How do these observations relate to the eigenvalues?
- How do these observations relate to the eigenvectors?

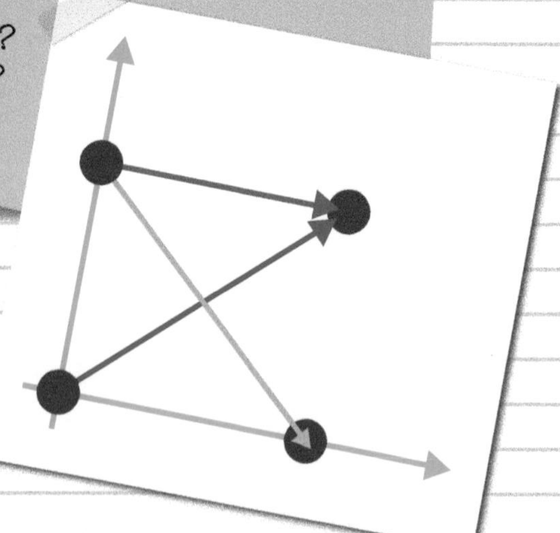

22 Assessment

1 Find the coordinates of the point of intersection between the three planes with equations

$2x+y-z=21$

$x+2y-3z=-14$

$5x-y+z=-7$ **[5 marks]**

2 A linear transformation is defined by the matrix $\mathbf{T}=\begin{pmatrix}1 & 0 & 0\\ 0 & 0.5 & -a\\ 0 & a & 0.5\end{pmatrix}$

When the transformation is applied to a cube, the volume of the image is the same as the volume of the image.

a Calculate the possible values of a **[3]**

b Describe the possible transformation represented by **T** **[2]**

3 Find the range of values of k for which the transformation $\begin{pmatrix}3-k & 2\\ k & k+1\end{pmatrix}$ involves a reflection. **[3]**

4 a Find the eigenvalues and corresponding eigenvectors of the matrix $\mathbf{A}=\begin{pmatrix}2 & -3\\ -8 & 4\end{pmatrix}$ **[5]**

b Verify that A satisfies its own characteristic equation. **[3]**

5 $\mathbf{M}=\begin{pmatrix}1 & k-1 & -k\\ k+2 & 3 & 0\\ -2k & 2 & 8\end{pmatrix}$

a Calculate the values of k for which **M** is singular. **[4]**

b Three planes have equations

$x-z=5$

$3x+3y=2$

$-2x+2y+8z=-1$

Explain how your answer to part **a** guarantees that the three planes have a unique point of intersection. **[3]**

6 Given that $\mathbf{A}=\begin{pmatrix}2 & 0 & 1\\ 0 & 1 & 0\\ 1 & 0 & 2\end{pmatrix}$, prove that $\mathbf{A}^n=\frac{1}{2}\begin{pmatrix}3^n+1 & 0 & 3^n-1\\ 0 & 2 & 0\\ 3^n-1 & 0 & 3^n+1\end{pmatrix}$ for all $n\in\mathbb{N}$ **[5]**

7 Use row and column operations to show that $\begin{vmatrix}2 & 1 & 2\\ 4 & 1 & 2\\ 1 & -1 & 1\end{vmatrix}=A\begin{vmatrix}0 & 1 & 0\\ 1 & 1 & 0\\ 1 & 0 & 1\end{vmatrix}$, where A is a constant to be found. Explain each step clearly. **[4]**

8 The 3×3 matrices **A** and **B** are such that $\mathbf{AB}=\begin{pmatrix}6 & 1 & -2\\ 14 & 2 & 4\\ 2 & 0 & -1\end{pmatrix}$ and $\mathbf{B}=\begin{pmatrix}a & 0 & 0\\ 0 & 2 & 2\\ 1 & 0 & 1\end{pmatrix}$

Find the matrix **A** in terms of a **[5]**

9 Given that the determinant of $\begin{pmatrix} a & b & c \\ d & e & f \\ g & h & i \end{pmatrix}$ is 30, write down the determinants of

a $\begin{pmatrix} a & b & 3c \\ 2d & 2e & 6f \\ g & h & 3i \end{pmatrix}$ b $\begin{pmatrix} g & h & i \\ d+a & e+b & f+c \\ a & b & c \end{pmatrix}$ **[2]**

10 a Use at least 2 row or column operations to fully factorise $\begin{vmatrix} a & b & c \\ a^2 & b^2 & c^2 \\ 1 & 1 & 1 \end{vmatrix}$ **[4]**

b Hence or otherwise, calculate the value of $\begin{vmatrix} k & k+1 & k-1 \\ k^2 & (k+1)^2 & (k-1)^2 \\ 1 & 1 & 1 \end{vmatrix}$ **[2]**

11 $\mathbf{T} = \begin{pmatrix} 1 & 2 & 0 \\ 0 & 0 & 3 \\ 2 & 0 & 1 \end{pmatrix}$

a Show that $\lambda = 3$ is the only real eigenvalue of **T** **[6]**

b Find the unit eigenvector corresponding to $\lambda = 3$ **[5]**

c Hence write down the vector equation of an invariant line under the transformation represented by matrix **T** **[2]**

12 Use row and column operations to solve the equation **[6]**

$$\begin{vmatrix} 1 & x+2 & x-1 \\ x^2 & 1 & 4 \\ x & -1 & 2 \end{vmatrix} = 0$$

13 Given that $\mathbf{M} = \begin{pmatrix} 0 & 0 & 2 \\ -2 & 1 & 0 \\ 2 & 0 & 0 \end{pmatrix}$

a Find matrix **P** and diagonal matrix **D** such that $\mathbf{M} = \mathbf{PDP}^{-1}$ **[8]**

b Hence find $\mathbf{M}^4$. You must show how you used your answer to part **a**. **[3]**

c Explain how you know from part **a** that the y-axis is a line of invariant points. **[2]**

14 Three planes have equations

$x + 2y - 3z = -2$

$4x - z = 3$

$2x - 4y + 5z = 1$

Identify the geometric configuration of the planes. Fully justify your answer. **[4]**

15 The matrix **A** can be written $\mathbf{A} = \begin{pmatrix} 1 & 0 & 1 \\ 1 & -1 & 2 \\ 0 & 1 & -2 \end{pmatrix} \begin{pmatrix} 3 & 0 & 0 \\ 0 & -7 & 0 \\ 0 & 0 & -7 \end{pmatrix} \begin{pmatrix} 1 & 0 & 1 \\ 1 & -1 & 2 \\ 0 & 1 & -2 \end{pmatrix}^{-1}$

Write down three eigenvectors of **A** and explain their geometrical significance in relation to the transformation represented by **A** **[4]**

23 Vectors 2

Your mobile phone may be a surprising thing to associate with the mathematics of vectors, but vectors are used extensively by the engineers who develop and design the technology and infrastructure of mobile phone networks. For example, the electromagnetic pulses that are used to carry the radio frequency signals between transmitters and your phone are analysed using such mathematics. Vectors are also used to analyse the circuitry and technology that underpins the design of the transmitter itself and your phone as a receiver/transmitter.

The use of vectors by engineers is widespread and of great importance in many aspects of our lives, particularly in the technology that we associate with modern living such as televisions, radios, microwave ovens and more. The development of our modern-day transportation systems including electric trains, electric cars and aircraft has all been achieved by our increasing knowledge and application of science, together with our use of vector mathematics.

Orientation

What you need to know	What you will learn	What this leads to
Ch6 Vectors 1	• To calculate the vector product of two vectors. • To use the vector product to find angles and areas. • To find the vector equation of a plane. • To calculate the distance between two lines. • To calculate the distance between a line and a plane.	**Careers** • Electrical engineering. • Mechanical engineering.

23.1 The vector product

Fluency and skills

Vectors can be multiplied by calculating the scalar product. Another sort of vector multiplication is the vector product.

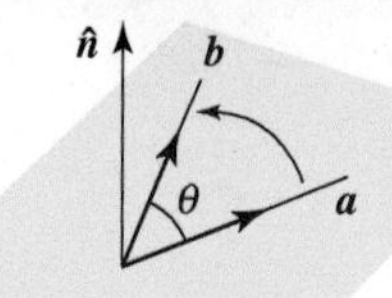

Key point

The **vector product** $\mathbf{a}\times\mathbf{b}$ of vectors **a** and **b** is defined as $\mathbf{a}\times\mathbf{b}=|\mathbf{a}||\mathbf{b}|\sin\theta\hat{\boldsymbol{n}}$ where θ is the angle between the vectors **a** and **b** and $\hat{\boldsymbol{n}}$ is a unit vector perpendicular to both **a** and **b**

The 'right-hand rule' can be used to determine the direction of the cross product. Use your index finger for vector **a** and middle finger for vector **b**; then your thumb will indicate the direction of $\mathbf{a}\times\mathbf{b}$

The vector product is sometimes called the **cross product**.

Suppose you have a plane containing the vectors **a** and **b**. An anticlockwise angle from vector **a** to vector **b** implies that $\hat{\boldsymbol{n}}$ is directed 'upwards,' whereas a clockwise angle implies that $\hat{\boldsymbol{n}}$ is directed in the opposite direction.

Therefore, you can see that the vector product is not commutative.

Example 1

The convention is to use 'right-handed' axes as shown here

Show that $\mathbf{i}\times\mathbf{j}=\mathbf{k}$ and $\mathbf{j}\times\mathbf{i}=-\mathbf{k}$

$\mathbf{i}\times\mathbf{j}=|1||1|\sin 90\mathbf{k}$

$=\mathbf{k}$

$\mathbf{j}\times\mathbf{i}=|1||1|\sin(-90)\mathbf{k}$

$=-\mathbf{k}$

Since **i**, **j** and **k** are unit vectors which are perpendicular to each other.

The angle between **i** and **j** is now 90° clockwise which you can write as –90° or +270°

Similarly, you can show that

Key point

$\mathbf{i}\times\mathbf{j}=\mathbf{k}$, $\mathbf{j}\times\mathbf{k}=\mathbf{i}$ and $\mathbf{k}\times\mathbf{i}=\mathbf{j}$

And that, in general

Key point

For any vectors **a**, **b** and **c**

- $\mathbf{a}\times\mathbf{b}=-\mathbf{b}\times\mathbf{a}$ (anticommutative property)
- $\mathbf{a}\times(\mathbf{b}+\mathbf{c})=\mathbf{a}\times\mathbf{b}+\mathbf{a}\times\mathbf{c}$ (distributive property)

From the definition $\mathbf{a}\times\mathbf{b}=|\mathbf{a}||\mathbf{b}|\sin\theta\,\hat{\mathbf{n}}$, it is clear that if vectors **a** and **b** are parallel, then $\mathbf{a}\times\mathbf{b}=\mathbf{0}$ since the angle between parallel vectors is either 0° or 180° and $\sin 0=\sin 180=0$

Key point

For any non-zero vectors **a** and **b**,

$\mathbf{a}\times\mathbf{b}=\mathbf{0}$ if and only if **a** and **b** are parallel

PURE

An important consequence of this fact is that $\mathbf{a}\times\mathbf{a}=\mathbf{0}$ for any vector **a**

Example 2

Given that **a** and **b** and **c** are non-parallel vectors and $\mathbf{b}\neq\mathbf{0}$

a Write $(\mathbf{a}+3\mathbf{b})\times(\mathbf{a}-\mathbf{b})$ in its simplest form,

b If $\mathbf{a}\times\mathbf{b}=\mathbf{b}\times\mathbf{c}$, show that $\mathbf{a}+\mathbf{c}=\lambda\mathbf{b}$ for some scalar λ

a $(\mathbf{a}+3\mathbf{b})\times(\mathbf{a}-\mathbf{b})=\mathbf{a}\times\mathbf{a}+\mathbf{a}\times(-\mathbf{b})+3\mathbf{b}\times\mathbf{a}+(3\mathbf{b})\times(-\mathbf{b})$

$=\mathbf{a}\times\mathbf{a}-\mathbf{a}\times\mathbf{b}+3\mathbf{b}\times\mathbf{a}-3\mathbf{b}\times\mathbf{b}$

$=-\mathbf{a}\times\mathbf{b}+3\mathbf{b}\times\mathbf{a}$ — Since $\mathbf{a}\times\mathbf{a}=\mathbf{0}$ and $\mathbf{b}\times\mathbf{b}=\mathbf{0}$

$=\mathbf{b}\times\mathbf{a}+3\mathbf{b}\times\mathbf{a}$ — Since $\mathbf{a}\times\mathbf{b}=-\mathbf{b}\times\mathbf{a}$

$=4\mathbf{b}\times\mathbf{a}$

b If $\mathbf{a}\times\mathbf{b}=\mathbf{b}\times\mathbf{c}$

Then $\mathbf{b}\times\mathbf{c}-\mathbf{a}\times\mathbf{b}=0$

$\mathbf{b}\times\mathbf{c}+\mathbf{b}\times\mathbf{a}=0$ — Use the anticommutative property.

Therefore $\mathbf{b}\times(\mathbf{a}+\mathbf{c})=0$ — Use the distributive property.

So $\mathbf{a}+\mathbf{c}$ must be parallel to **b**

Therefore $\mathbf{a}+\mathbf{c}=\lambda\mathbf{b}$ for some scalar λ — Since $\mathbf{b}\neq\mathbf{0}$ and $\mathbf{a}\neq-\mathbf{c}$ (since not parallel).

If you have two vectors $\mathbf{a}=a_1\mathbf{i}+a_2\mathbf{j}+a_3\mathbf{k}$ and $\mathbf{b}=b_1\mathbf{i}+b_2\mathbf{j}+b_3\mathbf{k}$ then you can find the vector product

$$\begin{aligned}\mathbf{a}\times\mathbf{b}&=(a_1\mathbf{i}+a_2\mathbf{j}+a_3\mathbf{k})\times(b_1\mathbf{i}+b_2\mathbf{j}+b_3\mathbf{k})\\&=a_1b_1(\mathbf{i}\times\mathbf{i})+a_1b_2(\mathbf{i}\times\mathbf{j})+a_1b_3(\mathbf{i}\times\mathbf{k})+a_2b_1(\mathbf{j}\times\mathbf{i})+a_2b_2(\mathbf{j}\times\mathbf{j})+a_2b_3(\mathbf{j}\times\mathbf{k})\\&\quad+a_3b_1(\mathbf{k}\times\mathbf{i})+a_3b_2(\mathbf{k}\times\mathbf{j})+a_3b_3(\mathbf{k}\times\mathbf{k})\\&=a_1b_2\mathbf{k}-a_1b_3\mathbf{j}-a_2b_1\mathbf{k}+a_2b_3\mathbf{i}+a_3b_1\mathbf{j}-a_3b_2\mathbf{i}\\&=(a_2b_3-a_3b_2)\mathbf{i}-(a_1b_3-a_3b_1)\mathbf{j}+(a_1b_2-a_1b_2)\mathbf{k}\end{aligned}$$

Since $\mathbf{i}\times\mathbf{j}=\mathbf{k}$, $\mathbf{j}\times\mathbf{k}=\mathbf{i}$ and $\mathbf{k}\times\mathbf{i}=\mathbf{j}$, and also $\mathbf{i}\times\mathbf{i}=\mathbf{j}\times\mathbf{j}=\mathbf{k}\times\mathbf{k}=0$

Key point

If you have two vectors $\mathbf{a}=a_1\mathbf{i}+a_2\mathbf{j}+a_3\mathbf{k}$ and $\mathbf{b}=b_1\mathbf{i}+b_2\mathbf{j}+b_3\mathbf{k}$, then the vector product is defined as $\mathbf{a}\times\mathbf{b}=(a_2b_3-a_3b_2)\mathbf{i}-(a_1b_3-a_3b_1)\mathbf{j}+(a_1b_2-a_1b_2)\mathbf{k}$

Example 3

Given $\mathbf{a}=\begin{pmatrix}1\\-3\\-2\end{pmatrix}$ and $\mathbf{b}=\begin{pmatrix}-2\\0\\4\end{pmatrix}$,

a Find a vector perpendicular to both **a** and **b**

b Calculate the acute angle between **a** and **b**

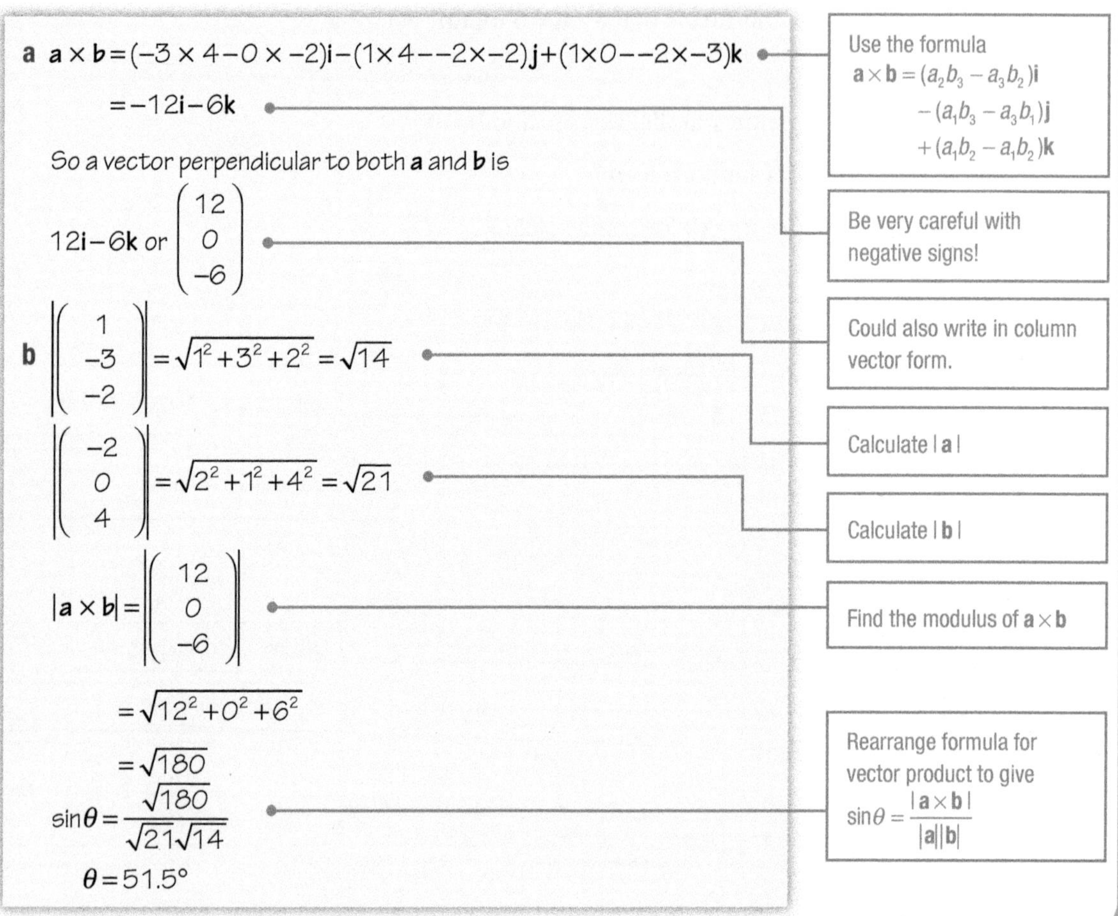

a $\mathbf{a}\times\mathbf{b}=(-3\times4-0\times-2)\mathbf{i}-(1\times4--2\times-2)\mathbf{j}+(1\times0--2\times-3)\mathbf{k}$

$=-12\mathbf{i}-6\mathbf{k}$

So a vector perpendicular to both **a** and **b** is

$12\mathbf{i}-6\mathbf{k}$ or $\begin{pmatrix}12\\0\\-6\end{pmatrix}$

b $\left|\begin{pmatrix}1\\-3\\-2\end{pmatrix}\right|=\sqrt{1^2+3^2+2^2}=\sqrt{14}$

$\left|\begin{pmatrix}-2\\0\\4\end{pmatrix}\right|=\sqrt{2^2+1^2+4^2}=\sqrt{21}$

$|\mathbf{a}\times\mathbf{b}|=\left|\begin{pmatrix}12\\0\\-6\end{pmatrix}\right|$

$=\sqrt{12^2+0^2+6^2}$

$=\sqrt{180}$

$\sin\theta=\dfrac{\sqrt{180}}{\sqrt{21}\sqrt{14}}$

$\theta=51.5°$

Use the formula $\mathbf{a}\times\mathbf{b}=(a_2b_3-a_3b_2)\mathbf{i}-(a_1b_3-a_3b_1)\mathbf{j}+(a_1b_2-a_2b_2)\mathbf{k}$

Be very careful with negative signs!

Could also write in column vector form.

Calculate $|\mathbf{a}|$

Calculate $|\mathbf{b}|$

Find the modulus of $\mathbf{a}\times\mathbf{b}$

Rearrange formula for vector product to give $\sin\theta=\dfrac{|\mathbf{a}\times\mathbf{b}|}{|\mathbf{a}||\mathbf{b}|}$

See Ch22.1

For a remind of how to calculate determinants.

From your study of determinants of 3×3 matrices, you will be able to use the fact that

$$\mathbf{a}\times\mathbf{b}=\left|\begin{pmatrix}\mathbf{i}&\mathbf{j}&\mathbf{k}\\a_1&a_2&a_3\\b_1&b_2&b_3\end{pmatrix}\right|$$

Example 4

Find the vector product of $2\mathbf{i}+\mathbf{j}-\mathbf{k}$ and $\mathbf{i}-3\mathbf{j}-2\mathbf{k}$

$\left|\begin{pmatrix}\mathbf{i}&\mathbf{j}&\mathbf{k}\\2&1&-1\\1&-3&-2\end{pmatrix}\right|=\mathbf{i}(-2-3)-\mathbf{j}(-4--1)+\mathbf{k}(-6-1)$

$=-5\mathbf{i}+3\mathbf{j}-7\mathbf{k}$

Use $\mathbf{a}\times\mathbf{b}=\left|\begin{pmatrix}\mathbf{i}&\mathbf{j}&\mathbf{k}\\a_1&a_2&a_3\\b_1&b_2&b_3\end{pmatrix}\right|$ with the coefficients of **i**, **j** and **k** you are given.

Calculator

Try it on your calculator

You can use a calculator to find the vector product.

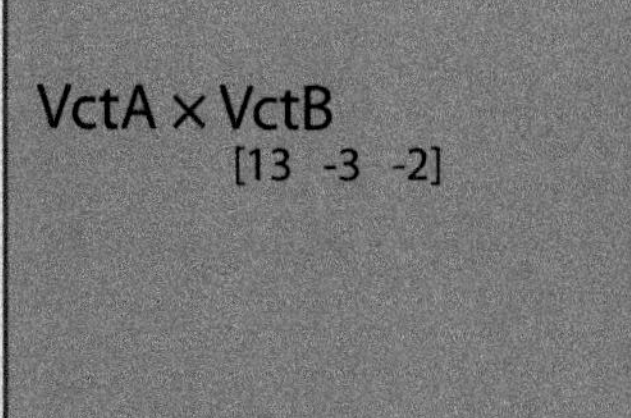

Activity

Find out how to work out $\begin{pmatrix}1\\5\\-1\end{pmatrix}\times\begin{pmatrix}0\\-2\\3\end{pmatrix}$ on your calculator.

The vector product can also be used to calculate areas. For example, the area of the triangle shown can be found using Area $=\frac{1}{2}|\mathbf{a}||\mathbf{b}|\sin\theta$ which, by definition, is $\frac{1}{2}|\mathbf{a}\times\mathbf{b}|$

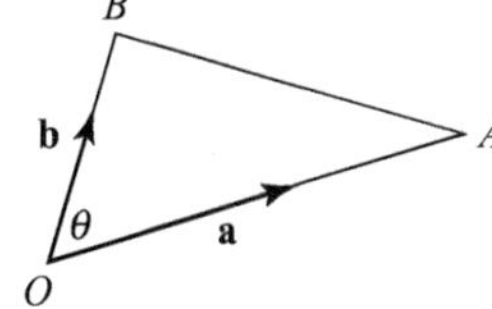

Key point

Area of triangle $OAB=\frac{1}{2}|\mathbf{a}\times\mathbf{b}|$

You may have to first work out vectors **a** and **b** from the coordinates of the vertices of the triangle. So, in general,

Key point

Area of triangle $ABC=\frac{1}{2}|\overrightarrow{AB}\times\overrightarrow{AC}|$

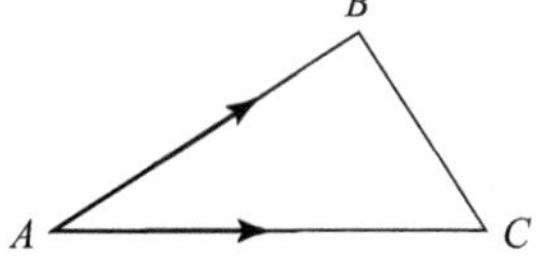

Example 5

Calculate the area of the triangle ABC with vertices at $A(2, 5, -1)$, $B(-4, 9, -6)$ and $C(3, -4, 8)$

$$\overrightarrow{AB}=\begin{pmatrix}-4\\9\\-6\end{pmatrix}-\begin{pmatrix}2\\5\\-1\end{pmatrix}=\begin{pmatrix}-6\\4\\-5\end{pmatrix}$$

$$\overrightarrow{AC}=\begin{pmatrix}3\\-4\\8\end{pmatrix}-\begin{pmatrix}2\\5\\-1\end{pmatrix}=\begin{pmatrix}1\\-9\\9\end{pmatrix}$$

Calculate $\overrightarrow{AB}$ and $\overrightarrow{AC}$ by subtracting position vectors, e.g. $\overrightarrow{AB}=\overrightarrow{OB}-\overrightarrow{OA}$

$$\text{Area of triangle}=\frac{1}{2}\left|\begin{pmatrix}-6\\4\\-5\end{pmatrix}\times\begin{pmatrix}1\\-9\\9\end{pmatrix}\right|$$

Use Area $=\frac{1}{2}|\overrightarrow{AB}\times\overrightarrow{AC}|$

$$=\frac{1}{2}\left|\begin{pmatrix}-9\\49\\50\end{pmatrix}\right|$$

$$=\frac{1}{2}\sqrt{(-9)^2+49^2+50^2}$$

Find magnitude of vector.

$= 35.3$ square units

Exercise 23.1A Fluency and skills

1 Simplify each of these vector products.

a $\mathbf{j}\times\mathbf{k}$ b $\mathbf{i}\times\mathbf{k}$

c $\mathbf{k}\times\mathbf{k}$ d $(\mathbf{i}+\mathbf{j})\times\mathbf{j}$

2 Fully simplify each of these expressions involving vectors **a**, **b** and **c**

a $\mathbf{a}\times(\mathbf{a}+\mathbf{b})$ b $\mathbf{a}\times(\mathbf{b}+2\mathbf{c})+\mathbf{c}\times\mathbf{a}$

c $(\mathbf{a}+\mathbf{b})\times(\mathbf{a}-\mathbf{b})$ d $(\mathbf{a}+\mathbf{b}+\mathbf{c})\times(\mathbf{a}-\mathbf{b}+\mathbf{c})$

3 Use the distributive and anticommutative properties of the vector product to show that

a $\mathbf{a}\times\mathbf{b}+\mathbf{b}\times\mathbf{c}+\mathbf{a}\times\mathbf{c}=(\mathbf{b}+\mathbf{a})\times(\mathbf{c}-\mathbf{a})$

b $\mathbf{b}\times(2\mathbf{a}+\mathbf{c})-\mathbf{c}\times(\mathbf{b}-\mathbf{c})=2\mathbf{b}\times(\mathbf{a}+\mathbf{c})$

4 The non-zero vectors **a** and **b** are such that $(\mathbf{a}+2\mathbf{b})\times(\mathbf{a}-3\mathbf{b})=0$

Deduce the possible sizes of the angle between **a** and **b**

5 For the vectors $\mathbf{a}=2\mathbf{i}-3\mathbf{j}+\mathbf{k}$, $\mathbf{b}=\mathbf{i}+5\mathbf{j}-2\mathbf{k}$ and $\mathbf{c}=-\mathbf{i}+3\mathbf{j}+4\mathbf{k}$, calculate

a $\mathbf{a}\times\mathbf{b}$ b $\mathbf{a}\times\mathbf{c}$ c $\mathbf{b}\times\mathbf{c}$

d $\mathbf{b}\times\mathbf{a}$ e $\mathbf{c}\times\mathbf{a}$ f $\mathbf{c}\times\mathbf{b}$

g $\mathbf{a}\times\mathbf{a}$ h $\mathbf{b}\times\mathbf{b}$

6 Calculate each of these vector products.

a $\begin{pmatrix}0\\2\\-5\end{pmatrix}\times\begin{pmatrix}3\\-1\\4\end{pmatrix}$

b $(2\mathbf{i}+3\mathbf{j}+\mathbf{k})\times(6\mathbf{i}-4\mathbf{j}+7\mathbf{k})$

c $(7\mathbf{j}-2\mathbf{k})\times(-4\mathbf{i}-9\mathbf{j}+3\mathbf{k})$

d $\begin{pmatrix}1\\-8\\-2\end{pmatrix}\times\begin{pmatrix}-4\\-11\\0\end{pmatrix}$

e $\begin{pmatrix}\sqrt{3}\\3\\1\end{pmatrix}\times\begin{pmatrix}0\\\sqrt{3}\\-\sqrt{3}\end{pmatrix}$

f $(a\mathbf{i}-2a\mathbf{j}-\mathbf{k})\times(a\mathbf{i}+\mathbf{j}-5a\mathbf{k})$

7 Find a vector which is perpendicular to both $\begin{pmatrix}1\\3\\-2\end{pmatrix}$ and $\begin{pmatrix}-1\\4\\7\end{pmatrix}$

8 Find a vector which is perpendicular to both $6\mathbf{i}+3\mathbf{j}-\mathbf{k}$ and $7\mathbf{i}+3\mathbf{k}$

9 Find a unit vector which is perpendicular to both $\begin{pmatrix}-5\\0\\0\end{pmatrix}$ and $\begin{pmatrix}-2\\1\\-3\end{pmatrix}$

10 Find a unit vector which is perpendicular to both $2\mathbf{i}-\mathbf{j}$ and $\mathbf{j}-2\mathbf{k}$

11 Use the vector product to show that the vectors $-2\mathbf{i}+6\mathbf{j}-4\mathbf{k}$ and $3\mathbf{i}-9\mathbf{j}+6\mathbf{k}$ are parallel.

12 Use the vector product to find the acute angle between these pairs of vectors.

a $\begin{pmatrix}9\\-2\\0\end{pmatrix}$ and $\begin{pmatrix}4\\1\\1\end{pmatrix}$

b $2\mathbf{i}-\mathbf{j}$ and $\mathbf{j}-6\mathbf{k}$

c $3\mathbf{i}+2\mathbf{j}+\mathbf{k}$ and $\mathbf{i}+\mathbf{k}$

d $\begin{pmatrix}5\\3\\\sqrt{2}\end{pmatrix}$ and $\begin{pmatrix}7\\-\sqrt{2}\\\sqrt{2}\end{pmatrix}$

13 Find, in surd form, the sine of the angle between $3\mathbf{i}+\mathbf{j}+2\mathbf{k}$ and $\mathbf{i}+2\mathbf{j}$

14 Find the values of *a*, *b* and *c* given that $\begin{pmatrix}1\\a\\-2\end{pmatrix}\times\begin{pmatrix}b\\0\\3\end{pmatrix}=\begin{pmatrix}12\\-13\\c\end{pmatrix}$

15 Find the values of *a*, *b* and *c* given that $\begin{pmatrix}3\\-4\\a\end{pmatrix}\times\begin{pmatrix}0\\2\\b\end{pmatrix}=\begin{pmatrix}10\\9\\c\end{pmatrix}$

16 A triangle *ABC* is such that $AB=2\mathbf{i}-\mathbf{j}+\mathbf{k}$ and $AC=7\mathbf{i}+\mathbf{j}-4\mathbf{k}$. Calculate the exact area of the triangle.

17 A triangle *DEF* is such that $ED=6\mathbf{i}+5\mathbf{j}-\mathbf{k}$ and $EF=-4\mathbf{i}-2\mathbf{j}+3\mathbf{k}$. Calculate the area of the triangle. Give your answer to 3 significant figures.

18 Calculate the exact area of the triangles with vertices at

a (0, 0, 0), (3, –1, 4) and (2, –3, –5)

b (1, 1, –2), (0, 1, –1) and (–2, 0, 1)

Reasoning and problem-solving

Two forms for the equation of a line in 3D are the Cartesian form $\frac{x-a_1}{b_1}=\frac{y-a_2}{b_2}=\frac{z-a_3}{b_3}$ and the vector form $\mathbf{r}=\mathbf{a}+\lambda\mathbf{b}$

A third possibility is the **vector product form**.

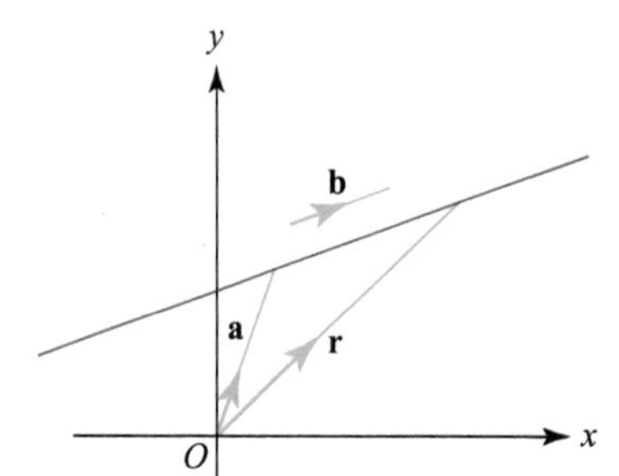

Key point

A straight line passing through the point with position vector **a** and parallel to the vector **b** has equation $(\mathbf{r}-\mathbf{a})\times\mathbf{b}=\mathbf{0}$

This follows from the fact that any vector on the line will be parallel to the vector **b** and the vector product of two parallel vectors is zero.

You can expand the brackets of $(\mathbf{r}-\mathbf{a})\times\mathbf{b}=\mathbf{0}$ to give $\mathbf{r}\times\mathbf{b}-\mathbf{a}\times\mathbf{b}=\mathbf{0}$ which rearranges to $\mathbf{r}\times\mathbf{b}=\mathbf{a}\times\mathbf{b}$

You should calculate the result of $\mathbf{a}\times\mathbf{b}$ using the vectors given.

Strategy

To find the equation of a line in vector product form

1. Find a vector parallel to the line by subtracting the position vectors of two points on the line.
2. Use the equation $(\mathbf{r}-\mathbf{a})\times\mathbf{b}=\mathbf{0}$ where **a** lies on the line and **b** is parallel to the line.
3. Expand the brackets and rearrange if required.

Example 6

Find the equation of the line through the points (2, 0, −1) and (−3, 1, 4) in the form $\mathbf{r}\times\mathbf{b}=\mathbf{c}$

A vector parallel to the plane is given by $\begin{pmatrix}2\\0\\-1\end{pmatrix}-\begin{pmatrix}-3\\1\\4\end{pmatrix}=\begin{pmatrix}5\\-1\\-5\end{pmatrix}$ — (1) Since these are the position vectors of two points on the line.

So the equation can be written $\left(\mathbf{r}-\begin{pmatrix}2\\0\\-1\end{pmatrix}\right)\times\begin{pmatrix}5\\-1\\-5\end{pmatrix}=\mathbf{0}$ — (2) Use $(\mathbf{r}-\mathbf{a})\times\mathbf{b}=\mathbf{0}$

$\mathbf{r}\times\begin{pmatrix}5\\-1\\-5\end{pmatrix}-\begin{pmatrix}2\\0\\-1\end{pmatrix}\times\begin{pmatrix}5\\-1\\-5\end{pmatrix}=\mathbf{0}$ — (3) Expand the brackets.

Which can be rearranged to give

$\mathbf{r}\times\begin{pmatrix}5\\-1\\-5\end{pmatrix}=\begin{pmatrix}2\\0\\-1\end{pmatrix}\times\begin{pmatrix}5\\-1\\-5\end{pmatrix}$

$\mathbf{r}\times\begin{pmatrix}5\\-1\\-5\end{pmatrix}=\begin{pmatrix}-1\\5\\-2\end{pmatrix}$ — By calculating the vector product on the right of the equation.

Exercise 23.1B Reasoning and problem-solving

1 Show that $\mathbf{a} \times (\mathbf{b}+\mathbf{c}) = \mathbf{a} \times \mathbf{b} + \mathbf{a} \times \mathbf{c}$ for any 3D vectors **a**, **b** and **c**

2 Show that $\mathbf{a} \times \mathbf{b} = -\mathbf{b} \times \mathbf{a}$

3 Find the equation of the line passing through the point (2, 5, 0) and parallel to the vector $\mathbf{i}+\mathbf{j}-\mathbf{k}$ in the form $(\mathbf{r}-\mathbf{a}) \times \mathbf{b} = \mathbf{0}$

4 Find the equation of the line that intercepts the y-axis at $y=3$ and is parallel to the vector **k** in the form $(\mathbf{r}-\mathbf{a}) \times \mathbf{b} = \mathbf{0}$

5 Find the equation of the line that intercepts the z-axis at $z=-2$ and is parallel to the vector $\mathbf{i}+\mathbf{k}$ in the form $\mathbf{r} \times \mathbf{b} = \mathbf{c}$

6 Find the equation of the line that passes through the points (1, 3, 5) and (4, 2, 7) in the form $(\mathbf{r}-\mathbf{a}) \times \mathbf{b} = \mathbf{0}$

7 Find two different versions of the equation of the line that passes through the points (0, 3, –2) and (1, 2, –1) in the form $\mathbf{r} \times \mathbf{b} = \mathbf{c}$

8 Convert each of these equations of lines to the form $(\mathbf{r}-\mathbf{a}) \times \mathbf{b} = \mathbf{0}$

a $\mathbf{r} = 2\mathbf{i} - \mathbf{k} + s(\mathbf{j}+5\mathbf{k})$

b $\mathbf{r} = \begin{pmatrix} 3 \\ 1 \\ -4 \end{pmatrix} + t\begin{pmatrix} 5 \\ 3 \\ -6 \end{pmatrix}$

c $\frac{x-1}{-2} = \frac{y+3}{4} = \frac{z}{-2}$

d $x = \frac{y}{-3} = \frac{z-1}{2}$

9 Show that these two equations represent the same line.

A: $\left(\mathbf{r} - \begin{pmatrix} 1 \\ 3 \\ 5 \end{pmatrix}\right) \times \begin{pmatrix} 4 \\ 3 \\ -1 \end{pmatrix} = \mathbf{0}$

B: $\mathbf{r} = \begin{pmatrix} 5 \\ 6 \\ 4 \end{pmatrix} + \lambda\begin{pmatrix} -8 \\ -6 \\ 2 \end{pmatrix}$

10 Do these equations represent the same line? Explain how you know.

A: $\frac{x-3}{-2} = \frac{y+1}{4} = \frac{z-5}{-6}$

B: $\left(\mathbf{r} - \begin{pmatrix} 4 \\ -3 \\ 8 \end{pmatrix}\right) \times \begin{pmatrix} -3 \\ 6 \\ -9 \end{pmatrix} = \mathbf{0}$

11 Find, in Cartesian form, a possible equation of the line given by $\mathbf{r} \times \begin{pmatrix} 1 \\ 0 \\ -1 \end{pmatrix} = \begin{pmatrix} -3 \\ 0 \\ -3 \end{pmatrix}$

12 Find, in the form $\mathbf{r} = \mathbf{a} + t\mathbf{b}$, an equation of the line given by $\mathbf{r} \times \begin{pmatrix} -2 \\ -1 \\ 0 \end{pmatrix} = \begin{pmatrix} 1 \\ -2 \\ 0 \end{pmatrix}$

13 The lines l_1 and l_2 have equations

$l_1: \mathbf{r} \times (2\mathbf{i}+\mathbf{j}-2\mathbf{k}) = (\mathbf{i}-2\mathbf{j}+\mathbf{k})$ and

$l_2: \mathbf{r} = (-3\mathbf{i}-11\mathbf{j}+5\mathbf{k}) + t(-\mathbf{i}+3\mathbf{j}+\mathbf{k})$

a Show that l_1 and l_2 intersect and find their point of intersection.

b Calculate the sine of the angle between l_1 and l_2

14 A line l_1 has equation $\left(\mathbf{r} - \begin{pmatrix} 0 \\ 4 \\ -2 \end{pmatrix}\right) \times \begin{pmatrix} 1 \\ -6 \\ 5 \end{pmatrix} = 0$

a Verify that the point $A(-2, 16, -12)$ lies on l_1

A second line, l_2 has equation

$\mathbf{r} = \begin{pmatrix} 2 \\ 0 \\ -3 \end{pmatrix} + \lambda\begin{pmatrix} 1 \\ 2 \\ -6 \end{pmatrix}$

The lines l_1 and l_2 intersect at the point B

b Find the coordinates of B

c Calculate the area of triangle OAB

15 The lines l_1 and l_2 have equations

$l_1: \mathbf{r} \times (5\mathbf{i}+\mathbf{k}) = (3\mathbf{j}-7\mathbf{k})$ and

$l_2: \mathbf{r} = (\mathbf{i}+\mathbf{j}-\mathbf{k}) + t(9\mathbf{i}+8\mathbf{j}+\mathbf{k})$

Find a unit vector which is perpendicular to both l_1 and l_2

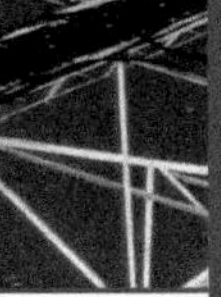

23.2 The equation of a plane

Fluency and skills

A plane is a 2D surface, extending infinitely far in both directions. You can write the equation of a plane in Cartesian, vector and scalar product form.

Suppose you wish to find the equation of the plane containing a point with position vector **a** and the non-parallel vectors **b** and **c**. Any other vector on the plane can be described as a multiple of **b** plus a multiple of **c**. Therefore the position vector of any point on the plane is given by $\mathbf{r} = \mathbf{a} + s\mathbf{b} + t\mathbf{c}$ for some values of s and t

Key point

The vector equation of the plane containing the point with position vector **a** and the non-parallel vectors **b** and **c** is $\mathbf{r} = \mathbf{a} + s\mathbf{b} + t\mathbf{c}$

Example 1

Find the equation of the plane containing the points (1, 4, –2), (0, 3, 2) and (–5, 0, 3)

$\mathbf{a} = \mathbf{i} + 4\mathbf{j} - 2\mathbf{k}$

A vector on the plane is $(\mathbf{i} + 4\mathbf{j} - 2\mathbf{k}) - (3\mathbf{j} + 2\mathbf{k}) = \mathbf{i} + \mathbf{j} - 4\mathbf{k}$

Another vector on the plane is $(\mathbf{i} + 4\mathbf{j} - 2\mathbf{k}) - (-5\mathbf{i} + 3\mathbf{k}) = 6\mathbf{i} + 4\mathbf{j} - 5\mathbf{k}$

So the vector equation of the plane is

$\mathbf{r} = \mathbf{i} + 4\mathbf{j} - 2\mathbf{k} + s(\mathbf{i} + \mathbf{j} - 4\mathbf{k}) + t(6\mathbf{i} + 4\mathbf{j} - 5\mathbf{k})$

Alternatively, you can write this in column vector form:

$$\mathbf{r} = \begin{pmatrix} 1 \\ 4 \\ -2 \end{pmatrix} + s\begin{pmatrix} 1 \\ 1 \\ -4 \end{pmatrix} + t\begin{pmatrix} 6 \\ 4 \\ -5 \end{pmatrix}$$

You could use the position vector of any of the three points.

Any combination of the position vectors of the three points can be used as long as they do not give parallel vectors.

Consider a plane containing the point with position vector **a** and with a perpendicular vector **n**. Any vector on this plane will be perpendicular to the vector **n**. Using the definition of the scalar product, any point on the plane with position vector **r**, satisfies $(\mathbf{r} - \mathbf{a}) \cdot \mathbf{n} = 0$. Expanding the bracket gives $\mathbf{r} \cdot \mathbf{n} - \mathbf{a} \cdot \mathbf{n} = 0$, which you can rearrange to give $\mathbf{r} \cdot \mathbf{n} = \mathbf{a} \cdot \mathbf{n}$

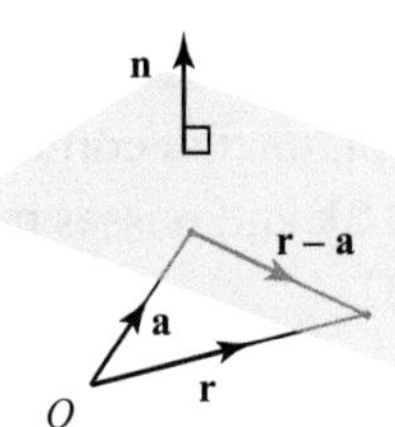

Key point

The scalar product equation of the plane perpendicular to the vector **n** and passing through the point with position vector **a** is $\mathbf{r} \cdot \mathbf{n} = p$, where $p = \mathbf{a} \cdot \mathbf{n}$

You can convert this into Cartesian form by writing the vector **r** in component form: $x\mathbf{i}+y\mathbf{j}+z\mathbf{k}$ and calculating the scalar product.

> **Key point**
> The vector perpendicular to a plane is called a **normal** to the plane.

Since any vectors on the plane will be perpendicular to the normal, you can use the vector product to find the normal vector.

The capital Greek letter Π is often used to denote a plane.

Example 2

The plane Π contains the vectors $\mathbf{a}=\mathbf{i}-\mathbf{j}+3\mathbf{k}$ and $\mathbf{b}=3\mathbf{i}+\mathbf{j}-2\mathbf{k}$ and the point (1, 0, –2). Find the equation of Π in

a Scalar product form, **b** Cartesian form.

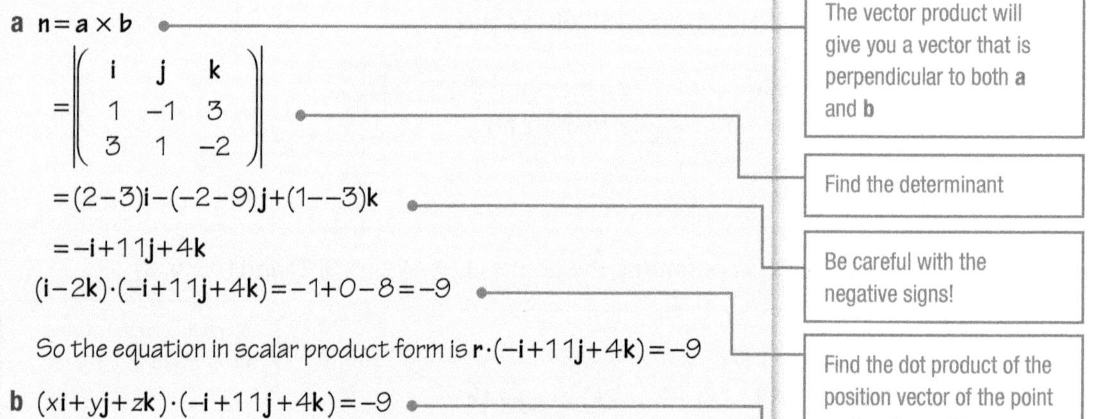

a $\mathbf{n}=\mathbf{a}\times\mathbf{b}$
$=\begin{vmatrix} \mathbf{i} & \mathbf{j} & \mathbf{k} \\ 1 & -1 & 3 \\ 3 & 1 & -2 \end{vmatrix}$
$=(2-3)\mathbf{i}-(-2-9)\mathbf{j}+(1--3)\mathbf{k}$
$=-\mathbf{i}+11\mathbf{j}+4\mathbf{k}$
$(\mathbf{i}-2\mathbf{k})\cdot(-\mathbf{i}+11\mathbf{j}+4\mathbf{k})=-1+0-8=-9$
So the equation in scalar product form is $\mathbf{r}\cdot(-\mathbf{i}+11\mathbf{j}+4\mathbf{k})=-9$
b $(x\mathbf{i}+y\mathbf{j}+z\mathbf{k})\cdot(-\mathbf{i}+11\mathbf{j}+4\mathbf{k})=-9$
So the equation in Cartesian form is $-x+11y+4z+9=0$
The vector product will give you a vector that is perpendicular to both **a** and **b**
Find the determinant
Be careful with the negative signs!
Find the dot product of the position vector of the point on the plane with **n**
Replace **r** with $x\mathbf{i}+y\mathbf{j}+z\mathbf{k}$

> **Key point**
> The Cartesian equation of a plane is $ax+by+cz+d=0$, where the vector $a\mathbf{i}+b\mathbf{j}+c\mathbf{k}$ is perpendicular to the plane.

Exercise 23.2A Fluency and skills

1 Find, in the form $\mathbf{r}=\mathbf{a}+s\mathbf{b}+t\mathbf{c}$, the equation of the plane that contains the vectors $\mathbf{i}+\mathbf{j}$ and $\mathbf{j}+2\mathbf{k}$ and passes through the point (1, 5, 2)

2 Find, in vector form, the equation of the plane that contains the vectors $\begin{pmatrix} 2 \\ 0 \\ -3 \end{pmatrix}$ and $\begin{pmatrix} 5 \\ -2 \\ 4 \end{pmatrix}$ and the point (0, 6, 2)

3 A plane contains the points (3, 1, 0), (2, 4, –2) and (–5, 0, 4)

Find the equation of the plane in

a Vector form,

b Scalar product form,

c Cartesian form.

4 A plane contains the points (5, 1, 1), (–2, 0, 0) and (6, 2, –5)

Find the equation of the plane in

a Vector form, b Scalar product form,

c Cartesian form.

5 A plane contains the points *A*, *B* and *C* with position vectors $\mathbf{a}=\mathbf{i}-2\mathbf{j}$, $\mathbf{b}=3\mathbf{j}+2\mathbf{k}$ and $\mathbf{c}=-\mathbf{i}+2\mathbf{j}-\mathbf{k}$ respectively. Find the equation of the plane in

a Vector form,

b Scalar product form,

c Cartesian form.

6 Find, in the form $\mathbf{r}\cdot\mathbf{n}=p$, the equation of the plane that passes through the point (3, –1, 4) and is perpendicular to the vector $\mathbf{i}+2\mathbf{k}$

7 Find, in the form $\mathbf{r}\cdot\mathbf{n}=p$, the equation of the plane that passes through the point (6, 0, –2) and has normal $3\mathbf{i}-\mathbf{j}+4\mathbf{k}$

8 Find the scalar product equation of the plane that passes through the point with position vector $\mathbf{i}+\mathbf{j}+\mathbf{k}$ and is perpendicular to the vector $2\mathbf{j}-3\mathbf{k}$

9 Find the Cartesian equation of the plane that passes through the point (0, 3, –3) and is perpendicular to the vector $5\mathbf{i}+4\mathbf{j}-2\mathbf{k}$

10 Find the Cartesian equation of the plane that passes through the point (1, –1, 2) and is perpendicular to the vector $2\mathbf{i}-5\mathbf{k}$

11 Find the Cartesian equation of the plane that passes through the point with position vector $3\mathbf{i}+7\mathbf{k}$ and is perpendicular to the vector $\mathbf{i}-\mathbf{j}+4\mathbf{k}$

12 The plane Π is perpendicular to the line with equation $\mathbf{r}=\mathbf{i}-\mathbf{j}+t(2\mathbf{i}+\mathbf{j}-3\mathbf{k})$ and passes through the point (4, –1, 3). Find the equation of Π in

a Scalar product form,

b Cartesian form.

13 The plane Π is perpendicular to the line with equation $\frac{x+1}{3}=\frac{y-4}{-2}=\frac{z}{5}$ and passes through the point (4, 3, –2). Find the equation of Π in

a Scalar product form,

b Cartesian form.

14 Convert each of these equations of planes into Cartesian form.

a $\mathbf{r}\cdot\begin{pmatrix}3\\5\\-1\end{pmatrix}=-2$

b $\mathbf{r}\cdot(7\mathbf{i}-\mathbf{j}+8\mathbf{k})=3$

c $\mathbf{r}=(2\mathbf{i}+\mathbf{j}-5\mathbf{k})+s(-\mathbf{i}+3\mathbf{j}+\mathbf{k})+t(6\mathbf{i}-2\mathbf{j}+7\mathbf{k})$

d $\mathbf{r}=\begin{pmatrix}0\\4\\-2\end{pmatrix}+s\begin{pmatrix}3\\-2\\5\end{pmatrix}+t\begin{pmatrix}1\\0\\4\end{pmatrix}$

15 Convert each of these equations of planes into scalar product form.

a $9x+3y-z=5$

b $2x-7y-15z+4=0$

c $\mathbf{r}=(5\mathbf{i}-2\mathbf{k})+s(-4\mathbf{i}+\mathbf{j})+t(8\mathbf{i}-3\mathbf{j}+4\mathbf{k})$

d $\mathbf{r}=\begin{pmatrix}6\\2\\-5\end{pmatrix}+s\begin{pmatrix}0\\1\\-3\end{pmatrix}+t\begin{pmatrix}5\\-8\\0\end{pmatrix}$

Reasoning and problem-solving

The vector equation of a plane is not unique, as you can chose any two vectors and any point on the plane to build the equation. In the case of the scalar product equation and the Cartesian equation, all vectors that are perpendicular to the plane are parallel to each other, so these forms of the equation will all simplify to a unique equation. You need to be able to tell if two equations represent the same plane.

Strategy 1

To check whether two equations represent the same plane

1. Check whether the normals are parallel.
2. Choose a point that you know is on one of the planes and check whether it also lies on the other plane.
3. Conclude whether the planes are the same, parallel or intersecting.

Example 3

Parallel planes will have parallel normals but will not have a point in common.

The planes Π_1 and Π_2 have equations $\mathbf{r}\cdot(\mathbf{i}-6\mathbf{j}+2\mathbf{k})=1$ and $\mathbf{r}=\mathbf{i}+\mathbf{j}+3\mathbf{k}+s(\mathbf{j}+3\mathbf{k})+t(2\mathbf{i}-\mathbf{k})$ respectively.

Show that Π_1 and Π_2 are, in fact, the same plane.

The normal to Π_2 is given by

$(\mathbf{j}+3\mathbf{k})\times(2\mathbf{i}-\mathbf{k})=-\mathbf{i}+6\mathbf{j}-2\mathbf{k}$ — Calculate the normal vector to Π_2

$-\mathbf{i}+6\mathbf{j}-2\mathbf{k}=-(\mathbf{i}-6\mathbf{j}+2\mathbf{k})$ — (1) So their normal vectors are parallel.

The point (1, 1, 3) lies on Π_2

$(\mathbf{i}+\mathbf{j}+3\mathbf{k})\cdot(\mathbf{i}-6\mathbf{j}+2\mathbf{k})=1-6+6$ — Substitute into the equation of Π_1

$=1$ — (2) So the point (1, 1, 3) lies on Π_1

Therefore Π_1 and Π_2 represent the same plane. — (3) Since their normals are parallel and they have a point in common.

Consider the intersection of a plane $\mathbf{r}\cdot\mathbf{n}=p$ and a line $\mathbf{r}=\mathbf{a}+\lambda\mathbf{b}$.

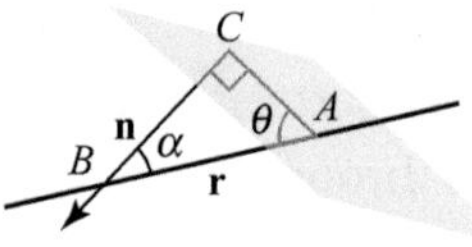

See Ch6.2
For a reminder of how to calculate the angle between two vectors.

You know that the acute angle between the vectors $\mathbf{n}$ and $\mathbf{b}$ is given by formula $\cos\alpha=\left|\frac{\mathbf{b}\cdot\mathbf{n}}{|\mathbf{b}||\mathbf{n}|}\right|$. Therefore, the angle θ is given by $\theta=90-\alpha$ which implies that $\sin\theta=\left|\frac{\mathbf{b}\cdot\mathbf{n}}{|\mathbf{b}||\mathbf{n}|}\right|$ since $\sin(90-\alpha)=\cos\alpha$

Key point

The acute angle, θ, between the plane $\mathbf{r}\cdot\mathbf{n}=p$ and the line $\mathbf{r}=\mathbf{a}+\lambda\mathbf{b}$ is given by $\sin\theta=\left|\frac{\mathbf{b}\cdot\mathbf{n}}{|\mathbf{b}||\mathbf{n}|}\right|$

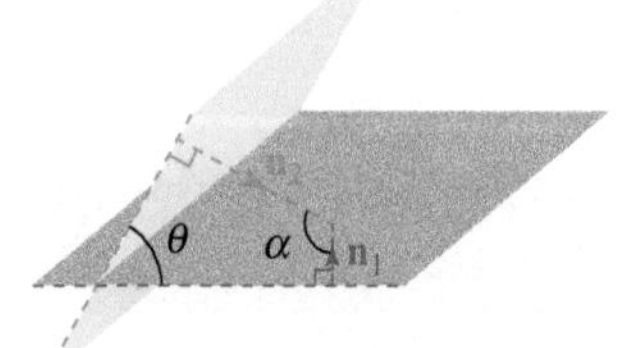

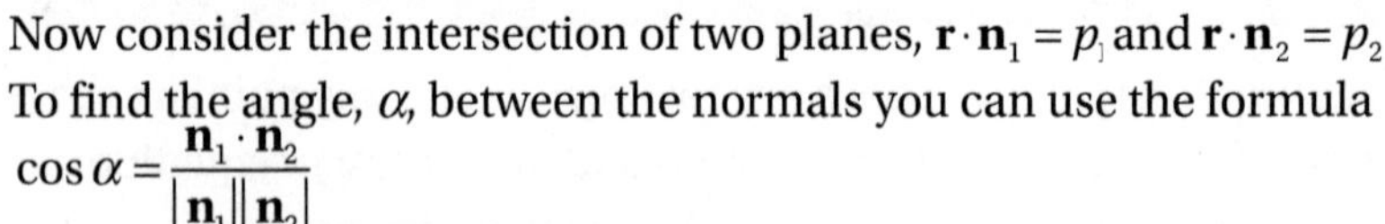

Now consider the intersection of two planes, $\mathbf{r}\cdot\mathbf{n}_1=p_1$ and $\mathbf{r}\cdot\mathbf{n}_2=p_2$

To find the angle, α, between the normals you can use the formula $\cos\alpha=\frac{\mathbf{n}_1\cdot\mathbf{n}_2}{|\mathbf{n}_1||\mathbf{n}_2|}$

Using the quadrilateral formed from the planes and their normals you can see that $\theta=180-\alpha$ which implies that $\cos\theta=\left|\frac{\mathbf{n}_1\cdot\mathbf{n}_2}{|\mathbf{n}_1||\mathbf{n}_2|}\right|$ since $|\cos(180-\alpha)|=|\cos\alpha|$

Key point

The acute angle, θ, between the planes $\mathbf{r}\cdot\mathbf{n}_1 = p_1$ and $\mathbf{r}\cdot\mathbf{n}_2 = p_2$ is given by $\cos\theta = \left|\frac{\mathbf{n}_1 \cdot \mathbf{n}_2}{|\mathbf{n}_1||\mathbf{n}_2|}\right|$

PURE

Strategy 2

To calculate the angle between two planes or between a line or a plane

1. Identify the normal vector/s.
2. Identify the direction vector of the line.
3. Use $\cos\theta = \left|\frac{\mathbf{n}_1 \cdot \mathbf{n}_2}{|\mathbf{n}_1||\mathbf{n}_2|}\right|$ to find the acute angle between two planes.
4. Use $\sin\theta = \left|\frac{\mathbf{b} \cdot \mathbf{n}}{|\mathbf{b}||\mathbf{n}|}\right|$ to find the acute angle between a line $r = \mathbf{a} + t\mathbf{b}$ and a plane.
5. Subtract from 180° to give an obtuse angle if necessary.

Example 4

Calculate the obtuse angle between the planes $\mathbf{r}\cdot(13\mathbf{i}-9\mathbf{j}+5\mathbf{k}) = 8$ and $\mathbf{r}\cdot(2\mathbf{j}+3\mathbf{k}) = 4$

The normal to the first plane is $\mathbf{n}_1 = 13\mathbf{i}-9\mathbf{j}+5\mathbf{k}$

Similarly the normal to the second plane is $\mathbf{n}_2 = 2\mathbf{j}+3\mathbf{k}$

(1) Identify the normals.

$$\cos\theta = \left|\frac{(13\mathbf{i}-9\mathbf{j}+5\mathbf{k})\cdot(2\mathbf{j}+3\mathbf{k})}{|13\mathbf{i}-9\mathbf{j}+5\mathbf{k}||2\mathbf{j}+3\mathbf{k}|}\right|$$

(3) Using $\cos\theta = \left|\frac{\mathbf{n}_1 \cdot \mathbf{n}_2}{|\mathbf{n}_1||\mathbf{n}_2|}\right|$

$$= \left|\frac{-3}{\sqrt{275}\sqrt{13}}\right|$$

$\theta = 87.1°$

So the obtuse angle is

$180 - 87.1 = 92.9°$

(5) Subtract from 180° since the question asked for an obtuse angle.

Example 5

Calculate the acute angle between the plane $2x+3y-z=1$ and the line $\frac{x}{4} = \frac{y+1}{-2} = \frac{z-2}{7}$

The normal to the plane is $\mathbf{n} = 2\mathbf{i}+3\mathbf{j}-\mathbf{k}$

(1) Since the Cartesian equation of a plane is $ax+by+cz=d$ where $\mathbf{n} = a\mathbf{i}+b\mathbf{j}+c\mathbf{k}$

The direction vector of the line is $\mathbf{b} = 4\mathbf{i}-2\mathbf{j}+7\mathbf{k}$

(2) Since the Cartesian equations of a line are $\frac{x-x_1}{b_1} = \frac{y-y_1}{b_2} = \frac{z-z_1}{b_3}$ where $\mathbf{b} = b_1\mathbf{i}+b_2\mathbf{j}+b_3\mathbf{k}$ is the direction vector of the line.

$$\sin\theta = \left|\frac{(4\mathbf{i}-2\mathbf{j}+7\mathbf{k})\cdot(2\mathbf{i}+3\mathbf{j}-\mathbf{k})}{|4\mathbf{i}-2\mathbf{j}+7\mathbf{k}||2\mathbf{i}+3\mathbf{j}-\mathbf{k}|}\right|$$

(4) Using $\sin\theta = \left|\frac{\mathbf{b} \cdot \mathbf{n}}{|\mathbf{b}||\mathbf{n}|}\right|$

$$= \left|\frac{-5}{\sqrt{69}\sqrt{14}}\right|$$

$\theta = 9.26°$

Exercise 23.2B Reasoning and problem-solving

1 Which of these planes does the point (–7, 10, 0) lie on? Show your working.

a $\mathbf{r}=2\mathbf{i}-3\mathbf{j}-7\mathbf{k}+s(-\mathbf{i}+5\mathbf{j}+3\mathbf{k})+t(7\mathbf{i}-3\mathbf{j}-\mathbf{k})$

b $\mathbf{r}=\begin{pmatrix}2\\0\\8\end{pmatrix}+s\begin{pmatrix}1\\3\\-2\end{pmatrix}+t\begin{pmatrix}-2\\4\\0\end{pmatrix}$

c $\mathbf{r}\cdot(\mathbf{i}-\mathbf{j}+3\mathbf{k})=17$

d $2x+3y-8z=16$

2 Verify whether each pair of equations represent the same plane.

a $2x+3y-z-2=0$ and $\mathbf{r}\cdot(2\mathbf{i}+3\mathbf{j}-\mathbf{k})=2$

b $\mathbf{r}=3\mathbf{i}-\mathbf{k}+s(4\mathbf{i}+\mathbf{j}+2\mathbf{k})+t(5\mathbf{i}+3\mathbf{j})$ and $\mathbf{r}=(\mathbf{i}-4\mathbf{j}+4\mathbf{k})+\lambda(2\mathbf{i}+4\mathbf{j}+5\mathbf{k})+\mu(5\mathbf{i}+3\mathbf{j})$

c $\mathbf{r}=\begin{pmatrix}1\\0\\-7\end{pmatrix}+s\begin{pmatrix}5\\-2\\4\end{pmatrix}+t\begin{pmatrix}0\\0\\1\end{pmatrix}$ and

$\mathbf{r}=\begin{pmatrix}-4\\2\\-7\end{pmatrix}+\lambda\begin{pmatrix}0\\1\\3\end{pmatrix}+\mu\begin{pmatrix}5\\-2\\0\end{pmatrix}$

d $\mathbf{r}=\begin{pmatrix}-2\\1\\0\end{pmatrix}+s\begin{pmatrix}4\\3\\2\end{pmatrix}+t\begin{pmatrix}-5\\0\\1\end{pmatrix}$ and

$\mathbf{r}\cdot\begin{pmatrix}6\\-2\\-1\end{pmatrix}=-14$

e $\mathbf{r}\cdot(2\mathbf{i}-7\mathbf{j}-3\mathbf{k})=-2$ and $\mathbf{r}=(5\mathbf{i}+4\mathbf{k})+\lambda(3\mathbf{j}-7\mathbf{k})+\mu(2\mathbf{i}+\mathbf{j}-\mathbf{k})$

f $x+3y+z=-1$ and $\mathbf{r}=(\mathbf{i}-2\mathbf{k})+\lambda(\mathbf{i}-9\mathbf{j}+6\mathbf{k})+\mu(4\mathbf{i}+3\mathbf{j}-3\mathbf{k})$

3 Calculate the acute angle between the line $\mathbf{r}=2\mathbf{i}-8\mathbf{j}+\mathbf{k}+t(\mathbf{j}+\mathbf{k})$ and the plane $\mathbf{r}\cdot(6\mathbf{i}-\mathbf{j}+5\mathbf{k})=9$

4 Calculate the obtuse angle between the line $\mathbf{r}=9\mathbf{i}-5\mathbf{j}+3\mathbf{k}+\lambda(\mathbf{i}-\mathbf{j}-4\mathbf{k})$ and the plane $6x-y+z=24$

5 Calculate the acute angle between the line $\frac{x-1}{5}=\frac{y-2}{-2}=\frac{z+3}{-2}$ and the plane $2x+7y-z=5$

6 Calculate the obtuse angle between the line $\mathbf{r}=\begin{pmatrix}-5\\0\\1\end{pmatrix}+\lambda\begin{pmatrix}7\\2\\-3\end{pmatrix}$ and the plane $\mathbf{r}\cdot\begin{pmatrix}-2\\4\\-9\end{pmatrix}=1$

7 Calculate the acute angle between the plane $\mathbf{r}\cdot(2\mathbf{i}+\mathbf{j}-3\mathbf{k})=2$ and the plane $\mathbf{r}\cdot(6\mathbf{i}-4\mathbf{j}-7\mathbf{k})=5$

8 Calculate the acute angle between the plane $\mathbf{r}\cdot(7\mathbf{i}-\mathbf{j}-5\mathbf{k})=9$ and the plane $2x-y+z=4$

9 Calculate the obtuse angle between the plane $x+y-7z-3=0$ and the plane $8x-3y+4z=15$

10 Find the scalar product equation of the plane that contains the vectors $\begin{pmatrix}1\\0\\-3\end{pmatrix}$ and $\begin{pmatrix}2\\6\\-1\end{pmatrix}$ and passes through the point (3, –1, 0)

11 Convert this equation of a plane into scalar product form.

$\mathbf{r}=\begin{pmatrix}0\\3\\-2\end{pmatrix}+\mu\begin{pmatrix}6\\-2\\2\end{pmatrix}+\lambda\begin{pmatrix}-1\\4\\7\end{pmatrix}$

12 Find the Cartesian equation of the plane that passes through the points (3, –5, 1), (2, 4, 5) and (1, 0, 3)

13 The plane Π has vector equation

$\mathbf{r}=\begin{pmatrix}3\\0\\1\end{pmatrix}+\lambda\begin{pmatrix}1\\-2\\4\end{pmatrix}+\mu\begin{pmatrix}6\\1\\-3\end{pmatrix}$

a Write the equation of Π in scalar product form.

b The line l passes through the point $P(1, 4, -2)$ and is perpendicular to Π

Find the vector product equation of l

23.3 Finding distances 2

Fluency and skills

A line will intersect any given plane unless the line and the plane are parallel.

You can find the point of intersection between a line and a plane by using the equation of the line to substitute for **r** in the equation of the plane.

Example 1

Find the point of intersection between the line with equation $\mathbf{r} = -4\mathbf{j} + 2\mathbf{k} + \lambda(3\mathbf{i} + \mathbf{j} + \mathbf{k})$ and the plane with equation $\mathbf{r} = \mathbf{i} - 2\mathbf{j} + s(\mathbf{j} + 3\mathbf{k}) + t(2\mathbf{i} + 2\mathbf{j} + \mathbf{k})$

At the point of intersection

$-4\mathbf{j} + 2\mathbf{k} + \lambda(3\mathbf{i} + \mathbf{j} + \mathbf{k}) = \mathbf{i} - 2\mathbf{j} + s(\mathbf{j} + 3\mathbf{k}) + t(2\mathbf{i} + 2\mathbf{j} + \mathbf{k})$ — Use the equation of the line to substitute for **r** in the equation of the plane.

$\mathbf{i}: 3\lambda = 1 + 2t$ — Consider the **i** components.

$\mathbf{j}: -4 + \lambda = -2 + s + 2t \Rightarrow \lambda = 2 + s + 2t$

Subtracting these two equations gives: $2\lambda = -1 - s$

Which can be rearranged to give $s = -1 - 2\lambda$

$\mathbf{k}: 2 + \lambda = 3s + t$

Substitute $s = -1 - 2\lambda$

$\Rightarrow 2 + \lambda = 3(-1 - 2\lambda) + t$

$\Rightarrow t = 5 + 7\lambda$

Solve this together with $3\lambda = 1 + 2t$

to give $\lambda = -1$ and $t = -2$

By substituting back into $s = -1 - 2\lambda$

$\Rightarrow s = 1$

Substitute $\lambda = -1$ into equation of the line.

$\mathbf{r} = -4\mathbf{j} + 2\mathbf{k} - 1(3\mathbf{i} + \mathbf{j} + \mathbf{k})$

$= -3\mathbf{i} - 5\mathbf{j} + \mathbf{k}$

So the point of intersection is $(-3, -5, 1)$ — Could also find by substituting $t = -2$ and $s = 1$ into equation of the plane.

Example 2

Find the point of intersection between the line with equation $\mathbf{r} = \mathbf{i} - \mathbf{k} + t(\mathbf{i} - 2\mathbf{j} + 3\mathbf{k})$ and the plane with equation $2x - y - 3z = 20$

At the point of intersection $2(1 + t) - (-2t) - 3(-1 + 3t) = 20$ — Substituting $x = 1 + t$, $y = -2t$, $z = -1 + 3t$ from the equation of the line.

$2 + 2t + 2t + 3 - 9t = 20$

$5t = -15$

$t = -3$

Substitute $t = -3$ into equation of the line.

So $\mathbf{r} = \mathbf{i} - \mathbf{k} - 3(\mathbf{i} - 2\mathbf{j} + 3\mathbf{k})$

$= -2\mathbf{i} + 6\mathbf{j} - 10\mathbf{k}$

So the point of intersection is $(-2, 6, -10)$

You can find the shortest distance from a point to a plane.

Suppose you have the plane Π with equation $\mathbf{r}\cdot\mathbf{n} = p$ and the point P The shortest distance from P to the plane will be perpendicular to the plane. Therefore you need the length of $\overrightarrow{PQ}$

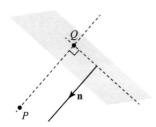

Q is the point of intersection of the plane and the line through P and Q. You can easily write down the equation of this line using the fact that it passes through P and has direction vector parallel to the vector **n** which is normal to the plane.

Once you've found the point of intersection Q you can find the length of the line segment PQ and hence the shortest distance from the point P to the plane Π

Example 3

Find the shortest distance from the plane Π with equation $\mathbf{r}\cdot\begin{pmatrix}1\\3\\-2\end{pmatrix} = -4$ to the point $P(4, 0, -7)$

The equation of the line though P and the plane is

$$\mathbf{r} = \begin{pmatrix}4\\0\\-7\end{pmatrix} + \lambda\begin{pmatrix}1\\3\\-2\end{pmatrix}$$

Since the line is perpendicular to the plane.

At Q, $\begin{pmatrix}4+\lambda\\3\lambda\\-7-2\lambda\end{pmatrix}\cdot\begin{pmatrix}1\\2\\-2\end{pmatrix} = -4$

Find the point of intersection between the plane and the line.

$$\Rightarrow 4+\lambda+2(3\lambda)-2(-7-2\lambda) = -4$$
$$\Rightarrow 11\lambda = -22$$
$$\Rightarrow \lambda = -2$$

Therefore, the position vector of Q is

$$\overrightarrow{OQ} = \begin{pmatrix}4\\0\\-7\end{pmatrix} - 2\begin{pmatrix}1\\3\\-2\end{pmatrix}$$

Substitute the value of λ found into the equation of the line.

$$\overrightarrow{PQ} = \overrightarrow{OQ} - \overrightarrow{OP}$$
$$= \begin{pmatrix}4\\0\\-7\end{pmatrix} - 2\begin{pmatrix}1\\3\\-2\end{pmatrix} - \begin{pmatrix}4\\0\\-7\end{pmatrix}$$
$$= \begin{pmatrix}-2\\-6\\4\end{pmatrix}$$

This is just the vector $\lambda\mathbf{n}$ where $\lambda = -2$

$$|\overrightarrow{PQ}| = \sqrt{(-2)^2+(-6)^2+4^2}$$
$$= 2\sqrt{14}$$

So the shortest distance from the point P to the plane Π is $2\sqrt{14}$

Notice that the distance of the point from the plane was given by $|\lambda\mathbf{n}|$

Exercise 23.3A Fluency and skills

1 Find the shortest distance between each point and plane.

a $(2, 5, 1)$ and $2x-4y+4z+5=0$

b $(3, 0, -4)$ and $3x+y-2z=6$

c $(-1, 5, 4)$ and $\mathbf{r}\cdot(\mathbf{i}-7\mathbf{j}+5\mathbf{k})=-3$

d $(6, 1, 4)$ and $\mathbf{r}\cdot\begin{pmatrix}0\\-8\\6\end{pmatrix}=12$

e $(-2, -5, -1)$ and $\mathbf{r}\cdot(\mathbf{i}-\mathbf{j}-3\mathbf{k})=1$

2 Show that the line $\mathbf{r}=\begin{pmatrix}-11\\-4\\10\end{pmatrix}+t\begin{pmatrix}4\\-2\\0\end{pmatrix}$ and the plane $\mathbf{r}=\begin{pmatrix}3\\-5\\-10\end{pmatrix}+\mu\begin{pmatrix}7\\-10\\0\end{pmatrix}+\lambda\begin{pmatrix}1\\-3\\4\end{pmatrix}$ intersect and find their point of intersection.

3 Find the point of intersection of the line with equation $\mathbf{r}=-6\mathbf{i}+2\mathbf{j}+16\mathbf{k}+\lambda(3\mathbf{i}-\mathbf{j}+\mathbf{k})$ and the plane with equation $\mathbf{r}=-7\mathbf{i}-9\mathbf{j}+\mathbf{k}+s(5\mathbf{i}+\mathbf{j}+6\mathbf{k})+t(-2\mathbf{i}+4\mathbf{j}+\mathbf{k})$

4 Find the point of intersection of the line with equation $\mathbf{r}=\begin{pmatrix}1\\-2\\5\end{pmatrix}+t\begin{pmatrix}-5\\1\\2\end{pmatrix}$ and the plane with equation $\mathbf{r}\cdot\begin{pmatrix}1\\6\\-5\end{pmatrix}=0$

5 Find the point of intersection of the line with equation $\frac{x}{5}=\frac{y+2}{-1}=\frac{3-z}{4}$ and the plane with equation $\mathbf{r}\cdot(5\mathbf{i}-\mathbf{j}+4\mathbf{k})=-6$

6 Find the point of intersection of the line with equation $\frac{x-3}{4}=\frac{y-2}{5}=\frac{z-1}{-3}$ and the plane with equation $\mathbf{r}=13\mathbf{i}+7\mathbf{j}-6\mathbf{k}+s(\mathbf{i}+2\mathbf{j}-5\mathbf{k})+t(-\mathbf{i}+\mathbf{j}+2\mathbf{k})$

7 The plane Π is perpendicular to the y-axis and passes through the point $(1, 2, 0)$

Calculate the shortest distance of this plane from the point $(3, 4, -1)$

8 Show that the shortest distance from the origin to the plane $\mathbf{r}\cdot\mathbf{n}=p$ is $\frac{p}{|\mathbf{n}|}$

9 For each pair of lines, decide whether they are parallel, skew or intersecting. If they are intersecting, find their point of intersection.

a $\mathbf{r}=\begin{pmatrix}7\\-6\\-2\end{pmatrix}+\lambda\begin{pmatrix}3\\0\\-1\end{pmatrix}$ and $\mathbf{r}=\begin{pmatrix}3\\4\\0\end{pmatrix}+\mu\begin{pmatrix}-2\\5\\1\end{pmatrix}$

b $\mathbf{r}=\mathbf{i}+2\mathbf{k}+\lambda(3\mathbf{i}+4\mathbf{j}-2\mathbf{k})$ and $\mathbf{r}=3\mathbf{i}+\mathbf{j}-\mathbf{k}+\mu(-6\mathbf{i}-8\mathbf{j}+4\mathbf{k})$

c $\frac{x+5}{-1}=\frac{y}{2}=\frac{z-1}{2}$ and $\frac{x-2}{1}=\frac{y+9}{3}=\frac{z-5}{-2}$

d $\frac{-4-x}{3}=\frac{y-2}{-1}=\frac{z}{5}$ and $\frac{x-6}{2}=\frac{y+8}{-2}=\frac{5-z}{-1}$

10 Two lines, L_1 and L_2, have equations $\frac{x+10}{5}=\frac{y-5}{-1}=\frac{z-5}{2}$ and $\mathbf{r}=\begin{pmatrix}3\\3\\5\end{pmatrix}+\lambda\begin{pmatrix}-3\\0\\4\end{pmatrix}$ respectively.

Given that L_1 and L_2 intersect at the point A, calculate the length of $\overrightarrow{OA}$

11 The points A, B, C are defined as $(2, -1, 4)$, $(0, -2, 4)$ and $(1, 0, 5)$ respectively.

Calculate the shortest distance from the origin to the plane containing points A, B and C

Reasoning and problem-solving

See Ch6.3

For a reminder of finding the distance between skew lines.

You can find the distance between lines by finding the length of the perpendicular to both lines. It is also possible to use the vector product to find this distance.

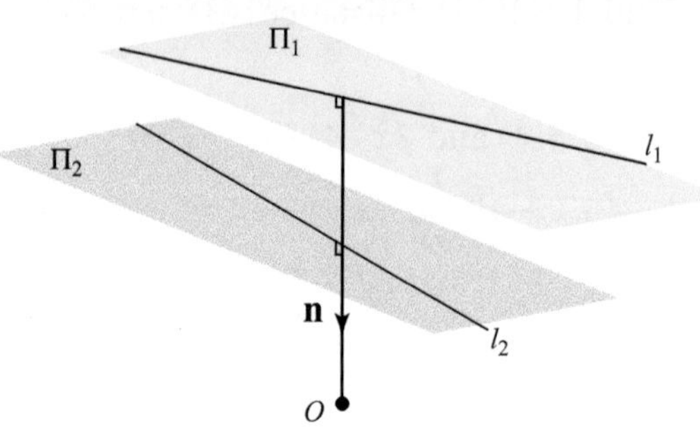

Consider the lines l_1 and l_2 with equations $\mathbf{r} = \mathbf{a} + \lambda\mathbf{b}$ and $\mathbf{r} = \mathbf{c} + \mu\mathbf{d}$ respectively. Since the lines do not intersect they must lie on parallel planes Π_1 and Π_2

The normal to the planes Π_1 and Π_2 will be the same since they are parallel and is given by the vector product of the two direction vectors: $\mathbf{n} = \mathbf{b} \times \mathbf{d}$

So the equations of the planes are $\mathbf{r} \cdot \mathbf{n} = \mathbf{a} \cdot \mathbf{n}$ and $\mathbf{r} \cdot \mathbf{n} = \mathbf{c} \cdot \mathbf{n}$

In Section 23.3A you saw how to find the distance of a point from a plane.

The equation of the line through the origin which is perpendicular to a plane $\mathbf{r} \cdot \mathbf{n} = p$ is $\mathbf{r} = \lambda\mathbf{n}$

Therefore, at the point of intersection, $\lambda\mathbf{n} \cdot \mathbf{n} = p$ which implies that $\lambda|\mathbf{n}|^2 = p$ so $\lambda = \dfrac{p}{|\mathbf{n}|^2}$

The distance of a point from a plane is given by $|\lambda\mathbf{n}|$. So, using $\lambda = \dfrac{p}{|\mathbf{n}|^2}$ gives

distance from plane to origin $= \left|\dfrac{p}{|\mathbf{n}|^2}|\mathbf{n}|\right| = \left|\dfrac{p}{|\mathbf{n}|}\right|$

The sign of p indicates on which side of the origin the plane lies.

You can then find the distance between the two planes, and hence the shortest distance between the lines, by adding or subtracting the two distances.

Strategy 1

To find the distance between skew lines equations with $\mathbf{r} = \mathbf{a} + \lambda\mathbf{b}$ and $\mathbf{r} = \mathbf{c} + \mu\mathbf{d}$

1. Find the vector perpendicular to both lines using $\mathbf{n} = \mathbf{b} \times \mathbf{d}$
2. Find parallel planes on which the lines lie: $\mathbf{r} \cdot \mathbf{n} = \mathbf{a} \cdot \mathbf{n}$ and $\mathbf{r} \cdot \mathbf{n} = \mathbf{c} \cdot \mathbf{n}$
3. Find the distance of each of these planes from the origin.
4. Find the distance between the planes and hence the shortest distance between the lines.

PURE

Example 4

Find the shortest distance between the lines l_1 and l_2 with equations

$l_1: \mathbf{r} = 2\mathbf{i} - \mathbf{j} + s(\mathbf{i} + \mathbf{j} - 3\mathbf{k})$ and $l_2: \mathbf{r} = 4\mathbf{i} + \mathbf{j} - \mathbf{k} + t(2\mathbf{i} - \mathbf{k})$

$\mathbf{n} = (\mathbf{i} + \mathbf{j} - 3\mathbf{k}) \times (2\mathbf{i} - \mathbf{k})$

$= -\mathbf{i} - 5\mathbf{j} - 2\mathbf{k}$

(1) Use $\mathbf{n} = \mathbf{b} \times \mathbf{d}$ to find a vector perpendicular to both lines.

Plane containing l_1 has equation

$\Pi_1 : \mathbf{r} \cdot (-\mathbf{i} - 5\mathbf{j} - 2\mathbf{k}) = (2\mathbf{i} - \mathbf{j}) \cdot (-\mathbf{i} - 5\mathbf{j} - 2\mathbf{k})$

(2) Use $\mathbf{r} \cdot \mathbf{n} = \mathbf{a} \cdot \mathbf{n}$

$\Rightarrow \mathbf{r} \cdot (-\mathbf{i} - 5\mathbf{j} - 2\mathbf{k}) = 3$

The distance of Π_1 to the origin is $\left|\frac{3}{\sqrt{1^2 + 5^2 + 2^2}}\right| = \frac{\sqrt{30}}{10}$

(3) Use distance to origin is $\frac{p}{|\mathbf{n}|}$

Plane containing l_2 has equation

$\Pi_2 : \mathbf{r} \cdot (-\mathbf{i} - 5\mathbf{j} - 2\mathbf{k}) = (4\mathbf{i} + \mathbf{j} - \mathbf{k}) \cdot (-\mathbf{i} - 5\mathbf{j} - 2\mathbf{k})$

(2) Use $\mathbf{r} \cdot \mathbf{n} = \mathbf{c} \cdot \mathbf{n}$

$\Rightarrow \mathbf{r} \cdot (-\mathbf{i} - 5\mathbf{j} - 2\mathbf{k}) = -7$

The distance of Π_2 to the origin is $\left|\frac{-7}{\sqrt{1^2 + 5^2 + 2^2}}\right| = \frac{7\sqrt{30}}{30}$

(3) Find distance of second plane to origin.

Therefore, the distance between the planes is $\frac{\sqrt{30}}{10} + \frac{7\sqrt{30}}{30} = \frac{\sqrt{30}}{3}$

(4) Add the distances since Π_1 and Π_2 are different sides of the origin.

Hence, the shortest distance between the lines is $\frac{\sqrt{30}}{3}$

Another application of the vector product is to find the image of a point reflected in a plane.

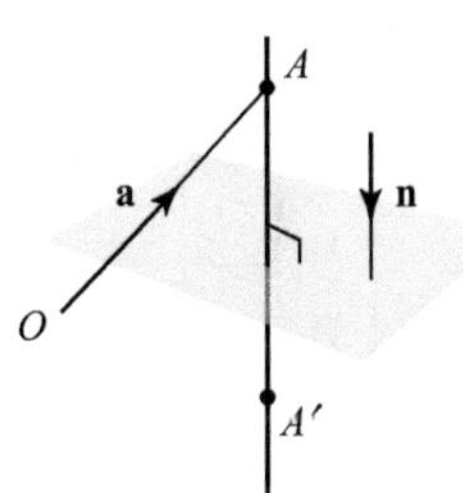

You know that the reflection will be along the line perpendicular to the plane and can use this fact to find the image.

You need to find a vector, **n**, which is perpendicular to the plane, then use this to write the equation of a line passing through the point you wish to reflect. The line will have equation $\mathbf{r} = \mathbf{a} + t\mathbf{n}$

By solving simultaneously, you can find the value of t where the line intersects the plane.

Then, since the image and the object are equidistant from the plane, the coordinates of the image can be found by doubling this value of t and substituting it back into the equation of the line.

Strategy 1

To find the reflection of a point in a plane

1. Find a vector, **n**, that is perpendicular to the plane.
2. Write the equation of a line perpendicular to the plane in the form $\mathbf{r} = \mathbf{a} + t\mathbf{n}$, where **a** is the point to be reflected.
3. Calculate the value of t where the line intersects the plane.
4. Multiply the value of t by two and substitute back into the equation to find the coordinates of the image.

Example 5

Find the image of the point $A(1, 0, 2)$ in the plane $\mathbf{r} = \begin{pmatrix}1\\0\\-2\end{pmatrix} + \mu\begin{pmatrix}-1\\3\\2\end{pmatrix} + \lambda\begin{pmatrix}0\\2\\1\end{pmatrix}$

$$\mathbf{n} = \begin{pmatrix}-1\\3\\2\end{pmatrix} \times \begin{pmatrix}0\\2\\1\end{pmatrix} = \begin{pmatrix}-1\\1\\-2\end{pmatrix}$$

$$\mathbf{r} = \begin{pmatrix}1\\0\\2\end{pmatrix} + t\begin{pmatrix}-1\\1\\-2\end{pmatrix}$$

$$\begin{pmatrix}1\\0\\2\end{pmatrix} + t\begin{pmatrix}-1\\1\\-2\end{pmatrix} = \begin{pmatrix}1\\0\\-2\end{pmatrix} + \mu\begin{pmatrix}-1\\3\\2\end{pmatrix} + \lambda\begin{pmatrix}0\\2\\1\end{pmatrix}$$

$1 - t = 1 - \mu$ (1)

$t = 3\mu + 2\lambda$ (2)

$2 - 2t = -2 + 2\mu + \lambda$ (3)

Substitute into equations (2) and (3)

$-t = \lambda$ and

$\lambda = 4 - 4t$

So $-t = 4 - 4t \Rightarrow t = \frac{4}{3}$

So

$$OA' = \begin{pmatrix}1\\0\\2\end{pmatrix} + \frac{8}{3}\begin{pmatrix}-1\\1\\-2\end{pmatrix}$$

$$= \begin{pmatrix}-\frac{5}{3}\\ \frac{8}{3}\\ -\frac{10}{3}\end{pmatrix}$$

So coordinates of A' are $\left(-\frac{5}{3}, \frac{8}{3}, -\frac{10}{3}\right)$

1. Use the vector product to find a vector perpendicular to the plane.

2. This is the equation of line through A which is perpendicular to the plane.

3. You need to solve these simultaneously to find the value of t

   Equation 1 gives $t = \mu$

4. $2 \times \frac{4}{3} = \frac{8}{3}$, substitute this for t in the equation of the line to find the position vector of A'

State the actual coordinates.

Exercise 23.3B Reasoning and problem-solving

1 The line l passes through the points $A(1, 4, -2)$ and $B(0, 2, -7)$

The plane Π has equation $5x-5y+z=3$

a Find the shortest distance from the plane to the point

i A ii B

b Does the line intersect the plane? Explain your answer.

2 a Find the perpendicular distance between each of these pairs of planes. Give your answers to 3 significant figures.

i $3x-y+9z=15$ and $6x-2y+18z-3=0$

ii $\mathbf{r}\cdot\begin{pmatrix}4\\1\\-2\end{pmatrix}=2$ and $\mathbf{r}\cdot\begin{pmatrix}-6\\-\frac{3}{2}\\3\end{pmatrix}=5$

iii $\mathbf{r}\cdot(\mathbf{i}+3\mathbf{j}-5\mathbf{k})=12$ and $2x-6y+10z+13=0$

b Explain why the planes in part **a** do not meet.

3 The plane Π is perpendicular to the x-axis and passes through the point $(5, 4, -2)$. The line L is parallel to the vector $\mathbf{i}-\mathbf{k}$ and passes through the point $(1, 5, 2)$. Π and L intersect at the point A. Calculate the length OA

4 The lines $\mathbf{r}=\begin{pmatrix}1\\0\\-2\end{pmatrix}+s\begin{pmatrix}5\\-2\\-4\end{pmatrix}$ and $\mathbf{r}=\begin{pmatrix}16\\10\\1\end{pmatrix}+t\begin{pmatrix}-1\\6\\6\end{pmatrix}$ intersect the plane $\mathbf{r}\cdot\begin{pmatrix}1\\-3\\2\end{pmatrix}=9$ at the points A and B respectively.

Calculate the area of triangle OAB

5 Three lines have equations as follows:

L_1: $\mathbf{r}=6\mathbf{i}-3\mathbf{j}+\lambda(\mathbf{i}+\mathbf{k})$,

L_2: $\mathbf{r}=s(3\mathbf{i}-\mathbf{j}+\mathbf{k})$ and

L_3: $\mathbf{r}=t(\mathbf{j}+2\mathbf{k})$

The lines L_1 and L_2 intersect at the point A, L_1 and L_3 intersect at the point B and L_2 and L_3 intersect at the point C, as shown.

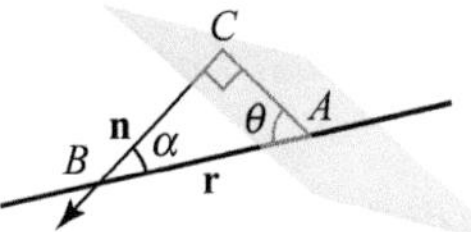

Calculate the exact area of triangle ABC

6 Find the shortest distance between these pairs of skew lines.

a $\mathbf{r}=2\mathbf{i}+\mathbf{j}+s(\mathbf{i}-2\mathbf{k})$ and $\mathbf{r}=8\mathbf{i}+2\mathbf{j}-3\mathbf{k}+t(\mathbf{i}-\mathbf{j})$

b $\mathbf{r}=\begin{pmatrix}2\\1\\1\end{pmatrix}+\lambda\begin{pmatrix}4\\0\\-3\end{pmatrix}$ and $\mathbf{r}=\begin{pmatrix}2\\7\\0\end{pmatrix}+\mu\begin{pmatrix}5\\3\\-1\end{pmatrix}$

c $\frac{x}{-1}=\frac{y+4}{0}=\frac{z-1}{2}$ and $\frac{x-2}{2}=\frac{y}{-1}=z-6$

d $\mathbf{r}=\begin{pmatrix}0\\0\\2\end{pmatrix}+\lambda\begin{pmatrix}-1\\1\\1\end{pmatrix}$ and $\frac{x-4}{2}=4-y=\frac{z+4}{-2}$

e $\mathbf{r}\times\begin{pmatrix}2\\0\\1\end{pmatrix}=\begin{pmatrix}2\\5\\-4\end{pmatrix}$ and $\mathbf{r}\times\begin{pmatrix}0\\-1\\0\end{pmatrix}=\begin{pmatrix}2\\0\\-4\end{pmatrix}$

7 Calculate the shortest distance between the x-axis and the line with equation

$\mathbf{r}=\begin{pmatrix}3\\-1\\-1\end{pmatrix}+\lambda\begin{pmatrix}2\\-1\\1\end{pmatrix}$

8 Find the image of the point $(1, 4, -7)$ when reflected in the plane $\mathbf{r}.\begin{pmatrix}1\\3\\2\end{pmatrix}=5$

9 The point $A(0, 2, -1)$ is reflected in the plane $\mathbf{r}=(2\mathbf{i}-\mathbf{k})+s(\mathbf{i}+\mathbf{j})+t(-3\mathbf{j}+\mathbf{k})$

Find the coordinates of the image of A

23 Summary and review

Chapter summary

- The vector product is defined as $\mathbf{a}\times\mathbf{b}=|\mathbf{a}||\mathbf{b}|\sin\theta\,\hat{\mathbf{n}}$, where θ is the angle between the vectors **a** and **b** and $\hat{\mathbf{n}}$ is a unit vector perpendicular to both **a** and **b**
- $\mathbf{i}\times\mathbf{j}=\mathbf{k}$, $\mathbf{j}\times\mathbf{k}=\mathbf{i}$ and $\mathbf{k}\times\mathbf{i}=\mathbf{j}$
- For any vectors **a** and **b**, $\mathbf{a}\times\mathbf{b}=-\mathbf{b}\times\mathbf{a}$
- For two vectors $\mathbf{a}=a_1\mathbf{i}+a_2\mathbf{j}+a_3\mathbf{k}$ and $\mathbf{b}=b_1\mathbf{i}+b_2\mathbf{j}+b_3\mathbf{k}$, the vector product is defined as $\mathbf{a}\times\mathbf{b}=(a_2b_3-a_3b_2)\mathbf{i}-(a_1b_3-a_3b_1)\mathbf{j}+(a_1b_2-a_1b_2)\mathbf{k}$. You can use the determinant to calculate
$$\mathbf{a}\times\mathbf{b}=\left|\begin{pmatrix}\mathbf{i} & \mathbf{j} & \mathbf{k}\\ a_1 & a_2 & a_3\\ b_1 & b_2 & b_3\end{pmatrix}\right|$$
- Area of triangle $ABC=\frac{1}{2}|AB\times AC|$
- A straight line passing through the point with position vector **a** and parallel to the vector **b** has equation $(\mathbf{r}-\mathbf{a})\times\mathbf{b}=\mathbf{0}$
- The vector equation of a plane is $\mathbf{r}=\mathbf{a}+s\mathbf{b}+t\mathbf{c}$ where **a** is the position vector of a point on the plane, and **b** and **c** are non-parallel vectors on the plane.
- The scalar product equation of a plane is $\mathbf{r}\cdot\mathbf{n}=p$ where **n** is perpendicular to the plane and $p=\mathbf{a}\cdot\mathbf{n}$ for **a** the position vector of a point on the plane.
- The Cartesian equation of a plane is $ax+by+cz=d$ where the vector $x\mathbf{i}+y\mathbf{j}+z\mathbf{k}$ is perpendicular to the plane.
- The acute angle, θ, between the plane $\mathbf{r}\cdot\mathbf{n}=p$ and the line $\mathbf{r}=\mathbf{a}+\lambda\mathbf{b}$ is given by $\sin\theta=\left|\frac{\mathbf{b}\cdot\mathbf{n}}{|\mathbf{b}||\mathbf{n}|}\right|$
- The acute angle, θ, between the planes $\mathbf{r}\cdot\mathbf{n}_1=p_1$ and $\mathbf{r}\cdot\mathbf{n}_2=p_2$ is given by $\cos\theta=\left|\frac{\mathbf{n}_1\cdot\mathbf{n}_2}{|\mathbf{n}_1||\mathbf{n}_2|}\right|$
- The shortest distance between the skew lines with equations $\mathbf{r}=\mathbf{a}+\lambda\mathbf{b}$ and $\mathbf{r}=\mathbf{c}+\mu\mathbf{d}$ is given by $\left|\frac{(\mathbf{a}-\mathbf{c})\cdot(\mathbf{b}\times\mathbf{d})}{|\mathbf{b}\times\mathbf{d}|}\right|$

Check and review

You should now be able to...	Try Questions
✔ Calculate the vector product of two vectors.	1
✔ Use the vector product to calculate the size of the angle between two vectors.	2
✔ Find a unit vector that is perpendicular to two other vectors.	3
✔ Use the vector product to calculate areas.	4

✔ Write the equation of a line using vector product form.	5, 6
✔ Write the equation of a plane in Cartesian, vector and scalar product form.	7, 8
✔ Find the point of intersection between a line and a plane.	9, 10
✔ Calculate the angle between a line and a plane.	9, 10
✔ Calculate the angle between two planes.	11, 12
✔ Calculate the shortest distance from a point to a plane.	13, 14
✔ Calculate the shortest distance between skew lines.	15

1 Calculate these vector products.

a $(\mathbf{i}+5\mathbf{j}-2\mathbf{k})\times(3\mathbf{i}-3\mathbf{j}-4\mathbf{k})$

b $(2\mathbf{i}-3\mathbf{j}+8\mathbf{k})\times(4\mathbf{i}-2\mathbf{k})$

2 Find the acute angle between the pairs of vectors in question **1**.

3 Find a unit vector which is perpendicular to both $\mathbf{a}=\mathbf{i}-3\mathbf{k}$ and $\mathbf{b}=-\mathbf{j}+5\mathbf{k}$

4 A triangle has vertices at (1, 3, –2), (2, 3, 0) and (4, 2, –3). Calculate its area.

5 A line is parallel to the vector $-5\mathbf{i}+3\mathbf{j}-\mathbf{k}$ and passes through the point with position vector $2\mathbf{j}+3\mathbf{k}$. Find the equation of the line in the form $\mathbf{r}\times\mathbf{a}=\mathbf{b}$

6 Write the equation of the line $\mathbf{r}=(\mathbf{i}-2\mathbf{j})+t(3\mathbf{j}-4\mathbf{k})$ in the form $\mathbf{r}\times\mathbf{a}=\mathbf{b}$

7 Write down the vector equation of the plane that contains the vectors $-2\mathbf{i}+3\mathbf{k}$ and $\mathbf{i}-4\mathbf{j}-\mathbf{k}$ and passes through the point (2, 0, –3)

8 A plane is perpendicular to the vector $7\mathbf{i}-5\mathbf{j}+\mathbf{k}$ and contains the point with position vector $2\mathbf{j}-5\mathbf{k}$. Write down the equation of the plane in

a The form $\mathbf{a}\cdot\mathbf{n}=p$ b Cartesian form.

9 A plane has equation $\mathbf{r}\cdot(\mathbf{i}-2\mathbf{j}+2\mathbf{k})=5$

The line with equation $\mathbf{r}=\mathbf{i}+t(2\mathbf{i}+3\mathbf{j}-\mathbf{k})$ intersects the plane at the point A

a Calculate the acute angle between the plane and the line at A

b Find the coordinates of A

c Calculate the length OA

10 The line with equation $\frac{x-1}{2}=y=\frac{z+2}{4}$ intersect the plane $8x+6y-3z-4=0$ at the point A

a Find the coordinates of A

b Calculate the obtuse angle between the line and the plane at point A

11 The plane Π_1 has equation $\mathbf{r}\cdot(\mathbf{i}-4\mathbf{k})=7$ and the plane Π_2 has equation $\mathbf{r}\cdot(-3\mathbf{i}+\mathbf{j}+2\mathbf{k})$

Given that the acute angle between the two planes is θ, find the exact value of $\cos\theta$

12 Calculate the size of the obtuse angle between the planes $x+y-5z=4$ and $2x-y+7z+1=0$

13 Calculate the shortest distance from the point (5, 2, 1) to the plane $\mathbf{r}\cdot(12\mathbf{i}+7\mathbf{j}-14\mathbf{k})=9$

14 Calculate the shortest distance from the point (–12, –7, 4) to the plane $2x-9y-12z=-2$

15 Calculate the shortest distance between the lines $\mathbf{r}=(2\mathbf{i}-\mathbf{k})+t(\mathbf{i}+\mathbf{j}-2\mathbf{k})$ and $\mathbf{r}\times(\mathbf{i}-\mathbf{k})=-2\mathbf{i}+\mathbf{j}-2\mathbf{k}$

Exploration

Going beyond the exams

Research

Galileo (1564–1642) was a major figure in the Scientific Renaissance, making significant contributions to a number of important areas in science. In order to investigate the motion of objects due to gravity, he rolled a ball along a sloping plane and recorded a notable result. The distance traversed *horizontally* by the ball was uniform, that is, in equal time intervals the ball travelled equal distances, but the distance travelled *vertically* was proportional to the square of the time taken.

Research Galileo's work in this field and use vectors to find an equation for the position of such an object measured from the point of projection.

History

Although mathematicians and scientists worked with a conceptual understanding of vectors prior to the 19th century, it was not until that time that vectors first began to be formalised. A number of mathematicians including Wessel, Argand and Gauss, explored complex numbers as vectors having two dimensions: a real and imaginary part. These new complex numbers provided many avenues to explore and helped mathematicians and scientists to think 'vectorially'.

In the mid-nineteenth century the mathematician Hamilton extended his thinking into four-dimensions developing a system that he called quaternions. Maxwell, a Scottish scientist, found that he could use this new way of thinking to develop understanding and insight into physical quantities that he classified as being either scalar or vector in nature.

Investigation

Consider geometrically how three planes in three-dimensional space might intersect each other.

Find a set of equations that illustrate each geometrical situation.

Try to solve your equations and consider how the algebra and geometry of the situation complement each other.

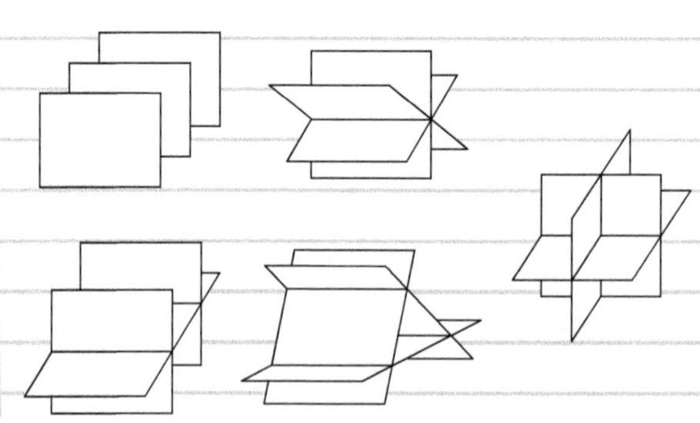

23 Assessment

1 a Calculate the vector product $(\mathbf{i}-\mathbf{j}-\mathbf{k})\times(\mathbf{i}+\mathbf{k})$ **[2]**

b Find a unit vector which is perpendicular to both $\mathbf{i}-\mathbf{j}-\mathbf{k}$ and $\mathbf{i}+\mathbf{k}$ **[2]**

2 The point (2, 7, –3) lies on the line with equation $\frac{x-a}{2}=\frac{y-b}{-3}=\frac{z-c}{1}$

a Write down the values of a, b and c **[1]**

b Find the equation of the line in

i Vector form, ii Vector product form. **[4]**

3 Give the equation of the straight line $\mathbf{r}=\mathbf{i}+2\mathbf{j}-3\mathbf{k}+t(2\mathbf{i}-\mathbf{k})$ in the form $(\mathbf{r}-\mathbf{a})\times\mathbf{b}=0$ **[2]**

4 Show that the point (3, 1, –2) lies on each of these planes.

a $\mathbf{r}\cdot\begin{pmatrix}0\\4\\-2\end{pmatrix}=8$ b $2x+3y-z=11$ **[4]**

5 A triangle has vertices at $A(3, 5, -1)$, $B(4, 0, 2)$ and $C(-3, 2, -4)$

Use the vector product to calculate the area of the triangle. **[5]**

6 A plane passes through the point with position vector $\mathbf{i}-5\mathbf{j}+2\mathbf{k}$ and contains the vectors $-3\mathbf{j}-\mathbf{k}$ and $-2\mathbf{i}+4\mathbf{j}$

a Find the equation of the plane in vector form. **[3]**

b Show that the vector $2\mathbf{i}+\mathbf{j}-3\mathbf{k}$ is perpendicular to the plane. **[4]**

c Find the equation of the plane in

i Scalar product form, ii Cartesian form. **[4]**

7 a Find the equation of the plane that passes through the points (7, 12, –14), (5, 4, –10) and (1, 9, –5) in the form $\mathbf{r}=\mathbf{a}+s\mathbf{b}+t\mathbf{c}$ **[3]**

b Does the point (–1, 0, 4) lie on the plane? Explain how you know. **[4]**

8 The lines l_1 and l_2 have equations l_1: $\mathbf{r}=(\mathbf{i}-\mathbf{j}+3\mathbf{k})+s(2\mathbf{i}-5\mathbf{k})$ and l_2: $\mathbf{r}=(3\mathbf{j}+\mathbf{k})+t(\mathbf{i}+3\mathbf{j})$

a Find the shortest distance between l_1 and l_2. Give your answer correct to 3 significant figures.

The plane Π has Cartesian equation $3x+y-z=2$

The line l_1 intersects the plane Π at the point A **[6]**

b Find the exact coordinates of A **[2]**

c Calculate the acute angle between l_1 and Π, giving your answer to 3 significant figures. **[2]**

9 The point A has coordinates (0, 3, –1). Find the image of A after reflection in the plane $\mathbf{r}\cdot(\mathbf{i}+\mathbf{j}-2\mathbf{k})=3$ **[6]**

10 Given L_1: $\mathbf{r}\times\begin{pmatrix}1\\0\\-2\end{pmatrix}=\begin{pmatrix}-4\\5\\-2\end{pmatrix}$ and L_2: $\mathbf{r}=\begin{pmatrix}5\\5\\2\end{pmatrix}+\lambda\begin{pmatrix}-8\\-6\\2\end{pmatrix}$

a Show that L_1 and L_2 intersect. **[3]**

b Find the point of intersection between L_1 and L_2

11 The plane Π_1 has equation $\mathbf{r}\cdot(2\mathbf{i}-3\mathbf{j})=4$ and the plane Π_2 has equation $\mathbf{r}\cdot(\mathbf{i}+2\mathbf{j}-\mathbf{k})=7$

Calculate the acute angle between planes Π_1 and Π_2 **[3]**

12 A plane Π_1 has equation $\mathbf{r}\cdot\begin{pmatrix}-1\\3\\-3\end{pmatrix}=-2$ and a line l_1 has equation $\mathbf{r}=\begin{pmatrix}5\\-7\\1\end{pmatrix}+t\begin{pmatrix}0\\-4\\2\end{pmatrix}$

a Calculate the acute angle between the plane Π_1 and the line l_1 **[3]**

b Find the point of intersection between the plane Π_1 and the line l_1 **[4]**

13 Find the perpendicular distance from the point (5, 0, –2) to the plane with equation $2x+y-3y=9$ **[3]**

14 Find the equation of the plane that contains the point (5, –2, 7) and the line with equation $\frac{x-2}{4}=\frac{y+1}{5}=\frac{z-1}{-2}$. Give your answer in the form $\mathbf{r}=\mathbf{a}+\lambda\mathbf{b}+\mu\mathbf{c}$ **[6]**

15 The planes Π_1 and Π_2 have equations $\mathbf{r}\cdot\begin{pmatrix}2\\0\\-3\end{pmatrix}=-11$ and $x-2y-z=1$ respectively.

The line with equation $\frac{x+2}{4}=\frac{y+1}{-3}=\frac{z-4}{1}$ intersects Π_1 at the point A and Π_2 at the point B

Calculate the length of AB **[9]**

16 The lines L_1 and L_2 have equations $\mathbf{r}=8\mathbf{i}-14\mathbf{j}+13\mathbf{k}+s(-4\mathbf{i}+7\mathbf{j}-6\mathbf{k})$ and $\frac{x}{2}=\frac{y-17}{5}=\frac{z+7}{-1}$ respectively. The plane Π contains both L_1 and L_2

a Find the vector equation of the plane. **[5]**

b Calculate the distance of the plane from the point (16, 11, –13) **[5]**

17 Find a vector equation of the line with vector product equation $\mathbf{r}\times\begin{pmatrix}3\\-1\\2\end{pmatrix}=\begin{pmatrix}2\\6\\1\end{pmatrix}$ **[5]**

18 Show that the line $\mathbf{r}=\begin{pmatrix}-11\\-4\\10\end{pmatrix}+t\begin{pmatrix}4\\-2\\0\end{pmatrix}$ and the plane

$\mathbf{r}=\begin{pmatrix}3\\-5\\-10\end{pmatrix}+\mu\begin{pmatrix}7\\-10\\0\end{pmatrix}+\lambda\begin{pmatrix}1\\-3\\4\end{pmatrix}$ intersect and find their point of intersection. **[5]**

19 The planes Π_1 and Π_2 have equations $\mathbf{r}=\begin{pmatrix}5\\3\\-2\end{pmatrix}+s\begin{pmatrix}4\\2\\-1\end{pmatrix}+t\begin{pmatrix}0\\-5\\1\end{pmatrix}$ and $\mathbf{r}\cdot\begin{pmatrix}3\\4\\20\end{pmatrix}=12$

a Show that Π_1 and Π_2 are parallel planes. **[6]**

b Find the perpendicular distance between Π_1 and Π_2 **[2]**

24 Circular motion 2

In sports such as cycling, it is important to consider the mechanics of circular motion to ensure both speed and safety. In a velodrome the circular parts of the track, at either end of two straight sections, are steeply banked. This banking helps cyclists to develop force towards the centre of their motion (modelled as being part of a circle) at speed. The maximum angle of banking in velodromes that are used to host world championship and Olympic events is forty-five degrees. This appears very steep when you ride a bike around such a track for the first time.

The same principles apply in the design of surfaces used by other vehicles, such as roads. Busy main roads must be designed to ensure that vehicles can take corners, and use roundabouts, safely. Engineers involved in road design, railway track engineering, and so on, use the principles of the mechanics of circular motion to make sure that their designs are safe under all conditions.

Orientation

What you need to know

Ch9 Circular motion 1
- Kinematics of circular motion.
- Horizontal circular motion.

What you will learn

- How to analyse circular motion.
- How to analyse the conical pendulum.
- How to analyse vertical circular motion.

What this leads to

Careers
Civil engineering.
Architecture.

24.1 Kinematics of circular motion 2

Fluency and skills

There are equations that describe the motion of a point mass moving in a horizontal circle at constant speed.

Key point

A body moving on a circular path of radius r with constant angular velocity ω about the centre has

- a constant linear speed $v = r\omega$ along the tangent
- an acceleration $a = r\omega^2 = \frac{v^2}{r}$ towards the centre
- a centripetal force $F = mr\omega^2$ or $\frac{mv^2}{r}$ towards the centre
- a time period $T = \frac{2\pi}{\omega}$ to make one revolution.

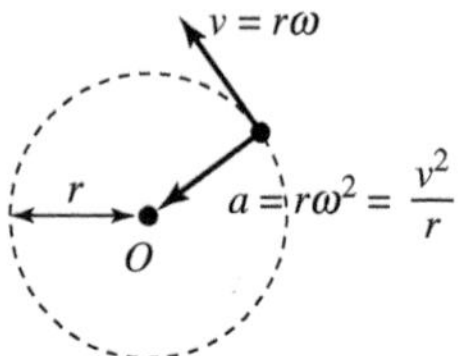

You can also use vectors to find some of these equations using a more mathematically elegant method.

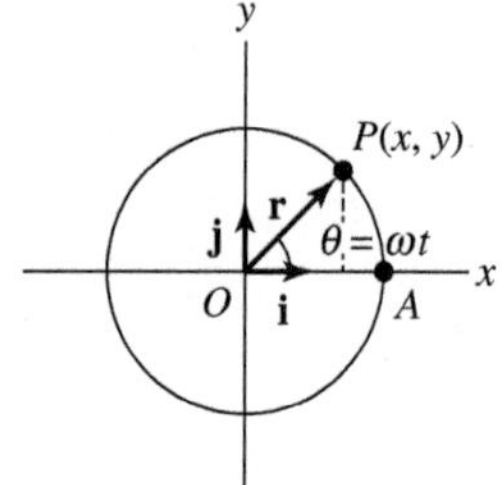

Let the circle have a radius r and a centre O

If the mass starts at point A and, after a time t seconds, it has reached point P, then $\angle AOP = \theta = \omega t$, where ω is the angular velocity in rad s^{-1}.

Let point P have coordinates (x, y) which, using trigonometry, can also be written as $(r\cos\omega t, r\sin\omega t)$.

So, the position vector of P is

$$\mathbf{r} = r\cos\omega t\mathbf{i} + r\sin\omega t\mathbf{j}$$

i and **j** are unit vectors parallel to the x-axis and y-axis respectively.

When you differentiate with respect to t, you get

$$\mathbf{v} = -r\omega\sin\omega t\mathbf{i} + r\omega\cos\omega t\mathbf{j}$$

When you differentiate again,

$$\mathbf{a} = -r\omega^2\cos\omega t\mathbf{i} - r\omega^2\sin\omega t\mathbf{j}$$

$$= -\omega^2(r\cos\omega t\mathbf{i} + r\sin\omega t\mathbf{j})$$

ω is a constant, so $\frac{\text{d}(\cos\omega t)}{\text{d}t} = -\omega\sin\omega t$ and $\frac{\text{d}(\sin\omega t)}{\text{d}t} = \omega\cos\omega t$

which you can write as

$$\mathbf{a} = -\omega^2\mathbf{r}$$

The negative sign in $\mathbf{a} = -\omega^2\mathbf{r}$ indicates that $\mathbf{a}$ is always *towards* the centre, in the opposite direction to $\mathbf{r}$.

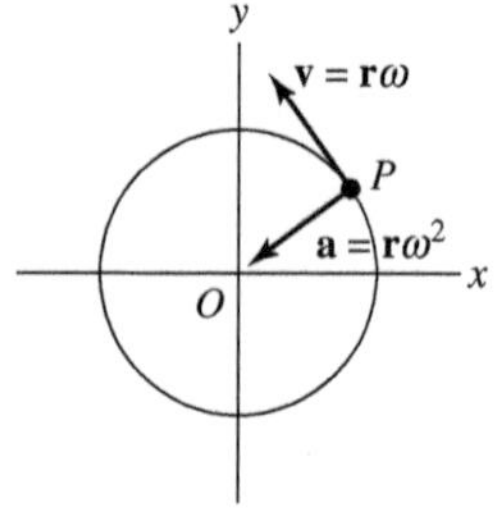

So, although there is no linear acceleration along the tangent, circular motion requires a linear acceleration towards the centre of the circle of magnitude $r\omega^2 = \frac{v^2}{r}$

You can think of the acceleration as changing the direction of the particle, rather than its speed.

Since there is an acceleration, Newton's 2nd law requires that there is a force $\mathbf{F} = m\mathbf{a} = -m\omega^2\mathbf{r}$ towards the centre. This force is called the **centripetal force**.

MECH

Example 1

A point mass of 5 kg has a position vector $\mathbf{r} = 8\sin\left(\frac{1}{2}t\right)\mathbf{i} + 8\cos\left(\frac{1}{2}t\right)\mathbf{j}$ metres at a time t seconds.

a Find the magnitude in newtons of the force **F** acting on the mass when $t = \frac{\pi}{3}$ seconds.

b Show that the force **F** and position vector **r** are parallel at all times.

a $\mathbf{r} = 8\sin\left(\frac{1}{2}t\right)\mathbf{i} + 8\cos\left(\frac{1}{2}t\right)\mathbf{j}$ [1]

$\mathbf{v} = \frac{d\mathbf{r}}{dt} = 8\times\frac{1}{2}\cos\left(\frac{1}{2}t\right)\mathbf{i} - 8\times\frac{1}{2}\sin\left(\frac{1}{2}t\right)\mathbf{j}$

Differentiate with respect to t

$= 4\cos\left(\frac{1}{2}t\right)\mathbf{i} - 4\sin\left(\frac{1}{2}t\right)\mathbf{j}$

$\mathbf{a} = \frac{d\mathbf{v}}{dt} = -4\times\frac{1}{2}\sin\left(\frac{1}{2}t\right)\mathbf{i} - 4\times\frac{1}{2}\cos\left(\frac{1}{2}t\right)\mathbf{j}$

Differentiate again with respect to t

$= -2\sin\left(\frac{1}{2}t\right)\mathbf{i} - 2\cos\left(\frac{1}{2}t\right)\mathbf{j}$

Newton's 2nd law gives

$\mathbf{F} = m\mathbf{a} = -10\sin\left(\frac{1}{2}t\right)\mathbf{i} - 10\cos\left(\frac{1}{2}t\right)\mathbf{j}$ [2]

When $t = \frac{\pi}{3}$

$\mathbf{F} = -10\left(\sin\frac{\pi}{6}\mathbf{i} + \cos\frac{\pi}{6}\mathbf{j}\right) = -5\mathbf{i} - 5\sqrt{3}\mathbf{j}$

Magnitude of $\mathbf{F} = |\mathbf{F}| = \sqrt{(-5)^2 + (-5\sqrt{3})^2}$

$= \sqrt{25+75} = 10\text{ N}$

b From [1] $\quad \mathbf{r} = 8\left[\sin\left(\frac{1}{2}t\right)\mathbf{i} + \cos\left(\frac{1}{2}t\right)\mathbf{j}\right]$

From [2] $\quad \mathbf{F} = -10\left[\sin\left(\frac{1}{2}t\right)\mathbf{i} + \cos\left(\frac{1}{2}t\right)\mathbf{j}\right]$

$= -10\times\frac{1}{8}\mathbf{r} = -1.25\mathbf{r}$

Hence, **F** and **r** are parallel but have opposite directions.

Exercise 24.1A Fluency and skills

1 A point mass of 10 kg has a position vector $\mathbf{r} = 2\sin 4t\mathbf{i} + 2\cos 4t\mathbf{j}$ metres at time t seconds.

Find vector expressions for

a The acceleration of the mass,

b The force acting on the mass.

2 A 2 kg particle P moves on a smooth horizontal plane containing x- and y-axes. Its velocity $\mathbf{v}$ is given by $\mathbf{v} = 6\cos 2t\mathbf{i} - 6\sin 2t\mathbf{j}\,\text{m s}^{-1}$.

When $t = 0$, P has the position vector $2\mathbf{i} + 4\mathbf{j}$.

a Find the position vector, $\mathbf{r}$, of P at time t

b Use the components of $\mathbf{r}$ to show that P moves on a circular path and find the equation of the path.

3 A mass, M, of 3 kg at the point (x, y) on coordinate axes moves on a circular path with the equation $x^2 + y^2 = 16$. The line OM rotates about the origin O with a constant angular speed of $0.5\,\text{rad s}^{-1}$, starting in line with the x-axis when $t = 0$

a Write expressions for

i The position vector $\mathbf{r}$ of M at time t

ii The acceleration of M as a vector at time t

b Find the time for M to make one revolution about O

c Calculate the magnitude of the force acting on M

4 The position vector of a particle at time t seconds is given by $\mathbf{r} = 2\sin 3t\mathbf{i} + 2\cos 3t\mathbf{j}$ metres.

a Express, as vectors, the velocity and acceleration of the particle.

b Show that the direction of the acceleration is

i Parallel to the direction of the position vector,

ii Perpendicular to the direction of the velocity.

5 A particle of mass 4 kg is acted on by a force $\mathbf{F}$ newtons and it moves in a horizontal plane with a velocity $\mathbf{v} = 4\cos 2t\mathbf{i} + 4\sin 2t\mathbf{j}\,\text{m s}^{-1}$ at a time t seconds.

a Find an expression for force $\mathbf{F}$ in terms of t and find its magnitude when $t = \pi$

b When $t = 0$, the particle is at a point with a position vector $3\mathbf{i} - 10\mathbf{j}$ metres. Find the position vector, $\mathbf{r}$ of the particle at time t seconds. Describe the path in which the particle moves.

6 A particle of mass m kg is acted upon by a force of $\mathbf{F}$ newtons acting through the origin, O. At time t seconds, the particle has position $\mathbf{r}$ in the x–y plane and velocity $\mathbf{v} = a\sin \omega t\mathbf{i} + b\cos \omega t\mathbf{j}\,\text{m s}^{-1}$.

Prove that $\mathbf{F} = k\mathbf{r}$ and find k in terms of the constants m and ω

Reasoning and problem-solving

Strategy

To solve problems involving motion in a horizontal circle

1. Draw and label a diagram showing all forces and other key variables.
2. Use Newton's 2nd law to write an equation of motion to the centre.
3. Write other equations involving forces, as necessary, and solve them.

Example 2

A car travels round a bend in a smooth road of radius 60 m which is banked at 10° to the horizontal.

a Find the only safe speed (in km h^{-1}) at which the car can travel.

b What assumption in this model is unrealistic?

a

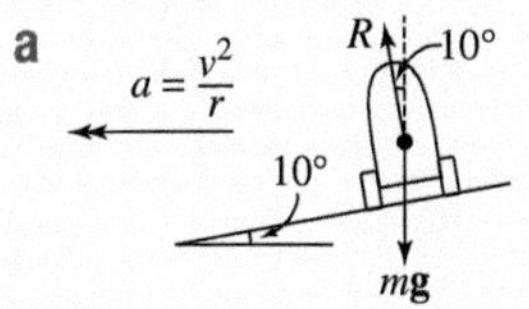

The car does not move vertically.

Resolve vertically

$$R\cos 10° = mg \qquad [1]$$

Horizontal equation of motion is

$$R\sin 10° = m \times \frac{v^2}{60} \qquad [2]$$

Divide [2] by [1] to eliminate R

$$\tan 10° = \frac{mv^2}{60} \div mg = \frac{v^2}{60g}$$

The only safe speed, $v = \sqrt{60g\tan 10°} = 10.1\text{ m s}^{-1} = 36.6\text{ km h}^{-1}$

If $v > 10.1\text{ m s}^{-1}$, the car skids up the slope.

If $v < 10.1\text{ m s}^{-1}$, the car skids down the slope.

b The model is unrealistic as it takes no account of friction acting along the slope and a speed of 10.1 m s^{-1} could not be maintained precisely.

1. Draw a labelled diagram. Let the mass, velocity and acceleration of the car be m, v and a. Let the normal reaction of the ground on the car be R

3. Vertical forces balance.

2. Use Newton's 2nd law.

MECH

Example 3

The bend in Example 2 is resurfaced so the coefficient of friction is μ

Find the least value of μ for the car to round the bend at a speed of 26 m s^{-1} without any side-slip up the slope.

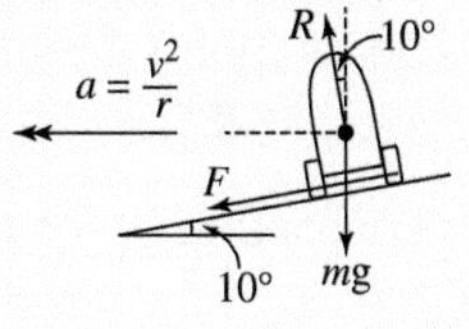

$a\sin 10°$, a, 10°, $a\cos 10°$

Equation of motion down the slope is

$$F + mg\sin 10° = \frac{m \times 26^2}{60} \times \cos 10°$$

1. At its greatest safe speed, the car is on the point of slipping up the slope, so friction acts down the slope.

2. Resolving parallel to the slope means that R does not appear in the equation, so it is easy to calculate F

(*Continued on the next page*)

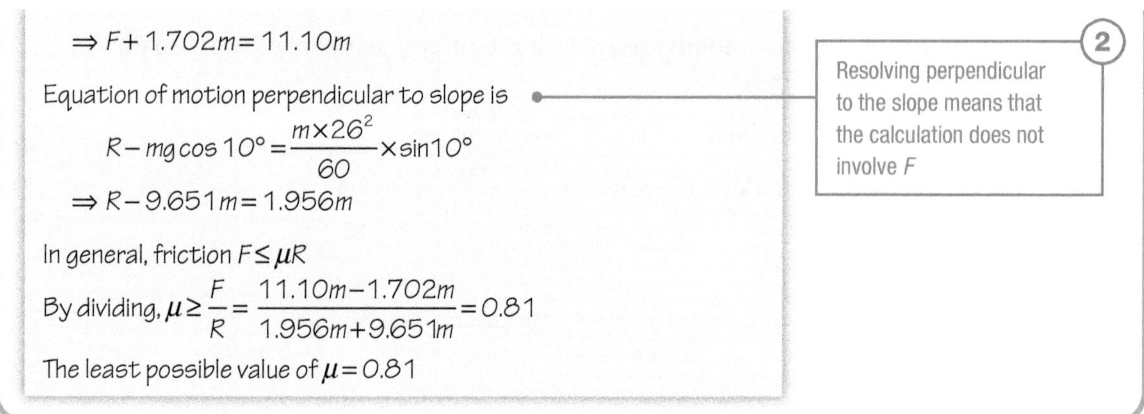

$\Rightarrow F+1.702m=11.10m$
Equation of motion perpendicular to slope is
$R-mg\cos 10°=\frac{m\times 26^2}{60}\times\sin 10°$
$\Rightarrow R-9.651m=1.956m$
In general, friction $F\leq\mu R$
By dividing, $\mu\geq\frac{F}{R}=\frac{11.10m-1.702m}{1.956m+9.651m}=0.81$
The least possible value of $\mu=0.81$
2
Resolving perpendicular to the slope means that the calculation does not involve F

Example 3 could also be solved by writing a horizontal equation of motion and by resolving vertically for vertical equilibirum (as in Example 2). F and R are then found by solving the equations simultaneously.

Exercise 24.1B Reasoning and problem-solving

1 The smooth road surface on a corner of radius 80 m is designed so that a car can travel without sideways movement when driven at a certain speed. If the road is banked at an angle of 15°, find the speed.

2 A lorry is driven round a bend in a smooth road of radius 100 m which is banked at an angle of 12°. At what speed should it be driven so that there is no sideways force on its tyres?

3 A smooth road is banked so that a car can travel round a circular bend without any skidding if its speed is 20 m s^{-1}. If the radius of the bend is 120 m, calculate the angle of the banking.

4 A circular bend in a road has a radius of 180 m and it is banked at 45° to the horizontal. The coefficient of friction between the road and a car is 0.5

a At what speed would a car rounding the bend have no sideways friction on its tyres?

b What are the car's maximum and minimum speeds that are possible without any sideways slipping?

5 A car is just on the point of skidding when it goes round a bend of radius 30 m at a speed of 54 km h^{-1} on a level racing track.

a At what angle to the horizontal should the track be banked so the car can round the same bend at 108 km h^{-1} without skidding up the slope?

b State an assumption you have made in your solution.

6 Find the maximum speed at which a car can be driven round a bend of radius 120 m if the coefficient of friction between the car and road is 0.4 and the road surface is

a Horizontal,

b Banked at an angle of 10°.

7 A bend in a road has a radius of 80 m and is banked at 14°. A car is on the point of sliding up the slope when it travels at 25 m s^{-1}

a Find the coefficient of friction, μ between the road and car at the bend.

b On an icy morning when $\mu=0.1$, what is the maximum speed of the car for it not to slide upwards?

8 Part of a railway track is a circular arc of radius 250 m. The track is banked so that there is no side-thrust on the flanges of the train's wheels when it travels at 50 km h^{-1}. Calculate the angle of the banking of the track.

9 A railway track curves with a radius of 60 m and it is banked so that an engine travelling at 15 m s^{-1} exerts no sideways force on the track. What sideways force would there be when a 50 tonne engine

a Stands at rest on the bend,

b Moves at 30 m s^{-1} round the bend?

10

a A motorcyclist rounds a curve of radius 12 m on a level road at 5 m s^{-1}. By considering the forces on the bike and the moments about the centre of mass G, find the angle at which he and his bike are inclined to the vertical and the least value of the coefficient of friction for the bike not to side-slip.

b Another corner on a level road has a radius of 30 m. Find the greatest speed in km h^{-1} at which he can round the corner if the coefficient of friction with the road is 0.4

11 A particle moves in a horizontal circle on the smooth inner surface of an inverted hollow cone. The circle is at a height h above the vertex of the cone. Prove that the particle's linear velocity is $\sqrt{hg}$

12 Four masses of 3 kg each are connected by four strings 10 cm long so they form a square with the strings as its sides. The square is placed with its centre at the centre of a turntable that is rotating with an angular speed of 1.5 rad s^{-1}. Find the tension in each of the strings.

13 A smooth bowl is formed from a segment of a sphere of radius $2x$ such that the bowl has a depth x in the middle. The bowl is rotated about its vertical axis with angular velocity ω such that a small particle is at rest relative to the bowl when placed just within the rim. Find the value of ω in terms of x

14 A car rounds a bend of radius r on a road banked at an angle θ to the horizontal, where the coefficient of friction $\mu = \tan\lambda$ Prove that the car can drive round the bend at a speed v without skidding provided that

$$\theta - \lambda < \arctan\left(\frac{v^2}{rg}\right) < \theta + \lambda$$

24.2 The conical pendulum

Fluency and skills

An ordinary pendulum swings to and fro with all the motion in the same vertical plane. A **conical pendulum**, however, has its bob moving in a horizontal circle so that the string of the pendulum traces out a hollow vertical cone.

Example 1

A conical pendulum has a bob P of mass 2 kg hanging at the lower end of a light, inextensible string of length 2 m. The upper end of the string is fixed to a point O. The bob moves in a horizontal circle, centre C and radius 1.2 m, at constant speed.

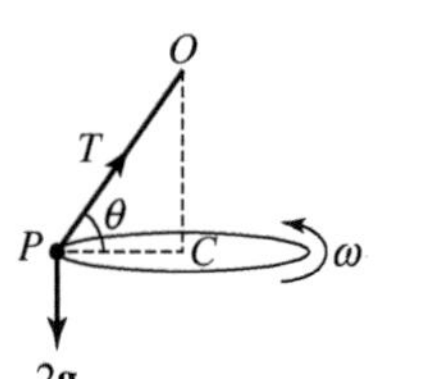

Find the tension in the string, the angular speed of the bob and the time taken for the bob to make one revolution.

In ΔOPC, $OC = \sqrt{2^2 - 1.2^2} = 1.6$ m

Resolve vertically $\quad T\sin\theta = 2g$ — P is in vertical equilibrium.

$$T \times \frac{1.6}{2} = 2 \times 9.8$$

$$\text{Tension, } T = \frac{4 \times 9.8}{1.6} = 24.5\text{ N}$$

Equation of motion of P towards C is

$$T\cos\theta = mr\omega^2$$ — Resolve horizontally and use Newton's 2nd law.

$$24.5 \times \frac{1.2}{2} = 2 \times 1.2 \times \omega^2$$

$$\omega^2 = 6.125$$

Angular speed, $\omega = 2.47$ rad s^{-1}

$$\text{Time for one revolution} = \frac{2\pi}{\omega} = \frac{2\pi}{2.47} = 2.5 \text{ seconds}$$ — Time $= \frac{\text{angular distance}}{\text{angular speed}}$

Exercise 24.2A Fluency and skills

In this exercise, all strings are light and inextensible.

1 A particle of mass 2 kg is attached to one end of a string 0.5 m long, the other end of which is fixed to a point O. The particle rotates in a horizontal circle about a point vertically below O so that the string makes an angle of 60° with the vertical. Find the tension in the string and the time taken for the particle to make one revolution.

2 The bob of a conical pendulum moves in a horizontal circle at a steady speed of 80 revolutions per minute (rpm). If the string of the pendulum is 0.25 m long, show that it makes an angle of approximately 56° with the vertical.

3 The string of a conical pendulum is 80 cm long and makes an angle of 60° with the vertical. Show that the bob of the pendulum makes about 8 revolutions every 10 seconds.

4 A mass of 0.5 kg is attached to a string of length 1.5 m. The mass acts as conical pendulum and moves in a horizontal circle at a steady speed of 60 rpm. Calculate the radius of the circle and the tension in the string.

5 A conical pendulum has a bob with a mass of 1.2 kg which makes a full rotation every second. If the length of the pendulum is 50 cm, calculate the tension in the string and the angle which the string makes with the vertical.

6 A small object is attached to a fixed point by a string of length l. It describes a horizontal circle with an angular speed of ω rad s^{-1}. Prove that its vertical distance below the fixed point is independent of the length of the string.

MECH

Reasoning and problem-solving

Strategy

To solve problems involving motion in horizontal circles

1. Draw a clear diagram showing all the forces acting on the moving parts.
2. Write equations of motion to the centre of the circle using Newton's 2nd law.
3. Write other equations involving forces, as necessary, and solve them simultaneously.

Example 2

A 0.5 kg mass P is tied to the midpoint of a 2 m string XY with X $\sqrt{3}$ m vertically above Y

P rotates in a horizontal circle so that the string stays taut.

Find the least possible angular velocity of P. Take g as 9.81 m s^{-2}.

X, 1 m, $\frac{\sqrt{3}}{2}$ m, T, θ, ω, P, U, 0.5g, Y

ΔXYP is isosceles, so $\cos\theta = \frac{\sqrt{3}}{2}$ and $\theta = 30°$

Vertical equilibrium:

$T\cos\theta = U\cos\theta + 0.5g$

giving $T - U = 5.664$ [1]

Horizontal equation of motion for P is

$T\sin\theta + U\sin\theta = mr\omega^2 = 0.5 \times 1 \times \sin\theta \times \omega^2$

$T + U = 0.5\omega^2$ [2]

The string stays taut if $U \geq 0$

From [1][2], $2U = 0.5\omega^2 - 5.664$

$U \geq 0$ if $\omega^2 \geq 11.33$,

so least angular velocity, ω is $\sqrt{11.33} = 3.4$ rad s^{-1}

1 Draw a labelled diagram. Let T and U be tensions, as shown

3 Write an equilibrium equation for vertical forces.

2 Use Newton's 2nd law for horizontal forces.

The string can only go slack between P and Y

3 Solve simultaneously to find an expression for U

If mass P were a smooth ring threaded onto string XY, then tensions T and U would be equal on either side of P. P will adjust its position and may not stay at the midpoint M

Example 3

Particle Y is connected to two light rods XY and YZ, both of length l, such that end X pivots about a fixed point.

End Z is attached to a ring which slides on a smooth vertical rod XZ. Y and Z have the same mass m

The system rotates so that Y performs horizontal circles with constant speed ω

Prove that, if $\angle ZXY = \theta$, then $\cos\theta = \frac{3g}{l\omega^2}$

ΔXYZ is isosceles. T and U are forces in the rods. R is the reaction of rod XZ on Z.

Z is in vertical equilibrium: $U\cos\theta = mg$ [1]

Y is in vertical equilibrium $T\cos\theta = U\cos\theta + mg$

$= 2mg$ [2]

Horizontal equation of motion for Y is

$T\sin\theta + U\sin\theta = m \times l\sin\theta \times \omega^2$ [3]

Substituting [1] and [2] into [3]

$$\frac{2mg}{\cos\theta} \times \sin\theta + \frac{mg}{\cos\theta} \times \sin\theta = m \times l\sin\theta \times \omega^2$$

Simplify to give $\cos\theta = \frac{3g}{l\omega^2}$

Write equations for Y and Z

Substitute from [1]

Divide by $m\sin\theta$ and rearrange.

Exercise 24.2B Reasoning and problem-solving

1 a A 2 kg mass is fixed to the midpoint of a string XY of length 10 m, with X 6 m vertically above Y. The mass describes a horizontal circle with an angular speed ω. Find the minimum value of ω so that both parts of XY remain taut.

b The mass in part **a** is changed to a smooth 2 kg ring R which can slide freely along the same string XY. Find the angular speed ω of the ring for it to rotate in a horizontal circle about Y with the horizontal distance $RY = 3.2$ m.

c Give two assumptions that you have made about this mathematical model.

2 A light, inextensible string PQ passes through a smooth fixed ring R in the ceiling. A 2 kg mass at Q hangs at rest vertically below R as a 1 kg mass at P acts as a conical pendulum moving in a horizontal circle with an unknown height and radius. What is the radius of the circle if the linear speed of the 1 kg mass is 7 m s^{-1}?

3 A light rod BC of length l is pivoted at a fixed point C and held horizontal by a string AB where A is a fixed point at a distance h vertically above C. A mass m attached to B rotates about C with angular velocity ω so that the rod BC remains horizontal. Find the force exerted on the mass by the rod in terms of m, l, h, ω and g.

4 A particle moves horizontally on a circular path of radius r inside a smooth hemispherical bowl of radius R. Find the period of its rotation in terms of r, R and g

5 The ends of an inextensible string of length 0.4 m are attached to a mass m on a smooth

horizontal table and a fixed point above the table. The mass moves in a horizontal circle of radius 0.2 m in contact with the table, with the string taut, at an angular speed ω. Find the reaction between the mass and table in terms of m and ω and show that the maximum value of ω to keep in contact with the table is 5.3 rad s^{-1}

6 The ends A and B of an inextensible string are fastened to a vertical rod with A above B A smooth ring R of mass 3 kg slides on the string and rotates with the string taut in a horizontal circle of radius 0.7 m about the rod below the level of B. If $\angle RAB = 28°$ and $\angle RBA = 144°$, find the angular speed of R.

7 An elastic string of natural length l and modulus of elasticity $2mg$ has its upper end fixed and its lower end attached to a particle of mass m which makes horizontal circles with angular velocity ω

a When the extension in the string is $\frac{3}{5}l$

i Show that $\omega^2 = \frac{3g}{4l}$

ii Find the exact value of the angle that the string makes with the vertical.

b State two assumptions you have made in this mathematical model.

8 A light, inextensible string has identical particles of mass m attached at each end. The string is threaded through a hole in the vertex V of a smooth, hollow cone of semi-vertical angle 30°. The particle inside the cone hangs at rest. The other particle describes horizontal circles on the outer surface of the cone at a distance of 0.5 metres from V. Show that the angular speed of the moving particle is 3.24 rad s^{-1}

9 A smooth vertical hoop of radius r rotates with angular speed ω about its vertical diameter. Two beads, A of mass m and B of mass M (with $M > m$), are threaded on the hoop with B vertically below A. They are joined by a light rigid rod of length r

a Show that the reaction R of the hoop on A is given by $R = mr\omega^2$

b If $m = 2$ kg, $M = 4$ kg and $r = 1.5$ m, find the value of ω

10 A particle P of mass m moves in a horizontal circle, centre C and radius 12 cm, on the inside surface of a smooth hemispherical bowl of radius 13 cm.

C is 8 cm vertically above the lowest point H of the bowl. A light, inextensible string attached to P passes through a small smooth hole at H with its lower end attached to an identical particle Q suspended at rest. Find the angular speed of P

11 Steam engines have a mechanism, called a governor, which rotates so that it opens or closes a valve to adjust the speed and so keep it constant.

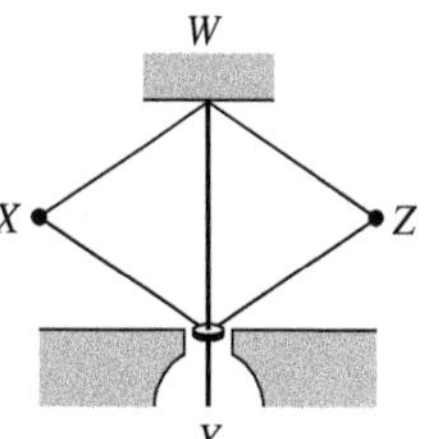

Four equal light rods are hinged to form a rhombus $WXYZ$ with W fixed. Y is a collar of mass m which slides smoothly on a vertical fixed rod WY. Two identical metal balls of mass M are attached at X and Z The mechanism rotates about rod WY at an angular speed ω

a Find the distance WY in terms of M, m and ω

b Describe three ways in which this model of a governor can be refined.

12 An elastic string of natural length l is fixed at one end. A mass m hangs from the other end and produces an extension e when in equilibrium. When the same mass acts as a conical pendulum and describes horizontal circles with angular velocity ω, the same string makes an angle θ with the vertical. Find an expression for $\cos\theta$ in terms of l, e, ω and g

24.3 Vertical circular motion

Fluency and skills

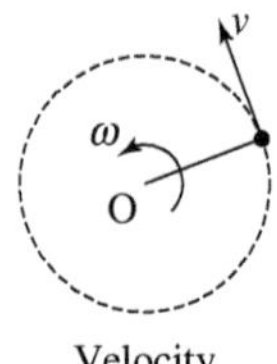

Velocity

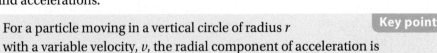

In the circle problems you have studied so far, angular velocity, ω, and tangential velocity, v, have both been constant. However, when circular motion is in a vertical plane, ω and v both vary, so $\frac{d\omega}{dt}$ and $\frac{dv}{dt}$ are not zero.

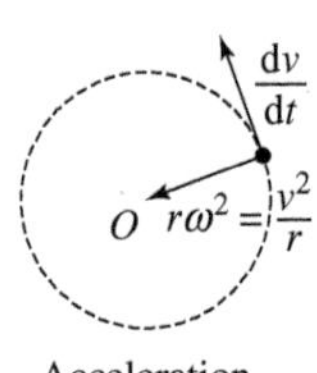

Acceleration

The acceleration of a particle now has two components:

- A **radial component** towards the centre of the circle, O, of magnitude $r\omega^2 = r\left(\frac{d\theta}{dt}\right)^2 = r\dot{\theta}^2 = \frac{v^2}{r}$

  where $\dot{\theta}$ is used as a shorthand for $\frac{d\theta}{dt}$
- A **tangential component** along the tangent of magnitude $\frac{dv}{dt} = r\frac{d\omega}{dt} = r\frac{d^2\theta}{dt^2} = r\ddot{\theta}$

  where $\ddot{\theta}$ is used as a shorthand for $\frac{d^2\theta}{dt^2}$

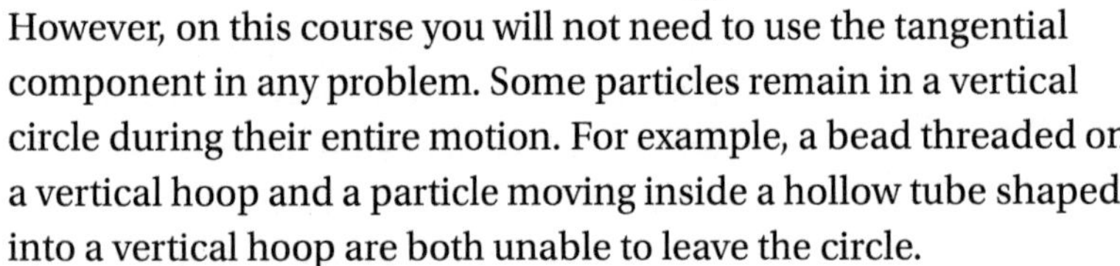

However, on this course you will not need to use the tangential component in any problem. Some particles remain in a vertical circle during their entire motion. For example, a bead threaded on a vertical hoop and a particle moving inside a hollow tube shaped into a vertical hoop are both unable to leave the circle.

High-energy particles in these, or similar, situations will be able to make complete circles, but low-energy particles will merely oscillate in the lower part of the circle.

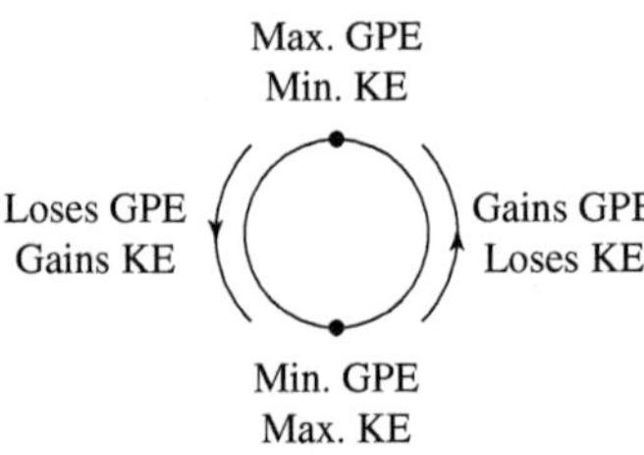

Consider a particle moving smoothly in a vertical circle under its own weight. It slows down as it travels towards the top of the circle and speeds up as it moves down towards the bottom. There are changes in its kinetic and gravitational potential energy.

An **energy equation** can state *either* that the total energy of the particle is constant *or* that a gain in one form of energy is balanced by a loss in another form.

The energy equation involves velocities and heights, but not forces.

You can also use Newton's 2nd law to write an equation of motion towards the centre of the circle. This equation will involve forces and accelerations.

Key point

For a particle moving in a vertical circle of radius r with a variable velocity, v, the radial component of acceleration is

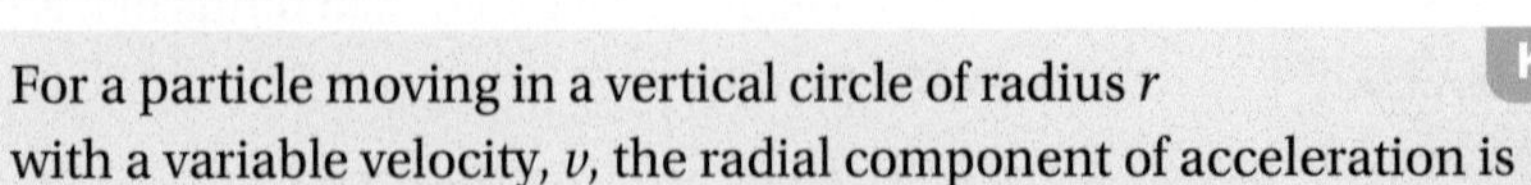

$$r\omega^2 = r\dot{\theta}^2 = \frac{v^2}{r}$$

Newton's 2nd law can be used for motion towards the centre of the circle. An energy equation can also be used.

Example 1

A 2 kg mass, P is attached to a string OP of length 1.2 m where O is a fixed point. P is given a horizontal velocity, u of 4 m s^{-1} when hanging vertically below O at point Q

When $\angle QOP = 60°$, find

a The velocity, v, of P

b The tension, T in the string.

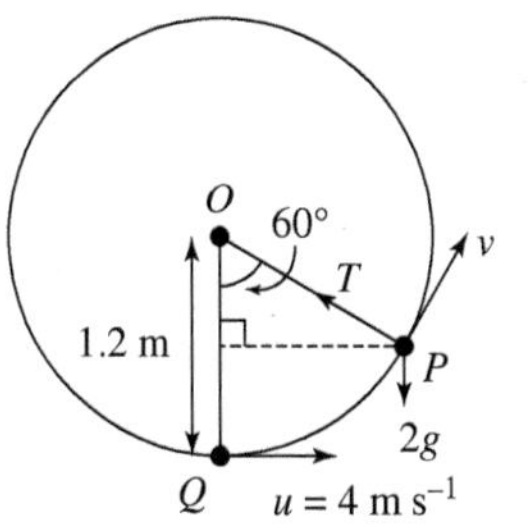

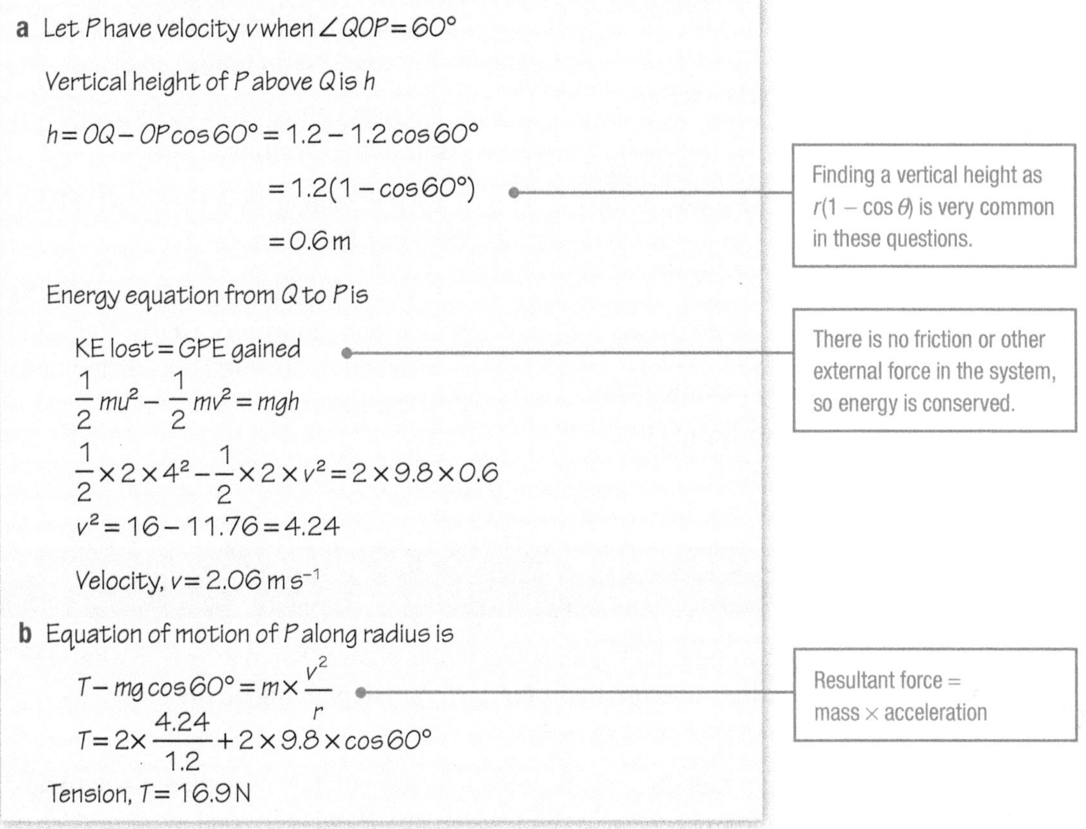

a Let P have velocity v when $\angle QOP = 60°$

Vertical height of P above Q is h

$h = OQ - OP\cos 60° = 1.2 - 1.2\cos 60°$

$= 1.2(1 - \cos 60°)$

$= 0.6\text{ m}$

Finding a vertical height as $r(1 - \cos\theta)$ is very common in these questions.

Energy equation from Q to P is

KE lost = GPE gained

$\frac{1}{2}mu^2 - \frac{1}{2}mv^2 = mgh$

$\frac{1}{2} \times 2 \times 4^2 - \frac{1}{2} \times 2 \times v^2 = 2 \times 9.8 \times 0.6$

$v^2 = 16 - 11.76 = 4.24$

Velocity, $v = 2.06\text{ m s}^{-1}$

There is no friction or other external force in the system, so energy is conserved.

b Equation of motion of P along radius is

$T - mg\cos 60° = m \times \frac{v^2}{r}$

$T = 2 \times \frac{4.24}{1.2} + 2 \times 9.8 \times \cos 60°$

Tension, $T = 16.9\text{ N}$

Resultant force = mass × acceleration

In Example 1, the energy equation gave you the velocity, v, which you then used in the equation of motion to the centre. This link between the two equations is a common strategy.

Exercise 24.3A Fluency and skills

In this exercise, use $g = 9.8\text{ m s}^{-2}$.

1 A 5 kg mass, P, is attached to a string OP of length 1.5 m where O is a fixed point. P is given a horizontal velocity, u, of 10 m s^{-1} when hanging vertically below O at point Q

Find the velocity of P and the tension in the string when

a $\angle QOP = 60°$,

b OP is horizontal.

2 A 10 kg mass P is attached to a light rigid rod OP of length 2 m such that the rod can rotate about the fixed point O

P is given a horizontal velocity, u, of 8 m s^{-1} when OP is vertical with P below O at point Q

Find the velocity of P and the force in the rod when

a $\angle QOP = 60°$, **b** $\angle QOP = 120°$

In each case, state whether the force in the rod is a tension or a thrust.

3 A 250 gram bead B is threaded on a fixed, smooth, vertical hoop of radius 2 m and centre O. It is projected horizontally from the lowest point A of the hoop with a velocity, u, of $10\,\text{m}\,\text{s}^{-1}$.

a Will it travel in complete circles of the hoop? Explain your answer.

b If its initial speed is reduced to $7\,\text{m}\,\text{s}^{-1}$, find the greatest value of $\angle AOB$ in the subsequent motion.

4 A particle P of mass 0.5 kg rotates on the inside of a smooth circular surface, in the vertical plane, of radius 1.4 m and centre O. It leaves the lowest point Q of the surface with a horizontal velocity of $6\,\text{m}\,\text{s}^{-1}$.

a When $\angle QOP = 45°$, find

i The velocity of P

ii The reaction between P and the surface.

b Will P rise to the same horizontal level as O? Explain your answer.

5 a A 5 kg particle P is attached to a string OP of length 1.5 m where O is a fixed point. P is given a horizontal velocity, u, of $8\,\text{m}\,\text{s}^{-1}$ when hanging vertically below O at point Q

i Find the velocity, v, of P and the tension, T in OP when $\angle QOP = 30°$.

ii Give a reason why the method used in part **i** might not apply when $\angle QOP$ is an obtuse angle.

b P is now connected to O by a light rod rather than a string.

i Will your answer to part **a i** still be valid?

ii Find the velocity, v, of P and the force in the rod when $\angle QOP = 150°$, stating whether the force is a tension or a thrust.

Reasoning and problem-solving

When a particle P rotates into the upper half of a vertical circle, centre O, it may not have enough energy to complete a full circle. The subsequent motion will vary depending on the particular situation. For example:

- If P is attached to O by a light rod, the rod will come to instantaneous rest when the particle's kinetic energy (and therefore its velocity) is reduced to zero. P will then oscillate to and fro on a circular arc, but never make a full rotation.
- If P is attached to O by a string, P will leave the circle if the tension in the string reduces to zero and the string becomes slack. P will fall inside the circle under gravity.
- If the particle P is sliding up the inside or down the outside of a circular surface, circular motion is broken when the reaction with the surface becomes zero. P will leave the circle and fall away under gravity.

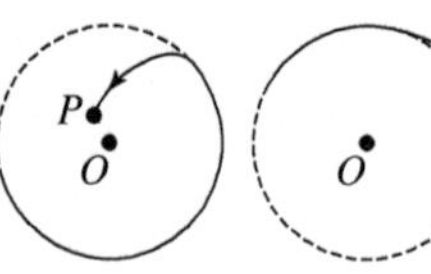

Strategy

To solve problems involving motion in vertical circles

1. Draw a clear diagram showing all relevant information.
2. Consider the conditions which apply for a complete circle or for when motion in a circle ceases.
3. Write energy equations and use Newton's 2nd law, as necessary.

Example 2

A particle P of mass m at point A hangs vertically from a fixed point O where OA is a light rod of length $r = 1.5$ m. P is given a horizontal speed, u, at A

Find the least value of u for P to perform a complete circle about O

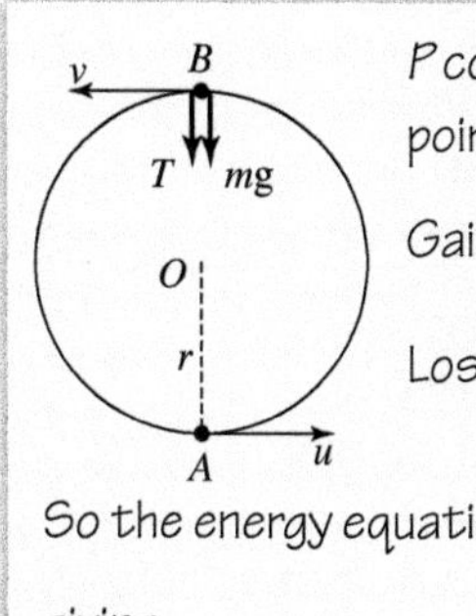

P completes a circle if it reaches the topmost point B with a speed $v \geq 0$

Gain in GPE from A to B is $m \times g \times 2r = 2mgr$

Loss of KE from A to B is $\frac{1}{2}mu^2 - \frac{1}{2}mv^2$

So the energy equation is $2mgr = \frac{1}{2}mu^2 - \frac{1}{2}mv^2$

giving $v^2 = u^2 - 4gr$

So $v \geq 0$ if $u^2 \geq 4gr$

Minimum speed, $u = \sqrt{4 \times 9.8 \times 1.5} = 7.67\ \text{m s}^{-1}$ for a complete circle for P

1 2 Draw a clear diagram.

State the condition for reaching the top.

3 Write the energy equation.

Example 3

The light rod OA in Example 2 is replaced by a string of the same length. Find the new least value of u for P to perform a complete circle about O

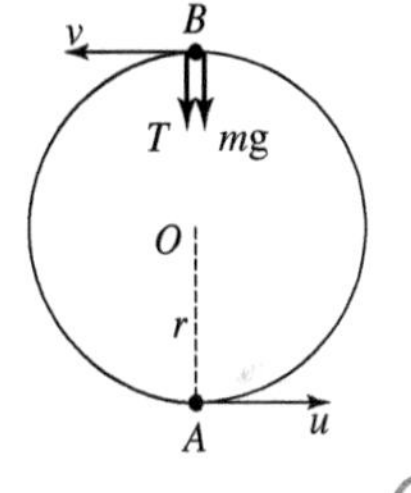

P now completes a circle if the string is still taut at B; that is, if its tension $T \geq 0$ at B

The energy equation is unchanged, so $v^2 = u^2 - 4gr$ [1]

From Newton's 2nd law, the equation of motion of P at B towards O is

$$T + mg = m \times \frac{v^2}{r} \Rightarrow T = \frac{mv^2}{r} - mg$$

So, $T \geq 0$ if $\frac{mv^2}{r} \geq mg$ or $v^2 \geq rg$ [2]

From [1] and [2], $u^2 - 4gr \geq rg \Rightarrow u^2 \geq 5gr$

Minimum speed, $u = \sqrt{5 \times 9.8 \times 1.5} = 8.57\ \text{m s}^{-1}$ for a complete circle for P

2 State the condition for reaching the top.

3 Write an energy equation and an equation of motion using Newton's 2nd law.

MECH

Example 4

A particle P of mass m is at the lowest point A on the inside of a thin, smooth sphere of radius 1.2 m with centre O. P is projected along the surface from A with a velocity, u, of 6 m s^{-1}. Find $\angle AOP$ when P is about to lose contact with the sphere.

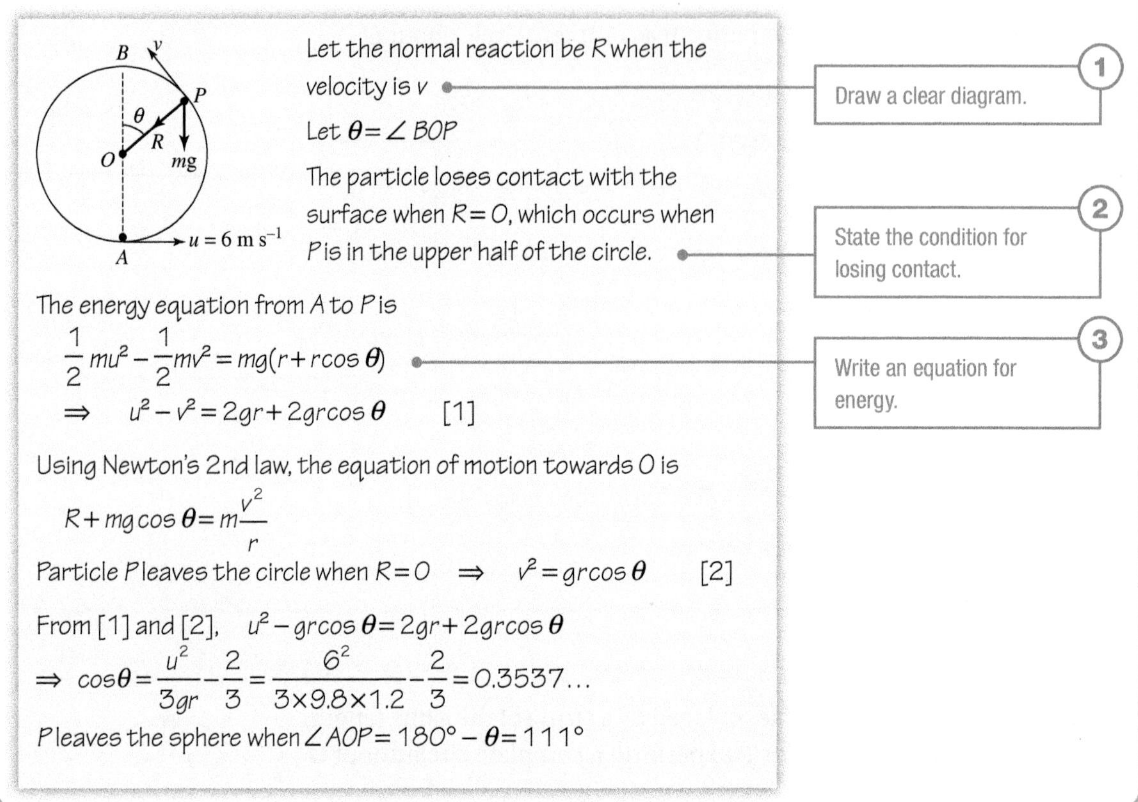
B
v
P
θ
R
O
mg
u = 6 m s⁻¹
A
Draw a clear diagram.
State the condition for losing contact.
Write an equation for energy.

Let the normal reaction be R when the velocity is v

Let $\theta = \angle BOP$

The particle loses contact with the surface when $R = 0$, which occurs when P is in the upper half of the circle.

The energy equation from A to P is

$$\frac{1}{2}mu^2 - \frac{1}{2}mv^2 = mg(r + r\cos\theta)$$

$$\Rightarrow \quad u^2 - v^2 = 2gr + 2gr\cos\theta \qquad [1]$$

Using Newton's 2nd law, the equation of motion towards O is

$$R + mg\cos\theta = m\frac{v^2}{r}$$

Particle P leaves the circle when $R = 0 \quad \Rightarrow \quad v^2 = gr\cos\theta \qquad [2]$

From [1] and [2], $\quad u^2 - gr\cos\theta = 2gr + 2gr\cos\theta$

$$\Rightarrow \cos\theta = \frac{u^2}{3gr} - \frac{2}{3} = \frac{6^2}{3 \times 9.8 \times 1.2} - \frac{2}{3} = 0.3537\ldots$$

P leaves the sphere when $\angle AOP = 180° - \theta = 111°$

Example 5

A particle P of mass 4 kg rests at the topmost point A of the surface of a smooth sphere of radius 1.5 m, fixed to a horizontal plane at point B. P is slightly disturbed from rest and moves down the surface of the sphere before leaving it at C. If P strikes the plane at D, find the distance BD

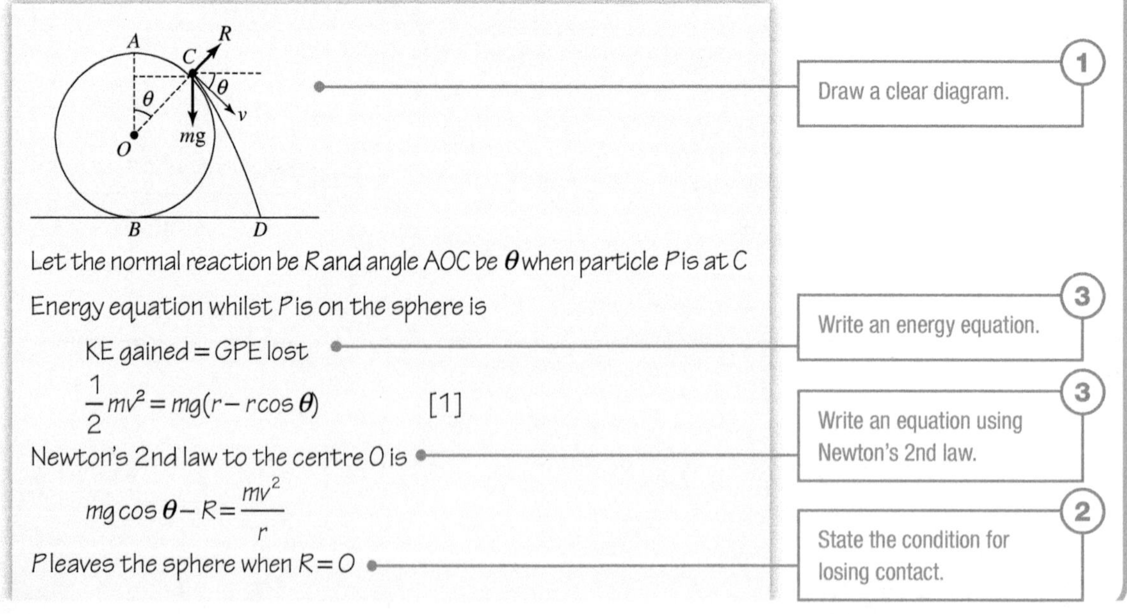
A
R
C
θ
θ
v
mg
O
B
D
Draw a clear diagram.
Write an energy equation.
Write an equation using Newton's 2nd law.
State the condition for losing contact.

Let the normal reaction be R and angle AOC be θ when particle P is at C

Energy equation whilst P is on the sphere is

KE gained = GPE lost

$$\frac{1}{2}mv^2 = mg(r - r\cos\theta) \qquad [1]$$

Newton's 2nd law to the centre O is

$$mg\cos\theta - R = \frac{mv^2}{r}$$

P leaves the sphere when $R = 0$

(*Continued on the next page*)

So $v^2 = gr\cos\theta$ [2]

[1] and [2] give $\frac{1}{2}mgr\cos\theta = mgr - mgr\cos\theta$

$\Rightarrow \cos\theta = \frac{2}{3}$ so $\theta = 48.2°$

$\Rightarrow v^2 = \frac{2gr}{3} = \frac{2\times9.8\times1.5}{3}$ and $v = 3.13\,\text{m s}^{-1}$

P leaves the sphere at a distance of $r + r\cos\theta = \frac{5r}{3} = 2.5\,\text{m}$ above the plane.

The vertical component of velocity is
$v\sin\theta = 3.13\times\sin48.2° = 2.33\,\text{m s}^{-1}$

The horizontal component of velocity is
$v\cos\theta = 3.13\times\cos48.2° = 2.09\,\text{m s}^{-1}$

Consider P as a projectile, taking a time t to reach the plane at D

$$s = ut + \frac{1}{2}at^2$$

$$2.5 = 2.33t + \frac{1}{2}\times9.8\times t^2$$

Take downwards as positive and write a kinematic equation to find t

$\Rightarrow 9.8t^2 + 4.66t - 5 = 0$

Solve the quadratic equation; t must be positive.

$\Rightarrow t = 0.515\,\text{s}\ (t > 0)$

Horizontal distance travelled in this time is

$$v\cos\theta\times t = 2.09\times0.515 = 1.08\,\text{m}$$

Horizontal velocity × time

Impact with plane occurs at a distance
$BD = BC + CD = 1.5\sin48.2° + 1.08 = 2.20\,\text{m}$

These problems emphasise that you should take care to state the correct condition for completing a full circle or breaking away from thc circle.

Exercise 24.3B Reasoning and problem-solving

In this exercise, assume that all strings are light and inextensible and use $\mathbf{g} = 9.8\,\text{m s}^{-2}$.

1 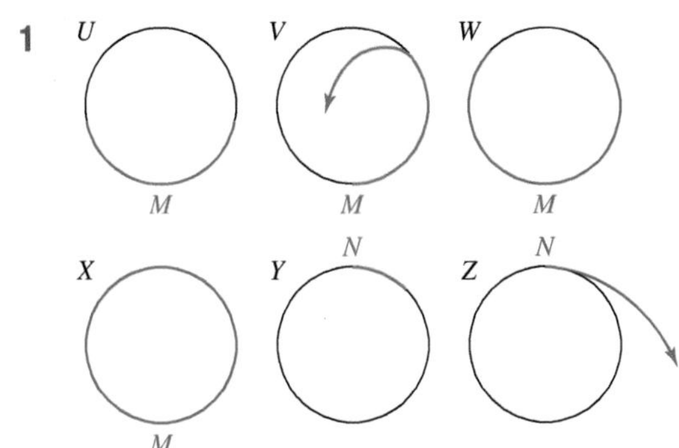

The loci in red show the possible paths traced out by particles or beads which move in a vertical circle centred at O after being given a starting velocity at point M or N. Match the possible paths U to Z with descriptions **a** to **f**. Each description may have more than one matching diagram.

a A bead threaded onto a smooth vertical hoop.

b A particle attached to a string OP

c A particle attached to a light rod OP

MECH

d A particle moving from the topmost point of a solid, smooth sphere.

e A particle moving from the topmost point of a smooth sphere.

f A particle moving on the inside of a smooth circular surface.

2 A 3 kg mass P hangs freely from a fixed point O on a string OP of length 2 m.

a Find the least velocity that P must be given so that it moves in a vertical circle about O

b Also find the radial component of its acceleration as the string OP passes through the horizontal, given the velocity calculated in part **a**.

3 A 2 kg particle P is placed at the lowest point inside the surface of a smooth hollow cylinder of radius 1.5 m and axis O. Find the velocity that P is given to move up the line of greatest slope if it

a Just reaches the horizontal through O

b Just makes a complete vertical circle about O

4 A particle P of mass m kg is at rest on the highest point of a smooth sphere of radius r and centre O. It is disturbed slightly so that it begins to move down the sphere's surface. Find the angle that OP makes with the vertical at the point where P leaves the surface. If a refined model takes air resistance into account, how would this affect your answer?

5 A 0.5 kg mass P hangs freely from a fixed point O on a string OP of length 2 m. It is given a horizontal velocity of 8 m s^{-1}. Find the angle through which OP has rotated when the string goes slack.

6 A hollow cylinder of radius r is fixed with its axis O horizontal. A particle of mass m is placed on the smooth inner surface of the cylinder and made to oscillate through 180°. Show that, when v is the speed of the particle, the reaction between the particle and the surface is $\frac{3mv^2}{2r}$

7 A road over a bridge has the shape of a circular arc of radius 20 m. Find the greatest speed in km h^{-1} at which a car can travel over the bridge without leaving the ground at the highest point of the road. State an assumption that you have made in your solution.

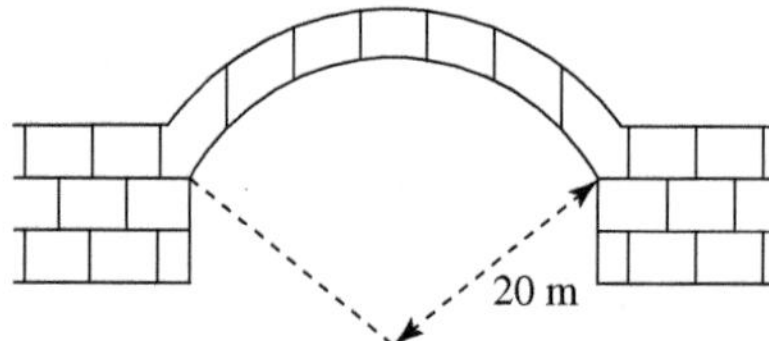

8 A 0.25 kg bead B is threaded onto a fixed vertical hoop with centre O and radius 2 m.

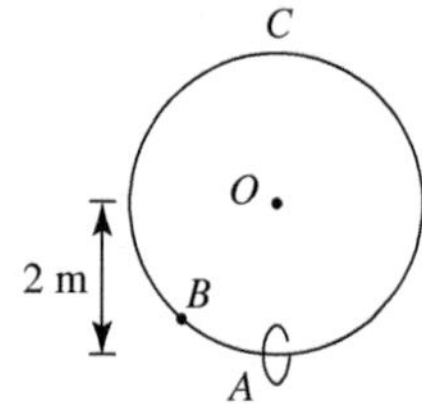

a B is given a horizontal velocity of 8 m s^{-1} at the lowest point, A, of the hoop. When $\angle AOB = 45°$, find

i The reaction between B and the hoop,

ii The radial component of its acceleration.

b If B is given a horizontal velocity of 8 m s^{-1} when at the highest point C of the hoop, find the radial component of the resultant acceleration of B when $\angle COB = 60°$

9 A 0.5 kg particle, P is at rest at the highest point inside a narrow hollow vertical circular tube of radius 2 m and centre O. It moves from rest and slides down inside the tube.

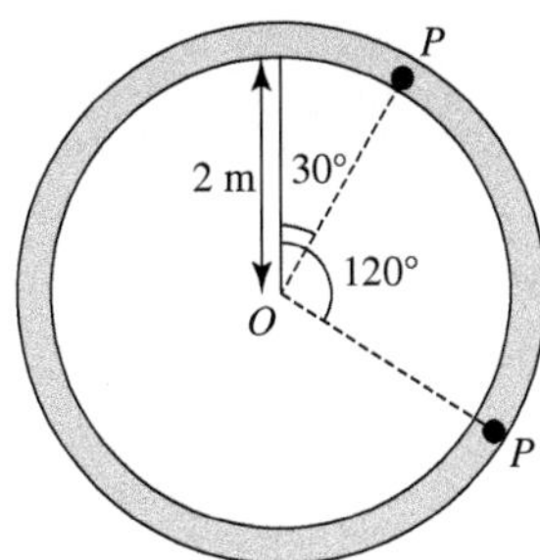

a Find the reaction R between P and the tube when OP has rotated through an angle of

i 30°, **ii** 120°.

b What angle must OP rotate through for the reaction R to be zero?

10 A smooth sphere of radius 2 m is fixed to a horizontal plane at point B. A 3 kg mass P at rest on the highest point A of the sphere is slightly disturbed and moves down the sphere's surface and leaves it. How far from B does P hit the plane?

11 The ends of a string OM of length 2 m are attached to a 3 kg mass M and a fixed point O. OM is held taut and horizontal and M is released from rest. When OM is vertical, it catches on a small fixed peg P which is a distance x below O

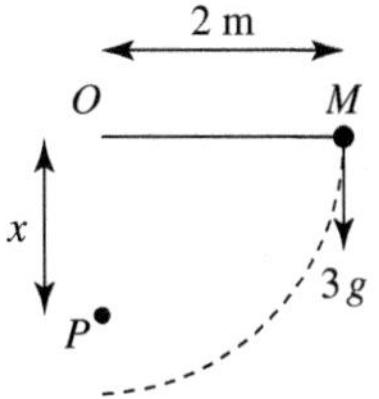

a Find x if M just completes a circle with P as centre.

b Find the ratio of the tensions in the string just before and just after making contact with P

12 A mass m, hanging at point A from a fixed point O on a string of length a, is projected from A with a horizontal velocity of $\sqrt{\frac{7ag}{2}}$

Show that the string becomes slack and that the mass, on leaving the circle, moves freely under gravity back to A

13 A 10 kg mass P hangs from a 2 m string attached to a fixed point O which is 3 m above point A on a horizontal plane.

P receives a horizontal impulse of 50 N s. When the string has rotated through 60° about O, it breaks. P now travels as a projectile and strikes the plane at B

Calculate the distance AB

24 Summary and review

Chapter summary

- For a particle moving in a circle, with centre O and radius r, at **constant** speed,
    - Its linear tangential speed, v, and angular speed, ω, are related by $v = r\omega = r\dot{\theta}$
    - Its acceleration, a, to the centre O is $r\omega^2 = \frac{v^2}{r}$
    - The centripetal force, F is $mr\omega^2 = \frac{mv^2}{r}$
    - The time, T, for one full revolution is $\frac{2\pi}{\omega}$
- A conical pendulum has a bob which moves in a horizontal circle at constant speed as its length traces out a hollow cone.
- For a particle moving in a vertical circle, centre O and radius r, at **variable** speed,
    - Its linear tangential speed, v, and angular speed, ω, are related by $v = r\omega = r\dot{\theta}$
    - Its acceleration has a radial component of $r\omega^2 = r\dot{\theta}^2 = \frac{v^2}{r}$
    - The centripetal force $F = mr\omega^2 = \frac{mv^2}{r}$
- Newton's 2nd law can give an equation of motion along the radius.
- An energy equation can be written in terms of *either* the total energy of the particle *or* its gains and losses of energy.

Check and review

You should now be able to...	Try Questions
✔ Use vectors to solve certain problems with circular motion.	1–4
✔ Solve problems when the motion is in a horizontal circle.	5–10, 14–15
✔ Solve problems when the motion is in a vertical circle.	11–13

1 A mass of 2 kg has a position vector $\mathbf{r} = 3\sin 2t\mathbf{i} + 3\cos 2t\mathbf{j}$ metres at a time t seconds.

a Calculate the magnitude of the force **F** acting on the mass when $t = \frac{\pi}{3}$ seconds.

b Find the equation of the circle on which the point moves in the x–y plane.

2 The position vector of a particle at time t seconds is $\mathbf{r} = (2\cos 3t + 5)\mathbf{i} + (2\sin 3t - 1)\mathbf{j}$ metres.

a What is the speed of the particle?

b Find the radius and centre of the circle on which the particle is moving.

3 A 2 kg mass M moves in a circle with the equation $x^2 + y^2 = 9$ at an angular speed of 2 rad s^{-1}, starting on the x-axis when $t = 0$. Find its position vector **r** at time t and the force **F** acting on M when $t = \frac{\pi}{4}$ seconds.

4 A particle P of mass 4 kg has a velocity $\mathbf{v} = 5\sin\left(\frac{1}{2}t\right)\mathbf{i} + 5\cos\left(\frac{1}{2}t\right)\mathbf{j}\text{ m s}^{-1}$ at a time t s. If its initial position is the point $(-2, 1)$, find its position vector at time t and the force acting on it when $t = \frac{\pi}{2}$ seconds.

5 A car of mass 900 kg rounds a bend of radius 40 m at 54 km h^{-1}. Find the smallest value of μ if the road is banked at an angle of 10° to the horizontal and there is no sideways slipping up the slope.

6 A motorcyclist goes round a curve of radius 20 m on a level road at 27 km h^{-1}. Find the angle at which she and her bike are inclined to the vertical and the least value of μ between the bike and the road for the bike not to side-slip.

7 A light, taut string connects particle P of mass m_1 and particle Q of mass m_2 as they lie on a rough horizontal surface, where the coefficient of friction for P is μ_1 and for Q is μ_2. P is given a velocity, u, perpendicular to PQ and rotates about Q which does not move. Find the greatest possible angle through which PQ can rotate in terms of m_1, m_2, μ_1 and μ_2.

8 A conical pendulum has a bob with a 2 kg mass making a full rotation every 2 seconds. If the pendulum's length is 1.5 m, calculate the tension in the string and the angle of the string to the vertical.

9 An elastic string has a natural length of 0.8 m and modulus of elasticity $\lambda = 100$ N. Its ends are attached to a fixed point O and to a 2 kg mass. Find the extension in the string when the mass rotates at 4 rad s^{-1} as a conical pendulum suspended from O

10 A ball of mass 2 kg rests on a smooth horizontal table and is attached to one end of a string of length 3 m. The other end is tied to a fixed point 1 m above the table. The ball moves in a horizontal circle while maintaining contact with the table. Show that the greatest linear speed for the ball to stay in contact with the table is $\sqrt{8g}$ m s^{-1}

11 A 1 kg mass slides down the surface of a smooth, fixed hemisphere of radius 2 m from being at rest at the topmost point. Find the distance it travels along the surface before contact is lost.

12 A 3 kg mass, P, is tied to the midpoint M of a string of length 2.6 m with its ends A and B fixed 2.4 m apart horizontally. Initially, P hangs at rest.

a Find the minimum speed it must be given to make complete vertical circles with the string taut.

b Find the maximum tension in the string during the motion.

13 A pendulum, P has a light rod OP 1 m in length which is fixed at O with a bob of 2 kg attached at P. When P is at rest vertically below O, the bob is struck so it has an initial velocity u. Find the angle through which the rod has turned when the bob reaches its highest point if

a $u = 3$ ms^{-1} b $u = 5$ ms^{-1}

14 Each end of a light, inextensible string is attached to a mass of 2 kg. The string passes through a hole in the vertex of a smooth, hollow cone of semi-vertical angle 45°, so that one mass hangs at rest inside the cone and the other mass moves in a horizontal circle on the outer surface of the cone. If the string is 2 m long and the two masses are at the same horizontal level, show that the angular velocity of the moving mass is $\sqrt{\frac{g}{2}}$ rad s^{-1}

15 a A bead of mass m is threaded on a smooth string QR of length 3 m where Q is fixed a distance of 2 m vertically above R. The bead rotates in a horizontal circle about QR. Find the angular velocity ω of the bead for it to position itself at point P on the string where $PQ = 2$ m.

b If the bead in part **a** is tied to the string at point P, find the minimum angular velocity, ω, which keeps both PQ and PR taut during the motion.

24 Exploration

Going beyond the exams

History

In the mid-nineteenth century, the French physicist Foucault realised that the plane of oscillation of a pendulum stays constant, even when the pivot point moves. This allowed him to demonstrate the rotation of the Earth because the plane of the swinging pendulum was seen to rotate over time. Many museums of science now include a 'Foucault pendulum' to illustrate the rotation of the Earth.

Research

The 'Wall of Death' motorcycle ride involves a motorcyclist riding a bike around a (near) vertical wooden circular wall inside a wooden cylinder. To do this safely, the motorcyclist needs to go at a speed that is sufficient to stop him or her from slipping down the wall and, because (s)he together with the motorcycle cannot be modelled as a particle, (s)he must also lean at an angle to the normal to the wall to ensure that the motorcycle remains in contact with the wall.

Carry out some research to find typical values of dimensions, mass and speed of a 'Wall of Death' ride and riders. Model and explore the situation.

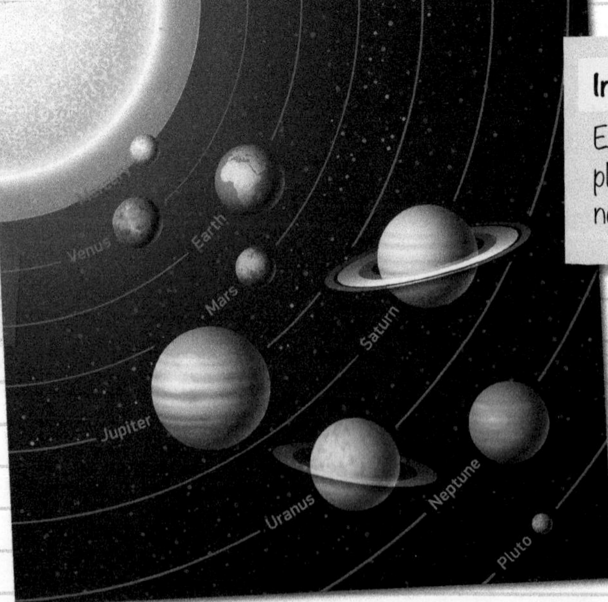

Investigation

Explore the forces of attraction acting on planets on their paths around the Sun. You will need to research planetary data to do this.

1 A particle of mass 2 kg is moving in a horizontal plane under the action of a force **F** N so that at time t s its position vector is given by

$$\mathbf{r} = (0.3\cos 10t\mathbf{i} + 0.3\sin 10t\mathbf{j})\text{ m}$$

a Show that the particle is moving in a circle. **[3]**

b Find an expression for **v**, its velocity vector, and hence show that it is moving with constant speed. **[4]**

c Find the magnitude of the force **F** and state its direction. **[4]**

2 A particle moves as a conical pendulum at the end of a light, inextensible string of length 40 cm. If the string makes an angle of 30° with the horizontal, find the angular speed of the particle. **[6]**

3 A car travels round a roundabout of radius 10 m at a uniform speed of 6 m s^{-1}. Taking the centre of the roundabout as the origin, the car is initially at the point with position vector 10**i** and is moving in an anticlockwise direction.

a Find an expression for the position vector **r** at time t **[3]**

b Find an expression for the acceleration of the car at time t and explain how you know that the acceleration is directed towards the origin. **[3]**

4 A car travels round a curve of radius 150 m on a track which is banked at 25° to the horizontal. The coefficient of friction between the wheels and the road is 0.4

a At what speed is the car travelling if there is no frictional force acting? **[5]**

b What is the maximum speed at which the car can travel without slipping up the slope? **[6]**

c What is the minimum speed at which the car can travel without slipping down the slope? **[5]**

5 A pendulum consists of a rod of length 1 m with a bob of mass 2 kg attached to one end. The rod is freely pivoted at the other end, O, so that it can rotate in a vertical circle. Initially, the bob is vertically below O when it is given an impulse so that it starts to move with speed 6.5 m s^{-1}. Assuming that the rod is light and that the bob can be modelled as a particle, calculate

a The speed of the bob, **[5]**

b The force in the rod, **[5]**

when the pendulum makes an angle of i 30°, ii 150° with the downward vertical.

6 A pendulum of length a has a bob of mass m. The speed of the bob at the lowest point of its path is U. Find the condition which U must satisfy for the bob to make complete revolutions if the length of the pendulum consists of

a A rod, **[4]**

b A string. **[4]**

7 A particle of mass m hangs at rest, suspended from a point, O, by a light, inextensible string of length a. The particle receives an impulse so that it starts moving with speed $\sqrt{3ga}$ m s^{-1}. Find the angle between the string and the upward vertical when the string goes slack. **[6]**

8 An engine of mass 60 tonnes is travelling on rails 1.5 m apart around a curve of radius 1500 m. The outer rail is raised a distance h m above the inner rail, so that at a speed of 50 km h^{-1} there is no sideways force on the rails.

a Find the value of h **[4]**

b Given that the maximum safe sideways force on the rails is 100 kN, what is the maximum safe speed of the engine round the curve? **[5]**

9 A ball of mass 3 kg is fastened to one end of a string of length 0.5 m. The other end of the string is fixed to a point A and the ball revolves as a conical pendulum at a rate of 5 rad s^{-1}.

a How far below A is the centre of the circle traced out by the ball? **[4]**

b What assumptions have you made in your answer? **[2]**

10 A mass of 0.5 kg, suspended by a light, inextensible string of length 1.5 m, revolves as a conical pendulum at 30 rev min^{-1}. Find the radius of the circle in which it travels and the tension in the string. **[6]**

11 A particle moves as a conical pendulum at the end of a light, inextensible string, which has its fixed end at point A. The angular speed of the particle is ω. The centre of the circle in which the particle moves is O

a Show that $AO = \dfrac{g}{\omega^2}$ **[5]**

b Explain why the string cannot be horizontal. **[2]**

12 A particle of mass 0.1 kg is attached by a string of length 1.5 m to a fixed point, and is made to travel in a vertical circle about that point.

a Find the minimum velocity the particle must have at the lowest point of the circle if it is to make complete revolutions. **[6]**

b For this velocity, find the tension in the string when the particle is at point A, a distance of 75 cm above the lowest point. **[5]**

13 A particle of mass 0.01 kg is placed on the topmost point, A, of a smooth sphere of centre O and radius 0.5 m. It is slightly displaced. When it reaches point B, it is about to leave the surface of the sphere. Calculate the angle AOB **[5]**

14 A ball of mass 1 kg is fastened to one end, B, of a light, inextensible string AB of length 1 m. The ball is placed on a smooth, horizontal table. The end A is attached to a fixed point above the table so that the particle moves as a conical pendulum whilst staying in contact with the table. The radius of the circle in which it travels is 0.5 m.

a If the speed of the particle is 1.5 m s^{-1}, what is the normal reaction between the ball and the table? **[4]**

b What is the maximum speed at which the ball could travel without lifting off the table? **[4]**

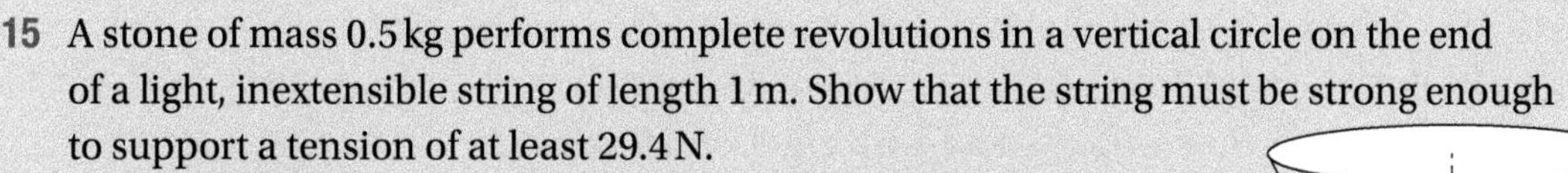

15 A stone of mass 0.5 kg performs complete revolutions in a vertical circle on the end of a light, inextensible string of length 1 m. Show that the string must be strong enough to support a tension of at least 29.4 N. **[6]**

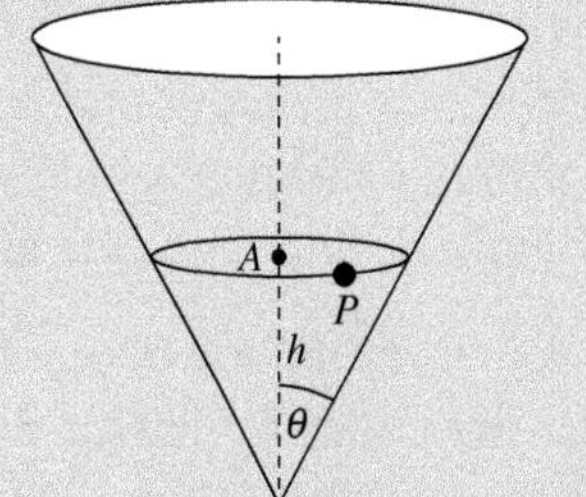

16 A smooth hollow cone with semi-vertical angle θ is fixed with its axis vertical and its vertex V at the bottom, as shown. A particle P travels with angular speed ω in a horizontal circle inside the cone about a point A on its axis, where $AV = h$. Show that

$$h = \frac{g}{\omega^2 \tan^2 \theta}$$ **[5]**

17 A bead of mass m is threaded onto a smooth, circular hoop of radius a, which is fixed in a vertical plane. The bead is displaced from rest at the top of the hoop. Find, in terms of g, the resultant acceleration of the bead when it has reached a point which is a vertical distance $\frac{3}{4}a$ below its starting point. **[9]**

18 A particle of mass m is projected horizontally with speed v from the topmost point, A, of a sphere of radius a and centre O. It remains in contact with the sphere until leaving the surface at point B. If angle AOB is 30°, find v **[6]**

19 A hemispherical bowl of radius 13 cm is fixed with its rim horizontal. A ball-bearing of negligible diameter is made to travel in a horizontal circle inside the bowl at a speed of 1.68 m s^{-1}. How far is the centre of the circle above the bottom of the bowl? **[8]**

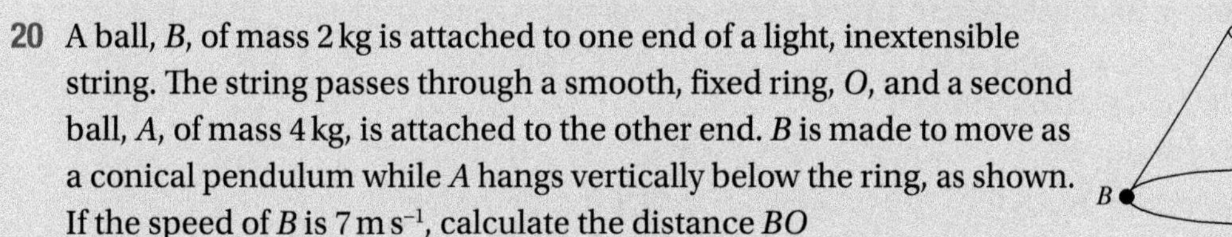

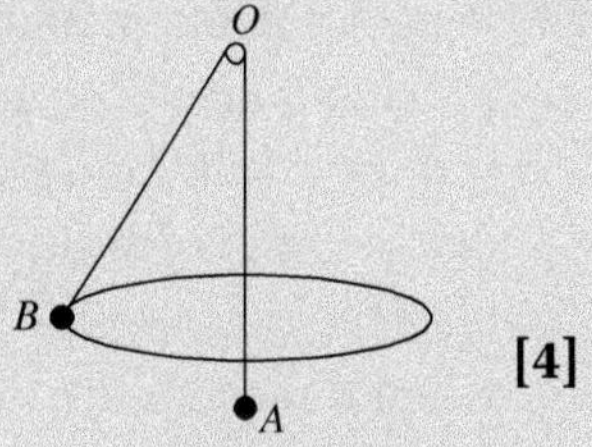

20 A ball, B, of mass 2 kg is attached to one end of a light, inextensible string. The string passes through a smooth, fixed ring, O, and a second ball, A, of mass 4 kg, is attached to the other end. B is made to move as a conical pendulum while A hangs vertically below the ring, as shown. If the speed of B is 7 m s^{-1}, calculate the distance BO **[4]**

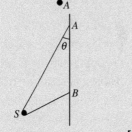

21 Two points, A and B, are on a vertical pole, 9 m apart with A above B, as shown. A rope of length 27 m is fastened at its ends to A and B. A smooth, heavy, metal ring, S, of mass m is threaded onto the rope. The ring is made to move in a horizontal circle about the pole. The upper section, AS, of the rope makes an angle θ with the vertical, as shown. Find the speed of the ring and the tension in the rope when

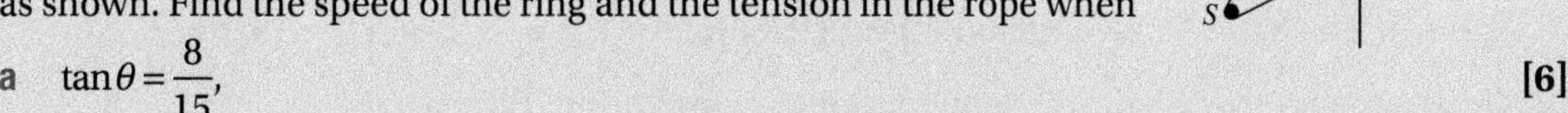

a $\tan\theta = \frac{8}{15}$, **[6]**

b $\tan\theta = \frac{4}{3}$ **[5]**

22 A smooth wire has a bead of mass 0.005 kg threaded onto it. The wire is bent to form a circular hoop of radius 0.2 m. The hoop is fastened in a horizontal position, whilst the bead travels round it at a constant speed of 1 m s^{-1}. Find the magnitude and direction of the reaction force between the hoop and the bead. **[6]**

23 A particle of mass m travels in complete vertical circles on the end of a light, inextensible string of length a. If the maximum tension in the string is three times the minimum tension, find the speed of the particle as it passes through the lowest point on the circle. **[6]**

24 A particle of mass 2 kg is attached to the end of a light, inextensible string of length 1 m, the other end of which is attached to a fixed point, O
The particle is held with the string taut and horizontal, and is released from rest. When the string reaches the vertical position, it meets a fixed pin, A, a distance x below O. Given that the particle just completes a circle about A, find the value of x **[8]**

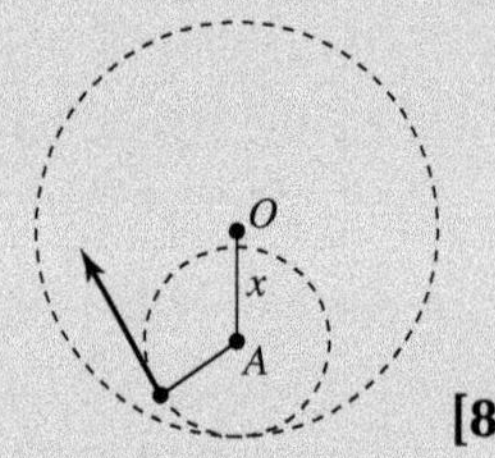

25 A pendulum bob of mass m is fastened to one end of a light, inextensible string of length r whose other end isfixed at a point O. The bob is at rest in its lowest position when it is set in motion with an initial speed of $\sqrt{\frac{7gr}{2}}$. As it swings upwards, the string meets a small, fixed peg, P, on the same level as O

The string then wraps round P. What is the closest that P can be to O so that the bob makes a complete revolution about P? **[8]**

26 A ring of mass 5 kg is threaded onto a rope of length 10 m, whose ends are attached to two fixed points 6 m apart and on the same level. The ring hangs at rest. It is then set in motion so that it travels on a circular path whose plane is perpendicular to the line joining the two ends of the rope. Given that the ring can just make complete revolutions, find the maximum tension in the rope. State any modelling assumptions that you have made in reaching your answer. **[10]**

27 Particles A and B, of masses m and $2m$, respectively, are connected by a light, inextensible string of length πa. The particles are placed symmetrically, and with the string taut, on the smooth outer surface of a cylinder of radius $3a$, as shown, and the system is released from rest. Find the reactions between the cylinder and the particles at the moment when A reaches the topmost point. **[9]**

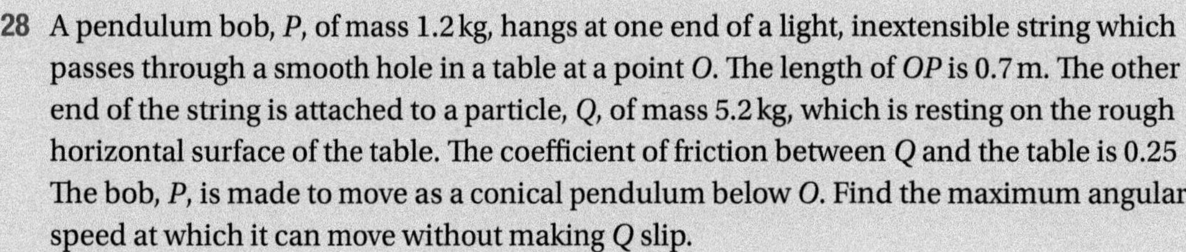

28 A pendulum bob, P, of mass 1.2 kg, hangs at one end of a light, inextensible string which passes through a smooth hole in a table at a point O. The length of OP is 0.7 m. The other end of the string is attached to a particle, Q, of mass 5.2 kg, which is resting on the rough horizontal surface of the table. The coefficient of friction between Q and the table is 0.25
The bob, P, is made to move as a conical pendulum below O. Find the maximum angular speed at which it can move without making Q slip. **[5]**

29 The diagram shows a loop-the-loop on a roller-coaster ride. The car approaches the loop on a horizontal track. The maximum speed at which the car can enter the loop is 80 km h^{-1}. What is the greatest radius with which the loop can be constructed if the car is not to leave the track? **[6]**

25 Centres of mass and stability

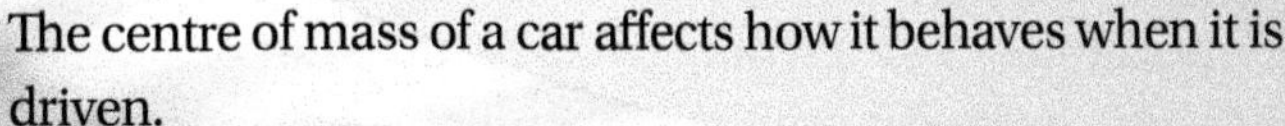

The centre of mass of a car affects how it behaves when it is driven.

This is important for all vehicles. It is particularly significant in racing cars, such as those used in Formula 1, where small differences in performance have a big effect on outcomes. Designers of racing cars make sure a car is stable by ensuring that the centre of mass of the car is as close to the ground as possible. When the brakes on a standard car are applied suddenly, the car might feel as though it is diving forwards towards the road. This happens when forces are unbalanced about the centre of mass. Such behaviour needs to be minimised in racing cars.

The position of the centre of mass of many other vehicles must also be considered at the design stages. For example, because of the uneven weight distribution of passengers in a double decker bus, the vehicle could quite easily become unstable when cornering on a banked section of road, unless its centre of mass is positioned correctly.

Orientation

What you need to know

Maths Ch19
- Moments.

What you will learn

- To calculate the moment about a point.
- How to find the centre of mass for point masses and laminas.
- How to find the centre of mass for solids.

What this leads to

Careers
Mechanical engineering.
Architecture.

25.1 Moments and couples

Fluency and skills

The magnitude of the turning effect or **moment** of a force F about a fixed point A depends on the perpendicular distance between F and A and its direction can be clockwise (negative) or anticlockwise (positive).

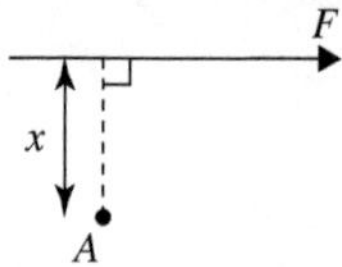

Key point

The moment of a force F about a point A is $F \times x$, where x is the perpendicular distance of the force from A

A moment is sometimes called a **torque**.

When several forces act on the same object at the same time, each force has its own turning effect on the object.

Key point

The **resultant moment** of several forces acting on an object equals the sum of the moments of the individual forces, taking into account their directions.

Moment and work have the same dimensions and thus the same units (N m). Whereas N m are called Joules for work, moments are always measured in N m.

For an object that can rotate about a fixed point A, it is in rotational equilibrium if the total moment acting on it is zero. (That is, when the clockwise moments balance the anticlockwise moments exactly.)

Key point

If a system of forces is in equilibrium, the resultant moment about any point is zero.

The fixed point A is called a **pivot**.

Example 1

A light rectangular lamina L, can rotate about point A

Find the resultant moment of the three forces about A

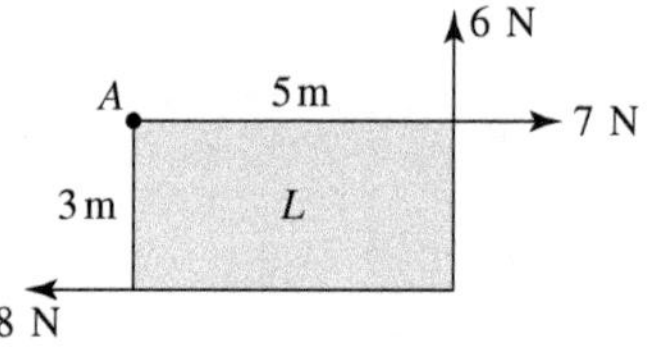

Total (anticlockwise) moment $= (6 \times 5) + (7 \times 0) - (8 \times 3)$

$= 30 - 24 = 6\text{ N m}$

The line of action of the 7 N force passes through A, so it produces a zero moment. The 8 N force produces a negative moment.

The net horizontal force on the object shown in the diagram is zero (as $F - F = 0$) and there is no vertical force. However, the object is not in equilibrium as the two parallel forces cause the object to rotate with a clockwise moment of $F \times d$. Two forces like these produce a **couple**.

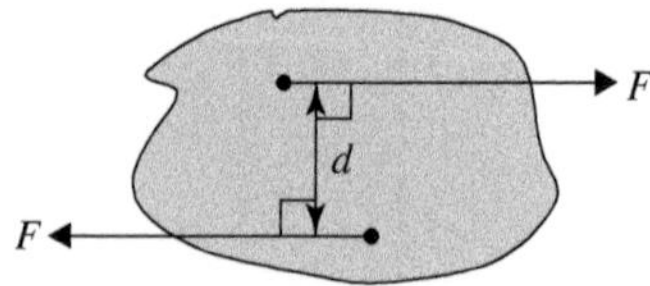

Key point

A couple comprises two equal and opposite forces which do not act in the same straight line. The moment of a couple, about any point, equals $F \times d$, where d is the perpendicular distance between the two forces. This is independent of the point about which the moment is taken.

The forces used to unscrew a bottle top or turn a doorknob cause rotation without translation. Such forces also form a couple which is measured by its moment.

Example 2

A cantilever is formed by clamping a uniform horizontal beam AB at A. The beam is 4 m long, weighs 20 N and has a load of 15 N hanging vertically at B

Find the force and couple which must act at A to keep the beam in equilibrium.

Let the force at A have components X and Y and the couple at A have a moment M

The components of the force at A prevent translation and the couple prevents rotation.

Resolve horizontally $\quad X=0$

Resolve vertically $\quad Y=20+15=35$

Take moments about A $\quad 20\times2+15\times4=M$

Clockwise moment = anti-clockwise moment

$\Rightarrow M=100\,\text{N m}$

The force at A is 35 N upwards and the couple at A is 100 N m anticlockwise.

Exercise 25.1A Fluency and skills

1 Find the total moment about point P of these systems of forces.

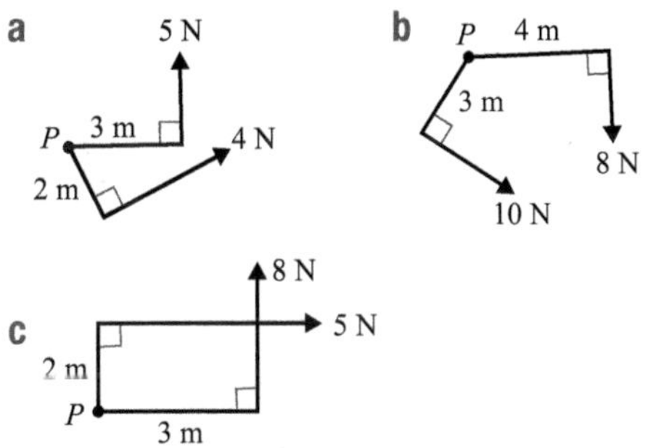

c

2 a Force $\mathbf{F}=2\mathbf{i}+3\mathbf{j}$ N acts at point (4, 3). Find its moment about point (2, 1).

b Force $\mathbf{F}=5\mathbf{i}-2\mathbf{j}$ N acts at point (2, 2). Find its moment about point (1, 4).

3 Rectangle $PQRS$ with sides 6 m by 4 m and centre O is a light lamina. Find the resultant moment of the forces about

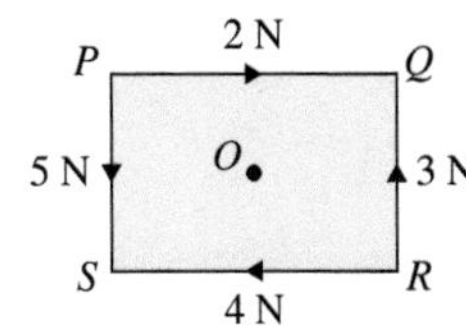

a Point Q b Point O

4 A uniform horizontal cantilever AB of mass 20 kg is 6 m long. It is clamped at A and carries a mass 10 kg at its midpoint. Find the force and couple acting at A to keep it in equilibrium.

5 A square $ABCD$ of side 2 m has forces of 10 N along AB and CD and forces of 6 N along CB and AD. Find the total moment acting on the square about A.

6 Find the resultant moment of the two forces about point A. Hence show that the moment of a couple is independent of the position of A

Reasoning and problem-solving

In situations where you apply a force F to an object at an angle θ which is not a right angle, you can use one of two methods to calculate its moment about a point, as illustrated in Example 3

The first method in Example 3 resolves force F into two components, $F\cos\theta$ and $F\sin\theta$, and finds the moment of each component about point A

The second method uses a right-angled triangle to find the perpendicular distance from the line of action of force F to point A

In each problem you meet, you can choose which method is more appropriate.

Strategy

To solve problems involving moments and couples

1. Draw a clear diagram, marking on all the forces and distances.
2. Take account of the directions of moments when finding their resultant.

Example 3

A force F of 10 N acts at end B of a light rod AB of length l, making an angle θ with the rod.

If $l = 3$ m and $\theta = 30°$, find the moment of F about A

First method

Moment of F about $A = F\cos\theta \times 0 + F\sin\theta \times l$

$= 0 + 10\sin 30° \times 3$

$= 15$ Nm

Second method

Moment of F about $A = F \times x$

$= F \times l\sin\theta$

$= 10 \times 3\sin 30° = 15$ Nm

(1) Draw a diagram and resolve the force F into components $F\cos\theta$ along AB and $F\sin\theta$ perpendicular to AB

(1) Draw a diagram and extend the line of the force to show the perpendicular distance, x from A to the line of action of the force.

Notice that both methods give the moment of F as $F \times l \times \sin\theta$

Example 4

A light square lamina $ABCD$ of side 10 cm can rotate about point A

The diagram shows three forces acting.

Find the magnitude of force F for the lamina to be in equilibrium.

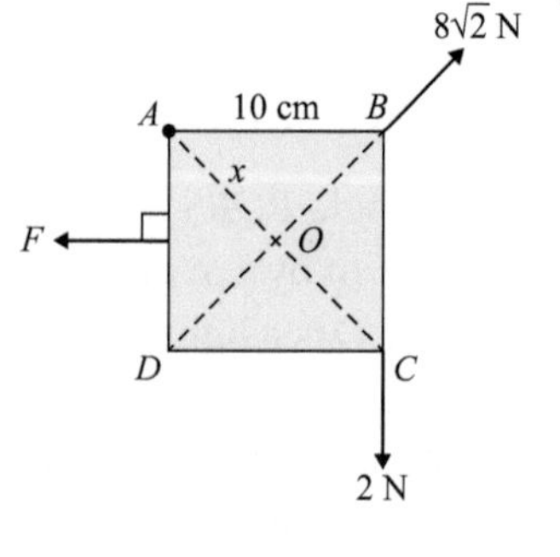

In ΔAOB, $x^2 + x^2 = 10^2$, so $x = \sqrt{50}$ cm $= 5\sqrt{2}$ cm

Take moments about point A

For equilibrium:

Anticlockwise moments = Clockwise moments

$8\sqrt{2} \times 5\sqrt{2} = (2 \times 10) + (F \times 5)$

$80 = 20 + 5F$

Force $F = 12$ N

(*Continued on the next page*)

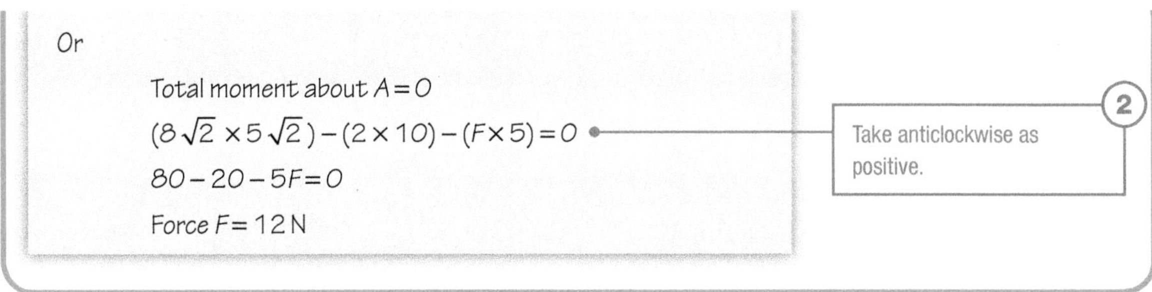

Or

Total moment about $A = 0$

$(8\sqrt{2} \times 5\sqrt{2}) - (2 \times 10) - (F \times 5) = 0$

$80 - 20 - 5F = 0$

Force $F = 12$ N

The reaction at point *A* is not on the diagram and it does not appear in the equation because moments are taken about *A*. This reaction exists because a balance of forces is needed for equilibrium as well as a balance of moments.

Exercise 25.1B Reasoning and problem-solving

1 Find the resultant moment about point *A* in each case.

a

b

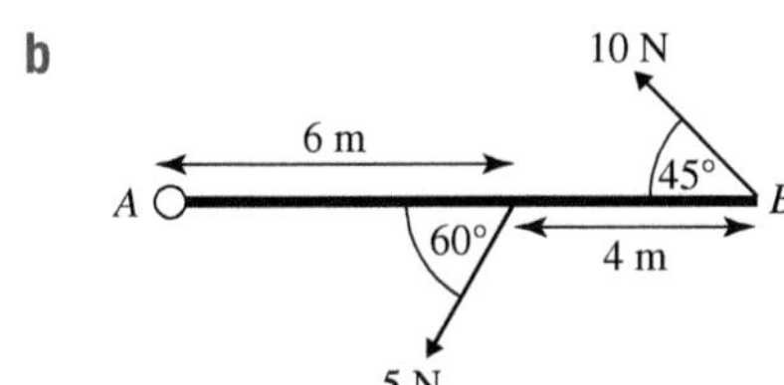

c

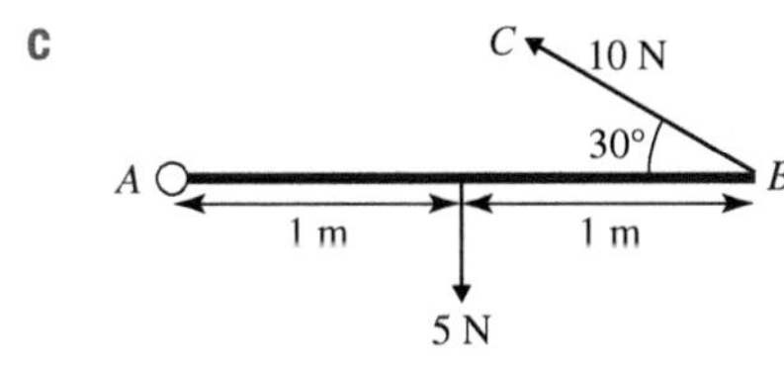

2 The equilateral triangle *ABC* of side 4 m and right-angled triangle *PQR* with $PQ = 3$ m and $QR = 4$ m are two light laminas. M_1 and M_2 are midpoints of sides *AB* and *PR*, respectively. Find the resultant moment of the given forces about points

a B b M_1 c Q d M_2

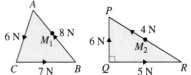

3 A uniform cantilever *AB* of length 2 m and weight 10 N is held horizontal by a tension of 45 N in a string *BC* when $\angle ABC$ is 30°. Find the resultant moment about point *A*

4 A uniform beam *AB* of length 8 m and weight 10 N is held by two strings *AC* and *BD* such that $\angle BAC = \angle ABD = 46°$. If the tension in the two strings is 7 N, find the total moment about *A* and also about a point on the beam 2 m from *A*

5 A rusty trapdoor of weight 40 N is made of a uniform material and has centre of mass at the point indicated on the diagram. It is held partially open in equilibrium by a vertical force of 30 N as shown.

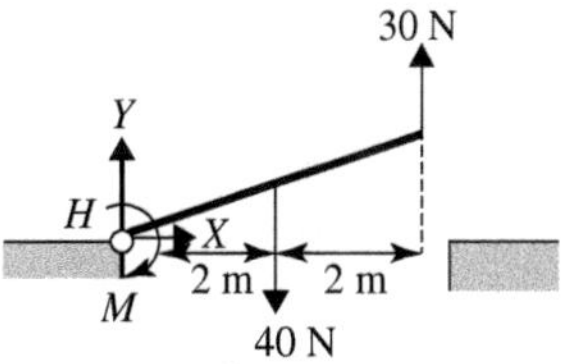

The reaction at the hinge *H* has components *X* and *Y*. The rust at the hinge provides a resisting couple of moment *M*. Find the values of *X*, *Y* and *M*

6 A horizontal uniform cantilever *YZ* of length 8 m, weight 30 N and midpoint *M* is fixed to a vertical wall *PYQ* with *P* above *Q* such that $PY = QY = 5$ m. Two light rods *PM* and *QZ* support *YZ* with a tension of 10 N in *PM* and a thrust of 12 N in *QZ*. Find the resultant moment about *Y*

25.2 Centre of mass for point masses and laminas

Fluency and skills

If you place n masses $m_1, m_2, m_3, \ldots, m_n$ along the x-axis at distances $x_1, x_2, x_3, \ldots, x_n$ from the origin O, the total mass M is given by

$$M = m_1 + m_2 + m_3 + \ldots = \sum_{i=1}^{n} m_i$$

This system of masses behaves as if its total weight Mg acts at a point, called the **centre of mass**, which is a distance $\bar{x}$ from O

To find $\bar{x}$, take moments about O and equate the moment of the total weight to the sum of the moments of the individual weights.

You can think of $\bar{x}$ as the weighted mean of the various x-values where the weight of each x-value is the mass at that point.

$$Mg \times \bar{x} = m_1 g \times x_1 + m_2 g \times x_2 + m_3 g \times x_3 + \ldots$$

giving $\quad M \times \bar{x} = \sum_{i=1}^{n} m_i x_i$

So, the centre of mass is at the point $(\bar{x}, 0)$ where $\bar{x} = \dfrac{\sum_{i=1}^{n} m_i x_i}{\sum_{i=1}^{n} m_i}$

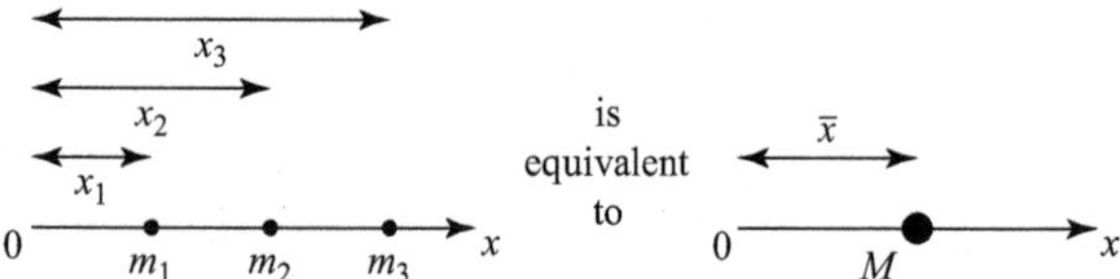

You can extend this method to two dimensions where m_1 is at the point (x_1, y_1), m_2 at (x_2, y_2), m_3 at (x_3, y_3), ... and the total mass M is at the point $(\bar{x}, \bar{y})$.

By taking moments about the x-axis, you get $M \times \bar{y} = \sum_{i=1}^{n} m_i \times y_i$

which can be written as $\bar{y} = \dfrac{\sum_{i=1}^{n} m_i \times y_i}{\sum_{i=1}^{n} m_i}$

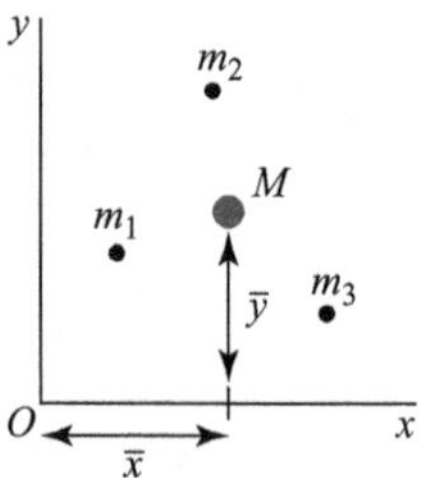

Key point

The centre of mass of a system of particles is at the point $(\bar{x}, \bar{y})$ where $\bar{x} = \dfrac{\sum_{i=1}^{n} m_i \times x_i}{\sum_{i=1}^{n} m_i}$ and $\bar{y} = \dfrac{\sum_{i=1}^{n} m_i \times y_i}{\sum_{i=1}^{n} m_i}$

Remember that $M = \sum_{i=1}^{n} m_i$

Example 1

Four masses of 3 kg, 2 kg, 4 kg and 1 kg lie at the points (2, 3), (5, 1), (8, 4) and (9, 6), respectively. Calculate the position of their centre of mass.

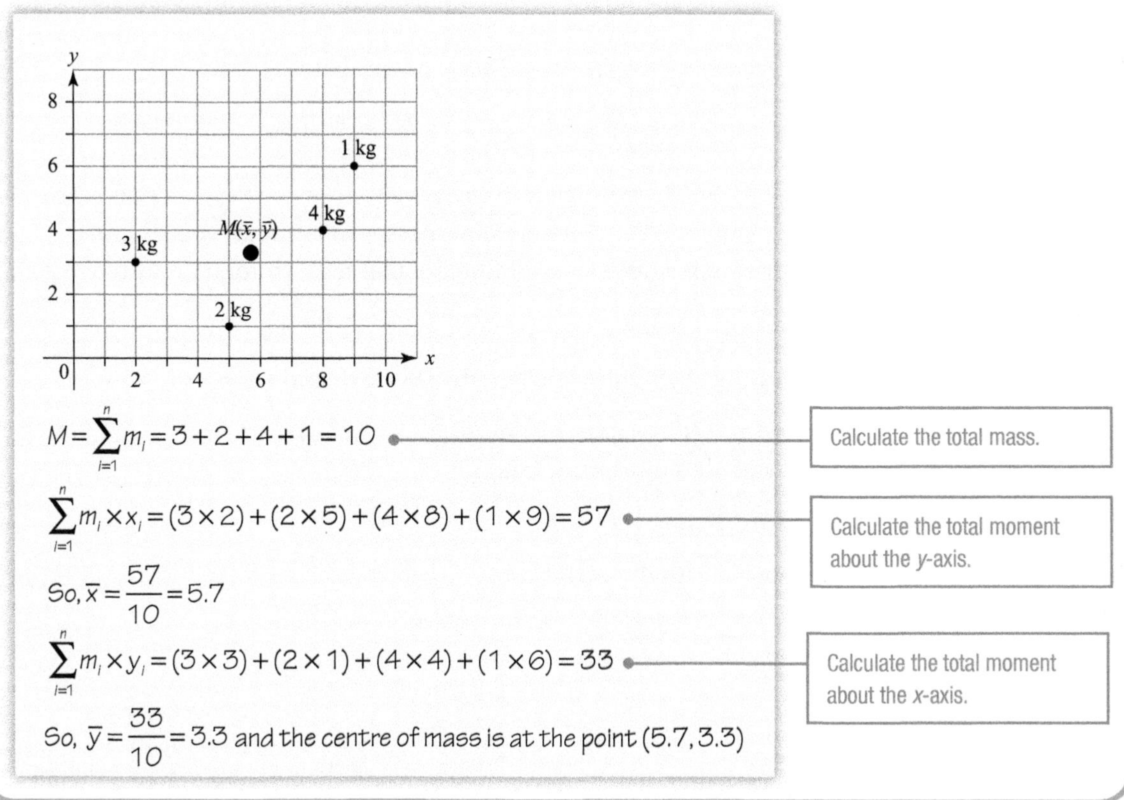

$M = \sum_{i=1}^{n} m_i = 3 + 2 + 4 + 1 = 10$ — Calculate the total mass.

$\sum_{i=1}^{n} m_i \times x_i = (3 \times 2) + (2 \times 5) + (4 \times 8) + (1 \times 9) = 57$ — Calculate the total moment about the y-axis.

So, $\bar{x} = \frac{57}{10} = 5.7$

$\sum_{i=1}^{n} m_i \times y_i = (3 \times 3) + (2 \times 1) + (4 \times 4) + (1 \times 6) = 33$ — Calculate the total moment about the x-axis.

So, $\bar{y} = \frac{33}{10} = 3.3$ and the centre of mass is at the point (5.7, 3.3)

Exercise 25.2A Fluency and skills

1 Find the coordinates of the centres of mass of these systems of masses.

a 2 kg at the point (4, 0), 3 kg at (6, 0) and 5 kg at (8, 0),

b 3 kg at the point (1, 0), 5 kg at (3, 0), 4 kg at (4, 0) and 8 kg at (6, 0),

c 4 kg at the point (0, 2), 6 kg at (0, 5) and 10 kg at (0, 4),

d 6 kg, 12 kg and 15 kg at the points (4, –4), (10, 1) and (0, –5), respectively.

2 Masses of 7 kg, 8 kg and 5 kg are placed at points with position vectors $3\mathbf{i} + 6\mathbf{j}$, $4\mathbf{i} - 2\mathbf{j}$ and $6\mathbf{i} + \mathbf{j}$, respectively. Find the position vector of their centre of mass.

3 Masses of 2 kg and 3 kg are placed at the points (6, 0) and (2, 0). Where should a mass of 4 kg be placed on the x-axis so that the centre of mass of all three masses is at the point (4, 0)?

4 Masses of 6 kg, 4 kg and 10 kg are placed at the points (5, 0), (0, 6) and (a, b), respectively. If their centre of mass is at the point (3.5, 2.2), find the values of a and b

5 Four masses of 0.5 kg, 1.5 kg, 1 kg and 2 kg are placed on coordinates axes at points with position vectors $4\mathbf{i} + 4\mathbf{j}$, $4\mathbf{i} + 8\mathbf{j}$, $10\mathbf{i} + 2\mathbf{j}$ and $a\mathbf{i} + b\mathbf{j}$, respectively. The position vector of their centre of mass is $6\mathbf{i} + 6\mathbf{j}$. Find the values of a and b

6 Four masses have a centre of mass at the point (0, –1). A 4 kg mass is at the point (5, 3), 9 kg is at (6, –2) and 6 kg is at (–1, 4). Find the position of the final mass of 5 kg.

Reasoning and problem-solving

Strategy

To solve problems involving a centre of mass

1. Draw a diagram showing the information and construct a table to summarise the data.
2. Take moments to calculate unknown values. Make sure you take distances from the correct axes when finding moments.

When working with any simple or composite system of point masses or laminas, you should consider the symmetry of both the shape and the distribution of masses. Symmetry may make your solution to a problem easier.

Example 2

A light, rectangular, rigid framework *OABC* has side lengths $OA = 3$ m and $OC = 2$ m.

D and *E* are the midpoints of *AB* and *BC*. Masses of 4 kg, 3 kg, 1 kg and 5 kg are fixed to *A*, *C*, *D* and *E*, respectively, and it is suspended by a string from *O*

a Find the position of the centre of mass, *G*

b Find the angle which *OA* makes with the vertical.

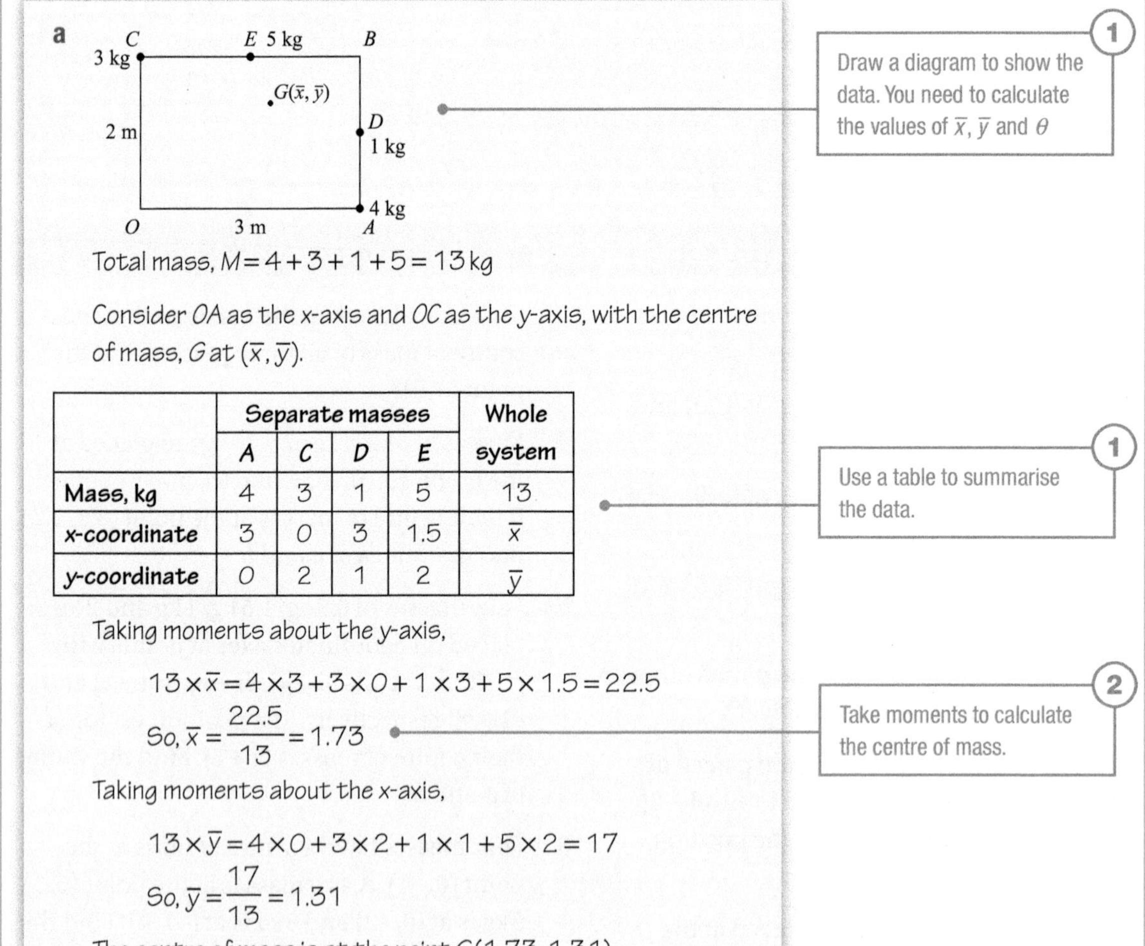

Draw a diagram to show the data. You need to calculate the values of $\bar{x}$, $\bar{y}$ and θ (1)

Total mass, $M = 4 + 3 + 1 + 5 = 13$ kg

Consider *OA* as the *x*-axis and *OC* as the *y*-axis, with the centre of mass, *G* at $(\bar{x}, \bar{y})$.

	Separate masses				Whole system
	A	C	D	E	
Mass, kg	4	3	1	5	13
x-coordinate	3	0	3	1.5	$\bar{x}$
y-coordinate	0	2	1	2	$\bar{y}$

Use a table to summarise the data. (1)

Taking moments about the *y*-axis,

$$13 \times \bar{x} = 4 \times 3 + 3 \times 0 + 1 \times 3 + 5 \times 1.5 = 22.5$$

$$\text{So, } \bar{x} = \frac{22.5}{13} = 1.73$$

Take moments to calculate the centre of mass. (2)

Taking moments about the *x*-axis,

$$13 \times \bar{y} = 4 \times 0 + 3 \times 2 + 1 \times 1 + 5 \times 2 = 17$$

$$\text{So, } \bar{y} = \frac{17}{13} = 1.31$$

The centre of mass is at the point $G(1.73, 1.31)$

(*Continued on the next page*)

b

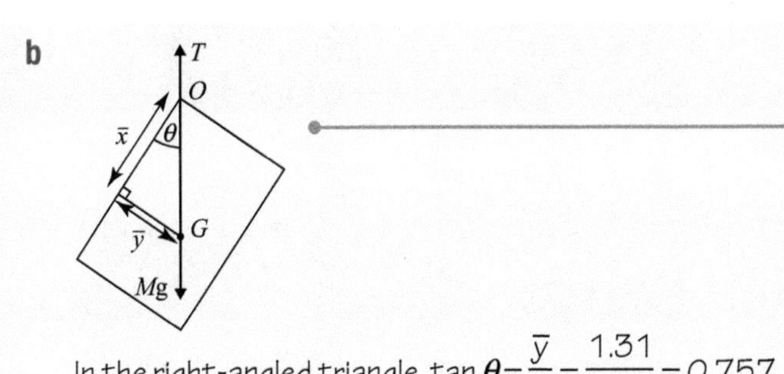

In the right-angled triangle, $\tan\theta = \frac{\bar{y}}{\bar{x}} = \frac{1.31}{1.73} = 0.757$

So, the angle θ that OA makes with the vertical is 37.1°

1 Draw a diagram of the suspended rectangle.

The line OG is vertical as the tension T and the weight Mg line up to create equilibrium.

If T and mg are not in line, they form a couple which rotates the system until it is in equilibrium with OG vertical.

Exercise 25.2B Reasoning and problem-solving

1. Masses of 2 kg, 4 kg, 6 kg and 9 kg are placed at the vertices of a light, rigid, rectangular framework $ABCD$, where $AB = 5$ m and $BC = 3$ m. Find the position of the centre of mass of the system from AB and AD

2. A light, rectangular lamina $OPQR$ has a mass of 5 kg fixed to P and 4 kg fixed to Q. A third mass of 8 kg is fixed to the centre C of the rectangle. $OP = 10$ m and $OR = 5$ m. Taking OP and OR as the x- and y-axes, find the position of the centre of mass of the system.

3. A rigid, square framework of light rods has particles fixed to its four corners and the centre of the square, as shown in the diagram.

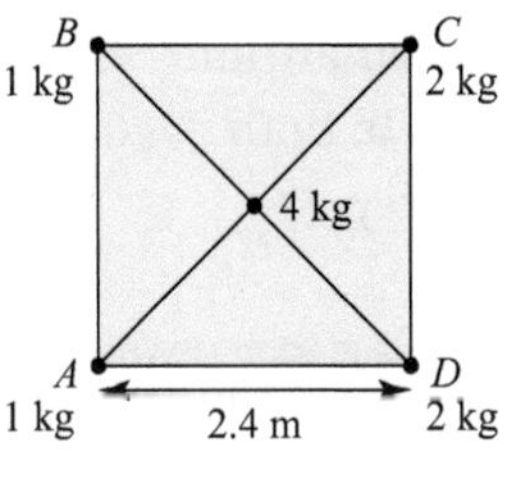

The square has sides 2.4 m long.

  a. Explain, without any calculation, why the distance of the centre of mass, G from side AD is 1.2 m.

  b. Find the distance of G from side AB and the angle of AB to the vertical when the framework is suspended from A

4. A light lamina $ABCDEF$ has the shape of a letter L made from two identical rectangles 10 cm by 4 cm and a square of side 4 cm.

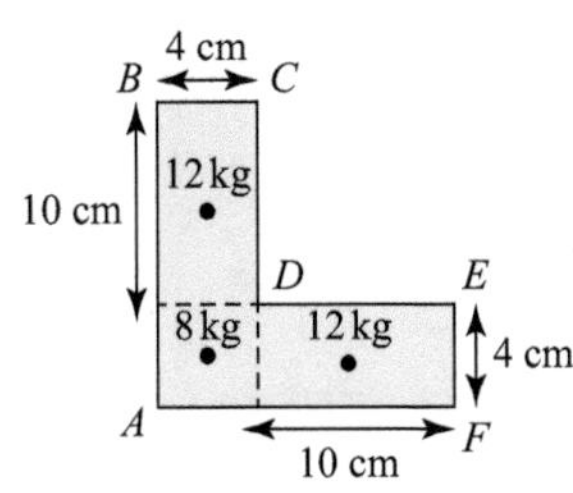

Three masses are attached to the lamina, two of 12 kg to the centre of each rectangle and one of 8 kg to the centre of the square.

  a. Explain why the centre of mass of the three masses must lie on the line AD

  b. Find the distances of the centre of mass from the sides AB and AF

5. A light, rectangular framework $ABCD$ has $AB = 4$ m and $BC = 3$ m. Masses of 5 kg, 4 kg, 2 kg and 3 kg are placed at A, B, C and D, respectively. A fifth mass m kg is placed at a point E on CD so that the centre of mass of the whole system is at the centre of the rectangle. Find the value of m and the position of E

6. A square, metal framework $OXZY$ of side 2 m is made from uniform rods of different densities. The density of OX is 1.2 kg per metre; the density of XZ and YZ is 2 kg per metre; and the density of OY is 1.8 kg per metre. Find the distance of the centre of mass from OX and OY

7. A uniform piece of wire has a mass of 10 grams per cm of length. It is bent into the shape of a triangle OXY, which is right-angled at O. If $OX = 24$ cm and $OY = 10$ cm, find the distance of the centre of mass from OX and OY

8. A rectangular framework $OXPY$ with $OX = YP = 1.2$ m is made from two bent uniform rods. The rods are both 2 m but rod OXP is twice as heavy as rod OYP. If OYP has a density of 4 kg per metre length, find the distance of the centre of mass from O

25.3 Centres of mass for laminas and solids

Fluency and skills

You can use symmetry to find the centres of mass of some common shapes, provided their mass is uniform over their area. Two simple cases are a rectangular lamina and a circular lamina. Symmetry gives their centres of mass G at their centre points.

Most triangular laminas are not symmetrical. But you can think of a triangle as being made up of an infinite number of uniform rods of negligible thickness which are parallel to one side. The symmetry of the rods gives their centres of mass at their midpoints G_1, G_2, G_3, ... so the centre of mass G of the whole triangle lies on the median AM When you repeat this method for the other two medians, you find point G where all the medians intersect.

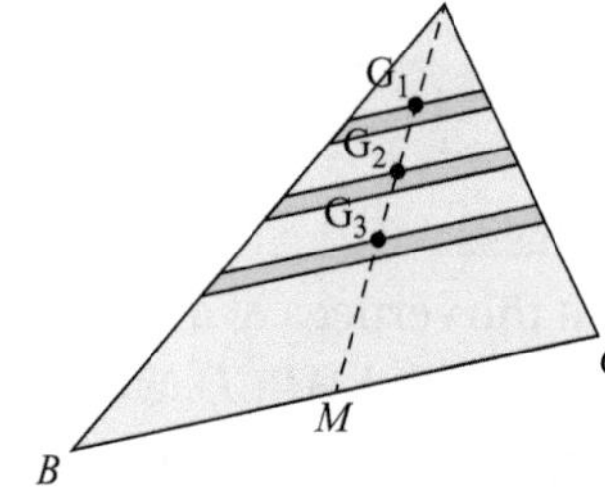

This point of intersection is called the **centroid** of the triangle.

Key point

The centre of mass, G, of a triangular lamina is at the point of intersection of its medians which is $\frac{2}{3}$ of the way along each median from the vertex.

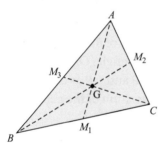

The formulae for the centre of mass for a uniform circular arc and for a sector of a circle are listed in your formula booket. In both cases, their centres of mass are on their lines of symmetry.

Key point

A circular arc of radius r and angle 2α at the centre has

$$OG = \frac{r\sin\alpha}{\alpha}$$

Key point

A sector of a circle of radius r and angle 2α at the centre has

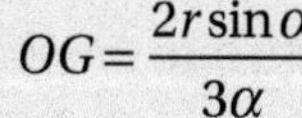

$$OG = \frac{2r\sin\alpha}{3\alpha}$$

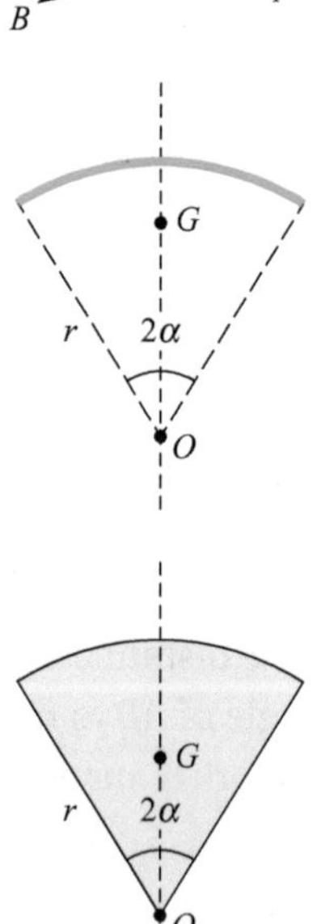

In both formulae, α must be in radians, not degrees.

Remember, for a sector with angle theta radians,

arc length, $s = r\theta$

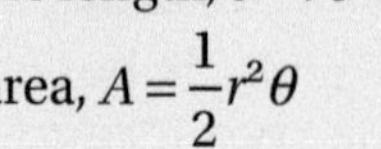

area, $A = \frac{1}{2}r^2\theta$

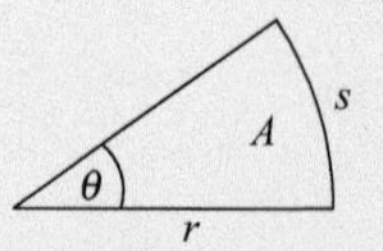

Example 1

A uniform wire of length 2.1 m is bent to make a circular arc of radius 1.4 m.

Find the position of its centre of mass.

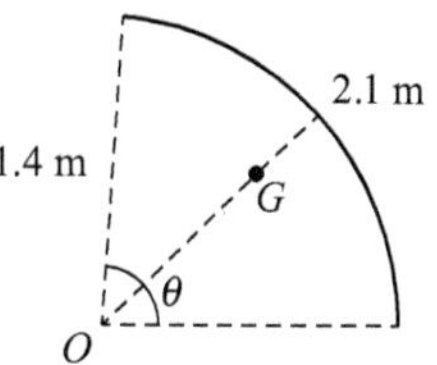

The length of a circular arc, $s = r\theta$

So angle θ at the centre $O = \frac{s}{r} = \frac{2.1}{1.4} = 1.5$ radians

As $\theta = 2\alpha$, angle $\alpha = 0.75$ radians

The centre of mass, G, is on the wire's line of symmetry such that

$$OG = \frac{r\sin\alpha}{\alpha} = \frac{1.4 \times \sin 0.75}{0.75} = 1.27\text{ m}$$

When working with any simple or composite system of laminas and point masses, you should consider the symmetry of both the shape and the distribution of any masses.

Exercise 25.3A Fluency and skills

1 Find the position of the centre of mass of a uniform wire bent into a circular arc with

a A radius of 2.4 m and a length of 3.6 m,

b A radius of 5 m and an angle at the centre of 135°,

c A semicircular shape and a diameter of 3.6 m.

2 Find the position of the centre of mass of a uniform sector of a circle with

a A radius of 2 m and an angle at the centre of 1.2 radians,

b A radius of 4 m and an area of 16 m^2,

c An area of 7.2 m^2 and an angle at the centre of 1.6 radians.

3 Find the position of the centre of mass of a uniform lamina which is

a A semicircle of diameter 1.6 m,

b A quadrant of a circle of radius 1.4 m.

4 Find the coordinates of the centre of mass of a uniform triangular lamina with its vertices at the points:

a $(1, 1)$, $(10, 1)$ and $(10, 7)$,

b $(0, 0)$, $(9, 0)$ and $(0, 6)$,

c $(2, 6)$, $(8, 6)$ and $(8, 0)$.

5 a A uniform lamina has the shape of an isosceles triangle ABC with $AB = AC$. If $AB = 13$ cm and $BC = 10$ cm, calculate the position of its centre of mass.

b Calculate the position of the centre of mass of an equilateral triangular lamina of side 10 cm.

6 A uniform triangular lamina OXY is right-angled at O with its two other vertices at $(a, 0)$ and $(0, b)$. If the triangle's centre of mass is at the point $(5, 4)$, find the values of a and b

7 Prove that the centre of mass of a uniform lamina in the shape of a parallelogram is at the intersection of the two lines which join the midpoints of pairs of opposite parallel sides.

Reasoning and problem-solving

Strategy

To solve problems involving the centre of mass of shapes

1. Draw a diagram showing the data and summarise the data in a table.
2. Use symmetry and standard results for the centre of mass of common shapes.
3. Take moments to find the centre of mass of a shape.

When working with laminas which have a mass, you can use the relationship *mass* = *area* × *density*, where the units of density ρ are kg m^{-2}. You need to be very careful as, in some problems, the area density across the shape may vary.

Example 2

a A uniform lamina *OABCD* consists of a rectangle, a semicircle and a right-angled triangle with the dimensions shown. Taking *OB* and *OD* as axes, find the point $G(\bar{x}, \bar{y})$ of the lamina's centre of mass.

b If the rectangle is made from a material with twice the density of that for the semicircle and triangle, find the new position $G(\bar{x}, \bar{y})$ of the lamina's centre of mass.

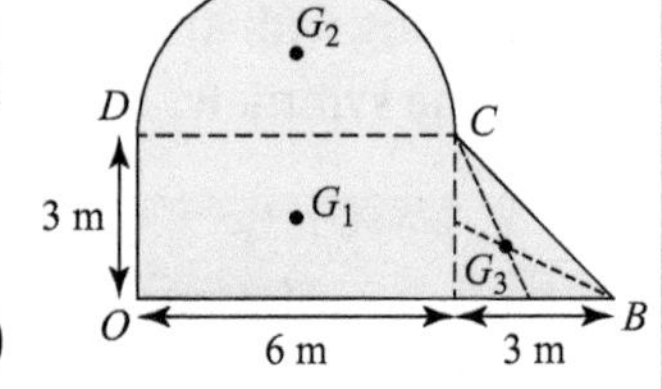

a Let the area density of the lamina be $\rho\,\text{kg m}^{-2}$.

	Area, m^2	Mass, kg	x-value	y-value
Rectangle	$3\times6=18$	18ρ	3	1.5
Semicircle	$\frac{1}{2}\pi\times3^2=14.1$	14.1ρ	3	$3+\dfrac{2\times3\sin\frac{\pi}{2}}{3\times\frac{\pi}{2}}=4.27$
Triangle	$\frac{1}{2}\times3\times3=4.5$	4.5ρ	$6+\frac{1}{3}\times3=7$	$\frac{1}{3}\times3=1$
Whole lamina	36.6	36.6ρ	$\bar{x}$	$\bar{y}$

Summarise the data in a table. (1)

Taking moments about *OD*

$36.6\rho\times\bar{x}=(18\rho\times3)+(14.1\rho\times3)+(4.5\rho\times7)$

$=127.8\rho$

giving $\bar{x}=3.49$

Taking moments about *OB*

$36.6\rho\times\bar{y}=(18\rho\times1.5)+(14.1\rho\times4.27)+(4.5\rho\times1)$

$=91.7\rho$

giving $\bar{y}=2.51$

Centre of mass is $G(3.49, 2.51)$

Take moments to find $\bar{x}$ and $\bar{y}$ (3)

(Continued on the next page)

b The density of the rectangle is now $2\rho\,\text{kg m}^{-2}$.

So the new total mass $= 18 \times 2\rho + 14.1\rho + 4.5\rho = 54.6\rho$

Taking moments about *OD* and *OB* gives

$54.6\rho \times \bar{x} = (18 \times 2\rho \times 3) + (14.1\rho \times 3) + (4.5\rho \times 7)$

$= 181.8\rho$

giving $\bar{x} = 3.33$

$54.6\rho \times \bar{y} = (18 \times 2\rho \times 1.5) + (14.1\rho \times 4.27) + (4.5\rho \times 1)$

$= 118.7\rho$

giving $\bar{y} = 2.17$

So the new centre of mass is $G(3.33, 2.17)$

Example 3

A uniform rectangular lamina *ABCD*, 4 m by 6 m, has a rectangular hole *PQRS*, 2 m by 1 m, removed, such that *S* coincides with the lamina's centre, G_1. Taking *AB* and *AD* as the *x*- and *y*-axes, find the distances $(\bar{x}, \bar{y})$ from *A* of the centre of mass of the remaining part.

Let the density of the lamina be $\rho\,\text{kg m}^{-2}$.

For *ABCD*

mass $= 6 \times 4 \times \rho = 24\rho$ with centre of mass G_1 at $(3, 2)$

For *PQRS*

mass $= 2 \times 1 \times \rho = 2\rho$ with centre of mass G_2 at $(3 + 1, 1 + 0.5) = (4, 1.5)$

For the remainder

mass $= 24\rho - 2\rho = 22\rho$ with centre of mass at $(\bar{x}, \bar{y})$

Taking moments about *AD*

Moment of *ABCD* = moment of *PQRS* + moment of remainder

$24\rho \times 3 = (2\rho \times 4) + (22\rho \times \bar{x})$

giving $72 = 8 + 22\bar{x}$

$\bar{x} = 2.91$

Taking moments about *AB*

$24\rho \times 2 = (2\rho \times 1.5) + (22\rho \times \bar{y})$

giving $48 = 3 + 22\bar{y}$

$\bar{y} = 2.05$

So, the centre of mass of the remaining part is (2.91, 2.05) metres from *A*

> **2** Work out the mass and centre of mass for each of the separate parts.

> **3** Take moments to find $\bar{x}$ and $\bar{y}$

MECH

The formulae booklet also provides some standard results for the centre of mass of 3D shapes which you can use without proof, as in Example 4

Example 4

A child's toy is made from a hollow, hemispherical shell attached to a hollow, conical shell of the same material and radius. The cone is 12 cm tall with a radius of 5 cm.

Find the position of the toy's centre of mass G from point O in the centre of the base of the hemisphere and cone.

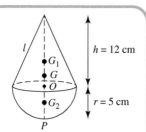

Let the area density be ρ and points G_1 and G_2 be the centres of mass of the two parts of the toy.

By symmetry, G lies on the line of symmetry through O

Using standard results, $OG_1 = \frac{1}{3} \times 12 = 4$ cm

and $OG_2 = \frac{1}{2} \times 5 = 2.5$ cm

> **2** The results for the centres of mass of a conical shell and a hemispherical shell are given in the formulae booklet.

Total mass = (area of cone + area of hemisphere) $\times \rho$

$= (\pi rl + \frac{1}{2} \times 4\pi r^2)\rho = (\pi \times 5 \times \sqrt{5^2 + 12^2} + 2 \times \pi \times 25)\rho$

$= (65\pi + 50\pi)\rho = 115\pi\rho$

1st method

Taking moments about point P

$115\pi\rho \times PG = 65\pi\rho \times PG_1 + 50\pi\rho \times PG_2$

$= 65\pi\rho \times (5 + 4) + 50\pi\rho \times (5 - 2.5) = 710\pi\rho$

$\Rightarrow PG = 6.2$ cm and $OG = 6.2 - 5 = 1.2$ cm

2nd method

Taking moments about point O

$115\pi\rho \times OG = 65\pi\rho \times OG_1 + 50\pi\rho \times OG_2$

$= 65\pi\rho \times 4 + 50\pi\rho \times (-2.5)$

$= 135\pi\rho$

$\Rightarrow OG = 1.2$ cm

> **3** Distances can be negative. Note the negative sign as G_2 is on the opposite side of O

The toy's centre of mass G is 1.2 cm from point O within the conical shell.

Exercise 25.3B Reasoning and problem-solving

1 Find the coordinates of the centres of mass of these uniform laminas.

a An L-shape (1, 7), (1, 9), (8, 9), (8, 3), (4, 3), (4, 7).

b A trapezium (2, 1), (8, 1), (8, 8), (2, 5).

c A rectangle (1, 1), (1, 6), (9, 6), (9, 1) with

i A semicircle attached to the side joining (1, 6) and (9, 6),

ii A circular hole, centre (5, 4) and radius 2 units, cut out.

2 For each uniform lamina, find the centre of mass taking O as the origin of x- and y-axes.

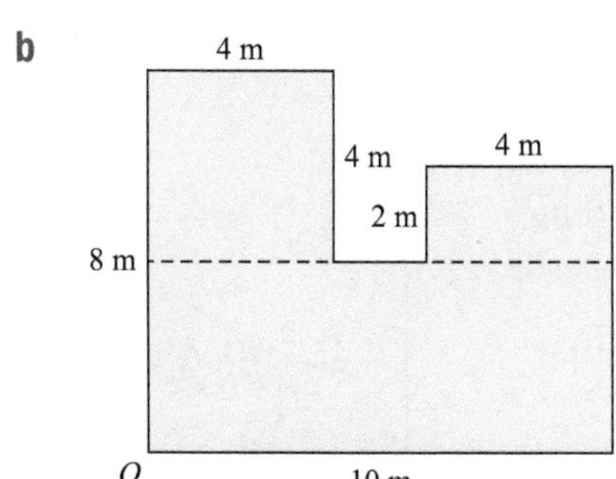

3 **a** A uniform wire of length 3 m is bent into a triangle ABC where $AB = 0.5$ m and $AC = 1.2$ m. Find the distances of the centre of mass from AB and AC

b The same wire is now bent to form the perimeter of a sector of a circle of radius 1 m. Find the distance of the centre of mass from the centre of the circle.

4 If the child's toy in Example 4 above were made from a solid cone and a solid hemisphere of the same material, use standard results to find the new position of the toy's centre of mass.

5 A circular hole of radius 3 cm is cut from a circular, metal disc of radius 9 cm. If the centre of mass of the remaining piece is 0.5 cm from the centre of the disc, find the position of the centre of the hole.

6 **a** A uniform square lamina of side 5 m has a square of side 2 m cut from one corner. Find the distance of the centre of mass of the remainder from that corner.

b A uniform semicircular lamina, diameter 24 cm, has a circular hole, diameter 6 cm, cut out of it. The hole's centre is 4 cm from its straight edge on the semicircle's line of symmetry. Find the exact position of the centre of mass of the remainder.

7 A child's building block is a prism with a symmetrical cross-section formed from a rectangle 10 cm by 5 cm with a semicircle of diameter 6 cm cut away. If the prism is uniformly dense, how far is its centre of mass from the baseline AB?

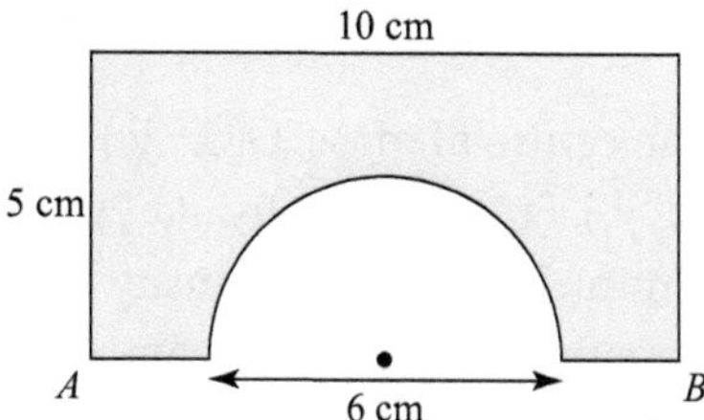

8 A uniform metal sheet $PQRST$ has the shape of a square $PQRT$ of side 6 cm joined to an isosceles triangle RST of height 4.5 cm with $SR = ST$. The triangle is twice the thickness of the square. How far is the centre of mass of the sheet from side PQ?

9 A toy rocket is made by joining the plane faces of a solid cone and a solid cylinder, both of radius 5 cm and height 9 cm. The density of the cylinder is twice that of the cone. How far is the rocket's centre of mass from the cone's vertex?

10 A frustum is made from a solid cone of height $6a$ and base radius $2a$ by removing its conical top so that the exposed plane surface has a radius of a. Find how far, in terms of a, the frustum's centre of mass is from the midpoint of the base.

11 An isosceles trapezium $ABCD$ has height h and parallel sides $AB = a$ and $CD = b$
Prove that its centre of mass is a distance of $\frac{h(a+2b)}{3(a+b)}$ from AB. Explain why this result applies to trapezia which are not isosceles.

MECH

Fluency and skills

You can find the centre of mass, G($\bar{x}$, $\bar{y}$) of n masses m_1, m_2, m_3, ... m_n at points (x_1, y_1), (x_2, y_2), (x_3, y_3), ... (x_n, y_n). You can apply the same method to a lamina with an area density ρ and mass M by dividing the lamina into n thin strips of mass $\delta m = \rho \times y\delta x$, where the base of each strip is on the x-axis so that the centre of mass of each strip is at the point $(x, \frac{1}{2}y)$.

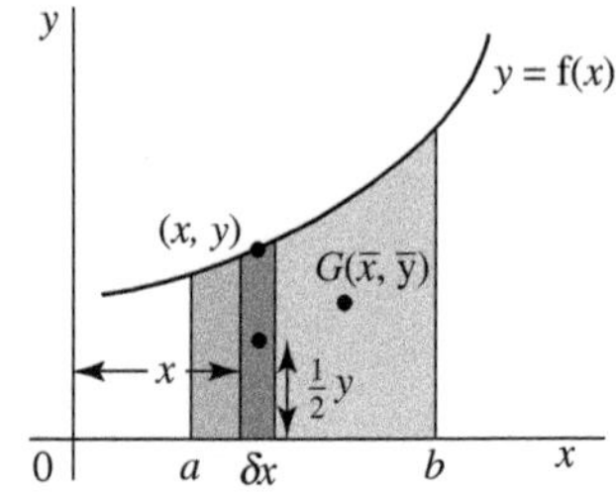

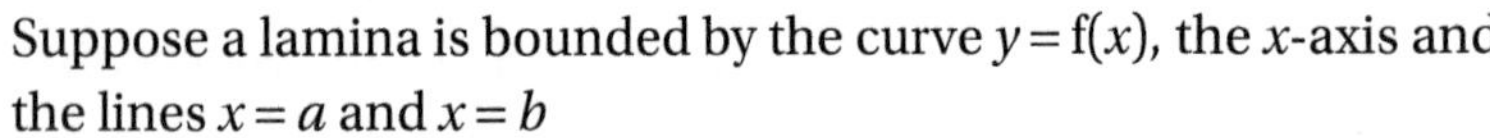

Suppose a lamina is bounded by the curve $y = f(x)$, the x-axis and the lines $x = a$ and $x = b$

The total mass $M = \sum_{x=a}^{b} \delta m = \sum_{x=a}^{b} \rho y \delta x$

By taking moments about the y-axis and the x-axis,

$$M \times \bar{x} = \sum_{x=a}^{b} \delta m \times x = \sum_{x=a}^{b} \rho xy\delta x$$

and $$M \times \bar{y} = \sum_{x=a}^{b} \delta m \times \frac{1}{2}y = \sum_{x=a}^{b} \frac{1}{2}\rho y^2 \delta x$$

As the number of strips increases, $\delta x \to 0$

In the limit, as $\delta x \to 0$, these summations combine strips across the whole area from $x = a$ to $x = b$ to give exact values for $\bar{x}$ and $\bar{y}$

So, in the limit,

$$M = \rho \int_a^b y \, dx$$

$$M \times \bar{x} = \rho \int_a^b xy \, dx$$

and $$M \times \bar{y} = \rho \int_a^b \frac{1}{2}y^2 \, dx$$

By substituting for M and cancelling ρ, you can find $\bar{x}$ and $\bar{y}$

Key point

A uniform lamina with an area bounded by the curve $y = f(x)$, the x-axis and the lines $x = a$ and $x = b$ has a centre of mass $G(\bar{x}, \bar{y})$, where

$$\bar{x} = \frac{\int_a^b xy \, dx}{\int_a^b y \, dx} \quad \text{and} \quad \bar{y} = \frac{\int_a^b \frac{1}{2}y^2 \, dx}{\int_a^b y \, dx}$$

You can *either* use the formulae for $\bar{x}$ and $\bar{y}$ as given in the key point, *or* set out your solution by taking moments as in Example 1. You have a choice.

Example 1

Find the centre of mass, G of the lamina formed from the area under the curve $y = \dfrac{12}{x}$ for $2 \le x \le 6$

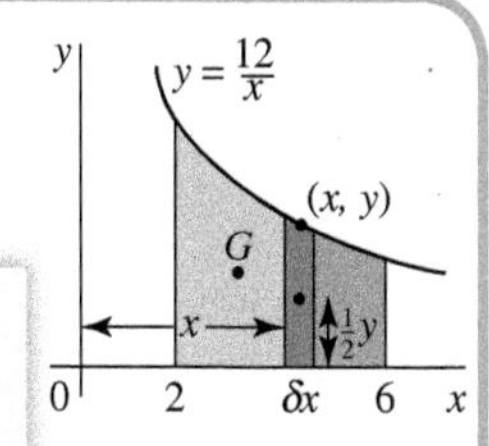

Let the area density be ρ and G be at the point $(\bar{x}, \bar{y})$

Total mass $M = \rho\int_2^6 y\,dx = \rho\int_2^6 \frac{12}{x}dx = 12\rho\int_2^6 \frac{1}{x}dx = 12\rho[\ln x]_2^6 = 12\rho\ln 3$

Taking moments about the y-axis

$$12\rho\ln 3 \times \bar{x} = \rho\int_2^6 xy\,dx = \rho\int_2^6 x\frac{12}{x}dx = 12\rho\int_2^6 1\,dx = 12\rho[x]_2^6 = 48\rho$$

$$\Rightarrow \bar{x} = \frac{48}{12\ln 3} = 3.64 \text{ to 3 sf}$$

Taking moments about the x-axis

$$12\rho\ln 3 \times \bar{y} = \rho\int_2^6 \frac{1}{2}y^2\,dx = \rho\int_2^6 \frac{1}{2}\left(\frac{12}{x}\right)^2 dx = 72\rho\int_2^6 \frac{1}{x^2}dx = 72\rho\left[-\frac{1}{x}\right]_2^6 = 24\rho$$

$$\Rightarrow \bar{y} = \frac{24}{12\ln 3} = 1.82 \text{ to 3 sf}$$

The centre of mass, G, is at the point (3.64, 1.82)

Note that the lamina in Example 1 has the x-axis as a boundary. Some laminas have a different shape. In the one shown here, the lamina is symmetrical about the x-axis, the x-axis is not a boundary and the vertical strips have a length $2y$, so you have to take moments about the y-axis to find the centre of mass, noting that, by symmetry $\bar{y} = 0$.

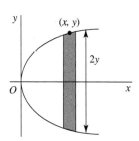

You can use a similar method for finding the centre of mass of a 3D solid with a volume density ρ. When the area under the curve $y = \text{f}(x)$ for $a \le x \le b$ is rotated about the x-axis, a thin strip of width δx generates a thin disc of mass $\delta m = \rho \times \pi y^2 \times \delta x$

In the limit as $\delta x \to 0$, a summation of these discs gives the total

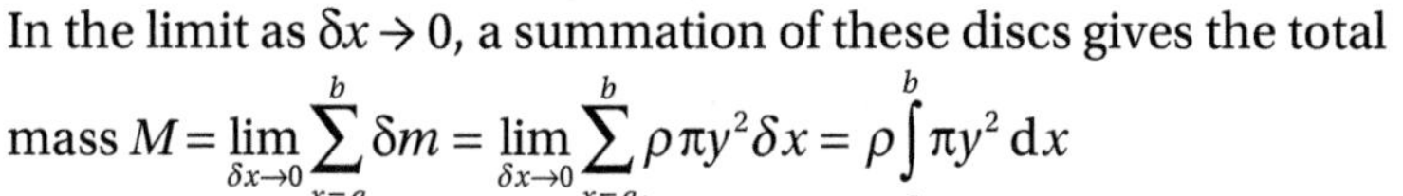

mass $M = \lim\limits_{\delta x \to 0}\sum\limits_{x=a}^{b}\delta m = \lim\limits_{\delta x \to 0}\sum\limits_{x=a}^{b}\rho\pi y^2\delta x = \rho\int_a^b \pi y^2\,dx$

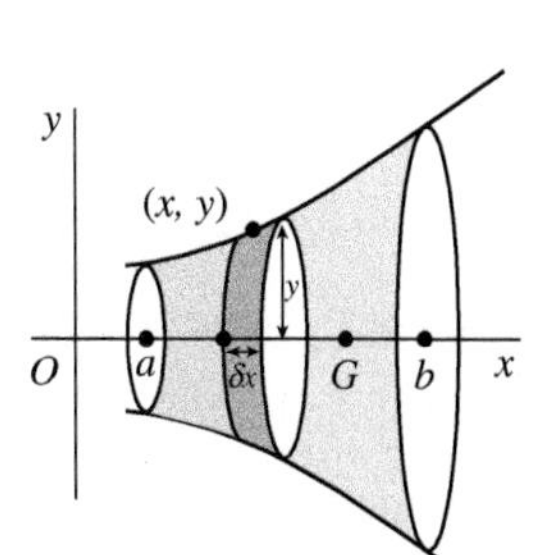

Taking moments about the y-axis, $M \times \bar{x} = \rho\int_a^b x \times \pi y^2\,dx$, which gives you the value of $\bar{x}$. Considering symmetry about the x-axis, you have $\bar{y} = 0$

MECH

Key point

A uniform solid of revolution formed by rotating the area between the curve $y = f(x)$, the x-axis and the lines $x = a$ and $x = b$ through 360° about the x-axis, has a centre of mass $G(\bar{x}, \bar{y})$, where

$$\bar{x} = \frac{\int_a^b xy^2\,dx}{\int_a^b y^2\,dx} \text{ and } \bar{y} = 0$$

As before, you can choose whether to use the formula given in this key point or set out your solution by taking moments about the y-axis, as in Example 2. You have a choice.

Example 2

Find the centre of mass, G, of the solid of revolution formed by rotating the area under the curve $y = \frac{12}{x}$ for $2 \le x \le 6$ through 2π radians about the x-axis. Give an exact answer.

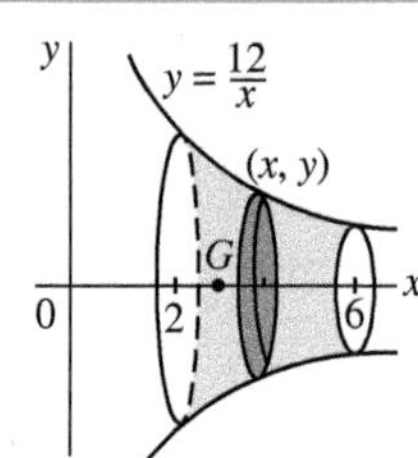

Let the volume density be ρ and G be at the point $(\bar{x}, \bar{y})$

Total mass $M = \rho\int_2^6 \pi y^2\,dx = \rho\pi\int_2^6 \frac{144}{x^2}dx = 144\rho\pi\left[-\frac{1}{x}\right]_2^6 = 48\rho\pi$

Taking moments about the y-axis

$48\rho\pi \times \bar{x} = \rho\int_2^6 x \times \pi y^2\,dx = \rho\pi\int_2^6 \frac{144}{x}dx = 144\rho\pi[\ln x]_2^6 = 144\rho\pi\ln 3$

$\Rightarrow \bar{x} = \frac{144\rho\pi\ln 3}{48\rho\pi} = 3\ln 3$

By symmetry, $\bar{y} = 0$

The solid's centre of mass, G, is at the point $(3\ln 3, 0)$

Exercise 25.4A Fluency and skills

1 Find the centre of mass of a uniform lamina defined by the area between the curve $y = f(x)$ and the x-axis for the given range, when

a $y = \frac{9}{x}$ for $1 \le x \le 3$

b $y = x^2$ for $0 \le x \le 2$

c $y = 1 + x^2$ for $0 \le x \le 1$

d $y = x(1 - x)$ for $0 \le x \le 1$

2 Find the centre of mass of a uniform lamina defined by the area between

a the curve $y^2 = x$ and the line $x = 4$

b the curve $y = 9 - x^2$ where $y \ge 0$

c the curve $y^2 = x^3$ and the line $x = 4$

3 Find the centre of mass of a uniform solid of revolution generated by rotating the area between the curve and the x-axis through 2π about the x-axis.

a $y = 2x$ for $0 \le x \le 4$

b $y = \frac{9}{x}$ for $1 \le x \le 3$

c $y = x^2$ for $0 \le x \le 2$

d $y^2 = x$ and the line $x = 4$, where $y \ge 0$

4 Find the centre of mass of a uniform semicircular lamina bounded by these curves.

a $x^2 + y^2 = 4$, where $x \ge 0$

b $x^2 + y^2 = r^2$, where $x \ge 0$

5 **a** A uniform solid hemisphere is generated by rotating a semicircle given by $x^2 + y^2 = 1$, where $x \geq 0$, through 180° about the x-axis. Find the coordinates of its centre of mass.

b Repeat for the hemisphere generated by $x^2 + y^2 = r^2$, where $x \geq 0$

6 Find the position of the centre of mass of a solid cone generated by a rotation of 360° about the x-axis of the straight line $y = mx$ for $0 \leq x \leq h$

MECH

Reasoning and problem-solving

Strategy

To solve problems involving the centre of mass of shapes

1. Draw a diagram showing the information you are given.
2. Work from first principles or, if you understand the method and can recall accurately, you may decide to quote a standard formula.
3. Take moments to find the centre of mass of a shape, using symmetry where possible.

A lamina can be suspended in equilibrium *either* by a string attached to its perimeter *or* from a nail or other smooth, horizontal axis through any point on its surface.

Only two forces act on the lamina: its weight and *either* the tension in the string *or* the reaction at the axis.

The lamina adjusts its position until the two forces act in the same vertical line and it is then in equilibrium. If the forces were not in the same vertical line, there would be a couple that would tend to rotate the suspended object. The same applies when a solid is suspended in equilibrium.

Tension | Reaction
G | G
Weight | Weight

Key point

When a lamina or solid is freely suspended in equilibrium, its centre of mass is vertically below its point of suspension.

Example 3

A uniform semicircular lamina of centre O with diameter $PQ = 2a$ is suspended from P and hangs freely. Calculate the angle θ that PQ makes with the vertical.

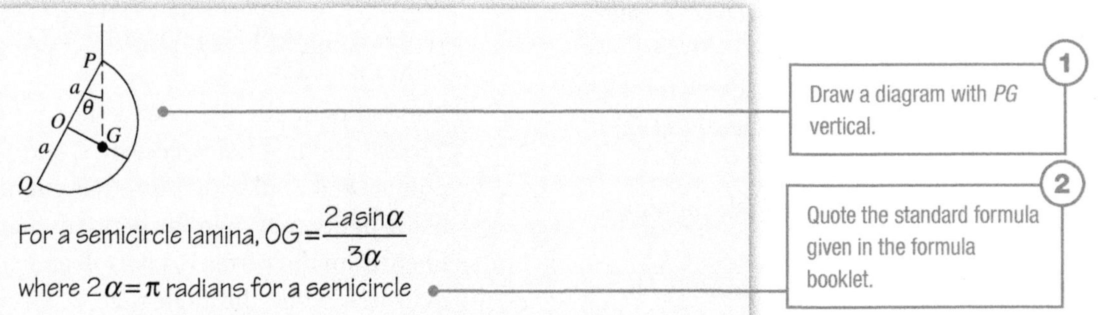

For a semicircle lamina, $OG = \dfrac{2a\sin\alpha}{3\alpha}$

where $2\alpha = \pi$ radians for a semicircle

1 Draw a diagram with PG vertical.

2 Quote the standard formula given in the formula booklet.

(*Continued on the next page*)

$$\Rightarrow OG = \frac{2a\sin\frac{\pi}{2}}{3\times\frac{\pi}{2}} = \frac{4a}{3\pi}$$

In $\Delta\ OPG$, $\tan\theta = \frac{OG}{OP} = \frac{4}{3\pi}$

$\Rightarrow$ Angle of PQ to vertical is $\theta = 23.0°$

Exercise 25.4B Reasoning and problem-solving

Where possible, you may quote standard results from the formula booklet for the centres of mass of uniform bodies.

1 Each of these uniform objects is suspended by a string from any point, P, on its circular edge, centre O. Find the angle that OP makes with the vertical.

   a A solid hemisphere, radius a

   b A solid cone, radius 3 cm, height 8 cm,

   c A hemispherical shell, radius a

   d A conical shell, radius 3 cm, height 8 cm.

2 A lamina in the shape of an isosceles triangle LMN with $MN = LN = 13$ cm, is suspended by a string to hang freely from L. If $LM = 10$ cm, find the angle that LM makes with the vertical.

3 A lamina in the shape of a sector of a circle, centre O and radius a with an arc PQ, is suspended to hang freely from point P. Find the angle that OP makes with the vertical when $\angle POQ = \frac{\pi}{2}$

4 One end of a solid cylinder is joined to the base of a solid cone of the same material. The object formed is hung from a point on the edge of the other end of the cylinder. The cylinder and cone have the same radius, r and height, h. Find the angle between the string and the object's axis in terms of r, and h

5 A uniform plane lamina is made by joining a semicircle, centre O and diameter $PQ = 2a$, to an equilateral triangle PQR

   a Find the exact distance of the lamina's centre of mass from O

   b The lamina is hung freely from P. Find the angle between OP and the vertical.

6 A vertical uniform rectangular laminar board $OABC$ of mass 2 kg is designed to hold boxes on its surface. Taking OA and OC as axes with $OA = 1$ m and $OC = 0.5$ m, model three boxes as particles A, B and C with masses 2 kg at point (0.2, 0.4), 1 kg at (0.4, 0.3) and 3 kg at (0.6, 0.4). Find the position G of the centre of mass. If the board is hung freely from a nail at the midpoint M of BC, find the angle that BC makes with the horizontal.

7 Two uniform semicircular laminas of masses m and $2m$ have the same radius r They combine to form a circle, centre O and centre of mass G. Calculate the distance OG The circle is suspended from one end P of their common diameter. Show that the angle that this diameter makes with the vertical is equal to $\tan^{-1}\left(\frac{4}{9\pi}\right)$

8

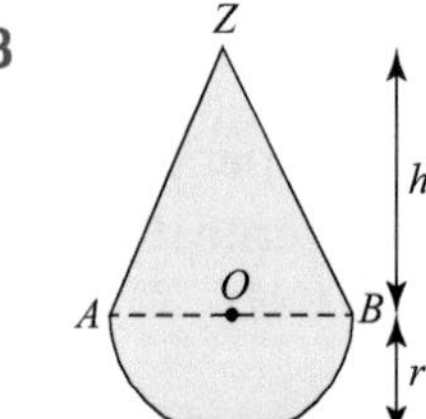

   a A pendant is made from a uniform lamina consisting of a semicircle, centre O and radius r, attached to an isosceles triangle of base $2r$ and height h. If the centre of mass of the pendant is at O, show that $h = r\sqrt{2}$

   b Another pendant of the same design has $2r = h = 15$ cm. A hole of centre C and radius $OC = 5$ cm with C on AB is cut out of the pendant. When the pendant hangs freely from the vertex Z, take OB and OZ as axes and find the angle that AZ makes with the vertical.

25.5 Equilibrium

Fluency and skills

When an object is subjected to a system of forces and couples which all act in the same plane, it will be in equilibrium only if

(a) The resultant of all forces must be zero.

(b) The total moment of all forces and couples must be zero.

The sum of the components of all forces must be zero in any two different directions to ensure that there is no linear movement. You can write three equations to describe the equilibrium and then solve them simultaneously.

Key point

For an object to be in equilibrium under a system of coplanar forces, these conditions must be met:

- The sum of the components of all forces must be zero in any *two different* directions
- The total moment of all forces and couples must be zero.

Coplanar means 'in the same plane'.

Sometimes you can use equivalent conditions: for example, resolve in one direction and take moments about two different points.

Take care when choosing the directions in which you resolve and also when choosing the point to take moments about. Try to keep the algebra as simple as possible.

Example 1

A uniform ladder AB weighing 120 N rests on a smooth floor with its upper end B on a smooth wall.
End A is held by a string AC fixed to the base of the wall at C
A weight of 80 N is attached three-quarters of the way up the ladder at D
If $AC = 4$ m and $BC = 6$ m, find the tension T in the string and the reaction S at the ground.

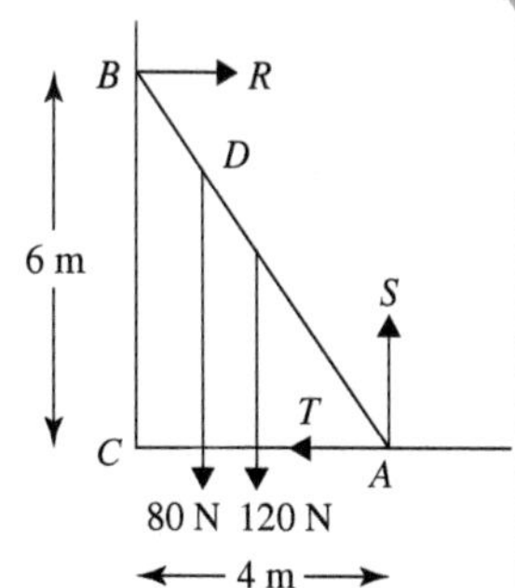

As both surfaces are smooth, the reactions R and S are normal to the wall and ground.

For vertical equilibrium,

resultant vertical force is zero: $S - 120 - 80 = 0$

or upward force balances downward force: $S = 120 + 80$ [1]

$\Rightarrow$ Reaction $S = 200$ N

(*Continued on the next page*)

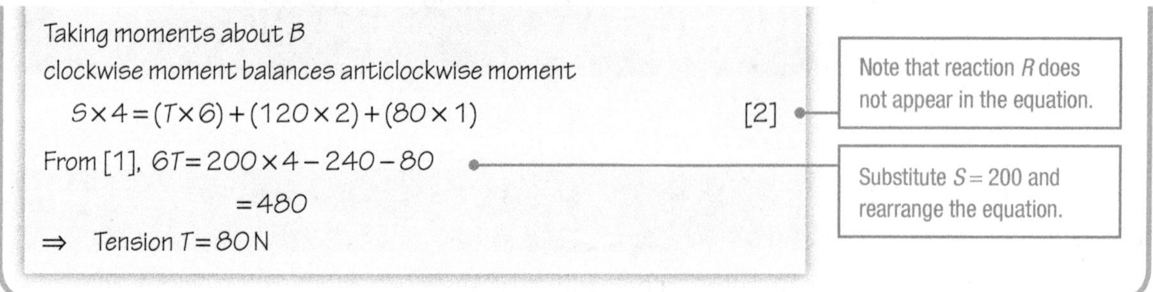

Taking moments about B

clockwise moment balances anticlockwise moment

$S \times 4 = (T \times 6) + (120 \times 2) + (80 \times 1)$ [2]

From [1], $6T = 200 \times 4 - 240 - 80$

$= 480$

$\Rightarrow$ Tension $T = 80$ N

Note that reaction R does not appear in the equation.

Substitute $S = 200$ and rearrange the equation.

The method in Example 1 used only two equations because it involved only two unknowns (T and S). Taking moments about B to give equation [2] was chosen because it did not involve R and so it made the algebra easier.

Instead of taking moments about B, you could have used the condition for horizontal equilibrium (giving $R = T$) and then taken moments about A. This method, however, involves three unknowns (R, T and S) and three equations. The algebra is not as easy.

You should think about your method before you start, so that you minimise the number of equations and unknowns.

Exercise 25.5A Fluency and skills

1 A uniform ladder PQ weighs 200 N. It stands on a smooth floor with its upper end Q against a smooth vertical wall. The lower end P is joined to the foot of the wall R by a string PR. If $PR = 6$ m and $QR = 8$ m, find the tension in the string and the reactions at the ground and at the wall.

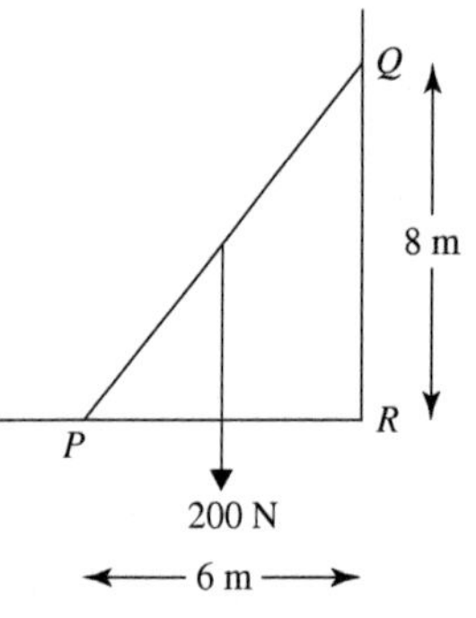

2 A uniform ladder AB weighing 100 N rests with end A on rough horizontal ground where the coefficient of friction, μ is $\frac{3}{4}$

End B rests against a smooth vertical wall. If A is 4 m from the foot of the wall, find the reaction at B and the frictional force at A if the ladder is on the point of slipping.

3 A uniform rod PQ weighs 8 N. It slopes up at 45° from a hinge at P and has a weight of 10 N hanging from Q. It is held in equilibrium by a horizontal string attached to Q so that Q is 4 m higher than P. Find the horizontal and vertical components of the reaction at the hinge and hence the total reaction at the hinge.

4 A uniform square lamina $EFGH$ of mass 4 kg and side 3 m is hinged at H with its upper edge GH horizontal. It is held in equilibrium by a horizontal string attached to F. Find the tension in the string and the exact value of the total reaction at the hinge.

H G E F T

5 A man of mass 75 kg climbs two-thirds of the way up a ladder AB of mass 20 kg which rests on a smooth floor at A and against a smooth wall at B. A string joins A to the foot of the wall at C such that $AC = 3$ m and $BC = 6$ m. Find the tension in the string in terms of g

6 A rectangular shop-sign $ABCD$, such that $AB = CD = 1$ m and $AD = BC = 0.5$ m, with a mass of 4 kg is hinged to a smooth wall at A. Its upper side AB is kept horizontal by corner D making contact with the wall. A 2 kg mass hangs from C. Find

a The reaction with the wall at D

b The total reaction at the hinge A

Reasoning and problem-solving

When an object is placed on an inclined plane, it may rest in equilibrium, slide down the plane or topple over. If there is enough friction to prevent sliding, there are three possible outcomes depending on the angle of inclination, θ

Consider a rectangular lamina $ABCD$ of weight W and centre of mass G.

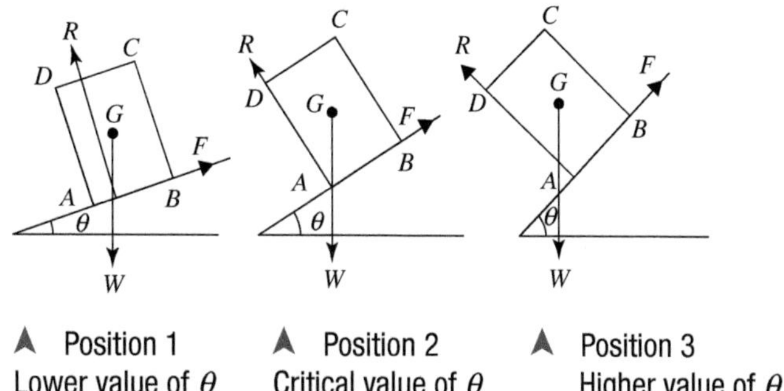

▲ Position 1 Lower value of θ ▲ Position 2 Critical value of θ ▲ Position 3 Higher value of θ

In position 1, the rectangle and plane make contact all along AB producing a normal reaction R. The moments of R and W about A are balanced and the rectangle is in equilibrium.

In position 2, θ has increased and the rectangle is about to topple. R, F and W all act through the only point of contact, A. The rectangle is in limiting equilibrium.

In position 3, θ has increased further. R and F still act through A, but W has an anticlockwise moment about A. The rectangle is not in equilibrium and it topples.

Key point

An object on an inclined plane is in equilibrium without toppling if the line of action of its weight falls within the side of the object that makes contact with the plane.

Strategy

To solve problems involving the equilibrium of objects

1. Draw a diagram showing the information you are given and the information you need to find.
2. Devise a strategy to minimise the number of unknowns and equations.
3. Write equations to satisfy the conditions for equilibrium, along with any other associated equations, and solve them simultaneously.

Example 2

Rectangle $ABCD$ with $AD = 6$ cm and $CD = 4$ cm rests on an inclined plane as shown. If it does not slide down the plane, find the greatest possible angle of inclination θ

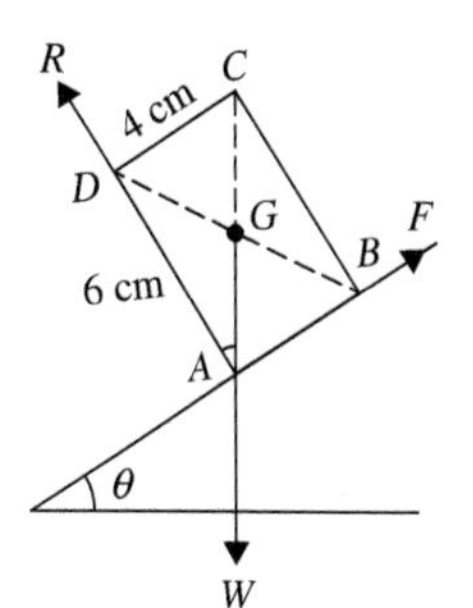

Let G be the centre of mass and the weight be W.

The rectangle is on the point of toppling when G is vertically above A

Geometry gives $\angle GAD = \theta$

In ΔACD, $\tan\theta = \frac{4}{6}$

So the greatest possible angle of inclination is $\theta = 33.7°$

Example 3

A cylindrical roller, centre O and radius 63 cm, is fixed to smooth, horizontal ground with a uniform plank resting against it at point C

One end, A, of the plank lies on the ground and the other end, B, projects beyond the roller. The plank is perpendicular to the horizontal axis of the roller.

A light string AO of length 1.05 m keeps the plank from slipping.

If the plank has weight W and length 90 cm, find the tension T in the string in terms of W.

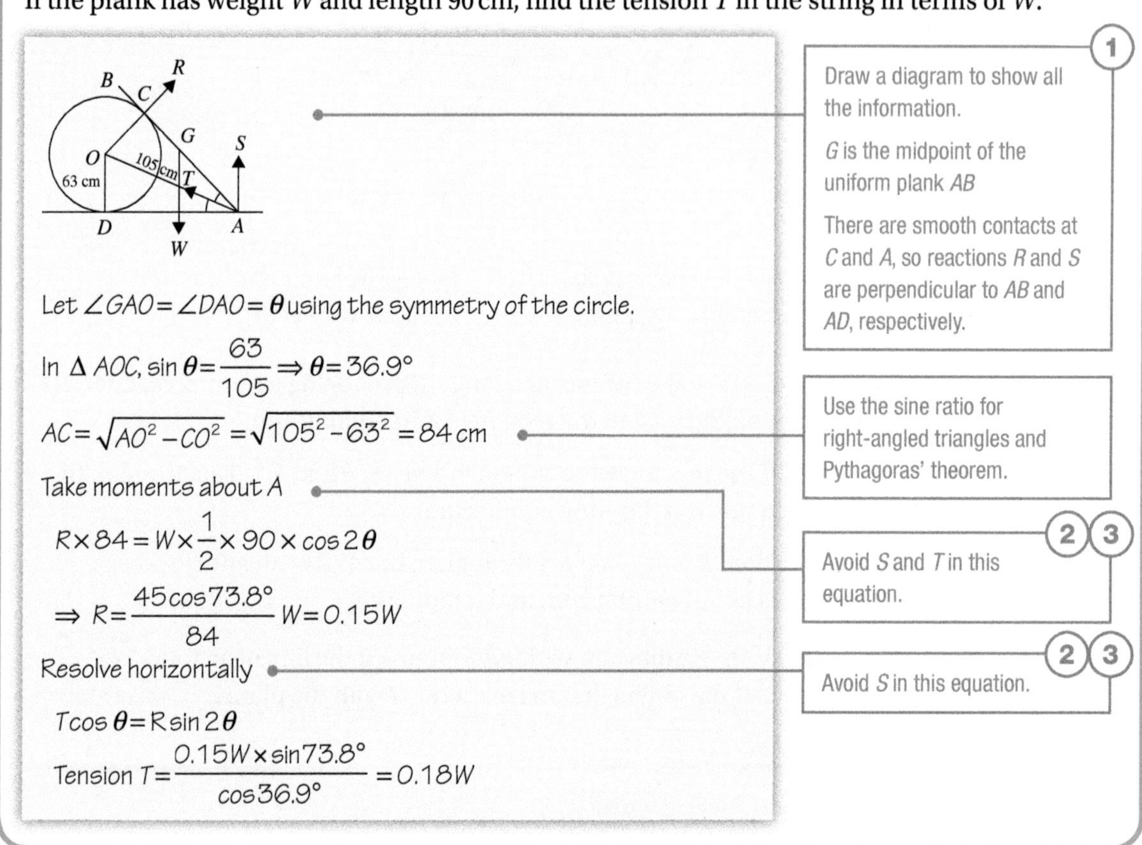

Draw a diagram to show all the information.

G is the midpoint of the uniform plank AB

There are smooth contacts at C and A, so reactions R and S are perpendicular to AB and AD, respectively.

Let $\angle GAO = \angle DAO = \theta$ using the symmetry of the circle.

In ΔAOC, $\sin\theta = \frac{63}{105} \Rightarrow \theta = 36.9°$

$AC = \sqrt{AO^2 - CO^2} = \sqrt{105^2 - 63^2} = 84\text{ cm}$

Use the sine ratio for right-angled triangles and Pythagoras' theorem.

Take moments about A

$R \times 84 = W \times \frac{1}{2} \times 90 \times \cos 2\theta$

$\Rightarrow R = \frac{45\cos 73.8°}{84} W = 0.15W$

Avoid S and T in this equation.

Resolve horizontally

$T\cos\theta = R\sin 2\theta$

Tension $T = \frac{0.15W \times \sin 73.8°}{\cos 36.9°} = 0.18W$

Avoid S in this equation.

A third equation (from resolving vertically) was not needed as you had to find only the tension T.

Example 4

A uniform rod AB of weight 10 N and length $2a$ is hinged at A Its end B is held by a string so that the rod and string are both inclined at 30° to the horizontal. Find the tension in the string and the total reaction at the hinge.

Resolve horizontally

$X = T\cos 30° \quad \Rightarrow \quad 2X = \sqrt{3}T \quad [1]$

Resolve vertically

$Y + T\sin 30° = 10 \quad \Rightarrow \quad 2Y + T = 20 \quad [2]$

(*Continued on the next page*)

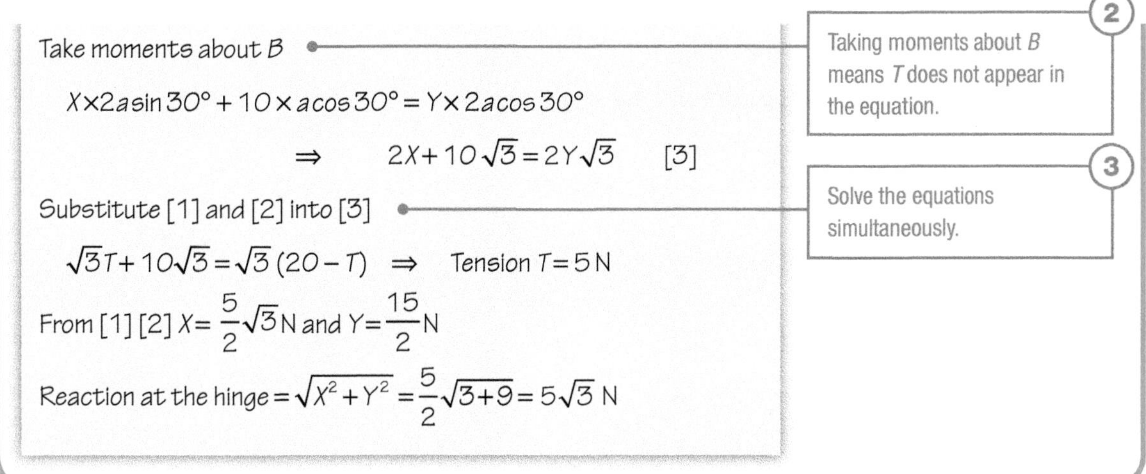

Take moments about B

$$X \times 2a\sin 30° + 10 \times a\cos 30° = Y \times 2a\cos 30°$$

$$\Rightarrow \quad 2X + 10\sqrt{3} = 2Y\sqrt{3} \quad [3]$$

Substitute [1] and [2] into [3]

$$\sqrt{3}T + 10\sqrt{3} = \sqrt{3}(20 - T) \quad \Rightarrow \quad \text{Tension } T = 5\text{ N}$$

From [1] [2] $X = \frac{5}{2}\sqrt{3}$ N and $Y = \frac{15}{2}$ N

Reaction at the hinge $= \sqrt{X^2 + Y^2} = \frac{5}{2}\sqrt{3+9} = 5\sqrt{3}$ N

2. Taking moments about B means T does not appear in the equation.

3. Solve the equations simultaneously.

The algebra could be reduced by taking moments about A with T resolved into two components. This means that X and Y do not appear in the equation.

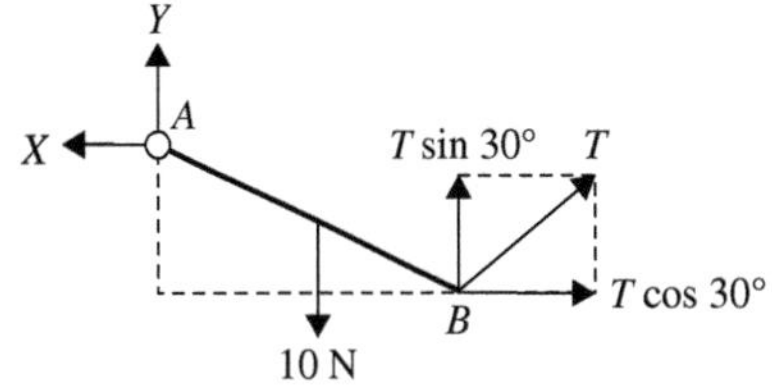

$$10 \times a\cos 30° = T\sin 30° \times 2a\cos 30° + T\cos 30° \times 2a\sin 30°$$

$$\Rightarrow T = 5\text{ N}$$

Using the vector product in mechanics

The vector product of two vectors $\mathbf{a}$ and $\mathbf{b}$ is given by $\mathbf{a} \times \mathbf{b} = |\mathbf{a}||\mathbf{b}|\sin\theta\hat{\mathbf{n}}$ and can be used to find the moment about the origin O of a force $\mathbf{F}$ acting at point R with position vector $\mathbf{r}$

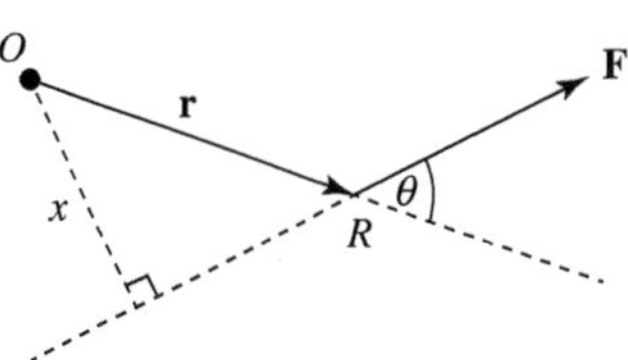

Let x be the shortest distance of O from the line of action of $\mathbf{F}$ and θ be the acute angle between the directions of $\mathbf{r}$ and $\mathbf{F}$

The moment of $\mathbf{F}$ about O has a magnitude of $\text{F} \times x = \text{F} \times \text{r}\sin\theta = |\mathbf{r}||\mathbf{F}|\sin\theta$

Using the definition of a vector product and the right-hand rule, you can express this moment as a vector $\mathbf{M}$ in the direction of the axis of rotation where $\mathbf{M} = \mathbf{r} \times \mathbf{F}$

The moment about O is actually the moment about an axis through O perpendicular to the plane containing O, R and $\mathbf{F}$

Key point

In general, a force $\mathbf{F}$ acting at point P has a moment $\mathbf{M}$ about point Q given by $\mathbf{M} = \mathbf{r} \times \mathbf{F}$, where $\mathbf{r} = \mathbf{QP}$

Example 5

Two forces $\mathbf{F}_1 = 2\mathbf{i} + 4\mathbf{j} - \mathbf{k}$ N and $\mathbf{F}_2 = 3\mathbf{i} - 3\mathbf{j} - 2\mathbf{k}$ N act at points $P(2, -2, 3)$ and $Q(0, 4, -3)$ respectively. Their resultant forms a couple with a third force $\mathbf{F}_3$ which acts at the origin O

Find

a The force $\mathbf{F}_3$ and its magnitude,

b The moment $\mathbf{M}$ of the couple and its magnitude.

(*Continued on the next page*)

a $\mathbf{F}_1 + \mathbf{F}_2 = (2\mathbf{i} + 4\mathbf{j} - \mathbf{k}) + (3\mathbf{i} - 3\mathbf{j} - 2\mathbf{k}) = 5\mathbf{i} + \mathbf{j} - 3\mathbf{k}$ N

As the overall resultant is a couple and not a single force,

Force $\mathbf{F}_3 = -(\mathbf{F}_1 + \mathbf{F}_2) = -5\mathbf{i} - \mathbf{j} + 3\mathbf{k}$

Magnitude of $\mathbf{F}_3 = \sqrt{(25+1+9)} = \sqrt{35}$N

b Moment **M** of couple = Total moment of all three forces about any point

Taking moments about O

Moment $\mathbf{M} = (\mathbf{p} \times \mathbf{F}_1) + (\mathbf{q} \times \mathbf{F}_2) + (\mathbf{O} \times \mathbf{F}_3)$

$$\text{Moment } \mathbf{M} = \begin{vmatrix} \mathbf{i} & \mathbf{j} & \mathbf{k} \\ 2 & -2 & 3 \\ 2 & 4 & -1 \end{vmatrix} + \begin{vmatrix} \mathbf{i} & \mathbf{j} & \mathbf{k} \\ 0 & 4 & -3 \\ 3 & -3 & -2 \end{vmatrix} + 0 = -27\mathbf{i} + 17\mathbf{j} + 24\mathbf{k}$$

Magnitude of the moment of the couple $= |\mathbf{M}| = \sqrt{(-27)^2 + 17^2 + 24^2} = 39.9$ Nm

Exercise 25.5B Reasoning and problem-solving

1 A uniform rectangular lamina $ABCD$ stands in a vertical plane with AB in contact with a rough plane inclined at an angle θ to the horizontal. Sides AD and BC are perpendicular to the slope with AD on the downside of BC. The rectangle does not slide. If $AD = 10$ cm and $CD = 5$ cm, find the value of θ when the lamina is on the point of toppling.

2 A lamina has the shape of an isosceles triangle ABC with $AC = BC = 12$ cm and $AB = 6$ cm. It stands in a vertical plane with AB in contact with a rough plane inclined at angle θ to the horizontal. Given that the lamina does not slide, find the greatest value of θ for it not to topple.

3 A triangular lamina PQR has $PR = 12$ cm, $PQ = 6$ cm and $\angle RPQ = 90°$. It rests in a vertical plane on a rough slope inclined at angle θ to the horizontal with PQ in contact with the slope and P lower than Q Given that the lamina does not slide, find the minimum value of θ for it to topple.

Also find the least value of the coefficient of friction, μ, for the lamina not to slide when it is about to topple.

4 A cone of height h and radius r rests in equilibrium with its plane face on a rough slope which makes an angle θ with the horizontal.

- **a** Calculate the maximum possible value of θ before the cone topples, without sliding, if
  - **i** The cone is solid,
  - **ii** The cone is hollow.
- **b** What must be the least value of the coefficient of friction, μ, between the hollow cone and the slope for the cone to topple before it slides?

5 **a** A uniform L-shaped lamina is made by removing the square $OCDE$ of side 2 cm from the square $ABDF$. If $AB = 4$ m, find the position of the centre of mass, G, of the lamina from AB and AF

b The lamina stands at rest with AB on a rough inclined plane. What is the greatest possible angle of inclination of the plane if the lamina is not to topple over for each of the two positions shown?

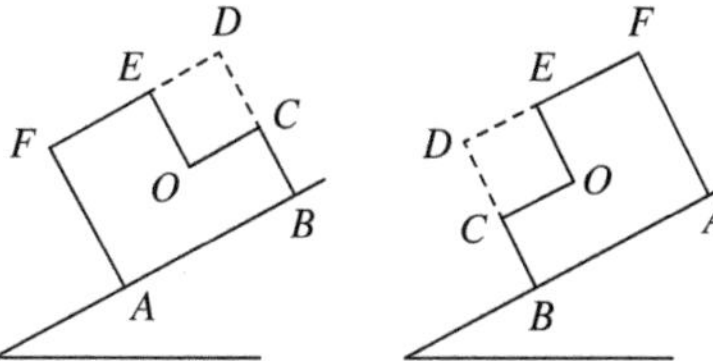

6 a A uniform horizontal beam PQ with mass 8 kg and length 2 m acts as a cantilever. It is clamped at P and has a 10 kg mass hanging from Q. Find the force and couple which acts at P to maintain the beam in equilibrium.

b The same beam and 10 kg mass are supported additionally by a sloping string QR fixed to a point R which is 2 m vertically above P. The string breaks if the tension in it exceeds 100 N. When the tension in the string is at its maximum, find the force and couple acting at P to keep the beam in equilibrium.

7 A light, triangular lamina ABC has $AB = 3$ cm, $BC = 4$ cm and $CA = 5$ cm. Forces of 15 N, 20 N and 25 N act along the sides AB, BC and CA, respectively. What force and couple, if any, must act at A to keep the lamina in equilibrium?

8 A uniform ladder of length $2a$ and weight W stands on a smooth floor against a smooth wall so that it is inclined at an angle α to the floor. A weight w is placed on the ladder at a distance x from the lower end. The ladder is held in equilibrium by a couple of moment M. Prove that $M = (Wa + wx)\cos\alpha$

9 A uniform rod PQ of mass 10 kg and length 4 m is hinged at P. A horizontal string attached to Q holds the rod at rest such that PQ makes 30° with the horizontal. Find the tension in the string if

a Q is higher than P

b Q is lower than P

10 A uniform rod AB of mass 8 kg and length 4 m rests at an angle θ to rough, level ground with A on the ground and B on a smooth peg 2 m above the ground. Find the total reaction at A and the least possible value of the coefficient of friction, μ, at A

11 a The area between the curve $y = f(x)$ and the x-axis for $0 \le x \le 2$ is rotated 360° about the x-axis to form a uniform solid. Find the exact coordinates of the centre of mass of the solid, when

i $f(x) = \frac{1}{2}x^2$

ii $f(x) = \sqrt{x}$

b Each solid is placed with its plane face on a rough plane inclined at an angle θ to the horizontal and it does not slide. Find, in each case, the exact value of θ when the solid is just about to topple.

12 a The area between the curve $y = \frac{12}{x}$ and the x-axis for $4 \le x \le 6$ is rotated through an angle of 2π about the x-axis to form a uniform solid. Find the exact coordinates of the centre of mass of the solid.

b The solid is placed with its smaller plane surface in contact with a rough plane that is inclined at an angle θ to the horizontal and has coefficient of friction, μ. As θ increases, find the value of θ at which it topples, but does not slide.

13 A uniform beam FH, of length 4 m and weight 50 N, is hinged to a vertical wall at H It is held at rest by a horizontal rope attached to the beam at F so that the beam slopes down at 20° from H to F. Find the tension in the rope and the magnitude and direction of the reaction at the hinge.

14 A horizontal light rod PQ of length 6 m is held in equilibrium by two strings PR and QS which slope upwards away from the rod such that $\angle QPR = 120°$ and $\angle PQS = 135°$. A weight of 10 N hangs from a point on the rod x m from P. Calculate the tension in each string and the value of x

MECH

15 A smooth cylinder, radius a, is fixed to rough, level ground at point C with its axis horizontal. A uniform rod AB of weight W rests at right angles to the axis with A on the ground and with B on the cylinder's surface such that $\angle BAC = 60°$. If the rod is about to slide, find the exact value of μ

16 A uniform ladder PQ of weight W and length $2a$ leans at an angle of 60° against a rough, vertical wall with its lower end Q on a rough level floor. The coefficient of friction at both ends is 0.8

What is the maximum weight of a person, in terms of W, who can climb to the top of the ladder without the ladder slipping?

17 A heavy, horizontal, uniform rod CD of mass 75 kg and length 90 cm is in contact with a rough vertical wall at C. A string DE holds the rod in limiting equilibrium, where E is on the wall 72 cm above C

A mass 100 kg hangs freely from D. Calculate the tension in the string, the thrust in the beam and the coefficient of friction at C

18 A uniform, square lamina $PQRS$ of weight W has its plane at right angles to a rough, vertical wall with P in contact with the wall and side PQ sloping upwards from P to Q at an angle θ to the vertical. It is held in equilibrium by a horizontal force W acting at R. Show that the coefficient of friction must be at least 1. Find the value of θ

19 A toy is made from two solid pieces with the same density ρ: a hemisphere and a cone with the same radius r. Their circular faces are joined together and have a common centre O. The toy stands with the hemisphere on a rough level surface.

a Show that the toy's centre of mass, G is a distance $\bar{x} = \dfrac{h^2 - 3r^2}{8r + 4h}$ from O, where h is the height of the cone.

b The toy stands with its axis at an angle to the vertical and with its hemisphere on a rough, level surface. By taking moments about its point of contact with the surface, explain what happens after it is released if G

i Lies within the hemisphere,

ii Lies within the cone,

iii Coincides with O

20 A uniform footbridge AB of weight 400 N and length 4 m is rigidly fixed at end A and hinged at end B. When a person of weight 800 N stands on it 1 m from B, the vertical force on the bridge at B is three times that at A. Find the magnitude and sense of the couple which acts at A to keep the bridge in equilibrium.

21 A uniform rod AB of weight W and length 4 m is acted on by two forces T and U as shown. If the three forces are equivalent to a couple when $U = \sqrt{2}$ N, find the rod's weight and the magnitude and sense of the couple.

22 A light rectangular lamina has vertices $A(6, 0)$, $B(6, 5)$, $C(0, 5)$ and the origin O, with distances in m. Three forces **F**, **G** and **H** act on the rectangle where $\mathbf{F} = 10\mathbf{i}$ N at the point (2, 1), $\mathbf{G} = -3\mathbf{j}$ N at (5, 4), $\mathbf{H} = 2\mathbf{i} + 8\mathbf{j}$ N at (0, 3). Show that this system is equivalent to

a A single force **R** acting along the line $12y - 5x = 31$. Find the magnitude of **R**

b A force **S** acting at the origin and a couple **C**. Find the magnitudes of **S** and **C**

23 Two equally rough planes are both inclined at 20° to the horizontal. Their lower edges meet in a horizontal line to form a V-shape in which an oil drum, of weight 800 N and radius 25 cm, rests symmetrically in equilibrium with its axis horizontal. If the coefficient of friction between the drum and planes is 0.5, what is the maximum external couple which can act on the drum in a vertical plane at right-angles to its axis if it is to stay in equilibrium?

24 A force $\mathbf{F} = 3\mathbf{i} - \mathbf{j} + 2\mathbf{k}$ newtons acts on an object at the point R(4, 3, –1). Find the magnitude of the moment of **F** about

a The origin,

b The point P(2, –2, 0).

25 Three forces $(4n\mathbf{i} + 2\mathbf{j} + 4\mathbf{k})$N, $(2n\mathbf{i} - 4\mathbf{j} + 6\mathbf{k})$N and $(4\mathbf{i} - 2n\mathbf{j} + 4n\mathbf{k})$N act on a light object at the points A(n, 0.5, 0), B(0.5, 1, 0) and C(–1, –0.5, 1) respectively, where n is a constant.

a Show that the total moment of the three forces about the origin O is independent of n

b Find the magnitude of the moment of the force $(4n\mathbf{i} + 2\mathbf{j} + 4\mathbf{k})$N about the point B when $n = 2$

26 Two forces $\mathbf{F}_1 = 2\mathbf{i} + \mathbf{j} - \mathbf{k}$ N and $\mathbf{F}_2 = \mathbf{i} - 2\mathbf{j} - 2\mathbf{k}$ N act at points P(1, –2, 3) and Q(0, 2, –3) respectively. The resultant of $\mathbf{F}_1$ and $\mathbf{F}_2$ forms a couple with a third force $\mathbf{F}_3$ which acts at the origin O

Find

a The resultant of $\mathbf{F}_1$ and $\mathbf{F}_2$

b The force $\mathbf{F}_3$ and its exact magnitude,

c The magnitude of the moment of the couple.

27 Three forces, in N, act in the x-y plane at the given points and they form a couple.

$\mathbf{F}_1 = \mathbf{i} + 2m\mathbf{j}$ at (3, 1)

$\mathbf{F}_2 = 4n\mathbf{i} + 11\mathbf{j}$ at (–2, –5)

$\mathbf{F}_3 = -2m\mathbf{i} + 8n\mathbf{j}$ at (4, –1)

a Find the constants m and n

b Calculate the couple and its magnitude.

28 The force $\mathbf{F} = 2\mathrm{i} + 5\mathrm{j} - \mathrm{k}$ newtons acts at the point A(3, –2, 1) and the force $\mathbf{G} = \mathrm{i} - \mathrm{j} - 2\mathrm{k}$ newtons acts at the point B(4, 2, –4). A third force **H** acts at the origin and forms a couple with the resultant of **F** and **G**. Find the exact values of

a The force **H** and its magnitude,

b The moment of the couple and its magnitude.

29 Three forces **F**, **G** and **H**, form a system of forces which is equivalent to a single force **Z** acting at the origin together with a couple of moment 96 N m. If $\mathbf{F} = 4\mathbf{i} + 2\mathbf{j}$, $\mathbf{G} = 10\mathbf{i} - 8\mathbf{j}$ and $\mathbf{H} = n\mathbf{j}$ in newtons act at the points (6, 8), (–4, –6) and (4, 0) respectively, find

a Two possible values of n

b The force **Z** in each case.

30 A rigid light rod AB has a force $\mathbf{F} = -\mathbf{i} + 3\mathbf{j} + 4\mathbf{k}$ N acting at its midpoint M

Given that A and B are at the points (–10, 8, 2) and (–2, –4, 6) respectively, find

a The vector **AM**

b The moment of **F** about A and its exact magnitude,

c The acute angle between the direction of **F** and the rod by using

i A vector product,

ii A scalar product.

MECH

25 Summary and review

Chapter summary

- The moment of a force F about a point A is $F \times d$, where d is the perpendicular distance of F from A. Moments are measured in N m. They can be positive (anticlockwise) or negative (clockwise).
- The resultant moment of several forces acting on an object equals the sum of the moments of the individual forces, taking into account their directions.
- A couple comprises two equal and opposite forces, F, which are not acting in the same straight line. The moment of a couple, about any point, equals $F \times d$
- When several masses $m_1, m_2, m_3, \ldots$ at the points $(x_1, y_1), (x_2, y_2), (x_3, y_3), \ldots$ have their centre of mass G at the point $(\bar{x}, \bar{y})$, then
    - Their total mass $M = \sum_{i=1}^{n} m_i$
    - Taking moments about the axes, $M \times \bar{x} = \sum_{i=1}^{n} m_i \times x_i$ and $M \times \bar{y} = \sum_{i=1}^{n} m_i \times y_i$
- You can use standard formulae to find the centre of mass of some laminas and solid bodies.
- When a uniform lamina with area density ρ has an area bounded by the curve $y = \mathrm{f}(x)$, the x-axis and the lines $x = a$ and $x = b$, it has a centre of mass G($\bar{x}$, $\bar{y}$), where

  $M = \rho \int_a^b y \, dx$, $M \times \bar{x} = \rho \int_a^b xy \, dx$ and $M \times \bar{y} = \rho \int_a^b \frac{1}{2} y^2 \, dx$
- When a uniform solid is formed by rotating the region between the curve $y = \mathrm{f}(x)$, the x-axis and the lines $x = a$ and $x = b$ through 360° about the x-axis, it has a centre of mass $G(\bar{x}, \bar{y})$, where

  $\bar{y} = 0$, $M = \rho \int_a^b \pi y^2 \, dx$ and $M \times \bar{x} = \rho \int_a^b x \times \pi y^2 \, dx$
- For an object to be in equilibrium under a system of coplanar forces
    - The sum of the components of all forces must be zero in any *two different* directions
    - The total moment of all forces and couples must be zero.
- When a plane shape is suspended from a point or can rotate about a horizontal axis, its centre of mass must be vertically below the point or the axis for it to be in equilibrium.
- An object on an inclined plane is in equilibrium without toppling if the line of action of its weight falls within the side of the object that makes contact with the plane.
- In general, a force **F** acting at point P has a moment **M** about point Q given by **M** = **r** × **F**, where **r** = **QP**

Check and review

You should now be able to...	Review Questions
✔ Find the centres of mass for systems of point masses, frameworks, laminas and solids of revolution.	1–4, 6
✔ Explore the stability of suspended objects and objects on inclined planes.	2, 4–7
✔ Use the conditions required for an object to be in equilibrium.	8–10
✔ Find forces and couples acting on a system	11–13

1 Masses of 2 kg, 4 kg, 6 kg and 8 kg are placed respectively at the vertices of a light, rigid rectangular framework *ABCD*, where $AB = 5$ m and $BC = 3$ m. Find the position of the centre of mass of the system from *AB* and *AD*

2 A uniform rectangular lamina *ABCD* with centre *O*, $AB = 12$ cm and $AD = 8$ cm has the rectangle *OMCN* cut away, where *M* and *N* are midpoints of *BC* and *CD*. Find the centre of mass, *G*, of the remaining piece and the angle that *AB* makes with the vertical when the piece is hung by a string from *A*

3 A uniform circular lamina, centre *O* and radius 12 cm, has a triangular hole *OMN* cut out of it, where *M* and *N* are midpoints of two radii with $\angle MON = 120°$. Find the position of the centre of mass of the remainder.

4 Find the centre of mass of a uniform lamina defined by the region between the curve $y^2 = x$ and the line $x = 4$. The lamina hangs freely by a thread from the point (4, 2). Find the angle between the thread and the *x*-axis.

5 A solid hemisphere of radius 24 cm is placed with its plane face on a rough slope inclined at an angle θ to the horizontal. It does not slide. Find the greatest value of θ for the hemisphere not to topple.

6 A graph is defined by

$y = x^2$ for $0 \le x \le 1$,
$y = 1$ for $1 \le x \le 3$

and $y = 0$ for all other values of *x*

The region between the graph and the *x*-axis is rotated through 360°

a Find the centre of mass of the solid generated.

b The solid is placed without sliding with its plane surface on a slope inclined at angle θ to the horizontal. Find the maximum value of θ for it not to topple.

7 A solid cone of height *h*, radius *r* and semi-vertical angle θ is cut to make a frustum by a plane parallel to its base at a distance of $\frac{1}{2}h$ from its vertex. The frustum is placed on its curved face on a horizontal plane. Prove that it will not topple if $\cos\theta \ge \sqrt{\frac{28}{45}}$

8 A uniform ladder of length 8 m and mass 24 kg rests on rough, level ground, where $\mu = 0.3$, against a smooth, vertical wall. The angle, θ, between the ladder and the ground is given by arctan 2. How far up the ladder can an object of mass 24 kg be placed without causing the ladder to slip?

9 A uniform rod *AB* of length 4 m and mass 10 kg is in equilibrium at 60° to the horizontal with its lower end *A* on a rough, level floor where the coefficient of friction, μ, is 0.5. The rod rests against a smooth peg *P* which moves along the length of *AB* so that $AP = 2 + x$ metres. Calculate the value of *x* for the rod to be on the point of moving.

10 A semicircular prism of radius *r* has its plane face fixed to a level floor. A uniform rod *PQ* of weight *W* and length 2*r* rests against the curved surface of the prism so that end *P* is on the floor and end *Q* extends beyond the point of contact with the prism. Both contacts are rough with the same coefficient of friction, μ. Prove that, if the rod is on the point of moving when inclined at an angle θ, $\sin^2\theta = \frac{\mu}{\mu^2 + 1}$

11 A uniform horizontal cantilever *AB* of mass 20 kg and length 10 m is in equilibrium rigidly fixed to a wall at *A*. End *B* is on a support with a vertical reaction of 30 N. Find the magnitude and direction of the force and the couple acting at *A*

12 A light, equilateral triangular lamina *OAB* lies in the first quadrant with *A*(5, 0) and *O* the origin. Forces of 6 N act along *OA*, 10 N along *AB* and 8N along *BO*. This system is equivalent to

a A single force *R*. Find the equation of its line of action.

b A force *S* acting at *O* and a couple *C*. Find the moment of the couple.

13 A light rod *OA* of length 6 m lies along the *x*-axis. Force $\mathbf{T} = a\mathbf{i} + b\mathbf{j}$ acts at the origin *O* at an angle of 30° with *OA*. Force $\mathbf{U} = -c\mathbf{j}$ acts at the point (4, 0). Force $\mathbf{V}$ of magnitude 4 N acts at *A* at an angle of 60° with *AO* This system is equivalent to a single couple $\mathbf{C}$. Find the values of *a*, *b* and *c* and the magnitude and sense of $\mathbf{C}$

25 Exploration

Going beyond the exams

Investigation

A castle drawbridge can be simplified and modelled as shown in the diagram.

Investigate how the tension in the rope varies as the drawbridge is raised from the horizontal.
Draw a graph that shows this.

Investigate how the tension in the rope varies as a person or vehicle crosses the drawbridge.
Again, draw a graph that shows this.

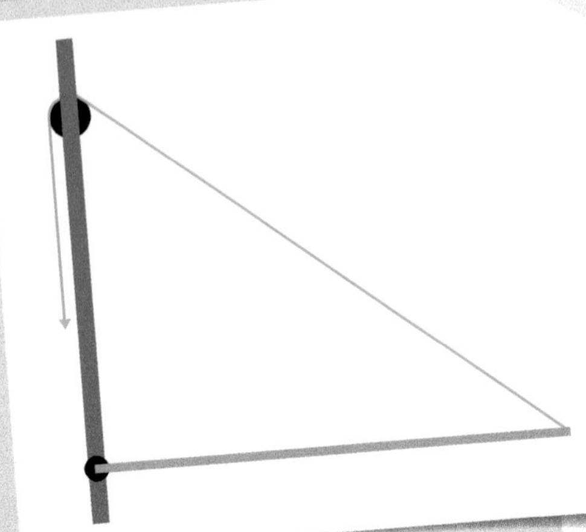

Applications

When designing buildings, an understanding of how forces act on, and within, frameworks is essential. Steel frames are often used when large buildings are built. For example, the Turning Torso skyscraper in Malmo, Sweden, has an external steel frame. It was designed by the famous Spanish architect Santiago Calatrava. The building is in nine segments made of irregular pentagons that twist, so that the top floor is rotated by 90° with respect to the ground floor. Calatrava has designed many iconic buildings around the world. Investigate some of these, and see his use of steel frames, which often add beauty to buildings as well as being an integral part of the structure.

Applications

Understanding the concept of centre of mass is important in field events in athletics, as well as in gymnastics. For example, the American high jumper Dick Fosbury developed the 'Fosbury flop'. In this technique, the athlete jumps over the bar with an arched back, passing as close as possible to the bar. Using this technique, the jumper's centre of mass passes under the bar by up to 20 centimetres.

Find out more about the technique and consider the advantages of a jump where the athlete's centre of mass is under the bar.

In what other sports is the centre of mass of a person important?

25 Assessment

1 Particles of mass 1 kg, 2 kg, 3 kg and 4 kg are attached in that order to a rod AB of length 1.5 m at distances of 0.3 m, 0.6 m, 0.9 m and 1.2 m from A

a Assuming that the rod is of negligible mass, find the distance from A of the centre of mass of the system. **[3]**

b In fact, the rod is uniform and of mass m kg. The centre of mass of the system is 0.85 m from A Find the value of m **[3]**

2 ABC is a triangle formed of three uniform rods. AB has length 4 m and mass 4 kg, AC has length 3 m and mass 2 kg and BC has length 5 m and mass 4 kg. Find the distance of the centre of mass of the triangle from

a AB **[3]**

b AC **[3]**

3 Masses of 2 kg, 4 kg, 6 kg and 9 kg are placed, respectively, at the vertices A, B, C and D of a light rectangular framework $ABCD$, where $AB = 5$ cm and $BC = 3$ m. Find the angle that AB makes with the vertical when the framework is suspended from A **[6]**

4 A uniform lamina consists of the region enclosed by $y = \sqrt{x}$, the x-axis and the lines $x = 1$ and $x = 4$

a Find the coordinates of the centre of mass of the lamina. **[8]**

b Find the coordinates of the uniform solid formed by rotating the region about the x-axis. **[5]**

5 A bridge on a model railway is a prism with a symmetrical cross-section in the form of a rectangle with a semicircular arch cut out of it. The whole structure is 28 cm high and 36 cm wide, and the radius of the arch is 10 cm.
Assuming that the bridge is made of a uniformly dense material, find the height of its centre of mass above the base. **[5]**

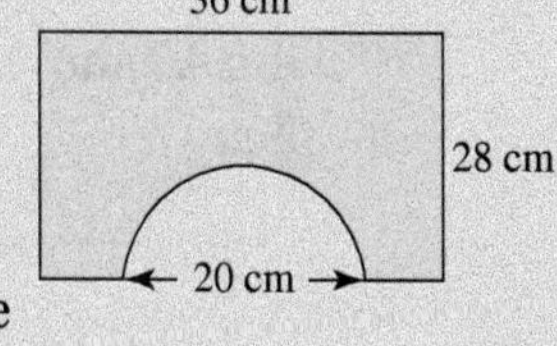

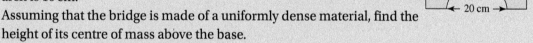

6 A uniform lamina is in the form of the minor segment of a circle of radius 60 cm cut off by a chord AB of length 60 cm. Find the distance of the centre of mass of the segment from the centre of the circle. **[8]**

7 Find the centre of mass of this uniform lamina relative to the origin O and the axes shown. **[6]**

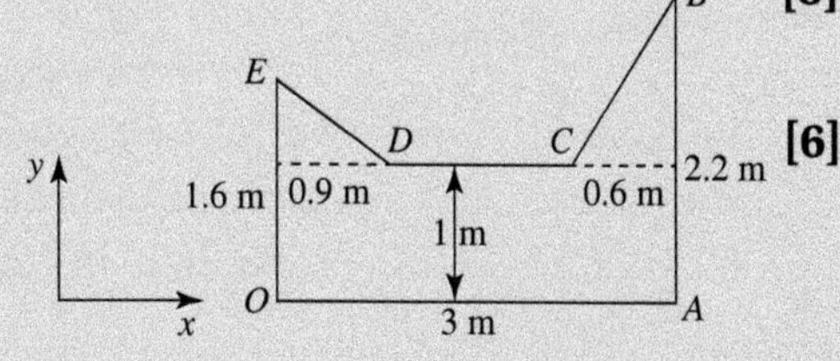

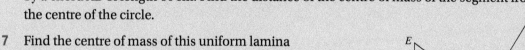

8 A uniform ladder AB of mass 20 kg and length 3 m rests with B against a smooth wall and A on smooth horizontal ground. A is attached to the base of the wall by a light inextensible string, 1 m long.

a Find the tension in the string. **[4]**

b If the breaking strain of the string is 250 N, find how far up the ladder a man of mass 80 kg can safely ascend. **[5]**

9 A uniform wire AB is bent to form the arc of a semicircle. The wire is then suspended freely from A. Find the angle between the diameter AB and the vertical. **[4]**

10 Masses of 5 kg, 4 kg, 2 kg and 3 kg are placed at A, B, C and D, respectively, on a light rectangular framework $ABCD$, where $AB = 4$ m and $BC = 3$ m. A mass m kg is placed at a point E on CD so that the centre of mass of the system is the centre of the rectangle. Find m and the position of E **[7]**

11 The diagram shows a carpenter's square, with a rectangular handle $PQUV$, of length 0.2 m and width 0.1 m, and a rectangular blade $RSTU$ of length 0.3 m and width 0.1 m. The blade is made from metal of density 1 kg m^{-2}, and the handle from material of density m kg m^{-2}.

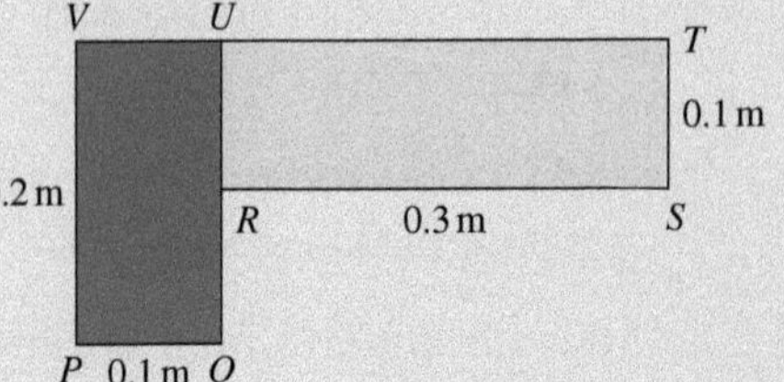

a Find the value of m if the centre of mass of the object lies on the line QU **[3]**

b For this value of m, find the distance of the centre of mass from PQ **[2]**

12 The triangular lamina with vertices at $(0, 0)$, $(h, 0)$ and (h, r) is rotated about the x-axis. Show, by integration, that the centre of mass of the resulting uniform cone is at $(0.75h, 0)$ **[6]**

13 Particles of mass 3 kg, 5 kg, m_1 kg and m_2 kg are placed, respectively, at points $(1, 4)$, $(4, 1)$, $(2, 1)$ and $(4, 2)$. Find the values of m_1 and m_2 if the centre of mass of the system is $(3, 2)$ **[8]**

14 $ABCD$ is a uniform rectangular lamina with $AB = 60$ cm and $BC = 30$ cm. E is the midpoint of CD The triangle BCE is removed from the lamina, and the remainder is suspended from E. Find the angle that the direction of AD makes with the vertical when the lamina hangs in equilibrium. **[7]**

15 A uniform rod of length 3 m is bent to form triangle ABC, with angle $BAC = 90°$, $AB = 1.2$ m and $AC = 0.5$ m.

a Taking AB and AC as the x- and y-axes, find the coordinates of the centre of mass of the triangle. **[4]**

b The triangle is freely suspended from B so that it hangs in equilibrium. Find the angle between side AB and the vertical. **[3]**

16 A uniform lamina consists of the region enclosed by the curve $y = e^x$, the axes and the line $x = 2$

a Find the exact coordinates of the centre of mass of the lamina. **[8]**

b Find the exact centre of mass of the uniform solid produced by rotating the region about the x-axis **[5]**

17 The diagram shows a rod AB of length 2 m and mass 4 kg. The rod is freely hinged at A to a point on a vertical wall and makes an angle of 50° with the wall. The rod is held in equilibrium by a horizontal force of T N applied at B

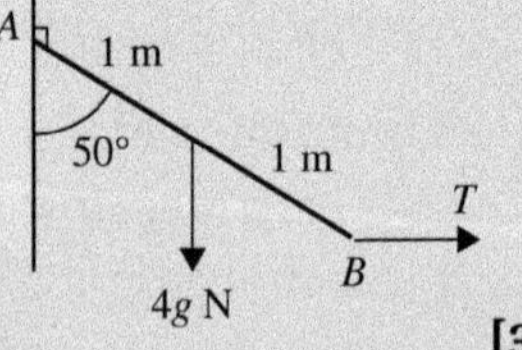

a Calculate the value of T. **[3]**

b Find the magnitude and direction of the reaction at the hinge. **[7]**

18 A uniform ladder AB has length 8 m and mass 20 kg. It leans against a rough wall with end A on rough, horizontal ground. The coefficient of friction at both contacts is 0.3 and the ladder makes an angle of 65° with the horizontal. A woman of mass 50 kg starts to climb the ladder.

a How far can she ascend up the ladder before the ladder is on the point of slipping? **[7]**

b Her daughter, of mass m kg, then stands on the ladder at A. This enables the woman to climb right to the top of the ladder. Find the least value of m **[6]**

19 A uniform rod, AB, of length 1.6 m and mass 6 kg, is freely hinged at A to a vertical wall. The rod rests in a horizontal position, supported by a light, inextensible string BC attached to the wall at the point C, which is 1.2 m vertically above A. The string has a breaking strain of 40 N, and equilibrium is maintained by a couple of moment P N m applied at A. Take $g = 9.8\,\text{m s}^{-2}$.

a Find the greatest and least possible values of P. **[6]**

b Find the magnitude and direction of the reaction force at the hinge in each case. **[8]**

20 A uniform right circular cone has height 1 m, base radius 0.6 m and density $1\,\text{kg m}^{-3}$. A uniform hemisphere has radius 0.6 m and density $\rho\,\text{kg m}^{-3}$. The two are joined together at their circular plane faces to form a symmetrical solid.

a Find the position of the centre of mass of the solid relative to O, the centre of the common plane face. Give your answer in terms of ρ **[7]**

b The solid is placed on a horizontal plane surface. The curved surface of the hemisphere is in contact with the horizontal plane and the solid remains in equilibrium whatever the angle between its axis and the vertical. Find the value of ρ **[4]**

21 The diagram shows a lamina in the form of a sector of a circle, centre O, of radius 6 cm. The angle of the sector is 60°. A uniform rod is bent so that it attaches exactly to the arc of the sector, as shown. The mass of the lamina is twice the mass of the rod. Show that the centre of mass of the combined object is $\frac{14}{\pi}$ cm from O **[7]**

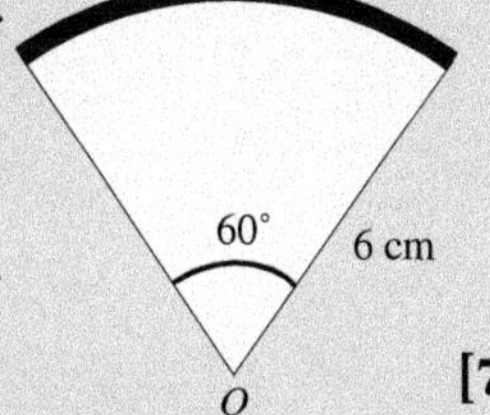

22 The diagram shows a sign outside a shop. It is made of a uniform lamina $ABCD$ in the shape of a square attached to a triangle. The mass of the lamina is 6 kg. The sign is mounted in a vertical plane with AD horizontal. It is smoothly hinged to the wall at A and is held in equilibrium by a horizontal string BE attached to the wall at E

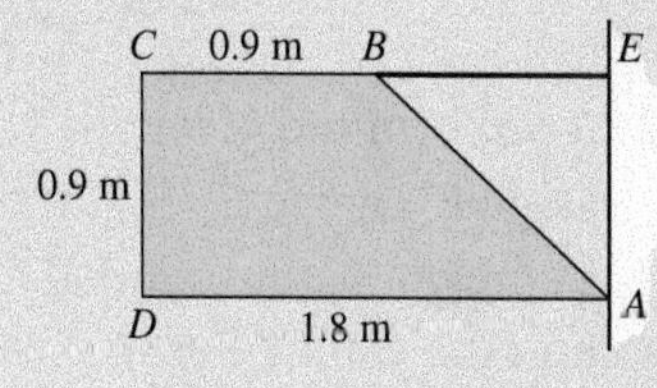

a Calculate the tension in the string. **[5]**

b Find the magnitude and direction of the reaction at the hinge. **[5]**

23 The diagram shows a lamina comprising a rectangle $ABDE$ of length 1.2 m and width 0.8 m together with an isosceles triangle BCD of height 0.6 m. F is the midpoint of AE. H is a point on CF, with $FH = 0.4$ m. A hole, centre H and radius r cm, is cut in the lamina. The centre of mass of the object is at G

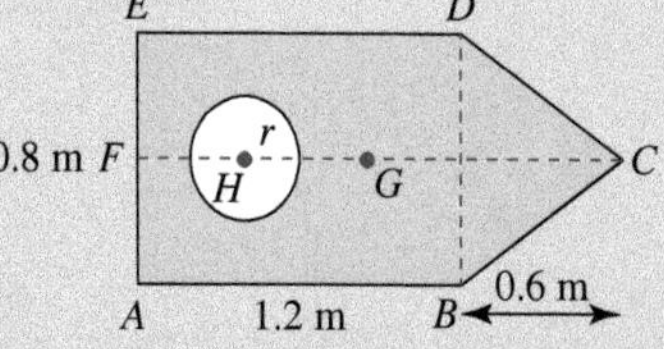

a Explain why G lies on CF **[1]**

b If $r = 0.2$, find the distance FG **[4]**

c If $FG = 1$ m, find the value of r **[3]**

d Explain why FG cannot be 1.1 m. **[2]**

24 A uniform ladder AB of length $2a$ and weight W rests with B against a smooth, vertical wall and A on rough, horizontal ground. The ladder makes an angle of 60° with the horizontal. A child of weight W can just climb to the top of the ladder without causing it to slip. Find how far up the ladder a man of weight $4W$ could safely climb. **[7]**

25 The diagram shows the cross-section of a prism consisting of a rectangle $ABCD$ of length 50 cm and width 30 cm from which a triangle BCE has been removed so that DE has length x cm. Find the minimum value of x for which the prism will stand on a horizontal surface with DE in contact with the surface. **[8]**

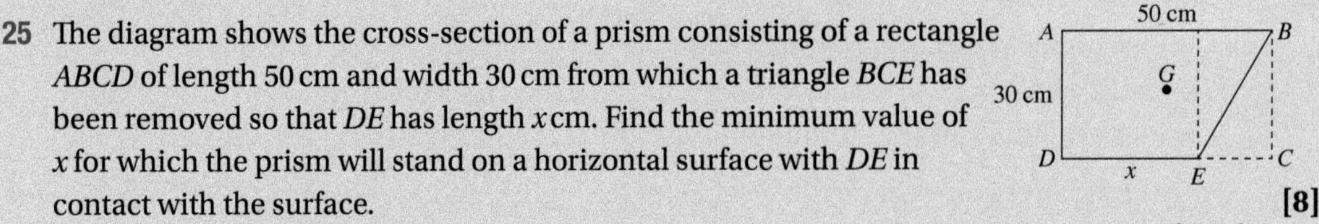

26 A uniform lamina comprises the region in the first quadrant enclosed by the ellipse $\frac{x^2}{a^2}+\frac{y^2}{b^2}=1$
The area of the ellipse is πab

a Find the coordinates of the centre of mass of the lamina. **[8]**

b Find the centre of mass of the uniform solid formed by rotating the region about the x-axis. **[5]**

27 The cross-section of a uniform prism is an isosceles trapezium of height 6 cm with parallel sides of length 4 cm and 10 cm. The prism is placed on a rough inclined plane, as shown. If the prism does not slip, but is on the point of toppling, find the angle of the sloping plane. **[7]**

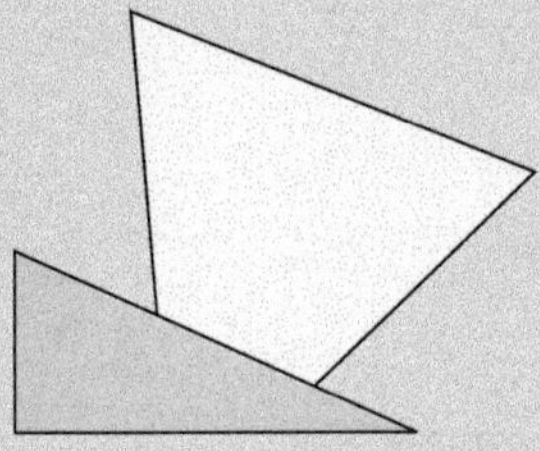

28 A uniform rod of length 1 m and mass 1 kg is bent to form the framework shown, where angle B = angle C = 90°. A particle of mass m kg is attached to the framework at D and the framework is suspended from A. If the point C hangs vertically below A, find the value of m **[9]**

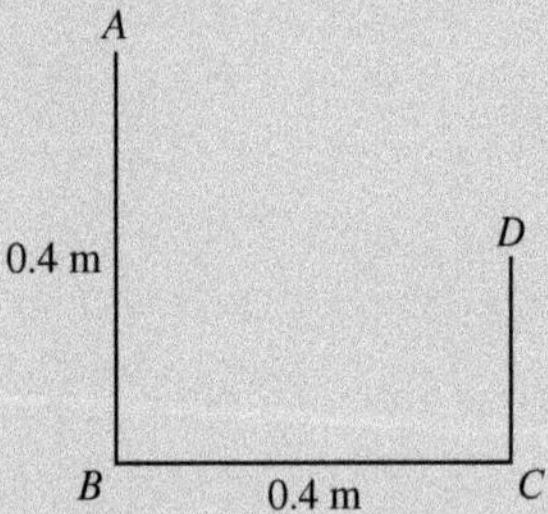

29 Two forces $\mathbf{F}_1 = 3\mathbf{i} + 2\mathbf{j} - \mathbf{k}$ N and $\mathbf{F}_2 = 2\mathbf{i} - \mathbf{j} - 2\mathbf{k}$ N act at points A(1, 0, 3) and B(0, 2, −3) respectively. Their resultant forms a couple $\mathbf{M}$ with a third force $\mathbf{F}_3$ which acts at the origin O. Find the force $\mathbf{F}_3$, the couple $\mathbf{M}$ and their magnitudes. **[4]**

26 Random processes

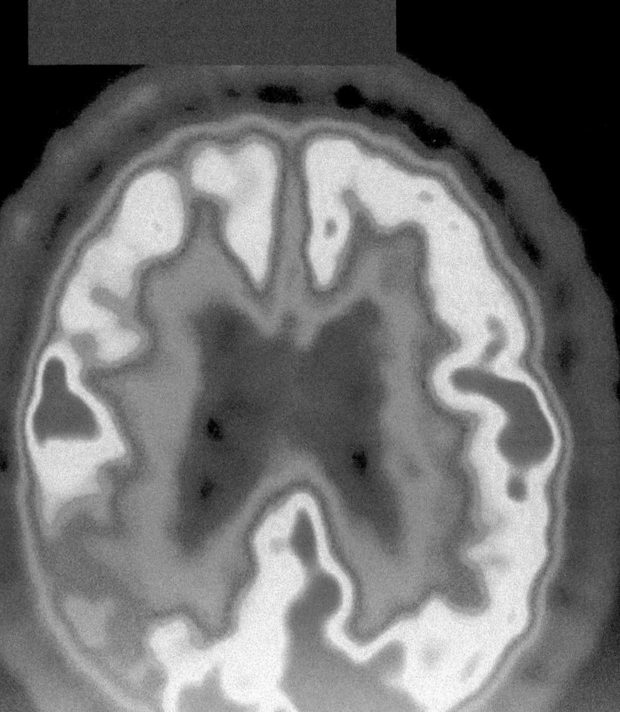

Radiation can help medical professionals investigate patients' bodies without resorting to invasive surgery. A radioactive tracer is ingested by a patient, and a radiographer detects the beta particles and gamma rays emitted by the patient's body. Due to the potentially harmful nature of the radiation, the amount used must be measured carefully to ensure that it will decay to a certain level after a short period of time. An exponential distribution provides a good model for radioactive disintegrations, and allows medical professionals to estimate the amount of radiation at any point in the process.

Continuous probability distributions are used to model a wide variety of continuous random variables, from waiting times on a public transport network to the frequency of flaws in a length of fibre-optic cable.

Orientation

What you need to know

Maths Ch15
- Differentiating exponential functions.

Maths Ch16
- Integrating exponential functions.

Ch10 Discrete and continuous random variables

What you will learn

- To calculate the cumulative distribution function and probability density function for a continuous random variable.
- To calculate the mean and variance of mixed random variables.
- To use rectangular and exponential probability models.

What this leads to

Careers
- Medicine.
- Quality assurance.
- Insurance.

26.1 Continuous distributions 2

Fluency and skills

See Ch10.3 For a reminder of continuous distributions.

If X is a continuous random variable with probability density function $f(x)$, then

$$P(a < X < b) = \int_a^b f(x)dx$$

The mean and variance of X are $\mu = E(X) = \int_{-\infty}^{+\infty} x f(x)dx$ and $\sigma^2 = Var[X] = E[X^2] - \mu^2 = \int_{-\infty}^{\infty} x^2 f(x)dx - \mu^2$

The cumulative distribution function, $F(x)$, of a random variable is useful when calculating probabilities as well as values of the median and other quartiles and percentiles. It is defined by the equation $F(x) = P(X < x)$

But, for a continuous random variable, $P(a < X < b) = \int_{x=a}^{x=b} f(x)dx$

Therefore

Key point

The cumulative distribution function of a continuous random variable with probability density function $f(t)$ is given by $F(x) = P(X < x) = \int_{-\infty}^{x} f(t)dt$

Example 1

A continuous random variable X has probability density function given by

$$f(x) = \begin{cases} 4x^a; & 0 < x < 1 \\ 0; & \text{otherwise} \end{cases}$$

Find:

a The value of a

b The cumulative distribution function $F(x)$

c The median value, M, of X

a $\int_0^1 4x^a\,dx = 1$ — Total probability equals 1

$\left[\frac{4x^{a+1}}{a+1}\right]_0^1 = 1 \rightarrow \frac{4}{a+1} = 1$ Therefore $a = 3$

b $F(x) = \int_0^x 4x^3\,dx = \left[x^4\right]_0^x = x^4$

c $\int_0^M 4x^3 dx = \frac{1}{2}$ — Definition of the median.

$M^4 = \frac{1}{2} \rightarrow M = 0.841$ (3dp)

Using the definition of integration as the inverse of differentiation, the probability density function of X can be found from the cumulative distribution function.

Key point

If $F(x)$ and $f(x)$ are the cumulative distribution function and the probability density function of X, respectively, then

$$f(x) = \frac{d}{dx}F(x)$$

Example 2

The cumulative distribution function of a random variable X is given by $F(x) = 1 - e^{-\lambda x}$

a Find the probability density function of X

b Write the modal value of X

c Prove that, when two X values are chosen at random, the probability that one is less than x_1 and the other is more than x_1 is given by $2e^{-\lambda x_1}(1 - e^{-\lambda x_1})$

a $f(x) = \frac{d}{dx}F(x) = \frac{d}{dx}(1 - e^{-\lambda x}) = \lambda e^{-\lambda x}$

Use the chain rule for differentiation.

b Mode $= 0$

The maximum value of $f(x)$ for $x \geq 0$ occurs when $x = 0$
Visualise the graph.

c $P(X < x_1) = F(x_1) = 1 - e^{-\lambda x_1}$. Therefore $P(X > x_1) = e^{-\lambda x_1}$

P(one x-value is less than x_1 and the other is more than x_1)
$= 2e^{-\lambda x_1}(1 - e^{-\lambda x_1})$

The outcome can occur in two ways.

Example 3

A continuous random variable X has a probability density function given by

$$f(x) = \begin{cases} ax^2; & 0 < x \leq 1 \\ \frac{-9x+15}{4}; & 1 < x < \frac{5}{3} \\ 0; & \text{otherwise} \end{cases}$$

a Show that $a = \frac{3}{2}$

b Find

i The cumulative distribution function $F(x)$

ii The median and the interquartile range of X

a $\int_1^{\frac{5}{3}} \frac{-9x+15}{4}dx = \left[\frac{1}{4}\left(\frac{-9x^2}{2} + 15x\right)\right]_1^{\frac{5}{3}} = \frac{1}{2}$

$\int_0^1 ax^2 dx = \left[\frac{ax^3}{3}\right]_0^1 = \frac{a}{3} = 1 - \frac{1}{2} = \frac{1}{2} \rightarrow a = \frac{3}{2}$

(*Continued on the next page*)

STATS

b i $F(x)=\begin{cases} \int_0^x \frac{3x^2}{2}dx; & 0<x\le 1 \\ \frac{1}{2}+\int_1^x \frac{-9x+15}{4}dx; & 1<x\le \frac{5}{3} \\ 1; & x>\frac{5}{3} \end{cases}$

Therefore

$$F(x)=\begin{cases} \frac{x^3}{2}; & 0<x\le 1 \\ \frac{-9x^2}{8}+\frac{15x}{4}-\frac{17}{8}; & 1<x\le \frac{5}{3} \\ 1; & x>\frac{5}{3} \end{cases}$$

ii Median = 1

If Q_1 and Q_3 are the first and third quartile, respectively, then

$F(Q_1)=\frac{1}{4}$ and $F(Q_3)=\frac{3}{4} \to$

$\frac{Q_1^3}{2}=\frac{1}{4} \to Q_1=\frac{1}{\sqrt[3]{2}}=0.79$ (2dp)

and

$\frac{-9Q_3^2}{8}+\frac{15Q_3}{4}-\frac{17}{8}=\frac{3}{4} \to Q_3=1.20$ (2dp)

IQR = 0.41

The rectangular distribution models many real-life situations where, within the domain of the function, all small intervals of width Δx are equally likely to occur when an experiment is performed. It is therefore characterised by having a probability density function equal to a constant over its entire domain. The value of the constant is chosen so that the total area under the function is equal to 1

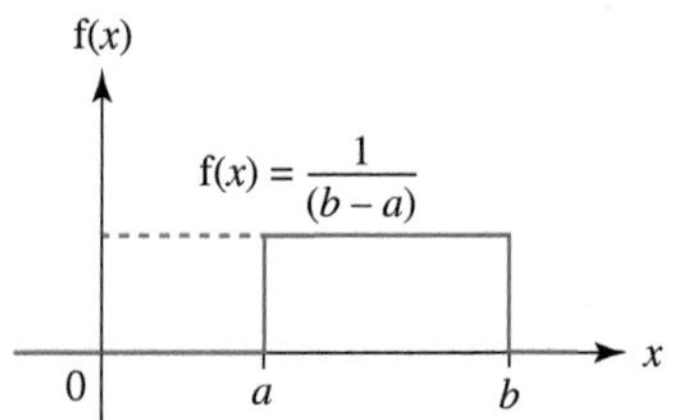

Key point

If R is a random variable with a rectangular probability distribution defined on the interval (a, b) (that is, $R \sim \text{Rectangular}(a,b)$), then its probability density function is given by $f(r)=\frac{1}{b-a}, a<r<b$

This probability density function can be used to find the mean and variance of the distribution.

$$E(R)=\int_{-\infty}^{+\infty} r f(r)dr=\int_a^b r\frac{1}{b-a}dr$$

$$=\frac{1}{b-a}\times\frac{b^2-a^2}{2}=\frac{a+b}{2}$$

$$\text{Var}(R) = \int_{-\infty}^{+\infty} r^2 \text{f}(r)\text{d}r - \{\text{E}(R)\}^2$$

$$= \int_{r=a}^{b} r^2 \frac{1}{b-a}\text{d}r - \left(\frac{a+b}{2}\right)^2$$

$$= \frac{1}{b-a} \times \frac{b^3-a^3}{3} - \left(\frac{a+b}{2}\right)^2$$

$$= \frac{1}{b-a} \times \frac{(b-a)(b^2+ab+a^2)}{3} - \left(\frac{a+b}{2}\right)^2$$

$$= \frac{(b^2+ab+a^2)}{3} - \left(\frac{a+b}{2}\right)^2$$

$$= \frac{4b^2+4ab+4a^2-3a^2-6ab-3b^2}{12}$$

$$= \frac{(b-a)^2}{12}$$

$$= \frac{(b-a)^2}{12}$$

Key point

If R has a rectangular probability distribution then the mean and variance of R are given by $\text{E}(X) = \frac{a+b}{2}$ and $\text{Var}(X) = \frac{(b-a)^2}{12}$

The rectangular distribution can be used to model 'round off' values as in Example 4

STATS

Exercise 26.1A Fluency and skills

1 A random variable, X, has a probability density function given by

$$\text{f}(x) = \begin{cases} \frac{1}{4}x; & 1 < x < k \\ 0; & \text{otherwise} \end{cases}$$

where k is a positive constant.

a Prove that $k = 3$

b Show that the cumulative distribution function is given by $\frac{1}{8}(x^2 - 1)$

c Find the median value of X

2 A random variable, X, has probability density function $\text{f}(x) = \begin{cases} ax^2(4x-1); & 0 < x < \frac{1}{4} \\ 0; & \text{otherwise} \end{cases}$

where a is a negative constant.

a Find the value of a and the cumulative distribution function $\text{F}(x)$

b Use your answer in part **a** to show that the median of X is approximately 0.15

3 A random variable, X, has probability density function $\text{f}(x) = \begin{cases} -kx(x-3); & 0 < x < 3 \\ 0; & \text{otherwise} \end{cases}$

where k is some positive constant.

a Find the value of k

b Find the cumulative distribution function of X and hence find $P(X > 2)$

4 a Sketch the probability density function of X, given by

$$\text{f}(x) = \begin{cases} 1-|x|; & -1 < x < 1 \\ 0; & \text{otherwise} \end{cases}$$

b Show that $P(S)=1$, where S is a sample space for X

c By finding the cumulative probability distribution $F(x)$, prove that the exact value of the upper quartile is given by $Q_3=\frac{\sqrt{2}-1}{\sqrt{2}}$

5 The random variable X has a cumulative probability distribution given by

$F(x)=\frac{1}{k}(x^2+x)$ for $0<x<5$

a Find the value of k

b Show that the lower quartile is 2.28 (2 dp) and find the upper quartile.

c Show that the probability density function of X is $f(x)=\frac{2x+1}{30}$ and hence write down the modal value of X

6 A random variable, X, has probability density function $f(x)=\begin{cases}\frac{1}{k}; & a<x<b\\ 0; & \text{otherwise}\end{cases}$

a Sketch the graph of the probability density function of X and show that $k=b-a$

b Find $E(X)$ and $Var(X)$ when $a=3$ and $b=5$

c Show that the cumulative probability distribution function of X is given by $F(x)=\frac{x-3}{2}$ and hence find the probability that $P(X<3.5)$

7 A random variable, X, has probability density function given by $f(x)=\begin{cases}\frac{1}{4}; & -1<x<a\\ 0; & \text{otherwise}\end{cases}$

a Find the value of a

b Sketch the probability density function of X and shade the area representing $P(X>1.5)$

c Use your diagram to write down the value of $P(X>1.5)$

d Find the cumulative distribution function of X and use it to show that the median value of X is 1

8 A continuous random variable X has probability density function given by

$$f(x)=\begin{cases}x; & 0<x\le 1\\ 2-x; & 1<x<2\\ 0; & \text{otherwise}\end{cases}$$

a Show that $f(x)$ is a valid probability density function.

b Find

i The cumulative probability distribution function $F(x)$

ii The median and the interquartile range of X

Reasoning and problem-solving

Strategy

To solve a problem using continuous distributions

1. Integrate the probability density function to find probabilities and the cumulative probability distribution function.
2. Recognise continuous variables that follow a rectangular distribution and identify values for the parameters *a* and *b*
3. Calculate the mean and variance of rectangularly distributed random variables.
4. Recognise random variables that are partly discrete and partly continuous and use appropriate formulae for mean and variance.

These ideas can be applied to the continuous rectangular distribution and to mixed distributions that are partly discrete and partly continuous.

Example 4

The length of a line segment is measured to the nearest millimetre. Calculate the mean and variance of the error in any recorded length.

X – a random variable for the error (mm) in any recorded length

$X \sim \text{Rectangular}(-0.5, 0.5)$

Defining the error as 'true' – recorded value.

Therefore mean is $\frac{0.5+(-0.5)}{2} = 0$ and variance is $\frac{(0.5-(-0.5))^2}{12} = \frac{1}{12}$

Example 5

A circle of radius R is drawn, where R is a rectangularly distributed random variable defined on the interval 1 to 5

The random variable A is the area of the circle. Find the cumulative distribution function of A and its mean.

If f(r) is the probability density function of R, then $f(r) = \frac{1}{4}, 1 < x < 5$

and $F(r) = \frac{r-1}{4}$

$$A = \pi R^2 \rightarrow R = \sqrt{\frac{A}{\pi}}$$

Therefore $\Phi(a) = P(A < a) = P(\pi R^2 < a) = P\left(R < \sqrt{\frac{a}{\pi}}\right) = \frac{\sqrt{\frac{a}{\pi}} - 1}{4}$

$$= \frac{1}{4}\left(\sqrt{\frac{a}{\pi}} - 1\right)$$

$$\varphi(a) = \frac{d}{da}\Phi(a) = \frac{d}{da}\frac{1}{4}\left(\sqrt{\frac{a}{\pi}} - 1\right) = \frac{1}{8\sqrt{a\pi}}$$

$\Phi(a)$ and $\varphi(a)$ are the cumulative distribution function and probability density function of A, respectively.

$$E(A) = \int_{\pi}^{25\pi} a\frac{1}{8\sqrt{a\pi}}\,da = \frac{1}{8}\int_{\pi}^{25\pi}\sqrt{\frac{a}{\pi}}\,da = \frac{1}{8\sqrt{\pi}}\int_{\pi}^{25\pi} a^{\frac{1}{2}}\,da = \left[\frac{1}{12\sqrt{\pi}}a^{\frac{3}{2}}\right]_{\pi}^{25\pi}$$

$$= \frac{\pi}{12}\left(25^{\frac{3}{2}} - 1\right) = \frac{31}{3}\pi$$

Some random variables are partly discrete and partly continuous. For example, queuing times may take the value zero exactly (discrete) if there is no one else queuing, or they may be a set of positive values (continuous).

Example 6

Supermarket checkout waiting times, T minutes, are found to have the following distribution: $P(T=0)=c$; $f(t)=kt(t-3)$, $0<t\le 3$

a Prove that $k=\dfrac{2(c-1)}{9}$

b If the probability of zero waiting time is 0.4, show that the expected waiting time is 54 seconds and find the probability that the waiting time exceeds one minute.

a $k\int_0^3 t(t-3)dt = 1-c$

Total probability in a sample space = 1

$$\left[k\left(\frac{t^3}{3}-\frac{3t^2}{2}\right)\right]_0^3 = 1-c \rightarrow k\left(9-\frac{27}{2}\right)=1-c \rightarrow k=\frac{2(c-1)}{9}$$

b $c=0.4 \rightarrow k=\dfrac{2(0.4-1)}{9}=-\dfrac{2}{15}$

$$E(T)=0\times 0.4-\frac{2}{15}\int_0^3 t^2(t-3)dt=\left[\frac{2}{15}\left(t^3-\frac{t^4}{4}\right)\right]_0^3=\frac{9}{10}=54\text{s}$$

$$P(T>1)=-\frac{2}{15}\int_1^3 t(t-3)dt=\left[\frac{2}{15}\left(\frac{3t^2}{2}-\frac{t^3}{3}\right)\right]_1^3=\frac{4}{9}$$

Exercise 26.1B Reasoning and problem-solving

1 Buses arrive at a bus stop on the hour and the half hour throughout the day. Isaac has no idea when the buses are due and arrives at the stop to catch a bus.

   a If T is a random variable for the number of minutes since the last bus, use an appropriate distribution to calculate the mean and variance of T

   b Using these results, find the mean and variance of the variable, S, the number of minutes Isaac has to wait for the next bus.

2 The random variable Y is part discrete, part continuous. The probability density function of Y is given by

$$f(y)=\begin{cases} c; & y=0 \\ k(y-1)(y-3); & 1<y<3 \\ 0; & \text{otherwise}\end{cases}$$

   a Find k in terms of c

   b Given that $c=0.1$, find the expected value of $\sqrt{Y}$

3 A random variable X takes the exact value 1 or any value in the interval $2\le x\le 4$
Its distribution is given by $P(X=1)=0.2$, $P(X<x)=k(-x^2+8x-11)$ for $2\le x\le 4$

   a Prove that $k=0.2$

   b Find the probability density function, $f(x)$, of X in the interval $2\le x\le 4$

c Find

i The expected value of X ii $P(X<3)$

4 A continuous random variable X, defined on the domain $x > 0$, has cumulative probability distribution function given by

$$F(x)=\begin{cases} ax; & 0<x\leq 2 \\ -2a+3ax-\frac{1}{2}ax^2; & 2<x\leq 3 \\ 1; & x>3 \end{cases}$$

a By calculating in terms of a an expression for F(3), find the value of a

b Find the value of F(2) and hence explain why the median, M, of X satisfies $M < 2$

c Find the median of X

d By differentiation, find the density function, f(x)

5 A continuous random variable X, defined on the domain $x > 0$, has cumulative probability distribution function given by

$$F(x)=\begin{cases} \frac{x^2}{k}; & 0<x\leq 3 \\ -3+2x-\frac{1}{4}x^2; & 3<x\leq 4 \\ 1; & x>4 \end{cases}$$

a Find the value of k

b By differentiation, find the density function, f(x) and hence sketch the density function.

c Find the median and show that $P(X > 3.5) = \frac{1}{16}$

STATS

26.2 Exponential distribution

Fluency and skills

The exponential distribution is used to model continuous random variables such as the waiting times between random events. These can include radioactive disintegrations, the failure of electronic components or even the lengths of certain telephone calls.

Key point

If X is a random variable with an exponential probability distribution with parameter λ, then its probability density function is given by $f(x) = \lambda e^{-\lambda x}, x > 0$

The probability density function has a positive skew and is defined for positive values of x

Its graph is shown for several values of λ

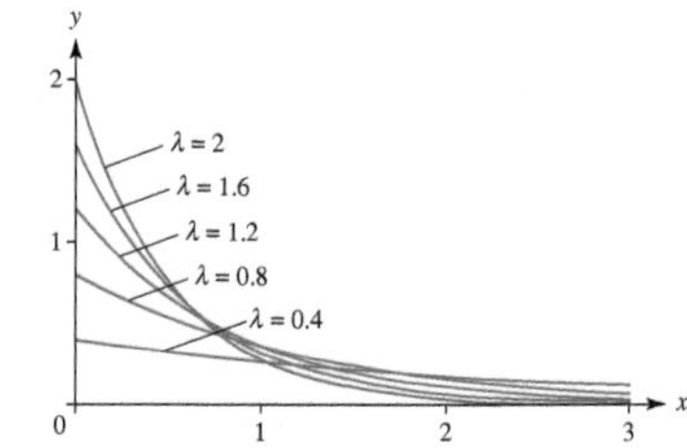

The exponential cumulative distribution function, F(x), can be found from the probability density function by integration:

$$F(x) = P(X < x) = \int_0^x f(t)\,dt$$

$$= \int_0^a \lambda e^{-\lambda t}\,dt = \left[\lambda \frac{e^{-\lambda t}}{-\lambda}\right]_0^x$$

$$= 1 - e^{-\lambda x}$$

Key point

If X is a random variable with an exponential probability distribution, then its cumulative probability distribution function is given by $F(x) = 1 - e^{-\lambda x}, x > 0$

Example 1

A random variable X has an exponential distribution with parameter $\lambda = \frac{1}{4}$

a Write down the cumulative distribution function and then use it to find

i $P(X < 1)$ **ii** $P(X > 3)$ **iii** $P(1 < X < 3)$

Use a calculator to check your answers.

b If three x-values are chosen at random, find the probability that one will be less than 1 and the other two will be greater than 1

a $F(x) = 1 - e^{(-\lambda x)}$

i $P(X < 1) = 1 - e^{-0.25} = 0.221$

ii $P(X > 3) = 1 - P(X < 3) = 1 - (1 - e^{-0.75}) = 0.472$

iii $P(1 < X < 3) = 1 - (0.221 + 0.472) = 0.31$ (2 dp)

b Y – number less than 1 out of 3 values. $Y \sim B(3, 0.221)$

$P(Y = 1) = 0.40$

The probability density function is used to calculate the mean value and variance of the distribution.

$$E(X) = \int_0^\infty x f(x) dx = \lambda \int_0^\infty x e^{-\lambda x} dx$$

$$= \lambda \left\{ \left[x \frac{e^{-\lambda x}}{-\lambda} \right]_0^\infty - \int_0^\infty \frac{e^{-\lambda x}}{-\lambda} dx \right\}$$

Using integration by parts.

$$= \lambda \left\{ 0 - \left[\frac{e^{-\lambda x}}{\lambda^2} \right]_0^\infty \right\} = \lambda \left(\frac{1}{\lambda^2} \right)$$

$$= \frac{1}{\lambda}$$

$$E(X^2) = \int_0^\infty x^2 f(x) dx = \lambda \int_0^\infty x^2 e^{-\lambda x} dx$$

$$= \lambda \left\{ \left[x^2 \frac{e^{-\lambda x}}{-\lambda} \right]_0^\infty + \int_0^\infty \frac{e^{-\lambda x}}{\lambda} 2x dx \right\}$$

Using integration by parts.

$$= \lambda \left\{ 0 - 2 \left(\left[x \frac{e^{-\lambda x}}{\lambda^2} \right]_0^\infty - \int_0^\infty \frac{e^{-\lambda x}}{\lambda^2} dx \right) \right\}$$

$$= \lambda \left\{ 0 - 2 \left(0 + \left[\frac{e^{-\lambda x}}{\lambda^3} \right]_0^\infty \right) \right\} = \lambda \frac{2}{\lambda^3} = \frac{2}{\lambda^2}$$

Using integration by parts.

$$\text{Var}(X) = E(X^2) - \{E(X)\}^2 = \frac{2}{\lambda^2} - \left(\frac{1}{\lambda} \right)^2 = \frac{1}{\lambda^2}$$

Key point

If X has a probability density function given by $f(x) = \lambda e^{-\lambda x}, x > 0$, then the mean and variance of X are given by $E(X) = \frac{1}{\lambda}$ and $\text{Var}(X) = \frac{1}{\lambda^2}$

The exponential distribution can be used in calculations on the waiting times between random events.

These events form a **Poisson process** with mean μ

$X \sim \text{Poisson}(\mu)$

Suppose that random events occur at a mean rate of μ per unit time. Let X be a random variable for the number of occurrences in a random unit interval and let T be the time between successive events.

$P(T < t) = 1 - P(T > t) = 1 - P(\text{no occurrences in interval } t) = 1 - e^{-\mu t}$, which is the cumulative distribution function of an exponential random variable, parameter μ and mean $\frac{1}{\mu}$

If the mean number of occurrences in unit time is Poisson (μ), the mean number in time t is Poisson (μt)

Key point

If T is a random variable for the time between successive random events that occur at a mean rate μ per unit time, then T follows an exponential distribution, parameter μ, with mean $\frac{1}{\mu}$ and variance $\frac{1}{\mu^2}$

Remember that μ is the mean of the underlying Poisson process, not the waiting time mean.

STATS

Example 2

a Write down the mean and variance of a random variable, X, that has an exponential distribution with $\lambda = 2$

b Find the probability that a randomly chosen value of X will be within 1 standard deviation of the mean.

a $X \sim \text{exponential}(\lambda = 2) \rightarrow \text{mean } \frac{1}{2}, \text{ variance } \frac{1}{2^2} = \frac{1}{4}$

b Standard deviation $\sqrt{\frac{1}{4}} = \frac{1}{2}$

$P\left(\frac{1}{2} - \frac{1}{2} < X < \frac{1}{2} + \frac{1}{2}\right) = P(0 < X < 1) = 0.865 \text{ (3dp)}$

Exercise 26.2A Fluency and skills

1 A random variable X has an exponential distribution with a mean of 3

Find the probability that X takes values

a Greater than 3 b Between 1 and 3

2 A random variable X has an exponential distribution with parameter $\lambda = 2$

a Using the probability density function of X, copy and complete this table

x	0	0.2	0.4	0.6	0.8	1.0	1.2	1.4
$f(x)$	2				0.40	0.27	0.18	0.12

b Sketch the probability density function of X for $0 < x < 1.4$

c Write the modal value of X

3 a Write the cumulative distribution function of a random variable X that follows an exponential distribution with parameter λ

b Hence prove that M, the median of X, satisfies the equation $M\lambda = \ln 2$

4 An exponentially distributed random variable X has parameter λ equal to 0.25

a Find the mean and standard deviation of X

b Using the definition of skewness, $\frac{\text{mean} - \text{mode}}{\text{standard deviation}}$, calculate the skewness of X

5 A random variable X has an exponential distribution, parameter $\lambda = 2.5$

a Write the probability density function of X

b Show that its cumulative distribution function is $F(x) = 1 - e^{-2.5x}, x > 0$

c Find the values of

i $P(X < 0.5)$

ii $P(X > 0.2)$

iii $P(0.30 < X < 0.65)$

d Using the general formulae for the mean and variance of a random variable, prove that the mean and variance of X are 0.4 and 0.16, respectively.

6 a State the condition for waiting times between events to follow an exponential distribution with parameter $\frac{1}{2}$

b Requests for cash withdrawals arrive at a bank as a Poisson process with mean 12 per second.

i What is the mean waiting time between requests?

ii What is the probability that, from a randomly chosen time, there is a gap of more than 0.2 s before the first request?

7 A random variable X has an exponential distribution with parameter $\lambda = 4$

a Write the cumulative distribution function of X and hence use differentiation to derive its probability density function.

b Sketch the probability density function of X

c Write the modal value of X

d Use your answer in part **a** to find $P\left(X > \frac{1}{2}\right)$

8 A random variable X has an exponential distribution, parameter $\lambda = 5$

a Write the probability density function of X and sketch this function, showing where it cuts the vertical axis.

b Using integration, show that the cumulative distribution function of X is $F(x) = 1 - e^{-5x}$

c Show that the mean and variance of X are 0.2 and 0.04, respectively. You may quote standard results for these quantities.

Reasoning and problem-solving

Strategy

To solve a problem using the exponential distribution

1. Ensure that the conditions for the exponential distribution apply to the problem.
2. Use a calculator, statistical tables or the exponential distribution function to find probabilities.
3. Where required, use standard results to find the mean and variance of the distribution.

Example 3

Random events occur at a rate of 4 per minute.

a Write the probability density function, $f(t)$, and the cumulative distribution function, $F(t)$, of the random variable T, the waiting time in minutes between events.

b Find the probability that, from the occurrence of one event, the waiting time until the next event will be greater than 15 seconds.

c Calculate the mean and variance of the waiting time.

a $f(t) = 4e^{-4t},\ t \geq 0$. $F(t) = 1 - e^{-4t},\ t \geq 0$

b $P(T > 0.25) = 1 - P(T < 0.25) = e^{-4 \times 0.25} = e^{-1} = 0.37$ (2dp)

c Mean $= \frac{1}{\mu} = \frac{1}{4}$; variance $= \frac{1}{\mu^2} = \frac{1}{16}$, where μ is the distribution parameter.

Example 4

Radioactive decays occur randomly in time with a mean of 3 per minute and T is a random variable for the waiting time in seconds between events.

a Specify fully the distribution of T

b Find the probability that, in the t_1 seconds after a clock is started, no decays occurred.

c Given that there were no decays in the first t_0 seconds after the clock started, find the probability that there were no decays in a further t_1 seconds.

d Comment on the results in parts **b** and **c**

(Continued on the next page)

a Exponential, parameter 0.05

b $P(\text{no decays in } t_1 \text{ s}) = P(T > t_1) = 1 - P(T < t_1) = 1 - (1 - e^{-0.05t_1}) = e^{-0.05t_1}$

c $P(\text{no decays in further } t_1 \text{ s} \mid \text{no decays in } t_0 \text{ s})$

$$= \frac{P(\text{no decays in } t_0 + t_1)}{P(\text{no decays in } t_0)} = \frac{e^{-0.05(t_0+t_1)}}{e^{-0.05t_0}} = \frac{e^{-0.05t_0} \times e^{-0.05t_1}}{e^{-0.05t_0}} = e^{-0.05t_1}$$

d The answers are equal. The probability that there were no decays in t_1 seconds is unaffected by the absence of decays during any previous time interval.

3 per minute is $\frac{3}{60} = 0.05$ per second

Or, substitute $x = 0$ into a Poisson distribution function, mean $0.05t_1$

Exercise 26.2B Reasoning and problem-solving

1 Random events occur at a mean rate of 6 per minute.

a Prove that the cumulative distribution function, F(t), of the random variable T, the waiting time in minutes between events, is given by $1 - e^{-6t}$

b By differentiation, find the probability density function, f(t)

c Show that the median waiting time between events, M_T, is 0.12 (2 dp)

A clock is started at some randomly chosen time.

d Find the probability that the waiting time until the next event will be greater than 20 seconds.

2 The mean number of customers who arrive at a supermarket checkout during a 6 minute period is 12

a Assuming that their arrivals constitute a Poisson process, what is the probability that a period of at least two minutes will occur without any customer appearing?

b Find the expected number of customers arriving during a 10-hour day and hence the mean number of gaps of at least 2 minutes between customers.

3 Telephone calls to a call centre arrive randomly with a mean of 4 calls per minute. The centre has 10 operators and calls last at least 4 minutes. Lines open at 9 am.

a Give the distribution of the waiting time between calls and hence find the probability that the first call arrives after 09:01

b Given that there are no calls by 09:01, with no further calculation, write down the probability that the operators wait until after 09:02 before the first call. Give a reason for your answer.

c Find the probability that at 09:03 at least half of the operators are busy.

4 A firm produces microchips and has found that the mean lifetime for these components is 2.1 years, with the exponential distribution providing a good model for the lifetime.

a Specify completely the distribution of the lifetime for a randomly chosen component and find the probability that its lifetime is less than one year.

The firm guarantees the components for one year. If a failure occurs within the first year, the component is replaced and a new guarantee for one year then applies to the new component. Subsequent replacements are not guaranteed.

b What is the probability that for a randomly chosen component, a buyer will apply for exactly one replacement under guarantee?

c Explain why it is unlikely that failure is generally due to wear and tear factors.

5 **a** T, the waiting times between random events, follows an exponential distribution with parameter $\frac{1}{3}$

Write the probability density function and the mean and variance of T

b Cars arriving at a traffic checkpoint can be considered as a Poisson process with mean 14 per hour.

i What is the mean waiting time between car arrivals?

ii What is the probability that, from the start of counting, there is a five-minute period during which no cars arrive?

6 In the production of fibre-optic cable, flaws occur at random with a mean of 0.21 per 10 metres.

a What is the probability that, from the start of production one day, a length of at least 80 metres will be produced without any flaws?

b In 10 000 metres of cable produced, find the mean number of flawless lengths of at least 80 metres.

7 Random events occur at a rate of 8 per hour.

a Write the probability density function, $f(t)$, and the cumulative distribution function, $F(t)$, of the random variable T, the waiting time in minutes between events.

b Find the probability that, from the occurrence of one event, the waiting time until the next event will be greater than 12 minutes.

c Calculate the mean and variance of the waiting time.

26 Summary and review

Chapter summary

- The cumulative distribution function of a continuous random variable with probability density function f(*t*) is given by $F(x) = P(X < x) = \int_{-\infty}^{x} f(t)\,dt$
- If F(*x*) and f(*x*) are the cumulative distribution function and the probability density function of *X*, respectively, then $f(x) = \frac{d}{dx}F(x)$
- A mixed distribution occurs when the corresponding random variable is partly discrete and partly continuous.
- If *R* is a random variable with a rectangular probability distribution defined on the interval (*a*, *b*) (that is, *R* ~ Rectangular(*a*,*b*)), then its probability density function is given by $f(r) = \frac{1}{b-a}; a < r < b$
- If *X* has a rectangular probability distribution, then the mean and variance of *X* are, respectively, given by $E(X) = \frac{a+b}{2}$ and $\text{Var}(X) = \frac{(b-a)^2}{12}$
- If *X* is a random variable with an exponential probability distribution, parameter λ, then its probability density function is given by $f(x) = \lambda e^{-\lambda x}; x > 0$
- If *X* is a random variable with an exponential probability distribution, parameter λ, then its cumulative probability distribution function is given by $F(x) = 1 - e^{-\lambda x}; x > 0$
- If *X* has a probability density function given by $f(x) = \lambda e^{-\lambda x}; x > 0$, then the mean and variance of *X* are, respectively, given by $E(X) = \frac{1}{\lambda}$ and $\text{Var}(X) = \frac{1}{\lambda^2}$
- If *T* is a random variable for the time between successive random events which occur at a mean rate μ, then *T* follows an exponential distribution, parameter μ, with mean $\frac{1}{\mu}$ and variance $\frac{1}{\mu^2}$

Check and review

You should now be able to...	Review Questions
✔ Calculate the cumulative distribution function and probability density function for continuous random variables.	1, 2, 3
✔ Recognise and calculate the mean and variance of mixed random variables.	4
✔ Apply rectangular and exponential probability models in different circumstances.	5, 6, 7, 8
✔ Prove and use formulae for the mean and variance of rectangular and exponential random variables.	3, 5, 6, 7, 8

1 A random variable, X, has a probability density function given by $f(x)=\begin{cases} ax(x-3)(x+1); & -1<x<0 \\ 0; & \text{otherwise} \end{cases}$.

a Find the value of a and the cumulative distribution function $F(x)$.

b Hence show that the median of the distribution is approximately -0.52

2 A continuous random variable X has pdf given by $f(x)=\begin{cases} \frac{(1-3x)}{2}; & a<x<\frac{1}{3} \\ 0; & \text{otherwise} \end{cases}$

a Show that $a=\frac{1-2\sqrt{3}}{3}$

b Find the cumulative distribution function of X

3 A random variable, X, has probability density function $f(x)=\begin{cases} \frac{1}{k}; & a<x<b \\ 0; & \text{otherwise} \end{cases}$

a Show that $k=b-a$

b Find $E(X)$ and $Var(X)$ when $a=-2$ and $b=2$

c Show that the cumulative distribution function of X is given by $F(x)=\frac{(x+2)}{4}$

4 A random variable X takes the exact value 0 or any value in the interval $2\leq x\leq 4$

Its distribution is given by $P(X=0)=0.1$, $P(2<X<x)=\left(\frac{k}{10}\right)(-x^2+10x-16)$ for $2<x<4$

a Prove that $k=1.125$

b Find the probability density function, $f(x)$, of X in the interval $2\leq x\leq 4$

c Find

i The expected value of X ii $P(X<3)$

5 Radioactive decay can be modelled as a Poisson process with mean 12 nuclei decaying per second.

a What is the mean waiting time between disintegrations?

b What is the probability that, from a randomly chosen time, there is a gap of more than 0.2 s before the first disintegration?

6 Cars pass a traffic checkpoint at a rate of 4 per minute.

a Find the probability that the interval between successive cars will be at least 20 seconds. You should state any distributional assumptions you make.

b Using the same assumptions as in part **a**, calculate how many cars you would expect to pass the checkpoint in 5 hours.

c How many intervals greater than 20 seconds would you expect during 5 hours?

7 Random events occur at a mean rate of 10 per minute.

a Prove that the cumulative distribution function, $F(t)$, of the random variable T, the waiting time in minutes between events, is given by $1 - e^{-10t}$

b Show that the median waiting time between events, M_T, is 0.069 (3 dp)

c Find the probability that the waiting time between successive events will be greater than 20 seconds.

8 On a quiet country road, cars pass a given point randomly in time with a mean of 6 every 10 minutes. Let T be a random variable for the waiting time in minutes between successive cars.

a Specify fully the distribution of T

An observer records car arrivals.

b Find the probability that she has to wait longer than 5 minutes before the first car arrives.

c Given that there were no cars in the first 2 minutes, use your answer to part **b** to find the probability that she has to wait more than a further 5 minutes until the first car. You should explain your answer.

d Given that observations start at 9 am, find the probability that the first car recorded arrives before 9:02 am and the second after 9:04 am.

26 Exploration

Going beyond the exams

Did you know?

The exponential distribution and the geometric distribution are the only probability distributions with the memoryless property:
$P(X > a + b | X > b) = P(X > a)$

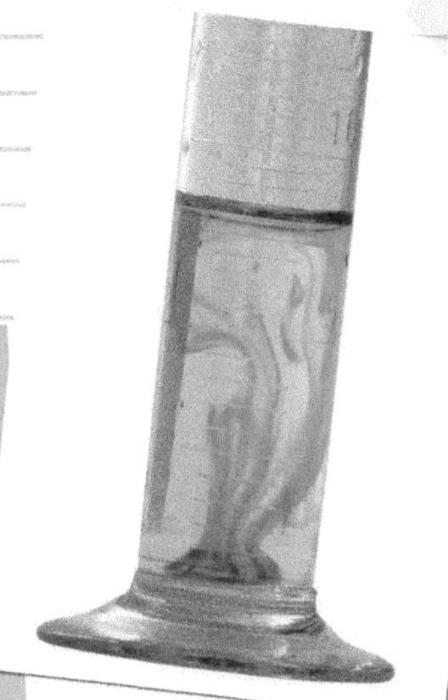

Research

The arcsine distribution has a cumulative distribution function of $F(x) = \frac{2}{\pi} \arcsin(\sqrt{x})$

Can you show that this is equivalent to

$F(x) = \frac{1}{\pi}\arcsin(2x - 1) + \frac{1}{2}$?

Can you find the corresponding probability distribution function through differentiating $F(x)$?

Find out about the three arcsine laws that cover one-dimensional random walks and Brownian motion.

Research

There are many applications of the exponential function that you could find out about. These include:

- Queuing theory
- The Barometric formula used in physics
- Daily rainfall calculations
- Dead time in particle detection analysis

26 Assessment

1 The continuous random variable X has probability density function given by

$$f(x) = \begin{cases} \frac{1}{4}(x-1); & 2 \le x \le 4 \\ 0; & \text{otherwise} \end{cases}$$

a Find $P(2.5 < X < 3)$ **[3]**

b Calculate the expectation and variance of X **[6]**

c Calculate $E(2X+1)$ and $Var(2X+1)$ **[4]**

2 The continuous random variable X has probability density function given by

$$f(x) = \begin{cases} k(1+3x^2); & 0 \le x \le 2 \\ 0; & \text{otherwise} \end{cases}$$

a Find the value of k **[2]**

b Sketch the probability density function of X **[2]**

c Write the mode of X **[1]**

d Show that the median, m, of X satisfies the equation $m^3 + m - 5 = 0$ **[3]**

3 The cumulative distribution function of the continuous random variable X is given by

$$F(x) = \begin{cases} 0; & x < 1 \\ k(x^3 - 1); & 1 \le x \le 2 \\ 1; & x > 2 \end{cases}$$

a Work out the value of k **[2]**

b Calculate $P(X < 1.5)$ **[2]**

c Work out the probability density function of X **[3]**

4 The continuous random variable U is uniformly distributed over the interval [1, 5]

a Sketch the probability density function of U **[2]**

b Write $P(X > 2)$ **[1]**

a Calculate the mean and variance of U **[4]**

5 The lifetime, X, of a component, measured in thousands of days, can be modelled by an exponential distribution with probability density function $f(x) = \begin{cases} \frac{1}{4}e^{-\frac{x}{4}}; & x > 0 \\ 0; & \text{otherwise} \end{cases}$

a State the values of the mean and the standard deviation of the lifetime. **[2]**

b Find the probability that the lifetime of the component is less than 2400 days. **[3]**

6 A random variable, R, is exponentially distributed with $\lambda = 5$

a Write the probability density function of R **[2]**

b Write the cumulative distribution function of R **[2]**

c Calculate the probability that R lies within one standard deviation of its mean. **[3]**

7 The continuous random variable X has probability density function given by

$$f(x) = \begin{cases} \frac{4}{3}(x^3 + x); & 0 \le x \le 1 \\ 0; & \text{otherwise} \end{cases}$$

a Find the cumulative distribution function of X **[3]**

b Calculate

i $E(5X - 3)$ ii $Var(5X - 3)$ **[8]**

8 The continuous random variable X has probability density function given by

$$f(x) = \begin{cases} \frac{x}{24} + \frac{1}{12}; & 2 \le x \le 6 \\ 0; & \text{otherwise} \end{cases}$$

a Find the cumulative distribution function of X **[4]**

b Find the exact value of the median of X **[3]**

c Calculate $P(\text{mean} < X < \text{median})$. **[5]**

9 The continuous random variable U is uniformly distributed over the interval $[4, 10]$

Work out the cumulative distribution function of U **[4]**

10 The duration, in minutes, of a phone call to a bank is modelled by a uniform distribution on the interval $[1, 10]$

a Use this model to calculate the probability that the length of a call is between 4 and 8 minutes. **[2]**

b Give a reason why this model may not be appropriate. **[1]**

It is decided instead to model the duration as an exponential distribution, using the mean of the uniform distribution as the parameter.

c Use this model to calculate the probability that a call is

i Between 4 and 8 minutes,

ii Over 15 minutes given it has already been over 10 minutes. **[5]**

11 Prove that the variance of an exponential random variable with parameter 2 is $\frac{1}{4}$ **[5]**

12 A continuous random variable, U, is uniformly distributed over the interval $[k, 2k]$

Prove that the variance of U is given by $\frac{1}{12}k^2$ **[4]**

13 The continuous random variable X has probability density function given by

$$f(x) = \begin{cases} \dfrac{1}{x}; & 1 \le x \le e \\ 0; & \text{otherwise} \end{cases}$$

a Work out the cumulative distribution function of X **[3]**

b Find the standard deviation of X **[6]**

14 The continuous random variable X has cumulative distribution function given by

$$F(x) = \begin{cases} 0; & x < 2\sqrt{2} \\ \dfrac{x^2}{8} - 1; & 2\sqrt{2} \le x \le 4 \\ 1; & x > 4 \end{cases}$$

a Calculate the median of X **[2]**

b Calculate the exact values of $E(2X^{-1})$ and $\text{Var}(2X^{-1})$. **[6]**

15 The continuous random variable X has probability density function given by

$$f(x) = \begin{cases} k; & -1 \le x < 1 \\ k(x-2)^2; & 1 \le x \le 2 \\ 0; & \text{otherwise} \end{cases}$$

a Sketch the probability density function of X **[4]**

b Calculate the value of k **[4]**

c Fully define the cumulative distribution function of X **[5]**

d Find the median of X **[3]**

e Calculate $E(4X)$ and $\text{Var}(4X)$ **[8]**

16 The rate of customers arriving at a bank is modelled by a Poisson distribution with a rate of 20 people per half-hour.

What is the probability that there is a gap of less than 30 seconds until the next customer? State clearly the distribution you are using. **[4]**

27 Hypothesis testing and the *t*-test

In 1972, researchers at Stanford University famously conducted the 'Marshmallow Experiment' to study whether deferred gratification is an indicator of success later in life. In the experiment, a sample of children aged between 4 and 6 were each given a marshmallow and asked to wait alone in a room for 15 minutes for the experimenter to return. The children were told that if the marshmallow was still uneaten when the experimenter returned, then they would be given a second marshmallow. One third of the children resisted eating the marshmallow for the full 15 minutes. Follow up experiments showed that the children who resisted eating the first marshmallow had higher SAT scores, thereby supporting the hypothesis that deferred gratification is an indicator of later success.

Hypothesis testing is used widely in psychology, and the ability to identify the risk of a Type I or Type II error is crucial to avoid incorrect conclusions from being widely accepted in the discipline.

Orientation

What you need to know

Ch11 Hypothesis testing and contingency tables

- Type I/II errors.

What you will learn

- To find the probability of a Type I error.
- To find the probability of a Type II error.
- Calculate the power of a hypothesis test.
- Use a *t*-test to evaluate a possible population mean for a Normal distribution.

What this leads to

Careers

- Scientific research.
- Quality control.
- Psychology.

27.1 Type II errors and power

Fluency and skills

See Ch11.1 For a reminder of Hypothesis testing.

In hypothesis testing, the null hypothesis is always assumed to be true when determining the critical region. The probability of a result in the critical region should be less than or equal to the test's significance level.

Key point

A **type I error** is when a null hypothesis which is true is rejected.

A **type II error** is when a null hypothesis which is false is accepted.

The probability of making a type I error is equal to the probability of a result in the critical region. It is commonly denoted by α

The probability of a type II error is more difficult to calculate as it depends on the true value of the population parameter being tested. If this was known already, then a test would be pointless. Taking some example true values of the parameter can provide an idea of the probability of making a type II error. The probability of making a type II error is commonly denoted by β

The probabilities of making type I and a type II errors are generally not independent; decreasing one is likely to increase the other. The probability of making a type I error is much easier to control via the significance level.

Example 1

A Poisson distribution is being tested with null hypothesis $\lambda = 1.9$ and alternative hypothesis $\lambda > 1.9$

The critical value at the 5% significance level is 5

a Find the probability of a type I error.

The true parameter value is $\lambda = 3.1$

b Find the probability of a type II error.

See Ch10.2 For a reminder of the Poisson distribution.

a The probability of a type I error is the probability of getting a result in the critical region. $P(X \geq 5) = 4.41\%$

The probabilities of the Poisson distribution are given by the formula $P(X = k) = \frac{\lambda^k e^{-\lambda}}{k!}$ where k is the number of events in the interval.

b The probability of a type II error is the probability of getting a result outside of the critical region when the parameter value is 3.1. $P(X \leq 4) = 79.8\%$

This is very large. This means that the situations $\lambda = 1.9$ and $\lambda = 3.1$ are difficult to distinguish. To overcome this, it is often a good idea to use a larger sample.

Key point

The **power** of a hypothesis test is the probability that a false null hypothesis is rejected. It is equal to $1-\beta$

This depends on the true parameter value and is rarely a fixed value. The larger the power, the better the test is at avoiding accepting a false null hypothesis by mistake.

Example 2

The mean of a Normal distribution with standard deviation 2.2 is being tested at the 10% level. The null hypothesis is $\mu=-0.8$ and the alternative hypothesis is $\mu\neq-0.8$

The true mean is $\mu=-3.2$

a Find the probability of a type II error.

b Calculate the power of the test.

a For this two-tailed test, the critical region is $X<-4.42$ and $X>2.82$

This was found using the Normal distribution with mean –0.8 and standard deviation 2.2. Each tail is 5% of the distribution.

When the mean is –3.2, the probability of obtaining a value outside that region is 70.7%

For $\mu=-3.2$,
$P(-4.42<X<2.82)$
$=1-P(X<-4.42)-P(X>2.82)$
$=1-0.2898-0.0031$

b The power is $1-0.707=0.293$

STATS

Exercise 27.1A Fluency and skills

1 With reference to the correct hypotheses, explain the difference between a type I and a type II error.

2 A Normal distribution with standard deviation 6.25 is being tested at the 5% significance level. The null hypothesis is H_0: $\mu=15.7$ and the alternative is H_1: $\mu>15.7$

a State the probability of a type I error.

b Find the probability of a type II error if actually $\mu=5.7$

c Find the probability of a type II error if actually $\mu=25.7$

3 A Poisson distribution is tested at the 5% level with hypotheses H_0: $\lambda=6.8$ and H_1: $\lambda>6.8$

a Find the critical region,

b Calculate the probability of a type I error.

The true value is $\lambda=9.7$

c Calculate the probability of a type II error,

d Calculate the power.

4 For a binomial distribution, the hypotheses H_0: $p=0.2$ and H_1: $p\neq0.2$ are tested at the 10% level. 20 trials are performed and the critical region is $X=0$ or $X>7$

Calculate the probability of a type II error and the power when the true value of p is

a 0.3 b 0.4 c 0.6

5 A Normal distribution is being tested with hypotheses H_0: $\mu = 0$ and H_1: $\mu > 0$
The significance level is 5%. The standard deviation is 4.5

a Calculate the power if the true mean is 12.1

b Explain why the power tends to 1 as the true mean tends to infinity.

6 A binomial distribution is being tested at the 5% level with the hypotheses H_0: $p = 0.4$ and H_1: $p > 0.4$

A sample of size 16 is taken.

a Find the critical region,

b Calculate the power if the true value of p is 0.65

7 A crisp manufacturer wants to check that the mean weight of a packet of crisps is not too different from the advertised weight. They perform a two-tailed test.

a Explain what a type I error would mean in context.

Actually, the mean weight is above the advertised weight but the crisp manufacturer believes that it is not.

b Decide whether or not they have made a type II error.

8 A binomial distribution $X \sim B(50, p)$ is tested at the 10% significance level to decide between the hypotheses H_0: $p = 0.4$ and H_1: $p < 0.4$

a Explain why the critical region is $X \leq 15$, showing supporting calculations,

b State the probability of making a type I error,

c Calculate the probability of making a type II error if actually $p = 0.35$

A common procedure to ease calculations is to approximate a binomial distribution with a Normal distribution.

d Explain why the approximating Normal distribution is $N(20, 12)$ and explain why the new hypotheses are H_0: $\mu = 20$ and H_1: $\mu < 20$

e Determine the new critical value at the 10% level,

f Determine the approximating distribution if actually $p = 0.35$, and hence calculate the probability of making a type II error.

9 Two hypothesis tests are considered for determining the mean of a distribution.
A statistician wishes to know which is likely to be a better test.

For both tests the probability of making a type I error is 5%

a Explain what a type I error is and hence why these probabilities might be made the same by the statistician to choose a better test.

For one test the probability of making a type II error is larger than for the other test.

b Explain what a type II error is and hence decide which test should be used by the statistician.

Reasoning and problem-solving

Strategy

1. When deciding on the significance level for a test you should consider whether it is more important to reduce the probability of making a type I error or of making a type II error.
2. The sample size should be large enough so that if the true value is very different from the null hypothesis, then it is likely to be inside the critical region.

Example 3

A Normal distribution with standard deviation 4 is being tested with hypotheses H_0: $\mu = 7.7$ and H_1: $\mu < 7.7$. Calculate the critical value and hence the probability of making a type II error if the true mean is 2.5 when the significance level is

a 10% **b** 5% **c** 1%

a The critical value is 2.57. $P(X < 2.57 \mid \mu = 2.5) = 40.3\%$

b The critical value is 1.12. $P(X < 1.12 \mid \mu = 2.5) = 63.5\%$

c The critical value is -1.61. $P(X < -1.61 \mid \mu = 2.5) = 84.8\%$

Cumulative probabilities for the Normal distribution can be found using your calculator.

Note that relatively small changes to the significance level have brought large changes to the probability of making a type II error.

Consider a Poisson distribution being tested with hypotheses H_0: $\lambda = 10$ and H_1: $\lambda > 10$

At the 5% significance level, the critical region is $X \geq 16$

The graph shows the power for a range of true parameter values.

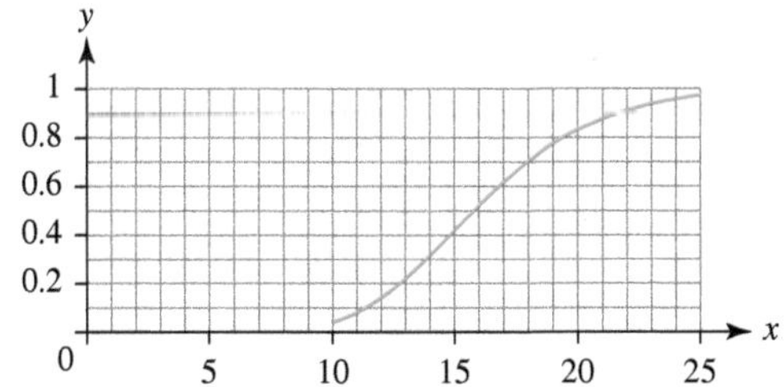

The power exceeds 0.5 near to 15.7, meaning that if the true value of the parameter is much smaller than the critical value of the test, then there is a better-than-evens chance that the false null hypothesis is accepted.

When the true value is much higher than the critical value, the power becomes close to 1. Note that the power isn't technically defined when the null hypothesis really is the true value.

This is because power is the probability that a false null hypothesis is rejected. A true null hypothesis is not false.

STATS

Example 4

A new medical test for detecting a rare illness is being trialled. The null hypothesis is that a person being tested does not have the illness.

a Explain why the probability of a type II error should be made as small as possible.

b This makes the probability of a type I error larger. Why might this not be a problem?

a A type II error is when a false null hypothesis is accepted. A medical test should aim to not overlook somebody who genuinely has the illness.

b A type I error is when a true null hypothesis is rejected. If the test claims that a person has the illness when they don't, then a simple repeat of the test will help to decide.

Exercise 27.1B Reasoning and problem-solving

1 A Normal distribution with standard deviation 0.6 is being tested with the hypotheses $H_0 : \mu = 0.3$ and $H_1 : \mu > 0.3$

Calculate the critical value, the probability of making a type II error if the true mean is 1.3 and the power when the significance level is

a 10%

b 5%

c 1%

2 A Poisson distribution is being tested with hypotheses $H_0: \lambda = 1.7$ and $H_1: \lambda \neq 1.7$ The true value of the mean is 2.3. For the significance level

a 10%

b 5%

calculate

i The critical region,

ii The probability of a type II error,

iii The power.

3 For a binomial distribution, the hypotheses $H_0: p = \frac{1}{3}$ and $H_1: p \neq \frac{1}{3}$ are tested at the 2% level. 20 trials are performed and the critical region is $X < 2$ or $X > 12$

The true value of p is 0.4

a State how many successes you would expect to see (using the true value of p),

b Calculate the probability of a type II error and the power.

Instead of 20 trials, 2000 trials are performed. The new critical region is $X < 618$ or $X > 716$

c State how many successes you would expect to see,

d Giving a reason, explain how you would expect the power of this test to compare with the 20-trial test.

4 A treasure hunter is configuring their metal detector to find buried valuables.

a Explain why the probability of a type II error should be made as small as possible,

b This makes the probability of a type I error larger. Why might this not be a problem?

5 The hypotheses H_0: $\lambda = 4.25$ and H_1: $\lambda < 4.25$ are tested at the 5% level, where λ is the parameter of a population Poisson distribution.

a Find the critical region,

b Find the probability of a type I error.

The true mean is 9.25

c Calculate the probability of a type II error and the power of the test.

6 A binomially distributed population is tested with the hypotheses H_0: $p = 0.2$ and H_1: $p > 0.2$ at the 5% level. Different-sized samples are taken to test this. The true value of p is 0.35. Calculate the critical region and hence the power when the sample size is

a 10

b 20

c 30

7 The number of goals scored by a team playing away from home in a football match can be modelled by a Poisson distribution with parameter $\lambda = 1.2$

A team has undergone a training regime that it is hoped will increase the number of goals they score away. The next 5 away games are played and the hypotheses H_0: $\lambda = 6$ and H_1: $\lambda > 6$ are tested at the 5% significance level.

a Explain why the hypotheses take this form,

b Find the critical region and calculate the probability of making a type I error.

After training, the number of goals scored away can now be modelled by a Poisson distribution with parameter $\lambda = 1.4$

c Calculate the probability of making a type II error,

d Explain why this error wouldn't be very concerning to a fan of the team,

e Explain why this error would be very concerning to a sports-betting company, who gamble on the results of matches.

27.2 *t*-tests

Fluency and skills

You can use a hypothesis test to investigate the mean of a Normal distribution with known variance. When the variance is unknown, more work is required.

Suppose you believe that a process follows a Normal distribution and expect the mean to be 12.0

You take a sample of size 10 and get the following results:

11.9	8.37	14.8	14.6	11.2
11.1	14.9	12.2	10.7	15.7

The mean of these values is 12.547, which could be close to 12.0 or it could be far from it, depending on the size of the variance. You need to know the distribution of the sample mean.

Key point

The **sample mean**, $\overline{X}$, itself has a probability distribution. The mean of this distribution (i.e. the mean sample mean) is the same as the population mean.

It is well worth stressing here that a particular sample's mean is very unlikely to be the same as that of the population, but the average of the means of many samples will be the same as the population mean.

Just as there is a standard Normal distribution that other Normal distributions can be related to, there are standard distributions for $\overline{X}$. These ***t*-distributions** describe the distribution of a sample mean for varying numbers of data points.

If you know the variance of the population distribution, σ, then the distribution of the test statistic, $\frac{\overline{x}-\mu_0}{\frac{\sigma}{\sqrt{n}}}$, is Normal with mean 0 and variance 1

If you do not know the variance of the population distribution, then you use the variance of the sample as an estimate but it will be inaccurate if the sample size is small.

For the sample of size 10 above, the sample variance is

$$\frac{1}{10-1}((11.9^2+8.37^2+\ldots+15.7^2)-10\times 12.547^2)=5.56$$

For a sample of size n the sample variance is

$$S^2=\frac{1}{n-1}\left(\sum X_i^2-n\overline{X}^2\right)=\frac{1}{n-1}\sum(X_i-\overline{X})^2$$

Key point

The ***t*-test statistic** is $t=\frac{\overline{x}-\mu_0}{\frac{S}{\sqrt{n}}}$

It follows the t-distribution with $n-1$ degrees of freedom.

The example t-test statistic is $\dfrac{12.547-12.0}{\dfrac{\sqrt{5.56}}{\sqrt{10}}}=0.734$

Using the t-distribution with $10-1=9$ degrees of freedom, the probability of getting a sample mean this big or larger is 24.1%

No significance level was stated, but it is reasonable to say that the sample has come from a Normal distribution with mean 12.0

Example 1

A population Normal distribution with unknown variance is being tested at the 10% significance level with hypotheses $H_0: \mu_0 = 3.2$ and $H_1: \mu_0 \neq 3.2$. A sample of size 8 is taken, which has a sample mean of 2.39 and a sample variance of 1.12

a Calculate the t-test statistic,

b State the number of degrees of freedom,

c Calculate the p-value of the statistic,

d Determine the conclusion of the test.

a $t=\dfrac{2.39-3.2}{\dfrac{\sqrt{1.12}}{\sqrt{8}}}=-2.165$

b There are $8-1=7$ degrees of freedom.

c $P(T<-2.165)=3.36\%$

Your calculator can find these cumulative probabilities from the t-test statistic.

d This is a two-tailed test and $3.36\% < 5\%$. There is significant evidence to reject the null hypothesis. You conclude that the sample is not likely to come from a population Normal distribution with mean 3.2

Key point

For a hypothesis test using the t-distribution to be valid, the population should be Normally distributed.
As with all hypothesis tests, the sample must be taken at random

Example 2

A student wants to test whether the average number of devices that can connect to the internet per UK household is five or greater. They ask seventeen friends to record their values and decide, due to the sample size, to use a t-test. Give two reasons why this test will not give a valid result.

1 The sample isn't random; it is likely to be biased due to the student asking friends.

2 The values aren't even approximately Normally distributed so a t-test is unsuitable.

Exercise 27.2A Fluency and skills

1 Calculate the test statistics for the following tests.

	Sample size	Sample mean	Sample variance	Hypothesis mean
a	16	2.51	1.08	1.92
b	4	−0.121	3.56	1.35
c	7	10.49	18.65	21.3
d	11	−21.7	5.16	−18.3

2 For the following data

a Calculate the sample mean,

b Calculate the sample variance,

c Calculate the test statistic if the hypothesis mean is 0.2

0.342	−0.348	0.692	3.845
−0.361	0.998	0.681	2.029

3 A population Normal distribution with unknown variance is being tested at the 5% level with hypotheses $H_0: \mu_0 = 1.61$ and $H_1: \mu_0 < 1.61$

A sample of size 12 is taken, which has sample mean -2.56 and sample variance 29.3

a Calculate the t-test statistic,

b State the number of degrees of freedom,

c Calculate the p-value of the statistic,

d Determine the conclusion of the test.

4 A population Normal distribution with unknown variance is being tested at the 5% significance level with hypotheses $H_0: \mu_0 = -2.1$ and $H_1: \mu_0 < -2.1$

A sample of size 17 is taken, which has sample mean −2.49 and sample variance 18.7

a Find the critical region,

b Calculate the t-test statistic,

c Determine the conclusion of the test.

5 The hypotheses $H_0: \mu_0 = 3.17$ and $H_1: \mu_0 \neq 3.17$ are tested at the 2% level for a population Normal distribution with unknown variance. A sample of size 14 is taken, which has a sample mean of 0.31 and a sample variance of 14.0

Showing your reasoning, determine the conclusion of the test.

6 The hypotheses $H_0: \mu_0 = 18.2$ and $H_1: \mu_0 < 18.2$ are tested at the 5% level for a population Normal distribution with unknown variance. The following sample is taken.

0.387	4.23	20.2	−15.3
22.9	26.5	−6.39	13.5

a Calculate the t-test statistic,

b State the number of degrees of freedom,

c Calculate the p-value of the statistic,

d Determine the conclusion of the test,

e State two assumptions made about the sample that are required for the test to be valid.

7 A population Normal distribution with unknown variance is being tested at the 5% significance level with hypotheses $H_0: \mu_0 = 129.1$ and $H_1: \mu_0 < 129.1$

This sample is taken.

153.9	26.53	120.8
183.5	75.95	

a Find the critical region,

b Calculate the t-test statistic,

c Determine the conclusion of the test.

8 A student wants to see whether the average height of a student in their school has changed over the last ten years. They obtain a list of all the students currently in the school and the previous average value.

a Could a t-test be suitable for this investigation?

b What would the student need to do with the sample to ensure that a valid test is performed?

Reasoning and problem-solving

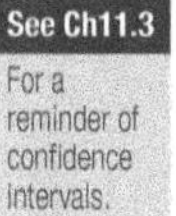
See Ch11.3 For a reminder of confidence intervals.

The t-distribution can also be used to generate confidence intervals for the mean of a Normal distribution when the sample size is small and the population variance is unknown. A rule of thumb is that a sample of size at least 30 is 'large enough' for the standard Normal distribution to be used. For a sample of size n taken from a Normal distribution $N(\mu, \sigma^2)$, the p%-confidence interval is

$$\bar{x} - t \times \frac{s}{\sqrt{n}} < \mu < \bar{x} + t \times \frac{s}{\sqrt{n}}$$

where t is drawn from the t-distribution with $n-1$ degrees of freedom, $\bar{x}$ is the sample mean and s^2 is the sample variance. For a p%-confidence interval $t = t_{n-1}^{-1}\left(\frac{1+p}{2}\right)$, you can work these out yourself using your calculator.

STATS

Example 3

A sample of size 16 is taken, whose mean is 13.6 and whose standard deviation is 20.4, in order to generate a 95% confidence interval for the mean of the population. Find the confidence interval.

For $p = 95\%$ and with $16 - 1 = 15$ degrees of freedom, $t_{15}^{-1}(0.975) = 2.13$

The interval $13.6 - 2.13 \times \frac{20.4}{\sqrt{16}} < \mu < 13.6 + 2.13 \times \frac{20.4}{\sqrt{16}}$

simplifies to $2.74 < \mu < 24.46$

Strategy

When using a t-test

1. Use a t-test if a sample comes from a Normal population with unknown variance.
2. For better estimates of population parameters, use a larger sample.

Key point

If the population variance is known, then a t-test is unsuitable. Instead, use a regular Normal distribution test.

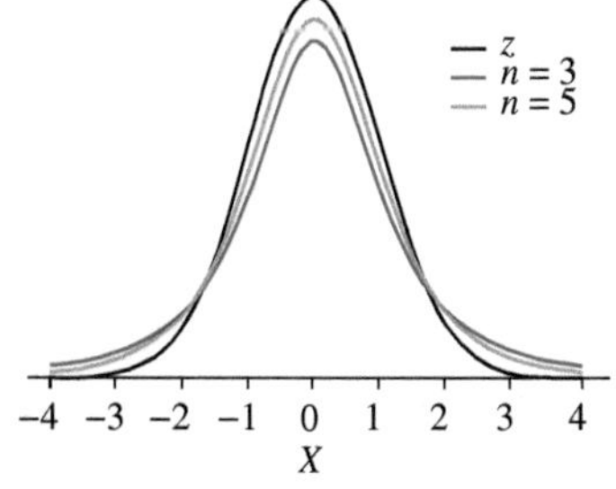

The t-distribution is used when only a small sample is available. As the sample size becomes larger, the distribution approaches the standard Normal distribution. In the limit as the number of degrees of freedom tends to infinity, the t-distribution does become the standard Normal distribution, as the sample approaches becoming the whole population.

A good rule of thumb is that if the sample size is larger than 30, then the t-distribution is numerically close enough to the standardised Normal distribution, Z, so that Z can be used instead.

Example 4

Two samples are taken from a Normal distribution with unknown mean and variance. The first sample is of size 12; it has sample mean 3.75 and sample variance 2.9. The second sample is of size 7; it has sample mean 5.1 and sample variance 4.3. The two samples are combined. The samples are tested to see whether or not the population mean is 4.5

a Calculate the t-test statistic for each sample,

b State the combined sample size,

c Calculate the combined sample mean,

d Calculate the combined sample variance,

e Calculate the t-test statistic for the combined sample.

a $t = \dfrac{3.75-4.5}{\dfrac{\sqrt{2.9}}{\sqrt{12}}} = -1.526$ for sample 1; $t = \dfrac{5.1-4.5}{\dfrac{\sqrt{4.3}}{\sqrt{7}}} = 0.766$ for sample 2

b 19

c $\dfrac{(12\times3.75+7\times5.1)}{19} = 4.25 = \overline{X}$

d The combined sample variance is $\dfrac{1}{19-1}\left(\sum X_i^2 - 19\times\overline{X}^2\right)$. To find $\sum X_i^2$ you need the contributions from each sample, which can be found by rearranging the variance formula. For the first sample it is $(12-1)\times2.9+12\times3.75^2 = 200.65$
For the second sample it is $(7-1)\times4.3+7\times5.1^2 = 207.87$. Their sum is 408.52 which, along with $\overline{X} = 4.25$, gives the combined variance as 3.63

e The combined sample t-test statistic is $t = \dfrac{4.25-4.5}{\dfrac{\sqrt{3.63}}{\sqrt{19}}} = -0.572$

Exercise 27.2B Reasoning and problem-solving

1 A sample of size 1000 is taken from a Normally distributed population with unknown mean and variance. The sample has mean 14.8 and variance 91.6

The hypotheses $H_0: \mu_0 = 12.1$ and $H_1: \mu_0 > 12.1$ are tested at the 5% level.

a Explain why a Normal test is required instead of a t-test,

b Calculate the test statistic,

c Determine the conclusion of the test.

2 A sample is taken from a Normally distributed population with unknown mean and variance. The hypotheses $H_0: \mu_0 = 13.4$ and $H_1: \mu_0 < 13.4$ are tested at the 10% level.

a Perform the test using the first row of five values,

b Perform the test using the first two rows (ten values total),

c Perform the test using all fifteen values.

−16.8	86.1	15.1	−149	96.8
−62.7	−10.3	−114	−52.5	−42.0
32.6	51.4	3.01	27.9	42.4

3 Two samples are taken from a Normal distribution with unknown mean and variance. The first sample is of size 8; it has sample mean 14.3 and sample variance 18.5

The second sample is of size 11; it has sample mean 12.7 and sample variance 22.3 The two samples are combined. The samples are tested with the hypotheses $H_0: \mu_0 = 9.1$ and $H_1: \mu_0 > 9.1$ at the 10% level.

a Calculate the combined sample mean,

b Calculate the combined sample variance,

c Calculate the t-test statistic for the combined sample.

4 The amount of sugar in grams found in a can of drink is measured. A sample of size 8 has mean 5.19 g and variance 2.34 g^2.

a Describe how to take such a sample,

b Give one assumption that must be made about the sample in order for a t-test to be appropriate,

c The intended sugar level is 4.26 g. Perform a t-test at the 5% level and state the conclusion of the test.

5 The distance at which a professional darts player throws a dart away from the bullseye is believed to have mean 0.31 cm. The hypotheses $H_0: \mu_0 = 0.31$ and $H_1: \mu_0 > 0.31$ are tested at the 5% level. They throw 12 darts and the sample has mean 0.68 cm and standard deviation 0.539 cm.

a Calculate the p-value of the statistic,

b Determine the conclusion of the test,

c Discuss the claim that the dart thrower's mean value being higher than the null hypothesis mean suggests that they throw inconsistently.

6 Two samples are taken from a Normal distribution with unknown mean and variance. The first sample is of size 10; it has sample mean 12.5 and sample variance 14.3

The second sample is of size 15; it has sample mean 12.3 and sample variance 27.3 The two samples are combined. The samples are tested with the hypotheses $H_0: \mu_0 = 14.7$ and $H_1: \mu_0 < 14.7$ at the 5% level.

a Calculate the t-test statistic for each sample and determine the conclusion of the test for each individual sample,

b Calculate the t-test statistic for the combined sample and determine the conclusion of the test for the combined sample.

7 The lengths of pipes produced by machines are Normally distributed with unknown variance. An engineer has two sets of pipes but has forgotten whether they came from the same machine. The machines are supposed to produce pipes with a mean length of 30.0 cm.

a The first sample of 14 pipes has sample mean 29.91 cm and sample variance 0.0576 cm^2. Test at the 10% level the hypotheses $H_0: \mu_0 = 30.0$ and $H_1: \mu_0 \neq 30.0$

b The second sample of 12 pipes has sample mean 30.07 cm and sample variance 0.0289 cm^2. Test at the 10% level the hypotheses $H_0: \mu_0 = 30.0$ and $H_1: \mu_0 \neq 30.0$

c Should the engineer conclude that the two sets of pipes have come from the same machine?

8 For each of these samples find a 90%-confidence interval for the population mean using an appropriate t-distribution.

a 11.6, 12.1, 12.5, 12.7, 13.0, 13.2, 14.8

b 3.2, 7.1, 8.8, 9.0, 9.5, 10.1, 10.9, 11.8, 12.1, 12.1, 12.7, 13.1, 15.3, 16.5

9 Find 99%-confidence intervals for samples with these summary statistics.

a $n = 9, \sum x = 83.5, \sum x^2 = 1296.9,$

b $n = 24, \sum x = 260, \sum x^2 = 3000$

10 Find the 95%-confidence interval for a sample with mean 25 and variance 25 when n takes the following values.

a $n = 5$ b $n = 10$ c $n = 25$

27 Summary and review

Chapter summary

- A **type I error** is when a null hypothesis that is true is rejected. The probability of such an error is denoted by α
- A **type II error** is when a null hypothesis that is false is accepted. The probability of such an error is denoted by β
- The **power** of a hypothesis test is the probability that a false null hypothesis is rejected. It is equal to $1 - \beta$
  This depends on the true parameter value and is rarely a fixed value.
- For a given sample size, decreasing the significance level will decrease α and increase β
- For a sample of size n taken from a Normally distributed population, the distribution of the sample mean follows a t-distribution with $n - 1$ degrees of freedom.
- The mean of the distribution of the sample mean is likely to be similar to the mean of the population, but the variance will be larger.
- A t-test may be used when the population is Normally distributed but only a small sample can be obtained; smaller than 30 is a good rule of thumb.
- The t-test determines whether a suggested value appears to be a suitable description of the population mean.

Check and review

You should now be able to...	Review Questions
✔ Calculate the probabilities of making type I and type II errors.	1, 2, 3
✔ Calculate the power of a hypothesis test.	1, 2
✔ Calculate the t-test statistic from a sample.	4, 5, 7
✔ Identify when a t-test is appropriate.	5
✔ Determine the conclusion of a t-test.	6, 7

1 A test is performed at the 5% significance level.

a Write down an inequality involving α, the probability of making a type I error.

The null hypothesis is H_0: $p = 15$ and the test has power P_1 when the true value is $p = k$, where $k > 15$

b Write down an inequality for the power P_2 of the test when the true value is $p = q$, where $q > k$

2 A Poisson distribution is tested at the 10% level with hypotheses H_0: $\lambda = 12.4$ and H_1: $\lambda > 12.4$

a Find the critical region,

b Calculate the probability of a type I error.

The true value is actually $\lambda = 18.3$

c Calculate

i The probability of making a type II error,

ii The power.

3 A Normal distribution with standard deviation 18.2 is being tested at the 2% significance level. The null hypothesis is $H_0: \mu = 30.1$ and the alternative is $H_1: \mu < 30.1$

a State the probability of a type I error,

b Find the probability of a type II error if actually $\mu = 15.2$

c Find the probability of a type II error if actually $\mu = -15.2$

4 Calculate the t-test statistics for the following samples. The mean is 26.7 in each case.

a 16.6 24.6 27.9 29.3 49.9

b 12.4 15.3 18.1 21.0 28.1 31.0 33.9

c −18.2 −8.83 6.13 23.0 24.8 41.7 47.3 49.1 60.4

5 A sample of size 23 is taken from a Normal distribution with sample mean 106.7 and sample variance 533

The hypotheses $H_0: \mu_0 = 104.5$ and $H_1: \mu_0 \neq 104.5$ are tested at the 5% level.

a Explain why a t-test is appropriate for this situation,

b Calculate the t-test statistic.

6 A population Normal distribution with unknown variance is being tested at the 5% level with hypotheses $H_0: \mu_0 = -5.14$ and $H_1: \mu_0 < -5.14$

For a sample of size 7, the t-test statistic is −2.87

a State the number of degrees of freedom,

b Calculate the p-value of the statistic,

c Determine the conclusion of the test.

7 Two samples are taken from a Normal distribution with unknown mean and variance. The first sample is of size 4; it has sample mean 71.6 and sample variance 23.8 The second sample is of size 6; it has sample mean 72.5 and sample variance 31.9

The two samples are combined. The samples are tested with the hypotheses $H_0: \mu_0 = 105$ and $H_1: \mu_0 < 105$ at the 5% level.

a Calculate the combined sample mean,

b Calculate the combined sample variance,

c Calculate the t-test statistic for the combined sample.

27

Exploration

Going beyond the exams

Did you know?

Student's t-distribution was developed by William Sealy Gosset. Gosset worked for Guinness who had banned publications by employees to prevent competing brewers from using their techniques. To get around this problem Gosset published under the pseudonym 'Student' – hence the name of the test.

Gosset was known for his modesty and once remarked that 'Fisher would have discovered it all anyway' in reference to his friend Ronald Aylmer Fisher.

Quotation

With regard to the ubiquity of the Normal curve, Poicare has been attributed as saying 'Everyone believes in it: experimentalists believing that it is a mathematical theorem, mathematicians believing that it is an empirical fact.'

Did you know?

You have studied type I and type II errors but you can also have type III errors. There are a number of interpretations for these errors of the third kind that you might like to look up. One popular interpretation is when you correctly reject a null hypothesis but for the wrong reasons!

27 Assessment

1 A Poisson distribution is being tested with null hypothesis $\lambda = 4.2$ and alternative hypothesis $\lambda > 4.2$

a Find the critical value for the test at a 5% significance level, **[1]**

b Find the probability of a type I error. **[1]**

The true parameter value is $\lambda = 8$

c Find the probability of a type II error. **[2]**

2 A biased coin is thrown 20 times and the number of heads noted. It is suggested that the probability of a head is 0.3

a State the null and alternative hypotheses in a test of this claim, **[1]**

b Find the critical region for the test using a significance level of 10% **[3]**

c Assuming that the probability of obtaining a head is actually 0.4, find the probability of making a type II error, **[2]**

d Write down the power of the test. **[2]**

3 The mean of a Normal distribution with standard deviation 12.1 is being tested at the 5% level. A sample of size 15 is used. The null hypothesis is $\mu = 24$ and the alternative hypothesis is $\mu > 24$. The true mean is 30

a Find the critical x-value for this test, **[7]**

b Find the probability of a type II error, **[4]**

c Calculate the power of the test. **[1]**

4 a Explain the meaning of the terms 'type II error' and 'the power of a hypothesis test'. **[2]**

b In the manufacture of soft drinks, a machine fills cans with cola. The nominal volume of cola in a can is 330 ml and it is known that the standard deviation of the volume per can is 4 ml. The manufacturer claims that the machine dispenses a mean volume greater than 330 ml. To test this claim, 10 cans are chosen at random and the mean volume calculated. You should assume that the quantity of drink in a randomly chosen can follows a Normal distribution.

i State the null and alternative hypotheses for this test, **[1]**

ii Find the set of sample mean values which would lead to the null hypothesis being accepted. **[6]**

It turns out that the population mean volume per can was 331.4 ml.

iii Find the probability of making a type II error, **[3]**

iv Write down the power of the test. **[1]**

5 A sample of size 12 is taken from a population modelled by a Normal distribution with unknown mean and variance. The sample values are:

54.1 50.9 46.3 49.2 56.3 51.1 50.6 48.3 47.2 51.4 52.0 50.3

The population mean is believed to be 52.0 and a 5% significance test is to be performed.

a Find the sample mean and variance, **[3]**

b Write the test statistic and calculate its value, **[3]**

c Calculate an appropriate p-value and state clearly your conclusion. **[2]**

6 a A sample of size 10 is taken from a population with unknown mean μ and variance σ^2

Explain why you should **not** assume that the test statistic $\dfrac{\overline{X}-\mu}{\dfrac{s}{\sqrt{n}}}$, where s is the sample standard deviation, follows a standard Normal distribution, **[1]**

b A sample of size 10 is taken from a Normally distributed population and has mean 12.1 and variance 4.4

The population mean was known to be 10.2 but is now thought to have increased. You wish to test this belief.

i Write down the null and alternative hypotheses for this test, **[1]**

ii Write down the test statistic and calculate its value, **[2]**

iii By calculating the appropriate p-value, perform the test at a significance level of 5% **[2]**

7 In a trial on the effectiveness of drugs designed to reduce cholesterol levels, 9 participants received a course of statins. The reduction in cholesterol level (mmol L^{-1}) for each person is as follows (– denotes an increase in level):

0.3 1.1 1.2 1.4 –0.3 0.5 –0.2 0.4 –0.5

a State the null and alternative hypotheses for a test of the effectiveness of the treatment, **[1]**

b Test at a 5% significance level whether the statin treatment affects cholesterol levels. **[7]**

You should state any assumptions you make.

c Comment on the confidence with which your final conclusion can be drawn. **[1]**

8 A sample of size 8 was taken from a Normal population and the values were as follows:

16.3 18.2 13.4 11.1 21.5 15.3 18.1 12.4

Find a 95% confidence interval for the mean. **[7]**

9 A sample of size 14 was taken from a Normal population and gave the following statistics:

$$\sum x = 94.2, \sum x^2 = 822.3$$

Find a 98% confidence interval for the mean. **[7]**

10 In an investigation into crop yield, 13 identical test beds are sown with a crop and the yield from each is noted. A 95% confidence interval for the population mean yield is (43.41, 44.99)

a State a necessary assumption related to the crop yield distribution, **[1]**

b Explain why it is not acceptable to use a Normal distribution to investigate the confidence interval for the population mean, **[1]**

c Find the sample mean and standard deviation. **[5]**

28 Graphs and networks 2

You will have noticed how transport systems, such as roads, get very congested at certain times of certain days: for example, during week-day mornings and late afternoons as large numbers of people travel to and from work. Equally, travel by train peaks at these times. Patterns also emerge over the course of the year as people take holidays in the summer months and travel to tourist destinations by the coast or in the countryside. The patterns underlying the movement of people need to be considered carefully by transport planners, who try to ensure that the infrastructure of roads and rail is adequate, as well as airports and passenger terminals for ferries at ports.

Transport planners use the mathematics of flow in such strategic planning to make sure that they can cater for the transportation expectations of workers, holidaymakers and other travellers. For example, it informs how many trains can travel on any section of the rail network and this, in turn, informs how many trains need to be bought. On the roads it informs the development of the 'smart motorway' network that seeks to optimise the flow of traffic along the busiest roads in the country.

Orientation

What you need to know	What you will learn	What this leads to
Ch12 Graphs and networks 1	• To construct the complement of a graph. • To identify isomorphic graphs. • To use Euler's formula. • To use Kuratowski's theorem. • To find the maximum flow through a network. • To solve problems involving multiple sources/sinks.	**Careers** • Transport industry. • Operational research.

28.1 Planar graphs and isomorphisms

Fluency and skills

See Ch12
For a reminder on graphs.

This section expands on some of the properties of graphs that were covered in the first year of this course. Recall that:

- A graph is a set of points (**vertices**) connected by a set of lines (**edges**).
- A graph is **connected** if, for every two vertices, there is a **trail** (a sequence of edges) joining one to the other.
- A graph may have loops (edges starting and ending at the same vertex) or multiple edges (two or more edges connecting a pair of vertices). A **simple graph** has no loops or multiple edges.
- A graph with 'one-way streets' is a **directed graph** or **digraph**.
- A simple graph with n vertices in which there is an edge connecting every pair of vertices is the **complete graph K_n** that has $\frac{n(n-1)}{2}$ edges.
- A graph with two distinct sets of vertices, in which every edge connects a vertex from one set with a vertex from the other, is a **bipartite graph**. If every vertex in a set of m vertices connects with every vertex in a set of n vertices, you have the **complete bipartite graph $K_{m,n}$** that has mn edges.
- The number of edges meeting at a vertex is its **degree**.
- The sum of the degrees = 2 × the number of edges (this is the **handshaking lemma** - each edge 'shakes hands' with two vertices).
- A trail which doesn't visit any vertex more than once (except perhaps the start/finish) is a **path**.
- A path which starts and ends at the same vertex is a **cycle**.
- A graph in which there is a closed trail that uses every edge of the graph only once is **traversable** or **Eulerian**. In an Eulerian graph, all vertices are of even degree.
- A cycle which visits every vertex of the graph once and only once and which returns to the starting vertex is a **Hamiltonian cycle**.
- A connected graph with no cycles is a **tree**. A **spanning tree** for a graph is a subgraph, which is a tree connecting all vertices. The spanning tree for a graph with n vertices has $(n - 1)$ edges.
- A **subgraph** of a graph G is a graph formed by some of the vertices and edges of G
- A **subdivision** of a graph G is a graph formed by adding vertices of degree 2 along edges of G

Consider Graph G

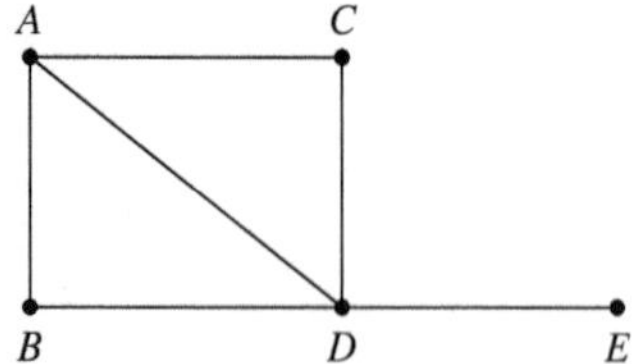

To describe G you can list the **vertex set** $\{A, B, C, D, E\}$ together with the **edge set** $\{AB, AC, AD, BD, CD, DE\}$ of G

Alternatively, you can construct the **adjacency matrix** of G. This shows the number of connections between each pair of vertices (an undirected loop would be recorded as 2).

	A	B	C	D	E
A	0	1	1	1	0
B	1	0	0	1	0
C	1	0	0	1	0
D	1	1	1	0	1
E	0	0	0	1	0

The **complement** of a simple graph G, denoted by G' (in some books by $\bar{G}$), and has the same vertices as G but contains only the edges which are *not* in G. It looks like this.

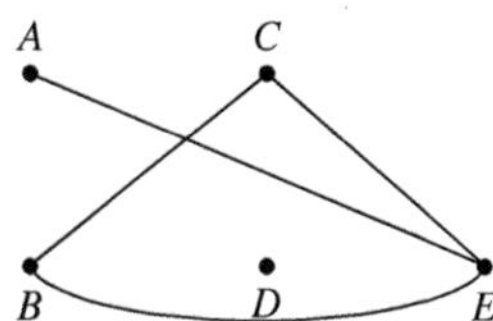

You find the adjacency matrix for G' by changing 1s to 0s and vice versa (ignoring the leading diagonal).

	A	B	C	D	E
A	0	0	0	0	1
B	0	0	1	0	1
C	0	1	0	0	1
D	0	0	0	0	0
E	1	1	1	0	0

If you combine G and G' you get the complete graph K_5

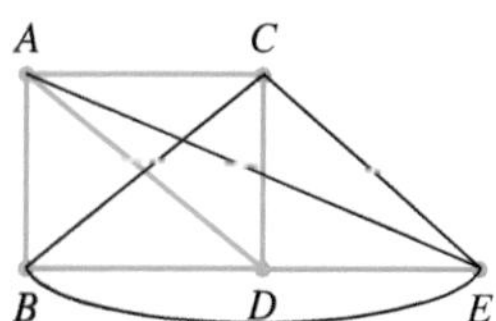

In general, if a simple graph has n vertices and you combine it with its complement, you get the complete graph K_n

Graphs G_1 and G_2 are **isomorphic** if they have the same structure. This means that for each vertex in G_1 there is a corresponding vertex in G_2, so that if an edge joins two vertices in G_1, then there is an edge between the corresponding vertices in G_2, and vice versa. Provided you list corresponding vertices in the same order, the adjacency matrices will look identical.

For example, look at these graphs.

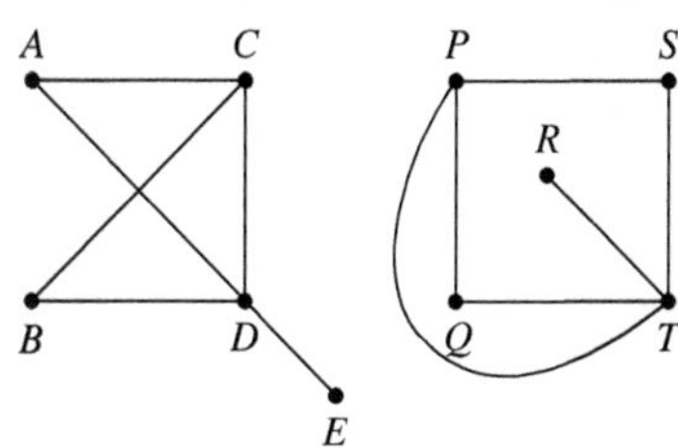

DISCRETE

If you pair vertices A, B, C, D and E with vertices Q, S, P, T and R, respectively, then the edge set $\{AC, AD, BC, BD, CD, DE\}$ of the first graph corresponds to the edge set $\{QP, QT, SP, ST, PT, TR\}$ of the second graph. The adjacency matrices are

	A	*B*	*C*	*D*	*E*
A	0	0	1	1	0
B	0	0	1	1	0
C	1	1	0	1	0
D	1	1	1	0	1
E	0	0	0	1	0

	Q	*S*	*P*	*T*	*R*
Q	0	0	1	1	0
S	0	0	1	1	0
P	1	1	0	1	0
T	1	1	1	0	1
R	0	0	0	1	0

When looking for corresponding vertices, it can help to use the degrees of the vertices – for example, in this case, you would expect to pair D with T because each has degree 4. However, just because two graphs have the same set of degrees you cannot assume they are isomorphic. For example, look at these graphs.

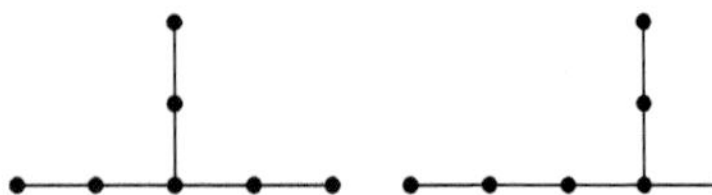

They each have one vertex of degree 3, three vertices of degree 2 and three of degree 1. However, they are clearly not isomorphic.

Example 1

The diagram shows the graph G

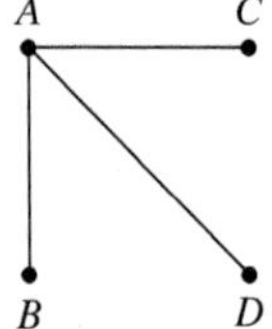

a Construct the adjacency matrix for G

b Find the adjacency matrix for the complement, G' and sketch G'

c Show that G and G' are not isomorphic.

d Find a graph H with 4 vertices such that H and H' are isomorphic.

a

	A	B	C	D
A	0	1	1	1
B	1	0	0	0
C	1	0	0	0
D	1	0	0	0

b

	A	B	C	D
A	0	0	0	0
B	0	0	1	1
C	0	1	0	1
D	0	1	1	0

Change 1s to 0s and 0s to 1s except on the leading diagonal.

(Continued on the next page)

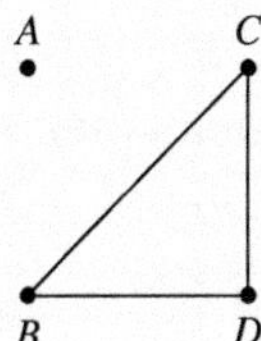

c G has vertices with degrees 3, 1, 1, 1. G' has degrees 2, 2, 2, 0, so they are not isomorphic.

d K_4 has $\frac{4\times3}{2} = 6$ edges and therefore the sum of the degrees of the vertices is 12

If H has 4 vertices and H and H' are isomorphic, then each has 3 edges, the degrees of their vertices are identical and each total 6. The degrees must be 2, 2, 1, 1 because the degrees of the complement would then be 1, 1, 2, 2

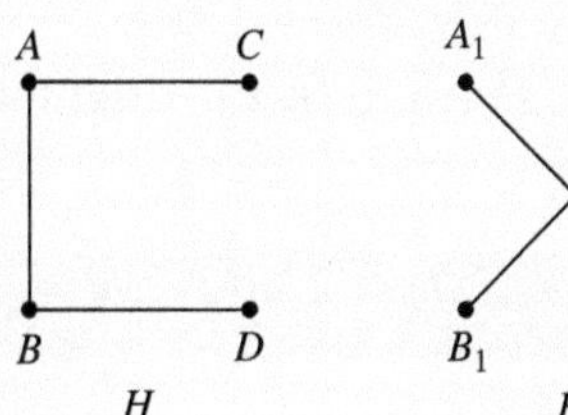

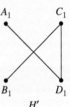

These are isomorphic. If you pair $\{C, A, B, D\}$ with $\{A_1, D_1, C_1, B_1\}$ in that order, then the edges $\{CA, AB, BD\}$ correspond to $\{A_1D_1, D_1C_1, C_1B_1\}$.

A graph is **planar** if it can be drawn without edges intersecting except at a vertex.

A face in a plane drawing is a region bounded by edges.

Key point

If a simple, connected graph drawn in the plane has V vertices, E edges and F faces (including the 'infinite face' surrounding the graph), then $\boldsymbol{V + F - E = 2}$ (Euler's formula). (If a graph is drawn in the plane then it is planar.)

Edges leading to vertices of degree 1, multiple edges and loops can all be temporarily removed from a graph and then added back once the graph has been drawn in the plane.

- K_n is non-planar if $n \geq 5$
- $K_{m,n}$ is non-planar if $m \geq 3$ and $n \geq 3$

There are some useful corollaries (consequences) of Euler's formula.

- If a graph drawn in the plane has no vertices of degree 1, loops or multiple edges then every edge borders two faces and every face has at least three edges.

  It follows that $2E \geq 3F$. Substituting this into Euler's formula gives $E \leq 3V - 6$. If $V = 5$, then $E \leq 9$. As K_5 has 10 edges, it follows that K_5 is non-planar.

- If you have a bipartite graph, every edge is associated with two faces and every face has at least four edges. It follows that $2E \geq 4F$. Substituting this into Euler's formula gives $E \leq 2V - 4$. If $V = 6$, then $E \leq 8$. As $K_{3,3}$ has 9 edges, it follows that $K_{3,3}$ is non-planar.

Notice that if $E \leq 3V - 6$, it does not mean that the graph is planar. However, if $E > 3V - 6$, then the graph is non-planar.

Key point

A graph is non-planar if and only if it has a subgraph which is a subdivision of K_5 or $K_{3,3}$ (**Kuratowski's theorem**).

For example:

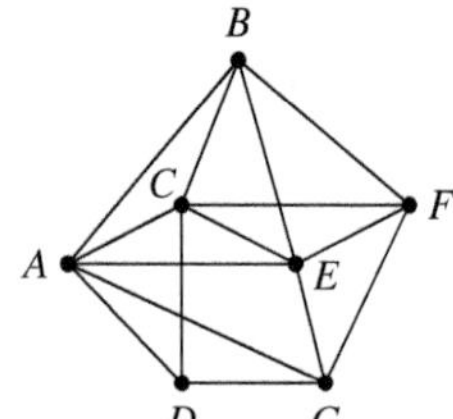

can be drawn as

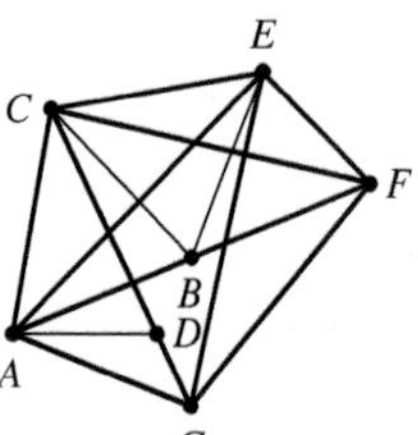

with subgraph

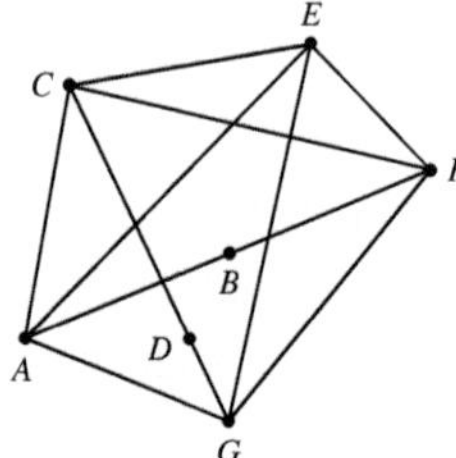

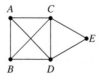

This is a subdivision (by vertices B and D) of the K_5 graph formed by A, C, E, F and G. The original graph is therefore non-planar.

Exercise 28.1A Fluency and skills

1

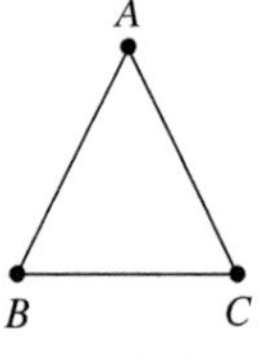

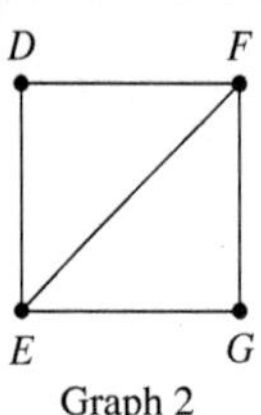

Identify

a A sub-graph of graph 2 which is isomorphic to Graph 1

b A sub-graph of graph 2 which is isomorphic to a subdivision of Graph 1

2 Identify a sub-graph of this graph which is a subdivision of the complete graph K_4

3 Identify a sub-graph of this graph which is isomorphic to the complete bipartite graph $K_{3,2}$

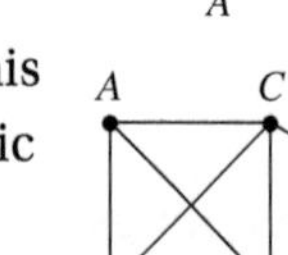

4 For graphs G_1, G_2, G_3 and G_4 shown, identify any pairs which are isomorphic. For each isomorphic pair you find, state a correspondence between their vertices and edges.

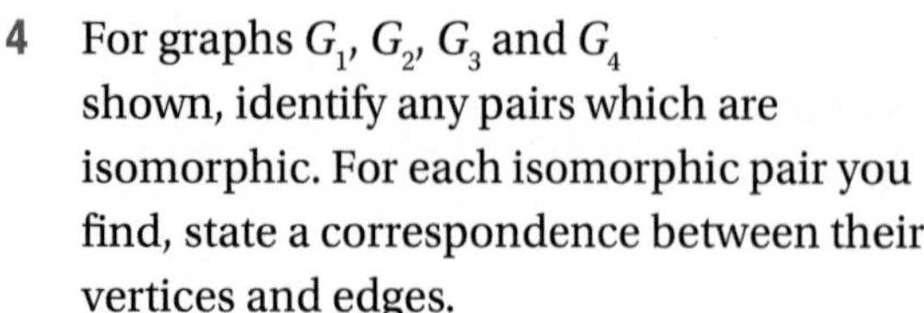

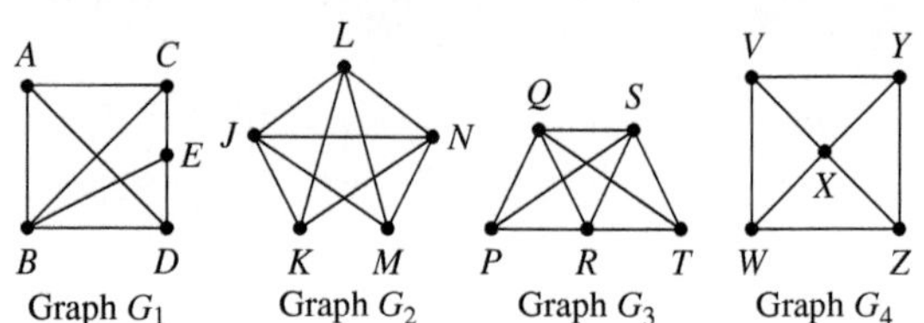

5 State whether each of these graphs is planar or non-planar. If the graph is planar, show this by redrawing it. If it is non-planar, explain using Kuratowski's theorem why this is the case.

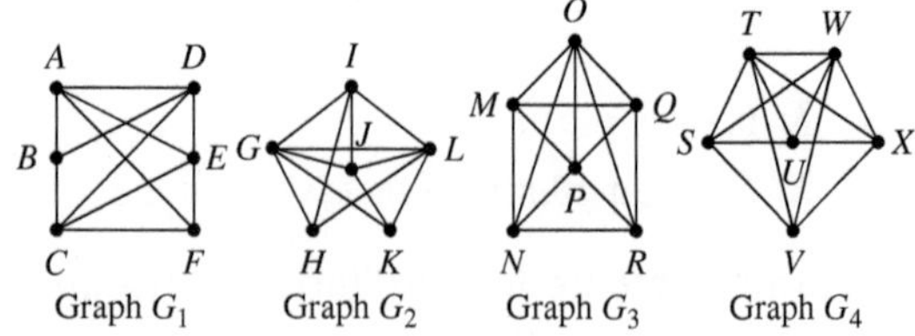

6 a Draw a graph to show that $K_{4,2}$ is planar.

b Draw the complement of $K_{4,2}$

7 This table is the adjacency matrix for the graph G

	A	B	C	D	E
A	0	0	1	0	1
B	0	0	0	1	1
C	1	0	0	0	0
D	0	1	0	0	1
E	1	1	0	1	0

a Construct the adjacency matrix of the graph G', the complement of G

b Draw the graphs G and G'

c State, with reasons, whether G and G' are isomorphic.

8 Show, by finding a sub-graph which is a subdivision of $K_{3,3}$, that this graph is non-planar.

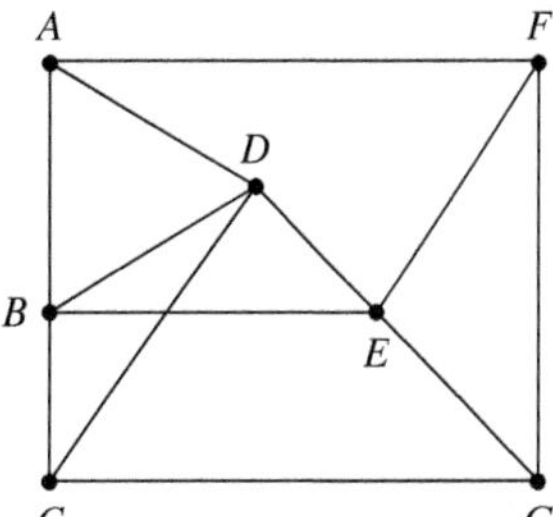

Reasoning and problem-solving

Strategy

To solve problems in graph theory

1. Draw clear, labelled diagrams.
2. Use the correct terminology.
3. Use theorems and rules to prove results.
4. Answer the question.

Example 2

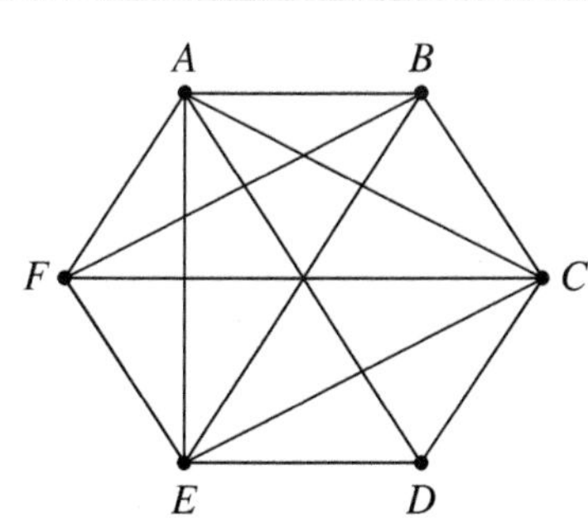

a By considering the number of edges and vertices, show that the graph is non-planar,

b Use Kuratowski's theorem to show that if CE is removed the graph is still non-planar.

(*Continued on the next page*)

DISCRETE

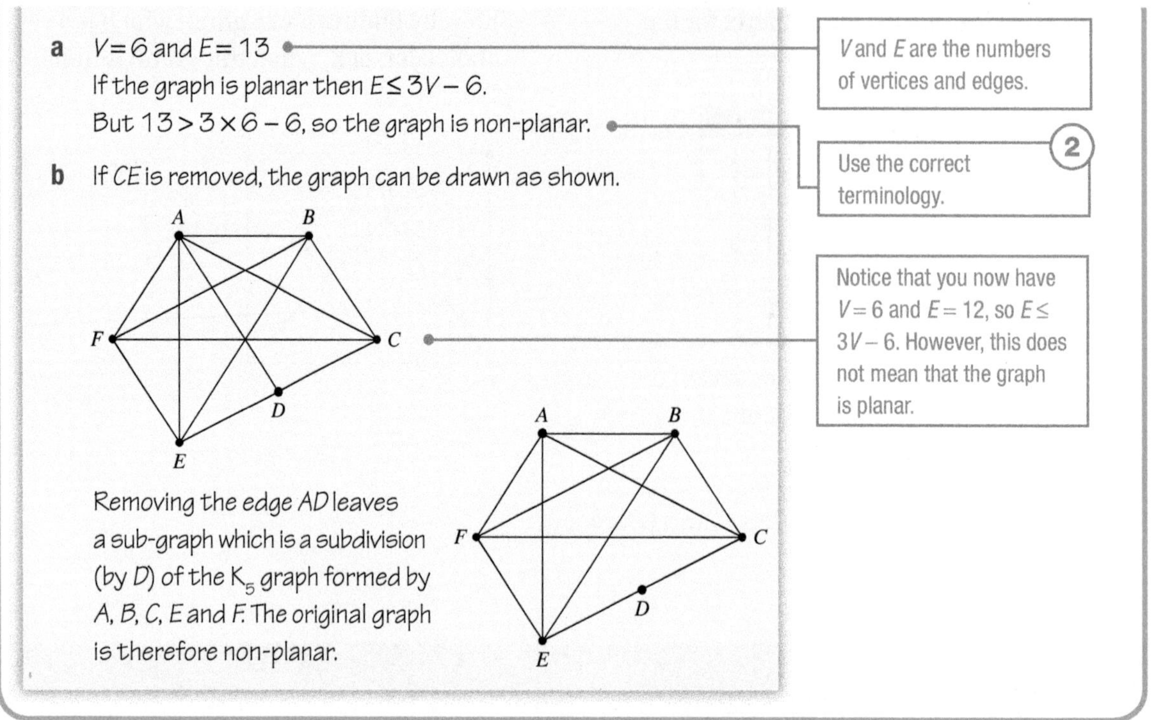

a V=6 and E=13
If the graph is planar then E≤3V−6.
But 13>3×6−6, so the graph is non-planar.
b If CE is removed, the graph can be drawn as shown.
A
B
F
C
D
E
Removing the edge AD leaves a sub-graph which is a subdivision (by D) of the K5 graph formed by A, B, C, E and F. The original graph is therefore non-planar.
V and E are the numbers of vertices and edges.
Use the correct terminology.
2
Notice that you now have V=6 and E=12, so E≤3V−6. However, this does not mean that the graph is planar.

Exercise 28.1B Reasoning and problem-solving

1 T is a spanning tree of K_n

a State the number of edges in T

b T' is the complement of T. For what value(s) of n could T and T' be isomorphic? Draw an example where this is true.

2 a Find the number of distinct, simple graphs with 3 vertices (ignoring isomorphisms).

b Find the number of distinct, simple, connected graphs with 4 vertices.

c A graph has 3 vertices, each of degree 4 (the graph cannot be simple).

i If the graph is connected, how many distinct graphs are possible?

ii How many more are there if they do not have to be connected?

3 A simple, connected graph has n vertices, each of degree d

a If the graph is non-Eulerian, what can you say about n?

b If $n = 6$, for what values of d is the graph planar?

4 A coding system uses three-digit binary codes $A(000)$, $B(001)$, $C(010)$, $D(011)$, $E(100)$, $F(101)$, $G(110)$ and $H(111)$.
A single-bit transmission error could change, for example, A into E

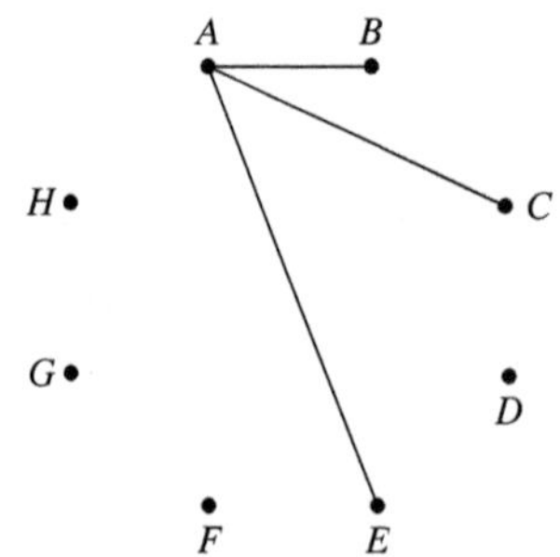

a Copy and complete this graph in which an edge connects codes which are just one error apart.

b How long is the shortest path from

i A to D

ii A to H?

c Show by drawing that the graph is planar.

5 The diagram shows the Petersen graph, which occurs in many mathematical topics related to graph theory.

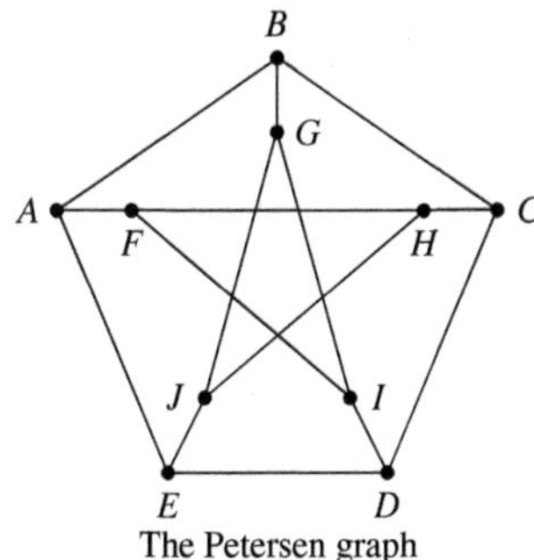

The Petersen graph

By considering the vertex sets $\{A, C, G\}$ and $\{B, E, F\}$, make a drawing to show that the Petersen graph has a subgraph which is a subdivision of $K_{3,3}$ and hence is non-planar.

6 a Draw a graph with four vertices which is isomorphic to its complement.

b Draw two distinct graphs with five vertices which are each isomorphic to their complement.

c A graph G has n vertices.

i If $n = 6$, explain why it is not possible for G and G' to be isomorphic.

ii For what values of n could G and G' be isomorphic?

7 The diagram shows a graph G which is not connected.

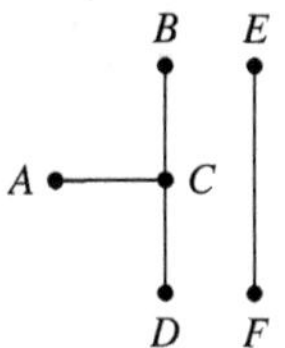

a Draw the complement, G'

b The vertices of a disconnected graph F are in two separate disconnected sets, $\{A_1, A_2, \ldots A_m\}$ and $\{B_1, B_2, \ldots, B_n\}$, each of which forms a connected subgraph of F

For what values of m and n is F' bound to be non-planar?

c In a, the complement of a disconnected graph was connected. Prove that this is always the case, that is, for any disconnected graph H, the complement H' is connected.

8 Use Kuratowski's theorem to show that this graph is non-planar.

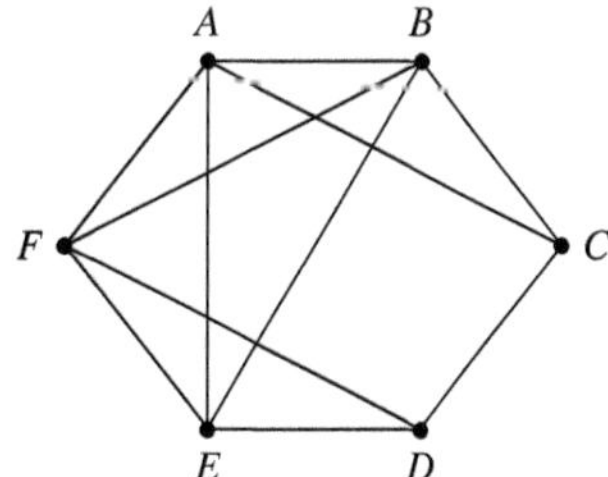

DISCRETE

28.2 Network flows 2

Fluency and skills

See Ch12.8
For a reminder on networks.

In Chapter 12 you met the idea of a commodity flowing through a directed network from a **source** (usually called S) to a **sink** (usually called T).

Each arc has a **capacity** and the **flow** along it cannot be more than this. (This is the **feasibility condition**.) Together these flows form the **flow in the network**.

If the flow in an arc equals its capacity you say it is **saturated**.

At every node of the network the total inflow equals the total outflow. (This is the conservation condition.)

The total outflow from S equals the total inflow at T. This quantity is the **value of the flow**. The **capacity of the network** is the value of the maximum possible flow.

You describe a **cut** either by listing the arcs in the cut (the cut set), or by listing the nodes in the source set X and in the sink set Y

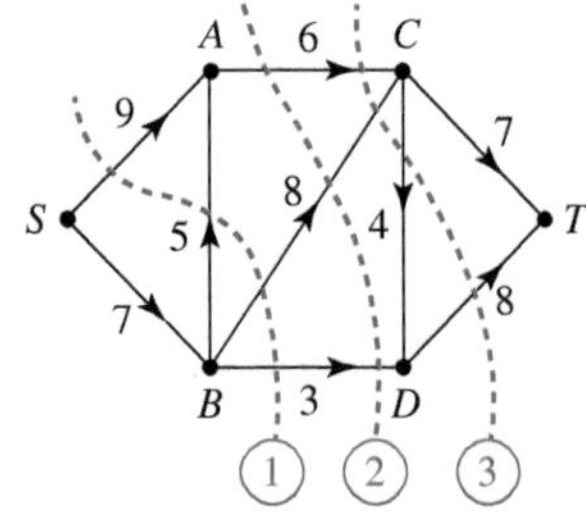

For this network, cut 2 has cut set = {AC, BC, BD}, source set X = {S, A, B} and sink set Y = {C, D, T}.

The capacity of a cut is the sum of the capacities of those arcs of the cut which are directed from X to Y. In this network, cut 1 has a capacity of 25, cut 2 a capacity of 17 and cut 3 a capacity of 22

The flow in the network cannot be more than the capacity of any cut. This gives the **maximum flow, minimum cut theorem**.

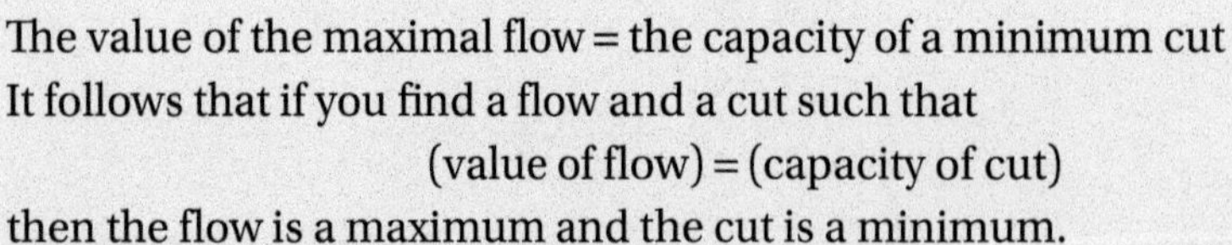

Key point

The value of the maximal flow = the capacity of a minimum cut
It follows that if you find a flow and a cut such that

(value of flow) = (capacity of cut)

then the flow is a maximum and the cut is a minimum.

You need a systematic way of finding the maximal flow. You start with an initial flow of some sort (which could just be zero, although you will usually do better than this) and look for ways to **augment** (increase) it. This **flow augmentation** process may happen several times until you have the maximal flow.

For example, look at this diagram.

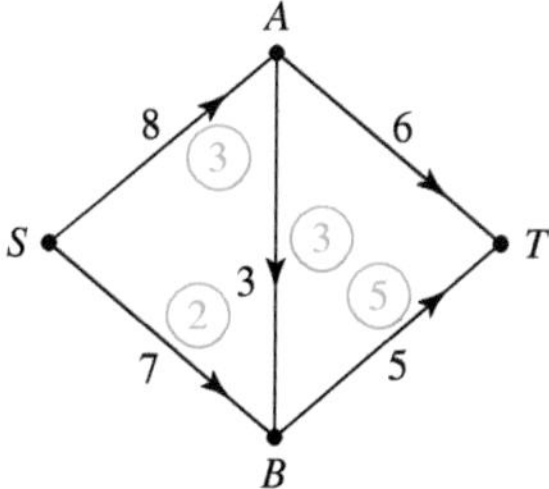

It shows a network with an initial flow of 3 along *SABT* and 2 along *SBT*. (The circled values show the flow in each arc.)

You subtract these flows from the capacities of the arcs to see how much spare capacity remains, as shown in this diagram. You can see that a flow of 5 is possible along *SAT*. *SAT* is called a **flow-augmenting path**.

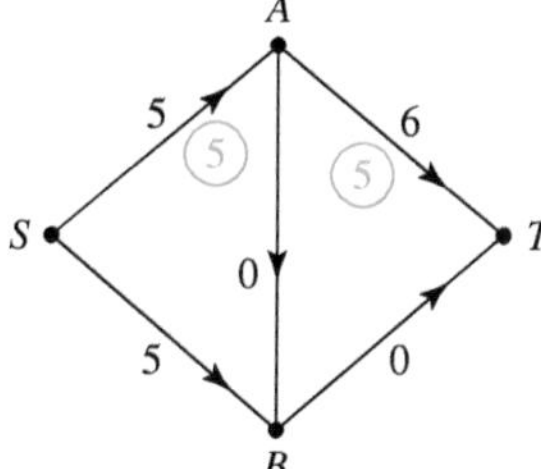

You again subtract this flow to see what capacity remains.

There is no obvious flow-augmenting path, but you have not yet reached the maximal flow.

If you increase the flow in *SB* and *AT* by 1, and **reduce** the flow in *AB* by 1, the overall flow increases by 1

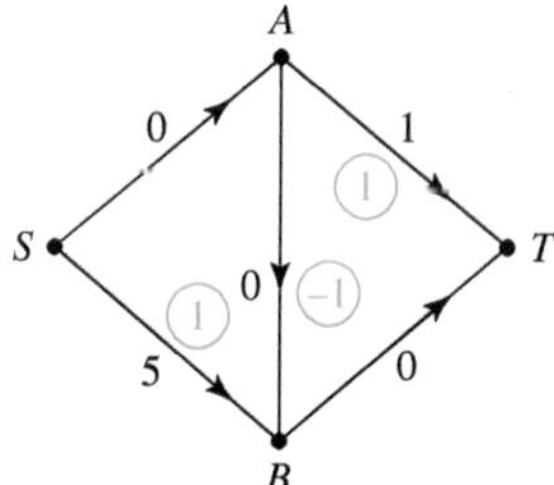

You now have an overall flow of 11, as shown.

The arcs *AT* and *BT* are saturated, so the flow is maximal.

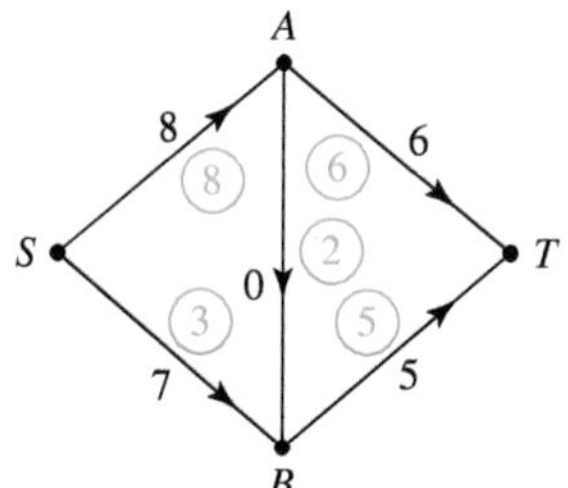

DISCRETE

It is not easy to spot flow-augmenting paths, especially if they involve reducing a flow. It is also tedious to keep redrawing the diagram. The **labelling procedure** overcomes these problems.

Suppose an arc *AB* of capacity 12 has a flow of 8 along it. So far you have shown this as

You could increase the flow by 4, this is the **potential flow**.

You could decrease the flow by 8, this is the **potential backflow**.

You show these as

The forward arrow represents the spare capacity while the backward arrow represents the actual flow. The total of the two gives the capacity of the arc.

To find a flow-augmenting path you look for a route from *S* to *T* where all the potential flows (arrows pointing forwards along the route) are greater than zero.

Key point

A flow-augmenting path is a route from *S* to T where the **arrows pointing forward** all have **non-zero values.**

The lowest of these values gives the possible extra flow along that path.

Example 1

Starting with an initial flow of 3 along the route *SABT* and 2 along *SBT*, find a maximal flow through this network.

Notice that the labelling procedure automatically reduces the flow along *AB* by 1 in the second stage of this solution.

This diagram shows the initial flow.

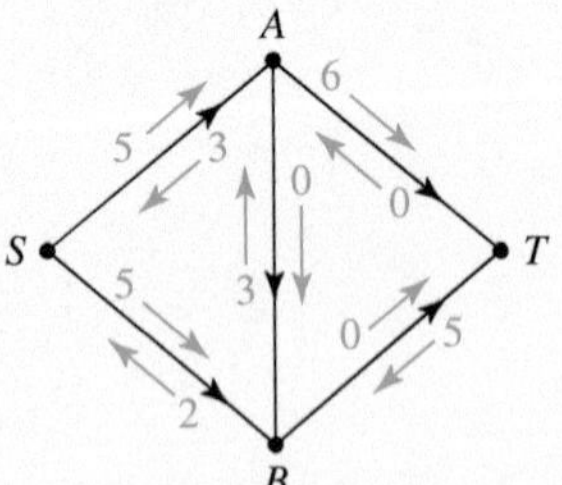

SAT is a flow-augmenting path with potential flow of 5

The arrows pointing forward along the route *SAT* are ≥ 5

(*Continued on the next page*)

Updating the labels gives this flow.

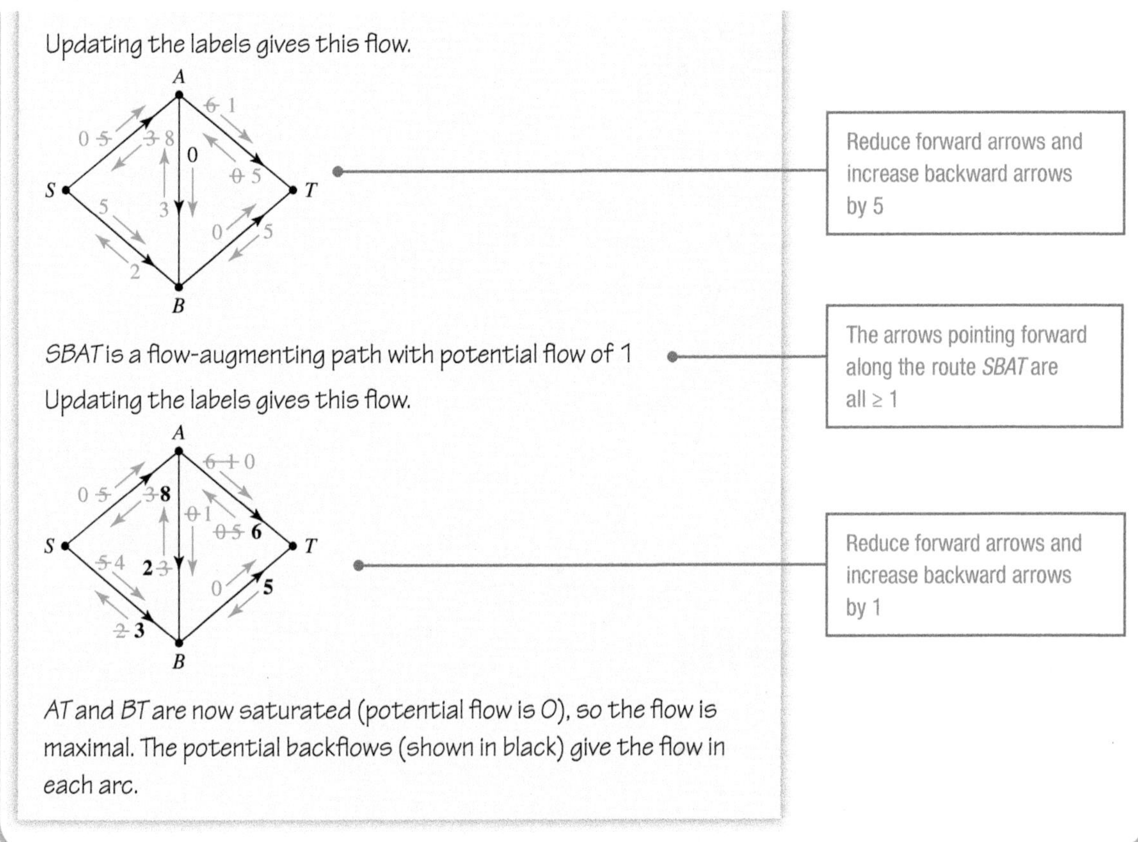

Reduce forward arrows and increase backward arrows by 5

SBAT is a flow-augmenting path with potential flow of 1

The arrows pointing forward along the route *SBAT* are all ≥ 1

Updating the labels gives this flow.

Reduce forward arrows and increase backward arrows by 1

AT and *BT* are now saturated (potential flow is 0), so the flow is maximal. The potential backflows (shown in black) give the flow in each arc.

Example 2

Taking a flow of 16 along *SACT* and 10 along *SBDT* as the initial flow:

a Use flow-augmenting paths to find a maximal flow for this network. Record your working in this table.

Augmenting path	Flow

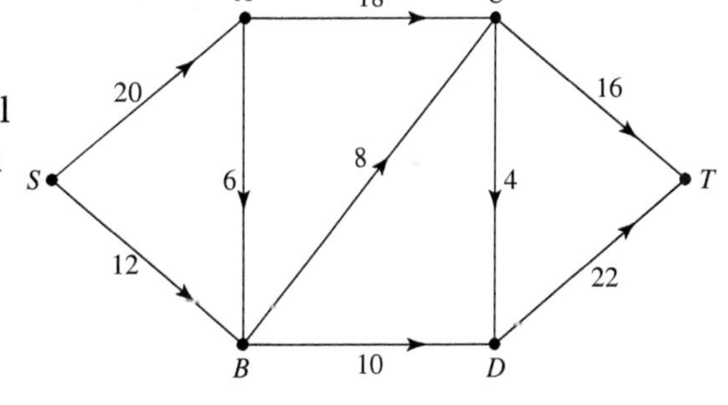

b Use the maximum flow, minimum cut theorem theorem to show that your flow is maximal.

a The initial flow is as shown.

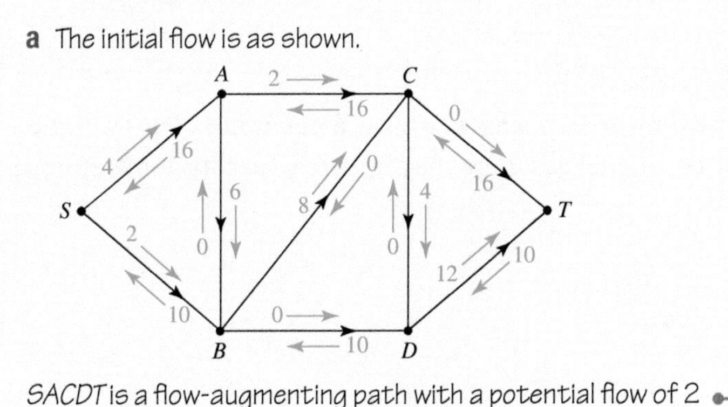

SACDT is a flow-augmenting path with a potential flow of 2

The arrows pointing forward along the route *SACDT* are ≥ 2

(*Continued on the next page*)

Augmenting path	Flow
SACDT	2

Update the labels to include this flow, as shown.

Reduce forward arrows and increase backward arrows by 2

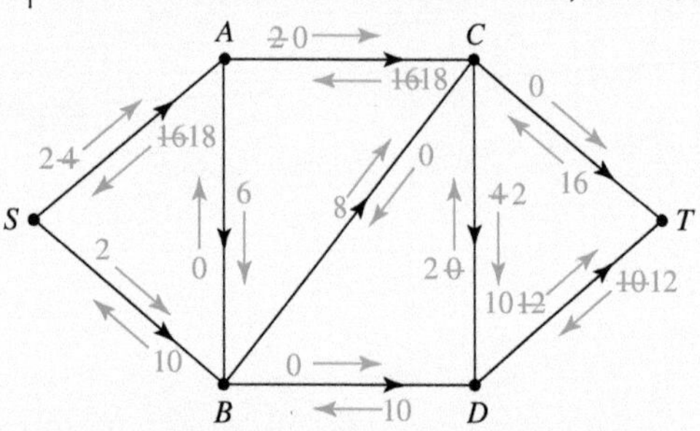

SBCDT is a flow-augmenting path with a potential flow of 2

Augmenting path	Flow
SACDT	2
SBCDT	2

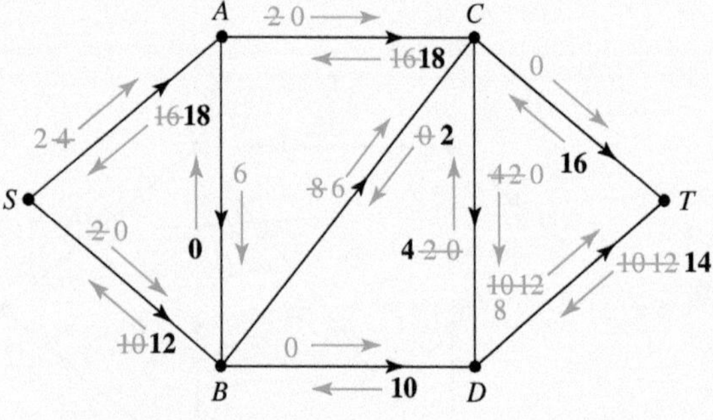

Update the labels to include this flow, as shown.

There is no flow-augmenting path available, so the flow (shown in black) is maximal.

The flow has been increased by a total of 4, so the value of the maximal flow is 30

In some circumstances arcs may have a **minimum capacity** as well as a maximum. For example, if the arcs represent pipelines carrying inflammable gas there may be safety reasons for keeping a certain amount of gas flowing.

The first task is to find an initial feasible flow. Previously you could start with an initial flow of zero, but this is not possible now. The flow through each node must be compatible with both the minimum and maximum total inflows and outflows.

Look at this partial network.

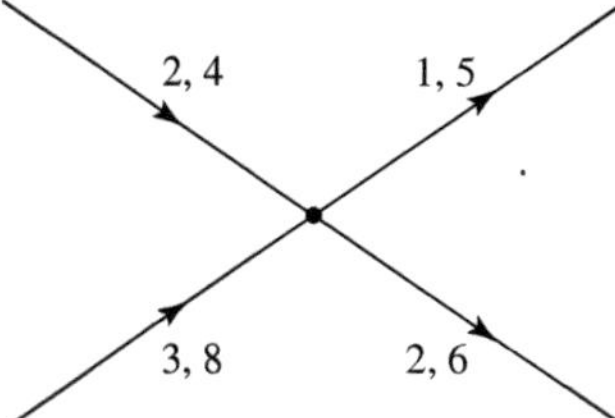

The values shown on each arc give the minimum and maximum permissible flow in that arc. At least 5 units must flow through this node because of the minimum incoming capacities. No more than 11 units can flow because of the maximum outgoing capacities.

In rare cases it is not possible to find a flow. This happens if the maximum outflow is less than the minimum inflow, or if the maximum inflow is less than the minimum outflow.

Once you have found an initial flow, you use the labelling procedure as before to find flow-augmenting paths. The only difference is that the potential flow and backflow in an arc must be compatible with the minimum and maximum flows in that arc.

For example, this diagram shows a flow of 6 units. The flow could be increased by 2 or decreased by 3

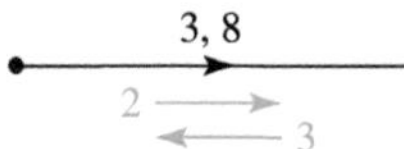

You can find the flow (6) from the diagram as (min capacity (3) + potential backflow (3)) or as (maximum capacity (8) – potential flow(2)).

To find the capacity of a cut you need to allow for the minimum flow for any arc directed from the sink set to the source set.

For example, $7 + 9 = 16$ could flow left to right through this cut, but at least 4 must flow right to left. The capacity of the cut is therefore $16 - 4 = 12$

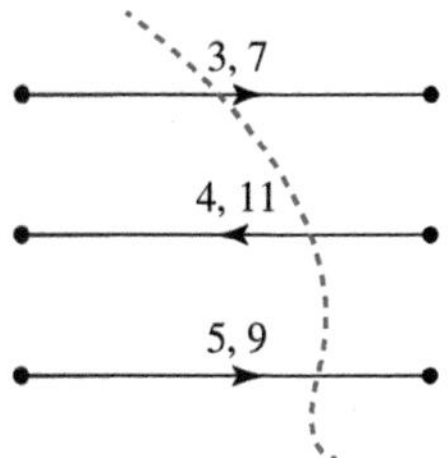

Key point

Capacity of cut = (maximum capacities from S to T) – (minimum capacities from T to S)

DISCRETE

Example 3

For this network, an initial feasible flow is *SAT* 3, *SABT* 4, *SBT* 5

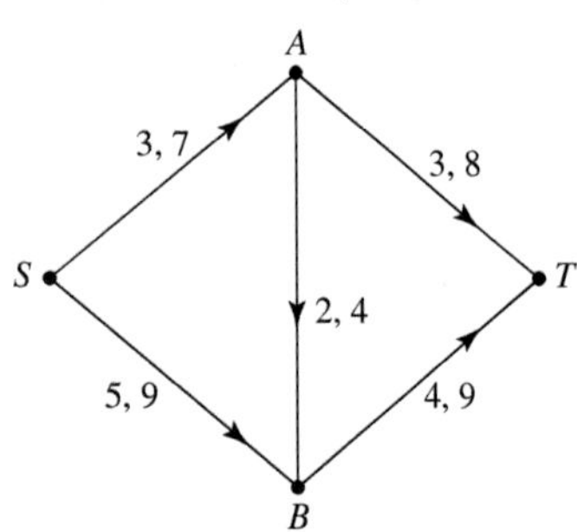

a Use flow augmentation to find a maximal flow,

b Confirm the result using the maximum flow, minimum cut theorem.

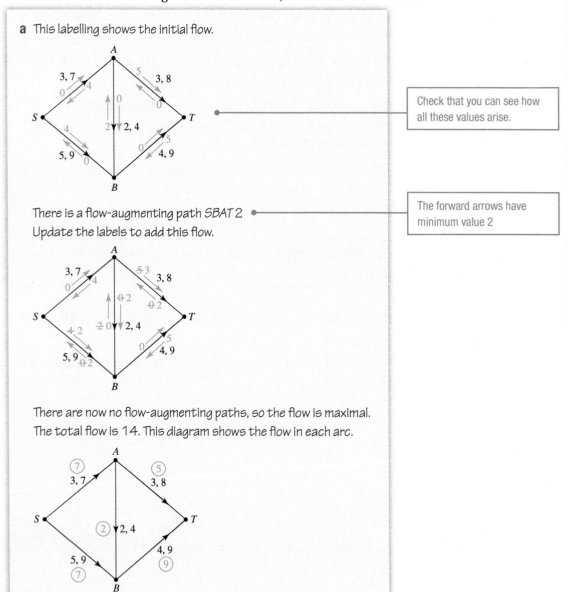

a This labelling shows the initial flow.

Check that you can see how all these values arise.

There is a flow-augmenting path *SBAT* 2
Update the labels to add this flow.

The forward arrows have minimum value 2

There are now no flow-augmenting paths, so the flow is maximal.
The total flow is 14. This diagram shows the flow in each arc.

(*Continued on the next page*)

b The cut {SA, AB, BT} shown in this diagram has capacity $7 + 9 - 2 = 14$

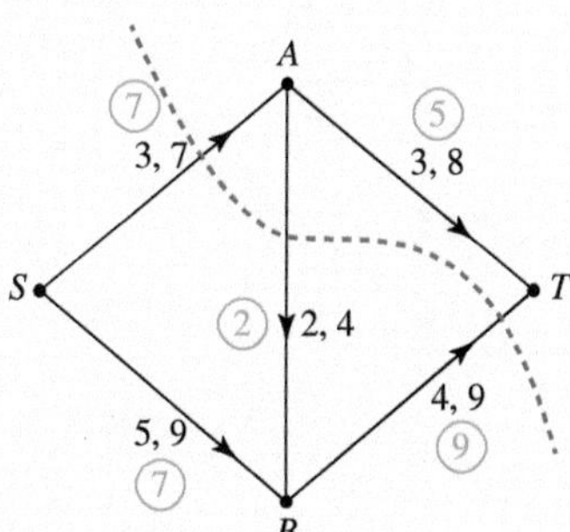

There is a flow of 14 and a cut of 14, so by the maximum flow, minimum cut theorem this flow is maximal.

Exercise 28.2A Fluency and skills

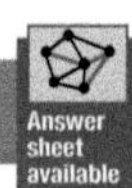

1 For this network, take an initial flow comprising a flow of 15 along *SAT*, 14 along *SBT* and 10 along *SCT*

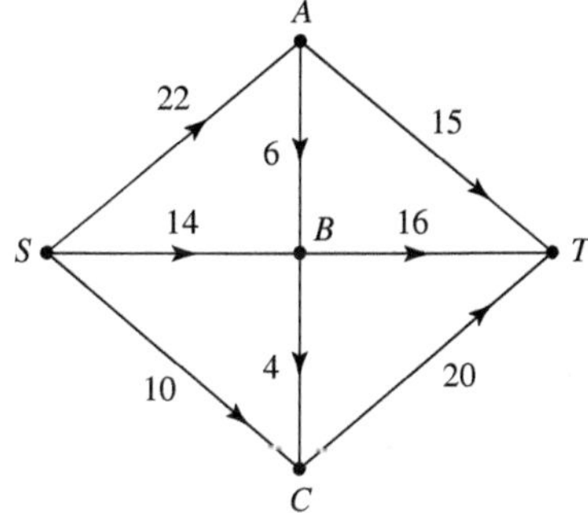

a Use the labelling procedure to augment the flow until a maximal flow is obtained.

b Confirm that your flow is maximal by using the maximum flow, minimum cut theorem.

2 For this network, take an initial flow comprising a flow of 20 along *SADT*, 20 along *SBCET* and 40 along *SBFT*

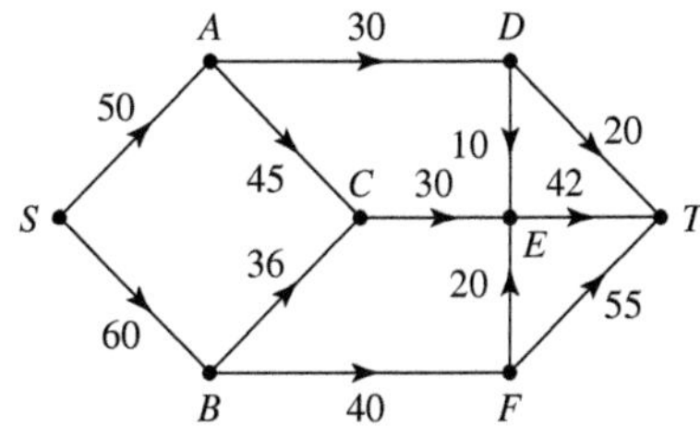

a Use the labelling procedure to augment the flow until a maximal flow is obtained.

b Confirm that your flow is maximal by using the maximum flow, minimum cut theorem.

3 Repeat question **2a** starting with a flow of 20 along *SADT*, 30 along *SACET* and 40 along *SBFT*

4 In this network, the maximum possible outflow from *S* and the maximum possible inflow to *T* are both 16

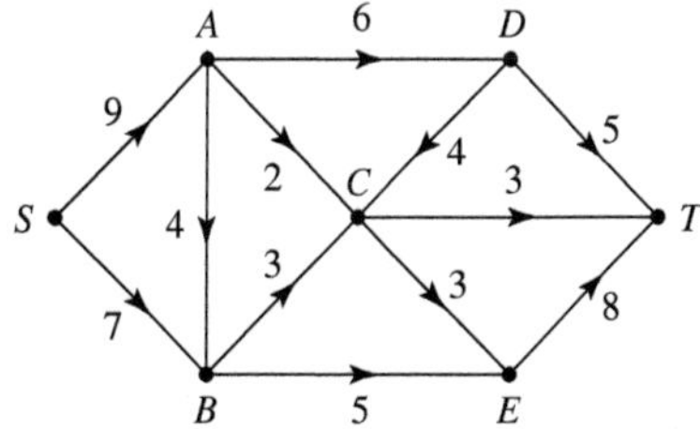

By starting with a flow of 5 along *SADT* and 5 along *SBET*, use flow augmentation to obtain a flow pattern with this maximum value.

DISCRETE

5 For this network

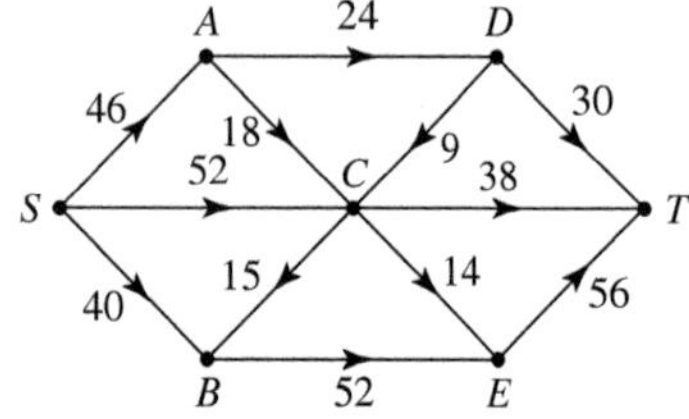

a Use an initial flow and flow augmentation to find the maximal flow.

b Use the maximum flow, minimum cut theorem to confirm that your flow is maximal.

6 Find a feasible flow, if one exists, for each of these networks.

a

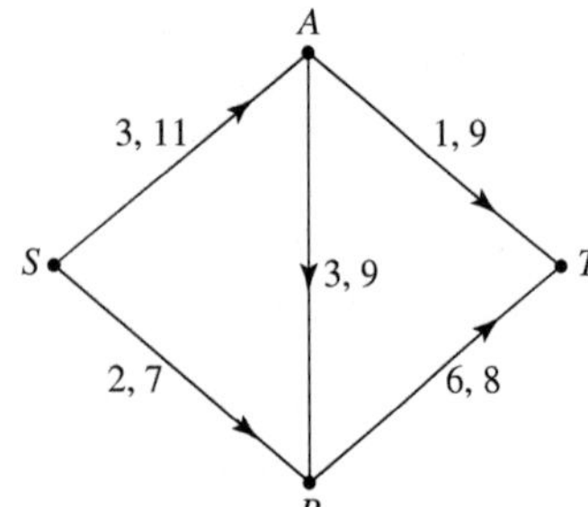

b

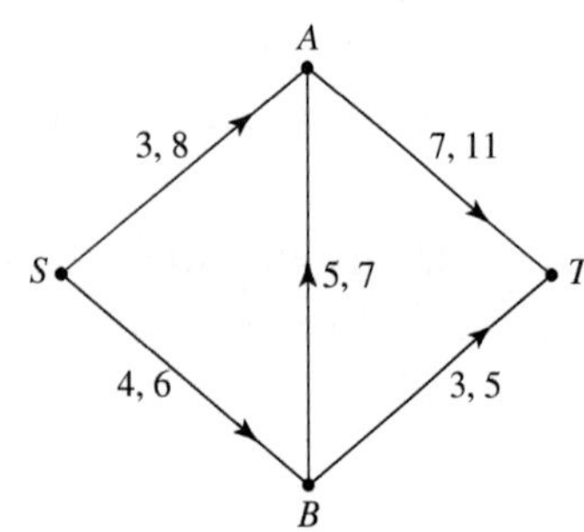

c

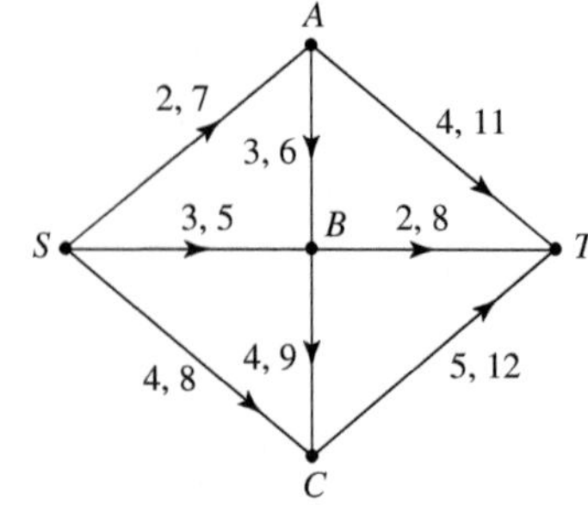

d

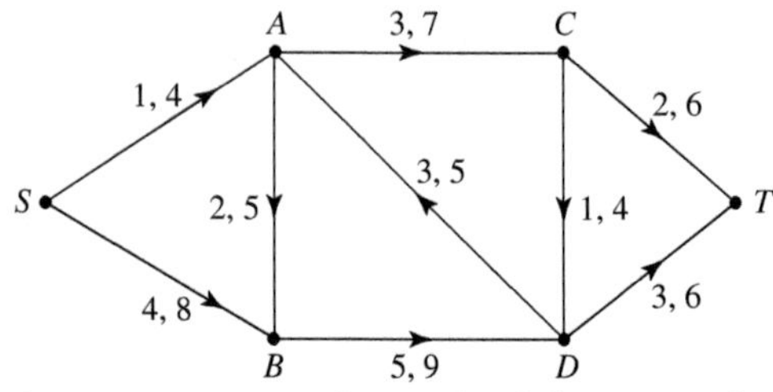

7 Find a minimum cut for each of the networks in question **6** for which a feasible flow exists.

8 This network shows the minimum and maximum flows along each arc and the circled values show a feasible flow.

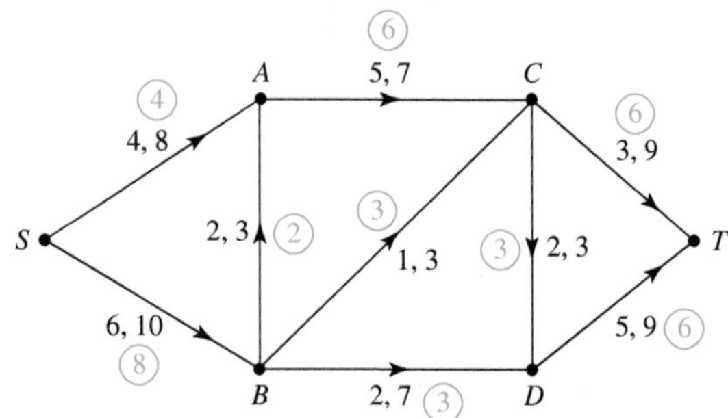

a Use the labelling procedure to find a maximum flow for the network.

b Confirm that it is a maximum by using the maximum flow, minimum cut theorem.

Reasoning and problem-solving

Strategy

To solve problems involving network flow

1. Use clear network diagrams.
2. Find an initial flow and label your diagram.
3. Find a flow-augmenting path and update the labels to include this flow.
4. Continue until there is no flow-augmenting path.
5. If required, use the maximum flow, minimum cut theorem to confirm your flow is maximal.
6. Answer the question.

If there is more than one source or sink, you introduce a dummy supersource or supersink.

Example 4

Use an initial flow and flow augmentation to find a maximal flow through this network.

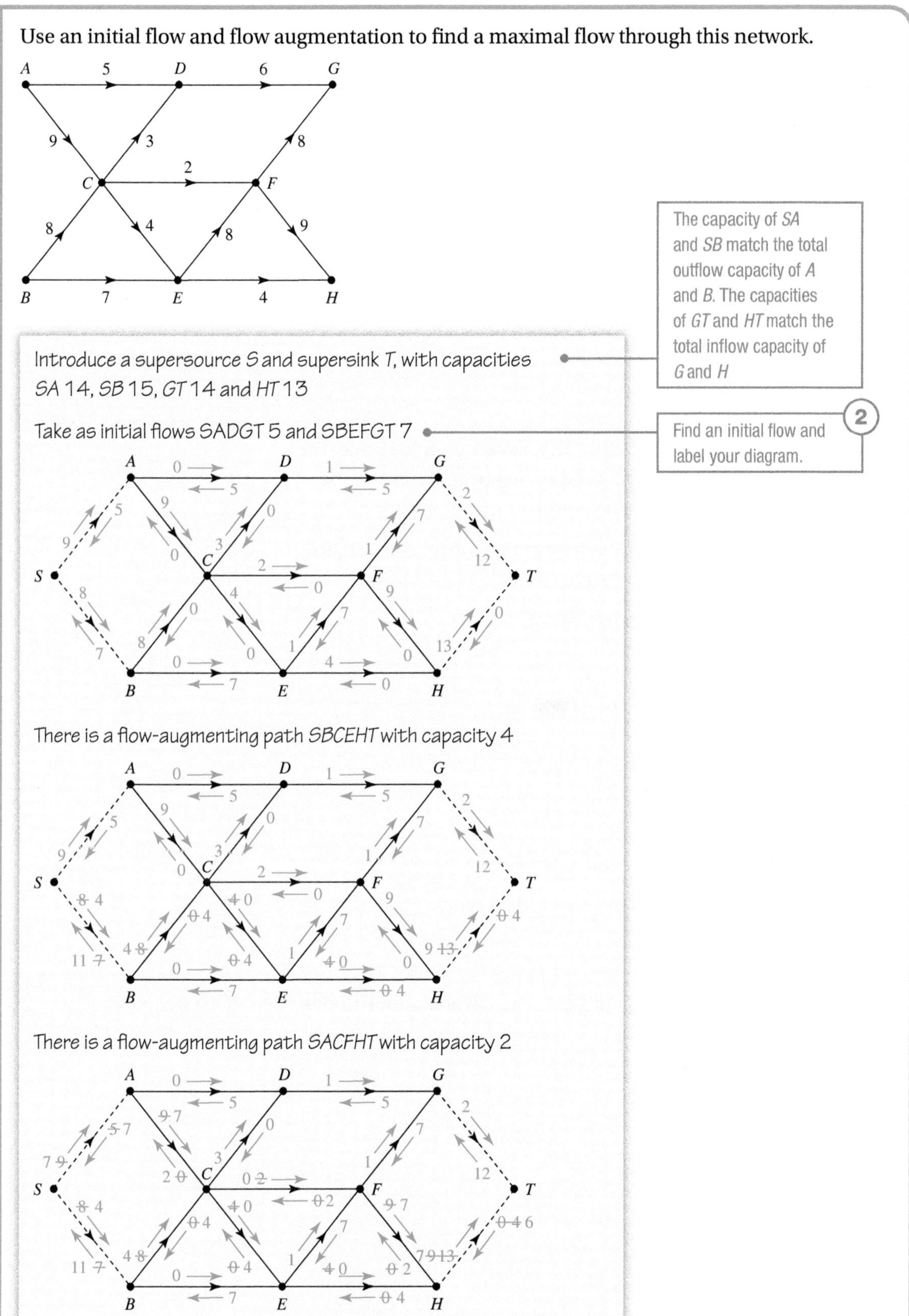

Introduce a supersource S and supersink T, with capacities SA 14, SB 15, GT 14 and HT 13
Take as initial flows SADGT 5 and SBEFGT 7
There is a flow-augmenting path SBCEHT with capacity 4
There is a flow-augmenting path SACFHT with capacity 2
The capacity of SA and SB match the total outflow capacity of A and B. The capacities of GT and HT match the total inflow capacity of G and H
2
Find an initial flow and label your diagram.

(Continued on the next page)

There is a flow-augmenting path *SBCDGT* with capacity 1

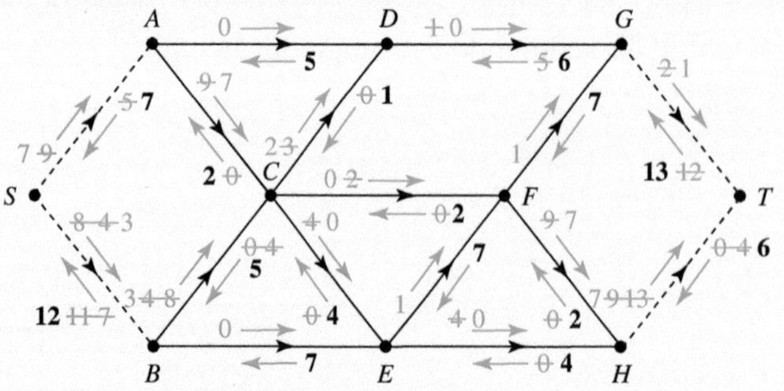

DG, *CF*, *CE* and *BE* are now saturated, so the flow is maximal.

The total network flow is 19

The flows on the arcs are shown in black.

> **4** Continue until there is no flow-augmenting path.

> **6** Answer the question.

> **4** You could now remove the dummy nodes and arcs to leave the solution for the original network.

There can be a restriction on the amount of flow through a node. For example, the capacity of a pumping station to pump water may be less than the capacity of the incoming and outgoing pipes.

To deal with this you draw a modified network, with an extra arc to represent the flow through the restricted node.

For example, if this node has a capacity of 4

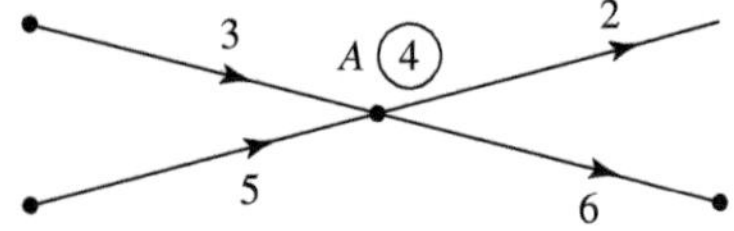

you redraw it as

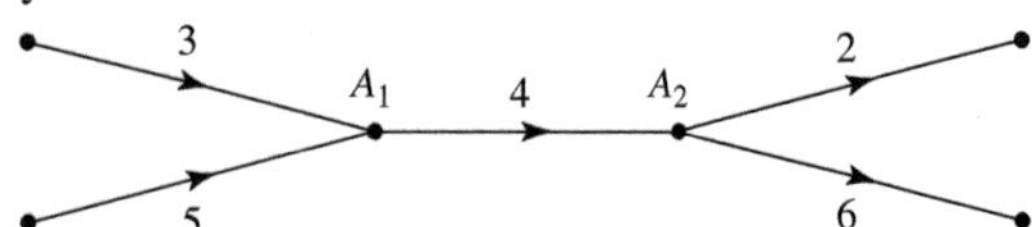

Notice that all inflows to A go to the first node A_1 and all outflows emerge from the second node A_2

Example 5

Find the maximal flow through this network, given that the nodes A and B have capacities of 5 and 4, respectively.

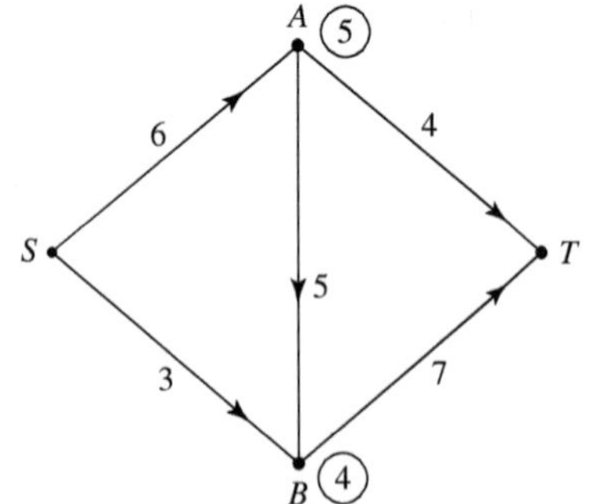

(*Continued on the next page*)

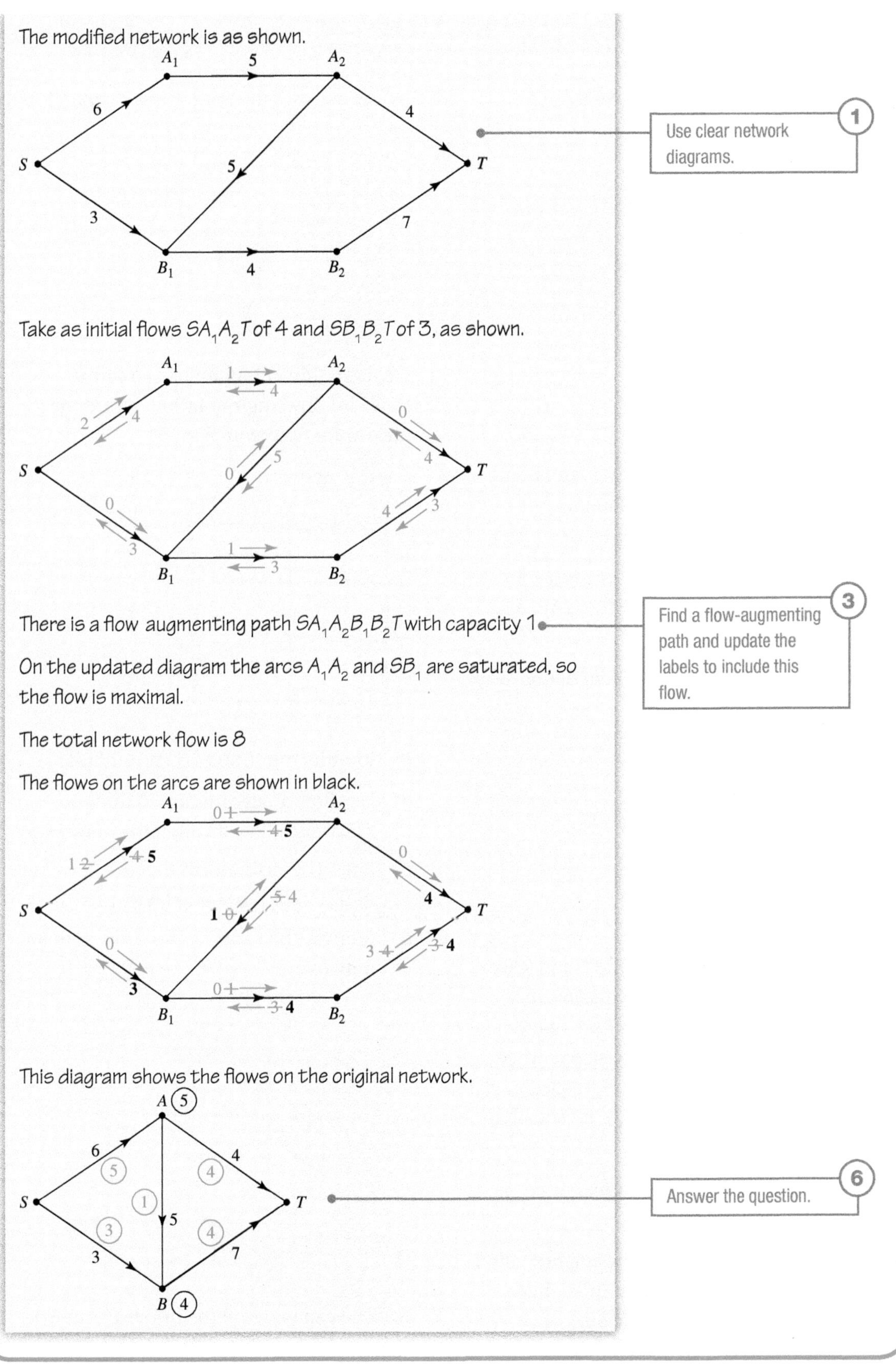

The modified network is as shown.

1 Use clear network diagrams.

Take as initial flows SA_1A_2T of 4 and SB_1B_2T of 3, as shown.

There is a flow augmenting path $SA_1A_2B_1B_2T$ with capacity 1

3 Find a flow-augmenting path and update the labels to include this flow.

On the updated diagram the arcs A_1A_2 and SB_1 are saturated, so the flow is maximal.

The total network flow is 8

The flows on the arcs are shown in black.

This diagram shows the flows on the original network.

6 Answer the question.

DISCRETE

Exercise 28.2B Reasoning and problem-solving

1 This network has a sink and two sources.

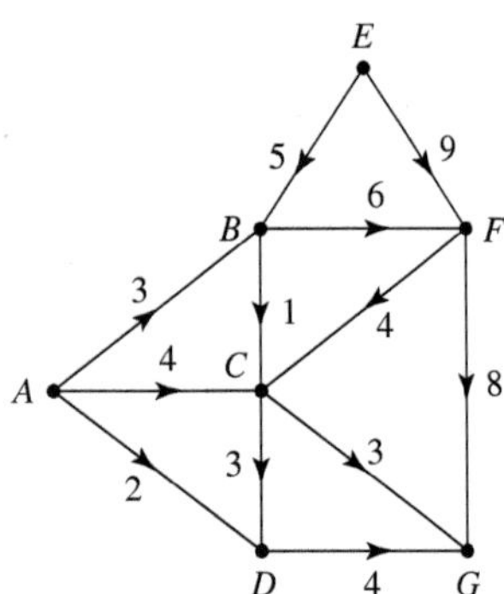

a Introduce a supersource S

b Use an initial flow and flow augmentation to find the maximal flow though the network.

c Draw the original network to show the flow you have found, together with a minimum cut.

2 This network represents a system of one-way streets.

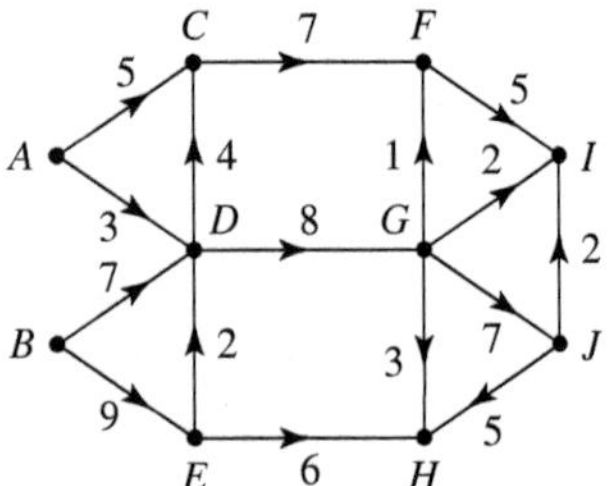

Traffic enters the system at A and B and leaves at I and H. The weights are the maximum traffic flows, in hundreds of cars per hour, which can safely pass along the streets. There is an environmental charge of £4 for all diesel vehicles using these roads, and it is estimated that 40% of vehicles are diesel.

a Draw a diagram with a supersource, S, and a supersink, T

b Use an initial flow and flow augmentation to find the maximum flow and hence estimate the maximum amount of environmental charge that could be collected in a peak hour.

3 In this network the node B has a capacity of 7

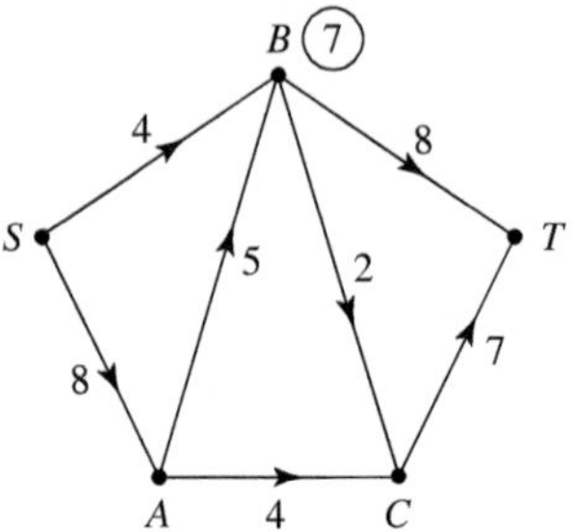

On a modified diagram use an initial flow and flow augmentation to find the maximum flow from S to T

4 For this network

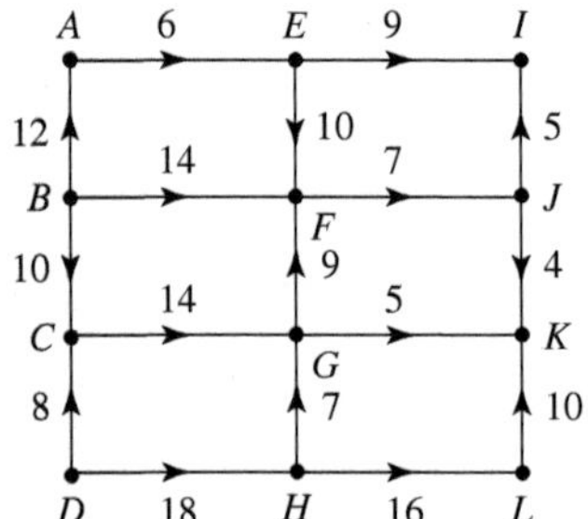

a Identify the source(s) and sink(s).

b Use flow augmentation to find the maximum flow through the network.

c Find a cut to confirm that the flow you found in part **b** is maximal.

5 a Identify the sources and sinks in this network.

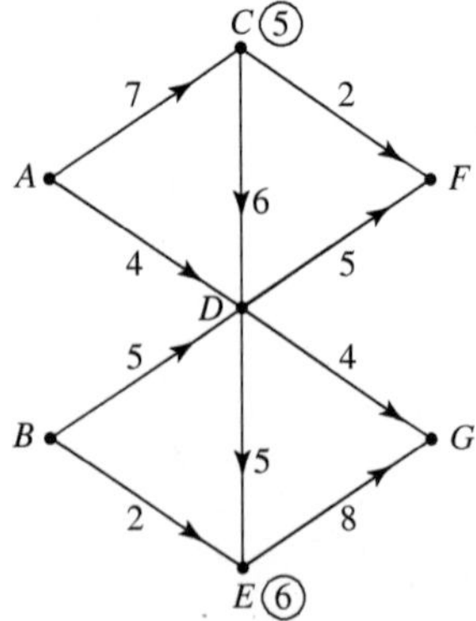

b The capacities at the nodes C and E are restricted to 5 and 6, respectively. Draw a modified network to allow for this and find an initial flow with a total capacity of 10

c Starting with your flow from part **b**, use flow augmentation to find a maximal flow through the network.

6 Find the maximum flow through this network and confirm your result using a minimum cut.

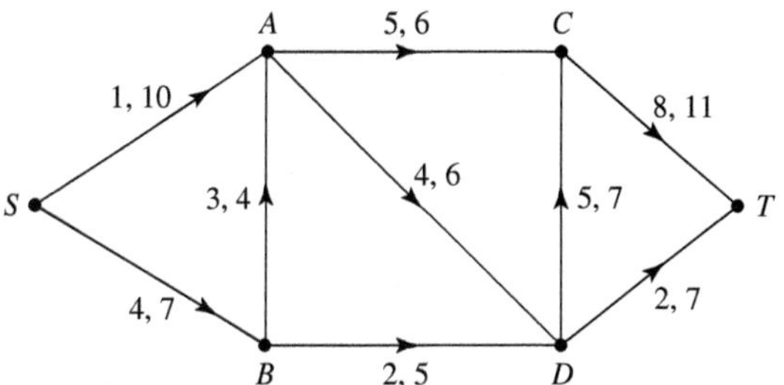

7 In a network of pipes used to deliver gas from S to T, each pipeline has a maximum flow, but for safety reasons it is also necessary to maintain a minimum flow in each pipe. This table shows both the minimum and the maximum number of units that can flow in each pipe.

		To				
		A	*B*	*C*	*D*	*T*
From	*S*	6,16	2,12	–	–	–
	A	–	3,4	4,12	1,2	–
	B	–	–	–	6,10	–
	C	–	–	–	2,8	4,12
	D	–	–	–	–	8,18

Find the maximum network flow consistent with these safety conditions and investigate whether it differs from the maximum flow possible if the safety restrictions were removed.

8 The diagram shows the minimum and maximum flows through the arcs of a network.

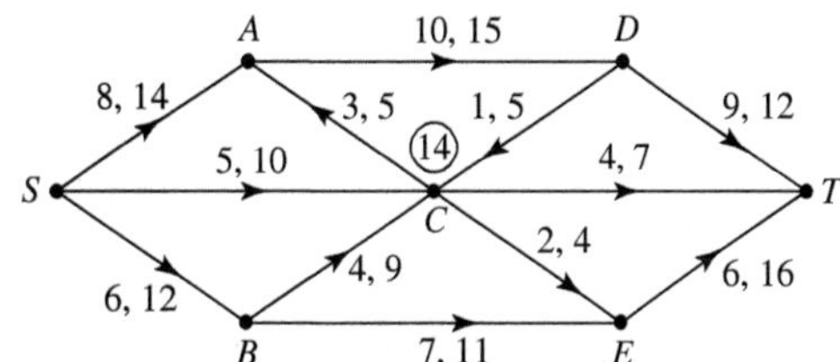

In addition, the node C has a maximum capacity of 14. Find the maximum flow through the network and confirm your result by finding a suitable cut.

DISCRETE

28 Summary and review

Chapter summary

- The complement of a simple graph G, denoted by G', has the same vertices as G but contains only the edges which are *not* in G
- You find the adjacency matrix for G' by changing 1s to 0s and vice versa (ignoring the leading diagonal).
- If a graph has n vertices and you combine it with its complement, you get the complete graph K_n
- Graphs G_1 and G_2 are isomorphic if, for each vertex in G_1, there is a corresponding vertex in G_2 so that if an edge joins two vertices in G_1, then there is an edge between the corresponding vertices in G_2, and vice versa. If corresponding vertices are listed in the same order, the adjacency matrices are identical.
- A subdivision of a graph G is a graph formed by adding vertices of degree 2 along edges of G
- A graph is planar if it can be drawn without edges intersecting except at a vertex.
- If a simple, connected, planar graph has V vertices, E edges and F faces (including the 'infinite face' surrounding the graph) then $\boldsymbol{V+F-E=2}$ (Euler's formula).
- It follows from Euler's formula that, for all planar graphs, $E \le 3V - 6$. (Be aware that this is also true for some non-planar graphs.) If $V = 5$, then $E \le 9$. As K_5 has 10 edges, it follows that K_5 is non-planar.
- For a bipartite graph, it follows from Euler's formula that $E \le 2V - 4$. If $V = 6$, then $E \le 8$. As $K_{3,3}$ has 9 edges, it follows that $K_{3,3}$ is non-planar.
- A graph is non-planar if and only if it has a subgraph which is a subdivision of K_5 or $K_{3,3}$ (Kuratowski's theorem).
- For a network flow from source S to sink T
  - The flow in each arc cannot be more than its capacity (feasibility condition).
  - At each node, the total inflow equals the total outflow (conservation condition).
  - A cut is a set of arcs whose removal disconnects the network into two parts X and Y, with X containing S and Y containing T
  - The value of the maximal flow = the capacity of a minimum cut.
  - If you have a flow and a cut such that (value of flow) = (capacity of cut), then the flow is a maximum and the cut is a minimum.
- Flow augmentation seeks to improve on an existing feasible flow.
  - The labelling procedure labels each arc with a forward arrow showing spare capacity and a backward arrow showing actual flow.
  - A flow-augmenting path is a route from S to T where the arrows pointing forwards along the route all have non-zero values. The lowest of these values gives the possible extra flow along that path.
- A network with more than one source or sink is modified by introducing a dummy supersource and/or supersink.
- If a node has a restricted capacity, you modify the network by replacing that node with an inflow node and an outflow node connected by an arc with the restricted capacity.

- If arcs have a minimum required flow, then the initial feasible flow must be consistent with these. You can still use the labelling procedure and flow augmentation.
- For a network with minimum and maximum capacities, capacity of cut = (maximum flows from S to T) – (minimum flows from T to S)

Check and review

You should now be able to ...	Review Questions
✔ Construct the complement of a graph given in the form of a drawing or an adjacency matrix.	1
✔ Decide whether two given graphs are isomorphic.	2
✔ Use Euler's formula in relation to planar graphs.	3
✔ Use Kuratowski's theorem to determine whether a graph is non-planar.	4
✔ Use the labelling procedure and flow-augmentation to find the maximum flow through a network.	5, 6, 7, 8
✔ Use the maximum flow, minimum cut theorem to show that a flow is maximal.	5, 8
✔ Work with multiple sources/sinks using a supersource/supersink.	6
✔ Solve problems in which there is a restricted flow through one or more nodes.	7
✔ Solve problems where arcs have a minimum flow as well as a maximum flow.	8

DISCRETE

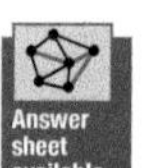

1 For the adjacency matrix shown:

	A	B	C	D	E	F
A	0	0	1	0	0	0
B	0	0	1	1	0	0
C	1	1	0	0	1	1
D	0	1	0	0	1	1
E	0	0	1	1	0	0
F	0	0	1	1	0	0

a Draw the corresponding graph, G

b Construct the adjacency table for the complement, G'

c Draw the graph G'

2 Two of these graphs are isomorphic. Identify these and list corresponding vertices.

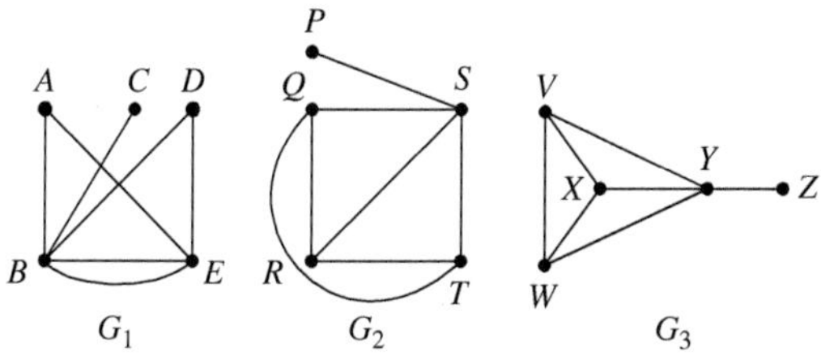

3 a A planar graph has 6 vertices and 9 edges. Calculate the number of faces in the graph and draw a graph which fits these facts.

b A graph has 7 vertices and 16 edges. Use a corollary of Euler's formula to show that the graph is non-planar.

4 Use Kuratowski's theorem to show that this graph is non-planar.

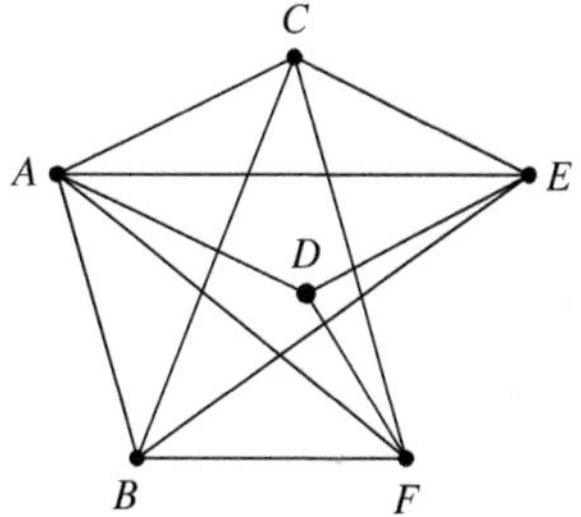

5 For this capacitated network:

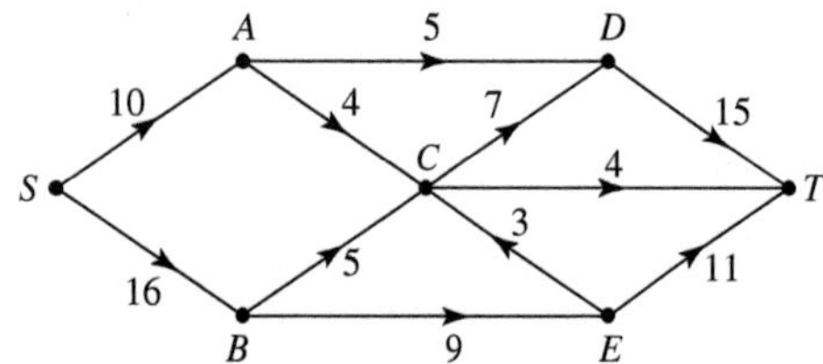

a Explain why a flow of 16 in *SB* is not possible.

b State the maximum flow along

i *SADT*

ii *SBET*

c Taking your results from part **b** as the initial flow, use the labelling procedure to find the maximum flow through the network.

d Prove that your flow is maximal.

6 For this capacitated network.

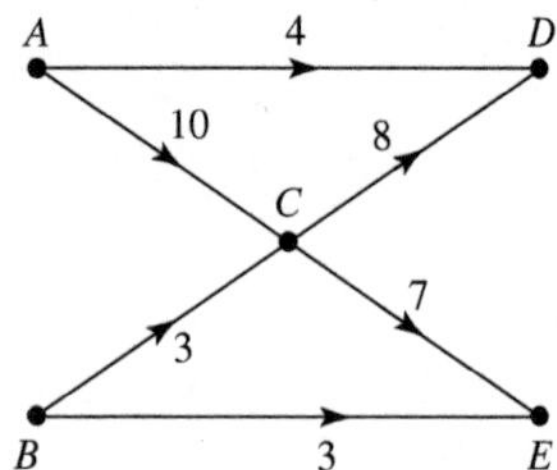

a Identify the source(s) and sink(s).

b Draw a diagram to include a dummy source and/or sink as necessary.

c Use the labelling procedure to find a maximal flow through the network.

7 In the network from question **6** the node *C* is then restricted to a flow of 8

a Draw a modified diagram to allow for this restriction.

b Use the labelling procedure to find a maximal flow through this network.

8

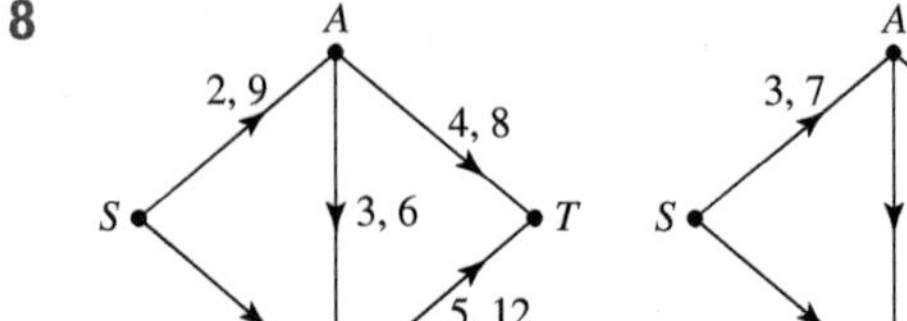
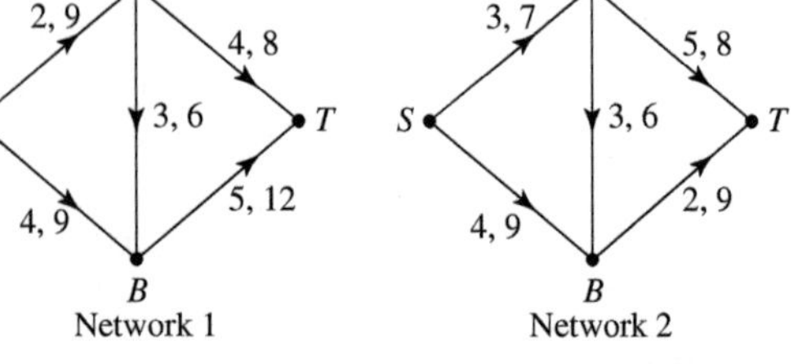

a For one of these networks, no feasible flow exists. Identify the network and explain why no flow is possible.

b Consider the other network.

i Find the capacity of the cut {*SA*, *AB*, *BT*}.

ii An initial feasible flow has 7 on *SA* and 4 on *SB*. Starting with this, use the labelling procedure to find the maximum flow.

iii Explain how you know it is a maximum.

28 Exploration

Going beyond the exams

Investigation

The complete bipartite graph on six vertices $K_{3,3}$ is also known as the utility graph. This comes from the puzzle of trying to connect three utility points to 3 buildings with no connection lines crossing.

$K_{3,3}$ is also a toroidal graph as it can be embedded on a torus without any edges crossing. This offers a potential solution for the utility problem – can you find it?

Research

Image segmentation is the process of segmenting a digital image to make it easier to analyse. Applications include video surveillance and magnetic resonance imaging. Research how the maximum flow problem can be used to assign pixels to the background or foreground of a picture.

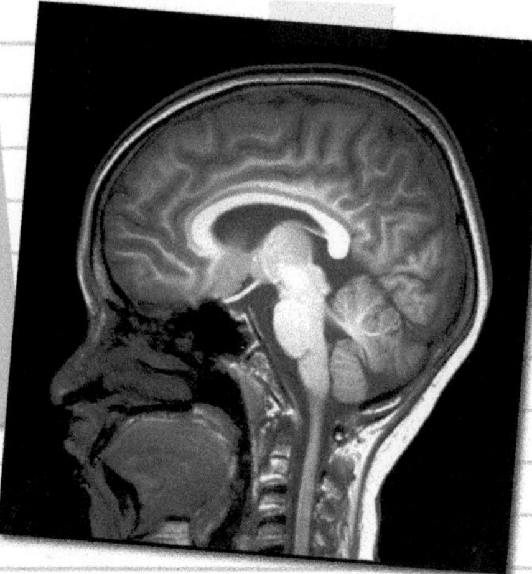

Did you know?

The first maximum flow problem was formulated in 1954 as a simplified model of Soviet railway traffic flow.

Research

Algorithms designed to find the maximum flow of a network include Ford-Fulkerson and push-relabel. The Ford-Fulkerson algorithm augments flows along paths from the source all the way to the sink. In contrast, the push-relabel algorithm gradually finds a maximum flow by moving flow locally between neighbouring nodes.

28 Assessment

Answer sheet available

1 Using Kuratowski's theorem, determine whether or not this graph is planar. [2]

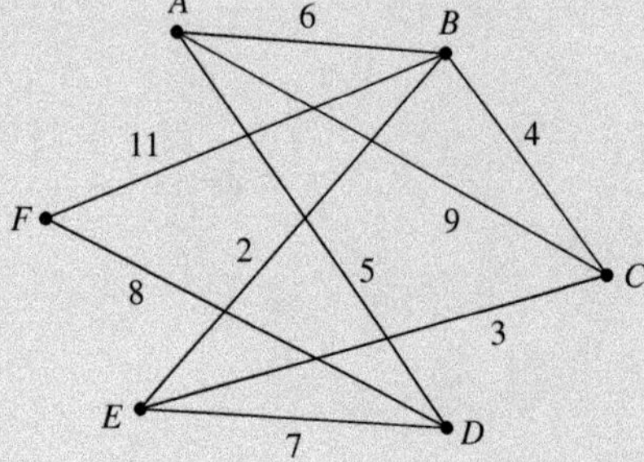

2

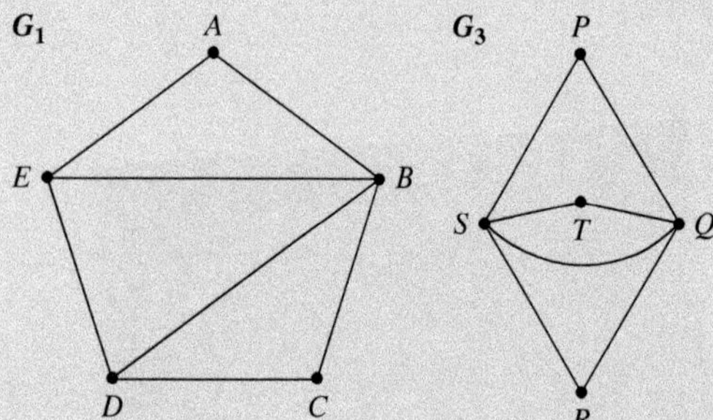

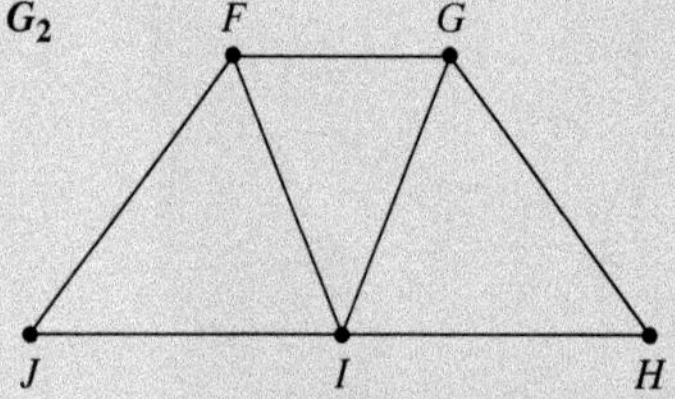

a Construct the adjacency matrix for the complement of G_1 [3]

b Explain whether or not each of G_2 and G_3 are isomorphic to G_1 [3]

3 a Using an initial flow of 7 along *SADT* and 9 along *SCFT*, augment the flow until a maximal flow is obtained. [5]

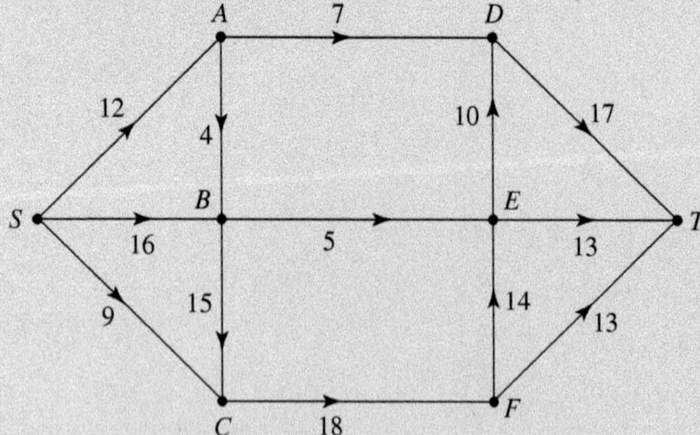

b Verify that your flow is maximal, stating the name of the theorem used. **[3]**

c Explain which edge in the network could be removed without affecting the maximum flow. **[2]**

d Which of the edges are saturated? **[2]**

4 The edges on this network represent road with the capacities shown.

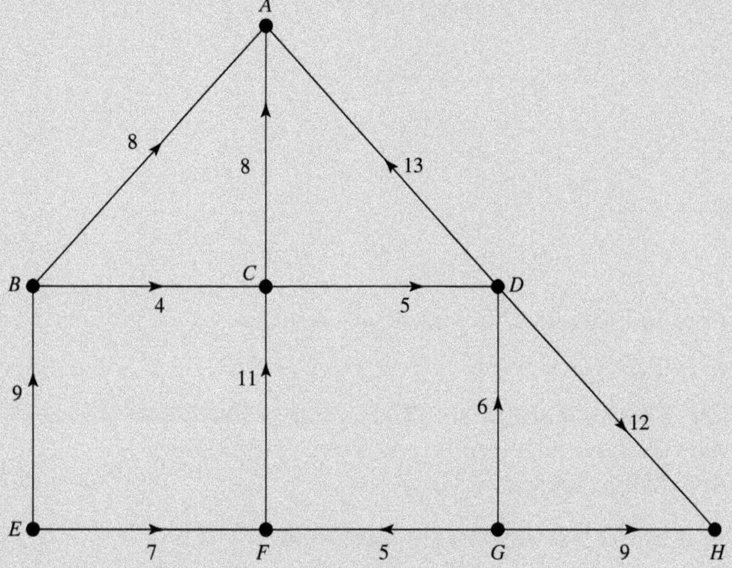

a Copy the network, introducing a supersource and/or supersink if needed. **[3]**

b Clearly stating your initial flows, use flow augmentation to obtain the maximal flow. **[5]**

c Draw your maximal flow onto the copy of the network. **[3]**

d Verify that the flow you have found is maximal. **[2]**

Additional capacity is to be added either at *EB* or *GF*

e Where would you suggest it is placed and how much additional capacity should bc added? Explain your answer fully. **[3]**

5 This network represents a system of pipes.

There is a blockage at *C* that restricts the flow to a maximum of 16

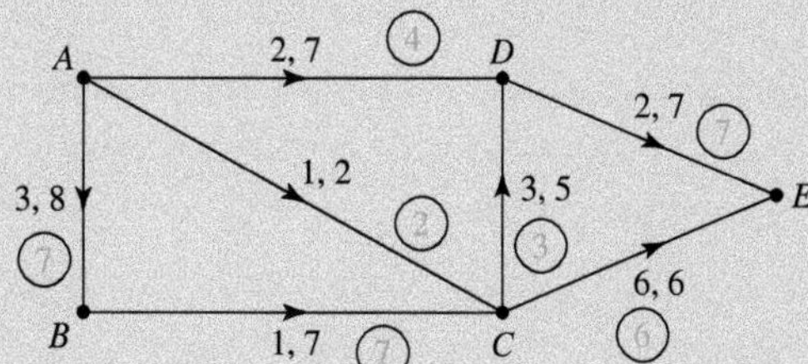

a Draw a modified network to account for this restricted node. **[3]**

b Find the maximal flow through this network and verify that it is a maximum. **[5]**

6 The numbers on the arcs of this network give the minimum and maximum permissible flow in that arc. A initial feasible flow is $ADE = 2$, $ACDE = 2$, $ABCE = 6$

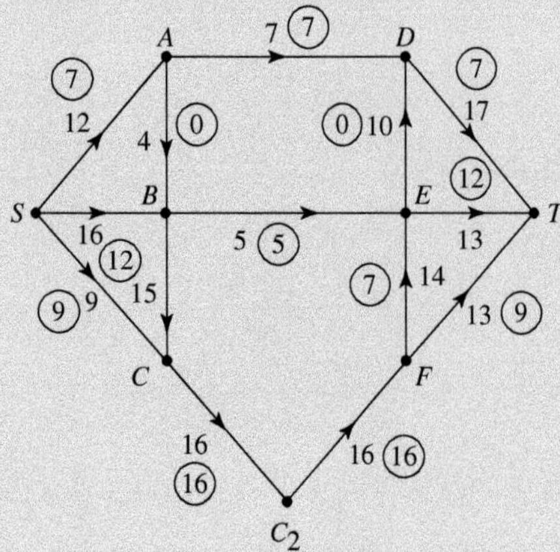

a Using arrows for potential flow and backflow, illustrate the initial flow on a copy of this diagram. **[3]**

b Use flow augmentation, starting with the initial flow given, to find a maximum flow for this network, and state the flow-augmenting paths you use. Clearly illustrate your flow on the copy of the network. **[4]**

c Use the cut $\{DE, DC, AC, BC\}$ to verify that the flow you have found is maximal. **[2]**

29 Critical path analysis 2

Natural disasters such as floods, volcanic eruptions, earthquakes and spread of a major disease, require a rapid response from disaster relief agencies. Once an incident occurs, such organisations are well prepared to move people to safety, arrange medical aid, transport food supplies and set up sanitation. The mathematics of critical path analysis is a crucial tool in ensuring that support reaches disaster victims as quickly as possible. Through this, agencies are able to coordinate large-scale operations using advance planning and variable factors, which can be adjusted using information from on-the-ground experts.

Critical path analysis is at the heart of project planning and implementation, from major building projects such as building a new school, airport terminal or retail distribution centre, through to holding a major event such as a summer festival or sports tournament.

Orientation

What you need to know

Ch13 Critical path analysis 1

What you will learn

- To calculate earliest start times and latest finish times for an activity network.
- To use Gantt charts.
- To use resource histograms to schedule activities.

What this leads to

Careers

- Project management.
- Construction.
- Event planning.

29.1 Gantt charts

Fluency and skills

See Ch13.1 For a reminder on activity networks and critical paths.

A project can be described by a precedence table showing each activity involved and the order in which they must be completed.

You draw an activity network and use a forward and backward pass to find the earliest possible start time and latest possible finish time for each activity. The time available for an activity is the gap between these. If the time available is greater than the duration of the activity, then the activity has float.

Key point

Float = (latest finish time – earliest start time) – duration

An activity with zero float is a critical activity. The critical activities form a critical path. Some projects may have more than one critical path.

With non-critical activities, you have some flexibility about when to start them. They will have a latest possible start time and an earliest possible finish time.

Key point

Earliest finish time = earliest start time + duration
Latest start time = latest finish time – duration

For example, look at this activity network.

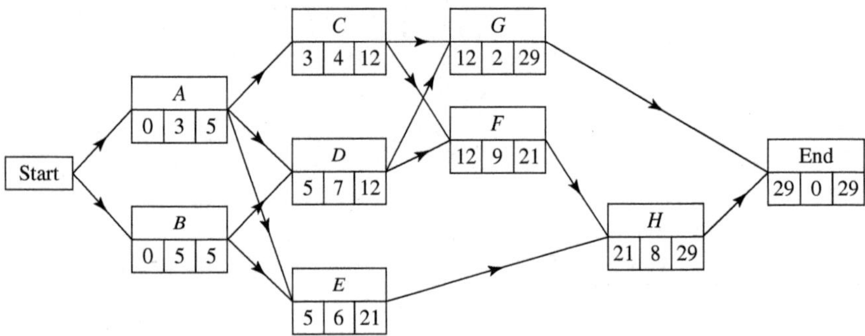

The table on the next page shows the complete information for this network.

For example, for activity C

Earliest start time = 3

Earliest finish time = 3 + duration of C = 7

Latest finish time = 12

Latest start time = 12 – duration of C = 8

Activity	Duration (days)	Start time		Finish time		Float
		Earliest	Latest	Earliest	Latest	
A	3	0	2	3	5	2
B	5	0	0	5	5	0
C	4	3	8	7	12	5
D	7	5	5	12	12	0
E	6	5	15	11	21	10
F	9	12	12	21	21	0
G	2	12	27	14	29	15
H	8	21	21	29	29	0

You can now illustrate this information on a **Gantt chart** or **cascade chart**. Activities are shown as bars against a time scale. The critical activities are fixed, so you insert those first.

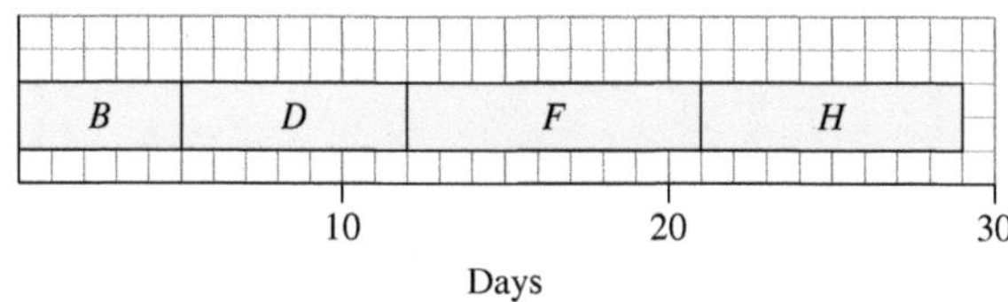

You now place each non-critical activity at its earliest start time, and show the boundaries within which it can be moved.

Activity A could start at 0 and has a latest finish of 5

Activity C has earliest start 3 and latest finish 12

Activity E has earliest start 5 and latest finish 21

Activity G has earliest start 12 and latest finish 29

Although C has 5 days float, 2 of these depend on when A starts. You can put a 'fence' on the chart to show that C cannot start until A is complete.

The finished Gantt chart looks like this.

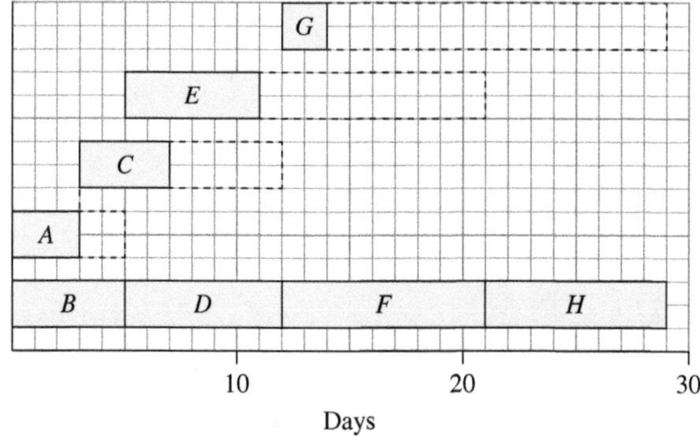

You may see variants of this layout. The most common is to put each activity on a separate line.

DISCRETE

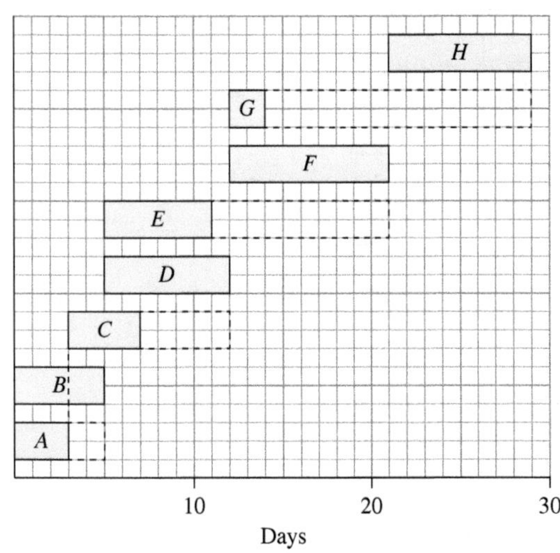

You will also see them drawn 'upside-down' with the early activities at the top of the diagram.

Example 1

Draw a Gantt chart corresponding to this activity network.

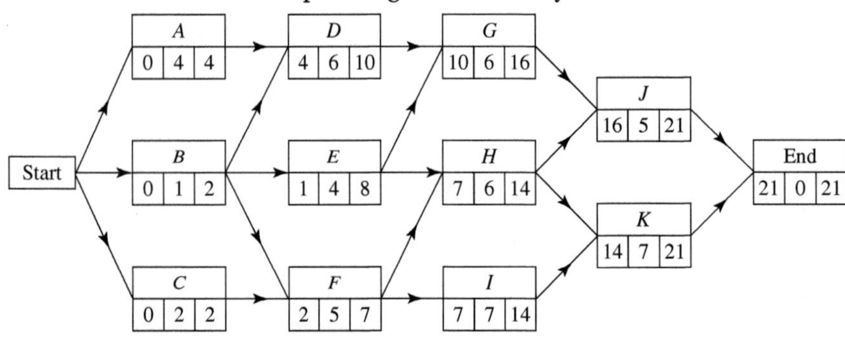

Activity	Duration (hours)	Start time		Finish time		Float
		Earliest	Latest	Earliest	Latest	
A	4	0	0	4	4	0
B	1	0	1	1	2	1
C	2	0	0	2	2	0
D	6	4	4	10	10	0
E	4	1	4	5	8	3
F	5	2	2	7	7	0
G	6	10	10	16	16	0
H	6	7	8	13	14	1
I	7	7	7	14	14	0
J	5	16	16	21	21	0
K	7	14	14	21	21	0

First, create a table to show the earliest and latest times.

The critical activities are *A*, *C*, *D*, *F*, *G*, *I*, *J*, *K*

They form two critical paths through the network – *ADGJ* and *CFIK*

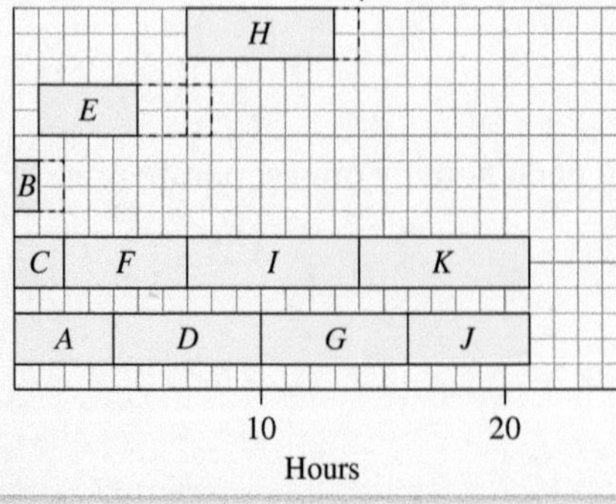

Draw the Gantt chart.

Exercise 29.1A Fluency and skills

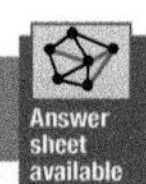

1 Complete the following table. Hence draw a Gantt chart to show the project.

Activity	Duration (hours)	Start		Finish		Float
		Earliest	Latest	Earliest	Latest	
A	6	0			6	
B	8	0			15	
C	14	6			22	
D	12	6			18	
E	3	8			18	
F	4	18			22	
G	3	22			25	

2 Draw a Gantt chart for each of these activity networks.

a

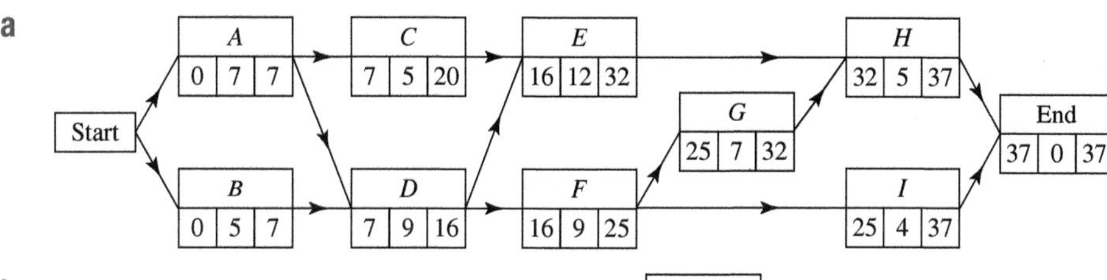

b

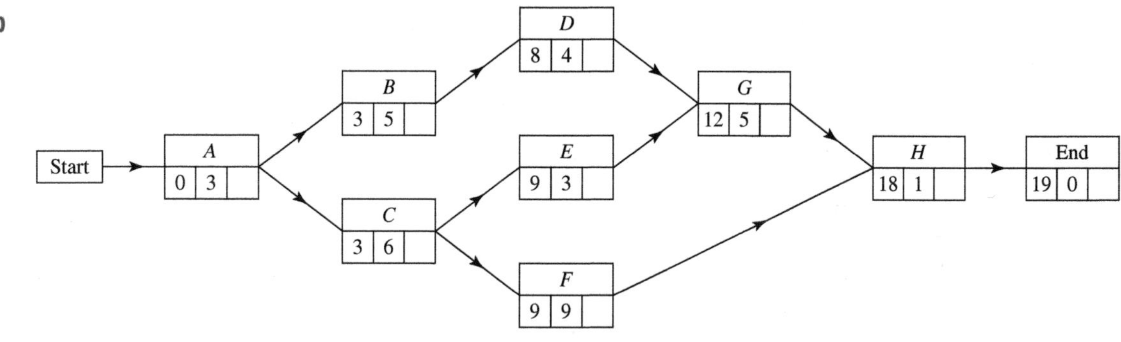

c

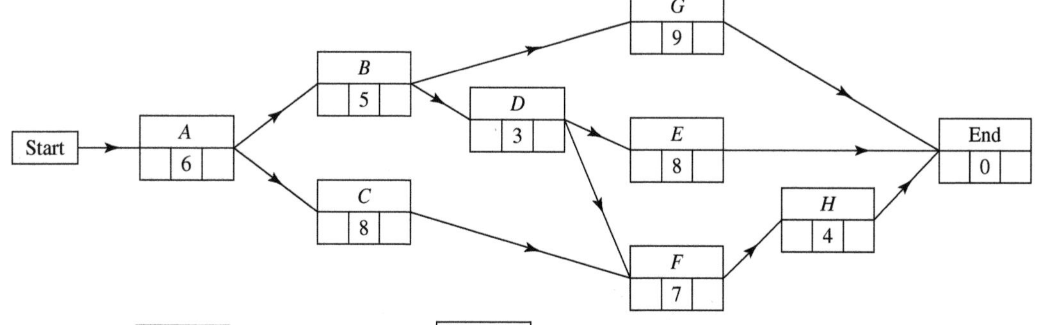

d

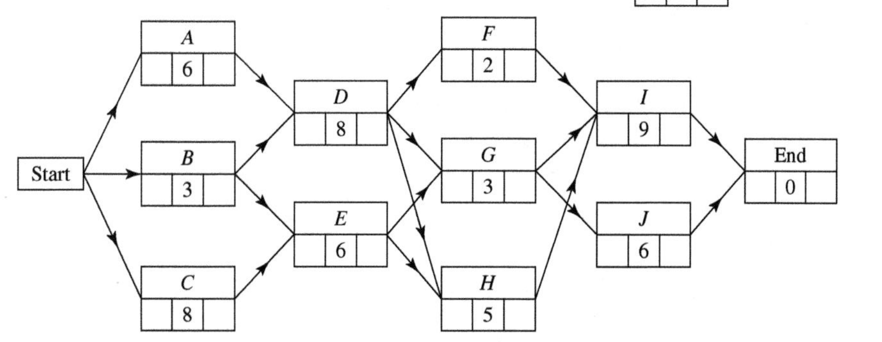

DISCRETE

Reasoning and problem-solving

Strategy

To solve problems involving Gantt charts

1. Find the earliest and latest start and finish times for each activity.
2. Identify the critical activities and critical path(s).
3. Draw the Gantt chart, entering critical activities first.
4. Answer the question.

Example 2

a Draw a Gantt chart to show this project.

Activity	Depends on:	Duration (days)
A	–	8
B	–	4
C	*A*	10
D	*B*	7
E	*C*	2
F	*C*, *D*	6
G	*E*, *F*	3

Explain what would happen if

b Activity *B* overran by 3 days,

c Activity *F* overran by 3 days.

Assume that, in each case, they started on time.

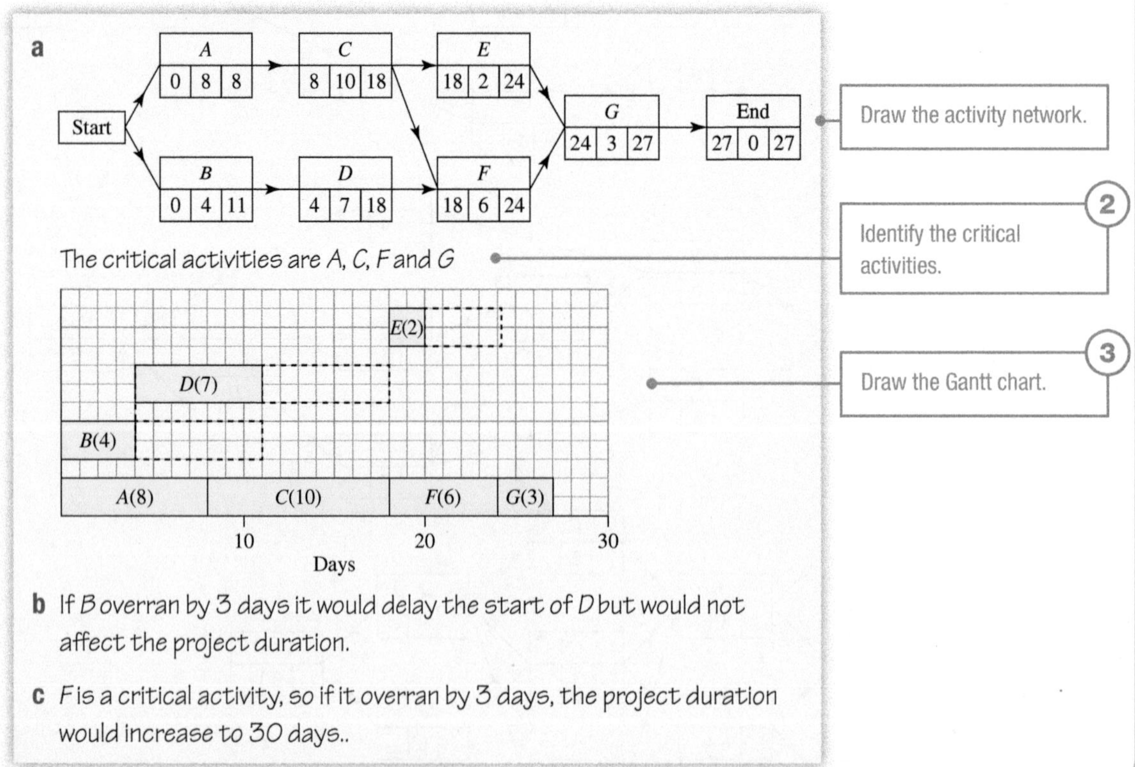

b If *B* overran by 3 days it would delay the start of *D* but would not affect the project duration.

c *F* is a critical activity, so if it overran by 3 days, the project duration would increase to 30 days..

Exercise 29.1B Reasoning and problem-solving

1 Draw a Gantt chart for the project described by this precedence table.

Activity	Depends on:	Duration (hours)
A	–	9
B	–	7
C	A	4
D	B	5
E	C, D	6
F	C, D	4
G	E	7
H	E, F	3

2 a Draw a Gantt chart for the project described by this precedence table.

Activity	Depends on:	Duration (days)
A	–	4
B	–	6
C	–	5
D	A, B	3
E	B, C	2
F	D	6
G	E	7
H	D, G	3
I	F, H	3
J	F	3
K	G	4

b Activities D and J overrun by 2 days each. What effect will this have on the project duration?

3 A gardener plans to make a pond. The table shows the tasks involved. Each task requires one worker.

	Activity	Depends on:	Duration (hours)
A	Clear site	–	2
B	Dig hole	A	10
C	Clear soil	B	3
D	Line hole	B	2
E	Fill pond	D	3
F	Install pump	D	2
G	Test pump	E, F	1
H	Put in fish	E	1
I	Put in plants	E	2
J	Landscape site	C, D	5

a How long will the project take if the gardener works alone?

b What is the least time needed for the complete project?

c Draw a Gantt chart for the project.

d Find the least workforce needed to finish the project in the minimum time.

4 The table shows a project for renovating a flat. Draw a Gantt chart for this project.

	Activity	Depends on:	Duration (hours)
A	Remove furniture	–	1
B	Remove old carpet	A	1
C	Remove curtains	–	1
D	Strip wallpaper	B, C	3
E	Sand down paintwork	B, C	2
F	Rewire	D	4
G	Install central heating	F	4
H	Repair plaster	G	2
I	Wash ceiling and walls	E, H	1
J	Paint ceiling	I	1
K	Paint walls	J	2
L	Paint woodwork	H	2
M	Install new carpet	K, L	1
N	Replace furniture	M	1
O	Replace curtains	K, L	1
P	Clean	N, O	2

29.2 Resourcing

Fluency and skills

You use the term **resource** to refer to the workforce needed for a project. The different activities may require different numbers of workers. You will need to

- decide how many workers are required at each stage if every activity starts as early as possible,
- reschedule activities to minimise the number of workers needed; this is known as **resource levelling**,
- extend the duration of the project to allow for the available number of workers.

To illustrate the number of workers required at each stage you use a **resource histogram**.

Example 1

This Gantt chart shows the activities in a project. The figures in brackets show the number of workers needed for the activity.

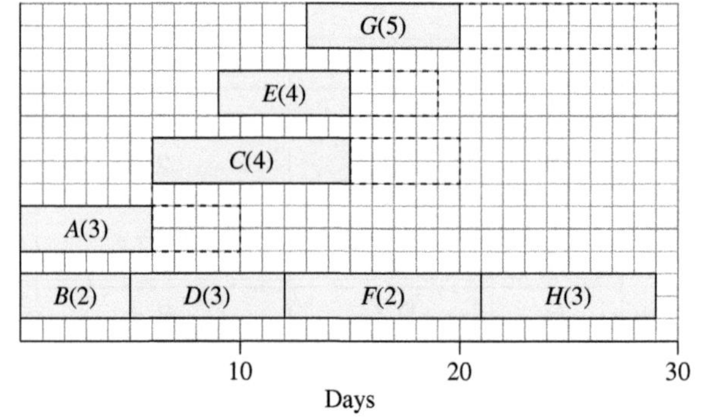

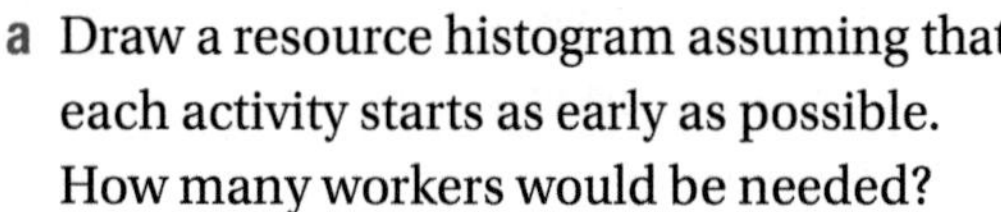

a Draw a resource histogram assuming that each activity starts as early as possible. How many workers would be needed?

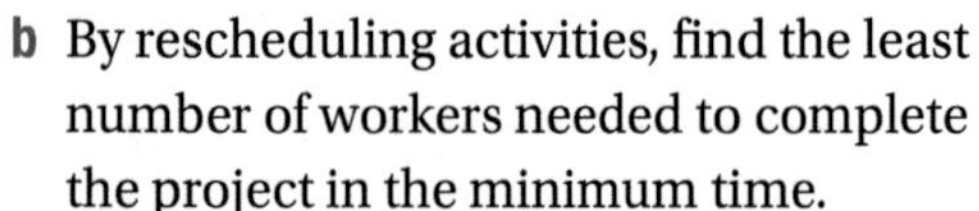

b By rescheduling activities, find the least number of workers needed to complete the project in the minimum time.

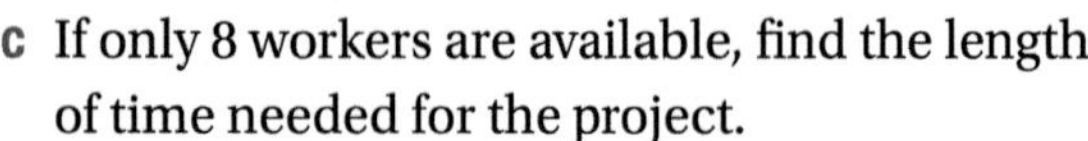

c If only 8 workers are available, find the length of time needed for the project.

a

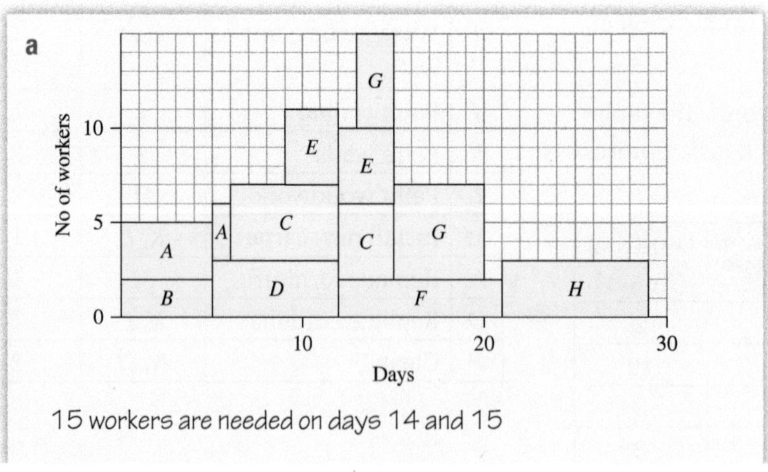

15 workers are needed on days 14 and 15

(*Continued on the next page*)

b There is enough float to delay the start of *E* until *D* has finished, and *G* until *E* has finished.

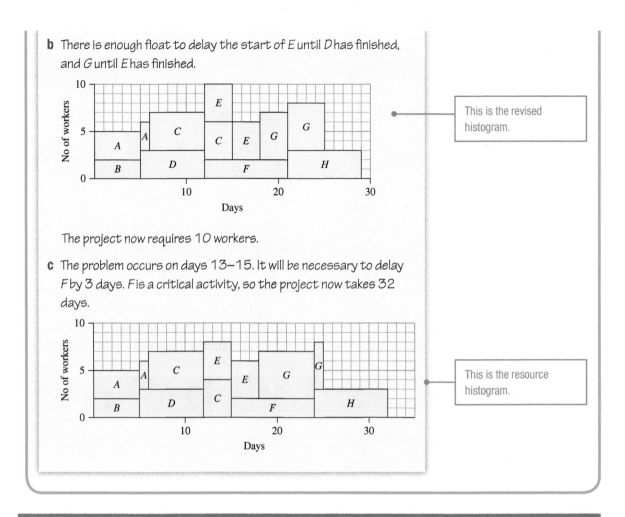

This is the revised histogram.

The project now requires 10 workers.

c The problem occurs on days 13–15. It will be necessary to delay *F* by 3 days. *F* is a critical activity, so the project now takes 32 days.

This is the resource histogram.

DISCRETE

Exercise 29.2A Fluency and skills

1 On each of these Gantt charts the values in brackets give the number of workers needed for the activity.

a Draw a resource histogram assuming that every activity starts as early as possible. How many workers are needed in this case?

b Use resource levelling to construct a revised resource histogram to minimise the workforce needed without extending the project duration. How many workers are now required?

i

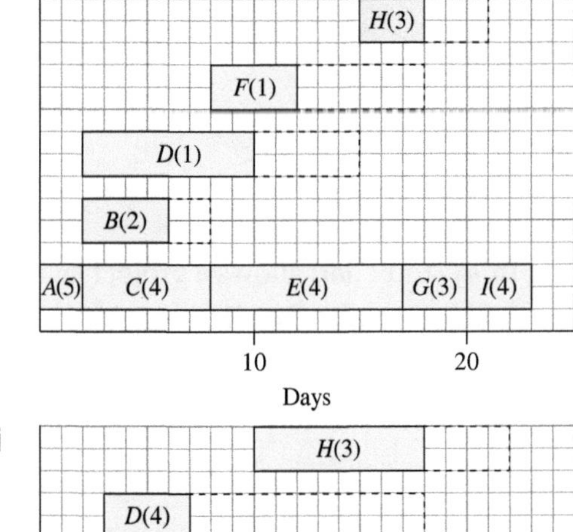

ii

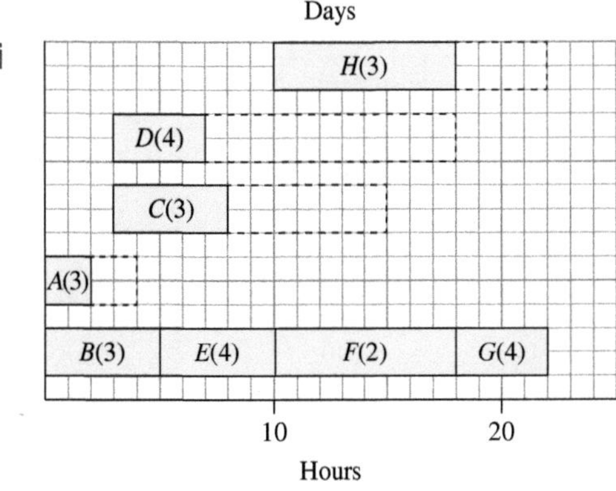

iii

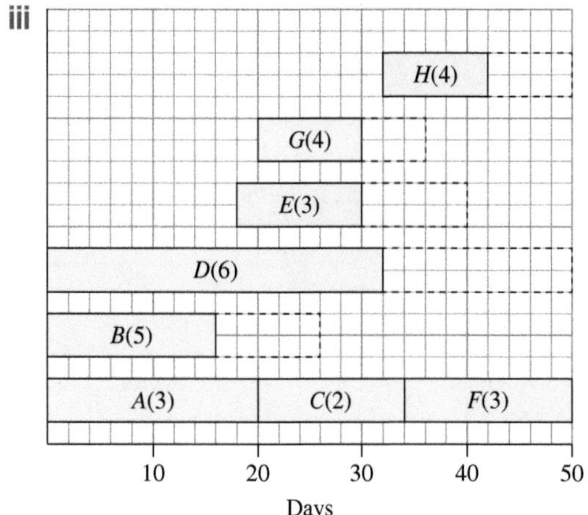

2 Draw a resource histogram for the project shown in this Gantt chart, assuming that every activity starts as early as possible.

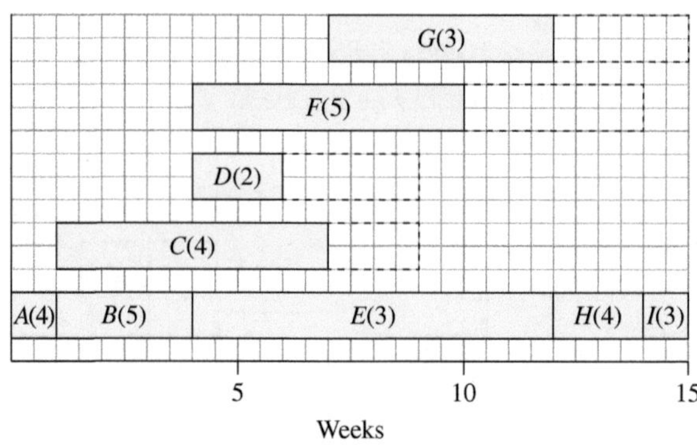

a Use resource levelling to show that the project can be completed in the minimum time using 12 workers.

b Show that, if each activity does not have to be completed in a single block of time, it is possible to complete the project in the minimum time using just 11 workers.

Reasoning and problem-solving

Strategy

To solve a problem involving resource levelling

1. If necessary, find the earliest start time, latest finish time and float for each activity.
2. Sketch a Gantt chart if one is not given.
3. Draw a resource histogram assuming that all activities start as early as possible.
4. Reschedule activities within their float times to reduce the size of the workforce needed.
5. If there is a given maximum number of workers available, it may be necessary to extend the duration of the project.
6. Answer the question.

Example 2

This Gantt chart shows a project lasting 27 days. The values in brackets give the number of workers needed for each activity.

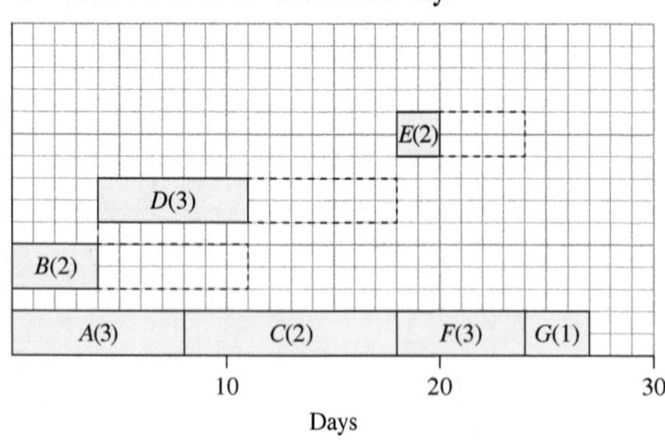

(*Continued on the next page*)

a Draw a resource histogram assuming that each activity starts as soon as possible. How many workers are needed?

b Show that the project can be completed on time if only 5 workers are available.

c Find the effect on the project duration if it has to be completed using just 4 workers.

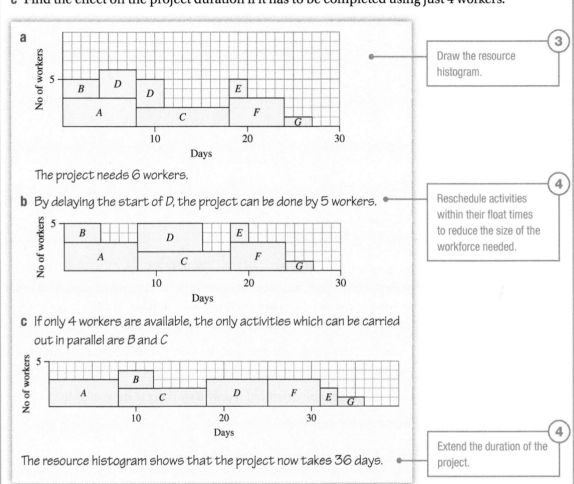

a

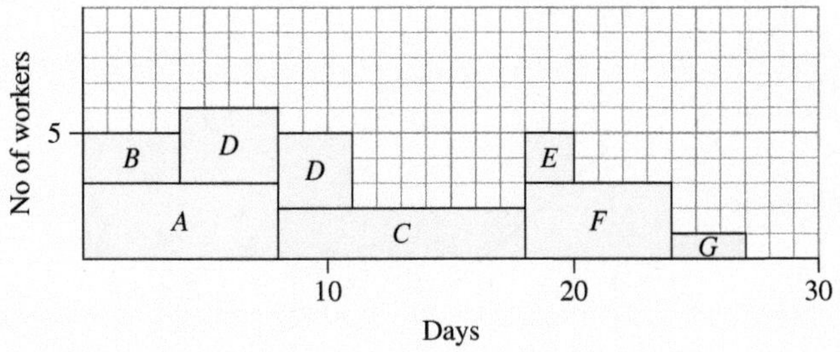

The project needs 6 workers.

> (3) Draw the resource histogram.

b By delaying the start of *D*, the project can be done by 5 workers.

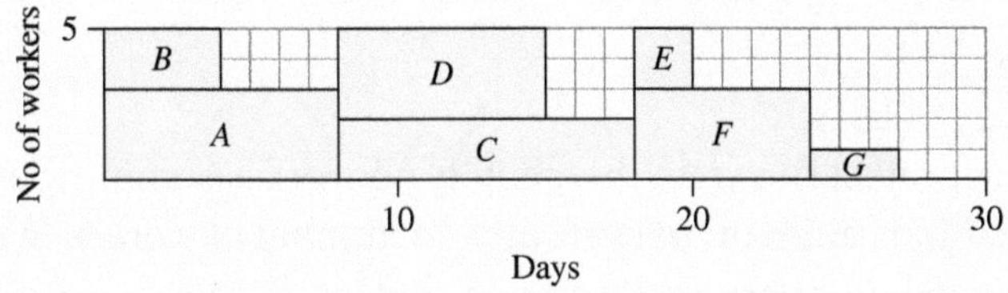

> (4) Reschedule activities within their float times to reduce the size of the workforce needed.

c If only 4 workers are available, the only activities which can be carried out in parallel are *B* and *C*

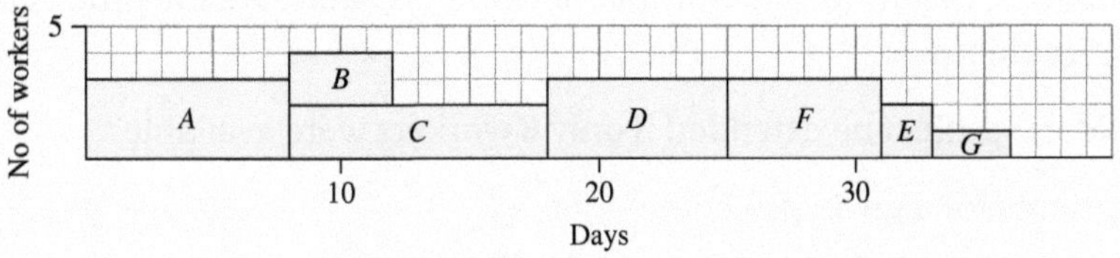

The resource histogram shows that the project now takes 36 days.

> (4) Extend the duration of the project.

Exercise 29.2B Reasoning and problem-solving

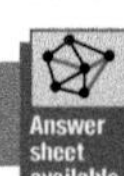

1 The table shows the requirements for a given project.

Activity	Preceded by	Duration (days)	Number of workers
A	–	5	3
B	*A*	3	5
C	*A*	4	2
D	*B*	6	4
E	*B, C*	4	4
F	*C*	2	3
G	*C*	5	5
H	*D, E*	2	4
I	*E, F*	2	2

a Draw a resource histogram for this project, assuming that all activities start as early as possible. State the number of workers needed.

b Show that the number of workers can be reduced by 3 without extending the duration of the project.

c How long will the project take if there are just 11 workers available?

2

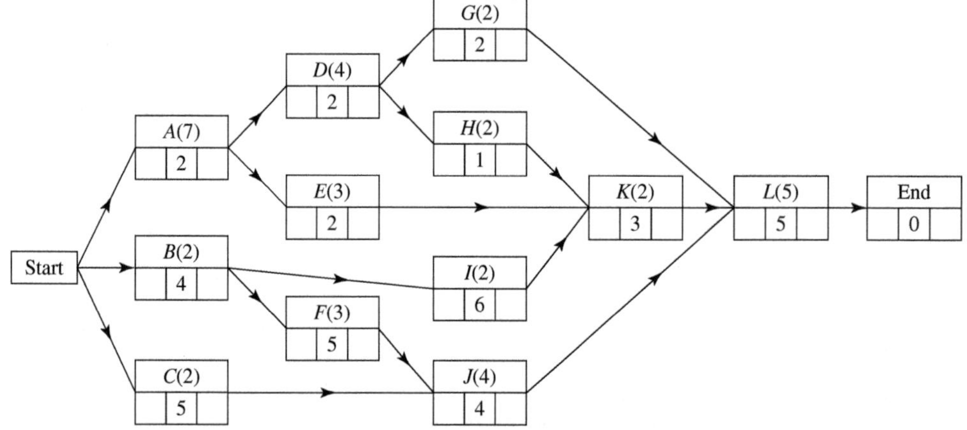

a Draw a resource histogram for the project shown in this activity network assuming that each activity starts as early as possible. (The values in brackets are the number of workers needed for the activity and the non-bracketed numbers are durations, in days, of each activity.) State the number of workers the project manager needs to be able to call on.

b By rescheduling activities, find the minimum number of workers required if the project is to be completed in the minimum time.

c By how many days would the project be extended if only 8 workers were available?

3 The table shows the requirements for a given project.

Activity	Preceded by	Duration (days)	Number of workers
A	–	3	2
B	*A*	5	1
C	*A*	3	2
D	*B*	3	3
E	*B*	4	3
F	*C*	3	2
G	*D*	2	4
H	*E*	4	2
I	*E*, *F*	5	2
J	*G*, *H*	3	1
K	*I*, *J*	3	4

a Draw a resource histogram for this project assuming that all activities start as early as possible. State the number of workers needed.

b Show, by resource levelling, that this number of workers can be reduced.

c By how many days would the project have to be extended if there were only

i 5 workers,

ii 4 workers available?

4

Activity	Preceded by	Duration (days)	Number of workers
A	–	2	5
B	*A*	4	2
C	*A*	5	4
D	*A*	7	2
E	*B*, *C*	8	3
F	*C*, *D*	4	2
G	*E*, *F*	2	3
H	*G*	2	3
I	*G*	4	4
J	*H*, *I*	2	2
K	*I*	2	2
L	*H*, *J*	1	3

a Draw a resource histogram for this project assuming that all activities start as early as possible. State the number of workers needed.

b In fact, there are only 6 workers available. Show how the project can be completed and state the number of extra days required.

29 Summary and review

Chapter summary

- From an activity network you find the earliest start time and latest finish time for each activity.
- Float = (latest finish time – earliest start time) – duration.
- Critical activities have zero float. They form one or more critical paths.
- Earliest finish time = earliest start time + duration.
  Latest start time = latest finish time – duration.
- You show this information on a Gantt chart or cascade chart. Activities are shown as bars against a time scale, starting as early as possible and with the float shown dotted.
- A resource histogram shows the number of workers needed at each stage of the project.
- Resource levelling involves rescheduling activities to reduce the number of workers needed.
- If the number of workers is limited, this may involve extending the duration of the project.

Check and review

You should now be able to...	Review Questions
✔ Calculate the earliest and latest start and finish times for an activity network.	1
✔ Draw a Gantt (cascade) chart for a project.	2
✔ Draw a resource histogram with every activity starting as early as possible.	3a
✔ Use resource levelling to reschedule activities to minimise the number of workers needed.	3b
✔ Use resource levelling to reschedule a project with extended duration when the number of workers is limited.	3c

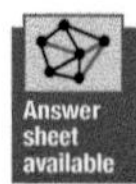

1 Construct a table showing the earliest and latest start and finish times for the activities in the project shown in this network.

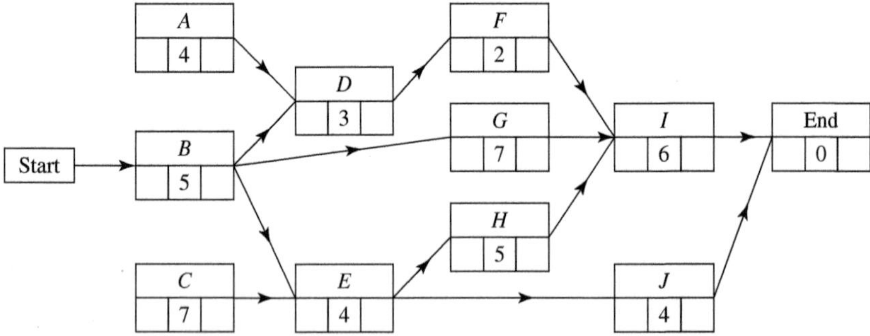

2 Draw a Gantt chart for the table you made in question **1**.

3

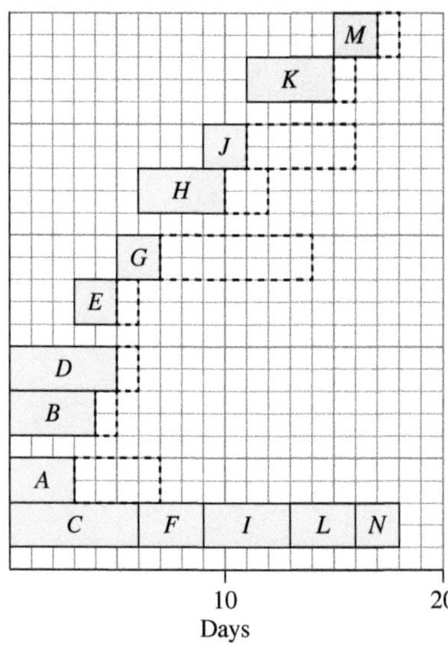

a All the activities in this Gantt chart require one worker, with the exception of $D(2)$, $G(3)$, $H(4)$, $I(2)$ and $K(2)$. Draw a resource histogram with every activity starting as early as possible. State the number of workers needed.

b Use resource levelling to reduce the number of workers needed. State the minimum workforce needed if the project is to be completed in 18 days.

c If the available workforce is 5 workers, show how the project can be completed with just one extra day.

DISCRETE

Exploration

Going beyond the exams

Research

Critical Path Drag is defined to be the amount of time that an activity on the critical path adds to the project duration. An important alternative way to understand this is to see it as the maximum amount of time that an activity can be shortened before it is no longer on the critical path.

For the activity network shown below: identify the floats, critical activities and hence the critical path drag.

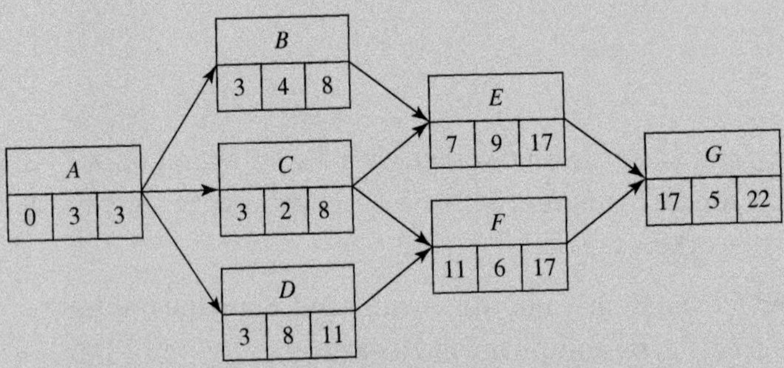

Can you explain why the critical path drag is equal to the minimum of either the remaining duration or (if there are parallel activities) the total float of the parallel activity that has the least total float?

Research

There are many criticisms (and offered solutions) of the critical path method that are worth finding out about.

These include:

1. The difficulty of estimating completion times – particularly for new projects.
2. Defining links for large projects becomes complicated very quickly.
3. The reality of resources (employees) shifting and changing the plan.
4. Identifying single critical paths.

Look into some suggested solutions to these problems by finding out a little more on probabilistic critical paths, critical chain project management and crash durations.

Did you know?

One modern critic of Gannt carts is Professor Edward Tufte. He argues that graphics should provide viewers with the greatest number of ideas in the shortest time, with the least ink, in the smallest space. He feels that the over simplistic Gannt charts have 'regressed to Microsoft mediocrity'.

29 Assessment

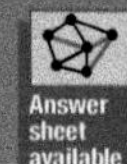

1 a Draw a Gantt chart for this activity network. **[5 marks]**

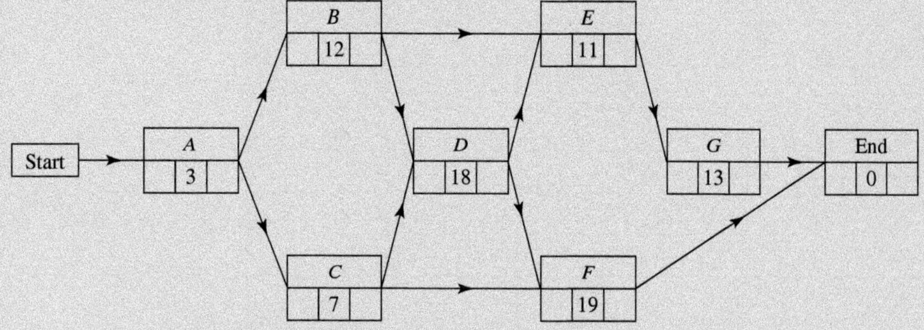

b Given that each activity requires 3 workers, how many workers are required to complete the project in the minimum time? **[1]**

2 The activity network for a project is shown below. The values given are the duration of each activity in hours.

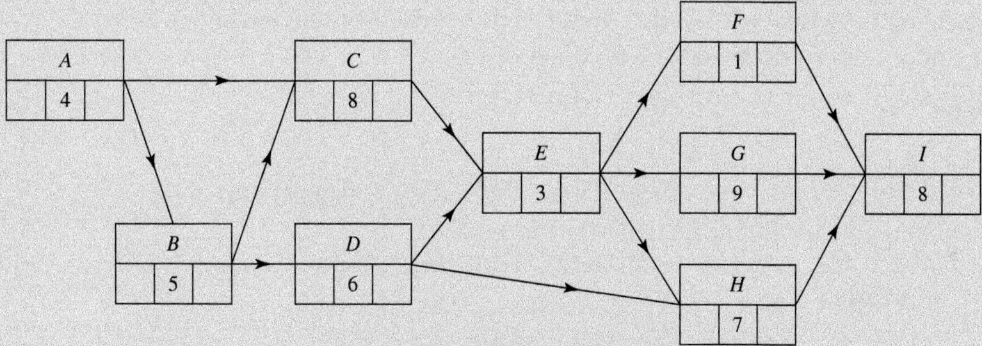

a Find the earliest start time and latest finish time for each activity and complete the activity network. **[4]**

b State the critical path. **[1]**

c Calculate the float of activity H **[2]**

d Explain how a delay of 3 hours for activity D will affect the minimum completion time for the project. **[2]**

3 The precedence table shows the dependencies and duration of the activities in a project.

Activity	Preceded by	Duration (days)
A	–	4
B	–	2
C	*A*	5
D	*A*, *B*	7
E	*B*	6
F	*E*	1
G	*D*, *E*	4
H	*F*, *G*	5
I	*F*	2
J	*C*, *H*, *I*	3
K	*I*, *J*	4

a Use the precedence table to draw an activity network. Include the duration, the earliest start time and the latest finish time for each activity. **[6]**

b Find the critical path for the network. **[1]**

c Explain how long each of these activities can be delayed without having an impact on the minimum completion time for the project.

i Activity *G* ii Activity *C* **[3]**

4 The table gives information about the duration, the earliest and latest start and finish times and the float for the activities in a project.

Activity	Duration (hours)	Start		Finish		Float
		Earliest	Latest	Earliest	Latest	
A	1.5	0			1.5	
B	2.4				3.9	0
C		3.9	3.9		10	0
D	1.8	3.9				4.6
E	0.3	10				0
F	3	3.9			10.3	
G	2.6				12.9	0
H	5.2	12.9			18.1	

a Copy and complete the table. **[8]**

b Draw a Gantt chart to show the project. **[3]**

c Use your Gantt chart to list which activities

i Are definitely taking place at time 2.5, **[1]** ii May be happening at time 9 **[3]**

5 A project taking place in a workshop is shown in the Gantt chart.

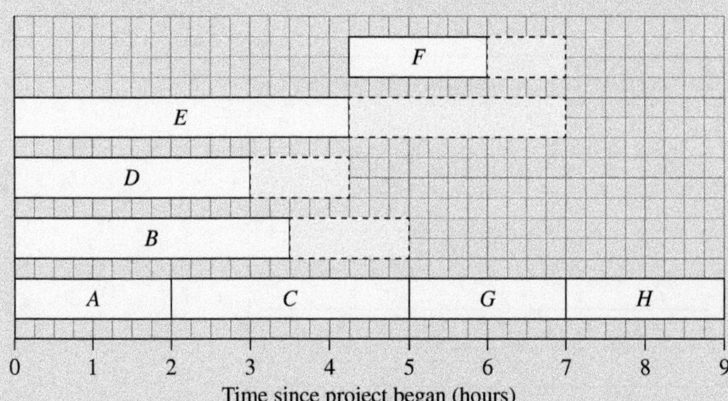

Work on the project begins at 8:00 am. An inspector visits at 11:45 am.

a Which activities will the inspector definitely see happening? **[1]**

Each activity requires two people to complete it and every person is capable of carrying out every activity.

b What is the lower bound for the number of people required to complete the project in the minimum time? **[2]**

c Draw a resource histogram to show how the project can be completed in the minimum time by 8 people. **[3]**

The machine needed for activity *F* is broken and will not be fixed until 2:30 pm.

d What is the earliest time at which the whole project can now end? **[2]**

6 On this Gantt chart, the values in brackets show the number of workers needed for the activity.

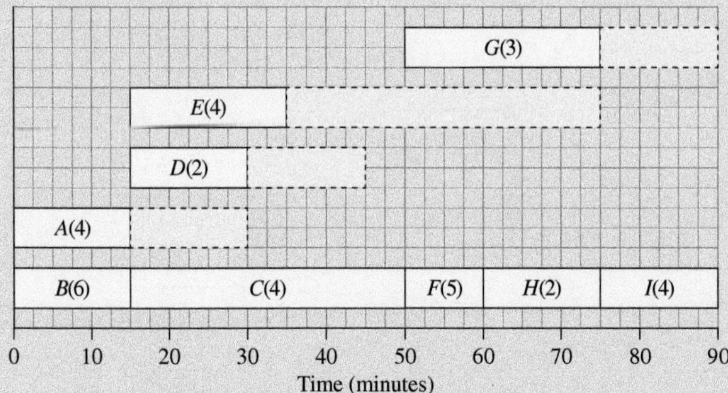

a Assuming that every activity starts as early as possible, draw a resource histogram and state the number of workers required. **[4]**

b Use resource levelling to show how the project can be completed in the minimum time by just 9 workers. **[3]**

Only 8 workers are now available.

c Draw a resource histogram and give the new minimum length of the project. **[4]**

7 The table gives the duration, dependencies and number of workers required for the activities in a project.

Activity	Depends on:	Duration (hours)	Number of workers
A	–	5	4
B	–	2	2
C	–	4	5
D	*A*, *B*	6	1
E	*A*, *B*, *C*	2	6
F	*B*, *C*	2	3
G	*F*	5	4
H	*E*, *F*	8	2
I	*G*	3	7
J	*I*	5	3
K	*J*	6	4
L	*I*	8	3

a Calculate a lower bound for the number of workers required to complete the project in 26 hours. **[2]**

b Draw a resource histogram for the project, assuming that all activities start as early as possible. **[4]**

It is possible for 4 workers to complete activity *C* in 5 hours.

c Would you recommend this approach? Fully explain your answer. **[3]**

8 This activity network has the number of workers required for each activity shown in brackets. The duration of each activity is given in days.

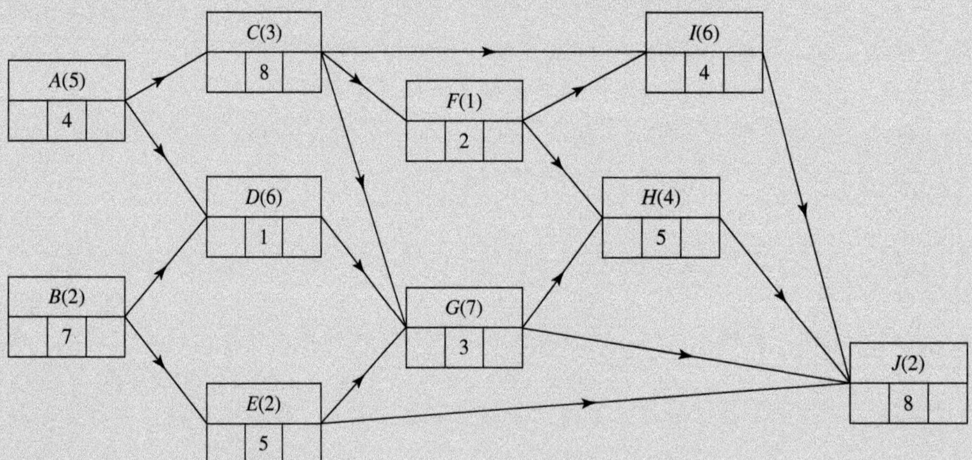

a Copy and complete the activity network with the earliest start and latest finish times. **[4]**

b Find the critical paths. **[2]**

c Draw a resource histogram, assuming that each activity starts as early as possible. State the number of workers required. **[5]**

d Explain how to reschedule the activities to find the minimum number of workers required to complete the project in the minimum time. **[2]**

e By how many days would the project have to be extended if only 10 workers were available? Explain your answer. **[2]**

30 Linear programming and game theory 2

Making decisions about situations that involve cooperation, or non-cooperation, with outcomes that are either advantageous or non-advantageous, is fundamental to **game theory**. Game theory has applications in a variety of areas including economics, business, political science and biology. The behaviours and decisions of humans, or animals, can lead to desirable or less desirable outcomes. Game theory can be used to model likely patterns of behaviour.

Until recently, computer programs have been able to take decisions only if every eventuality was pre-programmed. If certain conditions were not met, then the programming did not allow a decision to be made. Game theory has been used to advance programming for artificial intelligence (AI) systems. In future, AI systems may generate payoff matrices that are based on observed stimuli and experience. This will mean that, when decisions need to be made, the system responds more like a human.

Orientation

What you need to know

Ch14 Linear programming and game theory 1

What you will learn

- How to use the simplex algorithm.
- How to analyse zero-sum games.
- How to analyse mixed-strategy games.

What this leads to

Careers

- Business.
- Economics.
- Political science.
- Biology.
- Computer programming.

30.1 Simplex algorithm

Fluency and skills

In linear programming, you aim to find the best combination of a number of quantities (the decision variables) to maximise or minimise a given quantity (the objective function), subject to a number of constraints (usually inequalities).

To solve a problem with just two decision variables using a graph, you need to identify the feasible region. This is the set of points on the graph that satisfy all the constraints. You then draw an objective line. This is a line joining all points for which the objective function takes the same value.

By moving the objective line you can find the optimal value of the objective function for which the line intersects the feasible region. The optimal solution corresponds to a vertex of the feasible region.

If the problem has three or more decision variables, a graphical approach may be difficult or infeasible, so it is necessary to find a solution by algebraic methods. The simplex algorithm is an algebraic method that tests whether a vertex gives the optimal solution and, if not, moves systematically to a better vertex until the problem is solved.

You first write the problem using equations rather than inequalities. You do this by introducing **slack variables**.

Look at this linear programming formulation. This has only two variables, so you can compare the simplex method with the graphical method.

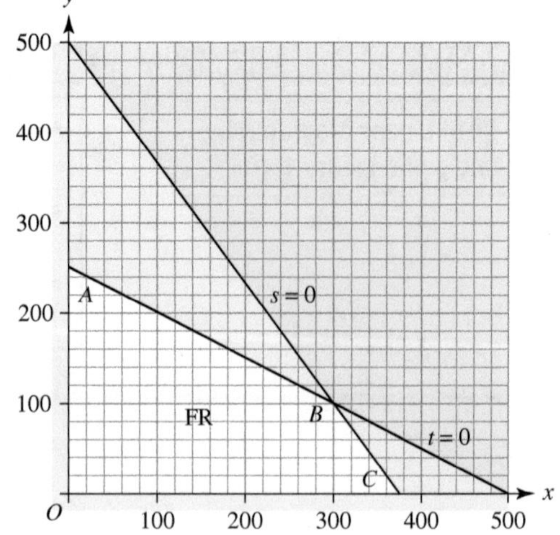

Maximise $P = 6x + 8y$

subject to $4x + 3y \leq 1500$

$x + 2y \leq 500$

$x \geq 0, y \geq 0$

You use slack variables s and t to show the difference (the slack) between the two sides of the inequalities. You write the formulation like this:

Maximise $P = 6x + 8y$

subject to $4x + 3y + s = 1500$

$x + 2y + t = 500$

$x \geq 0, y \geq 0, s \geq 0, t \geq 0$

The line $4x + 3y = 1500$ is equivalent to $s = 0$

and the line $x + 2y = 500$ to $t = 0$

At each of the vertices O, A, B and C of the feasible region, two of the variables x, y, s and t are zero.

The algebraic method effectively sets two of the variables to zero, finds the values of the other two and evaluates P

For example, setting $x = 0$, $t = 0$ (that is vertex A) gives

$3y + s = 1500$

$2y = 500$

Solving gives $y = 250$, $s = 750$ and $P = 6 \times 0 + 8 \times 250 = 2000$

The complete table is

Vertex	x	y	s	t	P
O	0	0	1500	500	0
A	0	250	750	0	2000
B	300	100	0	0	2600
C	375	0	0	125	2250
–	0	500	0	–500	–
–	500	0	–500	0	–

The optimal solution is $x = 300$, $y = 100$ $P = 2600$

In each row of the table, the non-zero variables form the basis of a possible solution. They are called the **basic variables** and the solution is a **basic solution**. Each of the first four rows is a **basic feasible solution** because it corresponds to a point of the feasible region.

The last two rows do not correspond to vertices of the feasible region. This is shown by the negative values taken by the variables.

You are going to use the **simplex algorithm** to search the basic solutions for the optimal solution.

You write each constraint as an equation, using a slack variable, in **standard form**. All the variables should be on the left-hand side and a non-negative value should be on the right. You also write the objective function in standard form, though the right-hand side can be negative.

You then enter the coefficients into a table called a **simplex tableau** (plural tableaux)

So for the above example

$P - 6x - 8y - 0s - 0t = 0$

$4x + 3y + s + 0t = 1500$

$x + 2y + 0s + t = 500$

and the simplex tableau is

Basic variable	P	x	y	s	t	Value	Row
	1	–6	–8	0	0	0	R1
s	0	4	3	1	0	1500	R2
t	0	1	2	0	1	500	R3

The basic variables, s and t, always have a coefficient of 1 in their row, and zeros elsewhere in their column. The 'Row' column is for recording your working. The left-hand column is not really needed, but you might find it helpful to keep track of the process.

This starting tableau corresponds to the origin O in the feasible region, with non-basic variables $x = 0$, $y = 0$, giving $P = 0$

The basic variables are $s = 1500$ and $t = 500$

Because $P - 6x - 8y = 0$, increasing either x or y will make P increase. In general, a negative number in the objective row tells you that the corresponding variable must be increased to reach the optimal solution.

Test for optimality

Key point

The table shows the optimal solution when the objective (top) row contains no negative coefficients.

You move to a different vertex of the feasible region. This is called **changing the basis** of the solution.

This means either x or y becomes a basic variable. It is said to **enter the basis** as you increase its value. In turn, either s or t must become a non-basic variable. It is said to **leave the basis** as you make it zero.

You first have to choose between x and y for the new basic variable. You could choose either, but with coefficients –6 and –8 it is clear that increasing y will have a greater effect on P than increasing x

Key point

Choose the variable with the 'most negative' entry in the objective row to enter the basis. The corresponding column is called the **pivot column.**

Next you choose which of s and t should become zero. You divide each *positive* entry in the pivot column into the corresponding entry in the 'Value' column. These ratios are called **θ-values.**

Key point

The row giving the smallest θ-value is the **pivot row.** The entry in the pivot row and pivot column is called the **pivot.** The basic variable in the pivot row is the one that will leave the basis.

In the example, the y-column is the pivot column.

The θ-value for $s = \dfrac{1500}{3} = 500$, and for $t = \dfrac{500}{2} = 250$

It is smaller for t, so this is the pivot row, and t will leave the basis.

The pivot is the 2 in the pivot column and pivot row.

Pivot column

Pivot row → t

Pivot

Basic variable	P	x	y	s	t	Value	Row
	1	−6	−8	0	0	0	R1
s	0	4	3	1	0	1500	R2
t	0	1	2	0	1	500	R3

You divide the pivot row by the pivot to give a 1 in the pivot position. In this case, you divide row R3 by 2

You now want to get zeros in the rest of the pivot column by combining the other rows with a suitable multiple of the pivot row.

In the y-column, change the 3 to a 0 by subtracting $3 \times$ R6

In the y-column, change the −8 to 0 by adding $8 \times$ R6

Remember that basic variables have a 1 in their row and zeros in the rest of their column. You need to make this true for y

Basic variable	P	x	y	s	t	Value	Row
	1	−2	0	0	4	2000	R4 = R1 + 8 × R6
s	0	$2\frac{1}{2}$	0	1	$-1\frac{1}{2}$	750	R5 = R2 − 3 × R6
y	0	$\frac{1}{2}$	1	0	$\frac{1}{2}$	250	R6 = $\frac{\text{R3}}{2}$

Use the 'Row' column to record what has happened, as shown. Always use fractions, not decimals, to avoid introducing rounding errors.

At this stage, $s = 750$ and $y = 250$, $x = 0$, $t = 0$, giving $P = 2000$

You apply the optimality test. There is −2 in the x-column, so you need to do another change of basis.

Remember that the solution is optimal if there are no negative coefficients in the objective row.

The x-column is the pivot column, so x will enter the basis.

The θ-value for $s = \frac{750}{2\frac{1}{2}} = 300$, and for $t = \frac{250}{\frac{1}{2}} = 500$

It is smaller for s, so this is the pivot row, and s will leave the basis.

Basic variable	P	x	y	s	t	Value	Row
	1	−2	0	0	4	2000	R4
s	0	$2\frac{1}{2}$	0	1	$-1\frac{1}{2}$	750	R5
y	0	$\frac{1}{2}$	1	0	$\frac{1}{2}$	250	R6

Divide the pivot row by $2\frac{1}{2}$ to give row R8. The basic variable will change from s to x

In the x-column on the y-row, change the $\frac{1}{2}$ to a 0 by subtracting $\frac{1}{2} \times$ R8

In the x-column on the objective row, change the −2 to 0 by adding $2 \times$ R8

DISCRETE

Basic variable	P	x	y	s	t	Value	Row
	1	0	0	$\frac{4}{5}$	$2\frac{4}{5}$	2600	$R7 = R4 + 2 \times R8$
x	0	1	0	$\frac{2}{5}$	$-\frac{3}{5}$	300	$R8 = \frac{R5}{2\frac{1}{2}}$
y	0	0	1	$-\frac{1}{5}$	$\frac{4}{5}$	100	$R9 = R6 - \frac{1}{2} \times R8$

You now have $x = 300$, $y = 100$, $s = 0$, $t = 0$, giving $P = 2600$

There are no negative coefficients in the objective row, so the solution is optimal.

Key point

The simplex algorithm

Step 1 Write the constraints and the objective function as equations in standard form, using slack variables.

Step 2 Transfer the data to a simplex tableau. At this stage the slack variables form the basis.

Step 3 Choose the column with the most negative coefficient in the objective row. This is the **pivot column.**

Step 4 If the **positive** numbers in the pivot column are $p_1, p_2, \ldots$ and the corresponding numbers in the 'Value' column are $v_1, v_2, \ldots$, calculate $\theta_1 = \frac{v_1}{p_1}, \theta_2 = \frac{v_2}{p_2}, \ldots$

The row giving the smallest θ-value is the **pivot row.** (If there is a 'tie', choose at random). The number in the pivot column and pivot row is the **pivot.**

Step 5 Divide the pivot row by the pivot. Replace the basic variable for that row by the variable for the pivot column.

Step 6 Combine suitable multiples of the new pivot row with the other rows to give zeros in the pivot column.

Step 7 If there are no negative coefficients in the objective row, the solution is optimal. Otherwise, go to Step 3

The main advantage of the simplex algorithm is that you can use it when there are three or more decision variables.

Example 1

Maximise $P = x + 2y + z$

subject to $x + 3y + 2z \le 60$

$2x + y \le 40$

$x + 3z \le 30$

$x, y, z \ge 0$

Step 1 In standard form with slack variables s, t and u

Maximise $P - x - 2y - z = 0$

subject to $x + 3y + 2z + s = 60$

$2x + y + t = 40$

$x + 3z + u = 30$

$x, y, z, s, t, u \ge 0$

Step 2

Basic variables.

B.V.	P	x	y	z	s	t	u	Value	Row
	1	−1	−2	−1	0	0	0	0	R1
s	0	1	3	2	1	0	0	60	R2
t	0	2	1	0	0	1	0	40	R3
u	0	1	0	3	0	0	1	30	R4

This is the simplex tableau.

Step 3 The y-column is the pivot column.

Step 4 The θ-values are: R2: $60 \div 3 = 20$, R3: $40 \div 1 = 40$

These give R2 as the pivot row and the 3 as the pivot, as shown.

Step 5 Divide R2 by 3

y becomes the basic variable.

Step 6

B.V.	P	x	y	z	s	t	u	Value	Row
	1	$-\frac{1}{3}$	0	$\frac{1}{3}$	$\frac{2}{3}$	0	0	40	R5 = R1 + 2 × R6
y	0	$\frac{1}{3}$	1	$\frac{2}{3}$	$\frac{1}{3}$	0	0	20	R6 = $\frac{\text{R2}}{3}$
t	0	$1\frac{2}{3}$	0	$-\frac{2}{3}$	$-\frac{1}{3}$	1	0	20	R7 = R3 − R6
u	0	1	0	3	0	0	1	30	R8 = R4

Combine the rows to get zeros in the y-column.

Step 7 The solution is not optimal because there is a negative value in the top row.

(*Continued on the next page*)

DISCRETE

Step 3 The x-column is the pivot column.

Step 4 The θ-values are: R6: $20 \div \frac{1}{3} = 60$

R7: $20 \div 1\frac{2}{3} = 12$

R8: $30 \div 1 = 30$

These give R7 as the pivot row and $1\frac{2}{3}$ as the pivot, as shown.

Step 5 Divide R7 by $1\frac{2}{3}$

x becomes the basic variable.

Step 6

B.V.	P	x	y	z	s	t	u	Value	Row
	1	0	0	$\frac{1}{5}$	$\frac{3}{5}$	$\frac{1}{5}$	0	44	$R9 = R5 + \frac{R11}{3}$
y	0	0	1	$\frac{4}{5}$	$\frac{2}{5}$	$-\frac{1}{5}$	0	16	$R10 = R6 - \frac{R11}{3}$
x	0	1	0	$-\frac{2}{5}$	$-\frac{1}{5}$	$\frac{3}{5}$	0	12	$R11 = \frac{R7}{1\frac{2}{3}}$
u	0	0	0	$3\frac{2}{5}$	$\frac{1}{5}$	$-\frac{3}{5}$	1	18	$R12 = R8 - R11$

Combine the rows to get zeros in the x-column.

Step 7 There are no negative numbers in the top row so the solution is optimal.

The optimal solution is $y = 16$, $x = 12$, $u = 18$, $z = 0$, $s = 0$, $t = 0$, giving $P = 44$

Exercise 30.1A Fluency and skills

1 Use the simplex method to solve the following linear programming problems.

a Maximise $P = x + y$

subject to $x + 2y \le 40$

$3x + 2y \le 60$

$x \ge 0, y \ge 0$

b Maximise $P = 2x + y$

subject to $4x + 3y \le 170$

$5x + 2y \le 160$

$x \ge 0, y \ge 0$

c Maximise $P = 4x + 5y$

subject to $2x + 3y \le 30$

$x + 3y \le 24$

$4x + 3y \le 48$

$x \ge 0, y \ge 0$

d Maximise $P = x + 2y$

subject to $x + y \le 6$

$2x + y \le 9$

$3x + 2y \le 15$

$x \ge 0, y \ge 0$

e Maximise $P = 3x + 4y + 2z$

subject to $8x + 5y + 2z \le 7$

$x + 2y + 3z \le 4$

$x \ge 0, y \ge 0, z \ge 0$

f Maximise $P = 7x + 6y + 4z$

subject to $2x + 4y - z \le 7$

$5x + 6y + 2z \le 16$

$7x + 7y + 4z \le 25$

$x \ge 0, y \ge 0, z \ge 0$

2

P	x	y	z	s	t	Value
1	−3	−1	−2	0	0	0
0	2	3	1	1	0	20
0	1	2	1	0	1	12

a Write the objective function and the constraints (in inequality form) corresponding to this simplex tableau.

b Perform one iteration of the simplex algorithm. Explain how you know that the solution is not yet optimal. State the values of the variables at this stage.

c Complete the solution of the problem, stating the final values of the variables.

3 Consider the following linear programming problem.

Maximise $P = x + 3y$

subject to $x + y \le 6$

$x + 4y \le 12$

$x + 2y \le 7$

$x \ge 0, y \ge 0$

a Write the problem in terms of equations with slack variables.

b Enter the data into a simplex tableau and perform one iteration of the simplex algorithm. Explain how you know that the optimal solution has not yet been reached.

c Perform a second iteration of the algorithm to obtain the optimal solution.

4 A linear programming problem is written as a simplex tableau as follows.

P	x	y	s	t	Value
1	−5	−6	0	0	0
0	3	3	1	0	40
0	1	2	0	1	25

a Write down the original linear programming formulation, stating the objective function and showing the constraints as inequalities. Explain the meaning of the variables s and t in the tableau.

b Perform one iteration of the simplex algorithm. State the values of x, y, s, t and P at this stage, and explain how you know that the solution is not optimal.

c Perform a second iteration of the algorithm to obtain the optimal solution.

DISCRETE

Reasoning and problem-solving

Strategy

To solve a problem using the simplex algorithm

1. Write the problem as a linear programming formulation.
2. Express the constraints as equations by using slack variables.
3. Apply the algorithm until there are no negative coefficients in the objective row.
4. Read off the values of the basic variables and the objective function.
5. State the solution in context, including units where appropriate.
6. If the problem is non-standard, modify it until you can apply the simplex algorithm.
7. Answer the question in context.

You will usually need to express a problem as a linear programming formulation before solving it.

Example 2

A farmer has 65 hectares of land on which to grow a mixture of wheat and potatoes. The costs and profits involved are shown in the table.

	Labour (man-hours per ha)	Fertiliser (kg per ha)	Profit (£ per ha)
Wheat	30	700	80
Potatoes	50	400	100

There are 2800 man-hours of labour and 40 tonnes of fertiliser available. Construct a simplex tableau and solve it to find the optimal planting scheme. Explain the meaning of slack variables in the final solution.

Plant x ha of wheat and y ha of potatoes.

Maximise $P = 80x + 100y$

subject to $30x + 50y \le 2800$ and so $3x + 5y \le 280$

$700x + 400y \le 40\,000$ and so $7x + 4y \le 400$

$x + y \le 65$

$x \ge 0, y \ge 0$

$P - 80x - 100y = 0$

$3x + 5y + s = 280$

$7x + 4y + t = 400$

$x + y + u = 65$

Write as equations.

P	x	y	s	t	u	Value	Row
1	−80	−100	0	0	0	0	R1
0	3	5	1	0	0	280	R2
0	7	4	0	1	0	400	R3
0	1	1	0	0	1	65	R4

Apply the simplex algorithm using the pivot shown.

P	x	y	s	t	u	Value	Row
1	−20	0	20	0	0	5600	$R5 = R1 + 100 \times R6$
0	$\frac{3}{5}$	1	$\frac{1}{5}$	0	0	56	$R6 = \frac{R2}{5}$
0	$4\frac{3}{5}$	0	$-\frac{4}{5}$	1	0	176	$R7 = R3 - 4 \times R6$
0	$\frac{2}{5}$	0	$-\frac{1}{5}$	0	1	9	$R8 = R4 - R6$

Not yet optimal, so apply the simplex algorithm using the pivot shown.

(*Continued on the next page*)

P	x	y	s	t	u	Value	Row
1	0	0	10	0	50	6050	$R9 = R5 + 20 \times R12$
0	0	1	$\frac{1}{2}$	0	$-1\frac{1}{2}$	$42\frac{1}{2}$	$R10 = R6 - R12 \times \frac{3}{5}$
0	0	0	$1\frac{1}{2}$	1	$-11\frac{1}{2}$	$72\frac{1}{2}$	$R11 = R7 - R12 \times \frac{23}{5}$
0	1	0	$-\frac{1}{2}$	0	$2\frac{1}{2}$	$22\frac{1}{2}$	$R12 = R8 \times \frac{5}{2}$

The solution is now optimal.

Maximum profit = £6050, by planting 22.5 ha of wheat and 42.5 ha of potatoes.

The other non-zero value is $t = 72.5$. This corresponds to 7250 kg of fertiliser left over.

Be careful to allow for any simplification made in the constraints before the slack variables were introduced.

You can use the standard simplex algorithm provided:

- you need to maximise the objective function
- every non-trivial constraint is an inequality using ≤
- all variables are ≥ 0
- the origin is a vertex of the feasible region.

Provided these conditions are met, you can write the problem as equations in standard form using slack variables, with the right-hand sides all being non-negative (except perhaps for the objective function).

You may need to deal with non-standard situations in which the aim is to minimise the objective function, or when one or more of the constraints involves an inequality using ≥

You convert a minimisation problem into a related maximisation problem.

Key point

Minimising the objective function C is equivalent to maximising the objective function $P = -C$

You can rewrite an inequality involving ≥ so that it uses ≤ by multiplying through by −1

If the resulting problem can be written in standard form, you can go ahead with the simplex algorithm.

Example 3

Minimise $C = 3x - 4y - 3z$

subject to $x + y - z \geq -2$

$x + 2y + z \leq 3$

$x \geq 0, y \geq 0, z \geq 0$

(*Continued on the next page*)

Maximise $P = -3x + 4y + 3z$

subject to $-x - y + z \leq 2$

$x + 2y + z \leq 3$

$x, y, z \geq 0$

First restate the problem.

$P + 3x - 4y - 3z = 0$

$-x - y + z + s = 2$

$x + 2y + z + t = 3$

$x, y, z, s, t \geq 0$

Write in standard form with slack variables.

P	x	y	z	s	t	Value	Row
1	3	−4	−3	0	0	0	R1
0	−1	−1	1	1	0	2	R2
0	1	2	1	0	1	3	R3

Complete the simplex tableau.

P	x	y	z	s	t	Value	Row
1	5	0	−1	0	2	6	R4 = R1 + 4 × R6
0	$-\frac{1}{2}$	0	$1\frac{1}{2}$	1	$\frac{1}{2}$	$3\frac{1}{2}$	R5 = R2 + R6
0	$\frac{1}{2}$	1	$\frac{1}{2}$	0	$\frac{1}{2}$	$1\frac{1}{2}$	$R6 = \frac{R3}{2}$

Apply the simplex algorithm with the pivot shown.

There is still a negative coefficient in the objective row, so apply the simplex algorithm again with the pivot shown.

P	x	y	z	s	t	Value	Row
1	$4\frac{2}{3}$	0	0	$\frac{2}{3}$	$2\frac{1}{3}$	$8\frac{1}{3}$	R7 = R4 + R8
0	$-\frac{1}{3}$	0	1	$\frac{2}{3}$	$\frac{1}{3}$	$2\frac{1}{3}$	$R8 = R5 \times \frac{2}{3}$
0	$\frac{2}{3}$	1	0	$-\frac{1}{3}$	$\frac{1}{3}$	$\frac{1}{3}$	$R9 = R6 - \frac{R8}{2}$

The solution is now optimal.

Maximum $P = 8\frac{1}{3}$ when $x = 0, y = \frac{1}{3}, z = 2\frac{1}{3}$

As $C = -P$, minimum $C = -8\frac{1}{3}$ when $x = 0, y = \frac{1}{3}, z = 2\frac{1}{3}$

An extra problem arises if you have a constraint such as $2x - 3y \geq 4$

The simplex method is based on the origin being a vertex of the feasible region, but clearly (0, 0) does not satisfy this inequality. Attempting to write the constraint as an equation with a slack variable would give $-2x + 3y + s = -4$, which if $x = 0$ and $y = 0$ violates the need to have $s \geq 0$

There are some methods beyond the scope of this syllabus for dealing with the situation, but it is sometimes possible to modify the simplex tableau to give a basic feasible solution (a vertex of the feasible region) from which to start the simplex algorithm. To do this you first pivot on a negative coefficient in the row with the negative value.

Example 4

Maximise $P = x + 2y$

subject to $2x + y \leq 6$

$x + y \geq 2$

$x, y \geq 0$

First write the second constraint as $-x - y \leq -2$

$P - x - 2y = 0$

$2x + y + s = 6$

$-x - y + t = -2$

Write the problem as equations with slack variables.

The last constraint is not in standard form.

P	x	y	s	t	Value	Row
1	−1	−2	0	0	0	R1
0	2	1	1	0	6	R2
0	−1	−1	0	1	−2	R3

Before starting the simplex algorithm, pivot on the cell shown to produce +2 in the value column.

P	x	y	s	t	Value	Row
1	0	−1	0	−1	2	R4 = R1 + R6
0	0	−1	1	2	2	R5 = R2 − 2 × R6
0	1	1	0	−1	2	R6 = − R3

There are negative coefficients in the objective row, so apply the simplex algorithm with the pivot shown.

This gives $x = 2$, $s = 2$ ($y = 0$, $t = 0$) as an initial feasible solution.

P	x	y	s	t	Value	Row
1	1	0	0	−2	4	R7 = R4 + R9
0	1	0	1	1	4	R8 = R5 + R9
0	1	1	0	−1	2	R9 = R6

The solution is not optimal, so apply the simplex algorithm with the pivot shown.

P	x	y	s	t	Value	Row
1	3	0	2	0	12	R10 = R7 + 2 × R11
0	1	0	1	1	4	R11 = R8
0	2	1	1	0	6	R12 = R9 + R11

This is now optimal.

Maximum $P = 12$ when $x = 0$, $y = 6$ ($t = 4$)

DISCRETE

Exercise 30.1B Reasoning and problem-solving

1 a Use the simplex method to minimise $C = x - y$

subject to $y \leq 2x$

$2y \geq x$

$2x + y \leq 12$

$x \geq 0, y \geq 0$

b Use the simplex method to minimise $C = 4x - 3y - 5z$

subject to $x + y - z \geq -2$

$x + 2y + z \leq 4$

$x \geq 0, y \geq 0, z \geq 0$

c Use the simplex method to minimise $C = x - 2y + z$

subject to $2x - y + z \geq 0$

$2x - 3y \geq -6$

$4y + 3z \leq 10$

$x \geq 0, y \geq 0, z \geq 0$

d Use the simplex method to maximise $P = 2x - 3y + z$

subject to $3y \geq x + 2z - 5$

$y - x - z \geq -8$

$x + y + z \geq -2$

$x \geq 0, y \geq 0, z \geq 0$

2 A situation involving variables x, y and z is subject to these constraints:

$x - 4y + z \geq -4$

$2x - 3y + z \leq 33$

$x \geq 10, y \geq 3, z \geq 8$

You need to minimise the cost function $C = x + 3y - 2z$

At present, the origin is not in the feasible region. Substitute $X = x - 10$, $Y = y - 3$ and $Z = z - 8$ and solve the resulting problem using the simplex method. Hence state the solution to the original problem.

3 A garden centre produces and sells two grades of grass seed, each a mixture of perennial ryegrass (PR) and creeping red fescue (CRF). Regular Lawn Mix is 70% PR and 30% CRF; Luxury Lawn Mix is 50% of each.

They buy PR at £4 per kg and CRF at £5 per kg. They sell regular mix at £6 per kg and luxury mix at £7 per kg. They have 8000 kg of PR and 6000 kg of CRF in stock.

Use the simplex algorithm to decide what quantities of regular and luxury mix they should make and state the profit they achieve.

4 An electric bicycle has two settings. On setting A the rider pedals, with the motor providing assistance. On setting B the motor does all the work. Setting A gives a speed of $4\,\text{m s}^{-1}$, using 6 J of battery energy per metre. Setting B gives a speed of $6\,\text{m s}^{-1}$ and uses 9 J per metre. The battery can store 45 000 J. The rider wishes to travel as far as possible in 20 minutes.

Given that the rider travels x metres on setting A and y metres on setting B, express the problem in linear programming terms and use the simplex algorithm to find the rider's furthest distance and the settings that should be used to achieve it.

5 A company makes three models of bicycle. In each week they make x bicycles of type A, y of type B and z of type C. The weekly profit, in £hundreds, is P. In a certain week there are constraints on resources and manpower. They need a production plan to maximise the profit. The simplex tableau corresponding to this problem is as shown.

P	x	y	z	s	t	Value
1	−4	−5	−2	0	0	0
0	5	6	3	1	0	90
0	2	4	1	0	1	42

a How much profit is made on a bicycle of type A?

b Write down the two constraints as inequalities.

c Use the simplex algorithm to solve this linear programming problem. Explain how you know when you have reached the optimal solution.

d Explain why the solution to part **c** is not practical.

e Find a practical solution which gives a profit of £7200

Verify that it is feasible.

6 Moltobuono Meals Ltd make three different pasta sauces. Each is a blend of tomato paste and onion (of which they have unlimited quantities), with the addition of varying amounts of garlic paste, oregano and basil. The table below gives details of the recipes (for 1 kg of sauce), together with the availability of ingredients and the profit on each type of sauce.

Sauce	Garlic (kg)	Oregano (kg)	Basil (kg)	Profit (£ per kg)
Assolato	0.03	0.02	0.01	0.60
Buona Salute	0.02	0.03	0.01	0.50
Contadino	0.03	0.04	0.02	0.90
Availability	50	60	40	

They wish to maximise their profit.

a Write the problem as a linear programming formulation, simplifying the constraints so they have integer coefficients.

b Use the simplex algorithm to find the most profitable production plan.

7 A manufacturer makes two fruit-based drinks. Froo-T is 50% grape juice, 30% peach juice and 20% cranberry juice. Joo-C is 40% grape juice, 20% peach juice and 40% cranberry juice. The profit made is 20p per litre for Froo-T and 25p per litre for Joo-C. The manufacturer has 2100 litres of grape juice, 1200 litres of peach juice and 1500 litres of cranberry juice in stock. Find the amount of each drink they should make to maximise their total profit, and the amount of raw materials left over.

8 A vegetable-box delivery scheme offers four different boxes, containing different proportions of baking potatoes, cabbages, swedes and butternut squash. The table below shows the proportions of these, the availability of the different vegetables and the profit made on the boxes.

Box	Potatoes	Cabbages	Swedes	Squash	Profit (p per box)
A	6	2	1	1	50
B	4	1	1	2	30
C	8	2	2	2	80
D	5	1	2	1	60
Availability	800	240	300	320	

a Express the problem as a simplex tableau and solve it to find the optimal solution. (Assume that all the boxes will be sold.)

b Explain why the result you obtained in **a** does not provide a solution to the original problem. Use your result to arrive at a solution to the problem.

9 **a** Use the simplex method to maximise $P = 3x + 2y$

subject to $2x + y \le 6$

$3x - 2y \ge 4$

$x, y \ge 0$

b Use the simplex method to minimise $C = x + 2y$

subject to $2x \ge 7$

$x + y \ge 4$

$x, y \ge 0$

10 Part of an area of land is being made into a garden, some of which will be covered with turf and the rest with paving slabs. The aim is to make as large a garden as possible for a maximum outlay of £2000

It costs £5 per m^2 to lay turf and £10 per m^2 to lay slabs. There should be at least 50 m^2 of slabs and at least half of the garden will be turf.

Let there be x m^2 of turf and y m^2 of slabs. Find the size of the garden and the areas of turf and slabs to be laid.

DISCRETE

11 A linear programming problem is stated as

Maximise $P = 5x + 2y - z$

subject to $x + 2y + z \leq 14$

$2x - y + 2z \leq 16$

$2x + y - z = 10$

$x, y, z \geq 0$

a Use the equality constraint to restate the problem as a two-variable problem. Solve it using the simplex method.

b Restate the original problem by replacing the equality constraint with the two constraints $2x + y - z \leq 10$ and $2x + y - z \geq 10$ and then solve it.

12 Three students, A, B and C, attempt to solve this linear programming problem using a simplex tableau:

Maximise $P = x + 2y$

subject to $x + y \leq 28$

$3x + y \geq 30$

$2x + 3y \geq 60$

$x, y \geq 0$

a Student A completes the exercise successfully. Show her solution and state the final result.

b Student B miscopies the 28 in the first constraint as 18, and then complains that he can't find a solution. Show how his solution breaks down and explain, by means of a sketch graph or otherwise, why there has been a problem.

c Student C miscopies the first constraint as $x - y \leq 28$

She also complains that there are problems with the solution. Show how her solution breaks down and explain, by means of a sketch graph or otherwise, why there has been a problem.

30.2 Games as linear programming problems

Fluency and skills

Here is a brief summary of two-player zero-sum games.

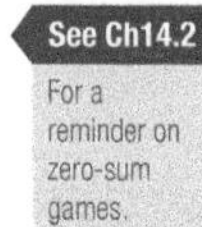

See Ch14.2
For a reminder on zero-sum games.

The table always shows the pay-offs for the row player. In this table, if player A plays A_1 they will lose 3 or gain 5 depending on what player B does.

		B			
		B_1	B_2	**Row minimum**	
A	A_1	−3	5	−3	
	A_2	3	−1	−1	← max. = −1
Column maximum		3	5		
		↑ min. = 3			

You find the play-safe strategy for A by finding the maximum of the row minima, and for B by finding the minimum of the column maxima. In this table, the play-safe strategies are A_2 and B_1

If they play a **pure strategy game** (both always play safe), A will win 3.

This is the value of the game.

If the solution is stable, there is no advantage in moving from the play-safe strategy.

Key point

A game has a stable solution if
maximum of row minima = minimum of column maxima

In the table shown the solution is not stable. It is best for each player to play each strategy for some proportion of the time, i.e play a **mixed strategy game**. They choose the proportions so as to maximise their expected pay-off.

Key point

If pay-offs $x_1, x_2, \ldots, x_n$ occur with probabilities $p_1, p_2, \ldots, p_n$, the **expectation** or **expected (mean) pay-off** $\mathrm{E}(x)$ is given by

$$\mathrm{E}(x) = x_1p_1 + x_2p_2 + \ldots + x_np_n = \sum_{i=1}^{n} x_ip_i$$

DISCRETE

If A plays A_1 with probability p and B plays B_1, the pay-offs and probabilities are as shown.

	A_1	A_2
Pay-off	-3	3
Probability	p	$(1-p)$

The expected pay-off is $(-3)\times p+3\times(1-p)=3-6p$

If B plays B_2, the possible pay-offs for A are as shown.

	A_1	A_2
Pay-off	5	-1
Probability	p	$(1-p)$

The expected pay-off is $5\times p+(-1)\times(1-p)=6p-1$

The value, v, of the game cannot be more than either of these expected pay-offs, so

$v\le 3-6p$ and $v\le 6p-1$

Player A needs to maximise v subject to $v\le 3-6p$ and $v\le 6p-1$

This is a linear programming problem that can be solved graphically if there are just two strategies available to A

If more than two strategies are involved, you need to use the simplex algorithm.

You will need to modify the pay-off matrix if any of the entries are negative, because the simplex method requires all variables to be non-negative including, in this case, the value v of the game. You do this by adding a constant to each entry in the matrix – once the problem has been solved, you then subtract this constant from the value you found.

Example 1

Use the simplex method to solve this zero-sum game to find the optimal strategy for players A and B and the value of the game.

		B	
		B_1	B_2
A	A_1	1	-2
	A_2	0	-1
	A_3	-1	0

Add 2 to each table entry.

		B	
		B_1	B_2
A	A_1	3	0
	A_2	2	1
	A_3	1	2

All entries must be non-negative.

(*Continued on the next page*)

Let A play A_1, A_2 and A_3 with probabilities p_1, p_2 and p_3

Maximise $\quad P = v$

subject to $\quad v \le 3p_1 + 2p_2 + p_3$

$v \le p_2 + 2p_3$

$p_1 + p_2 + p_3 \le 1$

$v, p_1, p_2, p_3 \ge 0$

$P - v = 0$

$v - 3p_1 - 2p_2 - p_3 + s = 0$

$v - p_2 - 2p_3 + t = 0$

$p_1 + p_2 + p_3 + u = 1$

A's expected profit, P, is the value, v, of the game.

You replace $p_1 + p_2 + p_3 = 1$ with the inequality shown so that you can introduce a slack variable for the simplex tableau. The slack variable will be found to be zero.

Introduce slack variables.

P	v	p_1	p_2	p_3	s	t	u	Value	Row
1	−1	0	0	0	0	0	0	0	R1
0	1	−3	−2	−1	1	0	0	0	R2
0	1	0	−1	−2	0	1	0	0	R3
0	0	1	1	1	0	0	1	1	R4
1	0	−3	−2	−1	1	0	0	0	R5 = R1 + R6
0	1	−3	−2	−1	1	0	0	0	R6 = R2
0	0	3	1	−1	−1	1	0	0	R7 = R3 − R6
0	0	1	1	1	0	0	1	1	R8 = R4
1	0	0	−1	−2	0	1	0	0	R9 = R5 + R11 × 3
0	1	0	−1	−2	0	1	0	0	R10 = R6 + R11 × 3
0	0	1	$\frac{1}{3}$	$-\frac{1}{3}$	$-\frac{1}{3}$	$\frac{1}{3}$	0	0	$R11 = \frac{R7}{3}$
0	0	0	$\frac{2}{3}$	$1\frac{1}{3}$	$\frac{1}{3}$	$-\frac{1}{3}$	1	1	R12 = R8 − R11
1	0	0	0	0	$\frac{1}{2}$	$\frac{1}{2}$	$1\frac{1}{2}$	$1\frac{1}{2}$	R13 = R9 + R16 × 2
0	1	0	0	0	$\frac{1}{2}$	$\frac{1}{2}$	$1\frac{1}{2}$	$1\frac{1}{2}$	R14 = R10 + R16 × 2
0	0	1	$\frac{1}{2}$	0	$-\frac{1}{4}$	$\frac{1}{4}$	$\frac{1}{4}$	$\frac{1}{4}$	$R15 = R11 + \frac{R16}{3}$
0	0	0	$\frac{1}{2}$	1	$\frac{1}{4}$	$-\frac{1}{4}$	$\frac{3}{4}$	$\frac{3}{4}$	$R16 = R12 \times \frac{3}{4}$

This is the simplex solution.

(*Continued on the next page*)

DISCRETE

This gives $v = 1.5$ when $p_1 = \frac{1}{4}$, $p_2 = 0$ and $p_3 = \frac{3}{4}$

So the value of the game is -0.5

Subtract the 2 you added initially.

A's optimal strategy is to play A_1 and A_3 with probabilities 0.25, 0.75 and never play A_2

The value of the game to B is 0.5. Suppose B plays B_1 and B_2 with probabilities q_1 and q_2

If A plays A_1, the value to B is $-q_1 + 2q_2$, so $\quad -q_1 + 2q_2 = 0.5 \quad$ [1]

Similarly, for A_2 $\quad q_2 = 0.5 \quad$ [2]

and A_3 $\quad q_1 = 0.5 \quad$ [3]

[1], [2] and [3] are all satisfied by $q_1 = 0.5$ and $q_2 = 0.5$, so B plays B_1 and B_2 with equal probability.

Of course, instead of using all three probabilities, you could rewrite the problem by replacing p_3 by $(1 - p_1 - p_2)$. The solution then looks like this.

Example 2

Maximise $\quad P = v$

subject to $\quad v \le 3p_1 + 2p_2 + (1 - p_1 - p_2) \quad \rightarrow \quad v \le 2p_1 + p_2 + 1$

$v \le p_2 + 2(1 - p_1 - p_2) \quad \rightarrow \quad v \le 2 - 2p_1 - p_2$

$v, p_1, p_2 \ge 0$

Introduce slack variables:

$P - v = 0$

$v - 2p_1 - p_2 + s = 1$

$v + 2p_1 + p_2 + t = 2$

Here is the simplex solution.

P	v	p_1	p_2	s	t	Value	Row
1	−1	0	0	0	0	0	R1
0	1	−2	−1	1	0	1	R2
0	1	2	1	0	1	2	R3

(*Continued on the next page*)

P	v	p_1	p_2	s	t	Value	Row
1	0	−2	−1	1	0	1	R4 = R1+R5
0	1	−2	−1	1	0	1	R5 = R2
0	0	4	2	−1	1	1	R6 = R3-R5
1	0	0	0	$\frac{1}{2}$	$\frac{1}{2}$	$1\frac{1}{2}$	R7 = R4+2 × R9
0	1	0	0	$\frac{1}{2}$	$\frac{1}{2}$	$1\frac{1}{2}$	R8 = R5 +2 × R9
0	0	1	$\frac{1}{2}$	$-\frac{1}{4}$	$\frac{1}{4}$	$\frac{1}{4}$	R9 = R$\frac{6}{4}$

This gives $v = 1.5 - 2 = -0.5$ when $p_1 = \frac{1}{4}$, $p_2 = 0$ and $p_3 = 1 - p_1 - p_2 = \frac{3}{4}$, as before.

Exercise 30.2A Fluency and skills

1 For each of these games, use the simplex method to find the optimal mixed strategy for each player and the value of the game.

a

		B	
		B_1	B_2
A	A_1	2	−2
	A_2	−2	3

b

		B	
		B_1	B_2
A	A_1	7	6
	A_2	5	8

c

		B	
		B_1	B_2
A	A_1	5	2
	A_2	−1	3

d

		B	
		B_1	B_2
A	A_1	−2	1
	A_2	3	0

2 For each of these zero-sum games, use the simplex method to find the optimal strategy for A and the value of the game.

a

		B		
		B_1	B_2	B_3
A	A_1	1	−1	0
	A_2	−2	0	−1

b

		B	
		B_1	B_2
A	A_1	1	−2
	A_2	−1	1
	A_3	0	−1

3 Consider each of the following games. If there is a stable solution, find the play-safe strategies and the value of the game. Otherwise use the simplex method to find the optimal strategy for A and the value of the game. (Simplify the table using dominance where possible.)

a

		B	
		B_1	B_2
A	A_1	−3	−3
	A_2	1	−3
	A_3	−2	1

b

		B	
		B_1	B_2
A	A_1	−1	2
	A_2	−3	4
	A_3	2	1

c

		B		
		B_1	B_2	B_3
A	A_1	−2	5	3
	A_2	2	3	2
	A_3	−1	0	1

d

		B		
		B_1	B_2	B_3
A	A_1	−4	−1	1
	A_2	0	−3	−1
	A_3	−1	2	0

e

		B			
		B_1	B_2	B_3	B_4
A	A_1	3	2	−1	3
	A_2	−1	0	3	1
	A_3	4	3	2	5
	A_4	2	−1	−2	1

f

		B			
		B_1	B_2	B_3	B_4
A	A_1	−3	−6	2	−11
	A_2	13	−10	−4	8
	A_3	5	8	4	7
	A_4	4	9	−5	0

Reasoning and problem-solving

Strategy

To solve a zero-sum game using the simplex algorithm

1. If necessary construct a pay-off matrix.
2. If possible, use dominance to reduce the table.
3. Check that the game does not have a stable solution.
4. Write the problem as a linear programming formulation.
5. Express the constraints as equations by using slack variables.
6. Apply the algorithm until there are no negative coefficients in the objective row.
7. Read off the values of the probabilities and the value of the game.
8. Answer the question in context.

Example 3

Show that this zero-sum game requires a mixed strategy, and find A's best strategy and the value of the game.

		B		
		B_1	B_2	B_3
A	A_1	0	−1	−1
	A_2	−1	0	1
	A_3	−2	1	2

Column 2 dominates column 3, so delete column 3

The revised table is

		B	
		B_1	B_2
A	A_1	0	−1
	A_2	−1	0
	A_3	−2	1

The row minima are −1, −1, −2, with a maximum value of −1

The column maxima are 0, 1, with a minimum of 0

These are not equal, so the solution is not stable.

Add 2 to all the pay-offs. Let the value of the revised game be v

		B	
		B_1	B_2
A	A_1	2	1
	A_2	1	2
	A_3	0	3

Suppose A plays A_1, A_2 and A_3 with probabilities p_1, p_2 and p_3

If B plays B_1 then A's expected pay-off is $2p_1 + p_2$

If B plays B_2 then A's expected pay-off is $p_1 + 2p_2 + 3p_3$

The linear programming formulation is

Maximise $P = v$

subject to $v \le 2p_1 + p_2$

$v \le p_1 + 2p_2 + 3p_3$

$p_1 + p_2 + p_3 \le 1$

$v, p_1, p_2, p_3 \ge 0$

2 If possible use dominance to reduce the table. $-1 = -1$, $0 < 1$, $1 < 2$ (Remember: these are A's gains.)

3 Check that the game does not have a stable solution.

So there are no negative entries.

4 Write the problem as a linear programming formulation.

(*Continued on the next page*)

Write as equations:

$P - v = 0$

$v - 2p_1 - p_2 + s = 0$

$v - p_1 - 2p_2 - 3p_3 + t = 0$

$p_1 + p_2 + p_3 + u = 1$

Express the constraints as equations by using slack variables. (5)

Enter these in a tableau and apply the simplex algorithm.

P	v	p_1	p_2	p_3	s	t	u	Value	Row
1	−1	0	0	0	0	0	0	0	R1
0	1	−2	−1	0	1	0	0	0	R2
0	1	−1	−2	−3	0	1	0	0	R3
0	0	1	1	1	0	0	1	1	R4
1	0	−2	−1	0	1	0	0	0	R5 = R1 + R6
0	1	−2	−1	0	1	0	0	0	R6 = R2
0	0	1	−1	−3	−1	1	0	0	R7 = R3 − R6
0	0	1	1	1	0	0	1	1	R8 = R4
1	0	0	−3	−6	−1	2	0	0	R9 = R5 + R11 × 2
0	1	0	−3	−6	−1	2	0	0	R10 = R6 + R11 × 2
0	0	1	−1	−3	−1	1	0	0	R11 = R7
0	0	0	2	4	1	−1	1	1	R12 = R8 − R11
1	0	0	0	0	$\frac{1}{2}$	$\frac{1}{2}$	$1\frac{1}{2}$	$1\frac{1}{2}$	R13 = R9 + R16 × 6
0	1	0	0	0	$\frac{1}{2}$	$\frac{1}{2}$	$1\frac{1}{2}$	$1\frac{1}{2}$	R14 = R10 + R16 × 6
0	0	1	$\frac{1}{2}$	0	$-\frac{1}{4}$	$\frac{1}{4}$	$\frac{3}{4}$	$\frac{3}{4}$	R15 = R11 + R16 × 3
0	0	0	$\frac{1}{2}$	1	$\frac{1}{4}$	$-\frac{1}{4}$	$\frac{1}{4}$	$\frac{1}{4}$	$R16 = \frac{R12}{4}$

This gives $v = 1.5$ and so the value of the original game is −0.5

Remember to subtract the 2 from v

The optimal strategy for A is to play A_2 and A_3 with probabilities 0.75 and 0.25

Example 4

Solve this zero-sum game to find the optimal strategy for player A and the value of the game.

		B		
		B_1	B_2	B_3
A	A_1	1	−1	0
	A_2	0	2	1
	A_3	1	1	−1

The game cannot be reduced using dominance.

Add 1 to all the entries.

Entries must be non-negative.

		B		
		B_1	B_2	B_3
A	A_1	2	0	1
	A_2	1	3	2
	A_3	2	2	0

Suppose A plays A_1, A_2, A_3 with probabilities p_1, p_2, p_3

Let the value of the revised game be v

The linear programming formulation is

Maximise $P = v$

subject to $v \le 2p_1 + p_2 + 2p_3$

$v \le 3p_2 + 2p_3$

$v \le p_1 + 2p_2$

$p_1 + p_2 + p_3 \le 1$

$v, p_1, p_2, p_3 \ge 0$

Introduce slack variables:

$P - v = 0$

$v - 2p_1 - p_2 - 2p_3 + s_1 = 0$

$v - 3p_2 - 2p_3 + s_2 = 0$

$v - p_1 - 2p_2 + s_3 = 0$

$p_1 + p_2 + p_3 + s_4 = 1$

Here is the simplex solution.

P	v	p_1	p_2	p_3	s_1	s_2	s_3	s_4	Value	Row
1	−1	0	0	0	0	0	0	0	0	R1
0	1	−2	−1	−2	1	0	0	0	0	R2
0	1	0	−3	−2	0	1	0	0	0	R3
0	1	−1	−2	0	0	0	1	0	0	R4
0	0	1	1	1	0	0	0	1	1	R5

(*Continued on the next page*)

DISCRETE

1	0	−2	−1	−2	1	0	0	0	0	R6 = R1 + R7
0	1	−2	−1	−2	1	0	0	0	0	R7 = R2
0	0	2	−2	0	−1	1	0	0	0	R8 = R3 − R7
0	0	1	−1	2	−1	0	1	0	0	R9 = R4 − R7
0	0	1	1	1	0	0	0	1	1	R10 = R5
1	0	0	−3	−2	0	1	0	0	0	R11 = R6 + 2 × R13
0	1	0	−3	−2	0	1	0	0	0	R12 = R7 + 2 × R13
0	0	1	−1	0	$-\frac{1}{2}$	$\frac{1}{2}$	0	0	0	$R13 = R\frac{8}{2}$
0	0	0	0	2	$-\frac{1}{2}$	$-\frac{1}{2}$	1	0	0	R14 = R9 − R13
0	0	0	2	1	$\frac{1}{2}$	$-\frac{1}{2}$	0	1	1	R15 = R10 − R13
1	0	0	0	$-\frac{1}{2}$	$\frac{3}{4}$	$-\frac{1}{4}$	0	$1\frac{1}{2}$	$1\frac{1}{2}$	R16 = R11 + 3 × R20
0	1	0	0	$-\frac{1}{2}$	$\frac{3}{4}$	$-\frac{1}{4}$	0	$1\frac{1}{2}$	$1\frac{1}{2}$	R17 = R12 + 3 × R20
0	0	1	0	$\frac{1}{2}$	$-\frac{1}{4}$	$\frac{1}{4}$	0	$\frac{1}{2}$	$\frac{1}{2}$	R18 = R13 + R20
0	0	0	0	2	$-\frac{1}{2}$	$-\frac{1}{2}$	1	0	0	R19 = R14
0	0	0	1	$\frac{1}{2}$	$\frac{1}{4}$	$-\frac{1}{4}$	0	$\frac{1}{2}$	$\frac{1}{2}$	$R20 = R\frac{15}{2}$
1	0	0	0	0	$\frac{5}{8}$	$\frac{1}{8}$	$\frac{1}{4}$	$1\frac{1}{2}$	$1\frac{1}{2}$	$R21 = R16 + R\frac{24}{2}$
0	1	0	0	0	$\frac{5}{8}$	$\frac{1}{8}$	$\frac{1}{4}$	$1\frac{1}{2}$	$1\frac{1}{2}$	$R22 = R17 + R\frac{24}{2}$
0	0	1	0	0	$-\frac{1}{8}$	$\frac{3}{8}$	$-\frac{1}{4}$	$\frac{1}{2}$	$\frac{1}{2}$	$R23 = R18 - R\frac{24}{2}$
0	0	0	0	1	$-\frac{1}{4}$	$-\frac{1}{4}$	$\frac{1}{2}$	0	0	$R24 = R\frac{19}{2}$
0	0	0	1	0	$\frac{3}{8}$	$-\frac{1}{8}$	$-\frac{1}{4}$	$\frac{1}{2}$	$\frac{1}{2}$	$R25 = R20 - R\frac{24}{2}$

This gives $v = 1.5$, so the value of the original game is 0.5

The optimal strategy is for A to play A_1 and A_2 each with a probability 0.5

Remember to subtract the 1 from v

Exercise 30.2B Reasoning and problem-solving

1 The table shows the pay-offs in a game between A and B.

		B		
		B_1	B_2	B_3
A	A_1	−1	2	1
	A_2	1	−1	−2
	A_3	0	−1	−1

a Show that A must play a mixed strategy in this game.

b Find A's optimal strategy and the value of the game.

c Find B's optimal strategy.

2 Analyse the games shown in these tables to find the optimal strategy for each player and the values of the games.

a

		B		
		B_1	B_2	B_3
A	A_1	0	2	−1
	A_2	2	1	3

b

		B	
		B_1	B_2
A	A_1	−3	1
	A_2	−1	−4
	A_3	0	−5

c

		B	
		I	II
A	I	−2	4
	II	−1	2
	III	4	1

d

		B		
		B_1	B_2	B_3
A	A_1	−8	−2	1
	A_2	−3	1	4
	A_3	0	−4	−2

e

		B		
		B_1	B_2	B_3
A	A_1	2	1	5
	A_2	1	−4	4
	A_3	5	2	1

f

		N			
		I	II	III	IV
M	I	−2	−1	3	−1
	II	2	−1	1	2
	III	−3	−2	1	−2
	IV	−3	−3	2	−1

3 You may be familiar with the game 'rock, paper, scissors' where the rock blunts the scissors, the scissors cut the paper and the paper wraps round the rock. Awarding 1 point for a win and 0 for a draw, construct a pay-off table for the game and analyse it to find the best strategy.

4 An entertainment company plans to run nine outdoor events in different localities on the same day. If the weather at a given locality is fine they expect to make £1200 for that event, but in a place where it rains they will lose £400

They could choose to take out onc of two possible insurance policies. The basic policy costs £500 and pays £1300 if it rains. The comprehensive policy costs £800 and pays £2400 if it rains.

a Their strategies are A_1 no insurance, A_2 basic, A_3 comprehensive. Analyse this as a 'game against the weather' (the weather's strategies are rain or no rain) and advise the company on their best strategy.

b On what assumptions is your advice based?

DISCRETE

30 Summary and review

Chapter summary

- A linear programming problem can be solved using the simplex algorithm if:
  - the aim is to maximise the objective function,
  - every constraint can be written in standard form, that is, as an equation with a slack variable ≥ 0 and a non-negative value on the right-hand side.
- To minimise an objective function, C, you maximise the objective function $P = -C$.

  You rewrite inequalities involving $\geq$ by multiplying through by -1, so that they involve $\leq$.

  If the problem can then be written in standard form, you use the simplex algorithm.
- If introducing a slack variable gives a negative value on the right-hand side, then the origin is not in the feasible region. You pivot on a negative coefficient in the row with the negative value, to generate a basic feasible solution (a vertex of the feasible region) from which to start the simplex algorithm.
- In a pure strategy game, each player always plays the option that maximises their pay-off. This corresponds to the row with the greatest minimum or the column with the least maximum entry.
- A game has a stable solution (a saddle point) if

  maximum of row minima = minimum of column maxima
- This gives the value of the game.
- For an unstable situation, each player should play a mixed-strategy game, that is, play each strategy some of the time.
- In a mixed-strategy game, you aim to maximise the expected pay-off, where for pay-offs $x_1, x_2, \ldots, x_n$ occuring with probabilities $p_1, p_2, \ldots, p_n$, the expected (mean) pay-off $\mathrm{E}(x)$ is given by

$$\mathrm{E}(x) = x_1p_1 + x_2p_2 + \ldots + x_np_n = \sum_{i=1}^{n} x_ip_i$$

- To solve a mixed-strategy game you assume that the row player plays their strategies with probabilities $p_1, p_2, \ldots$ and write inequalities connecting v, their expected value, with the ps for each of the column player's strategies. For a 2×2 game you can solve the resulting linear programming problem graphically, but for larger games you need to use the simplex algorithm.

Check and review

You should now be able to...	Review Questions
✔ Use slack variables to write a linear programming problem as a simplex tableau.	1
✔ Solve a standard maximisation problem using the simplex algorithm.	1
✔ Convert a minimisation problem to a maximisation problem and solve it using the simplex algorithm.	2
✔ Solve a non-standard linear programming problem by pivoting on negative coefficients to generate a starting feasible solution.	3
✔ Decide whether a game has a stable solution.	4
✔ Formulate a game as a linear programming problem, and solve it by using the simplex method to find the optimal mixed strategy.	4

1 A linear programming problem is stated as follows:

Maximise $P = x + 2y - z$

subject to $x + y + 2z \leq 14$

$2x + y - z \leq 8$

$x, y, z \geq 0$

a Rewrite the problem as equations in standard form using slack variables.

b Construct a simplex tableau and apply the simplex algorithm to solve the problem.

2 Use the simplex algorithm to solve this linear programming problem:

Minimise $C = 2x - y - 2z$

subject to $2x + y - 2z \geq -10$

$y \geq 2x$

$z \geq 3x + y$

$x, y, z \geq 0$

3 A linear programming problem is stated as follows:

Maximise $P = 5x + 4y$

subject to $3x + y \geq 10$

$4x + 3y \leq 24$

$x, y, z \geq 0$

a Explain why the problem cannot be written directly in standard form using equations and slack variables.

b Construct a tableau and pivot on a suitable coefficient to produce a basic feasible solution. State the values of the variables at this stage.

c Use the simplex algorithm to complete the solution of the problem.

4 The table shows the pay-offs in a game between A and B

		B	
		B_1	B_2
A	A_1	−3	2
	A_2	1	−2
	A_3	0	1

a Show that the game does not have a stable solution.

b Use the simplex method to find the value of this game and the optimal strategy for player A

Research

One of the most famous games analysed in game theory is that of the Prisoner's dilemma. It was formalised by Albert Tucker as follows:

Two members of a criminal gang are arrested and imprisoned. Each prisoner is in solitary confinement with no means of communicating with the other. The prosecutors lack sufficient evidence to convict the pair on the principal charge. They hope to get both sentenced to a year in prison on a lesser charge. Simultaneously, the prosecutors offer each prisoner a bargain. Each prisoner is given the opportunity either to betray the other by testifying that the other committed the crime, or to cooperate with the other by remaining silent. The offer is:

- *If A and B each betray the other, each of them serves 2 years in prison,*
- *If A betrays B but B remains silent, A will be set free and B will serve 3 years in prison (and vice versa),*
- *If A and B both remain silent, both of them will only serve 1 year in prison (on the lesser charge).*

Using this scenario, try to determine the dominant strategy, and then explain why this is considered a dilemma.

You could also find out about iterated versions of the game where the prisoners repeatedly play the game and are able to base their decisions on previous actions.

Did you know?

Le Her is a simple French card game based on a standard 52 card deck dating from the 16th century. A minimax mixed strategy game based around Le Her is described by James Waldergrave in a 1713 letter.

Research

The Klee-Minty cube is a system of linear inequalities for which the simplex algorithm has poor worst-case performance when initialised at the origin.

Try solving the following using the simplex algorithm.

Maximise $4x + 2y + z$

Subject to $x \leq 5,\ 4x + y \leq 25,\ 8x + 4y + z \leq 125,$
$x, y, z \geq 0$

Find out why this is called the Klee-Minty cube problem.

30 Assessment

1 A linear programming problem is defined as

Maximise $P = x + y$

subject to $3x + 4y \leq 120$

$4x + y \leq 80$

$x \geq 0, y \geq 0$

The constraints are rewritten as

$3x + 4y + s = 120$

$4x + y + t = 80$

$x \geq 0,\ y \geq 0,\ s \geq 0,\ t \geq 0$

a State the name given to s and t and rewrite the objective function as an equation involving s and t **[2]**

b Display the linear programming problem in a simplex tableau. **[2]**

The first pivot is chosen from the x-column.

c State a pivot and explain why this value is chosen. **[2]**

d Perform one iteration of the simplex algorithm. **[4]**

e Explain how you know that this solution is not optimal. **[1]**

f Find the optimal solution to this problem. **[5]**

2 Use the simplex method to solve this linear programming problem.

Maximise $P = 2x + y + 3z$

subject to $x + 2y + 3z \leq 10$

$3x + 4y + 2z \leq 22$

$5x + y + z \leq 15$

$x, y, z \geq 0$

Show your method clearly. **[10]**

3 Sarina and Jeremy play a zero-sum game. The pay-off matrix for Sarina is given.

Strategy	J_1	J_2	J_3
S_1	−4	9	−1
S_2	3	4	4
S_3	3	5	5

a Work out the play-safe strategies for Sarina and for Jeremy and state the value of the game. You must show your method. **[5]**

b Explain whether or not this is a stable solution. **[2]**

c Write down the pay-off matrix for Jeremy. **[2]**

4 Three types of drink, X, Y and Z, are to be produced. They all require orange, pineapple, grapefruit juice and water in different quantities. The amount of each type of juice (in litres) required per litre of each drink along with the availability of each juice and the profit per litre of each drink produced is shown in the table. You can assume that there is an unlimited supply of water.

	Orange	Pineapple	Grapefruit	Profit
X	0.4	0.1	0.1	£1
Y	0.6	0.2	0.1	£1.50
Z	0.5	0.3	0.05	£1.20
Availability	550	160	100	

a Given that you wish to maximise the profit, formulate this scenario as a linear programming problem. Ensure that you define your variables clearly. **[4]**

b Show the simplex tableau after one iteration of the simplex algorithm. **[6]**

c Write down the solution illustrated by this tableau and explain how you know whether or not this solution is optimal. **[3]**

A second iteration gives the solution

$P = £1300,\ x = 400,\ y = 600,\ z = 0,\ r = 30,\ s = 0,\ t = 0$

d Explain the significance of the values of r, s and t in this solution. **[2]**

5 The table shows a zero-sum game between two players, R and C

Strategy	C_1	C_2	C_3	C_4
R_1	7	2	−4	−5
R_2	−2	−3	2	3
R_3	5	4	−7	6
R_4	4	3	−3	5

a Use dominance arguments to explain why the matrix can be written as **[3]**

Strategy	C_2	C_3
R_2	−3	2
R_3	4	−7
R_4	3	−3

b Use a graphical method to find the optimal strategy for both players and work out the value of the game. **[10]**

6 A zero-sum game between two players, A and B, is represented by this pay-off matrix.

Strategy	B_1	B_2	B_3	B_4
A_1	−3	2	3	−1
A_2	6	−1	6	7
A_3	5	−3	4	−1

a Explain why player A should never play A_3 **[2]**

b Simplify the pay-off matrix as far as possible. **[3]**

c Verify that there is not a stable solution to the game. **[2]**

d Find the optimal mixed-strategy for both players. **[7]**

7 This optimal mixed-strategy problem for Alex in the zero-sum game shown in this pay-off matrix is to be solved using the simplex algorithm.

		Becky		
		A	*B*	*C*
Alex	I	3	−1	2
	II	−2	0	3
	III	0	1	−1

a Add 2 to every value in the pay-off matrix and then express the game as a linear programming problem for Alex. Write the constraints as equations involving slack variables. **[5]**

b Why was it necessary to add 2 to every value in the matrix? **[1]**

c Reduce the number of decision variables to three. **[3]**

d Construct a simplex tableau and carry out one iteration of the simplex algorithm. **[5]**

After a further two iterations, the tableau becomes

P	V	p_1	p_2	r	s	t	Value
1	0	0	0	$\frac{1}{6}$	$\frac{3}{5}$	$\frac{7}{30}$	$\frac{71}{30}$
0	0	0	1	$\frac{1}{6}$	0	$-\frac{1}{6}$	$\frac{1}{6}$
0	0	1	0	$-\frac{1}{6}$	$\frac{1}{5}$	$-\frac{1}{30}$	$\frac{7}{30}$
0	1	0	0	$\frac{1}{6}$	$\frac{3}{5}$	$\frac{7}{30}$	$\frac{71}{30}$

e State the optimal solution for Alex and the value of the game. **[4]**

8 Hayley and Sanjit play a zero-sum game. The pay-off matrix for the game is given.

	Sanjit			
	Strategy	S_1	S_2	S_3
Hayley	H_1	4	2	1
	H_2	1	1	3
	H_3	2	3	2

Hayley choses to play strategy H_i with probability p_i, $i = 1, 2, 3$

a Formulate the problem of finding the value, V, of the game as a linear programming problem.

Give the constraints as equations. **[4]**

b Perform two iterations of the simplex algorithm and indicate your pivots. **[8]**

9 Greg and Aidan play a zero-sum game. The pay-off matrix for the game is given.

	Aidan				
	Strategy	**W**	**X**	**Y**	**Z**
Greg	**A**	1	3	−2	−3
	B	2	−1	3	1
	C	3	1	0	2
	D	4	4	−1	3

a Verify that this game does not have a stable solution. **[3]**

b Reduce this 4×4 game to a 3×3 game. Justify your solution. **[3]**

c Formulate the 3×3 game as a linear programming problem, giving your constraints as inequalities. **[5]**

31 Group theory

Materials scientists use experimentation and the mathematical analysis of molecular symmetries to ensure that a material has particular properties. Symmetries of molecules are classified using **groups**, and the group provides an indication of the physical properties of its elements. The materials that are developed are used in products such as satellites. They may also be used to manufacture artificial limbs, and implants used in the human body.

Materials scientists are also involved in developing materials used in the manufacture of mobile phones. Mobile phones need screens that are made of a material that is resistant to breakages, sensitive to the touch, and has high levels of clarity. Inside the phone, the materials that are used in the complex circuitry must work efficiently, whilst being on a very small scale to ensure that the phone can be as small and powerful as possible

Orientation

What you need to know

KS4
- Sets.

Ch15 Abstract algebra
- Binary operations

What you will learn

- How to analyse groups mathematically.
- How to analyse subgroups.
- How to recognise and find isomorphisms.

What this leads to

Careers
- Materials science.

31.1 Groups

Fluency and skills

A group is a collection of mathematical objects that can be combined subject to a set of basic rules, or axioms.

An example of a group is the set of permutations of the numbers 1, 2 and 3

You can list these permutations: (1 2 3), (1 3 2), (2 1 3), (2 3 1), (3 1 2) and (3 2 1). You can also write the permutations using notation which indicates how the original order of the numbers changes under each permutation.

$$e=\begin{pmatrix}1&2&3\\1&2&3\end{pmatrix}, a=\begin{pmatrix}1&2&3\\1&3&2\end{pmatrix}, b=\begin{pmatrix}1&2&3\\2&1&3\end{pmatrix}, c=\begin{pmatrix}1&2&3\\2&3&1\end{pmatrix}, d=\begin{pmatrix}1&2&3\\3&1&2\end{pmatrix}, f=\begin{pmatrix}1&2&3\\3&2&1\end{pmatrix}$$

The six permutations are called **elements.** The first of these permutations leaves the order of the numbers unchanged. This element, e, is known as the **identity**.

In order to be a group, the set of elements must contain an identity element.

A second condition for the set of objects to be a group is that any combination of the elements is contained within the original set. This condition is known as **closure**.

If you apply permutation b and then permutation c, then under the permutation cb: 1 maps to 2 maps to 3, 2 maps to 1 maps to 2 and 3 maps to 3 maps to 1

Overall, the result is permutation f

All of the possible combinations of two permutations can be shown in a **Cayley table**.

		First permutation					
	followed by	e	a	b	c	d	f
Second permutation	e	e	a	b	c	d	f
	a	a	e	d	f	b	c
	b	b	c	e	a	f	d
	c	c	b	f	d	e	a
	d	d	f	a	e	c	b
	f	f	d	c	b	a	e

You can see from the Cayley table that for any combination of two permutations, the result is one of the original set of permutations.

In addition to this, you can see that every element has another element which, when combined, leads to the identity element e

For example, $aa = e$ and $cd = e$

These elements are known as **inverse** elements. Since a, b, e and f combine with themselves to produce the identity element they are known as **self-inverses**, while d is the inverse of c and vice versa.

The final condition for a set of elements to be a group is known as **associativity**. Three elements are associative if $a(bc) = (ab)c$, that is, bc followed by a gives the same result as c followed by ab

From the table on the previous page, you can see that $a(bc) = aa$ and $(ab)c = dc$ and both aa and dc give e

Hence a, b and c are associative. This can be shown for any other combination of the elements in the same way.

Since this final condition is met, you can say that the set of permutations of the numbers 1, 2 and 3 form a **group**.

Key point

The conditions under which a set of mathematical objects form a group under a given binary operation are known as the **axioms**. These are **closure**, the existence of an **identity element**, **associativity** and the existence of an **inverse** element for each member of the set.

Using formal notation, a group can be defined as follows.

Key point

A group $(S, \circledcirc)$ is a non-empty set S with a binary operation $\circledcirc$ such that:

$\circledcirc$ is closed in S

$\circledcirc$ is associative

there is an identity element such that $x \circledcirc e = e \circledcirc x = x$ for all x

each element has an inverse x^{-1} such that $x \circledcirc x^{-1} = x^{-1} \circledcirc x = e$

A binary operation is an operation that combines two elements of a set according to a given rule.

See Ch15
For a reminder on binary operations.

DISCRETE

The most difficult axiom to test for is associativity. However, there are certain binary operations that are known to be associative. For example, modular multiplication and addition are associative, as is matrix multiplication. Also, any binary operation that can be interpreted as the composition of mappings is associative. You can use these facts without proof.

Example 1

a Draw a Cayley table for the binary operation addition modulo 4 ($+_4$) on the set $S = \{0, 1, 2, 3\}$

b Is the set closed under $+_4$?

c State the element that is the identity element.

d For each element, state its inverse.

a

$+_4$	0	1	2	3
0	0	1	2	3
1	1	2	3	0
2	2	3	0	1
3	3	0	1	2

b Yes, because every combination of elements is in the original set of elements.

c 0

Combining 0 with any other element leaves that element unchanged.

d 0 and 2 are self-inverse, 3 is the inverse of 1 and 1 is the inverse of 3

There are several terms that you need to know.

> **Key point**
>
> The **order** of a group is equal to the number of elements in the group.
>
> The **period** (or order) of a particular element, x, of a group is the smallest non-negative integer n such that $x^n = e$, where e is the identity.

In the permutations example described at the start of this section, the order of the group is 6 since there are 6 different permutations.

Element a has period 2 since $a^2 = e$ and element c has period 3 since $c^3 = c(cc) = cd = e$

> **Key point**
>
> An **abelian** group is a group with the additional property of **commutativity** between the elements of the group.

Commutativity is the property that, for all elements of a group, $xy = yx$ under the given binary operation.

In the permutations example, $ba \neq ab$; therefore this group is not an abelian group.

In Example 1, $1 + 3 = 3 + 1$ and so on, so this group *is* an abelian group.

Abelian groups are characterised by a line of symmetry down the leading diagonal.

Exercise 31.1A Fluency and skills

1 a Draw a Cayley table for the binary operation multiplication modulo 5 ($\times_5$) on the set $S = \{0, 1, 2, 3, 4\}$

b Is the set closed under $\times_5$?

c State the element that is the identity element.

d For each element, write down its inverse.

e State, with reasons, whether or not S forms a group.

2 a Draw a Cayley table for the binary operation addition modulo 6 ($+_6$) on the set $S = \{0, 1, 2, 3, 4, 5\}$

b Is the set closed under $+_6$?

c State the element that is the identity element.

d Show that S forms a group under $+_6$

e Is the group formed an abelian group? Give a reason for your answer.

3 The binary operation $x \bullet y$ is defined as $|x - y|$

a Draw a Cayley table for the binary operation when applied to the set $S = \{0, 1, 2, 3\}$

b Is the set closed for $x \bullet y$?

c Identify the identity element.

d Fahima says that the set S forms a group under $\bullet$. Is Fahima correct? Give a reason for your answer.

4 Show that the set of matrices, S, of the form $\begin{pmatrix} 1 & p \\ 0 & 1 \end{pmatrix}$, $p \in \mathbb{Z}$ forms an abelian group under the operation of matrix multiplication.

5 Prove that the set of natural numbers, $\mathbb{N}$, does not form a group under the operation of subtraction.

6 The set S, defined as $S = \{1, 2, 3, 4, 5, 6\}$, forms a group G under the binary operation $\times_7$

a State the order of the group G

b Determine the period of the element 6 in the group G

7 a Show that the set $S = \{\mathbb{Z}\}$ forms a group under the binary operation $\bullet$ where $x \bullet y = x + y - 2$

b Explain why the set $T = \{\mathbb{Z}^+\}$ does not form a group under the same binary operation.

Reasoning and problem-solving

The set of all symmetries of a regular polygon also form a group.

These groups are called **dihedral groups** (denoted by $\boldsymbol{D_n}$ where n is the number of sides in the regular n-gon).

A **cyclic group** is formed if, for example, only rotational symmetries are considered. Cyclic groups are groups that can be generated by a single element.

Strategy

To show that a given set forms a group under a given binary operation

1. Write down the elements of the set.
2. Produce a Cayley table to show the combinations of each element under the given binary operation.
3. Check that the binary operation meets the axioms necessary to be a group.

DISCRETE

Example 2

a Show that the set of all symmetries of an equilateral triangle forms a group.

b Write down the cyclic group, C_3, represented by the set of rotational symmetries of an equilateral triangle and state an element that can be used as a generator.

a

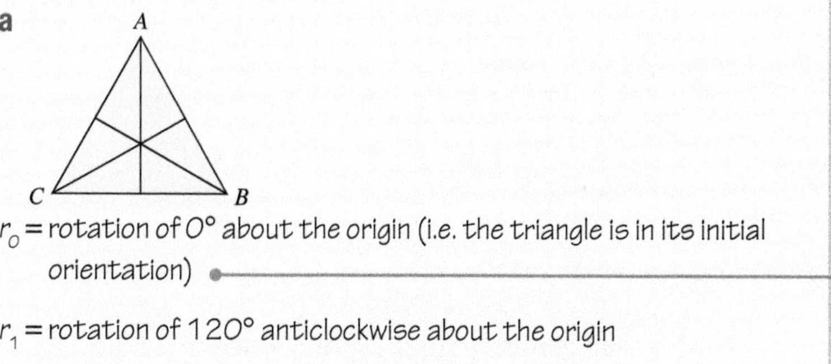

r_0 = rotation of 0° about the origin (i.e. the triangle is in its initial orientation)

r_1 = rotation of 120° anticlockwise about the origin

r_2 = rotation of 240° anticlockwise about the origin

(1) Define the point of intersection of the lines of symmetry as the origin and then list the symmetries in turn.

(*continued on the next page*)

m_1 = reflection in the mirror line through A

m_2 = reflection in the mirror line through B

m_3 = reflection in the mirror line through C

	r_0	r_1	r_2	m_1	m_2	m_3
r_0	r_0	r_1	r_2	m_1	m_2	m_3
r_1	r_1	r_2	r_0	m_2	m_3	m_1
r_2	r_2	r_0	r_1	m_3	m_1	m_2
m_1	m_1	m_3	m_2	r_0	r_2	r_1
m_2	m_2	m_1	m_3	r_1	r_0	r_2
m_3	m_3	m_2	m_1	r_2	r_1	r_0

2 Draw up a Cayley table to show the combinations of the different symmetries.

Axioms

Identity: r_0 is the identity element.

Inverses: r_0, m_1, m_2 and m_3 are self-inverses; r_1 is the inverse of r_2 and vice versa.

Closure: Every combination of symmetries is in the original set of symmetries.

3 Check that the binary operation representing the combination of two symmetries satisfies the axioms.

Associativity: Since the binary operation is a composition of mappings, the operation is associative.

Hence the set of symmetries of an equilateral triangle form a group (this is the dihedral group D_3)

b

	r_0	r_1	r_2
r_0	r_0	r_1	r_2
r_1	r_1	r_2	r_0
r_2	r_2	r_0	r_1

This is the Cayley table for the rotational symmetries.

The group C_3 is therefore $\{r_0, r_1, r_2\}$

The group can be generated by successively applying symmetry r_1

$r_1^{\,1} = r_1$

$r_1^{\,2} = r_2$

$r_1^{\,3} = r_0$

Hence r_1 is a generator of the group which can be denoted by $\langle r_1 \rangle$

r_2 is also a generator of the group since $r_2^{\,2} = r_1$ and $r_2^{\,3} = r_0$
Either r_1 or r_2 are suitable answers.

Exercise 31.1B Reasoning and problem-solving

1 a Show that the set of all symmetries of a square form a group and state the order of the group.

b Write the cyclic group, C_4, represented by the set of rotational symmetries of a square and state two different elements that are generators.

2 Jeremiah defines the cyclic group represented by the set of rotational symmetries of a regular hexagon as $C_6 = \{r_0, r_1, r_2, r_3, r_4, r_5\}$

a Using a Cayley table, or otherwise, find all of the generators of Jeremiah's group.

He claims that the group C_6 is abelian.

b Is Jeremiah correct?

3 a Give a geometric interpretation of the dihedral group D_5

b Explain why the order of D_5 is 10

Philomena claims that the order of any dihedral group D_n is equal to $2n$

c Explain why Philomena is correct.

4 The group G is defined as $G = (\langle 5 \rangle, \times_7)$

a Find the set of all elements contained within G

b Is G an abelian group?

5 The group R is defined as $R = (\langle 3 \rangle, +_7)$

a Draw a Cayley table for R

b Write the order of the group.

Joanna claims that 3 is not the unique generator of the group.

c Assess Joanna's claim.

6 The group G is defined as

$$G = \left(\left\langle \begin{pmatrix} 0 & 1 \\ -1 & 0 \end{pmatrix} \right\rangle, \text{matrix multiplication} \right)$$

a Find all of the elements of G and state its order.

b Give a geometrical interpretation of G

7 A cyclic group G under the binary operation • has the Cayley table

•	*a*	*b*	*c*	*d*
a	*a*	*b*	*c*	*d*
b	*b*	*a*	*d*	*c*
c	*c*	*d*	*b*	*a*
d	*d*	*c*	*a*	*b*

a Write the identity element.

b Find a generator of G

8 The set of symmetries of a circle form an infinite group, O_2

a Write the identity element.

b State why the associativity property holds for this group.

c Write the inverse of each:

i A rotation through 32.1°

ii A reflection in a line of symmetry inclined at $\theta°$

The table shows the effects of combining two symmetries, 'rot' (a rotation) and 'ref' (a reflection):

	rot	**ref**
rot	?	ref
ref	ref	rot

d What is represented by '?'
Give a reason for your answer.

e Use this table to explain why the set of symmetries of a circle has the closure property.

[Note: The group O_2 is known as the **orthogonal** group and the subscript '2' tells you that you are dealing with a two-dimensional object.]

31.2 Subgroups

Fluency and skills

Key point

A **subgroup** of a group G is any subset H of G such that H itself is also a group under the same binary operation as G

The **trivial subgroup** consisting of just the identity element is a subgroup of the (parent) group.

The group itself is also a subgroup of the group.

Key point

A **non-trivial subgroup** is any subgroup that is not the trivial subgroup.

A **proper subgroup** is any subgroup that is not the group itself.

You need to be able to show that a stated group H is a subgroup of another stated group G. You do this by checking four conditions:

- That H is non-empty
- That the identity element in G exists in H
- That H is closed under the binary operation for G
- That the inverse of each element of H belongs to H

$\mathbb{R}\setminus\{0\}$ denotes the set of real numbers *excluding zero.*

Example 1

a Show that $H=(\mathbb{Q},+)$ is a subgroup of $G=(\mathbb{R},+)$

b Show that the set $J=\{x\in\mathbb{R}:x\geq1\}$ is not a subgroup of $K=(\mathbb{R}\setminus\{0\},\times)$

a H is non-empty since there exist (an infinite number of) elements in H

The identity element of G under the binary operation addition is 0 and $0\in\mathbb{Q}$

H is closed since if $x, y\in\mathbb{Q}$, then $x+y\in\mathbb{Q}$

The inverse of an element x is x^{-1}, which is $-x$ under the binary operation of addition, and if $x\in\mathbb{Q}$, then $-x\in\mathbb{Q}$

Hence H is a subgroup of G

b J is non-empty since there exist (an infinite number of) elements in J

The identity element of K under the binary operation multiplication is 1, and 1 is in J

(*continued on the next page*)

J is closed since if *x* and *y* are in *J*, their product *xy* is in *J*

The inverse of an element *x* is x^{-1}, which is $\frac{1}{x}$ under the binary operation of multiplication, but if $x > 1$, $x^{-1} < 1$, so then *J* does not contain the inverse of *x*

Hence *J* is not a subgroup of *K*

If you can show that any given condition is false, you do not have to check all of the conditions, since a single counter-example is enough.

You also need to be able to find subgroups for a given group.

Example 2

A group *G* is formed by the set $S = \{a, b, c, d\}$ under the binary operation •

A Cayley table is drawn to show the outcome for each pair of elements under •

•	*a*	*b*	*c*	*d*
a	*a*	*b*	*c*	*d*
b	*b*	*a*	*d*	*c*
c	*c*	*d*	*b*	*a*
d	*d*	*c*	*a*	*b*

Find all the non-trivial subgroups of *G*

Since *a* is the identity element, the group $\{a\}$ is not a non-trivial subgroup, so this should not be included.

Since subgroups are groups under the same binary operation as the original group, all subgroups must contain the identity element *a*.

Hence the options are $\{a, b\}$, $\{a, c\}$, $\{a, d\}$, $\{a, b, c\}$, $\{a, b, d\}$, $\{a, c, d\}$ and $\{a, b, c, d\}$

Checking each one:

$\{a, b\}$ is non-empty, closed and since *a* and *b* are self-inverse, $\{a, b\}$ is a subgroup of *G*

$\{a, c\}$ is non-empty but since $c^2 = b$, it is not closed.

$\{a, d\}$ is non-empty but since $d^2 = b$, it is not closed.

$\{a, b, c\}$ is non-empty but since $bc = d$, it is not closed.

$\{a, b, d\}$ is non-empty but since $bd = c$, it is not closed.

$\{a, c, d\}$ is non-empty but since $cd = b$, it is not closed.

$\{a, b, c, d\}$ is the group itself so is, by definition, a non-trivial subgroup of *G*

Hence the non-trivial subgroups of *G* are $\{a, b\}$ and $\{a, b, c, d\}$

Exercise 31.2A Fluency and skills

1 Show that the group $H=(\mathbb{Z},+)$ is a subgroup of $G=(\mathbb{R},+)$

2 Show that the group of rotations of an equilateral triangle is a subgroup of the group of all symmetries of an equilateral triangle.

3 A group $G=(\{1, 2, 3, 4\}, \times_5)$. Find all the proper subgroups of G

4 A group $H=(\{0, 1, 2, 3\}, +_4)$. Find all the non-trivial subgroups of H

5 a Show that the group C_4 contains only one proper subgroup other than the trivial subgroup.

b Explain how group H in question **4** and C_4 are related.

6 The Klein group $V=(\{a, b, c, d\}, \bullet)$ has the corresponding Cayley table

•	***a***	***b***	***c***	***d***
a	*a*	*b*	*c*	*d*
b	*b*	*a*	*d*	*c*
c	*c*	*d*	*a*	*b*
d	*d*	*c*	*b*	*a*

Find all the proper subgroups of V

7 M is the set of all non-singular real 2 by 2 matrices such that

$$M=\left\{\begin{pmatrix} a & b \\ c & d \end{pmatrix}: a,b,c,d\in\mathbb{R}, ad-bc\neq 0\right\}$$

Given that M forms a group under the binary operation of matrix multiplication, show that the set $N=\left\{\begin{pmatrix} 1 & 0 \\ 0 & 1 \end{pmatrix}, \begin{pmatrix} -1 & 0 \\ 0 & -1 \end{pmatrix}\right\}$ is a subgroup of M

Reasoning and problem-solving

So far you have been identifying subgroups by checking all possible combinations of elements. If the number of elements is large, however, this becomes an arduous task. Imagine the group of all symmetries of a regular octagon. This would have order 16 and you would have to check a substantial number of cases.

Lagrange's theorem allows you to reduce the number of cases significantly.

Key point

Lagrange's theorem states that for any finite group G, the order of every subgroup of G divides the order of G

Also by Lagrange's theorem, the period of any element in the group G is a factor of the order of G

The consequence of this is that, for the example of the symmetries of the regular octagon, you would only need to look for subgroups of order 1 (the trivial subgroup containing just the identity which is always a subgroup), order 16 (the group itself), order 2, order 4 and order 8. For a group of order 16, this is still quite a few, but significantly fewer than if you had to check for groups of order 3, 5, 7, etc. as well.

Strategy

To find all the possible subgroups of a particular group

1. Use Lagrange's theorem to find the order of the possible subgroups.
2. Use a systematic process to identify the possibilities.
3. Check each possibility against the conditions for being a subgroup.

Example 3

The group $G = (\{0, 1, 2, 3, 4, 5\}, +_6)$

a State the order of G

b Write down the order of possible subgroups of G. Justify your answer.

c Use your answer to **b** to find all the proper subgroups of G

a There are six elements in G so the order is 6

b 1, 2, 3 and 6 are factors of 6 so these are the orders of possible subgroups.

1 Use Lagrange's theorem.

c List all the possibilities:

{0} – this is the trivial subgroup.

{0, 1, 2, 3, 4, 5, 6} – this is the group itself so is *not* a proper subgroup.

Other possibilities: {0, 1}, {0, 2}, {0, 3}, {0, 4}, {0, 5}, {0, 1, 2}, {0, 1, 3}, {0, 1, 4}, {0, 1, 5}, {0, 2, 3}, {0, 2, 4}, {0, 2, 5}, {0, 3, 4}, {0, 3, 5} and {0, 4, 5}

2 List the possibilities.

For the two-element options, the element other than the identity must be self-inverse.

$2 + 2 \neq 0 \pmod 6$, $3 + 3 = 0 \pmod 6$, $4 + 4 \neq 0 \pmod 6$, $5 + 5 \neq 0 \pmod 6$

Hence {0, 3} is a subgroup.

For the three-element options, both elements other than the identity must be self-inverse, or they must be inverse pairs.

Since 3 is the only self-inverse element other than the identity, {0, 1, 3}, {0, 2, 3}, {0, 3, 4} and {0, 3, 5} can be discounted.

$2 + 4 = 0 \pmod 6$ and $1 + 5 = 0 \pmod 6$, so {0, 1, 5} and {0, 2, 4} are also subgroups.

3 Check each possibility against the conditions. Checking for a closed group using inverses is a good way of doing this.

Exercise 31.2B Reasoning and problem-solving

1 Determine the only possible orders of the subgroups of a group which has an order of 90. Fully justify your answer.

2 Which of the following groups *cannot* be a subgroup of G, which has order 30

a A, order 5

b B, order 3

c C, order 12

d D, order 10

3 Prove that no subgroups of order 6 exist for a group of order 92. Fully justify your answer.

4 A group

$H = (\{0, 1, 2, 3, 4, 5, 6, 7, 8, 9, 10\}, +_{11})$

Explain why H has only two subgroups, the trivial subgroup and the group itself.

5 A group $G = (\{e, p, q, r, s, t, u, v\}, \bullet)$ has the following Cayley table.

$\bullet$	e	p	q	r	s	t	u	v
e	e	p	q	r	s	t	u	v
p	p	q	u	t	v	s	e	r
q	q	u	e	s	r	v	p	t
r	r	t	s	u	p	e	v	q
s	s	v	r	p	u	q	t	e
t	t	s	v	e	q	p	r	u
u	u	e	p	v	t	r	q	s
v	v	r	t	q	e	u	s	p

a State the order of G

b Write down the orders of possible subgroups of G. Justify your answer.

c Use your answer to **b** to find all the non-trivial proper subgroups of G

6 A group $J = (\{1, -1, i, -i\}, \times\}$. This is a subgroup of the complex numbers under multiplication.

Carmelita says that $(\{1, i, -i\}, \times)$ is a proper subgroup of J

a Explain why Carmelita is wrong.

b Identify all the proper subgroups of J

7 A group $G = (\{0, 1, 2, 3, 4, 5, 6, 7, 8\}, +_9)$. Explain why there are exactly three subgroups of G

8 The group, G, of all operations on a standard Rubik's Cube has 43 252 003 274 489 856 000 elements. Use Lagrange's theorem to work out the number of orders of possible subgroups of G

Hint: Try writing the number as the product of prime factors. You do not need to write out all the possible orders.

31.3 Isomorphisms

Fluency and skills

Isomorphism is an important concept in group theory. Suppose that you prove certain results for a group of order 4. If you encounter another group of order 4, your first thought is to start to prove these same results again for this different group. However, if you can show that the two groups are **isomorphic**, then the results proved for the first group automatically hold for the second group.

Key point

Two groups are **isomorphic** if there is a one-to-one mapping (an **isomorphism**) which associates each of the elements of one group with one of the elements of the other such that:

If p maps to a and q maps to b, then the result of combining p and q under the binary operation of the first set maps to the result of combining a and b under the binary operation of the second set.

DISCRETE

Example 1

A group G under the binary operation • has the Cayley table

•	p	q	r	s
p	p	q	r	s
q	q	p	s	r
r	r	s	q	p
s	s	r	p	q

A second group $H = (\{i, -i, 1, -1\}, \times)$

Show that H is isomorphic to G and state the corresponding elements in each group.

The Cayley table for H is

×	i	–i	1	–1
i	–1	1	i	–i
–i	1	–1	–i	i
1	i	–i	1	–1
–1	–i	i	–1	1

The identity element for H is 1 since $1 \times a = a \times 1 = a$

Elements 1 and –1 are self-inverse, so 1 corresponds to p and –1 to q

(continued on the next page)

Reorder the columns and rows in the Cayley table:

×	1	−1	i	−i
1	1	−1	i	−i
−1	−1	1	−i	i
i	i	−i	−1	1
−i	−i	i	1	−1

By reordering the columns and rows in the Cayley table, you can see that the pattern of entries here is the same as in the Cayley table for G

Hence i corresponds to r and −i corresponds to s

The groups are isomorphic.

The corresponding elements for each group are

$1 \leftrightarrow p$, $-1 \leftrightarrow q$, $i \leftrightarrow r$ and $-i \leftrightarrow s$

Note that this correspondence is not unique. Interchanging i and −i would also give the same pattern of entries as in G

The notation $H \cong G$ is used to denote isomorphism between groups.

Exercise 31.3A Fluency and skills

1 Show that the group formed by the set of rotational symmetries of a square is also isomorphic to the groups in Example 1

2 a Show that the groups $A = (\{1, 2, 3, 4\}, \times_5)$ and $B = (\{0, 1, 2, 3\}, +_4)$ are isomorphic.

b Are groups A and B also isomorphic to those in Example 1?

3 $M = \left(\left\{\begin{pmatrix} 1 & 0 \\ 0 & 1 \end{pmatrix}, \begin{pmatrix} 1 & 0 \\ 0 & -1 \end{pmatrix}, \begin{pmatrix} -1 & 0 \\ 0 & 1 \end{pmatrix}, \begin{pmatrix} -1 & 0 \\ 0 & -1 \end{pmatrix}\right\}, \right.$ matrix multiplication)

$N = (\{1, 3, 5, 7\}, \times_8)$

a Show that M and N are isomorphic.

b Are M and N also isomorphic to those in Example 1?

4 Show that the group formed by the set of rotational symmetries of an equilateral triangle is isomorphic to the group $R = (\langle 4 \rangle, \times_7)$

Reasoning and problem-solving

Showing that two groups of order 3 or 4 are isomorphic (or not) is quite straightforward since it is simple to rearrange the columns and rows of one of the Cayley tables to see if the pattern of entries matches the other table. For groups of a larger order, a systematic approach is advised.

Strategy 1

To show that two groups are isomorphic

1. Identify the identity element in each group.
2. Find the self-inverse elements in each group.
3. Identify the elements in each group that are not self-inverse and choose an arbitrary mapping.
4. Choose an arbitrary pairing of the self-inverse elements and write down a possible mapping.
5. Rewrite the Cayley table for one of the groups using the possible mapping and check that the pattern of entries is the same as for the Cayley table of the other group.

Example 2

Groups G and H have Cayley tables as shown.

G

	a	b	c	d	e	f
a	a	b	c	d	e	f
b	b	c	a	f	d	e
c	c	a	b	e	f	d
d	d	e	f	a	b	c
e	e	f	d	c	a	b
f	f	d	e	b	c	a

H

	P	Q	R	S	T	U
P	S	U	T	P	R	Q
Q	T	R	S	Q	U	P
R	U	S	Q	R	P	T
S	P	Q	R	S	T	U
T	Q	P	U	T	S	R
U	R	T	P	U	Q	S

Show that H is isomorphic to G

In G, the identity element is a

In H, the identity element is S

(1) Identify the identity element in each group.

The self-inverse elements in G are d, e and f

The self-inverse elements in H are P, T and U

(2) Identify the self-inverse elements in each group.

In G, b and c are not self-inverse.

In H, Q and R are not self-inverse.

Choose an arbitrary mapping, say $b \leftrightarrow Q$

This means that $c \leftrightarrow R$

(3) Identify the elements that are not self-inverse and choose an arbitrary mapping.

Choose an arbitrary mapping for the self-inverse elements:

Let $d \leftrightarrow P$; hence $e \leftrightarrow U$ and $f \leftrightarrow T$

A possible mapping is therefore $[b, c, d, e, f] \leftrightarrow [Q, R, P, U, T]$

(4) Choose an arbitrary pairing of the self-inverse elements and write down a possible mapping.

The Cayley table for H is now

	S	Q	R	P	U	T
S	S	Q	R	P	U	T
Q	Q	R	S	T	P	U
R	R	S	Q	U	T	P
P	P	U	T	S	Q	R
U	U	T	P	R	S	Q
T	T	P	U	Q	R	S

The pattern of entries in the redrawn Cayley table for H now matches that for the Cayley table for G

(5) Redraw the Cayley table for one of the groups using the possible mapping and check the pattern of entries.

Hence $H \cong G$

Strategy 2

To show that two groups are *not* isomorphic, you need to establish at least one of the following

(1) That there are a different number of self-inverse elements in the two groups.

(2) That some of the elements in one group do not have the same period as in the other.

(3) That one group is cyclic and the other is not.

Example 3

Group S has a Cayley table as shown.

Show that S is not isomorphic to the two groups in Example 1

	1	**2**	**3**	**4**	**5**	**6**
1	1	2	3	4	5	6
2	2	4	6	1	3	5
3	3	6	2	5	1	4
4	4	1	5	2	6	3
5	5	3	1	6	4	2
6	6	5	4	3	2	1

The identity element is 1

The only other self-inverse element in S is 6

G and H both have three self-inverse elements other than the identity.

Hence S is not isomorphic to G nor to H

(1) There are a different number of self-inverse elements.

You could also have shown that S has elements of period 6 (the elements 3 and 5) whereas G and H do not.

Exercise 31.3B Reasoning and problem-solving

1 Explain why all groups of order 2 are isomorphic to each other.

2 Explain why all groups of order 3 are isomorphic to each other.

3 Brice claims that every group of order 4 is isomorphic to exactly one of four particular groups, A, B, C, D of order 4.
Group A has three elements of period 2
Group B has two elements of period 2 and one of period 4
Group C has one element of period 2 and two of period 4
Group D has four elements of period 4
Explain why Brice's claim is incorrect and show that every group of order 4 is isomorphic to exactly one of two of the groups Brice has listed.

4 Leona says that if p is a prime number, all groups of order p are isomorphic, i.e. all groups of order 7, for example, are isomorphic to each other. Explain why Leona is correct in her assertion.

5 Explain why it is impossible for an abelian group to be isomorphic to a non-abelian group.

6 The group $G = (\{0, 1, 2, 3, 4, 5\}, +_6)$
The group H consists of the set of rotational symmetries of a regular hexagon.
Show that $G \cong H$

7 The group $S = (\{1, 2, 4, 7, 8, 11, 13, 14\}, \times_{15})$
Another group M has elements consisting of the set of rotations of angle $\frac{k\pi}{4}$, $k = 1, 2, 3, 4, 5, 6, 7, 8$, of a unit square about the origin.
Show that S is not isomorphic to M

8 Show that the group of matrices of the form $\begin{pmatrix} 1-a & a \\ -a & 1+a \end{pmatrix}$ under the operation matrix multiplication is isomorphic to the group $G = (\mathbb{Z}, +)$

Hint: Since the groups are infinite, you need to specify a mapping connecting each element in one group with an element in the other group.

Chapter summary

- The conditions under which a set of mathematical objects form a group under a given binary operation are known as the axioms.
- These are closure, the existence of an identity element, associativity and the existence of an inverse element for each member of the set.
- A group $(S, \circledcirc)$ is a non-empty set S with a binary operation $\circledcirc$ such that:
  - $\circledcirc$ is closed in S
  - $\circledcirc$ is associative
  - there is an identity element such that $x \circledcirc e = e \circledcirc x = x$ for all x
  - each element has an inverse, x^{-1}, such that $x \circledcirc x^{-1} = x^{-1} \circledcirc x = e$
- The order of a group is equal to the number of elements in the group.
- The period (or order) of a particular element, x, of a group is the smallest non-negative integer n such that $x^n = e$, where e is the identity.
- An abelian group is a group with the additional property of commutativity between the elements of the group.
- To show that a given set forms a group under a given binary operation:
  - Write down the elements of the set.
  - Produce a Cayley table to show the combinations of each element under the given binary operation.
  - Check that the binary operation meets the axioms necessary to be a group.
- A subgroup of a group G is any subset H of G such that H itself is also a group under the same binary operation as G
- A non-trivial subgroup is any subgroup that is not the trivial subgroup.
- A proper subgroup is any subgroup that is not the group itself.
- Lagrange's theorem states that, for any finite group G, the order of every subgroup of G divides the order of G
- To find all the possible subgroups of a particular group:
  - Use Lagrange's theorem to find the order of the possible subgroups.
  - Use a systematic process to identify the possibilities.
  - Check each possibility against the conditions for being a subgroup.
- Two groups are isomorphic if there is a one-to-one mapping (an isomorphism) which associates each of the elements of one group with one of the elements of the other such that if p maps to a and q maps to b, then the result of combining p and q under the binary operation of the first set maps to the result of combining a and b under the binary operation of the second set.

- To show that two groups are isomorphic:
  - Identify the identity element in each group.
  - Find the self-inverse elements in each group.
  - Identify the elements in each group that are not self-inverse and choose an arbitrary mapping.
  - Choose an arbitrary pairing of the self-inverse elements and write down a possible mapping.
  - Rewrite the Cayley table for one of the groups using the possible mapping and check that the pattern of entries is the same as for the Cayley table of the other group.
- To show that two groups are not isomorphic, you need to establish at least one of the following:
  - That there are a different number of self-inverse elements in the two groups.
  - That some of the elements in one group do not have the same period as in the other.
  - That one group is cyclic and the other is not.

Check and review

You should now be able to...	Review Questions
✔ Understand and use the language of groups including: order, period, subgroup, proper, trivial, non-trivial.	1, 3
✔ Understand and use the group axioms: closure, identity, inverses and associativity, including use of Cayley tables.	1, 2
✔ Recognise and use finite and infinite groups and their subgroups, including: groups of symmetries of regular polygons, cyclic groups and abelian groups.	1, 3, 4
✔ Understand and use Lagrange's theorem.	3
✔ Identify and use the generators of a group.	1
✔ Recognise and find isomorphism between groups of finite order.	4

1 $G = (\{1, 2, 4, 5, 7, 8\}, \times_9)$

a State the order of G

b Draw a Cayley table for G

c State the period of the element '5'

d Explain why your answer to **c** shows that G is cyclic.

e Is G an abelian group? Explain your answer.

2 It is suggested that the set $S = \{1, 5, 7, 11\}$ forms a group under $\times_{12}$

By drawing a Cayley table, or otherwise,

a Write down the identity element,

b Show that S is closed under $\times_{12}$

c Show that every element in S has an inverse,

d Give at least two examples of the axiom of associativity holding for S

3 The group, S, formed by the set of all symmetries of an equilateral triangle has Cayley table

	r_0	r_1	r_2	m_1	m_2	m_3
r_0	r_0	r_1	r_2	m_1	m_2	m_3
r_1	r_1	r_2	r_0	m_2	m_3	m_1
r_2	r_2	r_0	r_1	m_3	m_1	m_2
m_1	m_1	m_3	m_2	r_0	r_2	r_1
m_2	m_2	m_1	m_3	r_1	r_0	r_2
m_3	m_3	m_2	m_1	r_2	r_1	r_0

a State the order of possible subgroups of S

b Write down the trivial subgroup of S

c Find all proper subgroups of S

4 a Show that the groups $G = (\{1, 3, 7, 9\}, \times_{10})$ and $H = (\{0, 1, 2, 3\}, +_4)$ are isomorphic and write down a possible mapping.

b Are the groups cyclic? Explain your reasoning.

c Are the groups abelian? Explain your reasoning.

Exploration

Going beyond the exams

Did you know?

An automorphism is an isomorphism from a mathematical object to itself. One of the earliest group automorphisms was found by William Rowan Hamilton as he developed icosian calculus. This resulted in the development of the popular icosian game. The aim the game is to find a Hamiltonian cycle along the edges of a dodecahedron, visiting every vertex only once.

Research

The three general isomorphism theorems were formulated by Emmy Noether in 1927. Find out about them, particularly in the context of groups. You may also like to extend this research by looking at the lattice theorem and the Butterfly lemma.

Research

You can use the theories around zero knowledge proofs to discover whether two graphs are isomorphic.

Start by finding out about zero knowledge proofs. For example, look at the research paper *How to Explain Zero-Knowledge Protocols to Your Children* by Quisquater. In particular, well-known examples are the Ali Baba cave, and two-coloured balls with a colour-blind friend.

31 Assessment

1 The group G has the following Cayley table.

*	***a***	***b***	***c***	***d***
a	c	a	d	b
b	a	b	c	d
c	d	c	b	a
d	b	d	a	c

a State, with a reason, which element is the identity element. **[2]**

b Write down all of the elements that are self-inverse. **[1]**

c Explain why G is an abelian group. **[1]**

2 A finite group H has proper subgroups $\{a, b\}$, $\{b, d, e\}$ and $\{b, f, g, h, k\}$

a State, with a reason, which element of H is the identity. **[2]**

b State, with a reason, the smallest possible order of H **[2]**

3 The group $G = (\{1, 3, 5, 7\}, \times_8)$

The group $H = (\{1, 3, 7, 9\}, \times_{10})$

By drawing Cayley tables for G and H, or otherwise, show that G and H are not isomorphic. **[4]**

4 The cyclic group $C_n = (\langle 9 \rangle, \times_{64})$

a Explain what is meant by $\langle 9 \rangle$ **[1]**

b Find the value of n and justify your answer. **[3]**

c State the order of all of the possible subgroups, giving a reason for your answer. **[2]**

5 The binary operation $\circledcirc$ is defined as $a \circledcirc b = a + b + 3$ [mod 6], where $a, b \in \mathbb{Z}$

a Show that the set $\{0, 1, 2, 3, 4, 5\}$ forms a group G under $\circledcirc$ **[5]**

b Find all of the proper subgroups of G **[2]**

c Determine whether G is isomorphic to the group $K = (\langle 2 \rangle, \times_9)$ **[3]**

6 The group G has Cayley table

	a	**b**	**c**	**d**	**e**	**f**	**g**	**h**
a	c	e	b	f	a	h	d	g
b	e	c	a	g	b	d	h	f
c	b	a	e	h	c	g	f	d
d	f	g	h	a	d	c	e	b
e	a	b	c	d	e	f	g	h
f	h	d	g	c	f	b	a	e
g	d	h	f	e	g	a	b	c
h	g	f	d	b	h	e	c	a

a Explain why element e is the identity element. **[1]**

Emelia says that $\{e, d, g\}$ is a proper subgroup of G

b Explain, with a reason, whether Emilia is correct. **[2]**

c Show that G is cyclic and hence explain why G is not isomorphic to the group of symmetries of a square. **[5]**

7 The set P consists of the set of all integers under the binary operation $\bullet$ such that

$x \bullet y = x + y - 1$

a Show that $\bullet$ is associative. **[2]**

b Hence, show that P forms a group under $\bullet$ **[4]**

c State, with a reason, whether the set of positive integers forms a group under $\bullet$ **[2]**

Mathematical formulae to learn

For A Level Further Maths

Pure Mathematics

Quadratic equations

$ax^2 + bx + c = 0$ has roots $\dfrac{-b \pm \sqrt{b^2 - 4ac}}{2a}$

Laws of indices

$a^x a^y \equiv a^{x+y}$

$a^x \div a^y \equiv a^{x-y}$

$(a^x)^y \equiv a^{xy}$

Laws of logarithms

$x = a^n \Leftrightarrow n = \log_a x$ for $a > 0$ and $x > 0$

$\log_a x + \log_a y \equiv \log_a xy$

$\log_a x - \log_a y \equiv \log_a \left(\dfrac{x}{y}\right)$

$k \log_a x \equiv \log_a (x)^k$

Coordinate geometry

A straight-line graph, gradient m passing through (x_1, y_1) has equation $y - y_1 = m(x - x_1)$

Straight lines with gradients m_1 and m_2 are perpendicular when $m_1 m_2 = -1$

Sequences

General term of an arithmetic progression: $u_n = a + (n - 1)d$

General term of a geometric progression: $u_n = ar^{n-1}$

Trigonometry

In the triangle ABC

Sine rule: $\dfrac{a}{\sin A} = \dfrac{b}{\sin B} = \dfrac{c}{\sin C}$

Cosine rule: $a^2 = b^2 + c^2 - 2bc\cos A$

Area $= \dfrac{1}{2}ab\sin C$

$\cos^2 A + \sin^2 A \equiv 1$

$\sec^2 A \equiv 1 + \tan^2 A$

$\text{cosec}^2 A \equiv 1 + \cot^2 A$

$\sin 2A \equiv 2 \sin A \cos A$

$\cos 2A \equiv \cos^2 A - \sin^2 A$

$\tan 2A \equiv \dfrac{2 \tan A}{1 - \tan^2 A}$

Mensuration

Circumference and Area of circle, radius r and diameter d:

$C = 2\pi r = \pi d \qquad A = \pi r^2$

Pythagoras' Theorem:

In any right-angled triangle where a, b and c are the lengths of the sides and c is the hypotenus: $c^2 = a^2 + b^2$

Area of a trapezium $= \frac{1}{2}(a+b)h$, where a and b are the lengths of the parallel sides and h is their perpendicular separation.

Volume of a prism = area of cross section × length

For a circle of radius r, where an angle at the centre of θ radians subtends an arc of length s and encloses an associated sector of area A:

$s = r\theta \qquad A = \frac{1}{2}r^2\theta$

Complex numbers

For two complex numbers $z_1 = r_1 e^{i\theta_1}$ and $z_2 = r_2 e^{i\theta_2}$:

$$z_1 z_2 = r_1 r_2 \, e^{i(\theta_1 + \theta_2)}$$

$$\frac{z_1}{z_2} = \frac{r_1}{r_2} e^{i(\theta_1 - \theta_2)}$$

Loci in the Argand diagram:

$|z - a| = r$ is a circle radius r centred at a

$\arg(z - a) = \theta$ is a half line drawn from a at angle θ to a line parallel to the positive real axis.

Exponential form: $e^{i\theta} = \cos\theta + i\sin\theta$

Matrices

For a 2 by 2 matrix $\begin{pmatrix} a & b \\ c & d \end{pmatrix}$ the determinant $\Delta = \begin{vmatrix} a & b \\ c & d \end{vmatrix} = ad - bc$

the inverse is $\frac{1}{\Delta}\begin{pmatrix} d & -b \\ -c & a \end{pmatrix}$

The transformation represented by matrix **AB** is the transformation represented by matrix **B** followed by the transformation represented by matrix **A**.

For matrices **A**, **B**:

$(\mathbf{AB})^{-1} = \mathbf{B}^{-1}\mathbf{A}^{-1}$

Algebra

$$\sum_{r=1}^{n} r = \frac{1}{2}n(n+1)$$

For $ax^2 + bx + c = 0$ with roots α and β:

$$\alpha + \beta = \frac{-b}{a} \qquad \alpha\beta = \frac{c}{a}$$

For $ax^3 + bx^2 + cx + d = 0$ with roots α, β and γ:

$$\sum\alpha = \frac{-b}{a} \qquad \sum\alpha\beta = \frac{c}{a} \qquad \alpha\beta\gamma = \frac{-d}{a}$$

Hyperbolic functions

$$\cosh x \equiv \frac{1}{2}(e^{x} + e^{-x}) \qquad \sinh x \equiv \frac{1}{2}(e^{x} - e^{-x}) \qquad \tanh x \equiv \frac{\sinh x}{\cosh x}$$

Calculus and differential equations

Differentiation

Function	Derivative	Function	Derivative
x^n	nx^{n-1}	e^{kx}	ke^{kx}
$\sin kx$	$k\cos kx$	$\ln x$	$\frac{1}{x}$
$\cos kx$	$-k\sin kx$	$f(x) + g(x)$	$f'(x) + g'(x)$
		$f(x)g(x)$	$f'(x)g(x) + f(x)g'(x)$
		$f(g(x))$	$f'(g(x))g'(x)$

Integration

Function	Integral	Function	Integral
x^n	$\frac{1}{n+1}x^{n+1}+c,\ n\neq -1$	e^{kx}	$\frac{1}{k}e^{kx}+c$
$\cos kx$	$\frac{1}{k}\sin kx+c$	$\frac{1}{x}$	$\ln\lvert x\rvert+c,\ x\neq 0$
		$f'(x)+g'(x)$	$f(x)+g(x)+c$
$\sin kx$	$-\frac{1}{k}\cos kx+c$	$f'(g(x))g'(x)$	$f(g(x))+c$

Area under a curve $= \int_a^b y\,dx\ (y \geq 0)$

Volumes of revolution about the x and y axes:

$$V_x = \pi\int_a^b y^2\,dx \qquad V_y = \pi\int_c^d x^2\,dy$$

Simple Harmonic Motion: $\ddot{x} = -\omega^2 x$

Vectors

$$|x\mathbf{i}+y\mathbf{j}+z\mathbf{k}| = \sqrt{(x^2+y^2+z^2)}$$

Scalar product of two vectors $\mathbf{a} = \begin{pmatrix} a_1 \\ a_2 \\ a_3 \end{pmatrix}$ and $\mathbf{b} = \begin{pmatrix} b_1 \\ b_2 \\ b_3 \end{pmatrix}$ is

$$\begin{pmatrix} a_1 \\ a_2 \\ a_3 \end{pmatrix} \cdot \begin{pmatrix} b_1 \\ b_2 \\ b_3 \end{pmatrix} = a_1 b_1 + a_2 b_2 + a_3 b_3 = |\mathbf{a}|\,|\mathbf{b}|\cos\theta$$

where θ is the acute angle between the vectors **a** and **b**.

The equation of the line through the point with position vector **a** parallel to vector **b** is:

$$\mathbf{r} = \mathbf{a} + t\mathbf{b}$$

The equation of the plane containing the point with position vector **a** and perpendicular to vector **n** is:

$$(\mathbf{r}-\mathbf{a})\cdot\mathbf{n} = 0$$

Mechanics

Forces and equilibrium

Weight = mass $\times\, g$

Friction: $F \le \mu R$

Newton's second law in the form: $F = ma$

Kinematics

For motion in a straight line with variable acceleration:

$$v = \frac{\mathrm{d}r}{\mathrm{d}t} \qquad a = \frac{\mathrm{d}v}{\mathrm{d}t} = \frac{\mathrm{d}^2 r}{\mathrm{d}t^2}$$

$$r = \int v\,\mathrm{d}t \qquad v = \int a\,\mathrm{d}t$$

Statistics

The mean of a set of data: $\overline{x} = \dfrac{\sum x}{n} = \dfrac{\sum fx}{\sum f}$

The standard Normal variable: $Z = \dfrac{X - \mu}{\sigma}$ where $X \sim \mathrm{N}(\mu, \sigma^2)$

Mathematical notation
For A Level Further Maths

Set Notation

Notation	Meaning
$\in$	is an element of
$\notin$	is not an element of
$\subseteq$	is a subset of
$\subset$	is a proper subset of
$\{x_1, x_2, \ldots\}$	the set with elements $x_1, x_2, \ldots$
$\{x: \ldots\}$	the set of all x such that ...
$n(A)$	the number of elements in set A
$\varnothing$	the empty set
ε	the universal set
A'	the complement of the set A
$\mathbb{N}$	the set of natural numbers, $\{1, 2, 3, \ldots\}$
$\mathbb{Z}$	the set of integers, $\{0, \pm1, \pm2, \pm3, \ldots\}$
$\mathbb{Z}^+$	the set of positive integers, $\{1, 2, 3, \ldots\}$
$\mathbb{Z}_0^+$	the set of non-negative integers, $\{0, 1, 2, 3, \ldots\}$
$\mathbb{R}$	the set of real numbers
$\mathbb{Q}$	the set of rational numbers, $\left\{\frac{p}{q}: p \in \mathbb{Z},\ q \in \mathbb{Z}^+\right\}$
$\cup$	union
$\cap$	intersection
(x, y)	the ordered pair x, y
$[a, b]$	the closed interval $\{x \in \mathbb{R} : a \le x \le b\}$
$[a, b)$	the interval $\{x \in \mathbb{R} : a \le x < b\}$
$(a, b]$	the interval $\{x \in \mathbb{R} : a < x \le b\}$
(a, b)	the open interval $\{x \in \mathbb{R} : a < x < b\}$
$\mathbb{C}$	the set of complex numbers

Miscellaneous Symbols

Notation	Meaning
$=$	is equal to
$\neq$	is not equal to
$\equiv$	is identical to or is congruent to
$\approx$	is approximately equal to
∞	infinity
$\propto$	is proportional to
$\therefore$	therefore
$\because$	because
$<$	is less than
$\leqslant, \le$	is less than or equal to; is not greater than
$>$	is greater than
$\geqslant, \ge$	is greater than or equal to; is not less than
$p \Rightarrow q$	p implies q (if p then q)
$p \Leftarrow q$	p is implied by q (if q then p)
$p \Leftrightarrow q$	p implies and is implied by q (p is equivalent to q)

Notation	Meaning
a	first term for an arithmetic or geometric sequence
l	last term for an arithmetic sequence
d	common difference for an arithmetic sequence
r	common ratio for a geometric sequence
S_n	sum to n terms of a sequence
S_∞	sum to infinity of a sequence
$\cong$	is isomorphic to

Operations

Notation	Meaning
$a+b$	a plus b
$a-b$	a minus b
$a\times b$, ab, $a\cdot b$	a multiplied by b
$a\div b$, $\frac{a}{b}$	a divided by b
$\sum_{i=1}^{n} a_i$	$a_1+a_2+\ldots+a_n$
$\prod_{i=1}^{n} a_i$	$a_1\times a_2\times\ldots\times a_n$
$\sqrt{a}$	the non-negative square root of a
$\lvert a\rvert$	the modulus of a
$n!$	n factorial: $n!=n\times(n-1)\times\ldots\times 2\times 1$, $n\in\mathbb{N}$; $0!=1$
$\binom{n}{r}$, ${}^{n}C_r$, ${}_{n}C_r$	the binomial coefficient $\frac{n!}{r!(n-r)!}$ for $n, r\in\mathbb{Z}_0^+$, $r\le n$ or $\frac{n(n-1)\ldots(n-r+1)}{r!}$ for $n\in\mathbb{Q}$, $r\in\mathbb{Z}_0^+$
$a\times_n b$	multiplication modulo n of a by b
$a+_n b$	addition modulo n of a and b
$G=(<n>,*)$	n is the generator of a given group G under the operation*

Functions

Notation	Meaning
$\mathrm{f}(x)$	the value of the function f at x
$\mathrm{f}: x\mapsto y$	the function f maps the element x to the element y
f^{-1}	the inverse function of the function f
gf	the composite function of f and g which is defined by $\mathrm{gf}(x)=\mathrm{g}(\mathrm{f}(x))$
$\lim_{x\to a}\mathrm{f}(x)$	the limit of $\mathrm{f}(x)$ as x tends to a
Δx, δx	an increment of x
$\frac{\mathrm{d}y}{\mathrm{d}x}$	the derivative of y with respect to x
$\frac{\mathrm{d}^n y}{\mathrm{d}x^n}$	the nth derivative of y with respect to x
$\mathrm{f}'(x)$, $\mathrm{f}''(x)$,…, $\mathrm{f}^{(n)}(x)$	the first, second, …, nth derivatives of $\mathrm{f}(x)$ with respect to x
$\dot{x}$, $\ddot{x}$, …	the first, second, … derivatives of x with respect to t
$\int y\,\mathrm{d}x$	the indefinite integral of y with respect to x

Notation	Meaning
$\int_a^b y\,\mathrm{d}x$	the definite integral of y with respect to x between the limits $x = a$ and $x = b$

Exponential and Logarithmic Functions

Notation	Meaning
e	base of natural logarithms
e^x, $\exp x$	exponential function of x
$\log_a x$	logarithm to the base a of x
$\ln x$, $\log_\mathrm{e} x$	natural logarithm of x

Trigonometric Functions

Notation	Meaning
sin, cos, tan, cosec, sec, cot	the trigonometric functions
$\sin^{-1}$, $\cos^{-1}$, $\tan^{-1}$ arcsin, arccos, arctan	the inverse trigonometric functions
°	degrees
rad	radians
$\operatorname{cosec}^{-1}$, $\sec^{-1}$, $\cot^{-1}$, arccosec, arcsec, arccot	the inverse trigonometric functions
sinh, cosh, tanh, cosech, sech, coth	the hyperbolic functions
$\sinh^{-1}$, $\cosh^{-1}$, $\tanh^{-1}$ $\operatorname{cosech}^{-1}$, sech^{-1}, $\coth^{-1}$ arcsinh, arccosh, arctanh, arccosech, arcsech, arccoth, arsinh, arcosh, artanh, arcosech, arsech, arcoth	the inverse hyperbolic functions

Complex numbers

Notation	Meaning
i, j	square root of -1
$x + iy$	complex number with real part x and imaginary part y
$r(\cos\theta + i\sin\theta)$	modulus argument form of a complex number with modulus r and argument θ
z	a complex number, $z = x + iy = r(\cos\theta + i\sin\theta)$
$\mathrm{Re}(z)$	the real part of z, $\mathrm{Re}(z) = x$
$\mathrm{Im}(z)$	the imaginary part of z, $\mathrm{Im}(z) = y$
$\lvert z\rvert$	the modulus of z, $\lvert z\rvert = r = \sqrt{x^2 + y^2}$
$\arg(z)$	the argument of z, $\arg(z) = \theta$, $-\pi < \theta \le \pi$
z^*	the complex conjugate of z, $x - iy$

Matrices

Notation	Meaning
$\mathbf{M}$	a matrix $\mathbf{M}$
$\mathbf{0}$	zero matrix
$\mathbf{I}$	identity matrix
$\mathbf{M}^{-1}$	the inverse of the matrix $\mathbf{M}$
$\mathbf{M}^{\mathrm{T}}$	the transpose of the matrix $\mathbf{M}$
Δ, det $\mathbf{M}$ or $\lvert\mathbf{M}\rvert$	the determinant of the square matrix $\mathbf{M}$
$\mathbf{Mr}$	Image of the column vector $\mathbf{r}$ under the transformation associated with the matrix $\mathbf{M}$

Vectors

Notation	Meaning
$\mathbf{a}$, $\underline{a}$, $\underset{\sim}{a}$	the vector $\mathbf{a}$, $\underline{a}$, $\underset{\sim}{a}$ these alternatives apply throughout section 9
$\overrightarrow{AB}$	the vector represented in magnitude and direction by the directed line segment AB
$\hat{\mathbf{a}}$	a unit vector in the direction of $\mathbf{a}$
$\mathbf{i}, \mathbf{j}, \mathbf{k}$	unit vectors in the directions of the Cartesian coordinate axes
$\lvert\mathbf{a}\rvert$, a	the magnitude of $\mathbf{a}$
$\lvert\overrightarrow{AB}\rvert$, AB	the magnitude of $\overrightarrow{AB}$
$\begin{pmatrix} a \\ b \end{pmatrix}$, $a\mathbf{i} + b\mathbf{j}$	column vector and corresponding unit vector notation
$\mathbf{r}$	position vector
$\mathbf{s}$	displacement vector
$\mathbf{v}$	velocity vector
$\mathbf{a}$	acceleration vector
$\mathbf{a} \cdot \mathbf{b}$	the scalar product of $\mathbf{a}$ and $\mathbf{b}$

Differential equations

Notation	Meaning
ω	angular speed

Probability and statistics

Notation	Meaning
A, B, C, etc.	events
$A \cup B$	union of the events A and B
$A \cap B$	intersection of the events A and B
$\mathrm{P}(A)$	probability of the event A
A'	complement of the event A
$\mathrm{P}(A \mid B)$	probability of the event A conditional on the event B
X, Y, R etc.	random variables
x, y, r etc.	values of the random variables X, Y, R etc.
$x_1, x_2, \ldots$	values of observations

$f_1, f_2, \ldots$	frequencies with which the observations $x_1, x_2, \ldots$ occur
$\mathrm{p}(x)$, $\mathrm{P}(X=x)$	probability function of the discrete random variable X
$p_1, p_2, \ldots$	probabilities of the values $x_1, x_2, \ldots$ of the discrete random variable X
$\mathrm{E}(X)$	expectation of the random variable X
$\mathrm{Var}(X)$	variance of the random variable X
$\sim$	has the distribution
$\mathrm{B}(n, p)$	binomial distribution with parameters n and p, where n is the number of trials and p is the probability of success in a trial
q	$q=1-p$ for binomial distribution
$\mathrm{N}(\mu, \sigma^2)$	Normal distribution with mean μ and variance σ^2
$Z \sim \mathrm{N}(0, 1)$	standard Normal distribution
ϕ	probability density function of the standardised Normal variable with distribution $\mathrm{N}(0, 1)$
Φ	corresponding cumulative distribution function
μ	population mean
σ^2	population variance
σ	population standard deviation
$\bar{x}$	sample mean
s^2	sample variance
s	sample standard deviation
H_0	null hypothesis
H_1	alternative hypothesis
r	product moment correlation coefficient for a sample
ρ	product moment correlation coefficient for a population

Mechanics

Notation	**Meaning**
kg	kilograms
m	metres
km	kilometres
m/s, $\mathrm{m\,s^{-1}}$	metre(s) per second (velocity)
$\mathrm{m/s^2}$, $\mathrm{m\,s^{-2}}$	metre(s) per second square (acceleration)
F	Force or resultant force
N	newton
N m	newton metre (moment of a force)
t	time
s	displacement
u	initial velocity
v	velocity or final velocity
a	acceleration
g	acceleration due to gravity
μ	coefficient of friction

Answers

Chapter 16

Exercise 16.1A

1 **a** $5e^{i0.927}$ **b** $\sqrt{5}e^{-i0.464}$ **c** $10e^{i0}$

d $5e^{i\pi}$ **e** $2e^{i\frac{\pi}{2}}$ **f** $6e^{-i\frac{\pi}{2}}$

g $13e^{i1.97}$ **h** $4\sqrt{5}e^{-i2.03}$ **i** $2e^{i\frac{\pi}{6}}$ **j** $5\sqrt{2}e^{-i\frac{\pi}{4}}$

2 **a** $2e^{\frac{\pi}{12}i}$ **b** $4e^{-\frac{2\pi}{3}i}$ **c** $3e^{-\frac{5\pi}{6}i}$

d $6e^{-\frac{\pi}{7}i}$ **e** $e^{-\frac{3\pi}{5}i}$ **f** $\sqrt{2}e^{\frac{\pi}{8}i}$

g $\sqrt{3}e^{\frac{5\pi}{6}i}$ **h** $8e^{-\frac{7\pi}{12}i}$

3 **a** $2i$ **b** $\frac{7}{2}-\frac{7\sqrt{3}}{2}i$

c $1+i$ **d** $\frac{\sqrt{3}}{2}-\frac{1}{2}i$

e $-2\sqrt{2}$ **f** $-\frac{3}{2}+\frac{\sqrt{3}}{2}i$

4 **a** 6 **b** $\frac{2}{3}$ **c** 0 **d** $\frac{2\pi}{3}$

5 **a** 1 **b** 25 **c** $\frac{\pi}{7}$ **d** $\frac{3\pi}{7}$

6 **a** $3\sqrt{2}$ **b** $\sqrt{2}$ **c** $\frac{11\pi}{12}$ **d** $\frac{7\pi}{12}$

Exercise 16.1B

1 $e^{i\theta}=\cos\theta+i\sin\theta$

$e^{-i\theta}=\cos(-\theta)+i\sin(-\theta)=\cos\theta-i\sin\theta$

$e^{i\theta}-e^{-i\theta}=\cos\theta+i\sin\theta-(\cos\theta-i\sin\theta)$

$=2i\sin\theta$

$\sin\theta=\frac{e^{i\theta}-e^{-i\theta}}{2i}$ as required

2 $z=e^{\theta i}$

a $z^n+\frac{1}{z^n}=(e^{\theta i})^n+\frac{1}{(e^{\theta i})^n}$

$=e^{n\theta i}+e^{-n\theta i}$

$=2\cos(n\theta)$ as required

b $z^n-\frac{1}{z^n}=(e^{\theta i})^n-\frac{1}{(e^{\theta i})^n}$

$=e^{n\theta i}-e^{-n\theta i}$

$=2i\sin(n\theta)$ as required

3 **a** $z_1z_2=r_1e^{\theta_1 i}\times r_2e^{\theta_2 i}=r_1r_2e^{\theta_1 i}e^{\theta_2 i}=r_1r_2e^{(\theta_1+\theta_2)i}$

So $|z_1z_2|=r_1r_2=|z_1||z_2|$

and $\arg(z_1z_2)=\theta_1+\theta_2=\arg z_1+\arg z_2$

b $\frac{z_1}{z_2}=\frac{r_1e^{\theta_1 i}}{r_2e^{\theta_2 i}}=\frac{r_1}{r_2}e^{(\theta_1-\theta_2)i}$

So $\left|\frac{z_1}{z_2}\right|=\frac{r_1}{r_2}=\frac{|z_1|}{|z_2|}$

and $\arg\left(\frac{z_1}{z_2}\right)=\theta_1-\theta_2=\arg z_1-\arg z_2$

4 **a** $\text{RHS}=\left(\frac{e^{iA}-e^{-iA}}{2i}\right)\left(\frac{e^{iB}+e^{-iB}}{2}\right)+\left(\frac{e^{iB}-e^{-iB}}{2i}\right)\left(\frac{e^{iA}+e^{-iA}}{2}\right)$

$=\frac{\left(e^{i(A+B)}+e^{i(A-B)}-e^{-i(A-B)}-e^{-i(A+B)}\right)}{4i}$

$+\frac{\left(e^{i(A+B)}+e^{-i(A-B)}-e^{i(A-B)}-e^{-i(A+B)}\right)}{4i}$

$=\frac{2e^{i(A+B)}-2e^{-i(A+B)}}{4i}$

$=\frac{e^{i(A+B)}-e^{-i(A+B)}}{2i}=\sin(A+B)$ as required

b $\text{RHS}=\left(\frac{e^{iA}+e^{-iA}}{2}\right)\left(\frac{e^{iB}+e^{-iB}}{2}\right)-\left(\frac{e^{iA}-e^{-iA}}{2i}\right)\left(\frac{e^{iB}-e^{-iB}}{2i}\right)$

$=\frac{e^{i(A+B)}+e^{i(A-B)}+e^{-i(A-B)}+e^{-i(A+B)}}{4}$

$-\frac{e^{i(A+B)}-e^{i(A-B)}-e^{-i(A-B)}+e^{-i(A+B)}}{4i^2}$

$=\frac{e^{i(A+B)}+e^{i(A-B)}+e^{-i(A-B)}+e^{-i(A+B)}}{4}$

$+\frac{e^{i(A+B)}-e^{i(A-B)}-e^{-i(A-B)}+e^{-i(A+B)}}{4}$

$=\frac{2e^{i(A+B)}+2e^{-i(A+B)}}{4}$

$=\frac{e^{i(A+B)}+e^{-i(A+B)}}{2}=\cos(A+B)$ as required

5 **a** $\text{RHS}=\left(\frac{e^{ix}+e^{-ix}}{2}\right)^2-\left(\frac{e^{ix}-e^{-ix}}{2i}\right)^2$

$=\frac{e^{2ix}+2+e^{-2ix}}{4}-\frac{e^{2ix}-2+e^{-2ix}}{4i^2}$

$=\frac{e^{2ix}+2+e^{-2ix}}{4}+\frac{e^{2ix}-2+e^{-2ix}}{4}$

$=\frac{2e^{2ix}+2e^{-2ix}}{4}$

$=\frac{e^{2ix}+e^{-2ix}}{2}=\cos 2x$ as required

b $\text{LHS}=\left(\frac{e^{ix}+e^{-ix}}{2}\right)^2+\left(\frac{e^{ix}-e^{-ix}}{2i}\right)^2$

$=\frac{e^{2ix}+2+e^{-2ix}}{4}+\frac{e^{2ix}-2+e^{-2ix}}{4i^2}$

$=\frac{e^{2ix}+2+e^{-2ix}}{4}-\frac{e^{2ix}-2+e^{-2ix}}{4}$

$=\frac{4}{4}=1$ as required

6 a $\text{LHS} = \left(\frac{e^{i\theta}+e^{-i\theta}}{2} + i\left(\frac{e^{i\theta}-e^{-i\theta}}{2i}\right)\right)^2$

$= \left(\frac{e^{i\theta}+e^{-i\theta}}{2} + \frac{e^{i\theta}-e^{-i\theta}}{2}\right)^2$

$= \left(\frac{2e^{i\theta}}{2}\right)^2 = (e^{i\theta})^2$

$= e^{2i\theta}$

$\text{RHS} = \frac{e^{2i\theta}+e^{-2i\theta}}{2} + i\left(\frac{e^{2i\theta}-e^{-2i\theta}}{2i}\right)$

$= \frac{e^{2i\theta}+e^{-2i\theta}}{2} + \frac{e^{2i\theta}-e^{-2i\theta}}{2}$

$= \frac{2e^{2i\theta}}{2} = e^{2i\theta} = \text{LHS}$

so $(\cos\theta + i\sin\theta)^2 \equiv \cos 2\theta + i\sin 2\theta$

b $\text{LHS} = \left(\frac{e^{i\theta}+e^{-i\theta}}{2} + i\left(\frac{e^{i\theta}-e^{-i\theta}}{2i}\right)\right)^n$

$= \left(\frac{e^{i\theta}+e^{-i\theta}}{2} + \frac{e^{i\theta}-e^{-i\theta}}{2}\right)^n$

$= \left(\frac{2e^{i\theta}}{2}\right)^n = (e^{i\theta})^n$

$= e^{in\theta}$

$\text{RHS} = \frac{e^{in\theta}+e^{-in\theta}}{2} + i\left(\frac{e^{in\theta}-e^{-in\theta}}{2i}\right)$

$= \frac{e^{in\theta}+e^{-in\theta}}{2} + \frac{e^{in\theta}-e^{-in\theta}}{2}$

$= \frac{2e^{in\theta}}{2} = e^{in\theta} = \text{LHS}$

so $(\cos\theta + i\sin\theta)^n \equiv \cos(n\theta) + i\sin(n\theta)$

7 $zw = (4\times 3)\left(\cos\left(\frac{\pi}{9}+\frac{2\pi}{9}\right) + i\sin\left(\frac{\pi}{9}+\frac{2\pi}{9}\right)\right)$

$= 12\left(\cos\left(\frac{\pi}{3}\right) + i\sin\left(\frac{\pi}{3}\right)\right)$

$= 12\left(\frac{1}{2} + i\left(\frac{\sqrt{3}}{2}\right)\right)$

$= 6 + 6\sqrt{3}i$ as required

8 $z^2 = \left(8e^{i\frac{5\pi}{12}}\right)^2 = 64e^{i\frac{5\pi}{6}}$

$= 64\left(\cos\left(\frac{5\pi}{6}\right) + i\sin\left(\frac{5\pi}{6}\right)\right)$

$= 64\left(\left(-\frac{\sqrt{3}}{2}\right) + i\left(\frac{1}{2}\right)\right) = -32\sqrt{3} + 32i$

9 $|w| = \sqrt{1^2+1^2} = \sqrt{2}$ $\arg(w) = -\frac{\pi}{4}$

a $|zw| = |z||w| = \sqrt{2}k$

$\arg(zw) = \arg(z) + \arg(w) = \theta - \frac{\pi}{4}$

b $\left|\frac{z}{w}\right| = \frac{|z|}{|w|} = \frac{k}{\sqrt{2}} = \frac{\sqrt{2}}{2}k$

$\arg\left(\frac{z}{w}\right) = \arg(z) - \arg(w) = \theta + \frac{\pi}{4}$

Exercise 16.2A

1 a 1 **b** $-\frac{\sqrt{3}}{2} + \frac{1}{2}i$ **c** $\frac{1}{2} - \frac{\sqrt{3}}{2}i$ **d** $-i$

e $-\frac{\sqrt{2}}{2} - \frac{\sqrt{2}}{2}i$

2 a $\frac{81}{2}\sqrt{3} + \frac{81}{2}i$ **b** $\frac{\sqrt{2}}{1458} - \frac{\sqrt{2}}{1458}i$

3 a $4\sqrt{2} + 4\sqrt{2}i$ **b** $\frac{\sqrt{3}}{8} - \frac{1}{8}i$

c $64i$ **d** $\frac{1}{32} - \frac{\sqrt{3}}{32}i$

4 a -16 **b** $64i$

c $\frac{1}{4}i$ **d** $\frac{1}{16}$

5 a $-2-2i$ **b** $8+8i$

c $-\frac{1}{8} - \frac{1}{8}i$ **d** $-\frac{1}{8}i$

6 a -9 **b** $-\frac{1}{3}i$

c $\frac{1}{27}i$ **d** $\frac{1}{9}i$

7 a $8 - 8\sqrt{3}i$ **b** $-\frac{1}{8}i$

c $\frac{1}{8} + \frac{\sqrt{3}}{8}i$ **d** $-\frac{1}{8}$

Exercise 16.2B

1 a $2\cos^2\theta \equiv \frac{1}{2}(2\cos\theta)^2$

$\equiv \frac{1}{2}(e^{i\theta} + e^{-i\theta})^2$

$\equiv \frac{1}{2}(e^{2i\theta} + 2 + e^{-2i\theta})$

$\equiv \frac{1}{2}(2\cos 2\theta + 2)$

$\equiv \cos 2\theta + 1$

b $8\sin^3\theta \equiv \frac{1}{i^3}(2i\sin\theta)^3$

$\equiv \frac{1}{-i}(e^{i\theta} - e^{-i\theta})^3$

$\equiv \frac{-1}{i}(e^{3i\theta} - e^{-3i\theta} - 3e^{i\theta} + 3e^{-i\theta})$

$\equiv -\frac{1}{i}(2i\sin 3\theta - 6i\sin\theta)$

$\equiv 6\sin\theta - 2\sin 3\theta$

c $4\sin^4\theta \equiv \frac{1}{4}(2i\sin\theta)^4$

$\equiv \frac{1}{4}(e^{i\theta} - e^{-i\theta})^4$

$\equiv \frac{1}{4}(e^{4i\theta} + e^{-4i\theta} - 4e^{2i\theta} - 4e^{-2i\theta} + 6)$

$\equiv \frac{1}{4}(2\cos 4\theta - 8\cos 2\theta + 6)$

$\equiv \frac{1}{2}\cos 4\theta - 2\cos 2\theta + \frac{3}{2}$

2 **a** $\cos^5\theta \equiv \frac{1}{32}(2\cos\theta)^5$

$\equiv \frac{1}{32}(e^{i\theta} + e^{-i\theta})^5$

$\equiv \frac{1}{32}(e^{5i\theta} + e^{-5i\theta} + 5e^{3i\theta} + 5e^{-3i\theta} + 10e^{i\theta} + 10e^{-i\theta})$

$\equiv \frac{1}{32}(2\cos 5\theta + 10\cos 3\theta + 20\cos\theta)$

$\equiv \frac{1}{16}(10\cos\theta + 5\cos 3\theta + \cos 5\theta)$

So $A = \frac{1}{16}$

b $\frac{1}{16}\left(10\sin\theta + \frac{5}{3}\sin 3\theta + \frac{1}{5}\sin 5\theta\right) + c$

3 **a** $\sin^6\theta \equiv \frac{1}{-64}(2i\sin\theta)^6 \equiv -\frac{1}{64}(e^{i\theta} - e^{-i\theta})^6$

$\equiv -\frac{1}{64}(e^{6i\theta} + e^{-6i\theta} - 6e^{4i\theta} - 6e^{-4i\theta} + 15e^{2i\theta} + 15e^{-2i\theta} - 20)$

$\equiv -\frac{1}{64}(2\cos 6\theta - 12\cos 4\theta + 30\cos 2\theta - 20)$

$\equiv -\frac{1}{32}(15\cos 2\theta - 6\cos 4\theta + \cos 6\theta - 10)$

So $B = -\frac{1}{32}$

b $-\frac{1}{32}\left(\frac{15}{2}\sin 2\theta - \frac{3}{2}\sin 4\theta + \frac{1}{6}\sin 6\theta - 10\theta\right) + c$

4 **a** $2\sin^3\theta = \frac{1}{-4i}(2i\sin\theta)^3 = -\frac{1}{4i}(e^{i\theta} - e^{-i\theta})^3$

$= -\frac{1}{4i}\left(e^{3i\theta} - e^{-3i\theta} - 3e^{i\theta} + 3e^{-i\theta}\right)$

$= -\frac{1}{4i}(2i\sin 3\theta - 6i\sin\theta) = \frac{3}{2}\sin\theta - \frac{1}{2}\sin 3\theta$

b $\theta = \frac{\pi}{6}, \frac{5\pi}{6}$

5 **a** $5\cos^4\theta = \frac{5}{16}(2\cos\theta)^4 = \frac{5}{16}(e^{i\theta} + e^{-i\theta})^4$

$= \frac{5}{16}(e^{4i\theta} + e^{-4i\theta} + 4e^{2i\theta} + 4e^{-2i\theta} + 6)$

$= \frac{5}{16}(2\cos 4\theta + 8\cos 2\theta + 6)$

$= \frac{5}{8}\cos 4\theta + \frac{5}{2}\cos 2\theta + \frac{15}{8}$

So $A = \frac{5}{8}$, $B = \frac{5}{2}$ and $C = \frac{15}{8}$

b $\theta = \pm\frac{\pi}{4}, \pm\frac{3\pi}{4}$

6 **a** $\cos 2\theta + i\sin 2\theta = (\cos\theta + i\sin\theta)^2$

$= \cos^2\theta + 2i\cos\theta\sin\theta + i^2\sin^2\theta$

$= \cos^2\theta + 2i\cos\theta\sin\theta - \sin^2\theta$

Im : $\sin 2\theta = 2\cos\theta\sin\theta$

b $\cos 3\theta + i\sin 3\theta = (\cos\theta + i\sin\theta)^3$

$= \cos^3\theta + 3\cos^2\theta(i\sin\theta) + 3\cos\theta(i\sin\theta)^2 + (i\sin\theta)^3$

$= \cos^3\theta + 3i\cos^2\theta\sin\theta - 3\cos\theta\sin^2\theta - i\sin^3\theta$

Im : $\sin 3\theta = 3\cos^2\theta\sin\theta - \sin^3\theta$

$= 3(1 - \sin^2\theta)\sin\theta - \sin^3\theta$

$= 3\sin\theta - 3\sin^3\theta - \sin^3\theta$

$= 3\sin\theta - 4\sin^3\theta$

c Re : $\cos 3\theta = \cos^3\theta - 3\cos\theta\sin^2\theta$

$= \cos^3\theta - 3\cos\theta(1 - \cos^2\theta)$

$= \cos^3\theta - 3\cos\theta + 3\cos^3\theta$

$= 4\cos^3\theta - 3\cos\theta$

d $\cos 4\theta + i\sin 4\theta = (\cos\theta + i\sin\theta)^4$

$= \cos^4\theta + 4\cos^3\theta(i\sin\theta) + 6\cos^2\theta(i\sin\theta)^2 + 4\cos\theta(i\sin\theta)^3 + (i\sin\theta)^4$

$= \cos^4\theta + 4i\cos^3\theta\sin\theta - 6\cos^2\theta\sin^2\theta - 4i\cos\theta\sin^3\theta + \sin^4\theta$

Im : $\sin 4\theta = 4\cos^3\theta\sin\theta - 4\cos\theta\sin^3\theta$

$= 4\cos\theta\sin\theta(\cos^2\theta - \sin^2\theta)$

$= 4\cos\theta\sin\theta(1 - 2\sin^2\theta)$

$= 4\cos\theta\sin\theta - 8\cos\theta\sin^3\theta$

7 $\cos 6\theta + i\sin 6\theta = (\cos\theta + i\sin\theta)^6$

$= \cos^6\theta + 6i\cos^5\theta\sin\theta + 15i^2\cos^4\theta\sin^2\theta + 20i^3\cos^3\theta\sin^3\theta + 15i^4\cos^2\theta\sin^4\theta + 6i^5\cos\theta\sin^5\theta + i^6\sin^6\theta$

$= \cos^6\theta + 6i\cos^5\theta\sin\theta - 15\cos^4\theta\sin^2\theta - 20i\cos^3\theta\sin^3\theta + 15\cos^2\theta\sin^4\theta + 6i\cos\theta\sin^5\theta - \sin^6\theta$

a Re : $\cos 6\theta = \cos^6\theta - 15\cos^4\theta\sin^2\theta + 15\cos^2\theta\sin^4\theta - \sin^6\theta$

$= \cos^6\theta - 15\cos^4\theta(1 - \cos^2\theta) + 15\cos^2\theta(1 - \cos^2\theta)^2 - (1 - \cos^2\theta)^3$

$= \cos^6\theta - 15\cos^4\theta + 15\cos^6\theta + 15\cos^2\theta - 30\cos^4\theta + 15\cos^6\theta - 1 + 3\cos^2\theta - 3\cos^4\theta + \cos^6\theta$

$= 32\cos^6\theta - 48\cos^4\theta + 18\cos^2\theta - 1$

b Im : $\sin 6\theta = 6\cos^5\theta\sin\theta - 20\cos^3\theta\sin^3\theta + 6\cos\theta\sin^5\theta$

$= 2\sin\theta\cos\theta(3\cos^4\theta - 10\cos^2\theta\sin^2\theta + 3\sin^4\theta)$

$= 2\sin\theta\cos\theta(3(1 - \sin^2\theta)^2 - 10(1 - \sin^2\theta)\sin^2\theta + 3\sin^4\theta)$

$= 2\sin\theta\cos\theta(3 - 6\sin^2\theta + 3\sin^4\theta - 10\sin^2\theta + 10\sin^4\theta + 3\sin^4\theta)$

$= 2\sin\theta\cos\theta(16\sin^4\theta - 16\sin^2\theta + 3)$

8 **a** $\cos 5\theta + i\sin 5\theta = (\cos\theta + i\sin\theta)^5$

$= \cos^5\theta + 5\cos^4\theta(i\sin\theta) + 10\cos^3\theta(i\sin\theta)^2 + 10\cos^2\theta(i\sin\theta)^3 + 5\cos\theta(i\sin\theta)^4 + (i\sin\theta)^5$

$= \cos^5\theta + 5i\cos^4\theta\sin\theta - 10\cos^3\theta\sin^2\theta - 10i\cos^2\theta\sin^3\theta + 5\cos\theta\sin^4\theta + i\sin^5\theta$

$$\text{Re}: \cos 5\theta = \cos^5\theta - 10\cos^3\theta\sin^2\theta + 5\cos\theta\sin^4\theta$$
$$= \cos^5\theta - 10\cos^3\theta(1-\cos^2\theta) + 5\cos\theta(1-\cos^2\theta)^2$$
$$= \cos^5\theta - 10\cos^3\theta + 10\cos^5\theta + 5\cos\theta$$
$$-10\cos^3\theta + 5\cos^5\theta$$
$$= 16\cos^5\theta - 20\cos^3\theta + 5\cos\theta$$

b $x = 1, 0.309, -0.809$

9 a $\cos 4\theta + i\sin 4\theta = (\cos\theta + i\sin\theta)^4$
$$= \cos^4\theta + 4\cos^3\theta(i\sin\theta) + 6\cos^2\theta(i\sin\theta)^2$$
$$+ 4\cos\theta(i\sin\theta)^3 + (i\sin\theta)^4$$
$$= \cos^4\theta + 4i\cos^3\theta\sin\theta - 6\cos^2\theta\sin^2\theta$$
$$- 4i\cos\theta\sin^3\theta + \sin^4\theta$$
$$\text{Re}: \cos 4\theta = \cos^4\theta - 6\cos^2\theta(1-\cos^2\theta) + (1-\cos^2\theta)^2$$
$$= \cos^4\theta - 6\cos^2\theta + 6\cos^4\theta + 1 - 2\cos^2\theta + \cos^4\theta$$
$$= 8\cos^4\theta - 8\cos^2\theta + 1$$

b $x = \pm 0.966, \pm 0.259$

10 $\cos 2\theta + i\sin 2\theta = (\cos\theta + i\sin\theta)^2 = \cos^2\theta + 2i\cos\theta\sin\theta - \sin^2\theta$

$\text{Re}: \cos 2\theta = \cos^2\theta - \sin^2\theta$

$\text{Im}: \sin 2\theta = 2\cos\theta\sin\theta$

$$\tan 2\theta \equiv \frac{\sin 2\theta}{\cos 2\theta}$$
$$= \frac{2\cos\theta\sin\theta}{\cos^2\theta - \sin^2\theta}$$
$$= \frac{\frac{2\cos\theta\sin\theta}{\cos^2\theta}}{\frac{\cos^2\theta}{\cos^2\theta} - \frac{\sin^2\theta}{\cos^2\theta}}$$
$$= \frac{2\tan\theta}{1-\tan^2\theta}$$

11 $[r(\cos\theta + i\sin\theta)]^1 = r(\cos\theta + i\sin\theta)$

and $r^1(\cos(1\theta) + i\sin(1\theta)) = r(\cos\theta + i\sin\theta)$

So true for $n = 1$

Assume true for $n = k$

$$[r(\cos\theta + i\sin\theta)]^{k+1} = [r(\cos\theta + i\sin\theta)]^k[r(\cos\theta + i\sin\theta)]$$
$$= r^k(\cos k\theta + i\sin k\theta)[r(\cos\theta + i\sin\theta)]$$
$$= r^{k+1}(\cos k\theta\cos\theta + i\cos k\theta\sin\theta$$
$$+ i\cos\theta\sin k\theta + i^2\sin k\theta\sin\theta)$$
$$= r^{k+1}(\cos k\theta\cos\theta - \sin k\theta\sin\theta$$
$$+ i(\cos k\theta\sin\theta + \cos\theta\sin k\theta))$$
$$= r^{k+1}(\cos(k\theta + \theta) + i\sin(k\theta + \theta))$$
$= r^{k+1}[\cos(k+1)\theta + i\sin(k+1)\theta]$ as required

Since true for $n = 1$ and assuming true for $n = k$ implies true for $n = k + 1$ hence true for all positive integers n.

12 a $z^n = (\cos\theta + i\sin\theta)^n = \cos(n\theta) + i\sin(n\theta)$
$$\frac{1}{z^n} = z^{-n} = (\cos\theta + i\sin\theta)^{-n}$$
$$= \cos(-n\theta) + i\sin(-n\theta)$$
$$= \cos(n\theta) - i\sin(n\theta)$$

Therefore $z^n + \frac{1}{z^n} = \cos(n\theta) + i\sin(n\theta)$
$$+ (\cos(n\theta) - i\sin(n\theta))$$
$= 2\cos(n\theta)$ as required

b $4\cos\theta\sin^2\theta \equiv 4\cos\theta(1-\cos^2\theta)$
$$\equiv 4\cos\theta - 4\cos^3\theta$$
$$\equiv 4\cos\theta - 4\left(\frac{1}{2}\right)^3\left(z + \frac{1}{z}\right)^3$$
$$\equiv 4\cos\theta - \frac{1}{2}\left(z^3 + \frac{1}{z^3} + 3\left(z + \frac{1}{z}\right)\right)$$
$$\equiv 4\cos\theta - \frac{1}{2}(2\cos(3\theta) + 6\cos\theta)$$
$$\equiv \cos\theta - \cos(3\theta)$$

13 a $z^n = (\cos\theta + i\sin\theta)^n$
$$= \cos(n\theta) + i\sin(n\theta)$$
$$\frac{1}{z^n} = z^{-n} = (\cos\theta + i\sin\theta)^{-n}$$
$$= \cos(-n\theta) + i\sin(-n\theta)$$
$$= \cos(n\theta) - i\sin(n\theta)$$

Therefore $z^n - \frac{1}{z^n} = \cos(n\theta) + i\sin(n\theta)$
$$- (\cos(n\theta) - i\sin(n\theta))$$
$= 2i\sin(n\theta)$ as required

b $16\sin^3\theta\cos^2\theta \equiv 16\sin^3\theta(1-\sin^2\theta)$
$$\equiv 16(\sin^3\theta - \sin^5\theta)$$
$$\equiv 16\left(\frac{1}{2i}\right)^3\left(z - \frac{1}{z}\right)^3 - 16\left(\frac{1}{2i}\right)^5\left(z - \frac{1}{z}\right)^5$$
$$\equiv 2i\left(z^3 - \frac{1}{z^3} - 3\left(z - \frac{1}{z}\right)\right)$$
$$+ \frac{i}{2}\left(z^5 - \frac{1}{z^5} - 5\left(z^3 - \frac{1}{z^3}\right) + 10\left(z - \frac{1}{z}\right)\right)$$
$$\equiv 2i(2i\sin(3\theta) - 6i\sin\theta) + \frac{i}{2}(2i\sin(5\theta)$$
$$- 10i\sin(3\theta) + 20i\sin\theta)$$
$$\equiv -4\sin(3\theta) + 12\sin\theta - \sin(5\theta)$$
$$+ 5\sin(3\theta) - 10\sin\theta$$
$$\equiv 2\sin\theta + \sin(3\theta) - \sin(5\theta)$$

14 $\sum_{r=1}^{n}\cos(r\theta) + i\sin(r\theta) = \sum_{r=1}^{n}(\cos(r\theta) + i\sin(r\theta))^r$
$$= (\cos\theta + i\sin\theta) + (\cos\theta + i\sin\theta)^2 + \ldots$$
$$+ (\cos\theta + i\sin\theta)^n$$
$$= \frac{(\cos\theta + i\sin\theta)(1 - (\cos(\theta) + i\sin(\theta))^n)}{1 - (\cos(\theta) + i\sin(\theta))}$$
$$= \frac{(\cos\theta + i\sin\theta)(1 - (\cos(n\theta) + i\sin(n\theta)))}{1 - (\cos(\theta) + i\sin(\theta))}$$
$$= \frac{(\cos\theta + i\sin\theta)\left(2\sin^2\left(\frac{n\theta}{2}\right) - 2i\sin\left(\frac{n\theta}{2}\right)\cos\left(\frac{n\theta}{2}\right)\right)}{2\sin^2\left(\frac{\theta}{2}\right) - 2i\sin\left(\frac{\theta}{2}\right)\cos\left(\frac{\theta}{2}\right)}$$

$$= \frac{(\cos\theta + i\sin\theta)2\sin\left(\frac{n\theta}{2}\right)\left(\sin\left(\frac{n\theta}{2}\right) - i\cos\left(\frac{n\theta}{2}\right)\right)}{2\sin\left(\frac{\theta}{2}\right)\left(\sin\left(\frac{\theta}{2}\right) - i\cos\left(\frac{\theta}{2}\right)\right)}$$

$$= \frac{(\cos\theta + i\sin\theta)\sin\left(\frac{n\theta}{2}\right)\left(\sin\left(\frac{n\theta}{2}\right) - i\cos\left(\frac{n\theta}{2}\right)\right)\left(\sin\left(\frac{\theta}{2}\right) + i\cos\left(\frac{\theta}{2}\right)\right)}{\sin\left(\frac{\theta}{2}\right)\left(\sin\left(\frac{\theta}{2}\right) - i\cos\left(\frac{\theta}{2}\right)\right)\left(\sin\left(\frac{\theta}{2}\right) + i\cos\left(\frac{\theta}{2}\right)\right)}$$

$$= \frac{(\cos\theta + i\sin\theta)\left(\begin{array}{l}\sin\left(\frac{n\theta}{2}\right)\sin\left(\frac{\theta}{2}\right) + \cos\left(\frac{n\theta}{2}\right)\cos\left(\frac{\theta}{2}\right) \\ + i\left(\cos\left(\frac{\theta}{2}\right)\sin\left(\frac{n\theta}{2}\right) - \cos\left(\frac{n\theta}{2}\right)\sin\left(\frac{\theta}{2}\right)\right)\end{array}\right)\sin\left(\frac{n\theta}{2}\right)}{\sin\left(\frac{\theta}{2}\right)\left(\sin^2\left(\frac{\theta}{2}\right) + \cos^2\left(\frac{\theta}{2}\right)\right)}$$

$$= \frac{(\cos\theta + i\sin\theta)\left(\begin{array}{l}\cos\left(\frac{(n-1)\theta}{2}\right) \\ + i\sin\left(\frac{(n-1)\theta}{2}\right)\end{array}\right)\sin\left(\frac{n\theta}{2}\right)}{\sin\left(\frac{\theta}{2}\right)}$$

Considering only the real parts of this sum gives

$$\sum_{r=1}^{n}\cos(r\theta) = \frac{\left(\cos\theta\cos\left(\frac{(n-1)\theta}{2}\right) - \sin\left(\frac{(n-1)\theta}{2}\right)\sin\theta\right)\sin\left(\frac{n\theta}{2}\right)}{\sin\left(\frac{\theta}{2}\right)}$$

$$= \frac{\cos\left(\frac{(n+1)\theta}{2}\right)\sin\left(\frac{n\theta}{2}\right)}{\sin\left(\frac{\theta}{2}\right)}$$

Exercise 16.3A

1 a $1, -\frac{1}{2} + \frac{\sqrt{3}}{2}i, -\frac{1}{2} - \frac{\sqrt{3}}{2}i$

b

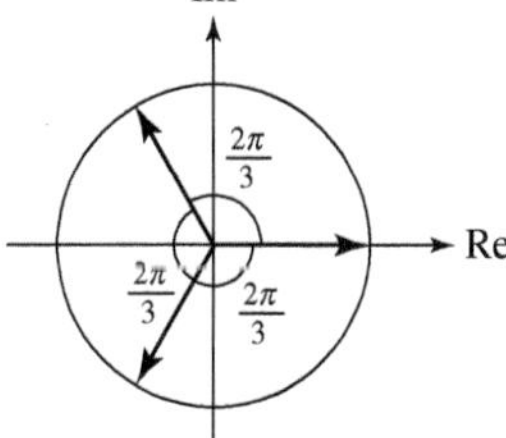

2 a $1, -1, i, -i$

b

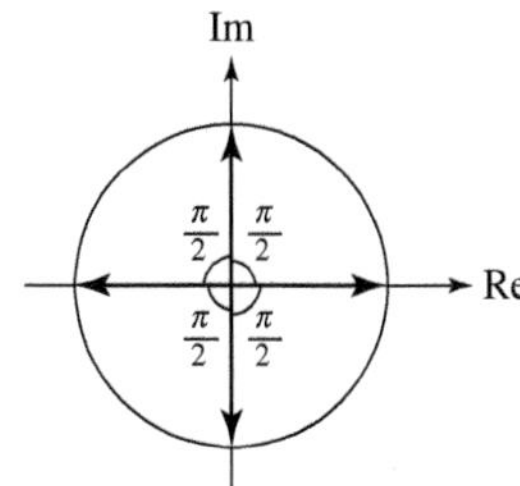

3 a $1, e^{\frac{2\pi i}{5}}, e^{\frac{4\pi i}{5}}, e^{\frac{6\pi i}{5}}, e^{\frac{8\pi i}{5}}$

b

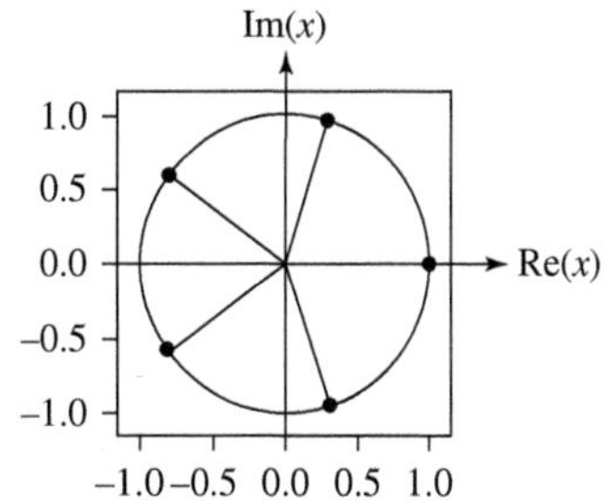

4 a $2, -1 + \sqrt{3}i, -1 - \sqrt{3}i$

b $\frac{\sqrt{3}}{2} + \frac{1}{2}i, -\frac{\sqrt{3}}{2} + \frac{1}{2}i, -i$

c $\frac{\sqrt{14}}{2} - \frac{\sqrt{14}}{2}i, \frac{\sqrt{14}}{2} + \frac{\sqrt{14}}{2}i, -\frac{\sqrt{14}}{2} + \frac{\sqrt{14}}{2}i,$
$-\frac{\sqrt{14}}{2} - \frac{\sqrt{14}}{2}i$

d $\frac{3\sqrt{3}}{2} - \frac{3}{2}i,\ 3i, -\frac{3\sqrt{3}}{2} - \frac{3}{2}i$

e $\frac{\sqrt{6}}{2} - \frac{\sqrt{2}}{2}i, \frac{\sqrt{6}}{2} + \frac{\sqrt{2}}{2}i, \sqrt{2}i, -\frac{\sqrt{6}}{2} + \frac{\sqrt{2}}{2}i, -\frac{\sqrt{6}}{2} - \frac{\sqrt{2}}{2}i,$
$-\sqrt{2}i$

5 a $2e^{-\frac{\pi}{8}i}, 2e^{\frac{3\pi}{8}i}, 2e^{\frac{7\pi}{8}i}, 2e^{-\frac{5\pi}{8}i}$

b $2e^{\frac{\pi}{10}i}, 2e^{\frac{\pi i}{2}}, 2e^{\frac{9\pi}{10}i}, 2e^{-\frac{7\pi}{10}i}, 2e^{-\frac{3\pi}{10}i}$

6 a $2e^{\frac{\pi}{12}i}, 2e^{\frac{3\pi}{4}i}, 2e^{-\frac{7\pi}{12}i}$

b $2e^{\frac{\pi}{4}i}, 2e^{\frac{11\pi}{12}i}, 2e^{-\frac{5\pi}{12}i}$

c $2e^{-\frac{\pi}{4}i}, 2e^{\frac{5\pi}{12}i}, 2e^{-\frac{11\pi}{12}i}$

d $2e^{-\frac{\pi}{12}i}, 2e^{\frac{7\pi}{12}i}, 2e^{-\frac{3\pi}{4}i}$

7 a $\sqrt{3}(\cos(0.18) + i\sin(0.18)), \sqrt{3}(\cos(1.75) + i\sin(1.75)),$
$\sqrt{3}(\cos(-1.39) + i\sin(-1.39)), \sqrt{3}(\cos(-2.96) + i\sin(-2.96))$

b $\sqrt{3}(\cos(0.60) + i\sin(0.60)),\ \sqrt{3}(\cos(2.17) + i\sin(2.17))$
$\sqrt{3}(\cos(-0.97) + i\sin(-0.97)),$
$\sqrt{3}(\cos(-2.54) + i\sin(-2.54))$

c $\sqrt{3}(\cos(-0.58) + i\sin(-0.58)), \sqrt{3}(\cos(1.00) + i\sin(1.00))$
$\sqrt{3}(\cos(2.57) + i\sin(2.57)), \sqrt{3}(\cos(-2.14) + i\sin(-2.14))$

d $\sqrt{3}(\cos(-0.21) + i\sin(-0.21)), \sqrt{3}(\cos(1.36) + i\sin(1.36))$
$\sqrt{3}(\cos(2.93) + i\sin(2.93)), \sqrt{3}(\cos(-1.78) + i\sin(-1.78))$

8 a $3e^{-\frac{\pi}{5}i}, 3e^{\frac{\pi}{5}i}, 3e^{\frac{3\pi}{5}i}, 3e^{-\frac{3\pi}{5}i}, 3e^{\pi i}$

b $\frac{1}{2}e^{-\frac{\pi}{5}i}, \frac{1}{2}e^{\frac{\pi}{5}i}, \frac{1}{2}e^{\frac{3\pi}{5}i}, \frac{1}{2}e^{-\frac{3\pi}{5}i}, \frac{1}{2}e^{\pi i}$

c $2e^{-\frac{\pi}{12}i}, 2e^{\frac{\pi}{4}i}, 2e^{\frac{7\pi}{12}i}, 2e^{\frac{11\pi}{12}i}, 2e^{-\frac{5\pi}{12}i}, 2e^{-\frac{3\pi}{4}i}$

d $1.06e^{-0.161i},\ 1.06e^{1.41i}, 1.06e^{2.98\,i}, 1.06e^{-1.73\,i}$

9 a $\frac{\sqrt{2}}{2}\left(1+\sqrt{3}+\left(\sqrt{3}-1\right)i\right), \frac{\sqrt{2}}{2}\left(1-\sqrt{3}+\left(\sqrt{3}+1\right)i\right)$
$\frac{\sqrt{2}}{2}\left(\sqrt{3}-1-\left(\sqrt{3}+1\right)i\right), \frac{\sqrt{2}}{2}\left(-1-\sqrt{3}+\left(-\sqrt{3}+1\right)i\right)$

b $\sqrt{3}+i, -1+\sqrt{3}i, -\sqrt{3}-i, 1-\sqrt{3}i$

10 $2-2\sqrt{3}\,i, -4, 2+2\sqrt{3}\,i$

11 a $\sqrt{6}+\sqrt{2}i, -\sqrt{2}+\sqrt{6}i, -\sqrt{6}-\sqrt{2}i, \sqrt{2}-\sqrt{6}i$

b

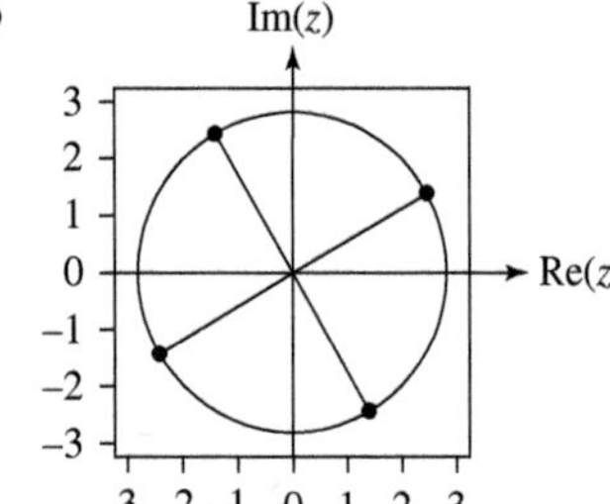

12 a $2e^{\frac{\pi i}{18}}, 2e^{\frac{13\pi i}{18}}, 2e^{-\frac{11\pi i}{18}}$

b $\sqrt[4]{6}e^{-\frac{\pi i}{16}}, \sqrt[4]{6}e^{\frac{7\pi i}{16}}, \sqrt[4]{6}e^{\frac{15\pi i}{16}}, \sqrt[4]{6}e^{-\frac{9\pi i}{16}}$

13 a $\sqrt{2}\left(\cos\left(-\frac{\pi}{9}\right)+i\sin\left(-\frac{\pi}{9}\right)\right), \sqrt{2}\left(\cos\left(\frac{5\pi}{9}\right)+i\sin\left(\frac{5\pi}{9}\right)\right),$
$\sqrt{2}\left(\cos\left(-\frac{7\pi}{9}\right)+i\sin\left(-\frac{7\pi}{9}\right)\right)$

b $\sqrt{2}\left(\cos\left(\frac{5\pi}{36}\right)+i\sin\left(\frac{5\pi}{36}\right)\right), \sqrt{2}\left(\cos\left(\frac{17\pi}{36}\right)+i\sin\left(\frac{17\pi}{36}\right)\right)$
$\sqrt{2}\left(\cos\left(\frac{29\pi}{36}\right)+i\sin\left(\frac{29\pi}{36}\right)\right),$
$\sqrt{2}\left(\cos\left(-\frac{31\pi}{36}\right)+i\sin\left(-\frac{31\pi}{36}\right)\right)$
$\sqrt{2}\left(\cos\left(-\frac{19\pi}{36}\right)+i\sin\left(-\frac{19\pi}{36}\right)\right),$
$\sqrt{2}\left(\cos\left(-\frac{7\pi}{36}\right)+i\sin\left(-\frac{7\pi}{36}\right)\right)$

Exercise 16.3B

1 a $1+\omega+\omega^2=\frac{(1-\omega^3)}{1-\omega}$
$=\frac{(1-1)}{1-\omega}=0$

b i 0 **ii** 1 **iii** −1 **iv** 0

2 a $1+\omega+\omega^2+\omega^3+\omega^4=\frac{(1-\omega^5)}{1-\omega}$
$=\frac{(1-1)}{1-\omega}=0$

b −1 **ii** 0 **iii** −1

3 a $\frac{3\sqrt{3}}{2}-\frac{3}{2}i, 3i, -\frac{3\sqrt{3}}{2}-\frac{3}{2}i$

b i $\frac{27\sqrt{3}}{4}$ square units **ii** $9\sqrt{3}$ units

4 a $z^3=125e^{(-\pi+2k\pi)i}$
$z=(125e^{(2k-1)\pi i})^{\frac{1}{3}}$
$=5e^{\frac{(2k-1)\pi i}{3}}$
$k=0: z=5e^{-\frac{\pi i}{3}}=\frac{5}{2}-\frac{5\sqrt{3}}{2}i$
$k=1: z=5e^{\frac{\pi i}{3}}=\frac{5}{2}+\frac{5\sqrt{3}}{2}i$
$k=2: z=5e^{\pi i}=-5$
$\text{Area}=\frac{1}{2}\times\left(5+\frac{5}{2}\right)\times\left(\frac{5\sqrt{3}}{2}+\frac{5\sqrt{3}}{2}\right)$
$=\frac{75\sqrt{3}}{4}$ square units $\left(k=\frac{75}{4}\right)$

b $15\sqrt{3}$ units

5 a Square **b** 4 square units

6 a $\frac{\sqrt{3}}{3}, \frac{\sqrt{3}}{3}i, -\frac{\sqrt{3}}{3}, -\frac{\sqrt{3}}{3}i$

b $A\left(\frac{\sqrt{3}}{3},0\right), B\left(0,\frac{\sqrt{3}}{3}\right), C\left(-\frac{\sqrt{3}}{3},0\right), D\left(0,-\frac{\sqrt{3}}{3}\right)$

gradient $AB=\frac{\frac{\sqrt{3}}{3}}{-\frac{\sqrt{3}}{3}}=-1$

gradient $BC=-\frac{\frac{\sqrt{3}}{3}}{-\frac{\sqrt{3}}{3}}=1$

$-1\times1=-1$ so a right-angle at B

gradient $CD=-\frac{\frac{\sqrt{3}}{3}}{\frac{\sqrt{3}}{3}}=-1$

$-1\times1=-1$ so a right-angle at C etc
Hence all angles are right angles.
The length of all four sides is given by

$\sqrt{\left(\frac{\sqrt{3}}{3}\right)^2+\left(\frac{\sqrt{3}}{3}\right)^2}=\frac{\sqrt{6}}{3}$

Hence all sides have the same length, and so the shape formed is a square.

c $\frac{2}{3}$ square units

7 a $n=6$

b $\left(-\frac{1}{2},\frac{\sqrt{3}}{2}\right)$ **c** $\frac{3\sqrt{3}}{2}$

Review exercise 16

1 a $3e^{-\frac{\pi}{2}i}$ **b** $\sqrt{2}e^{\frac{\pi}{4}i}$ **c** $5e^0$

d $2e^{-\frac{5\pi}{6}i}$ **e** $\sqrt{3}e^{-0.615i}$ **f** $2e^{\frac{2\pi}{3}i}$

2 a $3e^{\frac{\pi}{7}i}$ **b** $\sqrt{2}e^{\frac{\pi}{9}i}$ **c** $\sqrt{3}e^{-\frac{\pi}{8}i}$ **d** $5e^{\frac{\pi}{5}i}$

3 a $7i$ **b** $3+3\sqrt{3}i$ **c** $\frac{3}{2}-\frac{\sqrt{3}}{2}i$

d $-1+i$ **e** $-\frac{\sqrt{6}}{2}-\frac{3\sqrt{2}}{2}i$ **f** $-3+\sqrt{3}i$

4 a $|zw| = 15 \arg(zw) = \frac{12}{35}\pi$

b $|zw| = 2\sqrt{3} \arg(zw) = \frac{5}{24}\pi$

c $|zw| = 6\sqrt{2} \arg(zw) = \frac{\pi}{12}$

d $|zw| = 2 \arg(zw) = \frac{5}{9}\pi$

5 a $\left|\frac{z}{w}\right| = \frac{1}{2} \arg\left(\frac{z}{w}\right) = \frac{\pi}{4}$

b $\left|\frac{z}{w}\right| = \sqrt{3} \arg\left(\frac{z}{w}\right) = -\frac{5\pi}{8}$

c $\left|\frac{z}{w}\right| = \frac{5}{6}\sqrt{3} \arg\left(\frac{z}{w}\right) = -\frac{11\pi}{12}$

d $\left|\frac{z}{w}\right| = 8\sqrt{2} \arg\left(\frac{z}{w}\right) = -\frac{7\pi}{11}$

6 a -64 b $-512i$

7 $8+8\sqrt{3}i$

8 a
$$\begin{aligned}\cos 5\theta + i\sin 5\theta &= (\cos\theta + i\sin\theta)^5 \\ &= \cos^5\theta + 5\cos^4\theta(i\sin\theta) \\ &\quad +10\cos^3\theta(i\sin\theta)^2 \\ &\quad +10\cos^2\theta(i\sin\theta)^3 \\ &\quad +5\cos\theta(i\sin\theta)^4 + (i\sin\theta)^5 \\ &= \cos^5\theta + 5i\cos^4\theta\sin\theta - 10\cos^3\theta\sin^2\theta \\ &\quad -10i\cos^2\theta\sin^3\theta + 5\cos\theta\sin^4\theta + i\sin^5\theta\end{aligned}$$

$$\begin{aligned}\text{Im}: \sin 5\theta &\equiv 5\cos^4\theta\sin\theta - 10\cos^2\theta\sin^3\theta + \sin^5\theta \\ &= 5(1-\sin^2\theta)^2\sin\theta - 10(1-\sin^2\theta)\sin^3\theta + \sin^5\theta \\ &= 5\sin\theta - 10\sin^3\theta + 5\sin^5\theta - 10\sin^3\theta \\ &\quad +10\sin^5\theta + \sin^5\theta \\ &= 5\sin\theta - 20\sin^3\theta + 16\sin^5\theta\end{aligned}$$

b $x = 0, 0.588, 0.951$

9
$$\begin{aligned}\cos^3\theta &= \frac{1}{8}(2\cos\theta)^3 \\ &= \frac{1}{8}(e^{i\theta} + e^{-i\theta})^3 \\ &= \frac{1}{8}(e^{3i\theta} + 3e^{i\theta} + 3e^{-i\theta} + e^{-3i\theta}) \\ &= \frac{1}{8}(2\cos 3\theta + 6\cos\theta) \\ &= \frac{1}{4}(\cos 3\theta + 3\cos\theta) \text{ as required } (A = \frac{1}{4})\end{aligned}$$

10 a $1, \frac{\sqrt{2}}{2}+\frac{\sqrt{2}}{2}i, i, -\frac{\sqrt{2}}{2}+\frac{\sqrt{2}}{2}i, -1, -\frac{\sqrt{2}}{2}-\frac{\sqrt{2}}{2}i, -i, \frac{\sqrt{2}}{2}-\frac{\sqrt{2}}{2}i$

b

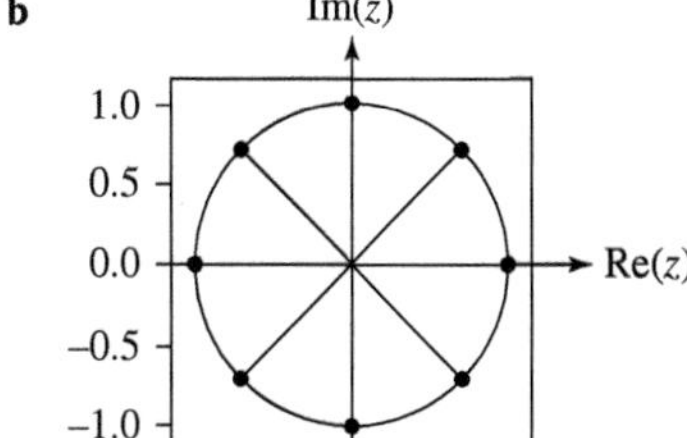

11 a $\sqrt{2}, 1+i, \sqrt{2}i, -1+i, -\sqrt{2}, -1-i, -\sqrt{2}i, 1-i$

b $\frac{\sqrt{3}}{2}+\frac{1}{2}i, -\frac{\sqrt{3}}{2}+\frac{1}{2}i, -i$

c $\frac{3\sqrt{2}}{2}-\frac{3\sqrt{2}}{2}i, -\frac{3\sqrt{2}}{2}+\frac{3\sqrt{2}}{2}i$

d $\frac{\sqrt{15}}{2}+\frac{\sqrt{5}}{2}i, \sqrt{5}i, -\frac{\sqrt{15}}{2}+\frac{\sqrt{5}}{2}i, -\frac{\sqrt{15}}{2}-\frac{\sqrt{5}}{2}i, -\sqrt{5}i, \frac{\sqrt{15}}{2}-\frac{\sqrt{5}}{2}i$

12 $\sqrt{2}e^{\frac{-3\pi i}{16}}, \sqrt{2}e^{\frac{5\pi i}{16}}, \sqrt{2}e^{\frac{13\pi i}{16}}, \sqrt{2}e^{\frac{-11\pi i}{16}}$

Assessment 16

1 a $2\sqrt{3}e^{-\frac{\pi}{6}i}$

b $144\left(\cos\left(-\frac{2\pi}{3}\right)+i\sin\left(-\frac{2\pi}{3}\right)\right)$

2 a $\text{mod}(z) = 2k \arg(z) = \frac{2\pi}{3}$

b $-128+128\sqrt{3}i$

3 a $e^{-\frac{\pi}{6}i}, e^{\frac{\pi}{2}i}, e^{-\frac{5\pi}{6}i}$

b

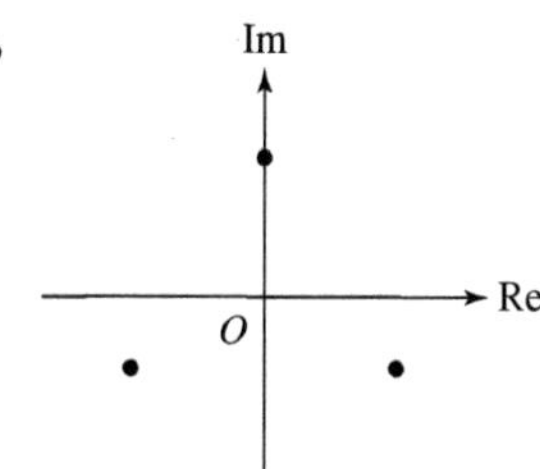

4 a $2\sqrt{2}\left(\cos\left(\frac{3\pi}{4}\right)+i\sin\left(\frac{3\pi}{4}\right)\right)$ b $\frac{1}{32}-\frac{1}{32}i$

5 $-8-8\sqrt{3}i$

6 $(\cos\theta - i\sin\theta)^n \equiv (\cos(-\theta)+i\sin(-\theta))^n$

since $\sin(-\theta) \equiv -\sin\theta$ and $\cos(-\theta) \equiv \cos\theta$

Therefore

$(\cos\theta - i\sin\theta)^n \equiv \cos(-n\theta)+i\sin(-n\theta)$ by de Moivre's theorem

$\equiv \cos(n\theta) - i\sin(n\theta)$

7 a
$$\begin{aligned}\cos(5\theta)+i\sin(5\theta) &\equiv (\cos\theta + i\sin\theta)^5 \\ &\equiv \cos^5\theta + 5i\cos^4\theta\sin\theta - 10\cos^3\theta\sin^2\theta \\ &\quad -10i\cos^2\theta\sin^3\theta + 5\cos\theta\sin^4\theta + i\sin^5\theta\end{aligned}$$

$$\begin{aligned}\sin(5\theta) &\equiv 5\cos^4\theta\sin\theta - 10\cos^2\theta\sin^3\theta + \sin^5\theta \\ &\equiv 5(1-\sin^2\theta)^2\sin\theta - 10(1-\sin^2\theta)\sin^3\theta + \sin^5\theta \\ &\equiv 5\sin\theta - 10\sin^3\theta + 5\sin^5\theta - 10\sin^3\theta \\ &\quad +10\sin^5\theta + \sin^5\theta \\ &\equiv 5\sin\theta - 20\sin^3\theta + 16\sin^5\theta\end{aligned}$$

b $\theta = 6°, 30°, 78°, 102°, 150°, 174°$

8 a $32\cos^5 x \equiv (2\cos x)^5$

$\equiv (e^{ix}+e^{-ix})^5$

$\equiv (e^{ix})^5 + 5(e^{ix})^4(e^{-ix}) + 10(e^{ix})^3(e^{-ix})^2 + 10(e^{ix})^2(e^{-ix})^3 + 5(e^{ix})(e^{-ix})^4 + (e^{-ix})^5$

$\equiv e^{5ix} + 5e^{3ix} + 10e^{ix} + \frac{10}{e^{ix}} + \frac{5}{e^{3ix}} + \frac{1}{e^{5ix}}$

$\equiv 2\cos(5x) + 10\cos(3x) + 20\cos(x)$

b $\frac{1}{80}\sin(5x) + \frac{5}{48}\sin(3x) + \frac{5}{8}\sin(x) + c$

9 a $\cos 3\theta + i\sin 3\theta \equiv (\cos\theta + i\sin\theta)^3$

$\equiv \cos^3\theta + 3i\cos^2\theta\sin\theta - 3\cos\theta\sin^2\theta - i\sin^3\theta$

$\text{Im}: \sin 3\theta \equiv 3\cos^2\theta\sin\theta - \sin^3\theta$ as required

b $\text{Re}: \cos 3\theta \equiv \cos^3\theta - 3\cos\theta\sin^2\theta$

$\tan 3\theta \equiv \frac{\sin 3\theta}{\cos 3\theta}$

$\equiv \frac{3\cos^2\theta\sin\theta - \sin^3\theta}{\cos^3\theta - 3\cos\theta\sin^2\theta}$

$\equiv \frac{3\tan\theta - \tan^3\theta}{1 - 3\tan^2\theta}$ as required

c $-\frac{18}{35}\sqrt{3}$

10 a 6

b $\sqrt{2}\left(\cos\left(-\frac{\pi}{6}\right) + i\sin\left(-\frac{\pi}{6}\right)\right)$,

$\sqrt{2}\left(\cos\left(\frac{\pi}{6}\right) + i\sin\left(\frac{\pi}{6}\right)\right)$,

$\sqrt{2}\left(\cos\left(\frac{\pi}{2}\right) + i\sin\left(\frac{\pi}{2}\right)\right)$,

$\sqrt{2}\left(\cos\left(\frac{5\pi}{6}\right) + i\sin\left(\frac{5\pi}{6}\right)\right)$,

$\sqrt{2}\left(\cos\left(-\frac{\pi}{2}\right) + i\sin\left(-\frac{\pi}{2}\right)\right)$,

$\sqrt{2}\left(\cos\left(-\frac{5\pi}{6}\right) + i\sin\left(-\frac{5\pi}{6}\right)\right)$

11 $\text{RHS} \equiv 1 - 2\left(\frac{(e^{ix} - e^{-ix})}{2i}\right)^2$

$\equiv 1 - \frac{2(e^{2ix} + e^{-2ix} - 2)}{4i^2}$

$\equiv 1 - \frac{e^{2ix} + e^{-2ix} - 2}{-2}$

$\equiv 1 + \frac{e^{2ix} + e^{-2ix} - 2}{2}$

$\equiv \frac{2 + e^{2ix} + e^{-2ix} - 2}{2}$

$\equiv \frac{e^{2ix} + e^{-2ix}}{2} \equiv \cos 2x$

12 a $\frac{3\pi}{4}$ **b** $\frac{\pi}{4}$ **c** $\frac{n}{4}\pi$ **d** $-\frac{3\pi}{4}$

13 a $\cos 7\theta + i\sin 7\theta \equiv (\cos\theta + i\sin\theta)^7$

$\text{Re}: \cos 7\theta \equiv \cos^7\theta - 21\cos^5\theta\sin^2\theta + 35\cos^3\theta\sin^4\theta - 7\cos\theta\sin^6\theta$

$\equiv \cos^7\theta - 21\cos^5\theta(1-\cos^2\theta) + 35\cos^3\theta(1-\cos^2\theta)^2 - 7\cos\theta(1-\cos^2\theta)^3$

$\equiv \cos^7\theta - 21\cos^5\theta + 21\cos^7\theta + 35\cos^3\theta - 70\cos^5\theta + 35\cos^7\theta - 7\cos\theta + 21\cos^3\theta - 21\cos^5\theta + 7\cos^7\theta$

$\equiv 64\cos^7\theta - 112\cos^5\theta + 56\cos^3\theta - 7\cos\theta$

$\equiv \cos\theta(64\cos^6\theta - 112\cos^4\theta + 56\cos^2\theta - 7)$

as required

b 0.975, 0.782, 0.434, –0.434, –0.782, –0.975

14 $\sqrt{2}e^{-\frac{5}{48}\pi i}, \sqrt{2}e^{\frac{7}{48}\pi i}, \sqrt{2}e^{\frac{19}{48}\pi i}, \sqrt{2}e^{\frac{31}{48}\pi i}, \sqrt{2}e^{\frac{43}{48}\pi i},$

$\sqrt{2}e^{-\frac{17}{48}\pi i}, \sqrt{2}e^{-\frac{29}{48}\pi i}, \sqrt{2}e^{-\frac{41}{48}\pi i}$

15 a $(\sin\theta)^4 \equiv \left(\frac{e^{i\theta} - e^{-i\theta}}{2}\right)^4$

$\equiv \frac{1}{16}\left[\begin{matrix}(e^{i\theta})^4 - 4(e^{i\theta})^3(e^{-i\theta}) + 6(e^{i\theta})^2(e^{-i\theta})^2 \\ -4(e^{i\theta})(e^{-i\theta})^3 + (e^{-i\theta})^4\end{matrix}\right]$

$\equiv \frac{1}{16}\left[e^{4i\theta} + e^{-4i\theta} - 4(e^{2i\theta} + e^{-2i\theta}) + 6\right]$

$\equiv \frac{1}{16}\left[2\cos(4\theta) - 8\cos(2\theta) + 6\right]$

$\equiv \frac{1}{8}\left(\cos(4\theta) - 4\cos(2\theta) + 3\right)$ as required

b $\frac{1}{4}\sin(4\theta) - 2\sin(2\theta) + 3\theta + c$

16 a $1.86\left(\cos\left(-\frac{\pi}{6}\right) + i\sin\left(-\frac{\pi}{6}\right)\right)$,

$1.86\left(\cos\left(\frac{\pi}{3}\right) + i\sin\left(\frac{\pi}{3}\right)\right)$,

$1.86\left(\cos\left(\frac{5\pi}{6}\right) + i\sin\left(\frac{5\pi}{6}\right)\right)$,

$1.86\left(\cos\left(-\frac{2\pi}{3}\right) + i\sin\left(-\frac{2\pi}{3}\right)\right)$

b $2.6 - 0.93i$ $1.9 + 1.6i$

$-0.61 + 0.93i$ $0.069 - 1.6i$

17 a $\frac{1}{z^n} \equiv z^{-n} \equiv (\cos\theta + i\sin\theta)^{-n}$

$\equiv \cos(-n\theta) + i\sin(-n\theta)$

$\equiv \cos(n\theta) - i\sin(n\theta)$

$z^n \equiv (\cos\theta + i\sin\theta)^n \equiv \cos(n\theta) + i\sin(n\theta)$

$z^n - \frac{1}{z^n} \equiv \cos(n\theta) + i\sin(n\theta) - (\cos(n\theta) - i\sin(n\theta))$

$\equiv 2i\sin(n\theta)$ as required

b $4\sin^3\theta \equiv \frac{1}{-2i}(2i\sin\theta)^3$

$\equiv \frac{1}{2}i\left(z-\frac{1}{z}\right)^3$

$\equiv \frac{1}{2}i\left[(z)^3 - 3(z)^2\left(\frac{1}{z}\right) + 3(z)\left(\frac{1}{z}\right)^2 - (z)^3\right]$

$\equiv \frac{1}{2}i\left[z^3 - \frac{1}{z^3} - 3\left(z - \frac{1}{z}\right)\right]$

$\equiv \frac{1}{2}i\left(2i\sin(3\theta) - 6i\sin\theta\right)$

$\equiv 3\sin\theta - \sin(3\theta)$ as required

c $\theta = \frac{7\pi}{6}, \frac{11\pi}{6}$

18 a

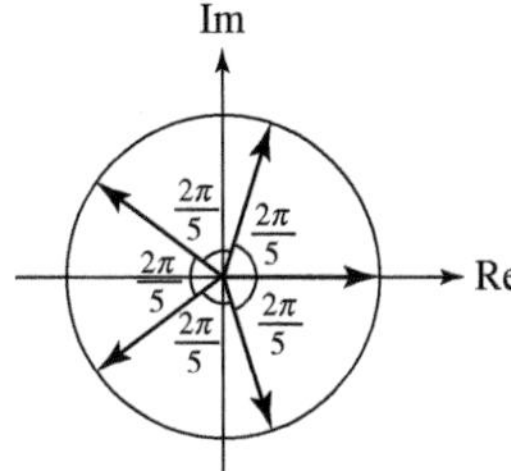

b **i** -1 **ii** 4

19 a $\cos(5\theta) + i\sin(5\theta) \equiv (\cos\theta + i\sin\theta)^5$

$\equiv \cos^5\theta + 5i\cos^4\theta\sin\theta$

$- 10\cos^3\theta\sin^2\theta$

$- 10i\cos^2\theta\sin^3\theta$

$+ 5\cos\theta\sin^4\theta + i\sin^5\theta$

$\text{Re}: \cos(5\theta) \equiv \cos^5\theta - 10\cos^3\theta\sin^2\theta + 5\cos\theta\sin^4\theta$

$\equiv \cos^5\theta - 10\cos^3\theta(1-\cos^2\theta)$

$+ 5\cos\theta(1-\cos^2\theta)^2$

$\equiv \cos^5\theta - 10\cos^3\theta + 10\cos^5\theta$

$+ 5\cos\theta - 10\cos^3\theta + 5\cos^5\theta$

$\equiv 16\cos^5\theta - 20\cos^3\theta + 5\cos\theta$
as required

b $\theta = \frac{\pi}{2}(2n-1)$ for $n \in \mathbb{Z}$

20 a $\omega^7 = \cos\left(\frac{2\pi}{7}\right) + i\sin\left(\frac{2\pi}{7}\right)$

$= \left(e^{\frac{2\pi i}{7}}\right)^7 = e^{2\pi i}$

$= \cos(2\pi) + i\sin(2\pi)$

$= 1$ as required

b $\omega^2, \omega^3, \omega^4, \omega^5, \omega^6$

c -1

21 a $\left(\frac{1}{2} + \frac{\sqrt{3}}{2}i - 1\right)^3 = \left(-\frac{1}{2} + \frac{\sqrt{3}}{2}i\right)^3$

$= \left(-\frac{1}{2}\right)^3 + 3\left(-\frac{1}{2}\right)^2\left(\frac{\sqrt{3}}{2}i\right)$

$+ 3\left(-\frac{1}{2}\right)\left(\frac{\sqrt{3}}{2}i\right)^2 + \left(\frac{\sqrt{3}}{2}i\right)^3$

$= -\frac{1}{8} + \frac{3\sqrt{3}}{8}i + \frac{9}{8} - \frac{3\sqrt{3}}{8}i$

$= 1$ as required

b $\frac{1}{2} - \frac{\sqrt{3}}{2}i$ $\quad 2$

c

Im; $\frac{\sqrt{3}}{2}$; z_1; z_3; Re; $\frac{1}{2}$; 2; $-\frac{\sqrt{3}}{2}$; z_2

d (1, 0), 1

22 a $\sqrt{2} + \sqrt{2}i$ $\quad \sqrt{2} - \sqrt{2}i$

$-\sqrt{2} - \sqrt{2}i$ $\quad -\sqrt{2} + \sqrt{2}i$

b 8 square units

23 a $\cos 5\theta + i\sin 5\theta = (\cos\theta + i\sin\theta)^5$

$= \cos^5\theta + 5i\cos^4\sin\theta - 10\cos^3\theta\sin^2\theta$

$- 10i\cos^2\theta\sin^3\theta + 5\cos\theta\sin^4\theta + i\sin^5\theta$

Imaginary parts give:

$\sin 5\theta = 5\cos^4\theta\sin\theta - 10\cos^2\theta\sin^3\theta + \sin^5\theta$

$= 5(1-\sin^2\theta)^2\sin\theta - 10\left(1-\sin^2\theta\right)\sin^3\theta + \sin^5\theta$

$= 5\sin\theta - 10\sin^3\theta + 5\sin^5\theta - 10\sin^3\theta$

$+ 10\sin^5\theta + \sin^5\theta$

$= 16\sin^5\theta - 20\sin^3\theta + 5\sin\theta$ as required

b $\sin\left(\frac{\pi}{10}\right), \sin\left(\frac{\pi}{2}\right), \sin\left(\frac{9\pi}{10}\right), \sin\left(\frac{13\pi}{10}\right), \sin\left(\frac{17\pi}{10}\right)$

24 a 3 $\quad \sqrt{3}i$ $\quad -\sqrt{3}i$

b

Im; $\sqrt{3}$; z_1; z_1; Re; 3; $-\sqrt{3}$; z_2

c $3\sqrt{3}$ square units

d $\frac{3}{2} + \frac{3\sqrt{3}}{2}i$ $\quad -\frac{3}{2} + \frac{\sqrt{3}}{2}i$

$\frac{3}{2} - \frac{\sqrt{3}}{2}i$

25 a $\left(\frac{\frac{1}{2}(1+i)-1}{\frac{1}{2}(1+i)}\right)^6 = \left(\frac{1+i-2}{1+i}\right)^6$

$= \left(\frac{i-1}{1+i}\right)^6$

$= \left(\frac{(i-1)(1-i)}{(1+i)(1-i)}\right)^6$

$= \left[-\frac{1}{2}(i-1)^2\right]^6$

$= \frac{1}{64}(i-1)^{12}$

$= \frac{1}{64}\left(\sqrt{2}\left(\cos\left(\frac{3\pi}{4}\right)+i\sin\left(\frac{3\pi}{4}\right)\right)\right)^{12}$

$= \frac{64}{64}\left(\cos\left(\frac{36\pi}{4}\right)+i\sin\left(\frac{36\pi}{4}\right)\right)$

$=(-1+0)$

$=-1$ as required

b $\frac{1}{2}+\frac{1}{2}i$

$\frac{1}{2}+\frac{2-\sqrt{3}}{2}i$

$\frac{1}{2}+\frac{-2+\sqrt{3}}{2}i$

$\frac{1}{2}-\frac{1}{2}i$

$\frac{1}{2}+\frac{-2-\sqrt{3}}{2}i$

c $\left(\frac{1}{4}\left(1+\sqrt{3}\right), \frac{1}{4}\left(-1+\sqrt{3}\right)\right)$

26 a $2e^{-\frac{\pi i}{9}}$ $\quad 2e^{\frac{5\pi i}{9}}$ $\quad 2e^{-\frac{7\pi i}{9}}$

b $3\sqrt{3}$ square units

c Sum of roots $= 0$

So sum of imaginary parts $= 0$

Therefore $\sin\left(-\frac{\pi}{9}\right)+\sin\left(\frac{5\pi}{9}\right)+\sin\left(-\frac{7\pi}{9}\right)=0$

$\Rightarrow -\sin\left(\frac{\pi}{9}\right)+\sin\left(\frac{5\pi}{9}\right)-\sin\left(\frac{7\pi}{9}\right)=0$

$\Rightarrow \sin\left(\frac{\pi}{9}\right)-\sin\left(\frac{5\pi}{9}\right)+\sin\left(\frac{7\pi}{9}\right)=0$

27 First consider the case when n is a positive integer.

Let $n=1$ then $(\cos\theta+i\sin\theta)^1=\cos(1\theta)+i\sin(1\theta)$ so true for $n=1$

Assume true for $n=k$ and consider when $n=k+1$

$(\cos\theta+i\sin\theta)^{k+1}=(\cos\theta+i\sin\theta)^k(\cos\theta+i\sin\theta)$

$=(\cos(k\theta)+i\sin(k\theta))(\cos\theta+i\sin\theta)$

$=\cos(k\theta)\cos\theta-\sin(k\theta)\sin\theta$

$\quad +i(\cos(k\theta)\sin\theta+\sin(k\theta)\cos\theta)$

$=\cos((k+1)\theta)+i\sin((k+1)\theta)$

So true for $n=k+1$

Hence, since true for $n=1$ and by assuming it is true for $n=k$ we have proved it is true for $n=k+1$, therefore it is true for all positive integers n.

When $n=0$, $(\cos\theta+i\sin\theta)^0=1$ and $\cos(0)+i\sin(0)=1$ so true for $n=0$

Consider when n is a negative integer, it can be written as $n=-m$ where m is a positive integer.

So $(\cos\theta+i\sin\theta)^n=(\cos\theta+i\sin\theta)^{-m}$

$=\frac{1}{(\cos\theta+i\sin\theta)^m}$

$=\frac{1}{(\cos(m\theta)+i\sin(m\theta))}$

since m is a positive integer and we have just proved de Moivre's holds for positive integers.

$=\frac{\cos(m\theta)-i\sin(m\theta)}{(\cos(m\theta)+i\sin(m\theta))(\cos(m\theta)-i\sin(m\theta))}$

$=\frac{\cos(m\theta)-i\sin(m\theta)}{\cos^2(m\theta)+\sin^2(m\theta)}$

$=\cos(m\theta)-i\sin(m\theta)$

$=\cos(-m\theta)+i\sin(-m\theta)$

$=\cos(n\theta)+i\sin(n\theta)$ as required

So de Moivre's theorem holds for all integers n

Chapter 17

Exercise 17.1A

1 a $\frac{2}{x+5}+\frac{4}{x-5}$ **b** $\frac{3}{x-3}-\frac{7}{x-7}$

c $\frac{9}{1-x}+\frac{2}{x+6}$ **d** $\frac{2}{x+2}+\frac{5}{x-2}-\frac{7}{x+3}$

2 a $f(x)-f(x+1)\equiv\frac{1}{x(x-1)}-\frac{1}{x(x+1)}\equiv\frac{(x+1)-(x-1)}{x(x-1)(x+1)}$

$\equiv\frac{2}{x(x-1)(x+1)}$

b $\frac{1}{2}\left[\frac{n^2+n-2}{2n(n+1)}\right]\equiv\left[\frac{(n+2)(n-1)}{4n(n+1)}\right]$

3 a $\left[\frac{5n^2+13n}{6(n+2)(n+3)}\right]$ **b** $\frac{2n}{(2n+1)}$ **c** $\left[\frac{8(n+1)(n+3)}{3(2n+3)(2n+5))}\right]$

4 a $\frac{n-1}{n}$ **b** $\frac{(n-1)}{2(n+1)}$

c $\frac{5n^2+13n}{12(n+2)(n+3)}$ **d** $\frac{(3n+2)(n-1)}{4n(n+1)}$

Exercise 17.1B

1 a $r(r+1)(r+2)-(r-1)r(r+1)$

$\equiv(r^3+3r^2+2r)-(r^3-r)\equiv 3r^2+3r\equiv 3r(r+1)$

b $\frac{n(n+1)(n+2)}{3}$

2 a $\frac{1}{2}\left[\frac{1}{2r-1}-\frac{1}{2r+1}\right]$ **b** $\frac{n}{(2n+1)}$ **c** $\frac{1}{2}$

3 $f(r)=\frac{1}{(r+1)(r+2)}$

a $f(r) - f(r+1) \equiv \frac{1}{(r+1)(r+2)} - \frac{1}{(r+2)(r+3)}$

$$\equiv \frac{(r+3)-(r+1)}{(r+1)(r+2)(r+3)} \equiv \frac{2}{(r+1)(r+2)(r+3)}$$

b $\frac{n(n+5)}{12(n+2)(n+3)}$ **c** $\frac{1}{12}$

4 **a** $4r^3$ **b** $\frac{1}{4}\left[n^2(n+1)^2\right]$

5 **a** $\frac{(3n+2)(n-1)}{n(n+1)}$ **b** $\frac{3}{4}$

6 $\frac{2r-1}{r(r+1)(r+2)} \equiv \frac{A}{r} + \frac{B}{(r+1)} + \frac{C}{(r+2)}$

$\Rightarrow 2r - 1 \equiv A(r+1)(r+2) + B(r)(r+2) + C(r)(r+1)$

When $r = 0, -1 = 2A$, so $A = -\frac{1}{2}$

When $r = -1, -3 = -B$, so $B = 3$

When $r = -2, -5 = 2C$, so $C = -\frac{5}{2}$

Hence $\frac{2r-1}{r(r+1)(r+2)} \equiv \frac{1}{2}\left[\frac{-1}{r} + \frac{6}{(r+1)} - \frac{5}{(r+2)}\right]$

$$\sum_1^n \frac{1}{2}\left[\frac{-1}{r} + \frac{6}{(r+1)} - \frac{5}{(r+2)}\right]$$

$$\equiv \frac{1}{2}\left[\frac{-1}{1} + \frac{6}{2} - \frac{5}{3}\right.$$
$$\frac{-1}{2} + \frac{6}{3} - \frac{5}{4}$$
$$\frac{-1}{3} + \frac{6}{4} - \frac{5}{5}$$
$$\frac{-1}{4} + \frac{6}{5} - \frac{5}{6}$$
$$- \dots$$
$$\frac{-1}{r-2} + \frac{6}{(r-1)} - \frac{5}{(r)}$$
$$\frac{-1}{r-1} + \frac{6}{(r)} - \frac{5}{(r+1)}$$
$$\left.\frac{-1}{r} + \frac{6}{(r+1)} - \frac{5}{(r+2)}\right]$$

Hence $\sum_1^n \frac{1}{2}\left[\frac{-1}{r} + \frac{6}{(r+1)} - \frac{5}{(r+2)}\right]$

$$\equiv \frac{1}{2}\left[-1 + 3 - \frac{1}{2} - \frac{5}{(r+1)} + \frac{6}{(r+1)} - \frac{5}{(r+2)}\right]$$

$$\equiv \frac{1}{2}\left[\frac{3}{2} + \frac{1}{(r+1)} - \frac{5}{(r+2)}\right]$$

Hence $\sum_1^\infty \frac{1}{2}\left[\frac{-1}{r} + \frac{6}{(r+1)} - \frac{5}{(r+2)}\right] = \frac{3}{4} + \frac{1}{\infty} - \frac{5}{\infty} = \frac{3}{4}$

7 **a** $\frac{x-1}{x} - \frac{x-2}{x-1} \equiv \frac{(x-1)^2 - x(x-2)}{x(x-1)}$

$$\equiv \frac{(x^2 - 2x + 1 - x^2 + 2x)}{x(x-1)} \equiv \frac{1}{x(x-1)}$$

b $\frac{n-2}{2n}$ **c** $\frac{1}{2}$

8 $\ln(n)$

9 **a** $\frac{-1}{(x-1)} + \frac{3}{(x)} - \frac{2}{(x+1)}$ **b** $\frac{1}{n} - \frac{2}{(n+1)}$

c $\sum_2^\infty \left[\left(\frac{-1}{(x-1)} + \frac{3}{(x)} - \frac{2}{(x+1)}\right)\right] = \lim_{n\to\infty} \frac{1}{n} - \frac{2}{(n+1)} = 0$

10 $\frac{1}{r(r+1)(r+2)(r+3)} \equiv \frac{A}{(r)} + \frac{B}{(r+1)} + \frac{C}{(r+2)} + \frac{D}{(r+3)}$

$\Rightarrow 1 \equiv A(r+1)(r+2)(r+3) + B(r)(r+2)(r+3) + C(r)(r+1)(r+3)$
$+ D(r)(r+1)(r+2)$

When $r = 0, 1 = 6A$ so $A = \frac{1}{6}$; When $r = -1, 1 = -2B$ so $B = -\frac{1}{2}$

When $r = -2, 1 = 2C$ so $C = \frac{1}{2}$; When $r = -3, 1 = -6D$ so $D = -\frac{1}{6}$

Hence $\frac{1}{r(r+1)(r+2)(r+3)} \equiv \frac{1}{(6r)} - \frac{1}{2(r+1)} + \frac{1}{2(r+2)} - \frac{1}{6(r+3)}$

Hence $\sum_1^n \frac{1}{r(r+1)(r+2)(r+3)} \equiv$

$$\frac{1}{6}\sum_1^n \frac{1}{(r)} - \frac{3}{(r+1)} + \frac{3}{(r+2)} - \frac{1}{(r+3)}$$

$$\equiv \frac{1}{6}\left[\frac{1}{1} - \frac{3}{2} + \frac{3}{3} - \frac{1}{4}\right.$$
$$+ \frac{1}{2} - \frac{3}{3} + \frac{3}{4} - \frac{1}{5}$$
$$+ \frac{1}{3} - \frac{3}{4} + \frac{3}{5} - \frac{1}{6}$$
$$+ \frac{1}{4} - \frac{3}{5} + \frac{3}{6} - \frac{1}{7}$$
$$+ \frac{1}{5} - \frac{3}{6} + \frac{3}{7} - \frac{1}{8}$$
$$+ \dots$$
$$+ \frac{1}{(n-3)} - \frac{3}{(n-2)} + \frac{3}{(n-1)} - \frac{1}{(n)}$$
$$+ \frac{1}{(n-2)} - \frac{3}{(n-1)} + \frac{3}{(n)} - \frac{1}{(n+1)}$$
$$+ \frac{1}{(n-1)} - \frac{3}{(n)} + \frac{3}{(n+1)} - \frac{1}{(n+2)}$$
$$\left.+ \frac{1}{(n)} - \frac{3}{(n+1)} + \frac{3}{(n+2)} - \frac{1}{(n+3)}\right]$$

Hence $\sum_1^n \frac{1}{r(r+1)(r+2)(r+3)}$

$$\equiv \frac{1}{6}\left[\frac{1}{1} - \frac{3}{2} + \frac{3}{3} + \frac{1}{2} - \frac{3}{3} + \frac{1}{3} - \frac{1}{(n+1)} + \frac{3}{(n+1)} - \frac{1}{(n+2)}\right.$$
$$\left.- \frac{3}{(n+1)} + \frac{3}{(n+2)} - \frac{1}{(n+3)}\right]$$

$$\equiv \frac{1}{6}\left[\frac{1}{3} - \frac{1}{(n+1)} + \frac{2}{(n+2)} - \frac{1}{(n+3)}\right]$$

Hence $\sum_1^\infty \frac{1}{r(r+1)(r+2)(r+3)} = \frac{1}{6}\left[\frac{1}{3} - \frac{1}{(\infty+1)} + \frac{2}{(\infty+2)}\right.$
$$\left.- \frac{1}{(\infty+3)}\right] = \frac{1}{18}$$

11 a $\frac{2}{3(1+2x)}+\frac{1}{3(1-x)}$

b $1-\frac{1}{2}x+\frac{19}{8}x^2-\frac{53}{16}x^3+\ldots$

Exercise 17.2A

1 a $\frac{(-x)^r}{r!}$ or $\frac{(-1)^r x^r}{r!}$

b $(-1)^{r+1}\frac{(-2x)^r}{r}$ or $-\frac{2^r x^r}{r}$

c $(-1)^r\frac{\left(\frac{x}{5}\right)^{2r+1}}{(2r+1)!}$ or $(-1)^r\frac{x^{2r+1}}{5^{2r+1}(2r+1)!}$

d $\frac{n(n-1)\ldots(n-1+r)(3x)^r}{r!}$ or $\frac{n(n-1)\ldots(n-1+r)(3^r)(x)^r}{r!}$

2 $\sin x \equiv x-\frac{x^3}{3!}+\frac{x^5}{5!}-\frac{x^7}{7!}+\ldots$

The general term is $(-1)^r\frac{x^{2r+1}}{(2r+1)!}$

3 $\cos x \equiv 1-\frac{x^2}{2!}+\frac{x^4}{4!}-\frac{x^6}{6!}+\ldots$

The general term is $(-1)^r\frac{x^{2r}}{(2r)!}$

4 $-4x-\frac{16x^3}{3}-\frac{64x^5}{5}-\ldots$

5 $1-x^2-\frac{x^4}{3}-\ldots$

6 $1-\frac{3x^2}{2}-\frac{9x^4}{8}-\ldots$

7 $x+2x^2+\frac{11x^3}{6}-\ldots$

8 $x+\frac{x^2}{2}-\frac{x^3}{3}-\ldots$

9 a 1 **b** −1 **c** 3 **d** $\frac{1}{2}$

e $-\frac{1}{4}$ **f** $\frac{1}{3}$ **g** −1 **h** −11

10 a 1.5 **b** $\frac{1}{3}$ **c** $-\frac{5}{8}$

d 2 **e** $\frac{\sqrt{2}}{2}$

11 a 1 **b** 1 **c** 0 **d** 1

e 2 **f** $\frac{29}{12}$ **g** $\frac{5}{2}$

Exercise 17.2B

1 a $4-16x+64x^2-256x^3+\ldots$

b $4x-8x^2+\frac{64}{3}x^3-64x^4+\ldots$

c $x-\frac{x^2}{2}+\frac{x^3}{3}-\frac{x^4}{4}+\frac{x^5}{5}\ldots$

So $\ln(1+4x)=4x-\frac{16x^2}{2}+\frac{64x^3}{3}-\frac{256x^4}{4}+\ldots$

$\equiv 4x-8x^2+\frac{64}{3}x^3-64x^4+\ldots$

2 a −6 **b** $\frac{1}{2}$ **c** 8 **d** −1

3 a 2 **b** 0 **c** 0

d $-\frac{1}{5}$ **e** $\frac{1}{4}$

4 a $x+\frac{x^2}{2}+\frac{5x^3}{6}+\frac{7x^4}{12}+\ldots$

b i $\sqrt{2}\left(1-\frac{9x}{4}+\frac{143x^2}{32}-\frac{1145x^3}{128}+\ldots\right)$

ii $\sqrt{2}$

When $x=0$, $\sqrt{2}\left(1-\frac{9x}{4}+\frac{9x^2}{2}-\frac{1145x^3}{128}+\ldots\right)$

$=\sqrt{2}\left(1-\frac{0}{4}+\frac{0}{2}-\frac{0}{128}+\ldots\right)=\sqrt{2}$

5 a $\frac{9}{(x+1)}-\frac{15}{(x+2)}$

b $\frac{3}{2}-\frac{21x}{4}+\frac{57x^2}{8}-\frac{129x^3}{16}+\ldots$

c $\frac{3}{2}$

When $x=0$, $\frac{3}{2}-\frac{21x}{4}+\frac{57x^2}{8}-\frac{129x^3}{16}+\ldots=\frac{3}{2}$

6 a −1

b $\lim_{x\to 0}\frac{e^x-e^{2x}}{x}=\frac{0}{0}\Rightarrow \lim_{x\to 0}\frac{e^x-2e^{2x}}{1}=-1$

7 a $e^2(2-3x+2x^2-\ldots)$; $2e^2$ **b** $2e^2$

8 a $1-\frac{x}{2}+\frac{x^2}{12}+\ldots$ 1

b $\lim_{x\to 0}\frac{x}{e^x-1}=\frac{0}{0}\Rightarrow \lim_{x\to 0}\frac{1}{e^x}=1$

9 a $2-x+\frac{x^2}{2}+\ldots$

b $x+\frac{9x^2}{8}$ **c** $\frac{1}{4}$

10 a $2x-\frac{4x^3}{3}+\frac{4x^5}{15}$

b $-\frac{2x^2}{3}-\frac{4x^4}{45}$ **c** $-\frac{2}{3}$

11 3

12 a $\ln 2+\frac{2}{3}x+\frac{3}{8}x^2+\frac{3}{8}x^3+\ldots$

b $\ln 2$

When $x=0$, $\ln 2+\frac{3}{2}x+\frac{3}{8}x^2+\frac{3}{8}x^3+\ldots=\ln 2$

13 a $\ln(\cos x)\equiv\ln(1+(\cos x-1))$

$\equiv\ln(1+[1-\frac{x^2}{2!}+\frac{x^4}{4!}-\frac{x^6}{6!}\ldots-1])$

$\equiv\ln\left(1+\left[-\frac{x^2}{2}+\frac{x^4}{24}-\frac{x^6}{6!}\right]\right)$

$\equiv\left[-\frac{x^2}{2}+\frac{x^4}{24}-\frac{x^6}{6!}\right]-\frac{1}{2}\left[-\frac{x^2}{2}+\frac{x^4}{24}-\frac{x^6}{6!}\right]^2$

$+\frac{1}{3}\left[-\frac{x^2}{2}+\frac{x^4}{24}-\frac{x^6}{6!}\right]^3+\ldots$

$\equiv-\frac{x^2}{2}+\frac{x^4}{24}-\frac{x^6}{6!}-\frac{1}{2}\left[\frac{x^4}{4}-\frac{x^6}{24}+\ldots\right]$

$+\frac{1}{3}\left[-\frac{x^6}{8}+\ldots\right]+\ldots$

$\equiv-\frac{x^2}{2}+\frac{x^4}{24}-\frac{x^4}{8}-\frac{x^6}{720}+\frac{x^6}{48}-\frac{x^6}{24}+\ldots$

$\equiv-\frac{x^2}{2}-\frac{x^4}{12}-\frac{x^6}{45}+\ldots$

b $\frac{1}{2}$

14 0

Review exercise 17

1 $\left(\frac{5}{(2x+9)}-\frac{4}{(3x-1)}\right)$

2 a $\frac{4}{x(x+4)} \equiv \frac{1}{x}-\frac{1}{x+4}$

b $\frac{25}{12}-\left[\frac{1}{n+1}+\frac{1}{n+2}+\frac{1}{n+3}+\frac{1}{n+4}\right]; \frac{25}{12}$

3 a $\frac{1}{r^2}-\frac{2}{(r+1)^2}+\frac{1}{(r+2)^2}$

$=\frac{(r+1)^2(r+2)^2-2r^2(r+2)^2+r^2(r+1)^2}{r^2(r+1)^2(r+2)^2}$

$=\frac{(r^2+2r+1)(r^2+4r+4)-2r^2(r^2+4r+4)+r^2(r^2+2r+1)}{r^2(r+1)^2(r+2)^2}$

$=\frac{6r^2+12r+4}{r^2(r+1)^2(r+2)^2}$

$=\frac{2(3r^2+6r+2)}{r^2(r+1)^2(r+2)^2}$

b $\frac{3}{4}+\frac{1}{(n+2)^2}-\frac{1}{(n+1)^2}$ c $\frac{3}{4}$

4 $2x+\frac{5x^3}{3}-2x^4\ldots$

5 a $\frac{x^r}{3^r r!}$ b $((-1)^r)\frac{x^{2r}}{r!}$

c $((-1)^r)\frac{(x)^{2r}}{(3)^{2r}(2r)!}$

d $((-1)^r)\frac{(4x+5)^{(2r+1)}}{(2r+1)!}$

e $\frac{n(n-1)(n-2)(n-3)\ldots(n-r+1)}{r!}\left(\frac{-x}{6}\right)^r$

f $\frac{x^{(r+1)}}{r!}$

6 a $-\sqrt{2}<x<\sqrt{2}$ b $-2\le x\le 2$

7 a $-6<x<6$ b $-\frac{1}{3}\le x<\frac{1}{3}$ c $-\frac{1}{3}\le x<\frac{1}{3}$

8 a $x+2x^2+\frac{11x^3}{6}+\ldots$

b $4-8x+16x^2+\ldots$

c $\frac{4}{\ln 2}+x\left(\frac{-2}{(\ln 2)^2}\right)+\frac{x^2}{2!}\left(\frac{1}{(\ln 2)^2}+\frac{2}{(\ln 2)^3}\right)+\ldots$

d $1-2x-2x^2+\ldots$

e $1+x\ln 3+\frac{(x\ln 3)^2}{2!}+\frac{(x\ln 3)^3}{3!}+\ldots$

f $\frac{2}{5}-\frac{2x}{25}-\frac{23x^2}{125}+\ldots$

9 a $\frac{4}{3}$ b ∞ c 4 d $\frac{1}{4}$

e 0 f $\frac{64}{3}$ g 2

Assessment 17

1 a $\frac{n}{2(n+2)}$ b 50

2 e.g. $(r+1)^3-(r-1)^3$

$=r^3+3r^2+3r+1-(r^3-3r^2+3r-1)$

$=6r^2+2$

$\sum_{r=1}^{n}(6r^2+2)=2^3-0^3$

$+3^3-1^3$

$+4^3-2^3$

$+5^3-3^3$

$+\ldots$

$+n^3-(n-2)^3$

$+(n+1)^3-(n-1)^3$

$=-1+n^3+(n+1)^3$

$=-1+n^3+n^3+3n^2+3n+1$

$=2n^3+3n^2+3n$

$=n(2n^2+3n+3)$

$\sum_{r=1}^{n}6r^2+2=6\sum_{r=1}^{n}r^2+\sum_{r=1}^{n}2$

Therefore, $\sum_{r=1}^{n}r^2=\frac{n(2n^2+3n+3)-2n}{6}$

$=\frac{n}{6}(2n^2+3n+3-2)$

$=\frac{n}{6}(2n^2+3n+1)$

$=\frac{n}{6}(n+1)(2n+1)$ as required

3 a $2x^2-\frac{4}{3}x^6+\frac{4}{15}x^{10}-\frac{8}{315}x^{14}+\ldots$

b Valid for all real values of x

c $\frac{1}{2}$

4 a $-5x-\frac{25}{2}x^2-\frac{125}{3}x^3-\frac{625}{4}x^4-\ldots$

b $-\frac{1}{5}\le x<\frac{1}{5}$ c $-\frac{1}{5}$

5 a $x+x^2+\frac{1}{3}x^3-\frac{1}{30}x^5+\ldots$

b 0.398 919

6 a $\frac{3n}{n+1}$

b $\sum_{r=n}^{2n}\frac{3}{r(r+1)}=\frac{3(2n)}{2n+1}-\frac{3(n-1)}{n-1+1}$

$=\frac{6n}{2n+1}-\frac{3(n-1)}{n}$

$=\frac{6n^2-3(n-1)(2n+1)}{n(2n+1)}$

$=\frac{6n^2-6n^2+3n+3}{n(2n+1)}$

$=\frac{3n+3}{n(2n+1)}$

$=\frac{3(n+1)}{n(2n+1)}$

c $\frac{30}{31}$

7 a $\sum_{r=2}^{n}\frac{2}{r^2-1}=1-\frac{1}{3}$

$+\frac{1}{2}-\frac{1}{4}$

$+\frac{1}{3}-\frac{1}{5}$

$+\frac{1}{4}-\frac{1}{6}$

$+\ldots$

$+\frac{1}{n-1-1}-\frac{1}{n-1+1}$

$+\frac{1}{n-1}-\frac{1}{n+1}$

$=1+\frac{1}{2}-\frac{1}{n}-\frac{1}{n+1}=\frac{3}{2}-\frac{n+1+n}{n(n+1)}$

$=\frac{3(n^2+n)-2(2n+1)}{2n(n+1)}$

$=\frac{3n^2-n-2}{2n(n+1)}=\frac{(3n+2)(n-1)}{2n(n+1)}$

So $\sum_{r=2}^{n}\frac{1}{r^2-1}=\frac{(3n+2)(n-1)}{2n(n+1)}\div 2$

$=\frac{(3n+2)(n-1)}{4n(n+1)}$ as required

b $\sum_{r=2}^{\infty}\frac{1}{r^2-1}=\lim_{n\to\infty}\frac{(3n+2)(n-1)}{4n(n+1)}$

$=\lim_{n\to\infty}\frac{3n^2-n-2}{4(n^2+n)}$

$=\lim_{n\to\infty}\frac{3-\frac{1}{n}-\frac{2}{n^2}}{4\left(1+\frac{1}{n}\right)}=\frac{3}{4}$

8 a $f(x)=\cos 3x \quad f(0)=1$

$f'(x)=-3\sin 3x \quad f'(0)=0$

$f''(x)=-9\cos x \quad f''(0)=-9$

$f'''(x)=27\sin x \quad f'''(0)=0$

$f^{iv}(x)=81\cos x \quad f^{iv}(0)=81$

So $\cos 3x=1+0x-\frac{9}{2!}x^2+0x^3+\frac{81}{4!}x^4+\ldots$

$=1-\frac{9}{2}x^2+\frac{27}{8}x^4-\ldots$

b $\frac{(-1)^{r-1}(3x)^{2r-2}}{(2r-2)!}$

9 a $1-\frac{3}{2}x-\frac{9}{8}x^2-\frac{27}{16}x^3+\ldots$ b $\frac{4}{3}$

10 $1+\frac{1}{2}x^2$

11 $1+2x+2x^2+\ldots$

12 a $\frac{2}{3}$

b $\lim_{x\to 0}2x=0$ and $\lim_{x\to 0}e^{3x}-1=1-1=0$

13 a $\frac{1}{3}$ b 2

14 a $\frac{1}{r!}-\frac{1}{(r+1)!}=\frac{(r+1)!-r!}{r!(r+1)!}$

$=\frac{r!(r+1)-r!}{r!(r+1)!}$

$=\frac{r!(r+1-1)}{r!(r+1)!}$

$=\frac{r}{(r+1)!}$ as required

b $\sum_{r=1}^{n}\frac{r}{(r+1)!}=1-\frac{1}{2!}$

$+\frac{1}{2!}-\frac{1}{3!}$

$+\frac{1}{3!}-\frac{1}{4!}$

$+\ldots$

$+\frac{1}{n!}-\frac{1}{(n+1)!}$

$=1-\frac{1}{(n+1)!}$

c $\frac{(2n+1)!-n!}{n!(2n+1)!}$

15 a $\frac{1}{r+2}-\frac{2}{r+1}+\frac{1}{r}$

b $\sum_{r=1}^{n}\frac{2}{r(r+1)(r+2)}=\frac{1}{3}-1+1$

$+\frac{1}{4}-\frac{2}{3}+\frac{1}{2}$

$+\frac{1}{5}-\frac{2}{4}+\frac{1}{3}$

$+\frac{1}{6}-\frac{2}{5}+\frac{1}{4}$

$+\ldots$

$+\frac{1}{n+1}-\frac{2}{n}+\frac{1}{n-1}$

$+\frac{1}{n+2}-\frac{2}{n+1}+\frac{1}{n}$

$=-1+1+\frac{1}{2}+\frac{1}{n+1}+\frac{1}{n+2}-\frac{2}{n+1}$

$=\frac{1}{2}+\frac{1}{n+2}-\frac{1}{n+1}$

$=\frac{(n+2)(n+1)+2(n+1)-2(n+2)}{2(n+2)(n+1)}$

$=\frac{n^2+3n+2+2n+2-2n-4}{2(n+2)(n+1)}$

$=\frac{n^2+3n}{2(n+2)(n+1)}$

$=\frac{n(n+3)}{2(n+2)(n+1)}$

So $\sum_{r=1}^{n}\frac{1}{r(r+1)(r+2)}=\frac{n(n+3)}{4(n+2)(n+1)}$ as required

c $=\frac{1}{4}$

16 $\lim_{x\to 0}\dfrac{3-3\cos x}{2x^2}=\lim_{x\to 0}\dfrac{3\sin x}{4x}$

Applicable since $\lim_{x\to 0}(3-3\cos x)=\lim_{x\to 0}2x^2=0$

$$=\lim_{x\to 0}\frac{3\cos x}{4}$$

Applicable since $\lim_{x\to 0}(3\sin x)=\lim_{x\to 0}4x=0=\dfrac{3}{4}$

17 $\lim_{x\to 0}x^2\ln(x^2)=\lim_{x\to 0}\dfrac{\ln(x^2)}{\dfrac{1}{x^2}}$

$\lim_{x\to 0}\ln(x^2)=-\infty$ and $\lim_{x\to 0}\dfrac{1}{x^2}=\infty$ so possible to use l'Hopital's rule

$$\lim_{x\to 0}\frac{\ln(x^2)}{\dfrac{1}{x^2}}=\lim_{x\to 0}\frac{\left(\dfrac{2x}{x^2}\right)}{\left(-\dfrac{2}{x^3}\right)}$$

$$=\lim_{x\to 0}-x^2$$

$$=0$$

18 $-\dfrac{5}{6}$

19 -1

20 $\dfrac{1}{r(r+1)(r+2)}=\dfrac{A}{r}+\dfrac{B}{r+1}+\dfrac{C}{r+2}$

$1=A(r+1)(r+2)+Br(r+2)+Cr(r+1)$

$r=-1\Rightarrow 1=-B\Rightarrow B=-1$

$r=0\Rightarrow 1=2A\Rightarrow A=\dfrac{1}{2}$

$r=-2\Rightarrow 1=2C\Rightarrow C=\dfrac{1}{2}$

Multiply through by 2

$\dfrac{2}{r(r+1)(r+2)}=\dfrac{1}{r}-\dfrac{2}{r+1}+\dfrac{1}{r+2}$

$$\sum_{r=1}^{n}\left(\frac{1}{r}-\frac{2}{r+1}+\frac{1}{r+2}\right)$$

$$=\cancel{1}-\cancel{1}+\cancel{\tfrac{1}{3}}$$
$$+\frac{1}{2}-\cancel{\tfrac{2}{3}}+\cancel{\tfrac{1}{4}}$$
$$+\cancel{\tfrac{1}{3}}-\cancel{\tfrac{2}{4}}+\cancel{\tfrac{1}{5}}$$
$$+\cancel{\tfrac{1}{4}}-\cancel{\tfrac{2}{5}}+\cancel{\tfrac{1}{6}}$$
$$+\ldots$$
$$+\cancel{\tfrac{1}{n-1}}-\cancel{\tfrac{2}{n}}+\frac{1}{n+1}$$
$$+\cancel{\tfrac{1}{n}}-\frac{2}{n+1}+\frac{1}{n+2}$$
$$=\frac{1}{2}-\frac{1}{n+1}+\frac{1}{n+2}$$
$$=\frac{n^2+3n}{2n^2+6n+4}$$

$$=\frac{1+\dfrac{3}{n}}{2+\dfrac{6}{n}+\dfrac{4}{n^2}}\to\frac{1}{2}\text{ as }n\to\infty$$

$$\sum_{r=1}^{\infty}\frac{1}{r(r+1)(r+2)}=\frac{1}{2}\div 2=\frac{1}{4}$$

21 $\dfrac{1}{5}$

Chapter 18

Exercise 18.1A

1 In each case the graph of f(x) is blue and either $\dfrac{1}{f(x)}$ or $|f(x)|$ is red.

a i

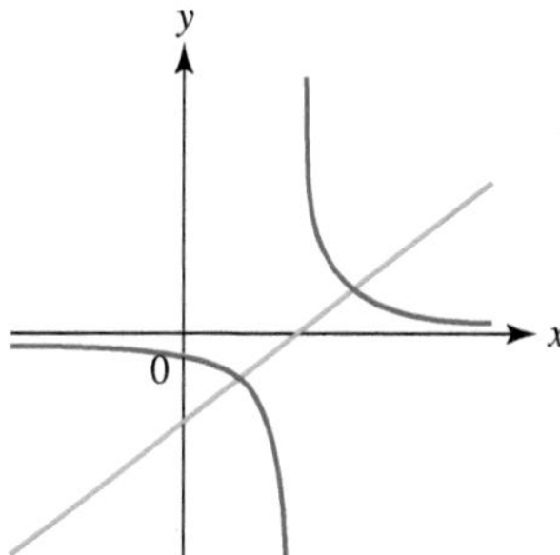

ii

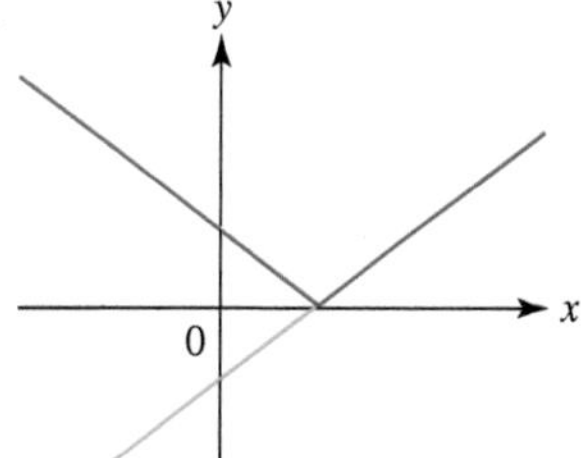

b i

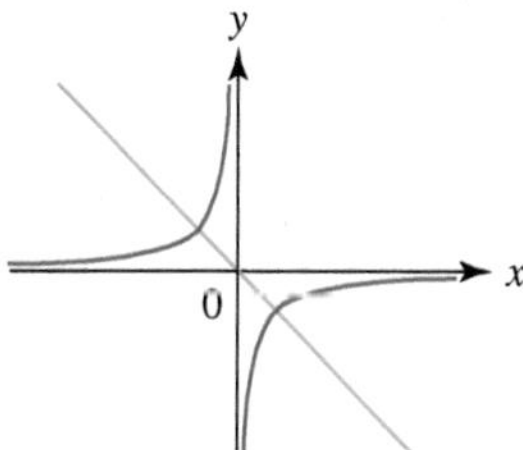

ii

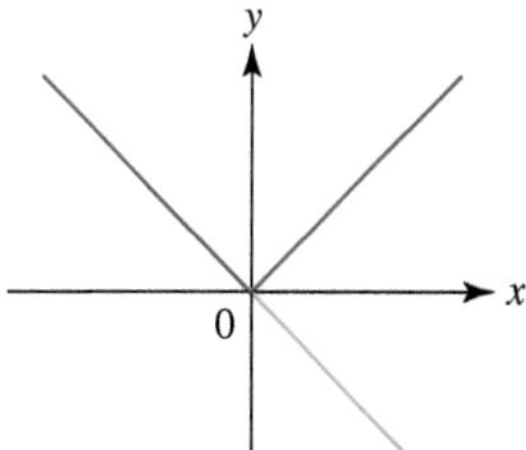

c i

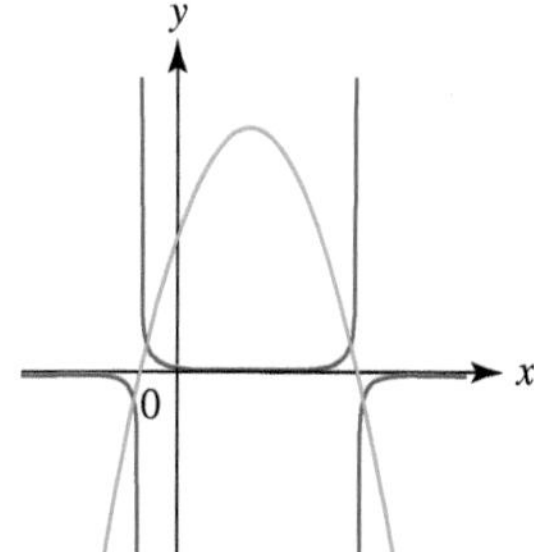

ii

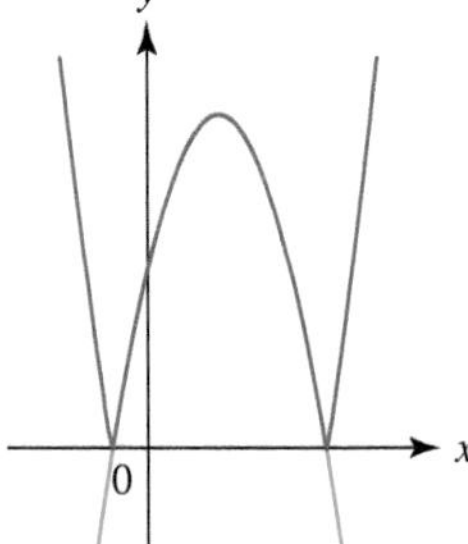

d i

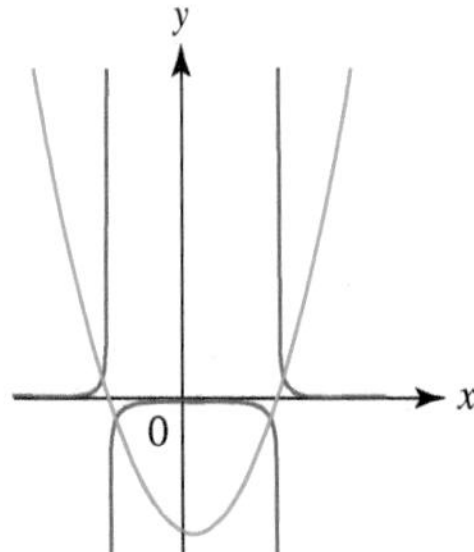

ii

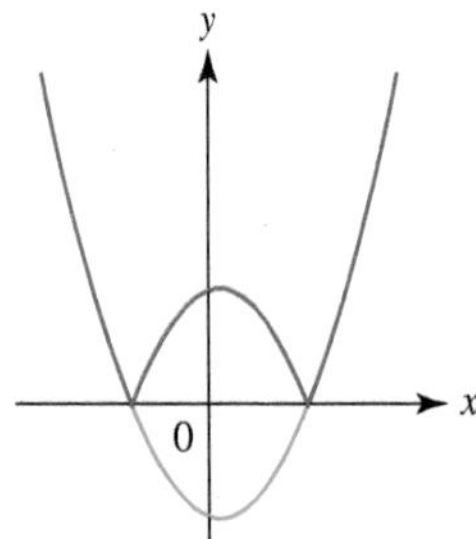

2 In each case the graph of f(x) is blue and $\frac{1}{f(x)}$ is red.

a

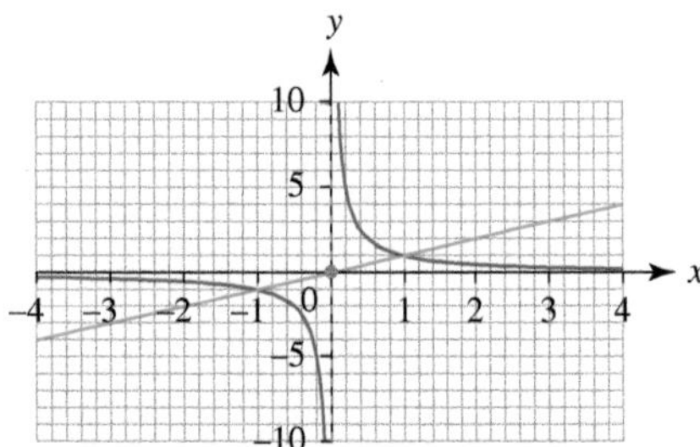

Asymptotes at $x = 0$ and at $y = 0$
Intersections at (−1, −1) and (1, 1)

b

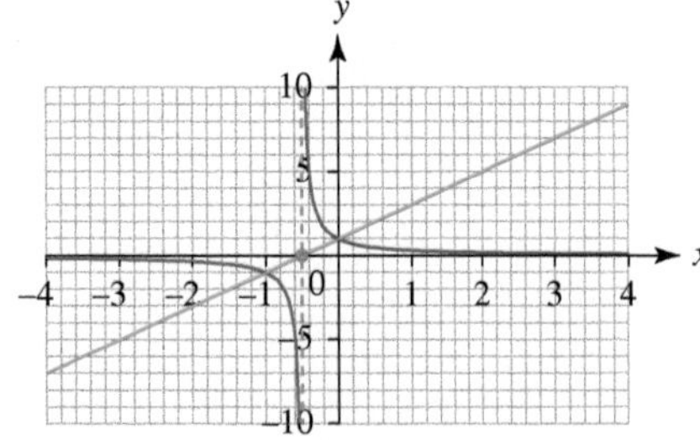

Asymptotes at $x = -\frac{1}{2}$ and at $y = 0$
Intersections at (−1, −1) and (0, 1)

c

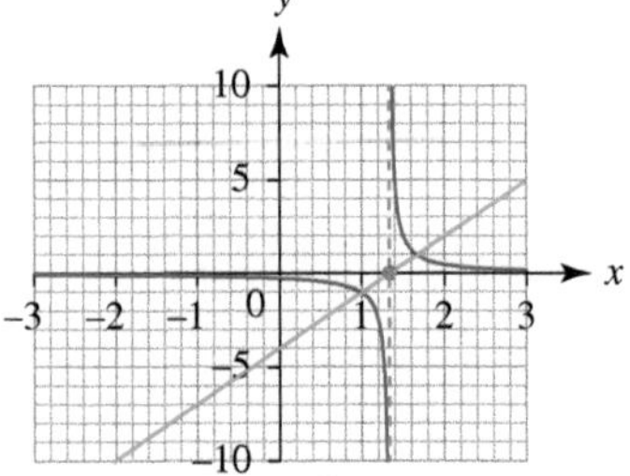

Asymptotes at $x = \frac{4}{3}$ and at $y = 0$
Intersections at (1, −1) and $(\frac{5}{3}, 1)$

d

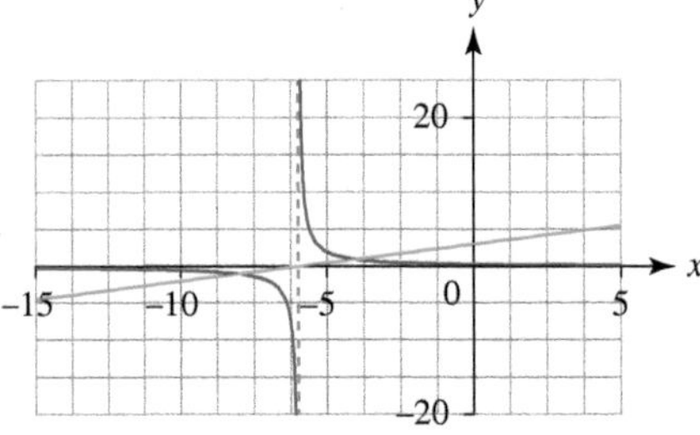

Asymptotes at $x = -6$ and at $y = 0$
Intersections at (−4, 1) and (−8, −1)

e

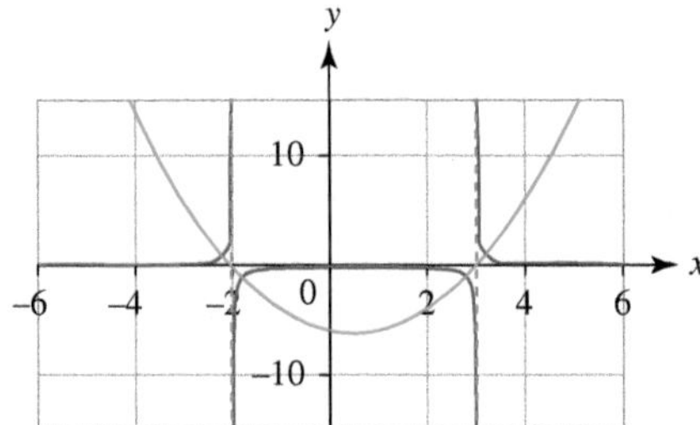

Asymptotes at $x = -2$, $x = 3$ and at $y = 0$
Intersections at (−2.193, 1) and (−1.791, −1) and (2.791, −1) and (3.193,1)

f

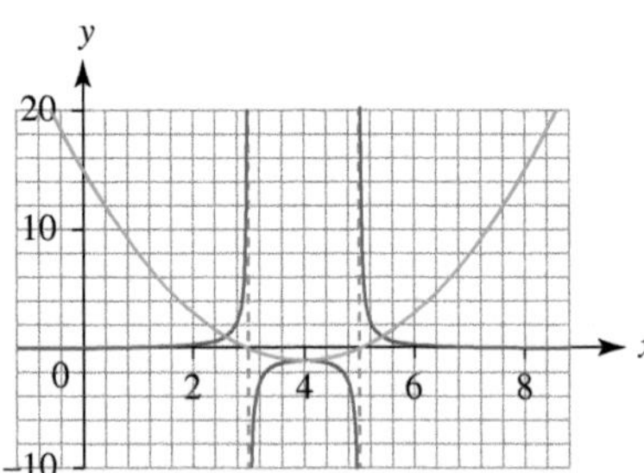

Asymptotes at $x = 3$, $x = 5$ and at $y = 0$
Intersections at (2.586, 1) and (4, −1) and (5.414, 1)

g

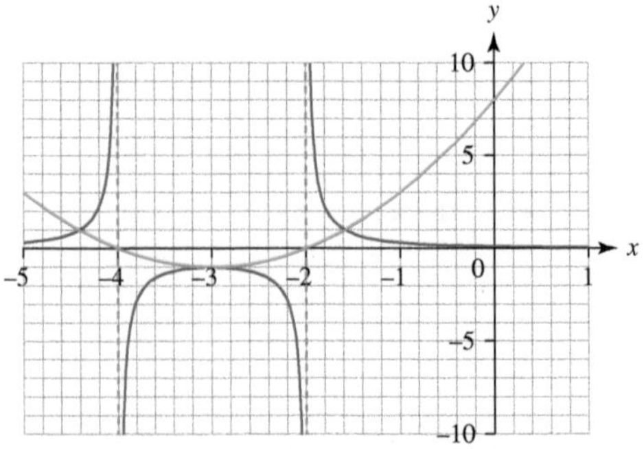

Asymptotes at $x = -4$, $x = -2$ and at $y = 0$
Intersections at (−4.414, 1) and (−3, −1) and (−1.586, 1)

h

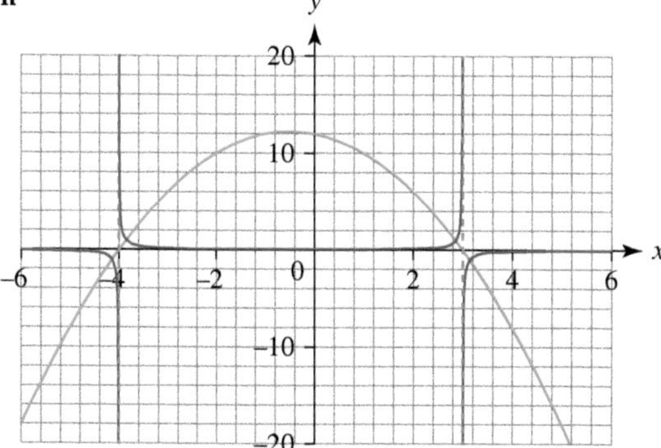

Asymptotes at $x = -4$, $x = 3$ and at $y = 0$
Intersections at $(-4.14, -1)$ and $(-3.854, 1)$ and $(2.854, 1)$ and $(3.14, -1)$

3 In each case the graph of $f(x)$ is blue and $|f(x)|$ is red.

a

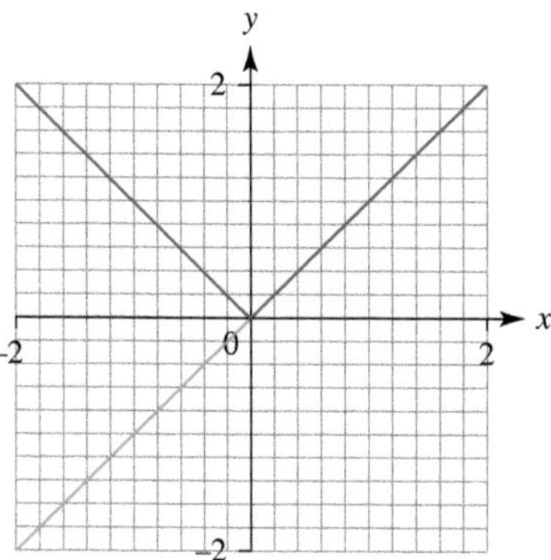

b

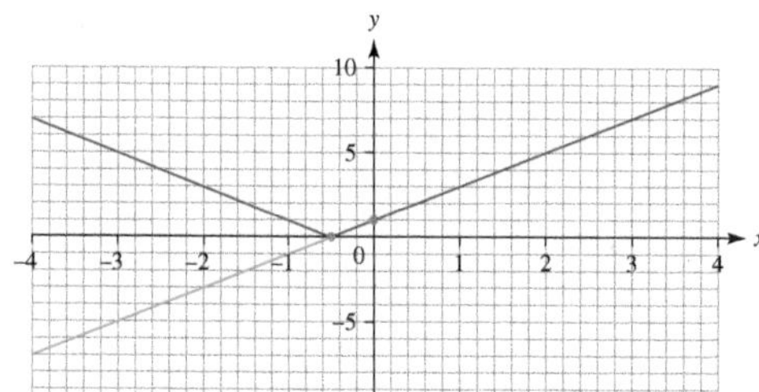

c

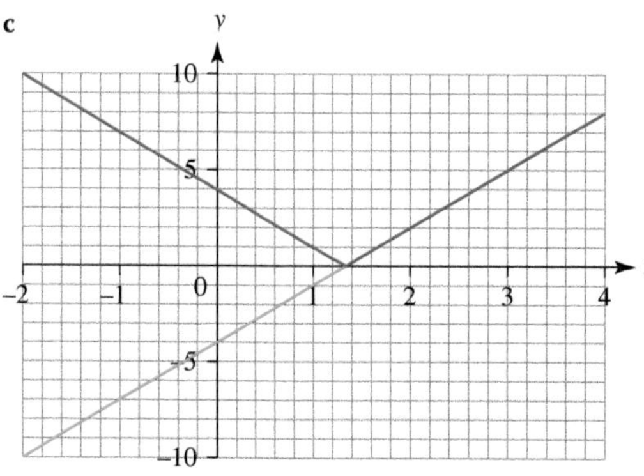

d

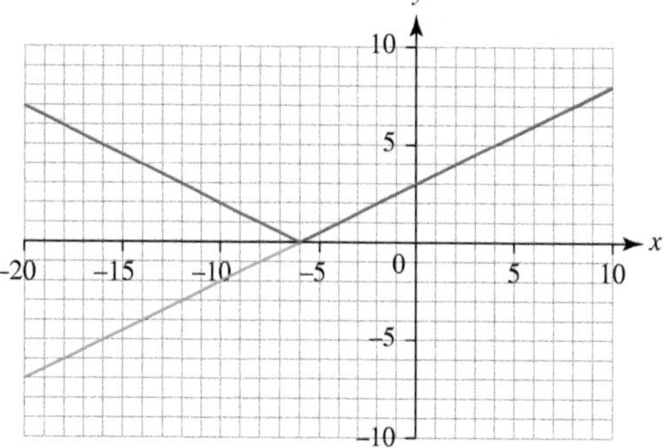

e

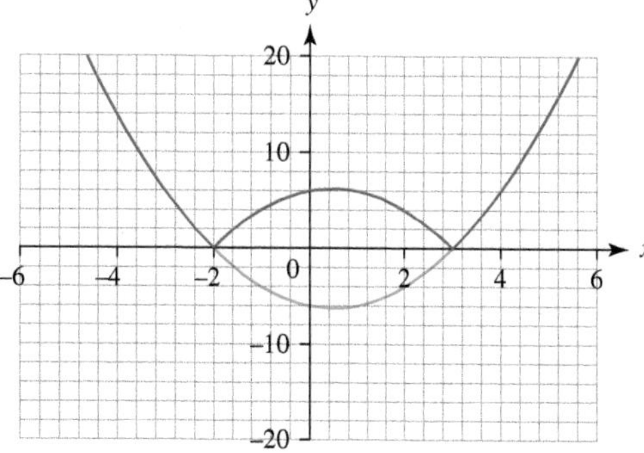

f

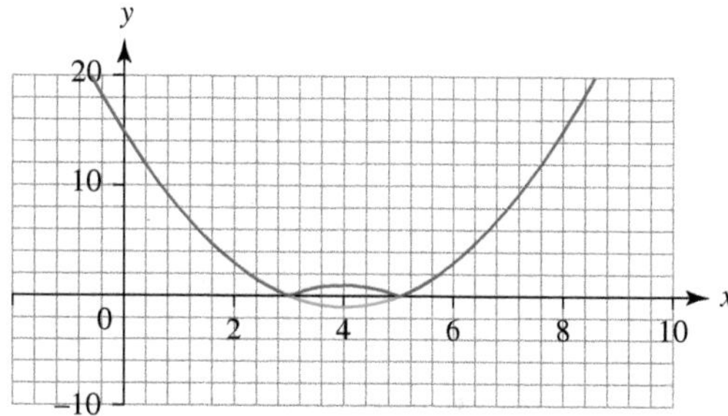

g

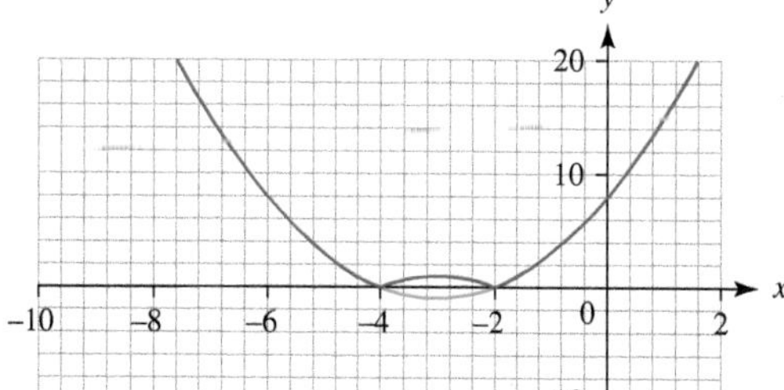

h

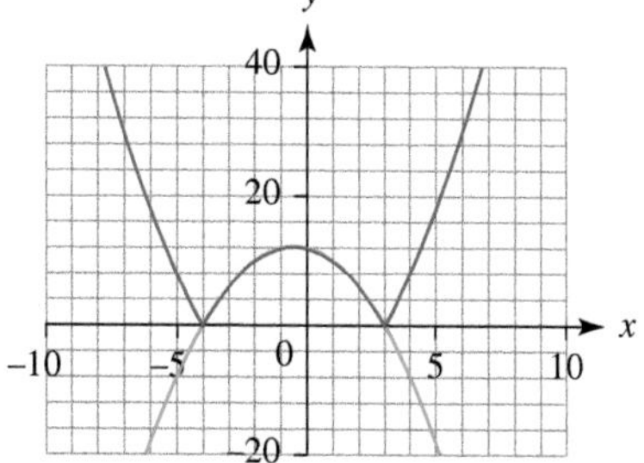

4 f(x) is blue and |f(x)| and $\frac{1}{f(x)}$ are red.

a i

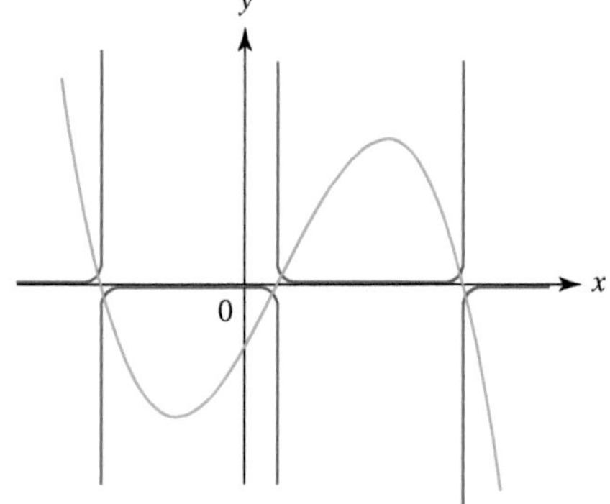

ii

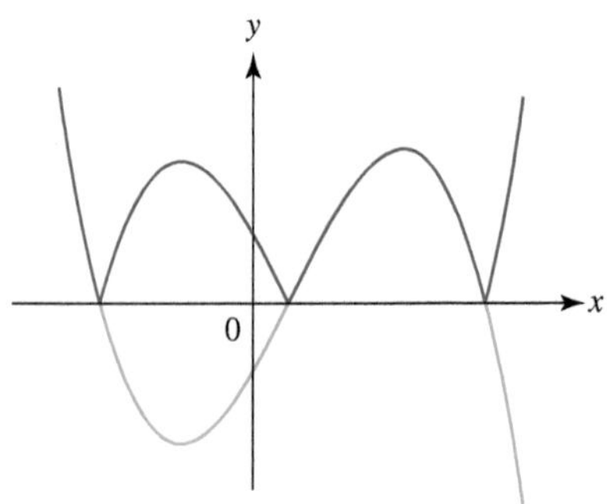

b i

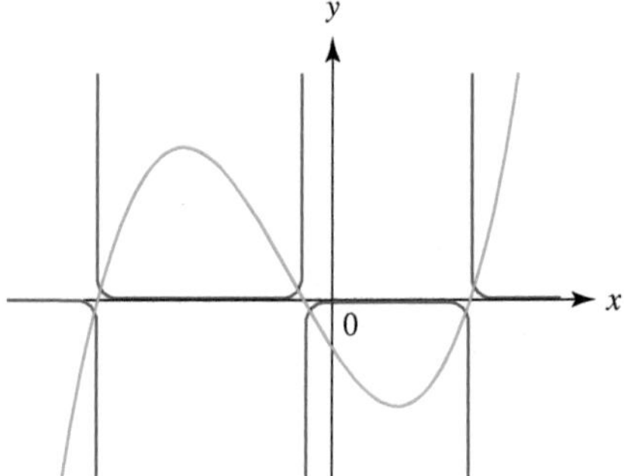

ii

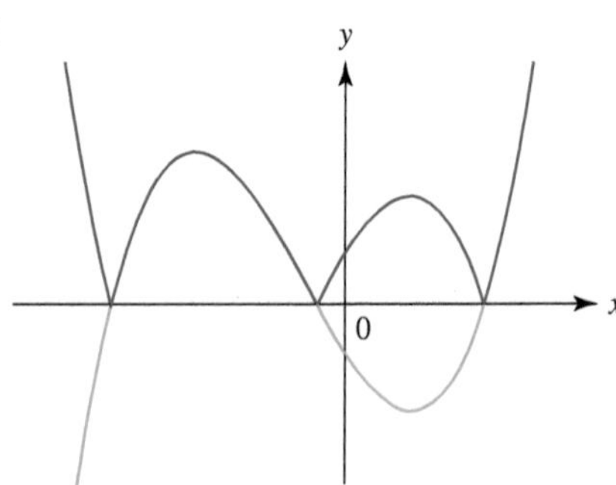

5 a i

ii

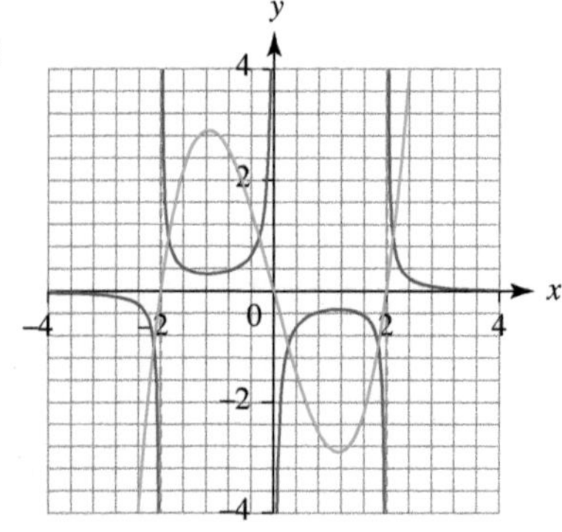

Asymptotes at $x = -2$, $x = 0$, $x = 2$ and at $y = 0$
Intersections at (−2.115, −1) and (−1.861, 1) and (−0.254, 1) and (0.254, −1) and (1.861, −1) and (2.115, 1)

iii

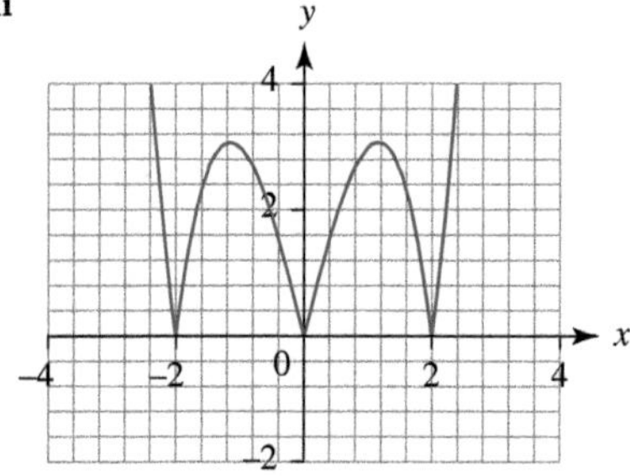

b i

ii

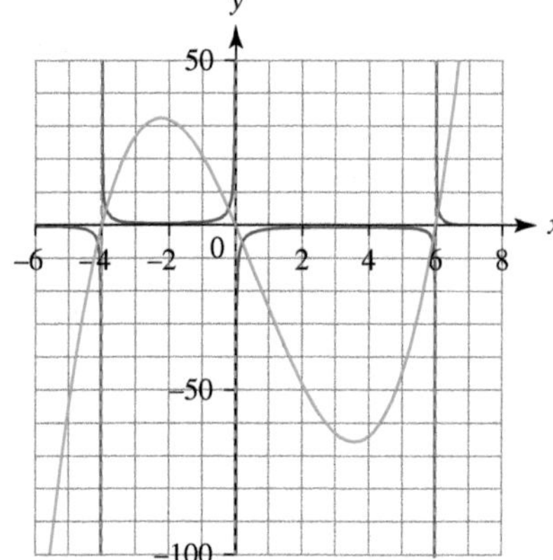

Asymptotes at $x = -4$, $x = 0$, $x = 6$ and at $y = 0$
Intersections at (−4.025, −1) and (−3.975, 1) and (−0.042, 1) and (0.042, −1) and (5.983, −1) and (6.017, 1)

iii

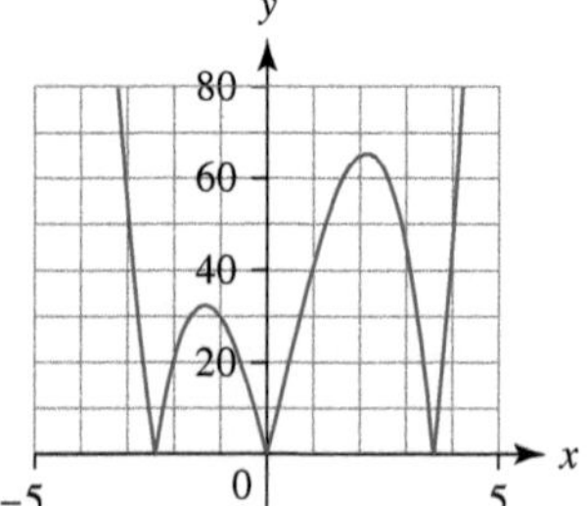

c i

ii

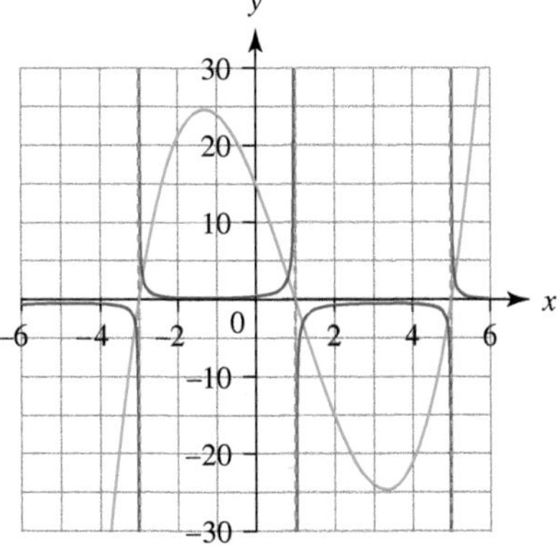

Asymptotes at $x = -3$, $x = 1$, $x = 5$

Intersections at (−3.031, −1) and (−2.968, 1) and (0.937, 1) and (1.063, −1) and (4.968, −1) and (5.031, 1)

iii

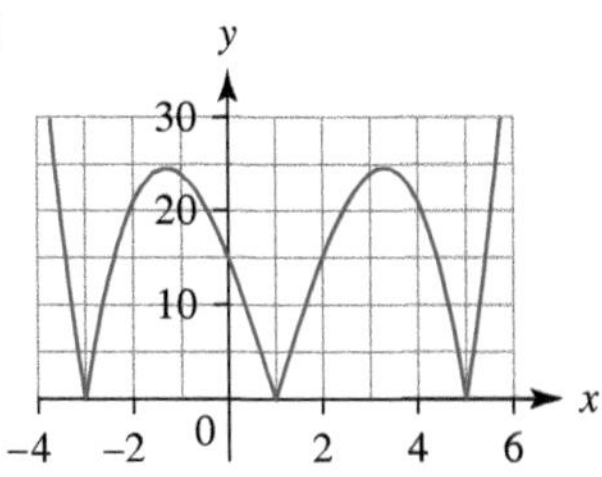

6

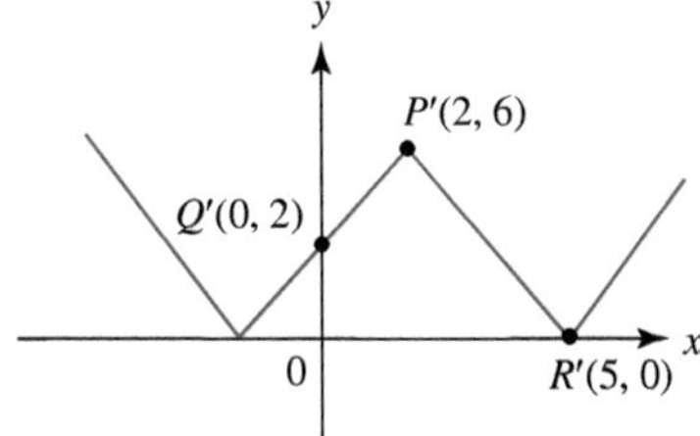

7

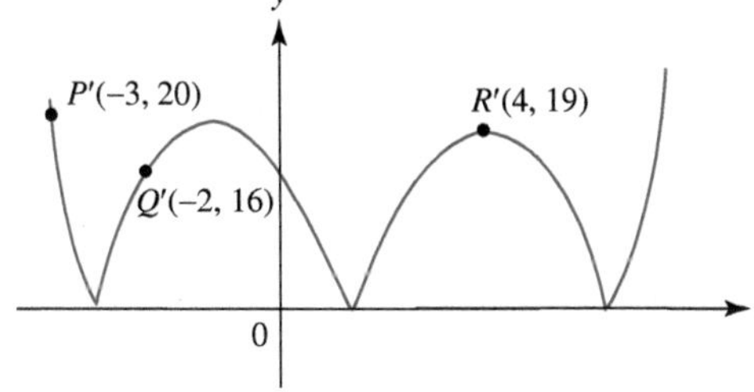

8

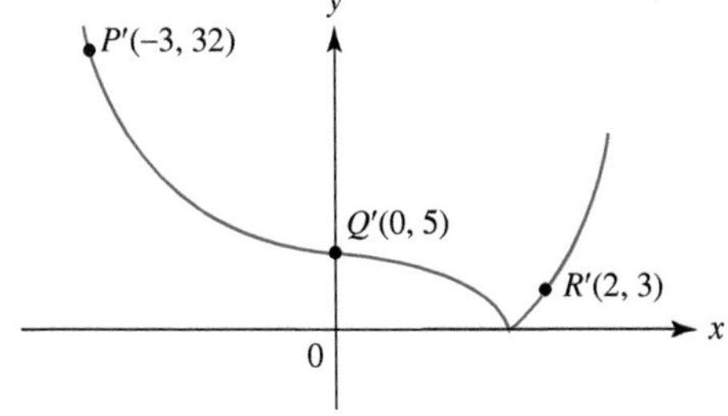

Exercise 18.1B

1 a i $y = \sin x$ and $y = \operatorname{cosec} x$

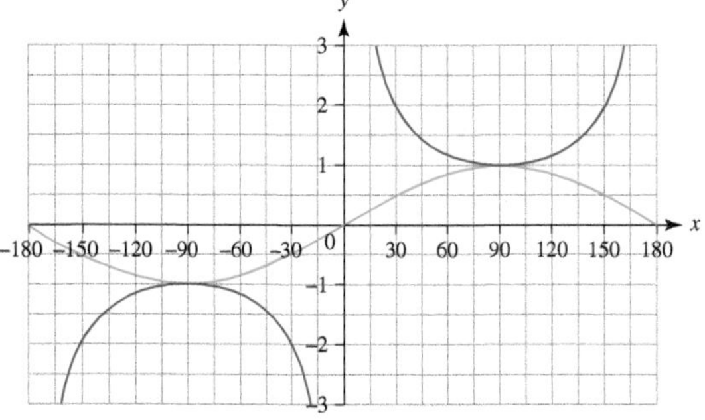

$\sin x$ is the blue graph and $y = \operatorname{cosec} x$ the red graph

ii $y = \cos x$ and $y = \sec x$

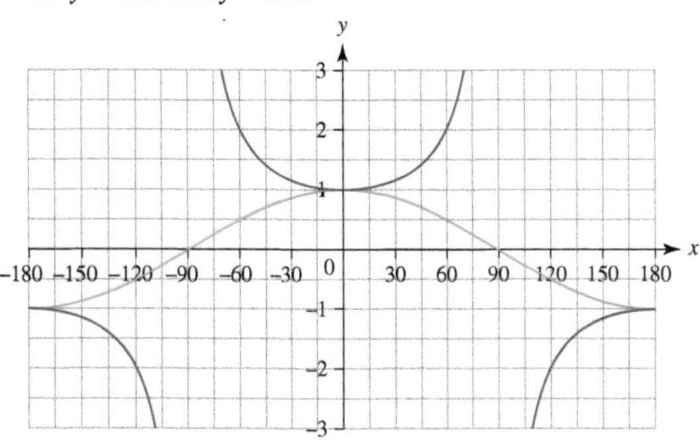

$\cos x$ is the blue graph and $y = \sec x$ the red graph

iii $y = \tan x$ and $y = \cot x$

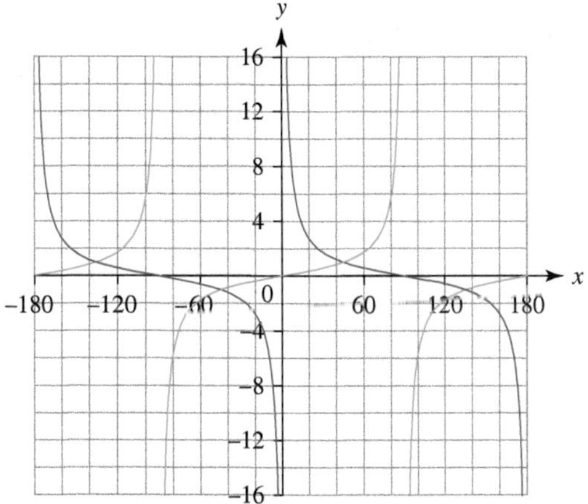

$\tan x$ is the blue graph and $y = \cot x$ the red graph

b i $0 < x < 180$ **ii** $-90 < x < 90$

iii $-180 < x < -90$ and $0 < x < 90$

2 a P becomes (0,3), Q becomes (2.5,8) and R stays as (6.5,0)

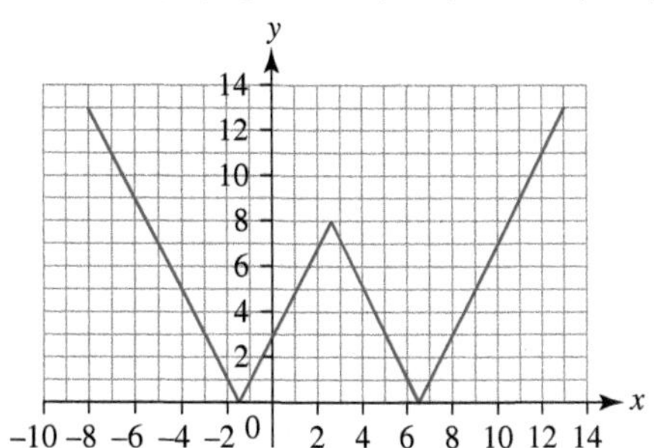

b $a = 2, b = -2.5$ and $c = -8$

3 a A' has coordinates $\left(-\frac{1}{2}, \frac{4}{25}\right)$ B' has coordinates $\left(3, -\frac{1}{6}\right)$

b $f(x)$ is the blue graph and $\frac{1}{f(x)}$ is the red graph

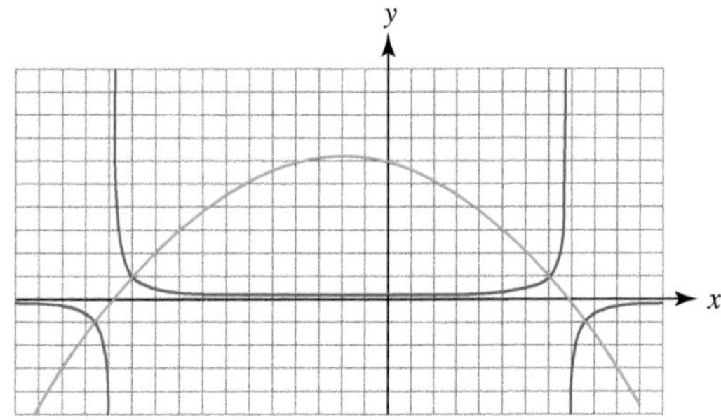

$\frac{1}{f(x)} > 0$ when x is between the values of the two roots of $f(x)$

4

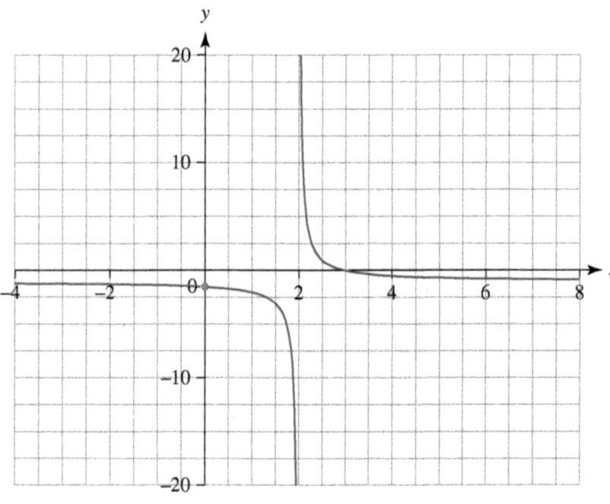

$x < 2$ or $x > 3$

5 a i $x = 1; y = 2; (0, 2); (2, 0)$

ii

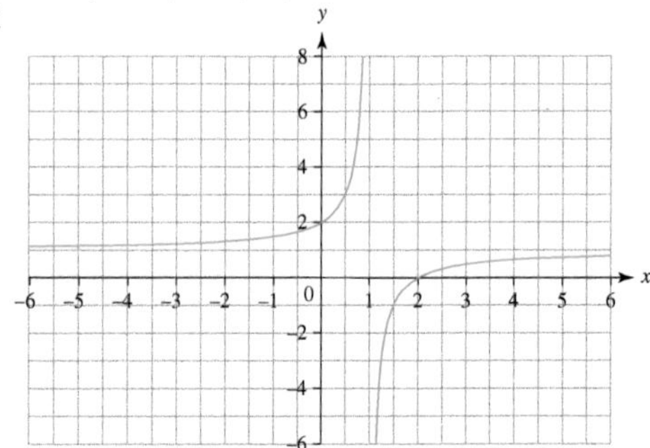

b

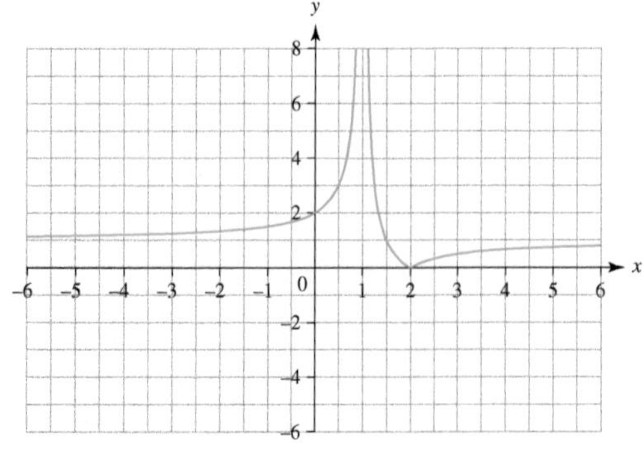

c

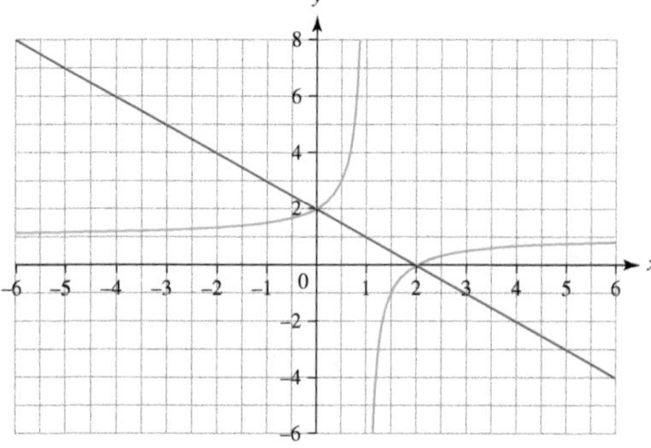

$x < 0$ or $1 < x < 2$

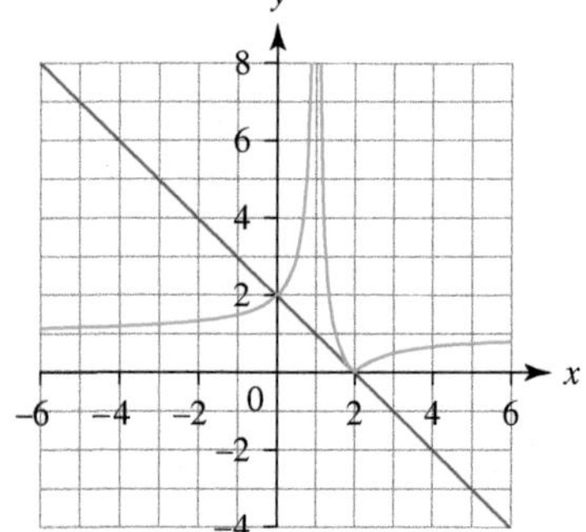

$x < 0$

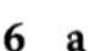

6 a

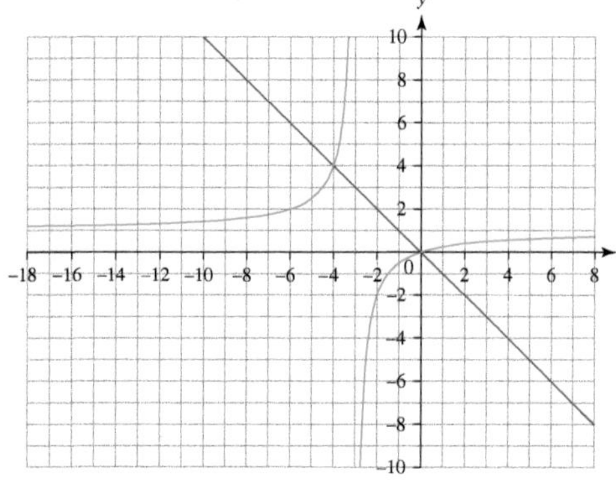

$x < -4$ or $-3 < x < 0$

b

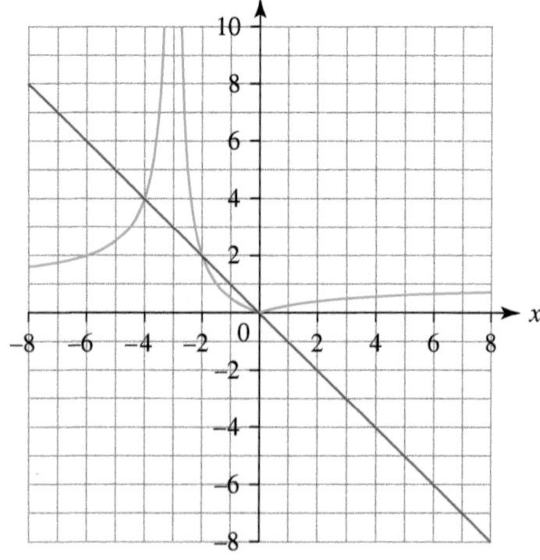

$x < -4$ or $-2 < x < 0$

Exercise 18.2A

1 a $\left(\frac{y}{2}\right)^2 = 12\left(\frac{x}{2}\right) \Rightarrow y^2 = 24x$

b $(3y)^2 = 12(3x) \Rightarrow y^2 = 4x$

c $(-x)^2 = 12y \Rightarrow x^2 = 12y$

d $(-y)^2 = 12(-x) \Rightarrow y^2 = -12x$

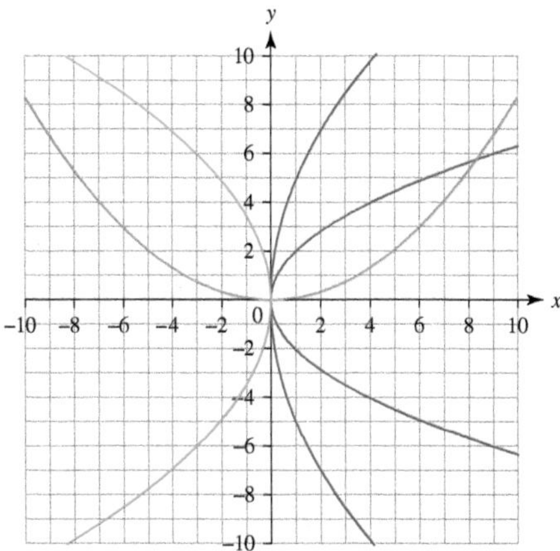

2 a $\frac{(2x)^2}{16} + \frac{(2y)^2}{25} = 1 \Rightarrow \frac{x^2}{4} + \frac{4y^2}{25} = 1$

Intercepts at (±2, 0) and $\left(0, \pm\frac{5}{2}\right)$

b $\frac{\left(\frac{x}{3}\right)^2}{16} + \frac{\left(\frac{y}{3}\right)^2}{25} = 1 \Rightarrow \frac{x^2}{144} + \frac{y^2}{225} = 1$

Intercepts at (±12, 0) and (0, ±15)

c $\frac{(-y)^2}{16} + \frac{x^2}{25} = 1 \Rightarrow \frac{x^2}{25} + \frac{y^2}{16} = 1.$

Intercepts at (±5, 0) and (0, ±4)

d $\frac{(-x)^2}{16} + \frac{(-y)^2}{25} = 1 \Rightarrow \frac{x^2}{16} + \frac{y^2}{25} = 1.$

Intercepts at (±4, 0) and (0, ±5)

Graphs: In order – red, purple, blue, green

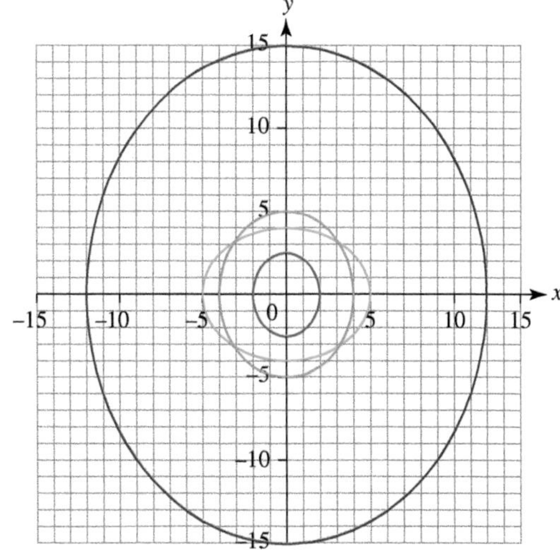

3 a $\frac{\left(\frac{x}{4}\right)^2}{9} - \frac{\left(\frac{y}{4}\right)^2}{4} = 1 \Rightarrow \frac{x^2}{144} - \frac{y^2}{64} = 1$

Asymptotes at $y = \pm\frac{2}{3}x$

b $\frac{\left(\frac{2x}{3}\right)^2}{9} - \frac{\left(\frac{2y}{3}\right)^2}{4} = 1 \Rightarrow \frac{4x^2}{81} - \frac{y^2}{9} = 1$

Asymptotes at $y = \pm\frac{2}{3}x$

c $\frac{(y)^2}{9} - \frac{(-x)^2}{4} = 1 \Rightarrow \frac{y^2}{9} - \frac{x^2}{4} = 1$ Asymptotes at $y = \pm\frac{3}{2}x$

d $\frac{(-y)^2}{9} - \frac{(x)^2}{4} = 1 \Rightarrow \frac{y^2}{9} - \frac{x^2}{4} = 1$ Asymptotes at $y = \pm\frac{3}{2}x$

Graphs: In order – red, blue, green

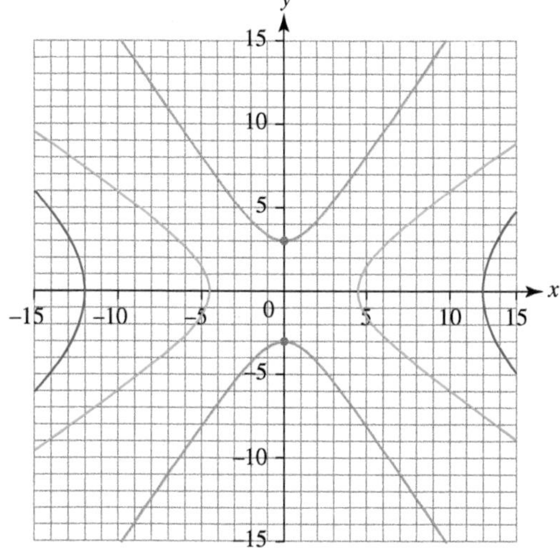

4 a $\left(\frac{3x}{2}\right)\left(\frac{3y}{2}\right) = 16 \Rightarrow 9xy = 64$

b $\left(\frac{3x}{4}\right)\left(\frac{3y}{4}\right) = 16 \Rightarrow 9xy = 256$

c $(-x)(-y) = 16 \Rightarrow xy = 16$

d $y(-x) = 16 \Rightarrow xy = -16$

Asymptotes in all cases are the lines $x = 0$ and $y = 0$
Graphs: In order – red, blue, purple, green

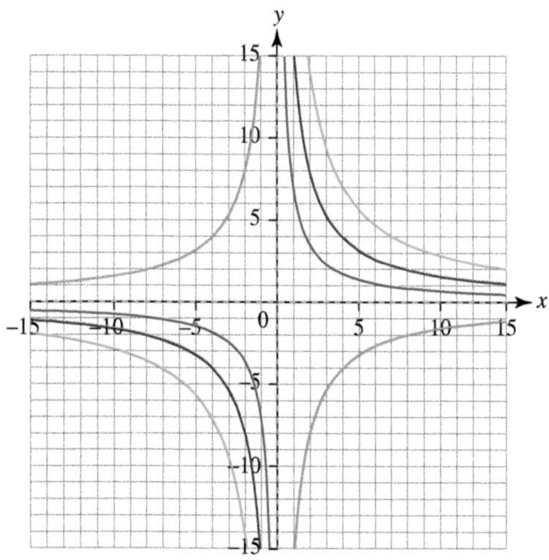

5 $\frac{2}{3}$ 6 $\frac{6}{5}$ 7 $\theta=\frac{3\pi}{2}$

Exercise 18.2B

1 a $9y^2-4x^2=324$ b Asymptotes: $y=\pm\frac{2}{3}x$

2 a $y^2+4y+12x=8$ b $(1, -2)$

c

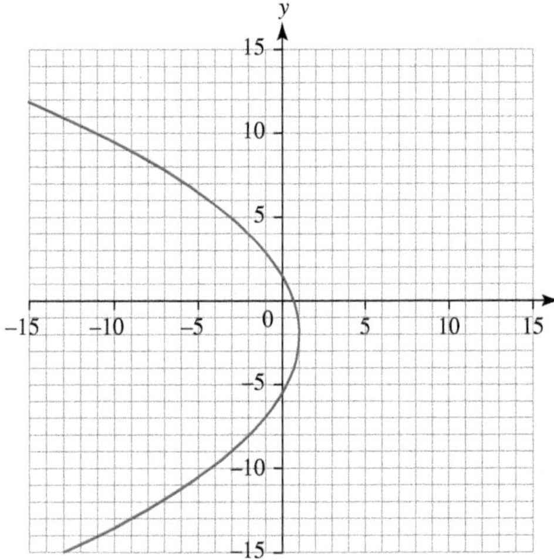

3 a Apply the translation: $\frac{(x-3)^2}{2}+\frac{(y+1)^2}{3}=1$

Apply the rotation: $\frac{(y-3)^2}{2}+\frac{(-x+1)^2}{3}=1$

Apply the stretch: $\frac{(y-3)^2}{2}+\frac{\left(-\frac{x}{2}+1\right)^2}{3}=1$

Expand and simplify:

$3(y^2-6y+9)+2\left(\frac{x^2}{4}-x+1\right)=6$

$3y^2-18y+27+\frac{1}{2}x^2-2x+2-6=0$

$x^2+6y^2-4x-36y+46=0$

b $3-\sqrt{2}<k<3+\sqrt{2}$

4 Translation with vector $\begin{pmatrix}2\\-2\end{pmatrix}$ followed by a stretch in the y-direction, scale factor $\frac{1}{2}$

5 Translated by vector $\begin{pmatrix}1\\3\end{pmatrix}$ and then enlarged by scale factor 3

6 a

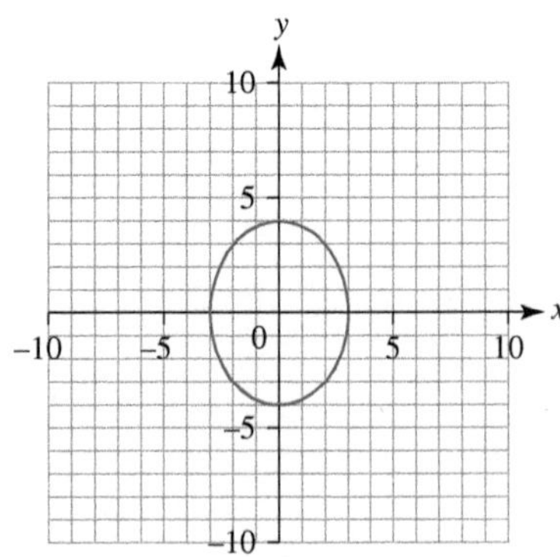

Intercepts at $(-3, 0)$ and $(3, 0)$ and $(0, 4)$ and $(0, -4)$

b $\frac{x^2}{9}+\frac{(c-x)^2}{16}=1$

$16x^2+9(c-x)^2=144$

$25x^2-18cx+9c^2-144=0$

Solve discriminant >0

$(-18c)^2-4\times25\times(9c^2-144)>0\Rightarrow576c^2<14\,400\Rightarrow c^2<25$

Hence $-5<c<5$

c $a=4$ $b=-1$ $d=121$

d $y=8-x$ and $y=-2-x$

7 a $x=2-\frac{3}{2}y$

Substitute into curve: $\frac{\left(2-\frac{3}{2}y\right)^2}{4}-\frac{y^2}{5}=1$

Expand and simplify:

$5\left(2-\frac{3}{2}y\right)^2-4y^2=20$

$\frac{45}{4}y^2-30y+20-4y^2=20$

$29y^2-120y=0$

$y(29y-120)=0$

Hence $y=0$ or $\frac{120}{29}$ giving coordinates $(2, 0)$ and $\left(\frac{-122}{29}, \frac{120}{29}\right)$

b $\frac{y^2}{4}-\frac{x^2}{5}=1$

$\frac{y^2}{4}-\frac{\left(2-\frac{3}{2}y\right)^2}{5}=1$

$5y^2-4\left(2-\frac{3}{2}y\right)^2=20$

$5y^2-9y^2+24y-16=20$

$y^2-6y+9=0$

$(y-3)^2=0$

Hence $y=3$. Repeated solution implies tangent.

Exercise 18.3A

1 a $\frac{4}{5}$ b $\frac{5}{12}$ c 3

d $\frac{32}{257}$ e $\frac{\sqrt{5}}{2}$ f $\frac{5}{4}$

2 a

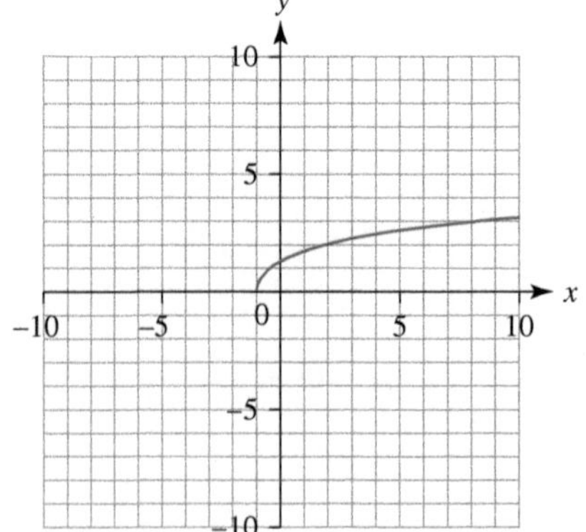

domain is $x\in\mathbb{R}$, $x\geq-1$; range is $y\in\mathbb{R}$

b

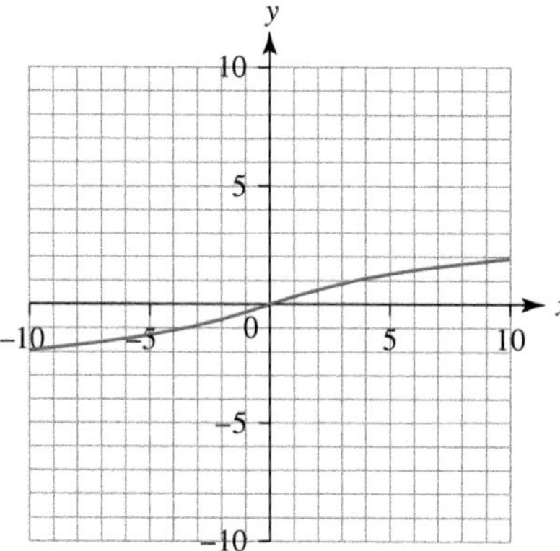

domain is $x \in \mathbb{R}$; range is $y \in \mathbb{R}$

c

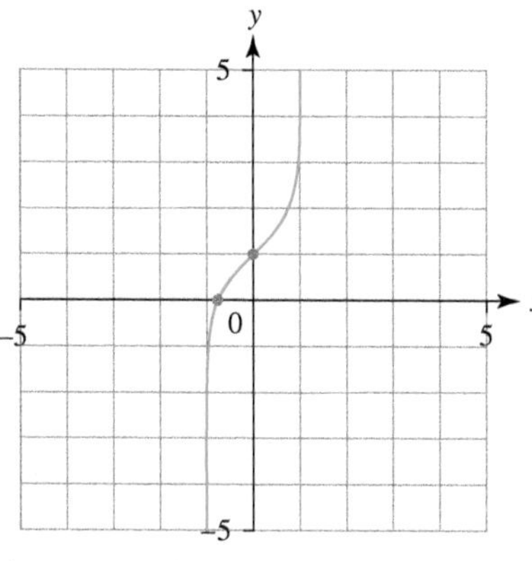

domain is $x \in \mathbb{R}$, $-1 < x < 1$; range is $y \in \mathbb{R}$

d

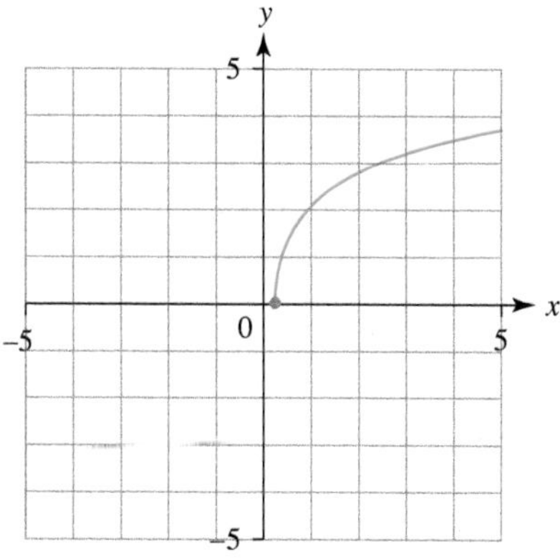

domain is $x \in \mathbb{R}$, $x \geq \frac{1}{4}$; range is $y \in \mathbb{R}$

e

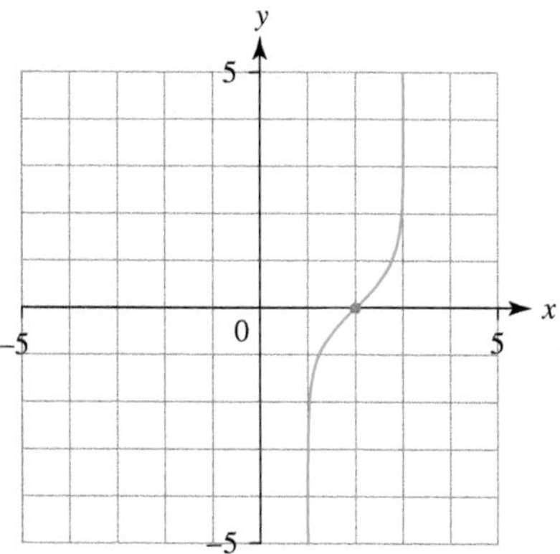

domain is $x \in \mathbb{R}$, $1 < x < 3$; range is $y \in \mathbb{R}$

f

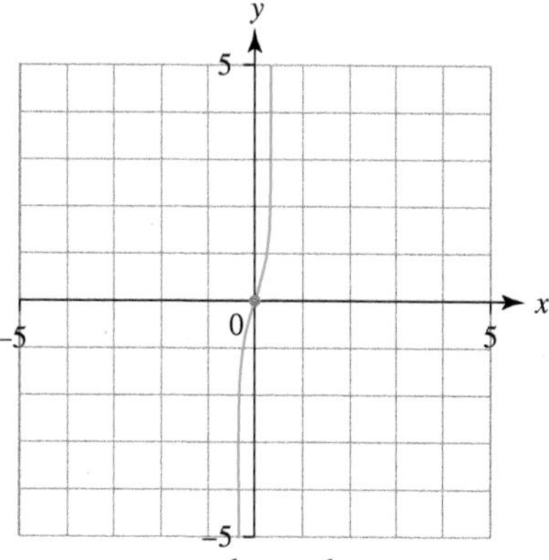

domain is $x \in \mathbb{R}$, $-\frac{1}{3} < x < \frac{1}{3}$; range is $y \in \mathbb{R}$

3 a

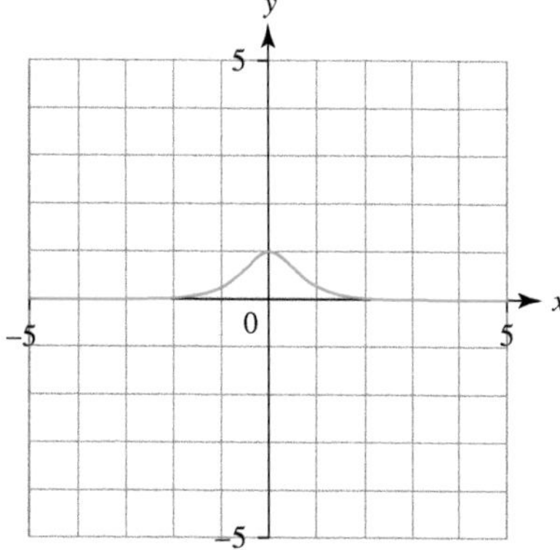

domain is $x \in \mathbb{R}$, range is $y \in \mathbb{R}$, $0 < y \leq 1$

b

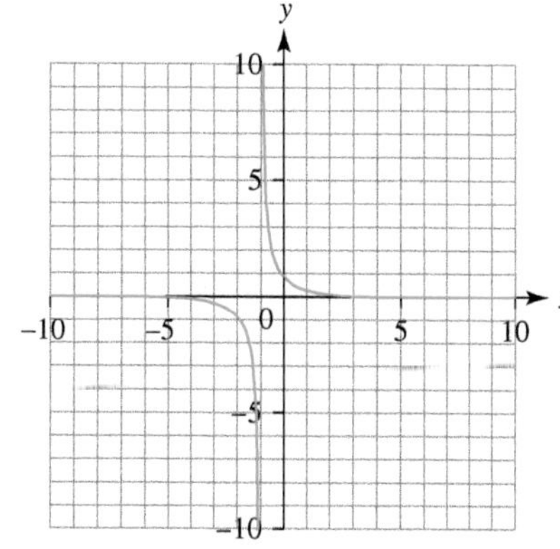

domain is $x \in \mathbb{R}$, $x \neq -1$; range is $y \in \mathbb{R}$, $y \neq 0$

c

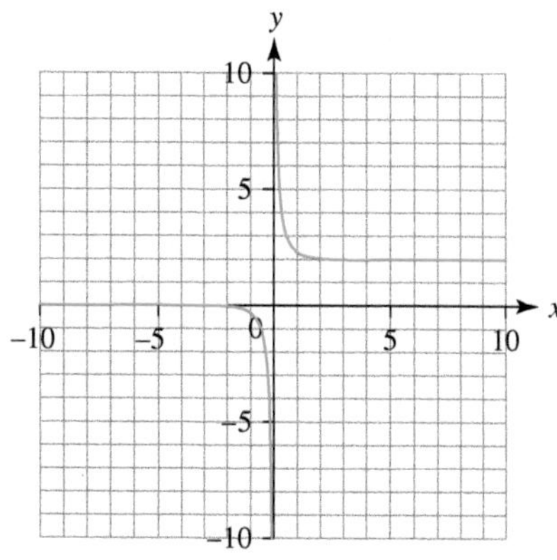

domain is $x \in \mathbb{R}$, $x \neq 0$; range is $y \in \mathbb{R}$, $y < 0$, $y > 2$

d

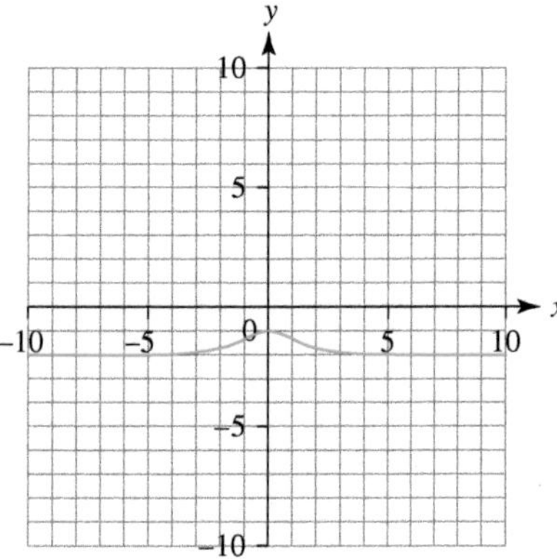

domain is $x \in \mathbb{R}$, range is $y \in \mathbb{R}$, $-2 < y \le -1$

e

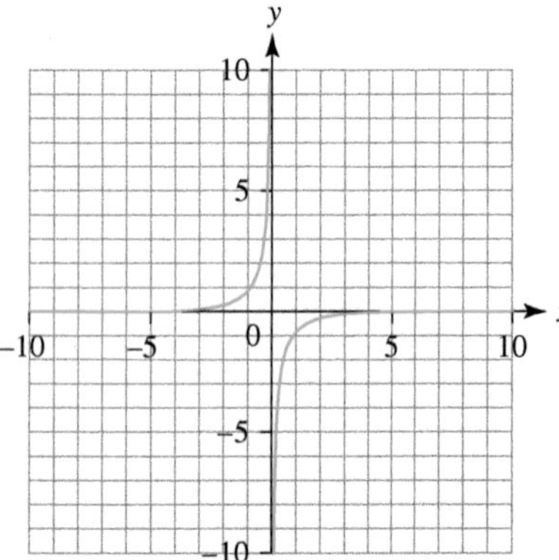

domain is $x \in \mathbb{R}$, $x \neq 0$; range is $y \in \mathbb{R}$, $y \neq 0$

f

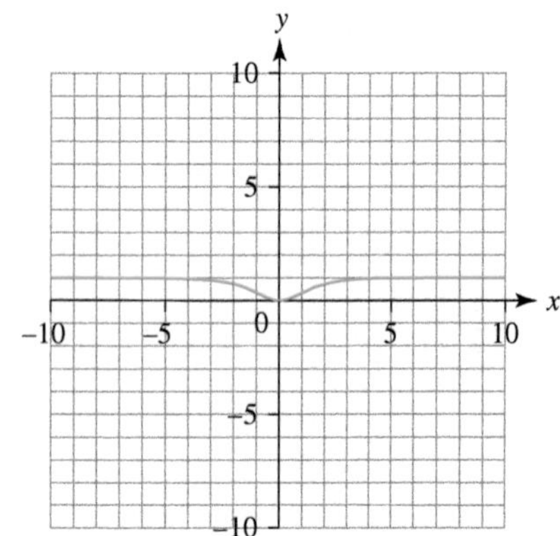

domain is $x \in \mathbb{R}$; range is $y \in \mathbb{R}$, $0 \le y < 1$

4 a $\ln\left(\frac{1+\sqrt{5}}{2}\right)$ **b** $\ln(\sqrt{2})$ **c** $\pm\ln(1+\sqrt{2})$

d $\ln\left(\sqrt{\frac{7}{3}}\right)$ **e** $\ln\left(\frac{\sqrt{17}-1}{4}\right)$ **f** $\pm\ln(2+\sqrt{3})$

5 a $\frac{7}{25}$ **b** $\frac{16}{63}$ **c** $\frac{2}{3}$ **d** $\frac{8}{85}$

6 a $\pm\ln(\sqrt{3})$ **b** $\ln(\sqrt{3}+2)$ **c** $\frac{1}{2}\ln(2\pm\sqrt{3})$

Exercise 18.3B

1 a $1+\operatorname{cosech}^2 x \equiv 1+\left(\frac{2}{e^x-e^{-x}}\right)^2$

$$\equiv 1+\frac{4}{(e^x-e^{-x})^2}$$

$$\equiv \frac{(e^x-e^{-x})^2+4}{(e^x-e^{-x})^2}$$

$$\equiv \frac{e^{2x}-2+e^{-2x}+4}{(e^x-e^{-x})^2}$$

$$\equiv \frac{e^{2x}+2+e^{-2x}}{(e^x-e^{-x})^2} \equiv \frac{(e^x+e^{-x})^2}{(e^x-e^{-x})^2}$$

$$\equiv \left(\frac{e^x+e^{-x}}{e^x-e^{-x}}\right)^2 \equiv \left(\frac{e^{2x}+1}{e^{2x}-1}\right)^2$$

$$\equiv \coth^2 x$$

b $\frac{1}{2}\operatorname{cosech} x \operatorname{sech} x \equiv \frac{1}{2}\left(\frac{2}{e^x-e^{-x}}\right)\left(\frac{2}{e^x+e^{-x}}\right)$

$$\equiv \frac{1}{2}\left(\frac{4}{(e^x-e^{-x})(e^x+e^{-x})}\right)$$

$$\equiv \frac{2}{e^{2x}-e^{-2x}} \equiv \operatorname{cosech} 2x$$

c $\frac{\coth^2\left(\frac{x}{2}\right)+1}{2\coth\left(\frac{x}{2}\right)} \equiv \frac{\left(\frac{e^x+1}{e^x-1}\right)^2+1}{2\left(\frac{e^x+1}{e^x-1}\right)}$

$$\equiv \frac{(e^x+1)^2+(e^x-1)^2}{2(e^x+1)(e^x-1)}$$

$$\equiv \frac{(e^{2x}+2e^x+1)+(e^{2x}-2e^x+1)}{2(e^{2x}-1)}$$

$$\equiv \frac{2(e^{2x}+1)}{2(e^{2x}-1)} \equiv \frac{e^{2x}+1}{e^{2x}-1}$$

$$\equiv \coth x$$

d $\frac{1-\operatorname{sech} x}{1+\operatorname{sech} x} \equiv \frac{1-\frac{2}{e^x+e^{-x}}}{1+\frac{2}{e^x+e^{-x}}}$

$$\equiv \frac{e^x+e^{-x}-2}{e^x+e^{-x}+2}$$

$$\equiv \frac{\left(e^{\frac{x}{2}}-e^{-\frac{x}{2}}\right)^2}{\left(e^{\frac{x}{2}}+e^{-\frac{x}{2}}\right)^2}$$

$$\equiv \left(\frac{e^{\frac{x}{2}}-e^{-\frac{x}{2}}}{e^{\frac{x}{2}}+e^{-\frac{x}{2}}}\right)^2$$

$$\equiv \tanh^2\left(\frac{x}{2}\right)$$

2 a $\cosh x+\frac{2}{\cosh x}=3$

$\cosh^2 x+2=3\cosh x$

$\cosh^2 x-3\cosh x+2=0$

b $x=0, \pm\ln(2+\sqrt{3})$

3 a $2\sinh x\cosh x = 2\left(\frac{e^x - e^{-x}}{2}\right)\left(\frac{e^x + e^{-x}}{2}\right)$

$= \frac{2(e^{2x} - e^{-2x})}{4}$

$= \frac{e^{2x} - e^{-2x}}{2} = \sinh(2x)$

b $\coth x - \tanh x = \frac{\cosh x}{\sinh x} - \frac{\sinh x}{\cosh x}$

$= \frac{\cosh^2 x - \sinh^2 x}{\sinh x\cosh x}$

$= \frac{1}{\frac{1}{2}\sinh(2x)} = \frac{2}{\sinh(2x)}$

$= 2\operatorname{cosech}(2x)$

c $x = \frac{1}{2}\ln(\sqrt{5} - 2)$

4 $x = \ln\sqrt{3}$

5 a $\sinh^2 x = \left(\frac{e^x - e^{-x}}{2}\right)^2$

$= \frac{e^{2x} - 2 + e^{-2x}}{4}$

$\frac{1}{2}(\cosh(2x) - 1) = \frac{1}{2}\left(\frac{e^{2x} + e^{-2x}}{2} - 1\right)$

$= \frac{1}{2}\left(\frac{e^{2x} + e^{-2x} - 2}{2}\right)$

$= \frac{e^{2x} - 2 + e^{-2x}}{4} = \sinh^2 x$

b $x = 0$ or $x = \ln\left(\frac{-1 + \sqrt{5}}{2}\right)$

6 a Let $y = \operatorname{arcosech} x$ then $x = \operatorname{cosech} y$

$x = \frac{2}{e^y - e^{-y}}$

$xe^y - xe^{-y} = 2$

$xe^{2y} - 2e^y - x = 0$

$e^y = \frac{2 \pm \sqrt{4 + 4x^2}}{2x}$

$e^y = \frac{1}{x} \pm \sqrt{\frac{1}{x^2} + 1}$

$y = \ln\left(\frac{1}{x} + \sqrt{\frac{1}{x^2} + 1}\right)$

$\frac{1}{x} - \sqrt{\frac{1}{x^2} + 1} < 0$ since $\sqrt{\frac{1}{x^2} + 1} > \frac{1}{x}$

so $\ln\left(\frac{1}{x} - \sqrt{\frac{1}{x^2} + 1}\right)$ is not a solution.

b $\operatorname{arcosech} x + \operatorname{arcosech}(-x)$

$= \ln\left(\frac{1}{x} + \sqrt{\frac{1}{x^2} + 1}\right) + \ln\left(\frac{1}{(-x)} + \sqrt{\frac{1}{(-x)^2} + 1}\right)$

$= \ln\left(\frac{1}{x} + \sqrt{\frac{1}{x^2} + 1}\right) + \ln\left(-\frac{1}{x} + \sqrt{\frac{1}{x^2} + 1}\right)$

$= \ln\left[\left(\frac{1}{x} + \sqrt{\frac{1}{x^2} + 1}\right)\left(-\frac{1}{x} + \sqrt{\frac{1}{x^2} + 1}\right)\right]$

$= \ln\left[-\frac{1}{x^2} + \left(\frac{1}{x^2} + 1\right)\right]$

$= \ln 1 = 0$

7 Let $y = \operatorname{arsech} x$ then $x = \operatorname{sech} y$

$x = \frac{2}{e^y + e^{-y}}$

$x(e^y + e^{-y}) = 2$

$xe^{2y} - 2e^y + x = 0$

$e^y = \frac{2 \pm \sqrt{4 - 4x^2}}{2x}$

$= \frac{1 \pm \sqrt{1 - x^2}}{x} = \frac{1}{x} \pm \sqrt{\frac{1}{x^2} - 1}$

$y = \ln\left(\frac{1}{x} \pm \sqrt{\frac{1}{x^2} - 1}\right)$

$= \pm\ln\left(\frac{1}{x} + \sqrt{\frac{1}{x^2} - 1}\right)$

since $\left(\frac{1}{x} - \sqrt{\frac{1}{x^2} - 1}\right)\left(\frac{1}{x} + \sqrt{\frac{1}{x^2} - 1}\right) = \frac{1}{x^2} - \left(\frac{1}{x^2} - 1\right) = 1$

so $\left(\frac{1}{x} - \sqrt{\frac{1}{x^2} - 1}\right) = \left(\frac{1}{x} + \sqrt{\frac{1}{x^2} - 1}\right)^{-1}$

$\ln\left(\frac{1}{x} - \sqrt{\frac{1}{x^2} - 1}\right) = \ln\left(\frac{1}{x} + \sqrt{\frac{1}{x^2} - 1}\right)^{-1}$

$= -\ln\left(\frac{1}{x} + \sqrt{\frac{1}{x^2} - 1}\right)$

8 a $\frac{\sqrt{6}}{2}$ **b** $\frac{\sqrt{3}}{3}$

9 a $\sqrt{2}$ **b** $\frac{\sqrt{6}}{2}$

10 a $\tanh(2x) = \frac{\sinh(2x)}{\cosh(2x)}$

$= \frac{2\sinh x\cosh x}{\cosh^2 x + \sinh^2 x}$

$= \frac{\frac{2\sinh x\cosh x}{\cosh^2 x}}{\frac{\cosh^2 x}{\cosh^2 x} + \frac{\sinh^2 x}{\cosh^2 x}}$

$= \frac{2\tanh x}{1 + \tanh^2 x}$

b $\coth(2x) = \frac{1 + \tanh^2 x}{2\tanh x}$

$= \frac{1}{2\tanh x} + \frac{\tanh^2 x}{2\tanh x}$

$= \frac{1}{2}(\coth x + \tanh x)$

c $\coth(2x) = 2$

$\tanh(2x) = \frac{1}{2}$

$2x = \frac{1}{2}\ln 3$

$x = \frac{1}{4}\ln 3$

11 a $\frac{1}{1+\coth x}+\frac{1}{1-\coth x}\equiv\frac{(1-\coth x)+(1+\coth x)}{(1+\coth x)(1-\coth x)}$

$$\equiv\frac{2}{1-\coth^2 x}$$

$$\equiv\frac{2}{-\operatorname{cosech}^2 x}$$

$$\equiv -2\sinh^2 x$$

b $x=\ln\left(\frac{3+\sqrt{13}}{2}\right)$ or $x=\ln\left(\frac{-3+\sqrt{13}}{2}\right)$

12 a

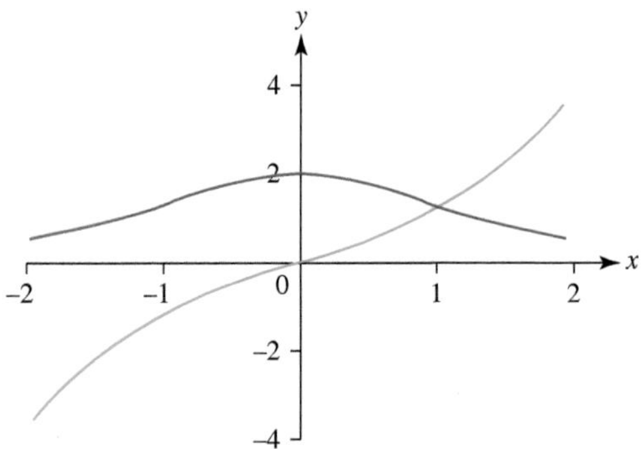

b one solution since they intersect once only

c $x=\frac{1}{2}\ln\left(4+\sqrt{17}\right)$

13 a

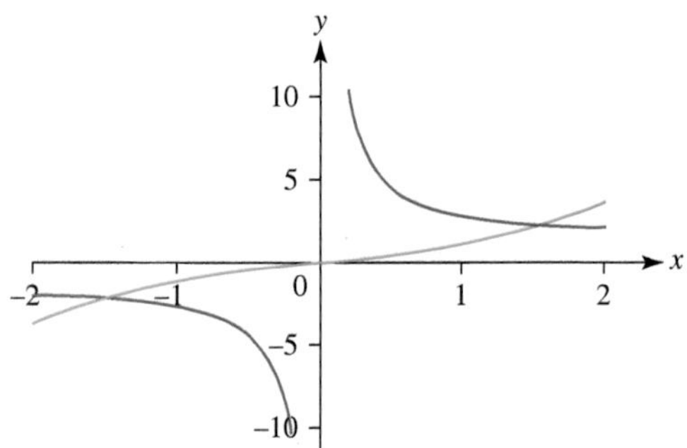

Asymptotes $y=\pm 2$ and $x=0$

b $x=\pm 1.53$

14 $x=\ln\left(1+\sqrt{2}\right)$, $\ln\left(\frac{-1+\sqrt{10}}{3}\right)$

15 $x=\pm\ln\left(2+\sqrt{3}\right)$

16 $x=\frac{1}{2}\ln 3$ Or $x=\frac{1}{2}\ln\frac{1}{5}$

17 $\ln\sqrt{5}$

18 $x=\frac{1}{2}\ln\left(\frac{3+\sqrt{5}}{2}\right)$ or $x=\frac{1}{2}\ln\left(\frac{3-\sqrt{5}}{2}\right)$

Exercise 18.4A

1 a $y=x+7$ and $x=5$ **b** $y=-2x-2$ and $x=1$

c $y=-3x-3$ and $x=1$ **d** $y=x+3$ and $x=2$

2 a 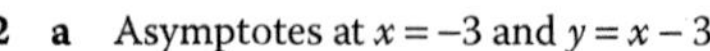 Asymptotes at $x=-3$ and $y=x-3$

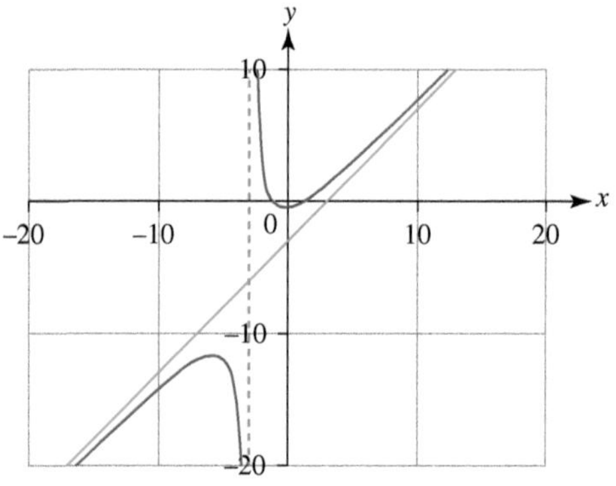

b Asymptotes at $x=-\frac{3}{2}$ and $y=x-\frac{3}{2}$

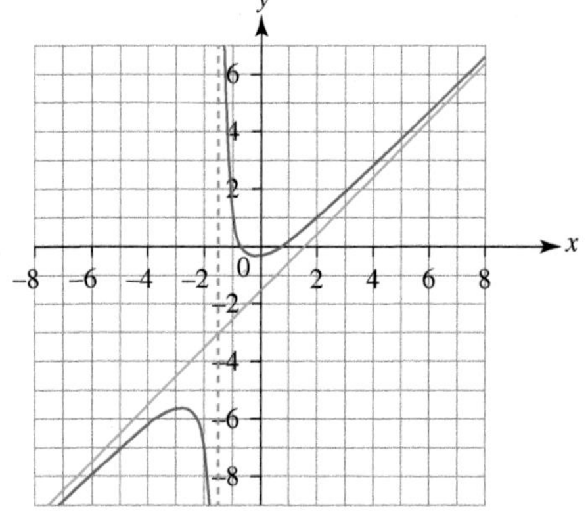

c Asymptotes at $x=4$ and $y=-x-2$

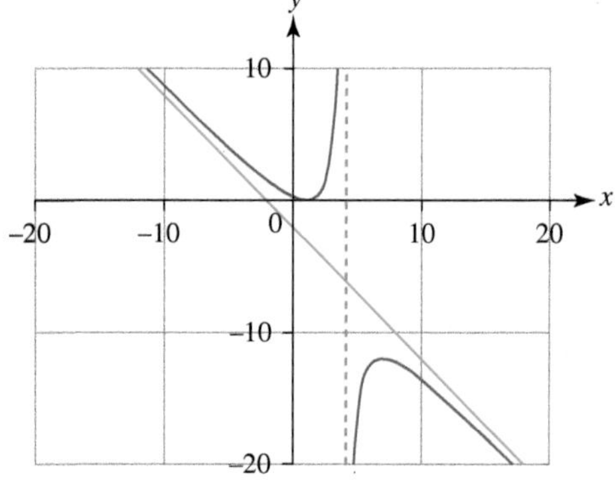

d Asymptotes at $x=\frac{3}{2}$ and $y=-\frac{1}{2}x+\frac{5}{4}$

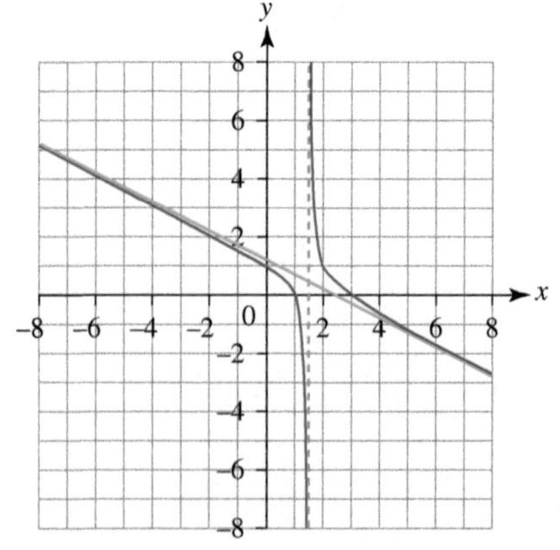

3 a Asymptotes at $x = \pm\frac{1}{\sqrt{2}}$ and $y = \frac{1}{2}x$

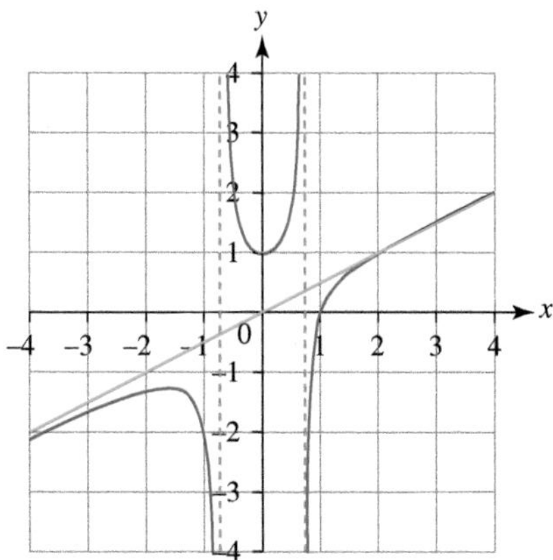

b Asymptotes at $x = \pm 2$ and $y = x$

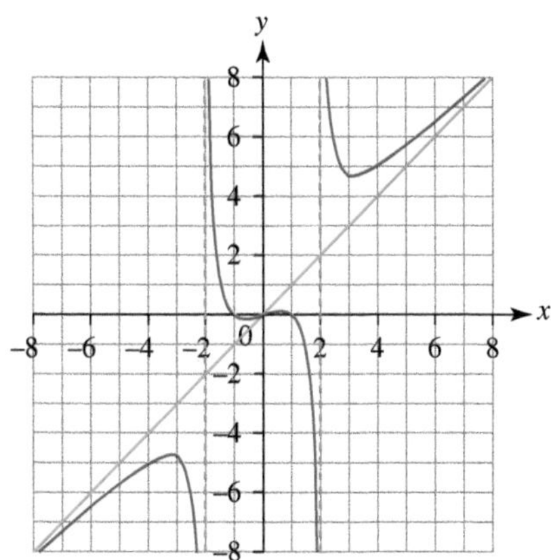

Exercise 18.4B

1 a $a = 4, b = 2$ **b** $x = -1$

c $\left(-\frac{1}{2}, -2\right), \left(-\frac{3}{2}, -10\right)$

d

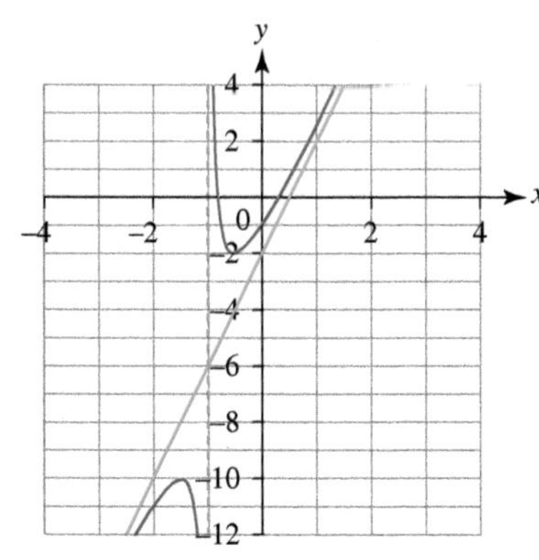

2 a $a = -1, b = -3$ **b** $x = 2$

c $(-1, -1), (5, -13)$

d

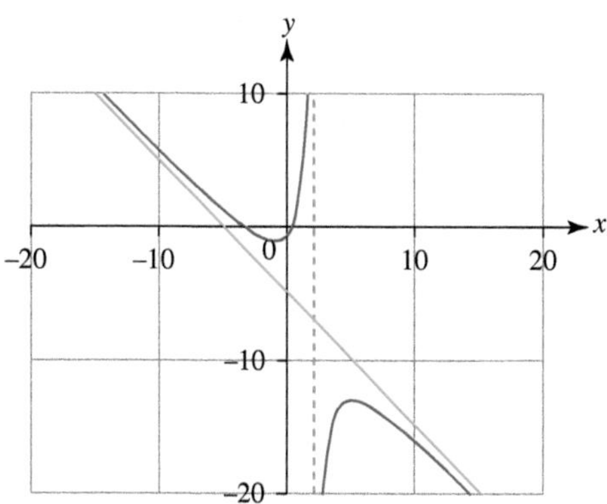

3 a $\frac{x^3 - 1}{2(x^2 - 1)} = \frac{(x-1)(x^2+x+1)}{2(x-1)(x+1)} = \frac{x^2+x+1}{2x+2} = \frac{1}{2}x + \frac{1}{2(x+1)}$

b $x = -1$ and $y = \frac{1}{2}x$ **c** $\left(0, \frac{1}{2}\right), \left(-2, -\frac{3}{2}\right)$

d

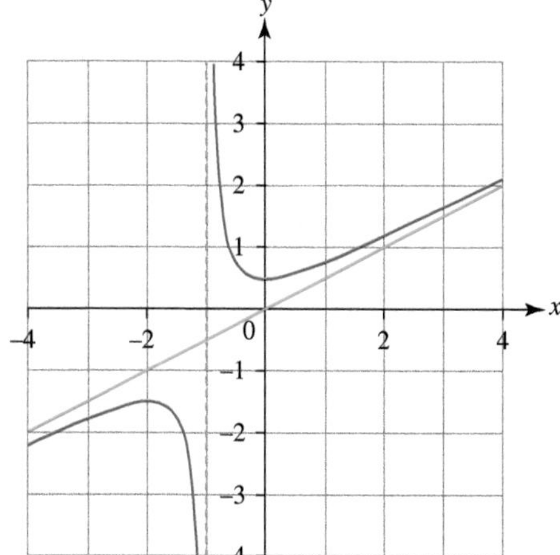

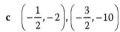

4 a $x^2 + x + 1 + \frac{1}{x-1}$

b $y = x^2 + x + 1$ and $x = 1$

c

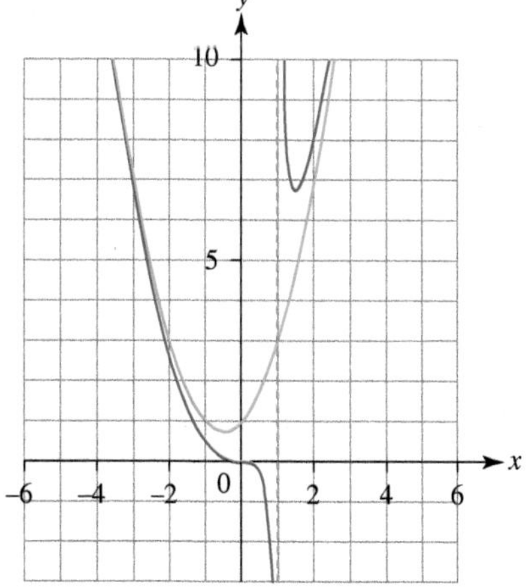

Review exercise 18

1

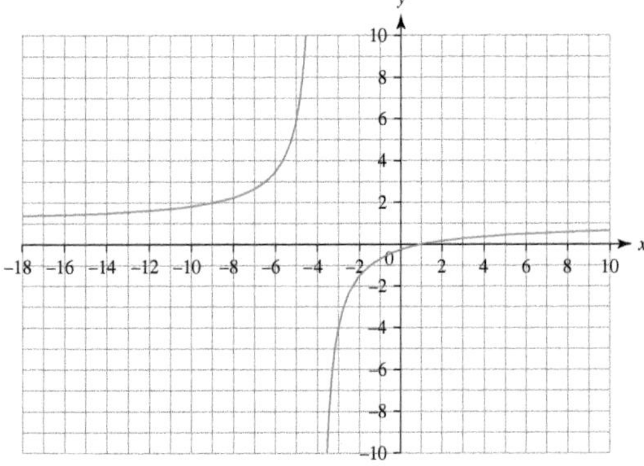

2 a

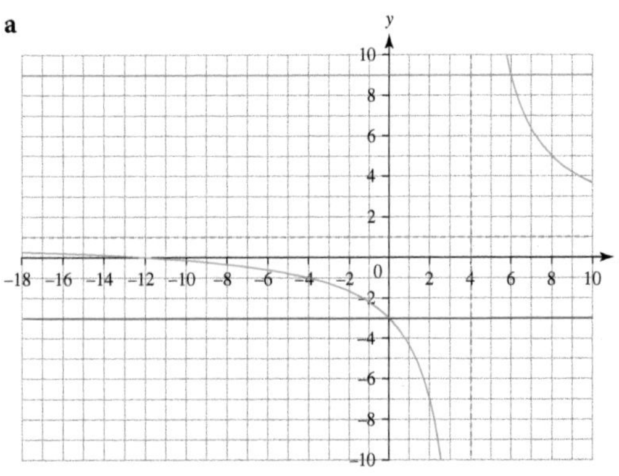

Intercepts are $(-12, 0)$ and $(0, -3)$

Asymptotes are $x = 4$ and $y = 1$

Solution to the inequality is $x < 0$ or $x > 6$

b

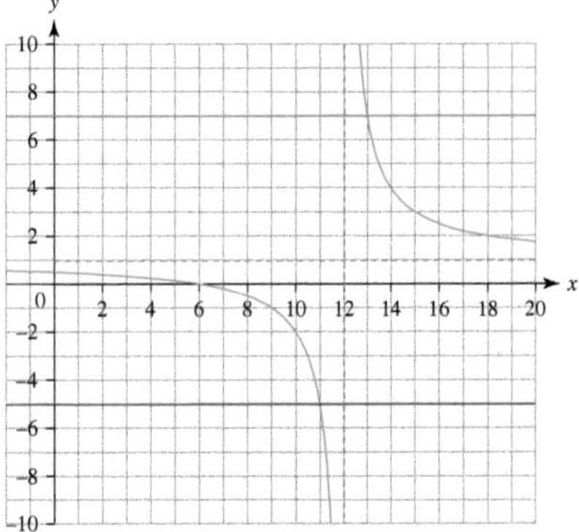

Intercepts are $(6, 0)$ and $(0, \frac{1}{2})$

Asymptotes are $x = 12$ and $y = 1$

Solution to the inequality is $x < 11$ or $x > 13$

3 a

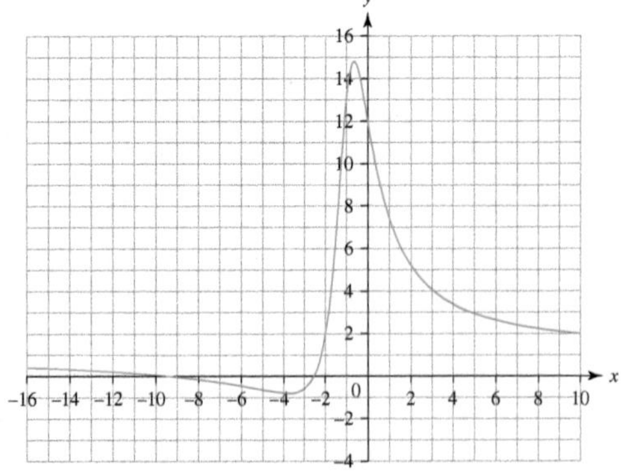

b

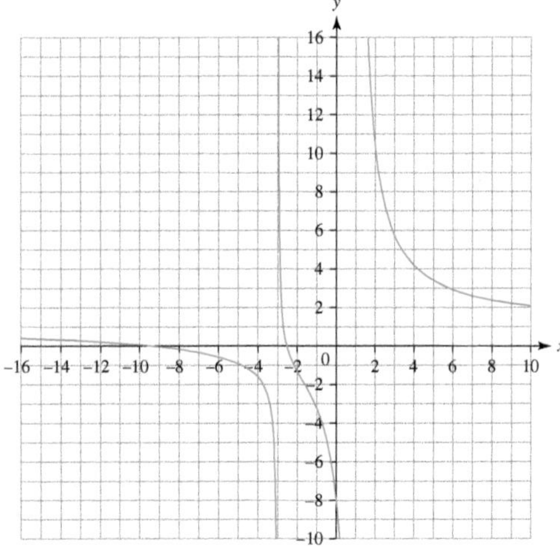

4 $\left(\frac{5}{2}, 5\right)$ Maximum value of $f(x)$ is infinite.

5

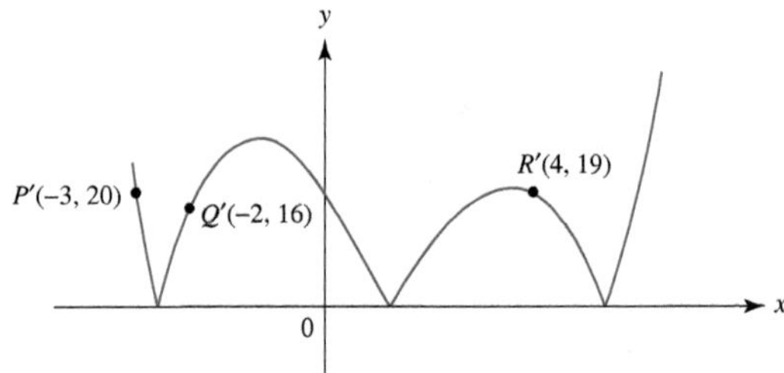

6 a

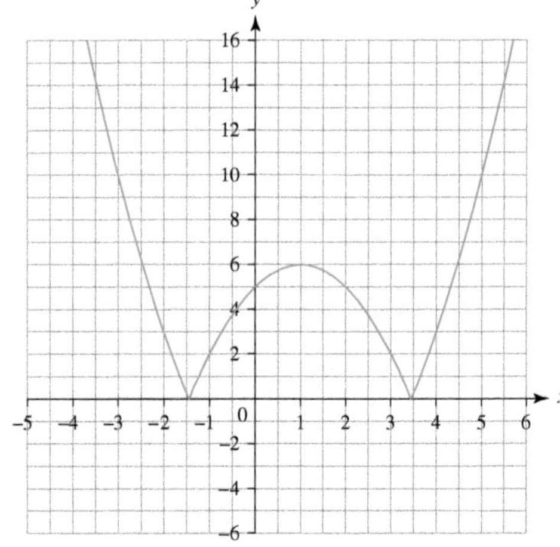

b

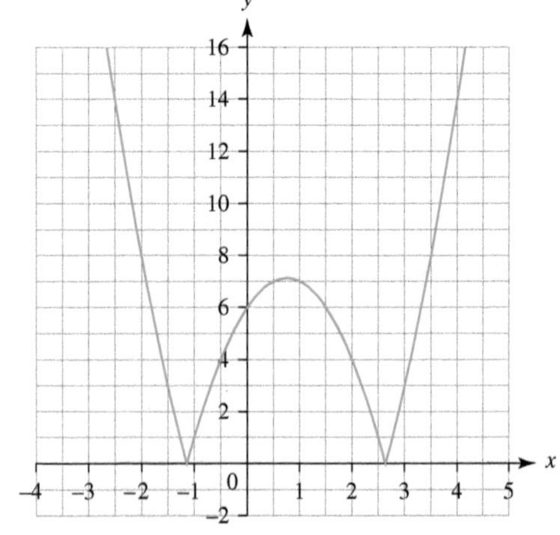

7 The blue line is f(x) and the red line is $\dfrac{1}{\mathrm{f}(x)}$

a

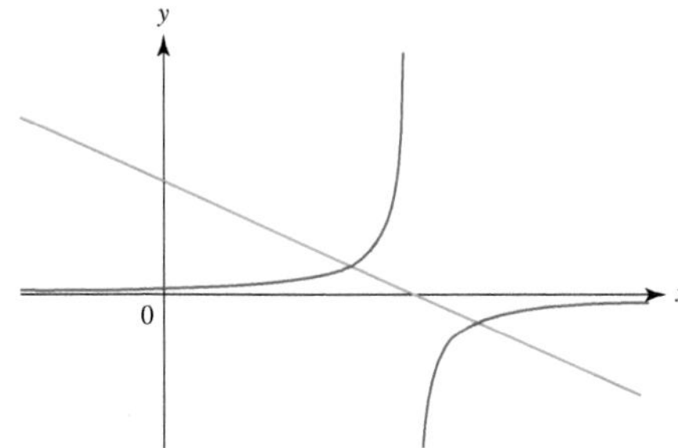

b

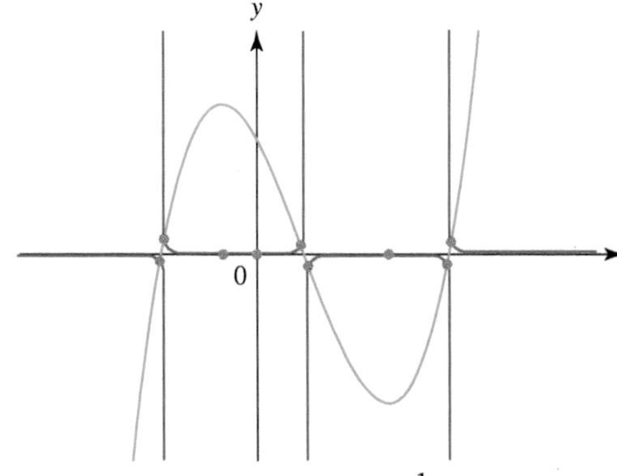

8 The blue line is f(x) and the red line is $\dfrac{1}{\mathrm{f}(x)}$

a

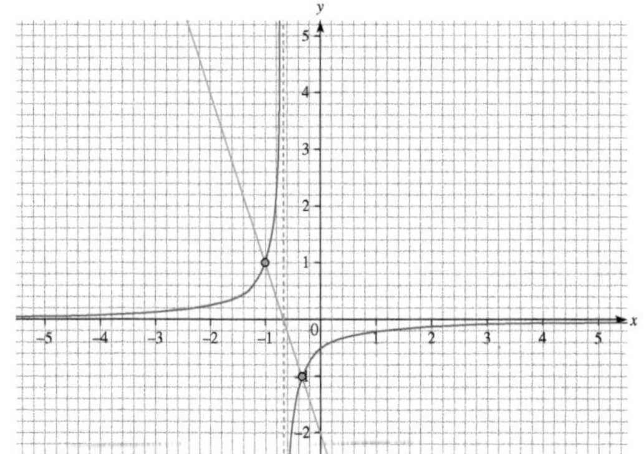

asymptotes: $x = -\dfrac{2}{3}$; $y = 0$

points of intersection: $(-1, 1)$, $(-\dfrac{1}{3}, -1)$

b asymptotes: $x = -5$; $x = -1$; $y = 0$

points of intersection: $(-3 - \sqrt{5}, 1)$, $(-3 + \sqrt{5}, 1)$, $(-3 - \sqrt{3}, -1)$, $(-3 + \sqrt{3}, -1)$

or (–5.24, 1), (–0.764, 1), (–4.73, –1), (–1.27, –1) to 3 sf

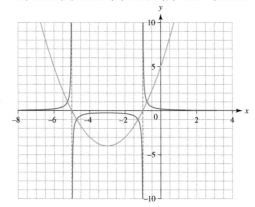

c asymptotes: $x = \sqrt[3]{-4}$ and $y = 0$

points of intersection: $(\sqrt[3]{-5}, -1)$, $(\sqrt[3]{-3}, 1)$ or (–1.71, –1), (–1.44, 1) to 3 sf

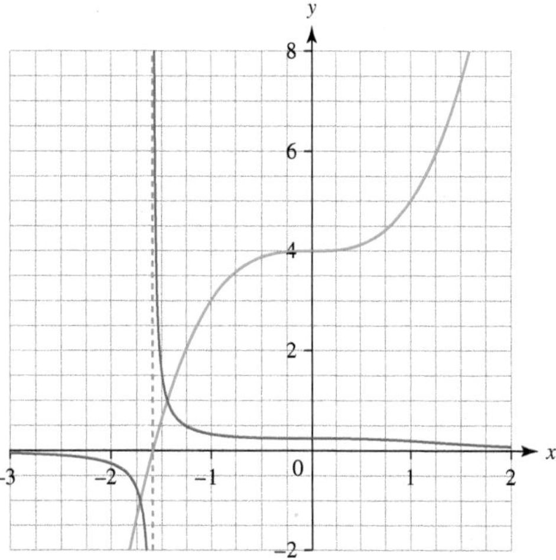

9 **a** **i** $x = 2$; $y = -1$; $(-4, 0)$, $(0, 2)$

ii

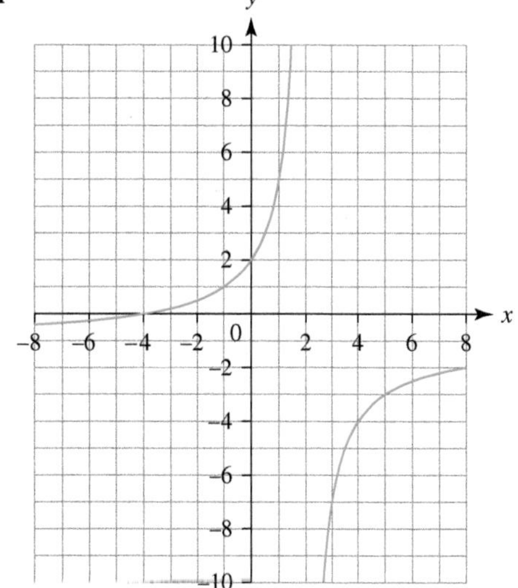

Intersections at (–4, 0) and (0, 2)

b

c

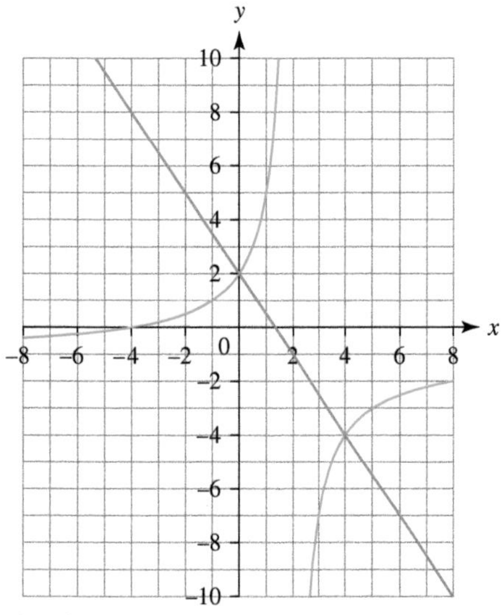

$\frac{(4+x)}{(2-x)} > 2 - \frac{3x}{2}$ when $0 < x < 2$ and when $4 < x$

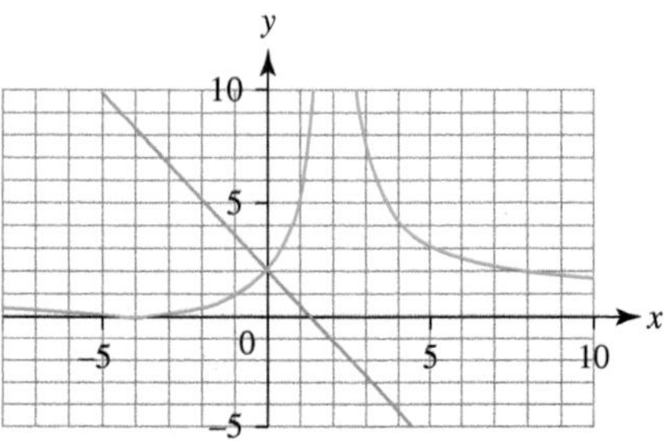

$0 < x < 2$ and $x > 2$

10 a $\frac{x^2}{45} + \frac{y^2}{81} = 1$

b

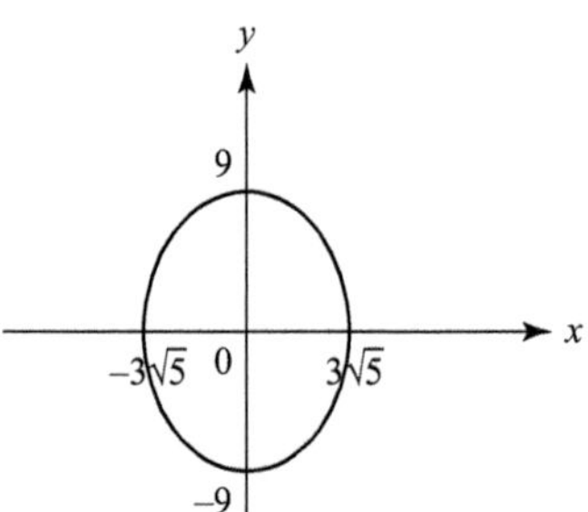

11 a $\frac{x^2}{16} - \frac{y^2}{9} = 1$

b asymptotes at $y = \pm\frac{3}{4}x$

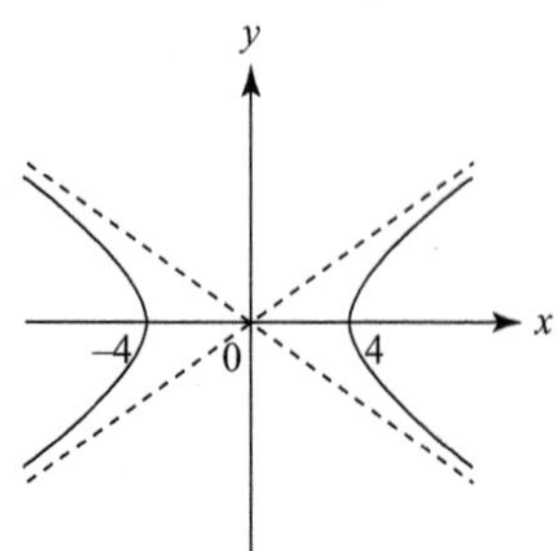

12 a $y = \frac{x^2}{18}$ **b** $(y-2) = \frac{1}{18}(x+1)^2$

c

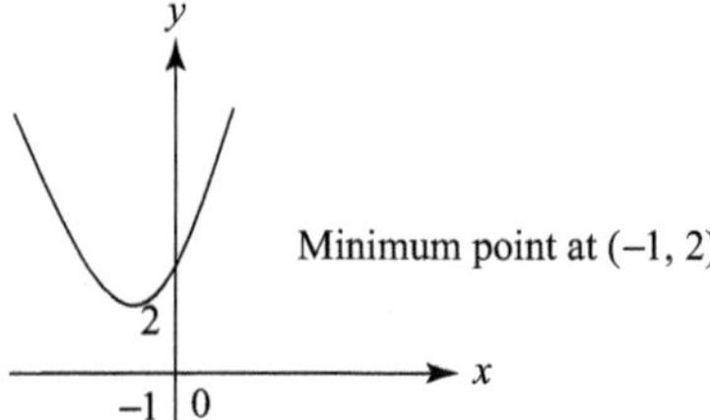

Minimum point at (−1, 2)

13 a $\frac{x^2}{4} + \frac{y^2}{9} = 1$

b $\frac{(x-2)^2}{4} + \frac{(y-3)^2}{9} = 1$

c

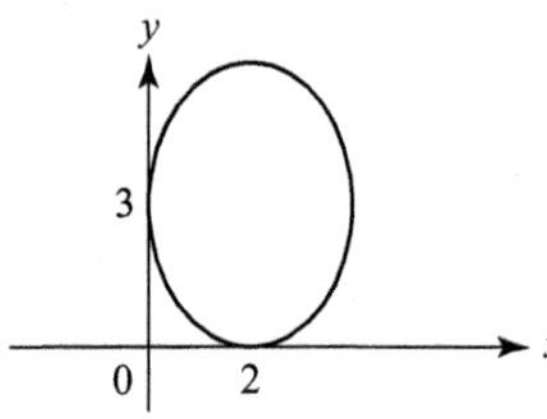

14 a $\frac{(x-1)^2}{100} + \frac{(y+4)^2}{75} = 1$

b

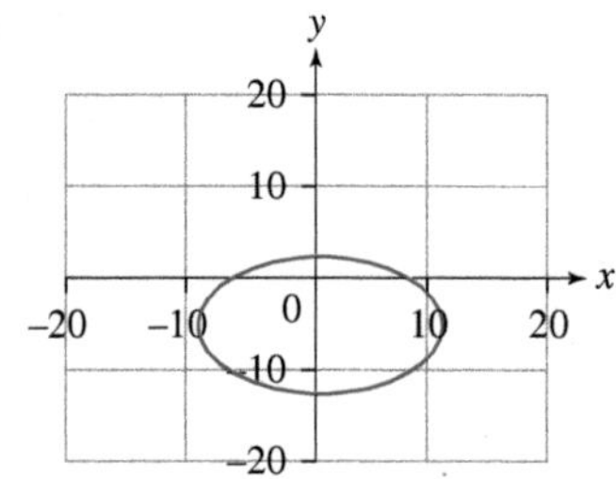

15 a $y^2 = 7x$

b

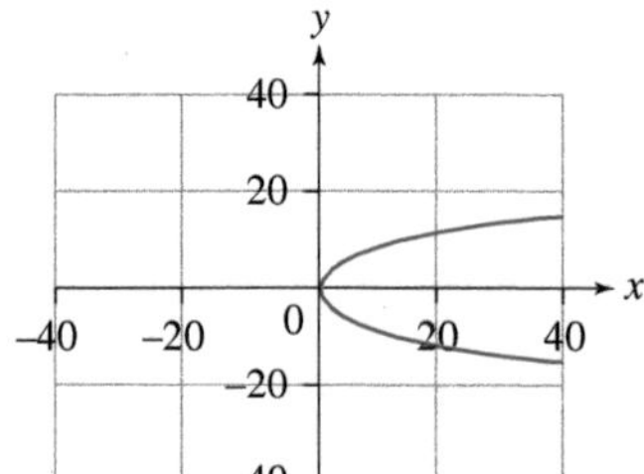

16 a

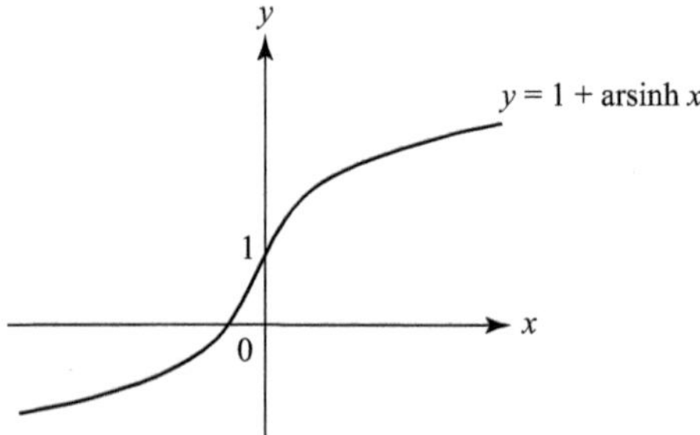

domain $x \in \mathbb{R}$; range $y \in \mathbb{R}$

b

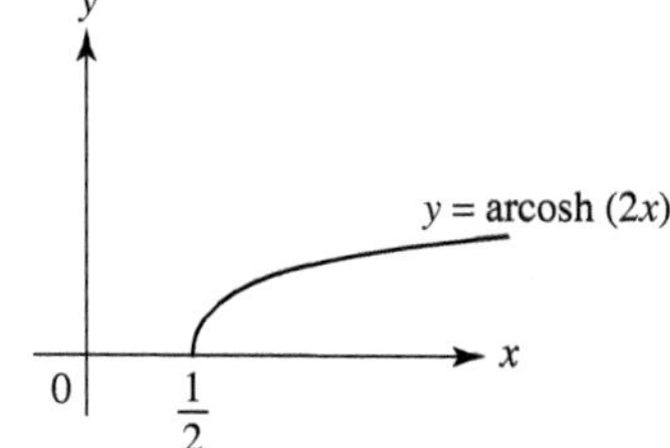

domain $x \in \mathbb{R}, x \geq \frac{1}{2}$; range $y \in \mathbb{R}, y \geq 0$

c

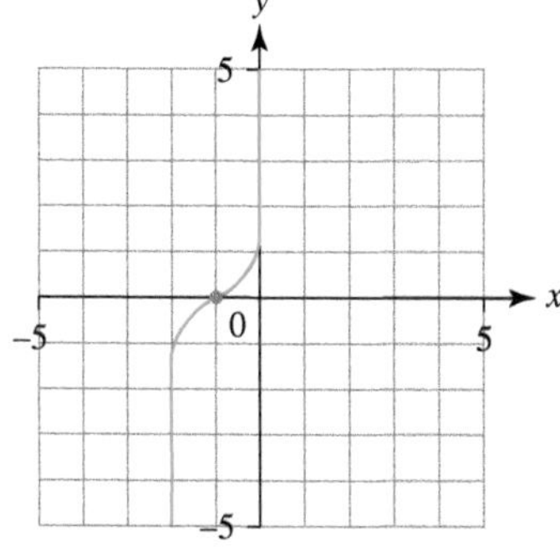

domain $x \in \mathbb{R}, -2 < x < 0$; range $y \in \mathbb{R}$

17

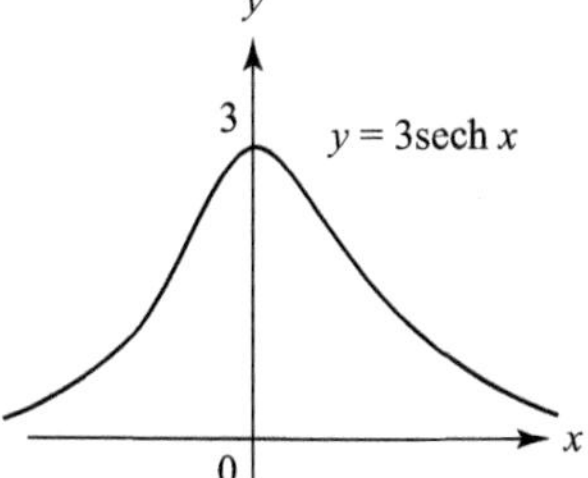

domain, $x \in \mathbb{R}$; range $y \in \mathbb{R}, 0 < y \leq 3$

18 a $\frac{1}{2}\ln(2)$ **b** $\frac{1}{2}\ln\left(\frac{1}{3}\right)$

c $\frac{1}{2}\ln(1+\sqrt{2})$

19 a $\text{cosech}(2x) = \frac{2}{e^{2x} - e^{-2x}}$

$$\frac{1}{2}\text{cosech}(x)\text{sech}(x) = \frac{1}{2}\left(\frac{2}{e^{x} - e^{-x}}\right)\left(\frac{2}{e^{x} + e^{-x}}\right)$$

$$= \frac{1}{2}\left(\frac{4}{e^{2x} - e^{-2x}}\right)$$

$$= \frac{2}{e^{2x} - e^{-2x}} = \text{cosech}(2x)$$

b $x = \frac{1}{2}\ln(-1+\sqrt{2})$

20 a $X = \ln(1+\sqrt{2})$ or $X = \ln\left(\frac{1}{4}(\sqrt{17}-1)\right)$

b $X = \frac{1}{2}\ln\left(\frac{3}{2}\right)$

21 $\tanh^2 x = 1 - \text{sech}^2 x$

$1 - \text{sech}^2 x + \text{sech}\, x + 5 = 0$

$\text{sech}^2 x - \text{sech}\, x - 6 = 0$

$\text{sech}\, x = 3, -2$

But the range of $y = \text{sech}\, x$ is $0 < y \leq 1$ so there are no solutions to the equation.

22 a $y = x + 3$ and $x = 5$

b

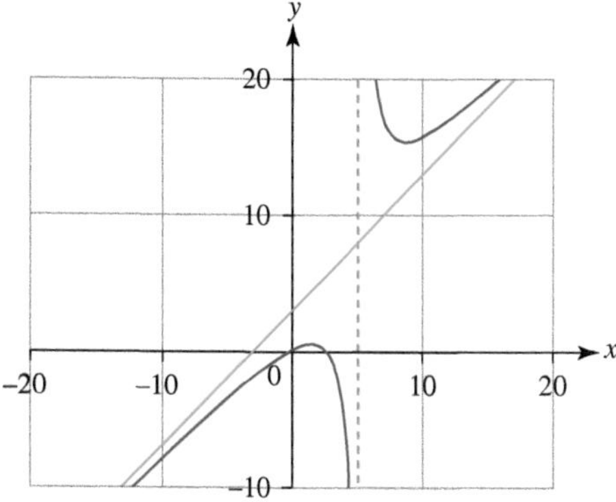

23 a $a = 3, b = -2$ **b** $x = \frac{1}{3}$

c

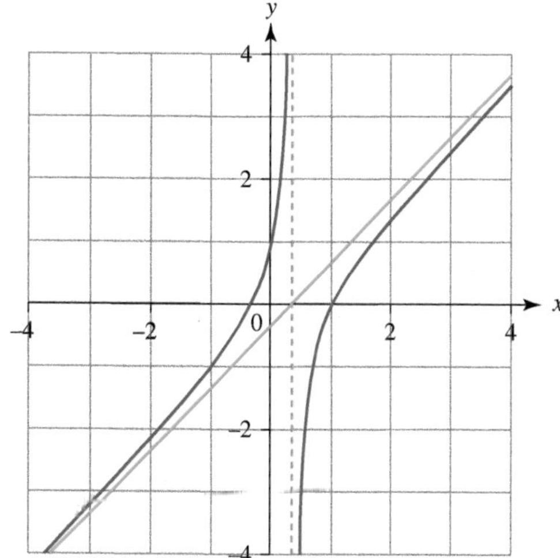

Assessment 18

1 a

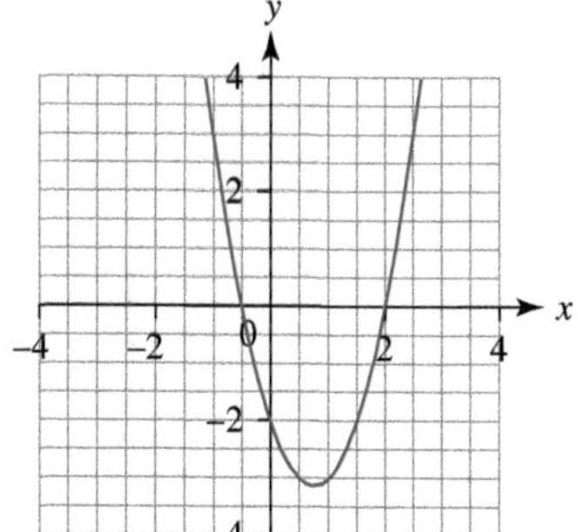

Intersections at (–0.5, 0), (2, 0) and (0, –2)

b

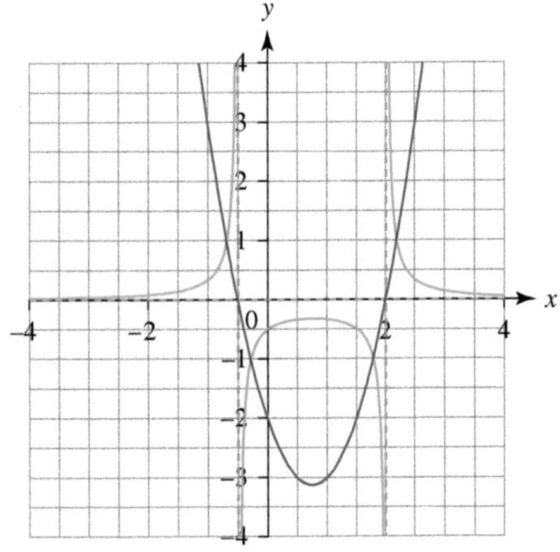

Asymptotes at $x = -0.5$, $x = 2$, $y = 0$

c $\left\{\left(\frac{3-\sqrt{33}}{4} < x < -\frac{1}{2}\right) \cup \left(\frac{3-\sqrt{17}}{4} < x < \frac{3+\sqrt{17}}{4}\right)\right.$

$\left.\cup\left(2 < x < \frac{3+\sqrt{33}}{4}\right)\right\}$

2 a

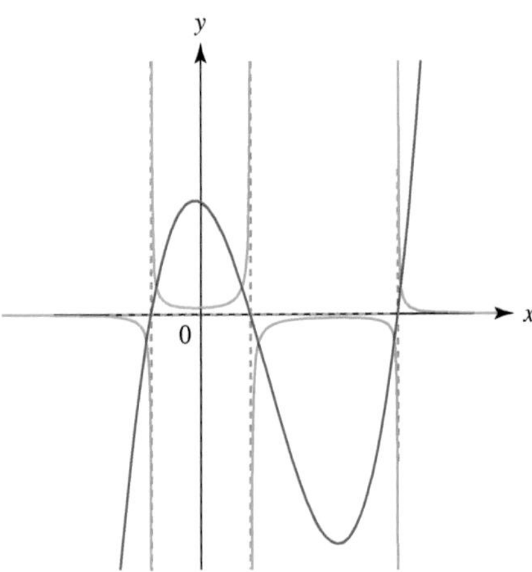

b

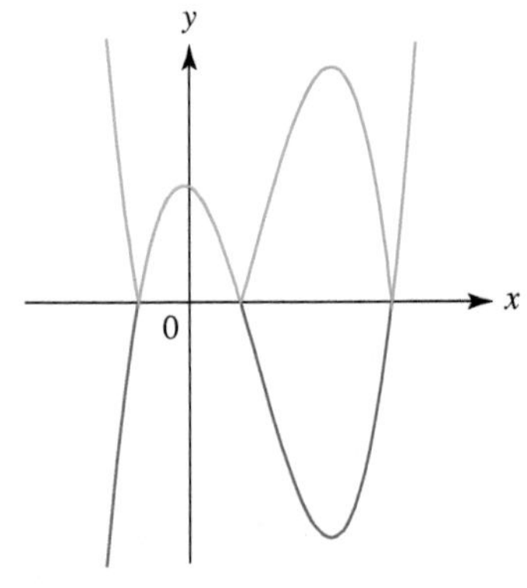

3 a $x=-1$ $y=-1$ $(0, 3)$, $(3, 0)$

b

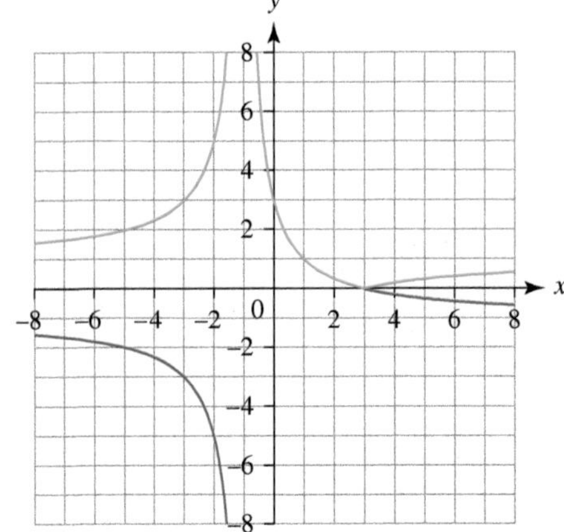

c $x < -3$, $x > 0$

4 a $\frac{\left(\frac{x}{2}\right)^2}{4} + \frac{\left(\frac{y}{2}\right)^2}{2} = 1$

$\frac{x^2}{16} + \frac{y^2}{8} = 1$

$\frac{(x-2)^2}{16} + \frac{(y+3)^2}{8} = 1$

$x^2 - 4x + 4 + 2(y^2 + 6y + 9) = 16$

$x^2 + 2y^2 - 4x + 12y + 6 = 0$ $(k = 6)$

b $y = -3 \pm 2\sqrt{2}$

5 $(x-1)^2 - 1 = -16y - 33$

$(x-1)^2 = -16(y+2)$

Recognise second transformation is translation vector $\begin{pmatrix} 1 \\ -2 \end{pmatrix}$

$x^2 = -16y \Rightarrow$ rotation $\frac{3\pi}{2}$ radians about $(0, 0)$

Must state: rotation *followed by* translation.

6 a $x = 3y - 3$ and substitute:

$\frac{(3y-3)^2}{9} - \frac{y^2}{4} = 1$

$4(9y^2 - 18y + 9) - 9y^2 = 36$

$3y^2 - 8y = 0$

$y = 0$ or $y = \frac{8}{3}$

Coordinates (both): $(-3, 0)$ and $\left(5, \frac{8}{3}\right)$

b

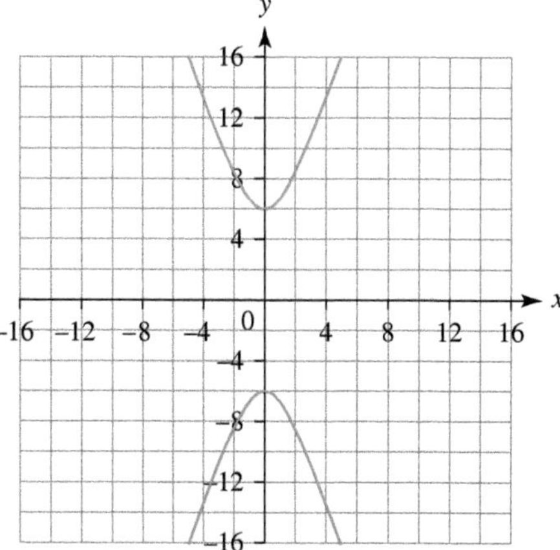

c $y = \pm 3x$

7 a $\left(\frac{e^x + e^{-x}}{2}\right)^2 - \left(\frac{e^x - e^{-x}}{2}\right)^2$

$= \frac{e^{2x} + 2 + e^{-2x}}{4} - \frac{e^{2x} - 2 + e^{-2x}}{4}$

$= \frac{2 - (-2)}{4} = 1$

b $x = \ln\frac{1}{2}$ or $x = -\ln 2$

8 a $2\cosh^2 x = 2\left(\frac{e^x + e^{-x}}{2}\right)^2$

$= \frac{e^{2x} + 2 + e^{-2x}}{2}$

$= \frac{e^{2x} + e^{-2x}}{2} + 1 = \cosh 2x + 1$

b $x = \ln(3 + \sqrt{8})$

9 a $\sinh x + \frac{3}{\sinh x} = 4 \Rightarrow \sinh^2 x + 3 = 4\sinh x$

$\sinh^2 x - 4\sinh x + 3 = 0$

b $x = \ln(3 + \sqrt{10})$ or $x = \ln(1 + \sqrt{2})$

10 a

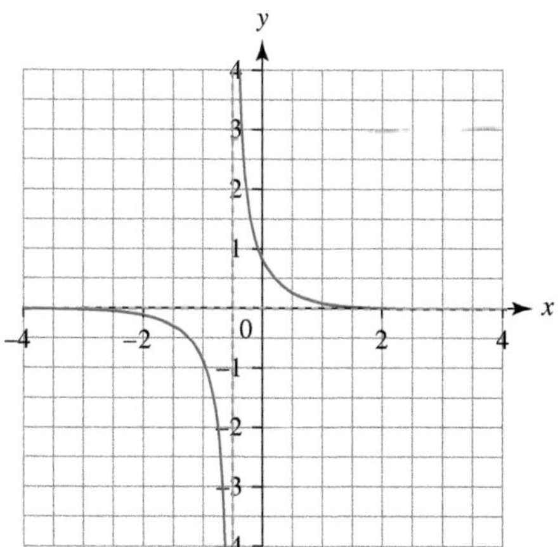

b Domain: $x \in \mathbb{R}, x \neq -\frac{1}{2}$

Range: $f(x) \in \mathbb{R}, f(x) \neq 0$

c $y = \operatorname{cosech} 1$

$\frac{2}{e^1 - e^{-1}} \times \frac{e^1}{e^1} = \frac{2e}{e^2 - 1}$

11 a $y = 2x - 2$ b $a = 4$ and $b = -6$

c

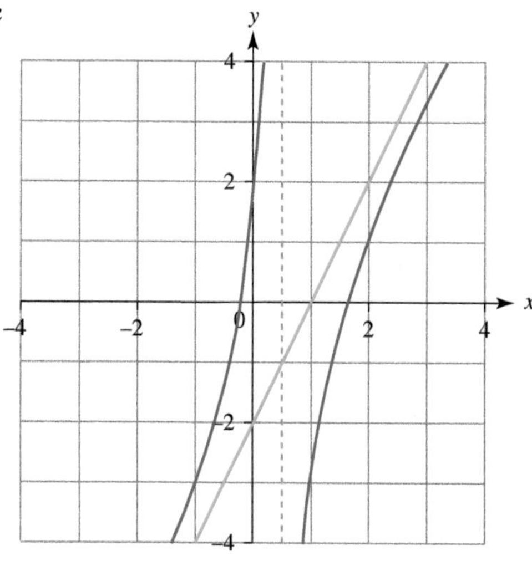

Chapter 19

Exercise 19.1A

1 a Not an improper integral, fully defined in interval and limits finite

b An improper integral since $\ln x$ undefined at $x = 0$

c An improper integral since one of the limits is ∞

d An improper integral since the limits are $\pm\infty$

e Not an improper integral, fully defined in interval and limits finite

f An improper integral since $\tan x$ undefined at $x = \frac{\pi}{2}$

2 a 1 b $\frac{1}{8}$ c 1

d $\frac{1}{6}$ e $\frac{1}{8}$ f $\frac{1}{3}$

g $\frac{1}{4}$ h 1

3 $\frac{1}{e^3}$

4 2

5 a 6 b $\frac{27}{2}$ c $2\sqrt{2}$

d 3 e −4 f 2

g 2 h $\sqrt{\frac{1}{2}}$

6 a $\frac{1}{4}e^2$ b $\frac{2}{9}e^3$

7 $\int_{1-e}^{a} \ln(1-x)\,dx = [-x + x\ln(1-x) - \ln(1-x)]_{1-e}^{a}$

$= (-a + a\ln(1-a) - \ln(1-a)) - (-(1-e) + (1-e)\ln e - \ln e)$

$= -a + (a-1)\ln(1-a) + (1-e) + e\ln e$

$= -a + (a-1)\ln(1-a) + (1-e) + e$

$= -a + (a-1)\ln(1-a) + 1$

$\to -1 + 1 = 0$ as $a \to 1$ since $-a \to -1$ and $(a-1)\ln(1-a) \to 0$ as $a \to 1$

8 $\frac{4}{9}e^3$

Exercise 19.1B

1 a $\int_a^1 x^{-4}\,dx = \left[-\frac{1}{3}x^{-3}\right]_a^1$

$= -\frac{1}{3}\left(1 - \frac{1}{a^3}\right)$

As $a \to 0$, $\frac{1}{a^3} \to \infty$ so integral does not converge.

b Integral converges and has value $\frac{1}{3}$

c Integral has value $\ln\sqrt{2}$

d $\int_0^a \tan x\,dx = [-\ln\cos x]_0^a$

$= -(\ln\cos a - \ln 1)$
$= -\ln\cos a$

As $a \to \frac{\pi}{2}$, $\cos a \to 0$ so $\ln\cos a \to \infty$

So integral does not converge.

e $\int_0^a \cos x\ dx = [\sin x]_0^a$

$= \sin a - \sin 0$
$= \sin a$

As $a \to \infty$, $\sin a$ does not converge so integral does not converge.

f $\int_1^a \frac{1}{x}\,dx = [\ln x]_1^a$

$= \ln a - \ln 1$
$= \ln a$

As $a \to \infty$, $\ln a \to \infty$ so integral does not converge.

g $\int_a^0 \frac{1}{3-x}\,dx = [-\ln(3-x)]_a^0$

$= -\ln 3 + \ln(3-a)$

As $a \to -\infty$, $\ln(3-a) \to \infty$ so integral does not converge.

h Integral converges and has value $2\sqrt{7}$

i $\int_0^7 (7-x)^{-2}\,dx = [(7-x)^{-1}]_0^a$

$= \frac{1}{7-a} - \frac{1}{7}$

As $a \to 7$, $\frac{1}{7-a} \to \infty$ so integral does not converge.

j Integral converges and has value $\ln\left(\frac{3}{2}\right)$

k $\int_1^a \frac{1}{x} - \frac{2x}{x^2+1}\,dx = [\ln x - \ln(x^2+1)]_1^a$

$= \left[\ln\frac{x}{x^2+1}\right]_1^a$

$= \ln\frac{a}{a^2+1} - \ln\frac{1}{2}$

$\frac{a}{a^2+1} = \frac{\frac{1}{a}}{1+\frac{1}{a^2}}$

As $a \to \infty$, $\frac{\frac{1}{a}}{1+\frac{1}{a^2}} \to 0$ so $\ln\frac{a}{a^2+1} \to -\infty$

So integral does not converge.

i Integral converges and has value $\ln\left(\frac{\sqrt{3}}{2}\right)$

2 a $\int \frac{x}{x^2+3} - \frac{2}{2x+3} = \frac{1}{2}\ln(x^2+3) - \ln(2x+3)$

$= \frac{1}{2}\ln(x^2+3) - \frac{1}{2}\ln(2x+3)^2$

$= \frac{1}{2}\ln\left(\frac{x^2+3}{(2x+3)^2}\right)$

$= \frac{1}{2}\ln\frac{x^2+3}{4x^2+12x+9}$

b $\int_0^a \frac{x}{x^2+3} - \frac{2}{2x+3} = \frac{1}{2}\left[\ln\frac{x^2+3}{4x^2+12x+9}\right]_0^a$

$= \frac{1}{2}\ln\frac{a^2+3}{4a^2+12a+9} - \frac{1}{2}\ln\frac{3}{9}$

$= \frac{1}{2}\ln\frac{1+\frac{3}{a^2}}{4+\frac{12}{a}+\frac{9}{a^2}} - \frac{1}{2}\ln\frac{1}{3}$

$\to \frac{1}{2}\ln\frac{1}{4} - \frac{1}{2}\ln\frac{1}{3}$ as $a \to \infty$ since

$\frac{1+\frac{3}{a^2}}{4+\frac{12}{a}+\frac{9}{a^2}} \to \frac{1}{4}$ as $a \to \infty$

$= \frac{1}{2}\ln\left(\frac{\frac{1}{4}}{\frac{1}{3}}\right)$

$= \frac{1}{2}\ln\left(\frac{3}{4}\right)$

$= \ln\left(\frac{\sqrt{3}}{2}\right) \qquad \left(k = \frac{\sqrt{3}}{2}\right)$

3 a $\int_a^0 \frac{6}{3x-2}-\frac{2x}{x^2+4}dx = [2\ln(3x-2)-\ln(x^2+4)]_a^0$

$$= \left[\ln\frac{(3x-2)^2}{x^2+4}\right]_a^0$$

$$= \ln 1 - \ln\frac{(3a-2)^2}{a^2+4}$$

$$= -\ln\frac{9a^2-12a+4}{a^2+4}$$

$$\frac{9a^2-12a+4}{a^2+4} = \frac{9-\frac{12}{a}+\frac{4}{a^2}}{1+\frac{4}{a^2}}$$

As $a \to -\infty$, $\frac{9-\frac{12}{a}+\frac{4}{a^2}}{1+\frac{4}{a^2}} \to 9$, so $-\ln\frac{9a^2-12a+4}{a^2+4} \to -\ln 9$

So integral converges and value is $-\ln 9$ or $\ln\frac{1}{9}$

b $\int\int_a^0 \frac{3-x}{x^2-6x+1}-\frac{8}{5-8x}\,dx$

$$= \left[-\frac{1}{2}\ln(x^2-6x+1)+\ln(5-8x)\right]_a^0$$

$$= \left[\frac{1}{2}\ln(5-8x)^2-\frac{1}{2}\ln(x^2-6x+1)\right]_a^0$$

$$= \left[\frac{1}{2}\ln\frac{(5-8x)^2}{x^2-6x+1}\right]_a^0$$

$$= \frac{1}{2}\ln 25 - \frac{1}{2}\ln\frac{(5-8a)^2}{a^2-6a+1}$$

$$\frac{(5-8a)^2}{a^2-6a+1} = \frac{25-80a+64a^2}{a^2-6a+1}$$

$$= \frac{\frac{25}{a^2}-\frac{80}{a}+64}{1-\frac{6}{a}+\frac{1}{a^2}} \to 64 \text{ as } a \to -\infty$$

So integral converges and is given by:

$$\frac{1}{2}\ln 25 - \frac{1}{2}\ln 64 = \frac{1}{2}\ln\left(\frac{25}{64}\right) = \ln\left(\frac{5}{8}\right)$$

4 Integral converges when $p<1$ and has value $\frac{1}{1-p}$

5 Integral converges when $p>1$ and has value $\frac{1}{p-1}$

Exercise 19.2A

1 a $-\frac{1}{\sqrt{1-x^2}}$

b $\frac{1}{1+x^2}$

c $\frac{2}{\sqrt{1-4x^2}}$

d $-\frac{5}{\sqrt{1-25x^2}}$

e $\frac{1}{x^2-2x+2}$

f $\frac{2}{\sqrt{1-x^2}}$

g $-\frac{3}{\sqrt{9-x^2}}$

h $-\frac{3}{\sqrt{4x-x^2-3}}$

i $-\frac{2x}{\sqrt{1-x^4}}$

j $\frac{x}{\sqrt{1-x^2}}+\arcsin x$

2 Let $y = \text{arcsec}\, x \Rightarrow x = \sec y = (\cos y)^{-1}$

$$\frac{dx}{dy} = -(\cos y)^{-2}(-\sin y)$$

$$= \frac{\sin y}{\cos^2 y}$$

$$= \tan y \sec y$$

$$= \sec y\sqrt{\sec^2 y - 1}$$

$$= x\sqrt{x^2-1}$$

$\frac{dy}{dx} = \frac{1}{x\sqrt{x^2-1}}$ as required

3 Let $y = \text{arccosec}\, x \Rightarrow x = \text{cosec}\, y = (\sin y)^{-1}$

$$\frac{dx}{dy} = -(\sin y)^{-2}(\cos y)$$

$$= -\frac{\cos y}{\sin^2 y}$$

$$= -\cot y\, \text{cosec}\, y$$

$$= -\text{cosec}\, y\sqrt{\text{cosec}^2 y - 1}$$

$$= -x\sqrt{x^2-1}$$

$\frac{dy}{dx} = -\frac{1}{x\sqrt{x^2-1}}$ as required

4 Let $y = \text{arccot}\, x \Rightarrow x = \cot y = (\tan y)^{-1}$

$$\frac{dx}{dy} = -(\tan y)^{-2}(\sec^2 y)$$

$$= -\frac{\sec^2 y}{\tan^2 y}$$

$$= -\text{cosec}^2 y$$

$$= -(1+\cot^2 y)$$

$$= -(1+x^2)$$

$\frac{dy}{dx} = -\frac{1}{1+x^2}$ as required

5 a $\frac{dy}{dx} = e^x \arctan x + \frac{e^x}{1+x^2}$

b $\frac{dy}{dx} = -\frac{1}{\sqrt{1-(3x^2-1)^2}}(6x)$

$$= -\frac{6x}{\sqrt{6x^2-9x^4}}$$

c $\frac{dy}{dx}=-\frac{2\sin x}{\sqrt{1-4x^2}}+\cos x\arccos 2x$

d $\frac{dy}{dx}=\frac{2(\arcsin x)}{\sqrt{1-x^2}}$

e $\frac{dy}{dx}=\frac{e^x}{\sqrt{1-e^{2x}}}$

f $\frac{dy}{dx}=\frac{2}{1+4x^2}e^{\arctan 2x}$

6 $x=\cos u\Rightarrow\frac{dx}{du}=-\sin u$

$$\int\frac{1}{\sqrt{1-x^2}}dx=\int\frac{1}{\sqrt{1-\cos^2 u}}(-\sin u)du$$
$$=-\int\frac{\sin u}{\sin u}du$$
$$=-\int 1du$$
$$=-u+c$$
$$=-\arccos x+c \text{ as required}$$

7 $x=\tan u\Rightarrow\frac{dx}{du}=\sec^2 u$

$$\int\frac{1}{1+x^2}dx=\int\frac{1}{1+\tan^2 u}(\sec^2 u)du$$
$$=\int\frac{\sec^2 u}{\sec^2 u}du$$
$$=\int 1du$$
$$=u+c$$
$$=\arctan x+c \text{ as required}$$

8 $\frac{1}{3}\arcsin 3x+c$

9 $-5\arccos\left(\frac{x}{5}\right)+c$

10 $\frac{1}{3}\arctan\left(\frac{x}{3}\right)+c$

Exercise 19.2B

1 a $\arcsin\left(\frac{x}{3}\right)+c$ b $\arcsin\left(\frac{x}{10}\right)+c$

c $-2\arcsin\left(\frac{x}{6}\right)+c$ d $-4\arcsin\left(\frac{x}{2\sqrt{2}}\right)+c$

e $\frac{1}{5}\arctan\left(\frac{x}{5}\right)+c$ f $\frac{1}{7}\arctan\left(\frac{x}{7}\right)+c$

g $-\frac{3\sqrt{2}}{2}\arctan\frac{\sqrt{2}}{2}x$ h $\sqrt{6}\arctan\left(\frac{\sqrt{6}}{6}x\right)+c$

i $\frac{1}{2}x\sqrt{1-x^2}+\frac{1}{2}\arcsin x+c$ j $\frac{1}{2}x\sqrt{16-x^2}+8\arcsin\left(\frac{x}{4}\right)+c$

2 a $8\arctan\left(\frac{x}{8}\right)+c$ b $5\arctan(5x)+c$

c $\sqrt{2}\arctan\left(\frac{\sqrt{2}}{4}x\right)+c$ d $\frac{2\sqrt{3}}{3}\arctan\left(\frac{\sqrt{3}}{3}x\right)+c$

e $3\arcsin\left(\frac{x}{3}\right)+c$ f $\frac{1}{2}\arcsin\left(\frac{x}{2}\right)+c$

g $-6\arcsin\left(\frac{x}{6}\right)+c$ h $\arcsin\left(\frac{\sqrt{2}}{2}x\right)+c$

i $\frac{1}{18}x\sqrt{81-x^2}+\frac{9}{2}\arcsin\left(\frac{x}{9}\right)+c$

j $\frac{3}{2}x\sqrt{4-x^2}+6\arcsin\left(\frac{x}{2}\right)+c$

3 a $\frac{1}{2}\arctan\left(\frac{x+1}{2}\right)+c$ b $\arctan(x-3)+c$

c $\frac{1}{2}\arctan\left(\frac{x-7}{2}\right)+c$ d $\sqrt{2}\arctan\left(\frac{x+6}{\sqrt{2}}\right)+c$

e $\arcsin\left(\frac{x+4}{6}\right)+c$ f $\arcsin\left(\frac{x-1}{\sqrt{2}}\right)+c$

g $-2\arcsin\left(\frac{x-4}{4}\right)+c$ h $\arcsin\left(\frac{x+2}{2\sqrt{2}}\right)+c$

i $\frac{x-2}{2}\sqrt{16-(x-2)^2}+8\arcsin\left(\frac{x-2}{4}\right)+c$

j $\frac{1}{2}(x+3)\sqrt{25-(x+3)^2}+\frac{25}{2}\arcsin\left(\frac{x+3}{5}\right)+c$

4 $\frac{\pi}{12}$

Exercise 19.3A

1 $\cos xh=\frac{1}{2}(e^x+e^{-x})$

So $\frac{d(\cosh x)}{dx}=\frac{1}{2}(e^x-e^{-x})$
$=\sinh x$

2 a $\tanh x=\frac{(e^x-e^{-x})}{e^x+e^{-x}}$

So $$\frac{d(\tanh x)}{dx}=\frac{(e^x+e^{-x})(e^x+e^{-x})-(e^x-e^{-x})(e^x-e^{-x})}{(e^x+e^{-x})^2}$$
$$=\frac{(e^{2x}+2+e^{-2x})-(e^{2x}-2+e^{-2x})}{(e^x+e^{-x})^2}$$
$$=\frac{4}{(e^x+e^{-x})^2}$$
$$=\left(\frac{2}{e^x+e^{-x}}\right)^2$$
$$=\text{sech}^2 x$$

b $\coth x=(\tanh x)^{-1}$

So $$\frac{d(\coth x)}{dx}=-(\tanh x)^{-2}\text{sech}^2 x$$
$$=-\frac{\text{sech}^2 x}{\tanh^2 x}$$
$$=-\frac{1}{\cosh^2 x}\div\frac{\sinh^2 x}{\cosh^2 x}$$
$$=-\frac{1}{\sinh^2 x}$$
$$=-\text{cosech}^2 x$$

3 $\text{cosech}\,x=\frac{2}{e^x-e^{-x}}=2(e^x-e^{-x})^{-1}$

$$\frac{d(\text{cosech}\,x)}{dx}=-2(e^x-e^{-x})^{-2}(e^x+e^{-x})$$
$$=-\frac{2(e^x+e^{-x})}{(e^x-e^{-x})^2}$$
$$=-\frac{2}{e^x-e^{-x}}\cdot\frac{e^x+e^{-x}}{e^x-e^{-x}}$$
$$=-\frac{2}{e^x-e^{-x}}\cdot\frac{e^{2x}+1}{e^{2x}-1}$$
$$=-\text{cosech}\,x\coth x$$

4 a $-3\coth x\,\text{cosech}\,x$ b $-\text{cosech}^2(x+1)$

c $2\cosh 2x$ d $\frac{1}{3}\sinh\left(\frac{x}{3}\right)$

e $\text{sech}^2(x-2)$ f $2x\cosh x^2$

g $2\sinh(2x-3)$ h $2\sinh x\cosh x$

i $\tanh x+x\,\text{sech}^2 x$ j $\sqrt{\text{sech}\,x}+\frac{1}{2}x\tanh x\sqrt{\text{sech}\,x}$

5 a $\tanh x+c$ **b** $-\operatorname{cosech} x+c$

c $\frac{1}{2}\sinh 2x+c$ **d** $3\cosh\left(\frac{x}{3}\right)+c$

e $\frac{1}{3}\sinh^3 x+c$ **f** $\frac{1}{4}\sinh 2x+\frac{1}{2}x+c$

g $\frac{1}{4}\ln\sinh 4x+c$ **h** $\frac{1}{5}\coth 5x+c$

6 a $\frac{9}{16}$ **b** $\frac{4}{3}$

c $\frac{27}{65}$ **d** $\frac{13}{240}$

7 Let $y=\operatorname{arsinh} x$ then $x=\sinh y$

$$\frac{dx}{dy}=\cosh y$$

$$\frac{dy}{dx}=\frac{1}{\cosh y}$$

$$=\frac{1}{\sqrt{1+\sinh^2 y}}$$

$$=\frac{1}{\sqrt{1+x^2}} \text{ as required}$$

8 Let $y=\operatorname{artanh} 2x$ then $x=\frac{1}{2}\tanh y$

$$\frac{dx}{dy}=\frac{1}{2}\operatorname{sech}^2 y$$

$$\frac{dy}{dx}=\frac{2}{\operatorname{sech}^2 y}$$

$$=\frac{2}{1-\tanh^2 y}$$

$$=\frac{2}{1-(2x)^2}$$

$$=\frac{2}{1-4x^2} \text{ as required}$$

9 Let $y=\operatorname{arcoth} x$ then $x=\coth y$

$$\frac{dx}{dy}=-\operatorname{cosech}^2 y$$

$$\frac{dy}{dx}=\frac{1}{-\operatorname{cosech}^2 y}$$

$$=\frac{1}{-(\coth^2 y-1)}$$

$$=\frac{1}{-(x^2-1)}$$

$$=\frac{1}{1-x^2} \text{ as required}$$

10 Let $y=\operatorname{arcosech} x$ then $x=\operatorname{cosech} y$

$$\frac{dx}{dy}=-\coth y\operatorname{cosech} y$$

$$\frac{dy}{dx}=\frac{1}{-\coth y\operatorname{cosech} y}$$

$$=\frac{1}{-\operatorname{cosech} y\sqrt{\operatorname{cosech}^2 y+1}}$$

$$=\frac{1}{-x\sqrt{x^2+1}} \text{ as required}$$

11 a $\frac{4}{\sqrt{16x^2-1}}$ **b** $\frac{1}{\sqrt{x^2+25}}$

c $\frac{2x}{\sqrt{x^4-1}}$ **d** $\frac{e^x}{\sqrt{e^{2x}+1}}$

e $\frac{\cosh x}{\sqrt{\sinh^2 x-1}}$ **f** $e^x\operatorname{arsinh}(x^2-1)+\frac{2xe^x}{\sqrt{x^4-2x^2+2}}$

Exercise 19.3B

1 a $\operatorname{arsinh}\left(\frac{x}{7}\right)+c$ **b** $\operatorname{arcosh}\left(\frac{x}{9}\right)+c$

c $\operatorname{arsinh}\left(\frac{x}{4}\right)+c$ **d** $-3\operatorname{arcosh}\left(\frac{x}{3}\right)+c$

e $2\operatorname{arsinh}\left(\frac{x+3}{4}\right)+c$ **f** $\operatorname{arsinh}(x-5)+c$

g $\operatorname{arcosh}\left(\frac{x+7}{5}\right)+c$ **h** $3\operatorname{arcosh}\left(\frac{x-12}{10}\right)+c$

2 a 0.398 **b** 23.7

3 a $\ln(1+\sqrt{2})$ **b** $\frac{1}{\sqrt{3}}\ln 2$

4 a $\frac{1}{9}\sqrt{16+9x^2}+\frac{1}{3}\operatorname{arsinh}\left(\frac{3x}{4}\right)+c$

b $6\operatorname{arcosh}\left(\frac{x}{\sqrt{12}}\right)-4\sqrt{\frac{x^2}{4}-3}+c$

c $\sqrt{8}\arcsin\left(\frac{x}{\sqrt{8}}\right)-\frac{1}{2}\sqrt{16-2x^2}+c$

d $\frac{35}{2}\ln(x^2+49)+2\arctan\left(\frac{x}{7}\right)+c$

5 $$\int \operatorname{arsinh} x\,dx=\int 1\times\operatorname{arsinh} x\,dx$$

$$=x\operatorname{arsinh} x-\int\frac{x}{\sqrt{x^2+1}}dx$$

$$=x\operatorname{ar\,sinh} x-\sqrt{x^2+1}+c$$

6 a $x\operatorname{arcosh} x-\sqrt{x^2-1}+c$

b $x\operatorname{arcoth} x+\frac{1}{2}\ln(1-x^2)+c$

7 Let $x=a\cosh u$, $\frac{dx}{du}=a\sinh u$

$$\int\frac{1}{\sqrt{x^2-a^2}}dx=\int\frac{1}{\sqrt{a^2\cosh^2 u-a^2}}a\sinh u\,du$$

$$=\int\frac{a\sinh u}{\sqrt{a^2\sinh^2 u}}du$$

$$=\int 1\,du$$

$$=u\ (+c)$$

$$=\operatorname{arcosh}\left(\frac{x}{a}\right)\ (+c)$$

A: $$\left[\operatorname{arcosh}\left(\frac{x}{a}\right)\right]_t^{2a}=\operatorname{arcosh}(2)-\operatorname{arcosh}\left(\frac{t}{a}\right)$$

$$=\ln(2+\sqrt{4-1})-\ln\left(\frac{t}{a}+\sqrt{\left(\frac{t}{a}\right)^2-1}\right)$$

As $t\to a$, $\frac{t}{a}\to 1$ so $\frac{t}{a}+\sqrt{\left(\frac{t}{a}\right)^2-1}\to 1$ so

$$\ln\left(\frac{t}{a}+\sqrt{\left(\frac{t}{a}\right)^2-1}\right)\to\ln 1=0$$

So the integral converges and is equal to $\ln(2+\sqrt{3})$

B: $$\left[\operatorname{arcosh}\left(\frac{x}{a}\right)\right]_{2a}^{t}=\operatorname{arcosh}\left(\frac{t}{a}\right)-\operatorname{arcosh}(2)$$

$= \ln\left(\frac{t}{a}+\sqrt{\left(\frac{t}{a}\right)^2-1}\right)-\ln(2+\sqrt{3})$

As $t\to\infty, \frac{t}{a}\to\infty$ so $\frac{t}{a}+\sqrt{\left(\frac{t}{a}\right)^2-1}\to\infty$ so

$\ln\left(\frac{t}{a}+\sqrt{\left(\frac{t}{a}\right)^2-1}\right)\to\infty$

So the integral does not converge.

Exercise 19.4A

1 a $5x+\ln(x-2)+3\ln(x+4)+c$

b $2x+3\ln(x+7)+\frac{4}{x+7}+c$

c $\frac{5}{2}x^2+x+2\ln(x+3)-3\ln(x-3)+c$

d $8x+3\ln(x+3)-2\ln(x+8)+3\ln x+c$

e $\frac{x^3}{3}+\frac{x^2}{2}-3x+\ln(x-9)-2\ln(x+1)+c$

f $\frac{9}{2}x^2+27x+\ln(x+1)+80\ln(x-2)-\frac{48}{x-2}+c$

g $-\frac{x^3}{3}-x+\frac{5}{3}\ln(3x+4)+\frac{3}{4}\ln(3-4x)+c$

2 a $3\ln\left(\frac{x+3}{\sqrt{x^2+1}}\right)+2\arctan x+c$

b $\ln\left(\frac{x-3}{\sqrt{x^2+25}}\right)+\arctan\left(\frac{x}{5}\right)+c$

c $\ln\left(\frac{\sqrt{x^2+36}}{6-x}\right)+\frac{5}{6}\arctan\left(\frac{x}{6}\right)+c$

d $\ln\left(\frac{x^2+2}{(1-x)^2}\right)-\frac{5}{\sqrt{2}}\arctan\left(\frac{x}{\sqrt{2}}\right)+c$

e $\ln(x+1)+2\ln(x-2)-\frac{3}{2}\ln(x^2+9)-\arctan\left(\frac{x}{3}\right)+c$

f $-2\ln(x-1)-\frac{3}{x-1}+\ln(x^2+16)-\frac{1}{4}\arctan\left(\frac{x}{4}\right)+c$

3 a $\ln x+\frac{3}{7}\arctan\left(\frac{x}{7}\right)-\frac{1}{2}\ln(x^2+49)+c$

b $-\frac{3}{2}\ln x+\frac{3}{4}\ln(x^2+64)+\frac{1}{16}\arctan\left(\frac{x}{8}\right)+c$

c $\ln(3x-1)+\frac{3}{2}\arctan\left(\frac{x}{2}\right)-\frac{1}{2}\ln(x^2+4)+c$

4 a $\frac{1}{x-3}+\frac{1}{(x-3)^2}-\frac{x+4}{x^2+5}$

b $\ln(x-3)-\frac{1}{x-3}-\frac{1}{2}\ln(x^2+5)-\frac{4}{\sqrt{5}}\arctan\left(\frac{x}{\sqrt{5}}\right)+c$

Exercise 19.4B

1 a $\ln\left(\frac{4}{3}\right)$

b $\ln\left(\frac{1}{6}\right)$

c $2+\ln\left(\frac{5}{2}\right)$

d $\frac{1}{2}\ln 10$

2 $\frac{-2x-12}{2x^3-x^2+6x-3}=\frac{Ax+B}{x^2+3}+\frac{C}{2x-1}$

$-2x-12=(Ax+B)(2x-1)+C(x^2+3)$

Let $x=\frac{1}{2}$ then $-13=\frac{13}{4}C\Rightarrow C=-4$

$x^2: 0=2A+C\Rightarrow A=2$

$1: -12=-B+3C\Rightarrow B=0$

A:

$\int_1^a \frac{2x}{x^2+3}-\frac{4}{2x-1}dx=[\ln(x^2+3)-2\ln(2x-1)]_1^a$

$=\left[\ln\left(\frac{x^2+3}{(2x-1)^2}\right)\right]_1^a$

$=\ln\left(\frac{a^2+3}{(2a-1)^2}\right)-\ln(4)$

$\frac{a^2+3}{(2a-1)^2}=\frac{a^2+3}{4a^2-4a+1}$

$=\frac{1+\frac{3}{a^2}}{4-\frac{4}{a}+\frac{1}{a^2}}$

$\frac{1+\frac{3}{a^2}}{4-\frac{4}{a}+\frac{1}{a^2}}\to\frac{1}{4}$ as $a\to\infty$ since $\frac{1}{a},\frac{1}{a^2}\to 0$

So $\ln\left(\frac{a^2+3}{(2a-1)^2}\right)\to\ln\left(\frac{1}{4}\right)=-\ln 4$

So integral converges and is $-2\ln 4$

B:

Point of discontinuity at $x=\frac{1}{2}$

$\frac{a^2+3}{(2a-1)^2}\to\infty$ as $x\to\frac{1}{2}$ since $2a-1\to 0$

So $\ln\left(\frac{a^2+3}{(2a-1)^2}\right)\to\infty$

Hence integral does not converge.

3 a $\frac{2-x}{x^2+a}+\frac{1}{x}$

b $\int_0^a \frac{1}{x}\,dx$ does not converge.

$\int_0^a \frac{2-x}{x^2+a}\,dx=\left[\frac{2}{\sqrt{a}}\arctan\left(\frac{x}{\sqrt{a}}\right)-\frac{1}{2}\ln(x^2+a)\right]_0^a$

$=\frac{2}{\sqrt{a}}\arctan(\sqrt{a})-\frac{1}{2}\ln(a^2+a)-0+\frac{1}{2}\ln(a)$

converges to a finite value.

Therefore $\int_0^a \frac{2x+a}{x^3+ax}dx$ does not converge.

4 $\frac{3-a}{(x+a)(x+3)}=\frac{1}{x+a}-\frac{1}{(x+3)}$

$\int \frac{3-a}{(x+a)(x+3)}dx=\int \frac{1}{x+a}-\frac{1}{(x+3)}dx$

$=ln(x+a)-ln(x+3)$

$=ln\left(\frac{x+a}{x+3}\right)$

a $\frac{1}{(x+3)}$ is discontinuous at $x=-3$

When $x \to -3$, $\frac{x+a}{x+3}$ does not converge (as the limit is different when approaching from the left and from the right). Hence $\ln\left(\frac{x+a}{x+3}\right)$ does not have a limit as $x \to -3$ so the integral is undefined for all values of a

b $\frac{1}{x+a}$ is discontinuous at $x=-a$, so if $a \le 0$ then this is between 0 and ∞ and as $x \to -a$ the limit of $\frac{1}{x+a}$ does not exist, hence the integral does not converge.

If $a > 0$ then there are no points of discontinuity in the specified range.

$\ln\left(\frac{x+a}{x+3}\right) \to \ln 1 = 0$ as $x \to \infty$

Hence $\int_0^t \frac{3-a}{(x+a)(x+3)} dx = \ln\left(\frac{t+a}{t+3}\right) - \ln\left(\frac{a}{3}\right) \to \ln\left(\frac{3}{a}\right)$

as $t \to \infty$

Thus the integral converges precisely when $a > 0$

Exercise 19.5A

1 a $I_n = \int x^n e^x dx = x^n e^x - \int nx^{n-1} e^x dx$

$= x^n e^x - nI_{n-1}$

b $x^4e^x - 4x^3e^x + 12x^2e^x - 24xe^x + 24e^x + c$

2 a $I_n = \int x^n e^{-\frac{x}{2}} dx = -2x^n e^{-\frac{x}{2}} + 2\int nx^{n-1} e^{-\frac{x}{2}} dx$

$= -2x^n e^{-\frac{x}{2}} + 2nI_{n-1}$

b $-2x^3 e^{-\frac{x}{2}} - 12x^2 e^{-\frac{x}{2}} - 48xe^{-\frac{x}{2}} - 96e^{-\frac{x}{2}} + c$

3 a $I_n = \int_0^1 x^n e^{3x} dx = \left[\frac{1}{3}x^n e^{3x}\right]_0^1 - \frac{1}{3}n\int_0^{\frac{1}{3}} x^{n-1} e^{3x} dx$

$= \frac{1}{3}(e^3 - 0) - \frac{1}{3}nI_{n-1}$

$= \frac{1}{3}e^3 - \frac{n}{3}I_{n-1}$

b $\frac{11}{81}e^3 - \frac{8}{81}$

4 a $\int x(\ln x)^n dx = \frac{x^2}{2}(\ln x)^n - n\int \frac{x^2}{2}\frac{1}{x}(\ln x)^{n-1} dx$

$= \frac{x^2}{2}(\ln x)^n - \frac{n}{2}\int x(\ln x)^{n-1} dx$

$= \frac{x^2}{2}(\ln x)^n - \frac{n}{2}I_{n-1}$

b $\frac{x^2}{2}(\ln x)^2 - \frac{x^2}{2}\ln x + \frac{x^2}{4} + c$

5 a $I_n = \int_0^1 \frac{x^n}{\sqrt{1-x}} dx = \left[-2x^n(1-x)^{\frac{1}{2}}\right]_0^1$

$- \int_0^1 -2nx^{n-1}(1-x)^{\frac{1}{2}} dx$

$= 0 + 2n\int_0^1 x^{n-1}(1-x)^{\frac{1}{2}} dx$

$= 2n\int_0^1 x^{n-1}(1-x)(1-x)^{-\frac{1}{2}} dx$

$= 2n\int_0^1 x^{n-1}(1-x)^{-\frac{1}{2}} - x^n(1-x)^{-\frac{1}{2}} dx$

$= 2nI_{n-1} - 2nI_n$

$I_n = \frac{2n}{1+2n}I_{n-1}$

b $\frac{2048}{3003}$

6 a $I_n = \int x^n \sin x\, dx = -x^n \cos x - \int -nx^{n-1}\cos x\, dx$

$= -x^n \cos x + n\int x^{n-1}\cos x\, dx$

$= -x^n \cos x + n\left(x^{n-1}\sin x - (n-1)\int x^{n-2}\sin x\, dx\right)$

$= -x^n \cos x + nx^{n-1}\sin x - n(n-1)I_{n-2}$

b i $-x^4\cos x + 4x^3\sin x + 12x^2\cos x - 24x\sin x$
$-24\cos x + c$

ii $-x^3\cos x + 3x^2\sin x + 6x\cos x - 6\sin x + c$

7 a $I_n = \left[\frac{1}{2}x^n \sin 2x\right]_0^{\frac{\pi}{4}} - \frac{n}{2}\int_0^{\frac{\pi}{4}} x^{n-1}\sin 2x\, dx$

$= \frac{1}{2}\left(\frac{\pi}{4}\right)^n - \frac{n}{2}\left(\left[-\frac{1}{2}x^{n-1}\cos 2x\right]_0^{\frac{\pi}{4}} + \frac{n-1}{2}\int_0^{\frac{\pi}{4}} x^{n-2}\cos 2x dx\right)$

$= \frac{1}{2}\left(\frac{\pi}{4}\right)^n - \frac{n}{2}\left(\frac{n-1}{2}\right)I_{n-2}$

$= \frac{1}{2}\left(\frac{\pi}{4}\right)^n - \frac{n(n-1)}{4}I_{n-2}$ as required

b i $\frac{1}{8192}\pi^6 - \frac{15}{1024}\pi^4 + \frac{45}{64}\pi^2 - \frac{45}{8}$

ii $\frac{1}{2048}\pi^5 - \frac{5}{128}\pi^3 + \frac{15}{16}\pi - \frac{15}{8}$

Exercise 19.5B

1 a $I_n = \int_0^{\frac{\pi}{2}} \cos x \cos^{n-1} x dx$

$= \left[\sin x \cos^{n-1} x\right]_0^{\frac{\pi}{2}} - (n-1)\int_0^{\frac{\pi}{2}} -\sin^2 x \cos^{n-2} x dx$

$= 0 + (n-1)\int_0^{\frac{\pi}{2}} (1-\cos^2 x)\cos^{n-2} x\, dx$

$= (n-1)\int_0^{\frac{\pi}{2}} \cos^{n-2} x - \cos^n x dx$

$= (n-1)(I_{n-2} - I_n)$

$I_n = \frac{n-1}{n}I_{n-2}$

b **i** $\frac{8}{15}$

ii $\frac{5}{32}\pi$

2 **a** $I_n = \int_{\frac{\pi}{2}}^{\pi} \sin x \sin^{n-1} x\,dx$

$$= \left[-\cos x \sin^{n-1} x\right]_{\frac{\pi}{2}}^{\pi} - (n-1)\int_{\frac{\pi}{2}}^{\pi} -\cos^2 x \sin^{n-2} x\,dx$$

$$= 0 + (n-1)\int_{\frac{\pi}{2}}^{\pi} (1 - \sin^2 x)\sin^{n-2} x\,dx$$

$$= (n-1)\int_{\frac{\pi}{2}}^{\pi} \sin^{n-2} x - \sin^n x\,dx$$

$$= (n-1)(I_{n-2} - I_n)$$

$$I_n = \frac{n-1}{n} I_{n-2}$$

b **i** $\frac{16}{35}$

ii $\frac{35}{256}\pi$

3 **a** $I_n = \frac{\tan^{n-1} x}{n-1} - I_{n-2}$

b **i** $\frac{1}{5}\tan^5 x - \frac{1}{3}\tan^3 x + \tan x - x + c$

ii $\frac{5}{12} - \frac{1}{2}\ln 2$

4 **a** $I_n = \int_0^1 x^n\sqrt{1+x^2}\,dx = \int_0^1 x^{n-1}x\sqrt{1+x^2}\,dx$

$$= \left[\frac{1}{3}x^{n-1}(1+x^2)^{\frac{3}{2}}\right]_0^1 - \frac{n-1}{3}\int_0^1 x^{n-2}(1+x^2)^{\frac{3}{2}}\,dx$$

$$= \left(\frac{1}{3}2\sqrt{2} - 0\right) - \frac{n-1}{3}\int_0^1 x^{n-2}(1+x^2)(1+x^2)^{\frac{1}{2}}\,dx$$

$$= \frac{2}{3}\sqrt{2} - \frac{n-1}{3}\int_0^1 x^{n-2}(1+x^2)^{\frac{1}{2}} + x^n(1+x^2)^{\frac{1}{2}}\,dx$$

$$= \frac{2}{3}\sqrt{2} - \frac{n-1}{3}(I_{n-2} + I_n)$$

$$3I_n = 2\sqrt{2} - (n-1)I_{n-2} - (n-1)I_n$$

$$I_n = \frac{2\sqrt{2} - (n-1)I_{n-2}}{2+n}$$

b **i** $\frac{22\sqrt{2} - 8}{105}$

ii $\frac{7\sqrt{2} + 3\ln(1+\sqrt{2})}{48}$

5 **a** $\int \frac{\cos(nx)}{\cos x}\,dx$

$$= \int \frac{\cos((n-1)x + x)}{\cos x}\,dx$$

$$= \int \frac{\cos(n-1)x\,\cos x - \sin(n-1)x\,\sin x}{\cos x}\,dx$$

$$= \int \cos(n-1)x - \frac{\sin(n-1)x\,\sin x}{\cos x}\,dx$$

$$= \frac{1}{n-1}\sin(n-1)x - \frac{1}{2}\int \frac{\cos(n-2)x - \cos nx}{\cos x}\,dx$$

$$= \frac{1}{n-1}\sin(n-1)x - \frac{1}{2}\int \frac{\cos(n-2)x}{\cos x} - \frac{\cos nx}{\cos x}\,dx$$

$$I_n = \int_0^{\pi} \frac{\cos(nx)}{\cos x}\,dx = \left[\frac{1}{(n-1)}\sin(n-1)x\right]_0^{\pi} - \frac{1}{2}I_{n-2} + \frac{1}{2}I_n$$

$$= -\frac{1}{2}I_{n-2} + \frac{1}{2}I_n$$

So $\frac{1}{2}I_n = -\frac{1}{2}I_{n-2}$, so $I_n = -I_{n-2}$ as required

b $-\pi$

c **i** $-\pi$

ii π

d Even values of n will reduce to I_0

$$I_0 = \int_0^{\pi} \frac{\cos 0}{\cos x}\,dx$$

$$= \int_0^{\pi} \frac{1}{\cos x}\,dx$$

Discontinuity at $x = \frac{\pi}{2}$

$$\int_0^a \frac{1}{\cos x}\,dx = \int_0^a \sec x\,dx$$

$$= \ln(\tan x + \sec x)$$

When $a \to \frac{\pi}{2}$ from the left, $\tan a \to \infty$ and $\sec a \to \infty$

When $a \to \frac{\pi}{2}$ from the right, $\tan a \to -\infty$ and $\sec a \to -\infty$

Hence the limit as $a \to \frac{\pi}{2}$ of $\tan x + \sec x$ does not exist, and hence nor does the limit of $\ln(\tan x + \sec x)$. Therefore the integral does not converge. So I_n will not exist for even values of n

Exercise 19.6A

1

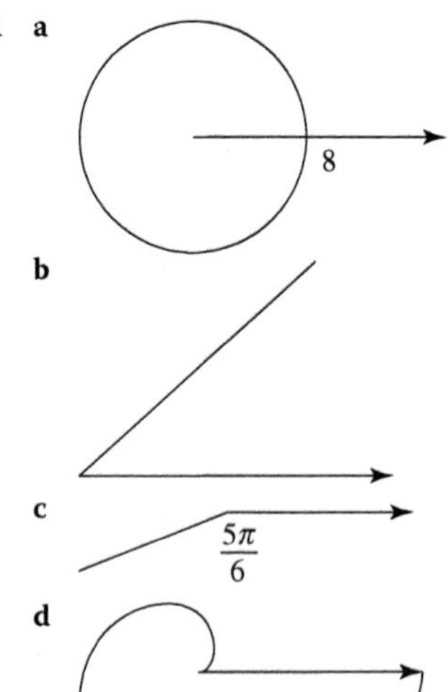

2 a max = 2, min = 0

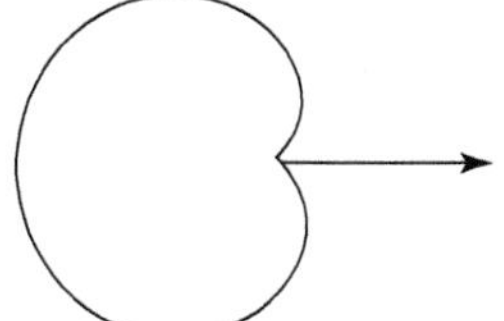

b max = 3, min = 1

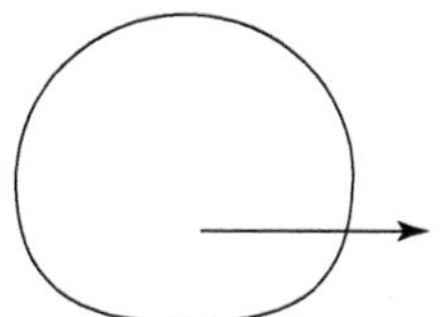

c max = 4, min = 0

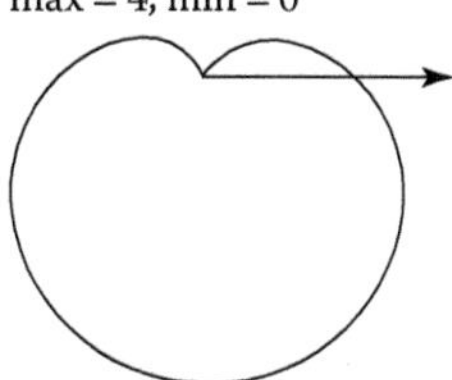

d max = 9, min = 3

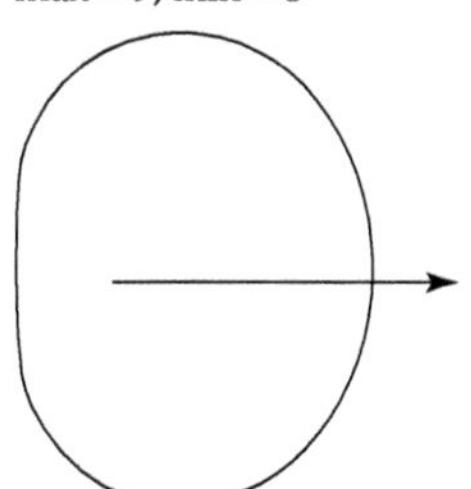

e max = 6, min = 4

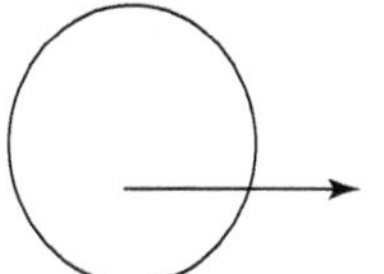

f max = 1, min = 0

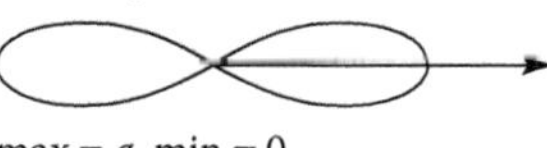

g max = a, min = 0

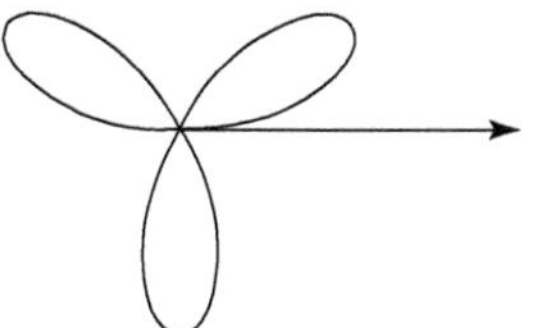

h max = b, min = 0

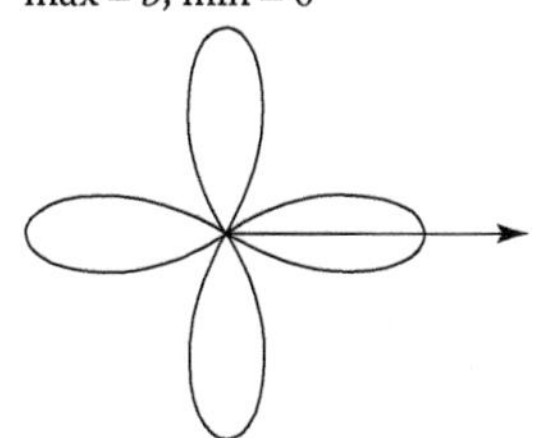

i max = c, min = 0

3 $\frac{4}{3}\pi+\sqrt{3}$

4 $\frac{\pi^3}{48}$

5 $\frac{\pi}{3}$

6 $2\sqrt{2}$

7 a $\frac{19}{2}\pi$ b 27π

c $\frac{3}{2}\pi$ d 11π

8 a $\frac{\pi}{4}$ b $\frac{\pi}{4}$

c 4π d $\frac{25}{4}\pi$

e 2 f 5π

Exercise 19.6B

1 a If there exists θ such that $\cos\theta+\sin\theta=2+\sin\theta$
Then $\cos\theta=2$ which has no solutions so they do not intersect

b 4π

2 a $(0, 0)$ and $\left(\frac{\pi}{3}, \frac{\sqrt{3}}{2}\right)$.

b $\frac{3\sqrt{3}}{16}-\frac{\pi}{12}$

3 $\frac{3\pi}{4}+\frac{3\sqrt{3}}{16}$

4 a $\frac{5}{24}\pi-\frac{\sqrt{3}}{4}$ b $\frac{9}{4}\pi-3\sqrt{3}$

c $\frac{7}{4}\pi$ d $\frac{\pi}{2}-\frac{3}{4}\sqrt{3}$ e $\frac{59}{3}\pi-\frac{19}{2}\sqrt{3}$

5 a 2.514 units b 8.473 square units

Exercise 19.7A

1 $\frac{dy}{dx}=x^{\frac{1}{2}}$

$$s=\int_0^1\sqrt{1+\left(x^{\frac{1}{2}}\right)^2}\,dx$$

$$=\int_0^1\sqrt{1+x}\,dx$$

$$=\left[\frac{2}{3}(1+x)^{\frac{3}{2}}\right]_0^1$$

$$=\frac{2}{3}(2\sqrt{2}-1)$$

$$=\frac{4}{3}\sqrt{2}-\frac{2}{3}$$

2 19.7

3 $\frac{9}{8}$

4 $\frac{15}{16}$

5 $\frac{56}{27}$

6 a $\frac{d(\ln(\sec x+\tan x))}{dx}=\frac{\sec x\tan x+\sec^2 x}{\sec x+\tan x}$

$=\frac{\sec x(\tan x+\sec x)}{\sec x+\tan x}$

$=\sec x$ as required

b $\ln(\sqrt{3})$

7 $\frac{8\pi}{3}(5\sqrt{5}-1)$

8 $\frac{61}{6}$

9 a $\frac{\pi}{56}$ b $\frac{\pi}{108}(10\sqrt{10}-1)$

10 a $\frac{\pi}{2}\ln 3+\frac{10\pi}{9}$ b $\pi\ln 3+\frac{20\pi}{9}$

Exercise 19.7B

1 a $\int\sqrt{1+16x^2}\,dx$

Let $x=\frac{1}{4}\sinh u$, then $\frac{dx}{du}=\frac{1}{4}\cosh u$

$$s=\int\sqrt{1+16\left(\frac{1}{16}\sinh^2 u\right)}\frac{1}{4}\cosh u\,du$$
$$=\frac{1}{4}\int\sqrt{1+\sinh^2 u}\cosh u\,du$$
$$=\frac{1}{4}\int\sqrt{\cosh^2 u}\cosh hu\,du$$
$$=\frac{1}{4}\int\cosh u\cosh u\,du$$
$$=\frac{1}{4}\int\cosh^2 u\,du$$
$$=\frac{1}{8}\int 1+\cosh 2u\,du$$
$$=\frac{1}{8}\left(u+\frac{1}{2}\sinh 2u\right)+c$$
$$=\frac{1}{8}(\text{arsinh }4x+\sinh(\text{arsinh }4x)\cosh(\text{arsinh }4x))+c$$
$$=\frac{1}{8}\text{arsinh}(4x)+\frac{1}{2}x\cosh(\text{arsinh }4x)+c$$

b $\frac{1}{8}\ln(1+\sqrt{2})+\frac{\sqrt{2}}{8}$

2 $\ln(1+\sqrt{2})+\sqrt{2}$

3 a $\frac{dy}{dx}=x^{-\frac{1}{2}}$

$$s=\int_0^4\sqrt{1+\left(x^{-\frac{1}{2}}\right)^2}\,dx$$
$$=\int_0^4\sqrt{1+\frac{1}{x}}\,dx$$
$$=\int_0^4\sqrt{\frac{x+1}{x}}\,dx \text{ as required}$$

b $\ln(2+\sqrt{5})+2\sqrt{5}$

4 $\frac{3}{2}$

5 a $\frac{3\pi}{2}$

b $\frac{dy}{dx}=-2e^{-x}$

$$S=2\pi\int_0^{\ln 2}2e^{-x}\sqrt{1+(-2e^{-x})^2}\,dx$$
$$=2\pi\int_0^{\ln 2}2e^{-x}\sqrt{1+4e^{-2x}}\,dx$$
$$=4\pi\int_0^{\ln 2}e^{-x}\sqrt{1+4e^{-2x}}\,dx$$

c $\pi(\ln(2+\sqrt{5})+2\sqrt{5}-\ln(1+\sqrt{2})-\sqrt{2})$

6 a $\frac{dx}{d\theta}=-r\sin\theta, \frac{dy}{d\theta}=r\cos\theta$

$$s=\int_0^{2\pi}\sqrt{(-r\sin\theta)^2+(r\cos\theta)^2}\,d\theta$$
$$=\int_0^{2\pi}\sqrt{r^2\sin^2\theta+r^2\cos^2\theta}\,d\theta$$
$$=\int_0^{2\pi}\sqrt{r^2}\,d\theta$$
$$=\int_0^{2\pi}r\,d\theta$$
$$=[r\theta]_0^{2\pi}$$
$$=2\pi r-0=2\pi r \text{ as required}$$

b $V=\pi\int_0^{\pi}(r\sin\theta)^2(-r\sin\theta)\,d\theta=\pi\int_0^{\pi}-r^3\sin^3\theta\,d\theta$

$$=\pi\int_0^{\pi}-r^3\sin\theta\sin^2\theta\,d\theta$$
$$=\pi\int_0^{\pi}-r^3\sin\theta(1-\cos^2\theta)\,d\theta$$
$$=-\pi r^3\int_0^{\pi}\sin\theta-\sin\theta\cos^2\theta\,d\theta$$
$$=-\pi r^3\left[-\cos\theta+\frac{1}{3}\cos^3\theta\right]_0^{\pi}$$
$$=-\pi r^3\left(1-\frac{1}{3}\right)-\pi r^3\left(-1+\frac{1}{3}\right)$$
$$=-\frac{4}{3}\pi r^3, \text{ so volume}=\frac{4}{3}\pi r^3 \text{ as required}$$

c $S=2\pi\int_0^{\pi}r\sin\theta\sqrt{(-r\sin\theta)^2+(r\cos\theta)^2}\,d\theta=2\pi\int_0^{\pi}r^2\sin\theta\,d\theta$

$$=2\pi r^2[-\cos\theta]_0^{\pi}$$
$$=2\pi r^2(1--1)$$
$$=4\pi r^2 \text{ as required}$$

7 $\frac{2\pi}{3}\left(\frac{13\sqrt{13}}{8}-1\right)$

8 a $\sqrt{1-\cos\theta}=\sqrt{2\sin^2\frac{\theta}{2}}$

$=\sqrt{2}\sin\frac{\theta}{2}$

b 4

c $\frac{32}{3}\pi$

Review exercise 19

1 A is an improper integral as integrand is undefined at $x=1$ which is within the limits
B is not an improper integral as it is fully defined within the limits
C is an improper integral as one of its limits is ∞

2 a 1
b 12

3 $\int_a^{2e} x^3\ln x\,dx=\left[\frac{1}{4}x^4\ln x\right]_a^{2e}-\int_a^{2e}\frac{1}{4}x^3$

$=\left[\frac{x^4}{4}\left(\ln x-\frac{1}{4}\right)\right]_a^{2e}$

$=\frac{(2e)^4}{4}\left(\ln 2+1-\frac{1}{4}\right)-\frac{a^4}{4}\left(\ln a-\frac{1}{4}\right)$

$\to e^4(\ln 16+3)$ as $a\to 0$ since $\ln a-\frac{1}{4}\to\frac{1}{4}$ and so $\frac{a^4}{4}\left(\ln a-\frac{1}{4}\right)\to 0$ as $a\to 0$

4 $\int_1^a \frac{2}{x}dx=[2\ln x]_1^a$

$=2\ln a-2\ln 1$

$\to\infty$ as $a\to\infty$ so integral does not converge.

5 Let $y=\arccos x$ then $x=\cos y\Rightarrow\frac{dx}{dy}=-\sin y$

$\frac{dy}{dx}=-\frac{1}{\sin y}$

$=-\frac{1}{\sqrt{1-\cos^2 y}}$

$=-\frac{1}{\sqrt{1-x^2}}$

6 a $2x\arctan x+\frac{x^2}{1+x^2}$ b $\frac{x}{\sqrt{1-\left(\frac{x^2}{2}\right)^2}}$ or $\frac{2x}{\sqrt{4-x^4}}$

7 a $\arcsin\left(\frac{x}{2}\right)+c$

b $\frac{1}{4}\arctan\left(\frac{x}{4}\right)+c$

8 a $\sinh A\cosh B-\sinh B\cosh A=\left(\frac{e^A-e^{-A}}{2}\right)\left(\frac{e^B+e^{-B}}{2}\right)$

$-\left(\frac{e^B-e^{-B}}{2}\right)\left(\frac{e^A+e^{-A}}{2}\right)$

$=\frac{1}{4}(e^{A+B}+e^{A-B}-e^{-(A-B)}-e^{-(A+B)})$

$-\frac{1}{4}(e^{A+B}+e^{-(A-B)}-e^{(A-B)}-e^{-(A+B)})$

$=\frac{1}{4}(2e^{A-B}-2e^{-(A-B)})$

$=\frac{1}{2}(e^{A-B}-e^{-(A-B)})$

$=\sinh(A-B)$ as required

b $\cosh^2 x=\left(\frac{e^x+e^{-x}}{2}\right)^2$

$=\frac{1}{4}(e^{2x}+2+e^{-2x})$

$=\frac{1}{2}+\frac{1}{4}(e^{2x}+e^{-2x})$

$=\frac{1}{2}\left(1+\frac{1}{2}(e^{2x}+e^{-2x})\right)$

$=\frac{1}{2}(1+\cosh 2x)$ as required

9 a $2\sinh 2x$ b $2x\sinh x+x^2\cosh x$

10 Let $y=\text{artanh}\,x$ then $x=\tanh y\Rightarrow\frac{dx}{dy}=\text{sech}^2 y$

$\frac{dy}{dx}=\frac{1}{\text{sech}^2 y}$

$=\frac{1}{1-\tanh^2 y}$

$=\frac{1}{1-x^2}$ as required

11 a $\frac{4x}{\sqrt{(2x^2)^2+1}}=\frac{4x}{\sqrt{4x^4+1}}$

b $\text{arcosh}(x-1)+\frac{x}{\sqrt{(x-1)^2-1}}$

12 a $\ln(1+\sqrt{2})$

b $\frac{\sqrt{2}}{2}\ln(3+2\sqrt{2})$

13 a $2x-3+\frac{4}{x-2}-\frac{1}{x+3}$

b $4+\ln\left(\frac{96}{7}\right)$

14 $\ln 640+3\arctan(3)$

15 a $I_n=\int x^n e^{-2x}dx=-\frac{1}{2}x^n e^{-2x}+\frac{n}{2}\int x^{n-1}e^{-2x}dx$

$=-\frac{1}{2}x^n e^{-2x}+\frac{n}{2}I_{n-1}$

b $-\frac{1}{8}e^{-2x}(4x^3+6x^2+6x+3)+c$

16 a $I_n=\int_0^{\pi} x^n\cos x\,dx=[x^n\sin x]_0^{\pi}-\int_0^{\pi} nx^{n-1}\sin x\,dx$

$=0-n\left([-x^{n-1}\cos x]_0^{\pi}-(n-1)\int_0^{\pi}-x^{n-2}\cos x\,dx\right)$

$=-n(-\pi^{n-1}(-1)-0+(n-1)I_{n-2})$

$=-n(\pi^{n-1}+(n-1)I_{n-2})=-n\pi^{n-1}-n(n-1)I_{n-2}$ as required

b **i** $-6\pi^5+120\pi^3-720\pi$

ii $-5\pi^4+60\pi^2-240$

17 a **i**

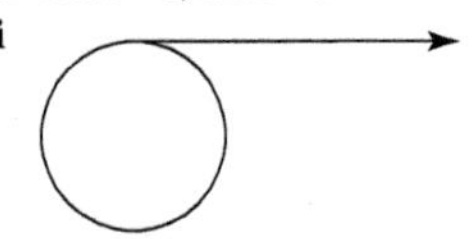

ii Max = 8, min = 0

b **i**

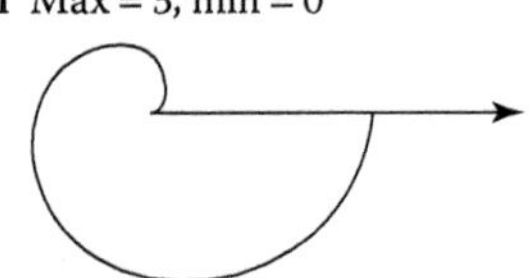

ii Max = 3, min = 0

c **i**

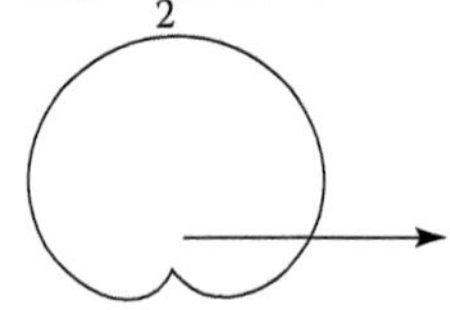

ii Max $= \frac{\pi}{2}$, min =0

d **i**

ii Max = 12, min = 2

18 a $\frac{19}{2}\pi$

b π

19 $\frac{7\pi}{4}$

20 $\frac{488}{27}$

21 $\ln(2+\sqrt{5})+2\sqrt{5}$

22 $\frac{\pi}{9}(2\sqrt{2}-1)$

23 $\frac{dx}{d\theta}=-\sin\theta, \frac{dy}{d\theta}=\cos\theta$

$$S=2\pi\int_{\frac{\pi}{2}}^{\frac{3\pi}{4}}\sin\theta\sqrt{(-\sin\theta)^2+(\cos\theta)^2}\,d\theta$$

$$=2\pi\int_{\frac{\pi}{2}}^{\frac{3\pi}{4}}\sin\theta\sqrt{\sin^2\theta+\cos^2\theta}\,d\theta$$

$$=2\pi\int_{\frac{\pi}{2}}^{\frac{3\pi}{4}}\sin\theta\,d\theta$$

$$=2\pi[-\cos\theta]_{\frac{\pi}{2}}^{\frac{3\pi}{4}}$$

$$=2\pi\left(\frac{\sqrt{2}}{2}-0\right)$$

$$=\sqrt{2}\pi$$

Assessment 19

1 $\ln\left(\frac{e^x-3}{e^x+3}\right)+c$

2 a $I_n=\int 1\times(\ln x)^n\,dx$

$v'=\frac{n(\ln x)^{n-1}}{x}$

$I_n=x(\ln x)^n-\int x\frac{n(\ln x)^{n-1}}{x}dx$

$I_n=x(\ln x)^n-nI_{n-1}$

b $4(\ln 4)^3-12(\ln 4)^2+24(\ln 4)-18$

3 a $P\left(\frac{3}{2},\frac{\pi}{3}\right)$ **b** $\frac{\pi}{2}-\frac{9\sqrt{3}}{16}$

4 a $\frac{\pi}{2}$ **b** $\cosh^{-1}2$

5 a $\frac{a^2}{3}$ **b** $\frac{2\pi a^3}{15}$

6 a $I_n=\int\tanh^n x\,dx=\int(1-\operatorname{sech}^2 x)\tanh^{n-2}x\,dx$

$=\int\tanh^{n-2}x\,dx-\int\operatorname{sech}^2 x\tanh^{n-2}x\,dx$

$=I_{n-2}-\frac{1}{n-1}\tanh^{n-1}x$

b $I_8=x-\tanh x-\frac{1}{3}\tanh^3 x-\frac{1}{5}\tanh^5 x-\frac{1}{7}\tanh^7 x\,(+c)$

7 a

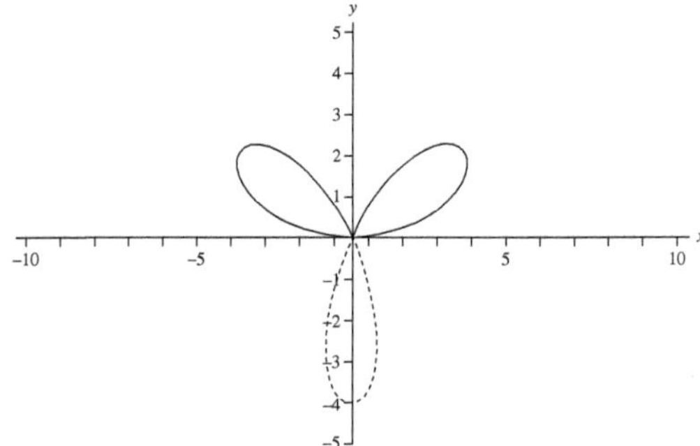

b $\frac{4\pi}{3}$

8 a $\sinh y=x$

$\cosh y\frac{dy}{dx}=1$

$\sqrt{1+\sinh^2 y}\,\frac{dy}{dx}=1$

$\frac{dy}{dx}=\frac{1}{\sqrt{1+x^2}}$

b $\int_0^2 1\times\sinh^{-1}x\,dx$

$=\left[x\sinh^{-1}x\right]_0^2-\int_0^2\frac{x}{\sqrt{1+x^2}}dx$

$=\left[x\sinh^{-1}x-\sqrt{1+x^2}\right]_0^2$

$2\sinh^{-1}2-(\sqrt{5})-(-1)$

$2\ln(2+\sqrt{5})+1-\sqrt{5}$

9 **a** $\frac{3a}{2}$ **b** $\frac{6\pi a^2}{5}$ square units

10 $x^2 - 2x + 10 \equiv (x-1)^2 + 9$

$$\int_1^4 \frac{2x+1}{\sqrt{x^2-2x+10}}\,dx = \int_1^4 \frac{2x+1}{\sqrt{(x-1)^2+9}}\,dx$$

$x - 1 = 3\sinh\theta$

$\frac{dx}{d\theta} = 3\cosh\theta$

Limits of 0 and arsinh 1

$$\int_0^{\sinh^{-1}1} \frac{6\sinh\theta+3}{\sqrt{9\sinh^2\theta+9}}\,3\cosh\theta\,d\theta$$

$$= \int_0^{\text{arsinh}\,1} 6\sinh\theta + 3\,d\theta$$

$$= \left[6\cosh\theta + 3\theta\right]_0^{\text{arsinh}\,1}$$

$$= \left[6\sqrt{1+\sinh^2\theta} + 3\theta\right]_0^{\text{arsinh}\,1}$$

$$= 6\sqrt{2} + 3\,\text{arsinh}1 - 6$$

$$= 6\sqrt{2} + 3\ln(1+\sqrt{2}) - 6$$

11 $\frac{\pi(\pi-2)}{2}$

12 **a** $\frac{1}{3}\tan^{-1}\left(\frac{x-2}{3}\right) + c$

b 2

13 **a** $\left(r = \frac{\sqrt{3}}{2}, \theta = \frac{\pi}{6}\right)$

b $\frac{1}{2}\int_0^{\frac{\pi}{6}} (\sin 2\theta)^2\,d\theta$

$$= \frac{1}{2}\int_0^{\frac{\pi}{6}} \frac{1-\cos 4\theta}{2}\,d\theta$$

$$= \left[\frac{\theta}{4} - \frac{\sin 4\theta}{16}\right]_0^{\frac{\pi}{6}}$$

$$= \frac{\pi}{24} - \frac{\sqrt{3}}{32}$$

$$\frac{1}{2}\int_{\frac{\pi}{6}}^{\frac{\pi}{2}} (\cos\theta)^2\,d\theta$$

$$= \frac{1}{2}\int_{\frac{\pi}{6}}^{\frac{\pi}{2}} \frac{1+\cos 2\theta}{2}\,d\theta$$

$$= \left[\frac{\theta}{4} + \frac{\sin 2\theta}{8}\right]_{\frac{\pi}{6}}^{\frac{\pi}{2}}$$

$$= \frac{\pi}{12} - \frac{\sqrt{3}}{16}$$

Total area $= \frac{\pi}{8} - \frac{3\sqrt{3}}{32}$

14 **a** $= 1 - \frac{1}{\sqrt{3}}$

b $\frac{3x-x^2}{(x+1)(x^2+3)} \equiv \frac{A}{x+1} + \frac{Bx+C}{x^2+3}$

$3x - x^2 \equiv A(x^2+3) + (Bx+C)(x+1)$

$-3 - 1 = 4A \Rightarrow A = -1$

$-1 = A + B \Rightarrow B = 0$

$0(x^2)$

$0 = 3A + C \Rightarrow C = 3$

$$\int_1^3 -\frac{1}{x+1} + \frac{3}{x^2+3}\,dx$$

$$\left[-\ln(x+1) + \frac{3}{\sqrt{3}}\tan^{-1}\left(\frac{x}{\sqrt{3}}\right)\right]_1^3$$

$$= -\ln 4 + \frac{3}{\sqrt{3}}\tan^{-1}\left(\sqrt{3}\right) - \left(-\ln 2 + \frac{3}{\sqrt{3}}\tan^{-1}\left(\frac{1}{\sqrt{3}}\right)\right)$$

$$-\ln\left(\frac{4}{2}\right) + \frac{3}{\sqrt{3}}\left(\frac{\pi}{3} - \frac{\pi}{6}\right)$$

$$= \frac{\pi}{2\sqrt{3}} - \ln 2$$

15 **a** $2\pi\int_0^2 y\sqrt{\left(\frac{dy}{dt}\right)^2 + \left(\frac{dx}{dt}\right)^2}\,dt$

$$= 2\pi\int_0^2 2at\sqrt{(2a)^2 + (2at)^2}\,dt$$

$$= 8\pi a^2\int_0^2 t\sqrt{1+t^2}\,dt$$

b $8\pi a^2\left(\frac{5\sqrt{5}-1}{3}\right)$

c $\int_0^2 \sqrt{\left(\frac{dy}{dt}\right)^2 + \left(\frac{dx}{dt}\right)^2}\,dt$

$$= \int_0^2 \sqrt{(2a)^2 + (2at)^2}\,dt$$

$$= 2a\int_0^2 \sqrt{1+t^2}\,dt$$

d $a(\sinh^{-1}2 + 2\sqrt{5})$

16 **a** $I_n = \int_0^{\frac{\pi}{2}} \sin^n x\,dx = \int_0^{\frac{\pi}{2}} \sin x \sin^{n-1} x\,dx,$

$v' = (n-1)\sin^{n-2} x\cos x$

$$\left[-\cos x\sin^{n-1} x\right]_0^{\frac{\pi}{2}} - \int_0^{\frac{\pi}{2}} -\cos x \times (n-1)\sin^{n-2} x\cos x\,dx$$

$$\left[-\cos x\sin^{n-1} x\right]_0^{\frac{\pi}{2}} - \int_0^{\frac{\pi}{2}} -(1-\sin^2 x) \times (n-1)\sin^{n-2} x\,dx$$

$$= 0 + (n-1)\int_0^{\frac{\pi}{2}} \sin^{n-2}x \, dx - (n-1)\int_0^{\frac{\pi}{2}} \sin^n x \, dx$$

$$I_n = (n-1)\, I_{n-2} - (n-1) I_n$$

$$n I_n = (n-1)\, I_{n-2}$$

b $\frac{3\pi}{16}$

c $\frac{35}{256}\pi^2$

17 a $\left(\frac{3}{2}, \frac{\pi}{3}\right)$ or $\left(\frac{3}{2}, -\frac{\pi}{3}\right)$

b

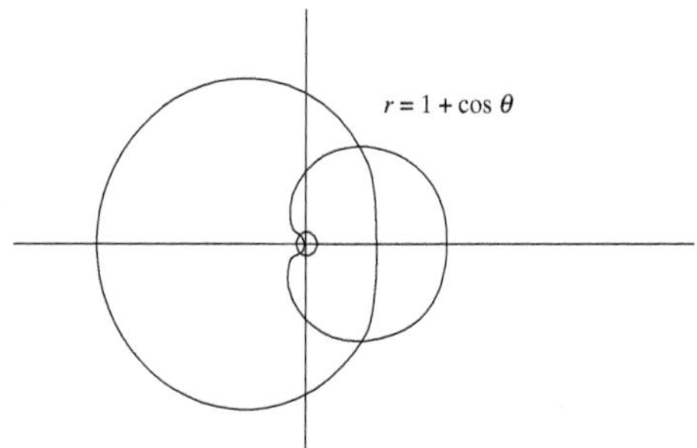

c $3\sqrt{3} - \pi$

18 a $\frac{1}{2}x^2 + \frac{1}{2}\ln(3x-2) - \frac{1}{2}\ln(3x+2) + c$

b Limits 0 and $\frac{\pi}{6}$

$dx = \cos\theta \, d\theta$

$$\int_0^{\frac{\pi}{6}} \frac{24\sin^2\theta}{\sqrt{1-\sin^2\theta}} \cos\theta \, d\theta$$

$$= \int_0^{\frac{\pi}{6}} 24\sin^2\theta \, d\theta$$

$$= \int_0^{\frac{\pi}{6}} 12 - 12\cos 2\theta \, d\theta$$

$$= \left[12\theta - 6\sin 2\theta\right]_0^{\frac{\pi}{6}}$$

$$= \frac{12\pi}{6} - \frac{6\sqrt{3}}{2} - (0)$$

$$= 2\pi - 3\sqrt{3}$$

19 a Consider

$$I_{n-1} - I_n = \int_0^{\frac{\pi}{4}} \frac{\sin^{2n-1}x}{\cos x} - \int_0^{\frac{\pi}{4}} \frac{\sin^{2n+1}x}{\cos x} dx$$

$$= \int_0^{\frac{\pi}{4}} \frac{(1-\sin^2 x)\sin^{2n-1}x}{\cos x} dx$$

$$= \int_0^{\frac{\pi}{4}} \frac{\cos^2 x \sin^{2n-1}x}{\cos x} dx = \int_0^{\frac{\pi}{4}} \cos x \sin^{2n-1} x dx$$

$$= \left[\frac{\sin^{2n}x}{2n}\right]_0^{\frac{\pi}{4}}$$

$$= \frac{1}{2n}\left(\frac{1}{\sqrt{2}}\right)^{2n} = \frac{1}{2^{n+1}n}$$

Hence $I_n = I_{n-1} - \frac{1}{2^{n+1}n}$

b $\frac{1}{2}\ln 2 - \frac{1}{3}$

20 a $P\left(4a, \frac{\pi}{3}\right)$ and $Q\left(4a, -\frac{\pi}{3}\right)$

b $\frac{1}{2}\int_0^{\frac{\pi}{3}} a^2(5 - 2\cos\theta)^2 \, d\theta$

$$= \frac{a^2}{2}\int_0^{\frac{\pi}{3}} 25 - 20\cos\theta + 4\cos^2\theta \, d\theta$$

$$= \frac{a^2}{2}\int_0^{\frac{\pi}{3}} 27 - 20\cos\theta + 2\cos 2\theta \, d\theta$$

$$\frac{a^2}{2}\left[27\theta - 20\sin\theta + \sin 2\theta\right]_0^{\frac{\pi}{3}}$$

$$= \frac{a^2}{2}\left(9\pi - 10\sqrt{3} + \frac{\sqrt{3}}{2}\right) = \frac{a^2}{2}\left(9\pi - \frac{19\sqrt{3}}{2}\right)$$

$$\frac{1}{2}\int_{\frac{\pi}{3}}^{\pi} a^2(3 + 2\cos\theta)^2 \, d\theta$$

$$\frac{a^2}{2}\int_{\frac{\pi}{3}}^{\pi} 9 + 12\cos\theta + 4\cos^2\theta \, d\theta$$

$$= \frac{a^2}{2}\int_{\frac{\pi}{3}}^{\pi} 11 + 12\cos\theta + 2\cos 2\theta \, d\theta$$

$$= \frac{a^2}{2}\left[11\theta + 12\sin\theta + \sin 2\theta\right]_{\frac{\pi}{3}}^{\pi}$$

$$= \frac{a^2}{2}\left(11\pi - \left(\frac{11\pi}{3} + 6\sqrt{3} + \frac{\sqrt{3}}{2}\right)\right) = \frac{a^2}{2}\left(\frac{22\pi}{3} - \frac{13\sqrt{3}}{2}\right)$$

$$\left(\frac{a^2}{2}\left(9\pi - \frac{19\sqrt{3}}{2}\right) + \frac{a^2}{2}\left(\frac{22\pi}{3} - \frac{13\sqrt{3}}{2}\right)\right)$$

$$\text{Total area} = 2 \times \left(\frac{a^2}{2}\left(9\pi - \frac{19\sqrt{3}}{2}\right) + \frac{a^2}{2}\left(\frac{22\pi}{3} - \frac{13\sqrt{3}}{2}\right)\right)$$

$$= a^2\left(\frac{49\pi}{3} - 16\sqrt{3}\right) = a^2\left(\frac{49\pi - 48\sqrt{3}}{3}\right)$$

21 a $(1,0)$

b $6y\left(\frac{dy}{dx}\right) = (1-x)^2 - 2x(1-x)$

$$\left(\frac{dy}{dx}\right) = \frac{(1-x)(1-3x)}{6y}$$

$$1+\left(\frac{dy}{dx}\right)^2 = \frac{36y^2+(1-x)^2(1-3x)^2}{36y^2}$$

$$= \frac{12x(1-x)^2+(1-x)^2(1-3x)^2}{12x(1-x)^2}$$

$$= \frac{(1-x)^2(1+6x+9x^2)}{12x(1-x)^2} = \frac{(1+3x)^2}{12x}$$

$$\text{So } 2\times\int_0^1 \sqrt{1+\left(\frac{dy}{dx}\right)^2}\,dx = 2\times\int_0^1 \sqrt{\frac{(1+3x)^2}{12x}}\,dx$$

$$= 2\times\int_0^1 \frac{1+3x}{2\sqrt{3x}}\,dx = \int_0^1 \frac{1+3x}{\sqrt{3x}}\,dx$$

c $= \frac{4\sqrt{3}}{3}$

d $2\pi\int_0^1 y\sqrt{1+\left(\frac{dy}{dx}\right)^2}\,dx$

$$= 2\pi\int_0^1 y\sqrt{\frac{(1-x)^2(1+6x+9x^2)}{36y^2}}\,dx$$

$$= \frac{\pi}{3}\int_0^1 (1-x)(1+3x)\,dx$$

$$= \frac{\pi}{3}\int_0^1 1+2x-3x^2\,dx$$

$$= \frac{\pi}{3}\left[x+x^2-x^3\right]_0^1 = \frac{\pi}{3}$$

22 a i $= \sin^{-1}\left(\frac{x+2}{4}\right)+c$

ii $x\operatorname{artanh}(2x)+\frac{1}{4}\ln(1-4x^2)+c$

b Let $x = 4\operatorname{cosech}u$, so $\frac{dx}{du} = -4\coth u\operatorname{cosech}u$

At $x=1$, $u = \operatorname{arsinh}4$; at $x=a$, $u=\operatorname{arsinh}\frac{4}{a}$

$$\int_1^a \frac{1}{x\sqrt{x^2+16}}\,dx$$

$$= \int_{\operatorname{arsinh}4}^{\operatorname{arsinh}\frac{4}{a}} \frac{1}{4\operatorname{cosech}u\sqrt{16\operatorname{cosech}^2u+16}}(-4\operatorname{cosech}u\coth u)\,du$$

$$= -\int_{\operatorname{arsinh}4}^{\operatorname{arsinh}\frac{4}{a}} \frac{1}{4}\,du$$

$$= \frac{1}{4}\operatorname{arsinh}4 - \frac{1}{4}\operatorname{arsinh}\frac{4}{a}$$

$$= \frac{1}{4}\ln\left(4+\sqrt{17}\right) - \frac{1}{4}\ln\left(\sqrt{\frac{16}{a^2}+1}+\frac{4}{a}\right)$$

As $a\to\infty$, $\sqrt{\frac{16}{a^2}+1}+\frac{4}{a}\to 1$ so $\frac{1}{4}\ln\left(\sqrt{\frac{16}{a^2}+1}+\frac{4}{a}\right)\to 0$

and so $\int_1^\infty \frac{1}{x\sqrt{x^2+16}}\,dx = \frac{1}{4}\ln 4+\sqrt{17}$

23 a $I_{2n+1} = \int_0^1 x^{2n}\,x\sqrt{1-x^2}\,dx$

$$= \left[-\frac{1}{3}x^{2n}\left(1-x^2\right)^{\frac{3}{2}}\right]_0^1 - \int_0^1 -\frac{2n}{3}x^{2n-1}\left(1-x^2\right)^{\frac{3}{2}}\,dx$$

$$= \int_0^1 \frac{2n}{3}x^{2n-1}(1-x^2)^{\frac{3}{2}}\,dx$$

$$= \frac{2n}{3}\int_0^1 x^{2n-1}(1-x^2)(1-x^2)^{\frac{1}{2}}\,dx$$

$$= \frac{2n}{3}I_{2n-1} - \frac{2n}{3}I_{2n+1}$$

So $(2n+3)I_{2n+1} = 2nI_{2n-1}$

ie $I_{2n+1} = \frac{2n}{2n+3}I_{2n-1}$

b $I_1 = \int_0^1 x\sqrt{1-x^2}\,dx$

$$= \left[-\frac{1}{3}(1-x^2)^{\frac{3}{2}}\right]_0^1 = \frac{1}{3}$$

$$I_{2n+1} = \frac{2n}{2n+3}I_{2n-1} = \frac{2n}{2n+3}\times\frac{2(n-1)}{2n+1}I_{2n-3}$$

$$= \frac{2n}{2n+3}\times\frac{2(n-1)}{2n+1}\times\frac{2(n-2)}{2n-1}\times\ldots\times\frac{2(1)}{5}I_1$$

$$= 2^n n!\times\frac{1}{(2n+3)\times(2n+1)\times(2n-1)\times\ldots\times5\times3\times1}$$

$$= 2^n n!\times\frac{1}{(2n+3)\times(2n+1)\times(2n-1)\times\ldots\times5\times3\times1}\times\frac{(2n+2)\times(2n)\times(2n-2)\times\ldots\times4\times2}{(2n+2)\times(2n)\times(2n-2)\times\ldots\times4\times2}$$

$$= 2^n n!\times\frac{2^{n+1}(n+1)!}{(2n+3)!}$$

$$= \frac{2^{2n+1}n!\,(n+1)!}{(2n+3)!}$$

24 a $Q\left(\frac{3}{2},\frac{\pi}{6}\right)$ and $P\left(\frac{3}{2},\frac{5\pi}{6}\right)$

b $\frac{1}{2}\int_{\frac{\pi}{6}}^{\frac{5\pi}{6}} (1+\sin\theta)^2\,d\theta$

$$= \frac{1}{2}\int_{\frac{\pi}{6}}^{\frac{5\pi}{6}} 1+2\sin\theta+\sin^2\theta\,d\theta$$

$$= \frac{1}{2}\int_{\frac{\pi}{6}}^{\frac{5\pi}{6}} 1+2\sin\theta+\frac{1-\cos2\theta}{2}\,d\theta$$

$$=\frac{1}{2}\left[\frac{3\theta}{2}-2\cos\theta-\frac{\sin 2\theta}{4}\right]_{\frac{\pi}{6}}^{\frac{5\pi}{6}}$$

$$=\frac{1}{2}\left(\frac{5\pi}{4}+\sqrt{3}+\frac{\sqrt{3}}{8}\right)-\frac{1}{2}\left(\frac{\pi}{4}-\sqrt{3}-\frac{\sqrt{3}}{8}\right)=\frac{4\pi+9\sqrt{3}}{8}$$

$$\frac{1}{2}\int_{\frac{\pi}{6}}^{\frac{5\pi}{6}}3^2(1-\sin\theta)^2\,d\theta$$

$$=\frac{9}{2}\int_{\frac{\pi}{6}}^{\frac{5\pi}{6}}1-2\sin\theta+\sin^2\theta\,d\theta$$

$$=\frac{9}{2}\int_{\frac{\pi}{6}}^{\frac{5\pi}{6}}1-2\sin\theta+\frac{1-\cos 2\theta}{2}\,d\theta$$

$$=\frac{9}{2}\left[\frac{3\theta}{2}+2\cos\theta-\frac{\sin 2\theta}{4}\right]_{\frac{\pi}{6}}^{\frac{5\pi}{6}}$$

$$=\frac{9}{2}\left(\frac{5\pi}{4}-\sqrt{3}+\frac{\sqrt{3}}{8}\right)-\frac{9}{2}\left(\frac{\pi}{4}+\sqrt{3}-\frac{\sqrt{3}}{8}\right)=\frac{36\pi-63\sqrt{3}}{8}$$

$$\frac{4\pi+9\sqrt{3}}{8}-\frac{36\pi-63\sqrt{3}}{8}$$

$$=9\sqrt{3}-4\pi$$

An alternative method is to work out

$$\frac{1}{2}\int_{\frac{\pi}{6}}^{\frac{5\pi}{6}}(1+\sin\theta)^2-9(1-\sin\theta)^2\,d\theta$$

25 a $\sinh^{-1}\left(\frac{x-1}{3}\right)+c$

b Let $x=\tan u$, so $\frac{dx}{du}=\sec^2 u$

At $x=a$, $u=\arctan a$

$$\int_a^b\frac{1}{1+x^2}dx=\int_{\arctan a}^{\arctan b}\frac{\sec^2 u}{1+\tan^2 u}du$$

$$=\int_{\arctan a}^{\arctan b}1\,du$$

$$=\arctan b-\arctan a$$

As $a\to-\infty\ \arctan a\to-\frac{\pi}{2}$, so $\int_{-\infty}^{b}\frac{1}{1+x^2}dx=\arctan b+\frac{\pi}{2}$

As $b\to\infty\ \arctan b\to\frac{\pi}{2}$, so $\int_{-\infty}^{\infty}\frac{1}{1+x^2}dx=\frac{\pi}{2}+\frac{\pi}{2}=\pi$

c The function $\frac{1}{(1+x)^2}$ is undefined at $x=-1$

In order to find $\int_{-\infty}^{\infty}\frac{1}{(1+x)^2}dx$ you need to work out

$\int_{-\infty}^{-1-\alpha}\frac{1}{(1+x)^2}dx+\int_{-1+\alpha}^{\infty}\frac{1}{(1+x)^2}dx$, and then let $\alpha\to 0$

26 a $A(2\pi,0)$, $B(4\pi,0)$, $C(6\pi,0)$

b $$\left(\frac{dx}{d\theta}\right)^2+\left(\frac{dy}{d\theta}\right)^2=(1-\cos\theta)^2+(\sin\theta)^2$$

$$=2(1-\cos\theta)$$

$$\int_0^{6\pi}\sqrt{\left(\frac{dx}{d\theta}\right)^2+\left(\frac{dy}{d\theta}\right)^2}\,d\theta=3\int_0^{2\pi}\sqrt{2(1-\cos\theta)}\,d\theta$$

$$=3\int_0^{2\pi}\sqrt{2\left(2\sin^2\left(\frac{\theta}{2}\right)\right)}\,d\theta$$

c 24

d $$2\pi\int_0^{6\pi}y\sqrt{\left(\frac{dx}{d\theta}\right)^2+\left(\frac{dy}{d\theta}\right)^2}\,d\theta$$

$$=2\pi\int_0^{6\pi}(1-\cos\theta)\sqrt{2(1-\cos\theta)}d\theta$$

$$=6\pi\int_0^{2\pi}(1-\cos\theta)\sqrt{2(1-\cos\theta)}d\theta$$

$$=6\pi\int_0^{2\pi}2\sin^2\frac{\theta}{2}2\sin\frac{\theta}{2}d\theta$$

$$=24\pi\int_0^{2\pi}\sin^3\frac{\theta}{2}d\theta$$

$$=24\pi\int_0^{2\pi}\sin\frac{\theta}{2}-\sin\frac{\theta}{2}\cos^2\frac{\theta}{2}d\theta$$

$$=24\pi\left[-2\cos\frac{\theta}{2}+\frac{2}{3}\cos^3\frac{\theta}{2}\right]_0^{2\pi}$$

$$=24\pi\left(2-\frac{2}{3}\right)-8\pi\left(-2+\frac{2}{3}\right)=64\pi$$

27 a $\frac{\pi}{4}$ **b** $\frac{\pi}{4}\ln\left(\frac{2+\sqrt{2}}{2-\sqrt{2}}\right)$

c $2\pi(2-\sqrt{2})$ cubic units

Chapter 20

Exercise 20.1A

1 a $y=\frac{e^{-x}}{2}+ce^{-3x}$ **b** $y=4+ce^{-x^2}$

c $y=\frac{\sin^3 x\cos x}{3}+c\cos x$

d $y=x\ln[A(x-5)]$ **e** $y=\frac{2}{5}x^{\frac{1}{2}}+cx^{-\frac{1}{3}}$

f $y=x^2\ln x-x^2+cx$

g $y=4\sec x\tan x+c\sec x$

h $y=\frac{1}{2}e^{(x^2-2x+1)}+ce^{-(x+1)^2}$

2 a $y=2xe^{3x}-2e^{3x}+e^{2x}$

b $y=\frac{2}{x}+\frac{3}{x^3}$ **c** $y=\frac{11-16\cos^4 x}{16\sin x}$

d $y=e^{\sin 2x}(2x+1)$ **e** $y=\frac{14x+31}{(x+1)^2}$

f $y=\frac{\sqrt{2}+4\cos^3 x}{2\sin x}$

g $y=\frac{x}{4}(2x^2\ln x-x^2+9)$

h $y=\frac{3+e^{5x}}{\cosh x}$

3 a $y\sec^2 x=\tan x-x+c$

b $y\sec^2 x=\tan x-x+\frac{\pi}{4}$

4 $y=\frac{x^3+2}{\cos x}$

5 $y=x^3\ln x-x^3+6x^2$

6 a $y=2\sin x-\cos x+ce^{-2x}$

b $y=2\sin x-\cos x+2e^{-2x}$

7 a By the chain rule $\frac{du}{dx}=\frac{du}{dy}\times\frac{dy}{dx}=-\frac{2}{y^3}\times\frac{dy}{dx}$

so $\frac{dy}{dx}=-\frac{y^3}{2}\frac{du}{dx}$

Substituting for $\frac{dy}{dx}$ gives

$-\frac{y^3}{2}\frac{du}{dx}=y+2xy^3\Rightarrow-\frac{1}{2}\frac{du}{dx}=\frac{1}{y^2}+2x$

or $-\frac{1}{2}\frac{du}{dx}=u+2x$ or $\frac{du}{dx}+2u=-4x$

b $y^2(1-2x+3e^{-2x})=1$

8 a $y=\cos^2 x(e^x+c)$ **b** $y=\cos^2 x(e^x+4)$

9 a $y\sin x=8\sin^3 x+c$

b $2\sin\left(\frac{\pi}{4}\right)=8\sin^3\left(\frac{\pi}{4}\right)+c$

$\sqrt{2}=2\sqrt{2}+c\Rightarrow c=-\sqrt{2}$

$y\sin x=8\sin^3 x-\sqrt{2}$

10 a $x^3e^x-3x^2e^x+6xe^x-6e^x+c$

b $x\frac{dy}{dx}+(x+2)y=2x^2$

i.e. $\frac{dy}{dx}+\left(1+\frac{2}{x}\right)y=2x$

Integrating factor $e^{\int\left(1+\frac{2}{x}\right)dx}=e^{x+2\ln x}=e^xe^{2\ln x}=x^2e^x$

So $x^2e^x\frac{dy}{dx}+e^x(x^2+2x)y=2x^3e^x$

$\frac{d}{dx}\left[x^2e^xy\right]=2x^3e^x$

$x^2e^xy=2\int x^3e^x\,dx$

$x^2e^xy=2e^x(x^3-3x^2+6x-6)+c$

or $x^2y=2(x^3-3x^2+6x-6)+ce^{-x}$

11 a $\int\sec^3 x\,dx=\int\sec x\sec^2 x\,dx=\sec x\tan x-$

$\int(\sec x\tan x)\tan x\,dx$

$=\sec x\tan x-\int\sec x(\sec^2 x-1)\,dx$

$=\sec x\tan x-\int\sec^3 x\,dx+\int\sec x\,dx$

$=\sec x\tan x-\int\sec^3 x\,dx+\ln(\sec x+\tan x)+c$

So $2\int\sec^3 x\;dx=\sec x\tan x+\ln(\sec x+\tan x)+c$

$\int\sec^3 x\;dx=\frac{1}{2}\left[\sec x\tan x+\ln(\sec x+\tan x)\right]+c$

b $x\frac{dy}{dx}+(1-x\tan x)y=2\sec^4 x$

i.e. $\frac{dy}{dx}+\left(\frac{1}{x}-\tan x\right)y=\frac{2\sec^4 x}{x}$

Integrating factor $e^{\int\left(\frac{1}{x}-\tan x\right)dx}=e^{(\ln x+\ln\cos x)}=x\cos x$

So $x\cos x\frac{dy}{dx}+(\cos x-x\sin x)y=2\sec^3 x$

$\frac{d}{dx}\left[y(x\cos x)\right]=2\sec^3 x$

$y(x\cos x)=\sec x\tan x+\ln(\sec x+\tan x)+c$

$y=\frac{\sec x\tan x+\ln(\sec x+\tan x)+c}{x\cos x}$

12 $x=\frac{e^{2t}+c}{\ln t}$

13 a $x=\frac{3\ln(1+t)-1}{9}+\frac{c}{(1+t)^3}$

b $\frac{8}{9}=-\frac{1}{9}+c\Rightarrow c=1$

$x=\frac{3\ln(1+t)-1}{9}+\frac{1}{(1+t)^3}$

14 a $y=z^2\Rightarrow\frac{dy}{dx}=2z\frac{dz}{dx}$

$\frac{dz}{dx}-3z\tan x=\frac{3}{2}$

b $y=\left(\frac{\frac{3}{2}\sin x-\frac{1}{2}\sin^3 x+c}{\cos^3 x}\right)^2$

15 a $\frac{dv}{dt}-\frac{1}{t}v=5t$

b $x=\frac{5}{8}t^4+ct^2+d$

Exercise 20.1B

1 a $\frac{dp}{dt}=20t-P$ so $\frac{dP}{dt}+P=20t$

$e^{\int F(t)dt}=e^{\int 1dt}=e^t$

$e^t\frac{dP}{dt}+e^tP=20te^t$

$\frac{d}{dt}\left[e^tP\right]=20te^t$

$e^tP=\int 20te^t\,dt$

$e^tP=20te^t-\int 20e^t\,dt=20te^t-20e^t+c$

$40=-20+c\Rightarrow c=60$

$e^tP=20te^t-20e^t+60$

$P=20t-20+60e^{-t}=20(t+3e^{-t}-1)$

b 260

c $\frac{dP}{dt}=20(1-3e^{-t})=0$

$1-3e^{-t}=0$ gives $t=\ln 3$

$P=20(\ln 3+1-1)$

By inspection this is a minimum, so the value of P never falls below $20\ln 3\approx 21.97$

2 a The integrating factor is $e^{\int 2\,dt} = e^{2t}$

$e^{2t}\frac{dI}{dt} + 2e^{2t}I = 6e^{2t}$

so $\frac{d}{dt}\left[e^{2t}I\right] = 6e^{2t}$

$e^{2t}I = 3e^{2t} + c$ or $I = 3 + ce^{-2t}$

$t = 0, I = 8$ gives $8 = 3 + c$, so $c = 5$

so $I = 3 + 5e^{-2t}$

b $e^{-2t} > 0$, so $I > 3$

c

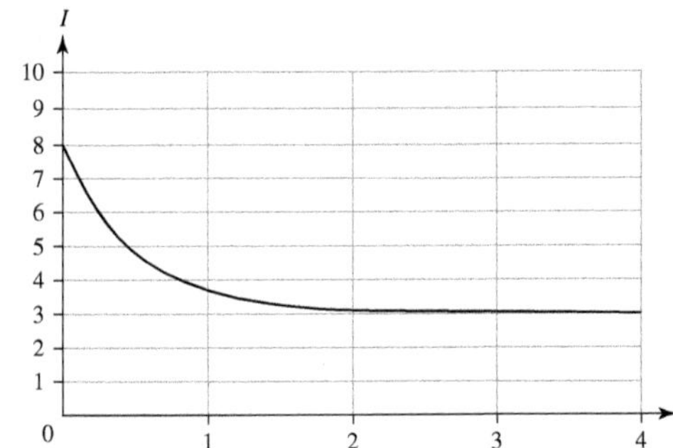

3 a $\frac{dx}{dt} = 0 - 8 + 4 = -4 < 0$ so population starts to decline initially

b $x = t + 3 + 5e^{-t}$ **c** 5.61 million (3 sf)

d Population continues to increase but at a constant rate in the long term.

4 a $v = \frac{10t(1+t)}{1+2t}$ **b** 4.4 s

c In this model, as $t \to \infty, v \to \infty$. In practice the hailstone will approach a terminal velocity.

5 a Integrating factor $e^{\int \frac{1}{20+t}dt} = e^{\ln(20+t)} = (20+t)$

$(20+t)\frac{dC}{dt} + C = 4(20+t)$

So $\frac{d}{dt}\left[(20+t)C\right] = 4(20+t)$

$(20+t)C = 2(20+t)^2 + K$

$t = 0, C = 10$ gives $200 = 800 + K$, so $K = -600$

so $(20+t)C = 2(20+t)^2 - 600$

or $C = 2(20+t) - \frac{600}{20+t}$

b 10 minutes

6 a $x = \frac{1}{2}e^{-\frac{1}{2}t}(4 - \sin t)$ **b** 0.204 m (3 sf)

7 a Multiplying throughout by $(1 + 2x)$ gives

$2v\frac{dv}{dx}(1+2x) + 2v^2 = -4(1+2x)$

Notice that $\frac{d}{dx}\left[v^2(1+2x)\right] = 2v\frac{dv}{dx}(1+2x) + 2v^2$

so $\frac{d}{dx}\left[v^2(1+2x)\right] = -4(1+2x)$

Integrating gives $v^2(1+2x) = -(1+2x)^2 + c$

$x = 0, v = 13$, gives $13^2 = -1 + c$, so $c = 170$

$v^2(1+2x) = 170 - (1+2x)^2$

b 6.02 m (3 sf)

8 a $\frac{dP}{dt} - \frac{P}{1+t} = 2$

Integrating factor is $e^{\int -\frac{1}{1+t}dt} = e^{-\ln(1+t)} = \frac{1}{1+t}$

$\frac{1}{1+t}\frac{dP}{dt} - \frac{1}{(1+t)^2}P = \frac{2}{1+t}$

$\frac{d}{dt}\left(\frac{P}{1+t}\right) = \frac{2}{1+t}$

$\frac{P}{1+t} = 2\ln(1+t) + c$

$10000 = 2\ln 1 + c \Rightarrow c = 10000$

$\frac{P}{1+t} = 2\ln(1+t) + 10000$

$P = 2(1+t)(5000 + \ln(1+t))$

b 70027

9 a Divide through by x to give:

$\frac{dy}{dx} + \frac{4}{x}y = x^3y^2$

which is in the form of a Bernoulli equation $\frac{dy}{dx} + py = qy^n$

with $p = \frac{4}{x}$ and $q = x^3$

b $y = \frac{1}{-x^4\ln x + \left(\ln 2 - \frac{1}{32}\right)x^4}$ (or equivalent)

Exercise 20.2A

1 a $y = Ae^{2x} + Be^{4x}$ **b** $y = (Ax + B)e^{-4x}$

c $y = e^{2x}(A\cos x + B\sin x)$

d $y = A + Be^{-3x}$ **e** $y = Ae^{3x} + Be^{-4x}$

f $y = A\cos x + B\sin x$

g $y = Ae^{\frac{1}{2}x} + Be^{-3x}$

h $y = e^{-\frac{1}{2}x}(A\cos x + B\sin x)$

2 a $y = Ae^{3x} + Be^{-5x} + 2x - 1$

b $y = (Ax + B)e^{3x} + 3e^{2x}$

c $y = e^{-x}(A\cos 4x + B\sin 4x) + \cos 3x$

d $y = Ae^{4x} + Be^{-4x} - 2x + 3$

e $y = Ae^{x} + Be^{-4x} - e^{-3x}$

f $y = (Ax + B)e^{-x} + x^2 + 5$

g $y = Ae^{-\frac{1}{3}x} + Be^{3x} - \frac{5}{7}e^{2x}$

h $y = (Ax + B)e^{-\frac{1}{2}x} + 5x - 2$

3 a $\frac{d^2y}{dx^2} - 2a\frac{dy}{dx} + a^2y = \frac{d^2y}{dx^2} - a\frac{dy}{dx} - a\frac{dy}{dx} + a^2y$

$= \frac{d}{dx}\left[\frac{dy}{dx} - ay\right] - a\left[\frac{dy}{dx} - ay\right]$

Let $u = \frac{dy}{dx} - ay$, then $\frac{du}{dx} - au = 0$ (1)

b Multiplying (1) throughout by the integrating factor of e^{-ax}:

$e^{-ax}\frac{du}{dx} - ae^{-ax}u = 0$

$\frac{d}{dx}\left[ue^{-ax}\right] = 0$

so $ue^{-ax} = A$

$u = Ae^{ax}$

c $u = \frac{dy}{dx} - ay$, so $\frac{dy}{dx} - ay = Ae^{ax}$ (2)

Multiplying (2) throughout by the integrating factor of e^{-ax}

$e^{-ax}\frac{dy}{dx} - ae^{-ax}y = A$

$\frac{d}{dx}\left[ye^{-ax}\right] = A$

$ye^{-ax} = Ax + B$

$y = (Ax + B)e^{ax}$

4 $Ae^{(m+in)x}+Be^{(m-in)x}=Ae^{mx}e^{inx}+Be^{mx}e^{inx}$
$=Ae^{mx}(\cos nx+i\sin nx)+Be^{mx}(\cos nx-i\sin nx)$
$=e^{mx}\left[(A+B)\cos nx+(Ai-Bi)\sin nx\right]$
Let $A+B=\alpha$, and $Ai-Bi=\beta$
Then $Ae^{(m+in)x}+Be^{(m-in)x}=e^{mx}(\alpha\cos nx+\beta\sin nx)$

Exercise 20.2B

1 a $y=2e^{x}+e^{6x}$ b $y=(4x+1)e^{-2x}$
c $y=3\cos 5x+6\sin 5x$
d $y=3e^{2x}-2e^{-3x}$ e $y=(5-12x)e^{3x}$
f $y=6(e^{4x}+e^{-3x})$ g $y=12e^{\frac{2}{3}x}$
h $y=2e^{\frac{1}{2}x}(3\cos x-2\sin x)$
2 a $y=15e^{2x}-5e^{5x}+3e^{x}$
b $y=(1-8x)e^{6x}+2$
c $y=e^{-2x}(3\cos x-2\sin x)+2(\cos x+\sin x)$
d $y=\frac{1}{6}(13e^{2x}+11e^{-4x})+3x-3$
e $y=6e^{\frac{2}{3}x}-9e^{-\frac{2}{3}x}-x^{2}-5$
f $y=e^{2x}(8\cos x+4\sin x)+\cos 2x$
g $y=3e^{\frac{3}{2}x}+7e^{-\frac{3}{2}x}+x^{2}-3x-2$
h $y=(x+7)e^{\frac{1}{4}x}+3e^{-\frac{1}{4}x}$
3 $a=-2,\ y=Ae^{4x}+Be^{-2x}-2xe^{-2x}$
4 $a=3,\ y=(Ax+B)e^{x}+3x^{2}e^{x}$
5 $y=2e^{x}-2e^{-6x}+3xe^{x}-2$
6 $y=(2x+3)e^{3x}+17x^{2}e^{3x}$
7 $y=3e^{2x}-e^{3x}+2e^{4x}$
8 a Given $x=e^{u}$, $\frac{dx}{du}=e^{u}=x$

By the chain rule $\frac{dy}{dx}=\frac{dy}{du}\times\frac{du}{dx}$

$$\frac{dy}{dx}=\frac{dy}{du}\times\frac{1}{e^{u}}=\frac{1}{x}\frac{dy}{du}$$
$$\frac{d^{2}y}{dx^{2}}=-\frac{1}{x^{2}}\frac{dy}{du}+\frac{1}{x}\frac{d^{2}y}{du^{2}}\times\frac{du}{dx}$$
$$\frac{d^{2}y}{dx^{2}}=-\frac{1}{x^{2}}\frac{dy}{du}+\frac{1}{x^{2}}\frac{d^{2}y}{du^{2}}$$

Substitute $\frac{dy}{dx}=\frac{1}{x}\frac{dy}{du}$ and $\frac{d^{2}y}{dx^{2}}=-\frac{1}{x^{2}}\frac{dy}{du}+\frac{1}{x^{2}}\frac{d^{2}y}{du^{2}}$ into

$$x^{2}\frac{d^{2}y}{dx^{2}}-4x\frac{dy}{dx}+6y=12$$
$$x^{2}\left(-\frac{1}{x^{2}}\frac{dy}{du}+\frac{1}{x^{2}}\frac{d^{2}y}{du^{2}}\right)-4x\left(\frac{1}{x}\frac{dy}{du}\right)+6y=12$$
$$-\frac{dy}{du}+\frac{d^{2}y}{du^{2}}-4\frac{dy}{du}+6y=12$$
$$\frac{d^{2}y}{du^{2}}-5\frac{dy}{du}+6y=12$$

b $y=7x^{2}-2x^{3}+2$

Exercise 20.3A

1 a $x=7\cos 5t+5\sin 5t$
b $x=-4\cos 2t+8\sin 2t$
c $x=-5\cos 10t+3\sin 10t$
d $x=-\cos t+9\sin t$
e $x=5\cos\frac{t}{2}-3\sin\frac{t}{2}$
f $x=6\sqrt{3}\cos\frac{t}{3}-4\sin\frac{t}{3}$
2 a $x=-3\cos 4t+2\sin 4t+3$
b $x=5\cos 5t+\sin 5t+4$
c $x=-5\cos 3t+4\sin 3t+4$
d $x=-6\cos 2t+4\sin 2t+3$
e $x=2\sqrt{2}\cos\frac{t}{4}+5\sqrt{2}\sin\frac{t}{4}+1$
f $x=-2\cos\frac{5t}{2}+6\sin\frac{5t}{2}-2$
3 $x=4\cos 5t+3\sin 5t$

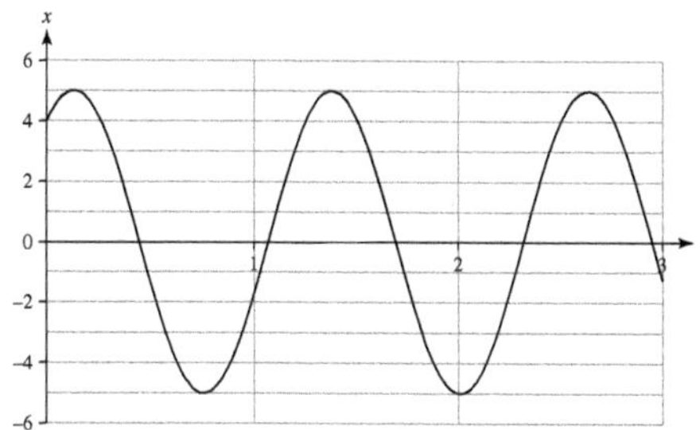

4 $x=5\cos 3t-12\sin 3t+2$

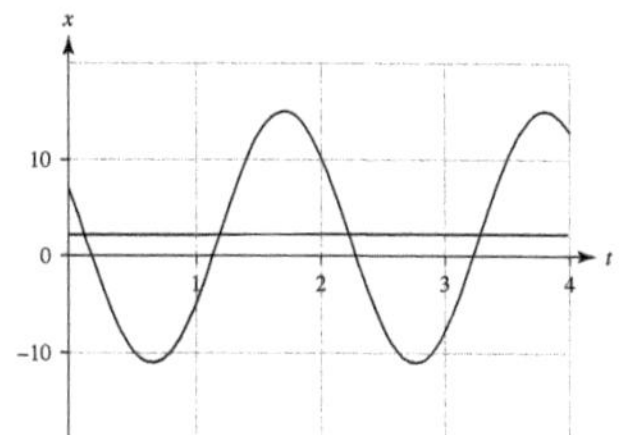

5 a $T=\frac{2\pi}{\omega}=\frac{2\pi}{3}$ s b $6\ \text{m s}^{-1}$
6 a $T=\frac{2\pi}{\omega}=\frac{2\pi}{2}=\pi$ s b $a=6$ metres
7 a $T=\frac{2\pi}{\omega}=\frac{2\pi}{0.2}=10\pi$ s
b $0.3=\sqrt{a^{2}-0.4^{2}}\Rightarrow a=0.5$ metres
c $v=\omega\sqrt{a^{2}-x^{2}}\Rightarrow v=0.2\sqrt{0.5^{2}-0.3^{2}}=0.08\ \text{m s}^{-1}$
8 a $v=\omega\sqrt{a^{2}-x^{2}}\Rightarrow v=2\sqrt{0.9^{2}-0^{2}}\Rightarrow v=1.8\ \text{m s}^{-1}$
b 0.446 s (3 sf)
9 a $v=\omega\sqrt{a^{2}-x^{2}}\Rightarrow v=4\sqrt{0.6^{2}-0^{2}}\Rightarrow v=2.4\ \text{m s}^{-1}$
b $0.6\sin 4t=0.3\Rightarrow\sin 4t=\frac{1}{2}\Rightarrow 4t=\frac{\pi}{6}\Rightarrow t=\frac{\pi}{24}$ s
10 $v=\omega\sqrt{a^{2}-x^{2}}\Rightarrow 4=8\sqrt{1.3^{2}-x^{2}}\Rightarrow x=1.2$ m
11 2.04 s
12 a $a^{2}=2.4^{2}+1\Rightarrow a=2.6$ m b $T=\frac{2\pi}{\omega}=\frac{2\pi}{3}$ s
13 a 50 cm b 100π s

Exercise 20.3B

1 a $-20x=5\frac{d^{2}x}{dt^{2}}$

$\frac{d^{2}x}{dt^{2}}=-4x$, which is SHM with $\omega=2$

b $T=\frac{2\pi}{\omega}=\frac{2\pi}{2}=\pi$ secs
c $v=\omega\sqrt{a^{2}-x^{2}}=2\sqrt{5^{2}-3^{2}}=8\ \text{m s}^{-1}$
2 a $-5x=0.2\frac{d^{2}x}{dt^{2}}$

$\frac{d^{2}x}{dt^{2}}=-25x$, which is SHM with $\omega=5$

b The auxiliary equation is $m^2 + 25 = 0$
$m^2 = -25$, so $m = \pm 5i$
So the general solution is $x = A\cos 5t + B\sin 5t$
Use $x = 0.3$, $\frac{dx}{dt} = 0$ at $t = 0$ to get $A = 0.3$, $B = 0$
$\Rightarrow x = 0.3\cos 5t$

c 15 cm **d** 1.3 m s^{-1}

3 a $-12x = \frac{d^2x}{dt^2}$
$\frac{d^2x}{dt^2} = -4x$, which is SHM with $\omega = 2$

b $T = \frac{2\pi}{\omega} = \frac{2\pi}{2} = \pi$ s

c $v = \omega\sqrt{a^2 - x^2} = 2\sqrt{0.2^2 - 0.16^2} = 0.24$ m s^{-1}

d No friction or air resistance. The body is a particle. The spring has no mass.

4 a $-10x = 0.1\frac{d^2x}{dt^2}$
$\frac{d^2x}{dt^2} = -100x$ which is SHM with $\omega = 10$

b 12 cm

c $v = \omega\sqrt{a^2 - x^2} = 10\sqrt{0.12^2 - 0.06^2} = \frac{3\sqrt{3}}{5} \approx 1.04$ m s^{-1}

5 a $-48x - 48x = 1.5\frac{d^2x}{dt^2}$
$\frac{d^2x}{dt^2} = -64x$, which is SHM with $\omega = 8$

b $T = \frac{2\pi}{8} = \frac{\pi}{4}$ s

c The auxiliary equation is $m^2 + 64 = 0$
$m^2 = -64$, so $m = \pm 8i$
So the general solution is $x = A\cos 8t + B\sin 8t$
And $\frac{dx}{dt} = -8A\sin 8t + 8B\cos 8t$
Use $x = 0$, $\frac{dx}{dt} = 4$ at $t = 0$ to get $A = 0$, $B = 0.5$
$\Rightarrow x = 0.5\sin 8t$
$0.5 = 0.5\sin 8t \Rightarrow \sin 8t = 1 \Rightarrow t = \frac{\pi}{16}$
(Or for a more elegant solution, notice that this is a quarter period.)

d $v = \omega\sqrt{a^2 - x^2} = 8\sqrt{0.5^2 - 0.3^2} = 3.2$ m s^{-1}

6 a $-0.32x = 0.5\frac{d^2x}{dt^2}$
$\frac{d^2x}{dt^2} = -0.64x$, which is SHM with $\omega = 0.8$

b The auxiliary equation is $m^2 + 0.64 = 0$
$m^2 = -0.64$, so $m = \pm 0.8i$
So the general solution is $x = A\cos 0.8t + B\sin 0.8t$
Use $x = 0.9$, $\frac{dx}{dt} = 0$ at $t = 0$ to get $A = 0.9$, $B = 0$
$\Rightarrow x = 0.9\cos 0.8t$

c 0.624 m s^{-1}

d If the body is pulled 120 cm in the direction OB, then one side of the string will be slack for part of the motion, and the model will need to be applied to two different situations – one when both parts of the string are taut, and one where one part of the string is slack.

7 a $-mx - 3mx = m\frac{d^2x}{dt^2}$
$\frac{d^2x}{dt^2} = -4x$, which is SHM with $\omega = 2$

b $T = \frac{2\pi}{2} = \pi$ s **c** 0.346 m s^{-1}

d No friction or air resistance, the box is a particle, the spring has no mass.

8 a $-\frac{2(x-2)}{5} = 0.1\frac{d^2x}{dt^2}$
$\frac{d^2x}{dt^2} + 4x = 8$

b The auxiliary equation is $m^2 + 4 = 0$
$m^2 = -4$, so $m = \pm 2i$
So the complementary function is $x = A\cos 2t + B\sin 2t$
By inspection the particular integral is $x = 2$
So the general solution is $x = 2 + A\cos 2t + B\sin 2t$
Use $x = 2.4$, $\frac{dx}{dt} = 0$ at $t = 0$ to get $A = 0.4$, $B = 0$
$x = 2 + 0.4\cos 2t$

c 0.8 m s^{-1}

d $\cos 2t = -1 \Rightarrow 2t = \pi \Rightarrow t = \frac{\pi}{2}$

9 a $-15(2x - 3) = 0.3\frac{d^2x}{dt^2}$
$\frac{d^2x}{dt^2} + 100x = 150$

b $x = 1.5 - 1.5\cos 10t + 2\sin 10t$

c $\frac{dx}{dt} = 15\sin 10t + 20\cos 10t$

d 25 m s^{-1}

10 a Apply $\mathbf{F} = m\mathbf{a}$ perpendicular to the string to get
$-mg\sin\theta = ml\frac{d^2\theta}{dt^2}$

b Use $\sin\theta \approx \theta$ to get
$-mg\theta \approx ml\frac{d^2\theta}{dt^2}$
i.e. $\frac{d^2\theta}{dt^2} \approx -\frac{g}{\ell}\theta$
This is the equation of SHM with $\omega = \sqrt{\frac{g}{\ell}}$
$T = \frac{2\pi}{\omega} = 2\pi\sqrt{\frac{\ell}{g}}$

Exercise 20.4A

1 a $x = -e^{-3t}(5\cos 2t + 6\sin 2t)$, light damping

b $x = 3e^{-2t} - e^{-4t}$, heavy damping

c $x = (3t - 1)e^{-2t}$, critical damping

d $x = e^{-5t}(2\cos 4t + \sin 4t)$, light damping

e $x = e^{-2.5t}(6\cos 0.5t + 14\sin 0.5t)$, light damping

f $x = 6t\,e^{-\frac{1}{3}t}$, critical damping

2 a $x = 3e^{-t} - e^{-9t} - 2$

b $x = (5t + 2)e^{-4t} + 3$

c $x = e^{-5t}(5\cos 2t + 2\sin 2t) + 1$

d $x = 4 - e^{-2t}\cos t$

e $x = (11t + 1)e^{-2t} + 3$

f $x = e^{-t}(\cos t + \sin t) + 5$

3 $x = 4e^{-t} - e^{-3t} + 2$

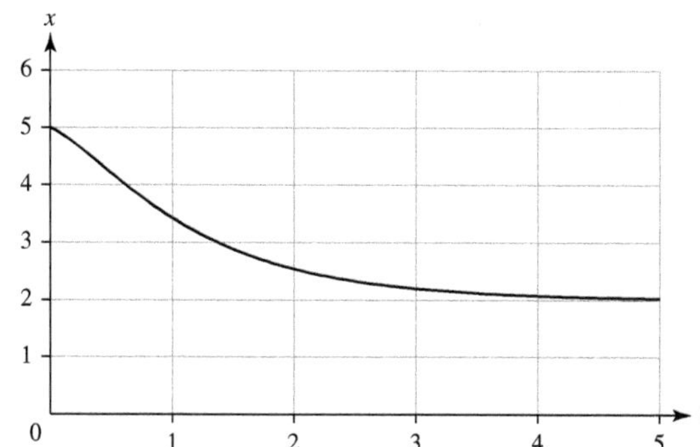

4 $x = 4 + e^{-t}\sin 2t$

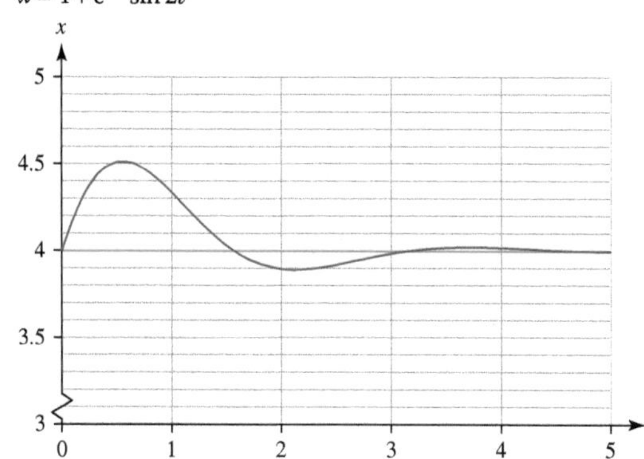

5 a $x = 0.4e^{-2t}\sin 2t$

b

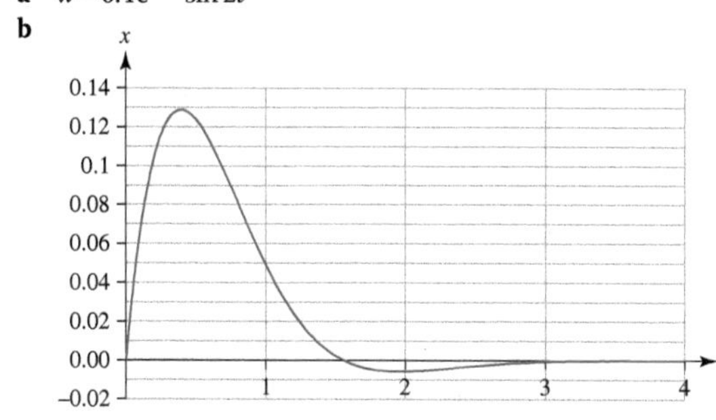

c This is light damping.

6 a $x = 4e^{-3t} - 3.5e^{-4t} + 0.5$ b 0.154 s (3 sf)

c

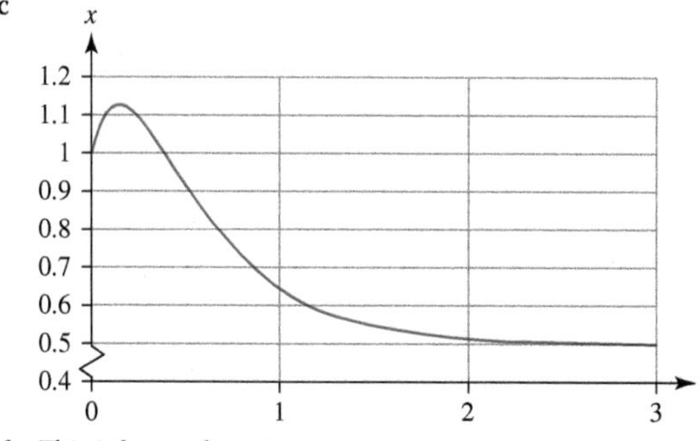

d This is heavy damping.

7 a $x = ute^{-4\omega t}$ b $t = \frac{1}{4\omega}$

c The damping is critical.

8 a $I = 2\sin 4t + 3\sin 2t$ b 1.21 s (3 sf)

9 a $\theta = \frac{\pi e^{-3t}}{8}(2\cos 2t + 3\sin 2t)$

b

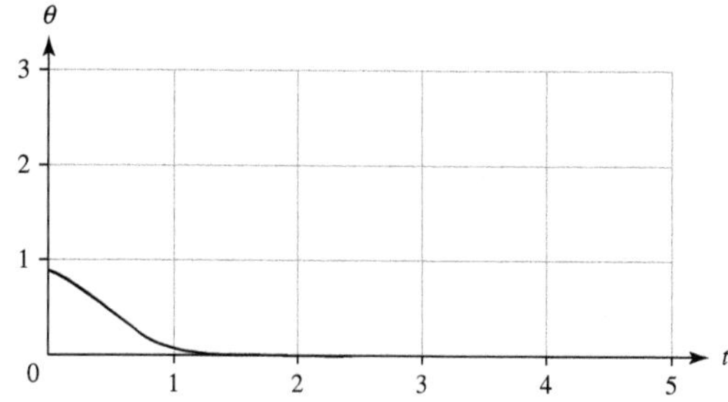

c As, $t \to \infty$, $\theta \to 0$, i.e. the door settles down towards a closed position.

10 a $x = e^{-\frac{1}{2}t}\left(3 + 2\cos\frac{1}{2}t + 2\sin\frac{1}{2}t\right)$

b $x = e^{-\frac{1}{2}t}\left(3 + 2\cos\frac{1}{2}t + 2\sin\frac{1}{2}t\right)$

$= e^{-\frac{1}{2}t}\left(3 + 2\sqrt{2}\sin\left(\frac{1}{2}t + \frac{\pi}{4}\right)\right) > 0$ for all values of t since $3 > 2\sqrt{2}$ and $\sin\left(\frac{1}{2}t + \frac{\pi}{4}\right) > -1$. So the particle never reaches the origin.

c 7.98 s

11 a $x = -\frac{4}{3}e^{-\frac{1}{2}t} + \frac{4}{3}e^{-2t} + 2\sin t$ b 1.35 cm above A

12 a $x = \frac{1}{4}(2t + 1 - \cos 2t - \sin 2t)$ b $t = \frac{3\pi}{4}$

c

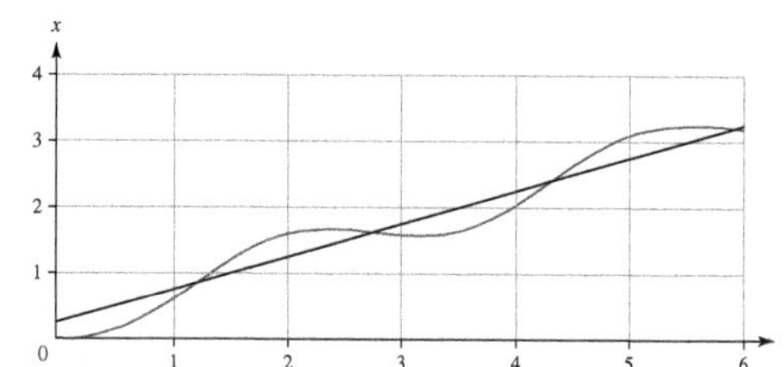

13 a $x = e^{-t}(\cos t + \sin t)$ b $t = \pi, 2\pi, 3\pi, \ldots$

c $x(\pi) = -e^{-\pi}, x(2\pi) = e^{-2\pi}, x(3\pi) = -e^{-3\pi}, \ldots$

So total distance $= 1 + 2(e^{-\pi} + e^{-2\pi} + e^{-3\pi} + \ldots)$

$= 1 + \frac{2e^{-\pi}}{1 - e^{-\pi}}$, summing an infinite GP

$= \frac{1 + e^{-\pi}}{1 - e^{-\pi}} = \frac{e^{\frac{\pi}{2}} + e^{-\frac{\pi}{2}}}{e^{\frac{\pi}{2}} - e^{-\frac{\pi}{2}}}$

$= \coth\left(\frac{\pi}{2}\right)$

Exercise 20.4B

1 a Using $\mathbf{F} = m\mathbf{a}$ gives $-36x - 45\frac{dx}{dt} = 9\frac{d^2x}{dt^2}$

So $\frac{d^2x}{dt^2} + 5\frac{dx}{dt} + 4x = 0$

b $x = e^{-t} - 3e^{-4t}$ c heavy damping

d 0.37 s

2 a Using $\mathbf{F} = m\mathbf{a}$ gives $-25mn^2x - km\frac{dx}{dt} = m\frac{d^2x}{dt^2}$

So $\frac{d^2x}{dt^2} + k\frac{dx}{dt} + 25n^2x = 0$

b $k = 10n$ c $x = (At + B)e^{-5nt}$

3 a $\frac{4m}{45}$ might be friction between the body and the table,

and $\frac{4m\frac{dx}{dt}}{3}$ might be air resistance.

b $9\frac{d^2x}{dt^2}+12\frac{dx}{dt}+4x=\frac{4}{5}$

c The auxiliary equation is $9m^2+12m+4=0$

$(3m+2)^2=0$, so $m=-\frac{2}{3}$ (repeated)

So the complementary function is $x=(At+B)e^{-\frac{2}{3}t}$

By inspection the particular integral is $x=\frac{1}{5}$

The general solution is $x=(At+B)e^{-\frac{2}{3}t}+\frac{1}{5}$

$x=\frac{2}{5}$ when $t=0$ gives $B+\frac{1}{5}=\frac{2}{5}$, so $B=\frac{1}{5}$

$\frac{dx}{dt}=Ae^{-\frac{2}{3}t}-\frac{2}{3}\left(At+\frac{1}{5}\right)e^{-\frac{2}{3}t}$

$\frac{dx}{dt}=0$ when t = 0 gives $A-\frac{2}{15}=0$ so $A=\frac{2}{15}$

Particular solution is $x=\left(\frac{2}{15}t+\frac{1}{5}\right)e^{-\frac{2}{3}t}+\frac{1}{5}$

$=\frac{1}{15}\left[(2t+3)e^{-\frac{2}{3}t}+3\right]$

d The damping is critical, so the body moves directly towards an equilibrium position in which the spring is stretched by 20 cm.

4 a Using $\mathbf{F}=m\mathbf{a}$ gives $24e^{-t}-15x-6\frac{dx}{dt}=3\frac{d^2x}{dt^2}$

So $\frac{d^2x}{dt^2}+2\frac{dx}{dt}+5x=8e^{-t}$

b $x=e^{-t}(2+3\cos 2t+4\sin 2t)$

c 1.45 s (3 sf)

5 a Using $\mathbf{F}=m\mathbf{a}$ gives $m\omega^2\ell e^{-\omega t}-5m\omega\frac{dx}{dt}-6m\omega^2\frac{dx}{dt}=m\frac{d^2x}{dt^2}$

So $\frac{d^2x}{dt^2}+5\omega\frac{dx}{dt}+6\omega^2x=\omega^2\ell e^{-\omega t}$

b The auxiliary equation is $m^2+5\omega m+6\omega^2=0$

$(m+2\omega)(m+3\omega)=0$, so $m=-2\omega$ or $m=-3\omega$

So the complementary function is $x=Ae^{-2\omega t}+Be^{-3\omega t}$

For the particular integral try $x=ae^{-\omega t}$

Then $a\omega^2e^{-\omega t}-5a\omega^2e^{-\omega t}+6a\omega^2e^{-\omega t}\equiv\omega^2\ell e^{-\omega t}$, so $a=\frac{\ell}{2}$

The particular integral is $x=\frac{\ell}{2}e^{-\omega t}$ and the general solution is $x=Ae^{-2\omega t}+Be^{-3\omega t}+\frac{\ell}{2}e^{-\omega t}$

$x=0$ when $t=0$ gives $A+B+\frac{\ell}{2}=0$ (1)

$\frac{dx}{dt}=-2A\omega e^{-2\omega t}-3B\omega e^{-3\omega t}-\frac{\ell}{2}\omega e^{-\omega t}$

$\frac{dx}{dt}=0$ when t = 0 gives $-2A\omega-3B\omega-\frac{\ell}{2}\omega=0$ ie

$-2A-3B-\frac{\ell}{2}=0$ (2)

Solving (1) and (2) gives $A=-\ell$ $B=\frac{\ell}{2}$

Particular solution is $x=-\ell e^{-2\omega t}+\frac{\ell}{2}e^{-3\omega t}+\frac{\ell}{2}e^{-\omega t}$

$=\frac{\ell}{2}e^{-\omega t}(1-2e^{-\omega t}+e^{-2\omega t})=\frac{\ell}{2}e^{-\omega t}(1-e^{-\omega t})^2$

c $\frac{dx}{dt}=-\frac{\ell}{2}\omega e^{-\omega t}(1-e^{-\omega t})^2+\ell\omega e^{-\omega t}(1-e^{-\omega t})$

$=\frac{\ell}{2}\omega e^{-\omega t}(1-e^{-\omega t})(-1+e^{-\omega t}+2e^{-\omega t})$

$=\frac{\ell}{2}\omega e^{-\omega t}(1-e^{-\omega t})(-1+3e^{-\omega t})$

So $\frac{dx}{dt}=0$ when $-1+3e^{-\omega t}=0$ ie when $e^{-\omega t}=\frac{1}{3}$

And $x=\frac{\ell}{2}\times\frac{1}{3}\left(1-\frac{1}{3}\right)^2=\frac{2\ell}{27}$

6 a Using $\mathbf{F}=m\mathbf{a}$ gives $-m\omega^2x-2mk\frac{dx}{dt}=m\frac{d^2x}{dt^2}$

So $\frac{d^2x}{dt^2}+2k\frac{dx}{dt}+\omega^2x=0$

b $x=Ute^{-\omega t}$ **c** critical

d $\frac{dx}{dt}=Ue^{-\omega t}-U\omega te^{-\omega t}=Ue^{-\omega t}(1-\omega t)$

Tub comes to rest when $1-\omega t=0\Rightarrow t=\frac{1}{\omega}$

$t<2\Rightarrow\frac{1}{\omega}<2\Rightarrow\omega>\frac{1}{2}$

e Tub is moving horizontally as it is brought to rest, and the spring has no mass.

7 a Using $\mathbf{F}=m\mathbf{a}$ gives $5g-25x-10\frac{dx}{dt}=5\frac{d^2x}{dt^2}$

So $5\frac{d^2x}{dt^2}+10\frac{dx}{dt}+25x=50$ or $\frac{d^2x}{dt^2}+2\frac{dx}{dt}+5x=10$

b $x=\frac{e^{-t}}{2}(\sin 2t-4\cos 2t)+2$

c 1.22 s (3 sf)

8 a Using $\mathbf{F}=m\mathbf{a}$ gives $80g-10x-40\frac{dx}{dt}=80\frac{d^2x}{dt^2}$

So

$80\frac{d^2x}{dt^2}+40\frac{dx}{dt}+10x=800$ or $8\frac{d^2x}{dt^2}+4\frac{dx}{dt}+x=80$

b $x=80-20e^{-\frac{1}{4}t}\left(4\cos\left(\frac{1}{4}t\right)+\sin\left(\frac{1}{4}t\right)\right)$

c 10.4 secs (3 sf)

d The man is a particle and the rope is light.

9 a i After t seconds the end A has moved a displacement of $3t$ metres, and box B has moved a displacement of y metres.

Since the spring is initially in equilibrium, the extension of the spring, x metres, is given by $x=3t-y$, ie $x+y=3t$

ii Differentiating $x+y=3t$ gives $\frac{dx}{dt}+\frac{dy}{dt}=3$ (1) and

$\frac{d^2x}{dt^2}+\frac{d^2y}{dt^2}=0$ (2)

Applying $\mathbf{F}=m\mathbf{a}$ to B gives $\frac{90x}{2}-30v=5\frac{d^2y}{dt^2}$

So $45x-30\frac{dy}{dt}=5\frac{d^2y}{dt^2}$ ie $9x-6\frac{dy}{dt}=\frac{d^2y}{dt^2}$

Substituting for $\frac{dy}{dt}$ and $\frac{d^2y}{dt^2}$ from (1) and (2):

$9x-6\left(3-\frac{dx}{dt}\right)=-\frac{d^2x}{dt^2}$

So $\frac{d^2x}{dt^2}+6\frac{dx}{dt}+9x=18$

iii Auxiliary equation is $m^2+6m+9=0$ so
$(m+3)^2=0 \Rightarrow m=-3$ (repeated)
Complementary function is $x=(At+B)e^{-3t}$
Particular integral is $x = 2$, so general solution is $x=(At+B)e^{-3t}+2$

$$\frac{dx}{dt}=A\,e^{-3t}-3(At+B)e^{-3t}$$

At $t=0$, $x=y=0$, and $\frac{dy}{dt}=0$. Using (1) gives $\frac{dx}{dt}=3$

$t=0, x=0$ and $\frac{dx}{dt}=3$ give $B+2=0$, and $A-3B=3$
$\Rightarrow B=-2$ and $A=-3$
So particular solution is $x=2-(2+3t)e^{-3t}$

b For large values of t the oscillations damp down, the box moves with a near constant velocity of 3 m s^{-1}, and the spring stretched by a near constant 2 metres.

Exercise 20.5A

1 a $y=Ae^{2t}+Be^{3t}$
b $y=A\cos 3t+B\sin 3t$
c $y=Ae^{\frac{16}{3}t}+Be^{t}$
d $y=Ae^{3t}+Be^{-2t}+\frac{4}{3}$
e $y=e^{-5t}(A\cos t+B\sin t)$
f $y=e^{-t}(A\cos 2t+B\sin 2t)+\frac{1}{2}\sin t-\frac{1}{2}\cos t$

2 a $y=2e^{3t}+e^{-2t},\ x=8e^{3t}-e^{-2t}$
b $y=e^{t}+3e^{-3t}-5,\ x=e^{t}+\frac{3}{5}e^{-3t}-4$
c $y=e^{-t}(3\cos t-2\sin t), x=e^{-t}(4\cos t-7\sin t)$
d $y=3t\,e^{-t}-5t+9,\ x=9t-11-\frac{3}{2}e^{-t}(2t+1)$
e $y=e^{-2t}(5\cos t+12\sin t), x=\frac{e^{-2t}}{2}(7\cos t-17\sin t)-1$
f $y=\left(2t+\frac{5}{3}\right)e^{t}+\frac{1}{3}e^{-2t}, x=3e^{t}(t+1)$

3 $x=e^{t}+2e^{3t}, y=-e^{t}+2e^{3t}$

4 $x=17-7t-(4t+9)e^{-t}, y=11(t-2)+(4t+11)e^{-t}$

5 a From (2): $\frac{d^2y}{dt^2}=\frac{dy}{dt}-\frac{dz}{dt}$,

so $\frac{d^2y}{dt^2}=\frac{dy}{dt}+x$ (4)

from (1): $\frac{d^2x}{dt^2}=\frac{dx}{dt}+2\frac{dy}{dt}$ (5)

differentiating (5): $\frac{d^3x}{dt^3}=\frac{d^2x}{dt^2}+2\frac{d^2y}{dt^2}$

from (4): $\frac{d^3x}{dt^3}=\frac{d^2x}{dt^2}+2\left(\frac{dy}{dt}+x\right)$

so $\frac{d^3x}{dt^3}=\frac{d^2x}{dt^2}+2\frac{dy}{dt}+2x$

substituting $2\frac{dy}{dt}=\frac{d^2x}{dt^2}-\frac{dx}{dt}$ from (5):

$$\frac{d^3x}{dt^3}=\frac{d^2x}{dt^2}+\frac{d^2x}{dt^2}-\frac{dx}{dt}+2x$$

$$\Rightarrow \frac{d^3x}{dt^3}-2\frac{d^2x}{dt^2}+\frac{dx}{dt}-2x=0$$

b $x=e^{2t}-\cos t+2\sin t$

Exercise 20.5B

1 a $\frac{dx}{dt}=-3y$

$\frac{dy}{dt}=x-4y$

b $y=20e^{-t}-20e^{-3t}$ $\quad x=60e^{-t}-20e^{-3t}$
c $t=\frac{1}{2}\ln 3$

2 a $\frac{dx}{dt}=3y-2x$ $\quad \frac{dy}{dt}=x-4y$
b $y=3e^{-t}+2e^{-5t}$ $\quad x=9e^{-t}-2e^{-5t}$
c $x-y=9e^{-t}-2e^{-5t}-(3e^{-t}+2e^{-5t})$
$=6e^{-t}-4e^{-5t}$
$=2e^{-5t}(3e^{4t}-2)$
For $t>0$, $3e^{4t}-2>0$ and $e^{-5t}>0$
$\Rightarrow x>y$, as required

3 a $\frac{dx}{dt}=3(x-y)$ $\quad \frac{dy}{dt}=5y-x$
$y=20e^{2t}+10e^{6t}$ $\quad x=60e^{2t}-10e^{6t}$
b X becomes extinct when
$t=\frac{1}{4}\ln 6$ months

4 a $\frac{dx}{dt}=y-4x+7$ $\quad \frac{dy}{dt}=6x-5y$
b $x=\frac{5e^{-2t}-4e^{-7t}+5}{2}$
$y=5e^{-2t}+6e^{-7t}+3$
c As $t\to\infty$, $e^{-kt}\to 0$ for $k>0$
$\Rightarrow x\to\frac{5}{2}$ and $y\to 3$
$\therefore x:y\to 5:6$

5 a $\frac{dx}{dt}=x-y$ $\quad \frac{dy}{dt}=3x+5y$
b $x=220e^{2t}-100e^{4t}$
$y=-220e^{2t}+300e^{4t}$ **c** 0.4 years

6 a $\frac{dx}{dt}=6x-y$ $\quad \frac{dy}{dt}=3x+2y$
$x=30e^{3t}+70e^{5t}$ $\quad y=90e^{3t}+70e^{5t}$

b $\frac{x}{y}=\frac{90e^{3t}+70e^{5t}}{30e^{3t}+70e^{5t}}$
$=\frac{90e^{-2t}+70}{30e^{-2t}+70}\to\frac{70}{70}=1$ as $t\to\infty$
$\therefore$ Over time, the number of foxes will be approximately the same as the number of rabbits.

7 $t=\frac{60}{7}=8\frac{4}{7}$ years

8 a $\frac{dx}{dt}=0.4x+0.1y$

$\frac{dy}{dt}=-0.2x+0.2y$

$\frac{d^2y}{dt^2}=-0.2\frac{dx}{dt}+0.2\frac{dy}{dt}$

$=-0.2(0.4x+0.1y)+0.2\frac{dy}{dt}$

$=0.4\left(\frac{dy}{dt}-0.2y\right)-0.02y+0.2\frac{dy}{dt}$

$10\frac{d^2y}{dt^2}-6\frac{dy}{dt}+y=0$

b $x=[4\cos(0.1t)+1004\sin(0.1t)]e^{0.3t}$
$y=[1000\cos(0.1t)-1008\sin(0.1t)]e^{0.3t}$
c 7.81 years (3 sf)
d Numbers of bears would grow without limit even after there are no fish left. This is unrealistic.

9 a $L=0$: $\frac{dH}{dt}=1.2H$ In the absence of lynx, hares increase at +120% = +1.2 per year

$L' \neq 0$: $\frac{dH}{dt}=1.2H-1.15L$ On average, each lynx eats 1.15 hares; this additional term reduces the rate of growth of the hare population.

b $\frac{dL}{dt}=0.05H$

$\frac{d^2H}{dt^2}=1.2\frac{dH}{dt}-1.15\frac{dL}{dt}$

$=1.2\frac{dH}{dt}-1.15\times 0.05H$

$\frac{d^2H}{dt^2}-1.2\frac{dH}{dt}+0.0575H=0$

c $L=20e^{1.15t}+40e^{0.05t}$ $\quad H=460e^{1.15t}+40e^{0.05t}$

d According to the model, the populations of both hares and lynx grow without limit. This is unrealistic since food and territory are both finite which would limit the growth.

10 a $\frac{dL}{dt}=0.2Z$ $\quad \frac{dZ}{dt}=1.3Z-1.1L$

$\frac{d^2Z}{dt^2}=1.3\frac{dZ}{dt}-1.1\frac{dL}{dt}=1.3\frac{dZ}{dt}-1.1\times 0.2Z$

$\frac{d^2Z}{dt^2}+1.3\frac{dZ}{dt}+0.22Z=0$

b $Z=1100e^{1.1t}-100e^{0.2t}$ $\quad L=200e^{1.1t}-100e^{0.2t}$

c $\frac{Z}{L}=\frac{1100e^{1.1t}-100e^{0.2t}}{200e^{1.1t}-100e^{0.2t}}$

$=\frac{1100-100e^{-0.9t}}{200-100^{-0.9t}}\to\frac{1100}{200}=\frac{11}{2}$ as $t\to\infty$

$\therefore$ For large values of t, $L:Z=2:11$

11 a The rate of change of X is $\frac{dx}{dt}$

It is decaying so $\frac{dx}{dt}<0$

The rate is 0.2 times the amount of X present: $\frac{dx}{dt}=-0.2x$

X changes into Y, so the $0.2x$ decrease in X is a $0.2x$ increase in Y.

In addition, Y decreases at a rate of 0.1 times the amount of Y present: $\frac{dy}{dt}=0.2x-0.1y$

Y changes into Z, so the $0.1y$ decrease in Y is a $0.1y$ increase in Z: $\frac{dz}{dt}=0.1y$

b $x=100e^{-0.2t}$

$y=-200e^{-0.2t}+200e^{-0.1t}$

$z=100e^{-0.2t}-200e^{-0.1t}+100$

c $x+y+z=100e^{-0.2t}-200e^{-0.2t}+200e^{-0.1t}+$

$100e^{-0.2t}-200e^{-0.1t}+100$

$=100$ as required

OR $\frac{d(x+y+z)}{dt}=-0.2x+0.2x-0.1y+0.1y$

$=0$

$\Rightarrow x+y+z=\text{constant}=100$

Review exercise 20

1 $y=\frac{-\frac{1}{4}\cos^4 x+1}{\sin x}$

2 a $S=\frac{t(300-t)}{600}$ **b** 37.5 kg

3 a $\frac{dT}{dt}=-kT$

$\int\frac{1}{T}dT=\int -k\,dt$

$\ln(cT)=-kt$

$t=0$, $T=100$ giving $\ln(100c)=0$, so $c=\frac{1}{100}$

$\ln\left(\frac{T}{100}\right)=-kt$

$T=100e^{-kt}$

b $T=25$ when $t=6$ gives $25=100e^{-6k}$

$\ln\left(\frac{1}{4}\right)=-6k$

$k=\frac{\ln 4}{6}=\frac{\ln 2}{3}$

c 13.0 minutes (3 sf)

4 a Using $\mathbf{F}=m\mathbf{a}$ gives $600-6v^2=120\frac{dv}{dt}\Rightarrow 100-v^2=20\frac{dv}{dt}$

b 2.94 seconds (3 sf)

c 16.6 metres (3 sf)

5 a $y=e^{-2x}(A\cos 3x+B\sin 3x)+2e^{4x}$

b $y=e^{-2x}(4\cos 3x-\sin 3x)+2e^{4x}$

6 a $y=-\frac{1}{8}\sin 4t+\frac{1}{2}t$ **b** $\frac{\pi}{2}$ seconds

c

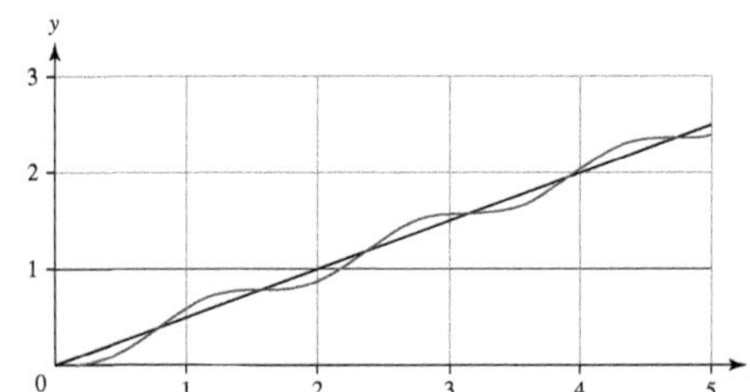

7 a Using $\mathbf{F}=m\mathbf{a}$ gives $-9x-12v=4a$

$\Rightarrow -9x-12\frac{dx}{dt}=4\frac{d^2x}{dt^2}$ or $4\frac{d^2x}{dt^2}+12\frac{dx}{dt}+9x=0$

b $4m^2+12m+9=0\Rightarrow(2m+3)^2=0$, so $m=-\frac{3}{2}$ (repeated)

$x=(At+B)e^{-\frac{3}{2}t}$

$\frac{dx}{dt}=-\frac{3}{2}e^{-\frac{3}{2}t}(At+B)+Ae^{-\frac{3}{2}t}$

$x=6$ and $\frac{dx}{dt}=-11$ when $t=0$ give $B=6$ and

$-\frac{3}{2}B+A=-11$,, so $A=-2$

The particular solution is $x=e^{-\frac{3}{2}t}(6-2t)$

c 3 s **d** 0.022 m s⁻¹ (3 dp)

8 a $\frac{dx}{dt}=x-y-3t$ $\quad \frac{dy}{dt}=y-4x-3$

$\frac{d^2x}{dt^2}=\frac{dx}{dt}-\frac{dy}{dt}-3$

$=\frac{dx}{dt}-y+4x+3-3$

$=\frac{dx}{dt}+\frac{dx}{dt}-x+3t+4x$

$\frac{d^2x}{dt^2}-2\frac{dx}{dt}-3x=3t$

b $x=\frac{1}{12}e^{3t}+\frac{5}{4}e^{-t}+\frac{2}{3}-t$

$y=-\frac{1}{6}e^{3t}+\frac{5}{2}e^{-t}+\frac{5}{3}-4t$

Assessment 20

1 $y = \ln \dfrac{2}{k - e^x(\cos x + \sin x)}$

2 $y = \ln \dfrac{1}{x \ln x - x + 2}$

3 a $y^2 = x^2 \ln x - \dfrac{x^2}{2} + k$

b $y^2 = x^2 \ln x - \dfrac{x^2}{2} + \dfrac{33}{2}$

4 a $P = 50 + ce^{-4t}$ b $P = 50 + 500e^{-4t}$

c The model predicts that the population will fall to 50, and will then remain constant.

d Because the exponential term is negative.
The model does not take into account external factors, so it is possible that the population will fall to zero, which this model does not predict.

5 $y = \dfrac{x^2}{4} + \dfrac{c}{x^2}$

6 $y = 4x^2 \tan x + 8x - 8\tan x + \sec x$

7 a $y = \ln x - 2 + \dfrac{c}{\sqrt{x}}$

b $y = \ln x - 2 + \dfrac{1}{\sqrt{x}}$

8 a $y = Ae^{-x} + Be^{-3x}$

b $y = Ae^{-2x} + Bxe^{-2x}$

c $y = e^{2x}(A\cos x + B\sin x)$

9 $y = 2e^{\frac{1}{2}x} + 3e^{-2x}$

10 $y = 6e^{\frac{x}{3}} + 2xe^{\frac{x}{3}}$

11 $y = e^{-x}(10\cos 2x + 6\sin 2x)$

12 $y = Ae^{3x} + Be^{-4x} - x - \dfrac{1}{6}$

13 $y = 4e^{2x}(\cos 3x - \sin 3x) + 3\sin x + \cos x$

14 a $0.5\dfrac{d^2x}{dt^2} = -\dfrac{7}{2}\dfrac{dx}{dt} - (5x - 6e^{-t})$

$\dfrac{d^2x}{dt^2} = -\dfrac{7}{2\times 0.5}\dfrac{dx}{dt} - \dfrac{1}{0.5}(5x - 6e^{-t})$

$\dfrac{d^2x}{dt^2} + 7\dfrac{dx}{dt} = -10x + 12e^{-t}$

$\dfrac{d^2x}{dt^2} + 7\dfrac{dx}{dt} + 10x = 12e^{-t}$

b $x = \dfrac{1}{3}e^{-5t} - \dfrac{10}{3}e^{-2t} + 3e^{-t}$

c For the auxiliary equation

$b^2 - 4ac = 7^2 - 4\times 1\times 10$

> 0 so the damping is heavy

d

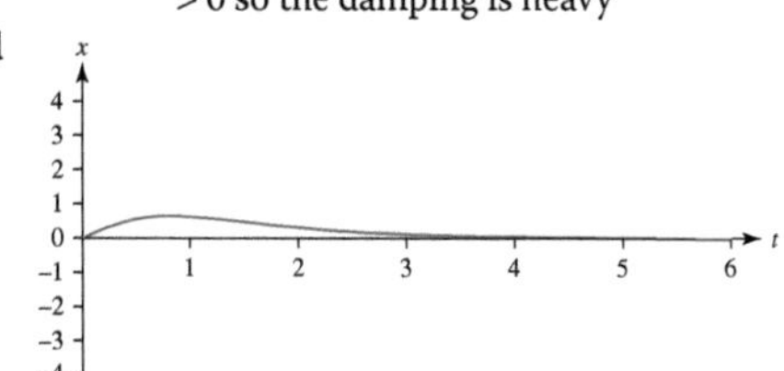

15 a $x = \sin 3t + \sin t$

b $\cos t = 0 \Rightarrow t = \dfrac{\pi}{2}, \dfrac{3\pi}{2}, \dfrac{5\pi}{2}, \ldots$

16 a $\dfrac{dA}{dt} = kA$; $A(0) = 1000$; $A(1) = 1005$

b $A = 1000\left(\dfrac{1005}{1000}\right)^t$

c The account has £1025.25 after 5 years.

17 a $V = RI + L\dfrac{dI}{dt}$

$\dfrac{V}{L} = \dfrac{RI}{L} + \dfrac{L}{L}\dfrac{dI}{dt}$

$\therefore \dfrac{dI}{dt} + \dfrac{RI}{L} = \dfrac{V}{L}$

b $I = \dfrac{V}{R} + \dfrac{V}{R}e^{-\frac{R}{L}t}$ $I = \dfrac{V}{R}\left(1 - e^{-\frac{R}{L}t}\right)$

c 0.787 amps (3 sf)

d The current reaches a steady value of 2 amps.

e

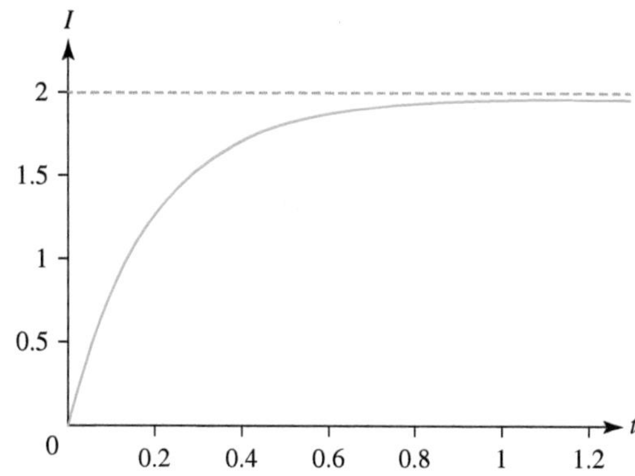

18 a $x = 4\sin \dfrac{1}{2}t$ b 2

19 a Auxiliary equation has equal roots so the damping is critical. The oscillations will die away quickly.

b $x = 0.5e^{-0.15t}(1 + t)$

c

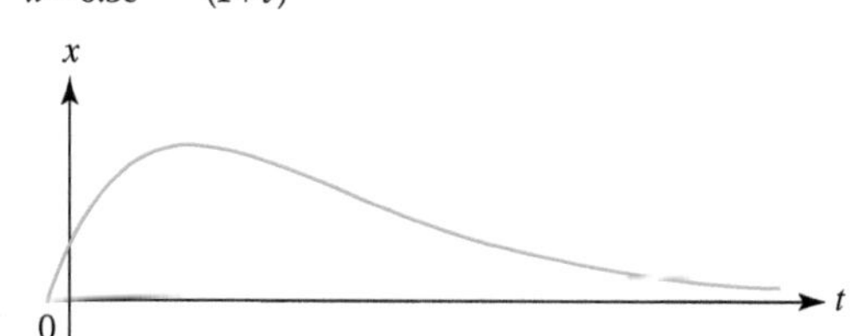

20 a $m\dfrac{d^2x}{dt^2} = -5m\dfrac{dx}{dt} - 6mx + 3m\sin 2t$

$\dfrac{d^2x}{dt^2} + 5\dfrac{dx}{dt} + 6x = 3\sin 2t$

b $x = Ae^{-2t} + Be^{-3t}$

c $x = -\dfrac{15}{52}\cos 2t + \dfrac{3}{52}\sin 2t$

d $x = \dfrac{7}{4}e^{-2t} - \dfrac{19}{13}e^{-3t} + \dfrac{3}{52}(\sin 2t - 5\cos 2t)$

21 a $x = 9(2e^{-t} - e^{-3t} - 1)$ $y = \dfrac{1}{2}(9e^{-3t} - 9e^{-t} + 2)$

b As $t \to \infty, e^{-t}$ and $e^{-3t} \to 0$ so $x \to -9$ and $y \to 1$ so the system settles around the point $(-9, 1)$

Chapter 21

Exercise 21.1A

1 a 12.9 square units
b π square units
c 4.07 square units
d 29.0 square units
e 0.654 square units
f 2.42 square units
g 3.71 square units
h 0.211 square units

2 a 12.99 square units (4 sf)
b 1.303 square units (4 sf)
c 1.764 square units (4 sf)
d 0.7751 square units (4 sf)
e 0.7819 square units (4 sf)
f 0.3227 square units (4 sf)
g 202.1 square units (4 sf)

3 a 165.333 square units
Area $\simeq$ 165.0 square units Relative error = 0.002016
b 2 square units
Area $\simeq$ 2.052 square units Relative error = 0.0262
c 114.583 square units
Area $\simeq$ 116.9 square units Relative error = 0.0200
d 32.47 square units (4 sf)
Area $\simeq$ 32.51 square units Relative error = 0.0011

4 a 3.343 square units (4 sf).
Area $\simeq$ 3.353 (4 sf) square units Relative error = 0.00299
b 8.959 square units (4 sf).
Area $\simeq$ 8.966 square units (4 sf) Relative error = 0.000781
c 6.220 square units(4 sf)
Area $\simeq$ 6.340 square units (4 sf) Relative error = 0.0193
d 36.92 square units (4 sf).
Area $\simeq$ 36.84 square units (4 sf) Relative error = 0.00217

Exercise 21.1B

1 1190 m (3 sf)

2 a $1+\frac{x^2}{2}+\frac{x^4}{24}+\ldots$
b 0.63665 (5 dp)

3 a $x-\frac{x^3}{3!}+\frac{x^5}{5!}-\frac{x^7}{7!}+\ldots$
b 0.4597 (4 d.p)

4 a $\int_0^{\frac{\pi}{4}} \tan x\,dx = \left[\ln|\sec x|\right]_0^{\frac{\pi}{4}} = \ln\left(\sec x\left(\frac{\pi}{4}\right)\right) - \ln(\sec x(0)) = \ln\sqrt{2}\ (=0.3466\ (4\text{ dp})).$
b Area $\simeq$ 0.3466 square units (4 dp) Relative error = 0

5 $(1+x^2)^{\frac{1}{3}} \equiv 1+\frac{1}{3}(x)^2+\frac{\left(\frac{1}{3}\right)\left(\frac{-2}{3}\right)}{2!}(x)^4+\frac{\left(\frac{1}{3}\right)\left(\frac{-2}{3}\right)\left(\frac{-5}{3}\right)}{3!}(x)^6+ \ldots\ldots \equiv 1+\frac{x^2}{3}-\frac{x^4}{9}+\frac{5x^6}{81}-\ldots$
a 0.6225 (4 dp)
b 0.6225 (4 dp)

6 Area $\simeq$ 26.61237 square units (5 dp)

7 a 2.68 square units (3 sf)
b 2.68 (3 sf)
c 0.000486 (3 sf) and 0.00000350 (3 sf)

8 1.09894 (6 sf)

9 $\bar{x} = \frac{\int xy\,dx}{\int y\,dx} = \frac{16}{9.3636} \simeq 1.709$ (3 sf) and $\bar{y} = \frac{\int \frac{y^2}{2}\,dx}{\int y\,dx} = \frac{12}{9.3636} \simeq 1.282$ (3 sf)

Exercise 21.2A

1

	$\frac{dy}{dx}$	x_0	y_0	h	End point
1	$2x+3y$	0	1	0.1	$y(0.1)$

A	B	C	D		
r	x	y(formula)	y	$y_{r+1}=y_r+hf(x_r,y_r)$	
0	0		1.0000000	f(x, y) =	$2x+3y$
1	0.1	(D3) + 0.1*(2(B3) + 3(D3))	1.3	h =	0.1

2

$\frac{dy}{dx}$	x_0	y_0	h	End point
$3+xy$	1	1	0.1	$y(1.2)$

A	B	C	D		
r	x	y(formula)	y	$y_{r+1}=y_r+hf(x_r,y_r)$	
0	1		1.0000000	f(x, y) =	$3+xy$
1	1.1	(D3) + 0.1*(3 + (B3)*(D3))	1.4	h =	0.1
2	1.2	(D4) + 0.1*(3 + (B4)*(D4))	1.854		

3

$\frac{dy}{dx}$	x_0	y_0	h	End point
x^2y	0	1	0.1	$y(0.3)$

A	B	C	D		
r	x	y(formula)	y	$y_{r+1}=y_r+hf(x_r,y_r)$	
0	0		1.0000000	f(x, y) =	x^2y
1	0.1	(D3) + 0.1*((B3)² *(D3))	1	h =	0.1
2	0.2	(D4) + 0.1*((B4)² *(D4))	1.001		
3	0.3	(D5) + 0.1*((B5)² *(D5))	1.005004		

4

$\frac{dy}{dx}$	x_0	y_0	h	End point
cos x	0	0	0.1	$y(0.5)$

A	B	C	D		
r	x	y(formula)	y	$y_{r+1}=y_r+hf(x_r, y_r)$	
0	0		0.0000000	f(x, y) =	cos$x$
1	0.1	(D3) + 0.1*(cos(B3))	0.1	h =	0.1
2	0.2	(D4) + 0.1*(cos(B4))	0.199500417		
3	0.3	(D5) + 0.1*(cos(B5))	0.297507074		
4	0.4	(D6) + 0.1*(cos(B6))	0.393040723		
5	0.5	(D7) + 0.1*(cos(B7))	0.485146823		

5

$\frac{dy}{dx}$	x_0	y_0	h	End point
$\sqrt{x}$	0	5	1	$y(10)$

A	B	C	D		
r	x	y(formula)	y	$y_{r+1}=y_r+hf(x_r, y_r)$	
0	0		5.0000000	f(x, y) =	$\sqrt{x}$
1	1	(D3) + 1*(√(B3))	5	h =	1
2	2	(D4) + 1*(√(B4))	6		
3	3	(D5) + 1*(√(B5))	7.414213562		
4	4	(D6) + 1*(√(B6))	9.14626437		
5	5	(D7) + 1*(√(B7))	11.14626437		
6	6	(D8) + 1*(√(B8))	13.38233235		
7	7	(D9) + 1*(√(B9))	15.83182209		
8	8	(D10) + 1*(√(B10))	18.4775734		
9	9	(D11) + 1*(√(B11))	21.30600053		
10	10	(D12) + 1*(√(B12))	24.30600053		

6

$\frac{dy}{dx}$	x_0	y_0	h	End point
$\ln(\sqrt{x}+\sqrt{y})$	1	2	1	$y(5)$

r	x	y (formula)	y	k_1 (formula)	k_1	k_2 (formula)	K_2
0	1		2				
1	2	(D2) + 0.5*((F3) + (H3))	3.008267	1*LN(SQRT (B2) +SQRT(D2))	0.881373587	1*(LN(SQRT(B3) + SQRT((D2) + (F3))))	1.135161001
2	3	(D3) + 0.5*((F4) + (H4))	4.245357	1*LN(SQRT (B3) +SQRT(D3))	1.146973564	1*(LN(SQRT(B4) + SQRT((D3) + (F4))))	1.32720539
3	4	(D4) + 0.5*((F5) + (H5))	5.648316	1*LN(SQRT (B4) +SQRT(D4))	1.333019411	1*(LN(SQRT(B5) + SQRT((D4) + (F5))))	1.472898258
4	5	(D5) + 0.5*((F6) + (H6))	7.181608	1*LN(SQRT(B5) +SQRT(D5))	1.4762764	1*(LN(SQRT(B6) + SQRT((D5) + (F6))))	1.590308326

7

$\frac{dy}{dx}$	x_0	y_0	h	End point
$\sin(x^2)$	1	1	0.1	$y(1.4)$

r	x	y (formula)	y	k_1 (formula)	k_1	k_2 (formula)	K_2
0	1		1				
1	1.1	(D2) + 0.5*((F3) + (H3))	1.088854	0.1*(SIN((B2)^2))	0.084147098	0.1*(SIN((B3)^2))	0.0935616
2	1.2	(D3) + 0.5*((F4) + (H4))	1.185208	0.1*(SIN((B3)^2))	0.0935616	0.1*(SIN((B4)^2))	0.099145835
3	1.3	(D4) + 0.5*((F5) + (H5))	1.284426	0.1*(SIN((B4)^2))	0.099145835	0.1*(SIN((B5)^2))	0.099290365
4	1.4	(D5) + 0.5*((F6) + (H6))	1.380332	0.1*(SIN((B5)^2))	0.099290365	0.1*(SIN((B6)^2))	0.092521152

8

$\frac{dy}{dx}$	x_0	y_0	h	End point
$\sin^2 x$	1	1	0.1	$y(1.4)$

r	x	y (formula)	y	k_1 (formula)	k_1	k_2 (formula)	K_2
0	1		1				
1	1.1	(D2) + 0.5*((F3) + (H3))	1.075116	0.1*(SIN(B2))^2	0.070807342	0.1*(SIN(B3))^2	0.079425056
2	1.2	(D3) + 0.5*((F4) + (H4))	1.158264	0.1*(SIN(B3))^2	0.079425056	0.1*(SIN(B4))^2	0.086869686
3	1.3	(D4) + 0.5*((F5) + (H5))	1.248121	0.1*(SIN(B4))^2	0.086869686	0.1*(SIN(B5))^2	0.092844438
4	1.4	(D5) + 0.5*((F6) + (H6))	1.343098	0.1*(SIN(B5))^2	0.092844438	0.1*(SIN(B6))^2	0.097111117

9

$\frac{dy}{dx}$	x_0	y_0	h	End point
$(x-y)e^{-x}$	0	5	1	$y(5)$

r	x	y (formula)	y	k_1 (formula)	k_1	k_2 (formula)	k_2
0	0		5				
1	1	(D2) + 0.5*((F3) + (H3))	2.68394	1*(((B2) – (D2))/ EXP(B2))	–5	1*(((B3) – ((D2) + (F3)))/EXP(B3))	0.367879441
2	2	(D3) + 0.5*((F4) + (H4))	2.369835	1*(((B3) – (D3))/ EXP(B3))	–0.619486803	1*(((B4) – ((D3) + (F4)))/EXP(B4))	–0.008722754
3	3	(D4) + 0.5*((F5) + (H5))	2.361742	1*(((B4) – (D4))/ EXP(B4))	–0.050051717	1*(((B5) – ((D4) + (F5)))/EXP(B5))	0.033865999
4	4	(D5) + 0.5*((F6) + (H6))	2.392342	1*(((B5) – (D5))/ EXP(B5))	0.031776991	1*(((B6) – ((D5) + (F6)))/EXP(B6))	0.029423725
5	5	(D6) + 0.5*((F7) + (H7))	2.415751	1*(((B6) – (D6))/ EXP(B6))	0.029445275	1*(((B7) – ((D6) + (F7)))/EXP(B7))	0.017371858

10

$\frac{dy}{dx}$	x_0	y_0	h	End point
$ye^{\tan x}$	0	5	0.25	$y(1)$

r	x	y (formula)	y	k_1 (formula)	k_1	k_2 (formula)	r
	0		5				
0	0.25	(D2) + 0.5*((F3) + (H3))	6.633518	0.25*((D2)*EXP(TAN(B2)))	1.25	0.25*(((D2) + (F3))EXP(TAN(B3)))	2.017035832
1	0.5	(D3) + 0.5*((F4) + (H4))	9.597921	0.25*((D3)*EXP(TAN(B3)))	2.140806933	0.25*(((D3) + (F4))EXP(TAN(B4)))	3.787999165
2	0.75	(D4) + 0.5*((F5) + (H5))	16.03014	0.25*((D4)*EXP(TAN(B4)))	4.143557165	0.25*(((D4) + (F5))EXP(TAN(B5)))	8.720887048
3	1	(D5) + 0.5*((F6) + (H6)	36.6637	0.25*((D5)*EXP(TAN(B5)))	10.17336459	0.25*(((D5) + (F6))EXP(TAN(B6)))	31.09374413

11

$\frac{dy}{dx}$	x_0	y_0	h	End point
$2x - 3y$	0	1	0.1	$y(0.3)$

r	x	y (formula)	y	$y_1 = y_0 + hf(x_0, y_0)$	
0	0		1	$y_{r+1} = y_{r-1} + 2hf(x_r, y_r)$	
1	0.1	(D2) + 0.1*(2*(B2) – 3*(D2))	0.7	$f(x, y) =$	$2x - 3y$
2	0.2	(D2) + 0.2*(2*(B3) – 3*(D3))	0.62	$h =$	0.1
3	0.3	(D3) + 0.2*(2*(B4) – 3*(D4))	0.408		

12

$\frac{dy}{dx}$	x_0	y_0	h	End point
$3x + xy - 4$	1	1	0.1	$y(1.3)$

r	x	y (formula)	y	$y_1 = y_0 + hf(x_0, y_0)$	
0	1		1.000000	$y_{r+1} = y_{r-1} + 2hf(x_r, y_r)$	
1	1.1	(D2) + 0.1*(3*(B2) + (B2)*(D2) – 4)	1.000000	$f(x, y) =$	$3x + xy - 4$
2	1.2	(D2) + 0.2*(3*(B3) + (B3)*(D3) – 4)	1.080000	$h =$	0.1
3	1.3	(D3) + 0.2*(3*(B4) + (B4)*(D4) – 4)	1.179200		

13

$\frac{dy}{dx}$	x_0	y_0	h	End point
x^2y^2	0	1	0.1	$y(0.3)$

r	x	y (formula)	y	$y_1 = y_0 + hf(x_0, y_0)$	
0	0		1.000000	$y_{r+1} = y_{r-1} + 2hf(x_r, y_r)$	
1	0.1	(D2) + 0.1*((B2)^2 + (D2)^2)	1.0	$f(x, y) =$	x^2y^2
2	0.2	(D2) + 0.2*((B3)^2 + (D3)^2)	1.002	$h =$	0.1
3	0.3	(D3) + 0.2*((B4)^2 + (D4)^2)	1.008032		

14

$\frac{dy}{dx}$	x_0	y_0	h	End point
tan$x$	0	0	0.1	$y(0.5)$

r	x	y (formula)	y	$y_1 = y_0 + hf(x_0, y_0)$	
0	0		0.000000	$y_{r+1} = y_{r-1} + 2hf(x_r, y_r)$	
1	0.1	(D2) + 0.1*(TAN(B2))	0.000000	f(x, y) =	tan$x$
2	0.2	(D2) + 0.2*(TAN(B3))	0.020067	h =	0.1
3	0.3	(D3) + 0.2*(TAN(B4))	0.040542		
4	0.4	(D4) + 0.2*(TAN(B5))	0.081934		
5	0.5	(D5) + 0.2*(TAN(B6))	0.125101		

15

$\frac{dy}{dx}$	x_0	y_0	h	End point
$x^{\frac{-1}{2}}$	1	5	1	$y(7)$

r	x	y (formula)	y	$y_1 = y_0 + hf(x_0, y_0)$	
0	1		5.000000	$y_{r+1} = y_{r-1} + 2hf(x_r, y_r)$	
1	2	(D2) + 1*(1/SQRT(B2))	6.000000	f(x, y) =	$\frac{1}{\sqrt{x}}$
2	3	(D2) +2*(1/SQRT(B3))	6.414214	h =	0.1
3	4	(D3) +2*(1/SQRT(B4))	7.154701		
4	5	(D4) +2*(1/SQRT(B5))	7.414214		
5	6	(D5) +2*(1/SQRT(B6))	8.049128		
6	7	(D6) +2*(1/SQRT(B7))	8.230710		

Exercise 21.2B

1 a

A	B	C	D		
r	x	y(formula)	y	$y_{r+1} = y_r + hf(x_r, y_r)$	
0	0		150.0000000	f(x, y) =	$2x - 1$
1	0.5	(D3) + 0.5*(2*(B3) −1)	149.5	h =	0.5
2	1	(D4) + 0.5*(2*(B4) −1)	149.5		
3	1.5	(D5) + 0.5*(2*(B5) −1)	150		
4	2	(D6) + 0.5*(2*(B6) −1)	151		
5	2.5	(D7) + 0.5*(2*(B7) −1)	152.5		
6	3	(D8) + 0.5*(2*(B8) −1)	154.5		
7	3.5	(D9) + 0.5*(2*(B9) −1)	157		
8	4	(D10) + 0.5*(2*(B10) −1)	160		

b $\frac{ds}{dt} = v = 2t - 1 \rightarrow s = \int (2t-1)dt = t^2 - t + c;$ $s = 150$ when $t = 0$, so $s = t^2 - t + 150$

Hence when $t = 4$, $s = 4^2 - 4 + 150 = 162$

2

A	B	C	D		
r	x	y (formula)	y	$y_{r+1} = y_r + hf(x_r, y_r)$	
0	0		0.0000000	f(x, y) =	10 – 0.5y
1	0.1	(D3) + 0.1*(10 – 0.5*(D3))	1.000000	h =	0.1
2	0.2	(D4) + 0.1*(10 – 0.5*(D4))	1.950000		
3	0.3	(D5) + 0.1*(10 – 0.5*(D5))	2.852500		
4	0.4	(D6) + 0.1*(10 – 0.5*(D6))	3.709875		
5	0.5	(D7) + 0.1*(10 – 0.5*(D7))	4.524381		
6	0.6	(D8) + 0.1*(10 – 0.5*(D8))	5.298162		
7	0.7	(D9) + 0.1*(10 – 0.5*(D9))	6.033254		
8	0.8	(D10) + 0.1*(10 – 0.5*(D10))	6.731591		
9	0.9	(D11) + 0.1*(10 – 0.5*(D11))	7.395012		
10	1	(D12) + 0.1*(10 – 0.5*(D12))	8.025261		

3

A	B	C	D		
r	x	y (formula)	y	$y_{r+1} = y_r + hf(x_r, y_r)$	
0	0		0.0000000	f(x, y) =	π(20x – x^2)
1	0.5	(D3) + 0.5*π*(20(B3) – (B3)^2)	0.000000	h =	0.5
2	1	(D4) + 0.5*π*(20(B4) – (B4)^2)	15.315251		
3	1.5	(D5) + 0.5*π*(20(B5) – (B5)^2)	45.160356		
4	2	(D6) + 0.5*π*(20(B6) – (B6)^2)	88.749918		

4 a

A	B	C	D		
r	x	y (formula)	y	$y_{r+1} = y_r + hf(x_r, y_r)$	
0	0		0.0000000	f(x,y) =	$3x^2 - 10x + 18$
1	1	(D3) + 1*(3*(B3)^2 – 10*(B3) + 18)	18.000000	h =	0.5
2	2	(D4) + 1*(3*(B4)^2 – 10*(B4) + 18)	29.000000		
3	3	(D5) + 1*(3*(B5)^2 – 10*(B5) + 18)	39.000000		
4	4	(D6) + 1*(3*(B6)^2 – 10*(B6) + 18)	54.000000		
5	5	(D7) + 1*(3*(B7)^2 – 10*(B7) + 18)	80.000000		
6	6	(D8) + 1*(3*(B8)^2 – 10*(B8) + 18)	123.000000		

b $\frac{ds}{dt} = 3t^2 - 10t + 18$. $s = 0$ when $t = 0 \rightarrow s = \int(3t^2 - 10t + 18)dt = t^3 - 5t^2 + 18t$

Hence when $t = 6$, $s = 6^3 - 5(6)^2 + 18(6) = 144$

5 a

A	B	C	D		
r	x	y (formula)	y	$y_{r+1} = y_r + hf(x_r, y_r)$	
0	1		4.0000000	f(x,y) =	$\frac{4}{\sqrt{t}}$
1	3	(D3) + 2*(4/sqrt(B3))	12.000000	h =	2
2	5	(D4) + 2*(4/sqrt(B4))	16.618802		
3	7	(D5) + 2*(4/sqrt(B5))	20.196511		
4	9	(D6) + 2*(4/sqrt(B6))	23.220227		

5	11	(D7) + 2*(4/sqrt(B7))	25.886893		
6	13	(D8) + 2*(4/sqrt(B8))	28.298941		
7	15	(D9) + 2*(4/sqrt(B9))	30.517785		
8	17	(D10) + 2*(4/sqrt(B10))	32.583376		
9	19	(D11) + 2*(4/sqrt(B11))	34.523661		
10	21	(D12) + 2*(4/sqrt(B12))	36.358987		

b $\frac{dv}{dt}=\frac{4}{\sqrt{t}}$, so $v=\int\frac{4+c}{\sqrt{t}}=8\sqrt{t}$ When $t=1$, $v=4$ so $c=-4$. Therefore $v=8\sqrt{t}-4$

so when $t=21$, $v=8\sqrt{21}-4=32.6606056$

relative error $=\frac{36.358987-32.6606056}{32.6606056}=0.1132$ (4sf) percentage error = 11.32% (4sf)

6 a

A	B	C	D		
r	*x*	*y* (formula)	*y*	$y_{r+1}=y_r+hf(x_r,y_r)$	
0	0		0.0000000	f(x, y) =	$\frac{x(60-3x)}{1000}$
1	1	(D3) + 1*((B3)*((60 – 3*(B3)))/1000	0.000000	h =	1
2	2	(D4) + 1*((B4)*((60 – 3*(B4)))/1000	0.057000		
3	3	(D5) + 1*((B5)*((60 – 3*(B5)))/1000	0.165000		
4	4	(D6) + 1*((B6)*((60 – 3*(B6)))/1000	0.318000		
5	5	(D7) + 1*((B7)*((60 – 3*(B7)))/1000	0.510000		
6	6	(D8) + 1*((B8)*((60 – 3*(B8)))/1000	0.735000		
7	7	(D9) + 1*((B9)*((60 – 3*(B9)))/1000	0.987000		
8	8	(D10) + 1*((B10)*((60 – 3*(B10)))/1000	1.260000		
9	9	(D11) + 1*((B11)*((60 – 3*(B11)))/1000	1.548000		
10	10	(D12) + 1*((B12)*((60 – 3*(B12)))/1000	1.845000		

b $\frac{dv}{dt}=\frac{t(60-3t)}{1000}=\frac{3t}{50}-\frac{3t^2}{1000}$. Therefore $v=\int\frac{3t}{50}-\frac{3t^2}{1000}dt$. When $t=10$, $v=\left[\frac{3t^2}{100}-\frac{t^3}{1000}\right]=3-1=2\,\mathrm{m\,s^{-1}}$

7 a

A	B	C	D		
r	*x*	*y* (formula)	*y*	$y_{r+1}=y_r+hf(x_r,y_r)$	
0	0		10.0000000	f(x, y) =	$-\frac{x}{y}$
1	1	(D3) + 1*((– (B3)/(D3)))	10.000000	h =	1
2	2	(D4) + 1*((– (B4)/(D4)))	9.900000		
3	3	(D5) + 1*((– (B5)/(D5)))	9.697980		
4	4	(D6) + 1*((– (B6)/(D6)))	9.388637		
5	5	(D7) + 1*((– (B7)/(D7)))	8.962590		

b $v\frac{dv}{dx}=-x\rightarrow\int v\,dv=\int -x\,dx$; $\rightarrow\frac{v^2}{2}=-\frac{x^2}{2}+C$; $\rightarrow$ Curve passes through (0, 10) so $\frac{10^2}{2}=-\frac{0^2}{2}+C\rightarrow C=50$

Hence $\frac{v^2}{2}=-\frac{x^2}{2}+50$; $\rightarrow v^2=100-x^2$; $\rightarrow$ When $x=5$, $v^2=100-5^2=75$; $\rightarrow v=\sqrt{75}=8.66$

8 a

r	x	y (formula)	y	k_1 (formula)	k_1	k_2 (formula)	k_2
0	0		3				
1	0.5	(D2) + 0.5*((F3) + (H3))	2.976563	0.5*(−9*(B2)/(16*(D2)))	0	0.5*(− 9*(B3)/(16*((D2) + (F3))))	−0.046875
2	1	(D3) + 0.5*((F4) + (H4))	2.904934	0.5*(−9*(B3)/(16*(D3)))	−0.047244094	0.5*(− 9*(B4)/(16*((D3) + (F4))))	−0.096012096
3	1.5	(D4) + 0.5*((F5) + (H5))	2.781408	0.5*(−9*(B4)/(16*(D4)))	−0.096818021	0.5*(− 9*(B5)/(16*((D4) + (F5))))	−0.150234158
4	2	(D5) + 0.5*((F6) + (H6)	2.59862	0.5*(−9*(B5)/(16*(D5)))	−0.151676759	0.5*(− 9*(B6)/(16*((D5) + (F6))))	−0.21390016
5	2.5	(D6) + 0.5*((F7) + (H7)	2.342808	0.5*(−9*(B6)/(16*(D6)))	−0.216461057	0.5*(− 9*(B7)/(16*((D6) + (F7))))	−0.295162942
6	3	(D7) + 0.5*((F8) + (H8))	1.986218	0.5*(−9*(B7)/(16*(D7)))	−0.300120643	0.5*(− 9*(B8)/(16*((D7) + (F8))))	−0.413058835

b Accurate value when $x = 3$: $3\sqrt{\left(1-\frac{3^2}{16}\right)} = 1.984313$ (6 dp) so relative error $= \frac{1.986218-1.984313}{1.984313} = 0.000960$ (3 sf) and percentage error = 0.0960% (3 sf)

9

r	t	v (formula)	v	k_1 (formula)	k_1	k_2 (formula)	k_2
0	1		4.000000				
1	3	(D2) + 0.5*((F3) + (H3))	7.386723	2*(2/(B2)^(1/3))	4.000000	2*(2/(B3)^(1/3))	2.773445
2	5	(D3) + 0.5*((F4) + (H4))	9.943052	2*(2/(B3)^(1/3))	2.773455	2*(2/(B4)^(1/3))	2.339214
3	7	(D4) + 0.5*((F5) + (H5))	12.158175	2*(2/(B4)^(1/3))	2.339214	2*(2/(B5)^(1/3))	2.091032
4	9	(D5) + 0.5*((F6) + (H6))	14.165191	2*(2/(B5)^(1/3))	2.091032	2*(2/(B6)^(1/3))	1.923014
5	11	(D6) + 0.5*((F7) + (H7))	16.025979	2*(2/(B6)^(1/3))	1.922999	2*(2/(B7)^(1/3))	1.798592
6	13	(D7) + 0.5*((F8) + (H8))	17.775849	2*(2/(B7)^(1/3))	1.798577	2*(2/(B8)^(1/3))	1.701176
7	15	(D8) + 0.5*((F9) + (H9))	19.437390	2*(2/(B8)^(1/3))	1.701161	2*(2/(B9)^(1/3))	1.621935
8	17	(D9) + 0.5*((F10) + (H10))	21.026172	2*(2/(B9)^(1/3))	1.621921	2*(2/(B10)^(1/3))	1.555659
9	19	(D10) + 0.5*((F11) + (H11))	22.553507	2*(2/(B10)^(1/3))	1.555644	2*(2/(B11)^(1/3))	1.499039
10	21	(D11) + 0.5*((F12) + (H12))	24.027939	2*(2/(B11)^(1/3))	1.499025	2*(2/(B12)^(1/3))	1.449855

10 a

A	B	C	D		
r	x	y (formula)	y	$y_{r+1} = y_r + hf(x_r, y_r)$	
0	1		2.0000000	f(x,y) =	$\frac{y}{(x^2+1)}$
1	1.2	(D3) + 0.2*((D3)/((B3)^2 + 1))	2.200000	h =	0.2
2	1.4	(D4) + 0.2*((D4)/((B4)^2 + 1))	2.380328		
3	1.6	(D5) + 0.2*((D5)/((B5)^2 + 1))	2.541161		
4	1.8	(D6) + 0.2*((D6)/((B6)^2 + 1))	2.683923		
5	2	(D7) + 0.2*((D7)/((B7)^2 + 1))	2.810523		

b $\frac{dy}{dx} = \frac{y}{x^2+1} \rightarrow \int \frac{dy}{y} = \frac{dx}{x^2+1} \rightarrow \ln(y) = \tan^{-1}(x) + C$; Curve passes through $(1, 2)$ so $\ln(2) = \tan^{-1}(1) + C \rightarrow C = \ln 2 - \frac{\pi}{4}$

$\rightarrow \ln(y) = \tan^{-1}(x) + \ln 2 - \frac{\pi}{4}; \rightarrow \ln\left(\frac{y}{2}\right) = \tan^{-1}(x) - \frac{\pi}{4}; \rightarrow y = 2e^{(\tan^{-1}(x) - \frac{\pi}{4})}$

Hence when $x = 2$, $y = 2e^{(\mathrm{Tan}^{-1}(2) - \frac{\pi}{4})} = 2.76$ (3 sf)

11 a

r	x	y(formula)	y
0	0		0.000000
1	1	(D2) + 1*((B2)^2 – 5*(B2) + 8)	8.000000
2	2	(D2) + 2*((B3)^2 – 5*(B3) + 8)	8.000000
3	3	(D3) + 2*((B4)^2 – 5*(B4) + 8)	12.000000
4	4	(D4) + 2*((B5)^2 – 5*(B5) + 8)	12.000000
5	5	(D5) + 2*((B6)^2 – 5*(B6) + 8)	20.000000

b $\frac{ds}{dt} = t^2 - 5t + 8$. $s = 0$ when $t = 0 \Rightarrow s = \int (t^2 - 5t + 8)\,dt$

$= \frac{1}{3}t^3 - \frac{5}{2}t^2 + 8t$

hence when $t = 5$, $s = \frac{1}{3}5^3 - \frac{5}{2}5^2 + 8 \times 5 = 19.17$ (2 dp)

Review exercise 21

1 a 5.72 square units (3 sf)
b Area ≈12.5 square units (3 sf)
c Area ≈ 6.80 square units (3 sf)

2 a Area ≃ 12.7 square units (3 sf)
b Area ≃ 0.687 square units (3 sf)
c Area ≃ 1.38 (3 sf)

3 a Relative error –0.0336364
Percentage error –3.3636364
b Relative error 0.01186
Percentage error 1.18644
c Relative error –0.77
Percentage error –77
d Relative error 0.01394
Percentage error 1.394

4 a

$\frac{dy}{dx}$	x_0	y_0	H	End point
10 – 3x	0	1	0.1	y(0.4)

r	x	y(formula)	y
0	0		1.0000000
1	0.1	(D2) + 0.1*(10 – 3*(B2))	2.000000
2	0.2	(D3) + 0.1*(10 – 3*(B3))	2.970000
3	0.3	(D4) + 0.1*(10 – 3*(B4))	3.910000
4	0.4	(D5) + 0.1*(10 – 3*(B5))	4.820000

b

$\frac{dy}{dx}$	x_0	y_0	H	End point
10 – 3y	1	1	0.1	y(1.5)

r	x	y(formula)	y
0	1		1.0000000
1	1.1	(D2) + 0.1*(10 – 3*(D2))	1.700000
2	1.2	(D3) + 0.1*(10 – 3*(D3))	2.190000
3	1.3	(D4) + 0.1*(10 – 3*(D4))	2.533000
4	1.4	(D5) + 0.1*(10 – 3*(D5))	2.773100
5	1.5	(D2) + 0.1*(10 – 3*(D6))	2.941170

c

$\frac{dy}{dx}$	x_0	y_0	H	End point
$x^2\sqrt{y}$	3	2	0.01	y(3.05)

r	x	y(formula)	y
0	3		2.0000000
1	3.01	(D2) + 0.01*((B2)^2)*SQRT(D2)	2.12727922
2	3.02	(D3) + 0.01* ((B3)^2)*SQRT(D3)	2.25942255
3	3.03	(D4) + 0.01* ((B4)^2)*SQRT(D4)	2.39651471
4	3.04	(D5) + 0.01* ((B5)^2)*SQRT(D5)	2.53864129
5	3.05	(D6) +0.01* ((B6)^2)*SQRT(D6)	2.68588876

5 a

$\frac{dy}{dx}$	x_0	y_0	h	End point
2x – y + xy	3	4	0.1	y(3.6)

r	x	y (formula)	y	k_1 (formula)	k_1	k_2 (formula)	k_2
0	3		4.000000				
1	3.1	(D2) + 0.5*((F3) + (H3))	5.577000	0.1*(2*(B2) – (D2) + (B2)*(D2))	1.400000	0.1*(2*(B3) – ((D2)+(F3)) + (B3)*((D2) + (F3)))	1.754000
2	3.2	(D3) + 0.5*((F4) + (H4))	7.603084	0.1*(2*(B3) – (D3) + (B3)*(D3))	1.789700	0.1*(2*(B4) – ((D3)+(F4)) + (B4)*((D3) + (F4)))	2.260997
3	3.3	(D4) + 0.5*((F5) + (H5))	10.229736	0.1*(2*(B4) – (D4) + (B4)*(D4))	2.31268	0.1*(2*(B5) – ((D4)+(F5)) + (B5)*((D4) + (F5)))	2.940625
4	3.4	(D5) + 0.5*((F6) + (H6))	13.665264	0.1*(2*(B5) – (D5) + (B5)*(D5))	3.01284	0.1*(2*(B6) – ((D5)+(F6)) + (B6)*((D5) + (F6)))	3.858218
5	3.5	(D6) + 0.5*((F7) + (H7))	18.198212	0.1*(2*(B6) – (D6) + (B6)*(D6))	3.95967	0.1*(2*(B7) – ((D6)+(F7)) + (B7)*((D6) + (F7)))	5.106232
6	3.6	(D7) + 0.5*((F8) + (H8))	24.231198	0.1*(2*(B7) – (D7) + (B7)*(D7))	5.24955	0.1*(2*(B8) – ((D7)+(F8)) + (B8)*((D7) + (F8)))	6.816419

b

$\frac{dy}{dx}$	x_0	y_0	h	End point
$\sqrt{xy}$	0	36	0.01	$y(0.04)$

r	x	y (formula)	y	k_1 (formula)	k_1	k_2 (formula)	k_2
0	0		36.000000				
1	0.01	(D2) + 0.5*((F3) + (H3))	36.003000	0.01*(SQRT((B2)*(D2)))	0.000000	0.01*(SQRT((B3)*((D2) + (F3))))	0.006000
2	0.02	(D3) + 0.5*((F4) + (H4))	36.010243	0.01*(SQRT((B3)*(D3)))	0.006000	0.01*(SQRT((B4)*((D3) + (F4))))	0.008486
3	0.03	(D4) + 0.5*((F5) + (H5))	36.019684	0.01*(SQRT((B4)*(D4)))	0.008486	0.01*(SQRT((B5)*((D4) + (F5))))	0.010395
4	0.04	(D5) + 0.5*((F6) + (H6))	36.030884	0.01*(SQRT((B5)*(D5)))	0.010395	0.01*(SQRT((B6)*((D5) + (F6))))	0.012005
5	0.05	(D6) + 0.5*((F7) + (H7))	36.043599	0.01*(SQRT((B6)*(D6)))	0.012005	0.01*(SQRT((B7)*((D6) + (F7))))	0.013424
6	0.06	(D7) + 0.5*((F8) + (H8))	36.057665	0.01*(SQRT((B7)*(D7)))	0.013425	0.01*(SQRT((B8)*((D7) + (F8))))	0.014709

c

$\frac{dy}{dx}$	x_0	y_0	h	End point
$\left(\frac{e^{x^2}}{y}\right)$	1	4	1	$y(5)$

r	x	y (formula)	y	k_1 (formula)	k_1	k_2 (formula)	k_2
0	1		4.000000				
1	2	(D2) + 0.5*((F3) + (H3))	10.173456	1*(EXP((B2)^2)/(D2))	0.679570	1*(EXP((B3)^2)/((D2) + (F3)))	11.66734223
2	3	(D3) + 0.5*((F4) + (H4))	273.570750	1*(EXP((B3)^2)/(D3))	5.366726	1*(EXP((B4)^2)/((D3) + (F4)))	521.4278625
3	4	(D4) + 0.5*((F5) + (H5))	14942.718275	1*(EXP((B4)^2)/(D4))	29.619701	1*(EXP((B5)^2)/((D4) + (F5)))	29308.67535
4	5	(D5) + 0.5*((F6) + (H6))	2332388.201382	1*(EXP((B5)^2)/(D5))	594.678315	1*(EXP((B6)^2)/((D5) + (F6)))	4634296.288

6 **a** 21.8 minutes.

b If $m = 10e^{-0.0318t}$, then $\frac{dm}{dt} \simeq -0.318e^{-0.0318t}$ hence using a starting point of $(0, 10)$ with a step length of 1, you get:

r	x	y (formula)	y	k_1 (formula)	k_1	k_2 (formula)	k_2
0	0		10.000000				
1	1	(D2) + 0.5*((F3) + (H3))	9.686977	1*(−0.318*EXP(−0.0318*(B2)))	−0.318000	1*(−0.318*EXP(−0.0318*(B3)))	−0.308046696
2	2	(D3) + 0.5*((F4) + (H4))	9.383751	1*(−0.318*EXP(−0.0318*(B3)))	−0.308047	1*(−0.318*EXP(−0.0318*(B4)))	−0.298404928
3	3	(D4) + 0.5*((F5) + (H5))	9.090016	1*(−0.318*EXP(−0.0318*(B4)))	−0.298405	1*(−0.318*EXP(−0.0318*(B5)))	−0.289064944
4	4	(D5) + 0.5*((F6) + (H6))	8.805475	1*(−0.318*EXP(−0.0318*(B5)))	−0.289065	1*(−0.318*EXP(−0.0318*(B6)))	−0.280017299
5	5	(D6) + 0.5*((F7) + (H7))	8.529840	1*(−0.318*EXP(−0.0318*(B6)))	−0.280017	1*(−0.318*EXP(−0.0318*(B7)))	−0.271252842
6	6	(D7) + 0.5*((F8) + (H8))	8.262832	1*(−0.318*EXP(−0.0318*(B7)))	−0.271253	1*(−0.318*EXP(−0.0318*(B8)))	−0.26276271
7	7	(D8) + 0.5*((F9) + (H9))	8.004181	1*(−0.318*EXP(−0.0318*(B8)))	−0.262763	1*(−0.318*EXP(−0.0318*(B9)))	−0.254538317
8	8	(D9) + 0.5*((F10) + (H10))	7.753627	1*(−0.318*EXP(−0.0318*(B9)))	−0.254538	1*(−0.318*EXP(−0.0318*(B10)))	−0.246571345
9	9	(D10) + 0.5*((F11) + (H11))	7.510914	1*(−0.318*EXP(−0.0318*(B10)))	−0.246571	1*(−0.318*EXP(−0.0318*(B11)))	−0.238853736
10	10	(D11) + 0.5*((F12) + (H12))	7.275798	1*(−0.318*EXP(−0.0318*(B11)))	−0.238854	1*(−0.318*EXP(−0.0318*(B12)))	−0.231377687
11	11	(D12) + 0.5*((F13) + (H13))	7.048042	1*(−0.318*EXP(−0.0318*(B12)))	−0.231378	1*(−0.318*EXP(−0.0318*(B13)))	−0.224135635
12	12	(D13) + 0.5*((F14) + (H14))	6.827414	1*(−0.318*EXP(−0.0318*(B13)))	−0.224136	1*(−0.318*EXP(−0.0318*(B14)))	−0.217120258
13	13	(D14) + 0.5*((F15) + (H15))	6.613691	1*(−0.318*EXP(−0.0318*(B14)))	−0.217120	1*(−0.318*EXP(−0.0318*(B15)))	−0.210324459

(Continued)

r	x	y (formula)	y	k_1 (formula)	k_1	k_2 (formula)	k_2
14	14	(D15) + 0.5*((F16) + (H16))	6.406658	1*(–0.318*EXP(–0.0318*(B15)))	–0.210324	1*(–0.318*EXP(–0.0318*(B16)))	–0.203741367
15	15	(D16) + 0.5*((F17) + (H17))	6.206106	1*(–0.318*EXP(–0.0318*(B16)))	–0.203741	1*(–0.318*EXP(–0.0318*(B17)))	–0.197364324
16	16	(D17) + 0.5*((F18) + (H18))	6.011830	1*(–0.318*EXP(–0.0318*(B17)))	–0.197364	1*(–0.318*EXP(–0.0318*(B18)))	–0.191186881
17	17	(D18) + 0.5*((F19) + (H19))	5.823635	1*(–0.318*EXP(–0.0318*(B18)))	–0.191187	1*(–0.318*EXP(–0.0318*(B19)))	–0.185202789
18	18	(D19) + 0.5*((F20) + (H20))	5.641331	1*(–0.318*EXP(–0.0318*(B19)))	–0.185203	1*(–0.318*EXP(–0.0318*(B20)))	–0.179405998
19	19	(D20) + 0.5*((F21) + (H21))	5.464732	1*(–0.318*EXP(–0.0318*(B20)))	–0.179406	1*(–0.318*EXP(–0.0318*(B21)))	–0.173790645
20	20	(D21) + 0.5*((F22) + (H22))	5.293662	1*(–0.318*EXP(–0.0318*(B21)))	–0.173791	1*(–0.318*EXP(–0.0318*(B22)))	–0.16835105
21	21	(D22) + 0.5*((F23) + (H23))	5.127945	1*(–0.318*EXP(–0.0318*(B22)))	–0.168351	1*(–0.318*EXP(–0.0318*(B23)))	–0.163081713
22	22	(D23) + 0.5*((F24) + (H24))	4.967416	1*(–0.318*EXP(–0.0318*(B23)))	–0.163082	1*(–0.318*EXP(–0.0318*(B24)))	–0.157977305
23	23	(D24) + 0.5*((F25) + (H25))	4.811911	1*(–0.318*EXP(–0.0318*(B24)))	–0.157977	1*(–0.318*EXP(–0.0318*(B25)))	–0.153032663

(using formulae in a spreadsheet)

The half life of Francium 223 is 22 minutes (nearest minute)

c 6.83 g (3 sf)

7 a

r	x	y (formula)	y	k_1 (formula)	k_1	k_2 (formula)	k_2
0	0		1.000000				
1	0.1	(D2) + 0.5*((F3) + (H3))	1.105556	0.1*((D2)/(1 – (B2)^2))	0.100000	0.1*(((D2) + (F3))/((1–(B3)^2)))	0.111111
2	0.2	(D3) + 0.5*((F4) + (H4))	1.224789	0.1*((D3)/(1 – (B3)^2))	0.111672	0.1*(((D3) + (F4))/((1–(B4)^2)))	0.126795
3	0.3	(D4) + 0.5*((F5) + (H5))	1.362886	0.1*((D4)/(1 – (B4)^2))	0.127582	0.1*(((D4) + (F5))/((1–(B5)^2)))	0.148612
4	0.4	(D5) + 0.5*((F6) + (H6))	1.527809	0.1*((D5)/(1 – (B5)^2))	0.149768	0.1*(((D5) + (F6))/((1–(B6)^2)))	0.180078
5	0.5	(D6) + 0.5*((F7) + (H7))	1.732729	0.1*((D6)/(1 – (B6)^2))	0.181882	0.1*(((D6) + (F7))/((1–(B7)^2)))	0.227959
6	0.6	(D7) + 0.5*((F8) + (H8))	2.001663	0.1*((D7)/(1 – (B7)^2))	0.231031	0.1*(((D7) + (F8))/((1–(B8)^2)))	0.306837

b 2

8 a

$\frac{dy}{dx}$	x_0	y_0	h	End point
2xy	1	1	0.1	y(1.4)

r	x	y(formula)	Y
0	1		1.000000
1	1.1	(D2) + 0.1*2*(B2)*(D2)	1.2
2	1.2	(D2) + 0.2*2*(B3)*(D3)	1.528
3	1.3	(D3) + 0.2*2*(B4)*(D4)	1.93344
4	1.4	(D4) + 0.2*2*(B5)*(D5)	2.5333888

b

$\frac{dy}{dx}$	x_0	y_0	h	End point
$\tan^2 x$	0	0	0.2	y(1)

r	x	y(formula)	Y
0	0		0.000000
1	0.2	(D2) + 0.2*(TAN(B2) * TAN(B2))	0.000000
2	0.4	(D2) + 0.4*(TAN(B3) * TAN(B3))	0.016437
3	0.6	(D3) + 0.4*(TAN(B4) * TAN(B4))	0.071502
4	0.8	(D4) + 0.4*(TAN(B5) * TAN(B5))	0.203654
5	1	(D5) + 0.4*(TAN(B6) * TAN(B6))	0.495564

c

$\frac{dy}{dx}$	x_0	y_0	h	End point
$2x + \ln(1 + x)$	1	2	0.01	y(1.05)

r	x	y(formula)	Y
0	1		2.000000
1	1.01	(D2)+0.01*(2*(B2) + LN(1+(B2)))	2.026931
2	1.02	(D2)+0.02*(2*(B3) + LN(1+(B3)))	2.054363
3	1.03	(D3)+0.02*(2*(B4) + LN(1+(B4)))	2.081793
4	1.04	(D4)+0.02*(2*(B5) + LN(1+(B5)))	2.109723
5	1.05	(D5)+0.02*(2*(B6) + LN(1+(B6)))	2.137652

9 a 12 383, and a baby on the way.

b

r	t	n(formula)	n
0	0		1000.000000
1	1	(D2) +700*1.5^(B2)	1700
2	2	(D2) +1400*1.5^(B3)	3100
3	3	(D3) +1400*1.5^(B4)	4850
4	4	(D4) +1400*1.5^(B5)	7825
5	5	(D5) +1400*1.5^(B6)	11937.5

Assessment 21

1 a 0.5775 square units (4 dp)
  b 0.5760 square units
2 a 3.307 square units (4 sf)
  b Use more strips
3 a 50.8 cm^2 (3 sf)
  b $\frac{25}{48}\pi^3 + 100\ln(\sqrt{2})$
  c Answers agree to 3 sf so approximation in part **a** is suitable, accurate, good, etc.
4 a 4.75
  b 5.5
  c 13.6% (3 sf)
5 £580.61
6 a $k_1 = 0.25 \times (2 - (1^2) \times 1) = ¼$
  $k_2 = 0.25 \times (2 - (1.25^2) \times (1 + 1/4)) = 3/256$
  $y_1 = 1 + (1/4 + 3/256)/2 = 1 + (67/256)/2 = 579/512$
  b 1.0755 (4dp)
7 a 6
  b 8.25
  c 0.273 (3 sf)
8 $v = 5.087$ m s^{-1}

Chapter 22
Exercise 22.1A

1 a 2 b $5a + b$ c 50 d $12b - 15ab$
2 a $x = -6$ b $x = 0, -5$ c $x = \frac{4}{3}$ d $x = \pm 2\sqrt{\frac{6}{19}}$
3 a i 21 square units ii 7 square units
  b i does not involve a reflection
  ii involves a reflection
4 a 1.5 square units
  b triangle has not been reflected
5 a 80 cube units
  b involves a reflection
6 a $\frac{1}{3}\begin{pmatrix} 0 & -3 \\ 1 & 2 \end{pmatrix}$ b $\frac{1}{a^2}\begin{pmatrix} a-2 & 2 \\ a & -a \end{pmatrix}$
  c $\frac{1}{6}\begin{pmatrix} -2 & 3 & -2 \\ -2 & -6 & 4 \\ 4 & -3 & 4 \end{pmatrix}$ d $\begin{pmatrix} -5 & 8 & 11 \\ 18 & -28 & -39 \\ 2 & -3 & -4 \end{pmatrix}$
  e $\begin{pmatrix} \frac{2}{a} & -\frac{3}{a} & \frac{4}{a} \\ 5 & -7 & 9 \\ -1 & 2 & -2 \end{pmatrix}$
  f $\frac{1}{9a^2 + 11a}\begin{pmatrix} 3a-7 & 6a+2 & -6-2a \\ 3a^2-9a & a+6a^2 & a^2-3a \\ 6a & 3a & 2a \end{pmatrix}$
7 $\begin{pmatrix} -\frac{1}{k^2} & \frac{k^2+1}{k^3} & -\frac{1}{k^3} \\ -\frac{1}{k^2} & \frac{1}{k^3} & \frac{k^2-1}{k^3} \\ \frac{1}{k} & -\frac{1}{k^2} & \frac{1}{k^2} \end{pmatrix}$
8 (1, −2, 0)

Exercise 22.1B

1 a (2, −5, −3)
  b $\left(\frac{1}{4}, -\frac{1}{2}, 1\right)$
2 $\begin{pmatrix} -3 & -1 & -4 \\ 0 & 0 & -5 \\ 2 & 3 & -1 \end{pmatrix}$
3 $\begin{pmatrix} -6 & 3 & 3 \\ 2 & -1 & 0 \\ 4 & -5 & -3 \end{pmatrix}$
4 $\begin{pmatrix} 17 & 9 & -1 \\ 22 & 0 & 4 \\ 15 & 6 & 2 \end{pmatrix}$
5 a If we multiply the first equation by 2 we get $6x - 4y + 2z = 14$ so this plane is parallel to the second as they are the same except the constant term. Therefore, the three planes do not meet.
  b Multiplying the first equation by −3 gives $6x - 9y + 3z = -12$ so this plane is parallel to the second as they are the same except the constant term. Therefore the three planes do not meet.
6 a $k > \frac{1+\sqrt{7}}{2}$ or $k < \frac{1-\sqrt{7}}{2}$
  b $k = -1$
7 a $\det\begin{pmatrix} -7 & -3 & 5 \\ 1 & 1 & 1 \\ 1 & 2 & 4 \end{pmatrix} = -7(4-2) + 3(4-1) + 5(2-1) = 0$

  Therefore they do not intersect at a unique point.
  b Let $x = 2\lambda + 1$ then second equation gives:
  $2\lambda + 1 + y + z = 0$
  Multiply by 2 to give $4\lambda + 2 + 2y + 2z = 0$
  Third equation becomes: $2\lambda + 2y + 4z = 0$
  Subtract to eliminate y: $2\lambda + 2 - 2z = 0 \Rightarrow z = \lambda + 1$
  Substitute back into second equation to give:
  $2\lambda + 1 + y + \lambda + 1 = 0 \Rightarrow y = -3\lambda - 2$
  So general point on line is $(2\lambda + 1, -3\lambda - 2, \lambda + 1)$
  Check first equation:
  $-7(2\lambda + 1) - 3(-3\lambda - 2) + 5(\lambda + 1) = 4$
8 a $\det\begin{pmatrix} 1 & 0 & 2 \\ -2 & 1 & 4 \\ 9 & -2 & 2 \end{pmatrix} = 1(12 - -8) + 2(4 - 9)$

  $= 0$ so not a unique solution
  First equation gives $x = 1 - 2z$
  Substitute into second equation to give
  $-2(1 - 2z) + y + 4z = 0 \Rightarrow y = 2 - 8z$
  Check in third equation:
  $9(1 - 2z) + 2(2 - 8z) + 2z = 5$
  So they form a sheaf (the line has equations such as $x = 1 - 2z, y = 2 - 8z$)
  b $\det\begin{pmatrix} 5 & -1 & 1 \\ 2 & 1 & -2 \\ 36 & 11 & -24 \end{pmatrix}$
  $= 5(-24 - -22) - -1(-48 - -72) + 1(22 - 36)$
  $= 0$ so not a unique solution
  Adding first two equations gives $7x - z = 7 \Rightarrow z = 7x - 7$
  Substitute into second equation to give
  $2x + y - 2(7x - 7) = 7 \Rightarrow y = 12x - 7$
  Check in third equation:
  $36x + 11(12x - 7) - 24(7x - 7) = 91$
  So they form a sheaf (the line had equations such as $z = 7x - 7, y = 12x - 7$)

9 $a = -3, 2$

10 a $\begin{pmatrix} 1 & 1 & 1 \\ 1 & -1 & 0 \\ 0 & 1 & -1 \end{pmatrix}\begin{pmatrix} r \\ h \\ f \end{pmatrix} = \begin{pmatrix} 17 \\ 1 \\ -10 \end{pmatrix}$

b 3 rabbits, 2 hamsters and 12 fish

11 a $\begin{pmatrix} 1 & 1 & 1 \\ 2 & 3 & 4 \\ 1 & 0 & -1 \end{pmatrix}\begin{pmatrix} x \\ y \\ z \end{pmatrix} = \begin{pmatrix} 12 \\ 36 \\ 0 \end{pmatrix}$

b $\det\begin{pmatrix} 1 & 1 & 1 \\ 2 & 3 & 4 \\ 1 & 0 & -1 \end{pmatrix}$

$= 1(-3-0) - 1(-2-4) + 1(0-3)$

$= 0$ so not a unique solution

c Equations are: $x + y + z = 12$

$2x + 3y + 4z = 36$

$x - z = 0$

Final equation gives $x = z$

Substitute into first equation to give

$z + y + z = 12 \Rightarrow y = 12 - 2z$

Check in second equation:

$2z + 3(12 - 2z) + 4z = 36$

So solutions lie on line $x = z$, $y = 12 - 2z$

But x, y, z must all be positive integers so possible solutions are:

(1, 10, 1), (2, 8, 2), (3, 6, 3), (4, 4, 4), (5, 2, 5)

So 5 possibilities (not 6 as (6, 0, 6) not a solution since has no three-bed houses)

12 a $\frac{16}{k}$ square units **b** $k < 0$

c $A\left(-\frac{6}{k}, 4\right), B(0,0), C\left(\frac{2}{k}, 4\right)$ **d** $k = -2$

13 a $\det\begin{pmatrix} k & -1 & -1 \\ 2 & k & -1 \\ 3 & 2 & -2 \end{pmatrix} = k(-2k+2) + (-4+3) - (4-3k)$

$= -2k^2 + 2k - 1 - 4 + 3k$

$= -2k^2 + 5k - 5$

$= -[2k^2 - 5k + 5]$

$- -\left[2\left(k - \frac{5}{4}\right)^2 + \frac{15}{8}\right] < -\frac{15}{8}$

therefore determinant $\neq 0$

So the planes intersect at a single, unique point for all possible values of k

b $\left(\frac{13}{3}, \frac{1}{3}, \frac{16}{3}\right)$

14 a $k = 4$

b $x = \frac{8-k}{k-4}$ $y = \frac{10+k}{4-k}$ $z = \frac{6}{k-4}$

15 a $a = 2$ **b** $b = 4$

Exercise 22.2A

1 a $\begin{vmatrix} 2 & 5 & 1 \\ 4 & 3 & -1 \\ -3 & 2 & 0 \end{vmatrix} = 1\begin{vmatrix} 4 & 3 \\ -3 & 2 \end{vmatrix} - -1\begin{vmatrix} 2 & 5 \\ -3 & 2 \end{vmatrix} + 0$

$= 1(8 - -9) + 1(4 - -15) = 36$

b $\begin{vmatrix} 0 & 14 & 3 \\ 0 & -1 & 2 \\ 8 & 3 & 1 \end{vmatrix} = 0 - 0 + 8\begin{vmatrix} 14 & 3 \\ -1 & 2 \end{vmatrix}$

$= 8(28 - -3) = 248$

c $\begin{vmatrix} 8 & 13 & -2 \\ 0 & 1 & 5 \\ 2 & 5 & 7 \end{vmatrix} = -0 + 1\begin{vmatrix} 8 & -2 \\ 2 & 7 \end{vmatrix} - 5\begin{vmatrix} 8 & 13 \\ 2 & 5 \end{vmatrix}$

$= 1(56 - -4) - 5(40 - 26)$

$= 60 - 70 = 10$

d $\begin{vmatrix} 12 & 4 & 6 \\ 8 & -2 & 5 \\ 0 & 9 & 0 \end{vmatrix} = 0 - 9\begin{vmatrix} 12 & 6 \\ 8 & 5 \end{vmatrix} + 0$

$= -9(60 - 48)$

$= -108$

e $\begin{vmatrix} 0 & b & 2a \\ a & a & 1 \\ 0 & b & 0 \end{vmatrix} = 0 - a\begin{vmatrix} b & 2a \\ b & 0 \end{vmatrix} + 0$

$= -a(0 - 2ab) = 2a^2b$

f $\begin{vmatrix} b & a & -b \\ 0 & -3 & 0 \\ a & 2 & a \end{vmatrix} = -0 - 3\begin{vmatrix} b & -b \\ a & a \end{vmatrix} - 0$

$= -3(ab - -ab) = -6ab$

2 a For example, $\begin{vmatrix} 2 & 1 & -3 \\ 4 & 2 & 1 \\ 1 & 1 & 2 \end{vmatrix} = \begin{vmatrix} 0 & 1 & -3 \\ 0 & 2 & 1 \\ -1 & 1 & 2 \end{vmatrix}$ C1 − 2C2

b −7

3 a For example, $\begin{vmatrix} 6 & 3 & -3 \\ 9 & 2 & -1 \\ -5 & -7 & -2 \end{vmatrix} = \begin{vmatrix} 0 & 3 & -3 \\ 5 & 2 & -1 \\ 9 & -7 & -2 \end{vmatrix}$ C1 − 2C2

$= \begin{vmatrix} 0 & 0 & -3 \\ 5 & 1 & -1 \\ 9 & -9 & -2 \end{vmatrix}$ C2 + C3

$= \begin{vmatrix} 0 & 0 & -3 \\ 6 & 1 & -1 \\ 0 & -9 & -2 \end{vmatrix}$ C1 + C2

$= \begin{vmatrix} 0 & 0 & -3 \\ 6 & 1 & 0 \\ 0 & -9 & -11 \end{vmatrix}$ C2 + C3

$= 162$

b 162

4 For example, $\begin{vmatrix} 5 & -2 & 12 \\ -2 & -8 & 24 \\ 4 & 1 & 16 \end{vmatrix} = \begin{vmatrix} 5 & -2 & 12 \\ -12 & -4 & 0 \\ 4 & 1 & 16 \end{vmatrix}$ R2 − 2R1

$= \begin{vmatrix} 11 & -2 & 12 \\ 0 & -4 & 0 \\ 1 & 1 & 16 \end{vmatrix}$ C1 − 3C2

5 a −128 since rows 1 and 2 swapped

b 128 since transpose

c 128 since row 2 added to row 1

d 128 since 4 × column 1 subtracted from column 3

e $128 \times 3 = 384$ since column 2 multiplied by 3

f $(128 \times -1) \div 2 = -64$ since column 2 divided by 2 and column 3 multiplied by −1

g 256 since columns 1 and 3 then columns 2 and 3 swapped and column 2 doubled

h $(128 \div 2) \times 4 = 256$ since row 3 divided by 2 and column 2 multiplied by 4

Exercise 22.2B

1 a
$$\begin{vmatrix} a & b & c \\ b & c & a \\ c & a & b \end{vmatrix} = \begin{vmatrix} a+b & b+c & c+a \\ b & c & a \\ c & a & b \end{vmatrix} \text{ R1+R2}$$

$$= \begin{vmatrix} a+b+c & b+c+a & c+a+b \\ b & c & a \\ c & a & b \end{vmatrix} \text{ R1+R3}$$

$$= (a+b+c)\begin{vmatrix} 1 & 1 & 1 \\ b & c & a \\ c & a & b \end{vmatrix}$$

$$= (a+b+c)[(bc-a^2)-(b^2-ac)+(ab-c^2)]$$

$$= (a+b+c)(ab+ac+bc-a^2-b^2-c^2)$$

as required

b
$$\begin{vmatrix} a^2 & b^2 & c^2 \\ bc & ca & ab \\ 1 & 1 & 1 \end{vmatrix} = \begin{vmatrix} a^2 & b^2-a^2 & c^2 \\ bc & ca-bc & ab \\ 1 & 0 & 1 \end{vmatrix} \text{ C2-C1}$$

$$= \begin{vmatrix} a^2 & b^2-a^2 & c^2-a^2 \\ bc & ca-bc & ab-bc \\ 1 & 0 & 0 \end{vmatrix} \text{ C3-C1}$$

$$= \begin{vmatrix} a^2 & (b-a)(b+a) & c^2-a^2 \\ bc & -c(b-a) & ab-bc \\ 1 & 0 & 0 \end{vmatrix}$$

$$= (b-a)\begin{vmatrix} a^2 & b+a & c^2-a^2 \\ bc & -c & ab-bc \\ 1 & 0 & 0 \end{vmatrix}$$

$$= (b-a)\begin{vmatrix} a^2 & b+a & (c-a)(c+a) \\ bc & -c & -b(c-a) \\ 1 & 0 & 0 \end{vmatrix}$$

$$= (b-a)(c-a)\begin{vmatrix} a^2 & b+a & c+a \\ bc & -c & -b \\ 1 & 0 & 0 \end{vmatrix}$$

$$= (b-a)(c-a)[-b(b+a)--c(c+a)]$$

$$= (b-a)(c-a)(-b^2-ab+c^2+ac)$$

$$= (b-a)(c-a)(c-b)(a+b+c) \text{ as required}$$

c
$$\begin{vmatrix} a+b & a+c & b+c \\ c & b & a \\ c^2 & b^2 & a^2 \end{vmatrix} = \begin{vmatrix} a+b+c & a+c+b & b+c+a \\ c & b & a \\ c^2 & b^2 & a^2 \end{vmatrix} \text{ R1+R2}$$

$$= (a+b+c)\begin{vmatrix} 1 & 1 & 1 \\ c & b & a \\ c^2 & b^2 & a^2 \end{vmatrix}$$

$$= (a+b+c)\begin{vmatrix} 0 & 1 & 1 \\ c-b & b & a \\ c^2-b^2 & b^2 & a^2 \end{vmatrix} \text{ C1-C2}$$

$$= (a+b+c)(c-b)\begin{vmatrix} 0 & 1 & 1 \\ 1 & b & a \\ c+b & b^2 & a^2 \end{vmatrix}$$

$$= (a+b+c)(c-b)\begin{vmatrix} 0 & 0 & 1 \\ 1 & b-a & a \\ c+b & b^2-a^2 & a^2 \end{vmatrix} \text{ C2-C3}$$

$$= (a+b+c)(c-b)(b-a)[(a+b)-(c+b)]$$

$$= (a+b+c)(c-b)(b-a)(a-c)$$

as required

d
$$\begin{vmatrix} a & -b & c \\ c-b & a+c & a-b \\ -bc & ac & -ab \end{vmatrix} = \begin{vmatrix} a+b & -b & c \\ -b-a & a+c & a-b \\ -bc-ac & ac & -ab \end{vmatrix} \text{ C1-C2}$$

$$= \begin{vmatrix} a+b & -b & c \\ -(a+b) & a+c & a-b \\ -c(a+b) & ac & -ab \end{vmatrix}$$

$$= (a+b)\begin{vmatrix} 1 & -b & b+c \\ -1 & a+c & -b-c \\ -c & ac & -ab-ac \end{vmatrix} \text{ C3-C2}$$

$$= (a+b)\begin{vmatrix} 1 & -b & b+c \\ -1 & a+c & -(b+c) \\ -c & ac & -a(b+c) \end{vmatrix}$$

$$= (a+b)(b+c)\begin{vmatrix} 1 & -b & 1 \\ -1 & a+c & -1 \\ -c & ac & -a \end{vmatrix}$$

$$= (a+b)(b+c)\begin{vmatrix} 0 & -b & 1 \\ 0 & a+c & -1 \\ -c+a & ac & -a \end{vmatrix} \text{ C1-C3}$$

$$= (a+b)(b+c)(c-a)\begin{vmatrix} 0 & -b & 1 \\ 0 & a+c & -1 \\ -1 & ac & -a \end{vmatrix}$$

$$= (a+b)(b+c)(c-a)[-1(b-(a+c))]$$

$$= (a+b)(b+c)(c-a)(a-b+c)$$

as required

2
$$\begin{vmatrix} 1 & 1 & 1 \\ a & a^2 & a^3 \\ b & b^2 & b^3 \end{vmatrix} = a\begin{vmatrix} 1 & 1 & 1 \\ 1 & a & a^2 \\ b & b^2 & b^3 \end{vmatrix}$$

$$= ab\begin{vmatrix} 1 & 1 & 1 \\ 1 & a & a^2 \\ 1 & b & b^2 \end{vmatrix}$$

$$= ab\begin{vmatrix} 1 & 1 & 1 \\ 0 & a-1 & a^2-1 \\ 1 & b & b^2 \end{vmatrix} \text{ R2-R1}$$

$$= ab\begin{vmatrix} 1 & 1 & 1 \\ 0 & a-1 & a^2-1 \\ 0 & b-1 & b^2-1 \end{vmatrix} \text{ R3-R1}$$

$$= ab(a-1)\begin{vmatrix} 1 & 1 & 1 \\ 0 & 1 & a+1 \\ 0 & b-1 & b^2-1 \end{vmatrix}$$

$$= ab(a-1)(b-1)\begin{vmatrix} 1 & 1 & 1 \\ 0 & 1 & a+1 \\ 0 & 1 & b+1 \end{vmatrix}$$

$$= ab(a-1)(b-1)[(b+1)-(a+1)]$$

$$= ab(a-1)(b-1)(b-a)$$

3 a $\begin{vmatrix} a & b & c \\ a^2 & b^2 & c^2 \\ a^3 & b^3 & c^3 \end{vmatrix} = a\begin{vmatrix} 1 & b & c \\ a & b^2 & c^2 \\ a^2 & b^3 & c^3 \end{vmatrix}$

$$= ab\begin{vmatrix} 1 & 1 & c \\ a & b & c^2 \\ a^2 & b^2 & c^3 \end{vmatrix}$$

$$= abc\begin{vmatrix} 1 & 1 & 1 \\ a & b & c \\ a^2 & b^2 & c^2 \end{vmatrix}$$

$$= abc\begin{vmatrix} 1 & 1 & 1 \\ a & b & c \\ 0 & b^2-ab & c^2-ac \end{vmatrix} \text{R3} - a\text{R2}$$

$$= abc\begin{vmatrix} 1 & 1 & 1 \\ 0 & b-a & c-a \\ 0 & b^2-ab & c^2-ac \end{vmatrix} \text{R2} - a\text{R1}$$

$$= abc\begin{vmatrix} 1 & 1 & 1 \\ 0 & b-a & c-a \\ 0 & 0 & c^2-ac-b(c-a) \end{vmatrix} \text{R3} - b\text{R2}$$

since $c^2 - ac - b(c-a) = c(c-a) - b(c-a)$
$= (c-b)(c-a)$

$$= abc\begin{vmatrix} 1 & 1 & 1 \\ 0 & b-a & c-a \\ 0 & 0 & (c-b)(c-a) \end{vmatrix}$$

as required

b **i** 48 **ii** –1680

4 a $(a-b)(a^2+ab+b^2) = a^3+a^2b+ab^2-a^2b-ab^2-b^3$
$= a^3 - b^3$ as required

b $-(a-b)(b-c)(c-a)(a+b+c)$

c Unique solution if: $a \neq b,\ a \neq c,\ b \neq c,\ a+b+c \neq 0$

5 a $(a-b)(b-c)(a-c)(ab+ac+bc)$

b $a \neq b$, $a \neq c$ and $b \neq c$

6 a $(a-b)(a+b)(a+b-2)$ **b** $a \neq b,\ a \neq -b,\ a+b \neq 2$

7 e.g. $\begin{vmatrix} a^2 & b^2 & a \\ 1 & c & b \\ a & ac & ab \end{vmatrix} = a\begin{vmatrix} a^2 & b^2 & a \\ 1 & c & b \\ 1 & c & b \end{vmatrix}$

$$= a\begin{vmatrix} a^2 & b^2 & a \\ 0 & 0 & 0 \\ 1 & c & b \end{vmatrix} \text{R2} - \text{R3}$$

$= a \times 0$
$= 0$ so singular

8 $x = 0, a, b$

9 $x = a, b, -a-b$

10 $\det \mathbf{A} = \begin{vmatrix} a & a^2-a & a^2+b^2+c^2 \\ b & b^2-b & a^2+b^2+c^2 \\ c & c^2-c & a^2+b^2+c^2 \end{vmatrix}$

$$= (a^2+b^2+c^2)\begin{vmatrix} a & a^2-a & 1 \\ b & b^2-b & 1 \\ c & c^2-c & 1 \end{vmatrix}$$

$$= (a^2+b^2+c^2)\begin{vmatrix} a & a^2-a & 1 \\ b-a & b^2-b-a^2+a & 0 \\ c-a & c^2-c-a^2+a & 0 \end{vmatrix}$$

subtract R1 from R2 & R3

$$= (a^2+b^2+c^2)\begin{vmatrix} a & a^2-a & 1 \\ b-a & (b-a)(b+a-1) & 0 \\ c-a & (c-a)(c+a-1) & 0 \end{vmatrix}$$

$$= (a^2+b^2+c^2)(b-a)(c-a)\begin{vmatrix} a & a^2-a & 1 \\ 1 & b+a-1 & 0 \\ 1 & c+a-1 & 0 \end{vmatrix}$$

$= (a^2+b^2+c^2)(b-a)(c-a)(c+a-1-b-a+1)$
$= (a^2+b^2+c^2)(b-a)(c-a)(c-b)$

$a \neq b \neq c$. so $(b-a)(c-a)(c-b) \neq 0$

a, b, c are distinct and real so $a^2+b^2+c^2 > 0$

Therefore $\det \text{A} \neq 0$ so **A** is non-singular for all real, distinct values of a, b and c.

Exercise 22.3A

1 a **i** $\lambda = 2, -4$, corresponding eigenvectors are $\begin{pmatrix} 1 \\ 0 \end{pmatrix}, \begin{pmatrix} 1 \\ -2 \end{pmatrix}$

ii $\lambda = 2, -1$, corresponding eigenvectors are $\begin{pmatrix} 1 \\ 1 \end{pmatrix}, \begin{pmatrix} 1 \\ 4 \end{pmatrix}$

iii $\lambda = 3, 7$, corresponding eigenvectors are $\begin{pmatrix} -4 \\ 3 \end{pmatrix}, \begin{pmatrix} 0 \\ 1 \end{pmatrix}$

iv $\lambda = -1, 5$, corresponding eigenvectors are $\begin{pmatrix} 1 \\ 1 \end{pmatrix}, \begin{pmatrix} -5 \\ 1 \end{pmatrix}$

b **i** Characteristic equation is $\lambda^2 + 2\lambda - 8 = 0$

Substitute matrix **A**:

$$\mathbf{A}^2 + 2\mathbf{A} - 8\mathbf{I} = \begin{pmatrix} 2 & 3 \\ 0 & -4 \end{pmatrix}^2 + 2\begin{pmatrix} 2 & 3 \\ 0 & -4 \end{pmatrix} - 8\begin{pmatrix} 1 & 0 \\ 0 & 1 \end{pmatrix}$$

$$= \begin{pmatrix} 4 & -6 \\ 0 & 16 \end{pmatrix} + \begin{pmatrix} 4 & 6 \\ 0 & -8 \end{pmatrix} - \begin{pmatrix} 8 & 0 \\ 0 & 8 \end{pmatrix}$$

$$= \begin{pmatrix} 0 & 0 \\ 0 & 0 \end{pmatrix} \text{ as required}$$

ii Characteristic equation is $\lambda^2 - \lambda - 2 = 0$
Substitute matrix **A**:

$$\mathbf{A}^2 - \mathbf{A} - 2\mathbf{I} = \begin{pmatrix} 3 & -1 \\ 4 & -2 \end{pmatrix}^2 - \begin{pmatrix} 3 & -1 \\ 4 & -2 \end{pmatrix} - 2\begin{pmatrix} 1 & 0 \\ 0 & 1 \end{pmatrix}$$
$$= \begin{pmatrix} 5 & -1 \\ 4 & 0 \end{pmatrix} - \begin{pmatrix} 3 & -1 \\ 4 & -2 \end{pmatrix} - \begin{pmatrix} 2 & 0 \\ 0 & 2 \end{pmatrix}$$
$$= \begin{pmatrix} 0 & 0 \\ 0 & 0 \end{pmatrix} \text{ as required}$$

iii Characteristic equation is $\lambda^2 - 10\lambda + 21 = 0$
Substitute matrix **A**:

$$\mathbf{A}^2 - 10\mathbf{A} + 21\mathbf{I} = \begin{pmatrix} 3 & 0 \\ 3 & 7 \end{pmatrix}^2 - 10\begin{pmatrix} 3 & 0 \\ 3 & 7 \end{pmatrix} + 21\begin{pmatrix} 1 & 0 \\ 0 & 1 \end{pmatrix}$$
$$= \begin{pmatrix} 9 & 0 \\ 30 & 49 \end{pmatrix} - \begin{pmatrix} 30 & 0 \\ 30 & 70 \end{pmatrix} + \begin{pmatrix} 21 & 0 \\ 0 & 21 \end{pmatrix}$$
$$= \begin{pmatrix} 0 & 0 \\ 0 & 0 \end{pmatrix} \text{ as required}$$

iv Characteristic equation is $\lambda^2 - 4\lambda - 5 = 0$
Substitute matrix **A**:

$$\mathbf{A}^2 - 4\mathbf{A} - 5\mathbf{I} = \begin{pmatrix} 4 & -5 \\ -1 & 0 \end{pmatrix}^2 - 4\begin{pmatrix} 4 & -5 \\ -1 & 0 \end{pmatrix} - 5\begin{pmatrix} 1 & 0 \\ 0 & 1 \end{pmatrix}$$
$$= \begin{pmatrix} 21 & -20 \\ -4 & 5 \end{pmatrix} - \begin{pmatrix} 16 & -20 \\ -4 & 0 \end{pmatrix} - \begin{pmatrix} 5 & 0 \\ 0 & 5 \end{pmatrix}$$
$$= \begin{pmatrix} 0 & 0 \\ 0 & 0 \end{pmatrix} \text{ as required}$$

2 a $\lambda = -3,\ 1,\ 5$, corresponding eigenvectors are

$$\begin{pmatrix} 5 \\ -3 \\ 4 \end{pmatrix}, \begin{pmatrix} 1 \\ -1 \\ 0 \end{pmatrix}, \begin{pmatrix} 0 \\ 1 \\ 0 \end{pmatrix}$$

b $\lambda = -2,\ 2,\ 3$, corresponding eigenvectors are

$$\begin{pmatrix} 1 \\ -1 \\ 0 \end{pmatrix}, \begin{pmatrix} 3 \\ 1 \\ -4 \end{pmatrix}, \begin{pmatrix} 3 \\ 2 \\ 0 \end{pmatrix}$$

c $\lambda = 5,\ 2,\ -4$, corresponding eigenvectors are

$$\begin{pmatrix} 1 \\ 0 \\ 0 \end{pmatrix}, \begin{pmatrix} 6 \\ 6 \\ -5 \end{pmatrix}, \begin{pmatrix} 0 \\ 0 \\ 1 \end{pmatrix}$$

d $\lambda = -7,\ 4,\ 9$, corresponding eigenvectors are

$$\begin{pmatrix} -44 \\ 19 \\ 11 \end{pmatrix}, \begin{pmatrix} 0 \\ 1 \\ 0 \end{pmatrix}, \begin{pmatrix} 0 \\ 1 \\ 5 \end{pmatrix}$$

3 a $\det\begin{pmatrix} 1-\lambda & 3 & -1 \\ -2 & 3-\lambda & 2 \\ 1 & 0 & 2-\lambda \end{pmatrix}$

$$= 1[6 - -1(3-\lambda)] + (2-\lambda)[(1-\lambda)(3-\lambda) - -6]$$
$$= 9 - \lambda + (2-\lambda)(\lambda^2 - 4\lambda + 9)$$
$$= 9 - \lambda + 2\lambda^2 - 8\lambda + 18 - \lambda^3 + 4\lambda^2 - 9\lambda$$
$$= -\lambda^3 + 6\lambda^2 - 18\lambda + 27$$

When $\lambda = 3$, $-\lambda^3 + 6\lambda^2 - 18\lambda + 27 = 0$ so 3 is an eigenvalue.

$-\lambda^3 + 6\lambda^2 - 18\lambda + 27 = (\lambda - 3)(-\lambda^2 + 3\lambda - 9)$

$-\lambda^2 + 3\lambda - 9 = 0$ has no real solutions since $b^2 - 4ac = 9 - 36 < 0$

An eigenvector is $\begin{pmatrix} 1 \\ 1 \\ 1 \end{pmatrix}$

b Characteristic equation is $-\lambda^3 + 6\lambda^2 - 18\lambda + 27 = 0$
Substitute matrix **A**:

$$-\mathbf{A}^3 + 6\mathbf{A}^2 - 16\mathbf{A} + 9\mathbf{I} = -\begin{pmatrix} 1 & 3 & -1 \\ -2 & 3 & 2 \\ 1 & 0 & 2 \end{pmatrix}^3$$
$$+6\begin{pmatrix} 1 & 3 & -1 \\ -2 & 3 & 2 \\ 1 & 0 & 2 \end{pmatrix}^2 - 18\begin{pmatrix} 1 & 3 & -1 \\ -2 & 3 & 2 \\ 1 & 0 & 2 \end{pmatrix} + 27\begin{pmatrix} 1 & 0 & 0 \\ 0 & 1 & 0 \\ 0 & 0 & 1 \end{pmatrix}$$
$$= -\begin{pmatrix} -27 & 18 & 36 \\ 0 & -9 & 36 \\ 0 & 18 & 9 \end{pmatrix} + 6\begin{pmatrix} -6 & 12 & 3 \\ -6 & 3 & 12 \\ 3 & 3 & 3 \end{pmatrix}$$
$$-18\begin{pmatrix} 1 & 3 & -1 \\ -2 & 3 & 2 \\ 1 & 0 & 2 \end{pmatrix} + 27\begin{pmatrix} 1 & 0 & 0 \\ 0 & 1 & 0 \\ 0 & 0 & 1 \end{pmatrix}$$
$$= -\begin{pmatrix} -27 & 18 & 36 \\ 0 & -9 & 36 \\ 0 & 18 & 9 \end{pmatrix} + \begin{pmatrix} -36 & 72 & 18 \\ -36 & 18 & 72 \\ 18 & 18 & 18 \end{pmatrix}$$
$$-\begin{pmatrix} 18 & 54 & -18 \\ -36 & 54 & 36 \\ 18 & 0 & 36 \end{pmatrix} + \begin{pmatrix} 27 & 0 & 0 \\ 0 & 27 & 0 \\ 0 & 0 & 27 \end{pmatrix}$$
$$= \begin{pmatrix} 0 & 0 & 0 \\ 0 & 0 & 0 \\ 0 & 0 & 0 \end{pmatrix} \text{ as required}$$

4 a $\det\begin{pmatrix} -2-\lambda & 6 & -3 \\ 3 & 2-\lambda & 6 \\ 2 & 6 & -3-\lambda \end{pmatrix}$

$= (-2-\lambda)\,[(2-\lambda)(3-\lambda) - 36] - 6[3(-3-\lambda) - 12]$
$-3[18 - 2(2-\lambda)]$
$= -\lambda^3 - 3\lambda^2 + 52\lambda + 168$

When $\lambda = -7$, $-\lambda^3 - 3\lambda^2 + 52\lambda + 168 = 0$ so -7 is an eigenvalue and $\lambda = 2 \pm 2\sqrt{7}$

b $\begin{pmatrix} 21 \\ -13 \\ 9 \end{pmatrix}$

c Characteristic equation is
$-\lambda^3 - 3\lambda^2 + 52\lambda + 168 = 0$
Substitute matrix **A**:
$-\mathbf{A}^3 - 3\mathbf{A}^2 + 52\mathbf{A} + 168\mathbf{I}$

$$= -\begin{pmatrix} -2 & 6 & -3 \\ 3 & 2 & 6 \\ 2 & 6 & -3 \end{pmatrix}^3 - 3\begin{pmatrix} -2 & 6 & -3 \\ 3 & 2 & 6 \\ 2 & 6 & -3 \end{pmatrix}^2$$
$$+52\begin{pmatrix} -2 & 6 & -3 \\ 3 & 2 & 6 \\ 2 & 6 & -3 \end{pmatrix} + 168\begin{pmatrix} 1 & 0 & 0 \\ 0 & 1 & 0 \\ 0 & 0 & 1 \end{pmatrix}$$

$$= -\begin{pmatrix} 16 & 366 & -309 \\ 120 & 98 & 357 \\ 80 & 294 & -105 \end{pmatrix} - 3\begin{pmatrix} 16 & -18 & 51 \\ 12 & 58 & -15 \\ 8 & 6 & 39 \end{pmatrix}$$

$$+52\begin{pmatrix} -2 & 6 & -3 \\ 3 & 2 & 6 \\ 2 & 6 & -3 \end{pmatrix} + 168\begin{pmatrix} 1 & 0 & 0 \\ 0 & 1 & 0 \\ 0 & 0 & 1 \end{pmatrix}$$

$$= -\begin{pmatrix} 16 & 366 & -309 \\ 120 & 98 & 357 \\ 80 & 294 & -105 \end{pmatrix} - \begin{pmatrix} 48 & -54 & 153 \\ 36 & 174 & -45 \\ 24 & 18 & 117 \end{pmatrix}$$

$$+\begin{pmatrix} -104 & 312 & -156 \\ 156 & 104 & 312 \\ 104 & 312 & -156 \end{pmatrix} + \begin{pmatrix} 168 & 0 & 0 \\ 0 & 168 & 0 \\ 0 & 0 & 168 \end{pmatrix}$$

$$= \begin{pmatrix} 0 & 0 & 0 \\ 0 & 0 & 0 \\ 0 & 0 & 0 \end{pmatrix} \text{ as required}$$

5 $P = \begin{pmatrix} 0 & 1 \\ 1 & 0 \end{pmatrix}$, $D = \begin{pmatrix} 7 & 0 \\ 0 & 2 \end{pmatrix}$

6 $B = \begin{pmatrix} -1 & 6 \\ 1 & 1 \end{pmatrix}\begin{pmatrix} -4 & 0 \\ 0 & 3 \end{pmatrix}\begin{pmatrix} -\frac{1}{7} & \frac{6}{7} \\ \frac{1}{7} & \frac{1}{7} \end{pmatrix}$

7 $P = \begin{pmatrix} 0 & 0 & -30 \\ -2 & 3 & 1 \\ 3 & 2 & 24 \end{pmatrix}$, $D = \begin{pmatrix} -7 & 0 & 0 \\ 0 & 6 & 0 \\ 0 & 0 & -4 \end{pmatrix}$

8 $T = \begin{pmatrix} -1 & 0 & 1 \\ 1 & 0 & 1 \\ 0 & 1 & 0 \end{pmatrix}\begin{pmatrix} 8 & 0 & 0 \\ 0 & -7 & 0 \\ 0 & 0 & 6 \end{pmatrix}\begin{pmatrix} -\frac{1}{2} & \frac{1}{2} & 0 \\ 0 & 0 & 1 \\ \frac{1}{2} & \frac{1}{2} & 0 \end{pmatrix}$

Exercise 22.3B

1 a $\begin{pmatrix} 1 & 0 \\ 189 & 64 \end{pmatrix}$

b $\begin{pmatrix} 1 & 0 \\ 3(2^n)-3 & 2^n \end{pmatrix}$

2 a $\begin{pmatrix} 5^4 & -2(5^4) \\ -2(5^4) & 4(5^4) \end{pmatrix}$

b $5^{n-1}\begin{pmatrix} 1 & -2 \\ -2 & 4 \end{pmatrix}$

3 a $\begin{pmatrix} -1 & 1 \\ 1 & 1 \end{pmatrix}, \begin{pmatrix} 3 & 0 \\ 0 & 7 \end{pmatrix}$

b $3^3 = 27$ and $7^3 = 343$

c $\begin{pmatrix} -1 \\ 1 \end{pmatrix}$ and $\begin{pmatrix} 1 \\ 1 \end{pmatrix}$

4 a $r = s\begin{pmatrix} 1 \\ -3 \end{pmatrix}$ and $r = t\begin{pmatrix} 2 \\ 1 \end{pmatrix}$

b Only eigenvalue is $7^2 = 49$, eigenvectors are $\begin{pmatrix} 1 \\ -3 \end{pmatrix}$ and $\begin{pmatrix} 2 \\ 1 \end{pmatrix}$

5 a Eigenvalues are $3^4 = 81$, $2^4 = 16$, $(-4)^4 = 256$

Eigenvectors are $\begin{pmatrix} 1 \\ 0 \\ 1 \end{pmatrix}, \begin{pmatrix} 2 \\ -3 \\ 0 \end{pmatrix}, \begin{pmatrix} 1 \\ 1 \\ 2 \end{pmatrix}$

b $r = s\begin{pmatrix} 1 \\ 0 \\ 1 \end{pmatrix}$, $r = t\begin{pmatrix} 2 \\ -3 \\ 0 \end{pmatrix}$ and $r = u\begin{pmatrix} 1 \\ 1 \\ 2 \end{pmatrix}$

6 a $\begin{pmatrix} 101 & -37 & 0 \\ 74 & -10 & 0 \\ 0 & 0 & 1 \end{pmatrix}$

b $\begin{pmatrix} 2(4^n)-3^n & 3^n-4^n & 0 \\ 2(4^n)-2(3^n) & 2(3^n)-4^n & 0 \\ 0 & 0 & 1 \end{pmatrix}$

7 a $x = -2y$ or $y = -\frac{1}{2}x$, $y = x$

b $y = x$ is a line of invariant points since corresponding eigenvalue is one.

8 $\lambda = a \pm b$, corresponding eigenvectors are $\begin{pmatrix} 1 \\ 1 \end{pmatrix}, \begin{pmatrix} 1 \\ -1 \end{pmatrix}$

9 a $\lambda = -3, 1$, corresponding set of eigenvectors are $\alpha\begin{pmatrix} 2 \\ 1 \end{pmatrix}$, $\beta\begin{pmatrix} 3 \\ 2 \end{pmatrix}$

b $\begin{pmatrix} 2 \\ 1 \end{pmatrix}$ gives the direction of an invariant line through the origin and $\begin{pmatrix} 3 \\ 2 \end{pmatrix}$ gives the direction of a line of invariant points.

10 a $\lambda = -1, 3, 1$

b Eigenvectors are of the form $\alpha\begin{pmatrix} 2 \\ 1 \\ -1 \end{pmatrix}, \beta\begin{pmatrix} 1 \\ 0 \\ 0 \end{pmatrix}, \gamma\begin{pmatrix} 1 \\ 0 \\ 1 \end{pmatrix}$

c The eigenvector $\begin{pmatrix} 1 \\ 0 \\ 1 \end{pmatrix}$ gives the direction of a line of invariant points through the origin.

The eigenvectors $\begin{pmatrix} 1 \\ 0 \\ 0 \end{pmatrix}$ and $\begin{pmatrix} 2 \\ 1 \\ -1 \end{pmatrix}$ gives the directions of two invariant lines through the origin.

11 a $\lambda = 2, 2, -1$

b Set of eigenvectors of the form $\alpha\begin{pmatrix} 0 \\ 1 \\ 2 \end{pmatrix} + \beta\begin{pmatrix} 1 \\ 0 \\ -1 \end{pmatrix}, \lambda\begin{pmatrix} 1 \\ 1 \\ -2 \end{pmatrix}$

c The set of eigenvectors $\alpha\begin{pmatrix} 0 \\ 1 \\ 2 \end{pmatrix} + \beta\begin{pmatrix} 1 \\ 0 \\ -1 \end{pmatrix}$ represent an invariant plane.

The eigenvector $\begin{pmatrix} 1 \\ 1 \\ -2 \end{pmatrix}$ gives the direction of an invariant line through the origin.

12 a $\det\begin{pmatrix}1-\lambda & 1 & 2\\ 0 & 1-\lambda & -1\\ 0 & 3 & -\lambda\end{pmatrix}=0$

$\Rightarrow (1-\lambda)(-\lambda(1-\lambda)+3)=0$

$\Rightarrow (1-\lambda)(\lambda^2-\lambda+3)=0$

$\Rightarrow 1-\lambda=0$ or $\lambda^2-\lambda+3=0$

$\lambda=1$ is a real eigenvalue

$\lambda^2-\lambda+3=0$ has no real solutions since discriminant $=(-1)^2-4(1)(3)=-11<0$

So only one real eigenvalue

b $\begin{pmatrix}1 & 1 & 2\\ 0 & 1 & -1\\ 0 & 3 & 0\end{pmatrix}\begin{pmatrix}x\\ y\\ z\end{pmatrix}=\begin{pmatrix}x\\ y\\ z\end{pmatrix}$

$x+y+2z=x\Rightarrow y=-2z$

$y-z=y\Rightarrow z=0$

$\Rightarrow y=0$

So eigenvector is $\begin{pmatrix}1\\ 0\\ 0\end{pmatrix}$

So, since eigenvalue is 1, $\mathbf{r}=s\begin{pmatrix}1\\ 0\\ 0\end{pmatrix}$ is a line of invariant points under the transformation represented by **M**, and this is the equation of the x-axis.

13 a $\mathbf{M}=\mathbf{UDU}^{-1}\Rightarrow \mathbf{M}^n=(\mathbf{UDU}^{-1})^n$

$=(\mathbf{UDU}^{-1})(\mathbf{UDU}^{-1})(\mathbf{UDU}^{-1})\ldots(\mathbf{UDU}^{-1})$

$=(\mathbf{UD})(\mathbf{U}^{-1}\mathbf{U})\mathbf{D}(\mathbf{U}^{-1}\mathbf{U})\mathbf{D}\ldots(\mathbf{U}^{-1}\mathbf{U})(\mathbf{DU}^{-1})$

$=(\mathbf{UD})\mathbf{IDID}\ldots\mathbf{I}(\mathbf{DU}^{-1})$

$=(\mathbf{UD})\mathbf{DD}\ldots(\mathbf{DU}^{-1})$

$=\mathbf{UD}^n\mathbf{U}^{-1}$ as required

b $\begin{pmatrix}0 & -\frac{1}{2}\\ 0 & 1\end{pmatrix}$

14 a $\lambda=\pm k$, corresponding eigenvectors are $\begin{pmatrix}k+3\\ k-3\end{pmatrix}, \begin{pmatrix}1\\ -1\end{pmatrix}$

b $\begin{pmatrix}k+3 & 1\\ k-3 & -1\end{pmatrix}^{-1}=\frac{1}{2k}\begin{pmatrix}1 & 1\\ k-3 & -k-3\end{pmatrix}$

$$\mathbf{M}=\frac{1}{2k}\begin{pmatrix}3+k & 1\\ k-3 & -1\end{pmatrix}\begin{pmatrix}k & 0\\ 0 & -k\end{pmatrix}\begin{pmatrix}1 & 1\\ k-3 & -k-3\end{pmatrix}$$

Therefore,

$$\mathbf{M}^{2n+1}=\frac{1}{2k}\begin{pmatrix}k+3 & 1\\ k-3 & -1\end{pmatrix}\begin{pmatrix}k & 0\\ 0 & -k\end{pmatrix}^{2n+1}\begin{pmatrix}1 & 1\\ k-3 & -k-3\end{pmatrix}$$

$$=\frac{1}{2k}\begin{pmatrix}k+3 & 1\\ k-3 & -1\end{pmatrix}\begin{pmatrix}k^{2n+1} & 0\\ 0 & (-k)^{2n+1}\end{pmatrix}\begin{pmatrix}1 & 1\\ k-3 & -k-3\end{pmatrix}$$

$$=\frac{k^{2n+1}}{2k}\begin{pmatrix}k+3 & 1\\ k-3 & -1\end{pmatrix}\begin{pmatrix}1 & 0\\ 0 & -1\end{pmatrix}\begin{pmatrix}1 & 1\\ k-3 & -k-3\end{pmatrix}$$

$$=\frac{k^{2n}}{2}\begin{pmatrix}k+3 & -1\\ k-3 & 1\end{pmatrix}\begin{pmatrix}1 & 1\\ k-3 & -k-3\end{pmatrix}$$

$$=\frac{k^{2n}}{2}\begin{pmatrix}(k+3)-(k-3) & (k+3)-(-k-3)\\ (k-3)+(k-3) & (k-3)+(-k-3)\end{pmatrix}$$

$$=\frac{k^{2n}}{2}\begin{pmatrix}6 & 2(k+3)\\ 2(k-3) & -6\end{pmatrix}$$

$$=k^{2n}\begin{pmatrix}3 & k+3\\ k-3 & -3\end{pmatrix}$$

$=k^{2n}\mathbf{M}$ as required

Review exercise 22

1 a i 200 square units **ii** 80 square units **iii** 8 square units **iv** $128a^2$ square units

b **ii, iv** determinants negative

2 a –63 **b** $-44a-21b$

3 a $x=-\frac{4}{3}$ **b** $x=\frac{-5\pm\sqrt{37}}{2}$

4 a i 4 cubic units **ii** $\frac{7}{4a}$ cubic units

b Determinant negative

5 a $\frac{1}{25}\begin{pmatrix}-5 & 5 & 10\\ 22 & -7 & -19\\ 4 & 1 & -8\end{pmatrix}$

b $\det\begin{pmatrix}6 & -2 & -3\\ 1 & 0 & -1\\ 0 & 2 & -3\end{pmatrix}=6(0--2)+2(-3-0)-3(2-0)=0$

therefore matrix is singular so inverse does not exist.

6 $x=2, y=-5, z=-10$

7 a $k=-\frac{5}{2}, 1$

b When $k=1$, the first and third planes are parallel hence do not intersect.

When $k=-\frac{5}{2}$, there is an inconsistency, hence the three planes form a triangular prism.

8 $\begin{vmatrix}3 & -2 & 5\\ -6 & 4 & 0\\ 5 & -1 & -3\end{vmatrix}=5(0-20)--1(0--30)+(12--12)$

$=-100+30+0=-70$

$\begin{vmatrix}3 & -2 & 5\\ -6 & 4 & 0\\ 5 & -1 & -3\end{vmatrix}=5(6-20)-0+3(12-2)$

$=-70+0=-70$

9 a –90 **b** –45 **c** 90 **d** $\frac{-45}{b}$

10 a $\begin{vmatrix}1 & -1 & 1\\ a & a^2 & -a^2\\ b & b^2 & -b^2\end{vmatrix}=\begin{vmatrix}1 & -1 & 1\\ a+1 & a^2-1 & 1-a^2\\ b+1 & b^2-1 & 1-b^2\end{vmatrix}$

$=(a+1)(b+1)\begin{vmatrix}1 & -1 & 1\\ 1 & a-1 & 1-a\\ 1 & b-1 & 1-b\end{vmatrix}$

$=(a+1)(b+1)\begin{vmatrix}1 & -1 & 1\\ 0 & a & -a\\ 0 & b & -b\end{vmatrix}$

$=ab(a+1)(b+1)\begin{vmatrix}1 & -1 & 1\\ 0 & 1 & -1\\ 0 & 1 & -1\end{vmatrix}$

$=ab(a+1)(b+1)\left(1\begin{vmatrix}1 & -1\\ 1 & -1\end{vmatrix}-0+0\right)$

$=ab(a+1)(b+1)\begin{vmatrix}1 & -1\\ 1 & -1\end{vmatrix}$

b 0

11 $(a-b)(c-a)(b-c)(ac+ab+bc)$

12 a $\lambda = 2, -7$, corresponding eigenvectors are $\begin{pmatrix}1\\1\end{pmatrix}, \begin{pmatrix}5\\-4\end{pmatrix}$

b $\lambda = -9, -27$, corresponding eigenvectors are $\begin{pmatrix}2\\1\end{pmatrix}, \begin{pmatrix}-1\\1\end{pmatrix}$

13 a $\lambda = 2, 1, -1$, corresponding eigenvectors are

$\begin{pmatrix}1\\8\\-5\end{pmatrix}, \begin{pmatrix}0\\1\\-1\end{pmatrix}, \begin{pmatrix}0\\1\\-2\end{pmatrix}$

b $\lambda = 1, 9, 10$, corresponding eigenvectors are

$\begin{pmatrix}-3\\0\\1\end{pmatrix}, \begin{pmatrix}13\\-8\\9\end{pmatrix}, \begin{pmatrix}3\\0\\2\end{pmatrix}$

14 a Characteristic equation is $\det(\mathbf{A} - \lambda\mathbf{I}) = 0$

So, in this case: $(a-\lambda)(2a-\lambda)+6=0$

$\Rightarrow \lambda^2 - 3a\lambda + 2a^2 + 6 = 0$ as required

b $\mathbf{A}^2 - 3a\mathbf{A} + (2a^2+6)\mathbf{I}$

$$= \begin{pmatrix} a & 3 \\ -2 & 2a \end{pmatrix}^2 - 3a\begin{pmatrix} a & 3 \\ -2 & 2a \end{pmatrix} + (2a^2+6)\begin{pmatrix} 1 & 0 \\ 0 & 1 \end{pmatrix}$$

$$= \begin{pmatrix} a^2-6 & 9a \\ -6a & 4a^2-6 \end{pmatrix} - \begin{pmatrix} 3a^2 & 9a \\ -6a & 6a^2 \end{pmatrix} + \begin{pmatrix} 2a^2+6 & 0 \\ 0 & 2a^2+6 \end{pmatrix}$$

$$= \begin{pmatrix} a^2-6-3a^2+2a^2+6 & 9a-9a+0 \\ -6a+6a+0 & 4a^2-6-6a^2+2a^2+6 \end{pmatrix}$$

$= \begin{pmatrix} 0 & 0 \\ 0 & 0 \end{pmatrix}$ as required

15 a $\det\begin{pmatrix} -1-\lambda & 0 & 2 \\ -1 & 2-\lambda & 0 \\ -4 & 0 & 3-\lambda \end{pmatrix} = (2-\lambda)[(-1-\lambda)(3-\lambda)--8]$

$(2-\lambda)(\lambda^2-2\lambda+5)=0 \Rightarrow \lambda = 2 \text{ or } \lambda^2-2\lambda+5=0$

Discriminant is $(-2)^2 - 4(5) = -16 < 0$ so no real solutions.

b $\begin{pmatrix}0\\1\\0\end{pmatrix}$

16 $\mathbf{P} = \begin{pmatrix} 0 & 4 & -1 \\ 0 & -10 & 0 \\ 1 & 1 & 1 \end{pmatrix}, \mathbf{D} = \begin{pmatrix} 0 & 0 & 0 \\ 0 & 2 & 0 \\ 0 & 0 & -3 \end{pmatrix}$

17 Eigenvalues are $\lambda = 1, 0, 3$

Eigenvectors are $\begin{pmatrix}0\\0\\1\end{pmatrix}, \begin{pmatrix}1\\-1\\2\end{pmatrix}, \begin{pmatrix}2\\1\\1\end{pmatrix}$

$$\mathbf{M}^n = \frac{1}{3}\begin{pmatrix} 0 & 1 & 2 \\ 0 & -1 & 1 \\ 1 & 2 & 1 \end{pmatrix}\begin{pmatrix} 1 & 0 & 0 \\ 0 & 0 & 0 \\ 0 & 0 & 3 \end{pmatrix}^n\begin{pmatrix} -3 & -3 & 3 \\ 1 & -2 & 0 \\ 1 & 1 & 0 \end{pmatrix}$$

$$= \frac{1}{3}\begin{pmatrix} 0 & 1 & 2 \\ 0 & -1 & 1 \\ 1 & 2 & 1 \end{pmatrix}\begin{pmatrix} 1 & 0 & 0 \\ 0 & 0 & 0 \\ 0 & 0 & 3^n \end{pmatrix}\begin{pmatrix} -3 & -3 & 3 \\ 1 & -2 & 0 \\ 1 & 1 & 0 \end{pmatrix}$$

$$= \frac{1}{3}\begin{pmatrix} 0 & 0 & 2(3^n) \\ 0 & 0 & 3^n \\ 1 & 0 & 3^n \end{pmatrix}\begin{pmatrix} -3 & -3 & 3 \\ 1 & -2 & 0 \\ 1 & 1 & 0 \end{pmatrix}$$

$$= \frac{1}{3}\begin{pmatrix} 2(3^n) & 2(3^n) & 0 \\ 3^n & 3^n & 0 \\ 3^n-3 & 3^n+3 & 3 \end{pmatrix}$$

$$= \begin{pmatrix} 2(3^{n-1}) & 2(3^{n-1}) & 0 \\ 3^{n-1} & 3^{n-1} & 0 \\ 3^{n-1}-1 & 3^{n-1}+1 & 1 \end{pmatrix}$$

18 a $\mathbf{r} = s\begin{pmatrix}3\\-1\end{pmatrix}$

b $\mathbf{r} = t\begin{pmatrix}2\\3\end{pmatrix}$

19 a $\lambda = -3, 2$ (repeated)

b Eigenvectors are of the form $\alpha\begin{pmatrix}-1\\2\\1\end{pmatrix}$ or the form $\beta\begin{pmatrix}2\\0\\-1\end{pmatrix} + \gamma\begin{pmatrix}1\\1\\0\end{pmatrix}$

c The eigenvector $\begin{pmatrix}-1\\2\\1\end{pmatrix}$ gives the direction of an invariant line through the origin.

Eigenvectors of the form $\beta\begin{pmatrix}2\\0\\-1\end{pmatrix} + \gamma\begin{pmatrix}1\\1\\0\end{pmatrix}$ define an invariant plane.

Assessment 22

1. (2, 67, 50)

2 a $\pm\frac{\sqrt{3}}{2}$ **b** Rotation of $\pm$ 60° around the x-axis

3 $k < -\sqrt{3}, k > \sqrt{3}$

4 a -2, 8; corresponding eigenvectors are $\begin{pmatrix}3\\4\end{pmatrix}, \begin{pmatrix}1\\-2\end{pmatrix}$

b $\begin{pmatrix} 2 & -3 \\ -8 & 4 \end{pmatrix} - 6\begin{pmatrix} 2 & -3 \\ -8 & 4 \end{pmatrix} - 16\begin{pmatrix} 1 & 0 \\ 0 & 1 \end{pmatrix}$

$$= \begin{pmatrix} 28 & -18 \\ -48 & 40 \end{pmatrix} - \begin{pmatrix} 12 & -18 \\ -48 & 24 \end{pmatrix} - \begin{pmatrix} 16 & 0 \\ 0 & 16 \end{pmatrix}$$

$= \begin{pmatrix} 0 & 0 \\ 0 & 0 \end{pmatrix}$ as required

5 a 1.25 and -2

b The equations are represented by the matrix **M** with $k = 1$. When $k \neq 1.25$ or -2, **M** is not singular. Therefore the equations have a unique solution so the planes have a unique point of intersection.

6 When $n=1$, $\mathbf{A}^n = \begin{pmatrix} 2 & 0 & 1 \\ 0 & 1 & 0 \\ 1 & 0 & 2 \end{pmatrix}^1 = \begin{pmatrix} 2 & 0 & 1 \\ 0 & 1 & 0 \\ 1 & 0 & 2 \end{pmatrix}$

and $\frac{1}{2}\begin{pmatrix} 3^1+1 & 0 & 3^1-1 \\ 0 & 2 & 0 \\ 3^1-1 & 0 & 3^1+1 \end{pmatrix} = \begin{pmatrix} 2 & 0 & 1 \\ 0 & 1 & 0 \\ 1 & 0 & 2 \end{pmatrix}$

So true for $n=1$

Assume true for $n=k$ and let $n=k+1$

$$\mathbf{A}^{k+1} = \begin{pmatrix} 2 & 0 & 1 \\ 0 & 1 & 0 \\ 1 & 0 & 2 \end{pmatrix}^{k+1}$$

$$= \begin{pmatrix} 2 & 0 & 1 \\ 0 & 1 & 0 \\ 1 & 0 & 2 \end{pmatrix}^{k}\begin{pmatrix} 2 & 0 & 1 \\ 0 & 1 & 0 \\ 1 & 0 & 2 \end{pmatrix}$$

$$= \frac{1}{2}\begin{pmatrix} 3^k+1 & 0 & 3^k-1 \\ 0 & 2 & 0 \\ 3^k-1 & 0 & 3^k+1 \end{pmatrix}\begin{pmatrix} 2 & 0 & 1 \\ 0 & 1 & 0 \\ 1 & 0 & 2 \end{pmatrix}$$

$$= \frac{1}{2}\begin{pmatrix} 2(3^k+1)+3^k-1 & 0 & 3^k+1+2(3^k-1) \\ 0 & 2 & 0 \\ 2(3^k-1)+3^k+1 & 0 & 3^k-1+2(3^k+1) \end{pmatrix}$$

$$= \frac{1}{2}\begin{pmatrix} 3(3^k)+1 & 0 & 3(3^k)-1 \\ 0 & 2 & 0 \\ 3(3^k)-1 & 0 & 3(3^k)+1 \end{pmatrix}$$

$$= \frac{1}{2}\begin{pmatrix} 3^{k+1}+1 & 0 & 3^{k+1}-1 \\ 0 & 2 & 0 \\ 3^{k+1}-1 & 0 & 3^{k+1}+1 \end{pmatrix}$$

So true for $n=k$

Since true for $n=1$ and assuming true for $n=k$ implies true for $n=k+1$, therefore true for all $n \in \mathbb{N}$

7 For example, add C3 to C2

$$\begin{vmatrix} 2 & 1 & 2 \\ 4 & 1 & 2 \\ 1 & -1 & 1 \end{vmatrix} = \begin{vmatrix} 2 & 3 & 2 \\ 4 & 3 & 2 \\ 1 & 0 & 1 \end{vmatrix}$$

multiply R3 by 2

$$= \frac{1}{2}\begin{vmatrix} 2 & 3 & 2 \\ 4 & 3 & 2 \\ 2 & 0 & 2 \end{vmatrix}$$

subtract R3 from R1

$$= \frac{1}{2}\begin{vmatrix} 0 & 3 & 0 \\ 4 & 3 & 2 \\ 2 & 0 & 2 \end{vmatrix}$$

divide C2 by 3

$$= \frac{3}{2}\begin{vmatrix} 0 & 1 & 0 \\ 4 & 1 & 2 \\ 2 & 0 & 2 \end{vmatrix}$$

divide C1 by 2

$$= 3\begin{vmatrix} 0 & 1 & 0 \\ 2 & 1 & 2 \\ 1 & 0 & 2 \end{vmatrix}$$

divide C3 by 2

$$= 6\begin{vmatrix} 0 & 1 & 0 \\ 2 & 1 & 1 \\ 1 & 0 & 1 \end{vmatrix}$$

subtract R3 from R2

$$= 6\begin{vmatrix} 0 & 1 & 0 \\ 1 & 1 & 0 \\ 1 & 0 & 1 \end{vmatrix}$$

8 $\begin{pmatrix} \frac{9}{a} & \frac{1}{2} & -3 \\ \frac{12}{a} & 1 & 2 \\ \frac{3}{a} & 0 & -1 \end{pmatrix}$

9 **a** 180 **b** −30

10 **a** $(b-a)(c-a)(c-b)$ **b** 2

11 **a** $\det\begin{pmatrix} 1-\lambda & 2 & 0 \\ 0 & -\lambda & 3 \\ 2 & 0 & 1-\lambda \end{pmatrix} = 0$

$(1-\lambda)(-\lambda)(1-\lambda) - 2(0-6) + 0 = 0$

$-\lambda^3 + 2\lambda^2 - \lambda + 12 = 0$

$(\lambda-3)(-\lambda^2-\lambda-4) = 0$

$\lambda = 3$ or $-\lambda^2 - \lambda - 4 = 0$

$b^2 - 4ac = (-1)^2 - 4(-1)(-4) = -15$

So no further real solutions.

Hence 3 is the only real eigenvalue.

b $\begin{pmatrix} \frac{1}{\sqrt{3}} \\ \frac{1}{\sqrt{3}} \\ \frac{1}{\sqrt{3}} \end{pmatrix}$ **c** $\mathbf{r} = s\begin{pmatrix} 1 \\ 1 \\ 1 \end{pmatrix}$

12 2, −1 (repeated)

13 **a** $\mathbf{P} = \begin{pmatrix} 1 & 3 & 0 \\ -2 & 2 & 1 \\ 1 & -3 & 0 \end{pmatrix}$ $\mathbf{D} = \begin{pmatrix} 2 & 0 & 0 \\ 0 & -2 & 0 \\ 0 & 0 & 1 \end{pmatrix}$

b $\mathbf{P}^{-1} = \frac{1}{6}\begin{pmatrix} 3 & 0 & 3 \\ 1 & 0 & -1 \\ 4 & 6 & 8 \end{pmatrix}$

$\mathbf{M}^4 = \mathbf{P}\mathbf{D}^4\mathbf{P}^{-1}$

$$= \begin{pmatrix} 1 & 3 & 0 \\ -2 & 2 & 1 \\ 1 & -3 & 0 \end{pmatrix}\begin{pmatrix} 2^4 & 0 & 0 \\ 0 & (-2)^4 & 0 \\ 0 & 0 & 1^4 \end{pmatrix}\frac{1}{6}\begin{pmatrix} 3 & 0 & 3 \\ 1 & 0 & -1 \\ 4 & 6 & 8 \end{pmatrix}$$

$$= \begin{pmatrix} 16 & 0 & 0 \\ -10 & 1 & -20 \\ 0 & 0 & 16 \end{pmatrix}$$

c The eigenvector $\begin{pmatrix} 0 \\ 1 \\ 0 \end{pmatrix}$ which represents the direction of the y-axis corresponds to an eigenvalue of 1

14 First equation × 2 gives: $2x+4y-6z=-4$

Add to third equation to give: $4x-z=-3$

But $4x-z=3$ so inconsistent.

Hence they do not intersect.

The three planes form a triangular prism.

15 $\begin{pmatrix}1\\1\\0\end{pmatrix}$ gives the direction of an invariant line, $\begin{pmatrix}0\\-1\\1\end{pmatrix}$, $\begin{pmatrix}1\\2\\-2\end{pmatrix}$ define an invariant plane.

Chapter 23

Exercise 23.1A

1 a $\mathbf{j}\times\mathbf{k}=\mathbf{i}$
 b $\mathbf{i}\times\mathbf{k}=-\mathbf{k}\times\mathbf{i}=-\mathbf{j}$
 c $\mathbf{k}\times\mathbf{k}=0$
 d $(\mathbf{i}+\mathbf{j})\times\mathbf{j}=\mathbf{i}\times\mathbf{j}+\mathbf{j}\times\mathbf{j}=\mathbf{k}$

2 a $\mathbf{a}\times\mathbf{b}$
 b $\mathbf{a}\times(\mathbf{b}+\mathbf{c})$
 c $-2\mathbf{a}\times\mathbf{b}$
 d $2\mathbf{b}\times(\mathbf{a}+\mathbf{c})$

3 a $\mathbf{a}\times\mathbf{b}+\mathbf{b}\times\mathbf{c}+\mathbf{a}\times\mathbf{c}=-\mathbf{b}\times\mathbf{a}+\mathbf{b}\times\mathbf{c}+\mathbf{a}\times\mathbf{c}$
 $=\mathbf{b}\times(\mathbf{c}-\mathbf{a})+\mathbf{a}\times\mathbf{c}$
 $=(\mathbf{b}+\mathbf{a})\times(\mathbf{c}-\mathbf{a})$ (since $\mathbf{a}\times\mathbf{a}=0$)
 b $\mathbf{b}\times(2\mathbf{a}+\mathbf{c})-\mathbf{c}\times(\mathbf{b}-\mathbf{c})=\mathbf{b}\times(2\mathbf{a})+\mathbf{b}\times\mathbf{c}-\mathbf{c}\times\mathbf{b}-\mathbf{c}\times(-\mathbf{c})$
 $=2\mathbf{b}\times\mathbf{a}+\mathbf{b}\times\mathbf{c}+\mathbf{b}\times\mathbf{c}+\mathbf{c}\times\mathbf{c}$
 $=2\mathbf{b}\times\mathbf{a}+2\mathbf{b}\times\mathbf{c}$
 $=2\mathbf{b}\times(\mathbf{a}+\mathbf{c})$

4 0° or 180°

5 a $\mathbf{i}+5\mathbf{j}+13\mathbf{k}$
 b $-15\mathbf{i}-9\mathbf{j}+3\mathbf{k}$
 c $26\mathbf{i}-2\mathbf{j}+8\mathbf{k}$
 d $-\mathbf{i}-5\mathbf{j}-13\mathbf{k}$
 e $15\mathbf{i}+9\mathbf{j}-3\mathbf{k}$
 f $-26\mathbf{i}+2\mathbf{j}-8\mathbf{k}$
 g 0
 h 0

6 a $\begin{pmatrix}3\\-15\\-6\end{pmatrix}$
 b $25\mathbf{i}-8\mathbf{j}-26\mathbf{k}$
 c $3\mathbf{i}+8\mathbf{j}+28\mathbf{k}$
 d $\begin{pmatrix}-22\\8\\-43\end{pmatrix}$
 e $\begin{pmatrix}-4\sqrt{3}\\3\\3\end{pmatrix}$
 f $\begin{pmatrix}10a^2+1\\5a^2-a\\2a^2+a\end{pmatrix}$

7 $\begin{pmatrix}29\\-5\\7\end{pmatrix}$

8 $9\mathbf{i}-25\mathbf{j}-21\mathbf{k}$

9 $\begin{pmatrix}0\\-\frac{3}{\sqrt{10}}\\-\frac{1}{\sqrt{10}}\end{pmatrix}$

10 $\begin{pmatrix}\frac{1}{\sqrt{6}}\\\frac{2}{\sqrt{6}}\\\frac{1}{\sqrt{6}}\end{pmatrix}$

11 $(-2\mathbf{i}+6\mathbf{j}-4\mathbf{k})\times(3\mathbf{i}-9\mathbf{j}+6\mathbf{k})$
 $=(6\times6--4\times-9)\mathbf{i}-(-2\times6--4\times3)\mathbf{j}+(-2\times-9-6\times3)\mathbf{k}$
 $=0\mathbf{i}-0\mathbf{j}+0\mathbf{k}$
 $=0$ so parallel as required

12 a 29.6°
 b 85.8°
 c 40.9°
 d $\theta=41.4$

13 $\frac{3}{\sqrt{14}}$

14 $a=4$
 $b=5$
 $c=-20$

15 $b=-3$
 $a=1$
 $c=6$

16 $\frac{3}{2}\sqrt{35}$ square units

17 10.4 square units

18 a $\frac{17\sqrt{3}}{2}$
 b $\frac{\sqrt{2}}{2}$

Exercise 23.1B

1 Let $\boldsymbol{a}=\begin{pmatrix}a_1\\a_2\\a_3\end{pmatrix}, \boldsymbol{b}=\begin{pmatrix}b_1\\b_2\\b_3\end{pmatrix}, \boldsymbol{c}=\begin{pmatrix}c_1\\c_2\\c_3\end{pmatrix}$

$$\mathbf{a}\times(\mathbf{b}+\mathbf{c})=\begin{pmatrix}a_1\\a_2\\a_3\end{pmatrix}\times\begin{pmatrix}b_1+c_1\\b_2+c_2\\b_3+c_3\end{pmatrix}$$

$$=\begin{pmatrix}a_2(b_3+c_3)-a_3(b_2+c_2)\\a_3(b_1+c_1)-a_1(b_3+c_3)\\a_1(b_2+c_2)-a_2(b_1+c_1)\end{pmatrix}$$

$$=\begin{pmatrix}(a_2b_3-a_3b_2)+(a_2c_3-a_3c_2)\\(a_3b_1-a_1b_3)+(a_3c_1-a_1c_3)\\(a_1b_2-a_2b_1)+(a_1c_2-a_2c_1)\end{pmatrix}$$

$$=\begin{pmatrix}a_2b_3-a_3b_2\\a_3b_1-a_1b_3\\a_1b_2-a_2b_1\end{pmatrix}+\begin{pmatrix}a_2c_3-a_3c_2\\a_3c_1-a_1c_3\\a_1c_2-a_2c_1\end{pmatrix}$$

$=\mathbf{a}\times\mathbf{b}+\mathbf{a}\times\mathbf{c}$ as required

2 $\mathbf{a}\times\mathbf{b}=\begin{pmatrix}a_1\\a_2\\a_3\end{pmatrix}\times\begin{pmatrix}b_1\\b_2\\b_3\end{pmatrix}$

$=\begin{pmatrix}a_2b_3-a_3b_2\\a_3b_1-a_1b_3\\a_1b_2-a_2b_1\end{pmatrix}$

$=-\begin{pmatrix}a_3b_2-a_2b_3\\a_1b_3-a_3b_1\\a_2b_1-a_1b_2\end{pmatrix}$

$=-\begin{pmatrix}b_1\\b_2\\b_3\end{pmatrix}\times\begin{pmatrix}a_1\\a_2\\a_3\end{pmatrix}$

$=-\mathbf{b}\times\mathbf{a}$ as required

3 $\left(\mathbf{r}-\begin{pmatrix}2\\5\\0\end{pmatrix}\right)\times\begin{pmatrix}1\\1\\-1\end{pmatrix}=0$

4 $\left(\mathbf{r}-\begin{pmatrix}0\\3\\0\end{pmatrix}\right)\times\begin{pmatrix}0\\0\\1\end{pmatrix}=0$

5 $\mathbf{r}\times\begin{pmatrix}1\\0\\1\end{pmatrix}=\begin{pmatrix}0\\-2\\0\end{pmatrix}$

6 $\left(\mathbf{r}-\begin{pmatrix}4\\2\\7\end{pmatrix}\right)\times\begin{pmatrix}3\\-1\\2\end{pmatrix}=0$ or $\left(\mathbf{r}-\begin{pmatrix}1\\3\\5\end{pmatrix}\right)\times\begin{pmatrix}3\\-1\\2\end{pmatrix}=0$

7 e.g. $\begin{pmatrix}0\\3\\-2\end{pmatrix}-\begin{pmatrix}1\\2\\-1\end{pmatrix}=\begin{pmatrix}-1\\1\\-1\end{pmatrix}$ (or could subtract other way round)

$\mathbf{r}\times\begin{pmatrix}-1\\1\\-1\end{pmatrix}=\begin{pmatrix}0\\3\\-2\end{pmatrix}\times\begin{pmatrix}-1\\1\\-1\end{pmatrix}$

$\mathbf{r}\times\begin{pmatrix}-1\\1\\-1\end{pmatrix}=\begin{pmatrix}-1\\2\\3\end{pmatrix}$

Alternatively: $\mathbf{r}\times\begin{pmatrix}1\\-1\\1\end{pmatrix}=\begin{pmatrix}1\\-2\\-3\end{pmatrix}$

8 a $\left(\mathbf{r}-\begin{pmatrix}2\\0\\-1\end{pmatrix}\right)\times\begin{pmatrix}0\\1\\5\end{pmatrix}=0$

b $\left(\mathbf{r}-\begin{pmatrix}3\\1\\-4\end{pmatrix}\right)\times\begin{pmatrix}5\\3\\-6\end{pmatrix}=0$

c $\left(\mathbf{r}-\begin{pmatrix}1\\-3\\0\end{pmatrix}\right)\times\begin{pmatrix}-2\\4\\-2\end{pmatrix}=0$

d $\left(\mathbf{r}-\begin{pmatrix}0\\0\\1\end{pmatrix}\right)\times\begin{pmatrix}1\\-3\\2\end{pmatrix}=0$

9 $\begin{pmatrix}-8\\-6\\2\end{pmatrix}=-2\begin{pmatrix}4\\3\\-1\end{pmatrix}$ so they are parallel

$\begin{pmatrix}1\\3\\5\end{pmatrix}$ is on line A so check B: $\begin{pmatrix}1\\3\\5\end{pmatrix}=\begin{pmatrix}5\\6\\4\end{pmatrix}+\lambda\begin{pmatrix}-8\\-6\\2\end{pmatrix}$

$1=5-8\lambda\Rightarrow\lambda=\frac{1}{2}$

$3=6-6\lambda\Rightarrow\lambda=\frac{1}{2}$

$5=4+2\lambda\Rightarrow\lambda=\frac{1}{2}$

Therefore (1, 3, 5) satisfies both equations so they do represent the same line.

10 $\begin{pmatrix}-3\\6\\-9\end{pmatrix}=\frac{3}{2}\begin{pmatrix}-2\\4\\-6\end{pmatrix}$ so they are parallel

(4, –3, 8) satisfies B so check A:

$\frac{4-3}{-2}=-\frac{1}{2}$

$\frac{-3+1}{4}=-\frac{1}{2}$

$\frac{8-5}{-6}=-\frac{1}{2}$ so (4, –3, 8) satisfies both equations therefore they represent the same line.

11 $\frac{x-1}{1}=\frac{y-3}{0}=\frac{z+1}{-1}$

12 $\mathbf{r}=\begin{pmatrix}2\\1\\1\end{pmatrix}+t\begin{pmatrix}-2\\-1\\0\end{pmatrix}$

13 a $\begin{pmatrix}1\\-2\\1\end{pmatrix}+s\begin{pmatrix}2\\1\\-2\end{pmatrix}=\begin{pmatrix}-3\\-11\\5\end{pmatrix}+t\begin{pmatrix}-1\\3\\1\end{pmatrix}$

$\mathbf{i}: 1+2s=-3-t\Rightarrow t=-4-2s$

$\mathbf{j}: -2+s=-11+3t\Rightarrow 3t=9+s$

$\Rightarrow 3(-4-2s)=9+s$

$\Rightarrow -12-6s=9+s$

$\Rightarrow 7s=-21$

$\Rightarrow s=-3$

$\Rightarrow t=-4-2(-3)=2$

$\mathbf{k}: 1-2s=1-2(-3)=7$

$5+t=5+2=7$ so they intersect when $t=2$

$\begin{pmatrix}-3\\-11\\5\end{pmatrix}+2\begin{pmatrix}-1\\3\\1\end{pmatrix}=\begin{pmatrix}-5\\-5\\7\end{pmatrix}$

They intersect at (–5, –5, 7)

b $\frac{7}{33}\sqrt{22}$

14 a $\left(\begin{pmatrix}-2\\16\\-12\end{pmatrix}-\begin{pmatrix}0\\4\\-2\end{pmatrix}\right)\times\begin{pmatrix}1\\-6\\5\end{pmatrix}=\begin{pmatrix}-2\\12\\-10\end{pmatrix}\times\begin{pmatrix}1\\-6\\5\end{pmatrix}$

$=\begin{pmatrix}60-60\\10-10\\12-12\end{pmatrix}=\begin{pmatrix}0\\0\\0\end{pmatrix}$

as required so point A lies on the line

b (1, −2, 3)

c $3\sqrt{21}$

15 $\frac{-2}{\sqrt{105}}\mathbf{i}+\frac{1}{\sqrt{105}}\mathbf{j}+\frac{10}{\sqrt{105}}\mathbf{k}$

Exercise 23.2A

1 $\mathbf{r}=\begin{pmatrix}1\\5\\2\end{pmatrix}+s\begin{pmatrix}1\\1\\0\end{pmatrix}+t\begin{pmatrix}0\\1\\2\end{pmatrix}$

2 $\mathbf{r}=\begin{pmatrix}0\\6\\2\end{pmatrix}+s\begin{pmatrix}2\\0\\-3\end{pmatrix}+t\begin{pmatrix}5\\-2\\4\end{pmatrix}$

3 a $\mathbf{r}=\begin{pmatrix}3\\1\\0\end{pmatrix}+s\begin{pmatrix}1\\-3\\2\end{pmatrix}+t\begin{pmatrix}8\\1\\-4\end{pmatrix}$

b $\mathbf{n}=\begin{pmatrix}1\\-3\\2\end{pmatrix}\times\begin{pmatrix}8\\1\\4\end{pmatrix}=\begin{pmatrix}10\\20\\25\end{pmatrix}=5\begin{pmatrix}2\\4\\5\end{pmatrix}$

$\begin{pmatrix}3\\1\\0\end{pmatrix}\cdot\begin{pmatrix}2\\4\\5\end{pmatrix}=6+4=10$

$\mathbf{r}\cdot(2\mathbf{i}+4\mathbf{j}+5\mathbf{k})=10$

c $2x+4y+5z-10=0$

4 a $\mathbf{r}=\begin{pmatrix}-2\\0\\0\end{pmatrix}+s\begin{pmatrix}7\\1\\1\end{pmatrix}+t\begin{pmatrix}8\\2\\-5\end{pmatrix}$

b $\mathbf{n}=\begin{pmatrix}7\\1\\1\end{pmatrix}\times\begin{pmatrix}8\\2\\-5\end{pmatrix}=\begin{pmatrix}-7\\43\\6\end{pmatrix}$

$\begin{pmatrix}-2\\0\\0\end{pmatrix}\cdot\begin{pmatrix}-7\\43\\6\end{pmatrix}=14$

$\mathbf{r}\cdot(-7\mathbf{i}+43\mathbf{j}+6\mathbf{k})=14$

c $-7x+43y+6z-14=0$

5 a $\mathbf{r}=\begin{pmatrix}0\\3\\2\end{pmatrix}+s\begin{pmatrix}1\\-5\\-2\end{pmatrix}+t\begin{pmatrix}-1\\-1\\-3\end{pmatrix}$

b $\mathbf{n}=\begin{pmatrix}1\\-5\\-2\end{pmatrix}\times\begin{pmatrix}-1\\-1\\-3\end{pmatrix}=\begin{pmatrix}13\\5\\-6\end{pmatrix}$

$\begin{pmatrix}0\\3\\2\end{pmatrix}\cdot\begin{pmatrix}13\\5\\-6\end{pmatrix}=3$

$\mathbf{r}\cdot(13\mathbf{i}+5\mathbf{j}-6\mathbf{k})=3$

c $13x+5y-6z-3=0$

6 $\mathbf{r}\cdot\begin{pmatrix}1\\0\\2\end{pmatrix}=11$

7 $\mathbf{r}\cdot\begin{pmatrix}3\\-1\\4\end{pmatrix}=10$

8 $\begin{pmatrix}1\\1\\1\end{pmatrix}\cdot\begin{pmatrix}0\\2\\-3\end{pmatrix}=0+2-3=-1$

$\mathbf{r}\cdot\begin{pmatrix}0\\2\\-3\end{pmatrix}=-1$

9 $5x+4y-2z=18$

10 $2x-5z=-8$

11 $x-y+4z=31$

12 a $\mathbf{r}\cdot\begin{pmatrix}2\\1\\-3\end{pmatrix}=-2$

b $2x+y-3z=-2$

13 a $r\cdot\begin{pmatrix}3\\-2\\5\end{pmatrix}=-4$

b $3x-2y+5z=-4$

14 a $3x+5y-z=-2$

b $7x-y+8z=3$

c $23x+13y-16z-139=0$

d $-8x-7y+2z+32=0$

15 a $\mathbf{r}\cdot\begin{pmatrix}9\\3\\-1\end{pmatrix}=5$

b $\mathbf{r}\cdot\begin{pmatrix}2\\-7\\-15\end{pmatrix}=-4$

c $\mathbf{r}\cdot\begin{pmatrix}4\\16\\4\end{pmatrix}=12$

d $\mathbf{r}\cdot\begin{pmatrix}-24\\-15\\-5\end{pmatrix}=-149$

Exercise 23.2B

1 a If there exist s and t such that

$\begin{pmatrix}2\\-3\\-7\end{pmatrix}+s\begin{pmatrix}-1\\5\\3\end{pmatrix}+t\begin{pmatrix}7\\-3\\-1\end{pmatrix}=\begin{pmatrix}-7\\10\\0\end{pmatrix}$

Then, taking **i**: $2-s+7t=-7\Rightarrow s=7t+9$

j: $-3+5s-3t=10\Rightarrow 5s=13+3t$

Substitute for s to give $5(7t+9)=13+3t$

$\Rightarrow 32t=-32$

$\Rightarrow t=-1,\ s=2$

k: $-7+2\times 3-1\times -1=0$ as required

So (−7, 10, 0) does lie on this plane.

b If there exist s and t such that

$$\begin{pmatrix}2\\0\\8\end{pmatrix}+s\begin{pmatrix}1\\3\\-2\end{pmatrix}+t\begin{pmatrix}-2\\4\\0\end{pmatrix}=\begin{pmatrix}-7\\10\\0\end{pmatrix}$$

Then, taking $\mathbf{k}: 8-2s=0 \Rightarrow s=4$

$\mathbf{j}: 3\times 4+4t=10 \Rightarrow t=-\dfrac{1}{2}$

$\mathbf{i}: 2+4-\dfrac{1}{2}\times -2=7$ not -7 so $(-7, 10, 0)$ not on the plane.

c $\begin{pmatrix}-7\\10\\0\end{pmatrix}\cdot\begin{pmatrix}1\\-1\\3\end{pmatrix}=-7\times 1+10\times -1+0\times 3=-17$

So $(-7, 10, 0)$ not on the plane.

d $\begin{pmatrix}-7\\10\\0\end{pmatrix}\cdot\begin{pmatrix}2\\3\\-8\end{pmatrix}=-7\times 2+10\times 3+0\times 8=16$ as required

So $(-7, 10, 0)$ does lie on this plane.

2 a $\begin{pmatrix}x\\y\\z\end{pmatrix}\cdot\begin{pmatrix}2\\3\\-1\end{pmatrix}=2x+3y-z=2$ so they represent the same plane.

b Check if $(3, 0, -1)$ satisfies second equation:

$$\begin{pmatrix}1\\-4\\4\end{pmatrix}+\lambda\begin{pmatrix}2\\4\\5\end{pmatrix}+\mu\begin{pmatrix}5\\3\\0\end{pmatrix}=\begin{pmatrix}3\\0\\-1\end{pmatrix}$$

$\mathbf{k}: 4-5\lambda=-1 \Rightarrow \lambda=-1$

$\mathbf{i}: 1-2+5\mu=3 \Rightarrow \mu=\dfrac{4}{5}$

$\mathbf{j}: -4-4+3\left(\dfrac{4}{5}\right)=-\dfrac{28}{5}$ (not 0) so not the same plane.

c Check if $(1, 0, -7)$ satisfies second equation:

$$\begin{pmatrix}-4\\2\\-7\end{pmatrix}+\lambda\begin{pmatrix}0\\1\\3\end{pmatrix}+\mu\begin{pmatrix}5\\-2\\0\end{pmatrix}=\begin{pmatrix}1\\0\\-7\end{pmatrix}$$

$\mathbf{i}: -4+5\mu=1 \Rightarrow \mu=1$

$\mathbf{j}: 2+\lambda-2=0 \Rightarrow \lambda=0$

$\mathbf{k}: -7+0+0=-7$ so on the same plane.

Check if normals are parallel.

Normal to first plane:

$$\begin{pmatrix}5\\-2\\4\end{pmatrix}\times\begin{pmatrix}0\\0\\1\end{pmatrix}=\begin{pmatrix}-2\\-5\\0\end{pmatrix}$$

Normal to second plane:

$$\begin{pmatrix}0\\1\\3\end{pmatrix}\times\begin{pmatrix}5\\-2\\0\end{pmatrix}=\begin{pmatrix}6\\15\\-5\end{pmatrix}$$

This is not a multiple of normal to first plane so not the same plane.

d Check if $(-2, 1, 0)$ satisfies second equation:

$\begin{pmatrix}-2\\1\\0\end{pmatrix}\cdot\begin{pmatrix}6\\-2\\-1\end{pmatrix}=-12-2=-14$

For first plane $\mathbf{n}=$

$$\begin{pmatrix}4\\3\\2\end{pmatrix}\times\begin{pmatrix}-5\\0\\1\end{pmatrix}=\begin{pmatrix}3\\-14\\15\end{pmatrix}$$

which is not equal to $\lambda\begin{pmatrix}6\\-2\\-1\end{pmatrix}$ for any λ

So not the same plane.

e Check if $(5, 0, 4)$ satisfies first equation:

$\begin{pmatrix}5\\0\\4\end{pmatrix}\cdot\begin{pmatrix}2\\-7\\-3\end{pmatrix}=10+0-12=-2$

For second plane

$$\mathbf{n}=\begin{pmatrix}0\\3\\-7\end{pmatrix}\times\begin{pmatrix}2\\1\\-1\end{pmatrix}=\begin{pmatrix}4\\-14\\-6\end{pmatrix}=2\begin{pmatrix}2\\-7\\-3\end{pmatrix}$$

so the planes are parallel and share a point so the equations represent the same plane.

f Check if $(1, 0, -2)$ satisfies first equation:

$\begin{pmatrix}1\\0\\-2\end{pmatrix}\cdot\begin{pmatrix}1\\3\\1\end{pmatrix}=1+0-2=-1$

Setting $\lambda=0$ and $\mu=1$gives point $(5, 3, -5)$ on second equation

Check if $(5, 3, -5)$ satisfies first equation:

$\begin{pmatrix}5\\3\\-5\end{pmatrix}\cdot\begin{pmatrix}1\\3\\1\end{pmatrix}=5+9-5=9$ (not -1)

Not the same plane.

3 21.1°

4 173°

5 2.72°

6 164.6°

7 39.5°

8 61.9°

9 110°

10 $\mathbf{r}\cdot\begin{pmatrix}18\\-5\\6\end{pmatrix}=59$

11 $\mathbf{r}\cdot\begin{pmatrix}-1\\-2\\1\end{pmatrix}=-8$

12 $2x+6y-13z=-37$

13 a $\mathbf{r}\cdot\begin{pmatrix}2\\27\\13\end{pmatrix}=19$

b $\left(\mathbf{r}-\begin{pmatrix}1\\4\\-2\end{pmatrix}\right)\times\begin{pmatrix}2\\27\\13\end{pmatrix}=0$

Exercise 23.3A

1 a $\dfrac{7}{6}$ units

b $\dfrac{11}{14}\sqrt{14}$ units

c $\dfrac{13}{15}\sqrt{3}$ units

d 0.4units

e $\frac{5}{11}\sqrt{11}$ units

2 $\begin{pmatrix}-11\\-4\\10\end{pmatrix}+t\begin{pmatrix}4\\-2\\0\end{pmatrix}=\begin{pmatrix}3\\-5\\-10\end{pmatrix}+\mu\begin{pmatrix}7\\-10\\0\end{pmatrix}+\lambda\begin{pmatrix}1\\-3\\4\end{pmatrix}$

$\mathbf{k}: 10=-10+4\lambda \Rightarrow \lambda=5$

$\mathbf{i}: -11+4t=3+7\mu+5$

$\Rightarrow 4t=19+7\mu$

$\mathbf{j}: -4-2t=-5-10\mu-15$

$\Rightarrow 2t=16+10\mu$

$\times 2$ to give $4t=32+20\mu$

Substitute into $4t=19+7\mu$ to give $32+20\mu=19+7\mu$

$\Rightarrow \mu=-1$

$\Rightarrow t=3$

$\begin{pmatrix}-11\\-4\\10\end{pmatrix}+3\begin{pmatrix}4\\-2\\0\end{pmatrix}=\begin{pmatrix}1\\-10\\10\end{pmatrix}$

and $\begin{pmatrix}3\\-5\\-10\end{pmatrix}-1\begin{pmatrix}7\\-10\\0\end{pmatrix}+5\begin{pmatrix}1\\-3\\4\end{pmatrix}=\begin{pmatrix}1\\-10\\10\end{pmatrix}$

So they intersect at the point (1, −10, 10)

3 (6, −2, 20)

4 (21, −6, −3)

5 (−10, 0, 11)

6 (11, 12, −5)

7 2 units

8 Distance $=\left|\frac{n_1\alpha+n_2\beta+n_3\gamma+d}{\sqrt{n_1^2+n_2^2+n_3^2}}\right|$

$=\left|\frac{n_1\times 0+n_2\times 0+n_3\times 0-p}{\sqrt{n_1^2+n_2^2+n_3^2}}\right|$

$=\left|-\frac{p}{|\mathbf{n}|}\right|$

$=\frac{p}{|\mathbf{n}|}$ as required

9 **a** So they intersect when $\lambda=0$
Which is point (7, −6, −2)

b Lines are parallel

c Skew lines

d Intersect at (−4, 2, 0)

10 $3\sqrt{10}$

11 $\frac{16}{\sqrt{14}}$

Exercise 23.3B

1 **a** **i** $\left|\frac{-20\sqrt{51}}{51}\right|$

ii $\left|\frac{-20\sqrt{51}}{51}\right|$

b Since A and B are the same distance from Π and are the same side of Π the line l must be parallel to the plane therefore it doesn't intersect it.

2 **a** **i** 1.42 units

ii 0.291 units

iii 3.13 units

b Each pair of planes is parallel, hence they do not meet and we can find the perpendicular distance between them since it is constant.

3 $\sqrt{54}$ (7.35)

4 11.67 square units

5 $=\frac{27}{2}\sqrt{6}$ square units

6 **a** $\frac{11}{3}$ (3.67)

b $\frac{78}{\sqrt{346}}$ (4.19)

c $\frac{29}{\sqrt{30}}$ (5.29)

d $\sqrt{2}$ (1.41)

e $\sqrt{5}$ (2.24)

7 $\sqrt{2}$ (1.41)

8 $\left(\frac{13}{7},\frac{46}{7},-\frac{37}{7}\right)$

9 $\left(\frac{8}{11},\frac{14}{11},-\frac{35}{11}\right)$

Review exercise 23

1 **a** $\begin{pmatrix}-26\\-2\\-18\end{pmatrix}$

b $\begin{pmatrix}6\\36\\12\end{pmatrix}$

2 **a** 82.8°

b 78.2°

3 $\pm\begin{pmatrix}\frac{3}{\sqrt{35}}\\\frac{5}{\sqrt{35}}\\\frac{1}{\sqrt{35}}\end{pmatrix}$

4 $\frac{3}{2}\sqrt{6}$ units²

5 $\mathbf{r}\times\begin{pmatrix}-5\\3\\-1\end{pmatrix}=\begin{pmatrix}-11\\-15\\10\end{pmatrix}$

6 $\mathbf{r}\times\begin{pmatrix}0\\3\\-4\end{pmatrix}=\begin{pmatrix}8\\4\\3\end{pmatrix}$

7 $\mathbf{r}=\begin{pmatrix}2\\0\\-3\end{pmatrix}+s\begin{pmatrix}-2\\0\\3\end{pmatrix}+t\begin{pmatrix}1\\-4\\-1\end{pmatrix}$

8 **a** $\mathbf{r}\cdot\begin{pmatrix}7\\-5\\1\end{pmatrix}=-15$

b $7x-5y+z=-15$

9 **a** 32.3°

b $\left(-\frac{1}{3},-2,\frac{2}{3}\right)$

c $\frac{\sqrt{41}}{3}$ (2.13)

10 **a** (−1, −1, −6)

b 168°

11 $-\frac{11}{\sqrt{238}}$

12 153°

13 2.59 units

14 0.463 units

15 $\frac{2\sqrt{3}}{3}$

Assessment 23

1 a $-\mathbf{i}-2\mathbf{j}+\mathbf{k}$

b $\frac{1}{\sqrt{6}}(-\mathbf{i}-2\mathbf{j}+\mathbf{k})$

2 a $a=2, b=7, c=-3$

b i $\mathbf{r}=\begin{pmatrix}2\\7\\-3\end{pmatrix}+t\begin{pmatrix}2\\-3\\1\end{pmatrix}$

ii $\left(\mathbf{r}-\begin{pmatrix}2\\7\\-3\end{pmatrix}\right)\times\begin{pmatrix}2\\-3\\1\end{pmatrix}=0$

3 $\left(\mathbf{r}-\begin{pmatrix}1\\2\\-3\end{pmatrix}\right)\times\begin{pmatrix}2\\0\\-1\end{pmatrix}=0$

4 a $\begin{pmatrix}3\\1\\-2\end{pmatrix}\cdot\begin{pmatrix}0\\4\\-2\end{pmatrix}=0+4+4$

$=8$ so lies on this plane

b $2(3)+3(1)-(-2)=11$

so lies on this plane

5 21.7 square units

6 a $\mathbf{r}=\begin{pmatrix}1\\-5\\2\end{pmatrix}+\mathbf{s}\begin{pmatrix}0\\-3\\-1\end{pmatrix}+\mathbf{t}\begin{pmatrix}-2\\4\\0\end{pmatrix}$

b $\begin{pmatrix}2\\1\\-3\end{pmatrix}\cdot\begin{pmatrix}0\\-3\\-1\end{pmatrix}=0-3+3=0$

So perpendicular to $\begin{pmatrix}0\\-3\\-1\end{pmatrix}$

$\begin{pmatrix}2\\1\\-3\end{pmatrix}\cdot\begin{pmatrix}-2\\4\\0\end{pmatrix}=-4+4+0=0$

So perpendicular to $\begin{pmatrix}-2\\4\\0\end{pmatrix}$

Hence perpendicular to plane

c i $\mathbf{r}\cdot\begin{pmatrix}2\\1\\-3\end{pmatrix}=-9$ ii $2x+y-3z=-9$

7 a $\mathbf{r}=\begin{pmatrix}7\\12\\-14\end{pmatrix}+s\begin{pmatrix}2\\8\\-4\end{pmatrix}+t\begin{pmatrix}2\\1\\-3\end{pmatrix}$

b If there exist s and t such that

$\begin{pmatrix}-1\\0\\4\end{pmatrix}=\begin{pmatrix}7\\12\\-14\end{pmatrix}+s\begin{pmatrix}2\\8\\-4\end{pmatrix}+t\begin{pmatrix}2\\1\\-3\end{pmatrix}$ then

$7+2s+2t=-1\Rightarrow s+t=-4$

$12+8s+t=0\Rightarrow 8s+t=-12$

$\Rightarrow s=-\frac{8}{7}$

$\Rightarrow t=-\frac{20}{7}$

Check for **k** component:

$-14-\frac{8}{7}(-4)-\frac{20}{7}(-3)=-\frac{6}{7}$ not 4

So does not lie on the plane

8 a 2.78

b $\left(\frac{17}{11},-1,\frac{18}{11}\right)$

c 0.664 rad (or 38.0°)

9 $\left(-\frac{2}{3},\frac{7}{3},\frac{1}{3}\right)$

10 a $\begin{pmatrix}5-8\lambda\\5-6\lambda\\2+2\lambda\end{pmatrix}\times\begin{pmatrix}1\\0\\-2\end{pmatrix}$

$=\begin{pmatrix}-2(5-6\lambda)-0\\(2+2\lambda)--2(5-8\lambda)\\0-(5-6\lambda)\end{pmatrix}$

$\begin{pmatrix}12\lambda-10\\12-14\lambda\\6\lambda-5\end{pmatrix}=\begin{pmatrix}-4\\5\\-2\end{pmatrix}$

$12\lambda-10=-4\Rightarrow\lambda=\frac{1}{2}$

$12-14\lambda=5\Rightarrow\lambda=\frac{1}{2}$

$6\lambda-5=-2\Rightarrow\lambda=\frac{1}{2}$

Therefore they intersect

b $\begin{pmatrix}1\\2\\3\end{pmatrix}$

11 63.1°

12 a 67.4°

b $\begin{pmatrix}5\\-1\\-2\end{pmatrix}$

13 $\frac{16}{\sqrt{14}}$

14 $\mathbf{r}=\begin{pmatrix}5\\-2\\7\end{pmatrix}+\lambda\begin{pmatrix}3\\-1\\6\end{pmatrix}+\mu\begin{pmatrix}-1\\-6\\8\end{pmatrix}$

15 $\frac{4}{9}\sqrt{26}$ or 2.27

16 a $\mathbf{r}=\begin{pmatrix}-4\\7\\-5\end{pmatrix}+\lambda\begin{pmatrix}-4\\7\\-6\end{pmatrix}+\mu\begin{pmatrix}2\\5\\-1\end{pmatrix}$

b 15.2 units

17 $\mathbf{r} = \begin{pmatrix} -3 \\ 1 \\ 0 \end{pmatrix} + \lambda \begin{pmatrix} 3 \\ -1 \\ 2 \end{pmatrix}$

18 $\begin{pmatrix} 3 \\ -5 \\ -10 \end{pmatrix} + \mu \begin{pmatrix} 7 \\ -10 \\ 0 \end{pmatrix} + \lambda \begin{pmatrix} 1 \\ -3 \\ 4 \end{pmatrix} = \begin{pmatrix} -11 \\ -4 \\ 10 \end{pmatrix} + t \begin{pmatrix} 4 \\ -2 \\ 0 \end{pmatrix}$

$\mathbf{k}: -10 + 4\lambda = 10 \Rightarrow \lambda = 5$

$\mathbf{i}: 3 + 7\mu + 5 = -11 + 4t \Rightarrow 7\mu + 19 = 4t$

$\mathbf{j}: -5 - 10\mu - 15 = -4 - 2t \Rightarrow 10\mu + 16 = 2t$

$\Rightarrow 20\mu + 32 = 7\mu + 19$

$\Rightarrow \mu = -1$

$\Rightarrow t = 3$

$\begin{pmatrix} -11 \\ -4 \\ 10 \end{pmatrix} + 3 \begin{pmatrix} 4 \\ -2 \\ 0 \end{pmatrix} = \begin{pmatrix} 1 \\ -10 \\ 10 \end{pmatrix}$

So they intersect at (1, –10, 10)

19 a $\begin{pmatrix} 4 \\ 2 \\ -1 \end{pmatrix} \times \begin{pmatrix} 0 \\ -5 \\ 1 \end{pmatrix} = \begin{pmatrix} -3 \\ -4 \\ -20 \end{pmatrix} = -1 \begin{pmatrix} 3 \\ 4 \\ 20 \end{pmatrix}$

Therefore planes are parallel

We know they aren't the same plane as (5, 3, –2) does not lie on Π_2, since $\begin{pmatrix} 5 \\ 3 \\ -2 \end{pmatrix} \cdot \begin{pmatrix} 3 \\ 4 \\ 20 \end{pmatrix} = 15 + 12 - 40 = -16$ (not 12).

b $\frac{5\sqrt{17}}{17}$ (1.23)

Chapter 24

Exercise 24.1A

1 a $\mathbf{a} = -32\sin 4t\mathbf{i} - 32\cos 4t\mathbf{j}\ \text{m s}^{-2}$

b $-320\sin 4t\mathbf{i} - 320\cos 4t\mathbf{j}$ N

2 a $(3\sin 2t + 2)\mathbf{i} + (3\cos 2t + 1)\mathbf{j}$ m

b $x = 3\sin 2t + 2$

$y = 3\cos 2t + 1$

$\cos^2 2t + \sin^2 2t = 1$

$\left(\frac{y-1}{3}\right)^2 + \left(\frac{x-2}{3}\right)^2 = 1$

$(y-1)^2 + (x-2)^2 = 9$

which is a circle, radius 3 and centre (2, 1).

3 a i $\mathbf{OM} = 4\cos\frac{1}{2}t\mathbf{i} + 4\sin\frac{1}{2}t\mathbf{j}$ m

ii $-\cos\frac{1}{2}t\mathbf{i} - \sin\frac{1}{2}t\mathbf{j}\ \text{m s}^{-2}$

b 4π seconds

c 3 N

4 a $\mathbf{v} = 6\cos 3t\mathbf{i} - 6\sin 3t\mathbf{j}\ \text{m s}^{-1}$

$\mathbf{a} = -18\sin 3t\mathbf{i} - 18\cos 3t\mathbf{j}\ \text{m s}^{-2}$

b i Gradient of $\mathbf{a} = \frac{-18\cos 3t}{-18\sin 3t} = \cot 3t$

Gradient of $\mathbf{r} = \frac{2\cos 3t}{2\sin 3t} = \cot 3t$

Gradients are equal, so **a** and **r** are parallel

Using vectors:

$\mathbf{a} = -9(2\sin 3t\mathbf{i} + 2\cos 3t\mathbf{j}) = -9\mathbf{r}$

So, **a** and **r** are parallel.

They are in opposite directions.

ii Gradient of $\mathbf{r} = \frac{-6\sin 3t}{6\cos 3t} = -\tan 3t$

As $\cot 3t \times (-\tan 3t) = -1$,

gradients of **a** and **v** multiply to –1.

So **a** and **v** are perpendicular.

Using vectors:

$\mathbf{a} \times \mathbf{v} = -18\begin{pmatrix} \sin 3t \\ \cos 3t \end{pmatrix} \times 6 \begin{pmatrix} \cos 3t \\ -\sin 3t \end{pmatrix}$

$= -108(\sin 3t\cos 3t - \cos 3t\sin 3t)$

$= 0$

So **a** and **v** are perpendicular.

5 a $-32\sin 2t\mathbf{i} + 32\cos 2t\mathbf{j}$, 32N

b $(2\sin 2t + 3)\mathbf{i} - (2\cos 2t + 8)\mathbf{j}$ m,

circle radius 2 and centre (3, –8)

6 $\mathbf{r} = \int \mathbf{v}\,dt = -\frac{a}{\omega}\cos t\mathbf{i} + \frac{b}{\omega}\sin\omega t\mathbf{j}$

$\mathbf{F} = m\mathbf{a} = m\frac{d\mathbf{v}}{dt} = m(a\omega\cos\omega t\mathbf{i} - b\omega\sin\omega t\mathbf{j})$

$= -m\omega^2\left(-\frac{a}{\omega}\cos\omega t\mathbf{i} + \frac{b}{\omega}\sin\omega t\mathbf{j}\right)$

$= -m\omega^2\mathbf{r}$

So $\mathbf{F} = k\mathbf{r}$ where $k = -m\omega^2$.

Exercise 24.1B

1 $14.5\ \text{m s}^{-1}$

2 $14.4\ \text{m s}^{-1}$

3 18.8°

4 a $42.0\ \text{m s}^{-1}$ **b** $24.2\ \text{m s}^{-1}$, $72.7\ \text{m s}^{-1}$

5 a 34.5°

b Assuming that $108\ \text{km h}^{-1}$ is the maximum speed around the bend.

Assume that the limiting friction force and hence the coefficient of friction remain the same.

6 a $21.7\ \text{m s}^{-1}$ **b** $27.0\ \text{m s}^{-1}$

7 a 0.46 (2 sf) **b** $16.8\ \text{m s}^{-1}$

8 4.50°

9 a 175 kN **b** 525 kN

10 a 12.0°, 0.21 **b** $39\ \text{km h}^{-1}$

11

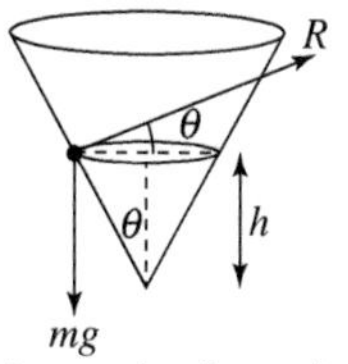

Let angle of cone be 2θ

Vertical equilibrium:

$R\sin\theta = mg$

Horizontal equation of motion:

$R\cos\theta = m \times \frac{v^2}{h\tan\theta}$

$v^2 = \frac{mg}{\sin\theta} \times \cos\theta \times \frac{h\tan\theta}{m}$

$= hg$

Velocity $v = \sqrt{hg}$

12 0.34 N

13 $\sqrt{\dfrac{g}{x}}$

14 Maximum speed occurs when on point of moving up slope.

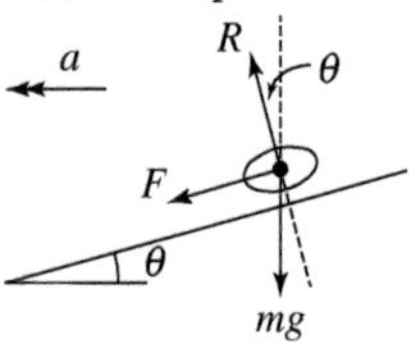

Equation of motion down slope:

$$F + mg\sin\theta = \frac{mv^2}{r}\cos\theta$$

Equation of motion $\perp$ slope:

$$R - mg\cos\theta = \frac{mv^2}{r}\sin\theta$$

Friction: $F \le \mu R$

$$\mu \ge \frac{F}{R} = \frac{mv^2\cos\theta - mgr\sin\theta}{mv^2\sin\theta + mgr\cos\theta}$$

$$= \frac{v^2 - gr\tan\theta}{r^2\tan\theta + gr}$$

$\mu v^2 \tan\theta + \mu gr \ge v^2 - gr\tan\theta$

and $\mu = \tan\lambda$

So $v^2(1 - \tan\lambda + \tan\theta) \le gr(\tan\lambda + \tan\theta)$

$$\frac{v^2}{gr} \le \frac{\tan\lambda + \tan\theta}{1 - \tan\lambda\tan\theta} = \tan(\theta + \lambda)$$

Minimum speed occurs when on point of moving down slope.

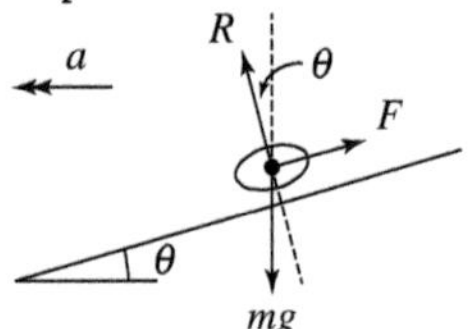

Equations are now:

$$mg\sin\theta - F = \frac{mv^2}{r}\cos\theta \text{ and}$$

$$R - mg\cos\theta = \frac{mv^2}{r}\sin\theta$$

Friction $F \le \mu R$

$$mg\sin\theta - \frac{mv^2}{r}\cos\theta \le \tan\lambda\left(mg\cos\theta + \frac{mv^2}{r}\sin\theta\right)$$

$$\frac{mv^2}{r}(\tan\lambda\sin\theta + \cos\theta) \ge mg(\sin\theta - \tan\lambda\cos\theta)$$

$$\frac{v^2}{rg}(\tan\lambda\tan\theta + 1) \ge \tan\theta - \tan\lambda$$

$$\frac{v^2}{rg} \ge \tan(\theta - \lambda)$$

So $\tan(\theta - \lambda) \le \dfrac{v^2}{rg} \le \tan(\theta + \lambda)$

or $\theta - \lambda \le \tan^{-1}\left(\dfrac{v^2}{rg}\right) \le \theta + \lambda$

Exercise 24.2A

1 $4g$ N, 1.0 s

2

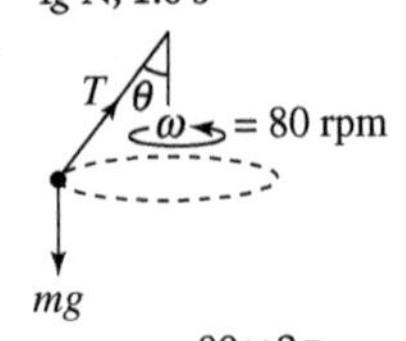

$$80\,\text{rpm} = \frac{80 \times 2\pi}{60} = 8.38\ \text{rad s}^{-1}$$

Vertical equilibrium: $T\cos\theta = mg$

Horizontal equation of motion:

$T\sin\theta = m \times 0.25\sin\theta \times \omega^2$

$$\Rightarrow \frac{mg}{\cos\theta} = m \times 0.25 \times 8.38^2$$

$\cos\theta = 0.559$

$\theta \approx 56°$

3

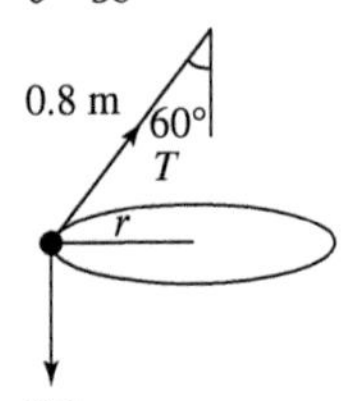

$T\cos 60° = mg$

$T\sin 60° = mrw^2$

$$\Rightarrow \tan 60° = \frac{rw^2}{9}$$

$$\omega^2 = \frac{g\sqrt{3}}{0.8 \times \sin 60°}$$

$\Rightarrow \omega = 4.95\ \text{rad s}^{-1}$

$= 0.788\ \text{rad s}^{-1}$

In 10 seconds, the number of revolutions $= 7.88 \approx 8$

4 1.48 m, 29.6 N

5 23. 7 N, 60.2°

6

Vertically: $T\cos\theta = mg$

Horizontally: $T\sin\theta = m \times l\sin\theta \times \omega^2$

Eliminating T: $ml\omega^2 \times \cos\theta = mg$

Height $h = l\cos\theta = \dfrac{g}{\omega^2}$ which is independent of l

Exercise 24.2B

1 **a** 1.8 rad s^{-1} **b** 2.26 rad s^{-1}

c Any two from:

String is light and inextensible.

There is no air resistance to slow the mass as it rotates.

Masses are modelled as point masses or particles.

2 2.89 m

3 $\dfrac{ml}{h}(\omega^2 h - g)$

4 $2\pi\sqrt{\dfrac{\sqrt{R^2 - r^2}}{g}}$ s

5 $mg - \frac{m\omega^2\sqrt{3}}{5}$

$R \geq 0$ implies $mg \geq \frac{m\omega^2\sqrt{3}}{5}$

$5g \geq \omega^2 \times \sqrt{3}$ So maximum $\omega^2 = \frac{5g}{\sqrt{3}}$

$\omega = \sqrt{28.29} = 5.3 \text{ rad s}^{-1}$

6 3.0 rad s^{-1}

7 a

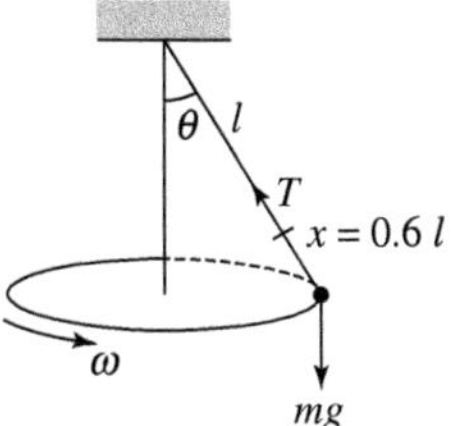

i Vertical equilibrium:
$T\cos\theta = mg$ (1)
Horizontal equation of motion:
$T\sin\theta = m \times 1.6l \sin\theta \times \omega^2$ (2)
Hooke's law:
$T = \frac{2mg}{l} \times 0.6l$

$T = 1.2\,mg$ (3)
From (2) and (3): $1.2\,mg = m \times 1.6l \times \omega^2$

$\omega^2 = \frac{1.2mg}{1.6ml} = \frac{3g}{4l}$

ii $\cos^{-1}\left(\frac{5}{6}\right)$

b String is light.
There is no air resistance to slow the mass as it rotates.

8

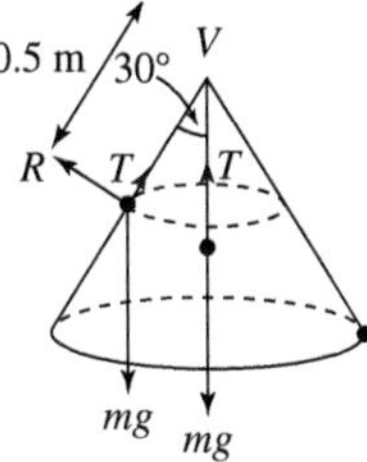

Cone is smooth, so tension T is constant along all the length.
For hanging mass in equilibrium: $T = mg$ (1)
Vertical equilibrium for moving mass:
$T\cos 30° + R\cos 60° = mg$ (2)
Horizontal equation of motion:
$T\sin 30° - R\sin 60° = mr\omega^2$ (3)
where $r = 0.5\sin 30° = \frac{1}{4}$ metre

From (1) and (2): $T\frac{\sqrt{3}}{2} + R \times \frac{1}{2} = mg$

$R = 2mg - 2 \times mg\frac{\sqrt{3}}{2}$

$R = 0.268\,mg$

Substitute into (3): $mg \times \frac{1}{2} - 0.268\,mg \times \sin 60° = m \times \frac{1}{4}\omega^2$

$\frac{1}{4}\omega^2 = 2.625$

Angular speed $\omega = 3.24 \text{ rad s}^{-1}$

9

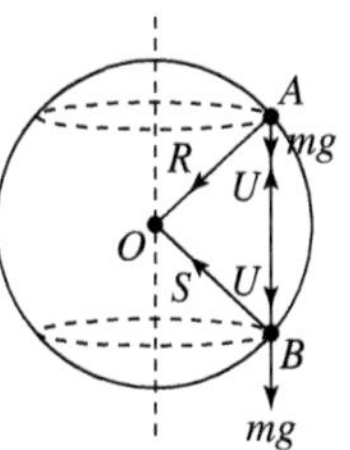

ΔOAB is equilateral with sides of length r
Assume directions of forces R, S, U as shown.

a For A, vertical equilibrium:
$R\cos 60° + mg = U$ (1)
Horizontal equation of motion:
$R\cos 30° = m \times r\sin 60° \times \omega^2$

$R \times \frac{\sqrt{3}}{2} = m \times r \times \frac{\sqrt{3}}{2} \times \omega$

$R = mr\,\omega^2$ (2)

b 6.3 rad s^{-1}

10 19.3 rad s^{-1}

11 a $\frac{2(M+m)g}{M\omega^2}$

b Let the rods have a mass that is not negligible.
Include friction at the contact between the collar and the rod.
Take air resistance into account.

12 $\frac{g - e\omega^2}{l\omega^2}$

Exercise 24.3A

1 a 9.24 m s^{-1}, 309 N b 8.40 m s^{-1}, 235 N

2 a 6.66 m s^{-1}, 271 N, tension b 2.28 m s^{-1}, −23 N, thrust

3 a Find out whether $\frac{1}{2}mu^2 > mgh$
$\frac{1}{2}u^2 > gh$ where $h = 2r = 4$ metres
Is $u^2 > 8g$?
$100 > 78.4$ so yes, it will complete the circle.

b 104°

4 a i 5.29 m s^{-1} ii 13.45 N

b P rises to level of O if $\frac{1}{2}mu^2 > mgr$
$\frac{1}{2}mu^2 = 9.0$ J
$mgr = \frac{1}{2} \times 9.8 \times 1.4 = 6.9$ J
P rises to level of O and goes higher.

5 a i 7.75 m s^{-1}, 243 N
ii The string may go slack, so that P leaves the circle.

b i Yes; the same equations apply.
ii 30.2 m s^{-1}, 12 N, a thrust

Exercise 24.3B

1 a UWX b UVX c UWX
d YZ e Z f UVX

2 a 10 m s^{-1} b −9.8 m s^{-2}

3 a 5.42 m s^{-1} b 8.57 m s^{-1}

4 48.2°
If air resistance is taken into account, the particle, P, would be slowed down and, travelling more slowly, it would leave the surface lower down, or at an angle to the vertical greater than 48.2°.

5 115°

6

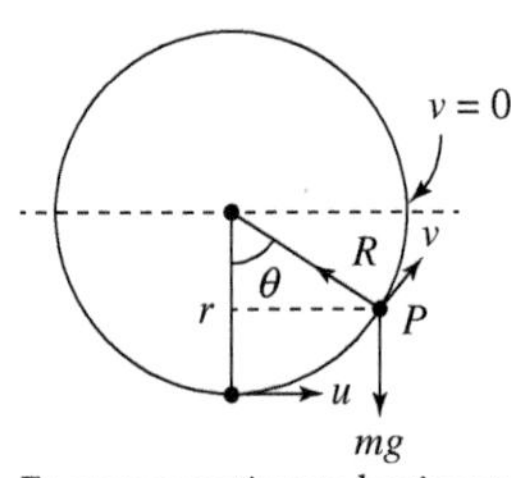

Energy equation to horizontal is

$\frac{1}{2}mu^2 - 0 = mgr$

$u^2 = 2gr$

Energy equation to point P is

$\frac{1}{2}\cancel{m}u^2 - \frac{1}{2}\cancel{m}v^2 = \cancel{m}g(r - r\cos\theta)$

$\cancel{gr} - \frac{v^2}{2} = \cancel{gr} - gr\cos\theta$

$\cos\theta = \frac{v^2}{2gr}$

Equation of motion at P towards centre is

$R - mg\cos\theta = m \times \frac{v^2}{r}$

Reaction $R = \frac{mv^2}{r} + m\cancel{g} \times \frac{v^2}{2\cancel{g}r}$

$= \frac{3mv^2}{2r}$

7 50.4 km h^{-1}

Car is a point mass, so no air resistance.

8 a i 8.3 N (3 sf) ii 26.3 m s^{-2}

b 41.8 m s^{-2}

9 a i 2.93 N ii 17.2 N

b 48.2°

10 2.92 m

11 a 1.2 m b 1 : 2

12

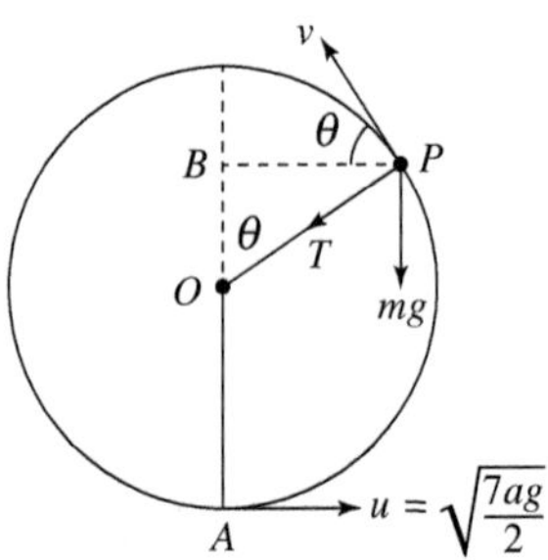

Energy equation for A to P: $\frac{1}{2}\cancel{m}u^2 - \frac{1}{2}\cancel{m}v^2 = \cancel{m}ga(1+\cos\theta)$

$\frac{7ag}{4} - \frac{v^2}{2} = ag(1+\cos\theta)$ (1)

Equation of motion along PO: $T + mg\cos\theta = m\frac{v^2}{a}$

Leaves circle when $T = 0$:

$v^2 = ag\cos\theta$ (2)

(1) and (2) give: $\frac{7ag}{4} - \frac{ag\cos\theta}{2} = ag + ag\cos\theta$

$\frac{3}{4}ag = \frac{3}{2}ag\cos\theta$

$\cos\theta = \frac{1}{2} \Rightarrow \theta = 60°$, so $v^2 = \frac{ag}{2}$

Time taken to travel PB horizontally

$= \frac{BP}{v\cos\theta} = \frac{a \times \frac{\sqrt{3}}{2}}{\sqrt{ag \times \frac{1}{2}} \times \frac{1}{2}} = \sqrt{\frac{6a}{g}}$

Using $s = ut + \frac{1}{2}at^2$ vertically from P:

$s = v\sin\theta \times t - \frac{1}{2} \times g \times t^2$

$= \sqrt{\frac{ag}{2}} \times \frac{\sqrt{3}}{2} \times \sqrt{\frac{6a}{g}} - \frac{1}{2} \times g \times \frac{6a}{g}$

$= \frac{3}{2}a - 3a$

$= -\frac{3a}{2}$, which is a distance of $\frac{3a}{2}$ below B

$AB = a + a\cos\theta = 3\frac{a}{2}$

So, mass passes through A

13 2.75 m

Review exercise 24

1 a 24 N b $x^2 + y^2 = 9$

2 a 6 m s^{-1} b 2 m, (5, −1)

3 $3\cos 2t\mathbf{i} + 3\sin 2t\mathbf{j}, -24\mathbf{j}$

4 $\left(8 - 10\cos\frac{t}{2}\right)\mathbf{i} + \left(1 + 10\sin\frac{t}{2}\right)\mathbf{j}, \frac{10}{\sqrt{2}}(\mathbf{i} - \mathbf{j})$

5 0.36

6 16°, 0.287

7 $\frac{\mu_2 m_2}{2\mu_1 m_1}$

8 29.6 N, 48.6°

9 0.28 m (28 cm)

10

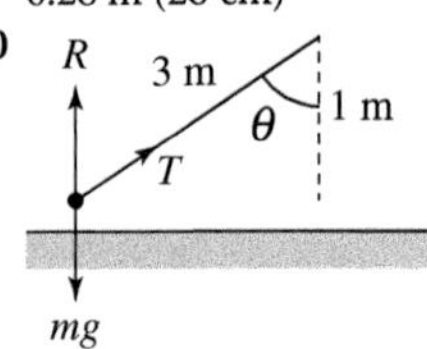

Vertical equation: $R + T\cos\theta = mg$

$R + \frac{T}{3} = mg$

Horizontal equation of motion:

$T\sin\theta = m \times 3\sin\theta \times \omega^2$

$T = 3m\omega^2$

Eliminating T

$R = mg - \frac{1}{3} \times 3m\omega^2 = mg - m\omega^2$

Contact occurs if $R > 0$

$mg > m\omega^2$

$\omega < \sqrt{g}$

Greatest linear velocity, $v = r\omega$

$= \sqrt{9-1} \times \omega = \sqrt{8g}$ m s^{-1}

11 1.68 m

12 a 4.95 m s^{-1} b 229 N

13 a 57° b 106°

14

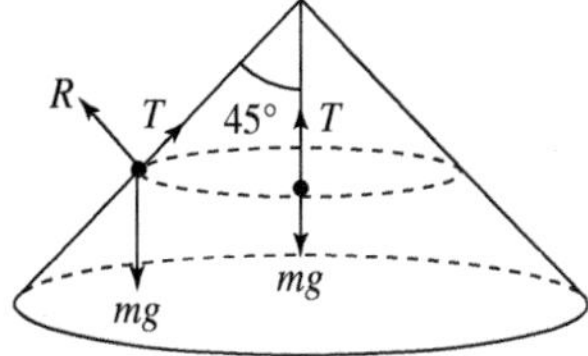

For hanging mass, $T = mg$
Tension constant as hole is smooth.
For outside mass,
vertical equilibrium means:
$T\cos 45° + R\cos 45° = mg$
$T + R = \sqrt{2} \times mg$ (1)
Horizontal equation of motion:
$T\sin 45° - R\sin 45° = m \times r\omega^2$
$T - R = \sqrt{2} \times mr\omega^2$ (2)
To find r

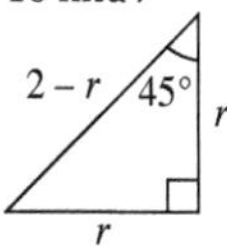

$\cos 45° = \frac{r}{2-r}$
$2 - r = \sqrt{2}r$
$r = \frac{2}{\sqrt{2}+1}$
Eliminate R from (1) and (2)
$2T = \sqrt{2}mg + \sqrt{2}m\omega^2 \times \frac{2}{\sqrt{2}+1}$
$\frac{2\sqrt{2}}{\sqrt{2}+1} m\omega^2 = 2mg - \sqrt{2}mg$
$= \sqrt{2}g(\sqrt{2}-1)$
$2\omega^2 = (\sqrt{2}-1)(\sqrt{2}+1)g = 1g$
Angular velocity $\omega = \sqrt{\frac{g}{2}}$

15 **a** 4.85 rad s^{-1} **b** 2.37 rad s^{-1}

Assessment 24

1 **a** $|\mathbf{r}| = \sqrt{(0.3\cos 10t)^2 + (0.3\sin 10t)^2}$
$= \sqrt{0.09(\cos^2 10t + \sin^2 10t)} = 0.3$
The particle moves at a constant distance from O
The path is a circle of radius 0.3 m about O
b $|\mathbf{v}| = \sqrt{(-3\sin 10t)^2 + (3\cos 10t)^2} = 3$
so speed is a constant 3 m s^{-1}
c 60 N directed towards O
2 7 rad s^{-1}
3 **a** $\mathbf{r} = 10\cos 0.6t\mathbf{i} + 10\sin 0.6t\mathbf{j}$
b $\mathbf{a} = -3.6\cos 0.6t\mathbf{i} - 3.6\sin 0.6t\mathbf{j}$
Hence $\mathbf{a} = -0.36\mathbf{r}$
$\mathbf{r}$ is directed away from the origin, so $\mathbf{a}$ is directed towards it.
4 **a** $v = 26.2$ m s^{-1} **b** $v = 39.6$ m s^{-1} **c** $v = 9.06$ m s^{-1}
5 **a i** $v = 6.29$ m s^{-1} **ii** $v = 2.38$ m s^{-1}
b i $T = 96.2$ N (tension) **ii** $T = -5.62$ N (thrust)
6 **a** $U \geq 2\sqrt{ga}$ **b** $U \geq \sqrt{5ga}$
7 70.5°
8 **a** 0.0193 m **b** 51.9 m s^{-1} or 187 km h^{-1}
9 **a** 0.392 m
b Assume ball is a particle, string is light and inextensible, no air resistance.
10 7.4 N, 1.12 m
11 **a** Let length of string be l, let AO be given by h
Resolve horizontally: $T\sin\theta = mr\omega^2$, where $r = l\sin\theta$
$T = ml\omega^2$
Resolve vertically: $T\cos\theta = mg$, where $\cos\theta = \frac{h}{l}$
$ml\omega^2 = \frac{h}{l} = mg \Rightarrow h = \frac{g}{\omega^2}$
b $\frac{g}{\omega^2} > 0$ for all values of ω, so $h > 0$ and string is not horizontal
or for a horizontal string there is no vertical component of T to balance the weight of the particle.
12 **a** 8.57 m s^{-1} **b** $T = 4.41$ N
13 48.2°
14 **a** 2.01 N **b** 1.68 m s^{-1}
15 Let speeds at bottom and top be u and v
Resolve vertically at top: $T + 0.5g = \frac{0.5v^2}{1}$
For a complete circle $T \geq 0$, so minimum $v^2 = g$
From conservation of energy: $\frac{1}{2}mu^2 = \frac{1}{2}mv^2 + 2mg$
$u^2 = v^2 + 4g$, so minimum $u^2 = 5g$
Resolve vertically at bottom: $T_1 - 0.5g = 0.5u^2 = 2.5g$
$T_1 = 3g = 29.4$ N
16 Radius of circle $= h\tan\theta$
Resolve vertically: $R\sin\theta = mg$ (1)
Resolve horizontally: $R\cos\theta = m\omega^2 h\tan\theta$ (2)
Divide (1) by (2): $\tan\theta = \frac{mg}{mh\omega^2\tan\theta} \Rightarrow h = \frac{g}{\omega^2\tan^2\theta}$
17 $\frac{g\sqrt{51}}{4}$ m s^{-2}
18 $v = \sqrt{\frac{ga(3\sqrt{3}-4)}{2}}$ m s^{-1}
19 8 cm
20 $3\frac{1}{3}$ m
21 **a** $\frac{85mg}{126}$ N, 8.20 m s^{-1} **b** $\frac{5mg}{3}$ N, 18.8 m s^{-1}
22 27.0°, 0.055N
23 $\sqrt{8ga}$
24 0.6 m
25 $0.5r$
26 183.75N
Assumes light, inextensible rope, ring is a particle, no friction/air resistance.
27 $R_A = \frac{mg(10-3\sqrt{3})}{6}$ and $R_B = \frac{mg(14-6\sqrt{3})}{6}$
28 3.89 rad s^{-1}
29 10.08 m

Chapter 25

Exercise 25.1A

1 **a** 23 m **b** −2 N m **c** 14 N m
2 **a** 2 N m **b** 8 N m
3 **a** 14 N m **b** 12 N m
4 $30g$ N, $90g$ N m anticlockwise
5 8 N m anticlockwise
6 Take moments about A clockwise.
Resultant clockwise moment $= F \times (d + x) - F \times x = F \times d$
which is independent of x and the position of point A

Exercise 25.1B

1 **a** 34.8 N m **b** 44.7 N m **c** 5 N m anticlockwise
2 **a** 20.8 N m **b** 22.5 N m **c** 9.6 N m **d** 4.5 N m
3 35 N m
4 0.010 47 N m, 0.005 24 N m
5 0 N, 10 N, −40 N m
6 −37.9 N m

Exercise 25.2A

1 **a** (6.6, 0) **b** (4.1, 0) **c** (0, 3.9) **d** (4.36, −2.64)
2 $\mathbf{r} = 4.15\mathbf{i} + 1.55\mathbf{j}$
3 (4.5, 0)
4 $a = 4, b = 2$
5 $a = 6, b = 7$
6 (−13.6, −8.4)

Exercise 25.2B

1

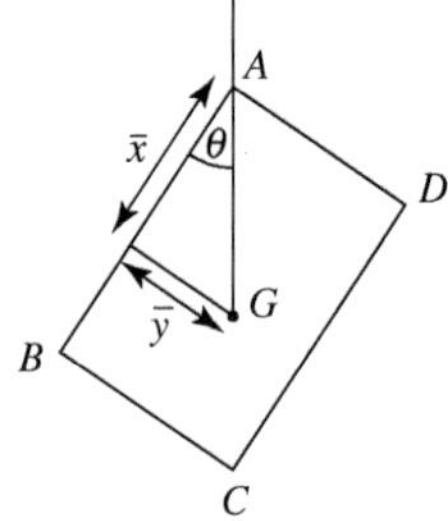

$\bar{x} = \frac{50}{21}, \bar{y} = \frac{45}{21}, \theta = 42.0°$

2 $\bar{x} = \frac{130}{17}, \bar{y} = \frac{40}{17}$

3 **a**

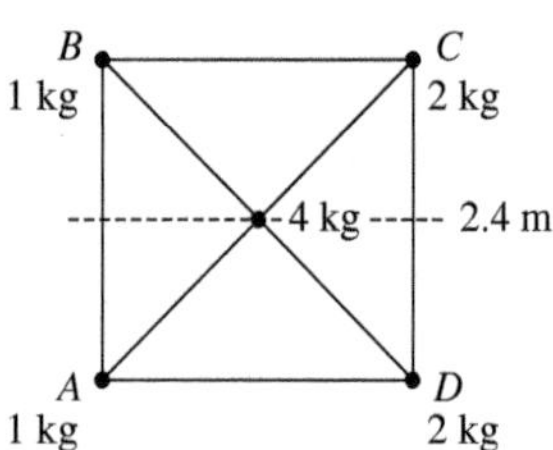

The masses are positioned, in position and in size, symmetrically about the dotted line.
So centre of mass, *G*, lies on this line.

b 1.44 m, 50.2°

4 **a**

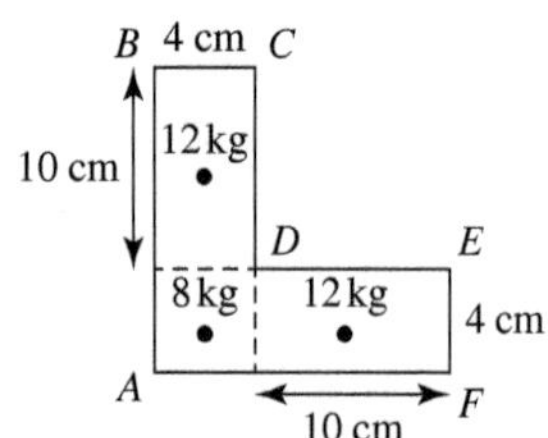

The line *AD* is a line of geometric symmetry and mass symmetry.
So centre of mass lies on *AD* (or *AD* extended).

b $\bar{x} = 4\frac{5}{8}$ cm, $\bar{y} = 4\frac{5}{8}$ cm

5 $m = 4$, $DE = 3$ metres

6 Centre of mass is 1.03 m from *OY* and 1.11 m from *OX*.
7 Centre of mass is 3 cm from *OX* and 10 cm from *OY*.
8 0.75 m

Exercise 25.3A

1 **a** *OG* is 2.18 m on line of symmetry of arc.
b *OG* is 3.9 m on line of symmetry of arc.
c *OG* is 1.15 m on line of symmetry of arc.
2 **a** *OG* is 1.24 m on line of symmetry.
b *OG* is 2.24 m on line of symmetry.
c *OG* is 1.79 m on line of symmetry.
3 **a** *OG* is 0.34 m on line of symmetry.
b *OG* is 0.84 m on line of symmetry.
4 **a** (7, 5) **b** (3, 2) **c** (6, 4)
5 **a** *AG* is 8 cm on line of symmetry.
b *AG* is $\frac{10\sqrt{3}}{3}$ cm on line of symmetry.
6 $a = 15, b = 12$
7

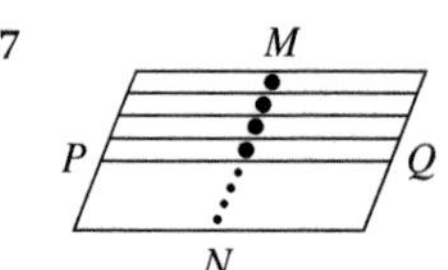

If *P*, *Q* are midpoints then mass of parallelogram above *PQ* equals mass below *PQ*, hence the centre of mass lies on *PQ*
Repeat the process with *MN* and the question is answered.

Exercise 25.3B

1 **a** (5.3, 6.4) **b** (5.27, 3.82)
c **i** (5, 5.12) **ii** (5, 3.27)
2 **a** (3.31, 1) **b** (4.625, 3.375)
3 **a** Centre of mass is 0.5 m from *AB* and 0.15 m from *AC*
b 0.612 m from centre on line of symmetry
4 0.78 cm above *O* on axis of symmetry
5

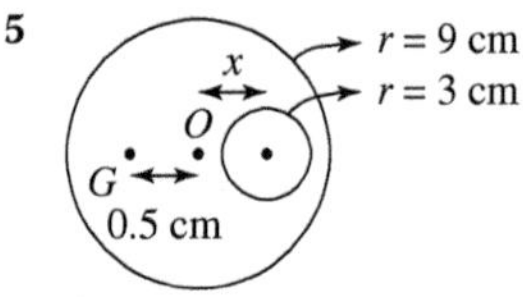

$x = 4$ cm

6 **a**

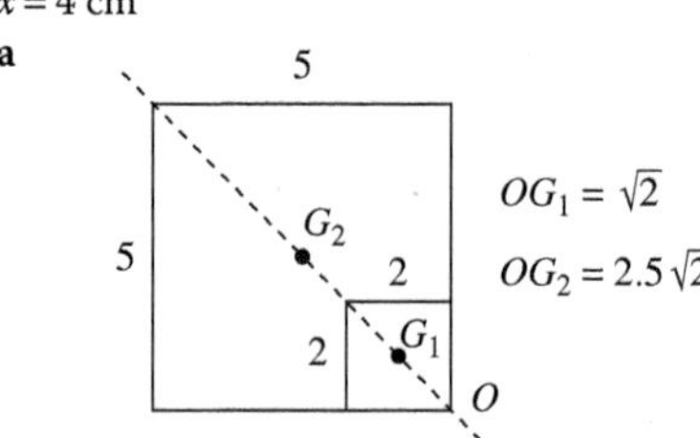

3.94 m from *O*

b 5.25 cm from the straight edge of the semicircle and along the line of symmetry of the lamina

7 2.98 cm
8 4.93 cm
9 12.5 cm
10 $\frac{33a}{28}$

11

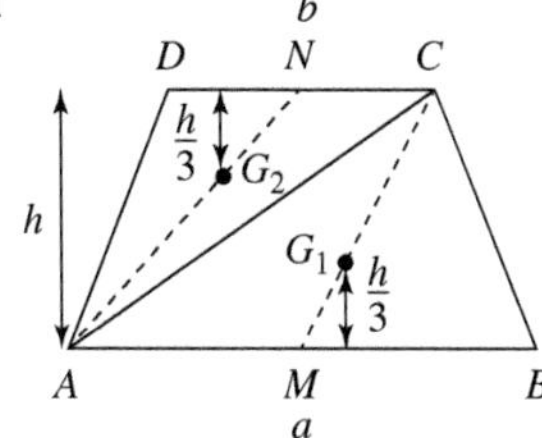

M, N are midpoints, so G_1 is $\frac{h}{3}$ from AB and G_2 is $h - \frac{h}{3} = \frac{2}{3}h$ from AB.

	Mass	C of M from AB
ΔABC	$\frac{1}{2}ah\rho$	$\frac{h}{3}$
ΔACD	$\frac{1}{2}bh\rho$	$\frac{2h}{3}$
Trapezium	$\frac{1}{2}h(a+b)\rho$	$\bar{y}$

Taking moments about AB

$$\frac{1}{2}h(a+b)\bar{y} = \frac{1}{2}ah \times \frac{h}{3} + \frac{1}{2}bh \times \frac{2}{3}h$$

$$(a+b)\bar{y} = (a+2b)\frac{h}{3}$$

Distance of centre of mass from $AB = \bar{y} = \frac{(a+2b)h}{3(a+b)}$

The symmetry of the isosceles trapezium has not been used in this proof, so the proof applies to all trapezia.

[NB: If the trapezium has been dissected like this, symmetry would been used.

The chosen method is more general.]

Exercise 25.4A

1 **a** $\left(\frac{2}{\ln 3}, \frac{3}{\ln 3}\right)$ **b** $\left(\frac{3}{2}, \frac{6}{5}\right)$

c $\left(\frac{9}{16}, \frac{7}{10}\right)$ **d** $\left(\frac{11}{8}, -18\frac{2}{5}\right)$

2 **a** (2.4, 0) **b** (1.8, 1.8) **c** $\left(2\frac{6}{7}, 0\right)$

3 **a** (3, 0) **b** $\left(\frac{3}{2}\ln 3, 0\right)$

c $\left(1\frac{2}{3}, 0\right)$ **d** $\left(2\frac{2}{3}, 0\right)$

4 **a** $\left(\frac{8}{3\pi}, 0\right)$ **b** $\left(\frac{4r}{3\pi}, 0\right)$

5 **a** $\left(\frac{3}{8}, 0\right)$ **b** $\left(\frac{3}{8}r, 0\right)$

6 $\left(\frac{3\pi}{4}, 0\right)$

Exercise 25.4B

1 **a** 20.6° **b** 33.7°

c 26.6° **d** 41.6°

2 38.7°

3 36.4°

4 $\tan^{-1}\left(\frac{16r}{11h}\right)$

5 **a** $\frac{2a}{3(\pi + 2\sqrt{3})}$ **b** 5.8°

6 (0.45, 0.35), 18.4°

7 Let G_1 and G_2 be centres of mass of the two semicircles.

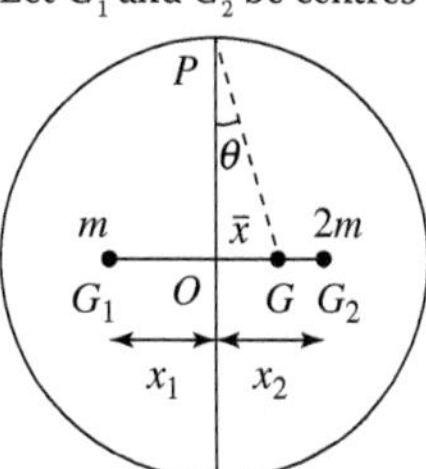

By symmetry, G lies on the line G_1G_2 which passes through O

Take moments about O

$3m_rOG = 2m_xx_2 - m_rx_1$

Move negative sign as G_1 is on opposite side of O

$$OG = \frac{2x_2 - x_1}{3} \text{ where } x_1 = \frac{2r\sin\frac{\pi}{2}}{3\frac{\pi}{2}} = \frac{4r}{3\pi} \text{ and } x_2 = x_1$$

So $OG = \frac{4r}{9\pi}$

When suspended from P, PG is vertical and the required angle θ is given by $\tan^{-1}\left(\frac{OG}{OP}\right) = \tan^{-1}\left(\frac{4}{9\pi}\right)$

8 **a** Take OB and OZ as x- and y- axes.

Centre of mass of pendant is at O, therefore taking moments about OB

$$0 = \frac{1}{2}(2r)h \times \frac{h}{3} - \frac{4r}{3\pi} \times \frac{1}{2}\pi r^2$$

$$\frac{h^2}{3} = \frac{2r^2}{3} \text{ therefore } h = r\sqrt{2}$$

b 47.6°

Exercise 25.5A

1 Reactions are 75 N, 200 N

Tension $T = 75$ N

2 Reaction at $B = S = 75$ N

Frictional force $F = 75$ N

3 Components $k = 14$ N

$y = 18$ N, 22.8 N at angle of 7.1° above rod

4 $2g$ N, $2\sqrt{5}g$ N

5 $30g$ N

6 **a** $8g$ N **b** 98.1 N

Exercise 25.5B

1 26.6°

2 37.8°

3 26.6°

4 **a** **i** $\tan^{-1}\left(\frac{4r}{h}\right)$ **ii** $\tan^{-1}\left(\frac{3r}{h}\right)$

b $\tan^{-1}\left(\frac{3r}{h}\right)$

5 **a** $\frac{5}{3}m$ **b** 45°, 54.5°

6 **a** $18g$ N, $28g$ N m anticlockwise

b 127.3 N, 35.2 N m

7 0 N, 60 N m clockwise

8

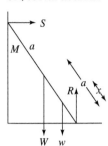

Let R and S be normal reactions at the two ends.

For equilibrium,

resolving horizontally: $S = 0$

resolving vertically: $R = W + w$

Take moments about lower end:

$Wa \cos\alpha + wx \cos\alpha = 0 + M$

Movement of couple, $M = (Wa + wx)\cos\alpha$

9 **a** 84.9.N **b** 84.9.N

10 $2\sqrt{7g}\ N$ at an angle 70.9°, $\frac{\sqrt{3}}{5}$

11 **a** **i** $\left(1\frac{2}{3}, 0\right)$ **ii** $\left(1\frac{1}{3}, 0\right)$

b **i** 80.5° **ii** 64.8°

12 **a** (12 ln 1.5, 0) **b** 60.4°

13 68.7.N, 85.0 N at 36° to horizontal

14 $T_1 = 7.32$ N and $T_2 = 5.18$ N, 2.2

15 $\frac{\sqrt{3}}{7}$

16 0.52 W (to 2 d.p.)

17 2160 N (3 s.f.), 1680 N, 0.218

18

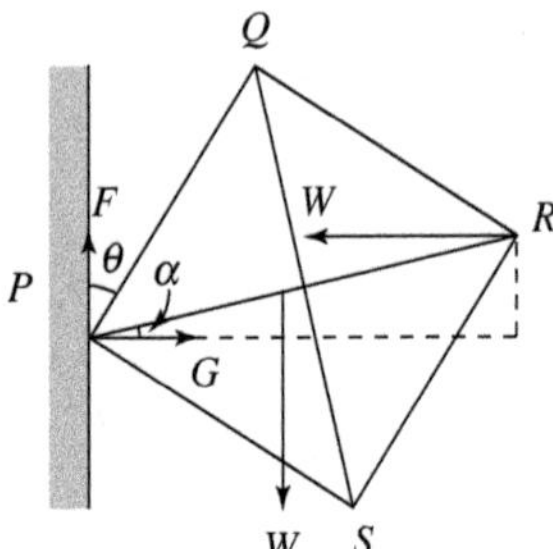

Let diagonals have length $2a$ and PR make angle α with the horizontal.

Let F and G be the friction and normal reaction at P

Resolving horizontally: $G = W$

Resolving vertically: $F = W$

At P, $F \leq \mu G$

$$\mu \geq \frac{F}{G} = \frac{W}{W}$$

So $\mu \geq 1$ and the coefficient must be at least 1

$\theta = 18.4°$

19 **a**

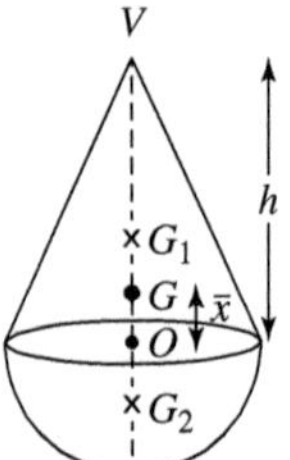

Centre of mass of cone is at G_1

Centre of mass of hemisphere is at G_2

Centre of mass of whole toy is at G.

Standard results give:

$OG_1 = \frac{1}{4}h, OG_2 = \frac{3}{8}r$

	Mass	Distance of C of M from O
Cone	$\frac{1}{3}\pi r^2 hp$	$\frac{1}{4}h$
Hemisphere	$\frac{1}{2}\left(\frac{4}{3}\pi r^2 p\right)$	$\frac{3}{8}r$
Whole	$\frac{1}{3}\pi r^2 p(h+2r)$	$\bar{x}$

Take moments about O

$$\frac{1}{3}\pi r^2 hp \times \frac{1}{4}h - \frac{1}{2} \times \frac{4}{3}\pi r^3 p \times \frac{3}{8}r$$

$$= \frac{1}{3}\pi r^2 \rho\,(h+2r) \times \bar{x}$$

Cancel $\frac{1}{3}\pi r^2 \rho$

$$\frac{1}{4}h^2 - \frac{3}{4}r^2 = (h+2r)\bar{x}$$

$$\bar{x} = \frac{h^2 - 3r^2}{4h + 8r}$$

b **i** G lies within the hemisphere.

$\bar{x}$ is negative.

$h^2 < 3r^2$

$h < r\sqrt{3}$

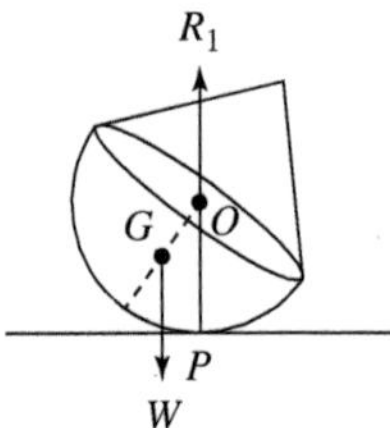

Normal reaction always passes through O

The anticlockwise moment about P returns the toy to the vertical position.

It is in stable equilibrium when it is vertical.

ii

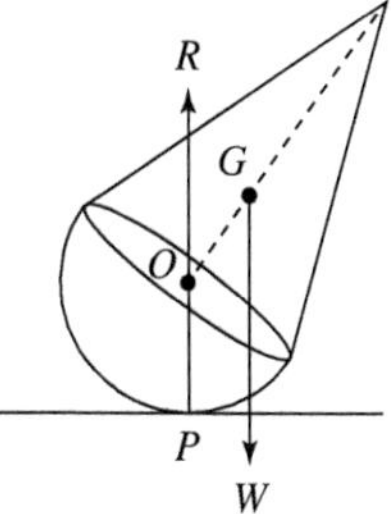

As before, R always passes through O. The clockwise moment about P turns the toy even further away from the vertical and the toy will always topple until the vertex of the cone lies on the surface.

The equilibrium when the toy is vertical is unstable for $h > r\sqrt{3}$

iii

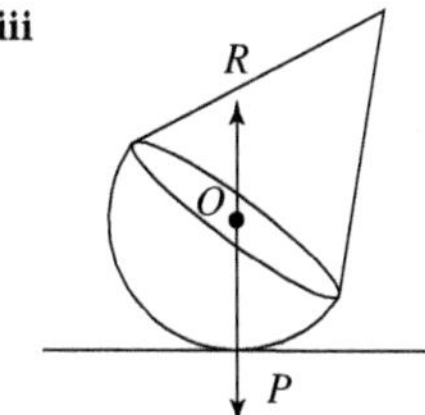

When G and O coincide, reaction R and weight W are equal and opposite.

There is no moment about P in any position. So the toy is in equilibrium in all positions (the equilibrium is said to be 'neutral') when $h = r\sqrt{3}$

20 400 N m clockwise

21 $1+\sqrt{3}\,N$, anticlockwise of magnitude $2\sqrt{3}-2$ N m

22 a 13 N **b** 13 N, 31 N m

23 142 N m

24 a 17.7 N m **b** 20.5 N m

25 a A total moment about O

$$= \begin{vmatrix} \mathbf{i} & \mathbf{j} & \mathbf{k} \\ n & \frac{1}{2} & 0 \\ 4n & 2 & 4 \end{vmatrix} + \begin{vmatrix} \mathbf{i} & \mathbf{j} & \mathbf{k} \\ \frac{1}{2} & 1 & 0 \\ 2n & -4 & 6 \end{vmatrix} + \begin{vmatrix} \mathbf{i} & \mathbf{j} & \mathbf{k} \\ -1 & -\frac{1}{2} & 1 \\ 4 & -2n & 4n \end{vmatrix}$$

$= (2\mathbf{i} - 4n\,\mathbf{j} + 0\mathbf{k})$

$+(6\mathbf{i} - 3\mathbf{j} + (-2-2n)\mathbf{k})$

$+(0\mathbf{i} - (4n-4)\mathbf{j} + (2n+2)\mathbf{k})$

$= 8\mathbf{i} + \mathbf{j} + 0\mathbf{k}$, which is independent of n

b 9.4.N m

26 a $\begin{pmatrix} 3 \\ -1 \\ -3 \end{pmatrix}$ N **b** $\begin{pmatrix} -3 \\ 1 \\ 3 \end{pmatrix}$ N, $\sqrt{19}$ N **c** 12.1 N m

27 a $m = -1\frac{1}{2}, n = -1$ **b** 81 N m anticlockwise

28 a $-3\mathbf{i} - 4\mathbf{j} + 3\mathbf{k}, \sqrt{34}$ N **b** $-11\mathbf{i} - 9\mathbf{j} + 13\mathbf{k}, \sqrt{371}$ Nm

29 a $n = 6$ or -42 **b** The force **Z** is either $\begin{pmatrix} 14 \\ 0 \end{pmatrix}$ N or $\begin{pmatrix} 14 \\ 48 \end{pmatrix}$ N

30 a $\begin{pmatrix} 4 \\ -6 \\ 2 \end{pmatrix}$ **b** $-30\mathbf{i} - 18\mathbf{j} + 6\mathbf{k}, 6\sqrt{35}$ N m

c i 68.5° **ii** 68.5°

Review exercise 25

1

y
D 8 kg
C 6 kg
• G ($\bar{x}$, $\bar{y}$)
3 m
x
A
5 m
B
2 kg
4 kg

(2.5, 2.1)

2 $\left(5, 3\frac{1}{3}\right)$ cm, 33.7°

3 Centre of mass of remainder is 0.07 cm from O on opposite side of O to the triangle, on the line of symmetry.

4 51.3°

5 69.4°

6 a (1.894, 0) **b** 42.1°

7

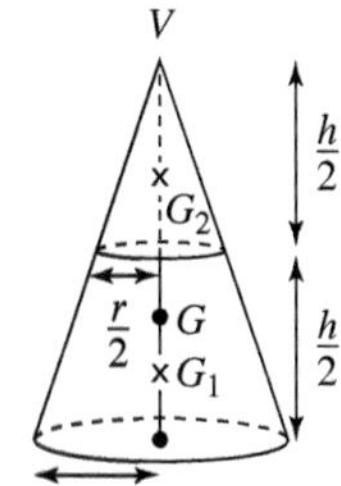

	Mass	Distance of C of M from vertex V
Cone removed	$\frac{1}{3}\pi\left(\frac{r}{2}\right)^2\left(\frac{h}{2}\right) = \frac{\pi r^2 h}{24}$	$VG_2 = \frac{3}{4}\left(\frac{h}{2}\right) = \frac{3h}{8}$
Whole cone	$\frac{1}{3}\pi r^2 h$	$VG_1 = \frac{3h}{4}$
Frustum	$\pi r^2 h\left(\frac{1}{3} - \frac{1}{24}\right) = \frac{7\pi r^2 h}{24}$	$\bar{x}$

Take moments about V

$$\frac{1}{\cancel{3}}\pi r^2 h \times \frac{\cancel{3}h}{4} - \frac{\pi r^2 h}{\cancel{24}_8} \times \frac{\cancel{3}h}{8} = \frac{7\pi r^2 h}{24} \times \bar{x}$$

Divide by $\pi r^2 h$

$$\frac{1}{4}h - \frac{1}{64}h = \frac{7}{24}\bar{x} \Rightarrow \bar{x} = \frac{15h}{64} \times \frac{24}{7} = \frac{45h}{56}$$

It is on the point of toppling when G is vertically above P

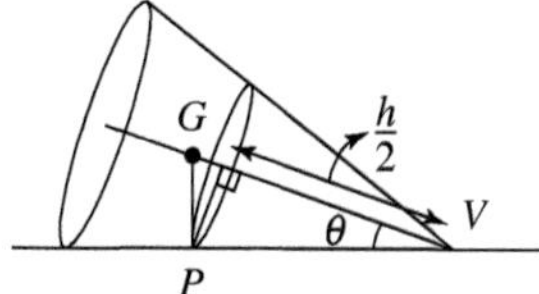

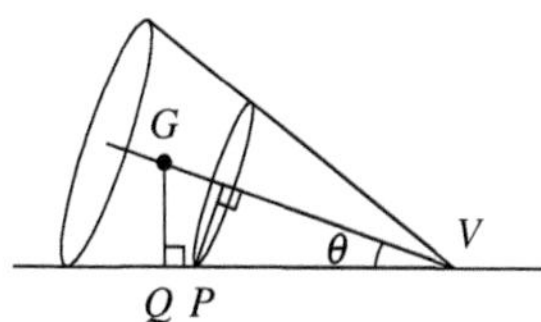

Will not topple if GQ is vertical such that $VQ > VP$

or $VG\cos\theta > \dfrac{\frac{h}{2}}{\cos\theta}$

$$\cos^2\theta > \frac{h}{2}\times\frac{1}{\bar{x}} = \frac{h}{2}\times\frac{56}{45h}$$

$$\cos^2\theta > \frac{28}{45}$$

$$\cos\theta > \sqrt{\frac{28}{45}}$$

8 5.6 m

9 0.23 m

10

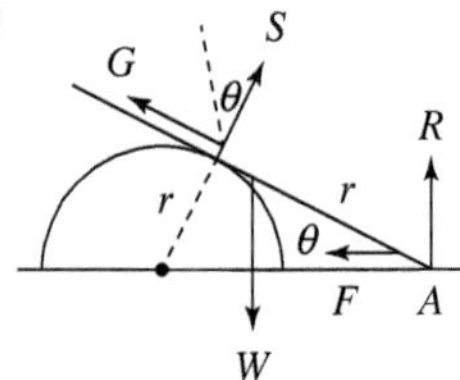

Resolving vertically:

$W = S\cos\theta + G\sin\theta + R$ (1)

Resolving horizontally:

$G\cos\theta + F = S\sin\theta$ (2)

Taking moments about A

$$S\times\frac{\not{r}}{\tan\theta} = W\times\not{r}\cos\theta$$

$S = W\sin\theta$ (3)

Friction is limiting.

$F = \mu R$ (4)

$G = \mu S$ (5)

[5 equations: 5 unknowns, R, F, S, G, θ]

Substitute from (4) and (5) into (2):

$\mu S\cos\theta + \mu R = S\sin\theta$

Substitute from (3):

$\mu R = W\sin^2\theta - \mu W\sin\theta\cos\theta$ (6)

Substitute from (6) and (5) into (1):

$$W = \cancel{W\sin\theta\cos\theta} + \mu W\sin^2\theta + \frac{W\sin^2\theta}{\mu} - \cancel{W\sin\theta\cos\theta}$$

$\mu = \mu^2\sin^2\theta + \sin^2\theta$

$$\sin^2\theta = \frac{\mu}{\mu^2+1}$$

11 166 N vertically downwards, 680 N m

12 a $\sqrt{3}y + x = 25$ **b** $25\sqrt{3}$ N m

13 $a = 2, b = \frac{2\sqrt{3}}{3}, c = \frac{8}{\sqrt{3}}, \frac{4}{\sqrt{3}}$ N m in an anticlockwise direction

Assessment 25

1 a 0.9 m **b** 5 kg

2 a 0.9 m **b** 1.6 m

3 42.0°

4 a (2.66, 0.804) **b** (2.8, 0)

5 15.8 cm

6 55.2 cm

7 (1.54, 0.641)

8 a 34.6 N **b** 2.33 m

9 32.5°

10 $m = 4$, $DE = 3$ m

11 a 4.5 kg m^{-2} **b** 0.1125 m from PQ

12 The region is bounded by the x-axis, the line $x = h$ and the line $y = \frac{rx}{h}$

By symmetry the centre of mass is on the x-axis.

$$\bar{x} = \frac{\int_0^h xy^2\,dx}{\int_0^h y^2\,dx} = \frac{\frac{r^2}{h^2}\int_0^h x^3\,dx}{\frac{r^2}{h^2}\int_0^h x^2\,dx} = \frac{\left[\frac{1}{4}x^4\right]_0^h}{\left[\frac{1}{3}x^3\right]_0^h} = 0.75h$$

13 $m_1 = 1, m_2 = 2$

14 21.8°

15 a (0.5, 0.15) **b** 12.1°

16 a $\left(\frac{e^2+1}{e^2-1}, \frac{1}{4e^2}+\frac{1}{4}\right)$ **b** $\frac{3e^4+1}{\frac{1}{2}(e^4-1)}$

17 a 23.4 N

b The resultant reaction has magnitude 45.6 N and makes an angle of 59.2° with the horizontal (into the wall).

18 a 5.94 m **b** 19.2 kg

19 a 47.04 N m, 8.64 N m

b P is a maximum when the reaction is $6g$ N vertically. 47.3 N at an angle of 47.4° to the horizontal

20 a $\bar{x} = \frac{25-27\rho}{100+120\rho}$ m

$\left(\text{or equivalent, for example, } \frac{0.03-0.0324\rho}{0.12+0.648\rho}\right)$

below point O on the line of symmetry

b $\frac{25}{27}$

21 In relation to the centre, the centres of mass are:

rod: $\frac{6\sin\frac{\pi}{6}}{\frac{\pi}{6}} = \frac{18}{\pi}$

lamina: $\frac{2\times6\sin\frac{\pi}{6}}{3\times\frac{\pi}{6}} = \frac{12}{\pi}$

Hence $2m\times\frac{12}{\pi} + m\times\frac{18}{\pi} = 3m\bar{x}$

This gives $\bar{x} = \frac{14}{\pi}$

22 a 71.9 N **b** 92.9 N at 39.3° to the horizontal

23 a G lies on CF because it is the line of symmetry.

b 0.802 m **c** 0.391 m

d Solving (1) when $\bar{x} = 1.1$ m gives $r = 0.43$ m, but as $FH = 0.4$ m the hole would overlap the edge of the lamina.

24 $\frac{13a}{8}$ m

25 18.3 cm

26 a $\left(\frac{4a}{3\pi}, \frac{4b}{3\pi}\right)$ **b** $\frac{3a}{8}$

27 30.3°

28 0.7 kg

29 $\begin{pmatrix}-5\\-1\\3\end{pmatrix}$ $\sqrt{35}$ N, $-13\mathbf{i}+4\mathbf{j}-2\mathbf{k}$, $\sqrt{189}$ N m

Chapter 26

Exercise 26.1A

1 a $\int_1^k \frac{1}{4}x\,dx = 1 \to \left[\frac{x^2}{8}\right]_1^k = 1$

$\frac{1}{8}(k^2-1) = 1 \to k = 3$

b $F(x) = \int_1^x \frac{1}{4}x\,dx = \left[\frac{x^2}{8}\right]_1^x = \frac{1}{8}(x^2-1)$

c $M = \sqrt{5}$

2 a $a = -768$

$F(x) = -768\left(x^4 - \frac{x^3}{3}\right)$

b $F(0.15) = -768\left(0.15^4 - \frac{0.15^3}{3}\right) = 0.48\text{ (2dp)} \approx 0.5$

Therefore $M \approx 0.15$

3 a $k = \frac{2}{9}$

b $F(x) = -\frac{2}{9}\left(\frac{x^3}{3} - 3\frac{x^2}{2}\right)$

$P(X > 2) = \frac{7}{27}$

4 a

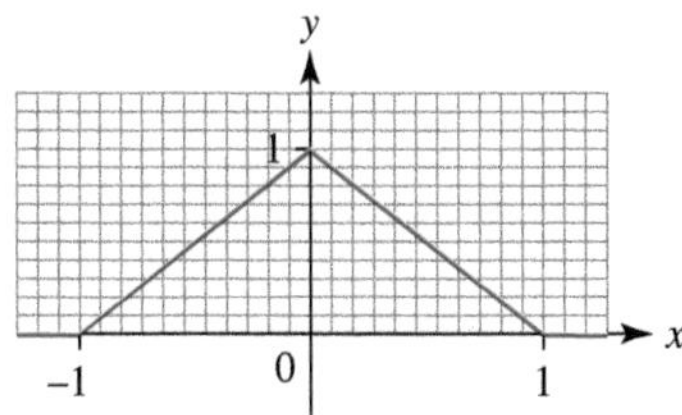

b Area under f(x) is $\frac{1}{2} \times 2 \times 1 = 1$

c $F(x) = \begin{cases} \frac{1}{2}(x+1)^2; & -1 < x < 0 \\ 1 - \frac{1}{2}(1-x)^2; & 0 < x < 1 \end{cases}$

$1 - \frac{1}{2}(1-Q_3)^2 = \frac{3}{4} \to Q_3 = \frac{\sqrt{2}-1}{\sqrt{2}}$

5 a $k = 30$

b $\frac{1}{30}(x^2+x) = \frac{1}{4}$

$2x^2 + 2x - 15 = 0 \to x = 2.28$ (2 d.p.)

$\frac{1}{30}(x^2+x) = \frac{3}{4}$

$2x^2 + 2x - 45 = 0 \to x = 4.27$ (2 d.p.)

c $f(x) = \frac{1}{30}\frac{d}{dx}(x^2+x) = \frac{2x+1}{30}$

From graph of f(x), modal value is 5

6 a

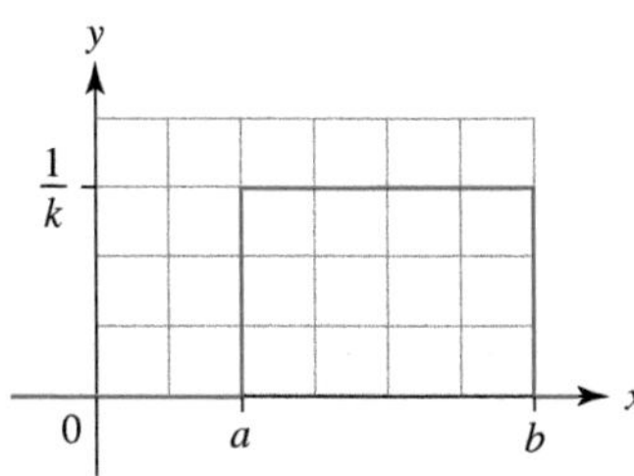

$(b-a) \times \frac{1}{k} = 1$

$k = b - a$

b $E(X) = 4$ and $Var(X) = \frac{1}{3}$

c $F(x) = P(X < x) = (x-3) \times \frac{1}{2} = \frac{x-3}{2}$

$P(X < 3.5) = (3.5 - 3) \times \frac{1}{2} = \frac{0.5}{2} = 0.25$

7 a $a = 3$

b

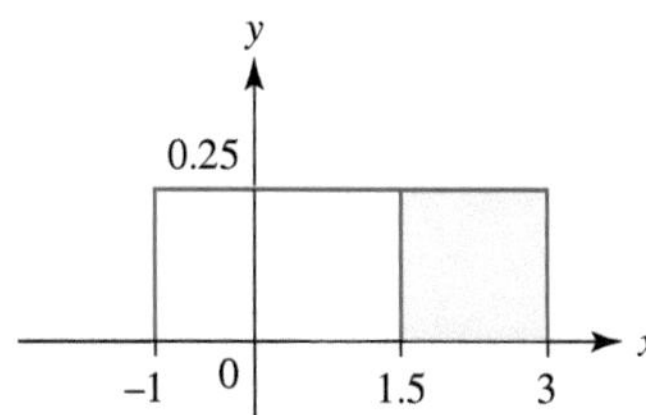

c $P(X > 1.5) = \frac{3}{8}$

d $P(X < x) = \frac{X+1}{4}$

$M = 1$

8 a $f(x) \geq 0$ for $0 < x \leq 2$ and area under $f(x) = \frac{1}{2} \times 1 + \frac{1}{2} \times 1 = 1$

b i $F(x) = \begin{cases} \int_0^x x\,dx; & 0 < x \leq 1 \\ \frac{1}{2} + \int_1^x 2 - x\,dx; & 1 < x \leq 2 \\ 1; & x > 2 \end{cases}$

Therefore:

$F(x) = \begin{cases} \frac{x^2}{2}; & 0 < x \leq 1 \\ -1 + 2x - \frac{x^2}{2}; & 1 < x \leq 2 \\ 1; & x > 2 \end{cases}$

ii Median = 1

IQR = 0.58

Exercise 26.1B

1 a $E[T] = 15$ and $Var[T] = 75$

b $E(S) = 30 - E(T) = 15$; $Var(S) = Var(T) = 75$

2 a $k = \frac{3(c-1)}{4}$ **b** $E(\sqrt{Y}) = 1.26$ (2 d.p.)

3 a $P(X \leq x) = k(-x^2 + 8x - 11)$ for $2 \leq x \leq 4$

$P(2 < X < 4) = P(X < 4) - P(X < 2)$

$= k(-4^2 + 8 \times 4 - 11) - k(-2^2 + 8 \times 2 - 11) = 4k = 1 - 0.2$

$k = 0.2$

b $f(x) = -0.4x + 1.6$

c i $E(X) = \frac{7}{3}$ **ii** $P(X < 3) = \frac{4}{5}$

4 a $a = \frac{2}{5}$

b $F(2) = 0.8$

$P(X < 2) = \frac{4}{5} > \frac{1}{2} \Rightarrow M < 2$

c $M = \frac{5}{4}$

d $f(x)=\begin{cases}\frac{2}{5}; & 0<x\le 2\\ \frac{6}{5}-\frac{2}{5}x; & 2<x<3\\ 0; & \text{otherwise}\end{cases}$

5 a $k=12$

b $f(x)=\begin{cases}\frac{x}{6}; & 0<x\le 3\\ 2-\frac{x}{2}; & 3<x\le 4\\ 0; & \text{otherwise}\end{cases}$

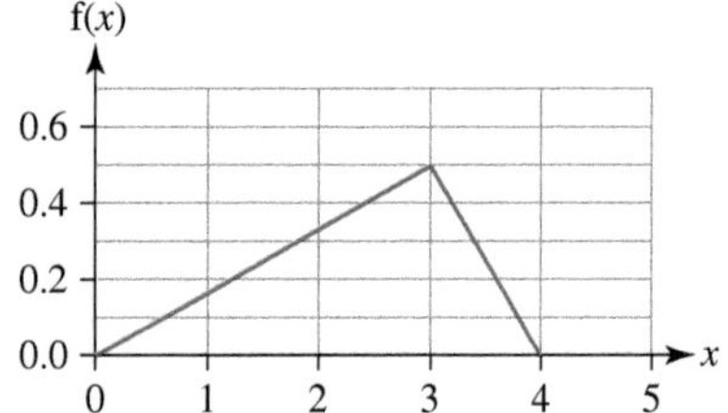

c $M=\sqrt{6}$

$P(X>3.5)=1-F(3.5)=1-\left(-3+7-\frac{49}{16}\right)=\frac{1}{16}$

Exercise 26.2A

1 a $P(X>3)=0.37$ **b** $P(1<X<3)=0.35$ (2 dp)

2 a

x	0	0.2	0.4	0.6	0.8	1.0	1.2	1.4
f(x)	2	1.34	0.90	0.60	0.40	0.27	0.18	0.12

b

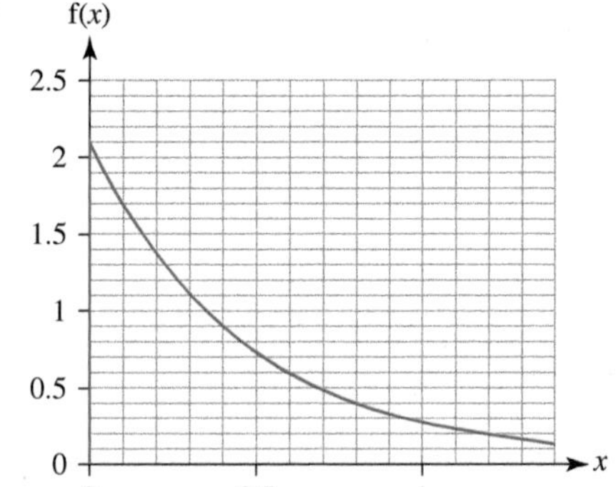

c $x=0$

3 a $F(x)=1-e^{-\lambda x}$

b $F(M)=\frac{1}{2}\Rightarrow 1-e^{-\lambda M}=\frac{1}{2}$

$e^{-\lambda M}=\frac{1}{2}\Rightarrow -\lambda M=\ln\frac{1}{2}$

$\lambda M=\ln 2$

4 a Mean = standard deviation = 4 **b** skewness = 1

5 a $f(x)=2.5e^{-2.5x}$ **b** $F(x)=1-e^{-2.5x}$

c i 0.71 **ii** 0.61 **iii** 0.28

d $E(X)=2.5\int_0^\infty xe^{-2.5x}dx=2.5\left\{\left[\frac{-xe^{-2.5x}}{2.5}\right]_0^\infty-\int_0^\infty\frac{-e^{-2.5x}}{2.5}dx\right\}$

$=2.5\left\{0-\left[\frac{e^{-2.5x}}{2.5^2}\right]_0^\infty\right\}=\frac{2.5}{2.5^2}=0.4$

$E(X^2)=\int_0^\infty x^2f(x)dx=2.5\int_0^\infty x^2e^{-2.5x}dx$

$=2.5\left\{\left[x^2\frac{e^{-2.5x}}{-2.5}\right]_0^\infty+\int_0^\infty\frac{e^{-2.5x}}{2.5}2x\,dx\right\}$

$=2.5\left\{0-2\left(\left[x\frac{e^{-2.5x}}{2.5^2}\right]_0^\infty-\int_0^\infty\frac{e^{-2.5x}}{2.5^2}dx\right)\right\}$

(Using integration by parts.)

$=2.5\left\{0-2\left(0+\left[\frac{e^{-2.5x}}{2.5^3}\right]_0^\infty\right)\right\}=2.5\frac{2}{2.5^3}=0.32$

$Var(X)=E(X^2)-\{E(X)\}^2=0.32-0.4^2=0.16$

6 a Events occur at random with mean 2 per unit time.

b i $\frac{1}{12}$ s **ii** 0.09

7 a $F(x)=1-e^{-4x}$

$f(x)=\frac{dF(x)}{dx}=-(-4e^{-4x})=4e^{-4x}$

b

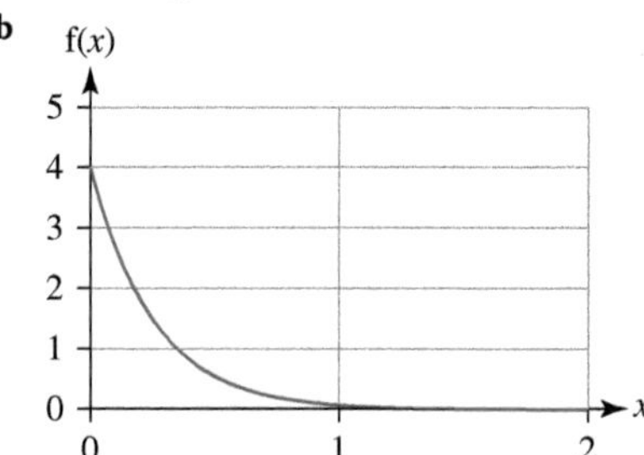

c 0 **d** 0.14

8 a $f(x)=5e^{-5x}$

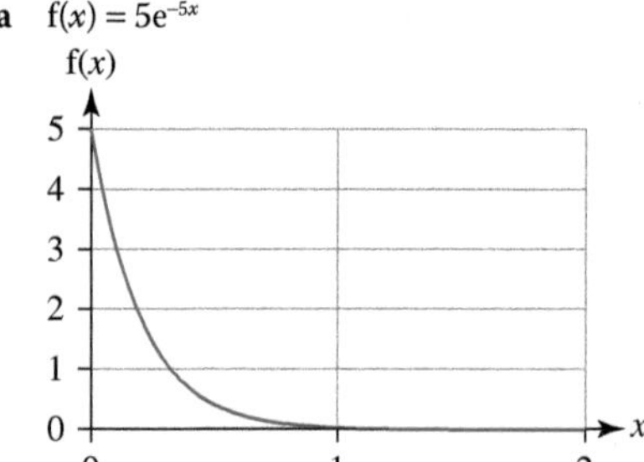

b $F(x)=\int_0^x 5e^{-5x}dx=[e^{-5x}]_x^0=1-e^{-5x}$

c Mean $=\frac{1}{\lambda}=\frac{1}{5}=0.2$; variance $=\frac{1}{\lambda^2}=\frac{1}{25}=0.04$

Exercise 26.2B

1 a $F(t)=P(T<t)=1-P(T>t)=1-P(\text{no events in }t)$
$=1-e^{-6t}$

b $f(t)=6e^{-6t}$

c $1-e^{-6M_T}=\frac{1}{2}\Rightarrow e^{-6M_T}=\frac{1}{2}$

$6M_T=\ln 2\Rightarrow M_T=\frac{1}{6}\ln 2=0.12$ (2 d.p.)

d 0.14

2 a 0.018 **b** 22.0

3 a $T\sim\text{Exp}(4)$

0.02

b 0.02. Exponential distribution is memoryless.

c 1.00

4 a $T \sim \text{Exp}\left(\frac{1}{2.1} = 0.476\ldots\right)$

0.38 (2 d.p.)

b 0.24

c Failure occurs at random, which is not a characteristic of wear and tear.

5 a $f(t) = \frac{1}{3}e^{-\frac{1}{3}t}$; 3, 9

b i $\frac{1}{14}$ hour **ii** 0.311

6 a 0.186 **b** 39.06

7 a Mean $\frac{2}{15}$ per minute

$f(t) = \frac{2}{15}e^{-\frac{2}{15}t};\ t \geq 0.\ F(t) = 1 - e^{-\frac{2}{15}t};\ t \geq 0.$

b 0.20

c Mean = 7.5; variance = 56.25

Review exercise 26

1 a $= \frac{12}{7}$

$F(x) = \frac{12}{7}\left\{\frac{x^4}{4} - \frac{2x^3}{3} - \frac{3x^2}{2} + \frac{7}{12}\right\}$

b $F(-0.52) = \frac{12}{7}\left\{\frac{0.52^4}{4} + \frac{2 \times 0.52^3}{3} - \frac{3 \times 0.52^2}{2} + \frac{7}{12}\right\} = 0.50$ (2 d.p.)

$M \approx -0.52$

2 $f(x) = \begin{cases} \frac{(1-3x)}{2}; & a < x < \frac{1}{3} \\ 0; & \text{otherwise} \end{cases}$

$\frac{1}{2}\int_a^{\frac{1}{3}} (1-3x)\,dx = 1$

$\frac{1}{2}\left[x - \frac{3x^2}{2}\right]_a^{\frac{1}{3}} = 1$

$\frac{1}{2}\left\{\left(\frac{1}{6}\right) - \left(a - \frac{3a^2}{2}\right)\right\} = 1$

$a = \frac{1 - 2\sqrt{3}}{3}$

b $F(x) = \left(\frac{x}{2} - \frac{3x^2}{4} + \frac{11}{12}\right)$

3 a $(b-a) \times \frac{1}{k} = 1$

$k = b - a$

b $E(X) = 0$ and $\text{Var}(X) = \frac{4}{3}$

c $F(x) = P(X < x) = (x+2) \times \frac{1}{4} = \frac{(x+2)}{4}$

4 a $P(2 < X < x) = \left(\frac{k}{10}\right)(-x^2 + 10x - 16)$ for $2 < x < 4$.

$P(2 < X < 4) = \left(\frac{k}{10}\right)(-16 + 40 - 16) = 1 - 0.1$

$\Rightarrow 0.8k = 0.9$

$k = 1.125$

b $f(x) = -0.225x + 1.125$

c i $E(X) = 2.55$

ii $P(X < 3) = 0.6625$

5 a $\frac{1}{12}$ s **b** 0.09

6 a Assume that cars arrive randomly and therefore constitute a Poisson process.

0.264

b 1200 **c** 316.8

7 a $F(t) = P(T < t) = 1 - P(T > t) = 1 - P(\text{no events in } t)$

$= 1 - e^{-10t}$

b $1 - e^{-10M_T} = \frac{1}{2} \rightarrow e^{-10M_T} = \frac{1}{2}$

$10M_T = \ln 2 \rightarrow M_T = \frac{1}{10}\ln 2 = 0.069$ (3 d.p.)

c 0.036

8 a $T \sim \text{Exp}(0.6)$

b 0.050

c 0.050 (3dp). Exponential distribution is memoryless; outcome of first two minutes is not 'remembered'.

d 0.11

Assessment 26

1 a $\frac{7}{32}$

b $E(X) = \frac{19}{6}$, $\text{Var}(X) = \frac{11}{36}$

c $E(2X+1) = \frac{22}{3}$, $\text{Var}(2X+1) = \frac{11}{9}$

2 a $k = \frac{1}{10}$

b Section of a positive quadratic curve between $x = 0$ and $x = 2$. y-intercept is 0.1

c 2

d $\frac{1}{10}\int_0^m (1 + 3x^2)\,dx = \frac{1}{2} \Rightarrow \left[x + x^3\right]_0^m = 5$

$m + m^3 - 0 = 5$

$m^3 + m - 5 = 0$ as required

3 a $k = \frac{1}{7}$ **b** $\frac{19}{56}$

c $f(x) = \begin{cases} \frac{3}{7}x^2; & 1 \leq x \leq 2 \\ 0, & \text{otherwise} \end{cases}$

4 a Rectangular distribution drawn on axes, with boundaries $x = 1$, $x = 5$, $y = 0.25$; graph continues along x-axis for $x \leq 1$ and $x \geq 5$

b $P(X > 2) = \frac{3}{4}$

c $E(X) = 3$, $\text{Var}(X) = \frac{4}{3}$

5 a Mean = 4000 days, standard deviation = 4000 days

b 0.451

6 a $f(x) = \begin{cases} 5e^{-5x}; & x > 0 \\ 0; & \text{otherwise} \end{cases}$

b $F(x) = \begin{cases} 0; & x < 0 \\ 1 - e^{-5x}; & x > 0 \end{cases}$

c 0.865

7 a $F(x) = \begin{cases} 0; & x < 0 \\ \frac{4}{3}\left(\frac{x^4}{4} + \frac{x^2}{2}\right); & 0 \leq x \leq 1 \\ 1; & x > 1 \end{cases}$

b $E(5X - 3) = \frac{5}{9}$, $\text{Var}(5X - 3) = \frac{101}{81}$

8 a $F(x)=\begin{cases} 0; & x<2 \\ \frac{x^2}{48}+\frac{1}{12}x-\frac{1}{4}; & 2\le x\le 6 \\ 1; & x>6 \end{cases}$

b $-2+2\sqrt{10}$ c $\frac{13}{486}$

9 $F(x)=\begin{cases} 0; & x<4 \\ \frac{x}{6}-\frac{2}{3}; & 4\le x\le 10 \\ 1; & x>10 \end{cases}$

10 a $\frac{4}{9}$

b Phone calls may last longer than 10 minutes.

c i 0.2497 ii 0.403

11 $\int_0^a 2x^2e^{-2x}dx=\left[\left(-x^2-x-\frac{1}{2}\right)e^{-2x}\right]_0^a$

$=\left(-a^2-a-\frac{1}{2}\right)e^{-2a}-\left(0-0-\frac{1}{2}\right)e^0$

$=-\left(a^2+a-\frac{1}{2}\right)e^{-2a}+\frac{1}{2}$

$\int_0^\infty 2x^2e^{-2x}dx=\lim_{a\to\infty}\left(\frac{1}{2}-\left(a^2+a-\frac{1}{2}\right)e^{-2a}\right)$

$=\frac{1}{2}$

$\int_0^a 2xe^{-2x}dx=\left[\left(-x-\frac{1}{2}\right)e^{-2x}\right]_0^a$

$=\left(-a-\frac{1}{2}\right)e^{-2a}-\left(-0-\frac{1}{2}\right)e^0$

$=-\left(a+\frac{1}{2}\right)e^{-2a}+\frac{1}{2}$

$\int_0^\infty 2xe^{-2x}dx=\lim_{a\to\infty}\left(\frac{1}{2}-\left(a+\frac{1}{2}\right)e^{-2a}\right)$

$=\frac{1}{2}$

$\mathrm{Var}(X)=\frac{1}{2}-\left(\frac{1}{2}\right)^2$

$=\frac{1}{4}$

12 pdf is $f(u)=\frac{1}{k}$ for $a\le u\le b$

$\mathrm{Var}(U)=\int_a^b\frac{u^2}{k}du-\left(\int_a^b\frac{u}{k}du\right)^2$

$=\left[\frac{u^3}{3k}\right]_k^{2k}-\left(\left[\frac{u^2}{2k}\right]_k^{2k}\right)^2$

$=\frac{8k^3-k^3}{3k}-\left(\frac{4k^2-k^2}{2k}\right)^2$

$=\frac{7k^2}{3}-\frac{9k^2}{4}$

$=\frac{28k^2-27k^2}{12}$

$=\frac{1}{12}k^2$ as required

13 a $F(x)=\begin{cases} 0; & x<1 \\ \ln x; & 1\le x\le e \\ 1; & x>e \end{cases}$

b Standard deviation is $\frac{1}{2}\sqrt{8e-2e^2-6}$

14 a $2\sqrt{3}$

b $E(2X^{-1})=2-\sqrt{2}$, $\mathrm{Var}(2X^{-1})=\ln 2-6+4\sqrt{2}$

15 a

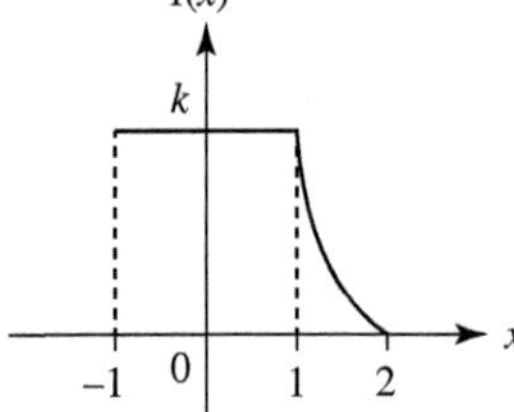

b $k=\frac{3}{7}$

c $F(x)=\begin{cases} 0; & x<-1 \\ \frac{3}{7}x+\frac{3}{7}; & -1\le x<1 \\ \frac{1}{7}(x-2)^3+1; & 1\le x\le 2 \\ 1; & x>2 \end{cases}$

d $\frac{1}{6}$

e $E(4X)=\frac{5}{7}$, $\mathrm{Var}(4X)=\frac{3039}{245}$

16 0.283

Exponential distribution with $\lambda=\frac{2}{3}$

Chapter 27

Exercise 27.1A

1 While both concern mistakes, type I errors occur when the null hypothesis is true and type II errors occur when the alternative hypothesis is true.

2 a 5% b 48.2% c 0.0588%

3 a $X\ge 12$ b 4.48% c 73.0% d 0.27

4 a 77.1%, 0.229 b 41.6%, 0.584 c 2.10%, 0.979

5 a 0.852

b As the true mean gets further from the hypothesis mean, the smaller the probability of obtaining an outcome not in the critical region, so the probability of making a type II error decreases.

6 a 11 or more b 0.490

7 a They would get a result inside the critical region and determine that the mean weight is not the advertised weight, when actually it is.

b They have accepted a null hypothesis which is false, so they have made a type II error.

8 a Because $P(X\le 15)=9.55\%<10\%$ and $P(X\le 16)=15.61\%>10\%$

b 9.55% c 71.99%

d $N(50\times0.4, 50\times0.4\times0.6)$. For a Normal distribution you can test the mean value, which corresponds to the expected number of outcomes ($n\times p$).

e 15.56

f $N(17.5,11.375).P(X\geq15.56\,|\,X\sim N(17.5,11.375))=71.74\%$

9 a A type I error occurs when a null hypothesis that is true is rejected. If both tests have the same probability of making such an error then a fair comparison can be made. It is easy to control the probability of making a type I error.

b A type II error occurs when a null hypothesis that is false is rejected. The statistician should use the test that has a lower probability of making such an error, since overall it produces a lower probability of error.

Exercise 27.1B

1 a Critical value = 1.07, P(type II error) = 35%, power = 0.650

b Critical value = 1.29, P(type II) = 49.1%, power = 0.509

c Critical value = 1.70, P(type II) = 74.5%, power = 0.255

2 a i $X\geq5$ **ii** 91.6% **iii** 0.0838

b i $X\geq6$ **ii** 97.0% **iii** 0.0300

3 a 8

b P(type II error) = 97.8%. Power = 0.022

c 800

d This expected number is inside the critical region. You should expect the power to be much closer to 1 than for the 20-trial test, due to the larger sample size.

4 a The treasure hunter does not want to miss any valuables.

b If the hunter digs a little and finds nothing valuable then they can keep trying other locations.

5 a $X=0$ **b** 1.43%

c P(type II error) = 99.99038%, power = 0.00961%

6 a $X\geq5$; 0.249 **b** $X\geq8$; 0.399 **c** $X\geq11$; 0.492

7 a Since 5 games are played, 1.2 goals per game leads to 6 goals per 5 games. The alternative is > 6 because the team thinks they might be scoring more goals. A two-tailed test would be inappropriate here as the team is not concerned with the probability of the training leading to fewer goals.

b $X\geq11$; 4.26% **c** 90.1%

d Most fans probably wouldn't notice an extra goal every 5 away matches, and it would have a small impact on overall performance, game by game or in the league.

e Bookmakers offer odds on specific scores in matches and this is a popular type of bet. An extra goal every 5 away games will change the odds enough that the company could lose a lot of money making bad predictions.

Exercise 27.2A

1 a 2.271 **b** −1.559 **c** −6.623 **d** −4.964

2 a 0.985 **b** 1.92 **c** 1.60

3 a −2.669 **b** 11 **c** 0.0109

d This is smaller than the significance level so there is sufficient evidence to reject the null hypothesis. You conclude that 1.61 is not a reasonable value for the mean of the parent population.

4 a $T<-1.75$ **b** $t=-0.372$

c This is not in the critical region; there is insufficient evidence to reject the null hypothesis. You conclude that −2.1 is a reasonable value for the mean of the parent population.

5 t-test statistic is −2.860. The p-value is 0.670%, or the critical region is $T<-2.65$ or $T>2.65$. The result is significant, so 3.17 is not a reasonable value for the mean of the parent population.

6 a $t=-1.88$ **b** 7 **c** 5.10%

d The p-value is larger than the significance level so there is insufficient evidence to reject the null hypothesis. You conclude that 18.2 is a reasonable value for the mean of the parent population.

e The population must be normally-distributed and the sample must be random.

7 a $T<-2.13$ **b** −0.608

c The t-test statistic is smaller than the critical value. There is insufficient evidence to reject the null hypothesis. You conclude that 129.1 is a reasonable value for the mean of the parent population.

8 a Yes as heights are reasonably approximated by a Normal distribution and the sample can be taken at random.

b The student would need to ensure that the sample is taken at random from the population, therefore avoiding possible bias.

Exercise 27.2B

1 a When the sample size is large, the distribution approaches a Normal distribution.
When the sample size is 30 the distribution is approximately a Normal distribution, so when it is 1000 the difference will be negligible and it is acceptable to use a Normal distribution.

b 8.92

c The critical value is 1.6463803
8.92 > 1.646... so reject H_0 in favour of H_1 and conclude that there is sufficient evidence at the 5% level of significance to suggest that the true mean is greater than 12.1

2 a The t-test statistic is −0.157, which has a p-value of 44.1%. This is larger than the significance level so there is insufficient evidence to reject the null hypothesis.

b The t-test statistic is −1.553, which has a p-value of 7.74%. This is smaller than the significance level so there is sufficient evidence to reject the null hypothesis.

c The t-test statistic is −1.095, which has a p-value of 14.6%. This is larger than the significance level so there is insufficient evidence to reject the null hypothesis.

3 a 13.4 **b** 20.2 **c** 4.14

4 a The cans should not have been produced together in the factory. Cans should be purchased from different locations if possible, and over a period of time.

b The sample must come from a Normal distribution for a t-test to be appropriate.

c The p-value of the sample is 6.46%. This is larger than the significance level so there is insufficient evidence to suggest that the sample has not come from a parent Normal distribution with mean 4.26 g.

5 a 1.83%

b The p-value is smaller than the significance level so there is sufficient evidence to reject the null hypothesis. You conclude that the dart player is throwing more than 0.31 cm away from the bullseye, on average.

c While the mean is higher, it could be that the dart thrower is consistently getting the darts to land 0.68 cm away at a specific point and not all around the target radially.

6 a For sample 1 the t-test statistic is −1.840. This has a p-value of 4.95%, which is smaller than the significance level. There is sufficient evidence with sample 1 to suggest that the true population mean is less than 14.7
For sample 2 the t-test statistic is −1.779. This has a p-value of 4.85%, which is smaller than the significance

level. There is sufficient evidence with sample 2 to suggest that 14.7 is not a suitable mean value of the population Normal distribution.

b For the combined sample the t-test statistic is –2.51. This has a p-value of 0.96%, which is smaller than the significance level. There is sufficient evidence with the combined sample to suggest that 14.7 is not a suitable mean value of the population Normal distribution.

7 a The t-test statistic is –1.403, which has a p-value of 9.20%. The significance level for this tail is 5%, so the result is not significant. There is insufficient evidence to reject the null hypothesis.

b The t-test statistic is 1.426, which has a p-value of 9.09%. The significance level for this tail is 5%, so the result is not significant. There is insufficient evidence to reject the null hypothesis.

c Each set of pipes could reasonably have come from a machine with mean 30.0 cm, so the engineer should believe that the two sets of pipes have come from the same machine. They might want larger samples to get a better idea though.

8 a (12.1, 13.6) **b** (9.28, 12.46)

9 a (0.241, 18.3) **b** (9.22, 12.5)

10 a (18.8, 31.7) **b** (21.4, 28.6) **c** (22.9, 27.1)

Review exercise 27

1 a $\alpha \le 5$ **b** $P_2 > P_1$

2 a $X \ge 18$ **b** 7.96%

c i 44.1% **ii** 0.559

3 a 2% **b** 89.2% **c** 33.2%

4 a 0.536 **b** –1.24 **c** –0.182

5 a The sample size isn't too large. **b** 0.457

6 a 6 **b** 0.0142

c Since the p-value is lower than the significance level there is sufficient evidence to reject the null hypothesis. You conclude that the mean is lower than –5.14.

7 a 72.14 **b** 25.87 **c** –20.43

Assessment 27

1 a 9 **b** 0.028 **c** 0.593

2 a $H_0: p = 0.3, H_1: p \neq 0.3$ **b** $X \le 2, X \ge 10$

c 0.751 **d** 0.249

3 a 29.14 **b** 0.39 **c** 0.61

4 a A type II error occurs when the null hypothesis is accepted when it is false. The power of a test is the probability of the complementary event when the null hypothesis is rejected when it is false.

b i $H_0: \mu = 330, H_1: \mu > 330$ **ii** $\bar{X} < 332.08$

iii 0.70 **iv** 0.30

5 a 7.7317 **b** –1.692

c p value = 0.1187, therefore do not reject null hypothesis

6 a Sample is not large enough to assume that sample and population standard deviations are approximately equal.

b i $\mu = 10.2, H_1: \mu > 10.2$ **ii** 2.8644

iii p value = 0.0093, therefore reject the null hypothesis.

7 a $\mu = 0, H_1: \mu > 0$

b X is normally distributed; $t = 1.8861$, p value = 0.048, therefore reject null hypothesis.

c Result is only just significant; small sample suggests further research would be a good idea.

8 (13.39, 18.18)

9 (4.03, 9.43)

10 a Crop yield is normally distributed.

b For a Normal distribution with unknown population variance, sample size must be large.

c 44.2 and 1.31

Chapter 28

Exercise 28.1A

1 a Either DEF or EFG. For example {AB, AC, BC} correspond to {DE, DF, EF}.

b The sub-graph formed by removing EF can be draw like this:

This is a subdivision (by G) of DEF

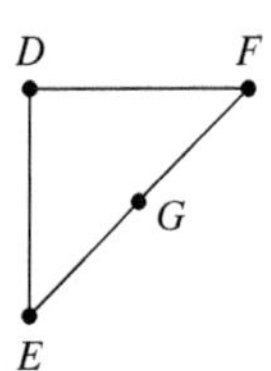

2 Removing the edge CD gives a subdivision (by C) of the K_4 graph $ABDE$

3 Removing edges AB and CD gives the complete bipartite graph joining the vertex sets {A, B, E} and {C, D}.

4 G_1 and G_4 are isomorphic – {A, B, C, D, E} corresponds to {V, X, Y, W, Z}

G_2 and G_3 are isomorphic – {J, K, L, M, N} corresponds to {Q, P, R, T, S}.

5 G_1 is planar:

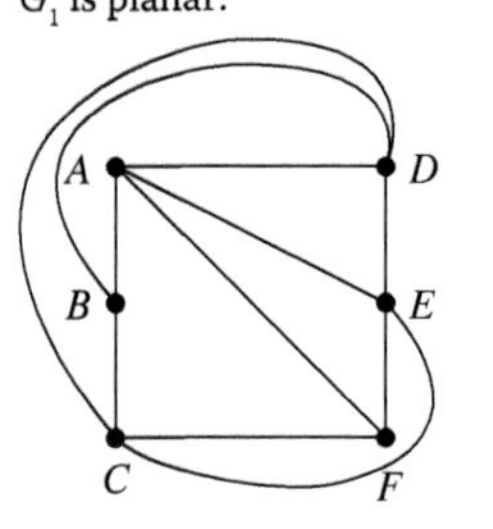

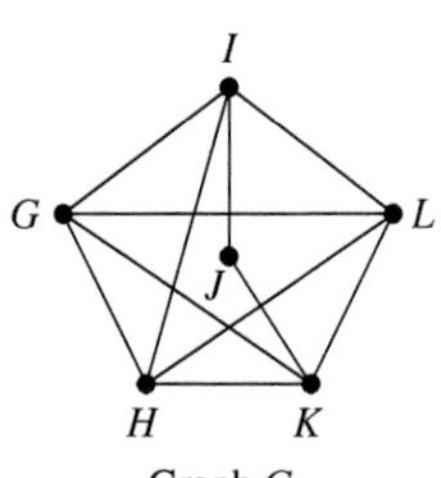

Graph G_2

G_2 contains a subdivision (by J) of the K_5 graph formed by G, H, I, K and L, so it is non-planar.

G_3 contains the complete bipartite graph $K_{3,3}$ joining {M, P, R} and {N, O, Q}, so is non-planar.

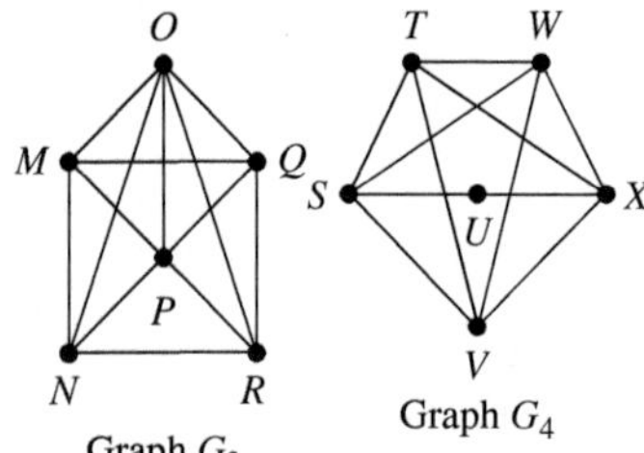

Graph G_3

Graph G_4

G_4 contains a subdivision (by U) of the K_5 graph formed by S, T, V, W and X, so it is non-planar.

6 a

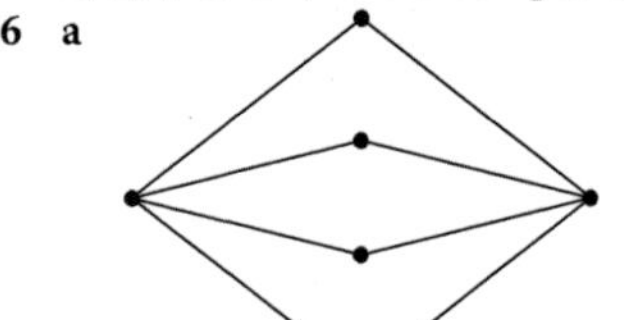

b

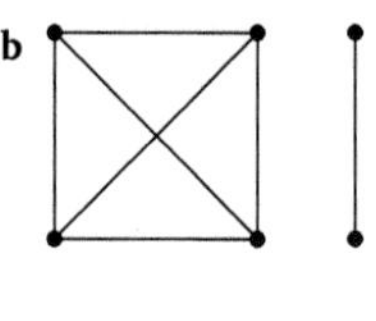

7 a

	A	B	C	D	E
A	0	1	0	1	0
B	1	0	1	0	0
C	0	1	0	1	1
D	1	0	1	0	0
E	0	0	1	0	0

b

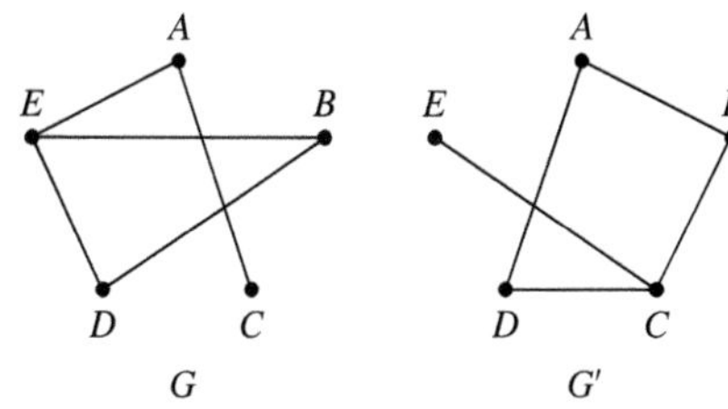

c Not isomorphic. For example, the vertex with degree 1 in G connects to a vertex of degree 2, but in G' it connects to a vertex with degree 3

8 The graph can be drawn as shown. Removing EG and BD gives a subdivision (by G) of $K_{3,3}$. Hence by Kuratowski's theorem the graph is non-planar.

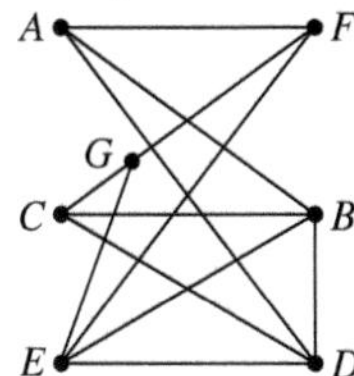

Exercise 28.1B

1 **a** $n-1$

b If T and T' are isomorphic they have the same number of edges.

K_n has $\frac{n(n-1)}{2}$ edges, so T and T' must each have $\frac{n(n-1)}{4}$

Hence $\frac{n(n-1)}{4}=n-1 \Rightarrow n^2-5n+4=0$, giving

$n=4$ (or $n=1$ which is trivial).

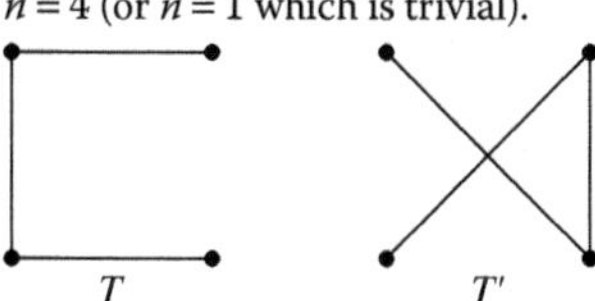

2 **a** There are 4

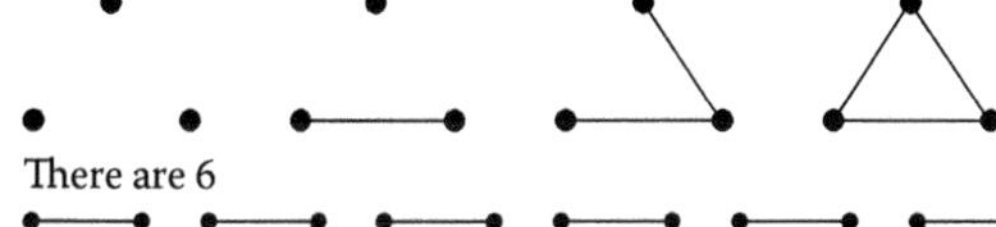

b There are 6

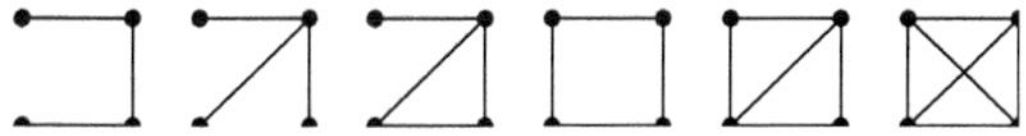

c **i** There are 4

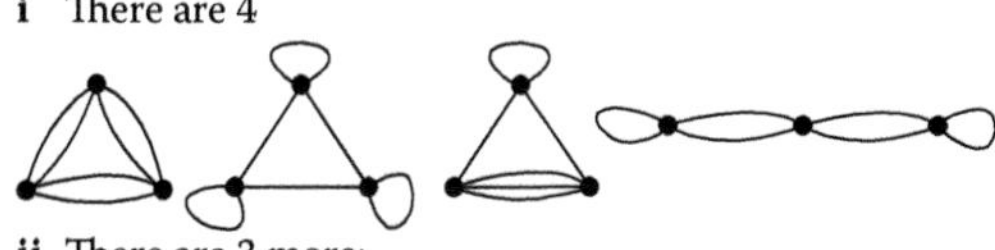

ii There are 3 more:

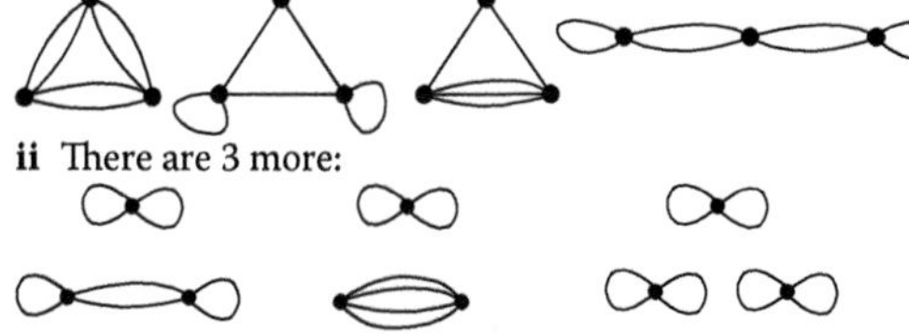

3 **a** d is odd but these odd vertices occur in pairs, so n is even.

b $d=2$ or 4 ($d=3$ gives $K_{3,3}$ and $d=5$ gives K_5 as a subgraph)

4 **a**

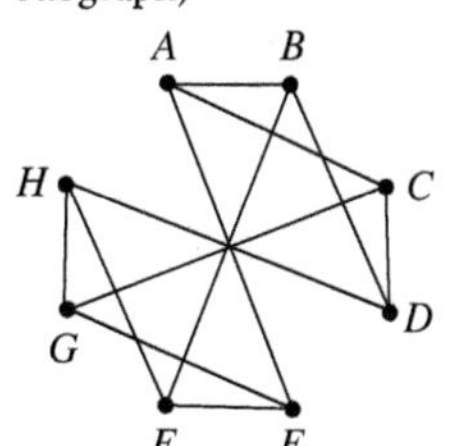

b **i** 2 **ii** 3

c AC and EG 'inside', BF and DH 'outside', so planar plane drawing:

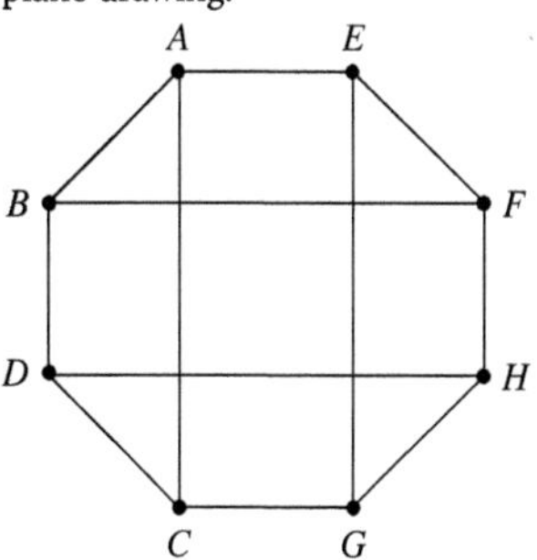

(The plane drawing is the cube graph, which you may have realised because the 'codes' are the 3D coordinates of the vertices of a unit cube.)

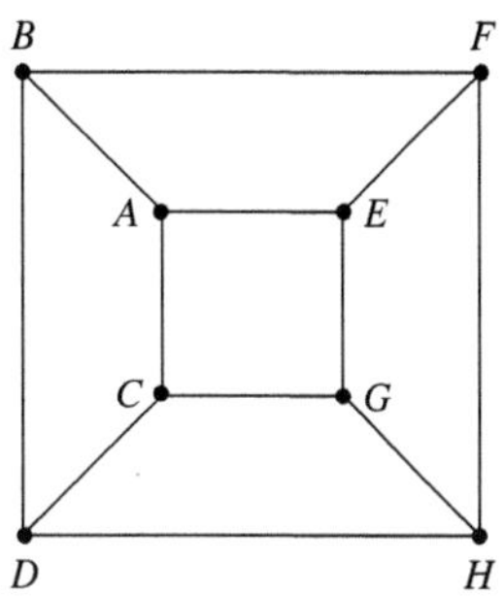

5

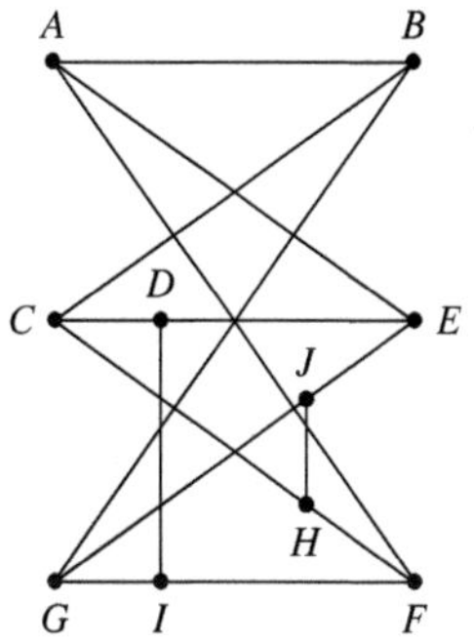

6 **a**

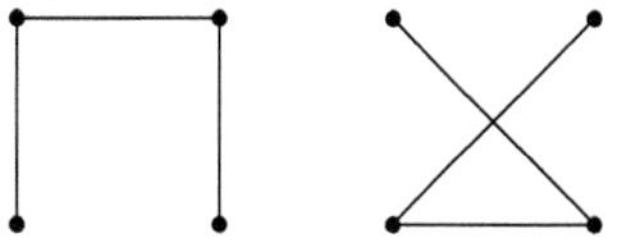

b

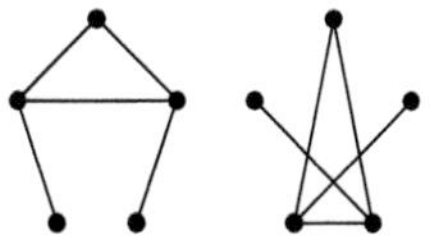

and

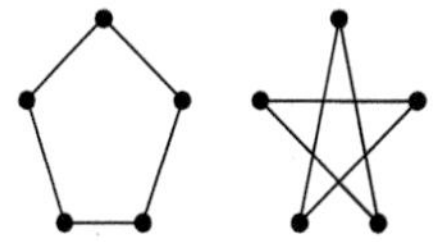

c **i** K_6 has 15 edges. As G and G' combine to form K_6, they must have different numbers of edges, so they are not isomorphic.

ii $\frac{n(n-1)}{2}$ must be even, so either n or $(n-1)$ is a multiple of 4

So $n=4m$ or $n=4m+1$

7 a

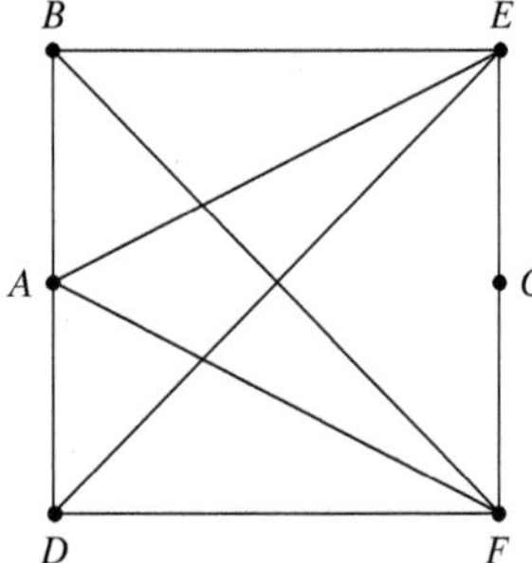

b $m \geq 3$ and $n \geq 3$

c Suppose the vertices of H are in two sets, $\{A_1, A_2, ...A_m\}$ and $\{B_1, B_2, ..., B_n\}$.
For vertices A_i and B_j there will be an edge A_iB_j in H'
Whether or not A_i and A_k were connected in H, there will be a trail $A_iB_jA_k$ between them in H'. Similarly, there will be a trail joining any two B vertices.
Hence there is a trail connecting any two vertices in H', so H' is a connected graph.
If H consists of more than two disjoint sets of vertices, the same argument applies between any two sets, so H' will be connected.

8 The graph does not contain a sub-division of K_5 (although you could generate one by "collapsing" C down onto D. The resulting graph is called a minor. Try Googling Wagner's theorem. A similar process is possible for the Petersen graph in question 5).
However, the graph does contain $K_{3,3}$ as is shown in this diagram, so the graph is non-planar.

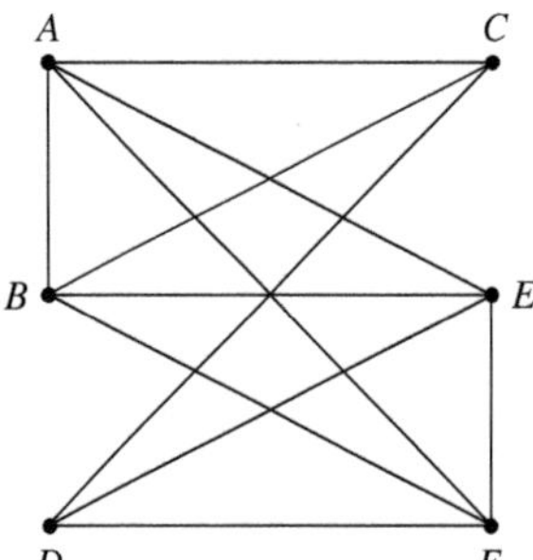

Exercise 28.2A

(Final flows are shown in black. There may be other possible flow patterns.)

1 a Flow-augmenting paths SABT 2 and SABCT 4
Maximum flow = 45

b There is a cut {SC, BC, BT, AT} with capacity 45. By the maximum flow-minimum cut theorem flow = cut = 45 means 45 is maximal.

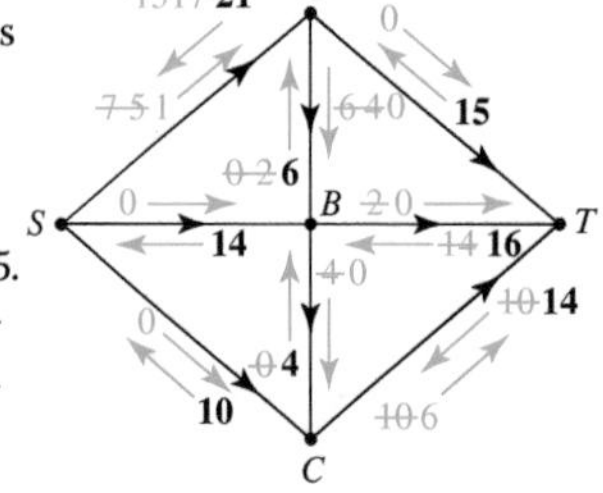

2 a Flow-augmenting paths *SADET* 10 and *SACET* 10
Maximum flow = 100

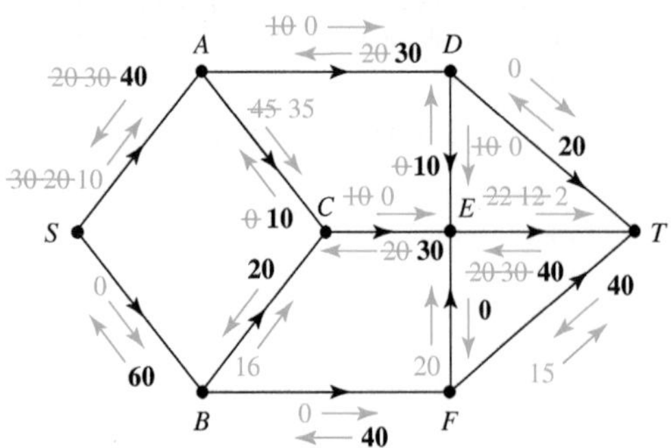

b Cut {*AD*, *CE*, *BF*} = 100
Flow = cut = 100 means 100 is maximal.

3 Flow-augmenting path *SBCADET* 10
Maximum flow = 100

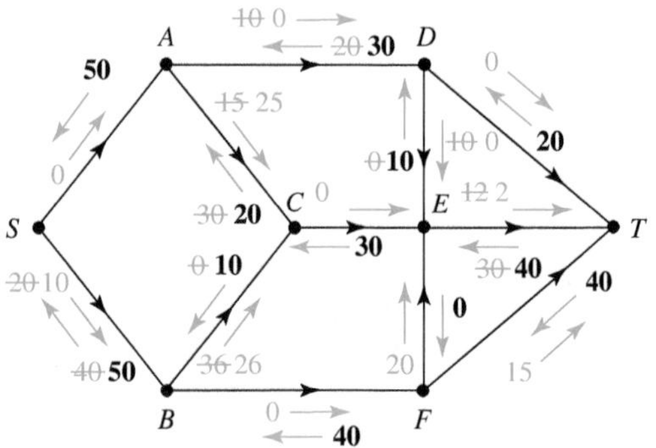

4 Flow-augmenting paths
SBCET 2, *SACT* 2, *SADCT* 1 and *SABCET* 1

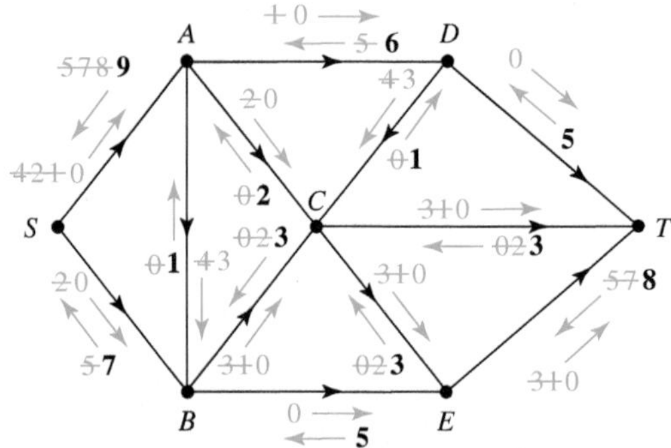

5 a Take, for example, initial flow *SADT* 24, *SCT* 38 and *SBET* 40
Flow augmenting paths *SCET* 14 and *SACBET* 2
Maximum flow = 118

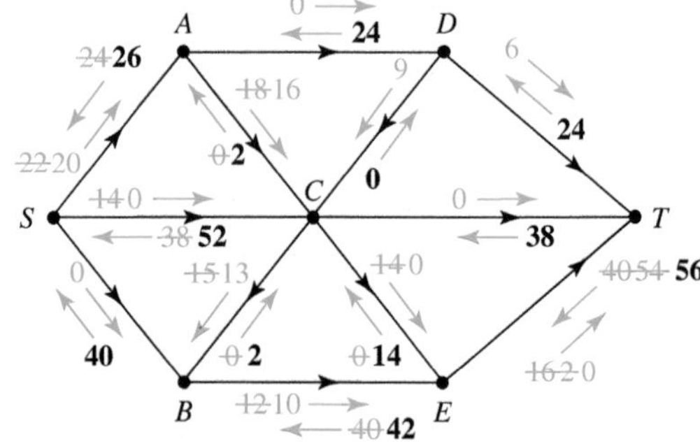

b Cut {*AD*, *CD*, *CT*, *ET*} = 118
Flow = cut = 118 means 118 is maximal.

6 These are just examples.

a

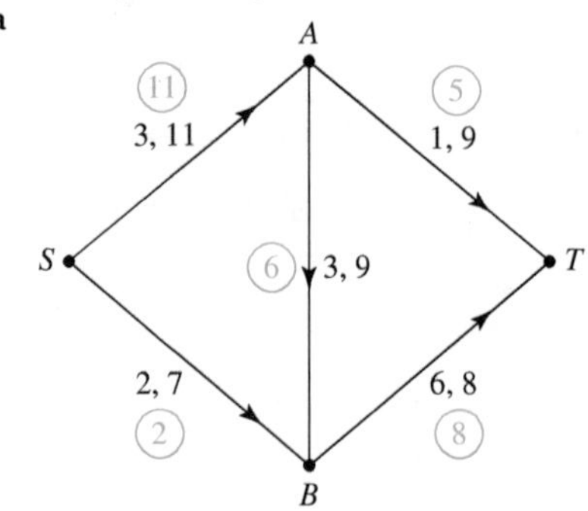

b No flow exists, because at B maximum inflow = 6 but minimum outflow = 5 + 3 = 8

c

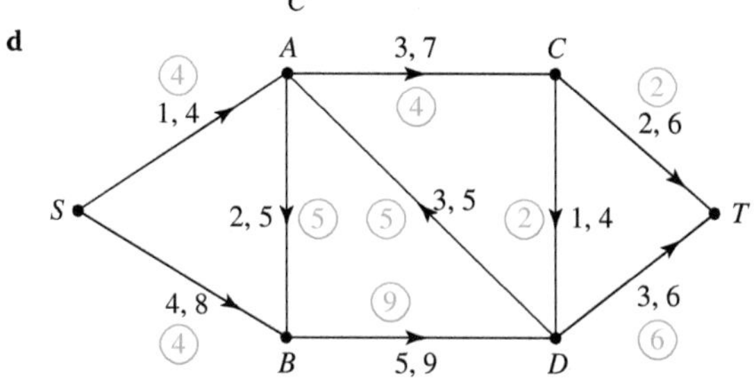

d

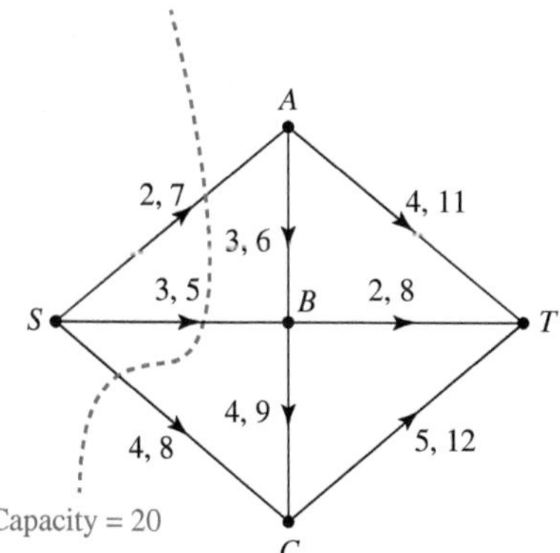

7 a

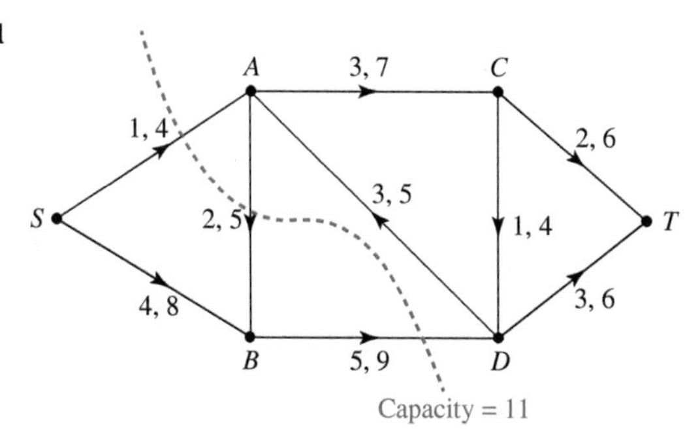

b No feasible flow

c

d

8 a Flow augmenting paths $SACT$ 1 and $SBDT$ 2
Maximum flow = 15
Final flows as in second diagram.

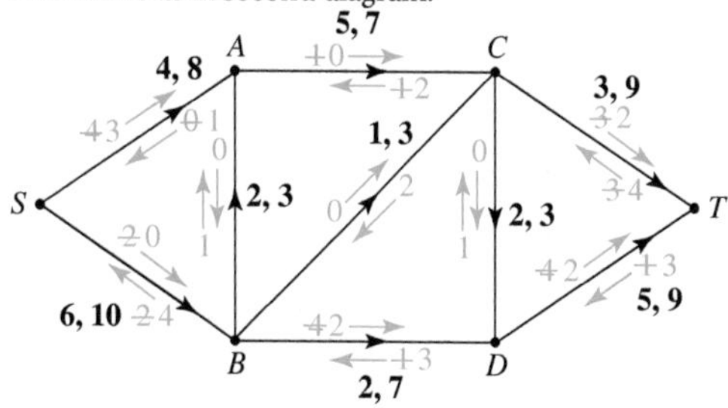

b Cut $\{SB, AB, AC\}$ = 15
Flow = cut, so maximal.

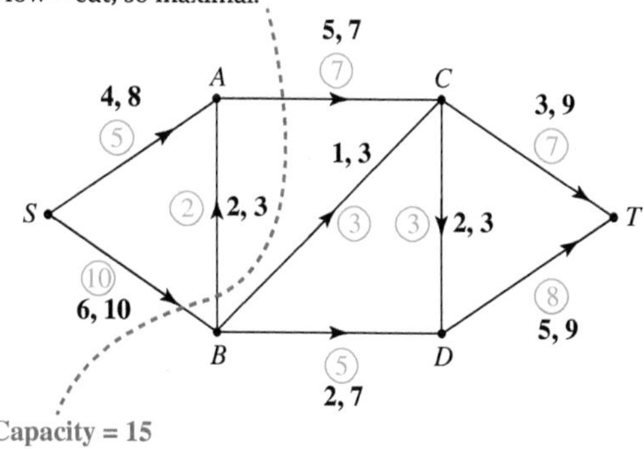

Exercise 28.2B

1 a, b Take, for example, initial flow $SEFG$ 8, $SACG$ 3 and $SADG$ 2
Flow-augmenting paths $SACDG$ 1 and $SABCDG$ 1
Maximum flow = 15

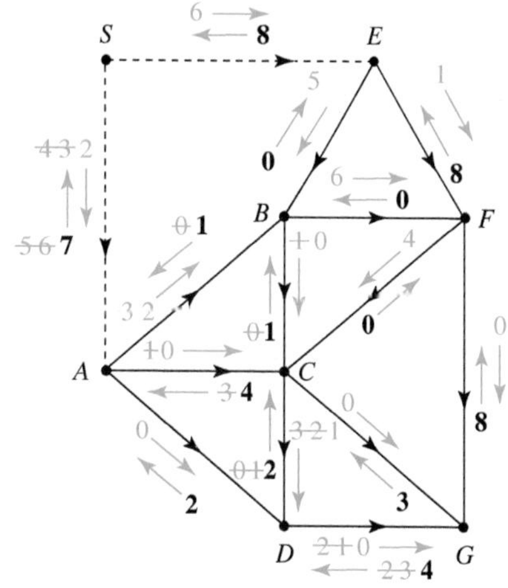

c

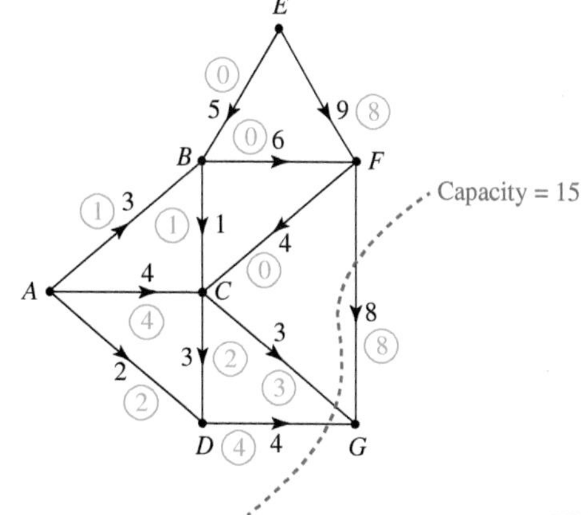

2 a, b Take, for example, initial flow *SACFIT* 5 and *SBEHT* 6
Flow-augmenting paths *SBDGJHT* 5, *SBDGIT* 2 and *SBEDGHT* 1

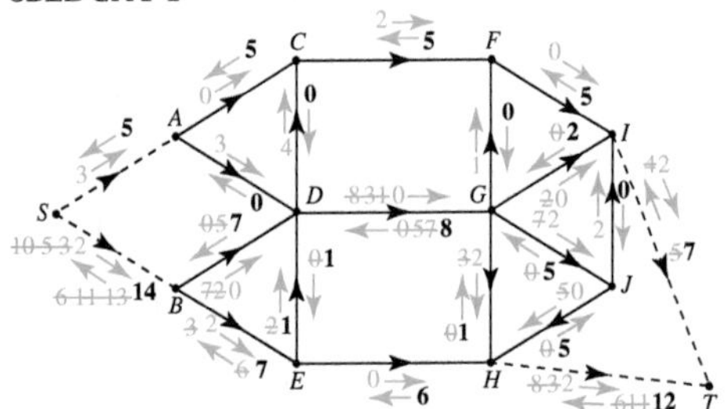

Maximum flow = 19
This is confirmed by cut {*EH*, *DG*, *FG*, *FI*} = 19
So expected revenue = 1900 × 0.4 × 4 = £3040

3 Convert *B* to B_1B_2 with capacity 7
Take, for example, initial flow SB_1B_2T 4 and *SACT* 4
Flow-augmenting path SAB_1B_2T 3
Maximum flow = 11
Confirmed by cut {B_1B_2, *AC*} = 11

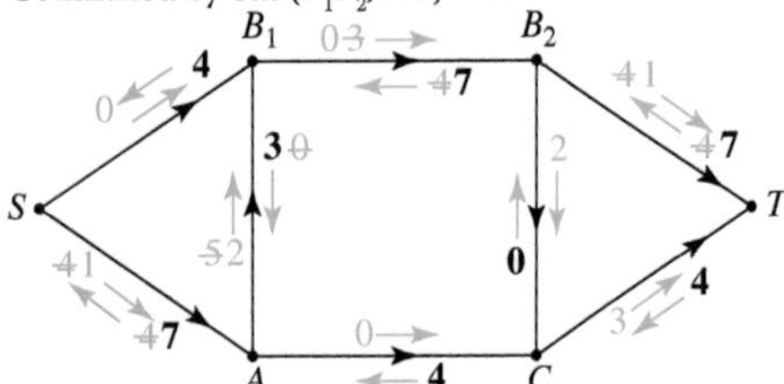

4 a Sources *B* and *D*, sinks *I* and *K*

b Take, for example, initial flow *SBAEIT* 6, *SBFJIT* 5, *SBCGKT* 5 and *SDHLKT* 10
Flow-augmenting path *SBFJKT* 2
Maximum flow = 28

c Cut {*AE*, *EF*, *FJ*, *GK*, *KL*} = 28

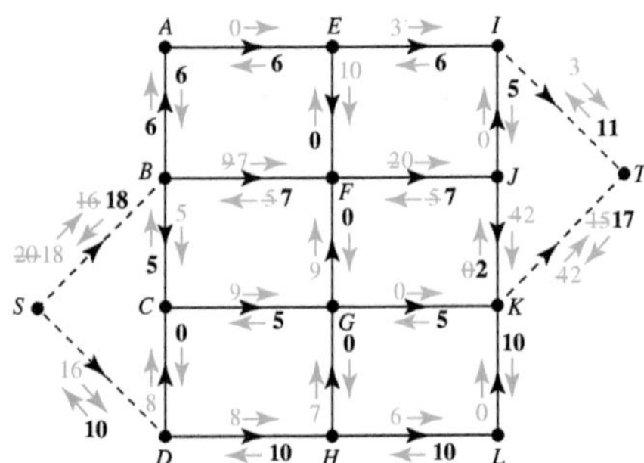

5 a Sources *A* and *B*, sinks *F* and *G*

b, c Take, for example, initial flow SAC_1C_2DFT 5 and $SBDE_1E_2GT$ 5
Flow-augmenting paths *SADGT* 4, SBE_1E_2GT 1 and SBE_1DC_2FT 1
Maximum flow = 16
Confirmed by cut (C_1C_2, *AD*, *BD*, BE_1} = 16

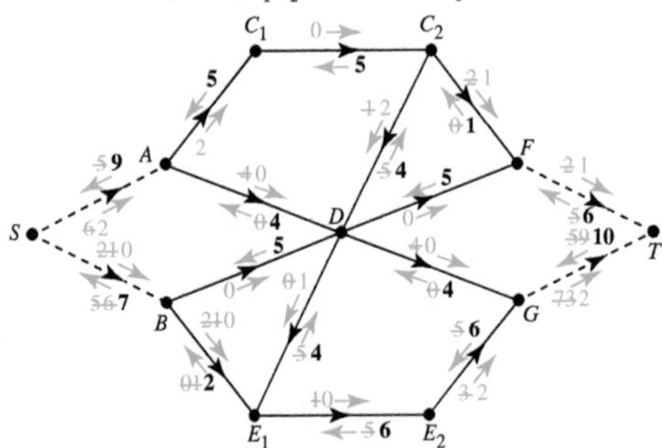

6 The first diagram shows working starting from a total initial flow of 13

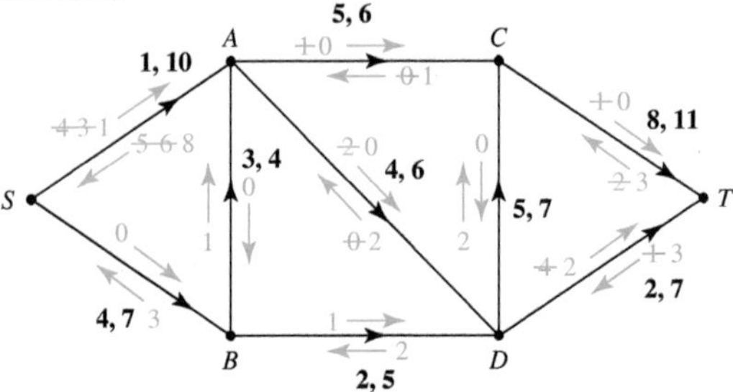

The second diagram shows final flows for a maximum flow of 16, with a cut of 16 to confirm this.

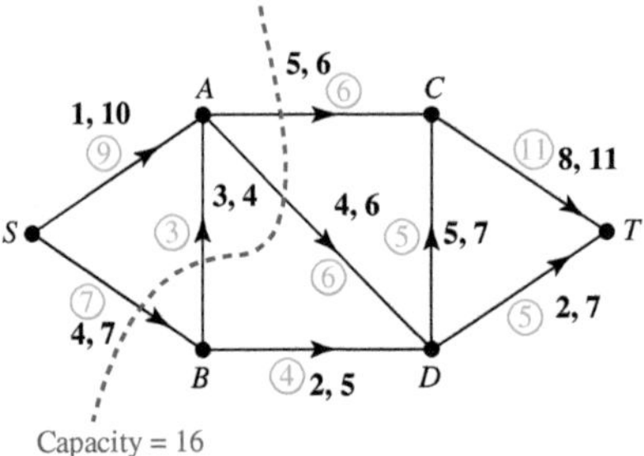

7 The first diagram shows the network with an initial feasible flow of 22

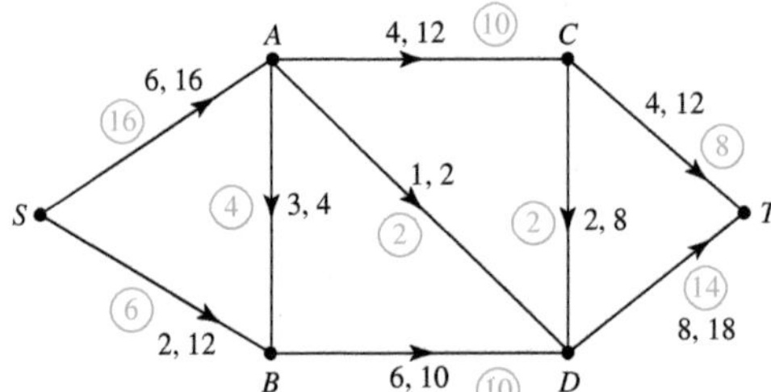

The second diagram shows the working using a flow-augmenting path *SBACT* 1. The flow (23) is then maximal. There is a cut {*SA*, *AB*, *BD*} with a capacity 23

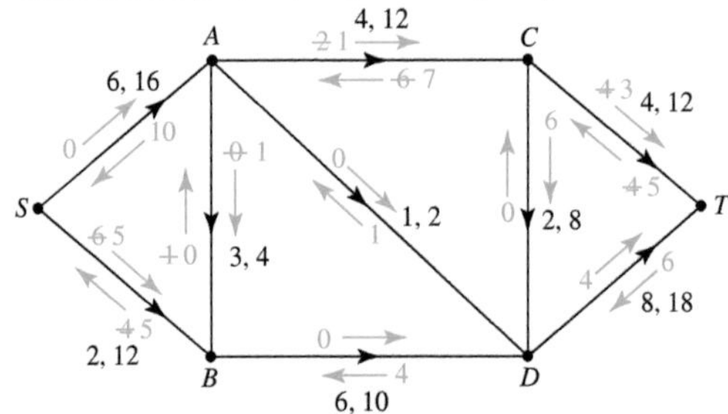

If there were no minimum requirements there is a flow of 24 (*SACT* 12, *SBDT* 10 *SADT* 2) and a cut of 24 {*AC*, *AD*, *BD*}, so the maximum flow is then 24

8 An initial feasible flow is as shown (others are possible).

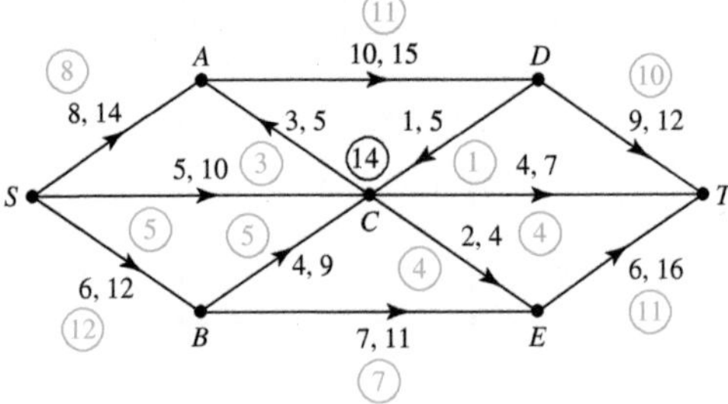

Split C into incoming node C_1 and outgoing node C_2. C_1C_2 has maximum capacity 14, and can be assumed to have minimum capacity 10 to be consistent with incoming values.

The second diagram shows the working using flow-augmenting paths SC_1C_2T 3, $SADT$ 2 and SC_1BET 1. This gives a total flow of 31, consistent with the cut shown.

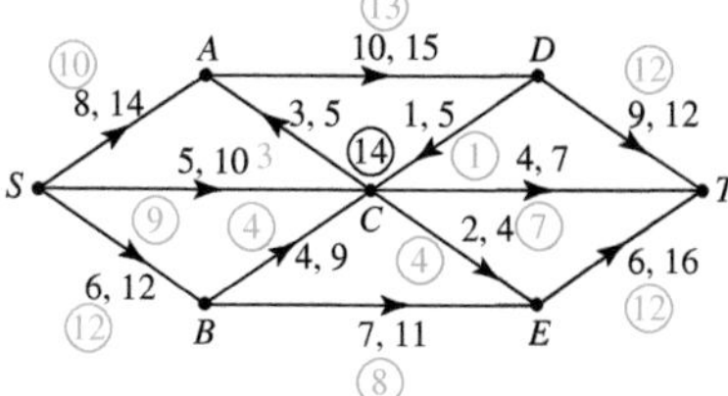

The third diagram shows the final flows.

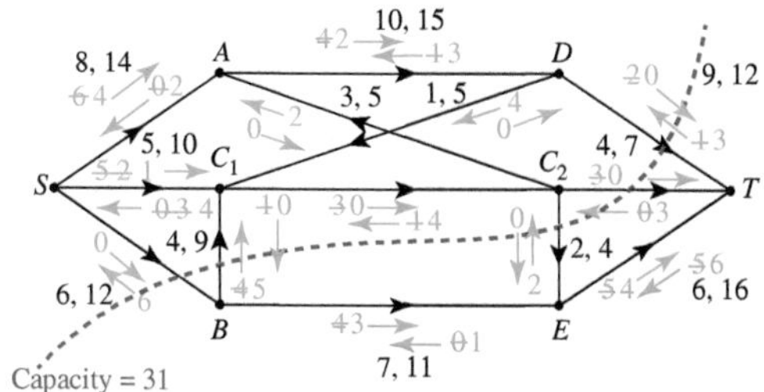

Review exercise 28

1 a

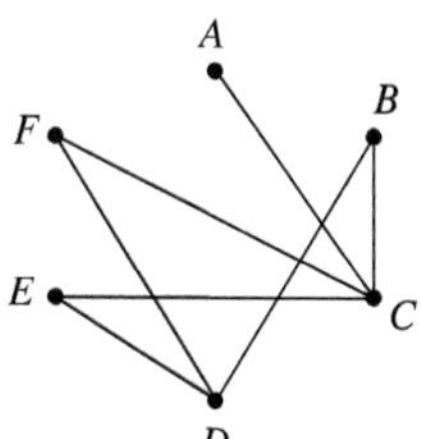

b

	A	B	C	D	E	F
A	0	1	0	1	1	1
B	1	0	0	0	1	1
C	0	0	0	1	0	0
D	1	0	1	0	0	0
E	1	1	0	0	0	1
F	1	1	0	0	1	0

c

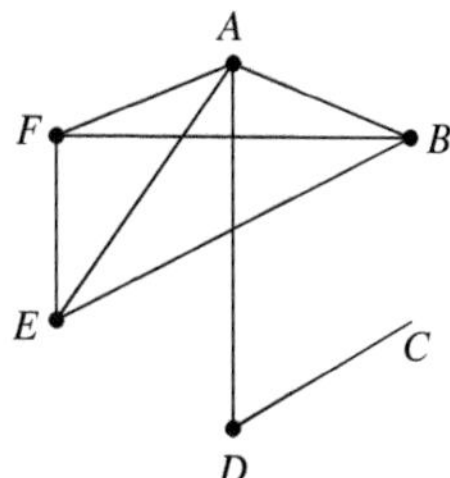

2 G_2 and G_3 are isomorphic. P corresponds to Z, S to Y and the remaining three can be in any order.

3 a $V = 6$ and $E = 9$. $V + F - E = 2$, so $F = 5$

These are two possible graphs.

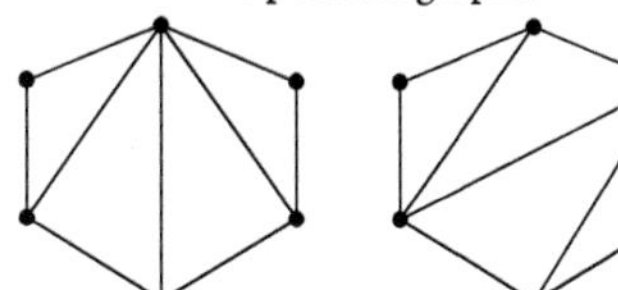

b $V = 7, E = 16$. From Euler a planar graph satisfies $E \leq 3V - 6$. Here $3V - 6 = 15$, so the graph is non-planar.

4 The graph contains a sub-graph which is a sub-division of K_5, as shown, so by Kuratowski's theorem it is non-planar.

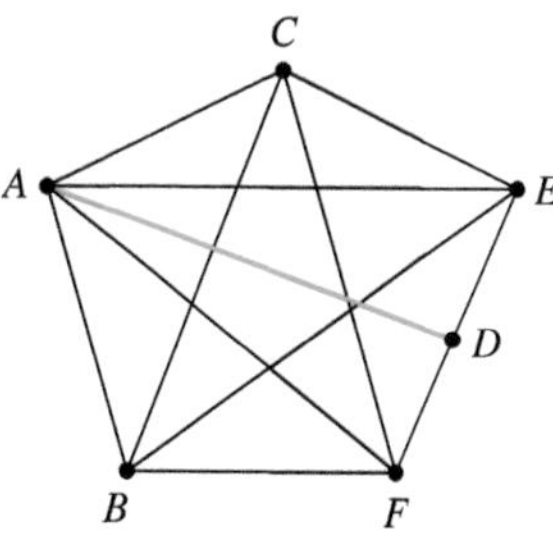

5 a The maximum outflow from B is 14, so the inflow ≤ 14

b i $SADT$ 5 **ii** $SBET$ 9

c Flow-augmenting paths $SBCT$ 4, $SACDT$ 4 and $SBCDT$ 1

Maximum flow = 23

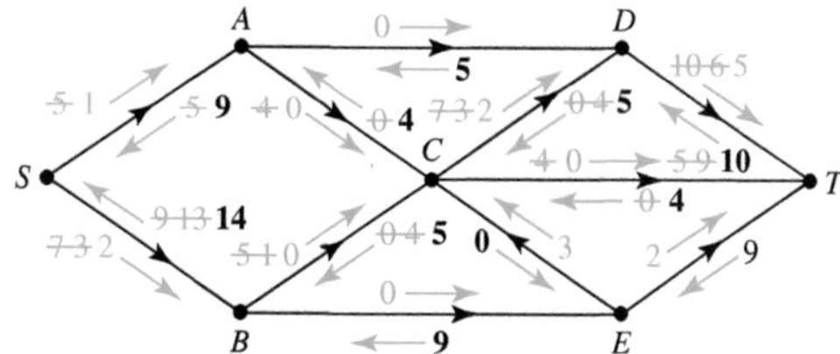

d Cut $\{AD, AC, BC, BE\} = 23$

Flow = cut = 23 means maximum flow is 23 (by the maximum flow-minimum cut theorem).

6 a Sources A and B, sinks, D and E

b, c Take, for example, initial flow $SADT$ 4 and $SBET$ 3

Flow-augmenting paths $SACDT$ 8, $SBCET$ 3 and $SACET$ 2

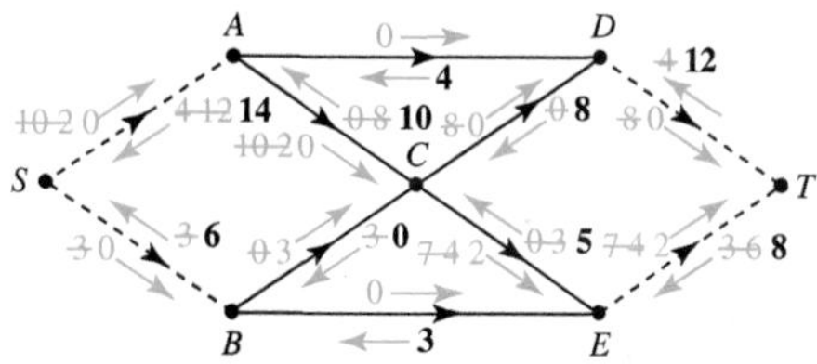

Maximum flow = 20

(You know because SA and SB are saturated.)

7 a, b Take, for example, initial flow $SADT$ 4 and $SBET$ 3

Flow-augmenting path SAC_1C_2DT 8

Maximum flow = 15

Confirmed by cut $(AD, C_1C_2, DE\} = 15$

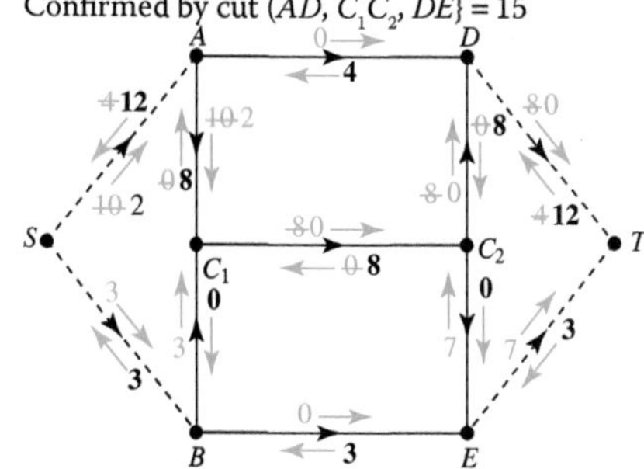

8 a There is no feasible flow for Network 2

Node A has a maximum inflow of 7, but needs a minimum outflow of 8

b i $9 + 12 - 3 = 18$

ii This is the initial flow pattern.

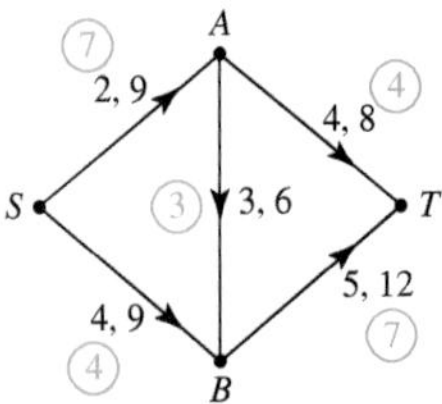

Using the labelling procedure with flow-augmenting paths SAT 2 and SBT 5 gives:

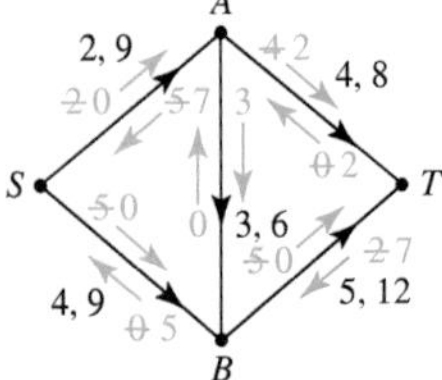

These are the final flows. Total flow = 18

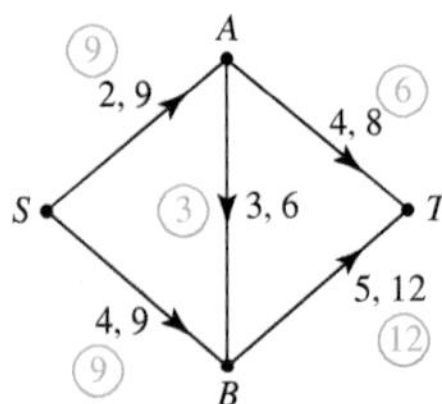

iii Flow = 18 and there is a cut of 18, so the flow is maximal by the maximum flow-minimum cut theorem.

Assessment 28

1 **a** The graph does not contain K_5 or $K_{3,3}$ as a subgraph therefore Kuratowski's theorem tells us it is planar.

2 **a**

	A	B	C	D	E
A	0	0	1	1	0
B	0	0	0	0	0
C	1	0	0	0	1
D	1	0	0	0	0
E	0	0	1	0	0

b **i** G_2 is isomorphic.
$J = A, F = E, I = B, G = D, H = C$

ii G_3 is not isomorphic: order of vertices is 2, 4, 2, 4, 2 not 2, 4, 2, 3, 3

3 **a** Flow-augmenting paths $SBCFES$ (9)
$SBEDS$ (5)
Final network:

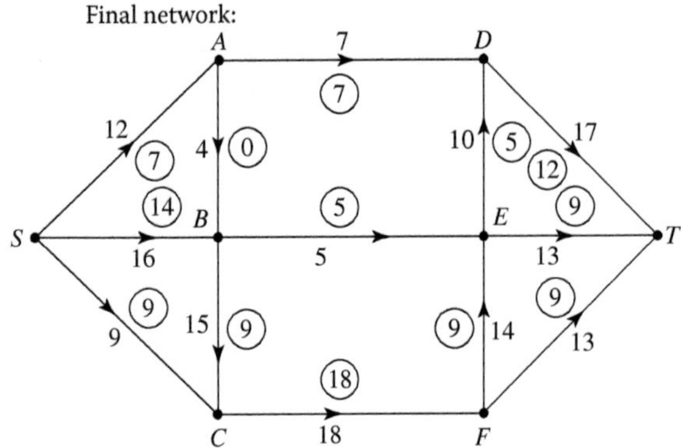

Maximum flow is 30

b A possible cut is {AD, BE, CF} of size 30 so flow of 30 is maximal using the maximum flow-minimum cut theorem.

c AB could be removed as no flow along this edge

d SC, CF, AD are saturated.

4 **a, c**

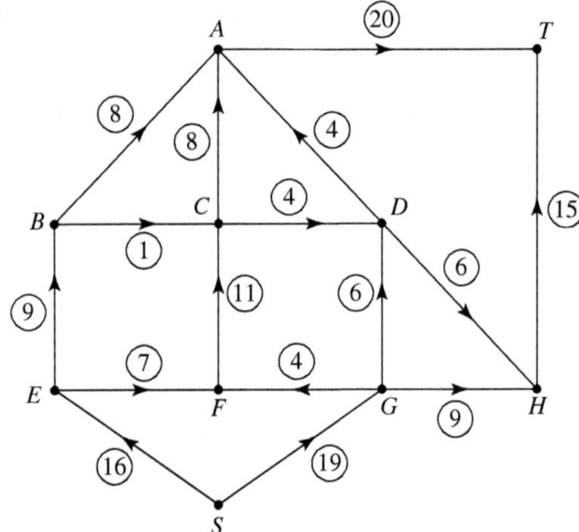

b For example, initial flow of 8 through $SEBA$
initial flow of 9 through $SGHT$
Flow-augmenting paths
$SEFCAT$ (7)
$SBCAT$ (1)
$SGFCDAT$ (4)
$SGDHT$ (6)

Maximum flow is 35

d Cut {GH, GD, FC, EB} has capacity 35
So 35 is maximum-flow using maximum flow-minimum cut theorem.

e EB as all edges from F are saturated
Only worth adding additional capacity of 1 as BA saturated so will go along BC then CA saturated so will go along CD which only has a spare capacity of 1. DH and DA both have plenty of spare capacity so can reach sink node.

5 **a**

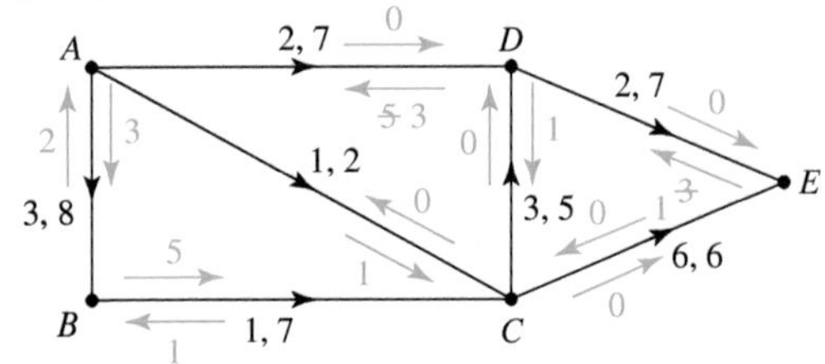

b Flow augmenting paths
ADE (+2)
and $ABCDE$ (+1)

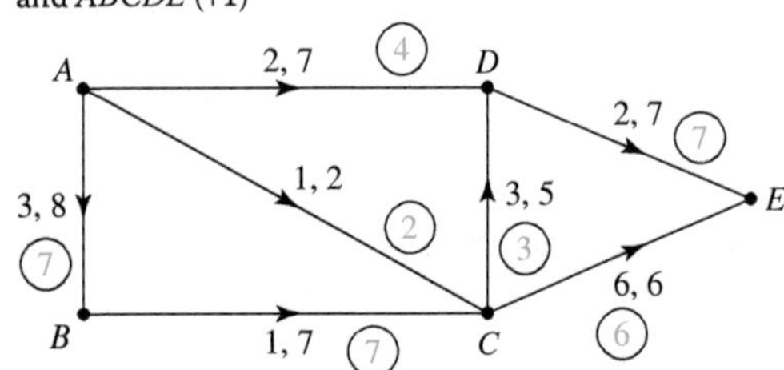

c Capacity of cut = 7 + 2 + 7 − 3 = 13
Therefore 13 is maximum-flow using maximum flow-minimum cut theorem.

6 a

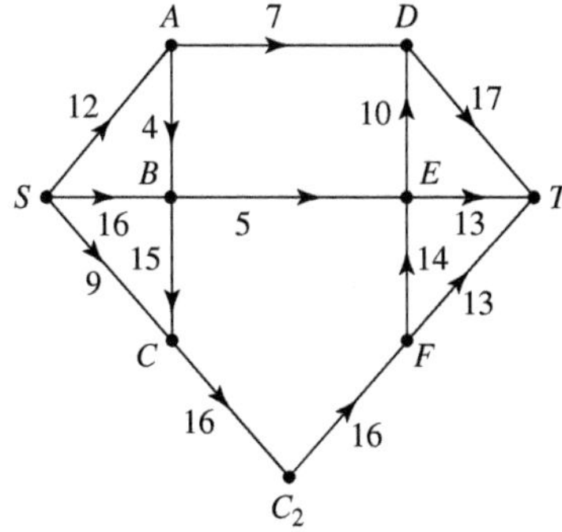

b e.g. initial feasible flow of 9 along SCC_2FT, 7 along $SADT$ and 5 along $SBET$

Flow augmenting path For example, $SBCC_2FET$ (+7)

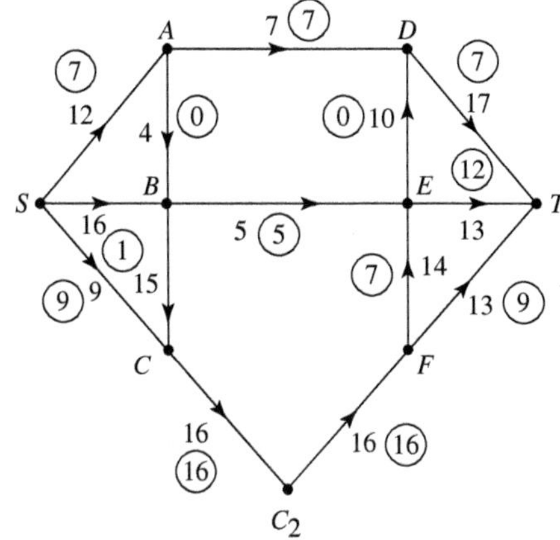

Min-cut e.g. $\{AD, BE, FE, FT\} = 7 + 5 + 7 + 9 = 28$

So flow of 28 is maximum by maximum flow-minimum cut theorem.

Chapter 29

Exercise 29.1A

1

Activity	Duration (hours)	Start		Finish		Float
		Earliest	Latest	Earliest	Latest	
A	6	0	0	6	6	0
B	8	0	7	8	15	7
C	14	6	8	20	22	2
D	12	6	6	18	18	0
E	3	8	15	11	18	7
F	4	18	18	22	22	0
G	3	22	22	25	25	0

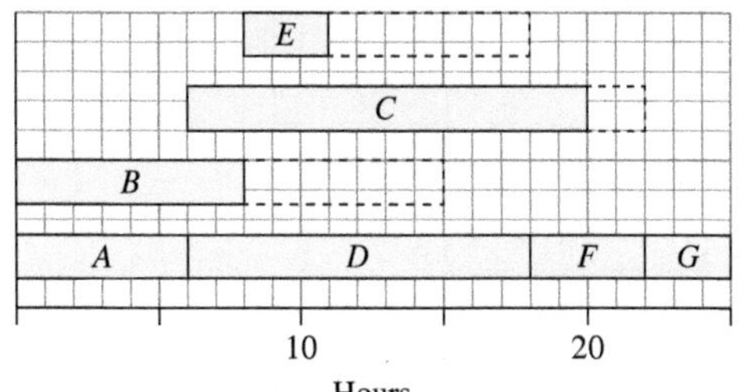

2 a

Activity	Duration	Start		Finish		Float
		Earliest	Latest	Earliest	Latest	
A	7	0	0	7	7	0
B	5	0	2	5	7	2
C	5	7	15	12	20	8
D	9	7	7	16	16	0
E	12	16	20	28	32	4
F	9	16	16	25	25	0
G	7	25	25	32	32	0
H	5	32	32	37	37	0
I	4	25	33	29	37	8

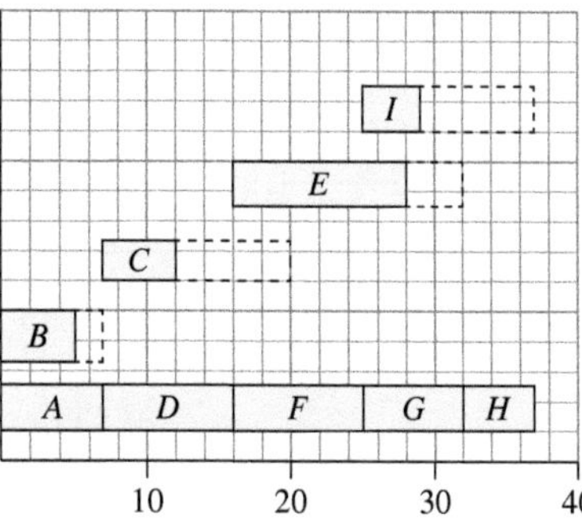

b

Activity	Duration	Start		Finish		Float
		Earliest	Latest	Earliest	Latest	
A	3	0	0	3	3	0
B	5	3	4	8	9	1
C	6	3	3	9	9	0
D	4	8	9	12	13	1
E	3	9	10	12	13	1
F	9	9	9	18	18	0
G	5	12	13	17	18	1
H	1	18	18	19	19	0

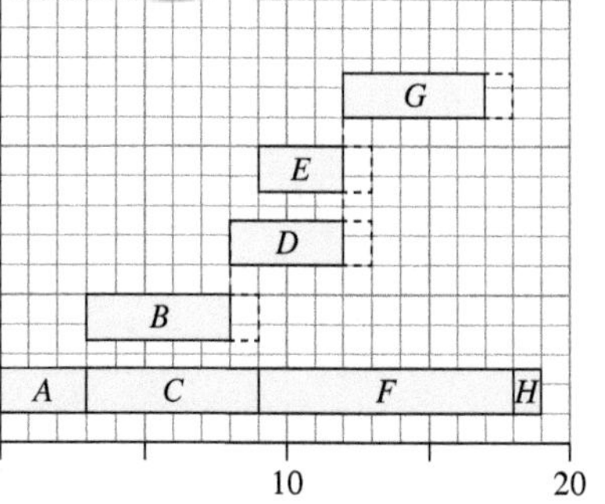

c

Activity	Duration	Start		Finish		Float
		Earliest	Latest	Earliest	Latest	
A	6	0	0	6	6	0
B	5	6	6	11	11	0
C	8	6	6	14	14	0
D	3	11	11	14	14	0
E	8	14	17	22	25	3
F	7	14	14	21	21	0
G	9	11	16	20	25	5
H	4	21	21	25	25	0

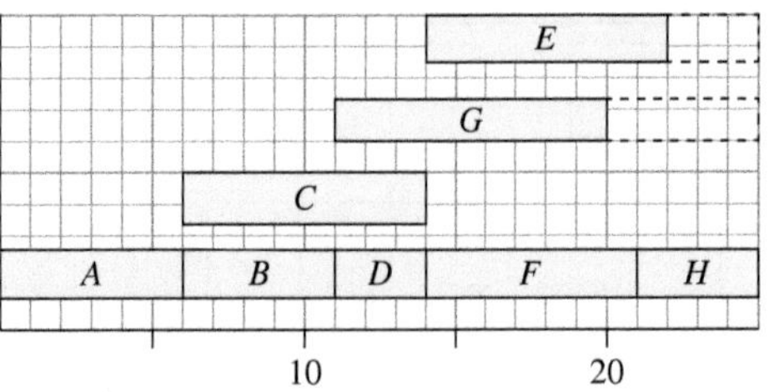

Activity	Duration	Start		Finish		Float
		Earliest	Latest	Earliest	Latest	
A	6	0	2	6	8	2
B	3	0	3	3	6	3
C	8	0	0	8	8	0
D	8	6	6	14	14	0
E	6	8	8	14	14	0
F	2	14	17	16	19	3
G	3	14	16	17	19	2
H	5	14	14	19	19	0
I	9	19	19	28	28	0
J	6	17	22	23	28	5

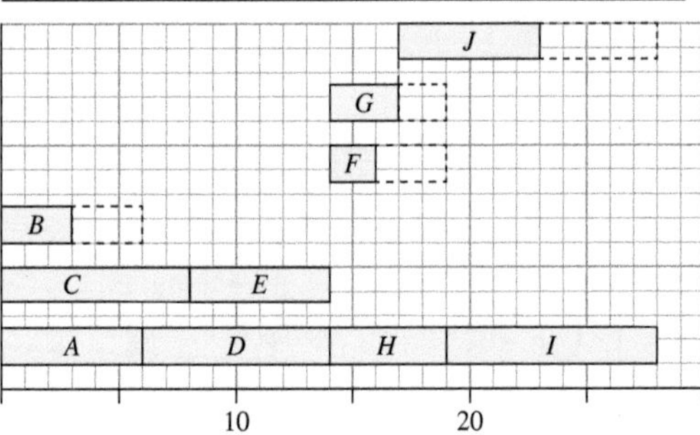

Exercise 29.1B

1

Activity	Duration (hours)	Start		Finish		Float
		Earliest	Latest	Earliest	Latest	
A	9	0	0	9	9	0
B	7	0	1	7	8	1
C	4	9	9	13	13	0
D	5	7	8	12	13	1
E	6	13	13	19	19	0
F	4	13	19	17	23	6
G	7	19	19	26	26	0
H	3	19	23	22	26	4

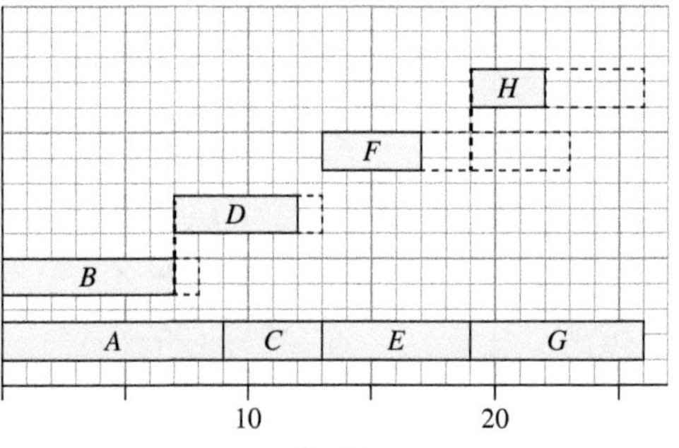

2 a

Activity	Duration	Start		Finish		Float
		Earliest	Latest	Earliest	Latest	
A	4	0	5	4	9	5
B	6	0	0	6	6	0
C	5	0	1	5	6	1
D	3	6	9	9	12	3
E	2	6	6	8	8	0
F	6	9	12	15	18	3
G	7	8	8	15	15	0
H	3	15	15	18	18	0
I	3	18	18	21	21	0
J	3	15	18	18	21	3
K	4	15	17	19	21	2

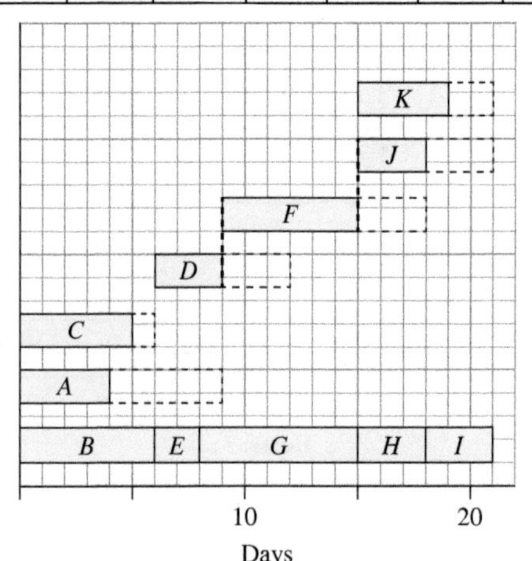

b *D* has 3 days of float. If *D* continues until 11 days, *F* cannot start until then, so *J* cannot start until 17 days. It can therefore be started on day 18 while *J* has 3 days float and will **still** finish at 20 days, so the project duration is not affected.

3 a 31 hours (the total of all the tasks)

b

Activity	Duration	Start		Finish		Float
		Earliest	Latest	Earliest	Latest	
A	2	0	0	2	2	0
B	10	2	2	12	12	0
C	3	12	12	15	15	0
D	2	12	13	14	15	1
E	3	14	15	17	18	1
F	2	14	17	16	19	3
G	1	17	19	18	20	2
H	1	17	19	18	20	2
I	2	17	18	19	20	1
J	5	15	15	20	20	0

Minimum project duration = 20 hours

c

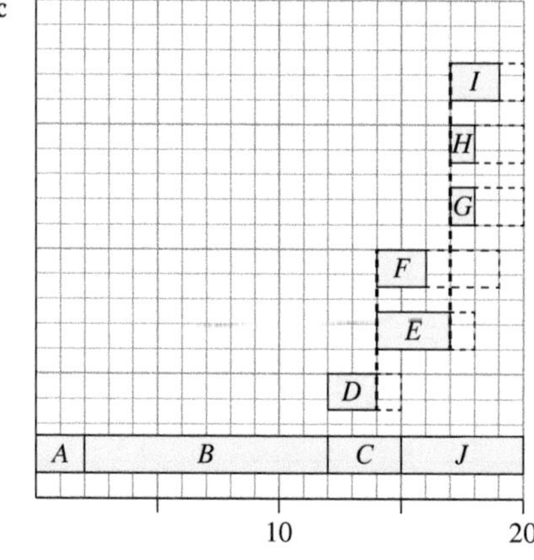

d At least 3 workers needed (once *D* is complete at least 2 of *E*, *F*, *G*, *H* and *I* must be taking place in addition to the critical tasks *C* and *J*).

4

Activity	Duration	Start		Finish		Float
		Earliest	Latest	Earliest	Latest	
A	1	0	0	1	1	0
B	1	1	1	2	2	0
C	1	0	1	1	2	1
D	3	2	2	5	5	0
E	2	2	13	4	15	11
F	4	5	5	9	9	0
G	4	9	9	13	13	0
H	2	13	13	15	15	0
I	1	15	15	16	16	0
J	1	16	16	17	17	0
K	2	17	17	19	19	0
L	2	15	17	17	19	2
M	1	19	19	20	20	0
N	1	20	20	21	21	0
O	1	19	20	20	21	1
P	2	21	21	23	23	0

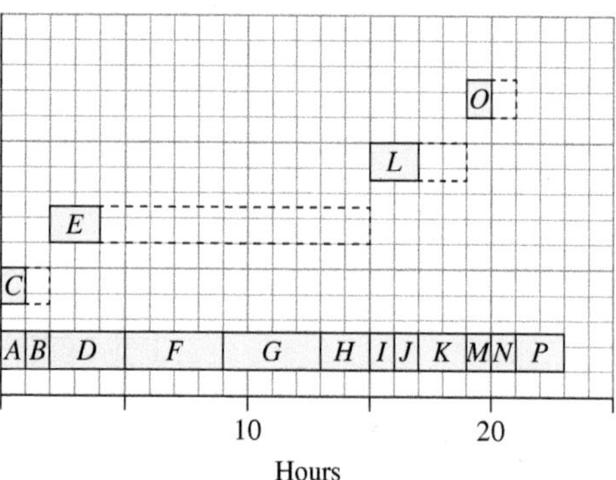

Exercise 29.2A

1 a i

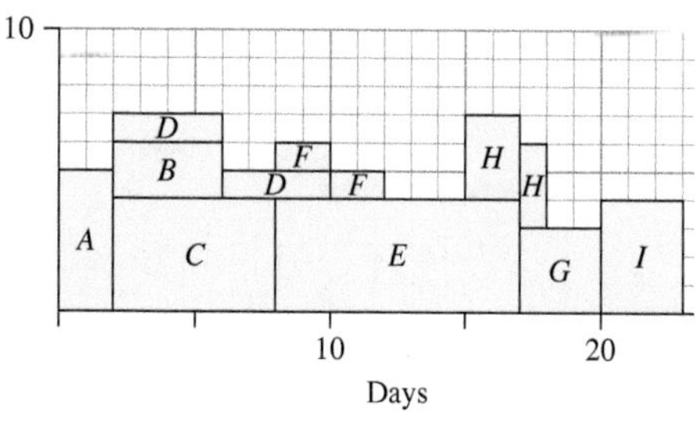

7 workers needed.

ii

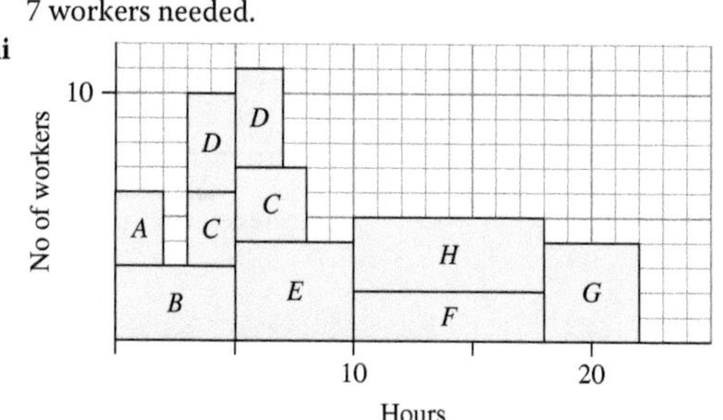

11 workers needed.

iii

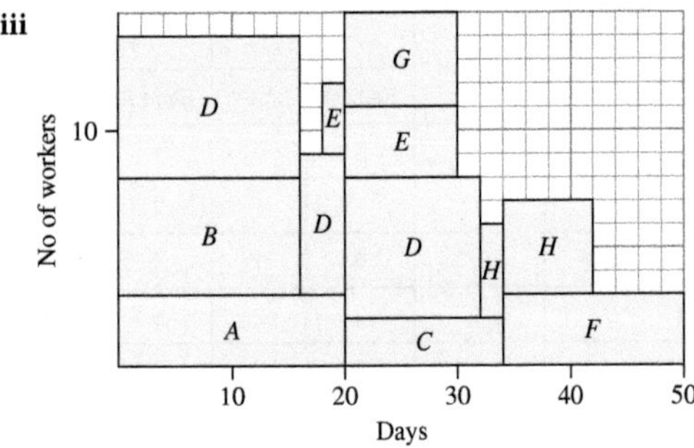

15 workers needed.

b i

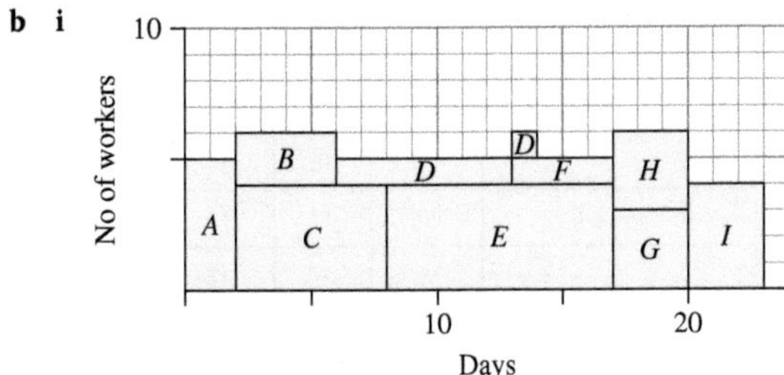

6 workers needed.

ii

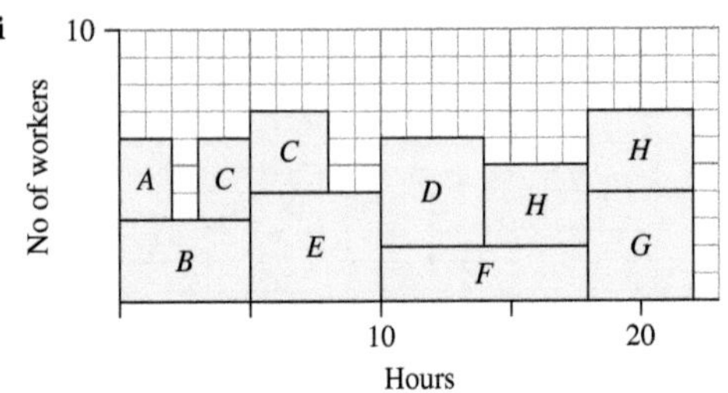

7 workers needed.

iii

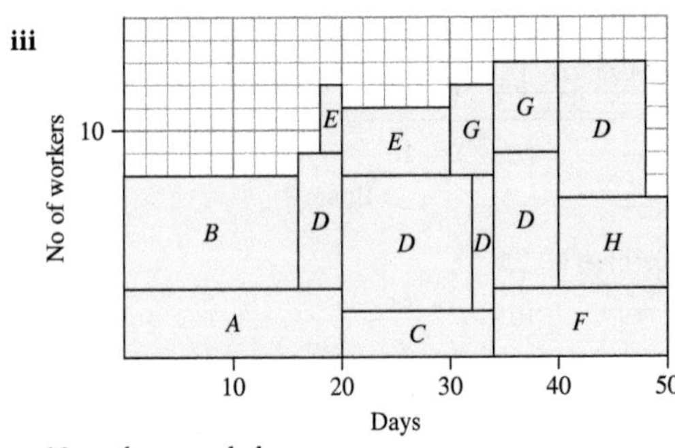

13 workers needed.

2

a

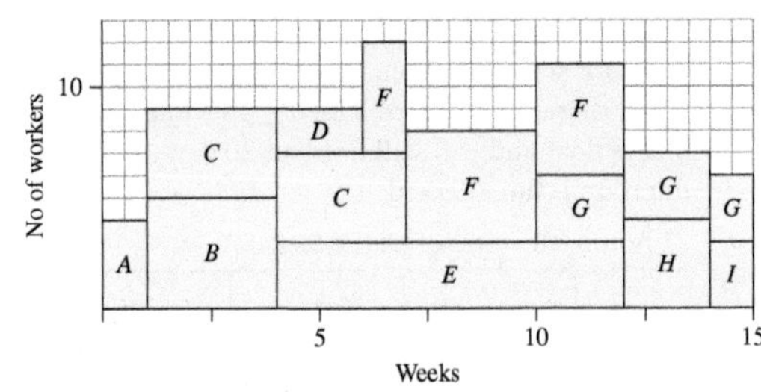

b

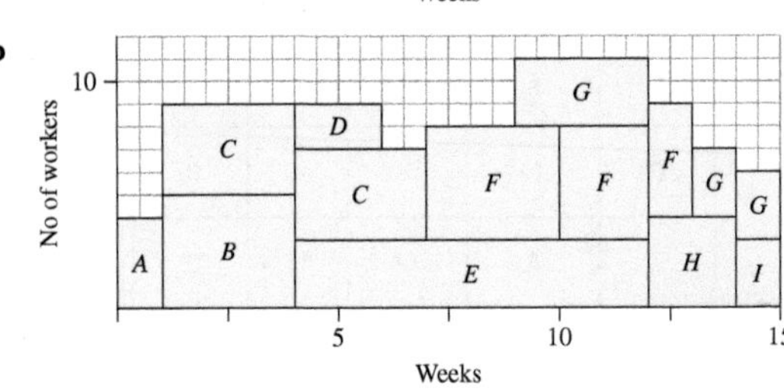

Exercise 29.2B

1 **a**

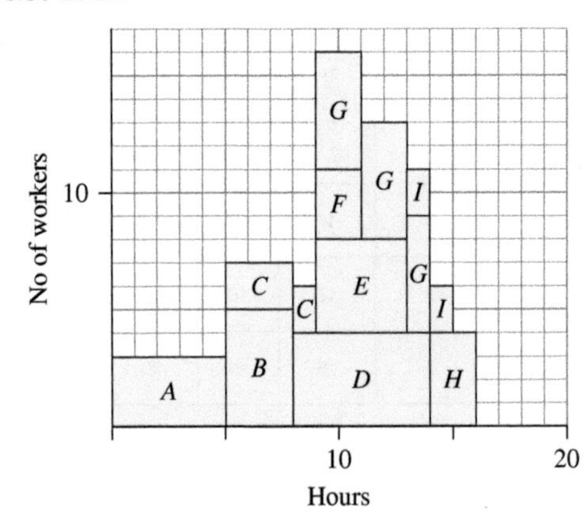

16 workers needed.

b

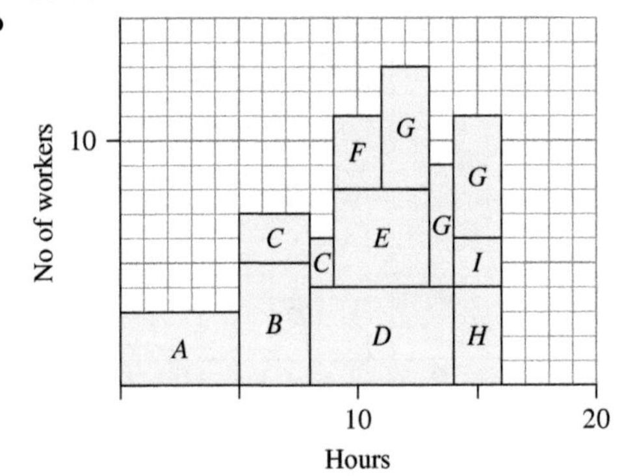

13 workers needed.

c

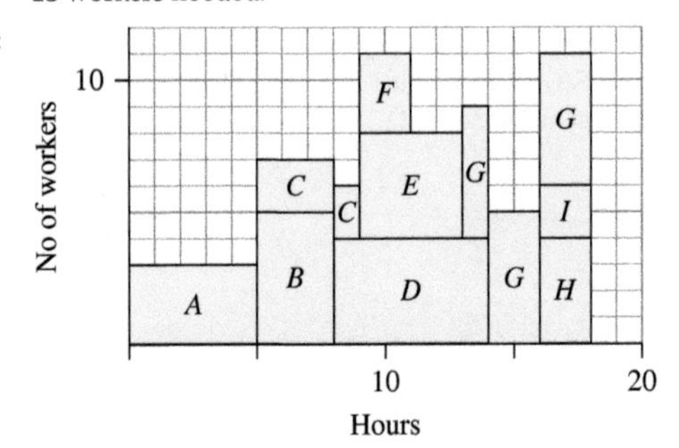

18 hours needed.

2 a

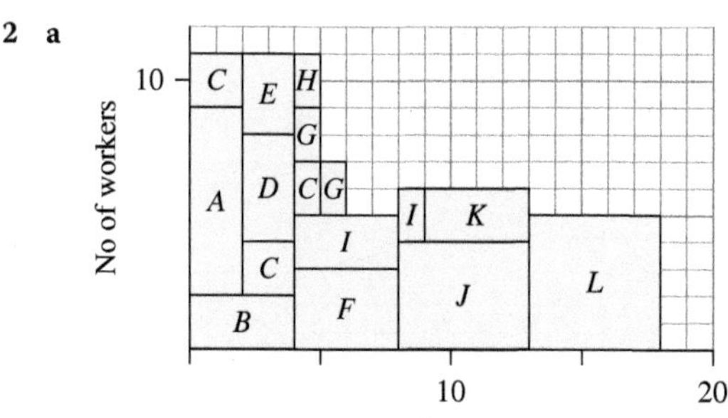

11 workers needed.

b

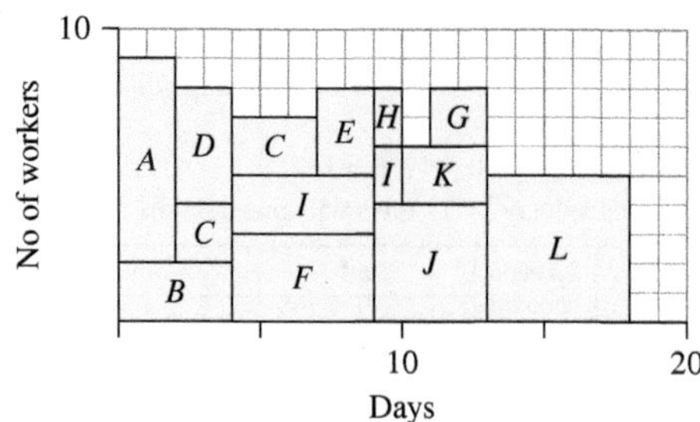

9 workers needed.

c

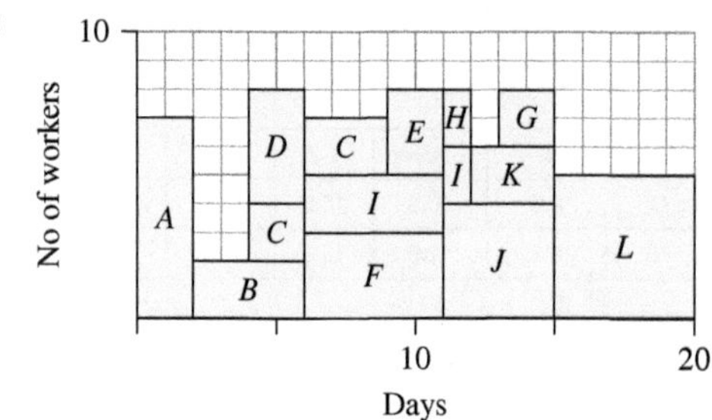

2 extra days needed.

3 a

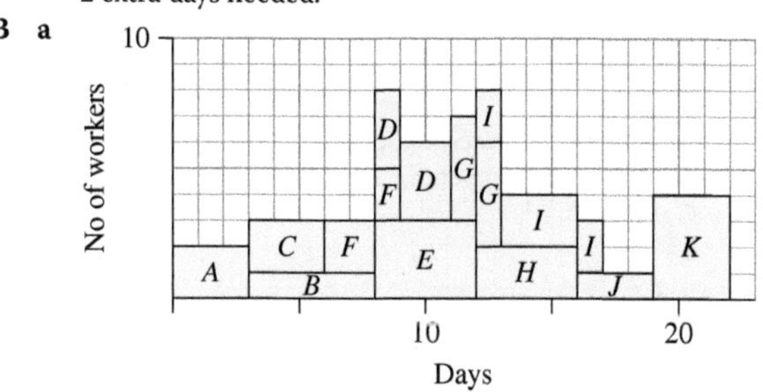

8 workers needed.

b

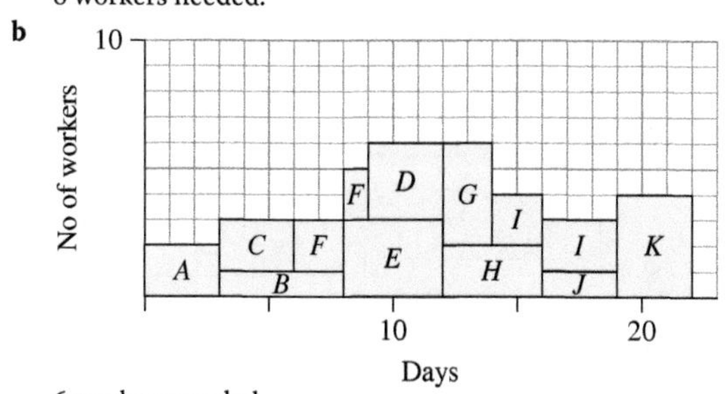

6 workers needed.

c i

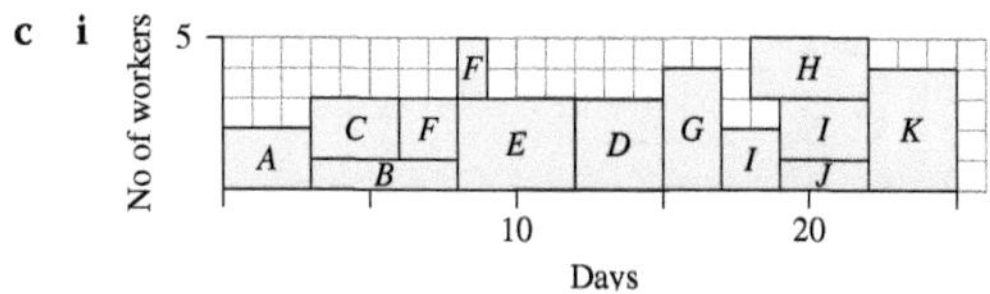

4 extra days

ii

No of workers

A C F B F E D I H G I J K

10 20 30

Days

4 a

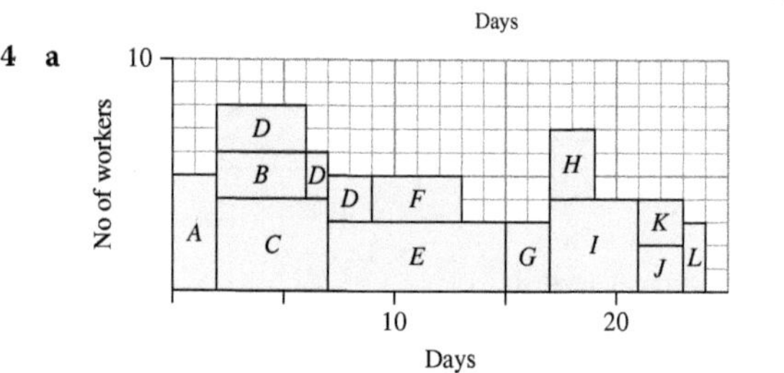

8 workers needed.

b

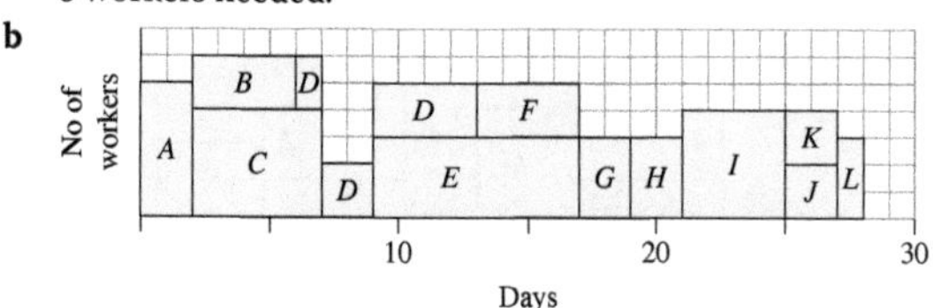

4 extra days needed.

Review exercise 29

1

Activity	Duration	Start		Finish		Float
		Earliest	Latest	Earliest	Latest	
A	4	0	7	4	11	7
B	5	0	2	5	7	2
C	7	0	0	7	7	0
D	3	5	11	8	14	6
E	4	7	7	11	11	0
F	2	8	14	10	16	6
G	7	5	9	12	16	4
H	5	11	11	16	16	0
I	6	16	16	22	22	0
J	4	11	18	15	22	7

2

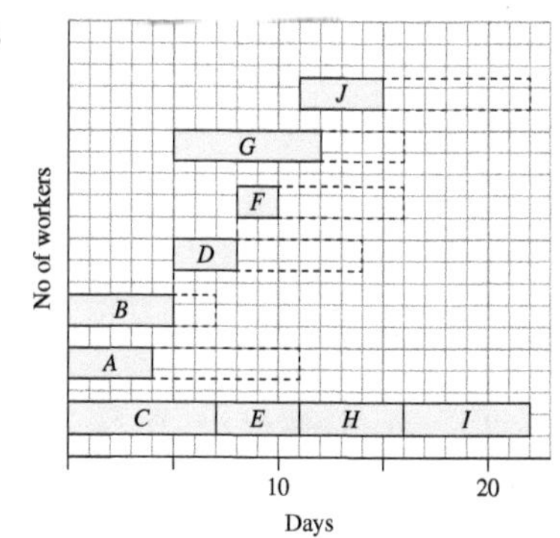

3 a

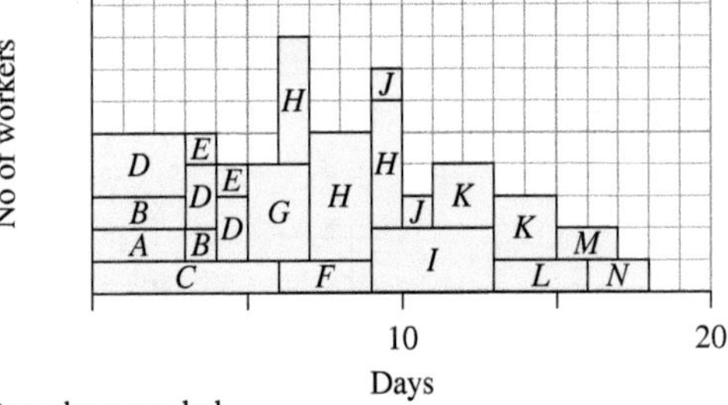

8 workers needed.

b

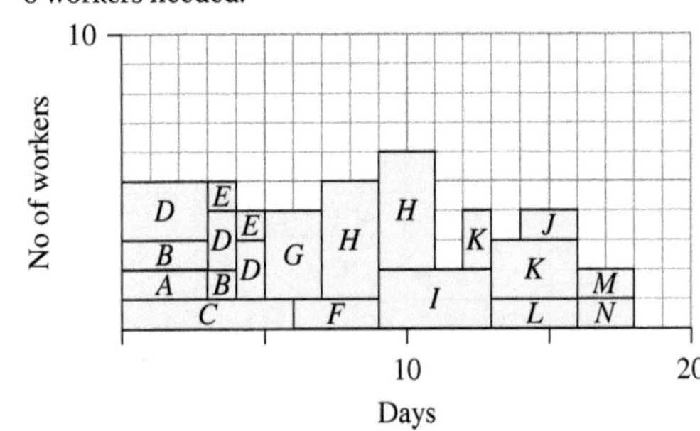

6 workers needed.

c

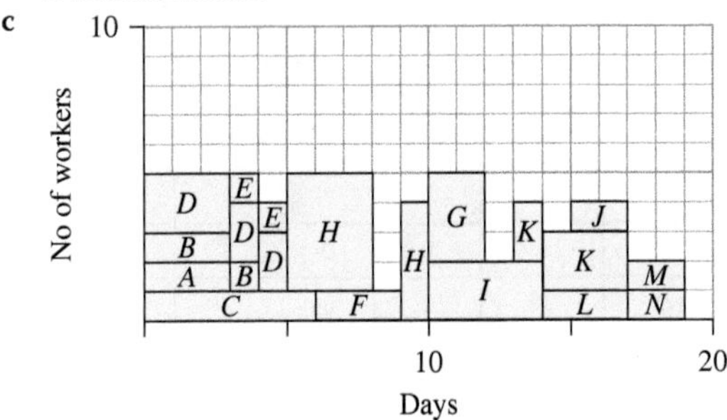

Assessment 29

1 a

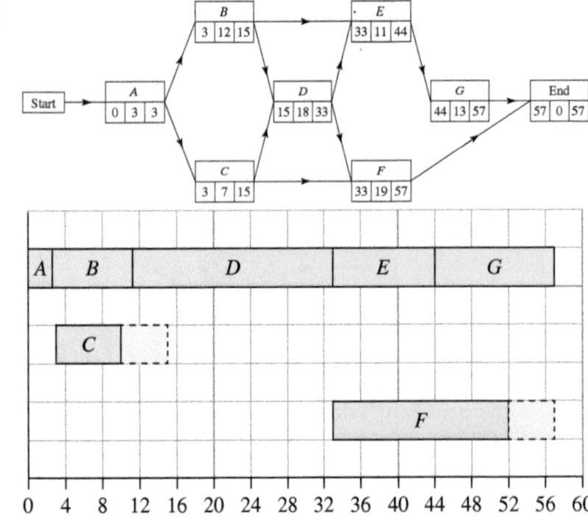

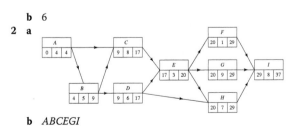

b 6

2 a

A 0 4 4; B 4 5 9; C 9 8 17; D 9 6 17; E 17 3 20; F 20 1 29; G 20 9 29; H 20 7 29; I 29 8 37

b *ABCEGI*

c $29 - 20 - 7$

$= 2$

d Float of *D* is $17 - 9 - 6 = 2$

Increase of 1 hour

3 a

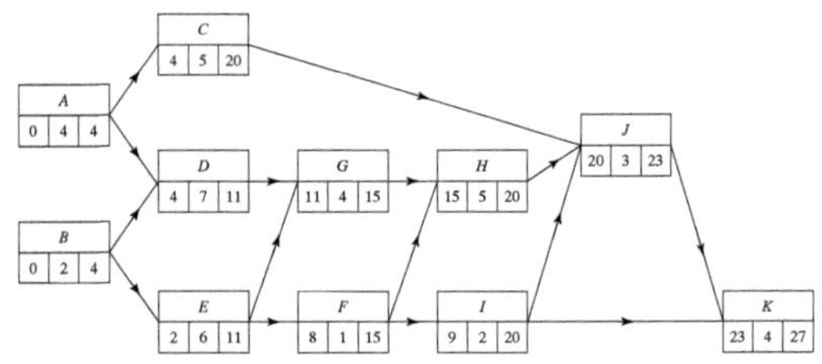

b *ADGHJK*

c **i** *G* is on the critical path so any delay will delay the project.

ii Float of *C* is $20 - 4 - 5 = 11$

So can delay by 11 days without delaying project.

4 a

Activity	Duration (hours)	Start		Finish		Float
		Earliest	Latest	Earliest	Latest	
A	1.5	0	0	1.5	1.5	0
B	2.4	1.5	1.5	3.9	3.9	0
C	6.1	3.9	3.9	10	10	0
D	1.8	3.9	8.5	5.7	10.3	4.6
E	0.3	10	10	10.3	10.3	0
F	3	3.9	7.3	6.9	10.3	3.4
G	2.6	10.3	10.3	12.9	12.9	0
H	5.2	12.9	12.9	18.1	18.1	0

b

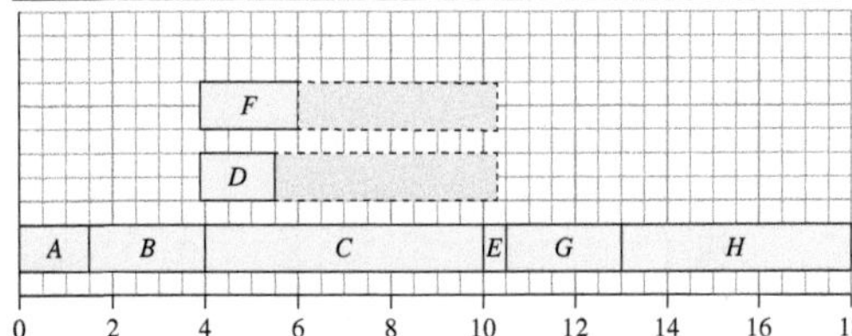

c **i** Only *B* is definitely happening.

ii *C* definitely happening, *D* and *F* may be happening.

5 a *C* and *E*

b 5

c

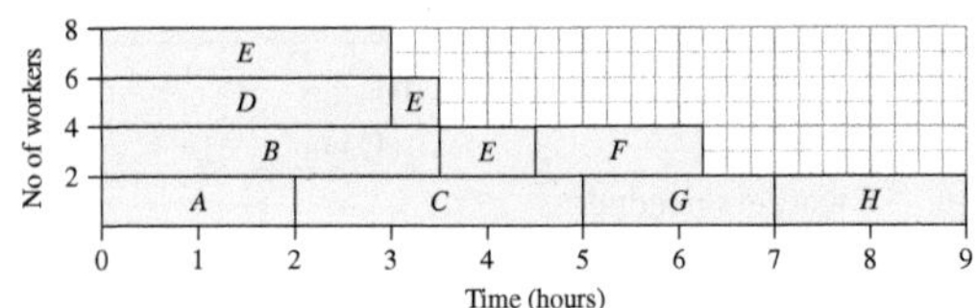

d Delay of 2.25 hours, float of 1 hour so delay of 1.25 hours.

6:15 pm

6 a

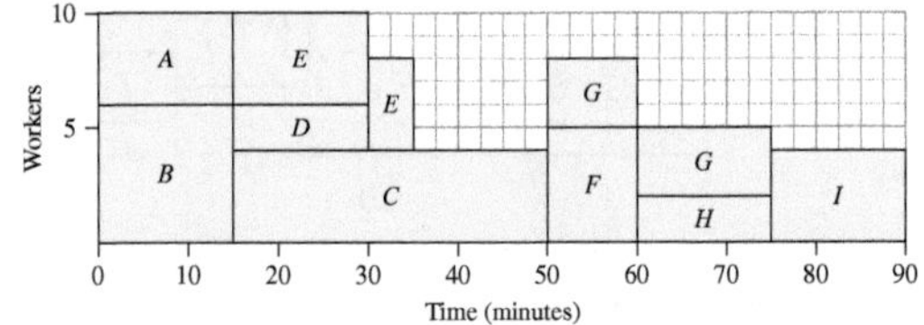

b For example:

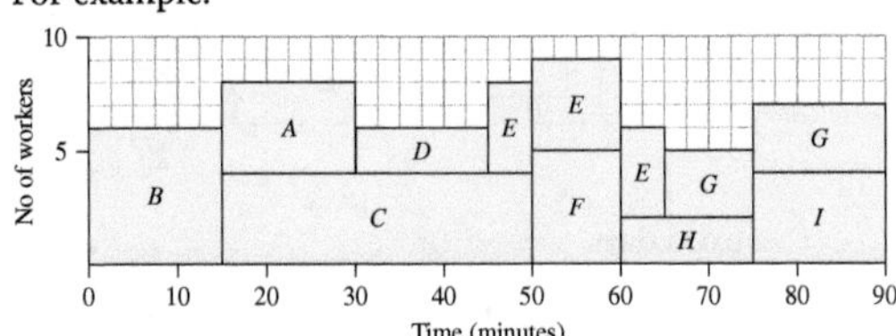

c 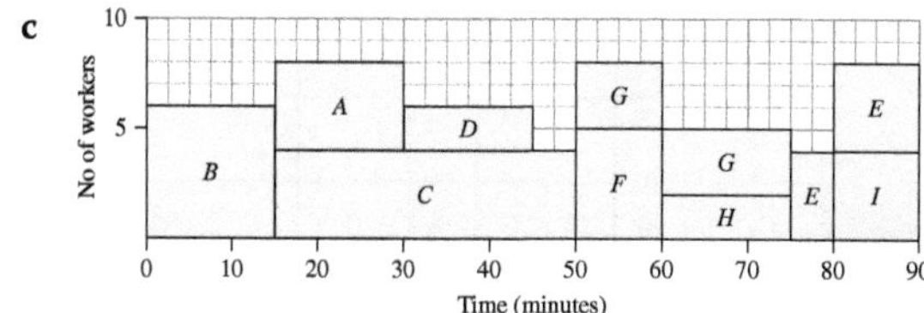

95 minutes

7 a $\frac{5\times4+2\times2+4\times5+6\times1+2\times6+3\times2+5\times4+8\times2+3\times7+5\times3+6\times4+8\times3}{26}$

$=7.23$

So 8 workers needed.

b 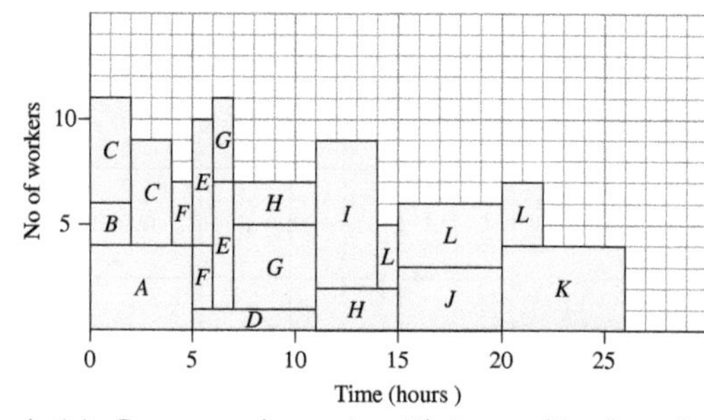

c Activity D can move (e.g. to time 15) since nothing depends on it. This will mean only 10 workers on the 7th day. Completing 4 workers completing C in 5 hours would mean only 10 workers required on 1st day. Activities G, I, L will be postponed by a day, but this will not increase the overall length of the project

So do recommend this approach as 10 workers will be required overall instead of 11

8 a 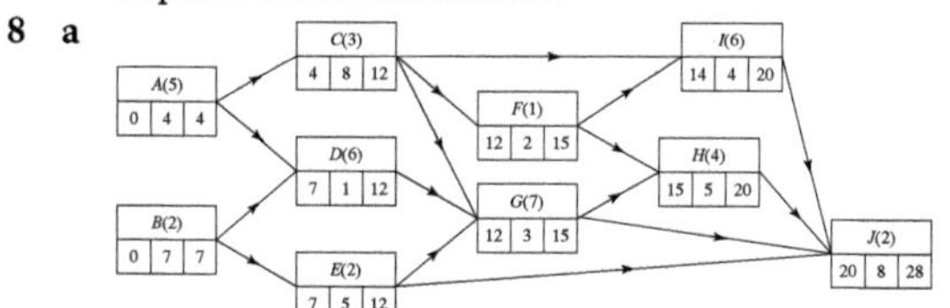

b Critical paths are: $ACGHJ$ and $BEGHJ$

c e.g.

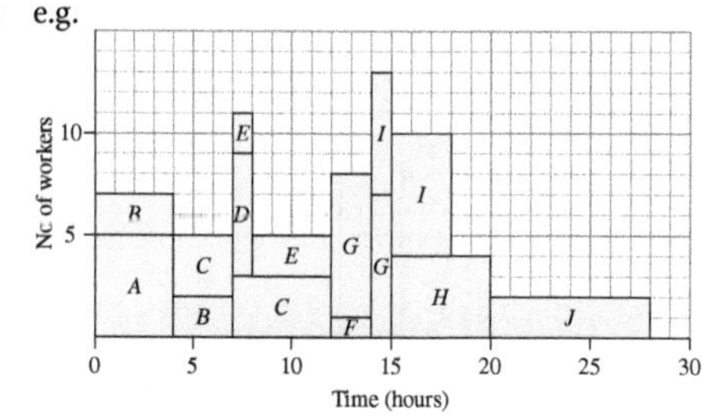

13 workers required.

d Activity I can start at time 15 instead without impacting the total time for the project.

So 11 workers required.

e Activity E will need to be postponed by 1 day so project will increase by 1 day since E critical.

(Could postpone D instead with same effect.)

Chapter 30

Exercise 30.1A

1 a Maximum $P = 25$, when $x = 10, y = 15$

b Maximum $P = 70$, when $x = 20, y = 30$

c Maximum $P = 56$, when $x = 9, y = 4, (t = 3)$

d Maximum $P = 12$, when $x = 0, y = 6, (t = 3, u = 3)$

e Maximum $P = 5\frac{9}{11}$ when $x = 0, y = 1\frac{2}{11}, z = \frac{6}{11}$

f Maximum $P = 25$ when $x = 2\frac{1}{3}, y = 0, z = 2\frac{1}{6}, \left(s = 4\frac{1}{2}\right)$

2 a $P = 3x + y + 2z, 2x + 3y + z \le 20, x + 2y + z \le 12, x \ge 0, y \ge 0, z \ge 0$

b

P	x	y	z	s	t	Value	Row
1	−3	−1	−2	0	0	0	R1
0	2	3	1	1	0	20	R2
0	1	2	1	0	1	12	R3
1	0	$3\frac{1}{2}$	$-\frac{1}{2}$	$1\frac{1}{2}$	0	30	R4 = R1 + 3 × R5
0	1	$1\frac{1}{2}$	$\frac{1}{2}$	$\frac{1}{2}$	0	10	R5 = $\frac{R2}{2}$
0	0	$\frac{1}{2}$	$\frac{1}{2}$	$-\frac{1}{2}$	1	2	R6 = R3 − R5

Not optimal because there is still a negatve entry in the objective row.

$P = 30, x = 10, y = 0, z = 0, s = 0, t = 2$

c Maximum $P = 32$, when $x = 8, y = 0, z = 4$

3 a $P - x - 3y = 0, x + y + s = 6, x + 4y + t = 12, x + 2y + u = 7$

b

P	x	y	s	t	u	Value	Row
1	−1	−3	0	0	0	0	R1
0	1	1	1	0	0	6	R2
0	1	4	0	1	0	12	R3
0	1	2	0	0	1	7	R4
1	$-\frac{1}{4}$	0	0	$\frac{3}{4}$	0	9	R5 = R1 + 3 × R7
0	$\frac{3}{4}$	0	1	$-\frac{1}{4}$	0	3	R6 = R2 − R7
0	$\frac{1}{4}$	1	0	$\frac{1}{4}$	0	3	R7 = $\frac{R3}{4}$
0	$\frac{1}{2}$	0	0	$-\frac{1}{2}$	1	1	R8 = R4 − 2 × R7

Not optimal because there is still a negative entry in the objective row.

c Maximum $P = 9\frac{1}{2}$, when $x = 2, y = 2\frac{1}{2}\left(s = 1\frac{1}{2}\right)$

4 a Maximise $P = 5x + 6y$

subject to $3x + 3y \le 40$

$x + 2y \le 25$

$x \ge 0, y \ge 0$ s, t are slack variables.

b

P	x	y	s	t	Value	Row
1	−5	−6	0	0	0	R1
0	3	3	1	0	40	R2
0	1	2	0	1	25	R3
1	−2	0	0	3	75	R4 = R1 + 6 × R6
0	$1\frac{1}{2}$	0	1	$-1\frac{1}{2}$	$2\frac{1}{2}$	R5 = R2 − 3 × R6
0	$\frac{1}{2}$	1	0	$\frac{1}{2}$	$12\frac{1}{2}$	R6 = $\frac{R3}{2}$

$P = 75, x = 0, y = 12.5, s = 2.5, t = 0$

Not optimal because there is still a negative entry in the objective row.

c Maximum $P = 78\frac{1}{3}, x = 1\frac{2}{3}, y = 11\frac{2}{3}$

Exercise 30.1B

1 a Maximum $P=3$, so minimum $C=-3$, when $x=3$, $y=6$, $(t=9)$

b Maximum $P=15\frac{1}{3}$, so minimum $C=-15\frac{1}{3}$, when $x=0$, $y=\frac{2}{3}$, $z=2\frac{2}{3}$

c Maximum $P=3\frac{3}{4}$, so minimum $C=-3\frac{3}{4}$, when $x=1\frac{1}{4}$, $y=2\frac{1}{2}$, $z=0$, $(t=1)$

d Maximum $P=14\frac{1}{2}$, when $x=9\frac{1}{2}$, $y=1\frac{1}{2}$ $z=0$, $(u=13)$

2 Maximum $P=97$, so minimum $C=-97$, when $X=0$, $Y=24$, $Z=86$, that is when $x=10$, $y=27$, $z=94$

3 Maximum profit = £31 000. Make 5000 kg Regular and 9000 kg Luxury.

4 Travel x m on setting A and y m on setting B.

Maximise $D=x+y$

subject to $3x+2y\le 14400$

$2x+3y\le 15000$

$x, y\ge 0$

Maximum $D=5880$ m. Travel 2640 m on setting A and 3240 m on setting B.

5 a £400

b $5x+6y+3z\le 90$

$2x+4y+z\le 42$

c Maximum $P=72.75$, when $x=13.5$, $y=3.75$, $z=0$ No negatives in the objective row.

d Cannot make fractions of a bicycle.

e 13 type A, 4 type B, no type C. Satisfies both constraints.

6 a Make x kg of A, y kg of B and z kg of C.

Maximise profit $P=0.6x+0.5y+0.9z$

subject to $0.03x+0.02y+0.03z\le 50$, so $3x+2y+3z\le 5000$

$0.02x+0.03y+0.04z\le 60$, so $2x+3y+4z\le 6000$

$0.01x+0.01y+0.02z\le 40$, so $x+y+2z\le 4000$

$x, y, z\ge 0$

b $P=1400$, $x=333\frac{1}{3}$, $y=0$, $z=1333\frac{1}{3}$, so make $333\frac{1}{3}$ kg Assolato and $1333\frac{1}{3}$ kg Contadino (and no Buona Salute), giving £1400 profit.

7 Maximum profit = £1087.50. Make 2000 litres of Froo-T and 2750 litres of Joo-C.

500 litres of peach juice left over.

8 a Maximum $P=9333\frac{1}{3}$ when $w=0$, $x=0$, $y=16\frac{2}{3}$, $z=133\frac{1}{3}$

b This is only an approximate answer, as the variables should be integers. Trial and error suggests that $y=16$, $z=134$, giving $P=9320$, is the best result.

So produce 16 box C and 134 box D, profit £93.20

9 a Maximum $P=9\frac{5}{7}$, when $x=2\frac{2}{7}$, $y=1\frac{3}{7}$

b Maximum $P=-4$, so minimum $C=4$, when $x=4$, $y=0$, $(s=1)$

10 Garden is 350 m^2, with 300 m^2 turf and 50 m^2 slabs.

11 a Substitute $z=2x+y-10$

$P=3x+y+10$, $x+y\le 8$, $6x+y\le 36$

$P-3x-y=10$, $x+y+s=8$, $6x+y+t=36$

Maximum $P=29.2$, when $x=5.6$, $y=2.4$, which gives $z=3.6$

b $P-5x-2y+z=0$, $x+2y+z+s=14$, $2x-y+2z+t=16$

$2x+y-z+u=10$. $-2x-y+z+v=-10$

Maximum $P=29.2$, when $x=5.6$, $y=2.4$, $z=3.6$

12 a

P	x	y	s	t	u	Value	Row
1	−1	−2	0	0	0	0	R1
0	1	1	1	0	0	28	R2
0	−3	−1	0	1	0	−30	R3
0	−2	−3	0	0	1	−60	R4
1	$\frac{1}{3}$	0	0	0	$-\frac{2}{3}$	40	R5 = R1 + 2 × R8
0	$\frac{1}{3}$	0	1	0	$\frac{1}{3}$	8	R6 = R2 − R8
0	$-2\frac{1}{3}$	0	0	1	$-\frac{1}{3}$	−10	R7 = R3 + R8
0	$\frac{2}{3}$	1	0	0	$-\frac{1}{3}$	20	$R8=-\frac{R4}{3}$
1	0	0	0	$\frac{1}{7}$	$-\frac{5}{7}$	$38\frac{4}{7}$	R9 = R5 − $\frac{R11}{3}$
0	0	0	1	$\frac{1}{7}$	$\frac{2}{7}$	$6\frac{4}{7}$	R10 = R6 − $\frac{R11}{3}$
0	1	0	0	$-\frac{3}{7}$	$\frac{1}{7}$	$4\frac{2}{7}$	R11 = −R7 × $\frac{3}{7}$
0	0	1	0	$\frac{2}{7}$	$-\frac{3}{7}$	$17\frac{1}{7}$	R12 = R8 − R11 × $\frac{2}{3}$
1	0	0	$2\frac{1}{2}$	$\frac{1}{2}$	0	55	R13 = R9 − R14 × $\frac{5}{7}$
0	0	0	$3\frac{1}{2}$	$\frac{1}{2}$	1	23	R14 = R10 × $\frac{7}{2}$
0	1	0	$-\frac{1}{2}$	$-\frac{1}{2}$	0	1	R15 = R11 − $\frac{R14}{7}$
0	0	1	$1\frac{1}{2}$	$\frac{1}{2}$	0	27	R16 = R12 − R14 × $\frac{3}{7}$

Maximum $P=55$, when $x=1$, $y=27$, $(u=23)$

b

P	x	y	s	t	u	Value	Row
1	−1	−2	0	0	0	0	R1
0	1	1	1	0	0	18	R2
0	−3	−1	0	1	0	−30	R3
0	−2	−3	0	0	1	−60	R4
1	$\frac{1}{3}$	0	0	0	$-\frac{2}{3}$	40	R5 = R1 + 2 × R8
0	$\frac{1}{3}$	0	1	0	$\frac{1}{3}$	−2	R6 = R2 − R8
0	$-2\frac{1}{3}$	0	0	1	$-\frac{1}{3}$	−10	R7 = R3 + R8
0	$\frac{2}{3}$	1	0	0	$-\frac{1}{3}$	20	$R8 = -\frac{R4}{3}$
1	0	0	0	$\frac{1}{7}$	$-\frac{5}{7}$	$38\frac{4}{7}$	$R9 = R5 - \frac{R11}{3}$
0	0	0	1	$\frac{1}{7}$	$\frac{2}{7}$	$-3\frac{3}{7}$	$R10 = R6 - \frac{R11}{3}$
0	1	0	0	$-\frac{3}{7}$	$\frac{1}{7}$	$4\frac{2}{7}$	$R11 = R7 \times \frac{(-3)}{7}$
0	0	1	0	$\frac{2}{7}$	$-\frac{3}{7}$	$17\frac{1}{7}$	$R12 = R8 - R11 \times \frac{2}{3}$

This can go no further because R10 has a negative right–hand side but no negative coefficients on the left–hand side on which to pivot.

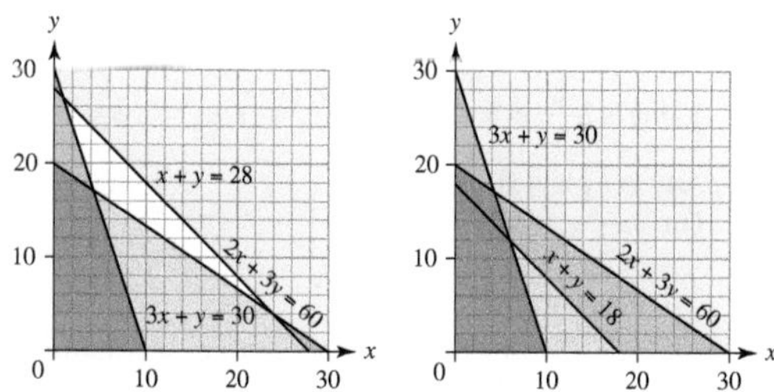

On the correct graph there is a feasible region, but putting 18 instead of 28 means that no point satisfies all three inequalities.

c

P	x	y	s	t	u	Value	Row
1	−1	−2	0	0	0	0	R1
0	1	−1	1	0	0	28	R2
0	−3	−1	0	1	0	−30	R3
0	−2	−3	0	0	1	−60	R4
1	$\frac{1}{3}$	0	0	0	$-\frac{2}{3}$	40	R5 = R1 + 2 × R8
0	$1\frac{2}{3}$	0	1	0	$-\frac{1}{3}$	48	R6 = R2 + R8
0	$-2\frac{1}{3}$	0	0	1	$-\frac{1}{3}$	−10	R7 = R3 + R8
0	$\frac{2}{3}$	1	0	0	$-\frac{1}{3}$	20	$R8 = -\frac{R4}{3}$
1	0	0	0	$\frac{1}{7}$	$-\frac{5}{7}$	$38\frac{4}{7}$	$R9 = R5 - \frac{R11}{3}$
0	0	0	1	$\frac{5}{7}$	$-\frac{4}{7}$	$40\frac{6}{7}$	$R10 = R6 - R11 \times \frac{5}{3}$
0	1	0	0	$-\frac{3}{7}$	$\frac{1}{7}$	$4\frac{2}{7}$	$R11 = R7 \times \frac{3}{7}$
0	0	1	0	$\frac{2}{7}$	$-\frac{3}{7}$	$17\frac{1}{7}$	$R12 = R8 - R11 \times \frac{2}{3}$
1	5	0	0	−2	0	60	R13
0	4	0	1	−1	0	58	R14
0	7	0	0	−3	1	30	R15 = R11 × 7
0	3	1	0	−1	0	30	R16

This breaks down because there are no positive values in the t column on which to pivot.

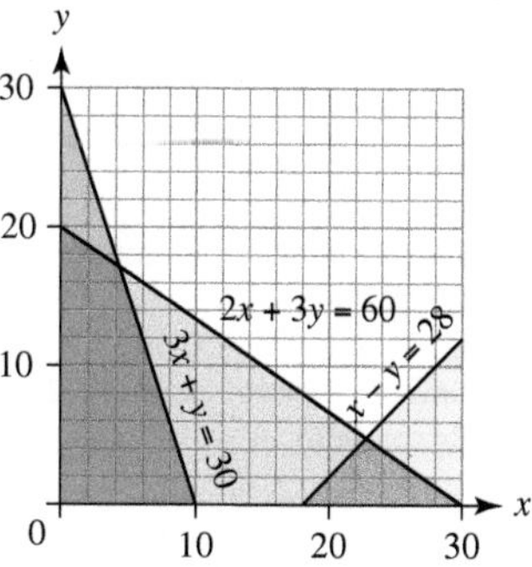

The feasible region is "open–ended" – there is no limit to how large P can be.

Exercise 30.2A

1 a Add 2 to all entries. A plays A_1, A_2 with probabilities p_1, p_2

$P = v$, $v \le 4p_1$, $v \le 5p_2$, $p_1 + p_2 \le 1$

$P - v = 0$, $v - 4p_1 + s = 0$, $v - 5p_2 + t = 0$, $p_1 + p_2 + u = 1$

Value is $2\frac{2}{9} - 2 = \frac{2}{9}$ when $p_1 = \frac{5}{9}, p_2 = \frac{4}{9}$

If B plays B_1, B_2 with probabilities q_1, q_2 then $4q_1 = 5q_2$ $= 2\frac{2}{9}$ so $q_1 = \frac{5}{9}, q_2 = \frac{4}{9}$

b $P - v = 0, v - 7p_1 - 5p_2 + s = 0, v - 6p_1 - 8p_2 + t = 0,$
$p_1 + p_2 + u = 1$
Value is 6.5 when $p_1 = 0.75, p_2 = 0.25$
For B, $7q_1 + 6q_2 = 5q_1 + 8q_2 = 6.5$ and solving gives $q_1 = q_2 = 0.5$

c Add 1 to all entries.
$P - v = 0, v - 6p_1 + s = 0, v - 3p_1 - 4p_2 + t = 0, p_1 + p_2 + u = 1$
Value $= 3\frac{3}{7} - 1 = 2\frac{3}{7}$ when $p_1 = \frac{4}{7}, p_2 = \frac{3}{7}$
For B, $6q_1 + 3q_2 = 4q_2 = 3\frac{3}{7}$ so $q_1 = \frac{1}{7}, q_2 = \frac{6}{7}$

d Add 2 to all entries.
$P - v = 0, v - 5p_2 + s = 0, v - 3p_1 - 2p_2 + t = 0,$
$p_1 + p_2 + p_3 + u = 1$
Value $= 2.5 - 2 = 0.5$ when $p_1 = p_2 = 0.5$
For B, $3q_2 = 5q_1 + 2q_2 = 2.5$, giving $q_1 = \frac{1}{6}, q_2 = \frac{5}{6}$

2 a Add 2 to all entries.
$P - v = 0, v - 3p_1 + s_1 = 0, v - p_1 - 2p_2 + s_2 = 0$
$v - 2p_1 - p_2 + s_3 = 0, p_1 + p_2 + s_4 = 1$
Value $= 1.5 - 2 = -0.5$ when $p_1 = p_2 = 0.5$

b Add 2 to all entries.
$P - v = 0, v - 3p_1 - p_2 - 2p_3 + s = 0, v - 3p_2 - p_3 + t = 0,$
$p_1 + p_2 + p_3 + u = 1$
Value $= 1.8 - 2 = -0.2$ when $p_1 = 0.4, p_2 = 0.6, p_3 = 0$

3 a Maximum of row minima $= -2$, minimum of column maxima $= 1$, no stable solution.
Row 2 dominates row 1, so table becomes

		B	
		B_1	B_2
A	A_2	1	–3
	A_3	–2	1

Add 3 to all entries.
$P - v = 0, v - 4p_1 - p_2 + s = 0, v - 4p_2 + t = 0, p_1 + p_2 + u = 0$
Value is $q_1 = \frac{1}{6}, q_2 = \frac{5}{6} \cdot 2\frac{2}{7} - 3 = -\frac{5}{7}$ when $p_1 = \frac{3}{7}, p_2 = \frac{4}{7}$.
$2\frac{2}{7} - 3 = -\frac{5}{7}$ when $p_1 = \frac{3}{7}, p_2 = \frac{4}{7}$

b Maximum of row minima $= 1$, minimum of column maxima $= 2$ so no stable solution.
Add 3 to all entries.
$P - v = 0, v - 2p_1 - 5p_3 + s = 0, v - 5p_1 - 7p_2 - 4p_3 + t = 0.$
$p_1 + p_2 + p_3 + u = 1$
Value is $4\frac{3}{8} - 3 = 1\frac{3}{8}$ when $p_1 = 0, p_2 = \frac{1}{8}, p_3 = \frac{7}{8}$

c Maximum of row minima = minimum of column maxima = 2, so stable solution.
Play-safe strategies are A_2 and B_1. The value = 2

d Maximum of row minima $= -2$, minimum of column maxima $= 0$, no stable solution.
Add 4 to all entries.
$P - v = 0, v - 4p_2 - 3p_3 + s_1 = 0, v - 3p_1 - p_2 - 6p_3 + s_2 = 0$
$v - 5p_1 - 3p_2 - 4p_3 + s_3 = 0, p_1 + p_2 + p_3 + s_4 = 1$
Value $= 3.5 - 4 = -0.5$ when $p_1 = 0, p_2 = p_3 = 0.5$

e Maximum of row minima $= 2$, minimum of column maxima $= 3$, no stable solution.
Row 3 dominates rows 1 and 4, then column 1 dominates column 4.
Table becomes

		B		
		B_1	B_2	B_3
A	A_2	–1	0	3
	A_3	4	3	2

Add 1 to all entries.
$P - v = 0, v - 5p_2 + s_1 = 0, v - p_1 - 4p_2 + s_2 = 0$
$v - 4p_1 - 3p_2 + s_3 = 0, p_1 + p_2 + s_4 = 1$
Value $= 3.25 - 1 = 2.25$ when $p_1 = 0.25, p_2 = 0.75$

f Maximum of row minima = minimum of column maxima = 4, so stable solution.
Play–safe strategies are A_3 and B_3. Value = 4.

Exercise 30.2B

1 a Maximum of row minima $= -1$, minimum of col maxima $= 1$, so no stable solution.

b Col 3 dominates col 2, so table becomes

		B	
		B_1	B_3
A	A_1	–1	1
	A_2	1	–2
	A_3	0	–1

Add 2 to all entries.
$P - v = 0, v - p_1 - 3p_2 - 2p_3 + s = 0, v - 3p_1 - p_3 + t = 0,$
$p_1 + p_2 + p_3 + u = 1$
Value $= 1.8 - 2 = -0.2$ when $p_1 = 0.6, p_2 = 0.4, p_3 = 0$

c A never plays A_3. If B plays B_1, B_3 with probs q_1, q_3 then $-q_1 + q_3 = q_1 - 2q_3 = -0.2$, which gives $q_1 = 0.6, q_3 = 0.4$

2 a Maximum of row minimums = 1, minimum of col maximums = 2, no stable solution.
Add 1 to all entries.
$P - v = 0, v - p_1 - 3p_2 + s_1 = 0, v - 3p_1 - 2p_2 + s_2 = 0$
$v - 4p_2 + s_3 = 0, p_1 + p_2 + s_4 = 1$
Value is $2\frac{1}{3} - 1 = 1\frac{1}{3}$ when $p_1 = \frac{1}{3}, p_2 = \frac{2}{3}$
A plays A_1, A_2 with probabilities $\frac{1}{3}, \frac{2}{3}$
If B plays B_1, B_2, B_3 with probabilities q_1, q_2, q_3
then $2q_2 - q_3 = 2q_1 + q_2 + 3q_3 = 1\frac{1}{3}$ and $q_1 + q_2 + q_3 = 1$.
Solving gives $q_1 = \frac{1}{3}, q_2 = \frac{2}{3}, q_3 = 0$

b Maximum of row minima $= -3$, minimum of column maxima $= 0$, no stable solution.
Add 5 to all entries.
$P - v = 0, v - 2p_1 - 4p_2 - 5p_3 + s = 0, v - 6p_1 - p_2 + t = 0,$
$p_1 + p_2 + p_3 + u = 1$
Value is $3\frac{1}{3} - 5 = -1\frac{2}{3}$ when $p_1 = \frac{5}{9}, p_2 = 0, p_3 = \frac{4}{9}$
A never plays A_2.
For B, $-3q_1 + q_2 = -5q_2 = -1\frac{2}{3}$
Solving gives $q_1 = \frac{2}{3}, q_2 = \frac{1}{3}$

c Maximum of row minima = 1, minima of column maxima = 4, no stable solution.
Add 2 to all entries.
$P - v = 0, v - p_2 - 6p_3 + s = 0, v - 6p_1 - 4p_2 - 3p_3 + t = 0,$
$p_1 + p_2 + p_3 + u = 1$
Value $= 4 - 2 = 2$ when $p_1 = \frac{1}{3}, p_2 = 0, p_3 = \frac{2}{3}$
A never plays II.
For B, $-2q_1 + 4q_2 = 4q_1 + q_2 = 2$
Solving gives $q_1 = \frac{1}{3}, q_2 = \frac{2}{3}$

d Maximum of row minima $= -3$, minimum of column maxima $= 0$, no stable solution.
Row 2 dominates row 1, column 2 dominates column 3.

Table becomes

		B	
		B_1	B_2
A	A_2	−3	1
	A_3	0	−4

Add 4 to all entries.

$P - v = 0, v - p_1 - 4p_2 + s = 0, v - 5p_1 + t = 0, p_1 + p_2 + u = 1$

Value is $2.5 - 4 = -1.5$ when $p_1 = p_2 = 0.5$

For B, $-3q_1 + q_2 = -4q_2 = -1.5$ which gives $q_1 = \frac{5}{8}, q_2 = \frac{3}{8}$

e Maximum of row minima = 1, minimum of column maxima = 2, no stable solution.

Row 1 dominates row 2, column 2 dominates column 1

Table becomes

		B	
		B_2	B_3
A	A_1	1	5
	A_3	2	1

$P - v = 0, v - p_1 - 2p_2 + s = 0, v - 5p_1 - p_2 + t = 0, p_1 + p_2 + u = 1$

Value is 1.8 when $p_1 = 0.2, p_2 = 0.8$

For B, $q_1 + 5q_2 = 2q_1 + q_2 = 1.8$, giving $q_1 = 0.8, q_2 = 0.2$

f Maximum of row minima = minimum of column maxima = −1, so stable solution.

M plays II, N plays II, value = −1

3

		B		
		R	P	S
A	R	0	−1	1
	P	1	0	−1
	S	−1	1	0

Add 1 to all entries.

$P - v = 0, v - p_1 - 2p_2 + s_1 = 0, v - p_2 - 2p_3 + s_2 = 0$

$v - 2p_1 - p_3 + s_3 = 0, p_1 + p_2 + p_3 + s_4 = 0$

Value is $1 - 1 = 0$ when all three strategies are equally probable. The same is true for player B. This accords with the symmetry of the situation.

4 a Allowing for the cost of insurance the pay-off matrix (in £00) is

		Weather	
		Rain	No rain
Acme	A_1	−4	12
	A_2	4	7
	A_3	12	4

Add 4 to all entries.

$P - v = 0, v - 8p_2 - 16p_3 + s = 0,$

$v - 16p_1 - 11p_2 - 8p_3 + t = 0, p_1 + p_2 + p_3 + u = 1$

Value is $10\frac{2}{3} - 4 = 6\frac{2}{3}$ when $p_1 = \frac{1}{3}, p_2 = 0, p_3 = \frac{2}{3}$

Advise them to avoid the basic insurance and to buy comprehensive insurance for a randomly chosen six of the nine events. They would expect to make a total of $9 \times 6\frac{2}{3} \times 100 = £6000$

b This assumes that the weather is independent at every location and it is equally likely to be wet at all the events.

Review exercise 30

1 a $P - x - 2y + z = 0, x + y + 2z + s = 14, 2x + y - z + t = 8$

b

P	x	y	z	s	t	Value	Row
1	−1	−2	1	0	0	0	R1
0	1	1	2	1	0	14	R2
0	2	1	−1	0	1	8	R3
1	3	0	−1	0	2	16	R4 = R1 + 2 × R6
0	−1	0	3	1	−1	6	R5 = R2 − R6
0	2	1	−1	0	1	8	R6 = R3
1	$2\frac{2}{3}$	0	0	$\frac{1}{3}$	$1\frac{2}{3}$	18	R7 = R4 + R8
0	$-\frac{1}{3}$	0	1	$\frac{1}{3}$	$-\frac{1}{3}$	2	R8 = $\frac{R5}{3}$
0	$1\frac{2}{3}$	1	0	$\frac{1}{3}$	$\frac{2}{3}$	10	R9 = R6 + R8

Maximum $P = 18$, when $x = 0, y = 10, z = 2$

2 Maximise $P = (-C) = -2x + y + 2z$

$P + 2x - y - 2z = 0, -2x - y + 2z + s = 10, 2x - y + t = 0,$

$3x + y - z + u = 0$

Maximum $P = 30$, so minimum $C = -30$, when $x = 0, y = 10,$ $z = 10, (t = 10)$

3 a As equations the problem is like this:

$P - 5x - 4y = 0, -3x - y + s = -10, 4x + 3y + t = 24$

This is not standard form because of the −10. The initial solution $x = 0, y = 0$ would give a negative value for s, which violates $s \geq 0$

b

P	x	y	s	t	Value	Row
1	−5	−4	0	0	0	R1
0	−3	−1	1	0	−10	R2
0	4	3	0	1	24	R3
1	0	$-\frac{1}{4}$	0	$1\frac{1}{4}$	30	R4 = R1 + 5 × R6
0	0	$1\frac{1}{4}$	1	$\frac{3}{4}$	8	R5 = R2 +3 × R6
0	1	$\frac{3}{4}$	0	$\frac{1}{4}$	6	R6 = $\frac{R3}{4}$

This is now a feasible solution with $x = 6, y = 0, s = 8, t = 0$

c Maximum $P = 31.6$, when $x = 1.2, y = 6.4$

4 a Maximum of row minima = 0, minimum of column maxima = 1, so solution is not stable.

b Add 3 to all entries.

$P - v = 0, v - 4p_2 - 3p_3 + s = 0, v - 5p_1 - p_2 - 4p_3 + t = 0,$

$p_1 + p_2 + p_3 + u = 1$

Value is $3.25 - 3 = 0.25$ when $p_1 = 0, p_2 = 0.25, p_3 = 0.75$

Assessment 30

1 a s and t are slack variables;

$13P + 8t - 2s = 1000$

b

P	x	y	s	t	Value
1	−1	−1	0	0	0
0	3	4	1	0	120
0	4	1	0	1	80

c The pivot is the 4 in the x–column;
since the θ-value of row 2 is $120 \div 3 = 40$
but the θ-value of row 3 is $80 \div 4 = 20$
$20 < 40$ so chose row 3/pivot is 4

d

P	x	y	s	t	Value
1	0	–0.75	0	0.25	20
0	0	3.25	1	–0.75	60
0	1	0.25	0	0.25	20

e Not optimal since one of the values in the objective row is negative

f $P_{max} = \frac{440}{13}$ when $x = \frac{200}{13}$, $y = \frac{140}{13}$

2 $P_{max} = \frac{25}{2}$, $x = \frac{5}{2}$, $y = 0$, $z = \frac{5}{2}$

3 a Play-safe strategy for Sarina is S_2 or S_3
Play-safe strategy for Jeremy is J_1
Value of game is 3

b Maximum of row minimums = minimum of column maxima, so a stable solution.

c

Strategy	S_1	S_2	S_3
J_1	4	–3	–3
J_2	–9	–4	–5
J_3	1	–4	–5

4 a x = number of litres of type X
y = number of litres of type Y
z = number of litres of type Z
Maximise $P = x + 1.5y + 1.2z$
subject to $0.4x + 0.6y + 0.5z \le 550$
$0.1x + 0.2y + 0.3z \le 160$
$0.1x + 0.1y + 0.05z \le 100$

b

P	x	y	z	r	s	t	Value
1	–0.25	0	1.05	0	7.5	0	1200
0	0.1	0	–0.4	1	–3	0	70
0	0.5	1	1.5	0	5	0	800
0	0.05	0	–0.1	0	–0.5	1	20

c $P = £1200$, $x = 0$, $y = 800$, $z = 0$, $(r = 70, s = 0, t = 20)$
Not optimal since negative value in objective function row

d There is $30l$ of orange juice remaining but no pineapple or grapefruit juice.

5 a C_2 dominates C_1 so delete C_1
R_4 dominates R_1 so delete R_1
C_3 dominates C_4 so delete C_4

b C should chose strategies C_2 and C_3 with probabilities $\frac{5}{11}$ and $\frac{6}{11}$ respectively
R should chose strategies R_2 and R_4 with probabilities $\frac{6}{11}$ and $\frac{5}{11}$ respectively
Value of game is $-\frac{3}{11}$ (to player C or $\frac{3}{11}$ to player R)

6 a Row 2 dominates row 3 since $6 > 5$, $-1 > -3$, $6 > 4$ and $7 > -1$. So player A should never play strategy A_3

b

Strategy	B_1	B_2
A_1	–3	2
A_2	6	–1

c Max of row minima = –1
Min of column maxima = 2
So no stable solution since $-1 \ne 2$

d A should play A_1 with probability $\frac{7}{12}$ and A_2 with probability $\frac{5}{12}$.
Value of the game is $-9 \times \frac{7}{12} + 6 = \frac{3}{4}$ to A

B should play B_1 with probability $\frac{1}{4}$ and B_2 with probability $\frac{3}{4}$
Value of the game is $-5 \times \frac{1}{4} + 2 = -\frac{3}{4}$ to B

7 a Let p_1 be the probability of playing strategy I, p_2 be the probability of playing strategy II, p_3 be the probability of playing strategy III,
Let V = value of the game
Maximise $P - V = 0$
Subject to $V - 5p_1 - 2p_3 + r = 0$
$V - p_1 - 2p_2 - 3p_3 + s = 0$
$V - 4p_1 - 5p_2 - p_3 + t = 0$
$p_1 + p_2 + p_3 = 1$
$v, p_1, p_2, p_3 \ge 1$

b Since the simplex algorithm requires decision variables to be non-negative

c For example:
$1 - p_1 - p_2 = p_3$
$V - 3p_1 + 2p_2 + r = 2$
$V + 2p_1 + p_2 + s = 3$
$V - 3p_1 - 4p_2 + t = 1$

d

P	V	p_1	p_2	r	s	t	Value
1	0	–3	–4	0	0	1	1
0	0	0	6	1	0	–1	1
0	0	5	5	0	1	–1	2
0	1	–3	–4	0	0	1	1

e Alex should chose strategy I $\frac{7}{30}$ of the time, II $\frac{1}{6}$ of the time and III $\frac{3}{5}$ of the time.
Value of game for Alex is $\frac{11}{30}$

8 a Maximise $P = V$
subject to: $V - 4p_1 - p_2 - 2p_3 + r = 0$
$V - 2p_1 - p_2 - 3p_3 + s = 0$
$V - p_1 - 3p_2 - 2p_3 + t = 0$
$p_1 + p_2 + p_3 + u = 1$
$p_1, p_2, p_3, r, s, t, u \ge 0$

b Set up table:

	V	P_1	P_2	P_3	r	s	t	u	Value
r	(1)	–4	–1	–2	1	0	0	0	0
s	1	–2	–1	–3	0	1	0	0	0
t	1	–1	–3	–2	0	0	1	0	0
u	0	1	1	1	0	0	0	1	1
P	–1	0	0	0	0	0	0	0	0

	V	P_1	P_2	P_3	s_1	s_3	s_3	s_4	Value
V	1	–4	–1	–2	1	0	0	0	0
s	0	(2)	0	–1	–1	1	0	0	0
t	0	3	–2	0	–1	0	1	0	0
u	0	1	1	1	0	0	0	1	1
P	0	–4	–1	–2	1	0	0	0	0

	V	P_1	P_2	P_3	s_1	s_2	s_3	s_4	Value
V	1	0	−1	−4	−1	2	0	0	0
P_1	0	1	0	−0.5	−0.5	0.5	0	0	0
t	0	0	−2	1.5	0.5	−1.5	1	0	0
u	0	0	1	1.5	0.5	−0.5	0	1	1
P	0	0	−1	−4	−1	2	0	0	0

9 a Maximum of row minima $= \max\{-3, -1, 0, -1\} = 0$
Minimum of column maxima $= \min\{4, 4, 3, 3\} = 3$
$0 \neq 3$ so no stable solution

b Strategy W is dominated by strategy Z and strategy A is dominated by strategy D.
So 3×3 pay-off matrix is

	Strategy	X	Y	Z
Greg	B	−1	3	1
	C	1	0	2
	D	4	−1	3

c

	Strategy	X	Y	Z
Greg	B	0	4	2
	C	2	1	3
	D	5	0	4

Greg plays B with probability p, C with probability q and D with probability r
Maximise $P = V - 1$
subject to: $V - 2p - 3q - 4r \leq 0$
$V - 2q - 5r \leq 0$
$V - 4p - q \leq 0$
$p + q + r \leq 1$
$V, p, q, r \geq 0$

Chapter 31

Exercise 31.1A

1 a

$\times_5$	0	1	2	3	4
0	0	0	0	0	0
1	0	1	2	3	4
2	0	2	4	1	3
3	0	3	1	4	2
4	0	4	3	2	1

b Yes c 1

d 1 and 4 are self-inverse, 2 is the inverse of 3 and vice versa. 0 has no inverse.

e S does not form a group since not every element has an inverse.

2 a

$+_6$	0	1	2	3	4	5
0	0	1	2	3	4	5
1	1	2	3	4	5	0
2	2	3	4	5	0	1
3	3	4	5	0	1	2
4	4	5	0	1	2	3
5	5	0	1	2	3	4

b Yes c 0

d S forms a group since modular addition is associative, every element has an inverse (e.g. the inverse of 5 is 1), and parts **b** and **c** give us that the operation is closed and has an identity element

e Yes: $x +_6 y = y +_6 x$ for all x, y. (The Cayley table has a line of symmetry down the leading diagonal.)

3 a

•	0	1	2	3
0	0	1	2	3
1	1	0	1	2
2	2	1	0	1
3	3	2	1	0

b Yes c 0

d No: inverse axiom is ok but $(2 \bullet 1) \bullet 3 \neq 2 \bullet (1 \bullet 3)$ so not associative.

4 Closed: $\begin{pmatrix} 1 & p \\ 0 & 1 \end{pmatrix}\begin{pmatrix} 1 & q \\ 0 & 1 \end{pmatrix} = \begin{pmatrix} 1 & p+q \\ 0 & 1 \end{pmatrix}$ which is also a member of the set S

Identity: $\mathbf{I} = \begin{pmatrix} 1 & 0 \\ 0 & 1 \end{pmatrix}$ is a member of the set S, and for any element s of S, $s\mathbf{I} = \mathbf{I}s = s$

Associativity: Matrix multiplication is associative.

Inverse: The inverse of $\begin{pmatrix} 1 & p \\ 0 & 1 \end{pmatrix}$ is $\begin{pmatrix} 1 & -p \\ 0 & 1 \end{pmatrix}$ which is also a member of set S

Commutativity: $\begin{pmatrix} 1 & p \\ 0 & 1 \end{pmatrix}\begin{pmatrix} 1 & q \\ 0 & 1 \end{pmatrix} = \begin{pmatrix} 1 & p+q \\ 0 & 1 \end{pmatrix}$ and

$\begin{pmatrix} 1 & q \\ 0 & 1 \end{pmatrix}\begin{pmatrix} 1 & p \\ 0 & 1 \end{pmatrix} = \begin{pmatrix} 1 & p+q \\ 0 & 1 \end{pmatrix}$ so the set forms an Abelian group under the operation of matrix multiplication.

5 For example, $1, 2 \in \mathbb{N}$ but $1 - 2 = -1 \notin \mathbb{N}$

6 a 6 b 2

7 a Closed: if x and y are integers then $x + y - 2$ is an integer.
Identity: $x \bullet e = x \Rightarrow x + e - 2 = x \Rightarrow e = 2$. Conversely, $2 \bullet x = 2 + x - 2 = x$
Associativity: $(x \bullet y) \bullet z = (x + y - 2) + z - 2 = x + y + z - 4$; $x \bullet (y \bullet z) = x + (y + z - 2) - 2 = x + y + z - 4$
Inverse: $x \bullet x^{-1} = e \Rightarrow x + x^{-1} - 2 = 2 \Rightarrow x^{-1} = 4 - x$ which is also an integer.

b $1 \in \mathbb{Z}^+$, but $1 \cdot 1 = 1 + 1 - 2 = 0 \notin \mathbb{Z}^+$. Hence T is not closed under •

Exercise 31.1B

1 a Define the symmetries:
Given that the centre of the square is the origin,
r_0 = rotation of 0° about the origin (i.e. the square is in its initial orientation)
r_1 = rotation of 90° anti-clockwise about the origin
r_2 = rotation of 180° anti-clockwise about the origin
r_3 = rotation of 270° anti-clockwise about the origin
m_1 = reflection in the x-axis
m_2 = reflection in the line $y = x$
m_3 = reflection in the y-axis
m_4 = reflection in the line $y = -x$

Draw up a Cayley table:

	r_0	r_1	r_2	r_3	m_1	m_2	m_3	m_4
r_0	r_0	r_1	r_2	r_3	m_1	m_2	m_3	m_4
r_1	r_1	r_2	r_3	r_0	m_2	m_3	m_4	m_1
r_2	r_2	r_3	r_0	r_1	m_3	m_4	m_1	m_2
r_3	r_3	r_0	r_1	r_2	m_4	m_1	m_2	m_3
m_1	m_1	m_4	m_3	m_2	r_0	r_3	r_2	r_1
m_2	m_2	m_1	m_4	m_3	r_1	r_0	r_3	r_2
m_3	m_3	m_2	m_1	m_4	r_2	r_1	r_0	r_3
m_4	m_4	m_3	m_2	m_1	r_3	r_2	r_1	r_0

Identity element: r_0

Inverses: The mirror lines are self-inverse, as are r_0 and r_2. r_1 is the inverse of r_3 and vice versa.

Associativity: Since the binary operation is a composition of mappings, the operation is associative.

Closure: Every combination of symmetries is in the original set of symmetries.

The order of the group is 8

b $C_4 = \{r_0, r_1, r_2, r_3\}$

Generators are r_1 and r_3

2 a

	r_0	r_1	r_2	r_3	r_4	r_5
r_0	r_0	r_1	r_2	r_3	r_4	r_5
r_1	r_1	r_2	r_3	r_4	r_5	r_0
r_2	r_2	r_3	r_4	r_5	r_0	r_1
r_3	r_3	r_4	r_5	r_0	r_1	r_2
r_4	r_4	r_5	r_0	r_1	r_2	r_3
r_5	r_5	r_0	r_1	r_2	r_3	r_4

Generators are r_1 and r_5

b Yes

3 a The set of all symmetries of a regular pentagon

b There is the identity element, r_0 plus four other rotations through multiples of 72° and five lines of symmetry, one through each vertex and the midpoint of its opposite side. Hence the group has at least 10 elements.

Conversely, select one vertex v of the pentagon, and consider an adjacent vertex w. Any symmetry of the pentagon maps v to one of 5 vertices of the pentagon, 5 possibilities.

Following this mapping, w must still be adjacent to v, either immediately clockwise, or immediately anticlockwise, so 2 possibilities.

These 5×2 possibilties uniquely determine the action of the symmetry on all vertices; hence there can be at most 10 elements of the group. Hence the group has 10 elements.

c For each group of symmetries there will be n rotations through multiples of $\frac{360°}{n}$ plus n lines of symmetry. For odd n, these will be through each vertex and the midpoint of its opposite side and for even n there will be $\frac{n}{2}$ lines through pairs of opposite vertices and $\frac{n}{2}$ lines through the midpoints of opposite sides. Hence the group has at least $2n$ elements.

Conversely, select one vertex v of the n-gon, and consider an adjacent vertex w. Any symmetry of the n-gon maps v to one of n vertices of the n-gon, so n possibilities.

Following this mapping, w must still be adjacent to v, either immediately clockwise or immediately anticlockwise, so two possibilities. These two possiilities uniquely determine the action of the symmetry on all vertices; hence there can be at most $2n$ elements of the group. Hence the group has exactly $2n$ elements.

4 a {1, 2, 3, 4, 5, 6} **b** Yes

5 a

$+_7$	0	1	2	3	4	5	6
0	0	1	2	3	4	5	6
1	1	2	3	4	5	6	0
2	2	3	4	5	6	0	1
3	3	4	5	6	0	1	2
4	4	5	6	0	1	2	3
5	5	6	0	1	2	3	4
6	6	0	1	2	3	4	5

b 7

c She is correct. Any element coprime to 7 is a generator under modular addition so in this case all of the elements except 0 are possible generators of the group.

6 a $\left\{\begin{pmatrix} 0 & 1 \\ -1 & 0 \end{pmatrix}, \begin{pmatrix} -1 & 0 \\ 0 & -1 \end{pmatrix}, \begin{pmatrix} 0 & -1 \\ 1 & 0 \end{pmatrix}, \begin{pmatrix} 1 & 0 \\ 0 & 1 \end{pmatrix}\right\}, 4$

b The group of rotational symmetries of the unit square.

7 a a **b** c or d

8 a A rotation through 0° about the centre of the circle

b Since the binary operation is a composition of mappings, the operation is associative.

c i A rotation through 327.9°

ii A reflection in a line of symmetry inclined at θ°

d *rot* since a combination of any two rotations is another rotation.

e Any combination of rotations and reflections leads to either another rotation or reflection and since the set of elements contains every rotation and every reflection, the resulting combination must be contained within the set.

Exercise 31.2A

1 H is non-empty since there exist (infinitely many) elements in H

The identity element of G under the binary operation addition is 0 and $0 \in \mathbb{Z}$

H is closed since if x and y are integers, then so is their sum $x + y$. Hence if $x, y \in \mathbb{Z}$ then $x + y \in \mathbb{Z}$

The inverse of an element x is x^{-1} which is $-x$ under the binary operation of addition. Since the negation of an integer is also an integer, it follows that if $x \in \mathbb{Z}$ then $x^{-1} \in \mathbb{Z}$

Hence H is a subgroup of G

2 The Cayley table of all symmetries of an equilateral triangle is:

	r_0	r_1	r_2	m_1	m_2	m_3
r_0	r_0	r_1	r_2	m_1	m_2	m_3
r_1	r_1	r_2	r_0	m_2	m_3	m_1
r_2	r_2	r_0	r_1	m_3	m_1	m_2
m_1	m_1	m_3	m_2	r_0	r_2	r_1
m_2	m_2	m_1	m_3	r_1	r_0	r_2
m_3	m_3	m_2	m_1	r_2	r_1	r_0

The group of rotations is given by:

	r_0	r_1	r_2
r_0	r_0	r_1	r_2
r_1	r_1	r_2	r_0
r_2	r_2	r_0	r_1

This contains the identity element r_0 and is non-empty. r_0 is self-inverse and r_1 and r_2 form an inverse pair. The set of rotations is closed since every combined element is one of the original elements, therefore the group of rotations is a subgroup of the group of all symmetries.

3 {1} and {1, 4}

4 {0, 2} and {0, 1, 2, 3}

5 **a** The Cayley table for C_4

	c_0	c_1	c_2	c_3
c_0	c_0	c_1	c_2	c_3
c_1	c_1	c_2	c_3	c_0
c_2	c_2	c_3	c_0	c_1
c_3	c_3	c_0	c_1	c_2

$\{c_0, c_1\}$, $\{c_0, c_3\}$, $\{c_0, c_1, c_2)$, $\{c_0, c_1, c_3\}$ and $\{c_0, c_2, c_3\}$ are not closed, e.g. $c_3c_2 = c_1$

$\{c_0, c_2\}$ is closed, contains the identity element c_0 and both elements are self-inverse so this is the only non-trivial proper subset of C_4

b The Cayley tables for **Q4** and **Q5** have exactly the same structure indicating that H is equivalent to C_4

6 $\{a\}$, $\{a, b\}$, $\{a, c\}$ and $\{a, d\}$

7 $N = \left\{\begin{pmatrix} 1 & 0 \\ 0 & 1 \end{pmatrix}\begin{pmatrix} -1 & 0 \\ 0 & -1 \end{pmatrix}\right\}$ contains the identity matrix and is non-empty.

$$\begin{pmatrix} 1 & 0 \\ 0 & 1 \end{pmatrix}\begin{pmatrix} -1 & 0 \\ 0 & -1 \end{pmatrix} = \begin{pmatrix} -1 & 0 \\ 0 & -1 \end{pmatrix}$$

$$\begin{pmatrix} -1 & 0 \\ 0 & -1 \end{pmatrix}\begin{pmatrix} 1 & 0 \\ 0 & 1 \end{pmatrix} = \begin{pmatrix} -1 & 0 \\ 0 & -1 \end{pmatrix}$$

$$\begin{pmatrix} -1 & 0 \\ 0 & -1 \end{pmatrix}\begin{pmatrix} -1 & 0 \\ 0 & -1 \end{pmatrix} = \begin{pmatrix} 1 & 0 \\ 0 & 1 \end{pmatrix}$$

Hence the set is closed under matrix multiplication. Both matrices are self-inverse, hence N is a subgroup of M

Exercise 31.2B

1 By Lagrange's theorem, only groups with order that is a factor of 90 can be subgroups so the only possibilities are: 1, 2, 3, 5, 6, 9, 10, 15, 18, 30, 45, 90

2 c C, order 12

Since 12 is not a factor of 30

3 6 does not divide 92, therefore, by Lagrange's theorem, 6 cannot be the order of a possible subgroup.

4 There are 11 elements in H and since 11 is prime, the only orders of possible subgroups are 1 and 11 The only subgroup of order 1 is the trivial subgroup, since any subgroup must contain the identity element, and the only subgroup of order 11 is the group itself, since it must contain every element. Hence the only subgroups are the trivial subgroup and the group itself.

5 **a** 8

b Orders of possible subgroups, by Lagrange's theorem, are 1, 2, 4, 8 (the factors of 8)

c $\{e, q\}$ and $\{e, q, p, u\}$

6 **a** There are three elements in Carmelita's proposed subgroup and since 3 is not a factor or 4, by Lagrange's theorem, Carmelita must be wrong. (Also accept that $-i \times -i = 1$ and 1 is not an element of Carmelita's proposed subgroup, so it's not closed.)

b {1} and {1, −1}

7 There are nine elements so the order of G is 9. Hence possible subgroups should be of orders 1, 3 or 9

Order 1: The trivial subgroup {0}

Order 9: The group itself, G

Order 3: There are no self-inverse elements under $+_9$ so a subgroup of order 3 will consist of the identity, an element and its inverse. There are four such sets: {0, 1, 8}, {0, 2, 7}, {0, 3, 6} and {0, 4, 5}. However, all except {0, 3, 6} are not closed under the group operation. Hence there is a single subgroup of order 3. Hence there are three subgroups in total.

8 $43\,252\,003\,274\,489\,856\,000 = 2^{27} \times 3^{14} \times 5^3 \times 7^2 \times 11$

$(27+1)(14+1)(3+1)(2+1)(1+1) = 10080$ so $43\,252\,003\,274\,489\,856\,000$ has 10080 factors and, by Lagrange's theorem, this gives the number of orders of possible subgroups.

Exercise 31.3A

1 Draw a Cayley table for the group:

	r_0	r_1	r_2	r_3
r_0	r_0	r_1	r_2	r_3
r_1	r_1	r_2	r_3	r_0
r_2	r_2	r_3	r_0	r_1
r_3	r_3	r_0	r_1	r_2

r_0 is the identity element. r_2 is self-inverse.

Rearranging the columns and rows gives

	r_0	r_2	r_1	r_3
r_0	r_0	r_2	r_1	r_3
r_2	r_2	r_0	r_3	r_1
r_1	r_1	r_3	r_2	r_0
r_3	r_3	r_1	r_0	r_2

Now it is clear to see that the pattern of entries is the same as for the groups in example 1, for instance, via the mapping $r_0 \mapsto 1$, $r_2 \mapsto -1$, $r_1 \mapsto i$, $r_3 \mapsto -i$, hence the groups are isomorphic.

2 **a** Write out the Cayley tables for each group:

$\times_5$	1	2	3	4
1	1	2	3	4
2	2	4	1	3
3	3	1	4	2
4	4	3	2	1

$+_4$	0	1	2	3
0	0	1	2	3
1	1	2	3	0
2	2	3	0	1
3	3	0	1	2

The self-inverse element in $+_4$ is 2 so rearrange the columns and rows:

$+_4$	0	1	3	2
0	0	1	3	2
1	1	2	0	3
3	3	0	2	1
2	2	3	1	0

Now it is clear to see the pattern of entries is the same for both groups, for instance via the mapping between A and B given by $1 \mapsto 0$, $2 \mapsto 1$, $3 \mapsto 3$, $4 \mapsto 2$ hence the groups are isomorphic.

b Yes

3 a For group M, let the four matrices be a, b, c and d respectively. Hence the Cayley table for M is

×	a	b	c	d
a	a	b	c	d
b	b	a	d	c
c	c	d	a	b
d	d	c	b	a

For group N, the Cayley table is

	1	3	5	7
1	1	3	5	7
3	3	1	7	5
5	5	7	1	3
7	7	5	3	1

The identity elements are a and 1 respectively and since all elements in both groups are self-inverse, the groups are isomorphic to each other, for instance via the mapping $a \mapsto 1, b \mapsto 3, c \mapsto 5, d \mapsto 7$.

b No

4 The Cayley table for the rotations of the equilateral triangle is

	r_0	r_1	r_2
r_0	r_0	r_1	r_2
r_1	r_1	r_2	r_0
r_2	r_2	r_0	r_1

For group R the Cayley table is

$\times_7$	4	2	1
4	2	1	4
2	1	4	2
1	4	2	1

Since the identity elements are r_0 and 1, rearrange the rows/columns in the table for $\times_7$

$\times_7$	1	4	2
1	1	4	2
4	4	2	1
2	2	1	4

Now it is clear to see the matching patterns of entries in the two tables, for instance via the mapping $r_0 \mapsto 1, r_1 \mapsto 4, r_2 \mapsto 2$, hence the two groups are isomorphic.

Exercise 31.3B

1 All groups of order 2 consist of the identity element and one other self-inverse element of period 2 so they must be isomorphic.

2 All groups of order 3 consist of the identity element and two other elements of period 3 (period 2 is not allowed since 2 does not divide into 3). The groups are therefore cyclic with elements that map to e, a and a^2.

3 Period 4 elements must occur in pairs, paired with their inverse. Hence one or three elements of period 4 cannot happen. So every group of order 4 is isomorphic to exactly one of two particular groups of order 4, Group A with all elements other than the identity of period 2 (the Klein 4-group) and Group C with two elements of period 4 (the C_4 group).

4 Groups of order p, where p is prime, have an identity element of period 1 and all remaining elements must have period p, since p has just two factors, 1 and p. Let G and H be two such groups of order p, and select an element $g \in G$ and $h \in H$, both of order p. Then the mapping $a^i \mapsto b^i$ gives a one-to-one mapping that respects the operations of G and H, hence giving an isomorphism. Hence all groups of order p are isomorphic to the cyclic group C_p

5 Let G be a non-abelian group, and H an abelian group. Since G is not abelian there are elements $a, b \in G$ such that $ab \neq ba$. Suppose there were an isomorphism between G and H with $a \mapsto x$, $b \mapsto y$. Then $ab \mapsto xy$, $ba \mapsto yx$, and since H is abelian $xy = yx$. This means that ab and ba map to the same element of H despite being different elements of G, so the mapping cannot have been one-to-one. Hence no isomorphism exists.

6 The Cayley table for G

$+_6$	0	1	2	3	4	5
0	0	1	2	3	4	5
1	1	2	3	4	5	0
2	2	3	4	5	0	1
3	3	4	5	0	1	2
4	4	5	0	1	2	3
5	5	0	1	2	3	4

The Cayley table for H

	r_0	r_1	r_2	r_3	r_4	r_5
r_0	r_0	r_1	r_2	r_3	r_4	r_5
r_1	r_1	r_2	r_3	r_4	r_5	r_0
r_2	r_2	r_3	r_4	r_5	r_0	r_1
r_3	r_3	r_4	r_5	r_0	r_1	r_2
r_4	r_4	r_5	r_0	r_1	r_2	r_3
r_5	r_5	r_0	r_1	r_2	r_3	r_4

It is evident that the two Cayley tables has the same pattern of entries, for instance via the mapping $i \mapsto r_i$, therefore the two groups are isomorphic.

7 M is a cyclic group since when $k = 1$, the period of the element is 8 (8 successive rotations of $\frac{\pi}{4}$ will map the unit square back to itself). In particular, the only self-inverse element of M is the rotation by π, so M only has one self-inverse element. However, in S the elements 4, 11 and 14 are all self-inverse. Hence the groups cannot be isomorphic.

8 0, which is the identity element of G

Consider a mapping between the sets, say a maps to $\begin{pmatrix} 1-a & a \\ -a & 1+a \end{pmatrix}$ for all $a \in \mathbb{Z}$. This means that p maps to $\begin{pmatrix} 1-p & p \\ -p & 1+p \end{pmatrix}$ and q maps to $\begin{pmatrix} 1-q & q \\ -q & 1+q \end{pmatrix}$. This mapping is clearly one-to-one.

You must now show that the product of these two matrices is $\begin{pmatrix} 1-(p+q) & (p+q) \\ -(p+q) & 1+(p+q) \end{pmatrix}$, i.e. $p+q$ maps to this matrix product.

$$\begin{pmatrix} 1-p & p \\ -p & 1+p \end{pmatrix}\begin{pmatrix} 1-q & q \\ -q & 1+q \end{pmatrix}$$

$$= \begin{pmatrix} (1-p)(1-q)-pq & (1-p)q+p(1+q) \\ -p(1-q)-q(1+p) & -pq+(1+p)(1+q) \end{pmatrix}$$

$$= \begin{pmatrix} 1-p-q & p+q \\ -p-q & 1+p+q \end{pmatrix} = \begin{pmatrix} 1-(p+q) & (p+q) \\ -(p+q) & 1+(p+q) \end{pmatrix}$$

as required.

Hence the two groups are isomorphic.

Review exercise 31

1 a 6

b

$\times_9$	1	2	4	5	7	8
1	1	2	4	5	7	8
2	2	4	8	1	5	7
4	4	8	7	2	1	5
5	5	1	2	7	8	4
7	7	5	1	8	4	7
8	8	7	5	4	7	1

c 6

d 5 has period equal to the order of the group; hence G is the group generated by 5 and hence the group is cyclic.

e Yes: The Cayley table has a line of symmetry down the leading diagonal. (Or G is a cyclic group and cyclic groups are abelian).

2 a 1

b Since all of the entries in the Cayley table are elements of S, the set is closed under the operation $\times_{12}$.

c All of the elements are self-inverse since for $a \in S$, $a^2 = 1$

d For instance:

$(5 \times 7) \times 11 = 11 \times 11 = 1 \pmod{12}$ and $5 \times (7 \times 11) = 5 \times 5 = 1 \pmod{12}$

$(11 \times 7) \times 5 = 5 \times 5 = 1 \pmod{12}$ and $11 \times (7 \times 5) = 11 \times 11 = 1 \pmod{12}$

3 a 1, 2, 3 and 6

b $\{r_0\}$

c $\{r_0\}$, $\{r_0, m_1\}$, $\{r_0, m_2\}$, $\{r_0, m_3\}$ and $\{r_0, r_1, r_2\}$.

4 a The Cayley tables are

$\times_{10}$	1	3	7	9
1	1	3	7	9
3	3	9	1	7
7	7	1	9	3
9	9	7	3	1

$\times_4$	0	1	2	3
0	0	1	2	3
1	1	2	3	0
2	2	3	0	1
3	3	0	1	2

The self-inverse element in $+_4$ is 2 so swapping the order of columns and rows gives

$+_4$	0	1	3	2
0	0	1	3	2
1	1	2	0	3
3	3	0	2	1
2	2	3	1	0

Now the pattern of elements in the two groups are the same, hence $G \cong H$.

A possible mapping is $[1, 3, 7, 9] \leftrightarrow [0, 1, 3, 2]$.

b Yes: 1 and 3 are generators of H and 3 and 7 are generators of G.

c Yes: The Cayley tables have a line of symmetry along the leading diagonal. Also the groups are cyclic, and cyclic groups are abelian.

Assessment 31

1 a b

Since $b * x = x * b = x$ for all elements x

b b, c

c Symmetric in the leading diagonal.

2 a b

Since it is the only element in all the subgroups.

b 30

Since it is the lowest common multiple of 2, 3 and 5

3

$\times_8$	1	3	5	7
1	1	3	5	7
3	3	1	7	5
5	5	7	1	3
7	7	5	3	1

$\times_{10}$	1	3	7	9
1	1	3	7	9
3	3	9	1	7
7	7	1	9	3
9	9	7	3	1

All elements in $\times_8$ are self-inverse, $\times_{10}$ has only two self-inverse elements, therefore not isomorphic.

4 a It means that 9 is a *generator* of the group.

b $9^2 = 17 \pmod{64}$

$9^3 = 25 \pmod{64}$

$9^4 = 33 \pmod{64}$

$9^5 = 41 \pmod{64}$

$9^6 = 49 \pmod{64}$

$9^7 = 57 \pmod{64}$

$9^8 = 1 \pmod{64}$

Since 1 is the multiplicative identity,

$n = 8$

c 1, 2, 4, 8

Since by Lagrange's theorem, order of subgroup must be a factor of the group order.

5 a

$\odot$	0	1	2	3	4	5
0	3	4	5	0	1	2
1	4	5	0	1	2	3
2	5	0	1	2	3	4
3	0	1	2	3	4	5
4	1	2	3	4	5	0
5	2	3	4	5	0	1

Since every element generated under $\odot$ is an element of the set, closure under $\odot$.

Identity element is 3

5, 1 and 4, 2 are inverse pairs and 0, 3 are self-inverse so every element has an inverse.

$(a \odot b) \odot c = (a + b + 3) + c + 3 = a + b + c + 6$

$a \odot (b \odot c) = a + (b + c + 3) + 3 = a + b + c + 6$

Therefore associative.

Since the four axioms of being a group are satisfied, G forms a group under $\odot$

b {3}, {1, 3, 5}, {0, 3}

c $G = (\langle 2 \rangle, \odot)$

or $K = (\{1, 2, 4, 5, 7, 8\}, \times_9)$

Produce a mapping of G to K, e.g.

$3 \mapsto 1$

$0 \mapsto 8$

$1 \mapsto 4$

$2 \mapsto 2$

$4 \mapsto 5$

$5 \mapsto 7$

As there is a one-to-one mapping between the elements of G and the elements of K which preserves the group operation, $G \cong K$

6 **a** Since $ex = xe = x$ for all x

b 3 does not divide into 8 so no, she is not correct.

c $d^2 = a$, $d^4 = c$,

$d^8 = e$

Since period of d is 8, d is a generator and group is cyclic. Rotations have period 2 or 4 and reflections have period 2, hence no element of order 8, hence not isomorphic.

7 **a** $(x \bullet y) \bullet z = (x + y - 1) + z - 1 = x + y + z - 2$

$x \bullet (y \bullet z) = x + (y + z - 1) - 1 = x + y + z - 2$

Therefore associative.

b $x \bullet 1 = x + 1 - 1 = x$, $1 \bullet x = x$ hence 1 is the identity.

$x \bullet x^{-1} = 1 \Rightarrow x + x^{-1} - 1 = 1 \Rightarrow x^{-1} = 2 - x$, and conversely $x \bullet (2 - x) = x + 2 - x - 1 = 1$.

$2 - x$ is an integer, therefore elements have inverse

$x + y - 1$ is an integer therefore closed.

Hence P forms a group under $\bullet$

c No.

Since inverse $2 - x$ is not a positive integer for all $x > 1$

For instance the inverse of 3 is −1, which is not a positive integer.

Index